XXIX International Conference on Photonic, Electronic and Atomic Collisions (ICPEAC 2015)

AA003360

Journal of Physics: Conference Series Volume 635

Toledo, Spain
22 – 28 July 2015

Editors:

Cristina Diaz
Ismanuel Rabadan
Gustavo Garcia
Luis Mendez
Fernando Martin

ISBN: 978-1-5108-1197-3
ISSN: 1742-6588

Printed from e-media with permission by:

Curran Associates, Inc.
57 Morehouse Lane
Red Hook, NY 12571

CURRAN ASSOCIATES INC. proceedings.com

Some format issues inherent in the e-media version may also appear in this print version.

Copyright© (2015) by the Institute of Physics
All rights reserved. The material featured in this book is subject to IOP copyright protection, unless otherwise indicated.

Printed by Curran Associates, Inc. (2015)

For permission requests, please contact the Institute of Physics at the address below.

Institute of Physics
Dirac House, Temple Back
Bristol BS1 6BE UK

Phone: 44 1 17 929 7481
Fax: 44 1 17 920 0979

techtracking@iop.org

Additional copies of this publication are available from:

Curran Associates, Inc.
57 Morehouse Lane
Red Hook, NY 12571 USA
Phone: 845-758-0400
Fax: 845-758-2634
Email: curran@proceedings.com
Web: www.proceedings.com

TABLE OF CONTENTS

HEAVY PARTICLE – ATOM AND ION

HEAVY PARTICLE – CONDENSED MATTER AND LARGE CLUSTER

HEAVY PARTICLE – OTHER

LEPTON – ATOM AND ION

LEPTON – CONDENSED MATTER AND LARGE CLUSTER

LEPTON – MOLECULE AND SMALL CLUSTER

LEPTON – OTHER

PHOTON – ATOM AND ION

PHOTON – CONDENSED MATTER AND LARGE CLUSTER

PHOTON – MOLECULE AND SMALL CLUSTER

PHOTON – OTHER

Preface

The 29th International Conference on Photonic, Electronic and Atomic Collisions (XXIX ICPEAC) was held at the Palacio de Congresos “El Greco”, Toledo, Spain, on 22–28 July, 2015, and was organized by the Universidad Autónoma de Madrid (UAM) and the Consejo Superior de Investigaciones Científicas (CSIC). ICPEAC is held biannually and is one of the most important international conferences on atomic and molecular physics. The topic of the conference covers the recent progresses in photonic, electronic, and atomic collisions with matter. With a history back to 1958, ICPEAC came to Spain in 2015 for the very first time. UAM and CSIC had been preparing the conference for six years, ever since the ICPEAC International General Committee made the decision to hold the XXIX ICPEAC in Toledo.

The conference gathered 670 participants from 52 countries and attracted 854 contributed papers for presentation in poster sessions. Among the latter, 754 are presented in issues 2-12 of this volume of the Journal of Physics Conference Series. In addition, five plenary lectures, including the opening one by the Nobel laureate Prof. Ahmed H. Zewail and the lectures by Prof. Maciej Lewenstein, Prof. Paul Scheier, Prof. Philip H. Bucksbaum, and Prof. Stephen J. Buckman, 62 progress reports and 26 special reports were presented following the decision of the ICPEAC International General Committee. Detailed write-ups of most of the latter are presented in issue 1 of this volume, constituting a comprehensive tangible record of the meeting.

On the occasion of the International Year of Light (IYL2015) and with the support of the Fundación Española para la Ciencia y la Tecnología (FECYT), the program was completed with two public lectures delivered by the Nobel laureate Prof. Serge Haroche and the Príncipe de Asturias laureate Prof. Pedro M. Echenique on, respectively, “Fifty years of laser revolutions in physics” and “The sublime usefulness of useless science”. Also a popularization initiative was held in parallel to the conference, consisting of a poster exhibition for the general public and an International Science Camp for children. Furthermore, before the conference, four distinguished scientists: Prof. Thomas F. Gallager, Prof. Jan Michael Rost, Prof. Joseph H. Macek, and Prof. Reinhard Dörner presented tutorial lectures.

During the conference, the 2015 IUPAP Young Scientist Prize was awarded to Dr. Gretchen K. Campbell from the University of Maryland. The Sheldon Datz Prize for an Outstanding Young Scientist Attending ICPEAC was awarded to Dr. Jayanta K. Saha from the Indian Association for the Cultivation of Science.

The Local Organizing Committee would like to express its gratitude to the Fundación Española para la Ciencia y la Tecnología (FECYT), the International Union for Pure and Applied Physics (IUPAP), all the conference sponsors for financial support, and, with particular emphasis, to the group of volunteers from UAM and CSIC for their hard and excellent work.

Finally, the Local Organizing Committee would like to thank all participants and authors of the proceedings for their support and contributions to the conference.

Content from this work may be used under the terms of the Creative Commons Attribution 3.0 licence. Any further distribution of this work must maintain attribution to the author(s) and the title of the work, journal citation and DOI.
Published under licence by IOP Publishing Ltd

Journal of Physics: Conference Series **635** (2015) 001001 doi:10.1088/1742-6596/635/0/001001

The editors,
Cristina Díaz, Universidad Autónoma de Madrid
Ismanuel Rabadán, Universidad Autónoma de Madrid
Gustavo García, Consejo Superior de Investigaciones Científicas
Luis Méndez, Universidad Autónoma de Madrid
Fernando Martín, Universidad Autónoma de Madrid

Journal of Physics: Conference Series **635** (2015) 001002 doi:10.1088/1742-6596/635/0/001002

Peer review statement

All papers published in this volume of *Journal of Physics: Conference Series* have been peer reviewed through processes administered by the proceedings Editors. Reviews were conducted by expert referees to the professional and scientific standards expected of a proceedings journal published by IOP Publishing.

Content from this work may be used under the terms of the Creative Commons Attribution 3.0 licence. Any further distribution of this work must maintain attribution to the author(s) and the title of the work, journal citation and DOI.
Published under licence by IOP Publishing Ltd

Journal of Physics: Conference Series **635** (2015) 022001 doi:10.1088/1742-6596/635/2/022001

Scattering of proton by Lithium atom with electron exchange

Elena Orlenko* [1], **Alexandr Evstafev*** [2]

[*] Departament for the Theoretical Physics, St.Petersburg State Polytechnic University, Polytechnicheskaya 29, 195251, St.Petersburg Russian Federation

Synopsis , We present formalism of Time-dependent Exchange Perturbation Theory (TDEPT) built to all orders of perturbation, for arbitrary time dependency of perturbation and apply it to the problem of proton scattering on Lithium atom where we calculate the differential and the total cross-sections

In this work, we develop a dedicated perturbation theory, named exchange perturbation theory (EPT), which takes into account the indistinguishability of electrons participating in multi-centre molecular / atomic collisions to any order of perturbation. It is worth noting that the development of EPT requires careful study of the wave functions associated with electrons assigned to different atomic centres. The overlapping of these wave functions is responsible for the exchange effects that play a crucial role in both adiabatic and dynamic scattering events.

In the following, we consider the collision associated with the redistribution of electrons, as for example the collisions of proton scattering on Lithium atom: $\mathrm{Li} + p \rightarrow \mathrm{Li}^{+} + \mathrm{H}$, accompanied by charge transfer.

The formalism of the time-dependent exchange perturbation theory is developed in an invariant form [1] . The obtained expression for the differential cross section is

$$\frac{d\sigma_{fi}}{d\Omega} = j^{-1}\frac{dw_{fi}}{d\Omega} = \frac{\mu_i \mu_f k_f}{\left(2\pi\hbar^2\right)^2 k_i}\left|\left\langle \Psi_f^0 \right| \hat{T} \left| \Phi_i^0 \right\rangle\right|^2 .$$

The transition operator $\hat{T}$ given in the form

$$\hat{T} = V_0^{\mathbb{N}} + V_0^{\mathbb{N}}\left(\frac{f_0^2}{P}\right)^{-1}\left(E_i - H + i\eta\right)^{-1} V_0^{\mathbb{N}},$$

where H is the total Hamiltonian of the system and where $V_0^{\mathbb{N}} = \left(\frac{f_0^2}{P}\right) V_0$ is a renormalized perturbation operator. The initial state is described by the vector $\left|\Phi_i^{0(0)}\right\rangle$, corresponding to the initial electron permutation between the centers, the antisymmetric vector $\left|\Psi_f\right\rangle$ of the final state is obtained by applying the normalized Young operator onto the wave function $\hat{A}\left|\Phi_f\right\rangle$ with respect to all electron permutations between the centers.

As the initial permutation ($p = 0$) we assume the following arrangement: three electrons of the Lithium atom and the incident proton. We label the electrons as 1, 2 and 3. Figure 1 illustrates the arrangement of the particles. $\vec{R}$ is vector originating at proton and pointing to Lithium atom.

The total cross section σ_{fi} is summarized in Table 1, where the cross section values, obtained in this work are compared to those obtained from the experiments.

Table 1. Summary of total cross section values

v(cm)	E (eV)	k *a.u.*	σ(cm²) Li-H⁺	σ(cm²) LiHe²⁺	σ (cm²)
$7.5\cdot10^{5}$	1.95	15	$7\cdot10^{-14}$	$6\cdot10^{-14}$	$6\cdot10^{-14}$
$2.5\cdot10^{6}$	21.6	50	$2\cdot10^{-14}$	$17\cdot10^{-15}$	$2.\cdot10^{-14}$

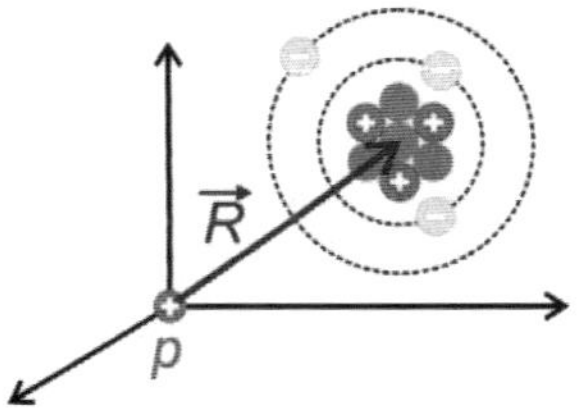

Figure 1. Arrangement of the particles in the proton - Lithium atom collision

References

[1] C. Orlenko EV, Evstafev AV, Orlenko FE 2015, *Journal of Experimental and Theoretical Physics* **147** 338 .

[1] E-mail: eorlenko@mail.ru

[2] E-mail: evstafev93@mail.ru

Content from this work may be used under the terms of the Creative Commons Attribution 3.0 licence. Any further distribution of this work must maintain attribution to the author(s) and the title of the work, journal citation and DOI.
Published under licence by IOP Publishing Ltd

Electron emission in collisions between atoms and dressed projectiles

M Purkait

Department of Physics, Ramakrishna Mission Residential College, Narendrapur, Kolkata - 700103, India

Keywords Ionization, collision, cross sections

Synopsis Double-differential cross sections (DDCS) for electron emission in ionization under the impact of dressed projectiles has been calculated within the framework of three-Coulomb wave (3CW) model. The interaction between the dressed projectile ion and the active electron is approximated by a model potential. In order to take into account of the effect of the passive electrons, we have also constructed a model potential which satisfies initial condition and the corresponding wave function has been calculated from the model hamiltonian of the active electron in the target.

The study of electronic interactions in collisions between partially stripped projectiles and atomic targets has received increasing interest in the past two decades. These systems provide a suitable framework in which to investigate the relative importance of the electron-electron interactions versus the electron-nucleus interactions. Therefore, detailed knowledge of the mechanisms of electron emission in collisions between fast partially striped ions with atoms is relevant in many scientific areas, such as astrophysics, plasma physics, radiation damage and in hadron therapy. The shape of electron emission spectrum observed in heavy ion-atom collisions is sensitive to the various ionization mechanisms. Such spectrum clearly identifies different processes, such as soft collision (SC), two-center electron emission (TCEE), electron capture to the continuum (ECC) and the binary encounter (BE). A large amount of experimental data for double differential cross sections (DDCS) are available in the literature involving bare [1-2] and partially striped projectile ions [3,4] impinging on atomic targets at intermediate and high collision energies. In the present investigation, we have focused our attention on the determination of DDCS for single ionization of partially striped lithium and carbon projectiles with helium atoms at intermediate and high energies. The transmission amplitude for the process may be written

$$T_{if} = <\psi_f^{3CW}|V_i|\psi_i>, \quad \text{where}$$

$$V_i = V_{TP} + V_{Pe} \quad V_{TP} = \frac{qz_T^{eff}}{R},$$

$$V_{pe} = -\frac{q}{r_p} - \frac{e^{-\lambda r_p}}{r_p}\{(Z_P - q) + br_p\},$$

where z_P and q are are respectively the number and asymptotic charge of the projectile ions. The final state wavefunction χ_f^{3CW} is approximated as

$$\chi_f^{3CW} = (2\pi)^{-3} e^{i\vec{k}_P\cdot\vec{R} + i\vec{k}_e\cdot\vec{r}_T} C_1(\alpha_{TP}, \vec{k}_R, R) \\ C_2(\alpha_{Te}, \vec{k}_e, \vec{r}_T) C_3(\alpha_{pe}, \vec{k}_{pe}, \vec{r}_p).$$

The initial state wavefunction is given by

$$\psi_i = \frac{1}{(2\pi)^{\frac{3}{2}}} e^{i\vec{k}_i\cdot\vec{R}} \phi(\vec{r}_T),$$

Here we used for $\phi(\vec{r}_T)$ an RHF initial wavefunction corresponding to the model potential taking into account the interaction of the electron with the remaining dressed target. We found that our calculated results are in good agreement with the available experimental and theoretical works.

References

[1] S.Jana, R.Samanta and M.Purkait 2012 *Eur.Phys. J. D* **66** 243
[2] S.Jana, R.Samanta and M.Purkait 2013 *Phys. Scr.* **88** 055301
[3] D. Fregenal *et al* 2014 *J. Phys. B: At. Mol. Opt. Phys.* **47** 155204
[4] J.M. Monti *et al* 2011 *J. Phys. B: At. Mol. Opt. Phys.* **44** 195206
E-mail:mpurkait_2007@rediffmail.com

Content from this work may be used under the terms of the Creative Commons Attribution 3.0 licence. Any further distribution of this work must maintain attribution to the author(s) and the title of the work, journal citation and DOI.
Published under licence by IOP Publishing Ltd

Angular differential cross sections for double and single capture in fast ion-atom collisions

A Mondal and M Purkait

Department of Physics, Ramakrishna Mission Residential College,

Narendrapur, Kolkata-700103, India

Keywords Collision, cross sections

Synopsis We investigate angular differential cross sections for double-electron and single-electron capture in ion-atom collisions for the intermediate and high energy region. We see the diffraction pattern in the spectrum of differential cross sections in the high energy range, i.e. another peak at 0.75 mrad in addition to Thomas peak at 0.47 mrad have been observed.

Electron capture by multi-charged ions from atom is of interest in many fields such as astrophysics, plasma production, planetary and upper-atmospheric reactions and irradiation of living matter. All mentioned areas need a full understanding of the mechanisms involves in ion-atom and molecular collisions. In classical picture, the H^+ projectile (P) scatters on one target electron (e) such that this electron (e) scatters on the target nucleus (N) and finally becomes bound to the projectile. Then, energy and momentum conservation requires that the electron (e), when initially at rest in the target frame, is scattered at 60° with respect to the projectile trajectory and this binary collision (p-e) thus gives the characteristic proton (p-e-N) Thomas projectile scattering angle of 0.47 mrad. Only two peaks at $\theta_P = 0$ mrad and $\theta_p = 0.47$ mard (θ_C) have been identified both experimentally and theoretically. Recently, third peak at $\theta_p = 0.75$ mrad for the 7.5 MeV $H^+ - He$ collision have been reported experimentally [1-2]. So our analysis for the multiple scattering events reveals a peak at $\theta_P \geq \theta_C$ only if two or more electrons are involved in the processes. However, such multiple scattering is explained by taking the higher order term in the transition amplitude, which may written as

$$T_{if} = \langle \psi_f^- | V_i | \phi_i \rangle + \langle \psi_f^- | U_f | \psi_i^+ \rangle,$$ where

ϕ_i is the asymptotic initial state wavefunction, V_i is the interaction potential between the projectile and the helium atom, ψ_f^- is the approximate final state wavefunction, U_f is the final state perturbation and ψ_i^+ is the exact initial sate wavefunction. The initial asymptotic state is given by

$$\phi_i = e^{i\vec{k}_i \cdot \vec{R}_i} \varphi_{He}(\vec{x}_1, \vec{x}_2).$$

The final state wavefunction is approximated by

$$\psi_f^- = e^{i\vec{k}_f \cdot \vec{R}_f} \phi_i(\vec{x}_2) C_{T-e_1}(\vec{x}_1) C_{P-e_2}(\vec{s}_2) C_{P-T}(\vec{R}_f),$$

where C_{P-e_2} represents the interaction between the projectile and the passive electron in the target ion, C_{T-e_1} represent the interaction between the target nucleus and the captured electron, and C_{P-T} is the scattered Coulomb wave between the projectile and the residual target. The exact initial state wavefunction is approximated as

$$\varphi_i^+ = e^{i\vec{k}_i \cdot \vec{R}_i} \phi_{He}(\vec{x}_1, \vec{x}_2) C_{P-e_1}(\vec{s}_1).$$

We will represent theoretical results for SC and DC cross sections for ion-atom collisions in the intermediate to high energy range, and compared them with the available experimental data.

References

[1] M. Gudmundsson *et al* 2010 *J. Phys. B: At. Mol. Opt. Phys.* **43** 185209

[2] M. Gudmundsson *et al* 2010 *J. Phys. B: At. Mol. Opt. Phys.* **44** 175201

E-mail: mpurkait_2007@rediffmail.com

Content from this work may be used under the terms of the Creative Commons Attribution 3.0 licence. Any further distribution of this work must maintain attribution to the author(s) and the title of the work, journal citation and DOI.
Published under licence by IOP Publishing Ltd

Journal of Physics: Conference Series **635** (2015) 022004 doi:10.1088/1742-6596/635/2/022004

3D numerical calculation of electron capture cross sections in $Be^{4+} + H$ collisions.

A. Jorge[*1] **J. Suárez**[*2] **Clara Illescas**[*3] **and L. Méndez**[*4]

[*] Departamento de Química, Módulo 13, Universidad Autónoma de Madrid, Cantoblanco 28049, Madrid, Spain

Synopsis Electron capture in Be^{4+} + H collisions is studied at collision energies between 1 and 100 keV/u, as a benchmark to discuss the accuracy of different theoretical approaches. Partial cross sections, obtained by solving numerically the time-dependent Schrödinger equation, are compared to the corresponding results from close-coupling and classical trajectory Monte Carlo calculations.

Collisions involving Be^{q+} ions are specially relevant because the first wall of ITER will contain Be, but experiments with these ions cannot be performed. In the present work, we have considered the electron capture reaction:

$$Be^{4+} + H(1s) \rightarrow Be^{3+}(n) + H^{+} \qquad (1)$$

in the energy range 1 keV/u $\leq E \leq$ 100 keV/u. We have calculated the partial cross sections by solving numerically the time-dependent Schrödinger equation with the Grid-TDSE package (GTDSE) [1]. At high velocities, we have also performed classical trajectory Monte Carlo (CTMC) calculations with the hydrogenic initial distribution, similar to those reported in [2].

In contrast to experimental data, which always include estimates of the uncertainties associated to the measurements, theoretical data do not present in general such uncertainties. In the last few years, several initiatives have encouraged data producers to include uncertainty estimates in their data; for instance, the International Atomic Energy Agency (e.g. [3]) has organized several workshops on this subject. In the present work, we have estimated the uncertainties of the GTDSE calculation by considering the convergence with the grid density and those of the CTMC calculation by the convergence of the cross sections with respect to the number of trajectories included in the calculation. At low energy, E = 1 keV/u, the molecular calculations of Harel *et al.* [4] indicate that the cross sections for populating the n = 3, 4 shells have converged with an absolute uncertainty of ± 0.3 Å^2, and our most precise GTDSE leads to cross sections within this range. As an illustration of the results at high E, we show in Fig. 1 the n-partial cross sections for reaction (1) at two collision energies, where it is clear the good agreement between the close-coupling atomic expansion (AOCC) and the GTDSE method at E = 30 keV/u. At E = 100 keV/u, the application of close-coupling methods is hindered by the need of representing transitions to very excited shells. The competition of the ionization reaction probably causes the unphysical increase of the AOCC cross section for $n > 5$, not shown by the GTDSE and CTMC results.

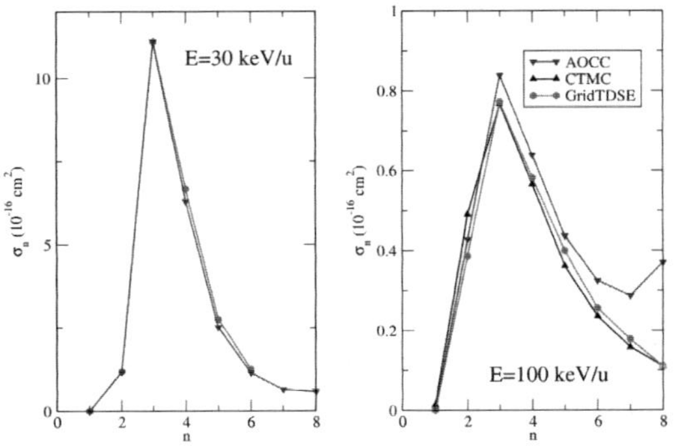

Figure 1. Comparison of state-selected capture cross sections calculated using an atomic expansion [5], the GTDSE method and the CTMC calculation at E =30 and 100 keV/u

References

[1] J. Suarez *et al* 2009 *Comput. Phys. Comm* **180**, 2025

[2] A. Jorge *et al* 2014 *Eur. Phys. J. D* **68**, 1

[3] Joint IAEA-ITAMP Technical Meeting, Cambridge 2014

[4] C. Harel *et al* 1998 *At. Data. Nucl. Data Tables* **68**, 279

[5] K. Igenbergs *et al* 2009 J. Phys. B **42**, 235206

[1] E-mail: alba.jorge@uam.es [2] E-mail: jaime.suarez@uam.es
[3] E-mail: clara.illescas@uam.es [4] E-mail: l.mendez@uam.es

Content from this work may be used under the terms of the Creative Commons Attribution 3.0 licence. Any further distribution of this work must maintain attribution to the author(s) and the title of the work, journal citation and DOI.
Published under licence by IOP Publishing Ltd

Journal of Physics: Conference Series **635** (2015) 022005 doi:10.1088/1742-6596/635/2/022005

Forward-angle electron spectroscopy in heavy-ion atom collisions studied at the ESR

P.-M. Hillenbrand[1,2 1], S. Hagmann[1,3 2], D. Banaś[4], K.-H. Blumenhagen[5], C. Brandau[1,2], W. Chen[1], E. De Filippo[6], A. Gumberidze[1], D. L. Guo[7], D. H. Jakubassa-Amundsen[8] M. Lestinsky[1], Yu. A. Litvinov[1], J. M. Monti[9] A. Müller[2], H. Rothard[10], S. Schippers[2], M. S. Schöffler[3], U. Spillmann[1], A. Surzhykov[5], S. Trotsenko[1,5], A. B. Voitkiv[11], X. L. Zhu[8], and Th. Stöhlker[1,5]

[1]GSI Darmstadt, Germany, [2]Universität Giessen, Germany, [3]Universität Frankfurt, Germany, [4]JKU Kielce, Poland, [5]Helmholtz Institut Jena, Germany, [6]INFN Catalania, Italy, [7]IMP Lanzhou, China, [8]LMU München, Germany, [9]CONICET Rosario, Argentina, [10]CIMAP-CIRIL-Ganil Caen, France, [11]Universität Düsseldorf, Germany

Synopsis For the collisions system $U^{88+} + N_2$ @ 90 MeV/u the emitted cusp electrons have been studied, and the three processes contributing to the electron cusp were compared to individual theoretical calculations.

In collisions of heavy highly-charged projectile ions with atomic targets, the energy distribution of the emitted electrons is a characteristic observable for the underlying elementary processes. At the experimental storage ring ESR of the heavy-ion accelerator facility GSI, a dedicated magnetic electron spectrometer was built downstream from the gas-jet target, which enabled the measurement of high-energetic electrons emitted in ion-atom collisions within a small cone in the forward direction. Using this electron spectrometer in combination with detectors for emitted x rays and charge-exchanged projectiles, the study of the collision system $U^{88+} + N_2$ @ 90 MeV/u revealed three processes resulting in the emission of electrons with a velocity similar to the projectile velocity, i.e., cusp electrons:

(a) The process of **electron loss to continuum** (ELC) corresponds to the ionization of an electron from the projectile into the projectile continuum during the collision with the target,

$$U^{88+} + N_2 \rightarrow U^{89+} + [N_2]^* + e^-.$$

For the ELC, the measured spectrum was compared to first-order perturbation theory using fully-relativistic Dirac wavefunctions [1].

(b) The process of **electron capture to continuum** (ECC) corresponds to the capture of a target electron into the projectile continuum, while the excess energy is carried away by the recoil of the generated target ion:

$$U^{88+} + N_2 \rightarrow U^{88+} + \left[N_2^+\right]^* + e^-.$$

For the ECC, the measured spectrum was compared to calculations in the impulse approximation using semi-relativistic Sommerfeld-Maue wavefunctions, and to calculations in the continuum-distorted-wave approach [2].

(c) The process of **radiative electron capture to continuum** (RECC) corresponds to the capture of a target electron into the projectile continuum, while the excess energy is carried away by a photon:

$$U^{88+} + N_2 \rightarrow U^{88+} + \left[N_2^+\right]^* + e^- + \gamma.$$

This latter process can be seen as the high-energy endpoint of bremsstrahlung studied in inverse kinematics. For the RECC, the measured spectra were compared to calculations in the impulse approximation using semi-relativistic Sommerfeld-Maue wavefunctions, and to calculations using fully-relativistic Dirac wavefunctions [3].

Within this study it was shown that each of the three processes is characterized by a unique electron cusp shape, being a clear messenger for the underlying charge-exchange mechanism.

References

[1] P.-M. Hillenbrand *et al.* 2014 *Phys. Rev. A* **90** 042713

[2] P.-M. Hillenbrand *et al.* 2015 *Phys. Rev. A* **91** 022705

[3] P.-M. Hillenbrand *et al.* 2014 *Phys. Rev. A* **90** 022707

[1]E-mail: P.M.Hillenbrand@gsi.de
[2]E-mail: S.Hagmann@gsi.de

Content from this work may be used under the terms of the Creative Commons Attribution 3.0 licence. Any further distribution of this work must maintain attribution to the author(s) and the title of the work, journal citation and DOI.
Published under licence by IOP Publishing Ltd

State-selected studies of electron capture by low-energy Ne^{q+} (q = 3 – 5) ions from CO_2 and H_2O molecules

E. Y. Kamber*,[1], O. Abu-Haija*,[2], J. Harris*, A. Hasan*,[3], and A. Kayani*

*Department of Physics, Western Michigan University, Kalamazoo, MI 49008, USA
[2] Physics Department, Tafila Technical University, P. O. Box 179, Zip Code 66110, Tafila, Jordan
[3] American University of Sharjah, Department of Physics, P. O. Box 26666, Sharjah, UAE

Synopsis Translational energy-gain spectroscopy has been used to measure energy spectra and absolute total cross sections for single- and double-electron capture processes in low-energy collisions of Ne^{q+} (q = 3- 5) ions with H_2O and CO_2 at collision energies between 15 and 500 qeV.

The study of single- and double-electron capture in low-energy collisions of multiply charged ions with molecules has grown tremendously in recent years, primarily due to the importance of these processes in many ionized media including plasmas and astrophysical environments. In addition, H_2O and CO_2 are important components of many planetary atmospheres, and also found in earth's atmosphere [1].

In the present work, energy-gain spectra and absolute total cross sections for single-and double-electron capture processes in collisions of Ne^{q+} (q = 3 – 5) recoil ions with CO_2 and H_2O targets at impact energies between 15 and 500 qeV have been studied using the translational energy-gain spectroscopy technique. The data have been obtained on a differential energy spectrometer, which has been described by Abu-Haija et al. [2].

Figure 1 shows the projectile charge state dependence of the cross section ratio of double to single-electron capture by Ne^{q+} (q = 3 – 6) ions from CO_2 and H_2O for two collision energies. The ratio increases slowly with charge state of the projectile on average from 0.1±0.04 for q = 3 to 0.55±0.18 for q = 6 in disagreement with the calculations based on an extension of the classical over-the-barrier model for multi-electron capture [3], which predicts that the ratio is independent of the charge state. This results may indicate a one-step mechanism for the double-electron capture in which the electrons are captured simultaneously in one step.

The energy dependence of total cross sections for single- and double-electron capture are also measured and found to slowly increase with increasing impact energy, a pattern that can be understood by the reaction window, which broadens with increasing energies and therefore increase the probability of capture channels with large Q-values.

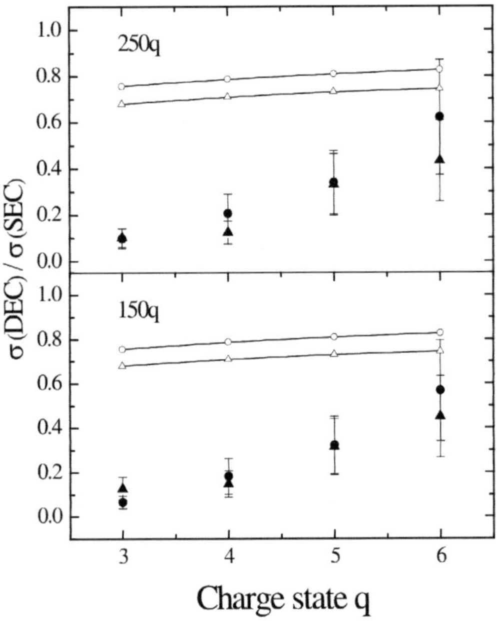

Figure 1. Projectile charge state dependence of the ratio of double-electron capture (DEC) cross sections to single-electron capture (SEC) cross sections in collisions of Ne^{q+} (q = 3 – 5) ions with H_2O and CO_2. Present work: (•) CO_2: (▲) H_2O. Extension of the classical over-the-barrier model: (○) CO_2; (Δ) H_2O.

References

[1] Cravens T E 1997 *Geophys. Res. Lett.* **25** 105.
[2] Abu-Haija O, Kamber E Y, Ferguson S M and Stolterfoht N 2005 *Phys. Rev. A* **72** 42701.
[3] Barany et al 1985 *Nucl. Instrum. Methods* **9** 421.

E-mail: emanuel.kamber@wmich.edu

Content from this work may be used under the terms of the Creative Commons Attribution 3.0 licence. Any further distribution of this work must maintain attribution to the author(s) and the title of the work, journal citation and DOI.
Published under licence by IOP Publishing Ltd

XXIX International Conference on Photonic, Electronic, and Atomic Collisions (ICPEAC2015) IOP Publishing
Journal of Physics: Conference Series **635** (2015) 022007 doi:10.1088/1742-6596/635/2/022007

Inelastic processes in low-energy Na^+ - Ar, K^+ - Ar and K^+ - He collisions

R.Lomsadze[1], M.Gochitashvili[1], N.Mosulishvili[1], D.Kuparashvili[1], and R.Kezerashvili[2, 3]

[1]Tbilisi State University, Department of Exact and Natural Sciences, Tbilisi, 0179, Republic of Georgia.
[2]New York City College of Technology, the City University of New York, Brooklyn, New York 11201, USA.
[3]The Graduate School and University Center, the City University of New York, New York, New York 10016, USA.

Synopsis The combined experimental approaches, that include a refined version of a condenser plate method, angle-and-energy-dependent collection of product ions, energy loss, and optical spectroscopy, are applied to measure the absolute value of the charge-exchange, ionization, stripping, and excitation cross sections for the collisions of Na^+, K^+ ions in the energy range of 0.7 – 10 keV with the He and Ar atoms. The correlation diagrams of the $(NaAr)^+$ and $(KHe)^+$ systems have been employed to discuss the mechanisms for these processes.

This work presents a multifaceted experimental study of collisions of Na^+, K^+ ions in the energy range 0.7-10 keV with He and Ar atoms. The combined experimental approaches, that include a refined version of a condenser plate method, angle-and-energy-dependent collection of product ions, energy loss, and optical spectroscopy, are applied to measure the absolute value of the charge-exchange, ionization, stripping, and excitation cross sections [1]. The correlation diagrams of the $(NaAr)^+$ and $(KHe)^+$ systems have been employed to discuss the mechanisms for these processes.

The charge – exchange processes, for all considered collision pairs, results from the capture of the electron to the ground state of the atom in the region of psevdocrossing of the potential curves of $^1\Sigma$ symmetry.

The primary ionization mechanism in K^+ - He collisions is the filling as a result of Σ- Π and Σ-Π-Δ transitions, of quasimolecular autoionization term and their decay in the region of the transition into the continuum (in the stage in which a quasimolecule exists).

Mechanism of ionization during Na^+ -Ar collisions is characterized by the liberation of predominantly slow electrons with energies E< 15 eV and connected with the two excited states of autoionization states of Ar ($3p^4$ (^1D) $4s^2$)) and with the decay in isolated atom.

Common mechanism of the population for the potassium ion and atom into 4p states in K^+ - He are explained.

The usage of the collision spectroscopy method has permitted to explore exceptionally high excited states of helium atom in the energy – loss spectrum and explain small value of cross section in case of Na^+ -Ar collision pair.

Reference

[1] R.A. Lomsadze et al 2013 *Phys. Rev. A***87**, 042710.

E-mail: lomsadze86@hotmail.com

Content from this work may be used under the terms of the Creative Commons Attribution 3.0 licence. Any further distribution of this work must maintain attribution to the author(s) and the title of the work, journal citation and DOI.
Published under licence by IOP Publishing Ltd

Enhancement of antihydrogen formation in antiproton collisions with excited-state positronium

A. S. Kadyrov[a, 1], C. M. Rawlins[a], A. T. Stelbovics[a], I. Bray[a], and M. Charlton[b]

[a] ARC Centre for Antimatter-Matter Studies, Curtin University, GPO Box U1987, Perth, WA 6845, Australia
[b] Department of Physics, College of Science, Swansea University, SA2 8PP, United Kingdom

Synopsis Antihydrogen formation in positronium scattering on antiprotons is investigated using the two-centre convergent close-coupling method. A several orders of magnitude enhancement in the formation of antihydrogen is found when positronium is in an excited state. The effect is greatest at the lowest energies considered which encompass those achievable in experiment. This suggests a practical approach to creating neutral antimatter for testing its interaction with gravity and for spectroscopic measurements.

Interactions of antiprotons with positronium can lead to antihydrogen formation via the reaction $\bar{p} + \mathrm{Ps}(nl) \to \bar{\mathrm{H}}(n'l') + e^-$. This process has been investigated by Humberston *et al* [1] for Ps($n = 1$) using the variational method. The study was limited to $\bar{\mathrm{H}}$ formation in the 1s state ($n' = 1$). There are few calculations involving Ps($n > 1$), and none as far as we are aware extend to low energies of practical interest for the current AEgIS and GBAR experiments [2]. Calculations in the unitarised Born approximation (UBA) [3] included Ps($n \leq 4$) initial states and extended to within 0.1 eV of threshold. However, such approximations are high-energy ones and, therefore, they cannot yield accurate results at such low energies.

We addressed the problem using the two-centre convergent close-coupling (CCC) method [4] and included all initial states with $n \leq 3$. The total wavefunction was expanded in both the atomic and positronium pseudostates with the resulting close-coupling equations solved in momentum space as coupled Lippmann-Schwinger equations. The pseudostates were generated by diagonalising the atomic and positronium Hamiltonians using the orthogonal Laguerre basis. We performed calculations with increasing basis size until the results converged to within ±5% at most energies. Such calculations have a total of 68 states, 34 for each center. The cross sections are obtained for Ps(nl) transitions to all open $\bar{\mathrm{H}}(n'l')$ states. We add the $\bar{\mathrm{H}}$-formation cross sections over $n'l'$, and present them for the summed over l Ps initial states. The results are shown in figure 1.

For the lowest energies presented we see several orders of magnitude cross section increase in going from Ps($n = 1$) to Ps($n = 2$) initial states. A further order of magnitude increase is obtained by starting with Ps($n = 3$). This indicates that antihydrogen creation via interaction of Ps in excited states with slow antiprotons is a very practical proposition. More details and more results will be presented at the conference.

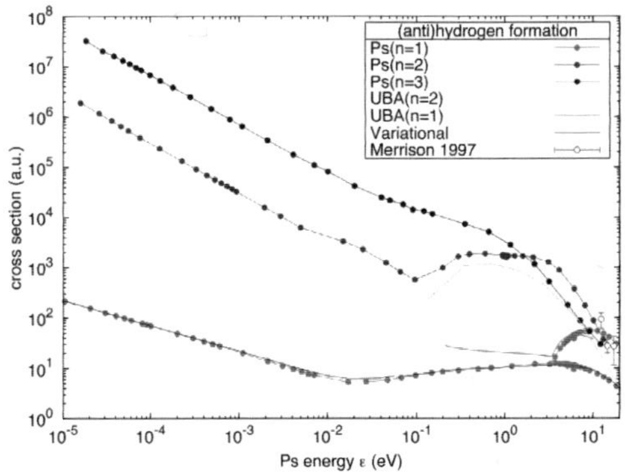

Figure 1. Total cross sections for $\bar{\mathrm{H}}$ formation in $\bar{p}$ collisions with Ps, in an initial state n, calculated using the CCC method. For Ps(1s), the variational calculations [1] are for $\bar{\mathrm{H}}$ formation in the 1s state only (corresponding CCC results are shown by unconnected points), while the UBA [3] and the CCC calculations are for $\bar{\mathrm{H}}$ formation in all open states. The experimental points are for H formation in p collisions with Ps [5].

This work was supported by the Australian Research Council and EPSRC (UK).

References

[1] J. W. Humberston *et al.* 1997 *J. Phys. B* **30** 2477; J. W. Humberston and P. Van Reeth 2015 (private communication)

[2] http://aegis.web.cern.ch and gbar.web.cern.ch

[3] J. Mitroy 1995 *Phys. Rev. A* **52** 2859

[4] A. S. Kadyrov and I. Bray 2002 *Phys. Rev. A* **66** 012710

[5] J. P. Merrison *et al.* 1997 *Phys. Rev. Lett.* **78** 2728

[1]E-mail: A.Kadyrov@curtin.edu.au

Content from this work may be used under the terms of the Creative Commons Attribution 3.0 licence. Any further distribution of this work must maintain attribution to the author(s) and the title of the work, journal citation and DOI.
Published under licence by IOP Publishing Ltd

A time-dependent analysis of transfer ionization in one-dimensional kinematics

Vladialv V. Serov[*1] **and Anatoli S. Kheifets**[†2]

[*] Department of Theoretical Physics, Saratov State University, 83 Astrakhanskaya, Saratov 410012, Russia
[†] Research School of Physical Sciences, The Australian National University, Canberra ACT 2601, Australia

Synopsis We analyze a transfer ionization (TI) reaction in the fast proton-helium collision $H^+ + He \to H^0 + He^{2+} + e^-$ by solving a time-dependent Schrödinger equation (TDSE) under the classical projectile motion approximation in one-dimensional kinematics. By comparing results of the TDSE and various perturbative Born series calculations, we conclude that the recent discrepancies of experimental and theoretical data may be attributed to deficiency of the Born models used by other authors. We demonstrate that the correct Born series for TI should include the momentum space overlap between the double ionization amplitude and the wave function of the transferred electron.

In a recent work [1], we reported calculations of transfer ionization in a fast proton scattering on the helium atom

$$H^+ + He \to H^0 + He^{2+} + e^-$$

for the paramenters of experiments [2, 3]. Our approach is based on the classical approximation for both a nucleus and a projectile and a solution of a time-dependent Schrödinger equation (TDSE) for two electrons in one-dimensional (1D) kinematics. The classical projectile motion approximation (CPA) is very accurate for the present problem because the proton wavelength $\lambda_p \sim 10^{-3}$ a.u. is much smaller than the atom size.

To gain a deeper physical insight into specific mechanisms of the TI reaction, we also performed perturbative 1D calculations which mimicked the plane-wave first Born calculations (PWFBA) performed by other authors [2, 3, 4]. We identified a strong effect of the non-orthogonality of the initial and final states which lead to a spurious unphysical term. This term may be responsible for poor performance of the PWFBA. The authors of Ref. [2, 3, 4] interpreted this term as a first-Born description of the second-Born binary encounter process.

Among various approximate perturbative TI schemes, the results that are closest to the non-perturbative CPA calculations are obtained by the overlap

$$A(k) = \sqrt{2}\int_{-\infty}^{\infty} u^*_{\mathrm{H}}(k_1 - v_H) A_{\mathrm{DI1B}}(k_1, k) dk_1$$

of the first-Born double ionization amplitude $A_{\mathrm{DI1B}}(k_1, k_2)$ with the momentum profile of the final state wave function $u_{\mathrm{H}}(k)$ of a fast hydrogen atom. This indicates that the most probable scenario of TI involves double ionization (TIDI) and subsequent capture of those of the ejected electrons which falls into the attractive potential well of the proton by matching its velocity.

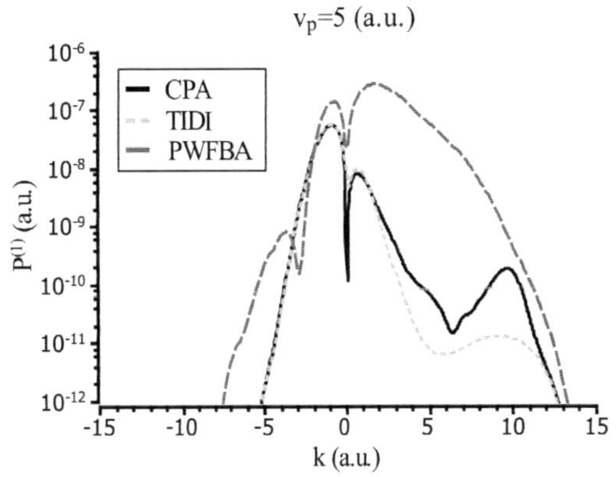

Figure 1. The TI probability density $P^{(1)}(k)$ as a function of the ejected electron momentum for the proton energy $E_p = 630$ keV. Various calculations are displayed with the following line styles: the CPA (black solid line), the PWFBA (red dashed line) and the TIDI (green dotted line).

References

[1] V. V. Serov and A. S. Kheifets 2014 *Phys. Rev. A* **90** 062706

[2] M. S. Schöffler *et al* 2013 *Phys. Rev. A* **87** 032715

[3] M. S. Schöffler *et al* 2013 *Phys. Rev. A* **88** 042710

[4] S. Houamer *et al* 2010 *Phys. Rev. A* **81** 032703

[1]E-mail: vladislav.serov@gmail.com
[2]E-mail: A.Kheifets@anu.edu.au

Content from this work may be used under the terms of the Creative Commons Attribution 3.0 licence. Any further distribution of this work must maintain attribution to the author(s) and the title of the work, journal citation and DOI.
Published under licence by IOP Publishing Ltd

Multiple ionization of Ar by alpha particles

C. C. Montanari* [1], **J. E. Miraglia*** [2]

* Instituto de Astronomía y Física del Espacio, Consejo Nacional de Investigaciones Científicas y Técnicas, Universidad de Buenos Aires, Ciudad Universitaria, C1428EGA, Buenos Aires, Argentina

Synopsis We present theoretical results on multiple ionization of Ar by He^{2+} by employing the continuum distorted-wave eikonal initial state together with photoionization branching ratios to include Auger-type contributions. We also analyze the inner-shell ionization cross sections (K and L-shells of Ar). The high energy experimental data is correctly described, showing differences with proton impact tendency in the post-collisional region. At intermediate impact energies (direct ionization region) the single and double ionization are described rather well, but we note certain overestimation for higher final charge states.

We present new results for single to quadruple ionization cross sections of Ar bombarded by He^{2+} ions in the intermediate and high energy range. Different experimental measurements involve the He-Ar system, i.e [1-2]. Also of interest are the very recent measurements on multiple ionization of Ar by Li^{+2} ions by the Rio de Janeiro group [3]. Our goal is to analyze the +2 charge projectiles in view of previous double ionization results by proton impact, which show an overestimation at high energies (post-collisional region). Our results are compared with the experimental data available considering only pure multiple-ionization cross sections (no capture contribution). Ionization cross sections of specific deep shells (L and M) are also evaluated, in order to test the values that contribute to the post-collisional ionization (PCI) via Auger processes.

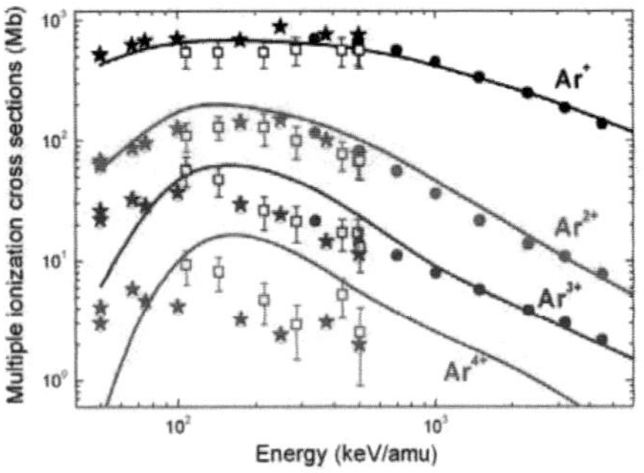

Figure 1. Multiple ionization of Ar by He^{2+}. Symbols: filled stars [1], filled circles [2], hollow squares [3] for Li^{+2} on Ar.

The comparison with the highly-charged ion production data requires the calculation of total ionization cross sections (direct plus PCI). We follow previous works [4] by employing the CDW-EIS [5], and including the PCI through the experimental branching ratios of single to multiple vacancy production.

In figure 1 we display single to quadruple ionization cross sections, together with the experimental data available. Only pure multiple-ionization cross sections are considered (no capture contribution). We also analyze the ionization of specific deep shells of Ar (K and L). In figure 2 we display these results and compare with the ECPSSR predictions [6] obtained with the ISICS11 code by Cipolla [7], and with the coupled-state results for L-shell ionization by Martir *et al* [8].

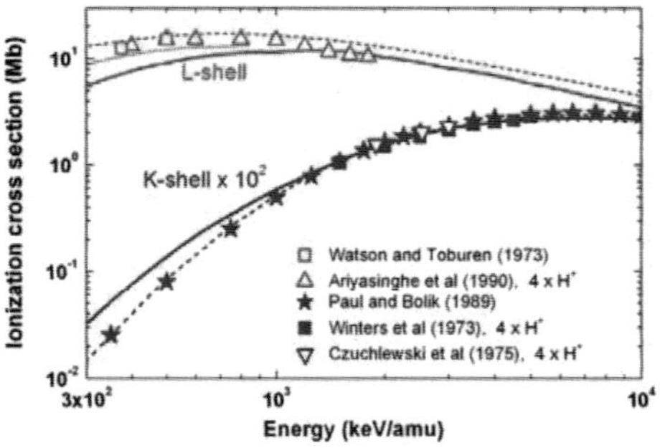

Figure 2. K and L-shell ionization cross sections of Ar by He^{2+}. Present results (solid lines), ECPSSR values [6-7] (dashed-lines), coupled state calculations by Martir et al [8] (dotted line). Details of the symbols in the inset.

References

[1] DuBois R D 1987, Phys. Rev A **36**, 2585
[2] Andersen *et al* 1987, Phys. Rev A **36**, 3612
[3] Losqui *et al* 2014, J. Phys. B **47** 045247
[4] Montanari C C *et al* 2012, J. Phys. Conf. Ser. **388** 012036
[5] Fainstein P D *et al* 1988, J. Phys. B **21** 287
[6] Lapicki G 2002, Nucl. Instrum. and Meth. B **189** 8
[7] Cipolla S J 2013, Comp. Phys. Comm **184** 2230.
[8] Martir M H *et al* 1982, J. Phys. B **15** 2405

[1] E-mail: mclaudia@iafe.uba.ar
[2] E-mail: miraglia@iafe.uba.ar

Content from this work may be used under the terms of the Creative Commons Attribution 3.0 licence. Any further distribution of this work must maintain attribution to the author(s) and the title of the work, journal citation and DOI.
Published under licence by IOP Publishing Ltd

Journal of Physics: Conference Series **635** (2015) 022011 doi:10.1088/1742-6596/635/2/022011

The use of Parr temperature within the Lindhard dielectric function

J. E. Miraglia * [1], C. C. Montanari * [2]

* Instituto de Astronomía y Física del Espacio, Consejo Nacional de Investigaciones Científicas y Técnicas, Universidad de Buenos Aires, Ciudad Universitaria, C1428EGA, Buenos Aires, Argentina

Synopsis The Parr distribution function is used within the Lindhard-Levine dielectric response function to calculate the energy loss of protons on the rare gases. This distribution is a function of the local temperature, which depends on the local kinetic energy. Present results compare quite well with the first Born approximation at high energies and open the way to use the Wigner distribution function.

In the well known local plasma approximarion (LPA), the term including the susceptibility is integrated on the momentum space with a weight given by the Thomas-Fermi (TF) distribution function

$$F_{TF}(\vec{r},\vec{k}) = \frac{2}{(2\pi)^3}\Theta(k_F(r)-k),$$

where $k_F = (3\pi\, n)^{1/3}$ is the Fermi velocity and $n=n(r)$ is the local electron density. The f-sum rule implies the normalization of the phase-space distribution, $\int d\vec{r}\, d\vec{k}\, F_{TF}(\vec{r},\vec{k}) = N$, with N being the number of electrons. Instead, in the present contribution we essay the Parr distribution function [1]

$$F_P(\vec{r},\vec{k}) = \frac{n(r)}{(2\pi\, k_B\, T(r))^{3/2}}\,\mathbf{exp}\left(-\frac{k^2}{2k_B\, T(r)}\right)$$

where k_B is the Boltzman constant and T(r) is the local temperature defined in terms of the local kinetics energy $t(r)$ given by

$$T(r) = \frac{2}{3\,n(r)\;k_B}\;t(r)\,.$$

The energy loss per atom due to ionization processes is calculated using the simplest local kinetics energy given by $t_{TF}(r) = 2.87\,n(r)^{5/3}$. The complexity of the calculation is equivalent to the space-independent Gaussian distribution developed by Kaneko [2]. The great advantage is that the Parr distribution makes possible to calculate probabilities as a function of the impact parameter and, in consequence, to estimate multiple process under the independent electron model. In Figure 1 we display present results for the energy loss per atom, for protons on the rare gases, using the Parr distribution within the Lindhard-Levine dielectric function [3]. We also display the values obtained by using the first Born approximation, and the shell-wise local plasma approximation (SLPA) [4] (TF distribution, and the atom in a shell to shell description). The use of Parr temperature does not necessarily improve the TF distribution, but opens the way to use the Wigner distribution function, which would provide the best attempt to deal with the local plasma approximation.

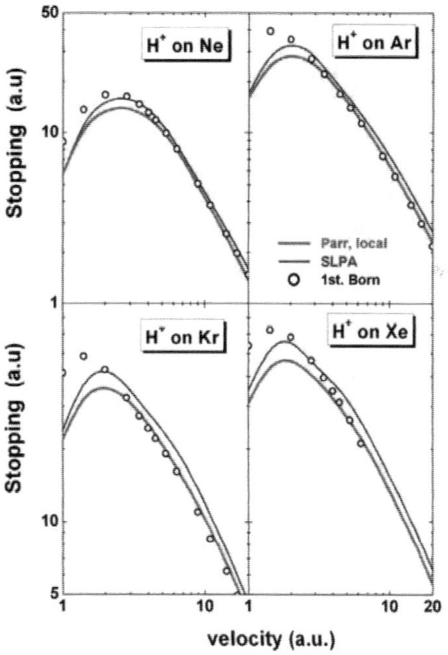

Figure 1. Stopping cross sections per atom for protons on the rare gases.

References

[1] R. Parr *et al*, Phys Rev Lett. **56**, 1555 (1986).
[2] T. Kaneko, Phys. Stat. Sol. **156**, 49 (1989).
[3] Z. H. Levine *et al*, Phys. Rev. B **25**, 6310 (1982).
[4] C.C. Montanari, J.E. Miraglia, Advances in Quantum Chemistry **65**, 165-202 (Ed. Dz. Belkic, Elsevier, Amsterdam, 2013).

[1]E-mail: miraglia@iafe.uba.ar
[2]E-mail: mclaudia@iafe.uba.ar

Content from this work may be used under the terms of the Creative Commons Attribution 3.0 licence. Any further distribution of this work must maintain attribution to the author(s) and the title of the work, journal citation and DOI.
Published under licence by IOP Publishing Ltd

Atomic Ionization by Dark Matter Particles

V. V. Flambaum*[1], V. A. Dzuba*, M. Pospelov †‡, A. Derevianko♯, B. Roberts*

* School of Physics, The University of New South Wales, Sydney NSW 2052, Australia
†Department of Physics and Astronomy, University of Victoria, Victoria, British Columbia, V8P IAI, Canada
‡Perimeter Institute for Theoretical Physics, Waterloo, Ontario, N2J 2W9, Canada
♯Department of Physics, University of Nevada, Reno, Nevada 89557, USA

Synopsis Using the relativistic Hartree-Fock approximation, we calculate the rates of atomic ionization by absorption of axions and other Dark matter candidates. We also calculate atomic ionization by scattering of heavy Dark matter particles on electrons. The results are used to interpret possible detection of Dark matter by DAMA collaboration.

Using the relativistic Hartree-Fock approximation, we [1, 2] calculate the rates of atomic ionization by absorption of pseudoscalar particles (called axions for short, although they may differ from QCD axions by their mass) in the mass range from 0 to $\sim$ 50 keV. We present numerical results for atoms relevant for the direct dark matter searches (*e.g.* Na, Ar, Ge, I and Xe), as well as the analytical formula which fits numerical calculations with few per cent accuracy and may be used for multi-electron atoms, molecules and condensed matter systems.

The results for ionization cross-section are presented in a form

$$\sigma_a(\epsilon_a) = \left(\frac{\epsilon_0}{f_a}\right)^2 \frac{c}{v} K(\epsilon_a) a_0^2,$$

where ϵ_a is the energy of a dark matter particle, ϵ_0 is an energy scale, $(1/f_a)$ is the axion-electron interaction strength constant, c is speed of light, v is the particle velocity in the laboratory frame, $a_0 = 0.52918\times10^{-8}$ cm is Borh radius, and $K(\epsilon_a)$ is dimensionless function of the energy which is found from the calculations.

Fig. 1 shows as an example the $K(\epsilon_a)$ function for Xe atom. Solid line represents the results of the Hartree-Fock calculations, dotted line - the result of analytical fitting of the calculations. The fitting is done for $18 < Z < 60$ and for dark matter particle energies from 10 to 50 keV.

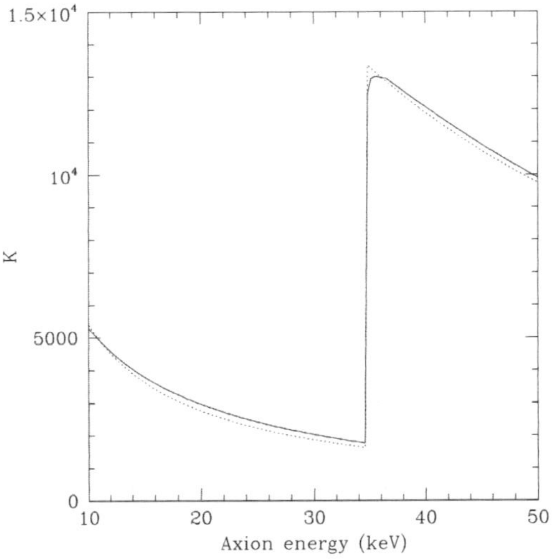

A possible detection of Dark matter by DAMA collaboration may be, in principle, explained by the ionization of Na and I atoms in NaI compound due to scattering of heavy Dark matter particles on electrons. We calculate the cross sections for these processes.

References

[1] V. A. Dzuba, V. V. Flambaum, M. Pospelov 2010 *Phys. Rev. D* **81** 103520.

[2] A. Derevianko, V. A. Dzuba, V. V. Flambaum, M. Pospelov 2010 *Phys. Rev. D* **82** 065006 .

[1]E-mail: v.flambaum@unsw.edu.au

Content from this work may be used under the terms of the Creative Commons Attribution 3.0 licence. Any further distribution of this work must maintain attribution to the author(s) and the title of the work, journal citation and DOI.
Published under licence by IOP Publishing Ltd

Journal of Physics: Conference Series **635** (2015) 022013   doi:10.1088/1742-6596/635/2/022013

Impact of Debye plasmas environments on He-like ions and He-atom collisions

Mukesh Kumar Pandey* [1], **Yew Kam Ho*** [2]

*Institute of Atomic and Molecular Sciences, Academia Sinica, P.O. Box 23-166, Taipei, Taiwan

Synopsis The impact of Debye plasmas on the electron capture and ionization process is investigated in multiply charged He-like ion (Li^+, Be^{2+}, B^{3+}, C^{4+}, N^{5+}, O^{6+}) and He atom collisions.

The effect of Debye plasmas on plasma-embedded atomic systems has received considerable attention from both theoretical and experimental research groups in recent years. Lately, we have reported the effect of Debye plasma conditions on charge transfer and ionization processes of fully stripped ion O^{8+} and He-like systems colliding with hydrogen atom [1, 2]. In the present work, we extend these calculations to a helium target. Moreover, as for a helium target very few works have been done in the environments of hot dense plasmas, especially for the models that take care of the plasma screening effects on many-electron systems, where Coulomb interactions between charged particles may be significantly screened. It is worthwhile to make an attempt to understand such interactions that affect the collision processes under Debye plasma environments.

In the present work, Classical Trajectory Monte Carlo (CTMC) method has been used. Here, we have considered the model potential suggested by Schweizer *et. al.* [3] representing the e - A^{q+} interaction in the form:

$$V_{\mathrm{mod}\,el}(r) = -\frac{1}{r}\left(\tilde{Z} + (Z - \tilde{Z})\exp(-a*r) + b*r\exp(-c*r)\right) \quad (1)$$

The basic idea of model potential $V_{\mathrm{mod}\,el}(r)$, with r being the radial coordinate, is to simulate the multi-electron core interaction with the valence electron by an analytic modification of the Coulomb potential. When the electron–ion interaction is described by the Debye Hückel potential (e is the unit charge)

$$V(r) = (V_{\mathrm{mod}\,el}(r))e^{-r/r_d}, \quad (2)$$

where the screening parameter r_d is related to the plasma temperature (T_e) and density (n_e) by $r_d = (K_B T_e / 4\pi e^2 n_e)^{1/2}$, with r_d being the Debye length and K_B is the Boltzmann constant.

The total charge exchange cross sections in the screened cases are smaller than the cross sections of the unscreened cases, and their magnitude decreases with decreasing Debye length.

[1] E-mail : pandey@pub.iams.sinica.edu.tw
[2] E-mail : ykho@pub.iams.sinica.edu.tw

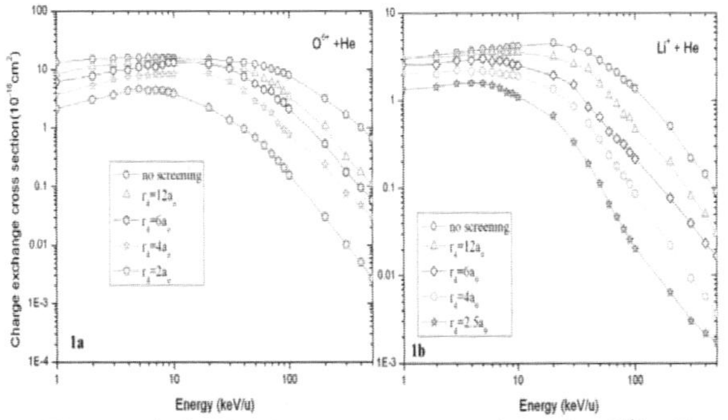

Figures 1a-b. Total capture cross sections for O^{6+}+ He, and Li^+ + He collisions systems, respectively, for screening parameter r_d = 12a$_0$, 6a$_0$, 4a$_0$, and 2a$_0$ in the energies ranging from 1 to 500 keV/u.

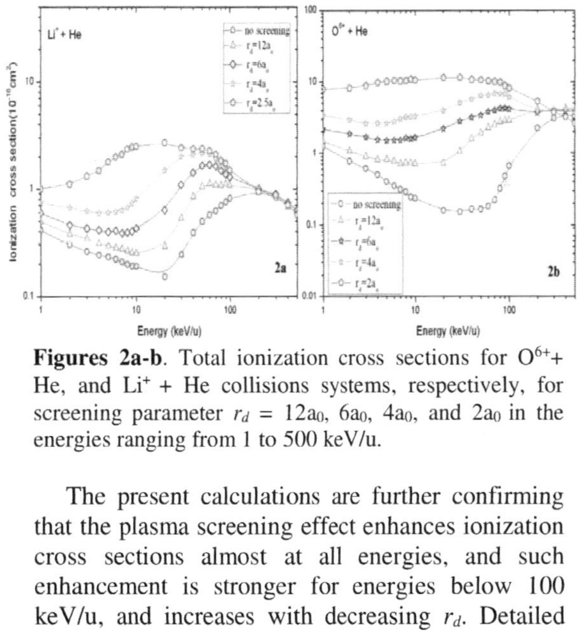

Figures 2a-b. Total ionization cross sections for O^{6+}+ He, and Li^+ + He collisions systems, respectively, for screening parameter r_d = 12a$_0$, 6a$_0$, 4a$_0$, and 2a$_0$ in the energies ranging from 1 to 500 keV/u.

The present calculations are further confirming that the plasma screening effect enhances ionization cross sections almost at all energies, and such enhancement is stronger for energies below 100 keV/u, and increases with decreasing r_d. Detailed results will be presented at the conference.

Acknowledgement

Work supported by Ministry of Science and Technology of Taiwan.

References:

[1] M. K. Pandey, Y.-C. Lin and Y. K. Ho 2013 *Phys. of Plasmas* **19** 062104
[2] M. K. Pandey, Y.-C. Lin and Y. K. Ho 2012 *Phys. of Plasmas* **20** 022104
[3] W. Schweizer *et. al.* 1999 *Atomic Data and Nuclear Data Table* **72** 33

Content from this work may be used under the terms of the Creative Commons Attribution 3.0 licence. Any further distribution of this work must maintain attribution to the author(s) and the title of the work, journal citation and DOI.
Published under licence by IOP Publishing Ltd

Theoretical study of very low-energy collision in Coulombic three-body systems

Y. Zhou[1], O. I. Tolstikhin[2], S. Watanabe[1] and T. Morishita[1,*]

[1] Department of Engineering Sciences, University of Electro-Communications, 1-5-1 Chofu-ga-oka, Chofu-shi, Tokyo 182-8585, Japan

2. Moscow Institute of Physics and Technology, Dolgoprudny 141700, Russia

Synopsis We present a systematic study for Coulombic three-body systems in very low-energy regime with mass ratio varying over several orders of magnitudes. Accurate phase shifts and scattering lengths are obtained by the Hyperspherical elliptic coordinate method. We identify the mass ratio dependence of the scattering lengths.

We present a theoretical study for very low-energy collisions in symmetric Coulombic three-body systems, where two of the constituting particles are identical. We solve the time-independent Schrödinger equation using the hyperspherical elliptic coordinate (HSE) method [1], and calculate phase shifts and extract scattering lengths for various systems. The HSE method is very powerful in studying the three-body problem. It is suitable to calculate general three-body systems with arbitrary masses just by treating the charges and masses as input parameters. In the HSE method the hyperspherical potentials and adiabatic channel functions in the internal region can be obtained efficiently using the quasi-separability among the hyperangles. The nonadiabatic couplings among the hyperspherical channels are treated with the slow/smooth variable discretization (SVD) method [2] in combination with the R-matrix propagation technique. In the asymptotic region, the solutions are obtained by using the Gailitis method [3]. The two-dimensional matching procedure is implemented to match the internal and asymptotic solutions in order to extract scattering information.

We compare the resulting phase shifts for the real physical systems, such as H^-, Ps^- and H_2^+, with previous data, and confirm the numerical accuracy. We also analyze the mass ratio dependence of the scattering length in symmetric Coulomb three-body systems. We find that the scattering length jumps from $-\infty$ to $+\infty$ at mass ratios where a new bound state appears and then it decreases monotonically between the two jumps (see Figure 1). This behavior is closely related to the Levinson's theorem. We will discuss more details at the conference.

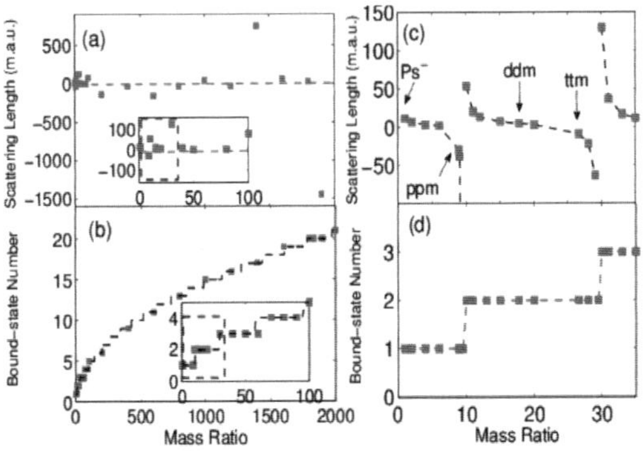

Figure 1. Scattering lengths (a) and number of bound states (b) as a function of mass ratio. (c) and (d) enlarge the left part of (a) and (b) [indicated by the blue dashed box].

References

[1] O. I. Tolstikhin *et al*, 2001, *Phys. Rev. A* **63**, 062705.
[2] O. I. Tolstikhin *et al*, 1996, *J. Phys. B* **29**, L389.
[3] M. Gailitis, 1976, *J. Phys. B* **9**, 843.

E-mail: toru@pc.uec.ac.jp

Content from this work may be used under the terms of the Creative Commons Attribution 3.0 licence. Any further distribution of this work must maintain attribution to the author(s) and the title of the work, journal citation and DOI.
Published under licence by IOP Publishing Ltd

Journal of Physics: Conference Series **635** (2015) 022015 doi:10.1088/1742-6596/635/2/022015

Analytical formulas for differential cross sections for ejection of electrons in ionization of water by protons in the PWBA and ECPSSR

Gregory Lapicki

Department of Physics, East Carolina University, Greenville, NC 27858, USA

Synopsis. For comparison with measured ionization cross sections as a function of the ejected electrons from water by protons, analytical formulas of the plane-wave Born approximation (PWBA) and the ECPSSR theory -- that accounting for the energy-loss (E) and Coulomb-deflection (C) protons and perturbed-stationary states (PSS) and relativistic (R) of electrons goes beyond the PWBA -- are evaluated analytically. As opposed to the PWBA that overestimates the data with the increasing energy of the ejected electrons, the ECPSSR is in good agreement with the measured spectra of the electrons ejected in with keV to MeV energies.

Single-differential cross sections (SDSC) with respect to the energy transfer in production of secondary electrons are required in detailed modeling of interaction of ionizing radiation with matter; in particular, in proton bombardment of biological materials of which water is a prime constituent. The existing SDCS calculations are typically developed in the first Born approach (FBA) of the plane-wave Born approximation (PWBA), done with semi-empirical corrections for the breakdown of the FBA with the decreasing proton energy and performed numerically [1,2].

Our analytical PWBA formulas based both on atomic and molecular composition of water are comprehensively compared with SDCS for electron ejection measured in ionization of water vapor by 0.15-4.2 MeV protons [3]. Analytical corrections of the ECPSSR theory [4] that goes beyond the FBA do not remove existing discrepancies with data obtained with slow protons and for the slowly ejected electrons. Predictions of the theory, however, are shown to result in increasingly smaller cross sections with the increasing energy of the ejected electrons and increasingly so in ionization by just a few hundred keV-protons. This could be a significant change as an input for Monte Carlo track structure codes in water and other biologically relevant materials.

As measurements with ice in conjunction with vapor data may support a bridge over troubled liquid water, new ideas that impose limitations on the range of momentum transfer in the derived analytical formulas for solid-state water will be discussed.

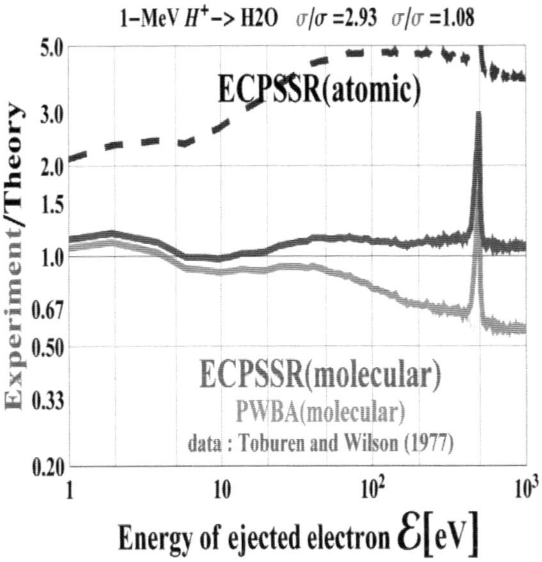

Fig.1. Ratios of experimental spectrum for SDSC in ionization of water vapor by 1-MeV protons [3] to the results of the plane-wave Born approximation [1,2] and the ECPSSR theory using atomic and molecular states for water [4].

References

[1] Miller, J.H., Wilson, W.E., Manson, S.T., 1985 Rad. Prot. Dos. **13**, 77.
[2] Dingfelder, M., Inokuti M., Paretzke, H.G., 2000 Rad. Phys. Chem. **59**, 255.
[3] Toburen, L.H., Wilson, W.E. 1977 J. Chem. Phys. **66**, 5202; Wilson W.E., Miller, J.H., Toburen, L.H., Manson, S.T., 1984 J. Chem. Phys. **80**, 5631.
[4] Brandt, W., Lapicki, G., 1981 Phys. Rev. A **23**, 1717.

Content from this work may be used under the terms of the Creative Commons Attribution 3.0 licence. Any further distribution of this work must maintain attribution to the author(s) and the title of the work, journal citation and DOI.
Published under licence by IOP Publishing Ltd

Journal of Physics: Conference Series **635** (2015) 022016 doi:10.1088/1742-6596/635/2/022016

Enhancement of the effects of space-time variation of the fundamental constants in collisions of cold atoms near Feshbach resonances

V. V. Flambaum*[1], C. Chin♯, J. Berengut*

* School of Physics, The University of New South Wales, Sydney NSW 2052, Australia
♯James Franck Institute and Department of Physics, The University of Chicago, Chicago, Illinois 60637, USA

Synopsis New results based on the quasar absorption data indicate variation of the fine structure constant alpha in space. These results are tested in different astrophysical measurements and laboratory experiments with atomic and nuclear clocks. Systems with enhanced effects of the variation may be especially important. Huge enhncement of the effects of space-time variation of the fundamental constants appears in collisions of cold atoms near Feshbach resonances.

Recent results for the variation of the fine structure constant alpha based on the quasar absorption spectra data indicate the variation of alpha in space [1]. The spatial variation can explain fine tuning of the fundamental constants which allows humans (and any life) to appear. We appeared in the area of the Universe where the values of the fundamental constants are consistent with our existence. There is an agreement between the results obtained using different telescopes and different redshifts. Also, now there are no contradictions between the results obtained by different groups. Other results and methods to search for the variation in astrophysics are also discussed. These astrophysical results may be used to predict the variation effects for atomic clocks which are very small and require improvement of the sensitivity by 1-2 orders of magnitude. This improvement may be achieved using 229Th nuclear clocks where the effect of the variation is hugely enhanced. There are also enhanced effects in multiply charged ions [2, 3], and certain atomic and molecular transitions.

There is a very strongly enhanced sensitivity to the fundamental constants in ultracold atomic and molecular systems near Feshbach resonances [4, 5]. Scattering length, which can be measured in Bose-Einstein condensate and Feshbach molecule experiments, is extremely sensitive to the variation of fundamental constants, in particular, the electron-to-proton mass ratio m_e/M_p. Based on single- and two-channel scattering models, we show how the variation of the mass ratio propagates to the scattering length. Our results suggest that effect of the variation of m_e/M_p may be enhanced up to 10^{12} times near a narrow magnetic or an optical Feshbach resonance. Derived formulas may also be used to estimate the isotopic shift of the scattering length.

References

[1] K. Webb, J. A. King, M. T. Murphy, V. V. Flambaum, R. F. Carswell, M. B. Bainbridge 2011 *Phys. Rev. Lett.* **107** 191101.

[2] J. C. Berengut, V. A. Dzuba and V. V. Flambaum, 2010 *Phys. Rev. Lett.* **105** 120801.

[3] J. C. Berengut, V. A. Dzuba, V. V. Flambaum, A.Ong 2011 *Phys. Rev. Lett.* **106** 210802.

[4] C. Chin, V.V. Flambaum 2006 *Phys. Rev. Lett.* **96** 96, 230801.

[5] K. Beloy, A. Borschevsky, V. V. Flambaum, and P. Schwerdtfeger 2011 *Phys. Rev. A* **84** 042117.

[1] flambaum@unsw.edu.au

Content from this work may be used under the terms of the Creative Commons Attribution 3.0 licence. Any further distribution of this work must maintain attribution to the author(s) and the title of the work, journal citation and DOI.
Published under licence by IOP Publishing Ltd

Possible Evidence of 3rd and 4th order Interactions Contributions to Double Ionization of Helium by Protons and Antiprotons

A. C. F. Santos*[1], R. D. DuBois†[2], and S. T. Manson#[3]

*Instituto de Física, Universidade Federal do Rio de Janeiro, PO 68528, 21941-972 Rio de Janeiro,RJ, Brazil
† Department of Physics, Missouri University of Science and Technology, Rolla, MO, 65409, USA
#Department of Physics & Astronomy, Georgia State University, Atlanta, GA, 30303, USA

SynopsisPublished cross sections for double ionization of helium by protons and antiprotons are analyzed in terms of the traditional interference between 1st and 2nd amplitudes. Between 3 and 10 a.u., the velocity dependence implies a dominant 2nd order interaction. For slower collisions it is shown that the velocity dependences are consistent with 3rd and 4th order interactions becoming important with these interactions having negative coefficients.

Recently, we analyzed a large data set of published cross sections for double ionization of helium [1] using the model proposed by McGuire [2] where three terms are considered, namely the 1st order shake off term, the 2nd order TS2 term, and an interference term. Since the interference term is of opposite sign for proton and antiproton impact, adding the double ionization cross sections for proton and antiproton impact removes the interference contribution. Under the assumption that the 1st order contribution is the same for proton and antiproton impact, the 2nd order TS2 contribution, σ_{2nd}, was isolated and found to have a $\ln(v)/v^4$ dependence at higher velocities. This was interpreted as implying sequential emission of two electrons via independent interactions. At lower velocities, σ_{2nd} reached a maximum and then decreased. These data and fit are shown by the black open squares and line in Fig. 1. Ref. 1 contains additional details plus references for the data used. Here we take a closer look at the lower velocity region and find that the velocity dependences in this region are consistent with 3rd and 4th order processes becoming increasingly important at lower impact en-ergies.

The absolute values for differences between the data and fit mentioned above are shown by the filled red circles with the red line illustrating these have a v^{-6} dependence above 3 a.u. The differences between the red data and line are shown by the filled blue stars which, as shown by the blue line, have a v^{-8} dependence. These velocity dependences suggest that for intermediate to high velocities, e.g., between ~ 3 and 10 a.u., the 2nd order TS2 mechanism dominates. But around the cross section maximum, the velocity de-pendence implies a 3rd order double ionization mech-anism having a negative coefficient becoming im-portant and eventually dominating below ~ 1.5 a.u. At still lower velocities, e.g., below 1 a.u., the v^{-8} de-pendence implies a 4th order double ionization mech-anism, also with a negative coefficient, becoming dominant.

Unfortunately, we do not know of any theoretical treatments of double ionization beyond 2nd order and can only state that in spite of uncertainties associated with error bars in the cross sections and several subtractions between data and fitted curves, the data in Figure 1 imply that some type of 3rd and 4th order double ionization mechanisms are present for low energy proton and antiproton impact. Finally, we do not believe our analysis is simply a fortuitous result of a polynomial fit to the data as this should yield both even and odd powers of v whereas we find only even powers to be present.

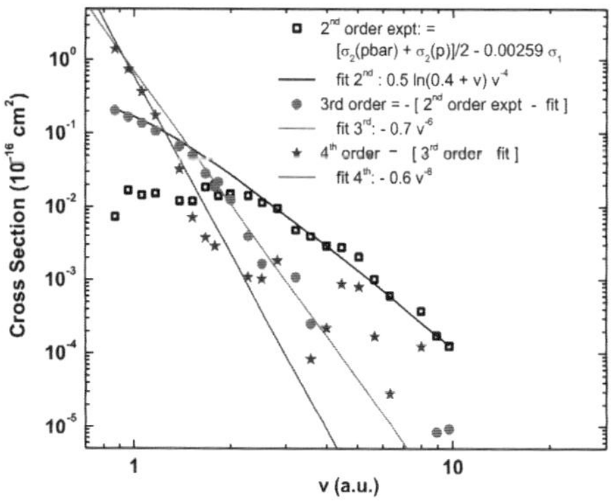

Figure 1. Possible 2nd, 3rd, and 4th order contributions to double ionization of helium by protons and antiprotons. See text for details and ref. 1 for sources of original data.

References

[1]R. D. DuBois, A. C. F. Santos, and S. T. Manson, 2014, Phys. Rev. A **90**, 052721.
[2] J. H. McGuire 1982, Phys. Rev. Lett. **49**, 1153.

[1]E-mail: toni@if.ufrj.br
#3E-mail: dubois@mst.edu
[3]E-mail: smanson@gsu.edu

Content from this work may be used under the terms of the Creative Commons Attribution 3.0 licence. Any further distribution of this work must maintain attribution to the author(s) and the title of the work, journal citation and DOI.
Published under licence by IOP Publishing Ltd

Journal of Physics: Conference Series **635** (2015) 022018 doi:10.1088/1742-6596/635/2/022018

On the reactivity of ion pairs into different diatomic systems

T. Launoy [1 §], N. Vaeck [1], X. Urbain [2], J. Liévin [1], J. Loreau [1], K. Béroff [3], M. Chabot [4]

[1] Laboratoire de Chimie Quantique et Photophysique (CQP), Faculté des Sciences, Université Libre de Bruxelles, 50, av. F. Roosevelt, CP160/09, 1050 Bruxelles, Belgium

[2] Institute of Condensed Matter and Nanosciences (IMCN), Université catholique de Louvain, chemin du cyclotron 2, 1348 Louvain-la-Neuve, Belgium

[3] Institut des Sciences Moléculaires d'Orsay (ISMO), Université Paris-Sud, F-91405 Orsay Cedex, France

[4] Institut de Physique Nucléaire d'Orsay (IPNO), Université Paris-Sud, F-91406 Orsay Cedex, France

Synopsis: Ion pair collisions are an important process in many astrophysical environments. We present *ab initio* calculations of highly excited states of C_2^+ and identify the ion pair channel C^{2+}/C^-. We use these results to interpret recent experiments on carbon cluster dissociation.

Carbon-based molecules have been detected in many astrophysical environments such as the interstellar medium (ISM) and circumstellar envelopes. In these environments, many processes can create molecules like neutral-neutral, ion-neutral or ion-ion collisions. There is a strong demand from the astrochemistry community for molecular data about these processes.

Our aim is to provide theoretical answers to experimental interrogations about ion pair relaxation of C_2^+ into C^{2+}/C^- in order to compare theoretical and experimental values.

We used multiconfigurational *ab initio* methods such as CASSCF/MRCI (+ Davidson correction) and different basis sets, e.g., ACV6Z for the lower dissociation limits or VTZ for the excited ones. With these methods, we were able to obtain the potential energy curves (PEC) for the fifty low-lying electronic states $^4\Sigma^-$ arising from various molecular dissociation channels of this system (see Fig. 1). Furthermore, based on the Mulliken population analysis of the excited states and with the use of C_{2v} symmetry into the Molpro package, we characterized the ionic channels of C^{2+} (^{1}S) / C^- (^{4}S°) witch interact with valence electronic states with a large number of avoided crossing. Based on that, we hope to provide some answers to the experimental branching ratio of carbon cluster dissociation into ion pair versus the dissociation process without anions [1].

With an optimized even-tempered basis set [2], we were able to obtain information on the reactivity of some Rydberg states of C_2^+. This has motivated us to study in a similar way the reactivity of Rydberg states for other systems.

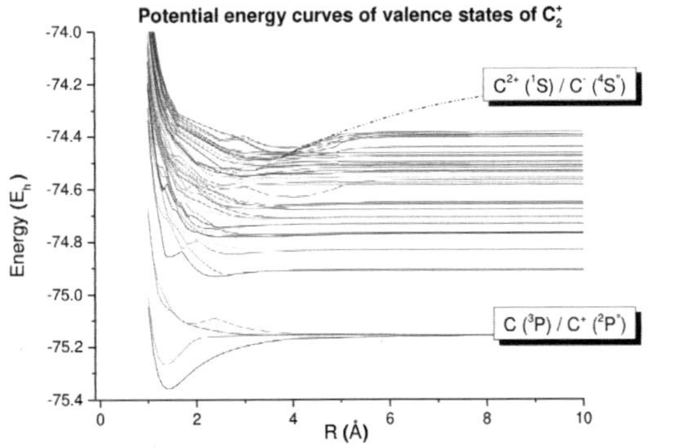

Figure 1. Potential energy curves of valence states of C_2^+ between the dissociation channel of C_2^+ into C (^{3}P)/ C^+ (^{2}P°) and the ionic channel C^{2+} (^{1}S)/C^- (^{4}S°)

We are currently working on the dynamics of the mutual neutralization of ion pairs in diatomic molecules such as LiH, BH and CH. We first compute the PEC of these systems and we subsequently investigate their dynamics using time-dependent quantum mechanical methods. Our goal is to obtain theoretical charge transfer cross sections to compare with experimental data that we will acquired with a merged-beam set up [3].

References

[1] Béroff, K. *et al* 2013 *J. Phys. B At. Mol. Opt. Phys. 46* 015201.

[2] Laruelle, F. *et al* 2009 *J. Phys. Chem. A 113* 13210–13220.

[3] Urbain, X. *et al* 2012 *J. Phys. Conf. Ser. 388* 092004.

§ E-mail : tlaunoy@ulb.ac.be

Content from this work may be used under the terms of the Creative Commons Attribution 3.0 licence. Any further distribution of this work must maintain attribution to the author(s) and the title of the work, journal citation and DOI.
Published under licence by IOP Publishing Ltd

Journal of Physics: Conference Series **635** (2015) 022019 doi:10.1088/1742-6596/635/2/022019

Fast electron emission in multiply charged ion-atom collisions at intermediate energies

A. Riascos Ochoa*, J.M. Monti†, J. Fiol*, P. Fainstein*, R. Rivarola† W. Wolff‡2, D. Fregenal*, G. Bernardi*, S. Suárez*1

* Centro Atómico Bariloche (CAB) Bustillo 9500, (8400) Bariloche – Argentina
† Inst. Fis. Rosario (UNR-CONICET) Avenida Pellegrini 250, 2000 Rosario, Argentina
‡ Instituto de Física, Universidade Federal de Río de Janeiro (IF-UFRJ), Brazil

Synopsis In the present work we show electron spectra from Si^{q+} on H_2 and He targets for projectile energies within the range 2.7 – 7 MeV for several emission angles. We discuss our results based on the picture of quantum interferences between the Short and Long range-projectile potential interactions in the emission of target electrons. Quantum mechanical CDW and CDW-EIS approximations are compared with present results.

Several experimental results have been reported concerning fast electron emission in ion-atom collisions [1]. In addition to the very well-known electron emission mechanisms, interference effects have been included to the discussion of electron spectra. These quantum interference effects were investigated by the end of the 1980s and the beginning of 1990s [2].

Many years after the first experimental evidence, a quantum mechanical calculation was presented by Monti et al [3] giving a more intuitive picture of the effect. The approach consisted in providing a potential for a projectile with bound electrons to produce a short range interaction, but charged, to ensure the long range potential. The projectile potential used in this calculation is the sum of the Coulomb or Long range term ($V_{LR} = q/s$), and that due to the screened projectile V_{SR} (V_p-V_{LR}) [3] or Short range contribution:

$$V_p = -\frac{1}{s}(Z_p - q)\left[H(e^{s/d} - 1) + 1\right]^{-1} - \frac{q}{s} \qquad (1)$$

The theoretical calculations showed the presence of two structures, near the binary emission due to the interference of Short and Long range-projectile potential interactions in the emission of target electrons, as experimentally observed. The calculations were able to reproduce well the spectra of fast electrons emitted in U^{21+} on He targets, and other systems, like multiply charged ions I^{7+}, I^{23+} and Au^{11} colliding with different targets [2].

Pursuing this search of quantum interferences with lighter projectiles at smaller incident energies, Al^{q+}, Si^{q+}, and P^{q+} with q = 3, 4, 5 (with Ne like electron configuration) were considered good candidates.

1 E-mail: suarez@cab.cnea.gov.ar 2 E-mail: wania@if.ufrj.br

We have measured electron Doubly Differential Cross Sections (DDCS) for several angles of emission from 0 to 50 degrees and covering a large range of emission energies, as shown in Figure 1, for example. In particular, we observe the region around the binary peak and down to the cusp due to electron transfer into the projectile continuum.

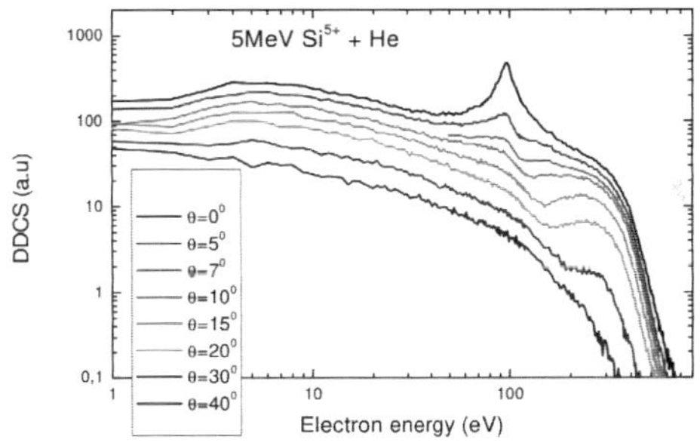

Figure 1. DDCS for electron emission from He targets by impact of 5 MeV Si^{5+} projectiles

Electron spectra from Si^{q+} on H_2 and He targets for projectile energies within the range 2.7 – 7 MeV are reported in the present work.

References

[1] N. Stolterfoht, R. D. DuBois, R. D. Rivarola, 1997 "Electron Emission in Heavy Ion-Atom Collisions" (Springer Series on Atoms + Plasmas, 20) Editors: G. Ecker, P. Lambropoulus, I. Sobel´man, H. Walther, H. Lotsch

[2] C. O. Reinhold et al 1991 Phys. Rev. Lett. 66 1842; W. Wolff et al 1992 J. Phys:B: Mol. Opt. Phys. 25 3683 and references therein

[3] J. M. Monti et al 2008 J. Phys.B: Mol. Opt. Phys. 41 201001

Content from this work may be used under the terms of the Creative Commons Attribution 3.0 licence. Any further distribution of this work must maintain attribution to the author(s) and the title of the work, journal citation and DOI.
Published under licence by IOP Publishing Ltd

Journal of Physics: Conference Series **635** (2015) 022020 doi:10.1088/1742-6596/635/2/022020

Fragmentation studies of Hydrogenated-Pyrene Polycyclic Aromatic Hydrocarbons in collisions with He

N. de Ruette* [1], **M. Gatchell***, **M. H. Stockett**†*, **L. Giacomozzi***, **T. Chen***, **E. K. Anderson***, **R. F. Nascimento***, **M. Wolf***, **J. D. Alexander***, **R. Delaunay**‡¶, **V. Vizcaino**‡¶, **P. Rousseau**‡¶, **L. Adoui**‡¶, **B. A. Huber**‡¶, **H. T. Schmidt***, **H. Zettergren*** and **H. Cederquist*** [2],

* Department of Physics, Stockholm University, Stockholm, SE-106 91, Sweden
† Department of Physics and Astronomy, Aarhus University, Ny Munkegade 120, DK 8000 Aarhus C, Denmark
‡ Centre de Recherche sur les Ions, les Matériaux et la Photonique (CIMAP),CEA-CNRS-ENSICAEN, Caen Cedex 05, France
¶Université de Caen Basse-Normandie, Caen, France

Synopsis We have studied the fragmentation of Pyrene or hydrogenated Pyrene cations ($C_{16}H^+_{10+x}$ with $x = 0$, 5, 6 and 16) in collision with He. We find that the carbon backbone fragmentation cross section increases with the number of H atoms added. This suggests that hydrogenation does not protect PAHs in, e.g., harsh interstellar environments.

Polycyclic Aromatic Hydrocarbons (PAHs) are an important component of interstellar dust and gas. In the 1980s, PAHs were proposed [1] to be the origin of the unidentified infrared (IR) emission bands. PAHs are known to emit strongly in the mid-IR through transitions between vibrational states [2]. Unfortunately, it is not easy to identify individual PAH species from these observations. The processes by which PAHs and other large molecules are formed and destroyed in the interstellar medium are not well understood. Collisions between PAHs and ions could, for example, be important destruction mechanisms. However, fragmentation cross-sections and related dissociation energies and barriers are largely unknown. Experiments on collisions between PAH ions and atoms in the 10-100 eV energy range, may elucidate the role of such collisions in the processing of interstellar carbon [3].

Reitsma et al [4] studied H-loss from hydrogenated coronene cations $C_{24}H^+_{12+m}$ (m = 0-7) following carbon K-shell X-ray absorption. They found that fewer of the twelve native H atoms were lost when the molecule was hydrogenated than when it was in its native form. This protective effect grew stronger as the degree of hydrogenation increased [4]. This was explained as being due to cooling of the excited molecular ions through evaporation of the loosely bound extra H atoms. Based on these results, Reitsma et al [4] suggested that PAH hydrogenation would also protect the carbon backbone and that this could explain the high PAH-abundance in interstellar space.

Here, we report experimental results on collisions between Pyrene or hydrogenated Pyrene cations ($C_{16}H^+_{10+x}$ with $x = 0$, 5, 6 and 16) and He atoms. We find that additional hydrogens do not protect the carbon backbone from fragmentation. Rather, the backbone fragmentation cross section increases with the degree of hydrogenation. This suggests that hydrogenation is not the reason for the high PAH abundance in interstellar media. Density Functional Theory and Molecular Dynamics simulations are used to aid the interpretation of the experimental results.

References

[1] A. Leger and J. L. Puget, 1984 *Astron. & Astrophys.* **137** L5-L8

[2] M. S. Povich, 2007 *Astrophys. J.* **660(1)** 346

[3] E. R. Micelotta *et al*, 2010 *Astron. Astrophys.* **510** A36

[4] Reitsma *et al*, 2014 *Phys. Rev. Lett.* **113** 053002

[1]E-mail: nathalie.deruette@fysik.su.se
[2]E-mail: cederq@fysik.su.se

Content from this work may be used under the terms of the Creative Commons Attribution 3.0 licence. Any further distribution of this work must maintain attribution to the author(s) and the title of the work, journal citation and DOI.
Published under licence by IOP Publishing Ltd

Main magnetic focus ion source, state-of-the-art technology for production of highly charged ions

V. P. Ovsyannikov* [1], A. V. Nefiodov⋆ [2], A. Borovik, Jr.† and A. Müller†

* Hochschulstr. 13, D-01069 Dresden, Germany
⋆ Petersburg Nuclear Physics Institute, 188300 Gatchina, St. Petersburg, Russia
† Institut für Atom- und Molekülphysik, Justus-Liebig-Universität Giessen, D-35392 Giessen, Germany

Synopsis An extremely efficient method for the production of highly charged ions has been realized. The method is based on the control of a local ion trap created by the rippled electron beam in a focusing magnetic field. Highly charged ions such as, for example, neon-like iridium can be prepared in the room-temperature hand-size device. This offers wide opportunities for laboratory studies of a variety of physical problems.

In electron beam ion sources and traps (EBIS/T), the ions confined in the electrostatic ion trap are produced by the electron beam via multiple sequential ionization. The ion trap is formed by the electron-optical system, which consists of the electron gun, a few (at least three) drift tubes with positive potentials applied to edge sections relative to the central electrode, and the electron collector. The ultimate current density of the long smooth electron beam does not exceed the Brillouin limit.

The rippled electron beam with its radius changing along the beam axis creates a sequence of local ion traps, the depths of which are determined by the maximum-to-minimum radius ratio. The electron current density in the local ion traps can reach the values, which are much higher than those predicted by the Brillouin conditions [1].

In a novel ion source classified as *main magnetic focus ion source* (MaMFIS), the local ion trap is formed in the crossover of the electron beam, which is rippled by a thick magnetic lens. The length of the local ion trap is very small, so that the MaMFIS can be small in comparison with the standard EBIS/Ts (see figure 1).

The electron optics was tested for electron beams with currents of up to 50 mA and energies of up to 10 keV. The relative magnitude of intercepted current is about 0.1% only. The electron current density is estimated to be larger than 10 kA/cm^2.

The experimental measurements demonstrate the extremely high efficiency of the local traps to produce highly charged ions. The radiation from the confined ions of such elements as Ir, Ce, Xe and Ar is detected by the method of x-ray spectroscopy. The analysis of the x-ray spectra allows one to identify the series of highly charged ions described by charge-state distributions with peaks near Ir^{60+}, Ce^{44+} and Ar^{16+}.

The ion extraction is realized by transformation of the rippled electron beam into a smooth one. In this case, the local ion trap disappears.

Figure 1. MaMFIS designed for operation with the electron energy of up to 10 keV.

References

[1] Ovsyannikov V P 2014 *Main Magnetic Focus Ion Trap, new tool for trapping of highly charged ions* arXiv e-print (arXiv:1403.2168)

[1] E-mail: v.ovsyannikov@yandex.com
[2] E-mail: anef@thd.pnpi.spb.ru

Content from this work may be used under the terms of the Creative Commons Attribution 3.0 licence. Any further distribution of this work must maintain attribution to the author(s) and the title of the work, journal citation and DOI.
Published under licence by IOP Publishing Ltd

Journal of Physics: Conference Series **635** (2015) 022024 doi:10.1088/1742-6596/635/2/022024

Reaction of NH^+ with atomic hydrogen at low temperatures – an experimental study

Radek Plašil*[1], Štěpán Roučka*, Dmytro Mulin*, Serhiy Rednyk*, Thuy Dung Tran*, Sunil S. Kumar† and Juraj Glosík*

* Charles University in Prague, Faculty of Mathematics and Physics, Prague, Czech Republic
† Universität Innsbruck, Institut für Ionenphysik und Agnewandte Physik, Innsbruck A-6020, Austria

Synopsis The reaction of NH^+ with H was studied at temperatures below 200 K. Using an rf-trapping technique, NH^+ ions were confined, cooled and exposed to a beam of thermalized atomic hydrogen.

Reactions of nitrogen and hydrogen are of importance in many environments. Ions N^+ and NH^+ play significant role in formation of ammonia and molecules with cyano group in interstellar space [1]. We studied formation of NH^+ at low temperatures in range of 10 K to 100 K [2]. We identified state specific rate coefficients of reaction:

$$N^+(^3P_{ja}) + H_2(J) \rightarrow NH^+ + H\,. \qquad (1)$$

The reaction is nearly thermoneutral, but the overall energy balance is not known. The lowest electronic state of NH^+, $X^2\Pi$, and the first electronically excited $a^4\Sigma$ state are separated by only a few hundred cm^{-1} which corresponds to temperature of 500 K [1,3]. This near-degeneracy makes it difficult to calculate the bond energy of NH^+ with accuracy needed for description of low-temperature reactions.

From our measurement we derived activation energy of reaction (1) as 220 K. It may be caused by endothermicity of reaction or by a barrier on reaction path.

To understand the NH_2^+ collision system we decided to study reverse reaction:

$$NH^+ + H \rightarrow N^+ + H_2\,. \qquad (2)$$

For the measurement we can employ the neutral beam of atomic hydrogen [4,5]. The unique source produces hydrogen atoms in an rf discharge and then thermalizes them in cooled tube. The temperature in the H atom beam is varied from 9 K to 180 K. The number density of H in the trap was determined by CO_2^+ + H reaction as 3.0×10^7 cm^{-3}. We combine the beam with NH^+ ions confined by 22-pole ion trap.

The measured rate coefficient of reaction (2), in order of 10^{-11} $cm^3\,s^{-1}$, is close to the detection limit of our experiment because of low number density of H. But from our measurements at temperatures from 10 K to 180 K we can find more about the reaction energetic. From a raw evaluation of measured rate coefficients, accounting for the lowest energy levels only, we can deduce that activation energy of reaction (2) is higher than ~ 150 K. Further evaluations and measurements are in progress.

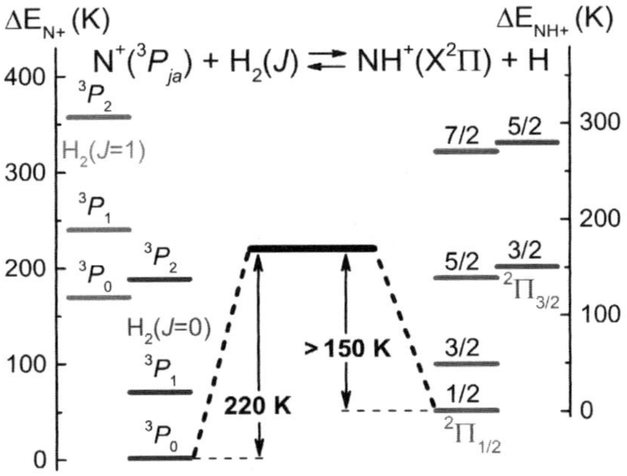

Figure 1. The lowest energy levels of reactants in reactions (1) and (2) using notation [1,2]. In the central part of the figure the activation energies are depicted by hypothetical barrier.

We thank the TU Chemnitz and the DFG for lending us the experimental apparatus and prof. D. Gerlich for help and consultations. This work was partially supported by GACR P209/12/0233, GACR 14-14715P and GAUK 572214.

References

[1] K. Beloy et al 2011 *Phys. Rev. A* **83** 062514
[2] I. Zymak at al 2013 *Astroph. J.* **768** 86
[3] R. Tarroni et al 1997 *J. Chem. Phys.* **106** 10265
[4] D. Gerlich et al 2011 *Zeitschrift Phys. Chemie* **252** 475
[5] R. Plasil et al 2011 *Astroph. J.* **737** 60

[1] E-mail: Radek.Plasil@mff.cuni.cz

Content from this work may be used under the terms of the Creative Commons Attribution 3.0 licence. Any further distribution of this work must maintain attribution to the author(s) and the title of the work, journal citation and DOI.
Published under licence by IOP Publishing Ltd

Journal of Physics: Conference Series **635** (2015) 022026 doi:10.1088/1742-6596/635/2/022026

Energy Spectra of Sputtering Ions Induced by $^{209}Bi^{25+, 28+}$ Ions bombardment on Al and Be

Yang Bian*, Ruan Fangfang*, Yu Deyang*, Chen Jing*, Wang Wei*, Wu Yehong*, Xue Yingli*, Cai Xiaohong* [1]

* Institute of Modern Physics, Chinese Academy of Sciences, Lanzhou 730000, China

Synopsis We have experimental investigated sputtering ions of ~MeV $^{209}Bi^{25+, 28+}$ collision with metal surfaces by 127° electrostatic coaxial cylindrical energy spectrometer, which enabled us to measure energy spectrum fast over energy of sputtering ions from 0 to 3.5 keV. The results show unprecedented details on the reaction dynamics.

The energy spectrum of sputtering ions is key information for understanding the mechanisms of sputtering and ionization processes of highly charged ions collision on surface[1]. Yield of sputtering ions is approximately less than 5% of all sputtering particles. Energy distribution of sputtering ions is wide in the range from several eV to keV. In addition, traditional step-by-step voltage scanning method has very low efficiency for sputtering ions measurement. Up to now, experimental investigation of energy distribution of sputtering ions induced by highly charged ions impact on surface is still rare[1].

We have experimentally investigated the energy spectra of sputtering ions produced by 2 MeV, 3.5 MeV, 6 MeV $^{209}Bi^{25+, 28+}$ collision on Al and Be surfaces. A 127° electrostatic coaxial cylindrical energy spectrometer equipped with a radial position sensitive detector[2] enabled us to perform segment of energy spectrum ($0.85\ E_c \leq E \leq 1.15\ E_c$) at single voltage (U), a wide range of energy spectrum is obtained with high efficiency. The experiment was carried out on 320 kV Multi-discipline Research Platform at the Institute of Modern Physics in Lanzhou. Schematic diagram of the experimental setup is in Ref. [3].

Figure 1 show segment spectra of sputtering ions induced by 2 MeV Bi^{28+} collision on Be at U = 95 V. The obtained energy spectra of Be^{+} is from dozens eV to more than 3.5 keV at 27 voltages. The shape of energy distribution of sputtering ions induced by 2 MeV Bi^{28+} has significant differences compared with that with collision energy region dominated by cascade collision. This information implies that thermal spike plays a dominant role in sputtering processes.

This work is supported in part by the Joint Funds of the National Natural Science Foundation of China (Grant No. U1332206) and the Young Scientists Funds of the National Natural Science Foundation of China (Grant No. 11204323).

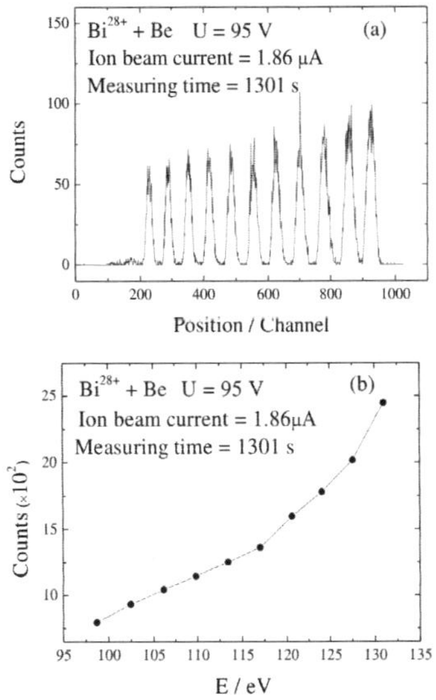

Figure 1. Segment spectra of sputtering ions induced by 2 MeV Bi^{28+} collision on Be.
(a) Radial position spectrum of sputtering ions,
(b) Energy spectrum of sputtering ions.

References

[1] R. Behrisch and W. Eckstein 2007 *Springer Berlin/ Heidelberg*
[2] F. Ruan *et al.* 2009 *Nucl. Phys. Rev.* 3 227 (in Chinese)
[3] F. Ruan 2009 *Ph.D Thesis* (in Chinese)

[1] E-mail: caixh@impcas.ac.cn

Content from this work may be used under the terms of the Creative Commons Attribution 3.0 licence. Any further distribution of this work must maintain attribution to the author(s) and the title of the work, journal citation and DOI.
Published under licence by IOP Publishing Ltd

XXIX International Conference on Photonic, Electronic, and Atomic Collisions (ICPEAC2015) IOP Publishing
Journal of Physics: Conference Series **635** (2015) 022027 doi:10.1088/1742-6596/635/2/022027

Systematic study of the K x-ray spectra for hollow Kr and Xe atoms

X Chen*, K Ma*, C Z Dong* [1], D H Zhang*, C J Shao†, D Y Yu† and X H Cai†

*Key Laboratory of Atomic and Molecular Physics & Functional Materials of Gansu Province, College of Physics and Electronic Engineering, Northwest Normal University, Lanzhou 730070, China

†Institute of Mordern Physics, Chinese Academy of Sciences, Lanzhou 730030, China.

synopsis In this work, the satellite and hypersatellite x-rays resulted from additional 1-7 holes in L-shell for Kr and Xe atoms have been calculated systematically by using the multiconfiguration-Dirac-Fock(MCDF) method. The present calculations are in good agreement with the latest experiment at the HIRFL-CSR in Lanzhou.

In collisions of atoms with the heavy ion, the intensities of x-ray satellites and hypersatellites will be significantly increased. Thanks to the fast development of accelerator techniques and the employment of heavy ions in many applications, the atom bombarded by heavy ions processes become accessible. Due to the strong Coulomb field of the fast projectile, several inner-shell electrons of the target atom can be ionized simultaneously, thus producing rich satellite and hypersatellite lines in the x-ray spectra. These satellite lines not only more or less strongly overlap each other, but also overlap the diagram lines.In order to interpret the spectra better, lots of precise calculations for concerning transition energies and rate data are necessary.

In the present work, the transition energies and rates of K_{α}, K_{β} diagrams satellites and hypersatellites for Kr and Xe atoms have been calculated systematically using the MCDF method with the inclusion of the Breit interaction and QED corrections. According to the calculated results, the relationship between the energy shifts for Kr and Xe satellites and hypersatellites and the number of spectator vacancies in L shell have been presented as shown in Figure 1. (for Xe only).

In order to compare with the latest experiments on Kr and Xe atoms collided with Xe^{54+} ions in Lanzhou Heavy-ion Research Facility (HIRFL), the present calculated results have been convoluted by using Gauss profile with the experimental resolution. Figure 2 shows the comparison between the simulation spectra and the experiment for Xe

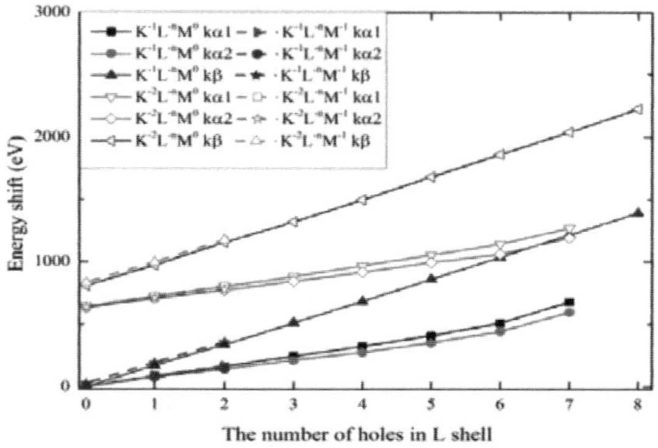

Figure 1. The energy shifts of K x-ray satellites and hypersatellites for Xe.

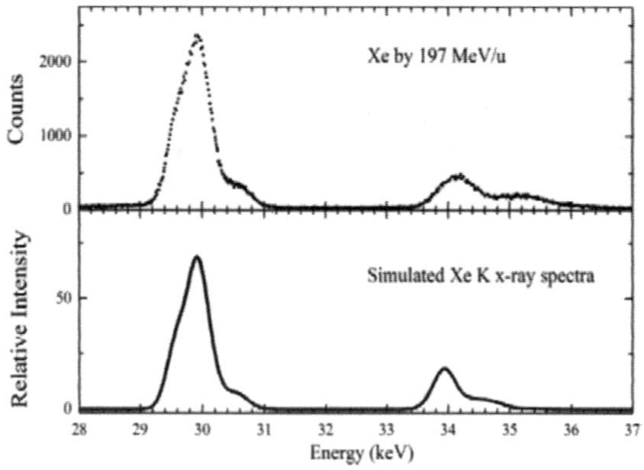

Figure 2. Comparison between the experimental specrta and the simulated one for Xe.

This work was supported by the National Natural Science Foundation of China (Grant nos. 11274254, U1332206 and 11464042).

References:

[1] M. Polasik, 2010 *Phys. Rev.* A **81** 064702.

[2] X L Wang, C Z Dong, M G Su, 2012 *Nucl Instr and Meth. B* **280** 9

[1] E-mail: dongcz@nwnu.edu.cn

Content from this work may be used under the terms of the Creative Commons Attribution 3.0 licence. Any further distribution of this work must maintain attribution to the author(s) and the title of the work, journal citation and DOI.
Published under licence by IOP Publishing Ltd

Journal of Physics: Conference Series **635** (2015) 022028 doi:10.1088/1742-6596/635/2/022028

Development and commissioning of RIken Cryogenic Electrostatic ring (RICE)

Y. Enomoto*[1], **Y. Nakano***, **T. Masunaga*** **and T. Azuma*** †

* Atomic, Molecular & Optical Physics Laboratory, RIKEN, Wako, Saitama 351-0198, Japan
† Department of Physics, Tokyo Metropolitan University, Hachioji, Tokyo, 192-0397, Japan

Synopsis We have been developing a cryogenic electrostatic ion storage ring to explore especially the cooling process and collision dynamics of the cold molecular ions in the specific vibrational and rotational state. Recently, we succeeded in storing of 15 keV Ne^+ ions in the ring stably.

We have been developing a cryogenic electrostatic ion storage ring (RIken Cryogenic Electrostatic ring, RICE) since 2010 to explore especially the cooling process and collision dynamics of the cold molecular ions in the specific vibrational and rotational state. Such cryogenic storage rings for atomic, molecular physics and chemistry are presently in operation in Sweden [1] and under construction in Germany[2].

Table 1 summarizes specifications of the RICE. The maximum energy of stored ions is 20 keV. Circumference is about 3 m, which is about a half or one third of typical room temperature electrostatic rings. Electrodes to confine ions in a race-track shaped orbit consist of two sets of 10 degree, 160 degree and 10 degree deflectors and four sets of quadrupole doublet lenses. In the straight sections, four sets of four-segmented pick-up electrodes are placed to monitor the beam position. A drift-tube-type buncher consisting of three cylindrical electrodes is also installed in the middle of one straight section. All of the electrodes are made of oxygen free copper and enclosed in a cryogenic inner vacuum chamber (IVC) which is covered by a radiation shield made of Al. The IVC and the radiation shield is cooled by three GM type cryocoolers whose cooling capacity is 3 W at 4.2 K in total. The system is enclosed in an outer vacuum chamber (OVC) for thermal isolation.

Following the assembly work, we tested cooling and vacuum performance of the ring. After 150 hours of cooling, Si diode temperature sensors attached to the IVC showed 4.7 K or less. A pressure gauge attached to the pumping station, which is connected to the IVC but is in the room temperature section, showed 1.8×10^{-11} Torr. Considering the pumping speed of the cryogenic chamber wall and conductance between cryogenic and room temperature sections, the pressure in the IVC is expected to be better than 10^{-13} Torr.

For the commissioning of the ring, an ECR ion source is connected via a transport beamline. 15 keV Ne^+ ions are injected into and stored in the ring. We confirmed stable confinement of the ions by monitoring the number of Ne atoms which are produced through neutralization by the collisions with residual gas and deviate from the trajectory in the ring. The maximum confinement time defined as a time constant assuming exponential decrease of the number of stored ions is 50 s so far.

Table 1. Design parameters of the RICE.

stored ion energy	20 keV
circumference	2965 mm
environmental temperature	≤ 5 K
background pressure	$\leq 10^{-13}$ Torr

References

[1] H. T. Schmidt *et al.* 2013 *Rev. Sci. Instrum.* **84** 0055115

[2] R. Von Hahn *et al.* 2011 *Nucl. Inst. Meth. B* **269** 2871

[1]E-mail: enomoto@riken.jp

Content from this work may be used under the terms of the Creative Commons Attribution 3.0 licence. Any further distribution of this work must maintain attribution to the author(s) and the title of the work, journal citation and DOI.
Published under licence by IOP Publishing Ltd

XXIX International Conference on Photonic, Electronic, and Atomic Collisions (ICPEAC2015) IOP Publishing
Journal of Physics: Conference Series **635** (2015) 022029 doi:10.1088/1742-6596/635/2/022029

Elastic scattering process for argon ions and oxygen atoms in 10-500 eV range of relative kinetic energies

A.A. Sycheva* [1], **A.P. Palov** [† 2]

* Lomonosov Moscow State University, Department of Physics, GSP-1, Leninskie gory, 119991, Moscow, Russian Federation

† Lomonosov Moscow State University, Skobeltsyn Institute of Nuclear Physics, GSP-1, Leninskie gory, 119991, Moscow, Russian Federation

Synopsis The interaction potential between oxygen atoms (^3P state) and argon ions (^2P state) has been calculated on *ab initio* level in the vast range of internuclear distances (0.8-100 a.u.). The elastic scattering process has been studied for relative kinetic energies of 10-500 eV.

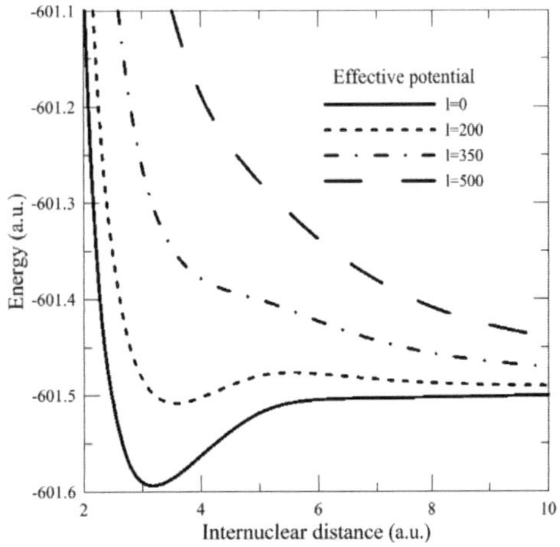

Figure 1. . Effective interaction potential as a function of the internuclear distance between Ar+ and O for orbital quantum numbers ***l*** = 0, 200, 350, 500.

In the field of modern microelectronics interaction between argon plasma and porous SiOCH low-K films is intently studied [1]. Thus, calculations of the O-Ar+ interaction potential and corresponding elastic scattering cross sections, is important for Monte Carlo and molecular dynamics simulations of sputtering processes.

We applied the MRCI method with the aug-cc-pV5Z basis set to obtain the exited $^2\Pi$ state of OAr+ ion. The following molecule is generated by argon ion (^2P) and oxygen atom (^3P) and seems to be a proper candidate of the elastic scattering process. Nine molecular orbitals were placed in the active configuration space. The calculated ground state was found to possess deeper well (2.58 eV) than that computed in previous studies (2.17 eV) [2].

To obtain elastic scattering cross section in the 10-500 eV range of relative kinetic energies phase shifts were computed. For the 10-50 eV range we have solved the Schrödinger equation and have implemented subsequent analysis of the solution in asymptotic region. In the 50-500 eV range the following approach was applied: the Schrödinger equation was solved for the orbital quantum numbers ***l*** less than 500 and WKB approximation was applied for larger ones. These results were employed to calculate differential, integral and transport scattering cross sections.As we have found the obtained OAr+ ground state potential supports 59 bound states oscillations one can expect the same amount of glory on integral cross section curve. Note shape resonances do not appear within the energy range given below because at these relative kinetic energy the effective potential does not possess quasi-stationary states.

The authors thank the Russian Scientific Fund (RSF) for financial support (Grant №14-12-01012).

Table 1. The integral elastic scattering cross section between argon ion and oxygen atom.

Relative kinetic energy, eV	Integral cross section values, Å²
10	7163.3
25	5375.6
50	2181.2
100	1372.2
200	370.6
500	341.1

References

[1] Keren J. Kanarik *et al* 2015 *J. Vac. Sci. Technol., A.* 33 020802

[2] Gernot Frenking *et al* 1989 *J. Phys. Chem.* 93 3410

[1] E-mail: sycheva.phys@gmail.com

[2] E-mail: a.palov@mics.msu.ru

Content from this work may be used under the terms of the Creative Commons Attribution 3.0 licence. Any further distribution of this work must maintain attribution to the author(s) and the title of the work, journal citation and DOI.
Published under licence by IOP Publishing Ltd

XXIX International Conference on Photonic, Electronic, and Atomic Collisions (ICPEAC2015) IOP Publishing
Journal of Physics: Conference Series **635** (2015) 022030 doi:10.1088/1742-6596/635/2/022030

Electron-positron pair production in low-energy heavy-ion collisions

I A Maltsev[*,†,1], **V M Shabaev**[*], **I I Tupitsyn**[*], **A I Bondarev**[*], **Y S Kozhedub**[*,†], **G Plunien**[‡], **and Th Stöhlker**[⋆]

[*] Department of Physics, St. Petersburg State University, Ulianovskaya 1, St. Petersburg 198504, Russia
[†] SSC RF ITEP of NRC "Kurchatov Institute", Bolshaya Cheremushkinskaya 25, Moscow 117218, Russia
[‡] Institut für Theoretische Physik, TU Dresden, Mommsenstraße 13, Dresden D-01062, Germany
[⋆] GSI Helmholtzzentrum für Schwerionenforschung GmbH, Planckstrasse 1, Darmstadt D-64291, Germany

Synopsis A method for calculations of pair production in low-energy collisions of heavy ions is developed. The approach is based on the numerical solving of the two-center time-dependent Dirac equation in the monopole approximation. Employing the developed method the positron energy spectra and the numbers of the created pairs are calculated for the symmetric collisions of bare nuclei with different values of the impact parameter and the nuclear charge. In order to investigate the role of the spontaneous pair production the different subcritical and supercritical scenarios are considered.

To date the effects of quantum electrodynamics (QED) have been tested with high accuracy in strong fields of highly charged ions [1]. However, the domain of the supercritical fields has not yet been probed in experiments. In the presence of such extremely strong fields the QED vacuum becomes unstable and can decay spontaneously via creation of electron-positron pair [2]. The supercritical regime can be achieved in collision of two heavy ions, if the total charge of their nuclei is larger than the critical value $Z_C \approx 173$. The future Facility for Antiproton and Ion Research (FAIR) will provide new opportunities for investigation of supercritical heavy-ion collisions [3].

For study of the supercritical effects one needs the theoretical methods for description of various processes in heavy-ion collisions. In the present work we develop the method which allows us to calculate pair production in the monopole approximation [4]. The approach is based on the expansion of the one-electron wave functions over the finite basis set. The basis functions are constructed from the B-splines using the dual-kinetic-balance technique which prevents the appearance of the nonphysical spurious states [5]. All the eigenstates of the initial Hamiltonian matrix are propagated via the numerical solving of the time-dependent Dirac equation. The pair-creation probabilities are obtained employing the second quantization technique [2].

Using the developed method the positron energy spectra and the numbers of the created pairs are calculated for symmetric collisions of bare nuclei with the Rutherford trajectory for different values of the impact parameter and the nuclear charge. In figure 1, we present the positron energy spectra calculated for the U−U collisions at energy $E_{\rm cm} = 740$ MeV. The results obtained for the U−U collisions are in a good agreement with the corresponding values from [6].

This work was supported by SPbSU (Grants No. 11.38.269.2014 and No. 11.38.654.2013), the Dynasty foundation, and FAIR-Russia Research Center.

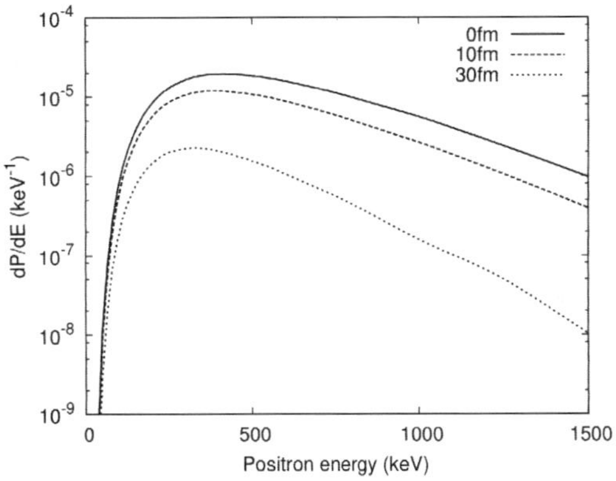

Figure 1. Positron energy spectrum for the U−U collision at energy $E_{\rm cm}$ =740 MeV for the different values of the impact parameter.

References

[1] Volotka A V *et al* 2013 *Ann. Phys. (Berlin)* **525** 636

[2] Greiner W, Müller B, and Rafelski J 1985 *Quantum Electrodynamics of Strong Fields* (Berlin: Springer-Verlag)

[3] Gumberidze A *et al* 2009 *Nucl. Instrum. Methods Phys. Res., Sect. B* **267** 248

[4] Maltsev I A *et al* 2015 arxiv:1412.5337

[5] Shabaev V M *et al* 2004 *Phys. Rev. Lett.* **93** 130405

[6] Müller U *et al* 1988 *Phys. Rev. A* **37** 1449

[1] E-mail: imaltsev@pcqnt1.phys.spbu.ru

Content from this work may be used under the terms of the Creative Commons Attribution 3.0 licence. Any further distribution of this work must maintain attribution to the author(s) and the title of the work, journal citation and DOI.
Published under licence by IOP Publishing Ltd

Journal of Physics: Conference Series **635** (2015) 022031 doi:10.1088/1742-6596/635/2/022031

Reaction rate measurements between sympathetically cooled ions and velocity selected polar molecules: search for the deuterium isotope effects

Yusuke Takada and Kunihiro Okada[1]

[*] Department of Physics, Sophia University, 7-1 Kioicho, Chiyoda, Tokyo 102-8554, Japan

Synopsis The rates of cold ion-polar molecule reactions are measured at a temperature lower than 10 K in order to search for the deuterium isotope effects.

Cold ion-molecule reactions at a temperature lower than 100 K are very important for astrochemistry [1]. However, there is a wide gap between the temperature required of the reaction rates in astrochemistry and the temperature of the available measurements. Actually most of such measurements were performed at the restricted range near room temperature [2]. In particular cold ion-polar molecule reactions have not been measured extensively at low temperatures owing to experimental difficulties, such as condensation or sublimation of polar gases. Actually the experimental information of the reactions at temperature lower than 77 K is missing. However, recent experimental developments combining a linear Paul trap for generating ion Coulomb crystals with a Stark velocity filter for producing cold polar molecules make it possible to measure the cold ion-polar molecule reactions under ultra-high vacuum conditions [3]. Recently we have demonstrated the reaction rate measurements between the sympathetically cooled N_2H^+ ions and cold CH_3CN molecules using the technique [4]. This potential technique possibly enables to find out quantum effects in cold ion-polar molecule reactions through systematic measurement.

In this work we search for the deuterium isotope effects in cold ion-polar molecule reactions. Up to now we have measured six reactions between cold acetonitrile isotopomers (CH_3CN, CD_3CN) and three sympathetically cooled ions (N_2H^+, N_2D^+ and Ne^+). As mentioned above, velocity-selected acetonitrile molecules were produced by the Stark velocity filter, which was previously described [4]. The translational temperature of about 6.5 K was estimated by combining the result of the time-of-flight measurement to the numerical simulations of the velocity filtering of the molecules. On the one hand, the average ion-kinetic energy, which is mainly determined by the micromotion energy, was evaluated by performing molecular dynamics simulations of two-component ion Coulomb crystals with the similar size as the observed crystals [4, 5]. As a result, the average-reaction temperature in the present reactions was estimated to be lower than 10 K, which corresponds to the energy lower than 1.2 meV. Thus, it is possible to detect the deuterium isotope effect if there is a reaction barrier larger than 1.2 meV in one of the pair of the present reactions.

Figure 1 (a) (b) show examples of the reaction rate measurements between velocity-selected acetonitrile molecules and sympathetically cooled Ne^+ ions. The graphs show plots of relative number of Ne^+ ions embedded in the Ca^+ ion Coulomb crystal as a function of the reaction time. The inset images give signification of LIF images of two-component Coulomb crystals consisting of Ca^+ and Ne^+ ions in the initial conditions of the measurements. We observed a difference in the reaction rates between (a) and (b) reactions. Details of the experiment, discussions and the results of the other reactions will be presented at the conference.

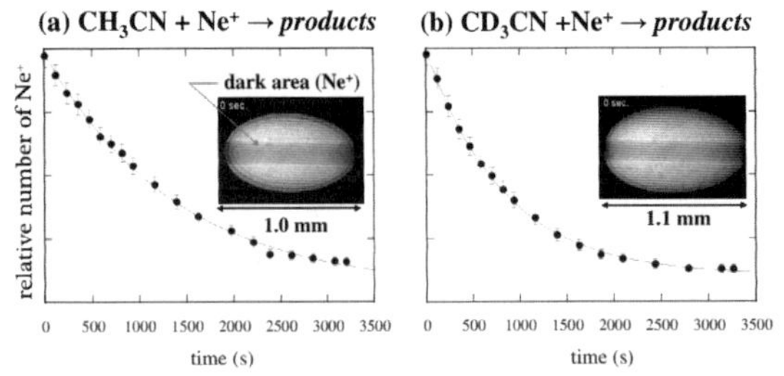

Figure 1. Reaction rate measurements between velocity-selected acetonitrile molecules and sympathetically cooled Ne^+ ions (see text).

References

[1] V. Wakelam *et al.*, Space Sci. Rev. **156**, 13
[2] J. Woodall *et al.*, A&A **466**, 1197
[3] S. Willitsch *et al.*, PRL **100**, 043203
[4] K. Okada *et al.*, Phys. Rev. A **87**, 043427.
[5] K. Okada *et al.*, submitted to Phys. Rev. Appl.

E-mail: okada-k@sophia.ac.jp

Content from this work may be used under the terms of the Creative Commons Attribution 3.0 licence. Any further distribution of this work must maintain attribution to the author(s) and the title of the work, journal citation and DOI.
Published under licence by IOP Publishing Ltd

XXIX International Conference on Photonic, Electronic, and Atomic Collisions (ICPEAC2015) IOP Publishing
Journal of Physics: Conference Series **635** (2015) 022032 doi:10.1088/1742-6596/635/2/022032

Ionisation of noble gas atoms and H_2O by antiproton impact

I. B. Abdurakhmanov[1], **A. S. Kadyrov, D. V. Fursa, S. K. Avazbaev, J. J. Bailey, and I. Bray**

Department of Physics, Astronomy and Medical Radiation Sciences, Curtin University, GPO Box U1987, Perth, WA 6845, Australia

Synopsis A time-dependent convergent close-coupling approach to antiproton collisions with noble gas atoms and H_2O has been developed using a fully multielectron treatment of the targets. Integrated cross sections for single ionisation have been calculated in a wide range of impact energies from 5 keV up to 2 MeV. Obtained results are in good agreement with experiment.

We investigate single ionisation of noble gas atoms and H_2O by antiprotons in the incident energy range from 5 keV to 2 MeV. Electron-electron correlations play a key role in the dynamics of these collisions. Presently available theories [1, 2] calculate single and multiple ionisation cross sections using effective one-electron approaches neglecting the electron-electron correlations [3].

We have developed a time-dependent convergent close-coupling (CCC) approach to antiproton collisions with multielectron targets which is capable of taking into account the electron-electron correlations. To this end we model the target structure as six p-electrons above an inert Hartree-Fock core. Only one-electron excitations from the outer p-shell are considered. The use of a Laguerre basis makes it possible to take into account all excitation and ionisation channels in a systematic manner. For H_2O we use a neon-isation method recently proposed by Montanari and Miraglia [4]. This method describes the ten-electron water molecule as a dressed Ne-like atom in a pseudospherical potential.

In figure 1 we present our calculated total single-ionisation cross section in comparison with the experiment and other calculations for the Ne atom. As one can see, the CCC results are in very good agreement with the experiment at all energies. Two types of independent-particle model calculations of Kirchner et al [1] and CDW-EIS calculations of Montanari and Miraglia [2] yield similar results, which are slightly higher than the experiment. The method has also been applied to the other noble gases and the simplified model of H_2O mentioned above [6]. Agreement between our results for the heavier noble-gas targets and experiment is slightly worse than for the lighter targets. For heavier targets it is likely that for a better agreement with experiment one needs a multiconfigurational treatment of the target structure.

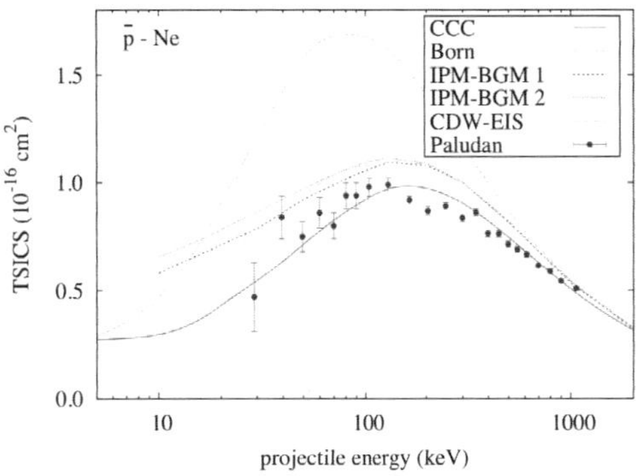

Figure 1. Integrated single-ionisation cross section for $\bar{p}$-Ne collisions. The present CCC and Born results are compared with the experimental measurements of Paludan et al. [5], independent-particle calculations of Kirchner et al. [1] with response (IPM-BGM 1) and with no response (IPM-BGM 2), and CDW-EIS calculations of Montanari and Miraglia [2].

References

[1] T. Kirchner, M. Horbatsch, and H. J. Lüdde, 2002 *Phys. Rev. A* **66** 052719

[2] C. C. Montanari and J. E. Miraglia 2012 *J. Phys. B* **45** 105201

[3] T. Kirchner and H. Knudsen 2011 *J. Phys. B* **44** 122001

[4] C. C. Montanari and J. E. Miraglia 2014 *J. Phys. B* **47** 015201

[5] K. Paludan *at al* 1997 *J. Phys. B* **30** 3951

[6] I. Abdurakhmanov *at al* 2015 *Phys. Rev. A* **91** 022712

[1]E-mail: Ilkhom.Abdurakhmanov@curtin.edu.au

Content from this work may be used under the terms of the Creative Commons Attribution 3.0 licence. Any further distribution of this work must maintain attribution to the author(s) and the title of the work, journal citation and DOI.
Published under licence by IOP Publishing Ltd

New Atomic Methods for Dark Matter Detection

B. M. Roberts*[1], Y. V. Stadnik*, V. A. Dzuba*, V. V. Flambaum*, N. Leefer†, and D. Budker†,‡,§

* School of Physics, University of New South Wales, Sydney 2052, Australia
† Helmholtz Institute Mainz, Johannes Gutenberg University, 55099 Mainz, Germany
‡ Department of Physics, University of California at Berkeley, Berkeley, CA 94720-7300, USA
§ Nuclear Science Division, Lawrence Berkeley National Laboratory, Berkeley, CA 94720, USA

Synopsis We calculate the parity and time-reversal violating effects that are induced in atoms, nuclei, and molecules by their interaction with various background cosmic fields, such as axion dark matter or dark energy.

We present our recent proposals to exploit the parity and time-reversal violating effects in atoms, nuclei, and molecules to search for evidence of various background cosmic fields, and to constrain the interaction strengths of these fields with fermions.

Candidates for such background fields include dark matter (e.g. axions) and dark energy, and also are motivated by various Standard Model extensions. Certain interactions of static and dynamic background cosmic fields would lead to the mixing of opposite parity states in atoms and molecules. This would, in turn, give rise to observable parity and time-invariance violating effects. The measurement of such effects would shed light on the interactions that gave rise to them.

Of particular interest is the interaction of a dynamic pseudoscalar field with atomic/molecular electrons and nuclei. Such a field could describe axions, a hypothetical pseudoscalar particle invoked to solve the strong CP problem from QCD, now also a leading cold dark matter candidate. We perform calculations of the parity and time-invariance violating effects that such a field would induce in atoms [1, 2], and demonstrate a potential very large enhancement of the effects in diatomic molecules.

Axion-induced nuclear Schiff moments were considered theoretically in [3], and a thorough experimental proposal to exploit this effect to search for axions was proposed in [4]. We note that nuclear magnetic quadupole moments are also generated via the interaction with axions [5]. Such moments may have significant advantages over nuclear Schiff moments.

Crucially, all the effects we consider here are linear in the small parameter that quantifies the interaction strength between the dark matter particles and ordinary matter particles; most current dark matter and axion searches rely on effects that are proportional to quadratic and higher powers of this parameter.

Oscillating electric dipole moments have the potential to be measured with very high accuracy, and experimental techniques in this field are evolving fast, making this a particularly exciting area for potential discovery in the near future. Pairs of closely spaced opposite parity levels that are found in diatomic molecules may also lead to a significant enhancement in these effects [2, 5, 6].

We are also interested in a possible explanation for the anomalous DAMA dark matter detection results. Our calculations may provide a possible mechanism for dark matter induced ionisation modulations that are not ruled out by other experiments. Alternatively, they could further reduce the available parameter space for certain dark matter models.

References

[1] Roberts B M, Stadnik Y V, Dzuba V A, Flambaum V V, Leefer N and Budker D 2014 *Phys. Rev. Lett.* **113** 081601

[2] Roberts B M, Stadnik Y V, Dzuba V A, Flambaum V V, Leefer N and Budker D 2014 Phys. Rev. D **90** 096005 (2014).

[3] Budker D, Graham P W, Ledbetter M P, Rajendran S and Sushkov A O 2014 *Phys. Rev. X* **4** 021030

[4] Graham P W and Rajendran S 2011 *Phys. Rev. D* **84** 055013

[5] Roberts B M, Stadnik Y V and Flambaum V V 2015 *In preparation*

[6] Stadnik Y V and Flambaum V V 2014 *Phys. Rev. D* **89** 043522

[1] E-mail: b.roberts@unsw.edu.au

Content from this work may be used under the terms of the Creative Commons Attribution 3.0 licence. Any further distribution of this work must maintain attribution to the author(s) and the title of the work, journal citation and DOI.
Published under licence by IOP Publishing Ltd

XXIX International Conference on Photonic, Electronic, and Atomic Collisions (ICPEAC2015) IOP Publishing
Journal of Physics: Conference Series **635** (2015) 022034 doi:10.1088/1742-6596/635/2/022034

Accurate stopping power calculations for antiprotons and protons

J. J. Bailey [1], A. S. Kadyrov, I. B. Abdurakhmanov, S. K. Avazbaev, D. V. Fursa, and I. Bray

Department of Physics, Astronomy and Medical Radiation Sciences, Curtin University, GPO Box U1987, Perth, WA 6845, Australia

Synopsis The convergent close-coupling method is applied to calculate antiproton and proton stopping cross sections for atomic and molecular targets. Excellent agreement with experimental measurements is obtained for antiprotons in helium while unexpectedly large disagreement is found for the hydrogen molecule, which is inconsistent with very good agreement between our ionisation cross section and the experiment.

Accurate knowledge of the stopping power (SP) for protons in atoms and molecules is important for proton therapy. Proton therapy is an advanced form of cancer treatment with greater benefits over the traditional x-ray therapy. Unlike x-rays, protons can deliver high doses of radiation very accurately to the site of a tumour with much less damage to surrounding healthy tissue. Knowledge of energy losses as particles travel through matter is a leading ingredient for depth-dose simulations required for treatment planning. Therefore energy loss data needs to be highly accurate. A theory of SP was developed by Bethe [1] using the first Born approximation. The Bethe theory is applicable only at high projectile velocities. Current depth-dose simulations use improved Bethe-type theories, however even with numerous improvements accurate SP data in the entire range of projectile velocities necessary for proton therapy cannot be obtained.

The convergent close coupling (CCC) method has been developed to be a highly accurate approach to ion-atom and ion-molecule collisions (see [2] and references therein). It couples all possible channels, including rearrangement ones, and gives a complete picture of the process. The method has recently been applied to calculate the stopping cross sections (SCS). In the CCC method the SCS is defined as

$$S(v) = \sum_f (\epsilon_f - \epsilon_i)\sigma_{fi}(v), \qquad (1)$$

where σ_{fi} is the scattering cross section for the transition of the target from an initial state i to some final pseudostate f and $\epsilon_f - \epsilon_i$ is the energy gained by the target during that transition. The SP is the SCS multiplied by the number of target atoms per unit volume.

We have calculated antiproton and proton SCS for H, He, H_2, Ne, Ar, Kr, Xe and $H_2$0. Figure 1 shows the results for antiproton in He and H_2 in comparison with experimental data [3]. One can see that agreement between the CCC results and the experiment is very good for He. However, for H_2 there is significant discrepancy between our calculations and the data from the same experiment [3]. This is unexpected. A puzzling aspect arises when we take the combined-nuclei limit (i.e., set the internuclear distance to 0) in our H_2 code we perfectly reproduce our He results. It is also important to note that our ionisation cross sections for H_2 are in excellent agreement with experiment [2]. It appears that these two facts, namely, large disagreement for SCS and very good agreement for ionisation cross sections are inconsistent with each other. More results will be presented at the conference.

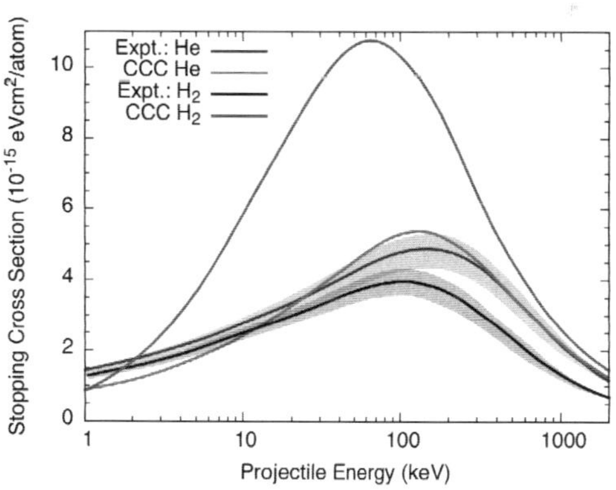

Figure 1. Stopping cross section for $\bar{p}$ in He and H_2. Experimental results [3] are shown by solid lines with shading representing the relevant error.

References

[1] H. Bethe 1930 *Annalen der Physik* **397** 325

[2] I. B. Abdurakhmanov *at al* 2013 *Phys. Rev. Lett.* **111** 173201

[3] M. Agnello *at al* 1995 *Phys. Rev. Lett.* **74** 371

[1]E-mail: jackson.bailey@postgrad.curtin.edu.au

Content from this work may be used under the terms of the Creative Commons Attribution 3.0 licence. Any further distribution of this work must maintain attribution to the author(s) and the title of the work, journal citation and DOI.
Published under licence by IOP Publishing Ltd

Journal of Physics: Conference Series **635** (2015) 022035 doi:10.1088/1742-6596/635/2/022035

Differential electron emission from multi-electronic targets under highly charged bare ion impact

Shubhadeep Biswas* [1], **D. Misra***, **J. M. Monti**†, **C. A. Tachino**†, **R. D. Rivarola**† **and L. C. Tribedi*** [2]

* Tata Institute of Fundamental Research, Mumbai 400005, India
† Institute de Fisica Rosario (CONICET-UNR), Universidad Nacional de Rosario, 2000

Synopsis We report a combined experimental and theoretical study of double differential electron emission in ionization of different multi-electronic atoms under fast highly charged bare ion impact. The main aim is to understand the electron dynamics for these targets which is quite different from one or two electron systems.

Study of doubly differential electron emission, in case of highly charged ion collision with atoms, can provide the microscopic details of the underlying electron dynamics. In this report, we provide a systematic study of the absolute double differential cross sections (DDCS) of electron emission in ionization of He, Ne and Xe atoms under the impact of 5-MeV/u and 6-MeV/u C^{6+} ions. An electrostatic hemispherical analyzer was used for detection of low energy (1-600 eV) electrons over a wide range of angles (20°-150°). The experimental data have been compared rigorously with the *post* and the *prior* forms of the CDW-EIS model. In general, the better agreement of the *prior* form compared to the *post* form indicates the role of the target passive electron dynamic screening effect for multi-electronic targets [1]. In case of the Xe target, several different features have been noticed. The energy distribution spectrum is dominated by different Auger transition lines right from the low energy region. Unlike He and Ne targets, the binary encounter peak in the angular distribution is almost absent. Besides, the nature of forward-backward angular asymmetry is also completely different from that of the He or Ne targets. As we approach the larger targets, the inner shell processes come into picture. This has been clearly seen from the estimates of sub-shell contributions to the total DDCS. In case of Ne, the electron emission spectrum is mostly dominated by the outer shell ionization process;

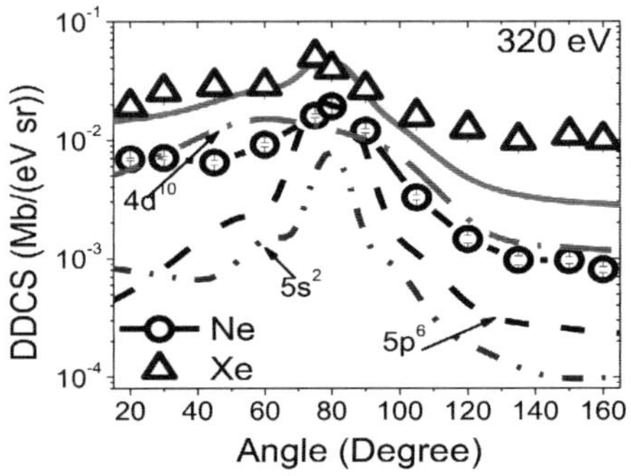

Figure 1. Angular distributions of electrons for Ne and Xe. The solid line represents total DDCS from CDW-EIS calculation. Other lines represent sub-shell contributions.

whereas for Xe, particularly in the higher energy region of the spectrum at extreme forward and backward angles, the inner shell ionization dominates over the outer shell contribution. This explains some of the observed differences to some extent. It is also realized that the inclusion of different characteristic excitation processes like Auger process, or some collective excitation processes is necessary to complete the theoretical description of the electron dynamics in case of the larger multi-electron targets [2].

References

[1] S. Biswas *et al* 2014 *Phys. Rev. A* **90** 052714

[2] S. Biswas *et al* 2015 *(J. Phys. B; accepted)*

[1] E-mail: shubhadeep@tifr.res.in

[2] E-mail: lokesh@tifr.res.in

Content from this work may be used under the terms of the Creative Commons Attribution 3.0 licence. Any further distribution of this work must maintain attribution to the author(s) and the title of the work, journal citation and DOI.
Published under licence by IOP Publishing Ltd

Journal of Physics: Conference Series **635** (2015) 022036 doi:10.1088/1742-6596/635/2/022036

Studies of atomic spin-orbit coupling synthesized with gradient magnetic fields

Xinyu Luo [1], Lingna Wu [1], Ruquan Wang [2,3], and L. You*[1,3]

[1]State Key Laboratory of Low Dimensional Quantum Physics, Department of Physics, Tsinghua University, Beijing 100084, China

[2]Beijing National Laboratory for Condensed Matter Physics, Institute of Physics, Chinese Academy of Sciences, Beijing 100080, China.

[3]Collaborative Innovation Center of Quantum Matter, Beijing, China

Synopsis: Atomic spin orbit coupling (SOC) can be synthesized with modulating gradient magnetic field (GMF). This is confirmed through studying collective dipole oscillations for a spin-1 atomic condensate in a harmonic trap after abruptly turning on SOC, and studying adiabatically adjusted equilibrium states when SOC strength is slowly ramped up. Further measurements reveal that SOC can be enhanced when the GMF modulation frequency approaches trapping frequency.

We report the observation of tunable spin-orbit coupling (SOC) in spin-1 atoms synthesized with periodically driven gradient magnetic fields (GMF) [1]. SOC represents an important type of synthetic gauge fields actively pursued in cold atom systems. Recently, a new scheme based on pulsed or periodically modulating GMF is proposed to synthesize one dimensional (1D) SOC in a spin-1 atomic Bose-Einstein condensate (BEC) [2,3]. Different from all previous experiments which demonstrate synthesized SOC with Raman coupling lasers, this new scheme employs only magnetic field, thus could avoid the heating from atomic spontaneous emission in the Raman scheme. The strength of SOC synthesized with GMF can be tuned simply through adjusting the impulse, or the time integrated area of GMF. This is confirmed based on the measurements of a) collective dipole oscillations of a BEC in a harmonic trap after abruptly turning on SOC, and b) its adiabatically adjusted equilibrium states when the strength of SOC is slowly ramped up.

The 1D SOC we synthesize is an equal weighted sum of the Rashba and Dresselhaus types. With SOC, the free atom dispersion from without SOC splits into multi sub-bands, with its original minimum centered at zero momentum shifted to $M_F k_{so}$ for spin component M_F. When SOC is turned on abruptly, atoms experience spin dependent restoring forces which induce both position and momentum oscillations for M_F =1,-1 components (Fig. 1). If SOC is slowly ramped up, atoms at the minimum will follow the ramp and stay at the shifted minimum (Fig. 2).

Our observations are in excellent agreement with theoretical predictions for both cases. This study also uncovers theoretical understanding and experimental confirmation that the strength of SOC can be enhanced making use of motional resonance associated with atomic center of mass in a harmonic trap. In addition to providing extra tunability and flexibility for GMF based scheme, this new finding also sheds light on experimental efforts towards synthesizing two-dimensional (2D) atomic SOC.

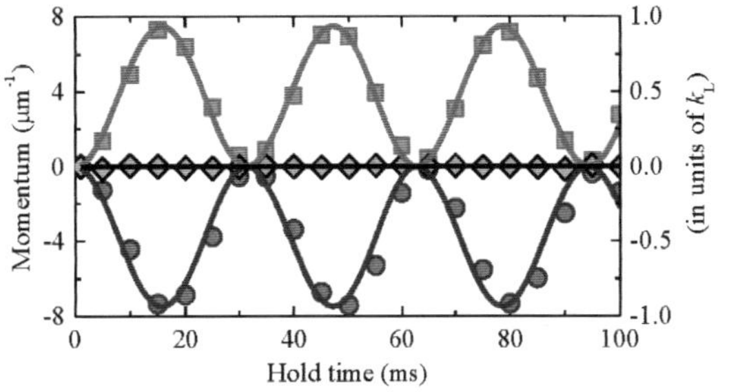

Figure 1. Measured momentum for the M_F=1, 0, -1 spin components respectively denoted by squares, diamonds, and disks as a function of hold time in a harmonic trap after abruptly turning on SOC.

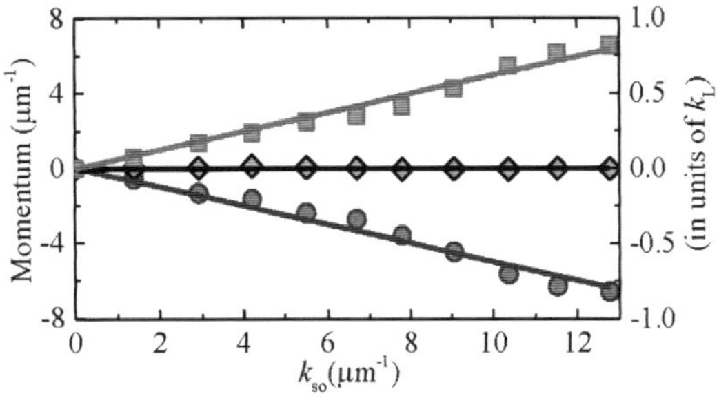

Figure 2. The same as in Fig. 1 but for the shifted minimum equilibrium momentums of the three spin components as a function of SOC strength k_{so}, after slowly ramping up SOC strength.

References

[1] X. Luo et al 2015 arXiv:1502.0709
[2] Z.-F. Xu et al 2013 *Phys. Rev. A* **87** 063634
[3] B. M. Anderson et al 2013 *Phys. Rev. Lett.* **111** 125301

* E-mail: lyou@mail.tsinghua.edu.cn

Content from this work may be used under the terms of the Creative Commons Attribution 3.0 licence. Any further distribution of this work must maintain attribution to the author(s) and the title of the work, journal citation and DOI.
Published under licence by IOP Publishing Ltd

Projectile coherence effects studied by *ab initio* calculations

F. Járai-Szabó*[1], S. Borbély*, and L. Nagy*

* Faculty of Physics, Babeş-Bolyai University, Kogălniceanu Street No. 1, 400084 Cluj, Romania

Synopsis Single ionization of helium produced by 100 MeV/u C^{6+} projectiles is studied theoretically. The ionization amplitudes are calculated by directly solving the two-electron Schrödinger equation in the framework of the semiclassical impact parameter approach. Projectile coherence effects are investigated through fully differential electron emission cross section ratios resulting from coherent and incoherent calculations based on the same amplitudes. The obtained results are compared to existing theories and experimental data, as well.

In the last few years, qualitative differences in case of ionization of helium by fast charged projectiles depending on the projectile beam coherence have been reported by experimental studies [1]. Fully differential cross sections (FDCS) for ionization of He by 3 MeV proton impact have been measured, with projectile beam of a coherence length larger than typical atomic scales. Compared to the incoherent 100 MeV/u C^{6+} projectile beam [2] pronounced effects have been observed.

In the present work we investigate theoretically this effect based on high precision $A(\vec{b})$ transition amplitudes calculated by using *ab initio* methods. In the employed approach [3] the ingoing projectile is treated as a classical particle moving on a straight line trajectory, while the two electrons of the target are treated quantum mechanically by exactly solving the time-dependent Schrödinger equation.

As described in [4] for calculating FDCS based on the transition amplitude $A(\vec{b})$, two possibilities arise. On one hand, by classical scattering theory impact parameter values b to certain momentum transfers q are assigned, and then the FDCSs are calculated [5]. By this impact parameter method each particle from the projectile beam is characterized by a single impact parameter value.Therefore, no coherence exists between different impact parameters. On the other hand, using the quantum scattering theory, the scattering matrix element is calculated by the inverse Fourier transform of the ionization amplitudes.The FDCSs are obtained as the square of the modulo of the scattering matrix element [6]. By these quantummechanical calculations the projectile beam is described as a coherent plane wave. Consequently, interference may occur between the ionization obtained for different impact parameters.

By calculating coherent and incoherent projectile FDCS ratios for single ionization of He by fast charged projectile impact, the significant differences between the two approaches are confirmed. Moreover, interference effects in the coherent projectile data are identified. As a result, the present calculations contribute to a better understanding of the kinematically complete experiments.

This work was supported by a grant of the Romanian National Authority for Scientific Research (CNCS UEFIS-CDI) project number PN-II-ID-PCE-2011-3-0192, by the Hungarian Scientific Research Fund OTKA No. K103917s.

References

[1] Wang X *et al.* 2012 *J. Phys. B: At. Mol. Opt. Phys.* **45** 211001.

[2] Schultz M *et al.* 2003 *Nature* **422** 48.

[3] S. Borbély *et al.* 2014 *Phys. Rev. A* **90** 052706.

[4] F. Járai-Szabó and L. Nagy 2015 *Eur. Phys. J. D* **69** 4.

[5] F. Járai-Szabó, L.Nagy 2009 *Nucl. Instrum. Methods B* **267** 292.

[6] R. D. Rivarola, P. D. Fainstein 2003 *Nucl. Instrum. Methods B* **205** 448.

[1]E-mail: jferenc@phys.ubbcluj.ro

Content from this work may be used under the terms of the Creative Commons Attribution 3.0 licence. Any further distribution of this work must maintain attribution to the author(s) and the title of the work, journal citation and DOI.
Published under licence by IOP Publishing Ltd

Journal of Physics: Conference Series **635** (2015) 022038 doi:10.1088/1742-6596/635/2/022038

Formation of molecular ion $LiHe^+$ by radiative association of metastable helium $He(2^3P)$ with lithium ions

M. Zámečníková*[1], L. Augustovičová*[2], W. P. Kraemer†[3], P. Soldán*[4]

* Charles University in Prague, Faculty of Mathematics and Physics, Department of Chemical Physics and Optics, Ke Karlovu 3, CZ-12116 Prague 2, Czech Republic

† Max-Planck-Institute of Astrophysics, Postfach 1371, D-85741 Garching, Germany

Synopsis Radiative association of metastable helium $He(2^3P)$ with lithium ions is theoretically investigated on the triplet-sigma manifold. Cross-sections and the corresponding rate coefficients for spontaneous and stimulated processes are calculated. Results are compared to other radiative and non-radiative processes in collisions of metastable helium with lithium ions.

Radiative association of metastable helium $He(2^3P)$ with lithium ions is investigated on the triplet-sigma manifold using a fully quantal approach [1]. Energy-dependent cross-sections for spontaneous and stimulated processes are calculated considering the association to rotational-vibrational states of the lower triplet-sigma electronic states $a^3\Sigma^+$, $b^3\Sigma^+$, and $c^3\Sigma^+$ from the continuum states of the $c^3\Sigma^+$ electronic state. The cross-sections for both the two-state processes ($c \to a$, $c \to b$) and one-state process ($c \to c$) are evaluated using highly accurate quantum calculations (see Table 1), following the methodology outlined by Augustovičová et al. [2] and making use of the potential energy curves, dipole moments, and transition dipole moments determined by Soldán and Kraemer [3]. All possible state-to-state transitions at thermal energies (for spontaneous association) or at higher background temperatures (stimulated association) are taken into account. The rate coefficients for the formation of the $LiHe^+$ molecular ion in the corresponding triplet-sigma electronic states are presented as functions of temperature. The calculations show that the rate coefficients for radiative association to the a and b states are several orders of magnitude larger than the one for the c state formation. On the other hand, only the one-state $c \to c$ process exhibits noticeable effects on the radiative association stimulated by black-body background radiation. The current results are compared with recent results obtained for the radiative association of lithium ions with metastable helium $He(2^3S)$ and $He(2^1S)$ [4, 5] and for the non-radiative inelastic processes in collisions of lithium ion with metastable helium [6]. Here it turned out that the radiative processes are by far the efficient ones in low temperature region (up to 2 000–8 000 K).

Table 1. Values of the cross-sections (in a_0^2) for $c \to a$, $c \to b$ and $c \to c$ processes for selected energies.

E [cm^{-1}]	$c \to a$	$c \to b$	$c \to c$
10	3.4×10^{-3}	2.7×10^{-5}	7.4×10^{-9}
100	7.2×10^{-4}	1.3×10^{-7}	7.2×10^{-10}
1 000	6.2×10^{-5}	1.1×10^{-14}	1.3×10^{-16}
10 000	4.6×10^{-7}	4.3×10^{-17}	5.9×10^{-17}

References

[1] B. Zygelman, A. Dalgarno 1990 *Astrophys. J.* **365** 239

[2] L. Augustovičová, V. Špirko, W. P. Kraemer, P. Soldán 2012 *Chem. Phys. Lett.* **531** 59

[3] P. Soldán, W. P. Kraemer 2012 *Chem. Phys.* **393** 135

[4] L. Augustovičová, W. P. Kraemer, P. Soldán 2014 *Astrophys. J.* **782** 46

[5] L. Augustovičová, W. P. Kraemer, P. Soldán 2014 *J. Quant. Spectros. Radiat. Transfer* **148** 27

[6] A. K. Belyaev, D. S. Rodionov, L. Augustovičová, P. Soldán, W. P. Kraemer 2015 *Mon. Not. R. Astron. Soc.*, in print

[1]E-mail: martina.zamecnikova@mff.cuni.cz
[2]E-mail: lucie.augustovicova@mff.cuni.cz
[3]E-mail: wpk@mpa-garching.mpg.de
[4]E-mail: pavel.soldan@mff.cuni.cz

Content from this work may be used under the terms of the Creative Commons Attribution 3.0 licence. Any further distribution of this work must maintain attribution to the author(s) and the title of the work, journal citation and DOI.
Published under licence by IOP Publishing Ltd

Journal of Physics: Conference Series **635** (2015) 022039 doi:10.1088/1742-6596/635/2/022039

Improving detection efficiency in a cryogenic environment - implications for DESIREE

E. K. Anderson*[1], S. Rosén*, P. Reinhed*, J. D. Alexander*, N. Sullivan†, A. Tremsin†, D. Gorelikov†, H. Zettergren*, H. Cederquist* and H. T. Schmidt*

* Department of Physics, Stockholm University, Alba Nova University Centre, SE-106 91 Stockholm, Sweden
† Arradiance, Inc.,142 North Road, Sudbury, MA 01776, USA

Synopsis The high electrical resistance of a microchannel plate at cryogenic temperatures presents a severe limitation on the detection count rate at low temperatures. We are currently investigating the detector properties of a novel atomic layer deposition activated MCP developed by Gorelikov *et al* [1]. This is of particular interest to us as MCPs are used for particle detection at the cryogenic DESIREE facility at Stockholm University.

Microchannel plate (MCP) detectors are widely used for position sensitive particle counting in a range of experimental systems. There are several cryogenic facilities, including the Double Electrostatic Storage Ring ExpEriment (DESIREE) [2] at Stockholm University that require MCP detectors. The electrical resistance of an MCP is temperature dependent, increasing strongly with decreasing temperature. At cryogenic temperatures the resistivity of the detectors is very high which severely limits the count rate as the deadtime becomes prohibitively long. Currently the best comercially available MCPs for use at low temperatures are extended dynamic range (EDR) MCPs which have a resistance in the order of 10^{11} Ω at ~ 14 K.

Gorelikov *et al* [1] report the development of atomic layer deposition (ALD) activated MCPs ideal for use in a cryogenic environment. Using ALD, conductive and secondary emission films are deposited on a borosilicate glass MCP substrate, producing an MCP with optimal resistance at cryogenic temperatures. It has not yet been shown that the ALD MCP can be successfully used for particle detection. If the ALD MCPs operation in a cryogenic environment is found to be superior to the currently used EDR MCPs, then it is our intention to use ALD MCPs in DESIREE.

Gorelikov and co-workers have provided a pair of ALD MCPs for testing. It is crucial to have a complete understanding of the ALD MCP function in a cryogenic environment if this kind of MCP is to be considered for installation. Investigations of the operation of the ALD MCPs are currently underway in the DESIREE test chamber. This is a double walled vacuum chamber which provides an UHV environment and a minimum temperature similar to that in DESIREE. The resistance of the ALD MCPs as a function of temperature and bias voltage has been measured by Gorelikov *et al* [1]. These measurements have been reproduced in our test system. The next step in the investigation is to use an alpha source to measure the count rate on the detector as a function of temperature and compare this to measured dark counts. The detectors will then be tested with an ion beam.

References

[1] D. Gorelikov *et al* 2014 *J. Vac. Sci. Technol. A.* **32** 020605

[2] H.T. Schmidt *et al* 2013 *Rev. Sci. Instrum.* **84** 055115

[1]E-mail: Emma.Anderson@fysik.su.se

Content from this work may be used under the terms of the Creative Commons Attribution 3.0 licence. Any further distribution of this work must maintain attribution to the author(s) and the title of the work, journal citation and DOI.
Published under licence by IOP Publishing Ltd

Journal of Physics: Conference Series **635** (2015) 022040 doi:10.1088/1742-6596/635/2/022040

Quasi-equilibrium in charge-state evolution for carbon, sulfur, and tungsten ions after passing through carbon foils

Alex M Imai* [1], **M. Sataka**[†‡], **M. Matsuda**[†], **S. Okayasu**[†], **K. Kawatsura**[§], **K. Takahiro**[‖], **K. Komaki**[#], **H. Shibata***[¶], **and K. Nishio**[†]

[*] Department of Nuclear Engineering, Kyoto University, Nishikyo, Kyoto 615-8540, Japan
[†] Japan Atomic Energy Agency (JAEA), Tokai, Ibaraki 319-1195, Japan
[‡] Research Facility Center for Science and Technology, University of Tsukuba, Ibaraki 305-8573, Japan
[§] Kansai Gaidai University, Hirakata, Osaka 573-1001, Japan
[‖] Kyoto Institute of Technology, Sakyo, Kyoto 606-8585, Japan
[#] RIKEN, Wako, Saitama 351-0198, Japan
[¶] The Institute of Scientific and Industrial Research, Osaka University, Ibaraki, Osaka 567-0047, Japan

Synopsis Charge-state evolution of 2.0 MeV/u carbon, sulfur, and 1.0 MeV/u tungsten ions after carbon-foil penetration has been studied experimentally in the foil thickness range covering both pre- and after the equilibrium. It was observed for the first time that mean charge-states for S^{q+} (q = 6–14) initial ions evolved to coincide in 12.3 at a target thickness of 6.9 μg/cm^2, showing a "quasi-equilibrium," and further evolved simultaneously to a real equilibrium value of 12.68, whereas those for S^{q+} (q = 15, 16) initial ions evolved directly to the real equilibrium, without through the "quasi-equilibrium," until the target thickness was increased to 100 μg/cm^2. Such kind of shell-effect was observed also for C-ions, and could be qualitatively reproduced by some simulations.

Both equilibrium and non-equilibrium charge-state distributions for 2.0 MeV/u S^{q+} (q = 6–16) and C^{q+} (q = 2–6) ions as well as 1.0 MeV/u W^{q+} (q = 13, 15, 28, 29, 30) ions after penetrating carbon foils have been investigated experimentally. Those wide ranges of the initial ion and charge-state have proved that charge-state distributions, mean charge-states, and distribution widths for projectile ions without K-shell holes, S^{q+} (q = 6–14) and C^{q+} (q = 2–4), once coincided at a target thickness of 6.9 and 5.7 μg/cm^2, respectively, showing a "quasi-equilibrium," and simultaneously evolved to establish real equilibrium when the foil thickness was increased further. Those for projectile ions with K-shell hole(s), $S^{15, 16+}$ and $C^{5, 6+}$, evolved differently and directly to the real equilibrium, established at a target thickness of around 100 μg/cm^2 or greater for S ions and at over 10 μg/cm^2 for C ions. The quasi- and real-equilibrium mean charge states for 2.0 MeV/u S ions were 12.3 and 12.68, respectively [1], whereas those for 2.0 MeV/u C ions were 5.48 and 5.57 [2]. The charge-state distributions, mean charge-states, and distribution widths for 1.0 MeV/u W ions established equilibrium at 10 μg/cm^2 in foil thickness. The mean charge state for 1.0 MeV/u W ions was 31.4, which was larger than the only existing empirical prediction [3]. Simulations using ETACHA code [4] and solution of simpler rate equations showed that the quasi-equilibrium was brought by a difference between the reaction-rates for K- and L-shell processes.

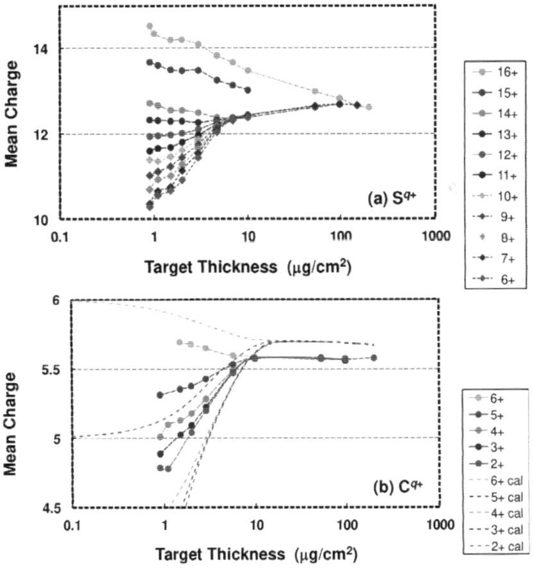

Figure 1. Mean charge sate evolution for 2.0 MeV/u (a) S^{q+} (q = 6–16) and (b) C^{q+} (q = 2–6) ions after penetrating C-foil targets.

References

[1] M. Imai *et al* 2009 *Nucl. Instr. and Meth.* B267 2675
[2] M. Imai *et al* 2015 *Nucl. Instr. and Meth.* B in press
[3] K. Shima *et al* 1992 *At. Data Nucl. Data Tables* 51 173
[4] J. P. Roset *et al* 1989 *J. Phys.* B22 33

[1] E-mail: imai@nucleng.kyoto-u.ac.jp

Content from this work may be used under the terms of the Creative Commons Attribution 3.0 licence. Any further distribution of this work must maintain attribution to the author(s) and the title of the work, journal citation and DOI.
Published under licence by IOP Publishing Ltd

XXIX International Conference on Photonic, Electronic, and Atomic Collisions (ICPEAC2015) IOP Publishing
Journal of Physics: Conference Series **635** (2015) 022041 doi:10.1088/1742-6596/635/2/022041

Collisions between spin polarized alkali atoms. Complex cross sections.

Victor Kartoshkin[1]

Ioffe Institute, Politechnicheskaya 26, 194021 St-Petersburg, Russia

Synopsis This work is devoted to the calculation of complex cross sections for the spin exchange collisions between polarized alkali atoms. This is of interest for two reasons. First, knowledge of the cross sections allows one to quantitatively analyze spin-exchange collisions. Second, the obtained results can be used in practice to create quantum electronic devices with optical orientation based on the spin-exchange principle.

Collisions of alkali-metal atoms in the ground state with the electron spin $\vec{S}=1/2$ are accompanied by the exchange of electron coordinates between the colliding particles, which leads to the polarization transfer between them, i.e., to the well-known phenomenon of spin exchange. Spin-exchange collisions between alkali atoms play an important role in quantum electronic devices, such as quantum magnetometers and time and frequency standards

The spin-exchange process can be described in terms of the complex spin-exchange cross sections

$$q^{AB} = \bar{q}^{AB} + i \cdot \bar{\bar{q}}^{AB}$$

The real part of the cross section ($\bar{q}^{AB}$) determines the transfer of orientation in a collision of particles, their relaxation, and the formation of higher polarization moments (the alignment and hyperfine polarization). The imaginary part ($\bar{\bar{q}}^{AB}$) of the cross section determines the magnetic resonance frequency shifts in the system of both Zeeman and hyperfine levels of atoms. Consequently, once the complex spin-exchange cross section is known, the processes that occur in spin-exchange collisions can be described.

In order to calculate the complex cross section one has to know potentials of the diatomic molecule, which is formed during the collision of alkali atoms in the ground state. This molecule might evolve on two terms – the singlet term V_s with the total spin $\vec{S}=0$ and the triplet term V_t with the total spin $\vec{S}=1$. Complex cross sections of spin exchange were calculated for collisions of alkali metal atoms based on the data of the singlet ($X^1\Sigma^+$) and triplet ($a^3\Sigma^+$) potentials that describe the interaction of these alkali atoms in the ground state. Calculations have been done for the following pairs of alkali atoms: K-Na [1], K-Rb [2], K-K, and Cs-Cs.

The complex spin-exchange cross section q^{AB} can be written in the form

$$q^{AB} = \frac{\pi}{k_{AB}^2} \sum_{l=0}^{\infty} (2l+1)\left[1 - T_0^{AB}(l) T_1^{AB}(l)^*\right].$$

Here $k_{AB}^2 = \mu_{AB} V_{AB} / h$ id the wave vector, μ_{AB} is the reduced mass of colliding atoms A and B, $V_{AB} = \left(8kT / \pi\mu_{AB} \right)^{1/2}$ is the average relative velocity of colliding atoms, and the sign * denotes the complex conjugate. The scattering matrix is: $T_S^{AB}(l) = \exp(2i\delta_S^{AB}(l))$.

In Fig.1 there are energy dependences of the real $\bar{q}^{AB}$ (1) and imaginary $\bar{\bar{q}}^{AB}$ (2) parts of complex cross section for Na_2 dimer.

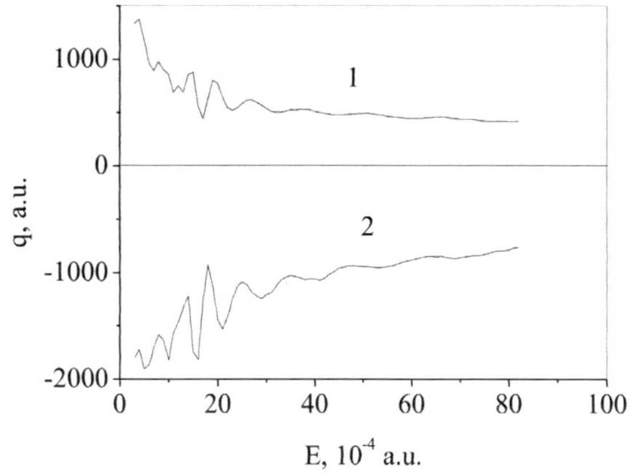

Figure 1. Real $\bar{q}^{AB}$ (1) and imaginary $\bar{\bar{q}}^{AB}$ (2) parts of complex cross section for Na_2 dimer

This work was supported in part by the RFBR (Grant № 14-02-004710)

References

[1] V.Kartoshkin *Opt.Spectrosc.* 2010 **109** 674
[2] V.Kartoshkin *Opt.Spectrosc.* 2011 **110** 665

[1] E-mail: victor.kart@mail.ioffe.ru
Content from this work may be used under the terms of the Creative Commons Attribution 3.0 licence. Any further distribution of this work must maintain attribution to the author(s) and the title of the work, journal citation and DOI.
Published under licence by IOP Publishing Ltd

Journal of Physics: Conference Series **635** (2015) 022042 doi:10.1088/1742-6596/635/2/022042

Observation of the spin polarization of Cs atoms during collisions with oriented metastable helium atoms

Victor Kartoshkin[1], **Sergei Dmitriev, Nikolai Dovator, George Klementiev**
Ioffe Institute, Politechnicheskaya 26, 194021 St-Petersburg, Russia

Synopsis We are the first to experimentally observe a magnetic resonance signal of $6^2S_{1/2}$ Cs atoms by absorption of light from a helium lamp that was used to optically orient metastable 2^3S_1 He atoms. The amplitude of the cesium signal proved to be almost three orders of magnitude lower than the amplitude of the magnetic resonance of 2^3S_1 He atoms. The theoretical explanation of the effect is presented.

Optical orientation of atoms by polarized radiation is a powerful tool for obtaining spin polarized atomic particles. Thus obtained polarized atomic particles have been used both to study various physical processes (e.g., to study atom–atom and atom–molecule collisions, to measure rate constants of collisional processes, etc.) and for practical purposes.

Upon optical orientation of atoms in a Cs–He mixture under gas discharge conditions, polarization transfer between particles can occur in the following collisions:

$$Cs(6^2S_{1/2})\uparrow + He(2^3S_1)\downarrow\downarrow \rightarrow Cs^+(2^1S_0) + He(1^1S_0) + e^-\downarrow \quad , (1)$$

$$Cs(6^2S_{1/2})\uparrow + He(2^3S_1)\downarrow\downarrow \leftrightarrow Cs(6^2S_{1/2})\downarrow He(2^3S_1)\uparrow\downarrow \quad , (2)$$

$$Cs(6^2S_{1/2})\uparrow + e^-\downarrow \leftrightarrow Cs(6^2S_{1/2})\downarrow + e^-\uparrow \quad , (3)$$

$$He(2^3S_1)\downarrow\downarrow + e^-\downarrow \leftrightarrow He(2^3S_1)\uparrow\downarrow + e^-\downarrow \quad . (4)$$

Here, the arrow denotes conventionally the direction of the electron spin of the particle.

As follows from (1)-(4), if the polarization transfer process between colliding particles involves the participation of metastable helium atoms, it is a result of collisional processes of two types—elastic and inelastic collisions. Thus, reactions (2)–(4) correspond to the elastic process, as a result of which the spin is transferred from one particle to the other, with the total momentum being preserved during collisions. At the same time, process (1) is an inelastic process. As a result of the collision of this type, the alkali atom is ionized (the Penning process) at the expense of the internal energy of the metastable helium atom (19.82 eV). As a result of collisions of this kind, the polarization is also transferred to an ensemble of alkali atoms.

Previously in works on optical orientation of atoms in alkali–helium plasma, indirect optical orientation of metastable helium atoms and electrons have been observed in the case in which Cs atoms were optically oriented by resonant radiation of a cesium lamp. It follows from reactions (1)–(4) that a pattern of this kind should also be observed in the case of an experiment in which metastable helium atoms are optically oriented in the 2^3S_1 state, whereas the polarizations of cesium atoms in the ground state and of electrons arise as a result of collisions with polarization transfer (by virtue of the symmetry of reactions (1)–(4) with respect to Cs and He atoms). However, experiment tally, this has not been detected. This paper describes our successful attempt to observe magnetic resonance signals of polarized Cs.

The explanation of a small value of the observed effect is the following. Since the values of the nuclear and electron spins of the cesium atom differ by seven times, upon the redistribution of the polarization between the electron and nuclear systems, the value of the electron polarization becomes considerably smaller compared to the polarization that was initially transferred upon collision. Subsequently, under the magnetic resonance conditions, the polarization of the hyperfine **F** level of the Cs atom is destroyed. In this case, a change in the electron polarization of the cesium atom that the helium atom "senses" upon collision will also be small, since, as was noted above, the electron spin of the cesium is considerably smaller than its nuclear spin. Therefore, as a result of these two processes, the change in the polarization transferred to cesium will affect helium more than an order of magnitude more weakly than in the case in which both the registration and the pumping are performed using the light from the cesium lamp.

References

[1] S.Dmitriev *et al 2014 Opt.Spectrosc.* **116** 216

[1]E-mail: victor.kart@mail.ioffe.ru

Content from this work may be used under the terms of the Creative Commons Attribution 3.0 licence. Any further distribution of this work must maintain attribution to the author(s) and the title of the work, journal citation and DOI.
Published under licence by IOP Publishing Ltd

Journal of Physics: Conference Series **635** (2015) 022043 doi:10.1088/1742-6596/635/2/022043

State-resolved measurements of mutual neutralization at subthermal collision energies

R. F. Nascimento*# [1], N. de Ruette*† , A. Dochain‡, M. Kaminska*, M. H. Stockett§*, H. T. Schmidt*, H. Cederquist*, and X. Urbain‡ [2]

* Department of Physics, Stockholm University, Stockholm, SE-106 91, Sweden
#Centro Federal de Educação Tecnológica Celso Suckow da Fonseca - 25620-003, Petrópolis, Brazil
† Columbia Astrophysics Laboratory, Columbia University, New York, NY 10027, U.S.A.
§ Department of Physics and Astronomy, Aarhus University, Ny Munkegade 120, DK 8000 Aarhus C, Denmark
‡ Institute of Condensed Matter and Nanosciences, Université catholique de Louvain, B-1348 Louvain-la-Neuve, Belgium

Synopsis We have measured the total cross sections and kinetic energy distributions of the neutral fragments for the mutual neutralization of C^- and O^- with C^+, N^+ and O^+ by using a merged beam apparatus. A fine resolution in the kinetic energy spectra of the product neutral atoms has allowed us to uniquely identify the atomic states which are participating in the mutual neutralization process. A detailed analysis of the branching ratios sheds new light on these charge transfer processes.

Mutual neutralization studies are generally limited to energies above a few eV, and do not specify the electronic state of the products, merely indicating a band of principal quantum numbers based on time-of-flight intervals [1]. This limitation of the merged beam technique results from the rapid degradation of the energy resolution with collision energy [2].

We recently upgraded the merged beam set-up in operation at UCLouvain (Louvain-la-Neuve, Belgium) to reach meV collision energies, and incorporated three-dimensional product imaging, as performed by a pair of position sensitive detectors located 3 to 5 meters downstream from the overlapping region. Besides providing clear coincidence signals, this technique gives unambiguous identification of the electronic states of the products via the measurement of kinetic energy release. The resolution (50 meV FWHM at 1 eV) is dominated by the uncertainty of the flight distance to the detectors, due to the finite length of the interaction region. Knowing the angular distribution of the products in the laboratory and center-of-mass frame, allows total and differential cross sections to be retrieved. They are put on an absolute scale by the simultaneous measurement of associative ionization products analyzed by a 180° magnet in combination with a 30° deflector.

Preliminary results for C^- and O^- in collision with C^+, N^+, and O^+, are presented, which reveal unexpected propensity rules regarding angular momentum and spin of the transferred electron. Further measurements are needed to disentangle the role of the electron affinity from that of the electronic configuration. Such studies will serve to benchmark the double electrostatic storage ring DESIREE now in operation at Stockholm University.

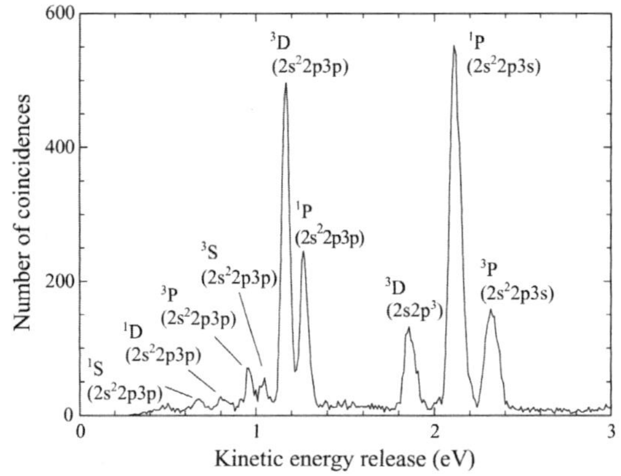

Figure 1. Kinetic energy release spectrum for the C^++O^- mutual neutralization. Terms and configurations refer to the final state of C in combination with $O(^3P)$.

References

[1] Terao M, Szücs S, Cherkani M, Brouillard F, Allan R J, Harel C, and Salin A 1986 *Europhys. Lett.* **1** 123

[2] Brouillard, F and Claeys, W 1983 in *Physics of Ion-Ion and Electron-Ion Collisions* (New York: Plenum Press) 415-459

[1]E-mail: rodrigo.nascimento@fysik.su.se
[2]E-mail: xavier.urbain@uclouvain.be

Content from this work may be used under the terms of the Creative Commons Attribution 3.0 licence. Any further distribution of this work must maintain attribution to the author(s) and the title of the work, journal citation and DOI.
Published under licence by IOP Publishing Ltd

Single and double electron capture cross sections in keV-collisions between fully stripped ions with helium atom

Aicha Ibaaz[1] **and Alain Dubois**[2]

Laboratoire de Chimie Physique-Matière et Rayonnement,
Université Pierre et Marie Curie - CNRS, 75005 Paris, France

Synopsis We present cross section calculations for single-, double-capture, and double capture to auto-ionizing states occurring in the course of collisions between fully stripped ions A^{q+} (q≤10) and helium atom at impact energies ranging from 0.25 to 625 keV/u. These calculations were performed by applying a semiclassical non-perturbative close coupling approach, based on the expansion of the scattering wave function into asymptotic bielectronic states with proper translational conditions and including both static and dynamic electronic correlation.

For the four last decades, the semiclassical nonperturbative close coupling formalism has been developed, improved and widely used to describe mainly mono-electronic processes occurring during collisions between atoms and molecules, neutral or charged. Such model has been proved to be reliable and unavoidable for intermediate impact energies, i.e. when the projectile-target relative velocity is of the same order of the (classical) velocity of the active electrons.

The objective of the present work is to apply this approach that has been recently improved to describe *ab initio* the dynamics of up to four electrons of the scattering system [1]. In the conference we shall report cross sections for collisions between He and fully stripped ions of elements of atomic number (Z_p) up to 10 in the 0.25-625 keV/u impact energy domain. We shall especially focus on single capture (SC) and double electron capture (TDC) which are the important ones for the considered energies. The analysis of these processes shows that they strongly depend upon Z_p and are highly final states selective. As an example, for Li^{3+}-He collisions, SC mainly populates n=2 states and our results show good agreement with other theoretical [2] and experimental [3-4] data. Concerning TDC processes which are described for the first time by a true two-electron non perturbative approach we can distinguish true double capture (DC), for which the projectile keeps the two captured electrons, from DC into autonionizing states (AIDC), where the projectile loses one electron. We shall present detailed results concerning the relative contribution of these two processes as function of the projectile charge and impact velocity: for low Z_p DC process is the dominant one while AIDC takes over for increasing Z_p.

Our results for He^{2+}-He and Li^{3+}-He systems are in good agreement with the experimental data [3, 5] but that is not the case for the other systems for which rare results are available.

References

[1] N. Sisourat, I. Pilskog and A. Dubois 2011 Phys. Rev. A **84** 052722
[2] L. F. Errea, F. Martin, L. Mendez, A. Riera and M. Yanez 1986 J. Chem. Phys. **84** 5422
[3] M. B. Shah and H. B. Gilbody 1985 J. Phys. B **18** 899
[4] J A. Guffey, L. D. Ellsworth and J. R. MacDonald 1977 Phys. Rev. A **15** 5
[5] M. E. Rudd, T. V. Guffey and A. Itoh 1985 Phys. Rev. A **32** 4

[1] E-mail: ibaaz.aicha@yahoo.com
[2] E-mail: alain.dubois@upmc.fr

Content from this work may be used under the terms of the Creative Commons Attribution 3.0 licence. Any further distribution of this work must maintain attribution to the author(s) and the title of the work, journal citation and DOI.
Published under licence by IOP Publishing Ltd

Journal of Physics: Conference Series **635** (2015) 022045 doi:10.1088/1742-6596/635/2/022045

Collision Induced Dissociation of PAHs and Biomolecules

L. Giacomozzi*[1], N. de Ruette*, M. Gatchell*, M. H. Stockett* †, H. T. Schmidt*, H. Zettergren* and H. Cederquist*

* Department of Physics, Stockholm University, Stockholm, SE-106 91, Sweden

† Department of Physics and Astronomy, Aarhus University, Ny Munkegade 120, DK 8000 Aarhus C, Denmark

Synopsis We have performed an extensive set of experiments with focus on collisions between PAH or biomolecular cations and He atoms at center-of-mass energies around 100 eV. In this energy range, non-statistical fragmentation processes are important.

Laboratory studies of the chemistry of planetary atmospheres and interstellar medium have historically focussed primarly on small and simple molecules, such as benzene. In more recent years, the attention has shifted towards larger molecules, mostly of organic nature. Among them, a special place is reserved for Polycyclic Aromatic Hydrocarbons (PAHs). PAHs are believed to be important components of the molecular inventory in space. However, the mechanism of collision between these molecules and other molecules or atoms is not yet very well understood [1]. In particular, the fragmentation and dissociation pathways and the corresponding absolute cross sections are parameters that are of particular interest in this context.

We have recently measured absolute cross sections for knocking out single carbon atoms in collision between PAH ions and noble gas atoms [2] [3] [4]. This gives highly reactive fragments which are not formed in statistical fragmentation processes and may be key intermediate steps for the formations of larger molecules in, e.g., the interstellar medium [5]. The knockout mechanism is expected to be important for any molecular system with a center of mass collision energy around 100 eV and down to the threshold energy for the process. PAHs are ideal for studying this process as their two-dimensional rigid structures makes further decay after the knockout less likely and the single carbon loss is strongly suppressed when the fragmentation is statistical [2] [3] [4].

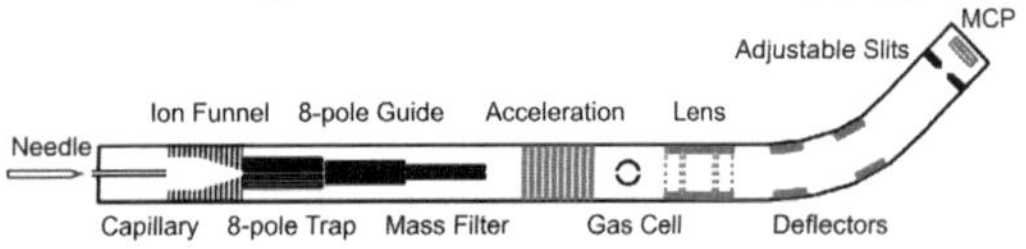

Figure 1. The Collision Induced Dissociation (CID) setup at Stockholm University.

[1]E-mail: linda.giacomozzi@fysik.su.se

However, it is more difficult to disentangle nonstatistical from statistical processes for three dimensional systems especially if they are also more fragile such as the biomolecules are.

Studies of collisions between molecular ions and atoms has been performed using a Collision Induced Dissociation (CID) type setup, which is a part of the DESIREE facility [6] [7] at Stockholm University. Briefly, this setup consists of an electrospray ionization source followed by a quadrupole mass filter and a collision gas cell. After the collisions, the fragment ions are analyzed by two pairs of electrostatic deflectors and recorded with a position-sensitive microchannel plate detector (see Fig. 1). This setup allows us to measure fragment mass spectra and absolute total fragmentation cross sections from attenuation measurements of the primary ion beam as a function of the gas pressure in the cell. Here we report our latest CID experiments on PAHs (e.g. Pyrene-based molecules) and biomolecules (e.g. Adenosine 5-monophosphate (AMP)) at center of mass energies of about 100 eV.

References

[1] E. Herbst. *et al* 2005 *J. Phys. Chem. A.* **109** 4017

[2] M. H. Stockett *et al* 2014 *Phys. Rev. A.* **89** 032701

[3] M. H. Stockett *et al* 2014 *Phys. Chem. Chem Phys.* **16** 21980

[4] M.Gatchell *et al* 2014 *Int. J. Mass Spectrom.* **365-366(0)** 260-265

[5] T.Chen *et al* 2014 *J. Chem. Phys.* **140(22)** 224306

[6] H. T. Schmidt *et al* 2013 *Rev. Sci. Instrum.* **84** 055115

[7] R. D. Thomas *et al* 2011 *Rev. Sci. Instrum.* **82** 065112

Content from this work may be used under the terms of the Creative Commons Attribution 3.0 licence. Any further distribution of this work must maintain attribution to the author(s) and the title of the work, journal citation and DOI.
Published under licence by IOP Publishing Ltd

XXIX International Conference on Photonic, Electronic, and Atomic Collisions (ICPEAC2015) IOP Publishing
Journal of Physics: Conference Series **635** (2015) 022046 doi:10.1088/1742-6596/635/2/022046

The magnetic toroidal sector: a broad-band electron-positron pair spectrometer

S. Hagmann[1,2 1], P.M. Hillenbrand[1,3 2], Yu. Litvinov[1], U. Spillmann[1], K. Stiebing[2]

[1] GSI Darmstadt, Germany , [2] Inst. f. Kernphysik, Univ. Frankfurt, Germany, [3] Univ. Giessen, Germany,

Synopsis We discuss the suitability of a toroidal magnetic spectrometer for free-free pair spectroscopy.

For heavy-ion atom collisions at near-relativistic collision velocities and above, electron-positron pairs from non-nuclear, atomic processes appear as significant reaction products. The underlying production mechanisms are a central topic of QED in strong fields. Kinematically complete experiments at the future relativistic storage-ring HESR at FAIR are focussing on momentum spectroscopy of emitted electrons and positrons and corresponding recoil ions. The first experiments will investigate capture pair production and negative continuum dielectronic recombination, in both cases the electron of the generated pair is captured into a bound atomic state and the positron emitted into the continuum in a narrow cone around the beam direction will be analyzed with a forward magnetic spectrometer in coincidence with the charge changed projectile[1,2].

In the related process free-free pair production both leptons, electron and positron, are emitted into the continuum. Theoretical calculations on the vector-momentum correlation of electron positron pairs beyond the elementary back-to-back emission in the emitter frame are not available and experimental benchmark data on the vector momentum correlation of electron positron pairs are highly desirable-for better elucidation of the details of the production mechanism[3].

For enabling a simultaneous spectroscopy of both leptons we are studying the previously described toroidal sector magnetic reaction microscope[4] for application as a free-free pair spectrometer in a storage ring and for electron/positron energies up to 1 MeV.

In this context we investigate the electron-optical properties of a modified 90^0 magnetic toroidal sector configuration for electrons and positrons suitable for implementation into a storage ring with a supersonic jet target and covering a wide range of emission into the forward hemisphere. The simulation calculations are performed using the OPERA-3d 17-64 code[5]. Our calculations use the TOSCA subpackage of the OPERA code for the calculation of the magnetic fields of the toroidal coils configuration and calculation of the trajectories of electrons and positrons in the magnetic field.

For electrons and positrons mapped from the target zone onto large 2D position sensitive detectors in the focal plane the instrument exhibits a large momentum dispersion of electrons and positrons in the upper and lower half-sphere, thus enabling a simul-taneous detection of both, electron and positron from a pair, over a large solid angle of emission. In the fig it is clearly visible how for a suitably cho-sen combination of magnetic field B and lepton ki-netic energy the desired lepton energies may be fo-cused on distinct locations in the focal plane.

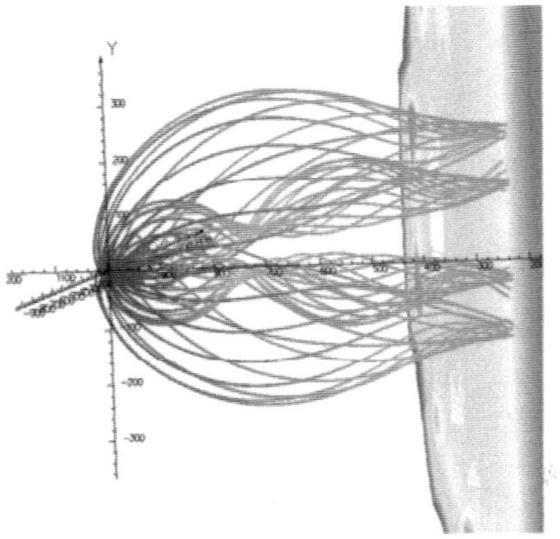

Figure 1. Trajectories of electrons (lower half-plane) and positrons (upper half-plane) of 12 keV and 50 keV energy, respectively, from the source at the jet target to the detector plane with 2D position sensitive detectors at the end of the toroidal sector.

References

[1] HESR-Techn. Design Report, GSI 2015
[2] P.M. Hillenbrand et al. Phys. Scr., submitted 2015
[3] A. Voitkiv, priv. communication, 2014
[4] X. Wang, S. Hagmann Meas. Sci. Technol. **18** (2007) 161-166
[5] OPERA-3d, Vector fields limited, Kidington, UK

[1] E-mail: s.hagmann@gsi.de

[2] E-mail: p.m.hillenbrand@gsi.de

Content from this work may be used under the terms of the Creative Commons Attribution 3.0 licence. Any further distribution of this work must maintain attribution to the author(s) and the title of the work, journal citation and DOI.
Published under licence by IOP Publishing Ltd

Measurement of Plasma Potential using Deceleration technique

T. Sairam*[1], Pragya Bhatt*, Ajit Kumar†, Herendra Kumar‡, J. Rajput‡, and C. P. Safvan*

* Inter University Accelerator Centre, Aruna Asaf Ali Marg, New Delhi-110067, India.
† Department of Physics, Jamia-Millia Islamia, New Delhi, 110025, India.
‡ Department of Physics and Astrophysics, University of Delhi, Delhi 110007, India.

Synopsis We report a novel technique to measure the plasma potential. In this technique, we extract an ion beam into the chamber with few keV of energy and then decelerate it using series of electrodes till the deceleration potential nullifies the extraction potential and measure the retaining energy of the beam using retarding plate analyzer (RPA) to determine the plasma potential.

To understand fundamental interaction between ions and atoms, the generation of variety of charged ions is a requisite and therefore different sources have been constructed for this cause. Among these, the Electron Cyclotron Resonance (ECR) source has a unique position in a way that the highly charged particles can be produced without utilizing any filament. As a result there is no requirement to replace the filament in the source and the source can operate for days by feeding only a small amount of gas. However, it is not easy to characterise the source as there are many parameters involved which can change the constitution of the plasma. The theoretical calculations of Drentje et al. [1] show that the plasma potential can be considered as a parameter which can reflect the dynamics with in the plasma. Different methods like Langmuir probe, retardation plate analyzer [2], decelerating lens [3] have been used earlier to measure the plasma potential. In this article, we report a novel technique where we have used decelerating electrodes and a retarding plate analyzer in unison to measure the plasma potential. This technique offers better resolution than the techniques when used independently. We have used six electrodes to decelerate a 15 keV/q beam where the last lens and the chamber has been floated to 15 kV. The voltages on the other five lenses have been optimized [4] to minimize the divergence of the beam. At the center of the chamber a Retarding Plate Analyzer (RPA) has been mounted to measure the energy of the beam. Since, the beam has been extracted and repelled by the same potential, the additional energy of the beam determines the plasma potential. A typical retardation curve for Ar^{7+} at 6 W microwave power and 3.4 x 10^{-8} Torr extraction pressure has been shown in fig.1. The extrapolation of the linear region in the retardation curve to the zero beam current gives us the plasma potential $\sim$ 9.7 eV. The subtraction of the extrapolated linear curve from the retarded curve provides an estimate of the ion energy spread inside plasma having FWHM $\sim$ 1.5 eV as shown in fig.1.

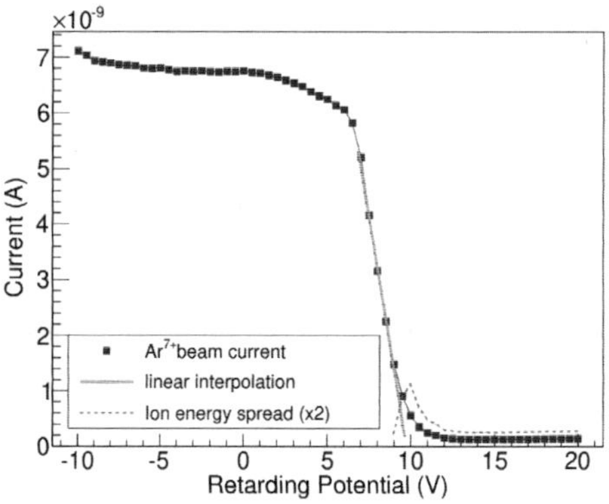

Figure 1. A typical retardation curve (legends are described inside the figure).

References

[1] A. G. Drentje *et al.* 2002 *Rev. Sci. Instrum.* **73** 516.

[2] O. Taravainen *et al.* 2004 *Rev. Sci. Instrum.* **75** 3138.

[3] P. R. Harris *et al.* 2010 *Rev. Sci. Instrum.* **81** 02A310.

[4] J. Rajput *et al.* 2010 *Rev. Sci. Instrum.* **81** 043301.

[1]E-mail: sairamtvv@gmail.com

Content from this work may be used under the terms of the Creative Commons Attribution 3.0 licence. Any further distribution of this work must maintain attribution to the author(s) and the title of the work, journal citation and DOI.
Published under licence by IOP Publishing Ltd

Journal of Physics: Conference Series **635** (2015) 022048 doi:10.1088/1742-6596/635/2/022048

A quantum mechanical analysis of the X-ray spectra from single-electron capture in Ne^{10+} collisions from He, Ne, and Ar atoms

Anthony C. K. Leung [1] **and Tom Kirchner** [2]

Department of Physics and Astronomy, York University, Toronto, Ontario, Canada

Synopsis The X-ray emission spectra due to single-electron capture in 4.54 keV/u Ne^{10+} collisions with He, Ne and Ar atoms are calculated using the two-center basis generator method. Present results of relative cross sections and X-ray spectra are compared across previous calculations and experimental measurements.

Experimental measurements of X-ray spectra from single-electron transfer in 4.54 keV/u Ne^{10+} collisions with He, Ne and Ar target atoms were first reported and compared with classical trajectory Monte Carlo (CTMC) calculations in [1]. Recent calculations using the two-center atomic-orbital close-coupling (TC-AOCC) method [2] were also compared with [1]. Results shown in these works indicate room for improvements.

In the present work, we calculate the state-selective single capture probabilities using the time-dependent two-center basis generator method (TC-BGM) [3]. We consider two variants of modeling the electron-electron interaction in terms of an effective single-particle potential: the no response approximation and a target response model to account for time-dependent screening effects [4]. We also consider two statistical perspectives for the capture probabilities: net capture and the exclusive single capture [5, 6]. The present relative cross sections using the exclusive single capture model show an overall good agreement with the experimental measurements and CTMC calculations in [1] across all target atoms. Differences in relative cross sections between the response and no response variants do not appear significant.

The calculated cross sections were then used in a radiative cascade model where they represent the initial population for a certain nl state [1]. Using the transition rates of the hydrogenlike Ne^{9+} ion we calculate the relative yields of the Lyman series, convolve the peaks to 126 eV full-width at half maximum and normalize them to the experimental areas [1]. Overall, the present X-ray spectra show reasonable agreement with the experimental measurements. We also compare the present X-ray peaks with CTMC and TC-AOCC calculations. In many cases, the present calculations of the X-ray peaks show better agreement than the CTMC and/ or TC-AOCC calculations; e.g, for single-electron capture to $n = 4$ in Ne^{10+}-He collisions (Fig. 1). Different calculation methods giving different X-ray peak heights can be understood by examining the distribution of the capture cross sections across the nl states.

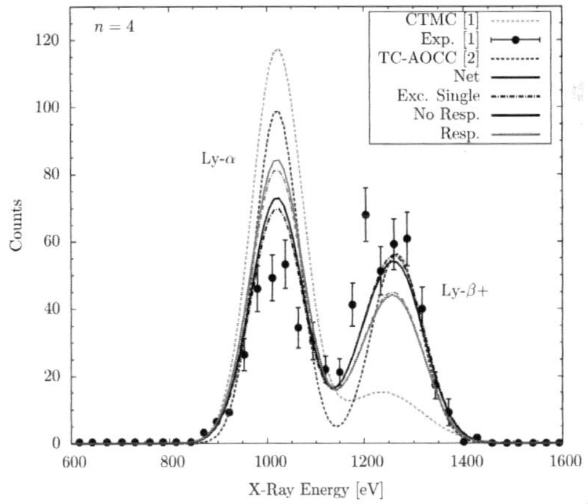

Figure 1. X-ray spectrum from single-electron capture to $n = 4$ in Ne^{10+}-He collisions.

This research was funded by the Ontario Graduate Scholarship and NSERC (Canada). High performance computing resources were provided through SHARCNET.

References

[1] R. Ali *et al* 2012 *Astrophys. J. Lett.* **716** L95

[2] L. Liu *et al* 2014 *Phys. Rev. A* **89** 012710

[3] M. Zapukhlyak *et al* 2005 *J. Phys. B* **38** 2353

[4] T. Kirchner *et al* 2000 *Phys. Rev. A* **62** 042704

[5] T. Kirchner *et al* 1999 *Phys. Rev. A* **61** 012705

[6] T. Kirchner *et al* 1999 *Phys. Scr.* **T80** 416

[1]E-mail: leungant@yorku.ca
[2]E-mail: tomk@yorku.ca

Content from this work may be used under the terms of the Creative Commons Attribution 3.0 licence. Any further distribution of this work must maintain attribution to the author(s) and the title of the work, journal citation and DOI.
Published under licence by IOP Publishing Ltd

XXIX International Conference on Photonic, Electronic, and Atomic Collisions (ICPEAC2015) IOP Publishing
Journal of Physics: Conference Series **635** (2015) 022049 doi:10.1088/1742-6596/635/2/022049

Electron emission spectra of U^{28+}-ions colliding with gaseous targets

P.-M. Hillenbrand[1,2 1], S. Hagmann[1,3 2], J. M. Monti[4], R. D. Rivarola[4], K.-H. Blumenhagen[5,6], C. Brandau[1,2], W. Chen[1], R. D. DuBois[7], A. Gumberidze[1], D. L. Guo[8], M. Lestinsky[1], Yu. A. Litvinov[1], A. Müller[2], M. S. Sanjari[1], S. Schippers[2], U. Spillmann[1], S. Trotsenko[1,5], G. Weber[5], and Th. Stöhlker[1,5,6]

[1]GSI Darmstadt, Germany, [2]Universität Giessen, Germany, [3]Universität Frankfurt, Germany, [4]CONICET Rosario, Argentina, [5]Helmholtz-Institut Jena, Germany, [6]Universität Jena, Germany, [7]Missouri University Rolla, USA, [8]IMP Lanzhou, China

Synopsis The electron-loss to continuum has been studied experimentally and theoretically for the collision systems $U^{28+} + H_2$ at a collision energy of 50 MeV/u, $U^{28+} + N_2$ at 30 MeV/u, and U^{28+} + Xe at 50 MeV/u.

Projectile ionization in ion-atom collisions has a significant relevance for understanding one of the fundamental processes of atomic physics. Its analysis has been shown to be of great importance for the beam physics of heavy-ion accelerators, since their beam intensities are limited by space-charge effects. Highest particle beam intensities can thus only be achieved by decreasing the charge state of the accelerated ions at the expense of an increased magnetic rigidity required for the same energy. Here, the ionization of heavy medium-charged ions by atoms of the rest gas plays a crucial role, since charge-exchanged ions are lost during the acceleration process. In the heavy-ion accelerators of FAIR, U^{28+}-ions are planned to be accelerated consecutively in the synchrotrons SIS18 and SIS100 from an energy of 11.4 MeV/u to 2.7 GeV/u. The charge state of the accelerated ions is far from the equilibrium charge state. For those charge states *a priori* calculations of projectile ionization have encountered substantial difficulties.

Complementary to studying absolute ionization cross sections [1], this motivates the study of the mechanisms of differential cross sections for the ionization of U^{28+} in collisions with different gaseous atomic targets, presented in this contribution. The measurement of the energy spectra of cusp electrons for the collision systems U^{28+}+H_2 at 50 MeV/u, U^{28+}+N_2 at 30 MeV/u, and U^{28+}+Xe at 50 MeV/u is shown. Unprecedentedly clean experimental conditions were facilitated by using the magnetic forward-angle electron spectrometer [2] installed in the experimental storage ring ESR of the heavy-ion accelerator facility GSI Helmholtzzentrum für Schwerionenforschung.

The experimental setup enabled the measurement of differential cross sections by studying the cusp-electron peak, i.e., the energy distribution of electrons emitted under an angle of $\vartheta_e \approx 0°$ with respect the projectile beam, having a velocity similar to the projectile velocity. These electrons are ionized from the projectile during the collision with the gaseous target atoms and form the electron-loss-to-continuum cusp. Remarkable deviations from a symmetric cusp shape measured in the laboratory frame-of-reference can be traced back to an anisotropic electron emission in the projectile frame, and asymmetries in the cusp shapes have been shown to be a clear messenger for the underlying charge-transfer mechanisms [2, 3, 4].

The measured spectra are compared with theoretical calculations based on the continuum-distorted-wave (CDW) approach, as a test for the description of ionization of heavy multi-electron projectiles in collisions with atomic targets [5]. The applied theoretical calculations required an extension of the approach used previously for the ionization of light neutral targets by dressed projectiles [6] towards the ionization of heavy dressed projectiles by neutral targets.

References

[1] G. Weber *et al.* 2015 *Phys. Rev. ST Accel. Beams*, accepted for publication

[2] P.-M. Hillenbrand *et al.* 2014 *Phys. Rev. A* **90** 042713

[3] P.-M. Hillenbrand *et al.* 2015 *Phys. Rev. A* **91** 022705

[4] P.-M. Hillenbrand *et al.* 2014 *Phys. Rev. A* **90** 022707

[5] P.-M. Hillenbrand *et al.*, in preparation

[6] J. M. Monti *et al.* 2011 *J. Phys. B* **44** 195206

[1]E-mail: P.M.Hillenbrand@gsi.de
[2]E-mail: S.Hagmann@gsi.de

Content from this work may be used under the terms of the Creative Commons Attribution 3.0 licence. Any further distribution of this work must maintain attribution to the author(s) and the title of the work, journal citation and DOI.
Published under licence by IOP Publishing Ltd

Journal of Physics: Conference Series **635** (2015) 022050 doi:10.1088/1742-6596/635/2/022050

Classical treatment of Li^{2+}+Ar and He^{2+}+Ar collisions

A. Jorge* [1], **Clara Illescas*** [2], **B. Pons**† [3]

* Departamento de Química, Módulo 13, Univ. Autónoma de Madrid, Cantoblanco, 28049 Madrid, Spain

† CELIA, Univ. Bordeaux - CNRS - CEA, 351 Cours de la Libération, 33405 Talence, France

Synopsis Classical Trajectory Monte Carlo calculations are carried out for Li^{2+}+Ar and He^{2+}+Ar collisions, motivated by recent experiments on these systems. Cross sections for electron capture, projectile electron loss and target multiple ionization processes are evaluated and compared to the experimental values in the 75-500 keV/amy impact energy range.

Absolute cross sections for electron capture and projectile electron loss, eventually accompanied by target ionization, have been measured recently in Li^{2+}+Ar collisions in the 75-500 keV/amu impact energy range [1]. We therefore performed calculations of these cross sections, in the framework of the impact-parameter Classical Trajectory Monte Carlo (CTMC) method [2]. Our calculations employ a two-active electron description, with one active electron on both target and projectile centres, in order to address dynamical processes involving both target and projectile electrons [3]. The interaction of these active electrons with the frozen target ionic core Ar^{+} is described by means of a (mean-field) model potential, similarly to what was performed in [4]. High-order processes, as target multiple ionization, are evaluated within the Independent Particle Model (IPM, [5]), where the multi-electron probabilities are obtained as products of one-electron probabilities.

The comparison of computed and measured cross sections enables to gauge the accuracy of the IPM and underlying model potential descriptions. We illustrate this in Figure 1 where are displayed the cross sections for multiple ionization of Ar accompanied by single electron loss of Li^{2+}. The agreement between theory and experiment is satisfactory, which signs the adequacy of the IPM to calculate high-order cross sections and of the classical collisional treatment to compute the underlying one-electron probabilities, from intermediate to high impact velocities.

As stated in [1], cross sections for Li^{2+}+Ar collisions can interestingly be compared to those for He^{2+}+Ar collisions [6], in order to assess the influence of projectile ionic cores on the dynamical processes. In the case of He^{2+} impact, we employed, alternatively to CTMC, a semi-classical monocentric approach [7] to obtain the one-electron probabilities entering the IPM. These probabilities are in good agreement with their classical counterparts, and the projectile ionic core is generally found to be of little importance in the impact energy range considered.

Details on the calculations and extensive comparison of theoretical and experimental results will be presented at the Conference.

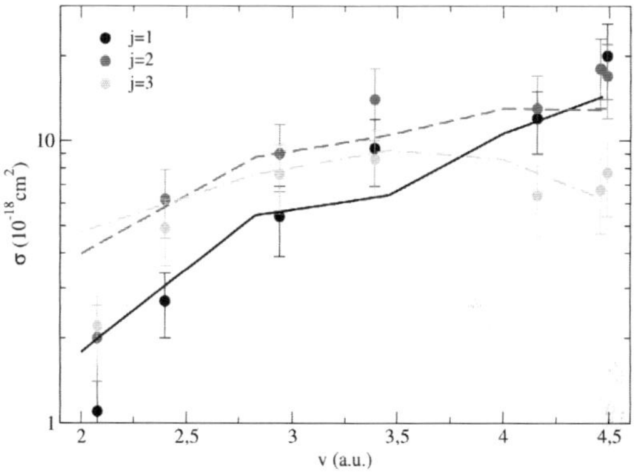

Figure 1. Cross sections for Li^{2+}+Ar → Li^{3+}+Ar^{j+}+$(j$+$1)$e. Symbols: experiments [1]; lines: present calculations.

References

[1] A. L. C. Losqui *et al.* 2014 *J. Phys. B.* **47** 045202

[2] C. Illescas and A. Riera 1999 *Phys. Rev. A* **60** 4546

[3] R. E. Olson 1979 *J. Phys. B.* **12** 109

[4] G. Labaigt *et al.* 2015 *J. Phys. B.* (in press)

[5] H. Lüdde and R. Dreizler 1985 *J. Phys. B.* **18** 107

[6] R. D. DuBois 1987 *Phys. Rev. A* **36** 2585

[7] B. Pons 2000 *Phys. Rev. Lett.* **84** 4569; 2001 *Phys. Rev. A* **63** 012704

[1]E-mail: alba.jorge@uam.es
[2]E-mail: clara.illescas@uam.es
[3]E-mail: pons@celia.u-bordeaux1.fr

Content from this work may be used under the terms of the Creative Commons Attribution 3.0 licence. Any further distribution of this work must maintain attribution to the author(s) and the title of the work, journal citation and DOI.
Published under licence by IOP Publishing Ltd

XXIX International Conference on Photonic, Electronic, and Atomic Collisions (ICPEAC2015) IOP Publishing
Journal of Physics: Conference Series **635** (2015) 022051 doi:10.1088/1742-6596/635/2/022051

Electron transfer processes in proton-atomic hydrogen excited states collisions

A. Taoutioui,[a,b,c 1] **A. Dubois,**[a 2] **and A. Makhoute** [b 3]

[a] Laboratoire de Chimie physique- Matière et Rayonnement, Pierre and Marie Curie University - CNRS, 75231 Paris Cedex 05, France

[b] Laboratory of Physics of Radiation and Laser-Matter Interactions, Faculty of Sciences. Moulay Ismail University, BP11201, Zitoune Meknes, Morocco

[c] Institut de Recherche pour le Développement, Bondy Nord Paris, France

Synopsis We have studied the collisions between protons and highly excited hydrogen atoms using the Classical Trajectory Monte Carlo method. We compare our numerical results for the electron transfer cross sections with those obtained by using semiclassical treatments. We present and discuss scaling laws for this process.

Proton-atom/molecule collisions have attracted considerable interest over the past few decades owing to their wide applications in plasma confinement, chemical reactive medium, ...and to the fundamental understanding of atomic structures and interactions [1]. The atomic processes of charge exchange (capture) in collisions between fully stripped ions and atomic hydrogen in its ground state have been extensively investigated both experimentally, e.g.[2,3], and theoretically, e.g.[4,5]. The case of H(n=2) initial state has also been analyzed by using Classical Trajectory Monte Carlo (CTMC)[6,8], semiclassical treatment [8] and Eikonal approximation (EA) [7].
The Classical Trajectory Monte Carlo has been proposed originally by Abrines and Percival [9] to study the collisions of protons with hydrogen atom in any excited state; these authors concluded that the CTMC method can be reliable for highly excited initial states. However this classical method has been used in numerous works, e.g.[4,5,6,7,8,9,10] to calculate ionization and electron transfer cross sections mainly for collisions involving ground state hydrogen and good agreement with experimental data were found in the intermediate energy range. Very recently CTMC calculations have also been performed on collisions between $N^{7+}$and $C^{6+}$with H(n=2) [10], with comparison with those obtained from semiclassical approaches, using atomic orbital (AOCC) and molecular orbitals close-coupling (MOCC) [11].

In the conference we present CTMC results, extending the study of $H^+ + H(n=1)$ (compared with experimental data and other recent CTMC results [12]), to collisions between proton and highly excited states of atomic hydrogen, focusing on total and state-to-state capture cross sections. The advent of supercomputers and parallel programming facilities allow the treatment of such processes that have been out of reach before due to extensive CPU time consumption. The aim of our work is to reveal the behaviour of the cross sections in the case of excited hydrogen and check different scaling laws, as those of Foster [13] and Janev [14].

References

[1] Dörner et al. Adv. At. Mol. Opt. Phys. **48**, 1(2002)
[2] J.E. Bayfield, Phys. Rev. **185**, 109 (1969)
[3]W.L. Fite et al., Phys. Rev. **112**, 1161 (1958)
[4] R. Abrines, andI.C Percival. Proc. Phys. Soc. *88*, 861. (1966)
[5]R. E. Olson, and A. Salop. Phys. Rev. A *16*, 531–541.(1977)
[6] C. O. Reinhold, and J. E. Miraglia. J. Phys. B: At. Mol. Phys. 20541-549 (1987)
[7] J. K. M. Eichler Phy. Rev. A **23** 498 (1981)
[8] M. J. Fitzpatrick, J. Fu , W. F. Smith, J. F. Reading, A. Dubois, and R. Gayet. Radiat. Phys. Chem. *76*, 426–430. (2007).
[9]R. Abrines, and I.C Percival. *Proc. Phys. Soc.* **88** 861(1966)
[10] A. Jorge, L.F. Errea, C. Illescas, and L . Méndez. Eur. Phys. J. D *68*, 1–11.(2014).
[11] K. Igenbergs, J. Schweinzer, A.Veiter, L. Perneczky, E. Frühwirth, M. Wallerberger, R.E. Olson, and F. Aumayr. J. Phys. B At. Mol. Opt. Phys. *45*, 065203.(2012)
[12] I. D. Kaganovich,., A. Shnidman, H. Mebane, and R.C. Davidson. Nucl. Instrum. Methods Phys. Res. Sect. Accel. Spectrometers Detect. Assoc. Equip. *606*, 196–204.(2009).
[13] A. Foster, Ph.D. thesis, University of Strathclyde, UK, (2008)
[14] R.K.Janev, D.Reiter, U.Samm: collision processes in Low-Temperature Hydrogen Plasmas JUEL 4105 (2003)

[1]E-mail: malek.taoutioui@gmail.com
[2]E-mail: alain.dubois@upmc.fr
[3]E-mail: makhoute@netcourrier.com

Content from this work may be used under the terms of the Creative Commons Attribution 3.0 licence. Any further distribution of this work must maintain attribution to the author(s) and the title of the work, journal citation and DOI.
Published under licence by IOP Publishing Ltd

Scaling for state-selective charge exchange due to collisions of multicharged ions with hydrogen

A. Jorge*[1], Clara Illescas*, J.E. Miraglia†, M.S. Gravielle†,

* Departamento de Química, Módulo 13, Universidad Autónoma de Madrid, Cantoblanco 28049, Madrid, Spain

† Instituto de Astronomía y Física del Espacio (IAFE, CONICET-UBA) Buenos Aires, Argentina.

Synopsis State-resolved charge exchange (CX) cross sections for Be^{4+}, C^{6+}, N^{7+} and O^{8+} projectiles colliding with atomic hydrogen are evaluated employing two different methods: the Classical Trajectory Monte Carlo (CTMC) and the eikonal impulse (EI) approximations. These cross sections are used to extend previously derived scaling laws for $n-$, $nl-$, and $nlm-$ distributions to highly excited final levels with $4 \leq n \leq 9$, covering energies in the range 50-2000 keV/amu. Present results are in agreement with available experimental and theoretical data for these collision systems, becoming a useful instrument for plasma research.

As reported in Ref. [1], it is possible to use the prior EI approach to derive a scaling rule for the following reaction:

$$A^{Z_P+} + H(1s) \rightarrow H^+ + A^{(Z_P-1)+}(n,l,m), \quad (1)$$

using that the corresponding CX cross section, σ_{nlm}, nearly satisfies:

$$\sigma^{EI}_{nlm} \simeq \tilde{z_P}^{-7} \left|C(Z_P/v)\right|^2 u_{nlm}(\tilde{W}), \quad (2)$$

where v is the impact velocity, $\tilde{z_P} = Z_P/v$, and $C(Z_P/v)$ is the Coulomb factor coming from the asymptotic conditions. In Eq. (2) the function u_{nlm} can be considered as an universal function of the scaled transferred momentum $\tilde{W} = (\tilde{v}^2 + \tilde{z_T}^2 - 1)/(2\tilde{v})$, with $\tilde{z_T} = Z_T/\tilde{z_P}$ (Z_T =1 for H) and $\tilde{v} = v/\tilde{z_P}$.
Two different approximations, EI [1] and CTMC [2], are used to evaluate state-selective CX cross sections for Be^{4+}, C^{6+}, N^{7+}, and O^{8+} projectiles impinging on H(1s), considering a different velocity range for each method: $v = 1-3$ a.u. for the CTMC approach and $v = 2-10$ a.u. for the EI one. The fact that there is a velocity region where both approaches can be successfully applied allows us to provide reliable scaling rules that cover a wide range of incidence energies.
In Fig. 1 we display scaled CX cross sections into the final level n, $S_n = \tilde{z_P}^7\sigma_n/|C(Z_P/v)|^2$, as a function of $\tilde{W}$, including CTMC and EI results for $n = 4-9$. The proposed scaling makes it possible to gather all the results in a relatively narrow band. Similar behavior is also obtained for the $nl-$distributions, defined as $P_{nl} = \sigma_{nl}/\sigma_n$, when they are plotted as a function of $\tilde{W}$, as shown in Fig. 2.

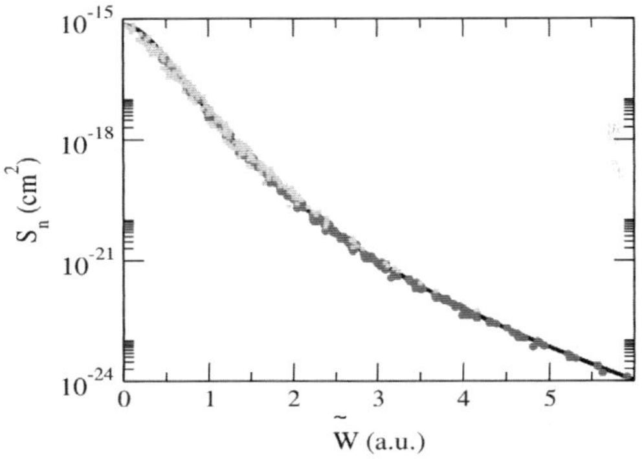

Figure 1. Scaled cross sections S_n in the range $n = 4-9$, evaluated with the considered approaches: ● EI, ■ CTMC.

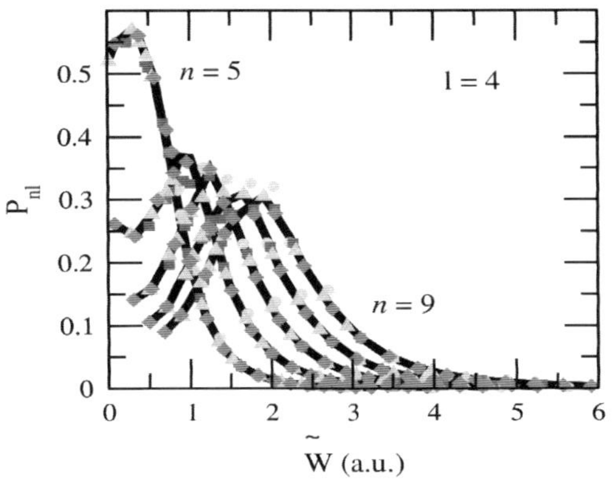

Figure 2. Joined CTMC and EI distributions for $l = 4$, as a function of $\tilde{W}$, considering different projectiles: ● Be^{4+}, ■ C^{6+}, ▲ N^{7+}, ♦ O^{8+}.

[1]E-mail: alba.jorge@uam.es

References

[1] M. S. Gravielle, J. E. Miraglia 1995 *Phys. Rev. A* **51** 2131–2139, ibidem **52** 851–854

[2] Clara Illescas, A. Riera 1999 *Phys. Rev. A* **60** 4546–4560

Content from this work may be used under the terms of the Creative Commons Attribution 3.0 licence. Any further distribution of this work must maintain attribution to the author(s) and the title of the work, journal citation and DOI.
Published under licence by IOP Publishing Ltd

Journal of Physics: Conference Series **635** (2015) 022053 doi:10.1088/1742-6596/635/2/022053

Single ionization of helium by fast proton impact: Searching for projectile coherence

H. Gassert*, O. Chuluunbaatar†,§, M. Waitz*, H.-K. Kim*, T. Bauer*, A. Laucke*, Ch. Müller*, J. Voigtsberger*, M. Weller*, J. Rist*, K. Pahl*, M. Honig*, M. Pitzer*, S. Zeller*, T. Jahnke*, L. Ph. H. Schmidt*, S. A. Zaytsev¶, A. A. Bulychev†, H. Schmidt-Böcking*, K. A. Kouzakov‡, R. Dörner*, M. S. Schöffler*[1], and Yu. V. Popov♭,†

*Institut für Kernphysik, University Frankfurt, Max-von-Laue-Str. 1, 60438 Frankfurt, Germany
†Joint Institute for Nuclear Research, Dubna, Moscow region 141980, Russia
§ School of Mathematics and Computer Science, National University of Mongolia, UlaanBaatar, Mongolia
¶Department of Physics, Pacific State University, Tikhookeanskaya 136, Khabarovsk 680035, Russia
‡Faculty of Physics, Lomonosov Moscow State University, Moscow 119991, Russia
♭Skobeltsyn Institute of Nuclear Physics, Lomonosov Moscow State University, Moscow 119991, Russia

Synopsis The fully differential cross section (FDCS) for single ionization $p + \mathrm{He} \to p + e + \mathrm{He}^+$ at proton energy of 1 MeV is studied both experimentally and theoretically. The 3D angular electron distribution is presented. The role of electron-electron correlations both in a trial helium ground-state wave function and in the final helium state is inspected.

A 3D angular distribution of the electron ejected in the reaction $p + \mathrm{He} \to p + e + \mathrm{He}^+$ was measured at an incident proton energy of 1 MeV. The momentum transfer q was fixed to 0.75 a.u. and the electron energy to 6.5 eV. The distribution has a shape of a pear with a narrow node in the P-plane, which is perpendicular to the scattering plane (C-plane) and crosses the latter along the proton-velocity axis. The first Born approximation (FBA) describes satisfactory all angle domains except that in the P-plane.

Other well-established approaches employed in our analysis, in particular, such as EWBA, SBA and J-matrix, also fail to reproduce the experimental data in the P-plane. We normalized all theories and experiment to the binary peak in the C-plane, thus obtaining the normalization coefficients for the P-plane. SBA and EWBA give practically no contribution to the FBA calculations, but we find a noticeable difference in the backward peak for the ground correlated [1] (blue curve in Fig. 1) and non-correlated [2] (green curve) helium wave functions, as well as for calculations within the J-matrix approach [3] (red curve), where final correlations between the electron and the helium ion are included. The results in the P-plane are presented in Fig. 1.

Possible explanations of the observed discrepancies will be discussed, including coherence/incoherence of a projectile beam (see the most recent paper [4] and references therein). Nevertheless, theoretical problems occur when FBA is close to zero in the P-plane. In such a case, both dynamical mechanisms beyond FBA and experimental uncertainties can manifest themselves.

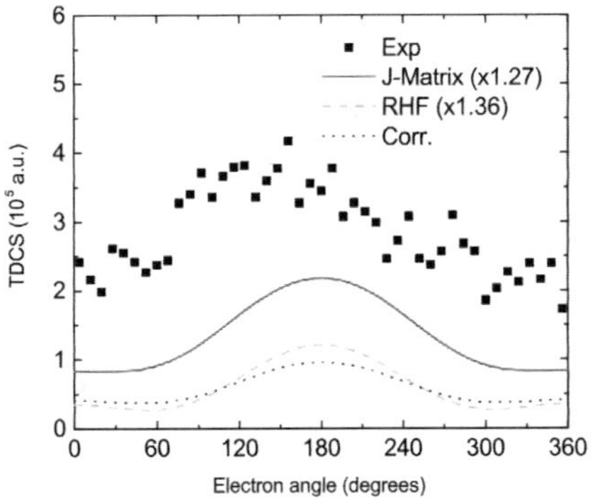

Figure 1. Experimental and theoretical electron angular distributions in the P-plane for singly ionizing 1-Mev $p+\mathrm{He}$ collisions. Blue curve represents FBA with a highly correlated helium wave function [1], green curve the same, but with a loosely correlated function [2], and red line the J-matrix calculations [3].

References

[1] O. Chuluunbaatar *et al* 2006 *Phys. Rev. A* **74** 014703

[2] E. Clementi and C. Roetti 1974 *At. Nucl. Data Tables* **14** 177

[3] S. A. Zaytsev *et al* 2007 *Phys. Rev. A* **75** 022718

[4] F. Járai-Szabóa and L. Nagy 2015 *Eur. Phys. J. D* **69** 4

[1]E-mail: schoeffler@atom.uni-frankfurt.de

Content from this work may be used under the terms of the Creative Commons Attribution 3.0 licence. Any further distribution of this work must maintain attribution to the author(s) and the title of the work, journal citation and DOI.
Published under licence by IOP Publishing Ltd

High energy resolution measurements of the radiative decay of double K-shell vacancies in $20 \leq Z \leq 29$ elements bombarded by fast C and Ne ions

J.-Cl. Dousse[*,1], Y.-P. Maillard[*], M. Berset[*], D. Castella[*], J. Hoszowska[*], M. Kavčič[†], M.Kobal[†], O. Mauron[*], A. Muehleisen[†], P.-A. Raboud[*], J. Rzadkiewicz[‡], K. Tökési[£] and B. Zorko[†]

[*] Department of Physics, University of Fribourg, CH-1700 Fribourg, Switzerland
[†] J. Stefan Institute, SI-1001 Ljubljana, Slovenia
[‡] National Centre for Nuclear Research, 05-400 Otwock, Swierk, Poland
[£] Institute of Nuclear Research of the Academy of Sciences (ATOMKI), H-4001 Debrecen, Hungary

Synopsis. We report on high energy resolution measurements of the Kα hypersatellite x-ray spectra of Ca, V, Fe and Cu induced by impact with 144 MeV C and 180 MeV Ne ions.

The double inner-shell ionization induced by photon and electron impact is rather weak. In contrast to that, x-ray satellites and hypersatellites with markedly higher intensities are observed in x-ray spectra resulting from collisions of atoms with heavy ions. Due to the strong Coulomb field of the projectile, several inner-shell electrons of the target atom can be indeed ionized simultaneously. As a consequence, x-ray spectra induced by heavy ion impact exhibit usually rich satellite and hypersatellite structures. High-resolution measurements of heavy-ion induced *K*-hypersatellites were found to be a sensitive tool for investigating the relativistic and quantum electrodynamics (QED) effects in atoms [1] and the dynamics of multiple ionization of atoms by heavy-ion impact [2].

In the present experiment the K-shell double ionization of Ca, V, Fe and Cu induced by impact with 12 MeV/amu C^{4+} ions and 9 MeV/amu Ne^{6+} ions was investigated. The measurements were performed at the Paul Scherrer Institute (PSI) in Villigen, Switzerland. The ions produced by a 10-GHz CAPRICE ECR source were accelerated to the final energies by the variable energy Philips cyclotron of PSI. The x-ray spectra were measured by means of high resolution x-ray spectroscopy, using the von Hamos curved crystal spectrometer of Fribourg [3]. The intensity of the beam which was in the range of 100–200 nA was monitored by observing the sample K x-ray fluorescence by means of a Si PIN photodiode. The fluorescence x-rays from the target were collimated by a 0.2 mm tantalum slit defining the effective source width. The Ca and V spectra were collected using a LiF (200) crystal, whereas for Fe and Cu a SiO_2 (2-23) crystal was employed. The diffracted x-rays were measured with a two-dimensional position-sensitive deep depleted CCD detector.

For illustration, the x-ray spectrum corresponding to the Ne-V collision is depicted in figure 1. As shown, L-satellites up to the 4th order were observed for the diagram and hypersatellite transitions.

From the fits of the high-resolution x-ray spectra, the centroid energies, linewidths and relative intensities of the transitions of interest were extracted and compared to theoretical predictions. A particular aim of this project was to determine the double K-shell photoionization cross sections for the investigated collisions. The cross sections were derived from the hypersatellite-to-diagram intensity ratios and compared to theoretical calculations performed within the SCA and CTMC models.

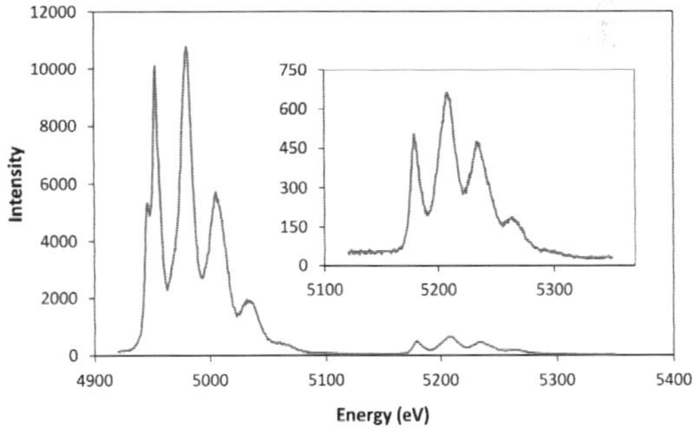

Figure 1. High-resolution K x-ray spectrum of V induced by impact with 180 MeV Ne ions. The first group of lines correspond to the $K\alpha L^n$ diagram and L-satellite transitions, the second one to the $K\alpha^h L^n$ hypersatellite transitions.

References

[1] M.H. Chen *et al* 1982 *Phys. Rev. A* **25** 391
[2] D.K. Olsen *et al* 1974 *Phys. Rev. Lett.* **33** 194
[3] J. Hoszowska *et al* 1996 *Nucl. Instrum. Phys. Res.* A **376** 1

[1]E-mail: jean-claude.dousse@unifr.ch

Content from this work may be used under the terms of the Creative Commons Attribution 3.0 licence. Any further distribution of this work must maintain attribution to the author(s) and the title of the work, journal citation and DOI.
Published under licence by IOP Publishing Ltd

Journal of Physics: Conference Series **635** (2015) 022055 doi:10.1088/1742-6596/635/2/022055

Study of ion-atom ionization using dressed ions and effective projectiles charge state

Shubhadeep Biswas* [1], **S. Kasthurirangan***, **J. M. Monti**†, **R. D. Rivarola**† **and L. C. Tribedi*** [2]

* Tata Institute of Fundamental Research, Mumbai 400005, India
† Institute de Fisica Rosario (CONICET-UNR), Universidad Nacional de Rosario, 2000

Synopsis In this report, we present a study of dressed ion projectile impact ionization of He and Ne atoms in the double differential cross section level. A large deviation from the q^2-scaling of the DDCS was observed even for large impact parameter collisions. Different q-dependence (i.e. q^n, n differs from 2) was found from which we calculated the effective projectile charge state. It also indicates the projectile screening mechanism which varies with impact parameter of the collision. For theoretical calculation we used GSZ model potential in the CDW-EIS framework, which gives an excellent agreement with the experimental data.

Double differential cross section (DDCS) measurement of electron emission in ionization of atoms or molecules by fast ions provides crucial information regarding the microscopic details of the underlying three body collision mechanisms. Most of the earlier measurements in the DDCS level involved the bare ions as projectile. But, in case of dressed projectile impact, apart from projectile nucleus–target active electron Coulomb interaction, residual projectile electrons play a crucial role in target ionization. Depending on the impact parameter of the collision, it can enhance or reduce the electron emission cross section.

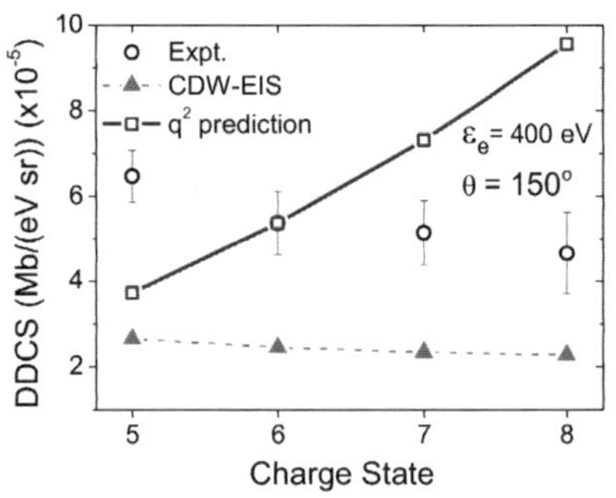

Figure 1. Charge state distribution of DDCS.

To investigate the influence of dressed projectile, we have measured the energy and angular distributions of the DDCS for electron emission in ionization of He and Ne atoms under the impact of fast highly charged bare and dressed oxygen ions with charge states q = 5, 6, 7 and 8, at different observation angles. We have also carried out the theoretical calculations based on Green-Sellin-Zachor (GSZ) model potential in the *prior* form of the state-of-the-art continuum distorted wave-eikonal initial state (CDW-EIS) framework [1]. This particular representation of the interaction potential suitably incorporates the effect of projectile dynamical screening by its residual electrons.

In general, the theoretical calculations provide an excellent agreement with the experimental data for all charge states and for all observation angles. It has been seen that the simple q^2-scaling of the electron DDCS is not applicable even at large impact parameter collisions. The obtained DDCS for dressed projectile impact is always higher than the expected DDCS from q^2-scalling law. This indicates towards an active ionization mechanism by direct projectile electron impact. From the obtained results, we derive a new q^n scaling law and estimated the value of effective projectile charge state for large impact parameter collisions which turned out to be observation angle dependent. The results also clearly indicate the projectile screening mechanism. In case of the dressed ion impact, with increasing electron energy, the DDCS gradually increases and approaches to the value of bare ion impact DDCS. This can only be explained if we consider the variation of projectile screening with impact parameter. In addition to the total DDCS, for dressed ion impact, we have also estimated the contributions from target ionization, projectile ionization and target-projectile simultaneous ionization. These show that, at higher electron energy, the projectile ionization and the target-projectile simultaneous ionization dominate over the target ionization [2, 3].

References

[1] J. M. Monti *et al* 2011 *J. Phys. B: At. Mol. Opt. Phys.* **44** 195206

[2] S. Biswas *et al* 2015 *Phys. Rev. A* **91** 022711

[3] S. Biswas *et al* 2015 *J. Phys. Conf. Ser.* **583** 012021

[1] E-mail: shubhadeep@tifr.res.in [2] E-mail: lokesh@tifr.res.in

Content from this work may be used under the terms of the Creative Commons Attribution 3.0 licence. Any further distribution of this work must maintain attribution to the author(s) and the title of the work, journal citation and DOI.
Published under licence by IOP Publishing Ltd

Multi-electron effects in fast ion collisions with Li atoms

M. D. Śpiewanowski*[1], L. Gulyás†[2], M. Horbatsch*[3], and T. Kirchner*[4]

* Department of Physics and Astronomy, York University, Toronto, Ontario, Canada M3J 1P3
† Institute for Nuclear Research, Hungarian Academy of Sciences (ATOMKI), H-4001 Debrecen, Hungary

Synopsis We investigate target electron removal in O^{8+}-Li and Li^{2+}-Li collisions at 1500 keV/amu and 2290 keV/amu, respectively. In the former case, we present doubly-differential cross sections for inner-shell electron removal showing the importance of multi-electron effects. In the latter case, the singly-differential cross-section results indicate that in outer-shell ionization a single-electron process plays the dominant part. The K-shell ionization results suggest only weak contributions from multi-electron processes, but a large discrepancy between experimental and single-particle theoretical results indicates a more complex nature of the process.

The recent development of the magneto-optical trap reaction microscope (MOTReMi) [1] has opened a new chapter for detailed investigations of charged-particle collisions from alkali atoms. In particular, lithium atoms received considerable attention. For example, it was shown that energy-differential cross sections (EDCS) for ionization from the outer shell in O^{8+}-Li collisions at 1500 keV/amu can be readily explained by the single-active electron (SAE) approximation [1, 2]. The understanding of K-shell ionization, however, requires the incorporation of many-electron effects, thus, going beyond the SAE approach. Independent-electron together with independent-event models proved to be useful for elucidating contributions from different ionization channels. It was shown that an excitation-ionization process is the dominant one in the measured EDCS [2]. Similarly, doubly-differential cross sections (DDCS) for inner-shell ionization cannot be understood within the SAE approach [3].

In this contribution we will first present a theoretical study of inner-shell electron removal in O^{8+}-Li collisions at 1500 keV/amu. To obtain DDCSs, we combined two-center basis generator method excitation amplitudes with amplitudes for electron removal obtained from the continuum distorted-wave with eikonal initial-state model. The combined amplitudes are subsequently Fourier transformed and integrated over the electron emission and the momentum transfer angles.

Secondly, we study target electron removal in Li^{2+}-Li collisions at 2290 keV/amu for ground and excited lithium target configurations [4]. The results indicate that in outer-shell ionization a single-electron process plays the dominant part, as in the O^{8+}-Li collisions at 1500 keV/amu. However, the K-shell ionization results are more difficult to interpret. Following the procedure in [2] we deduce that, according to our calculations, the process is strongly single-particle-like. We find a high resemblance between theoretical single-particle ionization and exclusive inner-shell ionization, and weak contributions from multi-electron processes. On the other hand, a large discrepancy between experimental and single-particle theoretical results indicates that multi-electron processes involving ionization from the outer-shell may be important for a complete understanding of the process.

References

[1] D. Fischer *et al* 2012 *Phys. Rev. Lett.* **109** 113202

[2] T. Kirchner, N. Khazai, and L. Gulyás 2014 *Phys. Rev. A* **89** 062702

[3] A. C. LaForge *et al* 2013 *J. Phys. B* **46** 031001

[4] M. D. Śpiewanowski *et al* 2015 accepted for publication in *J. Phys. Conf. Series*

[1]E-mail: maciekś@yorku.ca
[2]E-mail: gulyasl@atomki.mta.hu
[3]E-mail: marko@yorku.ca
[4]E-mail: tomk@yorku.ca

Content from this work may be used under the terms of the Creative Commons Attribution 3.0 licence. Any further distribution of this work must maintain attribution to the author(s) and the title of the work, journal citation and DOI.
Published under licence by IOP Publishing Ltd

Dissociative excitation and vibrational excitation of Cl_2^+

Mingwu Zhang* [1], Åsa Larson† [2], Ann E Orel††, Xiaohong Cai*

* Institute of Modern Physics, Chinese Academy of Sciences, Lanzhou, 730000, Gansu, China
† Department of Physics, Stockholm University, S-106 91 Stockholm, Sweden
†† Department of Chemical Engineering and Materials Science, University of California at Davis, Davis, California 95616, USA

Synopsis The cross sections for resonant dissociative excitation and vibrational excitation of Cl_2^+ molecular ion in the 0-10 eV region are calculated by solving a driven Schrödinger equation for the four lowest resonant states of all symmetries.

Molecular chlorine has attracted substantial attention for plasma processing and excimer lasers. It is also of atmospheric and environmental interest. In these media, electron collisions with chlorine molecules and molecular ions play a central role [1]. In our previous work [2], we have studied dissociative recombination (DR) of Cl_2^+, which is the process where an electron is captured by the molecular ion to form a doubly exited neutral state that dissociates to products. In addition, there are other competing processes occurring along with DR. After the formation of chlorine molecule in a resonant state, the incoming electron can autoionize, returning the target ion to its initial electronic state, either in a bound vibrational state or to the dissociation continuum. When a bound vibrational state is formed the process is resonant vibrational excitation (VE) (or de-excitation), and dissociation of the molecular ion results in resonant dissociative excitation – (DE).

With the potential energy curves and autoionization widths that have been computed by combining electron scattering calculations and *ab initio* structure calculations in our previous study [2], the reaction dynamics of DE and VE for collisions between electrons and chlorine molecular ions is investigated by solving driven time-independent Schrödinger equation. Cross sections for VE from initial v_i=0 vibrational state of Cl_2^+ to v_f=1, 2, …, 19 are computed. In Fig. 1, the sum of these cross sections is presented. Also the cross section for DE for ions in the v_i=0 is calculated.

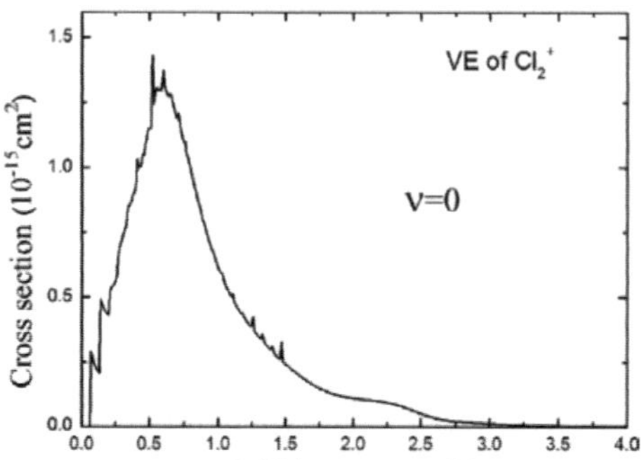

Figure 1. The vibrational excitation cross section of Cl_2^+ for the v_i=0 initial vibrational state of the target ion. This is the total cross section for excitation into all vibrational final states from v_f=1 up to v_f=19.

References

[1] E G Thorsteinsson *et al* 2010 *Plasma Sources Sci. Technol.* **19** 015001
[2] M Zhang *et al* 2011 *Phys. Rev. A.* **84** 052707

[1] E-mail: zhangmingwu@impcas.ac.cn
[2] E-mail: aasal@fysik.su.se

Content from this work may be used under the terms of the Creative Commons Attribution 3.0 licence. Any further distribution of this work must maintain attribution to the author(s) and the title of the work, journal citation and DOI.
Published under licence by IOP Publishing Ltd

XXIX International Conference on Photonic, Electronic, and Atomic Collisions (ICPEAC2015) IOP Publishing
Journal of Physics: Conference Series **635** (2015) 022059 doi:10.1088/1742-6596/635/2/022059

Coulomb crystallization of highly charged ions

L. Schmöger[1,2 1], O. O. Versolato[1,2], M. Schwarz[1,2], M. Kohnen[2], A. Windberger[1], B. Piest[1], S. Feuchtenbeiner[1], J. Pedregosa[3], T. Leopold[2], P. Micke[1,2], A. K. Hansen[4], T. M. Baumann[5], M. Drewsen[4], J. Ullrich[2], P. O. Schmidt[2,6], and J. R. Crespo López-Urrutia[1]

[1] Max-Planck-Institut für Kernphysik, Saupfercheckweg 1, 69117 Heidelberg, Germany
[2] Physikalisch-Technische-Bundesanstalt, , Bundesallee 100, 38116 Braunschweig, Germany
[3] Physique des Interactions Ioniques et Moléculaires, Aix-Marseille Université, 13397 Marseille, France
[4] Department of Physics and Astronomy, Aarhus University, 8000 Aarhus C, Denmark
[5] NSCL, Michigan State University, 640 S. Shaw Lane, East Lansing, Michigan, 48824, USA
[6] Institut für Quantenoptik, Leibniz Universität Hannover, Welfengarten 1, 30167 Hannover, Germany

Synopsis Sympathetic Coulomb crystallization of highly charged ions, retrapped in a cryogenic radiofrequency trap, is demonstrated by an over seven orders-of-magnitude decrease in motional temperature.

In highly charged ions (HCIs), the electronic wavefunction is much reduced in size. Subsequent advantages for precision spectroscopy are: higher sensitivity to electron-nucleus interactions and QED terms in general, and an extremely suppressed sensitivity to external field perturbations. Further, electric dipole forbidden optical transitions found near level crossings in HCIs are extremely sensitive to possible drifts in the fine structure constant α [1]. Thus, cold, strongly localized HCIs are of particular interest for bound-state QED studies (g-factor measurements), metrology (development of novel optical clocks) and the search for α variation. We report on Coulomb crystallization of highly-charged $^{40}Ar^{13+}$ ions through sympathetic cooling with co-trapped, laser-cooled $^{9}Be^{+}$ ions to final translational temperatures of about 200 mK or less [2]. The Ar^{13+} ions are produced in, and extracted from an electron beam ion trap (EBIT). They are decelerated and pre-cooled by means of two serrated interlaced pulsed drift tubes before they are injected into the cryogenic Paul trap CryPTEx [3]. Subsequently, they are forced to interact multiple times with a Coulomb crystal of laser-cooled Be^{+} ions, thereby losing enough energy to end up implanted as dark structures of spherical shape in the bright fluorescing Be^{+} crystal. The combination of an EBIT with a linear Paul trap operating at ~7 K facilitates not only the formation of mixed-species 3D Coulomb crystals, but also of 1D Coulomb crystals down to a single HCI cooled by a single Be^{+} ion (Fig. 1). This is a necessary step for future quantum logic spectroscopy at a potential 10^{-19} level accuracy. Our preparation technique of cold Ar^{13+} is readily applicable to a broad range of other highly charged elements and is thus a significant step forward for precision spectroscopy of HCIs.

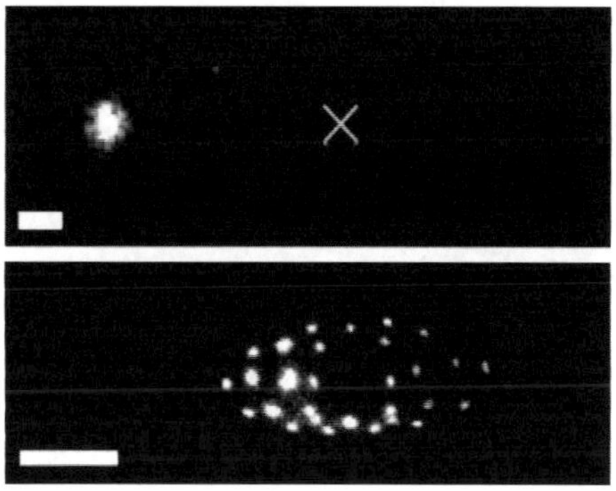

Figure 1. Top: A single Ar^{13+} ion (position marked by cross) sympathetically cooled by a single laser-cooled Be^{+} ion. White scale bar denotes 20 μm. Bottom: A single Ar^{13+} ion sympathetically cooled by several Be^{+} ions. White scale bar denotes 100 μm.

References

[1] J. C. Berengut *et al.* 2011 *Phys. Rev. Lett.* **106** 210802
[2] L. Schmöger *et al.* 2015 *Science* **347**, 1233-1236
[3] M. Schwarz. *et al.* 2012 *Rev. Sci. Instr.* **83** 083115

1 E-mail: lisa.schmoeger@mpi-hd.mpg.de

Content from this work may be used under the terms of the Creative Commons Attribution 3.0 licence. Any further distribution of this work must maintain attribution to the author(s) and the title of the work, journal citation and DOI.
Published under licence by IOP Publishing Ltd

Journal of Physics: Conference Series **635** (2015) 022061 doi:10.1088/1742-6596/635/2/022061

The development of the antihydrogen beam detector and the detection of the antihydrogen atoms for in-flight hyperfine spectroscopy

Y. Nagata[1 *], C. Sauerzopf[2], A. Capon[2], N. Kuroda[3], Y. Abo[4] M. Diermaier[2], P. Dupre[1], Y. Higashi[3], S. Ishikawa[3] C. Kaga[4], M. Leali[5], C. Malbrunot[2,6], V. Mascagna[5], D. Murtagh[1], B. Radics[1] M. C. Simon[2] M. Tajima[3], H. A. Torii[3], S. Van Gorp[1], J. Zmeskal[2], H. Breuker[6] H. Higaki[4], Y. Kanai[1], Y. Matsuda[3], S. Ulmer[7], L. Venturelli[5], E. Widmann[2], Y. Yamazaki[1]

[1]Atomic Physics Laboratory, RIKEN, Wako, Saitama 351-0198, Japan
[2]Stefan Meyer Institute for Subatomic Physics, Boltzmanngasse 3,1090 Vienna, Austria
[3]Graduate School of Arts and Sciences, The University of Tokyo, Komaba, Meguro, Tokyo 153-8902, Japan
[4]Graduate School of Advanced Sciences of Matter, Hiroshima University,1-3-1 Kagamiyama, Higashi-Hiroshima, Hiroshima 739-8530 Japan
[5]Dipartimento di Ingegneria dell' Informazione, Università di Brescia & Istituto Nazionale di Fisica Nucleare, Gruppo Collegato di Brescia, 25133 Brescia, Italy
[6]CERN, CH-1211 Geneva 23, Switzerland
[7]RIKEN Ulmer Initiative Research Unit, RIKEN, Wako, Saitama 351-0198, Japan

Synopsis We have been developing ground-state antihydrogen atomic beams to test CPT symmetry via in-flight hyperfine spectroscopy. A new antihydrogen beam detector has been developed. The overview of the experiment, the detail of the detector and latest results will be presented.

We have been developing ground-state antihydrogen ($\bar{\mathrm{H}}$) atomic beams to test CPT symmetry via in-flight hyperfine spectroscopy [1,2]. Figure 1 shows schematic view of the main part of our experimental setup which consists of the double cusp trap, a microwave cavity, a sextupole magnet, and a new $\bar{\mathrm{H}}$ beam detector.

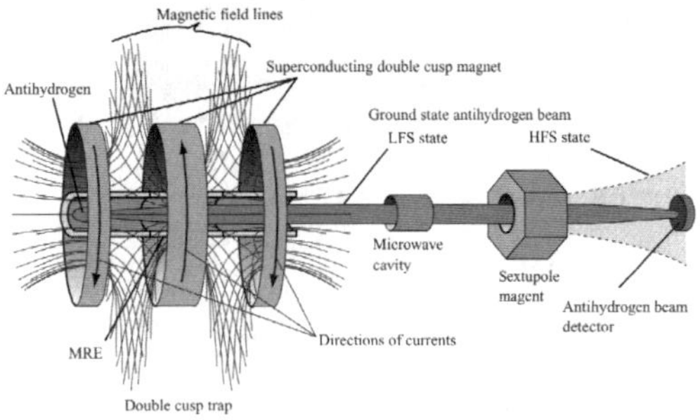

Figure 1. Schematic view of our experimental setup

We have been created $\bar{\mathrm{H}}$ atoms in the double cusp trap and tried extracted to the downstream of the trap. From our simulations, $\bar{\mathrm{H}}$ atoms were polarized to low field seeking (LFS) state in the magnetic field and transported as a beam well characterized by a focal length with a thin lens formula [3,4]. The polarized beam passes through the cavity and the sextupole magnet, and is focused on the $\bar{\mathrm{H}}$ beam detector. The $\bar{\mathrm{H}}$ atoms change their polarizations by spin-flip and are defocused on the detector when we apply the microwave tuned to the hyperfine frequency in the cavity. Therefore we can determine the hyperfine frequency by counting the number of $\bar{\mathrm{H}}$ atoms detected as a function of the microwave frequency.

The detector consisted of a BGO calorimeter and a hodoscope. The BGO crystal (9cmϕ $\times$ 0.5cm^t) was placed in UHV. The photons created by the charged particles in the BGO were detected by 4 multi-anode PMTs (256ch). We succeeded to measure the two-dimensional energy distribution by this setup without the segmentation of the BGO. This detector realize to be a two-dimensional calorimeter. The hodoscope surrounded the BGO calorimeter and was composed of 2 layers of 32 scintillator bars. The overview of the experiment, the detail of the detector and latest results will be presented.

References

[1] Y Yamazaki and S Ulmer 2013 Ann. Physik **525** 493

[2] N Kuroda et al 2014 Nat. Commun. **5** 3089

[3] A Mohri and Y Yamazaki 2003 Europhys. Lett. **63** 207

[4] Y Nagata and Y Yamazaki 2014 New. J. Phys. **16** 083026

*E-mail: nagata@radphys4.c.u-tokyo.ac.jp

Content from this work may be used under the terms of the Creative Commons Attribution 3.0 licence. Any further distribution of this work must maintain attribution to the author(s) and the title of the work, journal citation and DOI.
Published under licence by IOP Publishing Ltd

XXIX International Conference on Photonic, Electronic, and Atomic Collisions (ICPEAC2015) IOP Publishing
Journal of Physics: Conference Series **635** (2015) 022062 doi:10.1088/1742-6596/635/2/022062

The development of the superconducting double cusp magnet for intense antihydrogen beams

Y. Nagata[1] *, Y. Kanai[1], N. Kuroda[2], H. Higaki[3], Y. Matsuda[2], Y. Yamazaki[1]

[1] Atomic Physics Laboratory, RIKEN, Wako, Saitama 351-0198, Japan
[2] Graduate School of Arts and Sciences, The University of Tokyo, Komaba, Meguro, Tokyo 153-8902, Japan
[3]Graduate School of Advanced Sciences of Matter, Hiroshima University,1-3-1 Kagamiyama, Higashi-Hiroshima, Hiroshima 739-8530 Japan

Synopsis We have developed a superconducing double cusp magnet for the production of polarized intense antihydrogen atomic beams which are used to make microwave spectroscopy of a ground state hyperfine transition to test the CPT symmetry. The detail of specification and the focusing performance of the magnet will be presented.

Recently we have succeeded to produce a source of antihydrogen atoms [1]. We have developed a superconducing double cusp magnet to increase the intensity of polarized antihydrogen ($\bar{\mathrm{H}}$) atomic beams which are used to make in-flight microwave spectroscopy of a ground state hyperfine transition to test the CPT symmetry [2].

Previously we used a single cusp magnet of an anti-Helmholtz coil. The double cusp magnet consists of two sets of an anti-Helmholtz coil as shown in Fig. 1. The magnet creates a larger gradient of the magnetic field $\boldsymbol{B}$ than the single cusp magnet and the large force acting on $\bar{\mathrm{H}}$ atoms written by $\boldsymbol{\nabla}(\boldsymbol{\mu} \cdot \boldsymbol{B})$.

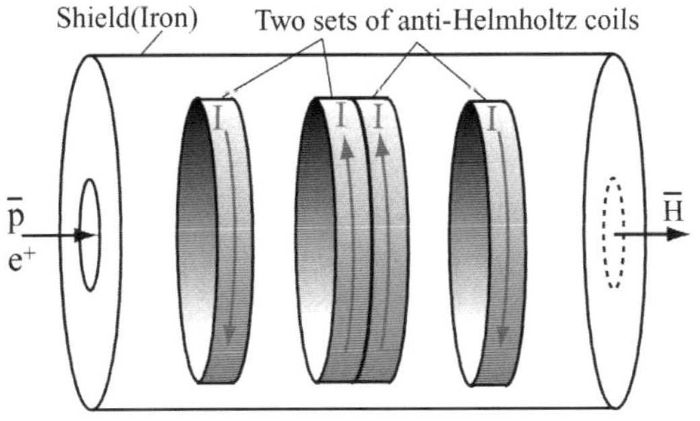

Figure 1. Schematic view of the double cusp magnet

Our simulations show that the cusp magnet is quite effective to produce an antihydrogen atomic beam polarized to low-field seeking states [3] and also a good magnetic lens characterized by a focal length with a thin lens formula. This has been explained using the momentum transfer dp from the magnetic field to the $\bar{\mathrm{H}}$ atom which is directly related to the deflection of the $\bar{\mathrm{H}}$ atoms [4]. Therefore we evaluated the focusing performance for the various parameters using dp.

Figure 2 shows the magnetic field of the double cusp magnet on axis. The magnetic field is slightly asymmetric by additional currents for the following reason. The magnetic field where antiprotons and positrons are confined is related to the cooling of them by the synchrotron radiation, hence, should be high (leftside negative peak). On the other hand, the leak field of downstream area (rightside) must not be high. The detail of the specification and the focusing performance of the magnet will be presented.

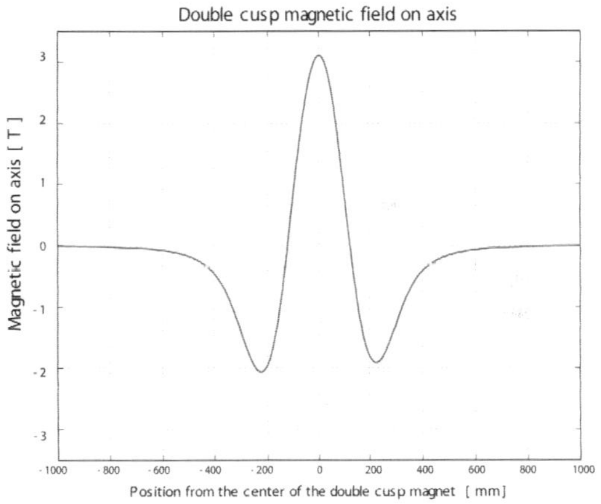

Figure 2. The magnetic field of the double cusp magnet

References

[1] N Kuroda et al 2014 Nat. Commun. **5** 3089

[2] Y Yamazaki and S Ulmer 2031 Ann. Physik **525** 493

[3] A Mohri and Y Yamazaki 2003 Europhys. Lett. **63** 207

[4] Y Nagata and Y Yamazaki 2014 New. J. Phys. **16** 083026

*E-mail: nagata@radphys4.c.u-tokyo.ac.jp

Content from this work may be used under the terms of the Creative Commons Attribution 3.0 licence. Any further distribution of this work must maintain attribution to the author(s) and the title of the work, journal citation and DOI.
Published under licence by IOP Publishing Ltd

Electron- and proton-impact excitation of He-like uranium

A. Gumberidze[a] [1], D.B. Thorn[b,c], C.J. Fontes[d], B. Najjari[e], H.L. Zhang[d] A. Surzhykov[f], A. Voitkiv[g], S. Fritzsche[f,h] D. Banaś[i], H. Beyer[a,f], W. Chen [a], F. Currell[j], R.D. DuBois[b,k], S. Geyer[a,l], R. Grisenti[a,l], S. Hagmann[a,l], M. Hegewald[a,l], P. Indelicato[m], C. Kozhuharov[a], R. Märtin[a,f], N. Petridis[a], R. Schuch[n], U. Spillmann[a], S. Trotsenko[a,f], A. Warczak[o] G. Weber[a,f], W. Wen[p], D.F.A. Winters[a], N. Winters[a], Z. Yin [a,q], Th. Stöhlker[a,f,h]

[a]GSI Helmholtzzentrum für Schwerionenforschung, 64291 Darmstadt, Germany; [b]ExtreMe Matter Institute EMMI and Research Division, GSI Helmholtzzentrum für Schwerionenforschung, 64291 Darmstadt, Germany; [c]FIAS Frankfurt Institute for Advanced Studies, 60438 Frankfurt am Main, Germany; [d]Computational Physics Division, Los Alamos National Laboratory, Los Alamos NM 87545, USA; [e]Institut Pluridisciplinaire Hubert-Curien, Groupe RaMsEs, 23 rue de Loess - BP 28 F-67037 Strasbourg Cedex 2, France; [f]Helmholtz-Institut Jena, D-07743 Jena, Germany; [g] Theoretische Physik I, Heinrich Heine Universität Düsseldorf, Universitätsstrae 1, 40225, Düsseldorf, Germany; [h]IOQ, Friedrich-Schiller-Universität Jena, 07743 Jena, Germany; [i]Institute of Physics, Jan Kochanowski University, PL-25-406 Kielce, Poland; [j]School of Mathematics and Physics, Queen's University, Belfast BT7 1NN, United Kingdom; [k]Missouri University of Science and Technology, Rolla, USA; [l]Institut für Kernphysik, Universität Frankfurt, 60486 Frankfurt am Main, Germany; [m]Laboratoire Kastler Brossel, École Normale Supérieure, CNRS, Sorbonne Universités, UPMC Univ. Paris 06, Case 74; 4, place Jussieu, 75252 Paris CEDEX 05, France; [n]Physics Department, Stockholm University, S-106 91 Stockholm, Sweden; [o]Institute of Physics, Jagiellonian University, PL-30-059 Cracow, Poland [p]Institute of Modern Physics, 730000 Lanzhou, China; [q] Max-Planck-Institut für Biophysikalische Chemie, DESY, Hamburg, Germany

Synopsis In this contribution, we present an experimental and theoretical study of the proton- and electron-impact excitation effects in helium-like uranium ions in relativistic collisions with different gaseous targets. The experiment was carried out at the experimental storage ring at GSI Darmstadt. Our experimental results are in agreement with predictions of the state-of-the art theory which treats both processes within the relativistic framework.

Electron-impact excitation (EIE) of bound electrons is a fundamental atomic processes which is also responsible for the vast majority of x-ray radiation produced in various kinds of laboratory and astrophysical plasmas.

Up to now, electron beam ion traps (EBITs) have been the preferred tool for studying the EIE [1]. Due to the small electron-impact ionization and excitation cross sections for heavy highly-charged ions, the focus of most of these EBIT studies has been confined to low and mid-Z systems.

In this contribution, we present an extension of our previous study [2] of the effect of electron-impact excitation in heavy highly-charged ions (HCI) undergoing collisions with neutral atoms. Namely, in a measurement carried out at the Experimental Storage Ring (ESR) we looked for electron- and proton (nucleus)-impact excitation (PIE) in relativistic collisions between Helium-like uranium ions and hydrogen and argon targets. Here, electron-electron correlation and relativistic effects can be addressed which are predicted to influence these processes significantly. By performing measurements with different targets as well as at different collision energies, we were able to gain access to both; PIE and EIE processes of the ground state in He-like uranium ions in the relativistic collisions. Combined calculations which treat both processes within the relativistic framework, provide a good agreement with the experimental data. Moreover, our experimental results clearly demonstrate the importance of including the effect of the generalized Breit interaction (GBI) in the EIE calculations [3].

References

[1] H. Chen and P. Beiersdorfer 2008 *Can. J. Phys.* **86** 55

[2] A. Gumberidze *et al.* 2013 *Phys. Rev. Lett.* **110**, 213201

[3] C. J. Fontes, D. H. Sampson, H. L. Zhang 1995 *Physical Reveiw A* **51** R12

[1]E-mail: a.gumberidze@gsi.de

Content from this work may be used under the terms of the Creative Commons Attribution 3.0 licence. Any further distribution of this work must maintain attribution to the author(s) and the title of the work, journal citation and DOI.
Published under licence by IOP Publishing Ltd

How include properly the dynamic-screening into the *post*-version of the Continuum Distorted Wave model

J M Monti*[1], O A Fojón*, J Hanssen *, R D Rivarola*

* Laboratorio de Colisiones Atómicas, Facultad de Ciencias Exactas, Ingeniería y Agrimensura and Instituto de Física Rosario, CONICET-UNR, Av. Pellegrini 250, 2000, Rosario, Argentina

Synopsis The *post*-version of the Continuum Distorted Wave (CDW) model neglects a part of the active electron-residual target interaction which has been shown to play a major role in obtaining the doubly differential cross sections (DDCS) for electron ionization. In this work we properly include the dynamic screening into the CDW model showing its influence over the ionization DDCS for several collision systems.

Among the many distorted wave models there are two that merit to be cited: the Continuum Distorted Wave (CDW) [1] and the Continuum Distorted Wave-Eikonal Initial State (CDW-EIS) [2]. These models were formulated to describe single ionization of monoelectronic targets by bare ion impact, then being extended to describe ionization of multielectronic targets [3]. In the CDW model the initial channel distortion is chosen as a Confluent Hypergeometric function (continuum factor) describing the interaction of the target active electron with the projectile, whereas in the exit channel the ionized electron evolves in a two center continuum approximated by the product of a plane wave and two continuum factors, one associated to the residual target and the other to the projectile. In the CDW-EIS models the initial channel distortion is chosen as an eikonal phase instead of the continuum factor.

In the distorted wave formalism, the action of the perturbative potentials can either be applied to the initial channel distorted wave function or to the final channel distorted one, giving place to the *prior* and *post*-versions of the transition matrix element, respectively. The *prior* version of the CDW transition amplitude was shown to have an intrinsic logarithmic divergence near the binary encounter structure which prohibit the cross section calculation [4]. Therefore its *post*-version has been used ever since. To facilitate the calculation of transition matrix elements, effective Coulomb potentials were chosen to represent the interaction between the active electron and the residual target in the exit channel. In such approximation a part of the dynamic screening is neglected (see [5] for the CDW-EIS case). In this work we show that including the non-Coulomb potentials in the *post*-version of the CDW model leads to the same divergences found in the *prior*- one. Therefore a hybrid CDW–CDW-EIS model is proposed in order to take into account the dynamic screening in the *post*-version of the CDW model without any divergence.

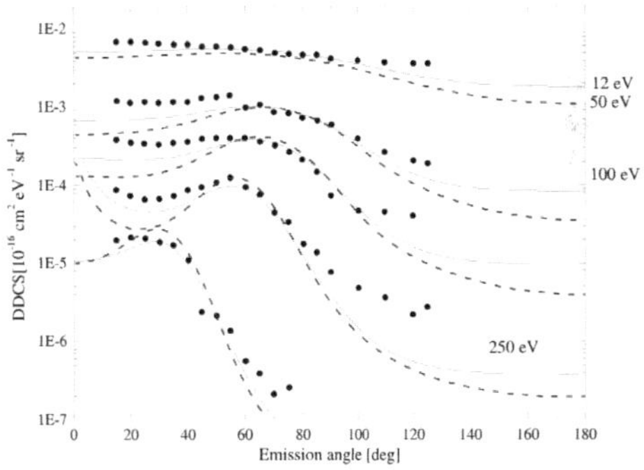

Figure 1. DDCS for single ionization of water vapour by 500 keV H^+ impact as a funtion of the emission angle for fixed electron energies. Dashed line: *post* CDW; full line: present *complete* CDW; dotted line: *prior* CDW-EIS. Experiments extracted from [6].

References

[1] Belkić Dž 1978 *J. Phys. B: At. Mol. Phys.* **11** 3529

[2] Crothers D S F and McCann J F 1983 *J. Phys. B: At. Mol. Phys.* **16** 3229

[3] Fainstein P D, Ponce V H and Rivarola R D 1988 *J. Phys. B: At. Mol. Opt. Phys.* **21** 287

[4] Brauner M and Macek J H 1992 *Phys. Rev. A* **46** 2519

[5] Monti J M, Fojón O A, Hanssen J and Rivarola R D 2010 *J. Phys. B: At. Mol. Opt. Phys.* **46** 145201

[6] Toburen H and Wilson W E 1977 *J. Chem. Phys.***66** 5202

[1]E-mail: monti@ifir-conicet.gov.ar

Content from this work may be used under the terms of the Creative Commons Attribution 3.0 licence. Any further distribution of this work must maintain attribution to the author(s) and the title of the work, journal citation and DOI.
Published under licence by IOP Publishing Ltd

Journal of Physics: Conference Series **635** (2015) 022065 doi:10.1088/1742-6596/635/2/022065

Relativistic effects in electron-capture to the continuum in 90 MeV/u U^{88+} $+N_2$ collisions

J M Monti* [1], **P-M Hillenbrand**[†,‡] [2], **S Hagmann**[†,§] [3], **R D Rivarola*** [4]

* Laboratorio de Colisiones Atómicas, Facultad de Ciencias Exactas, Ingeniería y Agrimensura and Instituto de Física Rosario, CONICET-UNR, Av. Pellegrini 250, 2000, Rosario, Argentina
† GSI Helmholtzzentrum für Schwerionenforschung, D-64291 Darmstadt, Germany
‡ Institut für Atom- und Molek ülphysik, Justus-Liebig-Universität Giessen, D-35392 Giessen, Germany
§ Institut für Kernphysik, Goethe Universität Frankfurt, D-60438 Frankfurt am Main, Germany

Synopsis The Continuum Distorted Wave (CDW) model is used to describe electron-capture to continuum cusp in 90 MeV/u U^{88+} + N_2 collisions. Relativistic effects are studied considering Lorentz transformations on the electron kinematics.

In this work we present experimental and theoretical results for doubly differential cross sections (DDCS) of electrons emitted from the N_2 target through the electron-capture to the continuum (ECC) process when colliding with a 90 MeV/u U^{88+} projectile. The ECC cusp is conformed by electrons captured into a projectile continuum state, whilst the excess energy is carried away by the recoil of the residual target. The experimental data was collected at the experimental storage ring ESR of the heavy-ion accelerator facility GSI using a dedicated magnetic electron spectrometer, enabling the measurement of high-energy electrons emitted within a small solid-angle in the forward direction (see [1]).
From the theoretical point of view, a semi-relativistic version of the Continuum Distorted Wave (CDW) model is introduced to describe the reaction. This model was developed to study the ionization of hydrogenic targets by swift bare projectile impact at non-relativistic collision energies [2]. Lately it was extended to multielectronic atomic targets [3], and more recently to the case of dressed-projectiles [4, 5].
In this work the molecular initial orbitals are described within a non-relativistic self-consistent field approximation, in which the the initial N_2 molecular orbitals are constructed as a linear combination of the atomic wave functions corresponding to the different orbitals of the N atom (MO-LCAO-SCF). The population analysis of the different orbitals is given by means of the neglect of differential overlap method (CNDO). Given the high collision energy, the projectile velocity is approximately $v_P \sim 0.41c$, therefore Lorentz transformations are taken into account. It is shown that the simultaneous relativistic description of the kinematics of both particles plays a fundamental role to give an adequate representation of the experimental data.
Further details about the calculations will be presented during the conference.

[1]E-mail: monti@ifir-conicet.gov.ar
[2]E-mail: p.m.hillenbrand@gsi.de
[3]E-mail: s.hagmann@gsi.de
[4]E-mail: rivarola@ifir-conicet.gov.ar

References

[1] P.-M. Hillenbrand *et al.* 2015 *Phys. Rev. A* **91** 022705

[2] Belkić Dž 1978 *J. Phys. B: At. Mol. Phys.* **11** 3529

[3] Fainstein P D, Ponce V H and Rivarola R D 1988 *J. Phys. B: At. Mol. Opt. Phys.* **21** 287

[4] Monti J M, Rivarola R D and Fainstein P D 2008 *J. Phys. B: At. Mol. Opt. Phys.* **41** 201001

[5] Monti J M, Rivarola R D and Fainstein P D 2011 *J. Phys. B: At. Mol. Opt. Phys.* **44** 195206

Content from this work may be used under the terms of the Creative Commons Attribution 3.0 licence. Any further distribution of this work must maintain attribution to the author(s) and the title of the work, journal citation and DOI.
Published under licence by IOP Publishing Ltd

Anion and cation emission via binary-encounters in OH^+ + Ar collisions: The role of dissociative excitation and statistical aspects

Z. Juhász[*1], E. Lattouf[†], J.-Y. Chesnel[†], S. T. S. Kovács[*], E. Bene[*], P. Herczku[*], B. A. Huber[†], A. Méry[†], J.-C. Poully[†], J. Rangama[†], and B. Sulik[*]

[*] Institute for Nuclear Research, Hungarian Academy of Sciences (MTA Atomki), H-4001 Debrecen, Hungary
[†] Centre de Recherche sur les Ions, les Matériaux et la Photonique (CIMAP), 6 Bd Mar. Juin, F-14000 Caen, France

Synopsis Emission of H^- and H^+ ions via nucleus-nucleus collisions were observed in OH^+ + Ar collisions in wide angular range. They show similar angular distributions, which indicate a statistical distribution of the final charge states. Non-vanishing emission at small angles show that dissociative excitation of the projectile has a crucial role in the process.

In a previous work, we observed that H^- ions can be formed from OH^+ ions when the H center is collisionally removed from it by the target atom [1-3]. It was found that the angular distribution of the so created H^- ions is proportional to the calculated one for H scattering on the target atomic center. The question arose whether it holds for the collisionally removed H^+ ions, too. So, molecular fragmentation leading to the formation of negatively and positively charged hydrogen ions in 7-keV OH^+ + Ar collisions was investigated experimentally at the Grand Accélérateur National d'Ions Lourds (GANIL) in Caen, France [4].

We have found that negative and positive hydrogen ions are indeed emitted with very similar angular dependences, which are proportional to simple H scattering cross sections (see figure 1). Also, the kinetic energy distribution of the H^+ fragment shows strong similarities with that of the ejected H^- ion. These findings indicate that the final charge state distribution of the emitted H center does not depend on how closely the atomic centers approach each other during the collision. Rather, it seems to follow simple statistical laws.

Below 5° scattering angle, one could expect a diminishing H^- formation, since the kinetic energy transferred to the H center is not enough to release it from its parent molecular ion. However, we found a rather high emission rate in this range, too. This is possible only if the OH^+ molecular ion gets into an excited state resulting in weak molecular bound or spontaneous dissociation. The later process leads to a kinetic energy release (KER) contributing to the energy of the emitted fragments. Indeed we found that the energy of the emitted H^- and H^+ centers is somewhat larger than expected in this angular range. We performed model calculations, which demonstrate the effects of KER.

The process is of general relevance where collisions of hydrogen containing ions or molecules are involved, because it is not specific on the target or projectile species as experiments on other collision systems have already shown. The created highly reactive species H^- and H^+ are relevant in astrophysics and life sciences.

This work was supported by the Hungarian National Science Foundation OTKA (Grant No. K109440), the French-Hungarian Cooperation Program PHC Balaton (No. 27860ZL/TéT_11-2-2012-0028), and the French-Hungarian CNRS-MTA Cooperation (ANION project).

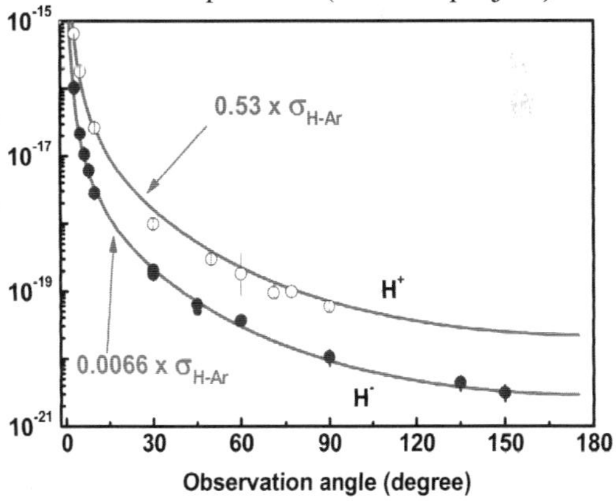

Figure 1. Single differential cross sections (SDCS) for H^- (full circles) and H^+ (open circles) emission as a function of the observation angle. Curves: calculated cross section for two-body elastic scattering of 412-eV H on Ar, multiplied by factors representing the fraction of the different charge state components.

References

[1] Z. Juhász *et al* 2012 *J. Phys. Conf. Ser.* **388** 102051
[2] Z. Juhász *et al* 2013 *Phys. Rev. A* **87** 032718
[3] Z. Juhász *et al* 2014 *J. Phys. Conf. Ser.* **488** 102024
[4] E. Lattouf *et al* 2014 *Phys. Rev. A* **89** 062721

1 E-mail: zjuhasz@atomki.mta.hu

Content from this work may be used under the terms of the Creative Commons Attribution 3.0 licence. Any further distribution of this work must maintain attribution to the author(s) and the title of the work, journal citation and DOI.
Published under licence by IOP Publishing Ltd

Effects on nitrogen-acetone ice induced by energetic heavy ion collisions

K. F. Alcantara[*1], **A. L. F. de Barros**[†], **E. F. da Silveira**[*], **H. Rothard**[¤], **P. Boduch**[¤]

[*]Departamento de Física, PUC- Rio, Rua Marquês de São Vicente 225, 22451-900, RJ, Brazil.

[†] Departamento de Física, CEFET-RJ, Av.Maracanã, 229, 20271-110 Rio de Janeiro, RJ, Brazil

[¤]CIMAP-CIRIL-Ganil, Boulevard Henri Becquerel, BP 5133, F-14070 Caen Cedex 05, France

Synopsis The radiolysis of a mixture of acetone and nitrogen condensed at 13 K by 40-MeV $^{58}Ni^{11+}$ ions was studied. The bombardment with heavy ions is highly efficient for inducing chemical reactions in ices. The dissociation rate of acetone mixed with nitrogen as a function of the fluence is determined from a sequence of infrared spectra. The formation of new molecular species due to the irradiation is investigated and their cross-sections are calculated

The N_2 molecule, formed by one of the most abundant elements in the Universe, is the most common constituent of Earth's modern atmosphere. It is also a major component of the atmosphere of Saturn's moon Titan and has been detected in trace amounts in the atmospheres of Venus and Mars. In acetone (CH_3COCH_3) the C=O link, which also occurs in several molecules of biological interest, is present. The mixture of these two molecular species, N_2:CH_3COCH_3, causes asymmetries in the nitrogen molecule, which enhances the IR absorption and allows its study by FTIR spectroscopy. Moreover, the cosmic ray - ice interaction can be simulated and analyzed by irradiating the N_2:CH_3COCH_3 ice in laboratory [1].

Therefore, N_2: CH_3COCH_3 ice (1:1) was irradiated with 40 MeV $^{58}Ni^{11+}$ ions, and FTIR spectra were obtained at certain fluences. Figure 1 shows spectra of N_2: CH_3COCH_3 before (dotted line) and after (solid line) irradiation. Daughter molecules resulting from the induced chemical reactions were identified; from the dependence of the molecular abundance, the evolution with fluence of destruction and compaction cross sections of the ice was determined. The average value of 2.2 ± 0.4 x 10^{-13} cm^2 was obtained for the average destruction cross-section determined via seven IR bands. At the beginning of irradiation, associated with ice compaction, the presence of N_2 accelerates the process of acetone destruction. However, for high fluences, the structural rearrangement caused by N_2 reduces the radiolysis rate compared to pure acetone [2].

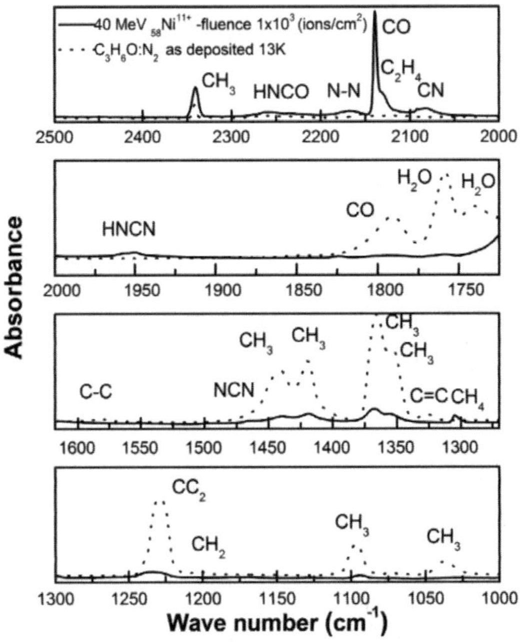

Figure 1. FTIR spectra of N_2: CH_3COCH_3 before (dotted line) and after (solid line). Some corresponding vibrational modes are indicated.

References

[1] A. L. F de Barros *et al 2014 MNRAS* **438**, 2026–2035

[2] Diana P. P. Andrade *et al 2014 MNRAS* **444**, 3792–3801

[1]E-mail: katianne.alcantara@gmail.com

Content from this work may be used under the terms of the Creative Commons Attribution 3.0 licence. Any further distribution of this work must maintain attribution to the author(s) and the title of the work, journal citation and DOI.
Published under licence by IOP Publishing Ltd

XXIX International Conference on Photonic, Electronic, and Atomic Collisions (ICPEAC2015) IOP Publishing
Journal of Physics: Conference Series **635** (2015) 022068 doi:10.1088/1742-6596/635/2/022068

The Tungsten Project: Dielectronic Recombination data for the International Thermonuclear Experimental Reactor (ITER)

S. P. Preval* [1], **N. R. Badnell*** [2], **M. O'Mullane*** [3]

* Department of Physics, University of Strathclyde, Glasgow, G4 0NG, United Kingdom

Synopsis The plasma facing components in ITER will be constructed using Tungsten. Sputtering of plasma against these walls introduces Tungsten impurities into the plasma, possibly causing quenching. As data for Tungsten in the relevant charge states is historically poor, this project aims to calculate this data, providing partial and total Dielectronic/Radiative Recombination rates for use in collisional-radiative modelling of the ITER plasma.

The International Thermonuclear Experimental Reactor (ITER) is thought to be the penultimate step in realizing commercial nuclear fusion power. ITER will dwarf the Joint European Torus in terms of physical size, core temperature, and plasma volume.

Tungsten has been chosen as a plasma facing metal for the ITER reactor. Prior to this, reactors such as the Joint European Torus (JET) used walls composed of Beryllium and/or Carbon, however, the former is extremely toxic to humans, making reactor decommissioning more difficult, while the latter is susceptible to high tritiation rates. Tungsten is a non toxic metal, and also has very low tritiation rates.

One disadvantage of using Tungsten for the reactor wall is its potential to quench the plasma. Tungsten has 74 electrons, and will hence not be completely ionized if introduced into the plasma. Therefore, it is necessary to model the effect of introducing Tungsten impurities into the plasma. This can be seen in Figure 1, where two sets of ionisation balance calculations are presented as a function of temperature in the low electron density limit. While both sets use ionisation rate data from [1], the red-solid curve uses recombination data from [2], and the blue-dashed curve uses data from [3]. This figure demonstrates the large uncertainty in the current recombination data used over the whole range of charge states that will arise in ITER.

The Tungsten Project is an effort to calculate ab initio the Dielectronic/Radiative Recombination (DR and RR respectively) rates for each ion. The DR and RR data is being calculated as described in [4] and [5] respectively, using the atomic collision package AUTOSTRUCTURE [6]. Results of the first phase of this program (H-Ar like) will be presented, covering the core component of the ITER plasma. We also provide partial recombination rates, enabling quantities such as the ionisation balance and power loss to be calculated at magnetic fusion densities where the low density limit is not valid.

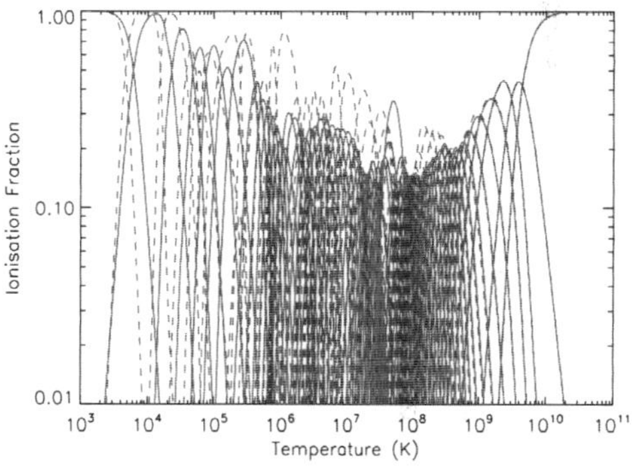

Figure 1. Comparison of calculated ionisation fractions for Tungsten using the Pütterich et al. [2] (red solid) and Foster [3] (blue dashed) recombination rates and Loch et al. [1] ionisation rates in the low electron density limit. From left to right, W^{0+}-W^{74+}.

References

[1] S. D. Loch *et al*, 2005, *Phys Rev A*, 72, 052716

[2] T. Pütterich *et al* 2008 *Plasma Physics and Controlled Fusion* **50** 085016

[3] A. Foster, 2008, *PhD Thesis*, University of Strathclyde

[4] N. R. Badnell 2003, *A&A*, 406, 1151

[5] N. R. Badnell 2006, *ApJS*, 167, 334

[6] N. R. Badnell, 2011, *Computer Physics Communications*, 182, 1528

[1] E-mail: simon.preval@strath.ac.uk
[2] E-mail: badnell@phys.strath.ac.uk
[3] E-mail: martin.omullane@strath.ac.uk

Content from this work may be used under the terms of the Creative Commons Attribution 3.0 licence. Any further distribution of this work must maintain attribution to the author(s) and the title of the work, journal citation and DOI.
Published under licence by IOP Publishing Ltd

Journal of Physics: Conference Series **635** (2015) 022069 doi:10.1088/1742-6596/635/2/022069

Ionization of Helium by neutronic impact

M.J. Ambrosio*[1], D.M. Mitnik*, G. Gasaneo†, and F.D. Colavecchia°

* Instituto de Astronomía y Física del Espacio (IAFE, CONICET–UBA), Buenos Aires, Argentina.

† Departamento de Física, Universidad Nacional del Sur and Consejo Nacional de Investigaciones Científicas y Técnicas, 8000 Bahía Blanca, Buenos Aires, Argentina.

° División Física Atómica, Molecular y Óptica, Centro Atómico Bariloche and Consejo Nacional de Investigaciones Científicas y Técnicas, 8400 S. C. de Bariloche, Río Negro, Argentina.

Synopsis We present fully differentiated cross sections for the Helium double ionization by neutronic impact. From our theoretical results we observe the underlying mechanisms that lead to the breakup of the target.

The ionization of two-electron atoms has been studied in numerous publications, both from chiefly theoretical and experimental standpoints. See Ref. [1] for a very complete review. In particular, a significant ammount of discussion revolved around the issue of the First Born Approximation validity even when fast projectiles. In the case of neutronic projectiles, the extremely short range of the projectile-target interaction makes additional Born orders completely irrelevant.

However, attempting to ionize an atom by quickly removing the core is not very efficient way, because the comparatively low electron mass allows them to react and follow their parent core. Large ammounts of momentum transfer are needed to accelerate the core to a velocity such that results in the electrons detachment [2].

In this contribution we wanted to delve into the ionization mechanisms present; this information can be extracted from a fully differentiated cross section. The interaction between the nucleus and the neutron is considered deltiform in the their relative vector $\mathbf{r}_n$: $V_{PT} = g\delta(\mathbf{r}_n)$ [2]. The resulting equation, which rules the behaviour of the target electrons is, within the First Born regime, as follows [2, 3]:

$$[E_a - h_{He}]\,\Phi^+_{sc}(\mathbf{r}_2, \mathbf{r}_3) = \frac{g}{(2\pi)^3}$$
$$\left(e^{-i\mathbf{q}\cdot\mathbf{r}_2/(m_\alpha+1)} + e^{-i\mathbf{q}\cdot\mathbf{r}_3/(m_\alpha+2)}\right)\Phi_i(\mathbf{r}_2, \mathbf{r}_3), \quad (1)$$

and we solved with the Generalize Sturmian Functions method [4]. There factor m_α is the mass of the Helium core in atomic units.

Two main structures (and their symmetric counterparts) can be observed in Fig. 1. One corresponds to a binary emission, implying that one electron collides with the other one, sending them in directions nearly orthogonal. These, however, are afterwards widened by the interelectronic repulsion.

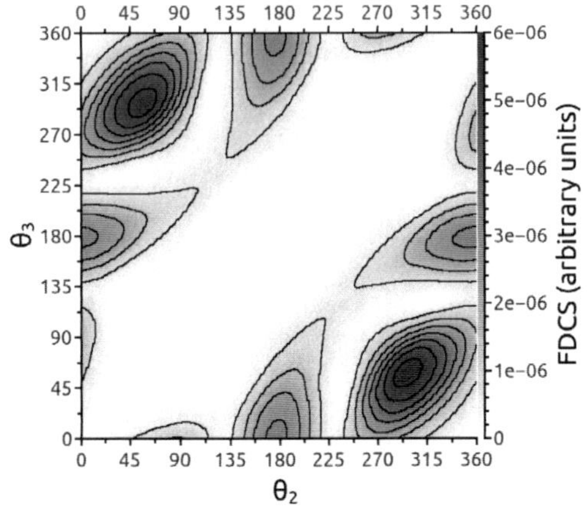

Figure 1. Fully differential cross section for the Helium double ionization by — eV neutrons. Momentum transfer q =9000 a.u. and $E_2 = E_3 = 50$ eV. The neutrons carry an incident energy of 150 keV and get deflected at 178°.

The other type of structure in Fig. 1 implies a back-to-back emission. The core is suddenly removed from between the electrons, and their repulsion gains the most relevance. However, this emission type is important only when one of the electrons is pulled towards the same direction as the momentum transfer vector, i.e., one electron is partially dragged by the ejected core, while the other is emmited in the opposite direction.

References

[1] J. Berakdar *et al* 2003 *Phys. Rep.* **374** 91-164

[2] J Berakdar 2002 *J. Phys. B* **35** L31

[3] M.J. Ambrosio *et al* 2015 *J. Phys. B* **48** 055204

[4] G. Gasaneo *et al* 2013 *Adv. Quant. Chem.* **67** 153-216

[1]E-mail: mja1984@gmail.com

Content from this work may be used under the terms of the Creative Commons Attribution 3.0 licence. Any further distribution of this work must maintain attribution to the author(s) and the title of the work, journal citation and DOI.
Published under licence by IOP Publishing Ltd

Fast calculation of cross sections using graphics processing units

F. D. Colavecchia[*,†1],

[*] División Física Atómica, Molecular y Óptica,
Centro Atómico Bariloche, Av. Bustillo 9500, 8400 San Carlos de Bariloche, Argentina
[†] Consejo Nacional de Investigaciones Científicas y Técnicas (Conicet)

Synopsis Continuum Distorted Wave methods are adapted to new heterogeneous computing architectures, exhibiting excellent speed ups in performance.

The theory of collisions of photons, leptons, ions or other species with atomic targets is devoted to compute a key observable of physics: the cross sections. Much of the work in this area in the last decades has been oriented to establish the formal mathematical foundations of different models and theories to represent the complex quantum-mechanical dynamics of the few particle problem. Besides, the calculation of cross sections involves the use of innovative numerical methods and becomes more and more intensive in the use of computational resources.

The first step in the computation of cross sections is the evaluation of the transition matrix element T_{if} between the initial and final states of the collision. The second step involves the (usually multidimensional) integration of the probabilities of transition to match the specific cross sections obtained in the experimental realization of the collision under scrutiny. While the details of the model determine the first step, this crude two-stages description is quite independent of the theory selected.

Theoretical models have become more and more advanced as the computational infraestructure evolved. Nowadays, there are several methods that solve three-body Schrdinger equation from first princiles, without approximations, but require the intensive use of clusters of computing machines[1, 2, 3]. However, the rise of heterogeneous computer architectures in the last years requires the development of new numerical methods and different approaches to the calculations: the computational capability of each piece of hardware as well as the vastly different bandwidths among them have to be carefully considered to port codes to contemporary computers. The ability to use graphics processing units (GPUs) as massivelly parallel processors with standard programming languages leads to big speed ups in optimized codes[4].

In this work the performance of the well-known Continuum Distorded Wave – Eikonal Initial State (CDW–EIS) method for the ionization of atoms by charged particles at intermediate to high energies[5] is explored. The code is completely rewritten to take advantage of heterogeneous hardware architectures with GPUs. Two main optimization points were developed: the calculation of partial waves (or hypergeometric functions for pure Coulomb potentials), and the multidimensional integration to compute total cross sections. Special purpose libraries are used to compute on the GPU[6, 7]. Workload balancing between CPU cores and GPU is an important factor affecting performance. Preliminary calculations show that the code runs at least ten times faster than the CPU only counterpart.

References

[1] McCurdy C W, Baertschy M and Rescigno T N 2004 *Journal of Physics B: Atomic, Molecular and Optical Physics* **37** R137–R187

[2] Bray I and Stelbovics A T 1995 *Computer Physics Communications* **85** 1–17

[3] Gasaneo G, Ancarani L, Mitnik D, Randazzo J, Frapiccini A and Colavecchia F 2013 *Adv. Quantum Chem* **67** 153

[4] Colavecchia F D 2014 *Computer Physics Communications* **185** 1955–1964

[5] Fainstein P D, Ponce V H and Rivarola R D 1991 *Journal of Physics B: Atomic, Molecular and Optical Physics* **24** 3091

[6] Merril D 2014 Cub template library version 1.3.2

[7] NVIDIA Corporation 2014 *NVIDIA CUDA C Programming Guide*

[1]E-mail: flavioc@cab.cnea.gov.ar

Content from this work may be used under the terms of the Creative Commons Attribution 3.0 licence. Any further distribution of this work must maintain attribution to the author(s) and the title of the work, journal citation and DOI.
Published under licence by IOP Publishing Ltd

Journal of Physics: Conference Series **635** (2015) 022071 doi:10.1088/1742-6596/635/2/022071

Double KK excited states in highly charged sulphur

P. Amaro[* 1], **J. P. Marques**[†], **P. Indelicato**[‡], **T. K. Mukherjee**[#], **J. K. Saha**[II], **L. C. Tribedi**[&], **J. P. Santos**[* 2]

[*] Laboratório de Instrumentação, Engenharia Biomédica e Física da Radiação (LIBPhys-UNL), Departamento de Física, Faculdade de Ciências e Tecnologia, FCT, Universidade Nova de Lisboa, 2829-516 Caparica, Portugal

[†] BioISI - Biosystems & Integrative Sciences Institute, Faculdade de Ciências da Universidade de Lisboa, Campo Grande, C8, 1749-016 Lisboa, Portugal

[‡] Laboratoire Kastler Brossel, École Normale Supérieure, CNRS, Université P. et M. Curie – Paris 6, Case 74; 4, place Jussieu, 75252 Paris CEDEX 05, France

[#] Narula Institute of Technology, Agarpara, Kolkata-700109, India

[II] Indian Association for the Cultivation of Science, Jadavpur, Kolkata-700032, India

[&] Department of Nuclear and Atomic Physics, Tata Institute of Fundamental Research, Colaba, Mumbai 400005, India

Synopsis Calculation of energies and radiative decays was performed for several highly charge states of sulphur in order to identify experimental lines associated to double excited KK atomic states.

Double KK excited states have several important applications in astrophysics as well as in plasma diagnostics [1, 2]. Besides its applicability, the fundamental interest of dominance of fluorescence over autoionization decay for some KK excited states has attract much attention, both from theoreticians and experimentalists [3, 4]. Recently, it was observed the $2p3d(^1\mathrm{P}^0) \rightarrow 1s3d(^1\mathrm{D}^e)$ in helium with sufficient accuracy to distinguish from the H-like K_α line, showing an enhanced fluorescence rate compared to the one assuming standard selection rules [5].In this work, we continue the investigation of KK excited states for sulphur by performing a systematic identification of experimental lines, which includes both single K and double KK excited states. The relativistic calculations were made within the MCDF approach using the general relativistic MCDF code (MDFGME) [6]. Experimental measurements were performed at the Pelletron accelerator facility in TIFR Mumbai.

In these preliminary calculations, all transition energies and probabilities were computed in a monoconfiguration approach. Almost all possible combinations of excitations for K_α, K_β and K_γ transitions of He through Be charge states were considered. The positions of the experimental lines are displayed in Fig. 1 along with the decay rates of the computed transitions. Transitions from KK excited states were also considered in both He- and Li-like states. It can observed in Fig. 1 that all experimental lines lines are unambiguously identified, with the exception of one line next to the He-like K_α line. Lines near the H-like K_α line can only be associated to the KK excited states of He or Li charge states. A precise identification can be done by calculating the respective autoionization rate.

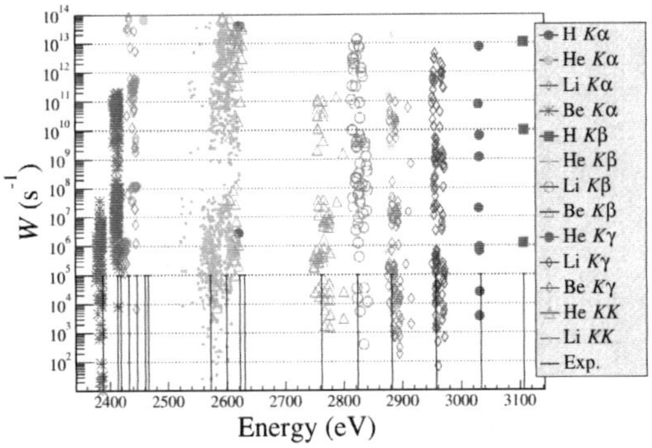

Figure 1. Comparison of obtained theoretical lines with experimental lines.

Referências

[1] A. B. C. Walker and Jr., H. R. Rugge, 1971 *Astrophys. J.* **164** 181

[2] T. Fujimoto and T. Kato, 1981 *Astrophys. J.* **264** 994

[3] T.W. Gorczyca *at al* 2000 *Phys.Rev.Lett.* **85** 1202

[4] J. G. Lambourne *at al* 2003 *Phys.Rev.Lett.* **90** 153004

[5] S. Kasthurirangan *at al* 2013 *Phys.Rev.Lett.* **24** 243201

[6] P. Indelicato and J. P. Desclaux 1990 *Phys.Rev.A* **42** 5139-5149

[1]E-mail: pdamaro@fct.unl.pt
[2]E-mail: jps@fct.unl.pt

Content from this work may be used under the terms of the Creative Commons Attribution 3.0 licence. Any further distribution of this work must maintain attribution to the author(s) and the title of the work, journal citation and DOI.
Published under licence by IOP Publishing Ltd

Magnetic Field Manipulation Applied to Feshbach Resonances in Ultracold Quantum Gases

Tianwei Zhou*[1], Tian Luan*, Chen Li*, Shizhe Li*, and Xuzong Chen*,

* School of Electronics Engineering and Computer Science, Peking University, Beijing 100871, China

Synopsis With a feedforward feedback compound control current servo loop, we present Bitter-type electromagnets featuring high magnetic field and fine electronic properties applied to Feshbach Resonances in Ultracold Quantum Gases.

The extraordinary degree of control that the ultracold quantum gases systems offer makes a great impact on present-day physics. In all applications, tuning two-body interactions via a Feshbach resonances is the experimental key to control collective phenomena in both atomic Bose-Einstein condensates and degenerate Fermi gases. The Feshbach resonance for alkali-metal atoms typically requires the magnetic field to approach 1000 G and the width of the resonance generally determines the magnetic field resolution required for detection. Many s-wave Feshbach resonances have widths larger than 1 G while high partial wave Feshbach resonances are typically much narrower [1].

In our experiment of ultracold atomic mixture ^{87}Rb-^{133}Cs-^{40}K, controlling the magnetic field is primarily required for manipulating atomic collision properties. The magnetic field needs to be stable to within a few milligauss to access certain narrow Feshbach resonances and has to switch fast enough to offer a sufficient time window for monitoring fast collision processes. We present Bitter-type electromagnets constructed from a stack of alternating layers consisting of copper arcs and insulating mica spacers. Efficient cooling of the copper is achieved via parallel cooling channel on the insulating layers and the parallel ducts between the layers ensuring low resistance for cooling water to flow.

We also built a feedforward feedback compound control current servo loop consisting of several paralleling power MOSFET connected with the magnetic coil as shown in Fig 1. To achieve good current stability, the feedback signal is provided by a current sensor through measuring the potential drop across a shunt resistor, connected in series with the magnetic coil. The closed loop bandwidth is 50 kHz. An independent Holzer current sensor monitors the long term current stability to be 10^{-5}. We used a IGBT module to achieve the fast switching of the magnetic field current in 1 ms.

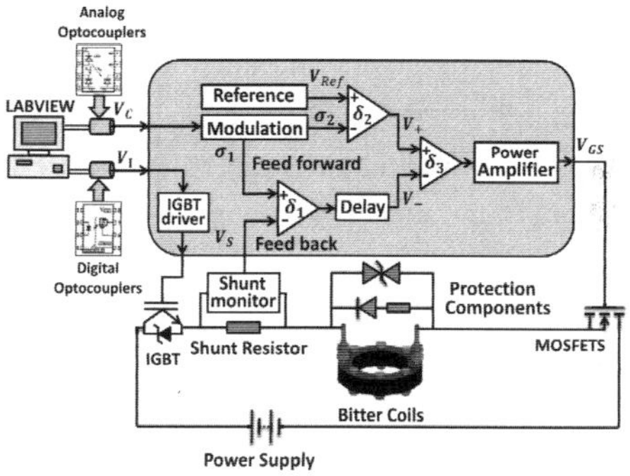

Figure 1. Schematic of current servo loop.

A peak magnetic field of 1000 G is generated in the center of the apparatus with a current of 230 A as shown in FIG 2. and the resolution of magnetic field is in the order of one hundred milligauss.

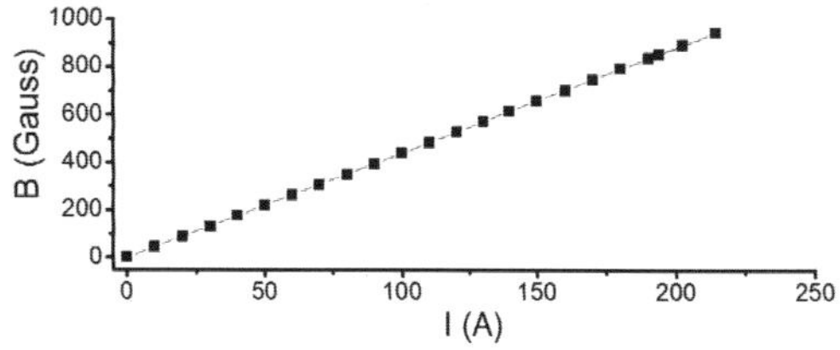

Figure 1. Magnetic field as a function of current through the electromagnets.

References

[1] Cheng Chin *at al* 2010 *Rev. Mod. Phys.* **82** 1225

[1]E-mail: ztwzkx@163.com

Content from this work may be used under the terms of the Creative Commons Attribution 3.0 licence. Any further distribution of this work must maintain attribution to the author(s) and the title of the work, journal citation and DOI.
Published under licence by IOP Publishing Ltd

Beryllium ionisation by ion impact

S. D. López [*,‡,1], **J. M. Randazzo** [†,‡] **R. Della Picca** [†,‡] **F. D. Colavecchia** [†,‡]

[*] Instituto de Astronomía y Física del Espacio, Casilla de Correo 67 Suc. 28, C1428ZAA Ciudad Autónoma de Buenos Aires, Argentina

[†] División Física Atómica, Molecular y Óptica, Centro Atómico Bariloche, Av. Bustillo 9500, 8400 Bariloche, Argentina

‡ CONICET, Consejo Nacional de Investigaciones Científicas y Técnicas, Argentina

Synopsis We study ionization of Beryllium atoms by ion impact with continuum distorted waves models. We calculate cross sections in the intermediate to high energy regime for state resolved and total ionization.

The knowledge of cross sections of atomic ionization and capture by charged particles plays a key role in the development of models for plasma dynamics in fusion reactors. Besides the constituents of the fuel, deuterium and tritium, and the byproducts of the reaction, alpha particles and neutrons, the plasma is contaminated by atomic species detached from the tokamak recipient. In this work we investigate the beryllium (Be) ionisation by ion impact, predicted to appear in ITER plasma since Be is the selected material for the walls [1].

From the theoretical point of view, Be ionisation presents several interesting features to be considered. The ground state has closed shells with zero angular momentum for the four electrons. This fact allows us to study the effect of the ground state by comparison between electron emission from $n = 1$ and $n = 2$ levels with the same angular symmetry.

In this work Doubly differential and total cross sections (DDCS) are calculated in the continuum-distorted-wave-eikonal-initial-state (CDW-EIS) approximation for ion-impact [2, 3].

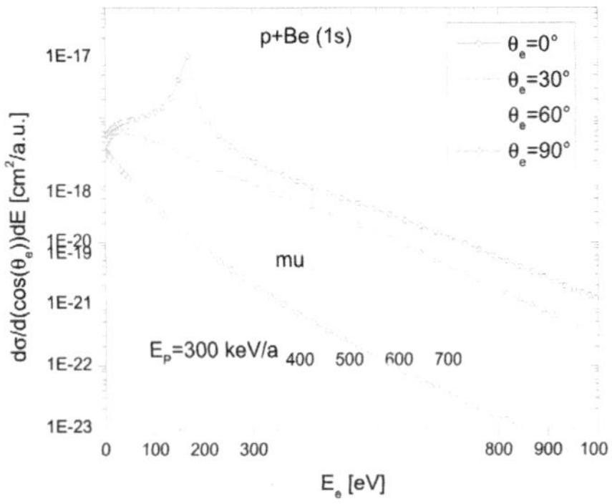

Figure 1. DDCS as a function ef the electron energy for single ionisation of Be(1s) by proton impact at 300 keV/amu for different emission angles.

The initial bounded and final electronic wave functions are obtained with the Optimized Potential Model (OPM) presented by Talman [4].

As an example, in Fig. 1 and 2 we present DDCS where we can observe the electron capture to the continuum (ECC) peak for emission at 0 degrees in both subshells ionisation. For higher emission angles this peak vanishes and for lower angles the 2s ground state displays a large binary peak, opposite to the case of 1s ground states. On the other hand, for 1s electrons the DDCS monotonically decreases with the emission angle, whereas for the 2s ground state the behaviour is very different. These observations will be discussed during the conference.

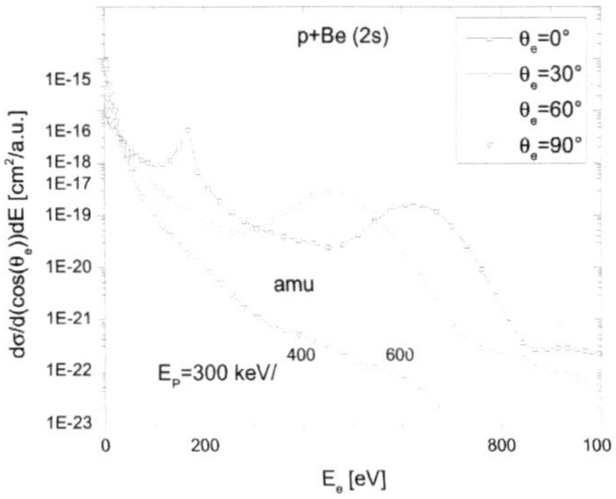

Figure 2. Same as Fig. 1, for Be(2s) ionization.

References

[1] T. Pütterich *et al* 2008 *Plasma Phys. Contr. F.* **50**, 085016

[2] L. Gulyas *et al* 1995 *J. Phys. B.* **28**, 245

[3] S. D. López *et al* 2015 *J Phys Conf Ser* **583**, 012035

[4] J. D. Talman 1989 *Comp. Phys. Comm.* **54**, 85

[1] E-mail: sebastlop@iafe.uba.ar

Content from this work may be used under the terms of the Creative Commons Attribution 3.0 licence. Any further distribution of this work must maintain attribution to the author(s) and the title of the work, journal citation and DOI.
Published under licence by IOP Publishing Ltd

XXIX International Conference on Photonic, Electronic, and Atomic Collisions (ICPEAC2015) IOP Publishing
Journal of Physics: Conference Series **635** (2015) 022074 doi:10.1088/1742-6596/635/2/022074

Molecular orbital perspective for inner shell couplings: Level diagrams

Punita Verma*[1], C.P. Safvan[2], A. Jhingan[2], T. Nandi[2], A. Mandal[2]

[1]Department of Physics, Kalindi College, University of Delhi, East Patel Nagar, 110008, New Delhi, India.
[2] Inter University Accelerator Centre, Aruna Asaf Ali Marg, New Delhi, 110062, India.

Synopsis Level diagrams drawn for the Ag-Gd, Ho and Yb collision systems allow a qualitative explanation for the experimentally observed enhanced target and projectile cross sections. Super heavy quasimolecules formed during these slow, heavy-ion, heavy-atom collisions act as a beacon of light to investigate the couplings of the inner shells and the dependence of inner shell ionization on projectile charge state..

Relatively slow, heavy ion-atom collisions with a collision velocity smaller than the orbital velocity of the innermost electrons lead to the formation of super-heavy atomic systems with united atomic number $Z_{UA}=Z_1+Z_2>100$ (Z_1 and Z_2 being the atomic numbers of the projectile and target respectively). These super-heavy quasi-atoms or quasi-molecules have aroused interest [1-3] due to their changed properties and accompanied enormous relativistic effects. The couplings of the inner shell levels become evident only for projectiles carrying vacancies in the collision and their subsequent filling leading to x-ray emission in the separated partners or in the quasi-molecule itself.

X-ray production and ionization cross sections of Gd, Ho and Yb have been measured due to a collision induced by Ag^{q+}-ions ($5<q<14$) between 0.4-1.7 MeV/u at the 15 MV tandem accelerator of Inter University Accelerator Centre (IUAC), New Delhi. Since $v_{ion}/v_{target}<1$ for these ions and consequently, η the adiabaticity parameter being also less than 1, the investigation of these slow, heavy ion-heavy-atom collisions have allowed a glance into the quasi-molecular phenomena. The results can thus be explained qualitatively using the molecular orbital picture. Formerly, inner shell couplings have been investigated at GeV projectile energies [4] to understand the role of projectile charge state in inner shell ionization and the present investigation facilitate the same at < 2 MeV/u energies.

Diabatic level diagrams [2,3] for the Ag-(Gd, Ho and Yb) collision systems have been drawn as per the binding energy values of Fricke [5] to develop an understanding of how the inner shells couple with each other and how do these couplings facilitate the transfer of vacancies from one partner to another. An approximate value of the average ionization limit of the projectile gives the validity of the level diagram for any collision system which is given by the $v_{ion}=v_{target}$ limit or the $\eta=1$ limit (i.e. the limit when the velocity of the projectile is equal to that of the electron in the Bohr orbit.

From the level diagrams it is clear that target (Gd, Ho and Yb) L x-ray emission is a closed channel problem. No vacancies are transferred to the $3d\sigma$ level of the united atom on the ingoing part of the collision and direct ionization out of this level has a high probability during the collision. However, on the outgoing part of the collision, vacancies existing in the $3d\sigma$ level of the united atom may be transferred to the $2p_{3/2}$ level of the Ag-ions. Certain vacancies might be transferred to the $2p_{3/2}$ level of the targets due to the sharing between the $2p_{3/2}$ level of the Ag-ion and the former levels thus influencing the target L x-ray emission. This process has the highest probability of occurrence.

A second process taking place on the ingoing part of the collision is the transfer of vacancies to the $4f\sigma$ level of the united atom. The level diagrams indicate that these vacancies can be transferred from the $4f\sigma_{UA}$ level to the $2p_{3/2}$ level of the Ag-ion projectile. Due to the sharing between the $2p_{3/2}$ level of the Ag-ion projectile and $2p_{3/2}$ level of the targets, certain vacancies may be transferred from the former to the latter thus influencing its L x-ray emission. Simultaneously a third process takes place which is of least consequence for its influence on target L x-ray emission. On the ingoing part of the collision no vacancies are transferred to the $2p_{UA}$ level and during the collision direct ionization out of the $2p_{UA}$ level takes place. However, it is known that the cross sections for the latter are negligible as compared to those for direct ionization out of the $3d\sigma_{UA}$ level. All the above possibilities lead to enhanced projectile (Ag-ion) and target (Gd, Ho and Yb) L x-ray cross sections compared to the predictions of the direct ionization theories. This enhancement has been indeed observed experimentally.

References

[1] D. Maor *et al* 1981 '*Inner Shell and X-ray Physics of Atoms and Solids*', ed. Fabien, H. Klienpoppen and L.M. Watson.
[2] M. Barat and W. Lichten, 1972 *Phys. Rev. A* **6** 211.
[3] P. Mokler *et al* 1971 *Phys. Rev. Lett.* **26** 811.
[4] P. Verma *et al*, 2011 Physica Scripta **T 144**, 014032.
[5] B. Fricke, University of Kassel, private commun.

*1 E-mail: p.verma.du@gmail.com

Content from this work may be used under the terms of the Creative Commons Attribution 3.0 licence. Any further distribution of this work must maintain attribution to the author(s) and the title of the work, journal citation and DOI.
Published under licence by IOP Publishing Ltd

Observation of singlet-scattering channels in long-range Rydberg dimers

Johannes Deiglmayr [1], **Heiner Saßmannshausen, Frédéric Merkt**

Laboratory of Physical Chemistry, ETH Zrich, CH-8093 Zurich

Synopsis We report on the observation of long-range Cs_2 Rydberg molecules consisting of a p-state Rydberg atom and a ground-state atom. Two types of molecular states are visible in the photoassociation spectra: states bound purely by triplet s-wave scattering and states bound by mixed singlet-triplet s-wave scattering. An effective Hamiltonian including s-wave scattering pseudopotentials, the hyperfine interaction of the ground-state atom, and the spin-orbit interaction of the Rydberg atom reproduces the experimental observations. The detailed analysis reveals the important role of singlet scattering in long-range Rydberg molecules and yields an effective singlet s-wave scattering length for low-energy e^--Cs collisions.

Long-range Rydberg dimers are molecules in which one of the atoms is in the ground state and the other one is in a highly excited electronic state, a Rydberg state. The binding of these molecules is caused by the elastic scattering of the Rydberg electron off the ground-state atom, which is located within the electron orbit. This type of molecular bond was initially proposed by Greene *et al.* [1] and has been observed experimentally in the photoassociation of ultracold rubidium and cesium atoms close to transitions to s, p and d Rydberg states [2–7].

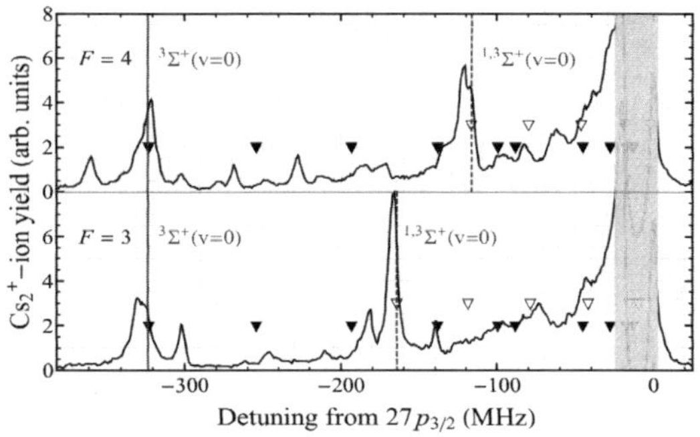

Figure 1. Detected Cs_2^+ ions after photoassociation in an ultracold gas of cesium atoms as a function of the laser frequency near the $27p_{3/2}+6s_{1/2}(F=3,4)$ asymptotes. The solid (dashed) red lines mark the calculated positions of the $v=0$ levels of the $^3\Sigma^+$ ($^{1,3}\Sigma^+$) states. Full and open triangles mark the calculated positions of higher vibrational levels of the $^3\Sigma^+$ and $^{1,3}\Sigma^+$ states, respectively. The strongly saturated atomic transition corresponds to the grey area.

Here we report on the formation of long-range Rydberg molecules correlated to np Rydberg states of cesium with principal quantum number n ranging from 26 to 34 [7]. The observed binding energies reach up to 400 MHz and the internuclear separations exceed 45 nm at n=26. As shown in Figure 1 for n=27, the photoassociation spectra reveal two types of molecular states recently predicted by Anderson *et al.* [4]: Bound states for which the binding results only from triplet s-wave scattering (labeled $^3\Sigma^+$), and more weakly bound states with contributions from both singlet and triplet s-wave scattering (labeled $^{1,3}\Sigma^+$). The experimental observations are well described by a model including s-wave scattering, the hyperfine interaction of the ground-state atom and the fine structure of the Rydberg atom.

By adjusting the values of the zero-energy singlet and triplet s-wave scattering lengths to $a_{S,0} = -3.5 \pm 0.4\ a_0$ and $a_{T,0} = -21.8 \pm 0.2\ a_0$, respectively, we are able to reproduce the experimental binding energies within the experimental uncertainties. This yields the first experimental determination of the zero-energy singlet s-wave scattering length, which is in reasonable agreement with an extrapolation from higher-energy electron-Cs scattering data [8].

References

[1] C. H. Greene, A. S. Dickinson and H. R. Sadeghpour 2000 *Phys. Rev. Lett.* **85** 2458
[2] V. Bendkowsky *et al.* 2009 *Nature* **458** 1005
[3] A. Krupp *et al.* 2014 *Phys. Rev. Lett.* **112** 143008
[4] D. A. Anderson, S. A. Miller and G. Raithel 2014 *Phys. Rev. Lett.* **112** 163201
[5] K. R. Overstreet *et al.* 2009 *Nat. Phys.* **5** 581
[6] J. Tallant *et al.* 2012 *Phys. Rev. Lett.* **109** 173202
[7] H. Saßmannshausen, F. Merkt and J. Deiglmayr 2015 *Phys. Rev. Lett. in press* (preprint at arXiv:1412.0846)
[8] C. Bahrim, U. Thumm and I. I. Fabrikant 2001 *J. Phys. B* **34** L195

[1] E-mail: j.deiglmayr@phys.chem.ethz.ch

Content from this work may be used under the terms of the Creative Commons Attribution 3.0 licence. Any further distribution of this work must maintain attribution to the author(s) and the title of the work, journal citation and DOI.
Published under licence by IOP Publishing Ltd

Measurements and Theoretical Predictions of Charge Exchange Cross Sections and Emission Spectra for O^{6+} with H_2O, CO, CO_2, CH_4, N_2, NO, N_2O and Ar

J. R. Machacek[1], D. P. Mahapatra[2, 3], D. R. Schultz[4 *], Yu. Ralchenko[5], A. Moradmand[2], M. O. A. El Ghazaly[2], and A. Chutjian[2 †]

[1] Positron Physics Group, The Australian National University, ACT 2601, Australia
[2] Astrophysics and Space Sciences Section, Jet Propulsion Laboratory/Caltech, Pasadena CA 91109
[3] Department of Physics, Utkal University, Bhubaneswar-751004, India
[4] Department of Physics, Univ. of North Texas, Denton, TX 76203
[5] Atomic Spectroscopy Group, National Institute of Standards and Technology, Gaithersburg, MD 20899-8422

Synopsis Relevant to modeling and understanding X-ray emission from cometary and planetary atmospheres, total charge-exchange cross sections for 1.17 and 2.33 keV/u O^{6+} ions colliding with H_2O, CO, CO_2, CH_4, N_2, NO, N_2O, and Ar have been measured and calculated for the processes of single, double, and triple exchanges. Synthetic emission spectra spanning the X-ray, UV, and visible ranges have also been calculated, based on theoretical treatment of the transfer of between one and six electrons from the target neutral to the projectile ion, followed by radiative and non-radiative decay of the resulting highly-excited projectile states.

The phenomenon of charge exchange (CE) between highly-charged ions (HCIs) and neutral molecules is observed throughout our Solar System, and is very likely occurring in more-distant stellar atmospheres. In such exchanges an HCI in (for example) the solar wind (SW) captures an electron from the neutral gas into a state with high principal quantum number n that subsequently de-excites by emitting one or more photons at visible–to–X-ray wavelengths [1]. The gaseous targets can be comet or planetary atmospheres, circumstellar clouds, and the interplanetary medium; while the HCIs arise from the SW, the Jovian magnetosphere and cosmic rays [2]. Since O^{6+} is the most abundant heavy ion in the SW, its interactions can be used to probe objects as distant as Pluto.

Reported herein are absolute CE cross sections for up to three exchanges of the O^{6+} ion with electrons from the comet/planetary gases H_2O, CO, CO_2, CH_4, N_2, NO, N_2O and Ar. The measurements are compared with results in calculations using the classical trajectory Monte Carlo (CTMC) method. With the measurements as benchmarks, synthetic emission spectra spanning the X-ray, UV, and visible range have been calculated based on theoretical treatment of the transfer of between one and six electrons from the target neutrals to O^{6+}, followed by radiative and non-radiative decay of the highly-excited ion states.

Experimental measurements were carried out using the HCI source at JPL. Details are given in [3] for results with He and H_2 targets; as well as details of the CTMC calculational approach. Extension of the theory for calculating cross sections and spectra for 1-6 captures using an *ab initio* formulation of the atomic structure and decay parameters will be described in [4, 5].

Shown in Table 1 are results of measurements and calculations for CO and H_2O, for three exchanges, at the lower energy of 1.17 keV/u [4]. Results will be presented for three exchanges in all the target gases at 1.17 and 2.33 keV/u, and simulated emission spectra for relaxation of the $O^{(5-3)+}$ ions to their lower electronic states.

Target	CE Channel	CE Cross Section (Exp.)	CE Cross Section (Theory)
CO	σ_{6-5}	5.41±0.35	4.51
	σ_{6-4}	1.10±0.08	0.58
	σ_{6-3}	0.37±0.03	0.05
H_2O	σ_{6-5}	4.73±0.31	5.50
	σ_{6-4}	0.83±0.06	0.63
	σ_{6-3}	0.37±0.03	0.06

Table 1. Experimental and CTMC results for CE of O^{6+} with CO and H_2O at an energy of 1.17 keV/u. Units are 10^{-15} cm^2, and errors are at the 2σ level.

References

[1] N. A. Schwadron, T. E. Cravens 2000 *ApJ* **554** 558
[2] A. Bhardwaj *et al* 2007 *Planet. Sp. Sci.* **55** 1135
[3] J. R. Machacek *et al* 2014 *Phys. Rev. A* **90** 052708
[4] J. R. Machacek *et al.* 2015 *ApJ*, in preparation
[5] D. R. Schultz and Yu. Ralchenko, in preparation

* E-mail: david.schultz@unt.edu † E-mail: ara.chutjian@jpl.nasa.gov

Content from this work may be used under the terms of the Creative Commons Attribution 3.0 licence. Any further distribution of this work must maintain attribution to the author(s) and the title of the work, journal citation and DOI.
Published under licence by IOP Publishing Ltd

The nucleus-nucleus scattering in transfer ionization of helium by He^{2+} impact

W. T. Feng*[1], X. Ma*†[2], X. L. Zhu*, D. L. Guo*, Y. Gao*†, S. C. Yan*, R. T. Zhang* and S. Xu*

* Institute of Modern Physics, Chinese Academy of Sciences, Lanzhou, 730000, China
† University of Chinese Academy of Sciences, Beijing, 100049, China

Synopsis We have performed kinematically complete measurement on transfer ionization of He by 50 keV/u He^{2+} impact. The data were analyzed in terms of Dalitz spectrum. We evaluated the role of three particles interactions and found that the main momentum exchange occurs between the projectile and the recoil ion. This result indicate that the nucleus-nucleus high order interaction could not be overlooked in transfer ionization collision dynamics of atom by intermediate energy ion impact.

The transfer ionization processes are particularly suitable to investigate the quantum many-body problem [1]. The advantage of transfer ionization is that the final state involves three particles (the projectile, the emitted electron and the recoil ion). Therein, the dynamics is more complex than in cases where the final state involves only two unbound particles. As a result, higher order effects tend to be more pronounced. This provides valuable information about the ion-atom collision dynamics.

The experiment was performed using a reaction microscope at the Institute of Modern Physics, CAS [2]. The momentum-analyzed recoil ions and emitted electrons were collected in triple coincidence with the projectiles. The Dalitz spectrum was presented in Figure 1. Instead of the relative energy in original Dalitz plot, the relative squared momenta $\pi_i = p_i^2 / \sum p_i^2$ (*I* denote electron, recoil ion and projectile) of the three collision fragments were plotted. For the projectile, the momentum transfer q was used instead of its final state momentum. It should be noted that all the data points were inside an inner circle because of the momentum conservation.

As seen in Figure 1, a large peak at the contact point of the internal circle with the emitted electrons triangle side e(r-p) clearly shows that a large momentum exchange in the collision occurs between the projectile and the recoil ion, suggesting the existence of high order interaction. At the right triangle side, a secondary maximum is seen near r(p-e), illustrating important contribution from binary interactions between the projectile and the emitted electron. However, few data points are seen near the lower triangle side p(r-e), indicating a weak momentum correlation between the emitted electron and the residual target. In contrast, these characteristics are different from those reported for pure target ionization of helium by fast ion impact [3], where the main contribution of cross section are resulting from the first order interaction between the projectile and the emitted electron.

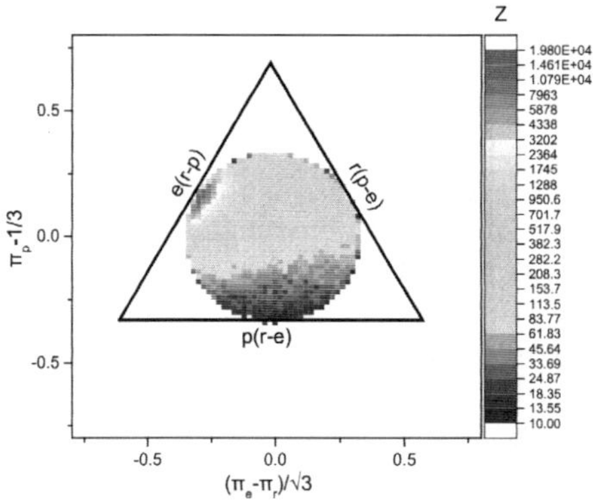

Figure 1. Dalitz plot in transfer ionization of helium by 50 keV/u He^{2+} impact.

This work was partially supported by the NSFC Grants No. 11004202

References

[1] M. Schulz *et al.*, *Phys. Rev. A.* **74** 042705
[2] X Ma *et al.*, *Sci. China. Ser. G.* **51** 755
[3] M. Schulz *et al.*, *J. Phys. B: At. Mol. Opt. Phys.* **37** 4055

[1] E-mail: fengwentian@impcas.ac.cn
[2] E-mail: x.ma@impcas.ac.cn

Content from this work may be used under the terms of the Creative Commons Attribution 3.0 licence. Any further distribution of this work must maintain attribution to the author(s) and the title of the work, journal citation and DOI.
Published under licence by IOP Publishing Ltd

Time-depending screening effects in relativistic calculations of electronic quantum dynamics in ion-atom collisions

Yury S. Kozhedub* [1], **Ilya I. Tupitsyn***, **Vladimir M. Shabaev***, **and Thomas Sthölker**†

* Department of Physics, St. Petersburg State University, Ulianovskaya 1, Petrodvorets, St. Petersburg 198504, Russia

† GSI Helmholtzzentrum für Schwerionenforschung, Planckstrasse 1, D-64291 Darmstadt, Germany

We present the results of non-perturbative relativistic calculations of electronic quantum dynamics in low-energy ion-atom collisions taking into account time-depending screening effects. Method of calculations employs an independent particle model with effective single-electron Dirac-Kohn-Sham operator [1]. For the inclusion of the time-depending screening effect we use a simple model, where a radially symmetric time-dependent term is added to the ground-state potential of atoms (ions) [2, 3]. Solving of the single-electron equations is based on coupled-channel approach with atomic-like Dirac-Sturm-Fock orbitals, localized at the ions (atoms) [4]. Many-particle probabilities are calculated in terms of single-particle amplitudes employing the formalism of inclusive probabilities [5]. The analysis of the post-collisional processes resulting in the x-ray and electron is based on the fluorescence yields, the radiation, and Auger decay rates, and allows one to derive intensities of the x-ray emission and compare them with experimental data. Calculations are performed for systems already studied experimentally and theoretically as well as for systems which experimental research is planed at FAIR (Darmstadt) in the nearest future [6, 7]. The role of relativistic effects is analyzed.

[1]E-mail: kozhedub@pcqnt1.phys.spbu.ru

References

[1] Kozhedub Y S *at al* 2014 *Phys. Rev. A* **90** 042709

[2] Kirchner T *at al* 2002 *Phys. Rev. A* **66** 052719

[3] Kirchner T *at al* 2000 *Phys. Rev. A* **62** 042704

[4] Tupitsyn I I *at al* 2010 *Phys. Rev. A* **82** 042701; 2012 *Phys. Rev. A* **82** 032712

[5] Lüdde H J and Dreizler R M 1985 *J. Phys.* **18** 107

[6] *htt ://www.gs .de/e /research/fa r.htm*

[7] *htt ://www.gs .de/s arc*

Content from this work may be used under the terms of the Creative Commons Attribution 3.0 licence. Any further distribution of this work must maintain attribution to the author(s) and the title of the work, journal citation and DOI.
Published under licence by IOP Publishing Ltd

XXIX International Conference on Photonic, Electronic, and Atomic Collisions (ICPEAC2015) IOP Publishing
Journal of Physics: Conference Series **635** (2015) 022079 doi:10.1088/1742-6596/635/2/022079

Experimental investigation of state selective single-electron capture in 120 keV Ar^{8+} - He collisions

R. T. Zhang* [1], **X. Ma*** [2], **X. L. Zhu***, **S. F. Zhang***, **W. T. Feng***, **D. L. Guo***, **Y. Gao***†, **S. C. Yan***, **D. B. Qian***, **P. J. Zhang***, **and S. Xu***

* Institute of Modern Physics, Chinese Academy of Sciences, Lanzhou 730000, China
† University of Chinese Academy of Sciences, Beijing 100190, China

Synopsis The longitudinal and the transversal momentum of recoil ions were measured for single electron capture in 120 keV Ar^{8+} - He collisions, using the reaction microscope techniques. The characteristics of the radial and the rotational coupling were identified for 1s to 4s and 1s to 4p sub-state selectivity.

Electron capture by multi-charged ions in collision with He is a fundamental process of intrinsic interest, because such studies are referred to electron-correlation, momentum exchange in many body coulomb field. Additionally, the studies of electron capture process play an important role in gaining recognition of plasma behavior and its diagnostics [1]. Recently, the experimental studies have been advanced to the point where measurement of multi-differential even fully differential cross sections became available, utilizing the momentum imaging techniques [2]. This fertilizes our understanding about the dynamical mechanism and the dynamical effects in capture processes.

In 1965, Fano and Lichten have firstly introduced the molecular orbital promotion (MO) into ion atom collisions, by studying energy transfer in 50 keV Ar^{+} - Ar collisions [3]. Since then, MO mechanism has been widely investigated, such as MO x ray emission in relativistic highly charged ion-atom collisions [4], MO electron emission in p - He [5] and He^{2+} - He collisions [6] etc. However, it is seldom reported for single capture in slow multi-charged ion on He collisions. Here, we investigate the single electron capture of 120 keV Ar^{8+} -He collision with reaction microscope [7], aiming at illustrating the contribution of quasi-molecular promotion to the sub-state selectivity. The preliminary results are presented in Figure 1.

As known from dynamical couplings, the theoretical basis of quasi-molecular promotion, radial coupling cannot change the angular momentum projection on the inter-nuclear vector and mainly occurs at large nuclear separation (qualitatively small transversal momentum of recoil ion), while rotational coupling could change the value 1 of angular momentum projection and mainly occurs at small inter-nuclear separation (qualitatively big transversal momentum of recoil ion) [1].

It is clearly seen that the 1s to 4p state selective capture process preferentially happened at the relatively large transversal momentum of recoil ions, compared to that for 1s to 4s state selective electron capture. Therefore, we come to the conclusion that 1s to 4p and 1s to 4s substate selective electron capture provide the indication of quasi-molecular promotion which was not investigated in reference [8].

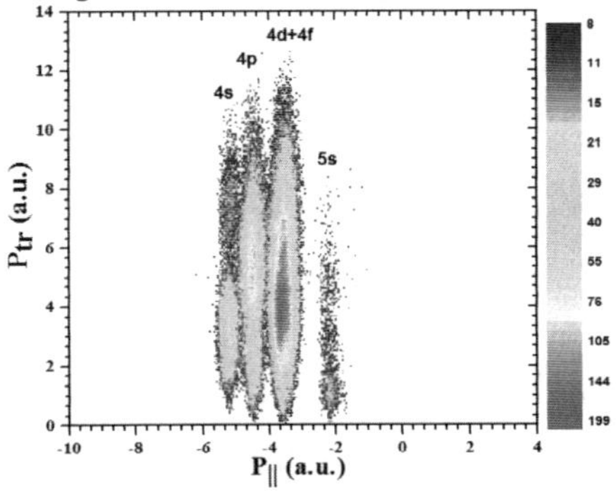

Figure 1. Two dimensional momentum distributions in 120 keV Ar^{8+} - He single electron capture process. The horizontal and the vertical axis represent the longitudinal and the transversal momentum of recoil ions.

References

[1] B. H. Bransden *et al* 1992 Charge exchange and the theory of ion-atom collisions, Oxford University press
[2] K. Schneider *et al* 2013 *Phys. Rev. Lett.* **110** 113201
[3] U. Fano *et al* 1965 *Phys. Rev. Lett.* **14** 627
[4] J. S. Greenberg *et al* 1974 *Phys. Rev. Lett.* **33** 473
[5] R. Dörner *et al* 1996 *Phys. Rev. Lett.* **77** 4520
[6] M. A. Abdallah *et al* 1998 *Phys. Rev. Lett.* **81** 3627
[7] X. Ma et al 2011 *Phys. Rev. A* **83** 052701
[8] M. A. Abdallah *et al* 1998 *Phys. Rev. A* **57** 4373

[1] E-mail: zhangrt@impcas.ac.cn
*[2] E-mail: x.ma@impcas.ac.cn

Content from this work may be used under the terms of the Creative Commons Attribution 3.0 licence. Any further distribution of this work must maintain attribution to the author(s) and the title of the work, journal citation and DOI.
Published under licence by IOP Publishing Ltd

Journal of Physics: Conference Series **635** (2015) 022080 doi:10.1088/1742-6596/635/2/022080

Progress of laser cooling of $^{12}C^{3+}$ ions at the CSRe

H. B. Wang [*,†,1], **X. Ma** [*,2], **W. Q. Wen**[*], **Z. K. Huang**[*,†], **D. C. Zhang**[*], **B. Hai**[*,†], **X. L. Zhu**[*], **D. M. Zhao**[*], **J. Li**[*], **X. M. Ma**[*], **T. L. Yan**[*], **R. S. Mao**[*], **T. C. Zhao**[*], **J. X. Wu**[*], **J. C. Yang**[*], **Y. J. Yuan**[*], **J. W. Xia**[*], **M. Loeser**[&], **M. Siebold**[&], **U. Schramm**[&], **O. Boine-Frankenheim**[$], **L. Eidam**[$], **D. Winters**[$], **G. Birkl**[#], **B. Rein**[#], **Th. Walther**[#], **and M. Bussmann**[&]

[*] Institute of Modern Physics, Chinese Academy of Sciences, 730000 Lanzhou, China
[†] University of Chinese Academy of Sciences, 100049 Beijing, China
[$] GSI Helmholtzzentrum für Schwerionenforschung GmbH, 64291 Darmstadt, Germany
[#] Institut für Angewandte Physik, Technische Universität Darmstadt, 64289 Darmstadt, Germany
[&] Helmholtz-Zentrum Dresden-Rossendorf, 01328 Dresden, Germany

Synopsis We have performed a test run for laser cooling experiments with $^{12}C^{3+}$ ion beams at an energy of 122 MeV/u at the CSRe with a pulsed laser. During this beamtime a lot of progress have been made. This was the first time we could successfully separate the $^{12}C^{3+}$ ions and $^{16}O^{4+}$ ions in the Schottky spectrum with the help of electron cooling. And our newly installed CPM detector worked well during the experiment. We tried to see effects from the interaction of the pulsed laser light with the stored $^{12}C^{3+}$ ion beams, but no cooling effects could yet be observed.

Laser cooling is the most promising method to achieve high phase-space densities or even crystalline beams for relativistic heavy ion at storage rings [1]. In order to achieve laser cooling at the CSRe in IMP, we performed a test run for a laser cooling experiment with $^{12}C^{3+}$ ion beams at an energy of 122 MeV/u at the CSRe with a pulsed laser in September 2014.

During this beamtime we have made a lot of progress and managed to get a large part of the experiment under control. Firstly, we could successfully separate $^{12}C^{3+}$ ions and $^{16}O^{4+}$ ions in the Schottky spectrum with the help of electron cooling for the first time (Figure 1). Secondly, the new CPM (Channeltron photomultiplier) detector which was used to detect the fluorescence from the (laser excited) $2p_{1/2}$ and the $2p_{3/2}$ states was tested systematically. With the laser off, the CPM detector was also used to determine the lifetime by recording fluorescence due to charge-changing collisions between $^{12}C^{3+}$ ions and the residual gas. The lifetime was about 20 seconds, which is much longer than it was during the last beam time. Thirdly, we tried to see effects from laser cooling with the pulsed laser, and the beam and laser widths were measured using the scrapers and a wire scanner to make sure they were overlapped. Moreover, the dynamics of the electron-cooled and RF-bunched ion beams were investigated systematically by longitudinal modulation of sinusoidal, square and barrier bucket waveforms.

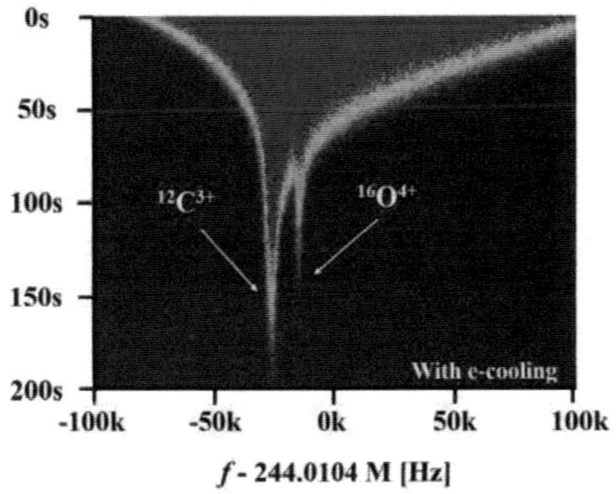

Figure 1. Schottky spectrum of $^{12}C^{3+}$ and $^{16}O^{4+}$.

We tried to see effects from the interaction of the pulsed laser with the stored $^{12}C^{3+}$ ion beams. But unfortunately, we did not find any evidence for effects from the interaction in the Schottky spectrum or the count rate of the CPM detector. We attribute this to the background and the low repetition rate of the laser. Therefore, we will repeat the experiment in 2015 using a laser system with a higher repetition rate.

This work is supported by the NSFC No. 10921504 and the External Cooperation Program of the CAS, Grant No. GJHZ1305. We acknowledge Tobias Beck for his help for preparing the beam time.

References

[1] U. Schramm *et al* 2004 *Progress in Particle and Nuclear Physics* **53** 583-67

[1] E-mail: wanghanbing@impcas.ac.cn
[2] E-mail: x.ma@impcas.ac.cn

Content from this work may be used under the terms of the Creative Commons Attribution 3.0 licence. Any further distribution of this work must maintain attribution to the author(s) and the title of the work, journal citation and DOI.
Published under licence by IOP Publishing Ltd

Double K-shell ionization of Kr induced by swift Xe^{54+} ions

Caojie Shao*, †, §, Deyang Yu*, 1, Wei Wang*, Yingli Xue*, Mingwu Zhang*, Bian Yang*, Zhangyong Song*, Rongchun Lu*, Junliang Liu*, Yipan Guo*, §, and Xiaohong Cai*, 2

* Institute of Modern Physics, Chinese Academy of Science, Lanzhou 730000, P.R. China
† School of Nuclear Science and Technology, Lanzhou University, Lanzhou 730000, P.R. China
§ University of Chinese Academy of Sciences, Beijing 100049, P.R. China

Synopsis Kα satellite and hypersatellite lines of Kr induced by 52-, 94-, 146-, and 197-MeV/u Xe^{54+} ions were measured. Experimental values were compared with theoretical ones obtained within the independent electron model (IEM) employing single electron probabilities calculated with the semi-classical approach (SCA). This comparison suggests that first order perturbation approximation is consistent well with the experimental result at 197-MeV/u projectile collision while breaks down for the lower energies.

Double K-shell ionization of atoms by collisions with charged ions is one of typical two-electron processes and attracts considerable attention both in term of basic theory and experiment. Radiative de-excitation of the double K-shell vacancy states of atoms leads to the emission of so called K x-ray hyper-satellites ($K\alpha^h$, $K\beta^h$ ...) [1], which gives us the insight into the decay modes of multiply ionized ions as well as the ionization processes during ion-atom collisions. Contrary to the long-winded and difficult experiments with heavy target due to the low detection efficiency of K x-ray hyper-satellites with crystal spectrometers [2-4], the bulk of knowledge concerning double K-shell ionization in ion-atom collisions has been obtained for light target.

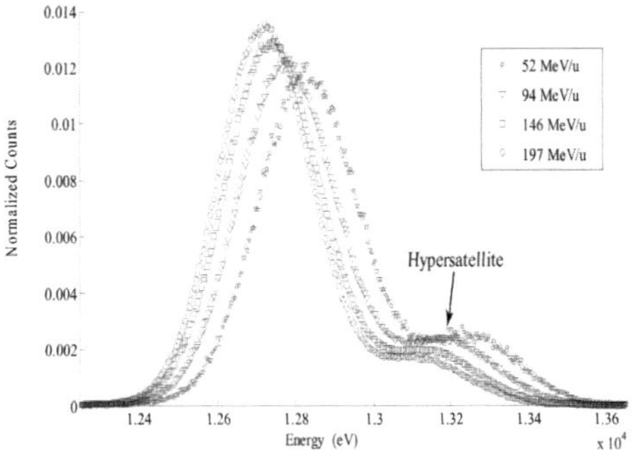

Figure 1. Kα X-ray spectra of Kr atom induced by 52-197 MeV/u Xe^{54+}

In this work we report on the double- to single-*K*-shell ionization ratios of Kr by collisions with 52-, 94-, 146-, and 197-MeV/u Xe^{54+} ions. Taking advantage of the storage ring and its internal target facility where secondary effects could be negligible and charge state could keep single, the double- to single-*K*-shell ionization ratios for the four collisions were investigated by detecting corresponding Kα line. The measured Kα X-ray energy spectra of Kr obtained at 90° observe angle with a Si (Li) detector are presented in Fig.1. The Kα X-ray hypersatellites were resolved by Si (Li) detector with energy resolution of 250 eV at 12.6 keV. The peaks position and counts of hyper-satellite could be extracted by double Gaussian function fitting.

Experimental values were compared with theoretical ones obtained within the independent electron model (IEM) [5] employing single electron probabilities calculated with the semi-classical approach (SCA) [6]. This comparison suggests that first order perturbation approximation is consistent well with the experimental result at 197 MeV/u projectile collision while breaks down for the lower energies.

This work is supported by the National Natural Science Foundation of China (Nos.11179017, 11105201, and U1332206).

References

[1] J. P. Briand 1971 *Phys. Rev. Lett.* 27 777
[2] P. Richard, W. Hodge, and C. F. Moore 1972 *Phys. Rev. Lett.* 29 393
[3] D. Olsen and C. F. Moore 1974 *Phys. Rev. Lett.* 33 194
[4] P. Richard, D. K. Olsen, R. Kauffman, and C. F. Moore 1973 *Phys. Rev. A* 7 1437
[5] R.L. Becker, A.L. Ford and J.F. Reading 1983 *Nucl. Instr. and Meth.* 214 49
[6] J.M. Hansteen 1975 *Adv. At. Mol. Phys.* 11 299

1 E-mail: d.yu@impcas.ac.cn
2 E-mail: caixh@impcas.ac.cn

Content from this work may be used under the terms of the Creative Commons Attribution 3.0 licence. Any further distribution of this work must maintain attribution to the author(s) and the title of the work, journal citation and DOI.
Published under licence by IOP Publishing Ltd

Single and double electron capture by fast Xe^{54+} from Kr and Xe

Deyang Yu*†1, Caojie Shao*, Rongchun Lu*, Wei Wang*, Junliang Liu*, Yingli Xue*, Zhangyong Song*, Mingwu Zhang*, Bian Yang*, Fangfang Ruan* and Xiaohong Cai*

* Institute of Modern Physics, Chinese Academy of Sciences, Lanzhou, 730000, China

Synopsis Decay X-ray spectrum of the scattered ions in 52–197 MeV/u Xe^{54+} collision with Kr and Xe atoms are measured. The transitions form the single capture ion Xe^{53+*} and from the double capture ion Xe^{52+*} are observed, both in a single collision. The single capture is the dominate channel at high energies while the double capture becomes dominate at lower energies.

When an energetic highly charged ion (HCI) collide with an atom, the target electrons may be captured by the projectile ion, either radiatively or non-radiatively. During a radiative electron capture (REC), a target electron is transferred to the projectile accompanying with a photon emission, which carries away the excess energy and momentum. During a non-radiative electron capture (NRC), the energy and momentum conservations are ensured by the target nucleus[1]. If the captured electrons are populated in excited states, photons maybe emitted during the following stabilization processes, and therefore the X-ray spectrum can provide information about the initial population. However, if a solid target is employed, the single-collision condition cannot be ensures when considering the capture processes, as well as strong background will be produced[2]. With the development of heavy ion cooling storage rings, the experimental luminance is enhanced by the strong ion beams when a gaseous target is prerequisite[3].

In this work, 52–197 MeV/u Xe^{54+} ions are stored and electron-cooled in the HIRFL-CSR (Heavy Ion Research Facility at Lanzhou - Cooling Storage Ring[4]). Then they are employed to collide with krypton and xenon atoms at the internal target. The X-ray are measured from 35°, 60°, 90°, 120° and 145°, respect to the beam direction.

The decay X-ray spectrum of the scattered ions in 94 MeV/u Xe^{54+} collision with Kr and Xe atoms, which observed from 35°, is shown in **Fig. 1**. The transitions form the single capture ion Xe^{53+*} and from the double capture ion Xe^{52+*} are observed, both in a single collision. The ratio between double and single capture channel of the Xe target is higher than the Kr target.

Although the single capture is the dominate channel at high energies, when the ion energy is lower (e. g, 52 MeV/u), the double capture becomes dominate, accompany with that the NRC instead of the REC becomes the dominate mechanism[1, 5]. Therefore, it is interesting that in a double capture process of Xe^{54+} with Kr and Xe, the channels of double REC, double NRC and REC plus NRC are comparable with each other at some energies between 52–197 MeV/u.

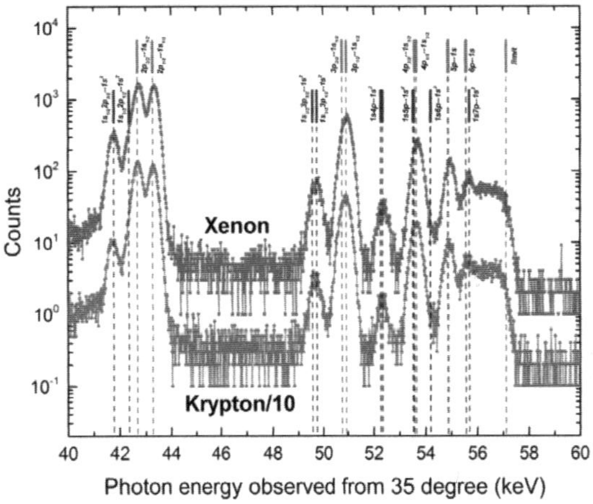

Figure 1. Decay X-ray spectrum of the scattered Xe^{53+*} and Xe^{52+*} in 94 MeV/u Xe^{54+} collision with Kr and Xe atoms, observed from 35°. The counts of the Kr target are divided by 10 to shift the plot.

This work was partially supported by the NSFC Grants Nos. 11275240 and U1332206.

References

[1] J. Eichler *et al 2007 Phys. Rep.* **439** 1
[2] R. Anholt *et al 1984 Phys. Rev. A.* **30** 2234
[3] D.Yu *et al 2011 Nucl. Instr. Meth. B.* **269** 692
[4] J. Xia *et al 2002 Nucl. Instr. Meth. A.* **488** 11
[5] J. Eicher and W. Meyerhof *1995 Relativistic Atomic Collisions, Academic, New York*

1 E-mail: d.yu@impcas.ac.cn

Content from this work may be used under the terms of the Creative Commons Attribution 3.0 licence. Any further distribution of this work must maintain attribution to the author(s) and the title of the work, journal citation and DOI.
Published under licence by IOP Publishing Ltd

High-resolution dielectronic recombination experiments at the upcoming CRYRING@ESR facility

C. Brandau[*,◇ 1], **M. Lestinsky**[◇ 2], **C. Kozhuharov**[◇], **A. Müller**[*], **S. Schippers**[*], **Th. Stöhlker**[◇,○,•]

[*] Justus-Liebig-Universität, 35392 Gießen, Germany
[◇] GSI-Helmholtzzentrum für Schwerionenforschung, 64291 Darmstadt, Germany
[○] Friedrich-Schiller-Universität, 07743 Jena, Germany
[•] Helmholtz-Institut, 07743 Jena, Germany

Synopsis The heavy ion storage ring CRYRING formerly located at the Manne-Siegbahn-Laboratory (MSL) in Stockholm, Sweden was dismantled and transferred to the GSI-Helmholtzzentrum für Schwerionenphysik (GSI) in Darmstadt, Germany and is presently being rebuilt. At its new location at GSI's storage ring ESR, the ultracold electron beam of the CRYRING electron cooler will allow for a new generation of high-resolution dielectronic recombination experiments with highly charged heavy ions and radioisotopes.

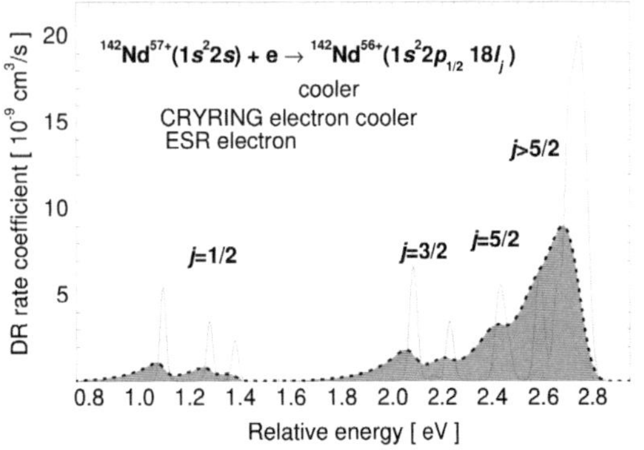

Figure 1. Comparison of the experimental conditions for DR experiments at the ESR (dashed line, gray area; $k_BT_\parallel = 175\mu$eV / $k_BT_\perp = 120$ meV) and at CRYRING@ESR (full line; $k_BT_\parallel = 50\mu$eV / $k_BT_\perp = 1.5$ meV) for the example of Nd^{57+}.

At the heavy ion storage ring CRYRING, formerly located at the Manne-Siegbahn-Laboratory in Stockholm, Sweden, pioneering experiments with an adiabatically expanded ultracold electron beam were performed, e.g. [1–3]. In particular, the resonant electron-ion collision process of dielectronic recombination (DR) benefits from a very cold electron beam since its two temperatures in the longitudinal and in the transversal direction $T_\parallel$ and $T_\perp$ determine the experimental resolution in DR studies at cooler storage rings [4]. The ring is presently being reinstalled as a Swedish in-kind contribution to the upcoming Facility for Antiproton and Ion Research (FAIR) at the site of the present GSI [5, 6]. CRYRING will be located behind the present storage ring ESR and will be store precooled highly charged ions (HCI) up to bare uranium and few-electron artificially synthesized radioisotopes at low energies [6].

The combination of an ultracold electron beam and a very capable injector that can deliver high intensities of HCI and radioisotopes is world-unique and will open up new pathways for precision studies of quantum electrodynamics in strong fields and for the investigation of nuclear properties such as charge radius, magnetic moment and nuclear spin using DR collision spectroscopy [4, 7]. Figure 1 illustrates the improved experimental conditions at CRYRING@ESR for the example of DR of Li-like Nd^{57+} using the temperatures typically obtained at the ESR electron cooler and using the ones measured at the CRYRING.

In this contribution we will introduce the project and will present the opportunities for next generation high-resolution DR studies with stable and unstable highly charged ions.

Work is supported by BMBF (contracts 06GI911I and 05P12R6FAN) and by the state of Hessen through the LOEWE funding scheme (project ELCH).

References

[1] H. Danared *et al* 1994 Phys. Rev. Lett. **72** 3775

[2] S. Mannervik *et al* 1998 Phys. Rev. Lett. **81** 313

[3] E. Lindroth *et al* 2001 Phys. Rev. Lett. **86** 5027

[4] C. Brandau and C. Kozhuharov 2012 in: *Atomic Processes in Basic and Applied Physics* ed. V. Shevelko and H. Tawara, p. 283ff

[5] http://www.fair-center.org

[6] M. Lestinsky *et al* 2015 Phys. Scr. T submitted; M. Lestinsky *et al.* 2015, this conference

[7] C. Brandau *et al* 2013 Phys. Scr. T **156** 014050

[1] E-mail: c.brandau@gsi.de [2] E-mail: m.lestinsky@gsi.de

Content from this work may be used under the terms of the Creative Commons Attribution 3.0 licence. Any further distribution of this work must maintain attribution to the author(s) and the title of the work, journal citation and DOI.
Published under licence by IOP Publishing Ltd

Journal of Physics: Conference Series **635** (2015) 022085 doi:10.1088/1742-6596/635/2/022085

A continuous, broadband data acquisition for Schottky signals in atomic and nuclear physics experiments at heavy-ion storage rings

C. Trageser[*,†,1], **C. Brandau**[*,†,‡,2], **C. Kozhuharov**[†], **Yu.A. Litvinov**[†,◇], **A. Müller**[*], **F. Nolden**[†], **S. Sanjari**[†], **Th. Stöhlker**[†,○,•]

[*] Justus-Liebig-Universität, 35392 Gießen, Germany
[†] GSI-Helmholtzzentrum für Schwerionenforschung, 64291 Darmstadt, Germany
[‡] ExtreMe Matter Institute, EMMI, and GSI Research Division, 64291 Darmstadt, Germany
[◇] Ruprecht-Karls-Universität, 69120 Heidelberg, Germany
[○] Friedrich-Schiller-Universität, 07743 Jena, Germany
[•] Helmholtz-Institut, 07743 Jena, Germany

Synopsis At heavy-ion storage rings Schottky noise techniques belong to the established non-destructive diagnostics methods. In addition, Schottky mass spectrometry is used to observe atomic masses and nuclear decays down to the level of a single stored ion without destruction of the circulating ion beam. In this contribution a new broadband, continuous data acquisition for Schottky signals and first experimental results obtained with this system during several commissioning runs at the storage ring ESR, Darmstadt are presented. Special emphasis will be given to present and future applications in atomic and nuclear physics experiments at storage rings.

Schottky probes are a common non-destructive tool at synchrotrons and storage rings to record the noise spectrum of ions in the ring. From the frequency spectrum of this Schottky noise a large variety of parameters of the ion-beam can be deduced, e.g., revolution frequency, beam intensity and momentum spread [1]. Besides application in beam-diagnostics this method is also exploited for experiments in atomic and nuclear physics in storage rings: the so-called (time-resolved) Schottky mass spectrometry (SMS) allows for a determination of atomic masses and for the investigation of decay processes with long half-lives [2–4]. The sensitivity of SMS reaches down to a single stored ion. The Schottky technique is also an indispensable tool in atomic physics experiments such as dielectronic-recombination measurements with artificially synthesized radioisotopes. This has been demonstrated and further developed at the Experimental Storage Ring (ESR) at the GSI-Helmholtzzentrum für Schwerionenforschung in Darmstadt, Germany [5, 6].

A decisive improvement has been obtained at the ESR in 2011 with the installation of a new resonant Schottky probe with high a quality factor, increased sensitivity and with a time resolution on the level of 10 ms [7]. The resonator typically operates at frequencies around 245 MHz, corresponding to about the 125th harmonic of the circulating ions. In order to fully exploit this new Schottky probe a new data acquisition system (DAQ) was developed and commissioned at the ESR which allows for a continuous and broadband recording of the radio frequency (RF) Schottky-signals of the stored ions [8]. The new DAQ is capable of persistently recording up to $3.5 \cdot 10^7$ pairs of inphase and quadrature samples per second (data rate: 140 MByte/s), thus allowing more than 10 harmonics of the ion beam to be recorded simultaneously.

In this contribution we report on the commissioning and first experimental results using this new advanced Schottky DAQ. In addition, we will discuss potential applications of this next generation Schottky set-up with emphasis on atomic physics experiments at the storage rings of GSI and of the upcoming Facility for Antiproton and Ion Research (FAIR) [9].

Work is supported by BMBF (contracts 06GI911I and 05P12R6FAN), by the Helmholtz Alliance HA216/EMMI, by the Helmholtz graduate school HGS-HIRe and by the Helmholtz-CAS Joint Research Group HGJRG-108 .

References

[1] D. Boussard 1993 in: *CERN Accelerator School* , ed. S. Turner, CERN 95-06, 749ff

[2] B. Franzke *et al* 2008 Mass Spectr. Rev. **27** 428

[3] F. Bosch *et al* 2013 Progr. Part. Nucl. Phys. **73** 84

[4] Y.A. Litvinov and F. Bosch 2011 Rep. Prog. Phys. **74** 016301

[5] C. Brandau *et al* 2009 J. Phys.: Conf. Ser. **194** 012023

[6] C. Brandau *et al* 2013 Phys. Scr. T **156** 014050

[7] F. Nolden *et al* 2011 Nucl. Intr. Meth. A **659**, 69

[8] C. Trageser *et al* 2015 Phys. Scr. T, in print

[9] http://www.fair-center.org

[1] E-mail: c.trageser@gsi.de [2] E-mail: c.brandau@gsi.de

Content from this work may be used under the terms of the Creative Commons Attribution 3.0 licence. Any further distribution of this work must maintain attribution to the author(s) and the title of the work, journal citation and DOI.
Published under licence by IOP Publishing Ltd

XXIX International Conference on Photonic, Electronic, and Atomic Collisions (ICPEAC2015) IOP Publishing
Journal of Physics: Conference Series **635** (2015) 022087 doi:10.1088/1742-6596/635/2/022087

A lepton spectrometer for studies of fundamental atomic processes at HESR at FAIR

K. E. Stiebing*[1], F. King*, P.-M. Hillenbrand†[2], S. Hagmann*,†, Yu. A. Litvinov†, S. Schippers‡, and Th. Stöhlker†,&,§

*Goethe Universität Frankfurt, Germany, †GSI Darmstadt, Germany, ‡Justus-Liebig-Universität Gießen, Germany, &Friedrich-Schiller-Universität Jena, Germany, §Helmholtz-Institut Jena, Germany

Synopsis A lepton spectrometer for the study of atomic processes in relativistic collisions of very heavy highly charged ions has been designed for experiments at the HESR storage ring at the FAIR facility.

Projectile energies of up to 5 GeV/u for very heavy highly ionized ions (up to e.g. U^{92+}) enable the study of electron-positron pair production processes in extremely strong transient electric and magnetic fields. Under these conditions, those reaction channels, where the electron of the dynamically created e^+/e^--pair is captured into the vacant K-shell of the projectile, become interesting channels of observation. For the specific case of collisions of U^{92+} on Ar, accessible at HESR, two fundamental processes can be contrasted here [1]:

(a) Bound-free pair production **(BFPP)**:

$$U^{92+} + Ar \rightarrow U^{91+} + Ar^{*} + e^{+}$$ and

(b) Negative continuum dielectronic recombination **(NCDR)**:

$$U^{92+} + Ar \rightarrow U^{90+} + [Ar^{+}]^{*} + e^{+}.$$

While in BFPP the electron is directly captured into the projectile bound state, the *hitherto unobserved* NCDR process results in the capture of two electrons (one from the e^+/e^-pair and one from the excited target atom) into the projectile. Therefore, both processes can be separated by their ejectile's charge states. The relativistic kinematics facilitates experimental studies by essentially focusing the emitted e^+ into a small forward cone. In fig.1, the expected positron-spectrum of NCDR at an ion energy of 2,37 GeV/u is shown. The simulation bases on calculations by Artemyev et al. [2], using the argon Compton profiles of Biggs et al. [3].

A (Q-2D-Q) spectrometer, consisting of two magnetic dipole sectors (60°) combined with magnetic quadrupole singlets at the entrance and the exit has been designed and simulated by 3D-FE calculations [4] (see fig. 2). Position resolving stacks of ΔE-E scintillation detectors will be installed in the focal plane. The e^+/e^--annihilation radiation is detected in an array of photon scintillation detectors.

[1] E-mail: stiebing@em.uni-frankfurt.de
[2] E-mail: p.m.hillenbrand@gsi.de

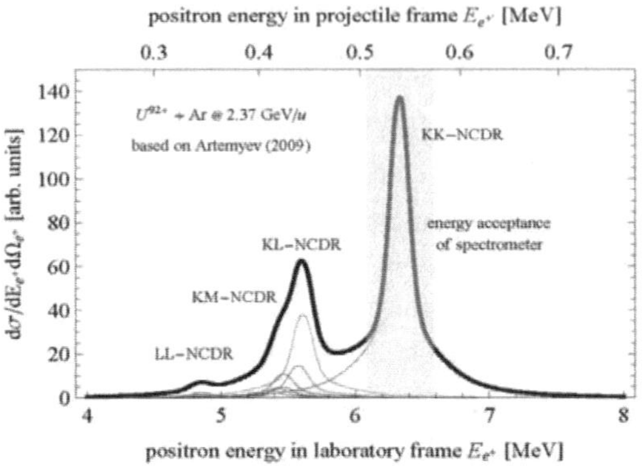

Figure 1: Simulated positron energy spectrum for NCDR emitted at an angle of **θ** = 0° for 2.37 GeV/u U^{92+}+Ar collisions.

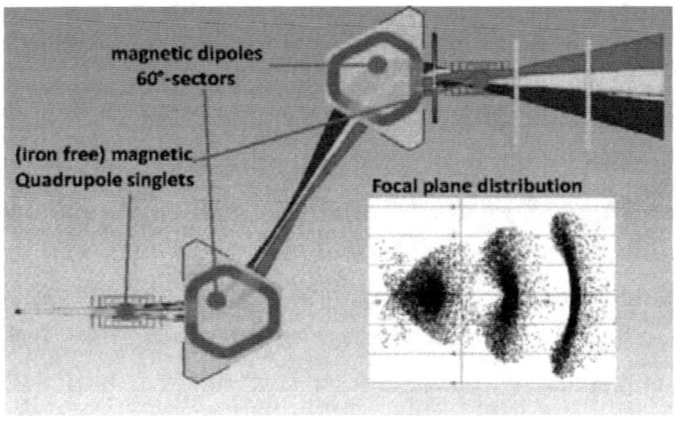

Figure 2: 3D Transport simulations [4] of leptons (three kinetic energies centered at 6.5 MeV±4%, with an emission angular spread of Δθ= 2.5°).

References

[1] P.-M. Hillenbrand *et al.*, Experimental concepts of positron spectroscopy at HESR, Phys. Scr., *submitted* (2015).
[2] Artemyev A N, Shabaev V M, Stöhlker T and Surzhykov A S; Phys. Rev. A79(3), 032713 (2009).
[3] Biggs F, Mendelsohn L and Mann J; Atomic Data Nuclear Data Tables 16(3), 201(1975).
[4]Computer simulation Technology, www.cst.com

Supported by Helmholtz-CAS Joint Research Group HCJRG-108 and by the Federal Ministry of Education and Research, FRG, (Contract No. 05P12RFFAH).

Content from this work may be used under the terms of the Creative Commons Attribution 3.0 licence. Any further distribution of this work must maintain attribution to the author(s) and the title of the work, journal citation and DOI.
Published under licence by IOP Publishing Ltd

Two-electron-one-photon transitions in Cu^{9+}-Zn collisions

Yipan Guo* #, Zhihu Yang* 1, Shubin Du†, Qiumei Xu* #, Hongwei Chang† and Xiaohong Cai*

* Institute of Modern Physics, Chinese Academy of Sciences, Lanzhou, 730000, P.R. China
† Department of Nuclear Physics, China Institute of Atomic Energy (CIAE), Beijing, 102413, P.R. China
# University of Chinese Academy of Sciences, Beijing, 100049, P.R. China

Synopsis The two-electron-one-photon Kαα x-rays emissions from 80 MeV Cu^{9+}-Zn collisions have been measured. These Kαα x-rays originate from the projectile ions and target atoms. It indicates that double K-vacancies appear in both Cu^{9+} ions and Zn atoms.

Two-electron-one-photon transitions (TEOP) have been reported in several ion-atom collision systems [1-3]. During these collisions, two vacancies in the K-shell of the same partner can be created first. Second, two outer- shell electrons jump into the empty K-shells. Meanwhile, a single photon which carries somewhat larger than the sum of the two independent electron transition energies emits. The study of TEOP transitions demonstrates that electron-electron correlation effects make contributions to the transition probability, and it also proves the limitations of the independent-particle model.

The experiment was performed using HI-13 tandem accelerator at China Institute of Atomic Energy (CIAE). In the experiment, Cu^{9+} ions were accelerated to the energy of 80 MeV first. Then, the beams entered the chamber and bombarded the target surface at 45°. The bombardment time for each measurement is seven hours. The emitted x-rays were detected with a Si (Li) detector mounted at 90° relative to the beam incident direction. The target chamber was sealed up with a Be window about 50 μm thick. Between the Be window and the detector, an aluminum absorber with a thickness of 960 μm is utilized to suppress the pile-up effect of the Kα and Kβ photons. In addition, the target has a thickness of 900 $\mu g/cm^2$ and a polished surface.

Figure 1 exhibits the total spectrum obtained from 80 MeV Cu^{9+}-Zn collisions with a 960 μm Al absorber. The red box shows the TEOP peaks. Fig. 2 is a larger version of the red zone in Fig. 1. On the basis of the energetics [1], the weak peaks (A) and (B) originate from Cu^{9+} ions Kαα x-rays and Zn Kαα x-rays, respectively. It indicates that both the projectile ions and target atoms appear the double K-vacancies.

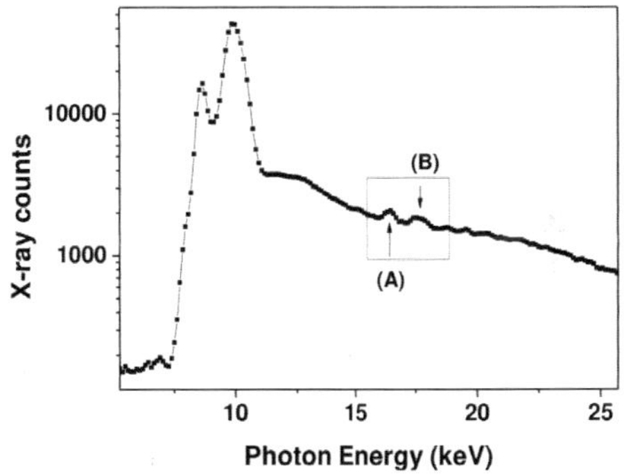

Figure 1. X-ray spectrum obtained with 80 MeV Cu^{9+} ions on Zn target (960 μm Al absorber). The red box shows the energy range of the TEOP transitions.

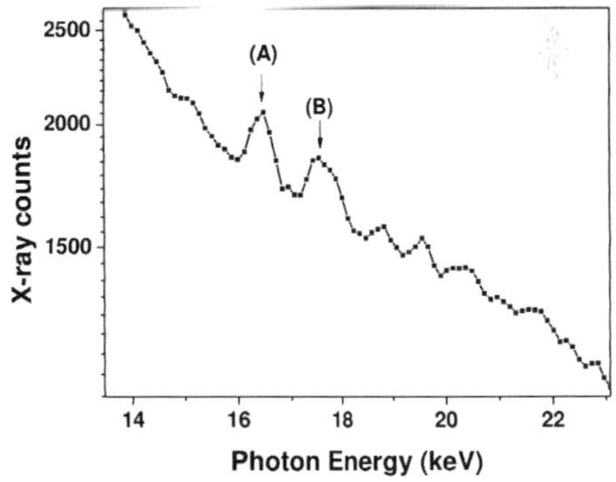

Figure 2. The TEOP transitions from 80 MeV Cu^{9+}-Zn collisions (red box in Fig.1). (A): Cu Kαα x-ray (16.48 keV); (B): Zn Kαα x-ray (17.54 keV).

References

[1] W. Wölfli *et al* 1975 *Phys. Rev. Lett.* **35** 656
[2] D. Nagel *et al* 1976 *Phys. Rev. Lett.* **36** 164
[3] C. Stoller *et al* 1977 *Phys. Rev. A* **15** 990

[1]E-mail: z.yang@impcas.ac.cn

Content from this work may be used under the terms of the Creative Commons Attribution 3.0 licence. Any further distribution of this work must maintain attribution to the author(s) and the title of the work, journal citation and DOI.
Published under licence by IOP Publishing Ltd

Electron loss in collisions of dressed projectiles with atoms and molecules

D. Fregenal†[1], J. Fiol†, G. Bernardi†, J.M. Monti§, S.G. Suárez†, P.D. Fainstein†, L.M. Rodríguez† R.D. Rivarola§ R. Schuch‡

† Centro Atómico Bariloche, Comisión Nacional de Energía Atómica, San Carlos de Bariloche (Río Negro) and Consejo Nacional de Investigaciones Científicas y Técnicas (CONICET), Argentina

§ Lab. de Colisiones Atómicas, Inst. de Fís. Rosario (CONICET-UNR) and Facultad de Ciencias Exactas, Ingeniería y Agrimensura, Universidad Nacional de Rosario, Rosario, Argentina

‡ Department of Physics, Stockholm University, AlbaNova University Center, SE-106 91 Stockholm, Sweden

Synopsis Energy distributions of electrons emitted with energies corresponding to the region of the electron-loss-cusp (ELP) were measured for collisions of several neutral projectiles with atoms and small molecules. Besides the well known projectile ELP cusp, a second structure is observed for Li and Na projectiles. Comparing the experimental data with four-body CTMC calculations we analise the origin of these structures in terms of nuclear-electron and electron-electron interactions.

Electrons emitted from a dressed projectile produce a peak structure in the forward direction when the electron energy matches the initial velocity that is known as electron loss peak (ELP). This structure, which is superimposed to the target electron emission, is generated either by a direct single ionization or as a result of simultaneous projectile and target ionizations. ELP dependencies on projectile charge state and initial electronic state have been extensively studied [1]. For negative incident ions were found two peaks, one corresponding to electrons moving at approximately the projectile velocity and one at a slightly lower energy [2]. The origin of the second structure was interpreted as the result of target ionization by the projectile electron [3].

In this work we present theoretical and experimental results for angular and energy distributions of electrons emitted in the ELP energy region due to collisions of dressed projectiles with atoms and molecules. Measurements for different incident projectiles (B, C, Li and Na) colliding with atoms and molecules (He, H_2, Ne, Ar) with energies between 50 keV/u and 450 keV/u, were done for several angles between 0 and 30 degrees. The experimental data is compared with theoretical calculations based on a four-body Classical Trajectory Monte Carlo (CTMC) approximation and three-body Continuum Distorted Wave (CDW) and Continuum-Distorted-Wave -Eikonal-Initial-State (CDW-EIS) models. The results include the contributions to the ELP from target ionization, projectile ionization and simultaneous ionization channels.

While charged-projectile data show a single cusp ELP as expected, a second structure at lower energies is observed for data associated with certain neutral projectiles (Li and Na). The energy position of this structure depends on the target species. Experimental results for Li, Na and C are displayed in Figure 1.

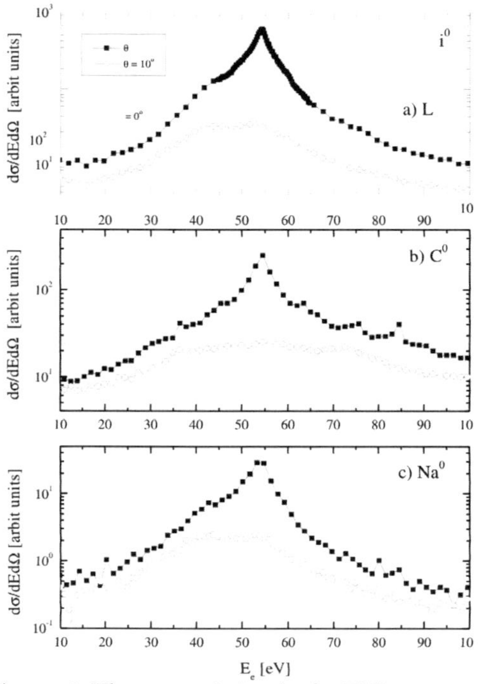

Figure 1. Electron emission in the ELP energy region for impact of 100 keV/u a) Li, b) C and c) Na incident on H_2. Emission angles are indicated.

References

[1] D. Schneider *et al* 1983 *Phys. Rev. A* **28** 649. O. Heil *et al* 1992 *Phys. Rev. A* **45** 2850

[2] M.G. Menendez and M.M. Duncan 1979 *Phys. Rev. A* **20** 2327.

[3] J. Sorensen, L.H. Andersen and L.B. Nielsen 1988, *J. Phys. B: At. Mol. Opt. Phys.* **21** 847

[1]E-mail: fregenal@cab.cnea.gov.ar

Content from this work may be used under the terms of the Creative Commons Attribution 3.0 licence. Any further distribution of this work must maintain attribution to the author(s) and the title of the work, journal citation and DOI.
Published under licence by IOP Publishing Ltd

CDW and CDW-EIS calculations of simultaneous ionization in collisions between dressed projectiles and atomic targets

J M Monti[†][1], D Fregenal[§][2], P D Fainstein[§], G Bernardi[§], R Schuch[‡], S G Suárez[§], R D Rivarola[†], J Fiol[§]

[§] Centro Atómico Bariloche, Comisión Nacional de Energía Atómica, 8400 San Carlos de Bariloche (Río Negro) and Consejo Nacional de Investigaciones Científicas y Técnicas (CONICET), Argentina
[†] Laboratorio de Colisiones Atómicas, Instituto de Física Rosario (CONICET-UNR) and Facultad de Ciencias Exactas, Ingeniería y Agrimensura, Univ. Nac. de Rosario, Av. Pellegrini 250, 2000 Rosario, Argentina
[‡] Department of Physics, Stockholm University, AlbaNova University Center, SE-106 91 Stockholm, Sweden

Synopsis The role of simultaneous electron emission in collision between dressed ions and atoms is discussed. Methods of calculations within quantum-mechanical Continuum-Distorted- Wave (CDW) and Continuum-distorted Wave–Eikonal-Initial-State (CDW-EIS) models are revisited and improved within a semiclassical probabilistic approach.

In collisions of atoms or dressed-ions with atoms and molecules, simultaneous ionization (SI) of both projectile and target accounts for an important fraction of the observed electron emission spectra. In recent publications, it has been shown that simultaneous ionization plays a major role in collision systems in which target and projectile have bound electrons with similar ionization energies, i.e. in symmetric collisions [1, 2, 3]. In certain cases, comparison of experimental data with Classical Trajectory Monte Carlo (CTMC) and quantum-mechanical CDW and CDW-EIS theories has shown that SI is the most important ionization mechanism for electrons emitted in the forward direction with velocities that match the initial projectile velocity [1, 2, 4].

In previous works, using either CDW or CDW-EIS models, the doubly differential cross sections (DDCS) for simultaneous ionization was approximated as a weighted average of target and projectile ionization DCCSs, where the weight factors are proportional to the total ionization cross section of each aggregate. This simple estimation of a two-electron process in terms of single-electron cross-sections has shown to provide sound results in a few systems but lacks real justification in more general cases.

In this work we revisit the calculation of simultaneous ionization cross-sections within a probabilistic approach, in order to obtain a more realistic and precise description of this process. In a semiclassical approach we calculate the ionization probability as a function of the impact parameter in the CDW and CDW-EIS approximations and obtain the SI probability by considering the ionization from each center as independent events. The cross-sections are finally evaluated by summing over all possible impact parameters. Details of the calculations, comparison with experimental data and with four-body CTMC results for specific systems will be presented at the conference.

[1]E-mail: monti@ifir-conicet.gov.ar
[2]E-mail: fregenal@cab.cnea.gov.ar

References

[1] J M Monti *et al.* 2013 *Phys. Scr.* **T156** 014031

[2] D Fregenal *et al.* 2014 *J. Phys. B: At. Mol. Opt. Phys.* **47** 155204

[3] D Fregenal *et al.* 2014 *J. Phys.: Conf. Ser.* **488** 082014

[4] D Fregenal *et al.* 2015 *J. Phys.: Conf. Ser.* **583** 012013

Content from this work may be used under the terms of the Creative Commons Attribution 3.0 licence. Any further distribution of this work must maintain attribution to the author(s) and the title of the work, journal citation and DOI.
Published under licence by IOP Publishing Ltd

XXIX International Conference on Photonic, Electronic, and Atomic Collisions (ICPEAC2015) IOP Publishing
Journal of Physics: Conference Series **635** (2015) 022092 doi:10.1088/1742-6596/635/2/022092

Ion trap study of the charge transfer and associative detachment reactions of $D^- + H$

Štěpán Roučka* [1], **Dmytro Mulin***, **Pavol Jusko***, **Radek Plašil***, **Juraj Glosík***

* Department of Surface and Plasma Science, Faculty of Mathematics and Physics, Charles University in Prague, Czech Republic

Synopsis The thermal rate coefficients of the associative detachment and charge transfer reactions of $D^- + H$ have been measured at temperatures below 160 K. The measurements have been carried out using a cryogenic 22-pole ion trap, combined with a beam of atomic hydrogen – the AB-22PT instrument.

The associative detachment reaction

$$H^- + H \xrightarrow{k_{AD}^{H}} H_2 + e^- \quad (1)$$

has recently been investigated in several experimental works [1, 2, 3] due to its relevance for astrochemistry [4]. Certain effort has also been dedicated to studying the isotopic effects in this system by observing the interaction of $D^- + D$ [5]. However, no experiment has yet been able to study the low temperature behavior of the isotopic system $D^- + H$, which is the subject of the present study. This system is especially interesting due to its asymmetry, which allows us to directly measure the rate coefficient of the charge transfer reaction

$$D^- + H \xrightarrow{k_{CT}^{D}} H^- + D \quad (2)$$

in addition to the rate coefficient of the associative detachment reaction

$$D^- + H \xrightarrow{k_{AD}^{D}} H_2 + e^- . \quad (3)$$

The presented work was carried out using the AB-22PT instrument (atomic beam combined with 22-pole trap) [6]. In this experiment, a cold beam of atomic hydrogen is formed by passing the atomic hydrogen from RF discharge in hydrogen through a nozzle, which can be cooled to temperatures in the range 8 K to 300 K. Ions are produced in a storage ion source by electron impact ionization of D_2 precursor gas. Selected ions are then trapped in the 22-pole ion trap [7] and cooled by collisions with H_2 buffer gas to temperatures between 10 K and 300 K. The time evolution of the numbers of trapped D^- and H^- ions due to interaction with H atoms is then monitored. Reactions (1), (2), and (3) occur in the trap and their relative rate coefficients (see Fig. 1) can be directly obtained from the observed time evolutions. Absolute rate coefficients k_{CT}^{D} and k_{AD}^{D} can be obtained by calibrating the number density of H atoms in the trap.

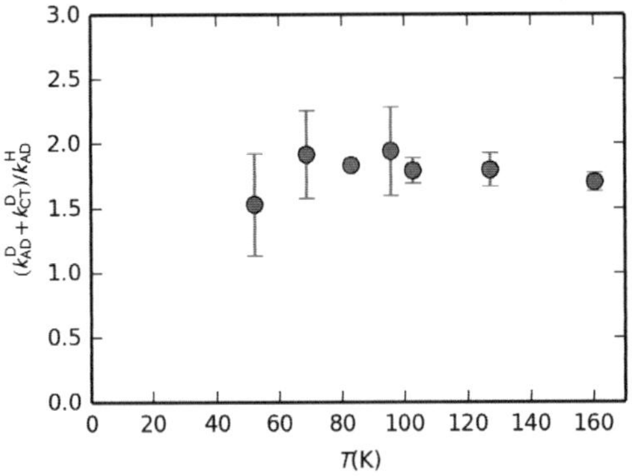

Figure 1. Ratios of the total reaction rate coefficient of D^-+H to the H^-+H reaction rate coefficient as a function of $D^- + H$ translational temperature.

Acknowledgment We thank prof. Dieter Gerlich for ideas and discussions regarding the presented experiment. We thank Chemnitz University of Technology and the DFG for lending the AB-22PT to the Charles University. This work was partly supported by GACR Grant No. P209/12/0233, 14-14715P, GAUK Grant No. 572214.

References

[1] O. Martinez *et al* 2009 *Astrophys. J. Lett.* **705** L172

[2] H. Kreckel *et al* 2010 *Science* **329** 69

[3] D. Gerlich *et al* 2012 *Astrophys. J.* **749** 22

[4] S. C. Glover *et al* 2006 *Astrophys. J.* **640** 553

[5] K. Miller *et al* 2012 *Phys. Rev. A* **86** 032714

[6] G. Borodi *et al* 2009 *Int. J. Mass. Spectrom.* **280** 218

[7] D. Gerlich 1992, Inhomogeneous RF fields: a versatile tool for the study of processes with slow ions *Adv. Chem. Phys.* **82** 1

[1]E-mail: Stepan.Roucka@mff.cuni.cz

Content from this work may be used under the terms of the Creative Commons Attribution 3.0 licence. Any further distribution of this work must maintain attribution to the author(s) and the title of the work, journal citation and DOI.
Published under licence by IOP Publishing Ltd

Journal of Physics: Conference Series **635** (2015) 022093 doi:10.1088/1742-6596/635/2/022093

Laser spectroscopy on lithium-like ions at the HESR

Rodolfo M. Sánchez Alarcón* [1]

* GSI Helmholtzzentrum für Schwerionenforschung GmbH, Planckstr. 1, 64291 Darmstadt, Germany

Synopsis Possibilities of studying D1 and D2 transitions in lithium-like ions by laser spectroscopy at the HESR at FAIR are presently being considered. First experiments could already be performed using conventional lasers to excite these transitions in medium-Z lithium-like ions. Hyperfine structure and isotope shift could be studied and nuclear moments and charge radii being extracted. However, challenges connected with these experiments will have to be resolved first.

One of the facilities of the upcoming FAIR project in Darmstadt is the High-Energy Storage Ring (HESR). In this ring, ion beams will be stored and cooled at relativistic energies with γ-values ranging from 2 to 6 [1]. This opens up opportunities of performing laser spectroscopy to effectively excite high-energy transitions in highly charged ions. In combination with the Doppler blue-shift in counter-propagating excitation, especially the $1s^2\ 2s\ ^2\mathrm{S}_{1/2} \to 1s^2\ 2p\ ^2\mathrm{P}_{1/2}$ (D1) and $1s^2\ 2s\ ^2\mathrm{S}_{1/2} \to 1s^2\ 2p\ ^2\mathrm{P}_{3/2}$ (D2) transitions in medium-Z lithium-like ions are accessible. Transitions between 20 eV and 60 eV [2] can be already reached with 5.5 eV photons, produced with off-the-shelf frequency-doubled cw or pulsed laser systems.

The theoretical values for the transition frequencies given in [2] do not include radiative corrections, therefore the experimental determination of these lines can already trigger more precise calculations including QED contributions, which have already been evualuated for some systems [3, 4]. If radioactive lithium-like ions are injected into the HESR, isotope shift measurements might provide information on their nuclear charge radii. For ions with nonvanishing nuclear spin I, hyperfine spectroscopy will provide access to nuclear magnetic and electric quadrupole moments (only D2) of these nuclei. The three-electron structure of these ions will have less systematic uncertainties from the calculation of hyperfine fields and might therefore serve as calibration data for studies on more exotic isotopes.

However fluorescence detection represents a challenge with these highly relativistic beams. To detect the fluorescence that is preferably emitted in a narrow cone in forward direction, a suitable detector has to be built and implemented into the beamline under ultra-high vacuum conditions. This Doppler-shifted fluorescence will be emitted with energies ranging between 75 eV and 750 eV under angles smaller than 5° with respect to the ion beam direction. Moreover, there is limited space in the HESR tunnel as well as in the beamline to install the laser setup and the detector, respectively.

References

[1] Th. Stöhlker *et al* 2014 *Hyperfine Interact.* **227** 45.

[2] W. R. Johnson *et al* 1996 *Atom. Data Nucl. Data* **64** 279.

[3] Y. S. Kozhedub *et al* 2010 *Phys. Rev. A* **81** 042513.

[4] J. Sapirstein and K. T. Cheng 2011 *Phys. Rev. A* **83** 012504.

[1]E-mail: R.Sanchez@gsi.de

Content from this work may be used under the terms of the Creative Commons Attribution 3.0 licence. Any further distribution of this work must maintain attribution to the author(s) and the title of the work, journal citation and DOI.
Published under licence by IOP Publishing Ltd

XXIX International Conference on Photonic, Electronic, and Atomic Collisions (ICPEAC2015) IOP Publishing
Journal of Physics: Conference Series **635** (2015) 022094 doi:10.1088/1742-6596/635/2/022094

Contribution of higher multipole terms of Coulomb interaction to ionization in low-energy heavy-ion collisions

A. I. Bondarev*[1], I. V. Ivanova*, I. I. Tupitsyn*, Y. S. Kozhedub*, G. Plunien†

* Department of Physics, St. Petersburg State University, St. Petersburg 198504, Russia

† Institut für Theoretische Physik, Technische Universität Dresden, D-01062 Dresden, Germany

Synopsis The two-center Dirac equation is solved in spherical coordinates using the multipole decomposition of the electron-nucleus interaction. The total and differential ionization probabilities in collisions of a hydrogenlike ion with a bare nucleus are calculated for different impact parameters. The contribution of the higher multipole terms of the interaction to these probabilities is analyzed. The influence of relativistic effects is examined as well.

Recent progress in development of experimental facilities mainly storage rings and accelerators which should yield new data on processes accompanying slow collisions of heavy ions amplifies theoretical studies in this area. In the 1970s a formation of the charged vacuum in such collisions, provided that the total nuclear charge is larger than the critical one $Z_c = 173$, has been predicted, and the following appropriate calculations were mostly carried out in the monopole approximation of the interaction [1]. In the 2010s interest to this problem has been renewed, and the results obtained beyond this approximation were reported [2, 3, 4].

In this contribution we implement the multipole expansion of the two-center potential to study the angular distribution of ionized electrons in the $U^{91+}(1s)$-U^{92+} collision. In order to evaluate firstly the total ionization probability in this collision, one above all has to propagate the electron wave function in time, that is to solve the two-center time-dependent Dirac equation (TDDE). Since the collision is slow, non-perturbative methods for solving TDDE are required.

Here the electron wave function is expanded over a finite basis set, and the resulting system of the first order differential coupled-channel equations is then solved numerically. We consider the collision in the spherical rotating reference frame where target is at the origin and the z-axis is collinear to the internuclear vector at each time moment. The latter facilitates the angular integration in the matrix elements, which can be carried out analytically, while numerical procedure is employed to perform the radial one. When the solution of the system of coupled-channel equation is found, one obtains the set of the expansion coefficients of the wave function over the basis set, which consists information about various electron transitions, including ionization.

The size of the angular part of the basis set, namely, the number N of channels with different symmetry (relativistic quantum number κ) defines the number $N = 2L_{\max} + 1$ of the multipole expansion terms with $L \leqslant 2L_{\max}$ taken into account.

Table 1. Total ionization probability in the $U^{91+}(1s)$-U^{92+} collision. A straight-line trajectory with impact parameter b is assumed for the point-like projectile with energy $E = 6$ MeV/u. The spherical charge model with $r_{\rm n} = 5.8569$ fm is used for the target.

b (fm)	$N = 1$	$N = 3$	$N = 5$
30	0.087	0.073	0.070
35	0.063	0.052	0.049
40	0.046	0.037	0.035
45	0.034	0.027	0.025
50	0.025	0.019	0.018

As one can see from Table 1, the total ionization probability does depend on accounting for the higher multipole terms of the interaction.

To obtain the differential ionization probabilities, we use the distorted plain wave in the field of one nucleus as a final electron state, and project the solution of the two-center TDDE onto this state. The corresponding transition amplitude and probability, in contrast to the calculation in the monopole approximation, depend on angles.

References

[1] U. Müller-Nehler and G. Soff 1994 *Phys. Rep.* **246** 101

[2] I. I. Tupitsyn *et al* 2010 *Phys. Rev. A* **82** 042701

[3] A. Marsman and M. Horbatsch 2011 *Phys. Rev. A* **84** 032517

[4] S. R. McConnell *et al* 2013 *Phys. Scr.* **T156** 014055

[1]E-mail: bondarev@pcqnt1.phys.spbu.ru

Content from this work may be used under the terms of the Creative Commons Attribution 3.0 licence. Any further distribution of this work must maintain attribution to the author(s) and the title of the work, journal citation and DOI.
Published under licence by IOP Publishing Ltd

Journal of Physics: Conference Series **635** (2015) 022095 doi:10.1088/1742-6596/635/2/022095

Production of N_3 from irradiation of the N_2-rich ices using soft X-rays and cosmic rays analog

Fredson de Araujo Vasconcelos * [1], Sergio Pilling * [2]

* Research and development institute (IP&D), Universidade do Vale do Paraiba - UNIVAP, Sao Jose dos Campos-SP/Brazil

Synopsis N_2-rich ices may exist in various environments in the solar system and interstellar medium. These ices are exposed to ultraviolet photons and cosmic rays, which may cause changes in the molecular structure, thus leading to formation of new species. In this study we investigated experimentally the production of N_3 from the irradiation of N_2-rich ices, using soft X-rays and cosmic rays analog. The results of the experiments using these sources of radiation showed that the azide radical, N_3, was observed in ices that have been bombarded by ions and in those that have been photolyzed (range of vacuum ultraviolet and soft X-rays).

Astrophysical ices are constantly processed by ionizing agents that may be in the form of radiation or particles (ions, x-rays and gamma, photons, electrons, neutrons). Significant amounts of molecular nitrogen (N_2) are found in the constitution of some of these ices. The ionizing agents deposit high doses of energy on the ice, inducing changes in their structures and a series of chemical reactions that can be verified in the laboratory. The result is the appearance of various molecular species.

In this study we investigated experimentally the production of N_3 from the irradiation of N_2-rich ices, using soft X-rays and cosmic rays analog.

The experiments with ions were performed in the Grand Accélérateur National d'Ions Lourds (GANIL) located at Caen, France. For photons (range of vacuum ultraviolet and soft X-rays), the experiments have been performed using a high vacuum portable chamber from the Laboratório de Astroquimica e Astrobiologia (LASA/UNIVAP) coupled to the the Brazilian Synchrotron Light Source (LNLS) located at Campinas, Brazil.

The results indicate that during the sample processing by radiation, the infrared spectra have presented several daughter species, including the N_3 (1657 cm^{-1}). This band of N_3 also was identified elsewhere [1], which showed that the irradiation of solid N_2 and N_2-rich ices with 0.8 MeV protons produces the N_3(azide) radical. In the experiment employing ions the species N_2O was also observed (2235 cm^{-1}) due to the implantantion of Oxygen atoms in the frozem sample. The destruction cross section of the parental species was determined, as well as, the formation cross section of N_3 and N_2O. Detaisl of this works is given elsewhere [2].

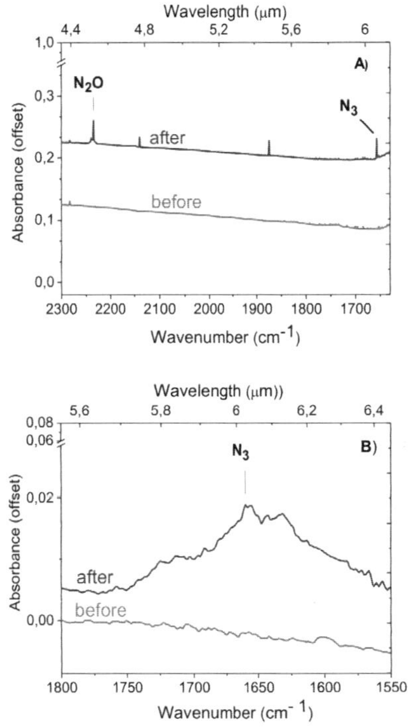

Figure 1. A) Infrared spectra of the N_2-rich ices at 16 K before (*lower*) and after (*upper*) irradiation using ions. The vibration 2235 cm^{-1} corresponds to the N_2O. B) Infrared spectra of the $N_2 + CH_4$ ice at 12 K before (*lower*) and after (*upper*) irradiation using photons (vacuum ultraviolet and soft X-rays). Other peaks in the infrared expectros are associated with other molecules.

References

[1] R.L. Hudson & M.H. Moore 2002 Astrophys. J. 568, 1095–1099.

[2] F. A. Vasconcelos et al 2015 A&A (in preparation).

[1] E-mail: fred.vasco@hotmail.com * [2] E-mail: sergiopiling@yahoo.com.br

Content from this work may be used under the terms of the Creative Commons Attribution 3.0 licence. Any further distribution of this work must maintain attribution to the author(s) and the title of the work, journal citation and DOI.
Published under licence by IOP Publishing Ltd

XXIX International Conference on Photonic, Electronic, and Atomic Collisions (ICPEAC2015) IOP Publishing
Journal of Physics: Conference Series **635** (2015) 022096 doi:10.1088/1742-6596/635/2/022096

First observation of coherence in a highly charged ion

S. Tashenov*[1], D. Banaś†, H. Beyer‡, C. Brandau♮,♯, S. Fritzsche♭,÷, A. Gumberidze♮,♯, S. Hagmann‡, P.-M. Hillenbrand‡, H. Jörg*, I. Kojouharov‡, Ch. Kozhuharov‡, M. Lestinsky‡, Yu.A. Litvinov*,‡, A.V. Maiorova∅, H. Schaffner‡, V.M. Shabaev∅, U. Spillmann‡, Th. Stöhlker♭,∇, A. Surzhykov♭, S. Trotsenko‡,♭

*Physikalisches Institut der Universität Heidelberg, 69120 Heidelberg, Germany
†Institute of Physics, Jan Kochanowski University, 25-406 Kielce, Poland
‡GSI Helmholtzzentrum für Schwerionenforschung GmbH, 64291 Darmstadt, Germany
♮ExtreMe Matter Institute EMMI and Research Division, GSI Helmholtzzentrum für Schwerionenforschung, 64291 Darmstadt, Germany
♯FIAS Frankfurt Institute for Advanced Studies, 60438 Frankfurt am Main, Germany
♭Helmholtz-Institut Jena, 07743 Jena, Germany
÷Theoretisch-Physikalisches Institut, Universität Jena, 07743 Jena, Germany
∅St. Petersburg State University, 198504 St. Petersburg, Russia
∇Institut für Optik und Quantenelektronik, Universität Jena, 07743 Jena, Germany

Synopsis We report the first observation of correlated x rays emitted by a heavy highly charged ion such as hydrogenlike uranium in the process of radiative recombination. The angular correlations between these x rays for the first time revealed a coherence of the populated state in this ion. The results also indicate a strong contribution of the spin–orbit interaction in radiative recombination.

Radiative recombination (RR) into highly charged ions offers a unique opportunity to study its time-reverse the fundamental process of the photoelectric effect in the otherwise inaccessible regime of hard x rays and strong Coulomb fields. Using the experimental storage ring ESR in GSI Darmstadt we have studied the RR into the excited $2p_{3/2}$ state of hydrogenlike uranium ions. This state decays to the ground state $1s_{1/2}$ by emitting a Lyα_1 x ray. In the experiment the RR and the Lyα_1 x rays were detected in time-coincidences using a setup of large germanium detectors, representing the first observation of the correlated x rays emitted by heavy highly charged ions [1].

We observed that the angular distribution the Lyα_1 x rays is strongly correlated with the emission direction of the RR x ray. To interpret this observation we consider a non statistical population of the magnetic sublevels (alignment) of the $2p_{3/2}$ state. When the RR x ray is not observed, as was the case in all previous alignment experiments with highly charged ions, the populated state is axially symmetric around the collision direction, and, therefore, its magnetic sublevels are populated incoherently. In contrast, in the present experiment, this state is populated *coherently*. Therefore, it loses its symmetry with respect to the collision axis. Figure 1 shows the effect of the coherence on the electric charge density distribution of the $2p_{3/2}$ state. The alignment axis of this distribution attains a finite angle γ with respect to the electron propagation direction. This angle is neither 0° nor 90° which manifests the coherence of the populated state. We extracted the alignment angle of the charge cloud from the measured angular distribution of the Lyα_1 x rays revealing for the first time the coherence in a heavy highly charged ion. Moreover, the experimental results indicate a strong contribution of the spin-orbit interaction to the RR process and, therefore, to the fundamental process of the photoelectric effect.

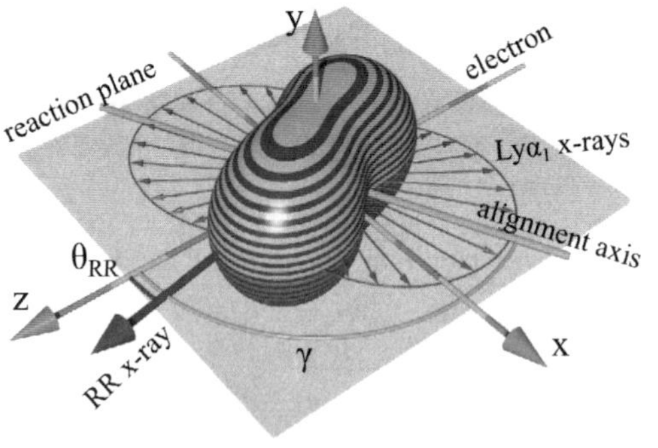

Figure 1. The angular density distribution of the charge cloud of the RR-populated $2p_{3/2}$ state when the RR x ray is observed.

References

[1] S. Tashenov *at al* 2014 *Phys. Rev. Lett.* **113** 113001

[1]E-mail: tashenov@physi.uni-heidelberg.de

Content from this work may be used under the terms of the Creative Commons Attribution 3.0 licence. Any further distribution of this work must maintain attribution to the author(s) and the title of the work, journal citation and DOI.
Published under licence by IOP Publishing Ltd

XXIX International Conference on Photonic, Electronic, and Atomic Collisions (ICPEAC2015) IOP Publishing
Journal of Physics: Conference Series **635** (2015) 022097 doi:10.1088/1742-6596/635/2/022097

Electron transfer induced by collision of low-energy doubly charged carbon ions with tetrahydrofuran

Ewa Erdmann[1*] **and Marta Łabuda**[2*]

[*] Faculty of Applied Physics and Mathematics, Gdańsk University of Technology, Narutowicza 11/12, 80233 Gdańsk, Poland

Synopsis In this communication, we present results of recent investigations of charge transfer process in the collision of C^{2+} ions with heterocyclic molecule of tetrahydrofuran.

Theoretical study of electron transfer processes involving multiply charged ions in collision with atomic and molecular targets are in our interest since several years [1-4]. Therefore, in our recent calculations we have selected tetrahydrofuran (THF), as a simple organic heterocyclic molecule, which serves as a perfect prototype model of deoxyribose sugar in the DNA helix. Such a model is important in the investigations of the action of ionizing radiation in biological material as well as in the synthesis of new compounds which are particularly useful in the medicinal chemistry. It is also known, that therapy with heavy carbon ions is proven to be the best choice for selectively irradiating deep-seated tumors. In this purpose, approaches at the molecular level are crucial to model biological effects of radiation in order to improve medical treatment.

In this contribution, we present results of recent investigations of electron transfer mechanism in the collision of C^{2+} ions with THF. As a first step, to determine the structure of the molecule, the results of the electronic properties and lowest excited states obtained by *ab initio* and density functional theory methods will be presented. The accuracy of the methods is estimated in good agreement with experimental and theoretical results given in literature [5-7]. Secondly, we show the lowest potential energy curves as well as nonadiabatic couplings as a function of the internuclear distance between the incident C^{2+} and target molecule. Such calculations are carried out at Complete Active Space Self Consistent Field level of theory by using different active spaces in which the most important orbitals of THF and C ion are included. Additionally, the anisotropic effects taking into account the different directions of the collision and dynamics of the process by semi-classical approach in the range of eV energy will be presented. In the Figure 1 the potential energy curves show avoided crossings around nuclear distance of 2Å-3.5Å. Such region corresponds to a strong interaction between electronic states and can be investigated by radial couplings.

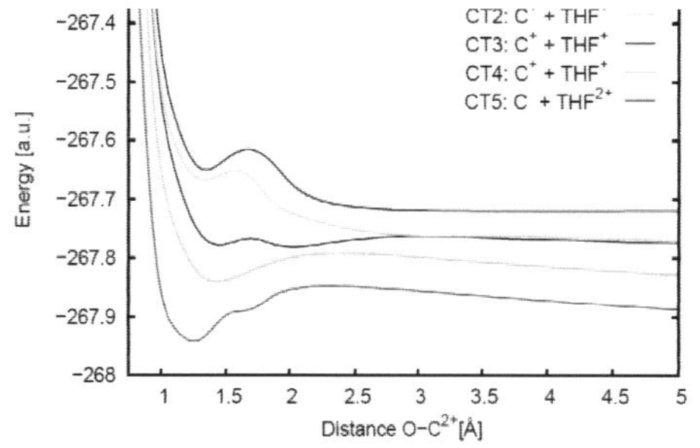

Figure 1. Adiabatic potential energy curves of the of C^{2+} + THF planar collision.

In this case both single and double electron transfer from the THF molecule toward the doubly charged ion can be noticed. Each potential energy curve in its asymptotic limit represents following excitation(s):

CT1: p_z on O $\rightarrow 2p_z$ on colliding carbon ion C(5)
CT2: p_y on O $\rightarrow 2p_y$on C(5)
CT3: $\pi^x_{C3C4} \rightarrow$ p_y on C(5)
CT4: $\pi^y_{C3C4} \rightarrow$ $p_x$on C(5)
CT5: p_z on O$\rightarrow 2p_x$ on C(5), $\pi^z_{C3C4} \rightarrow$ p_y on C(5)

Acknowledgements
This work has been partially supported by the COST Actions Nano-IBCT and XLIC. The calculations have been performed at the Academic Computer Centre (TASK) in Gdańsk .

References

[1] M.C.Bacchus-Montabonel, M.Łabuda *et al*, *Phys. Rev. A* **72**, 052706 (2005)
[2] M.C. Bacchus-Montabonel *et al*, *Int. J. Quantum Chem* **107**, (14) 2672 (2007)
[3] M. Łabuda *et al*, *Phys. Chem. Chem. Phys.* **12**, 5439 (2010)
[4] M. Łabuda *et al*, *Chemical Physics* **400**, 165 (2012)
[5] B. Cadioli *et al*, *J. Phys. Chem.* **97**, 7844 (1993)
[6] D.G. Melnik *et al*, *J. Chem. Phys.* **118**, 3589 (2003)
[7] V.M. Rayón, J.A. Sardo, *J. Chem. Phys.* **122**, 204303 (2005)

[1] E-mail: eerdmann@mif.pg.gda.pl [2]E-mail: marta@mif.pg.gda.pl

Content from this work may be used under the terms of the Creative Commons Attribution 3.0 licence. Any further distribution of this work must maintain attribution to the author(s) and the title of the work, journal citation and DOI.
Published under licence by IOP Publishing Ltd

XXIX International Conference on Photonic, Electronic, and Atomic Collisions (ICPEAC2015) IOP Publishing
Journal of Physics: Conference Series **635** (2015) 022099 doi:10.1088/1742-6596/635/2/022099

Cross section set and transport properties of Ne^+ in CF_4

Z. Raspopović[*], Ž. Nikitović[*], S. Tošić and V. Stojanović [*1]

[*]Institute of Physics Belgrade, University of Belgrade, P.O. Box 57, 11000 Zemun Serbia

Synopsis Cross section set for scattering Ne^+ ions in CF_4 is assessed by using available experimental data for charge transfer cross sections. Monte Carlo method is used to calculate transport properties of Ne^+ ions in CF_4.

Charge transfer reactions of ions with molecules are unavoidable elementary processes in modeling kinetics in terrestrial, industrial, and astrophysical plasmas. In selected cases charge transfer reactions are known to represent the most significant part of a cross section set.
Line spectra of excited atoms obtained in spectrometric measurements in CF_4 indicate that the charge transfer reaction is dominant process in collisions with inert gas ions. Thus, in this work we assessed cross section set for Ne^+ in CF_4 by using existing experimental data [1] for charge transfer collisions producing radical ions of CF_4.
Since no direct information is found in the literature how mobility of high recombination energy ions such as Ne^+ ions behaves in CF_4 we also calculated transport parameters by using Monte Carlo simulation technique [2].
The cross sections presented by Fisher *et al* [1] were used to determine the elastic momentum transfer cross section ("elastic" in Fig. 1) assuming the total momentum transfer cross section σ_{mt} is known. At low energies we assumed that σ_{mt} is Langevin's cross section and elastic momentum transfer cross section is determined by deducing all reactive cross sections.
Average polarizability of CF_4 is not well established [1] and may produce discrepancy for calculated mobility of ions in CF_4 [3] and thus affect plasma parameters prediction in modeling. We adopted value of 3.86 A^3 used by Stojanović *et al* [3] who found excellent agreement between experimental and calculated mobility of CF_3^+ ions in CF_4.

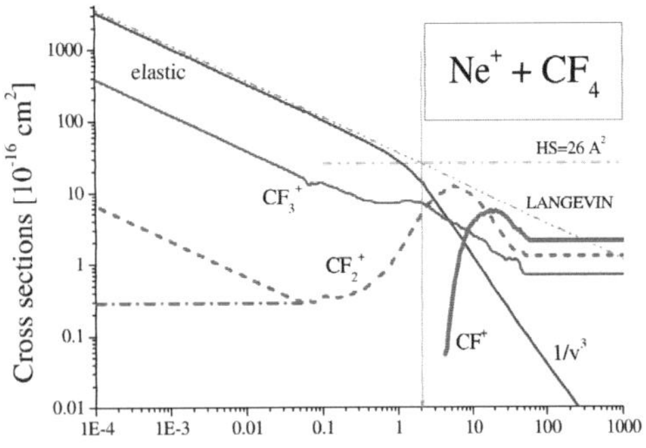

Figure 1. Cross section set for $Ne^+ + CF_4$.

Further, extrapolation of elastic momentum transfer cross section trend beyond crossing point of Langevin's and hard sphere (HS) cross section [1] is done by smoothly connecting to $1/v^3$ trend [4] where v is the center-of-mass velocity (see Fig. 1).
At all ion kinetic energies above 50 eV reactive cross sections are extrapolated by constant values.
Effect of various extrapolations (short dot-dashed or dashed line in Fig. 1) of unusual behavior at low energy, observed with measurements of the cross section leading to formation of CF_2^+ (where irrespective of the Ne^+ spin state exothermic behavior of reaction is expected) is found negligible on mobility. Reduced mobility for Ne^+ ions as a function of E/N(E-electric field, N-gas density) compared with Langevin's value is shown in Fig. 2.

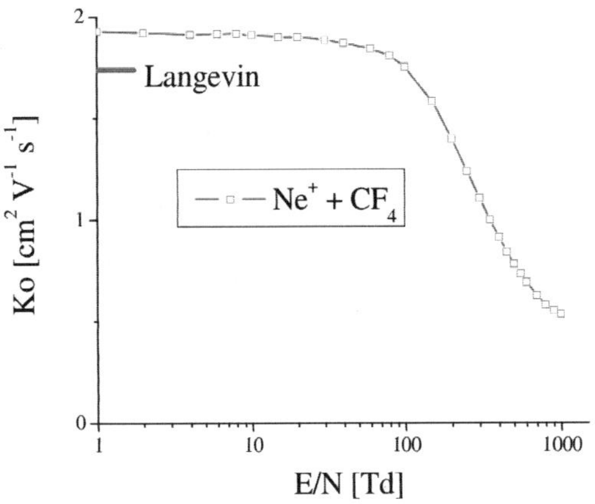

Figure 2. Reduced mobility for Ne^+ in CF_4 at 300 K.

References

[1] E R Fisher *et al.* 1990, *J.Chem.Phys.* **92** 2296
[2] Z Lj. Petrović Z. Lj *et al.* 2009 *J. Phys. D: Appl. Phys.* **42** 194002.
[3] V Stojanović *et al.* 2014 *J. Phys. Conf. Ser.* **514** 012059
[4] P S Krstić and D R Schultz 2009 *J. Phys. B: At. Mol. Opt. Phys.* **42** 065207

[1]E-mail: stoyanov@ipb.ac.rs

Content from this work may be used under the terms of the Creative Commons Attribution 3.0 licence. Any further distribution of this work must maintain attribution to the author(s) and the title of the work, journal citation and DOI.
Published under licence by IOP Publishing Ltd

XXIX International Conference on Photonic, Electronic, and Atomic Collisions (ICPEAC2015) IOP Publishing
Journal of Physics: Conference Series **635** (2015) 022100 doi:10.1088/1742-6596/635/2/022100

Fully quantum-mechanical treatment of proton-hydrogen scattering

I. B. Abdurakhmanov[1], **A. S. Kadyrov, S. K. Avazbaev, and I. Bray**

Department of Physics, Astronomy and Medical Radiation Sciences, Curtin University, GPO Box U1987, Perth, WA 6845, Australia

Synopsis A fully quantum-mechanical convergent close-coupling approach to proton collisions with atomic hydrogen has been developed. Cross sections for target ionisation and electron capture by the projectile have been calculated in the energy range from 20 keV to 1 MeV. Calculated electron capture cross sections are in good agreement with the experiment, however for ionisation discrepancies between theory and experiment at intermediate energies still remain.

Accurate knowledge of charge exchange, excitation and ionisation processes occurring during collisions between positive ions and various atomic and molecular targets is important in a number of fields including astrophysics, plasma physics, atmospheric modelling and hadron therapy. Proton-hydrogen scattering is the simplest prototype of such collisions. Even for this simplest three-body system there is no agreement between theory and experiment when ionisation is concerned.

A new fully quantum-mechanical convergent close-coupling approach has been developed and applied to the proton-hydrogen scattering problem. It is based on the solution of Lippmann-Schwinger (LS) integral equations emerging from the two-center expansion of the total scattering wave function. For the first time, the integral over the off-shell momentum in the LS equations has been evaluated analytically. This procedure allowed to greatly reduce the computational time and memory requirements for solving the scattering problem.

Here we present the results of our fully symmetric calculations which include s, p and d states both for the target and projectile. In figure 1 we show our calculated total ionisation (TICS) and electron capture (TECCS) cross sections in comparison with the experiment and other calculations for p-H collisions. As one can see, the present results for electron capture are in very good agreement with the experiment at all energies. For ionisation the discrepancies with the experiment at intermediate energies seen in the other semiclassical calculations [8–10] still remain. The present results are convergent in terms of the included s, p and d states. Results with larger angular momentum states will be reported later.

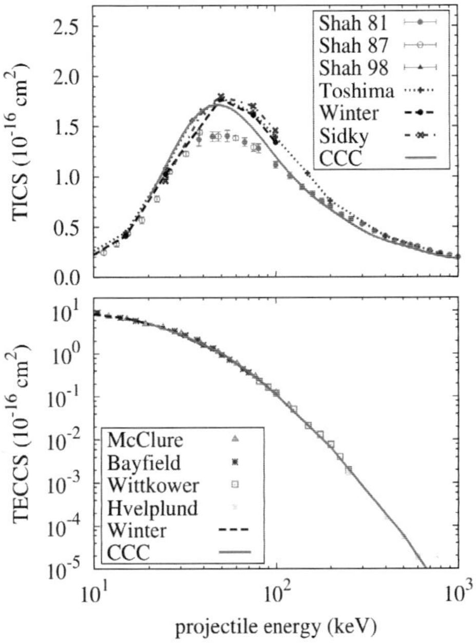

Figure 1. Total ionisation and electron capture cross sections for p-H collisions. The present CCC results are compared with the experimental data [1–7] and two-centre close-coupling calculations of Toshima [8], Winter [9], and Sidky and Lin [10].

References

[1] G. W. McClure, 1966 *Phys. Rev.* **148** 47
[2] J. J. Bayfield 1969 *J. Phys. B* **185** 105
[3] A. Wittkower *at al* 1966 *Proc. Phys. Soc.* **89** 541
[4] P. Hvelplund *at al* 1982 *Phys. Scr.* **26** 375
[5] M. B. Shah *at al* 1981 *J. Phys. B* **14** 2361
[6] M. B. Shah *at al* 1987 *J. Phys. B* **20** 2481
[7] M. B. Shah *at al* 1998 *J. Phys. B* **31** L757
[8] N. Toshima *at al* 1999 *Phys. Rev. A* **59** 1981
[9] T. G. Winter *at al* 2009 *Phys. Rev. A* **80** 032701
[10] E. Sidky and C. Lin 2001 *Phys. Rev. A* **65** 2001

[1]E-mail: Ilkhom.Abdurakhmanov@curtin.edu.au

Content from this work may be used under the terms of the Creative Commons Attribution 3.0 licence. Any further distribution of this work must maintain attribution to the author(s) and the title of the work, journal citation and DOI.
Published under licence by IOP Publishing Ltd

Studies of surface nanostructure formation due to swift heavy ion irradiation under grazing incidence

E. Gruber* [1]**, E. Lattouf**[†]**, L. Bergen*, P. Salou**[†]**, C. Grygiel**[†]**, Y. Y. Wang*, O. Ochedowski**[#]**, A. Benyagoub**[†]**, D. Levavasseur**[†]**, J. Rangama**[†]**, H. Lebius**[†]**, B. Ban d'Etat**[†]**, M. Schleberger**[#]**, and F. Aumayr***

* Institute of Applied Physics, TU Wien, 1040 Vienna, Austria, EU
† CIMAP, Blvd. Henri Bequerel, 14070 Caen Cedex 5, France, EU
# Experimentelle Physik und CENIDE, Universität Duisburg-Essen, 47048 Duisburg, Germany, EU

Synopsis In this contribution we present new experimental results for swift heavy ion irradiation under grazing incidence. This particular collision geometry forces the track formation to a region close to the surface, sometimes visible as a chain of individual nanodots, whose length can be controlled by the angle of incidence. For irradiation of mica surfaces a new track form is observed.

Swift heavy ion (SHI) irradiation of solid targets can lead to permanent structural modifications in the bulk and at the surface (see e.g. [1-5] and refs. therein). In most studies the irradiation is performed under normal incidence with respect to the surface plane of the target. The impact of each individual projectile induces a long, straight nanometric track consisting of amorphous or otherwise modified target material. At the impact site of the ion usually a hillock- or crater-type nanostructure can be observed.

Performing SHI irradiation under grazing incidence offers new possibilities for the modification of surface and bulk properties on the nanometer scale. Due to this special collision geometry, the tracks are formed in a region close to the surface, sometimes visible as a chain of individual nanodots. The length of these chains can be controlled by the angle of incidence [2, 5]. So far, the formation of these chain-nanostructures have been observed for insulating materials like polymers, oxides and ionic crystals and their formation has been linked to a critical electronic energy loss dE/dx of the incident projectiles [2, 5].

By using high resolution atomic force microscopy, comparative studies of the chain formation for different materials, i.e. $SrTiO_3$, TiO_2, CaF_2 and mica, have been performed at the IRRSUD line at GANIL. These targets have been irradiated with Xe and Pb ions at kinetic energies between 90 and 100 MeV under very small grazing incidence angles between 0.2° and 2°. Imaged under AFM the track structures observed are strongly material dependent.

For mica, e.g. surprisingly "double tracks" seem to be formed which eventually combine to a single track (see. Fig. 1).

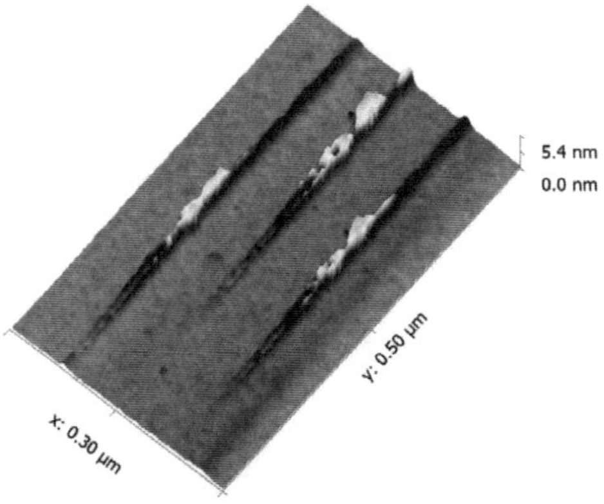

Figure 1. AFM image of surface tracks formed on a muskovit mica surface after 0.71 MeV/u $^{129}Xe^{23+}$ ion irradiation at 0.9°. The projectile is approaching the mica surface from the bottom-left opening first a groove, whose walls are visible as a "double track". After the projectile ion has entered deeper layers the double track is eventually combined into a single track.

This work has been supported by the French-Austrian collaboration SIISU (ANR project ANR-12-IS04-0005-01 and FWF project I 1114-N20).

References

[1] H. Dammak *et al* 1995 *Phys. Rev. Lett.* **74**, 1135
[2] E. Akcöltekin *et al* 2007 *Nature Nanotechnology* **2**, 290
[3] M. Toulemonde *et al* 2007 *Mat. Fys. Med. Dan. Vid. Selsk.* **52**, 263
[4] N. Khalfaoui *et al* 2005 *Nucl Instrum Meth B* **240**, 819
[5] F. Aumayr *et al* 2011 *J. Phys.: Cond.Mat.* **23**, 393001

[1] E-mail: egruber@iap.tuwien.ac.at

Content from this work may be used under the terms of the Creative Commons Attribution 3.0 licence. Any further distribution of this work must maintain attribution to the author(s) and the title of the work, journal citation and DOI.
Published under licence by IOP Publishing Ltd

Interaction of multiply charged ions with single layer graphene Part I: charge exchange and energy loss

V. Smejkal*, E. Gruber* [1], M. Kralik*, R. A. Wilhelm†, R. Heller†, S. Facsko†, and F. Aumayr* [2]

* Institute of Applied Physics, TU Wien, Wiedner Hauptstr. 8-10/E134, 1040 Vienna, Austria
† Helmholtz-Zentrum Dresden-Rossendorf, Inst. of Ion Beam Physics & Materials Research, 01328 Dresden, Germany

Synopsis The exit charge state distribution and the energy loss of slow multiply charged ions transmitted through single layers of graphene and 1 nm thick carbon nanomembranes is analyzed.

Due to its unique electronic properties, single layer graphene (SLG) has attracted enormous attention since its discovery [1] as an excellent candidate for future nanoelectronics. SLG can be considered as the ultimately thin membrane made of sp^2-hybridized carbon atoms. Disorder as induced e.g. by collisions with energetic electrons or ions affects the electronic structure and opens the possibility to modify and tailor the properties of this true 2D material. Collision studies between ions and free-standing SLG are also of fundamental interest, because they bridge the gap between atomic collisions in gaseous and those in solid targets.

For the transmission of slow highly charged ions through slightly thicker 1 nm carbon nanomembranes (CNMs) we have recently observed two distinct exit charge state distributions accompanied by a strong charge state dependent energy loss [2] as well as the creation of nano-sized pores for Xe^{q+} projectiles in charge states $q \geq 25$ mainly by deposition of their potential energy [3].

To learn more about the microscopic interaction mechanism we have extended our studies to the case of slow multiply charged Ar^{q+} ions transmitted through SLG (and CNMs in comparison). In a new experimental setup in Vienna, slow Ar^{q+} ions ($q \leq 9$) are extracted from an ECR-ion source [4], mass analyzed and collimated before traversing a free standing SLG or CNM target mounted on a TEM grid. The ions exiting the target foil are analyzed in a spherical electrostatic energy analyzer (COMSTOCK AC-903B) equipped with a channelplate detector. A typical spectrum (intensity vs. analyzer voltage) recorded for 144 eV/amu Ar^{4+} ions through a CNM is shown in figure 1.

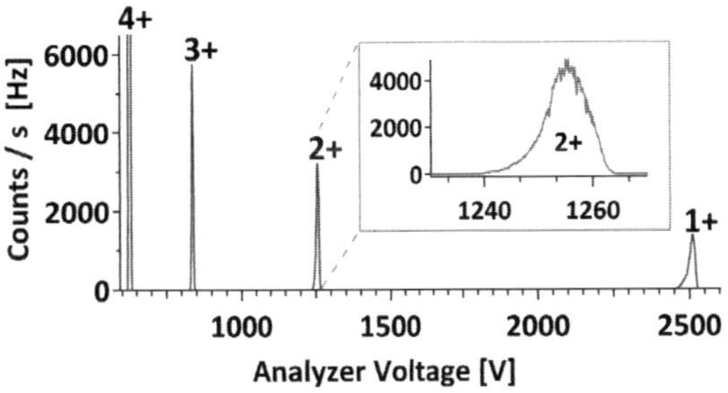

Figure 1. Charge state and energy spectrum (e.g. see insert for ions exiting as Ar^{2+}) of 5.76 keV Ar^{4+} projectile ions after transmission through a 1 nm thick carbon nanomembrane.

Such spectra give information on both the exit charge state distribution (i.e. charge exchange between multiply charged ions and SLG) as well as the mean value of the energy loss and the energy loss distribution (see the insert of fig. 1) as a function of the charge state and kinetic energy of the primary projectile. E.g., figure 1 clearly shows that charge equilibrium of the projectile has by far not been reached within the 1 nm thick foil and therefore non-equilibrium effects can be studied. Further information on the interaction is derived from the comparison between results for SLG and CNMs.

Work supported by Austrian FWF (Proj.No. I 1114-N20).

References

[1] A. K. Geim and K. S. Novoselov 2007 *Nature Mater.* **6**, 183
[2] R. A. Wilhelm, et al. 2014 *Phys. Rev. Lett.* **112** 153201
[3] R. Ritter, et al. 2013 *Appl. Phys. Lett.* **102** 06311
[4] E. Galutschek et al. 2007 *J. Phys. Conf. Ser.* **58** 395

[1] E-mail: egruber@iap.tuwien.ac.at
[2] E-mail: aumayr@iap.tuwien.ac.at

Content from this work may be used under the terms of the Creative Commons Attribution 3.0 licence. Any further distribution of this work must maintain attribution to the author(s) and the title of the work, journal citation and DOI.
Published under licence by IOP Publishing Ltd

Interaction of multiply charged ions with single layer graphene Part II: electron emission

A. Fuchs-Fuchs, J. Schwestka, E. Gruber, Y. Y. Wang, and F. Aumayr[1]

Institute of Applied Physics, TU Wien, Wiedner Hauptstr. 8-10/E134, 1040 Vienna, Austria

Synopsis A new setup is used to measure the number statistics of electrons emitted during the transmission of slow multiply charged ions through single layers of graphene and 1 nm thick carbon nanomembranes.

Single layer graphene (SLG) is an ultimately thin membrane made of sp^2-hybridized carbon atoms with unique electronic properties. Since its discovery [1] SLG has been considered as an excellent candidate for future nanoelectronics. Disorder, as induced e.g. by collisions with energetic electrons or ions, effects the electronic structure and opens the possibility to modify and tailor the properties of this true 2D material. Collision studies between ions and free-standing SLG are also of fundamental interest, because they bridge the gap between atomic collisions in gaseous and those in solid targets.

For the transmission of slow highly charged ions through slightly thicker 1 nm carbon nanomembranes (CNMs) we have recently observed two distinct exit charge state distributions accompanied by a strong charge state dependent energy loss [2] as well as the creation of nanosized pores for Xe^{q+} projectiles in charge states $q \geq 25$ mainly by deposition of their potential energy [3].

To learn more about the microscopic interaction mechanism we have built a new experimental setup (figure 1), which allows us to determine the numbers statistics of electrons emitted during the transmission of slow multiply charged Ar^{q+} ions through SLG (and CNMs in comparison). Slow Ar^{q+} ions ($q \leq 11$) are extracted from an ECR-ion source [4], mass analyzed and collimated before traversing a free standing SLG or CNM target (mounted on a TEM grid which is fixed onto a rotatable manipulator) and being registered by a position sensitive multichannel plate detector (MCP). Electrons emitted from the interaction region are extracted by a weak electric field through a highly transparent grid and accelerated onto a surface barrier type detector biased at +30 kV, where their statistical distribution is measured in the usual way (see, e.g. [5, 6]). By changing the orientation of the target film from 45° to 135° with respect to the incident beam direction, electrons emitted from the front (= ion entrance side) and back side (= ion exit side) can be distinguished. To separate emission events originating from ion transmission through SLG from ion impact on the (solid) TEM grid support, only electrons registered in coincidence with an ion signal on the MCP are counted.

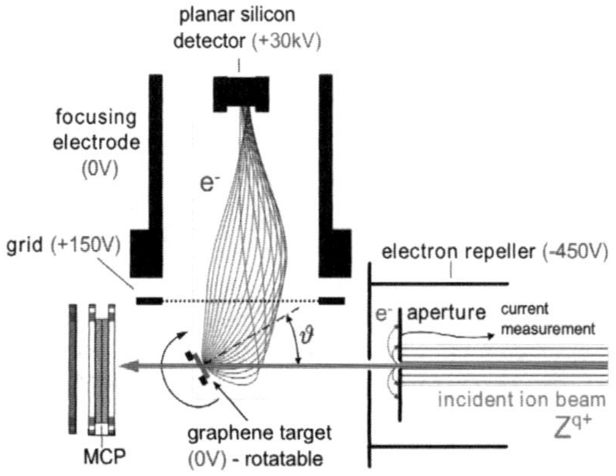

Figure 1. Experimental setup (c.f. text).

Since electron emission and projectile charge exchange are closely related, in a next step it is planned to register the electrons in coincidence with a particular projectile exit charge state.

Supported by Austrian FWF (Proj.No. I 1114-N20).

References

[1] A. K. Geim and K. S. Novoselov 2007 *Nature Mater.* **6**, 183
[2] R. A. Wilhelm, et al. 2014 *Phys. Rev. Lett.* **112** 153201
[3] R. Ritter, et al. 2013 *Appl. Phys. Lett.* **102** 06311
[4] E. Galutschek et al. 2007 *J. Phys. Conf. Ser.* **58** 395
[5] F. Aumayr et al. 1991 *Appl. Surf. Sci.* **47** 139
[6] K. Töglhofer et al. 1993 *Surf. Sci.* **281** 143

[1]E-mail: aumayr@iap.tuwien.ac.at

Content from this work may be used under the terms of the Creative Commons Attribution 3.0 licence. Any further distribution of this work must maintain attribution to the author(s) and the title of the work, journal citation and DOI.
Published under licence by IOP Publishing Ltd

Theoretical study of noble gases diffraction from Ru(0001) using van der Waals DFT-based potentials

M. del Cueto* [1], A. S. Muzas*, F. Martín*† and C. Díaz*

* Departamento de Química, Módulo 13, Universidad Autónoma de Madrid, Cantoblanco 28049, Madrid, Spain
† Insituto Madrileño de Estudios Avanzados en Nanociencia (IMDEA-Nanociencia), Cantoblanco 28049, Madrid, Spain

Synopsis This study aims to analyze the role of van der Waals forces in the diffraction process of noble gases from a metal surface. We made use of different vdW implementations to rationalize the effect of dispersion forces on the corrugation of the system, the resulting scattering patterns and on the eventual diffraction results.

Motivated by recent experimental studies about He, Ne and Ar diffraction from a Ru(0001) surface [1], where the last two cases display anticorrugation effects, i.e. corrugation inversion, and by the possibility of using van der Waals (vdW) DFT functionals which allow one to simulate accurately the mentioned experiments, we have studied these systems from a theoretical point of view, aiming to determine the role of vdW interactions on the anticorrugation phenomena observed in the mentioned diffraction processes.

To achieve this task, we built the 3D potential energy surfaces (PESs) based on periodic DFT calculations performed by means of the Vienna ab initio simulation package (VASP) [2].

We tested different vdW functionals during these calculations in order to simulate the dispersion forces. Finally, the diffraction spectra are being computed using both classical and quantum dynamics simulations. Although diffraction is a 'pure' quantum phenomenon and quantum dynamics is, in principle, unavoidable to study our system, a classical analysis based on parallel momentum binning [3] has already been revealed as a very useful tool to mimic and analyze diffraction spectra [4,5].

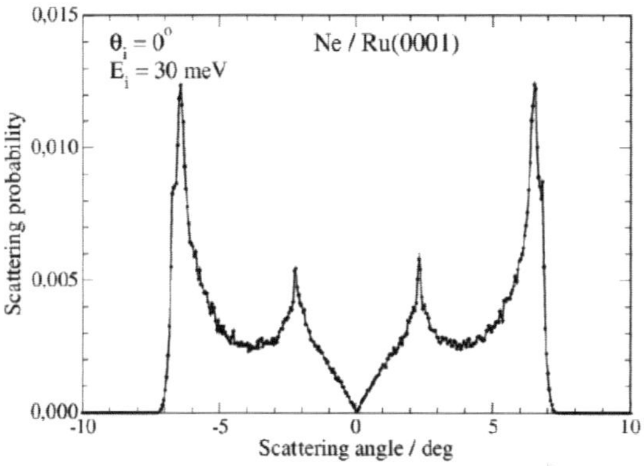

Figure 1. Scattering probability of Ne over Ru(0001) at normal incidence. The scattering distribution presents a classical rainbow, which is proportional to the surface corrugation and will therefore determine the diffraction distribution.

References

[1] M. Minniti *et al* 2012, *J.Phys. Condens. Matter* **24** 354002
[2] G. Kresse and J. Hafner 1993, *Phys. Rev. B* **47** 558
[3] C. Ray and J. Bowman 1975, *J. Chem. Phys.* **63** 5231
[4] D. Farías *et al* 2004, *Phys. Rev. Lett.* **93** 24104
[5] C. Díaz *et al* 2012, *J. Phys. Chem. C* **116** 13671

[1] E-mail: marcos.cueto@estudiante.uam.es

Content from this work may be used under the terms of the Creative Commons Attribution 3.0 licence. Any further distribution of this work must maintain attribution to the author(s) and the title of the work, journal citation and DOI.
Published under licence by IOP Publishing Ltd

Hillock formation on CaF_2, Al_2O_3, c-SiO_2 and MgO single crystal surfaces by ion impact - From potential energy deposition to electronic energy loss

Y. Y. Wang*,†1, C. Grygiel#, I. Monnet#, E. Gruber*, H. Lebius#, B. Ban d'Etat#, J. R. Sun†, Y. T. Zhao†, Z. G. Wang†, G. Q. Xiao†, F. Aumayr*

* Institute of Applied Physics, TU Wien, 1040 Vienna, Austria
† Institute of Modern Physics, Chinese Academy of Sciences, Lanzhou 730000, China
# CIMAP-GANIL, CEA-CNRS-ENSICAEN-University of CAEN, F-14070 Caen, France

Synopsis For impact of very slow highly charged ions on single crystal surfaces the appearance of hillocks can be linked to a threshold in potential energy while for swift heavy ions a minimum electronic energy loss per unit length is necessary. Recent investigations on CaF_2 in the medium energy range bridge the gap between these two extreme cases. These investigations are now extended to other target materials.

Great interest has been drawn by the use of ions as possible tools for lithographic modifications or "nanostructuring" of surfaces. Investigations have shown that swift heavy ions (SHI) with kinetic energies in the MeV to GeV range can induce severe structural modifications at the surface [1] and in the bulk [2]. Individual slow highly charged ions (HCI) in the keV energy regime are also able to induce surface modifications on a nanometric scale [3]. These modifications result from the deposition of potential energy carried by HCI in the topmost surface layers or from electronic energy loss of SHI along its trajectory. In both cases several keV/nm^3 are deposited within a few femtoseconds into the target's electronic system and subsequently transferred to the lattice atoms via electron-phonon coupling. This similarity between HCI and SHI was pointed out [3, 4] in order to establish a qualitative link between surface modifications by potential energy and electronic energy loss.

In the medium energy range (few MeV) hillock formation on single crystal surfaces induced by highly charged heavy ions has recently been studied. This study has shown an additive effect between depositions of kinetic energy and potential energy for surface nanostructure formation on CaF_2 [5].

We have now extended these investigations to other insulating materials including Al_2O_3, c-SiO_2 and MgO. As a first step we have irradiated these materials by 100 MeV Pb ions delivered by the IRRSUD beamline at GANIL. Because the electronic energy deposition depends on the SHI velocity [6], the impact energy of the SHI was varied by Al foil degradation.

The aim of these investigations is to find the S_e threshold for the SHI case. Unfortunately on several crystals the surface roughness or present surface contaminations prevented an unambiguous identification of ion-induced nanostructures. In these cases irradiation were also conducted under grazing angles of incidence. Surface tracks usually take the form of a chain of hillocks [7] and are therefore more easily identified than single ion-induced hillocks. Our results for SHI irradiation of Al_2O_3, c-SiO_2 and MgO will be presented at the conference.

This work has been supported by the French-Austrian collaboration SIISU, co-financed by ANR (France, ANR-12-IS04-0005-01) and FWF (Austria, Project No. I 1114-N20), as well as by the French-Chinese cooperation project IiN by PICS-CNRS (France) and NSFC (China, No. 11411130112).

References

[1] R. Neumann 1999 *Nucl. Instr. Meth. B* 151 42
[2] M. Toulemonde *et al* 1988 *Phil. Mag. A* 58, 799
[3] F. Aumayr *et al* 2011 *J. Phys. Cond. Matt.* 23 393001
[4] A. El Said *et al* 2008 *Phys. Rev. Lett.* 100 237601
[5] Y.Y. Wang *et al* 2014 Sci. Reps. 4 5742
[6] A. Meftah *et al* 1993 *Phys. Rev. B* 48, 920
[7] E. Akcöltekin *et al* 2007 *Nature Nanotechnology* **2**, 290

†1 E-mail: ywang@iap.tuwien.ac.at

Content from this work may be used under the terms of the Creative Commons Attribution 3.0 licence. Any further distribution of this work must maintain attribution to the author(s) and the title of the work, journal citation and DOI.
Published under licence by IOP Publishing Ltd

Journal of Physics: Conference Series **635** (2015) 032006 doi:10.1088/1742-6596/635/3/032006

Fast atom diffraction of H and He from a LiF(001) surface: semi-quantum approach

M.S. Gravielle* [1], J.E. Miraglia*, and G.A. Bocan†

* Instituto de Astronomía y Física del Espacio (IAFE, CONICET-UBA), Buenos Aires, Argentina
† Centro Atómico Bariloche and CONICET, S.C. de Bariloche, Río Negro, Argentina

Synopsis For fast atom diffraction (FAD) from crystal surfaces we introduce a distorted wave method, based on the semi-quantum approach of the quantum evolution operator, in order to improve the description of rainbow effects given by the Surface Eikonal (SE) model. The theory, named Surface Initial Value Representation (SIVR) approximation, is applied to H and He atoms colliding with a LiF(001) surface, using pair-wise additive and DFT–derived potentials. The SIVR approach provides a very good representation of the experimental distributions, even in the angular region around classical rainbow angles where the SE method diverges. In addition, the influence of the incident wave-packet is analyzed.

In spite of the successful performance of the SE approach for the simulation of FAD patterns [1], a weakness of the theory is its deficient description of the rainbow effect, which affects the intensity of the outermost diffraction maxima when these maxima are close to the classical rainbow angles, i.e. the extreme deflection angles of the classical projectile distribution. Such a deficiency is a well-known characteristic of the *classical* representation of the dynamics, which introduces a singularity at rainbow angles. Within a proper *quantum* treatment, these sharp rainbow peaks are replaced by smooth maxima that display an exponentially decaying behavior outside classical rainbow angles, just in the region of classically forbidden transitions.

The goal of this work is to develop a *semi-quantum* approximation for FAD, based on the Initial Value Representation (IVR) method by Miller [2], which can solve the drawback of the SE model without losing the simple description of the interference process in terms of classical scattering trajectories. For this purpose we extend the IVR method by using the corresponding semi-quantum time evolution operator in the frame of a time-dependent distorted-wave formalism. The proposed approach, named Surface-Initial Value Representation (SIVR) approximation [3], is applied to evaluate FAD patterns for H and He atoms grazingly impinging on a LiF(001) surface. The surface potentials were derived from two different ways: accurate density functional theory (DFT) calculations using the "Quantum Espresso" code (Fig. 1) and a pairwise additive model including non local contributions (Fig. 2). Furthermore, the effect of the shape of the incident wave-packet is investigated in the light of Ref. [6].

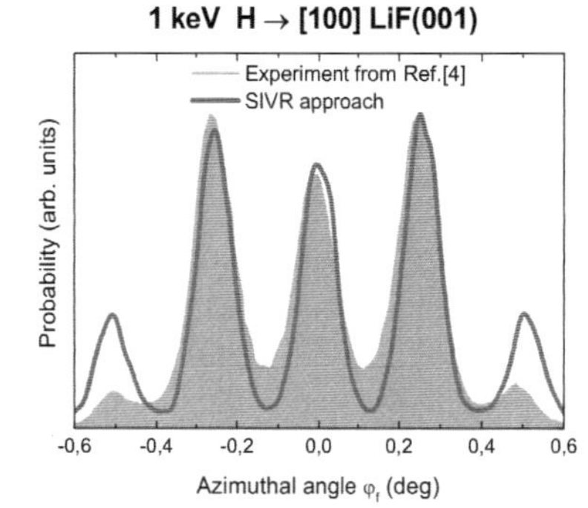

Figure 1. Angular projectile distribution, as a function of φ_f, for H atoms impinging on LiF(001) with the incidence angle θ_i=0.99 deg.

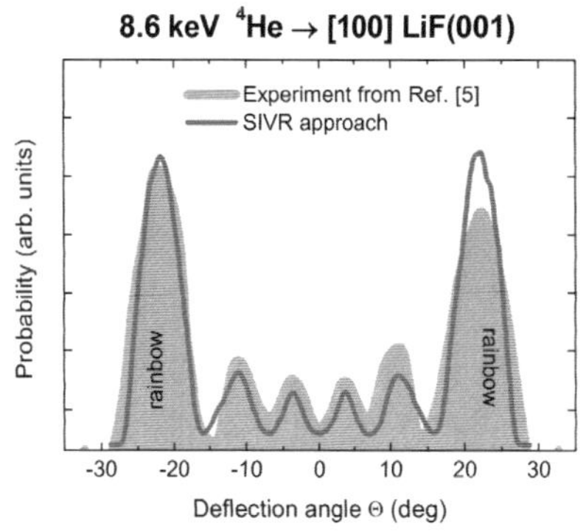

Figure 2. Similar to Fig. 1, as a function of $\Theta=\arctan(\varphi_f/\theta_f)$, for He projectiles with θ_i=0.71 deg.

References

[1] Gravielle *et al* 2008 *Phys. Rev. A* **78** 022901
[2] Miller 2001 *J. Phys. Chem. A* **105** 2942
[3] Gravielle *et al* 2014 *Phys. Rev. A* **90** 052718
[4] Rousseau *et al* 2008 *J. Phys. Conf. S.* **133** 012013
[5] Schüller *et al* 2008 *Phys. Rev. Lett.* **100** 097602
[6] Seifert *et al* 2014 *Phys. Procedia* to be published

E-mail: msilvia@iafe.uba.ar

Content from this work may be used under the terms of the Creative Commons Attribution 3.0 licence. Any further distribution of this work must maintain attribution to the author(s) and the title of the work, journal citation and DOI.
Published under licence by IOP Publishing Ltd

Journal of Physics: Conference Series **635** (2015) 032007 doi:10.1088/1742-6596/635/3/032007

Understanding the rotational excitation in scattering of D_2 from CH_3-Si(111)

**C. Díaz*[1], A. S. Muzas*, M. del Cueto*, T. J. Frankcombe‡ F. Martín*°
Z. M. Hund†, K. J. Nihill†, S. J. Sibener†**

* Departamento de Química, Módulo 13, Universidad Autónoma de Madrid, Cantoblanco 28049, Madrid, Spain

° Instituto Madrileño de Estudios Avanzado en Nanociencia (IMDEA-Nanociencia), Cantoblanco 28049, Madrid, Spain

‡ Research School of Chemistry, Australian National University, ACT 0200, Australia

† The James Franck Institute and Department of Chemistry, The University od Chicago, 929 East 57th Street, Chicago Illinois 60637, USA

Synopsis We have studied the origin of the striking rotational excitation probability, found experimentally, for D_2 upon scattering from a organic-terminated Si(111) surface.

Organic-terminated Si surfaces are the subject of a great interest in surface science. Such surfaces exhibit improved oxidative and electrochemical stability, relative to hydrogen-terminated silicon, for practical applications such as biosensing electronics [1] or photoelectrodes in electrochemical cells [2]. Silicon surfaces functionalized with methyl termination have emerged as the best alternative to hydrogen termination [3], giving suppression of surface reconstruction and resistance to oxidation.

This system has been investigated with a range of experimental techniques. Recent experimental results show [4] a noticeable rotational excitation for D_2 (see Fig.), which is not observed in the diffraction spectra measured for H_2. In order to understand the origin of this behavior, we have simulated these experimental spectra to determine the mechanism that explain this difference. In a first step we have construct a suitable potential energy surface (PES), using a Modified Shepard interpolation method [5]. In a second step we have tested the accuracy of our PES by means of classical dynamics simulations.

First dynamical results already show that our six-dimensional PES reproduces accurately the anisotropy of the system. Our classical dynamics calculations show rotational excitation probabilities of the order of 22% for D_2 and of 8% for H_2, always within the incidence energy experimental range. Finally, quantum calculations have been used to further assess the appropriateness of the classical analysis [6]

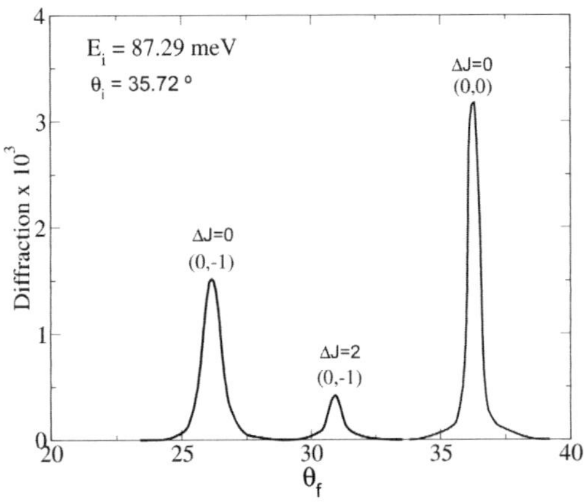

Figure 1. Experimental diffraction spectrum of D_2 scattered from CH_3-Si(111), along the incidence direction ΓM. Incidence energy 87.29 meV, polar angle 35.72 deg.

[1] E-mail: cristina.diaz@uam.es

References

[1] T. L. Lasseter *et al* 2004 *J. Am. Chem. Soc.* **126**, 10220.

[2] X. Shen *et al* 2010 *ACS Nano* **4**, 5869.

[3] R. D. Brown *et al* 2013 *Phys. Rev. Lett.* **110**, 156102.

[4] Z. M. Hund *et al* to be published

[5] T. J. Frankcome *et al* 2012 *J. Chem. Phys.* **137**, 144701.

[6] D. Farías *et al* 2004 *Phys. Rev. Lett.* **93**, 246104.

Content from this work may be used under the terms of the Creative Commons Attribution 3.0 licence. Any further distribution of this work must maintain attribution to the author(s) and the title of the work, journal citation and DOI.
Published under licence by IOP Publishing Ltd

XXIX International Conference on Photonic, Electronic, and Atomic Collisions (ICPEAC2015) IOP Publishing
Journal of Physics: Conference Series **635** (2015) 032008 doi:10.1088/1742-6596/635/3/032008

Structural changes in water and Ar-water clusters under high pressure

R. Prosmiti[† 1], **A. Vítek**[*], **D.J. Arismendi-Arrieta**[†], **R. Rodriguez-Cantano**[†], **P. Villarreal**[†], **R. Kalus**[*], **G. Delgado-Barrio**[†]

[†] Institute of Fundamental Physics, IFF-CSIC, Serrano 123, 28006 Madrid, Spain

[*] IT4Innovations National Supercomputing Center, and Department of Applied Mathematics, VSB–Technical University of Ostrava, 70833 Ostrava, Czech Republic

Synopsis Specific size gas-water clusters are currently receiving considerable attention, as models for inclusion compounds of different type of clathrate hydrates. As model microsolutions they retain many characteristics of the bulk, are theoretically tractable, and can be used to probe the relevant guest/host interactions, as well as to derive and to test intermolecular potentials that can be also used under different thermodynamic conditions.

A number of spectroscopic and diffraction studies have been conducted to explore at which conditions bulk rare-gas hydrates may be formed or decomposed and what types of them may be obtained at such conditions [1]. Theoretical attempts to describe structural transitions in these clathrate hydrates could provide useful information, and could serve to settle disagreement between different sets of experiments.

Thus, we performed [2] parallel tempering Monte Carlo simulations in the isothermal-isobaric (NPT) ensemble, aiming to provide a complete description of the cluster phase behavior under various temperature–pressure conditions. The TIP4P/ice model was employed for the water-water interactions, while both semiempirical and *ab initio*-based potentials were used to model the interaction between the rare-gas atom and water molecules (see Fig. 1). We constructed heat capacity phase diagrams for specific size clusters, such as the small polyhedral cages of the sI, sII and sH clathrate structures. By analyzing the heat capacity landscape and the Pearson correlation coefficient profile for the interaction energy and volume, structural changes were detected [2] (see Fig. 1). For the pure water $(H_2O)_{20}$ cluster, at low pressures such transitions correspond from the "all-surface" structures to the cage-1 like structures, while at higher pressures transitions to the cage-1 isomeric structure. For the $Ar(H_2O)_{20}$ cluster, changes at high pressure correspond to solid-to-solid transitions from outside-to-inside structures, and have been related to clathrate-like cages around the Ar atom. It is also shown that the formation and thermodynamic stability of the regular 5^{12} structures are determined by the intermolecular interaction between the Ar atom and host water network [2].

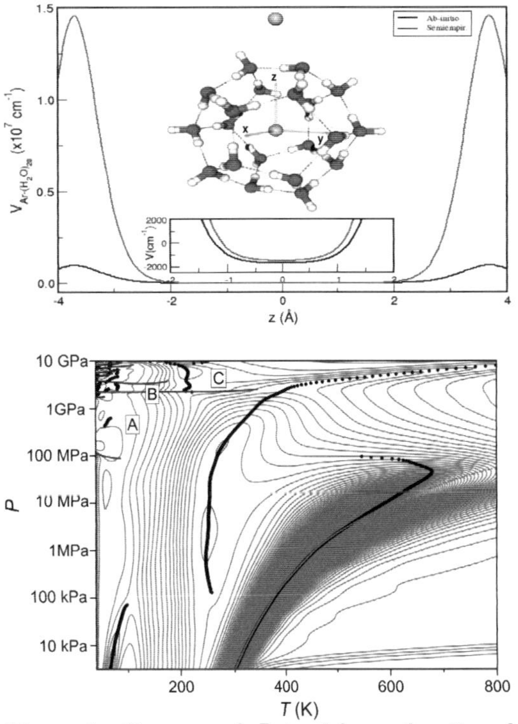

Figure 1. Upper panel: Potentials as a function of the distance between the Ar atom and the center of mass of the 5^{12} cage formed by 20 H_2O molecules (see inset plot). Lower panel: Phase diagrams of the $Ar(H_2O)_{20}$ cluster. Black dot lines correspond to the maxima of the heat capacity, blue/red dot lines to the min/max of the Pearson correlation coefficient, and A, B, and C to the different solid–solid transitions.

References

[1] J.S. Loveday *et al* 2008 *Phys. Chem. Chem. Phys.* **10** 937

[2] A. Vitek *et al* 2015 *Phys. Chem. Chem. Phys.* DOI: 10.1039/c4cp04862h

[1]E-mail: rita@iff.csic.es

Content from this work may be used under the terms of the Creative Commons Attribution 3.0 licence. Any further distribution of this work must maintain attribution to the author(s) and the title of the work, journal citation and DOI.
Published under licence by IOP Publishing Ltd

XXIX International Conference on Photonic, Electronic, and Atomic Collisions (ICPEAC2015) IOP Publishing
Journal of Physics: Conference Series **635** (2015) 032009 doi:10.1088/1742-6596/635/3/032009

Modeling the water-halide ion interactions

D.J. Arismendi-Arrieta[†1], **R. Prosmiti**[†], **F. Paesani**[*]

[†] Institute of Fundamental Physics, IFF-CSIC, Serrano 123, 28006 Madrid, Spain

[*] Department of Chemistry and Biochemistry, University of California, San Diego, La Jolla, CA 92093, USA

Synopsis Ions play an important role in many heterogeneous processes, and understanding specific ion effects is a key point in studying solvation mechanisms in liquids as well as reactivity of aerosol particles. In this regard, a crucial issue to be resolved is the extent to which ions influence the structural and dynamical properties of the surrounding hydrogen bond network and, at the same time, to provide a direct molecular-level interpretation of the experimental measurements.

So far, water-halide ion systems have attracted the interest of both experimentalists and theorists [1]. On the experimental side diffraction techniques and x-ray absorption spectroscopy have been used to understand the hydration of ions, while for their interpretation rather simplified models have been employed in the post-processing of raw data [2]. On the theoretical side, computational simulations have been limited by the accuracy of the models in describing weak interactions and hydrogen bonds, as well as due to the computational cost for an effective sampling of the molecular configurations.

Thus, we developed first-principles water-halide ion potentials [3], namely i-TTM (ion-Thole-type), by combining a newly constructed intermolecular ion-water one with the water-water MB-pol model [4]. The potentials include anisotropy in both short- and long-range interactions, and their parameters were determined by a two-step fitting procedure to electronic structure calculations for the X^-H_2O (X=F, Cl, Br, I) complex. In particular, the dispersion coefficients were evaluated first from DFT computations employing the XDM model, imposing the correct asymptotic behaviour, while for the electrostatic part the reparameterized TTM4-F variant of the Thole-type model was used, with the anion polarizabilites and the interaction energies calculated at the CCSD(T) and CCSD(T)-F12 level of theories, respectively.

The quality of the i-TTM model is validated through comparisons with high level *ab initio* calculations at optimized configurations of low-lying stationary points of each X^-H_2O system, as well as for small $X^-(H_2O)_n$ clusters (see Fig. 1). The obtained data suggest that the polarizable two-body i-TTM potentials can indeed be used to describe the halide ion-water gas phase interactions, while in conjuction with the MB-pol water-water model can be also used in molecular simulations of condensed-phase systems.

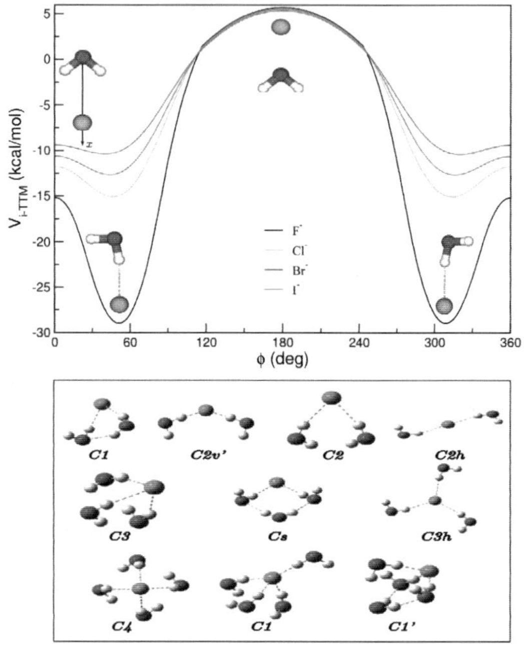

Figure 1. Upper panel: Minimun energy paths as a function of the azimuthal angle for the indicated X^-H_2O system, together with the configurations of the corresponding stationary points. Lower panel: Optimal structures for the indicated $X^-(H_2O)n$ clusters from the i-TTM model.

References

[1] W.H. Robertson *et al* 2003 *Ann. Rev. Phys. Chem.* **54** 173

[2] V. Migliorati *et al* 2014 *J. Chem. Phys.* **141** 044509

[3] D.J. Arismendi-Arrieta *et al* 2015 *J. Chem. Theor. Comput.* to be submitted

[4] V. Babin *et al* 2013 *J. Chem. Theor. Comput.* **9** 5395

[1]E-mail: d.arismendi@iff.csic.es

Content from this work may be used under the terms of the Creative Commons Attribution 3.0 licence. Any further distribution of this work must maintain attribution to the author(s) and the title of the work, journal citation and DOI.
Published under licence by IOP Publishing Ltd

Journal of Physics: Conference Series **635** (2015) 032010 doi:10.1088/1742-6596/635/3/032010

Confining CO_2 inside the nanocavities of the sI clathrate: a quantum dynamics study

R. Prosmiti† [1], **Á. Valdés***, **D.J. Arismendi-Arrieta**†

† Institute of Fundamental Physics, IFF-CSIC, Serrano 123, 28006 Madrid, Spain

* Dept de Física, Universidad Nacional de Colombia, Calle 26, Cra 39, Edificio 404, Bogotá, Colombia

Synopsis Clathrate hydrates have been found to occur naturally, and have been extensively studied due to their important industrial applications, such as the storage of the CO_2 from the atmosphere and control climate change. Thus, quantitative understanding of physical and chemical properties, as well as the factors that control the formation of CO_2 hydrates, on both macroscopic and microscopic levels is essential in several areas of physical science.

Carbon dioxide clathrate hydrates of sI type are essentially pure, and have been studied experimentally by X-ray, powder X-ray and neutron diffraction methods [1] while their spectroscopic characterization has been reported via Raman, solid-state NMR, and the infrared (IR) [2] spectroscopy. These experimental studies motivated us to perform [3] quantum calculations of the T-R-V dynamics of the CO_2 molecule in both 5^{12} and $5^{12}6^2$ cages of the sI clathrate hydrate, within the MCTDH framework. The present study had two objectives. The first was to investigate the quantum T-R dynamics of the enclathrated CO_2 molecule on a semiempirical PES checking its quality by direct comparison with experimental X-ray results. The second, and indeed the computationally most challenging objective, was to compute specific vibrational states of the CO_2 in both small and large cages of the sI clathate structure, in order to compare with recent FTIR measurements on fundamental and combination stretching modes. We found that in the small cage the CO_2 molecule is located almost at its center, while in the large one its center of mass is displaced by at least 0.2 Å out off the center of the equatorial plane of the cage (see Fig. 1). This finding is consistent with data available from X-ray diffraction measurements. Also, we show that the quantum dynamics of the rotational and translational excited states of the CO_2 in small and large cages is remarkably different, with the density of states to be higher and the T-R levels to be highly coupled in the large cage, even at low-energies, than in small one. The specific T-R-V levels, corresponding to ν_1 and ν_3 excitations, were calculated for the three different $^{12}CO_2$, $^{13}CO_2$, and ^{18}OCO isotopes. We observe that the fundamental ν_1 and ν_3 frequencies of the CO_2 in the small and large cage are shifted by about 16 cm^{-1}, as well as their overtones and $\nu_1+\nu_3$ combination modes. This is in close agreement with the characteristic double-peak profiles observed in the IR spectra of the CO_2 clathrate hydrate at 5.6 K (see Fig. 1), supporting the CO_2 single occupation of small and large cavities of the sI clathrate. Such good agreement, with the data from recent spectroscopic studies of carbon dioxide clathrate hydrate at low temperatures, indicates a single occupancy of the sI structure cages, allowing an assessment of the theoretical approaches employed.

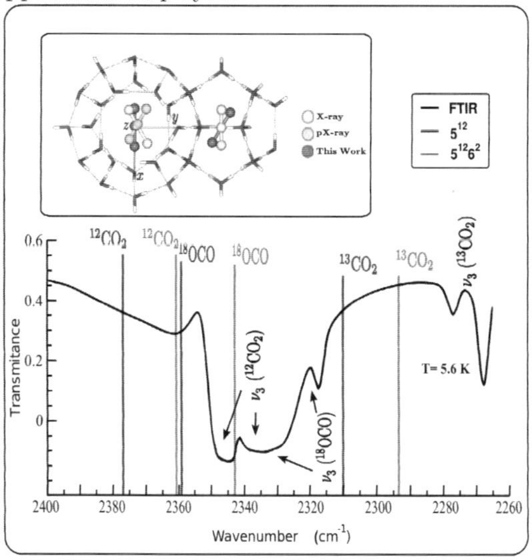

Figure 1. Comparison of the FTIR spectra with the 7D MCTDH vibrational energy levels for the indicated isotopes, and orientation of the CO_2 (see inset plot) in the small and large cages of the sI structure.

References

[1] S. Takeya *et al* 2010 *J. Am. Chem. Soc.* **132** 524

[2] E. Dartois *et al* 2009 *A&A* **504** 869

[3] Á. Valdés *et al* 2015 *J. Phys. Chem. C* **119** 3945

[1] E-mail: rita@iff.csic.es

Content from this work may be used under the terms of the Creative Commons Attribution 3.0 licence. Any further distribution of this work must maintain attribution to the author(s) and the title of the work, journal citation and DOI.
Published under licence by IOP Publishing Ltd

Threshold and Efficiency for Perforation of 1nm Thick Carbon Nano-membranes with Slow Highly Charged Ions

R.A. Wilhelm* [1], E. Gruber†, R. Ritter†, R. Heller*, S. Facsko*, F. Aumayr†

* Helmholtz-Zentrum Dresden-Rossendorf, Inst. of Ion Beam Physics and Materials Research, Bautzner Landstr. 400, 01328 Dresden, Germany
† Institute of Applied Physics, TU Wien, Wiedner Hauptstr. 8-10/E134, 1040 Vienna, Austria

Synopsis Slow highly charged ion induced nano-structure formation on 1nm thick carbon nano-membranes is investigated with respect to perforation yield and precise threshold determination.

Ion collision with surfaces leads to nano-structure formation usually by cumulative effects. Large ion fluencies are needed to lead to observable topographic surface features. For special classes of ions, i.e. swift heavy ions or slow highly charged ions (HCI) already single ion impacts may give rise to structural changes with a lateral extent of a few nm. Desired surface-only modification, however, can solely be achieved by HCI due to the very localized release of their potential energy.

Different kinds of surface structures have been observed on insulating [1] and semi-conducting surfaces [2]. All these studies have in common that a certain ion charge state or potential energy threshold has to be exceeded in order to form nano-hillocks, nano-pits, nano-craters, or even sub-surface damage only visible after chemical etching [3]. For high ion charge states the efficiency for nano-structure formation is considered to be 100%, i.e. every ion produces a structure upon impact. However, an accurate fluence measurement necessary to determine the structuring yield for HCI irradiation is challenging, because of very low ion currents (fA-range). Also it is an open question how well defined the potential energy threshold is.

Here we show that transmission of HCI through 1nm thick carbon nanomembranes (CNM) and subsequent determination of their charge state distribution and energy loss [4] allows us to measure the ion-induced nano-structure formation yield much more accurately. Since these insulating membranes are a surface-only material and HCI energy deposition is limited typically to the surface of a solid we consider this as a model system for insulating surfaces.

Figure 1 shows the efficiency for pore formation by single HCI impact as function of the incident charge state. It is obvious that a threshold exists at about $Q = 28$, where the yield increases from 0 to about 70%. For even higher charge states the efficiency increases to about 100% for $Q = 40$. Thus, only for sufficiently highly charged ions the structure formation yield may be considered to be 100%.

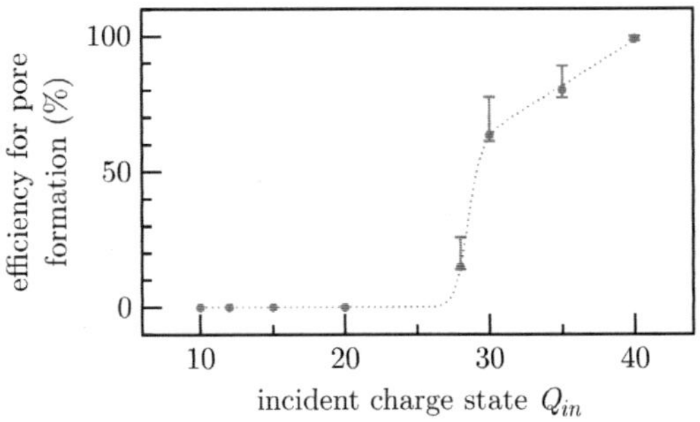

Figure 1. Pore formation yield due to single ion impacts as function of incident charge state. The kinetic ion energy is in all cases 4.4kV x Q.

References

[1] Aumayr, F. *et al.* 2011 *J. Phys. Condens. Matter* **23**, 393001
[2] Tona, M. *et al.* 2007 *Surf. Sci.* **601**, 723–727
[3] El-Said, A. S. *et al.* 2012 *Phys. Rev. Lett.* **109**, 117602
[4] Wilhelm, R.A. *et al.* 2014 *Phys. Rev. Lett.* **112**, 153201

[1] E-mail: r.wilhelm@hzdr.de

Content from this work may be used under the terms of the Creative Commons Attribution 3.0 licence. Any further distribution of this work must maintain attribution to the author(s) and the title of the work, journal citation and DOI.
Published under licence by IOP Publishing Ltd

XXIX International Conference on Photonic, Electronic, and Atomic Collisions (ICPEAC2015) IOP Publishing
Journal of Physics: Conference Series **635** (2015) 032012 doi:10.1088/1742-6596/635/3/032012

Six-dimensional theoretical study of H_2 scattering from LiF(001): From thermal to high incidence energies

A. S. Muzas*, **C. Díaz*** [1], **F. Martín***†

* Departamento de Química, Módulo 13, Universidad Autónoma de Madrid, Cantoblanco 28049, Madrid, Spain

† Instituto Madrileño de Estudios Avanzado en Nanociencia (IMDEA-Nanociencia), Cantoblanco 28049, Madrid, Spain

Synopsis We have computed an accurate potential energy surface (PES) describing the electronic structure of the system H2/LiF(001). Based on this PES, we have performed an analysis of the dynamics of the system trying to understand recent experimental results obtained at fast grazing incidence conditions.

Interaction of molecules with insulating surfaces has received little attention in surface science, mainly due to the fact that surface scientists have been very much focused on systems and phenomena relative to heterogeneous catalysis. However, recent experiments showing diffraction of atoms and light molecules under fast (100 eV-1 keV) grazing incidence (0.5°-2°) conditions upon scattering from insulating surfaces [1], have renewed the interest on this kind of surfaces.

We have focused on molecular projectiles because diffraction patterns are expected to be richer, due to the internal degree of freedom. In this case, theory is called to pay a key role in understanding experimental results. And therefore, accurate 6D PES describing the electronic structure, and accurate dynamics methods describing the dynamics of the system are required.

Here, we present, to our knowledge, the first 6D PES describing the electronic interaction between a diatomic molecule and an insulating surface ($H_2(D_2)$/LiF(001)), based on density functional theory calculations [2]. In order to test the accuracy of our PES, we have carried out dynamics calculations of elastic and inelastic scattering at low incidence energy (< 1 eV), and we have compared our results with previous experimental and theoretical ones [3, 4]. From the excellent agreement obtained, we have concluded that our PES is accurate enough to perform a complete analysis of the dynamics of the system, from very low to very high incidence energy. In fact, our first results at fast grazing incidence already show that the tools we have used are able to reproduce experimental results at these extreme conditions (see Fig.)

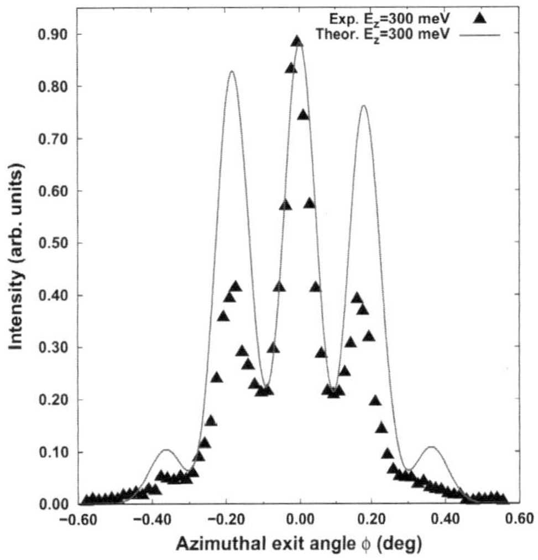

Figure 1. Experimental and simulated diffraction spectrum of H_2 scattered from LiF(001), along the incidence direction [100]. Experimental data from [5].

[1] E-mail: cristina.diaz@uam.es

References

[1] H. Winter *at al* 201 *Prog. Surf. Sci.* **86**, 169.

[2] A. S. Muzas *at al to be published.*

[3] E. Pijper *et al* 1998 *Phys. Rev. Lett.* 80, 488.

[4] M. F. Bertino *et al* 1998 *Phys. Rev. Lett.* 81, 5608.

[5] P. Rousseau *et al* 2008 *J. Phys. Conf.: Series* 133, 012013.

Content from this work may be used under the terms of the Creative Commons Attribution 3.0 licence. Any further distribution of this work must maintain attribution to the author(s) and the title of the work, journal citation and DOI.
Published under licence by IOP Publishing Ltd

Energy-loss contribution for fast He atoms axially scattered off a silver surface

C.A. Ríos Rubiano*[1], G.A. Bocan[†], J.I. Juaristi[#,‡], and M.S. Gravielle*

[*] Instituto de Astronomía y Física del Espacio (IAFE, CONICET-UBA), Buenos Aires, Argentina
[†] Centro Atómico Bariloche and CONICET, S.C. de Bariloche, Río Negro, Argentina
[#] Depto. de Física de Materiales, Facultad de Químicas, UPV/EHU, 20018 San Sebastian, Spain
[‡] Donostia International Physics Center (DIPC) and Centro de Física de Materiales CFM/MPC (CSIC-UPV/EHU), 20018 San Sebastian, Spain

Synopsis Angle- and energy-loss- resolved distributions for He atoms axially scattered off a Ag(110) surface are calculated in order to study the influence of inelastic processes on fast atom diffraction (FAD) spectra. Final projectile distributions are evaluated within a semi-classical formalism that includes dissipative effects due to electronic excitations. For monoenergetic beams the model predicts the presence of multiple peaks in energy-loss spectra. But these structures disappear when the experimental conditions are included, giving rise to an energy-loss distribution in good agreement with available experimental data. The inelastic contribution to FAD patterns is estimated.

Grazing-incidence fast-atom diffraction (GIFAD or FAD) on crystal surfaces has recently emerged as a powerful surface analysis method that allows one to probe surface interactions with an exceptional sensitivity. It is possible to observe FAD patterns for a wide variety of materials, even for metals where inelastic electronic processes are expected to play an important role against coherence. The aim of this work is to study the energy lost by swift He atoms after grazingly impinging on a Ag(110) surface, under the same conditions for which FAD has been observed [1].

We build the three-dimensional potential energy surface and charge density of the He-Ag(110) system from DFT calculations using the Quantum Espresso code. Angle- and energy-loss-resolved distributions of inelastically scattered projectiles are evaluated within a semi-classical formalism that includes dissipative effects due to electron-hole excitations through a friction force [2].

For incidence along different axial channels with a given impact energy, the model predicts the presence of multiple maxima in the energy-loss spectrum [3]. These structures are related to trajectories that probe different regions of the surface electron density. Such energy-loss peaks are however completely washed out when the experimental dispersion of the incident beam is taken into account, as shown in Fig. 1.

Present results (Fig. 2) suggest that the incoherent electronic excitations do not affect the relative intensities of internal FAD maxima, even though the total inelastic contribution to the momentum distribution is estimated around 75% for incidence along the [1-10] channel [3].

E-mail: cariosr@iafe.uba.ar

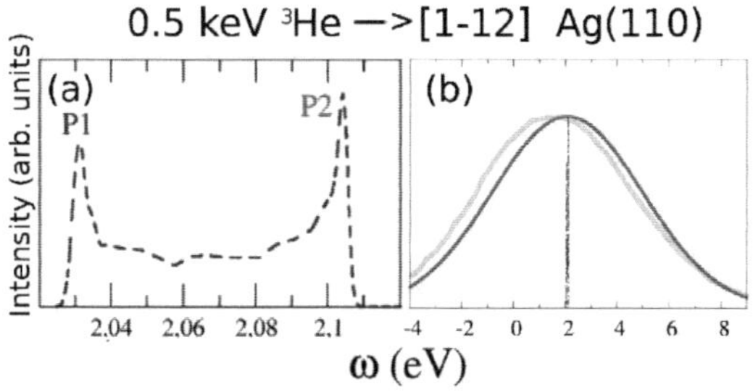

Figure 1. Energy-loss spectrum, as a function of the lost energy ω: (a) blue dashed line, results for a monoenergetic incident beam; (b) red solid line, results including the experimental energy spread; gray solid line, experimental data from Ref. [1].

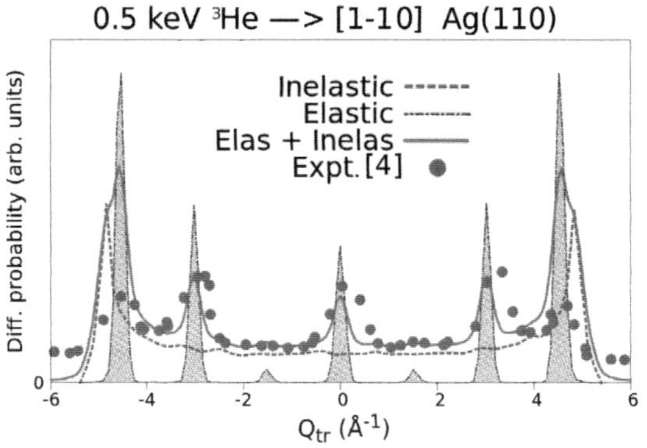

Figure 2. Momentum distribution, as a function of the transverse transferred momentum Q_{tr}, for the incidence angle $\theta_i = 0.75^{\circ}$.

References

[1] Bundaleski *et al.* 2011 *NIMB* **269** 1216
[2] J.I. Juaristi *et al.* 2008 *Phys. Rev. Lett.* **100** 116102
[3] C. Ríos Rubiano *et al.* 2013 *Phys. Rev. A* **87** 012903; idem. 2014 *NIMB* **340** 15
[4] Bundaleski *et al.* 2008 *Phys. Rev. Lett.* **101** 177601

Content from this work may be used under the terms of the Creative Commons Attribution 3.0 licence. Any further distribution of this work must maintain attribution to the author(s) and the title of the work, journal citation and DOI.
Published under licence by IOP Publishing Ltd

XXIX International Conference on Photonic, Electronic, and Atomic Collisions (ICPEAC2015) IOP Publishing
Journal of Physics: Conference Series **635** (2015) 032014 doi:10.1088/1742-6596/635/3/032014

Attachment of Water and Alcohol Molecules onto Water and Alcohol Clusters

Isabelle Braud* † [1], Julien Boulon* †, Sébastien Zamith * †, and Jean-Marc L'Hermite * †

* Laboratoire Collisions Agrégats Réactivité, IRSAMC, Université de Toulouse, UPS, F-31062 Toulouse, France

† CNRS, UMR 5589, F-31062 Toulouse, France

Synopsis We measured the absolute attachment cross-sections of water and alcohol molecules colliding with protonated water and alcohol clusters. Dynamical arguments are used to explain the lower than geometrical cross-sections at small sizes.

By studying the collision between clusters and molecules we can have a better understanding of the growth (nucleation) process. We focused on H-bonded systems such as water, methanol and ethanol that are solvent involved in many physical and chemical processes, in particular as nucleation seeds in atmospheric physics [1].

In the present study we report the measurement of absolute attachment cross-sections of single molecules M (M = water, methanol, or ethanol) onto positively charged mass-selected clusters X_nH^+ (X = water, methanol, or ethanol). Cluster sizes range from tens to hundreds of molecules and the center-of-mass collision energies vary from 0.1 to $\sim$ 1 eV [2].

The attachment cross-sections converge towards the geometrical (hard-sphere) cross-sections at large cluster sizes, whereas they are noticeably lower for small sizes as shown in figure 1.

For homogeneous attachment reactions $X_nH^+ + X \rightarrow X_{n+1}H^+$, this decrease in the cross-section is well explained by a dynamical collisional model. Indeed when the collision time is short compared to the main vibrational period of the cluster's surface vibration mode, the cluster cannot absorb all the collision energy. From our previous results on water [3] and using a mass-scaling, the experimental results are well reproduced for both methanol and ethanol clusters.

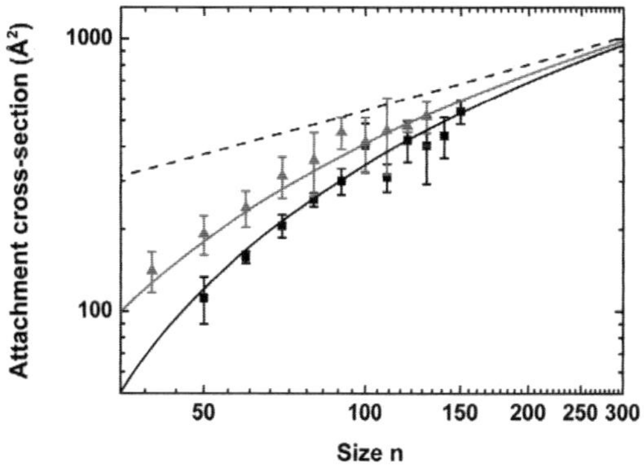

Figure 1. Log-log plot of the attachment cross-sections for methanol molecules onto protonated methanol clusters for two kinetic energies in the laboratory frame [2]. Red triangles: 12 eV. Black squares: 22 eV. Dashed line: geometrical cross-section. Full lines: predictions from the dynamical model.

For inhomogeneous attachment reactions $X_nH^+ + Y \rightarrow X_nYH^+$ (X, Y = water, ethanol, or methanol), however, this simple dynamical model does not satisfactorily account for the measured data, although the global size dependences are qualitatively reproduced.

References

[1] J. Lengyel *et al* 2012 *J Chem Phys* **137** 034304

[2] I. Braud *et al* 2015 *J Phys Chem A* doi: 10.1021/jp511854r

[3] S. Zamith *et al* 2010 *J Chem Phys* **133** 154305

[1]E-mail: isabelle.braud@irsamc.ups-tlse.fr

Content from this work may be used under the terms of the Creative Commons Attribution 3.0 licence. Any further distribution of this work must maintain attribution to the author(s) and the title of the work, journal citation and DOI.
Published under licence by IOP Publishing Ltd

Journal of Physics: Conference Series **635** (2015) 032015 doi:10.1088/1742-6596/635/3/032015

Experimental and theoretical studies of 4.5-keV Ar^{7+} guided through a conical glass macrocapillary

N. Stolterfoht[1], E. Gruber[2], P. Allinger[2], S. Wampl[2], Y. Wang[2,3], M. J. Simon[4] and F. Aumayr[2]

[1]Helmholtz-Zentrum Berlin für Materialien und Energie, 14109 Berlin, Germany
[2]Institute of Applied Physics, TU Wien, 1040 Vienna, Austria
[3]Institute of Modern Physics, Chinese Academy of Sciences, Lanzhou, 730000, Gansu, China
[4]Labor für Ionenstrahlphysik, ETH Zürich, 8093 Zürich, Switzerland

Synopsis Experimental and theoretical work was performed to investigate the guiding of 4.5- keV Ar^{7+} ions through a conical macrocapillary. The bulk conductivity was varied by changing the capillary temperature from 24 – 110 °C. A minimum in the transmitted ion intensity was experimentally observed at forward direction, which was reproduced and interpreted by the simulations.

The guidance of keV ions through insulating nanocapillaries [1] has received considerable attention during the past decade [2]. Ions deposited at the capillary wall produce a repulsive electric field that is capable to deflect the following ions at relatively large distances. Thus, the ions are guided along the capillary axis maintaining their incident charge state. Ion guiding is based on a self-organizing process, which governs the charge deposition and ion transmission through the capillaries.

In this work, experimental and theoretical studies were performed to investigate guiding of 4.5-keV Ar^{7+} ions through a conical glass macrocapillary [3]. In the experiments we used a capillary whose bulk conductivity could be varied by changing its temperature from 24 – 110 °C [4]. The tilt angle of the capillary axis relative to the incident beam was varied within 0° - 2°.

As shown in Fig. 1, a minimum in the transmitted ion intensity was observed at forward direction for lower temperatures. After strongly increasing the electrical conductivity of the capillary by increasing its temperature to 110 °C, all deposited charges are depleted from the capillary surface and the transmission profile becomes geometrical without a forward minimum.

The experimental data were compared with theoretical results, which are based on simulations previously developed for a straight macrocapillary [5]. Both the surface and bulk conductivities were implemented into the calculations providing clear evidence that the bulk conductivity is dominant. The calculated charge distributions show a maximum near the capillary exit and a minimum just before that. Evidently this overfocusing effect provides explanation for the minima in the forward direction, observed experimentally and theoretically.

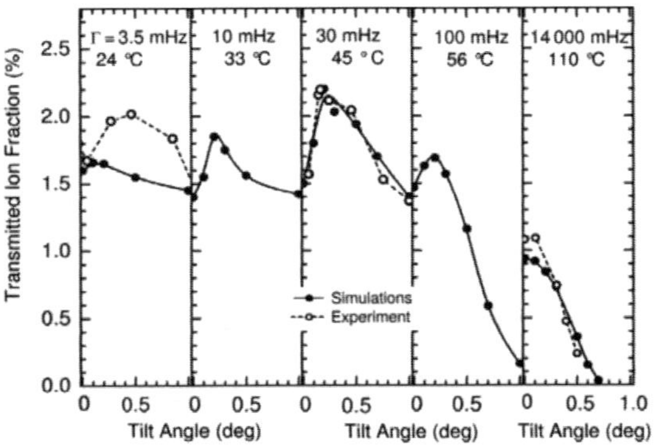

Figure 1. Transmitted ion fraction obtained in experiments (open circles) and simulations (full circles). Discharge rates are 3.5, 10, 30, 100, and 14000 mHz as indicated. The corresponding temperatures are also shown.

It should be emphasized that this overfocusing involves deflection angles significantly smaller than 1°. Thus, the scenario that the ions are fully blocked due to a repelling electrostatic field is inapplicable. The maximum potential created in the capillary is about 10 V. which is by far smaller than that required for a complete electrostatic blocking of keV ions.

References

[1] N. Stolterfoht *et al* 2002 Phys. Rev. Lett. **88** 133201
[2] C. Lemell, J. Burgdörfer, and F. Aumayr 2013 Progr. Surface Science **88** 237
[3] N. Stolterfoht *et al* 2015 Phys. Rev. A **91** 032705
[4] E. Gruber *et al.* 2012 Phys. Rev. A **86** 062901
[5] N. Stolterfoht 2014 Phys. Rev. A **89** 062706

E-mail: nico@stolterfoht.com

Content from this work may be used under the terms of the Creative Commons Attribution 3.0 licence. Any further distribution of this work must maintain attribution to the author(s) and the title of the work, journal citation and DOI.
Published under licence by IOP Publishing Ltd

Enhancing the energy resolution of resonant coherent excitation using the cooled U^{89+} beam extracted from the ESR

Y. Nakano[A 1], **A. Ananyeva**[B,C], **S. Menk**[A], **A. Bräuning-Demian**[B], **H. Bräuning**[B], **C. leffner**[B], **Th. Stöhlker**[B,D,E], **T. Azuma**[A,F]

[A]Atomic, Molecular & Optical Physics Laboratory, RIKEN , Saitama 351-0198, Japan
[B]GSI Helmholtzzentrum für Schwerionenforschung, D-64291 Darmstadt, Germany
[C]Johann Wolfgang Goethe Universität, Frankfurt, Germany, 60438 Frankfurt am Main, Germany
[D]Helmholtz-Institut Jena, D-07743 Jena, Germany
[E]Friedrich-Schiller University Jena, 07737 Jena, Germany
[F]Department of Physics, Tokyo Metropolitan University, Tokyo 192-0397, Japan

Synopsis We report on the resonant coherent excitation (RCE) of the $2s$-$2p_{3/2}$ transition in Li-like U^{89+} with an enhanced energy resolution, which was achieved by reducing the projectiles momentum spread. The kinetic temperature of the beam was decreased by electron cooling in the ESR, and the collisional momentum broadening in the target was suppressed by the use of thin crystal (1.0 and 2.5 μm-thick). The resonance width was observed to be ~1.4 eV in FWHM, which is three-times narrower than that from the previous work.

Energetic ions traveling through a crystalline target can be resonantly excited through the interaction with the periodic crystal eld. This process is called resonant coherent excitation (RCE), and expected to be a versatile tool in the atomic physics experiments in the short wavelength regime. Towards the high-resolution spectroscopy of highly-charged heavy ions, we have recently demonstrated the experimental proof-of-principle of the RCE of high-Z ions using a 7 μm-thick Si crystal and the 190 MeV/u Li-like U^{89+} beam from the SIS at GSI [1]. The RCE of $2s$-$2p_{3/2}$ transition was observed under the (220) planar channeling condition at the transition energy of 4459 eV, and the resonance width (FWHM) was 4.4 eV (Figure 1(a)) . The resonance width was predominantly limited by the momentum spread of the beam, which eventually determines the energy resolution of the RCE-based spectroscopy.

In autumn 2014, we carried out an upgraded experiment dedicated to the reduction of the momentum spread of the beam. Instead of getting the U^{89+} beam directly from the SIS, an U^{90+} beam from SIS was transported to the storage ring ESR. The beam was cooled by the electron cooler in the ESR, and slowly extracted as continuous U^{89+} beam using the electron capture at the cooler. The crystal thickness was also reduced to 1.0 and 2.5 μm to suppress the momentum spread caused by collisions in the crystal.

Figure 1(b) shows the $2s$-$2p_{3/2}$ RCE spectra observed in this work, exhibiting a sharp resonance width of ~1.4 eV. The successful reduction of the beam momentum spread led to the improvement of the energy resolution of RCE by a factor of 3. Moreover, the resonance shape observed in this work can be well tted to a Lorentzian curve as shown in the solid curve in Figure 1(b). This allows the determination of the peak position with a signi cantly better accuracy.

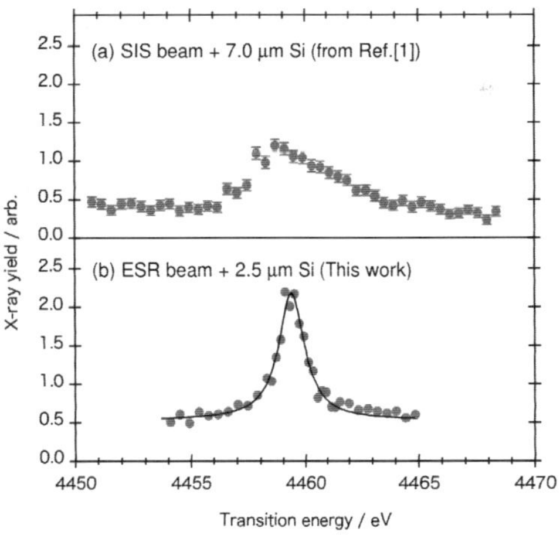

Figure 1. The RCE spectra of $2s$-$2p_{3/2}$ transition in the 190 MeV/u Li-like U^{89+} beam from (a) SIS and (b) ESR .

References

[1] Y. Nakano, Y. Takano, T. Ikeda, Y. Kanai, S. Suda, T. Azuma, H. Bräuning, A. Bräuning-Demian, D. Dauvergne, Th. Stöhlker and Y. Yamazaki, Phys. Rev. A **87** (2013) 060501(R).

[1]E-mail: nakano-y@riken.jp

Content from this work may be used under the terms of the Creative Commons Attribution 3.0 licence. Any further distribution of this work must maintain attribution to the author(s) and the title of the work, journal citation and DOI.
Published under licence by IOP Publishing Ltd

XXIX International Conference on Photonic, Electronic, and Atomic Collisions (ICPEAC2015) IOP Publishing
Journal of Physics: Conference Series **635** (2015) 032017 doi:10.1088/1742-6596/635/3/032017

New mechanism of ion desorption from rare gas solids by multiply-charged ion impact

eisuke Ban, Masaki Akiwa, Atsumi Sumita, Yuta Nidaira, Yuki uwabara, Takayuki Tachibana, and Takato Hirayama[1]

Department of Physics, Rikkyo University, Toshima-ku, Tokyo 171-8501 JAPAN

Synopsis We have measured the absolute ion desorption yield from the surface of solid Ne and Ar by low-energy multiply-charged ion impact. Present results could not be explained by well-known Coulomb explosion model. We here propose a new mechanism of ion desorption.

Potential sputtering is the desorption mechanism which is induced by transfer of the potential energy from incident ions to the solid surface. We have reported previously the relative ion yields of potential sputtering from solid Ne [1], and demonstrated that the sputtering ion yields are proportional to the potential energy of the incident ion. Here, we report the experimental results of the *absolute* sputtering yields, and propose a new mechanism of ion sputtering.

Figure 1 shows the absolute sputtering ion yields as a function of the potential energy of the incident ion. Incident ions are Ne^{q+}, Ar^{q+}, and Kr^{q+}, and the target is solid Ne (circles) and solid Ar (triangles). The thickness of the marks denotes the kinetic energy of the incident ions (200 2000 eV), the thicker color, the higher energy. It is clearly shown that the sputtering ion yields linearly increase with the potential energy of the incident ion, which is consistent with our previous work [1]. It is found that our results cannot be explained by the Coulomb explosion model, where the ion sputtering yields should be proportional to the number of ion pairs created in the solid.

In order to explain our results, we propose a new desorption mechanism, where ions are created in the solid by the charge transfer with the incident multiply-charged ions, and part of them desorb with surrounding neutral atoms. In this model the ion sputtering yield Y_{ion} should depend on a scaling parameter $\Gamma \equiv (\text{PE}/\text{IP}) \cdot \sigma \cdot Y_{\text{total}}$, where PE and IP are the potential energy of the incident ions, and the ionization energy of the solid atom, and σ and Y_{total} are the charge transfer cross section [2] and total desorption yields [3], respectively.

Figure 2 shows the scaling plot based on our new model. As can be seen in the figure, our model well describes the experimental results regardless of the incident ion species, incident energy, charge state, or target material. Detailed discussion will be given at the conference.

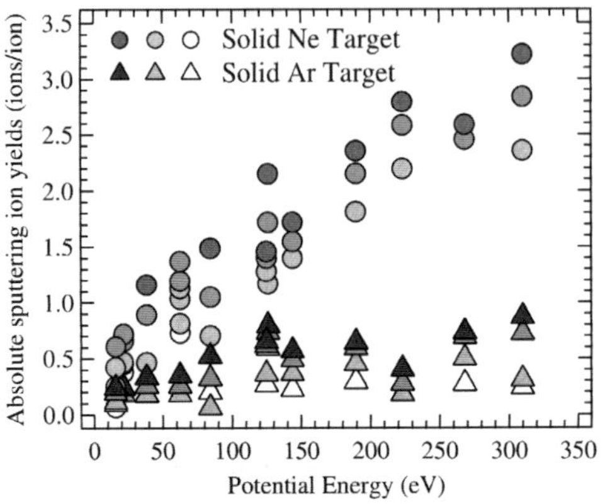

Figure 1. Absolute sputtering ion yields as a function of the potential energy of the incident ion. Incident ions are Ne^{q+}, Ar^{q+} and Kr^{q+}, and the target is solid Ne (circles) and solid Ar (triangles).

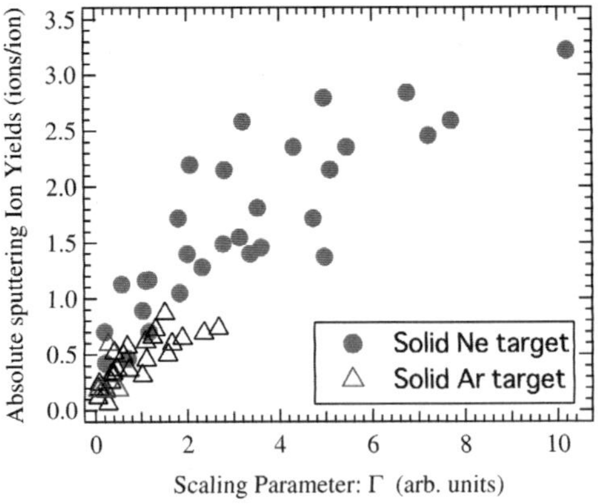

Figure 2. Scaling plot based on our new model.

References

[1] K. Fukai et al., 2010 *J. Phys. Cond. Matt.* **22**, 084007

[2] C. Cocke et al., 1981 *Phys. Rev. Lett.* **46**, 1671

[3] S. Fujita et al., 2009 *J. Phys. Conf. Ser.* **163**, 012083

[1]E-mail: hirayama@rikkyo.ac.jp

Content from this work may be used under the terms of the Creative Commons Attribution 3.0 licence. Any further distribution of this work must maintain attribution to the author(s) and the title of the work, journal citation and DOI.
Published under licence by IOP Publishing Ltd

Journal of Physics: Conference Series **635** (2015) 032018 doi:10.1088/1742-6596/635/3/032018

Positive and negative ion emission from microdroplets by MeV energy ions

T. Majima* [1], K. Kitajima†, Y. Oonishi†, H. Ueda†, H. Tsuchida*† and A. Itoh*†

* Quantum Science and Engineering Center, Kyoto University, Uji 611-0011, Japan
† Department of Nuclear Engineering, Kyoto University, Kyoto 615-8540, Japan

Synopsis We have developed a new experimental setup that allowed us to study collision interactions between fast ion beams and liquid droplets under a vacuum condition. Droplets of water and ethanol are irradiated with 0.4-1.5 MeV H^+ and 2.0 MeV C^{2+} ions. The droplet diameter is estimated from energy loss measurements of projectile ions penetrating through droplets. Time-of-flight mass spectra of positive and negative secondary ions exhibit a series of cluster ions generated via protonation and deprotonation.

Study on the interaction of fast ions in liquids is important to understand the initial physicochemical processes of radiation effects in living cells. Conventional experiments using liquid targets were limited mostly to irradiation under atmospheric pressure because of high vapor pressure in vacuum. Our group already succeeded in ion irradiation of liquid targets in vacuum by applying a liquid-jet technique and reported mass-spectrometric studies of secondary ions from liquid-jet surfaces [1]. In the present work, we have developed a new experimental setup using "liquid droplets" as an additional option to perform collision experiments between energetic ions and liquid targets under a higher vacuum condition. We obtained mass spectra of positive and negative secondary ions from water and ethanol microdroplets.

The experiment was performed by using the 2 MV tandem Pelletron accelerator at Quantum Science and Engineering Center, Kyoto University. Droplets of water or ethanol produced in atmosphere were transferred into a vacuum chamber through a differential pumping system and crossed with an ion beam of 0.4–2.0 MeV H^+ or 2.0 MeV C^{2+} ions from the accelerator. The collision chamber was kept the pressure around 10^{-4}–10^{-3} Pa during the measurements. To deduce size distributions of the droplets at the collision region, distributions of the energy loss of incident ions passing through water droplets were measured. As a result, the droplet sizes were found to be distributed around a few micrometers or less in diameter. Positive and negative secondary ions ejected from droplet surfaces induced by collisions were analyzed by time-of-flight mass spectrometry.

Figure 2 shows an example of mass spectra of negative secondary ions from ethanol droplets induced by 2.0 MeV C^{2+} ions. We observe fragment ions generated by deprotonation ($C_2H_5O^-$) and additional loss of two or four hydrogen atoms ($C_2H_3O^-$, C_2HO^-). Negative cluster ions which are also generated by deprotonation and additional loss of hydrogen atoms, $(EtOH)_n(C_2H_mO^-)$ (m=5,3,1), are observed up to n ~ 10, although the three peaks with different number of hydrogen atoms (m) are not resolved at higher n due to a limited mass resolution. It was found that negative cluster ions seem to be produced more effectively from droplet targets than from an ice target. This might be due to different charge delocalization time scale depending on the phase and/or temperature.

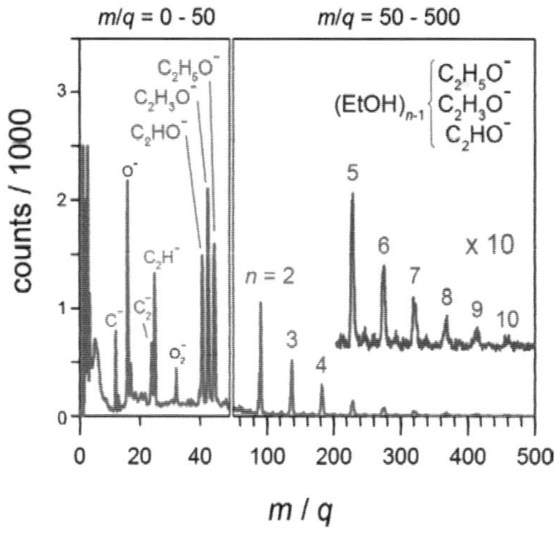

Figure 1. Mass spectrum of negative secondary ions from ethanol droplets induced by 2.0 MeV C^{2+}.

One of the authors (T.M.) would like to acknowledge the support of Toyota Physical & Chemical Research Institute Scholars.

References

[1] M. Kaneda *et al* 2010 *J. Chem. Phys.* **132** 144502.

[1] E-mail: majima@nucleng.kyoto-u.ac.jp

Content from this work may be used under the terms of the Creative Commons Attribution 3.0 licence. Any further distribution of this work must maintain attribution to the author(s) and the title of the work, journal citation and DOI.
Published under licence by IOP Publishing Ltd

Journal of Physics: Conference Series **635** (2015) 032019 doi:10.1088/1742-6596/635/3/032019

Formation of HCN^+ in heterogeneous surface reactions

Martina Harnisch*[1], Nikolaus Weinberger*, Paul Scheier* [2], Zdenek Herman† [3]

* Institute for Ion Physics and Applied Physics, University of Innsbruck, Technikerstr. 25, 6020 Innsbruck, Austria

† J. Heyrovsky Institute of Physical Chemistry, Academy of Sciences of the Czech Republic, Dolejskova 3, 18223 Prague 8, Czech Republic

Synopsis Mass spectrometric studies of ion-surface collisions between low-energy (20-100 eV) ions of N_2^+, N^+ [1] and NH^+ and hydrocarbon-covered room-temperature and heated surfaces of tungsten, carbon-fiber-composite, and beryllium showed significant increase of the ion yield at m/z 27 that was not observed in analogous collisions of Ar^+. This increase is ascribed to formation of HCN^+ in heterogeneous reactions between the nitrogen-containing projectile ions and the surface hydrocarbons.

Motivation

The atmosphere of Titan is composed mostly of molecular nitrogen (above 95%) with 2-5% of methane [2] and small amounts of more complex hydrocarbons, traces of nitriles and oxygen-containing compounds. Data from the Cassini-Huygens mission to Titan revealed a rich chemical composition of the ionosphere [3] and led to emphasis of the importance of nitrogen chemistry in the upper atmosphere. While data on gas phase processes relevant to the atmosphere of Titan are plentiful, information on possible heterogeneous processes of ions and neutrals on surfaces is scarce. To our best knowledge this is the first experimental report on a heterogeneous ion-surface reaction that could have some relevance to the subject.

Experimental

The experimental setup BESTOF is described in detail in previous publications [4]. It consists of two mass spectrometers arranged in tandem geometry. A beam of m/z-selected projectile ions is focused on a heatable surface sample and interacts with the target at well-defined incident energies ranging from 15 eV up to 100 eV. Ions produced during this interaction are analyzed with a linear time-of-flight mass spectrometer.

Details of the experimental setup of the He-droplet experiment can be found in [5]. In this experiment doped helium droplets are ionized by electron bombardment and the resulting positively or negatively charged clusters are analyzed by a time-of-flight mass spectrometer.

Results

Bombardment of hydrocarbon-covered surfaces with nitrogen-containing projectile ions shows chemical reactions forming HCN^+. The hydrocarbon fragments on the surface had a chain length of about C_8 and could be compared to a self-assembled hydrocarbon monolayer in their collisional behavior [6]. While the formation of HCN^+ in the reaction with N^+ is an exothermic reaction with no activation barrier, the reaction with N_2^+ was only observed at incident energies above 50 eV suggesting an activation barrier or reaction endothermicity of the heterogeneous reaction of about $3 - 3.5$ eV [1].

In a complementary experiment we aimed at investigating another mechanism to create a CN-bond under low-energy and low-temperature conditions. Helium nano droplets provide an ultracold environment (0.37 K) which represents an ideal inert medium to study chemical reactions, also if the temperature lies below the one of Titan as explained in [7]. He droplets doped with CH_4 and N_2 showed preferential formation of a CN^--group attached to different cluster species that were formed.

References

[1] M. Harnisch *et al.*, 2013 *J. Phys. Chem. A* **117** 9653

[2] M.G. Tomasko *et al.* 2005 *Nature* **438** 765

[3] T.E. Cravens *et al.*, *et al.* 2006 *Geophys. Res. Lett.* **33** L07105

[4] A. Keim *et al.*, 2011 *Int. J. Mass. Spectrom.*, **306** 204

[5] L. An der Lan *et al.* 2011 *J. Chem. Phys.* **135** 044309

[6] J. Zabka *et al.*, 2002 *J.Phys.Chem. A*, **106** 10861

[7] S. Ptasinska *et al.* 2012 *RSC Advances* **2** 10492

[1]E-mail: Martina.Harnisch@uibk.ac.at
[2]E-mail: Paul.Scheier@uibk.ac.at
[3]E-mail: zdenek.herman@jh-inst.cas.cz

Content from this work may be used under the terms of the Creative Commons Attribution 3.0 licence. Any further distribution of this work must maintain attribution to the author(s) and the title of the work, journal citation and DOI.
Published under licence by IOP Publishing Ltd

Journal of Physics: Conference Series **635** (2015) 032021 doi:10.1088/1742-6596/635/3/032021

Impacts of highly charged ions as seeds in a magneto-structural phase transition of magnetocaloric thin films

S. Cervera[1], M. Trassinelli[2], M. Marangolo, L. Bernard-Carlsson, M. Eddrief, V.H. Etgens, V. Gafton, S. Hidki, E. Lacaze, E. Lamour, C. Prigent, J.-P. Rozet, S. Steydli, Y. Zheng and D. Vernhet

CNRS, UMR 7588, Institut des NanoSciences de Paris (INSP), 4 Place Jussieu, 75005 Paris, France
Sorbonne Universités, UPMC Univ. Paris 06, INSP, UMR 7588, F-75005 Paris, France,

Synopsis Investigation on modifications of structural and magnetic properties of magnetocaloric thin films induced by slow highly charged ions bombardment under well-controlled conditions is presented. The ions induce defects/constraints that facilitate nucleation of one phase with respect to the other in the first-order magneto-structural MnAs, with a consequent suppression of thermal hysteresis, but without any significant perturbation on the other structural and magnetic properties.

Modifications of material properties by ion impact, and in particular changes in magnetic features, have been extensively studied in the last decades. Nevertheless, previous studies on the effect of ion impact were focused mostly on materials exhibiting a second-order magnetic transition and with singly charged ions whilst, here, we investigate thin film presenting a first-order magneto-structural transition, namely MnAs, irradiated with slow highly charged ions, i.e. 90 keV (v=0.4 a.u.) Ne^{9+} [1, 2]. Indeed, MnAs is one of the most promising candidates for developing magnetic refrigeration since it exhibits giant magnetocaloric effect (GMCE) associated to its phase transition close to room temperature (at T_C=40°). Up to now, the practical use of GMCE materials for real refrigerator systems is blocked by the presence of a large thermal hysteresis in the magnetization cycle, which is typical of first-order transition materials (fig. 1, solid lines). Differently from other technics applied previously to reduce the thermal hysteresis (like doping, external strains, ...) [3–5], ion collisions induce defects that act as seeds for the nucleation of one phase with the other during the transition. Consequently, the thermal hysteresis is entirely suppressed (fig. 1, dashed lines) whereas other structural and magnetic properties are only slightly affected. In particular, the large refrigeration power of MnAs thin films related to GMCE is preserved. At present, we extend our investigations using different collision conditions and other samples that are relevant for the magnetic refrigeration (FeRh, compounds based on Fe_2P, etc.). The goal is to provide new insights on the fundamental processes that play a major role in the suppression of the thermal hysteresis in order to get a full understanding on this effect.

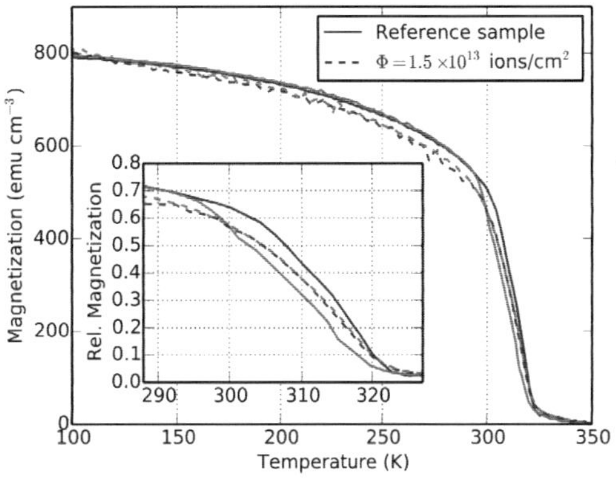

Figure 1. (color online) Magnetization as a function of temperature for the reference (solid lines) and for the irradiated samples (dashed lines). Data obtained by a temperature increase (from colder temperatures) and decrease (from hotter temperatures) are presented in blue and red, respectively.

This work was partially supported by French state funds managed by the ANR within the Investissements d'Avenir programme under reference ANR-11-IDEX-0004-02, and more specifically within the framework of the Cluster of Excellence MATISSE led by Sorbonne Universités.

References

[1] M. Trassinelli, M. Marangolo, M. Eddrief *et al.*, Appl. Phys. Lett. **104**, 081906 (2014)
[2] M. Trassinelli, V.E. Gafton, M. Eddrief *et al.*, Nucl. Instrum. Meth. B, **317**, 154 (2013)
[3] B.G. Shen, J.R. Sun, F.X. Hu, H.W. Zhang, and Z.H. Cheng, Adv. Mater. **21**, 4545-4564 (2009)
[4] J. Liu, T. Gottschall, K.P. Skokov, J.D. Moore, and O. Gutfleisch, Nature Mater. **11**, 620-626 (2012)
[5] X. Moya, L.E. Hueso, F. Maccherozzi *et al.*, Nature Mater. **12**, 52-58 (2013)

[1]E-mail: sophie.cervera@insp.jussieu.fr
[2]E-mail: martino.trassinelli@insp.jussieu.fr

Content from this work may be used under the terms of the Creative Commons Attribution 3.0 licence. Any further distribution of this work must maintain attribution to the author(s) and the title of the work, journal citation and DOI.
Published under licence by IOP Publishing Ltd

Improvement of the ETACHA Code towards low velocities and many-electron ions

E. Lamour[* 1], P D Fainstein[#], M Galassi[¥], C Prigent[*], C A Ramirez[¥], R D Rivarola[¥], J.-P. Rozet[*], M. Trassinelli[*] and D. Vernhet[* 2]

[*] Sorbonne Universités, UPMC, Institut des NanoSciences de Paris (INSP), CNRS-UMR 7588, F-75005 Paris, France
[#] Centro Atómico Bariloche, CNEA, 8400 San Carlos de Bariloche (Río Negro), Argentina
[¥] Laboratorio de Colisiones Atómicas, Instituto de Física Rosario (CONICET-UNR) Rosario, Argentina

Synopsis Knowledge of the detailed evolution of the whole charge state distribution of projectile ions colliding with targets is required in several fields of research such as material science, atomic, plasma and nuclear physics, in particular in regards of the several foreseen large scale facilities. Starting from the previous ETACHA model [1], we present extension of its validity domain towards lower velocities and larger distortions. Moreover, the system of rate equations is now able to take into account ions with up to 60 electrons.

Ab initio calculations of charge state distributions of fast ions at the exit of solid targets are useful in many circumstances, as in the context of energy loss in matter and for the design, or analysis, of atomic, plasma or nuclear physics experiments. Several empirical or semi empirical laws can be used to predict the mean charge state at equilibrium [2], however, in many cases, the detailed evolution of the whole charge state distribution as a function of the solid target thickness or even the evolution of the $n\ell$ substate populations are needed. The initial version of the ETACHA code [1] intended to calculate charge state distributions of fast few-electron ions with at most 28 electrons. It was therefore based on calculations of cross sections for monoelectronic atomic collision processes in the independent electron approximation and at high velocities. However, in the present days, many experiments take place in the non-perturbative regime, where previously used Born1 type calculations break down. Moreover, more states have to be included.

Several improvements have been made. More refined calculations are performed for ionization and excitation, namely, the Continuum Distorted Wave-Eikonal Initial State and Symmetric Eikonal approaches respectively. In parallel, studying carefully the evolution of cross sections of projectile electron-gain, -loss and –excitation versus the principal quantum number n, we also extend the applicability of the code up to dressed ions with (in principle) 60 electrons, i.e. with full n=4 shell.

Comparison with experiments for systems such as 28.9 MeV/u Pb^{56+} ions in carbon (see Fig.1) shows

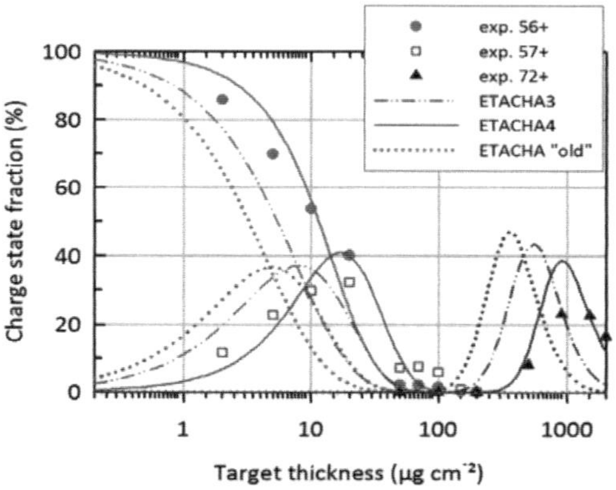

Figure 1. Evolution of selected charge states for 28.9 Mev u^{-1} Pb^{56+} with carbon target thickness. Symbols: experiment [2]; dotted lines: ETACHA "old" [1]; dashed-double-dotted lines: ETACHA3 (up to n=3); full lines: ETACHA4 (up to n=4).

very good agreement with the new fully *ab initio* calculations of the ETACHA code. It demonstrates that even for a projectile initially dressed with 26 electrons, i.e. Pb^{56+}, it is mandatory to properly take into account n=4, because here the excitation from n=3 to n=4 is as large as the ionization of n=3. This new version of the ETACHA code provides also accurate predictions for other systems like 0.05 – 0.5 MeV/u C ions in aluminum which is of interest in warm dense plasma experiments [3].

References

[1] Rozet J P, Stéphan C, and Vernhet D 1996 Nucl. Instrum. Meth. B **107** 67-70
[2] Leon A *et al.* 1998 Atomic Data and Nuclear Data Tables **69**, 217-238 (and references therein)
[3] Gauthier M *et al.* 2013 Phys. Rev. Lett. **110**, 13500

[1] E-mail: emily.lamour@insp.jussieu.fr
[2] E-mail: dominique.vernhet@insp.jussieu.fr

Content from this work may be used under the terms of the Creative Commons Attribution 3.0 licence. Any further distribution of this work must maintain attribution to the author(s) and the title of the work, journal citation and DOI.
Published under licence by IOP Publishing Ltd

XXIX International Conference on Photonic, Electronic, and Atomic Collisions (ICPEAC2015) IOP Publishing
Journal of Physics: Conference Series **635** (2015) 032023 doi:10.1088/1742-6596/635/3/032023

Cryogenic micro-calorimeters for mass spectrometry of keV neutral atoms and molecules

O Novotný*,† 1, S Allgeier‡, C Enss‡, A Fleischmann‡, L Gamer‡, D Hengstler‡, S Kempf‡, C Krantz*, A Pabinger‡, C Pies‡, D W Savin†, D Schwalm*,◇, A Wolf* 2

* Max Planck Institute for Nuclear Physics, D-69117 Heidelberg, Germany
† Columbia Astrophysics Laboratory, Columbia University, New York, NY 10027, USA
‡ Kirchhoff Institute for Physics, Heidelberg University, D-69120 Heidelberg, Germany
◇ Faculty of Physics, Weizmann Institute of Science, Rehovot 76100, Israel

Synopsis We demonstrate the capability of micro-calorimeters to detect and mass-resolve neutral atoms and molecules at $\sim$ keV energies, reaching single H-atom resolution.

Mass spectrometric identification of fragmentation products is one of the most important tools for molecular collision physics. Most molecular mass-spectrometry methods employ electromagnetic fields to determine the mass-to-charge ratio of ionized particles. Two major limitations arise from this approach. First, the ratios may be ambiguous, e.g., singly charged monomers are indistinguishable from doubly charged dimers. Second, neutral fragments cannot be directly analyzed and either a "neutral loss" must be accepted or an additional ionization stage must be used, which results in additional ambiguities.

As an alternative method, the mass of a neutral particle can be deduced from its kinetic energy if the particle velocity is known. This calorimetric approach has been successfully used in high-energy physics experiments by employing, for example, surface barrier detectors [1]. These detectors, however, are very restricted in their use at the $\sim$ keV energies relevant for low-energy collision experiments, as their energy resolution becomes poor and they often are insensitive to the low-energy incident particles.

A promising approach to solve the needs for low kinetic energy measurements is the use of cryo-detectors, such as micro-calorimeters. With these detectors the particle kinetic energy is transformed into heat and sensitive temperature monitoring then provides the measure of the kinetic energy deposited. The operation at $\sim$ 10 mK limits the thermal noise and thus enables resolving low measured energies.

We have systematically investigated the energy resolution of magnetic micro-calorimeters (MMC) [2] for atomic and molecular projectiles at impact energies ranging from $E \approx 13$ keV to 150 keV. For atoms we obtained relative energy resolutions down to $\Delta E/E \approx 10^{-3}$ and absolute energy resolutions down to $\Delta E \approx 120$ eV. We also studied in detail the MMC response function for molecular projectiles of masses up to 56 amu. We have demonstrated the capability of identifying neutral fragmentation products of these molecules by calorimetric mass spectrometry (see Figure 1) with single H-atom resolution. We have modeled the MMC response function for molecular projectiles and concluded that backscattering is the dominant source of the observed energy spread at the impact energies studied. Furthermore, we have successfully demonstrated the use of a detector absorber coating to minimize such spreads.

Our findings [3] are of general interest for mass spectrometric techniques with special advantage for neutral-particle mass measurements.

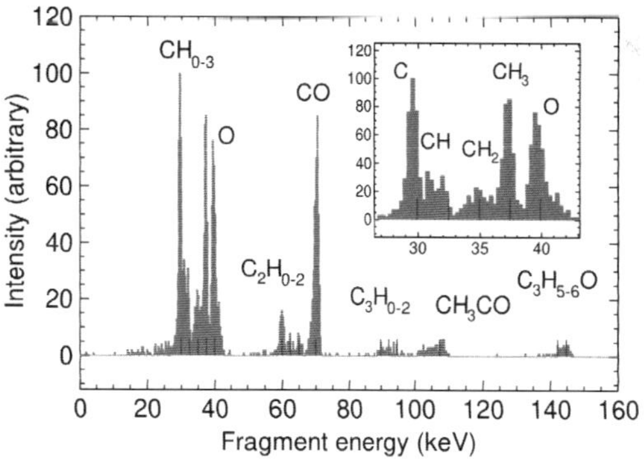

Figure 1. MMC energy spectrum of neutral products arising from collisions of 150 keV acetone radicals ($C_3H_6O^+$, 56 amu) with residual gas.

References

[1] H. Buhr *at al* 2010 *Phys. Rev. A* **81** 062702

[2] A. Fleischmann, C. Ens, and G.M. Seidel 2005 in *Cryoge c Part cle Detect o* pp. 151-216

[3] O. Novotný *et al*, in preparation

[1]E-mail: oldrich.novotny@mpi-hd.mpg.de
[2]E-mail: andreas.wolf@mpi-hd.mpg.de

Content from this work may be used under the terms of the Creative Commons Attribution 3.0 licence. Any further distribution of this work must maintain attribution to the author(s) and the title of the work, journal citation and DOI.
Published under licence by IOP Publishing Ltd

Journal of Physics: Conference Series **635** (2015) 032024 doi:10.1088/1742-6596/635/3/032024

Interference in the plasmon field excited by a diatomic molecule on metallic cylindrical nanostructures

S. Segui*‡ [1], J. L. Gervasoni*†‡ [2], N. R. Arista*† [3]

* Centro Atómico Bariloche (CNEA) Av. Bustillo 9500, 8400 S. C. de Bariloche, Río Negro, Argentina.
† Instituto Balseiro (CNEA - UNCuyo) Av. Bustillo 9500, 8400 S. C. de Bariloche, Río Negro, Argentina.
‡ Consejo Nacional de Investigaciones Científicas y Técnicas (CONICET) Argentina.

Synopsis In this work we analyze the fluctuations of the electronic density $\Delta\rho$ and the average number of the excited plasmons, for the case of the diatomic charged molecule traveling near or passing through a metallic nanocylinder at different positions, and compare the results with the case of a single particle.

The interaction of structured projectiles with polarizable media has gained visibility in recent years due to the interest of the scientific community in the adsorption of molecules on surfaces and nanoparticles. For instance, the detection and analysis of the composition of molecules or structures adsorbed on a metallic surface can be accomplished using surface plasmon resonance techniques [1]. It is therefore desirable to have a complete description of the coupling of plasmon fields with such systems. The purpose of this communication is to study the interferences that can occur when a charged diatomic molecule interacts with the plasmon field induced in a cylindrical nanostructure, such as a wire or a tube.

Plasmons represent perturbations of the electronic charge distribution of the solid; in an infinite medium, they are spatially localized in an average distance of the order of the velocity of the incident projectile divided by the characteristic plasma frequency of the material.

In this work we use a coherent states formulation to depict the interaction of the incident particles with the plasmon field, based on the formalisms developed for cylindrical geometry [2, 3]. Within this framework, plasmon modes are considered as a set of harmonic oscillators with characteristic frequencies $\omega_\nu(k)$. The unperturbed plasmon field is described by its corresponding hamiltonian, and the temporal evolution of the plasmon field interacting with the external particles can be obtained in an exact quantum mechanical way as coherent states.

We consider the incident molecule (external charge distribution) as formed by two point particles of charge Ze, separated by a fixed distance d and traveling along a rectilinear trajectory with constant mass center velocity. With these elements, we evaluate the wake potential, the variation of electronic density and other relevant quantities for different configurations of dimers traveling in the proximity or through tubes and wires. As an example, figure 1 shows the induced charge density on the surface of a metallic wire (aluminum, with plasmon frequency $\omega_P = 0.55$ a.u.) due to the action of a diatomic Hydrogen molecule ($Z = 1$) oriented with its axis perpendicular to the velocity and travelling on an external trajectory, parallel to the wire's axis.

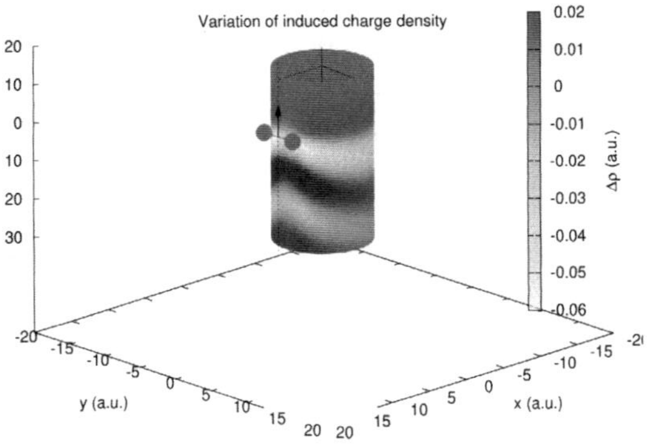

Figure 1. Variation of the induced charge density on an aluminum wire of radius $a = 5$ a.u. by a diatomic molecule with charges $Z = 1$. Velocity of the center of mass $v = 2$ a.u.; interatomic distance $d = 4$ a.u.; distance from the trajectory to the wire's axis $\rho_0 = 6$ a.u.

References

[1] L. S. Jung *et al.* 1998 *Langmuir* **14** 5636.

[2] J. L. Gervasoni and N. R. Arista 2003 *Phys. Rev. B* **68** 235302.

[3] D. J. Mowbray *et al.* 2010 *Phys. Rev. B* **82** 035405.

[1]E-mail: segui@cab.cnea.gov.ar
[2]E-mail: gervason@cab.cnea.gov.ar
[3]E-mail: arista@cab.cnea.gov.ar

Content from this work may be used under the terms of the Creative Commons Attribution 3.0 licence. Any further distribution of this work must maintain attribution to the author(s) and the title of the work, journal citation and DOI.
Published under licence by IOP Publishing Ltd

Journal of Physics: Conference Series **635** (2015) 032025 doi:10.1088/1742-6596/635/3/032025

Electron capture imaging of two-dimensional materials

Alain Dubois* [1]**, Gabriel Labaigt*, and Jan Petter Hansen**[† 2]

[*] Laboratoire de Chimie Physique-Matière et Rayonnement, UPMC-CNRS UMR 7614, F-75005, Paris, France
[†] Department of Physics and Technology, University of Bergen, N-5007, Bergen, Norway

Synopsis The principles of an imaging technique for 2D materials are presented. It is based on the measurement of differential cross sections for electron capture induced by impact of keV-proton on such materials. We illustrate these principles for graphene target and demonstrate the selectivity of this technique upon materials, geometries and surface defects through the highly coherent phenomenon giving rise to the imaging pattern.

Two-dimensional materials, e.g. graphene, designed with spectacular properties, possess an important potential for applications. Aside their macroscopic features, the present possibility to suspend them on boundaries [1] opens rather new lines of studies based on their quantum properties at the atomic level, such as tunneling, specific reactivity in a molecular environment, ... The dynamics response of graphene to strong and ultrafast perturbations is one of these lines and is still largely unexplored nowadays. Concerning laser field interaction theoretical investigations have predicted strong high harmonics generation and formation of focused beam of relativistic electrons [2, 3]. Concerning high-energy ion irradiation attention has been focused on defect formation and hardness properties, e.g. [4, 5]. For low-energy ion-impact a recent investigation [6] simulated images of graphene using the ionization process induced by 30-keV He^+ impact on single suspended sheet. Based on time-dependent density functional theory a large part of the ultrafast electronic dynamics during the scattering event was ignored.

During the conference, we shall present the principle of a new imaging technique based on electron transfer from the sample to the projectile ion. Our predictions stem from a newly developed approach based on scattering theory state-of-the-art semiclassical non perturbative treatment [7] where the periodicity of the target is taken into account. We shall show that contrasted - atom/site selective - images (cf. Figure 1) can be formed due to coherent mechanisms working on each node of the lattice in a rather narrow domain of impact energies. State-to-state probabilities, energy dependence and geometrical selectivity of the process will be also presented.

A surprising contrast of the images produced by our calculations is observed when comparing to the ion-atom counterpart. A simple and illuminating model will be presented in order to explain the coherent multicenter effect giving rise to this behavior, opening the way to generalize this imaging principle to any periodic two-dimensional materials.

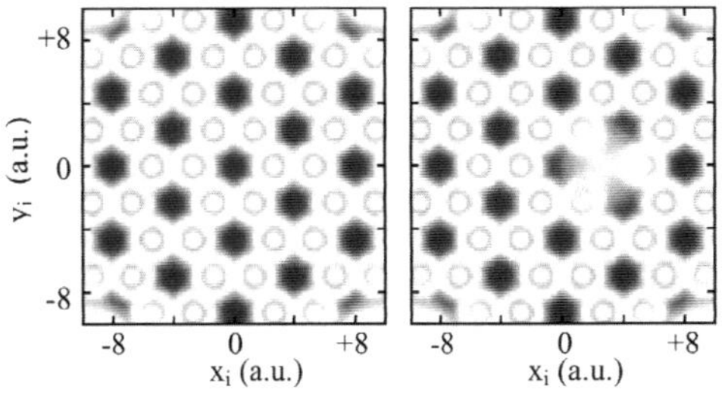

Figure 1. Left: Total electron capture probability as function of impact positions of H^+ on a semi-infinite graphene surface, perpendicular to the projectile beam. Collision energy is 7.5 keV. Right: same as left but with a single vacancy (missing carbon atom in the lattice).

References

[1] K. Yu *et al* 2011 *J. Phys. Chem. Lett.* **2** 537
[2] S.A. Sørngård, S.I. Simonsen, and J.P. Hansen 2013 *Phys. Rev. A* **87** 053803
[3] V.V. Kulagin *et al* 2009 *Phys. Rev. E* **80** 016404
[4] S. Mathew *et al* 2011 Carbon **49** 1720
[5] G. Compagnini *et al* 2009 *Carbon* **47** 3201
[6] H. Zhang, Y. Miyamoto, and A. Rubio 2012 *Phys. Rev. Lett.* **109** 265505
[7] G. Labaigt, A. Dubois, and J.P. Hansen 2014 *Phys. Rev. B* **89** 245438

[1] E-mail: alain.dubois@upmc.fr
[2] E-mail: jan.hansen@uib.no

Content from this work may be used under the terms of the Creative Commons Attribution 3.0 licence. Any further distribution of this work must maintain attribution to the author(s) and the title of the work, journal citation and DOI.
Published under licence by IOP Publishing Ltd

Journal of Physics: Conference Series **635** (2015) 032027 doi:10.1088/1742-6596/635/3/032027

Conductivity mechanism probed by ion transmission through nanocapillaries during the discharging process

P. Herczku* [1], **Z. Juhász*, S. T. S. Kovács*, R. Rácz*, S. Biri*, B. Sulik*** and **N. Stolterfoht**[†]

[*] Institute for Nuclear Research, Hungarian Academy of Sciences, Bem tér 18/c, H-4026 Debrecen Hungary

[†] Helmholtz-Zentrum Berlin für Materialien und Energie, Glienickerstr. 100, D-14109 Berlin, Germany

Synopsis We studied the decrease of the guided transmission due to discharging of nanocapillary walls in polyethylene-terephthalate (PET) foil. After developing the stable guided transmission, the ion beam was switched off and the transmission was tested time to time by short pulses. The transmission monotonically decreased but its time dependence significantly deviated from a simple exponential decay. Our results suggest a non-linear connection between the guiding field and the depleting current of the deposited charge.

Guiding properties of nanocapillaries in insulating materials has gained increasing interest in the last decade [1-4]. The root cause for this process is the self-organizing electrostatic charging up. The incident ions collide with the inner walls of the capillary, deposit their charge and form a repulsive electrostatic field. The subsequently incoming ions are deflected in this field and transmitted through the capillary.

In a regular case the transmission of incident ions monotonously increases until a stable transmission is reached. By switching off the incident beam, the guiding field slowly decays due to the discharge of the capillary walls. The subsequent decrease of the transmission was measured in two early works [1,5]. Later it was recognized [6,7] that it can be a probe for studying the motion of charges at insulator surfaces.

In the present work, we measured the transmission of 3-keV $^{22}Ne^{7+}$ ions through nanocapillaries (~200 nm diameter, $3x10^8/cm^2$ density) formed in a 12μm thick PET foil. After charging up the sample, we switched off the incident beam, and studied the decay of the transmission. From time to time, the transmission rate was tested by three consecutive short pulses. It is noted that the test pulses may recharge the capillary. Since the transmission was practically the same for all the three test pulses, we concluded that this was not the case.

The transmission rate was a monotonically decreasing function of time (see fig. 1.). At the beginning the transmission dropped quickly but later the decay slowed down significantly. In a linear-logarithmic plot the decay function clearly deviates from a straight line, demonstrating that the decay is not exponential, which would be expected in a linear system with field independent conductivities.

We assume that without incident beam, the time development of the transmission is determined only by the discharge current. As a first attempt, for the field dependence of the conductivity we apply the already considered nonlinear Frenkel-Poole model [3,8], which results in a good agreement with the experimental data.

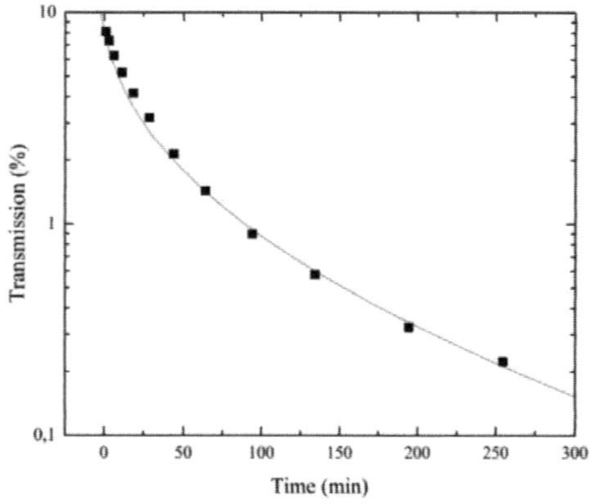

Figure 1. Experimentally measured decay of the transmission at Ψ=5.7° tilt angle as a function of time (squares). The curve is a fitting of the data by the nonlinear conductivity model.

This work was supported by the Hungarian National Science Foundation (OTKA-K83886) and by the TÁMOP-4.2.2.A-11/1/KONV-2012-0036 project, co-financed by the EU and the European Social Fund.

References

[1] N. Stolterfoht *et al* 2002 *Phys Rev Lett* **88** 133201
[2] N. Stolterfoht *et al* 2005 *Surf. & Coat. Techn.* **196** 389
[3] N. Stolterfoht *et al* 2013 *Phys. Rev. A* **87** 012902
[4] K. Schiessl *et al* 2005 *Phys. Rev. A* **72** 062902
[5] N. Stolterfoht *et al* 2004 Vacuum **73** 31
[6] G.P. Pokhill and K.A. Vokhmyania, *J. Surf. Invest. X-ray+* **2** 237
[7] P. Herczku *et al* 2013 *Acta Phys. Debr.* **XLVII** 67
[8] J. Frenkel 1938 *Phys. Rev.* **54** 647

[1] E-mail: herczku.peter@atomki.mta.hu

Content from this work may be used under the terms of the Creative Commons Attribution 3.0 licence. Any further distribution of this work must maintain attribution to the author(s) and the title of the work, journal citation and DOI.
Published under licence by IOP Publishing Ltd

Journal of Physics: Conference Series **635** (2015) 032028 doi:10.1088/1742-6596/635/3/032028

Blocking effect on transmission of Ne^{7+} ions through nanocapillaries

P. Herczku[1], Z. Juhász, S. T. S. Kovács, R. Rácz, S. Biri and B. Sulik

Institute for Nuclear Research (Atomki), Hungarian Academy of Sciences, Bem ter 18/c, H-4026 Debrecen Hungary

Synopsis Systematic investigation was carried out on the guided transmission of $^{22}Ne^{7+}$ ions through nanocapillaries in polycarbonate (PC) foil from 3 to 6-keV kinetic energy range. At 3-keV energy a strong blocking effect was observed contrary to the 6-keV case, where a stable transmission was developed.

Ion guiding by nanocapillaries attracted increasing interest in the last decade [1-8]. This capability is due to the self-organizing electrostatic charging up of the inner walls. Initially the incident ions collide with the surface and deposit their charge. Do to this repulsive electrostatic field is developed, by which the subsequently incoming ions are deflected and guided through the capillary.

Originally, the time development of the transmission of the incident ions was found be monotonically increasing until a stable transmission was developed for polyethylene-terephthalate (PET) capillaries [1,2]. Later, decaying transmission was observed for polycarbonate (PC) capillaries after a maximum transmission was reached [3,4]. At higher ion energies, however, the transmission was stable for PC too [5]. At the other end, blocking was also found for some PET samples [6]. The blocking effect for capillary samples can be attributed to the over-charge of capillary interior [6].

In this work, we systematically measure the transmission of $^{22}Ne^{7+}$ ions through nanocapillaries in a PC foil in the kinetic energy range of 3-6-keV. At 3-keV incident energy a strong blocking effect was observed (see Fig. 1.). After a short sudden increase, the development of the transmission stops slightly above 0.2%. Then it suddenly decreases and drops almost to zero. At 6-keV incident energy the increase is slower but monotonous (see Fig. 1). A stable transmission rate is being approached at the end.

At 4-keV kinetic energy, the blocking effect is still observable. However the transmitted portion of the beam is higher than at 3-keV, so the capillaries are only partially blocked.

The relatively small ion-energy range of the transition from blocked to non-blocked transmission suggests that there exists a characteristic voltage for the sample depending on its geometry and the electric properties of the insulating material. If the ion beam energy is lower than this critical voltage, the capillaries are blocked, while above it regular transmission develops. Partial blocking may be due to the fact that the potential distribution is not uniform along the radius of the beam spot at the sample.

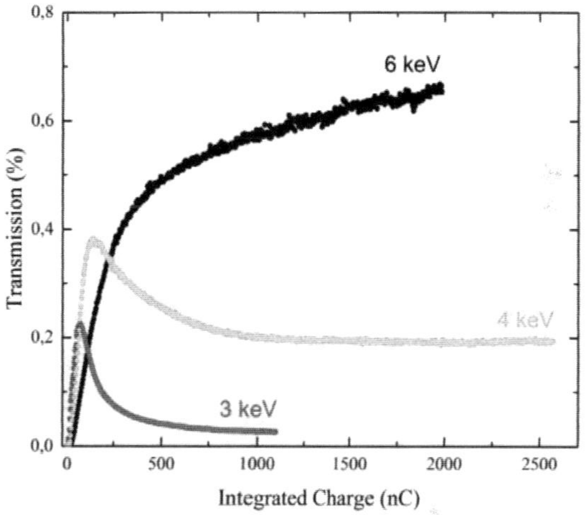

Figure 1. The measured transmission at around Ψ=4.2° tilt angle as a function of integrated charge at 3, 4 and 6-keV kinetic energies, respectively.

This work was supported by the Hungarian National Science Foundation (OTKA, Grant No. K83886) and by the TÁMOP-4.2.2.A-11/1/KONV-2012-0036 project, which is co-financed by the European Union and the European Social Fund.

References

[1] N. Stolterfoht *et al* 2002 *Phys. Rev. Lett.* **88** 133201
[2] N. Stolterfoht *et al* 2005 *Surf. & Coat. Techn.* **196** 389
[3] N. Stolterfoht *et al* 2013 *Phys. Rev. A* **83** 062901
[4] Z. Juhasz *et al* 2012 *Nucl. Instr. and Meth. B* **279** 177
[5] N. Stolterfoht *et al* 2013 *Phys. Rev. A* **79** 042902
[6] N. Stolterfoht *et al* 2013 *Phys. Rev. A* **88** 032902
[7]N. Stolterfoht *et al* 2013 *Phys. Rev. A* **87** 012902
[8] C. Lemell, F. Aumayr, J. Burgdörfer 2013 *Prog. Surf. Sci.* **88** 237

[1] E-mail: herczku.peter@atomki.mta.hu

Content from this work may be used under the terms of the Creative Commons Attribution 3.0 licence. Any further distribution of this work must maintain attribution to the author(s) and the title of the work, journal citation and DOI.
Published under licence by IOP Publishing Ltd

Journal of Physics: Conference Series **635** (2015) 032030 doi:10.1088/1742-6596/635/3/032030

X-ray spectroscopy as a tool to enlighten the growth of Van der Waals nanoparticles in a supersonic jet.

L. Bernard Carlsson[1], C. Prigent, E. Lamour[2], S. Macé, J.-P. Rozet, S. Steydli, M. Trassinelli, and D. Vernhet

Sorbonne Universités, UPMC, Institut des NanoSciences de Paris (INSP), CNRS-UMR 7588, F-75005 Paris, France

Synopsis The clustering thermodynamics in a supersonic jet of argon is studied using x-ray emission induced by keV electrons and slow highly charged ions (HCIs). The high sensitivity of the HCI interaction dynamics allows to probe the very first steps of the argon bunch in a supersonic jet.

Since many decades, studies of collision processes of electrons, photons and heavy particles interacting with matter (from gaseous to solid) are an important tool in physics to understand the internal structure of matter at atomic level (or at nanoscale) as well as at very short time scale (down to fs). For example, recently, pump-probe experiments with IR and XFEL photons allowed probing the morphological structure of an isolated Van der Waals cluster by time resolved imaging [1].

Taking advantage of x-ray spectroscopy, which acts also as a very short time probe, we have investigated the μs temporal structure of the cluster condensation in high-density rare gas expansion (supersonic beams). Briefly summarized, we have previously demonstrated that the x-ray emission allows to determine: i) the absolute total atomic density when the cluster jet is submitted to keV electron impact; ii) the relative cluster density profile when interacting with an intense IR femtosecond laser pulse; iii) and finally the free atom density when irradiated by slow highly charged ions. These first experiments led to the determination of a high degree of condensation (close to 100 %) of the clusters in the supersonic beam when using a skimmer [2]. These results have also paved the way towards new questions: what is the temporal evolution of the thermodynamic growing of clusters? What is the saturation time, needed for having stationary flow conditions in the beam?, etc.

To obtain deeper information on the growth of Van der Waals nanoparticles in a supersonic jet, we are performing new experiments with our set-up at the SIMPA facility (French acronym for "Highly Charged Ion facility in Paris") using Ne^{9+} of 90 keV on argon. Figure 1 shows a comparison of the temporal x-ray signal when the supersonic jet interacts with 10 keV electrons and Ne^{9+} ions for a backing pressure of around 20 bar upstream a conical nozzle with a 300 μm aperture diameter and using a skimmer of 500 μm. Clearly, the x-ray signal starts before with HCIs compared to electron impact. It is a clear signature of the high sensitivity of HCIs to probe very low free atomic density when the cluster begins growing, i.e. at a time scale that is not reachable with "traditional" techniques (like optical measurements). More systematic measurements are under progress and a complete set of results varying the cluster size will be presented during the ICPEAC conference. They will provide new insights on the thermodynamics of a supersonic beam and on the cluster formation.

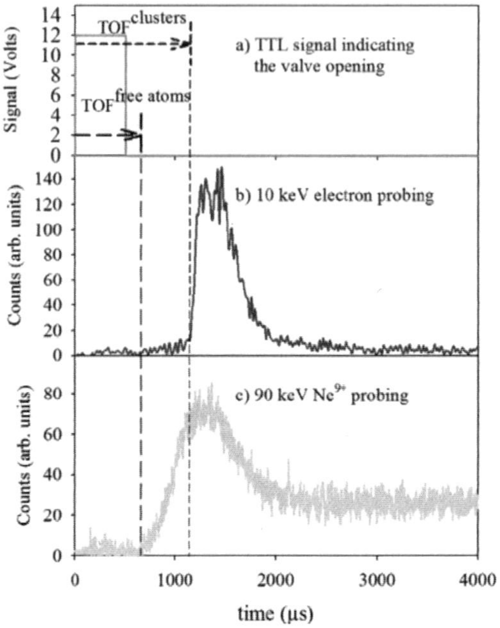

Figure 1. Preliminary results of temporal clusters' x-ray emission in case of b) 10 keV electrons and c) 90 keV Ne^{9+} ions.

References

[1] T. Möller *et al.*, PRL **108** (2012) 245005
[2] M. Trassinelli *et al.*, J.Phys. Conf. Ser **388** (2012) 082009

[1] E-mail: bernardc@insp.jussieu.fr
[2] E-mail: emily.lamour@insp.jussieu.fr

Content from this work may be used under the terms of the Creative Commons Attribution 3.0 licence. Any further distribution of this work must maintain attribution to the author(s) and the title of the work, journal citation and DOI.
Published under licence by IOP Publishing Ltd

A study of electron production in proton-biomolecule collisions using additivity rules

Sergio Paredes*, **Clara Illescas***[1] **and L. Méndez***[2]

* Departamento de Química, Módulo 13, Universidad Autónoma de Madrid, Cantoblanco 28049, Madrid, Spain

Synopsis Additivity rules are employed to estimate electron production cross sections for proton collisions with nucleobases and amino acids at collision energies 10 keV$\leq E \leq$2 MeV. Total and differential cross sections agree with the available theoretical and experimental cross sections, and the method can be easily applied to estimate electron production cross sections for collisions with other biomolecules.

Electron production (EP) in proton-molecule collisions takes place through the reactions:

$$\mathrm{H}^+ + \mathrm{M} \rightarrow ne^- + \mathrm{H}^{(1-m)+} + \mathrm{M}^{(n+m)+}, \quad (1)$$

where M is the target molecule, $m = 0, 1$ and $n \geq 1$. These processes are relevant in cancer proton therapy (see e.g., [1]), where high energy protons give rise to the ionization and ensuing fragmentation of the biomolecules.

In this work we present a semiempirical treatment to estimate the EP total and differential cross sections by employing additivity rules that combine the cross section for atoms and small molecules. This simple approach yields results in good agreement with existing experimental and theoretical values in a wide collision energy range that includes low collision energies, where perturbative calculations are not appropriate. The method can be easily applied to many collisions, provided that the cross sections for the small molecules involved in the additivity rules are available. As an illustration, we show in Fig. 1 our estimates of the single differential, in electron energy, cross sections (SDCS), σ^ϵ, for proton collisions with uracil and adenine. We have used the additivity rules:

$$\begin{aligned}\sigma^\epsilon(\mathrm{Base}) &= a_1\sigma^\epsilon(\mathrm{CO_2}) + a_2\sigma^\epsilon(\mathrm{O_2}) \\ &+ a_3\sigma^\epsilon(\mathrm{N_2}) + a_4\sigma^\epsilon(\mathrm{H_2}) \quad (2)\end{aligned}$$

where the SDCS for collisions with CO_2, O_2, N_2 and H_2 are the experimental values [2], and the coefficients a_i are determined by the molecular formulae of the nucleobases. One can note the excellent agreement between our values and those from the CB1 calculations. The disagreement between theoretical and experimental results at $\epsilon < 7$ eV remains unexplained. The "humps" in the experimental curves might arise from interference effects between collisions with different atoms, although a similar decrease of the experimental SDCS has not been observed in proton collisions with small molecules.

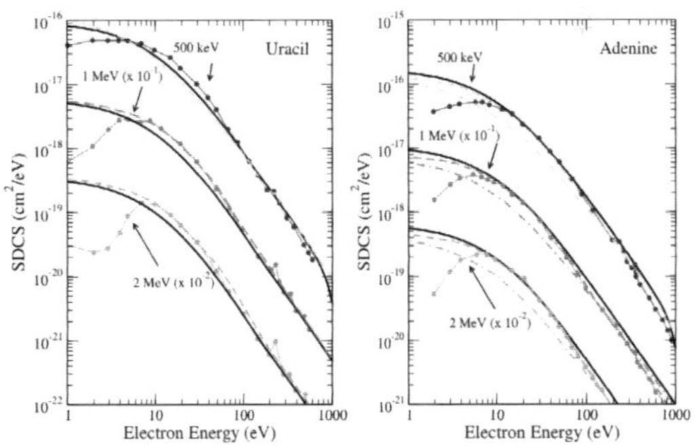

Figure 1. SDCS for EP in proton collisions with uracil and adenine, as functions of the energy of the emitted electron. The full lines are the present results from eq. (2). Dashed lines, first Born approximation with corrected boundary conditions (CB1) calculations [3, 4]; dashed-dotted lines in the right panel, continuum distorted wave-eikonal initial state (CDW-EIS) approximations [4]. Lines with circles, experimental results of references [3] (H^+ + uracil) and [5] (H^+ + adenine).

References

[1] D. Schardt *et al* 2010 *Rev. Mod. Phys.* **82** 383

[2] M. E. Rudd *et al* 1992 *Rev. Mod. Phys.* **64** 441

[3] A. Itoh *et al* 2013 *Phys. Rev. A* **88** 052711

[4] M. E. Galassi *et al* 2012 *Phys. Med. Biol* **57** 2081

[5] Y. Iriki *et al* 2011 *Phys. Rev. A* **84** 052719

[1]E-mail: clara.illescas@uam.es
[2]E-mail: l.mendez@uam.es

Content from this work may be used under the terms of the Creative Commons Attribution 3.0 licence. Any further distribution of this work must maintain attribution to the author(s) and the title of the work, journal citation and DOI.
Published under licence by IOP Publishing Ltd

Journal of Physics: Conference Series **635** (2015) 032034 doi:10.1088/1742-6596/635/3/032034

Direct Measurement of the Isomerization Barrier of the Isolated Retinal Chromophore

Jonathan Dilger[A,B], **Lihi Musbat**[C], **Mordechai Sheves**[D], **Anastasia V. Bochenkova**[E], **David E. Clemmer**[B] **and Yoni Toker**[C, 1]

[A] Spectrum Warfare Systems Department, NSWC Crane Division Crane, IN 47522 (USA)
[B]Department of Chemistry, Indiana University, Bloomington, Indiana 47405, USA
[C]Institute of Nanotechnology and Advanced Materials, Bar-Ilan University, Ramat-Gan 290002, Israel
[D]Chemistry Department, Weizmann Institute of Science, Rehovot 978007, Israel
[E]Department of Physics and Astronomy, Aarhus Universi-ty, DK-8000 Aarhus C, Denmark

Synopsis Energy barrier Heights for isomerization of the isolated retinal chromophore were measured using two stages of ion mobility spectroscopy (IMS-IMS).

Ion mobility spectroscopy (IMS) allows one to differentiate between different isomers of a given molecular ion according to their collisional cross-section. Using two stages of IMS (IMS-IMS) one can select a specific isomer, collisionally heat it and follow its isomerization pathways (See Fig. 1). Recently it has been shown that this technique allows one to determine internal energy barrier for isomerization [1].

Here we apply the technique to the important case of the retinal protonated Schiff base (RPSB) [2]. Photoisomerization of the RPSB is the primary in animal vision. We find that the energy barrier for a single *cis-trans* isomerization is 0.64±0.05 eV, which is significantly lower than that observed for the reaction within opsin proteins. Thus the protein has a significant role in increasing the barrier energy for thermal isomerization relative to the gas phase which lacks interaction with the RPSB counterion and steric constraints. High barrier energy is mandatory for efficient vision processes, otherwise thermal noise would overwhelm the signal originating from the photochemical isomerization.

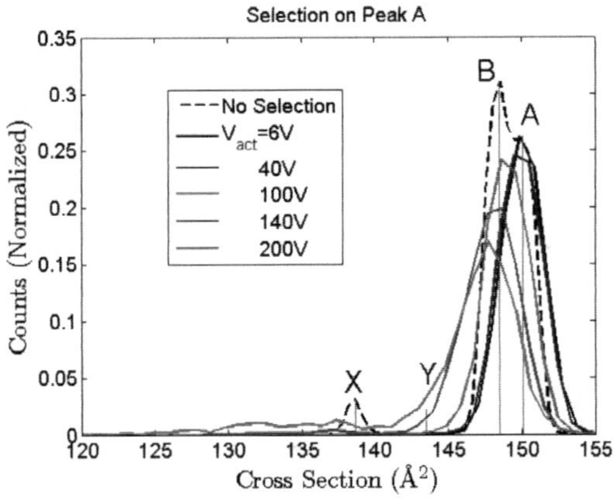

Figure 1. Results of selection and activation when the selection is applied to peak A (the *all-trans* isomer), and activation is performed for different activation voltages, V_{act}. The dashed line corresponds to the IMS of the RPSB with no selection.

References

[1] N. A. Pierson, S. J. Valentine, D. Clemmer 2015 Int. J. Mass. Spectrom, **377**, 646-654.
[2] J. Dilger. Y. Toker *et al.* 2015 Ang. Chemie Int. Ed. **127**, 4830-4834.

[1]E-mail: yonitoker@gmail.com

Content from this work may be used under the terms of the Creative Commons Attribution 3.0 licence. Any further distribution of this work must maintain attribution to the author(s) and the title of the work, journal citation and DOI.
Published under licence by IOP Publishing Ltd

Journal of Physics: Conference Series **635** (2015) 032035 doi:10.1088/1742-6596/635/3/032035

Double impulse effects in ion-molecule collisions at hyperthemal energies

Masato Nakamura*[1] and Atsushi Ichimura†[2]

* College of Science and Technology, Nihon University, 2748501, Fuhabashi, Japan
† Institute of Space and Astronautical Science, JAXA, Sagamihara 2298510, Japan

Synopsis The possibility of double impulse phenomenon in a single collision event is studied by the hard-shell model. Such an effect would be observed if the projectile is as heavy as target and the interaction potential has a strong anisotropy.

Short range repulsive force dominates the dynamics in the close collision between ions with molecules at hyperthermal energies. Generally, such a collision would be impulsive and the hard-shell model [1] and the hard-potential model [2, 3] are applicable to estimate the energy and angular momentum transfers. In these models, an impulse is exerted at a point on the hard-shell, which is given by the equipotential surface at the collision energy. As a result, the energy-loss spectrum at a fixed angle presents a double-peaked structure. If the angular momentum transfer and/or energy transfer from projectile to target is large in the first impulse, the projectile may be largely decelerated and it is plausible that the second impulse [4] takes place. In fact, Tanuma *et al.* [5] have observed collapse of the double-peaked structure in the energy-loss spectra in $Na^+ - N_2$ and concluded that it is due to the 'double collisions' between the projectile and the target through the classical trajectory calculation.

Here we discuss the possibility of *double m-ulse* effect in a single collision event within a framework of the hard-shell model. Let $\mathbf{l}_i$ be the initial orbital angular momentum of relative motion and $\mathbf{J}$ the rotational angular momentum of the molecule after the first impulse. The direction of angular momentum transfer can be categorized into the following to cases; $\mathbf{l}_i \cdot \mathbf{J} > 0$ (normal torque) and $\mathbf{l}_i \cdot \mathbf{J} < \mathbf{0}$ (anomalous torque) [6]. The double impulse effect would be observed in the latter case.

In Fig.1, we present the results of model calculation for $K^+ - N_2$. The hard-shell is given by a sum of monopole and quadrupole deformations as $r_s(\gamma) = r_0(1+\beta_2 P_2(\cos\gamma))$ with the anisotropy parameter $\beta = 0.3$ (0.4) in the left (right) panel. Trajectories of a projectile ion scattered by the hard-shell is shown in a molecular-fixed frame. We observe that the projectile contacts the shell again in the right panel, while it does not in the left panel. We find that the double impulse effect likely occurs in the collisions where the projectile mass is heavier than the target and the potential anisotropy is large.

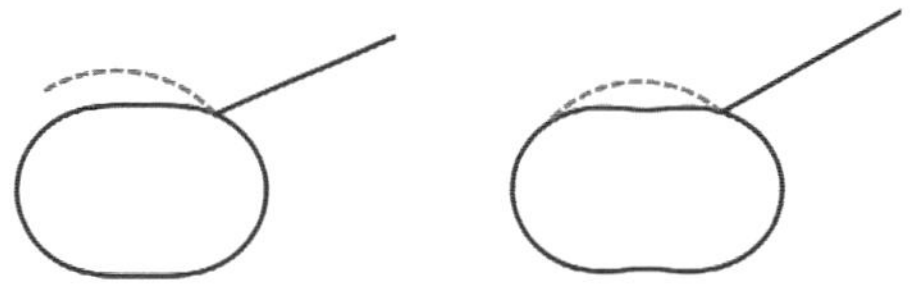

Figure 1. Trajectories of projectile ion scattered by the hard-shell in the molecular fixed frame for $K^+ - N_2$; incident and scattered trajectories are shown by solid lines and dashed curves, respectively. The anisotropy parameter $\beta_2 = 0.3(0.4)$ in the left (right) panel.

References

[1] D. Beck *et al*, Z. Phys. A **293** 107 (1979); ibid **299** 97 (1981).

[2] A. Ichimura and M. Nakamura, Phys. Rev. A **69** 0227116 (2004).

[3] M. Nakamura and A. Ichimura, Phys. Rev. A **71** 062701 (2005).

[4] D. Beck, Chem. Phys. **126** 19 (1988).

[5] H. Tanuma *et al.* Phys. Rev. A A**38** 5053 (1988)

[6] M. Bouard, B, Hornung and F. J. Aoiz, Phys. Rev. Lett. **111** 183202 (2013)

[1]E-mail: mooming@phys.ge.cst.nihon-u.ac.jp
[2]E-mail: ichimura@isas.jaxa.jp

Content from this work may be used under the terms of the Creative Commons Attribution 3.0 licence. Any further distribution of this work must maintain attribution to the author(s) and the title of the work, journal citation and DOI.
Published under licence by IOP Publishing Ltd

Journal of Physics: Conference Series **635** (2015) 032037 doi:10.1088/1742-6596/635/3/032037

Unusual hydrogen and hydroxyl migration in the fragmentation of excited doubly-positively-charged amino acids in the gas phase

Sergio Díaz-Tendero*[1], Dariusz G. Piekarski*, Manuel Alcamí*[†], Fernando Martín*[†], Sylvain Maclot[‡§], Rudy Delaunay[‡§], Alicja Domaracka[‡], Patrick Rousseau[‡§], Lamri Adoui[‡§], and Bernd A. Huber[‡]

[*] Departamento de Química, Módulo 13, Universidad Autónoma de Madrid, Cantoblanco 28049, Madrid Spain
[†] Instituto Madrileño de Estudios Avanzados en Nanociencias (IMDEA- Nanociencia), Cantoblanco 28049 Madrid,
[‡]CIMAP (UMR 6252) - CEA,CNRS,ENSICAEN,Blvd. Henri Becquerel, BP 5133, 14070 Caen cedex 5, France
[§]Université de Caen Basse-Normandie, Esplanade de la Paix, CS 14032, 14032 Caen cedex 5, France

Synopsis We present a combined experimental and theoretical study of the fragmentation of doubly-positively-charged amino acids in the gas phase. The combination of *ab initio* molecular dynamics simulations with ion-molecule collisions followed by multiple-coincidence mass spectrometric techniques, allows us to obtain a complete picture of the fragmentation dynamics. In addition to the expected Coulomb explosion, we have found evidence of hydrogen and hydroxyl-group migration processes, which leads to unusual fragmentation products.

Hadron therapy is radiation therapy using strongly interacting particles [1]. A better depth dose profile of the energetic ion beam (Bragg peak) has proven its superiority over gamma radiation for killing cancer cells selectively [2]. With the advent of these ion-beam cancer treatments [3,4], the interaction of biomolecules with ionising particles (X- rays, electrons, ions) in the gas phase has become a fundamental technique to investigate the radiation damage of biological tissues at the molecular level [5-7]. In biological tissues the damages produced in the biomolecules are not only caused directly by the particle-matter collision but also by radicals and secondary particles created after the fragmentation of different chemical species along the ionisation path [8,9]. This underlines the importance of a proper description of the fragmentation mechanisms after electron removal. Thus, studying the behavior of the amino acids after interaction with highly charged ions gives valuable information for the hadron therapy treatment based on ion beams.

In this communication we present recent results on fragmentation of small lineal amino acids, doubly-positively charged in the gas phase, $NH_2-(CH_2)_n-COOH$: n=1 glycine [10]; n=2 β−alanine [11] and n=3 γ−aminobutyric acid GABA [12]. Experimentally, we obtain the data in the gas phase for neutral molecules in collisions with low-energy highly charged ions. State-of-the-art multi-coincidence detection mass spectrometric techniques are used to determine the charge state of the molecule before fragmentation. The experimental data are analyzed by means of quantum chemistry calculations. In particular, *ab initio* molecular dynamics simulations of the excited and charged species provide valuable information on the fragmentation mechanisms; further exploration of the potential energy surfaces with the density functional theory calculations allows us to obtain energetic and structural information on the most populated dissociation channels.

Our results [10-12] have shown that in competition with the expected Coulomb explosion, the doubly positively charged lineal amino acids present several de-excitation mechanisms. In particular, hydrogen migration and hydroxyl group migration that appear in the femtosecond timescale and lead to unusual fragmentation products.

References

[1] Petti and Lennox, 1994, Hadronic Radiotherapy, *Ann. Rev. Nuclear & Particle Science* **44** 154
[2] H. Kaur *et al* 2013 *AIP Conf. Proc.* **205** 1530
[3] D. Schardt *et al* 2010 *Rev. Mod. Phys.* **82** 383
[4] M. Durante and J.S. Loeffler 2010 *Nat. Rev. Clin. Oncol.* **7** 37
[5] P.Swiderek 2006 *Angew.Chem.Int.Ed.* **45** 4056
[6] B.Boudaïffa *et al* 2000 *Science* **287** 1658
[7] M.H. Holzscheiter *et al* 2012 *Radiother.Oncol.* **105** 1
[8] L. Sanche 2005 *Eur. Phys. J. D* **35**, 367
[9] C. V. Parast *et al* 1995 *Biochemistry* **34** 2393
[10] S. Maclot *et al* 2013 *J.Phys.Chem.Lett.* **4** 3903
[11] D.G. Piekarski *et al* (submitted)
[12] M. Capron *et al* 2012 *Chem. Eur. J.* **18** 9321

[1] E-mail: sergio.diaztendero@uam.es

Content from this work may be used under the terms of the Creative Commons Attribution 3.0 licence. Any further distribution of this work must maintain attribution to the author(s) and the title of the work, journal citation and DOI.
Published under licence by IOP Publishing Ltd

Journal of Physics: Conference Series **635** (2015) 032038 doi:10.1088/1742-6596/635/3/032038

Argonization method to deal with inelastic processes in SiH_4, PH_3 and SH_2 by impact of protons

J. E. Miraglia * [1], C. C. Montanari * [2]

[*] Instituto de Astronomía y Física del Espacio, Consejo Nacional de Investigaciones Científicas y Técnicas, Universidad de Buenos Aires, Ciudad Universitaria, C1428EGA, Buenos Aires, Argentina

Synopsis We describe the eighteen-electron molecules SiH_4, PH_3 and SH_2 with a central potential forming an Argon-type atom. Ionization cross sections and stopping power are calculated using the full first Born approximation and the continuum-distorted-wave eikonal-initial-state approximation.

In a previous article [1] we introduced a method to deal with hydrides of the second line of the periodic table. Using the fact that CH_4, NH_3, H_2O have ten electrons, we assimilated these molecules with a Neon-type atom, and so it was called *neonization* model. It proved to be quite effective in the description of most of the inelastic collisional magnitudes of experimental interest: differential and total ionization cross sections, stopping power, energy-loss straggling and mean excitation energy [1].

In this work we deal with *argonization*, that is we extend the same concepts to the hydrides of the third line of the periodic table, SiH_4, PH_3 and SH_2. We have done a shortcut in relation with our previous work: First, the molecular wave functions of SiH_4, PH_3 and SH_2 reported by Moccia [2] were spherically averaged. Then, the central potentials for each *nlm*-state were directly determined, describing so an Argon-type atom.

Using this potential the ionization cross sections (upper figure) and inelastic (ionization) stopping power cross sections (lower figure) were rigorously calculated using the first Born approximation and the Continuum-Distorted-Wave Eikonal-Initial-state (CDW-EIS). As expected the CDW-EIS converge to the first Born approximation, which holds at high impact proton energies.

The interesting point to note here is that the theoretical treatment of these targets is especially necessary since these molecular hydrides are very difficult to manage in the laboratory because they are corrosive or explosive.

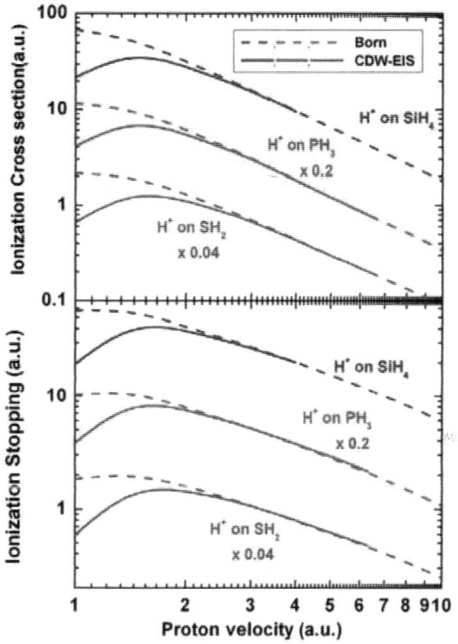

Figure 1. Ionization (upper) and stopping (lower) cross sections in first Born approximation (dashed-lines) and CDW-EIS approximation (solid-lines).

References

[1] C.C. Montanari and J.E. Miraglia, J. Phys. B **47**, 015201 (2014).
[2] R. Moccia , J. Chem. Phys. **40**, 2164, 2176 and 2186 (1964).

[1]E-mail: miraglia@iafe.uba.ar
[2]E-mail: mclaudia@iafe.uba.ar

Content from this work may be used under the terms of the Creative Commons Attribution 3.0 licence. Any further distribution of this work must maintain attribution to the author(s) and the title of the work, journal citation and DOI.
Published under licence by IOP Publishing Ltd

Journal of Physics: Conference Series **635** (2015) 032039 doi:10.1088/1742-6596/635/3/032039

Ionization and fragmentation of pyrimidine and pyridazine induced by proton and electron impact

W. Wolff* [1], H. Luna*, L. Sigaud † and E. C. Montenegro*

* Instituto de Física, Universidade Federal do Rio de Janeiro, 21941-972, Rio de Janeiro, RJ, Brazil
† Instituto de Física, Universidade Federal Fluminense, 24210-346, Niterói, RJ, Brazil

Synopsis The ionization and fragmentation of pyrimidine and pyridazine have been studied experimentally, by recording the time of flight mass spectra, and theoretically, by calculating the valence shell MOs ionization cross sections. The absolute total non-dissociative and dissociative cross section has been measured between 70 to 400 eV for electron impact on pyrimidine and from 250 to 2500 keV for proton impact on both molecules. The partial yields and the absolute cross sections measured as a function of the energy combined with the model calculation allows us to suggest the vacancy population of the valence MOs from which several sets of fragment ions are produced.

The study of the light and heavy particles interactions that drive the ionization and fragmentation of pyrimidine and pyridazine are of prime importance including in radiation therapy, since these biomolecules constitute key building blocks of many important natural and synthetic bio-active molecules, as the nucleobases and as core of drugs against disorders belonging to almost all therapeutic classes. Also, the large abundance of electrons and protons in regions of the interstellar space medium and the existence of nitrogen containing molecules that form subunits of hetero-aromatic molecules may produce pyrimidine.

In this context, the present work appears as one of the first combined theoretical and experimental study providing (i) measurements of the absolute total, non-dissociative and dissociative cross sections of the electron- and proton-induced ionization for pyrimidine [1] and pyridazine for energies ranging from 70 to 400 eV for electron impact and from 125 up to 2500 keV for proton impact and (ii) the percentage contribution of the valence molecular orbitals ionization in the production of the ionic parent molecule and fragments. The overlapping and velocity range of the electron and proton measurements allowed the determination of absolute cross sections for proton impact based on effusive gas target and the use of the Born Approximation to estimate the distribution of primary vacancies in the molecular orbitals.

Results, such as those shown in Figure 1, for pyrimidine losing two up to all atoms of the ring as a function of the proton impact energy will be presented and discussed for both molecules. A detailed analysis of the experimental partial dissociation cross sections as function of the impact energy shows characteristic changes in the ion production compared to the model calculations based on single vacancies: in most cases, the cross sections are well described from 200 keV upwards, but in some others, a steep increase is observed downwards in energy. This indication enables us to discriminate rather directly which process is active, a single step, where most species are dominated by a single vacancy, or a two-step process, that becomes important in multiple particle fragmentation.

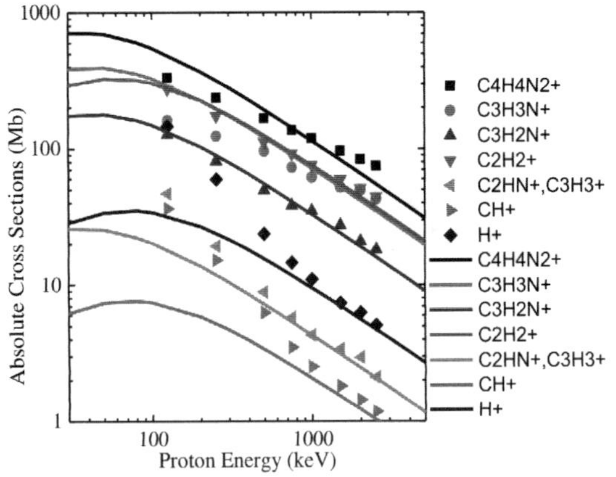

Figure 1. Absolute cross sections for the parent molecule and several fragments of pyrimidine: closed symbols and lines representing experimental results and model calculations, respectively.

This work show a simple and effective strategy to study the fragmentation of rather complex molecules in the gas phase based on standard time of flight mass spectroscopy and a simple semi-classical framework.

Reference

[1] W. Wolff *et al* 2014 *J. Chem. Phys.* **140**, 064309

E-mail: wania@if.ufrj.br

Content from this work may be used under the terms of the Creative Commons Attribution 3.0 licence. Any further distribution of this work must maintain attribution to the author(s) and the title of the work, journal citation and DOI.
Published under licence by IOP Publishing Ltd

Improved description of the interference effects in the ionization of H_2 by fast charged projectiles

L. Nagy [1], **L. Czipa and K. Nagy-Póra**

Faculty of Physics, Babeş-Bolyai University, 400084 Cluj, Romania

Synopsis We present a new approach for the calculation of differential cross sections for the ionization of diatomic molecules. The transition amplitudes are calculated separately for each atom and they are added coherently. The interference term in this method is separated and can be calculated with more precision than in our previous approach.

Interference effects in the ionization of diatomic molecules due to the two-center character of the target have been identified in the energy spectra of the ejected electrons [1, 2]. Right after the experimental observation of the phenomenon we have provided a simple and intuitive explanation of it, and have predicted the angular dependence of the interference pattern [3]. Another group has performed CDW-EIS calculations for the same process [4, 5]. Now we are improving our previous calculations by applying transition amplitudes calculated with more accuracy. The results are compared with the experimental data measured for the ionization of H_2 by 5 MeV/u bare fluorine ion projectiles [2].

Our previous model [3] has assumed several approximations for the calculation of the transition amplitude, such as the peaking approximation and using plain waves for the ejected electron. This had the advantage of obtaining an analytical formula for the amplitude, but did not provide proper values for the calculated cross sections, mainly for low velocities of the ejected electron.

In the present approach we calculate the atomic transition amplitude with a good precision applying the first order perturbational approximation and impact parameter method, suitable for projectiles of high energy. It can be shown, that due to the interference effects between the electron waves ejected from the two atomic centers, the differential ionization cross section for the H_2 molecule may be written as

$$\frac{d^2\sigma(\mathbf{D})}{d\Omega_k dk} = 2\frac{d^2\sigma_H}{d\Omega_k dk} + 2\cos\left[(k_\parallel - q)D_\parallel + \mathbf{k}_\perp \mathbf{D}_\perp\right] \times \int d\mathbf{b}\, a_H^*(\mathbf{b}_+)\, a_H(\mathbf{b}_-).$$

Here $\mathbf{k}$ is the wave vector of the ejected electron, $k_\parallel$ and $\mathbf{k}_\perp$ are its parallel and perpendicular components relative to the beam direction, $D_\parallel$ and $\mathbf{D}_\perp$ the corresponding components of the molecular axis, and q is the minimum momentum transfer. σ_H and a_H stand for the atomic cross section and amplitude, respectively, while $\mathbf{b}_+$ and $\mathbf{b}_-$ represent the impact parameters relative to the two atomic centers. The cross section above, dependent on the orientation of the molecular axis, has to be integrated over all possible orientations in order to be compared with the experimental data.

Using the method above we have calculated differential cross sections for the ionization of H_2 by 5 MeV/u bare fluorine ions, and also σ_{H_2}/σ_H cross section ratios in order to better visualize the interference effects.

This work was supported by a grant of the Romanian National Authority for Scientific Research, CNCS UEFISCDI, project number PN-II-ID-PCE-2011-3-0192.

References

[1] N. Stolterfoht *et al* 2001 *Phys. Rev. Lett.* **87** 023201

[2] S. Chatterjee *et al* 2010 *J. Phys. B: At. Mol. Opt. Phys.* **43** 125201

[3] L. Nagy *et al* 2002 *J. Phys. B: At. Mol. Opt. Phys.* **35** L453

[4] M. E. Galassi *et al* 2002 *Phys. Rev. A* **66** 052705

[5] M. E. Galassi *et al* 2004 *Phys. Rev. A* **70** 032721

[1] E-mail: lnagy@phys.ubbcluj.ro

Content from this work may be used under the terms of the Creative Commons Attribution 3.0 licence. Any further distribution of this work must maintain attribution to the author(s) and the title of the work, journal citation and DOI.
Published under licence by IOP Publishing Ltd

Journal of Physics: Conference Series **635** (2015) 032041 doi:10.1088/1742-6596/635/3/032041

Orientation and alignment effects in ion-induced fragmentation of isolated water molecules

C P Safvan*[1], Jyoti Rajput†[2]

* Inter University Accelerator Centre, Aruna Asaf Ali Marg, New Delhi 110067, India

† Department of Physics and Astrophysics, University of Delhi, Delhi 110007, India

Synopsis We report the angular distribution of the fragment ions formed in the collision of Xe^{9+} ions with isolated water molecules. A significant anisotropy, and strong alignment effects have been seen. It is observed that there is a strong forward - backward asymmetry in the emission of protons and oxygen ions in the three body dissociation of multiply charged water molecules.

With the advent of recoil ion momentum spectrometers a detailed study of ion-molecule collision processes has become possible. One of the important parameters in such collisions is frequently represented by $\kappa = q/v$ where q is the charge on the projectile and v is its velocity. For very fast ($v >> 1\ a.u.$) projectiles, ionization is dominant while for slower ions ($v < 1\ a.u.$) electron capture is enhanced.

For collisions of fast projectiles with molecules, the interaction time is much smaller than the rotational or vibrational time scales. Hence a vertical Frank-Condon transition is expected and the nuclear conformations of the molecule can be considered to be stationary during the collision. The formation of the multiply charged molecule and its dissociation can be considered independent steps. For slow collisions, not only does the incoming slow projectile influence the electron cloud of the neutral molecule, the outgoing projectile also influences the recoil fragments as they move apart. For molecules (as opposed to atoms) the non spherical nature of the target therefore also leads to alignment and orientation effects that are dependent on the projectile velocity.

Ion impact excitation and fragmentation of water molecules has been addressed in earlier studies. Singles measurements using electrostatic analyser have reported anisotropy in proton emission [1, 2, 3] and coincidence measurements [4, 5] which have focussed on the channel ($D^+ + D^+ + O^{q+}$) have shown the inability of a simple coulomb model to explain the energies and geometry of the dissociating molecular ion.

We present here, the angular distributions of recoil fragment momentum vectors with respect to the incoming projectile direction in the fragmentation of water molecules on multiple ionization. From these measurements, made in multi-particle coincidence mode, we note that there is a strong forward backward asymmetry in the D^+ and O^{q+} fragments. We also show a signature of non-planar dissociation of the water molecule.

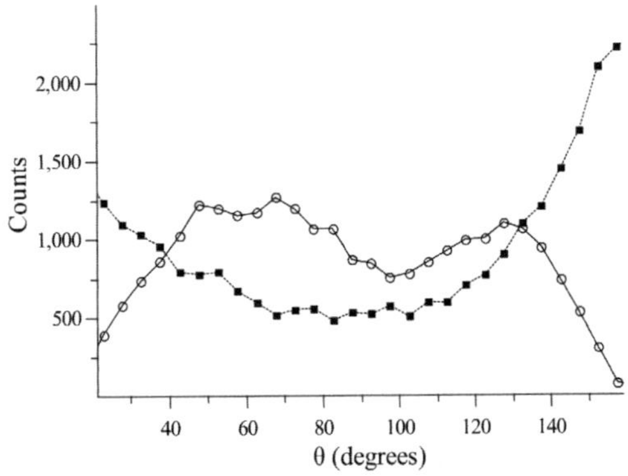

Figure 1. Angle between individual momentum vectors and beam direction for the dissociation channel $[D_2O]^{4+} \rightarrow D^+ + D^+ + O^{2+}$. Hollow circles are for D^+ fragment and full squares are for O^{2+}.

References

[1] Pesic *et al.* 2009 *Journal of Physics B: Atomic, Molecular and Optical Physics* **42** 235202

[2] Sobocinski *et al.* 2006 *Journal of Physics B: Atomic, Molecular and Optical Physics* **39**,927

[3] Pesic et al. 2004 *Journal of Physics B: Atomic, Molecular and Optical Physics* **37**, 1405

[4] Werner *et al.* 1995 *Nuclear Instruments and Methods in Physics Research Section B: Beam Interactions with Materials and Atoms* **98**, 385

[5] Werner *et al.* 1995 *Phys. Rev. Lett.* **74**, 1962

[1]E-mail: safvan@iuac.res.in
[2]E-mail: jyotirajput2803@gmail.com

Content from this work may be used under the terms of the Creative Commons Attribution 3.0 licence. Any further distribution of this work must maintain attribution to the author(s) and the title of the work, journal citation and DOI.
Published under licence by IOP Publishing Ltd

Journal of Physics: Conference Series **635** (2015) 032042 doi:10.1088/1742-6596/635/3/032042

Observation of Electronic States of Rb_2 in the Energy Range 20300-21900 cm^{-1}

J. Yang[1], Z. Zhou, W. Zhao, X. Han, X. Dai[2]

State Key Laboratory for Low-Dimensional Quantum Physics, Department of Physics, Tsinghua University, Beijing 100084, China

Synopsis Thousands of energy levels of Rb_2 along with dozens of collisional lines are observed in the energy range 20300 – 21900 cm^{-1} using double resonance technique. They are assigned as the $2^3\Pi_{0g}$, $4^1\Sigma_g^+$, and $3^1\Sigma_g^+$ states. Strong perturbations among the three states are observed.

Diatomic alkali molecules play an important role in the research of cold atoms and molecules, coherent control, etc [1-4]. So there is an obvious need for accurate knowledge of their electronic structures, especial for diatomic rubidium molecules.

Using Infrared-Infrared Double Resonance (IIDR) high resolution spectroscopy, we have observed the $2^3\Pi_{0g}$, $4^1\Sigma_g^+$, and $3^1\Sigma_g^+$ electronic states of Rb_2 in the 20300-21900 cm^{-1} energy range (see **Figure 1**) [5-6].

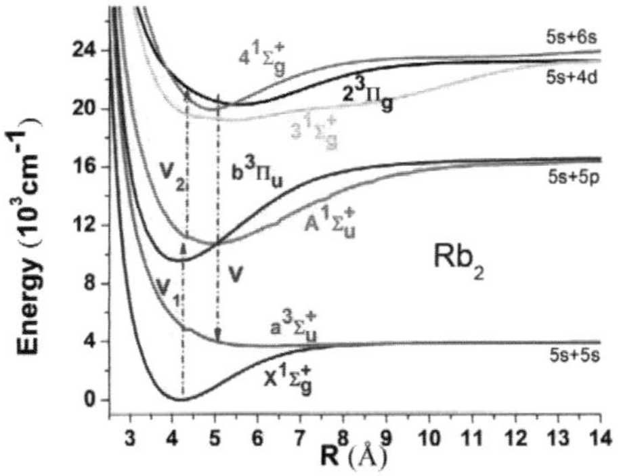

Figure 1. Ab *initio* potential energy functions for several states of Rb_2. The dotted line arrows v_1, v_2 and v represent the process of Infrared-Infrared Double Resonance (IIDR).

In our experimental setup, the temperature of the rubidium vapor was about 550 K and 100 Pa Argon was filled as a buffer gas. Two Toptica DL 100 single-mode cw diode lasers were employed as the pump laser and the probe laser, respectively. The lasers overlap at the center of the heat pipe oven. We collect the fluorescence by a photomultiplier with an optical filter.

The energy levels of the middle states $A^1\Sigma_u^+$ and $b^3\Pi_{0u}$, which are strongly mixed and have both singlet and triplet characteristics, were used as the intermediate levels. So both the upper singlet and triplet states can be accessed.

For each state, we obtained abundant energy levels. The absolute vibrational quantum number assignment is based on the isotope effect since the states of both the $^{85}Rb^{85}Rb$ and $^{85}Rb^{87}Rb$ isotopomers are observed.

In addition, we have detected dozens of lines due to collisional transfer and their assignments are all confirmed. Based on the observed term values of each state, we obtain their molecular constants and RKR potential curves.

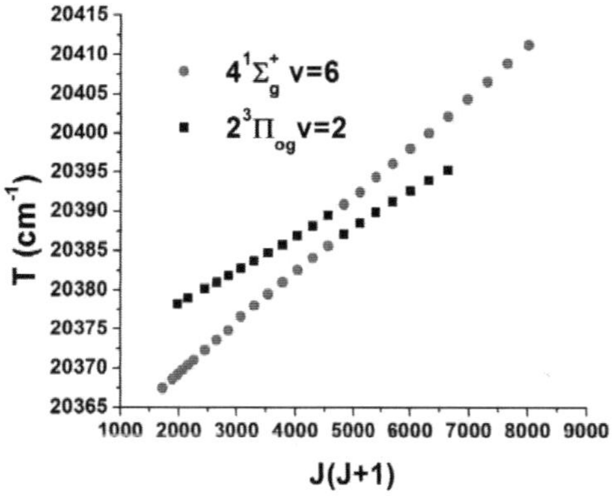

Figure 2. Observed avoided crossing between $2^3\Pi_g$ state v=2 and $4^1\Sigma_g^+$ state v=6 energy bands.

By comparing the observed term values and the calculated term values from the molecular constants, strong perturbations between the $2^3\Pi_{0g}$ and $4^1\Sigma_g^+$ states are clearly demonstrated (see **Figure 2**), as well as between the $4^1\Sigma_g^+$ and $3^1\Sigma_g^+$ states.

References

[1] T. Takekoshi *et al* 2011 *Phys. Rev. A* **83** 062504
[2] P. S. Zuchowski *et al* 2010 *Phys. Rev. A.* **81** 060703
[3] G. Auboeck *et al* 2010 *J. Chem. Phys.* **132** 054304
[4] H. E. L. Martay *et al* 2009 *J. Chem. Phys.* **80** 033403
[5] Y. Guan *et al* 2013 *J. Chem. Phys.* **139** 144303
[6] X. Han *et al* 2014 *Chem. Phys. Lett.* **601** 124058

[1]E-mail: yangjx12@mails.tsinghua.edu.cn
[2]E-mail: xingcandai@mail.tsinghua.edu.cn

Content from this work may be used under the terms of the Creative Commons Attribution 3.0 licence. Any further distribution of this work must maintain attribution to the author(s) and the title of the work, journal citation and DOI.
Published under licence by IOP Publishing Ltd

Journal of Physics: Conference Series **635** (2015) 032043 doi:10.1088/1742-6596/635/3/032043

Molecular dynamics studies of impulse driven reactions in molecules and molecular clusters

M. Gatchell*[1], R. Delaunay[‡,¶], M. H. Stockett[*,†], P. Rousseau[‡,¶], A. Domaracka[¶], Y .Wang[♭,♮] T. Chen*, L. Adoui[‡,¶], M. Alcamí[♭,♮], F. Martín[♭,♮,♯], H. T. Schmidt*, B. A. Huber[‡,¶], H. Zettergren*, and H. Cederquist*

* Department of Physics, Stockholm University, SE-106 91 Stockholm, Sweden
‡ Centre de Recherche sur les Ions, les Matériaux et la Photonique (CIMAP),CEA-CNRS-ENSICAEN, Bd Henri Becquerel, F-14070 Caen Cedex 05, France
¶ Université de Caen Basse-Normandie, Esplanade de la Paix, F-14032 Caen, France
† Department of Physics and Astronomy, Aarhus University, Ny Munkegade 120, DK 8000 Aarhus C, Denmark
♭ Departamento de Química, Módulo 13, Universidad Autónoma de Madrid, 28049 Madrid, Spain
♮ Instituto Madrilenõ de Estudios Avanzados en Nanociencias (IMDEA-Nanociencia), Cantoblanco, 28049 Madrid, Spain
♯ Condensed Matter Physics Center (IFIMAC), Universidad Autónoma de Madrid, 28049 Madrid, Spain

Synopsis We have used molecular dynamics methods to study collisions between atoms and molecules or molecular clusters. We find that these collisions may yield highly reactive molecular fragments and efficient intracluster covalent bond-forming reactions in the cluster. There is a good agreement between our simulations and various experimental results.

Recent studies have shown that collisions between atoms/ion and molecules, in the energy range of a few tens of eV to a few tens of keV, can result in fragmentation pathways not typically found when photons or electrons are used to excite molecular systems. An example of this is the knockout process, where individual atoms may be ejected from a molecule in Rutherford-like scattering processes [1]. This mechanism has also been found to result in the production of highly reactive fragments, leading to a significant reduction in barrier heights for forming covalent bonds between molecules, for example within clusters of PAH- or fullerene molecules [2, 3].

Classical and ab initio molecular dynamics simulations have played an important role in our understanding of the mechanisms that occur in collisions between atoms and isolated molecules or their clusters [2, 3, 4, 5, 6]. Overall we have found a good agreement between the simulations and the experimental results, both qualitatively and quantitatively. Examples of this include studying reaction barriers between molecules and fragments [2, 5, 6], determining reaction cross sections [3, 4], identifying molecular species that are formed in the reactions [2, 3, 5], and simulating fragment mass spectra [3].

At the conference we will present an overview of the simulation methods that we have developed as well as an update on recent results.

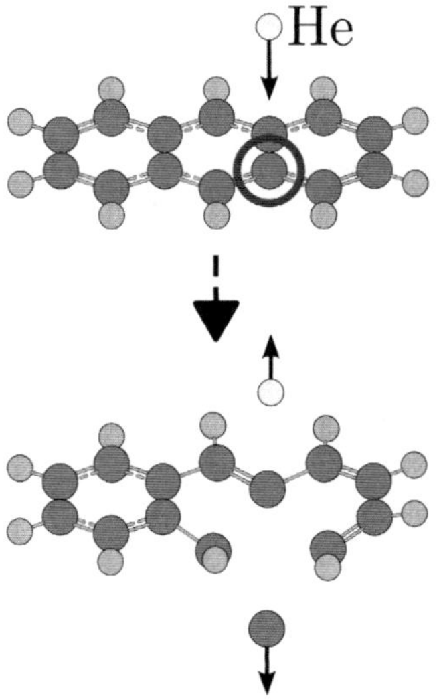

Figure 1. Two frames from a simulation of He colliding with anthracene ($C_{14}H_{10}$) where a single C-atom is knocked out from the molecule in a scattering process.

References

[1] M. Gatchell *et al* 2014 *IJMS* **365-366** 260
[2] H. Zettergren *et al* 2013 *PRL* **110** 185501
[3] R. Delaunay *et al* 2015 *submitted to JPCL*
[4] M. H. Stockett *et al* 2014 *PCCP* **16** 21980
[5] Y. Wang *et al* 2014 *PRA* **89** 062708
[6] M. Gatchell *et al* 2014 *PRA* **90** 022713

[1]E-mail: gatchell@fysik.su.se

Content from this work may be used under the terms of the Creative Commons Attribution 3.0 licence. Any further distribution of this work must maintain attribution to the author(s) and the title of the work, journal citation and DOI.
Published under licence by IOP Publishing Ltd

Journal of Physics: Conference Series **635** (2015) 032044 doi:10.1088/1742-6596/635/3/032044

Ionization and electron capture cross sections for single and multiple electron removal from H_2O by Li^{3+} impact

H. Luna[†1], **W. Wolff** [†], **A. C. Tavares***, **E C Montenegro**[† 2], **T. Pausz**[$] , **M. Murakami**[§] , **H. J. Lüdde**[$] , **M. Horbatsch**[§] , **and T. Kirchner**[§3]

[*] Departamento de Física, Pontifícia Universidade Católica do Rio de Janeiro, 22451-900, Rio de Janeiro, RJ, Brazil
[†] Instituto de Física, Universidade Federal do Rio de Janeiro, 21941-972, Rio de Janeiro, RJ, Brazil
[$] Institut für Theoretische Physik, Goethe-Universität, D-60438 Frankfurt, Germany
[§] Department of Physics and Astronomy, York University, Toronto, Ontario, Canada M3J 1P3

Synopsis We report experimental and theoretical ionization and single electron capture cross sections for single, double and triple electron removal from H_2O by Li^{3+} with energies ranging from 0.75 to 5.8 MeV. Single electron removal is dominant for ionization, while double electron removal is dominant for electron capture. For the one- and two-electron removal cases, the calculations based on time-dependent orbital propagation in density functional theory show good agreement with experiment for the two collision channels studied.

The quality of theoretical calculations of ionization and charge exchange processes when multiply charged ions collide with molecules cannot be assessed based on measurements of the produced ions alone, as is the case for atomic targets. This follows since the relationship between the observed ionic fragments and the distribution of primary vacancies, either single or multiple, is not straightforward. A better check is achieved from a finer selection of the measured ionic fragments by discriminating the collision events that produced either one, two or more ions. If this measurement is differential in the final projectile charge state, a quite stringent comparison with theory can be made.

In this work we report experimental and theoretical ionization and single-electron capture cross sections for single, double and triple electron removal from water by Li^{3+} with energies ranging from 0.75 to 5.8 MeV. The experiment was carried out selecting both the final charge state of the projectile and the ejected fragments in coincidence to obtain cross sections associated with ionization and electron capture channels [1]. The ionic fragments and the emitted electrons produced under single-collision conditions were collected by a time-of-flight spectrometer with single-hit (e.g. $OH^+ + H^0$) and double-hit events (e.g. $OH^+ + H^+$) properly discriminated.

The theoretical framework is a continuation of previous work [2]. The molecular orbitals from a self-consistent field calculation are represented in an atomic oxygen basis and propagated with the basis generator method. This can be achieved for a limited number of molecular orientations, but should be sufficient for the generation of reliable total cross sections.

The measured and calculated cross sections for single, double, and triple electron removal from water are shown in Figure 1.

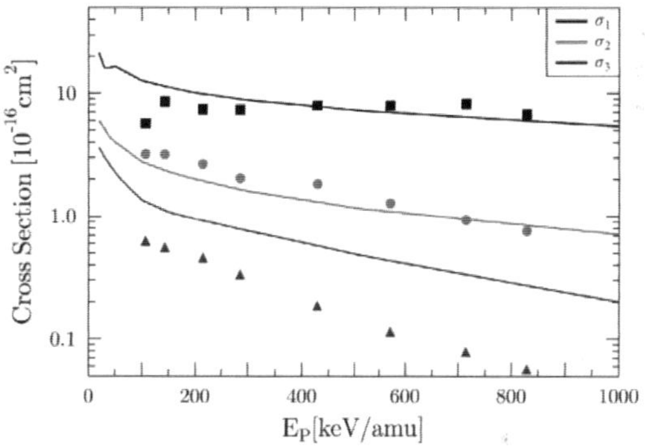

Figure 1. Cross sections for single, double, and triple electron removal from water by Li^{3+} impact.

These cross sections σ_j for removal of j electrons represent a relatively inclusive property, and should be predicted reliably by the theoretical method. We observe that for j=1,2 and impact energies above 100 keV/amu the agreement between theory and experiment is good. For j=3 the theory overestimates the measured data by a factor of two and more. Given that j matches the initial charge state of the projectile this overestimation is perhaps unexpected. Also unexpected is the different fall-off with projectile energy which would indicate a possible breakdown of the independent-electron model for this charge state. More detailed comparisons on the level of final projectile-charge-state-specific cross sections will be shown at the conference.

References

[1] J S Ihani *et al* 2013 *J. Phys. B: At. Mol. Opt. Phys.* **46** 115208
[2] M Murakami *et al* 2012 *Phys. Rev. A* **85** 052704

[1] E-mail: : hluna@if.ufrj.br
[2] E-mail: : montenegro@if.ufrj.br
[3] E-mail: tomk@yorku.ca

Content from this work may be used under the terms of the Creative Commons Attribution 3.0 licence. Any further distribution of this work must maintain attribution to the author(s) and the title of the work, journal citation and DOI.
Published under licence by IOP Publishing Ltd

XXIX International Conference on Photonic, Electronic, and Atomic Collisions (ICPEAC2015) IOP Publishing
Journal of Physics: Conference Series **635** (2015) 032047 doi:10.1088/1742-6596/635/3/032047

Double differential ionization cross section of tetrahydrofuran for proton impact

Ticia Buhr* [1], **Sándor Ricz**†, **Levente Ábrók**†, **Ákos Kövér**†, **Mingjie Wang***, **Gerhard Hilgers*** **and Hans Rabus***

* Physikalisch-Technische Bundesanstalt (PTB), D-38116 Braunschweig, Germany
† Institute for Nuclear Research, Hungarian Academy of Sciences (Atomki), H-4001 Debrecen, Hungary

Synopsis Double differential (in angle and energy) ionization cross sections were experimentally determined for the collisions of protons with a tetrahydrofuran gas target at impact energies between 100 keV and 1910 keV in the angular range between 15° and 165° relative to the beam direction. The derived absolute double differential cross sections will be included in the BioQuaRT [1] multi-scale track structure simulation program.

Within the framework of the EMRP Project "BioQuaRT" [1] the double differential cross sections (DDCS) for ionization of tetrahydrofuran (THF) by proton impact were measured in the projectile energy range between 100 keV and 1910 keV.

The experiments were carried out at the lately developed 150 kV electrostatic and the 3.75 MV Van de Graaff accelerators at PTB (Braunschweig, Germany). The emitted electrons were measured with a recently upgraded ESA-22-type electrostatic electron spectrometer developed at Atomki (Debrecen, Hungary). A detailed description of this type of analyzer is presented in Ref. [2]. The electrons were detected simultaneously at polar angles between ±15° and ±165° (except ±90°) in 15° steps relative to the beam direction.

The absolute normalization was performed by collecting electron spectra from collisions of 100 keV, 300 keV and 2 MeV protons with argon and comparing them with the experimental data of Rudd *et al* [3]. Because of the cylindrical symmetry of the collision system (the tetrahydrofuran molecules were randomly oriented), the experimental double differential cross sections were averaged for the symmetric emission angles relative to the projectile momentum vector.

Figure 1 shows preliminary results of the measured double differential cross sections for the ionization of tetrahydrofuran by 100 keV protons. The absolute values were derived by the comparison of our measurement and the experimental data from Ref. [3] for the collision of 100 keV protons with argon. The reference values for emission angles and secondary electron energies missing in Ref. [3] were determined by linear interpolation. The error bars denote the standard errors of the mean.

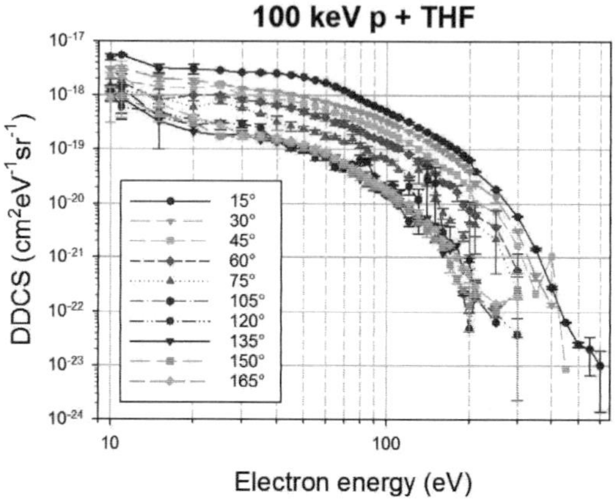

Figure 1. Absolute double differential ionization cross sections (DDCS) for the ejection of secondary electrons by collision of 100 keV protons with tetrahydrofuran (THF) at different emission angles (see legend) relative to the proton beam direction as a function of the electron energy. The error bars indicate the standard errors of the mean.

The assistance of the staff of the PTB accelerators and of the machine shop is gratefully acknowledged. The authors are especially thankful to Andreas Pausewang for his valuable support. This work was carried out within the EMRP Joint Research Project SIB06 BioQuaRT. The EMRP is jointly funded by the EMRP-participating countries within EURAMET and the European Union. This work was also supported by the PTB guest scientist program and the Hungarian Scientific Research Foundation (OTKA Grant No. K104409).

References

[1] BioQuaRT project website: http://www.ptb.de/emrp/bioquart.html
[2] S. Ricz *et al* 2002 *Phys. Rev A* **65** 042707
[3] M.E. Rudd *et al* 1979 *At. Data Nucl. Data Tables* **23** 405

[1] E-mail: ticia.buhr@ptb.de

Content from this work may be used under the terms of the Creative Commons Attribution 3.0 licence. Any further distribution of this work must maintain attribution to the author(s) and the title of the work, journal citation and DOI.
Published under licence by IOP Publishing Ltd

Journal of Physics: Conference Series **635** (2015) 032048 doi:10.1088/1742-6596/635/3/032048

Fragmentation of tryptophan molecule by low-energy electron-impact

A.V. Papp§, J. Tamuliene*, L.G. Romanova§, V.S. Vukstich§, S.S. Demesh§, A.V. Snegursky§

*Vilnius University, Institute of Theoretical Physics and Astronomy, Vilnius, Lithuania
§Institute of Electron Physics, Ukr. Nat. Acad. Sci., Uzhgorod, Ukraine

Synopsis The fragmentation of a gas-phase tryptophan molecule by a low-energy (<70 eV) electron impact was studied both experimentally and theoretically. Various positively charged fragments have been observed and analyzed. A special attention is paid to the energy characteristics of the ionic fragment yield.

This paper deals with the studies of the low-energy (<150 eV) electron interaction with the tryptophan ($C_{11}H_{12}N_2O_2$) molecule belonging to the essential amino acids in order to probe the intrinsic properties of the molecule and trace its chemical structure change(s) under the electron impact. Tryptophan is the metabolic precursor of serotonin, which plays an important role in a sleeping process and is necessary for normal growth and nitrogen balance in humans.

The experimental technique used in this work was described in detail in a number of our previous papers (see, e.g. [1]). A crossed-beam technique with the mass separation of the collision products by means of a magnetic mass-spectrometer was applied.

The tryptophan molecule mass-spectrum measured by us is (see Fig. 1) generally close to that for the electron-impact ionization quoted in the NIST database [2] and that for the case of photoionization [3].

Structurally, tryptophan, as the aromatic amino acid, could be presented as the indole-3-alanine or the 3-methyleneindole bound at the C_α position to a residue of a simple amino acid (glycine). The ion fragment $C_9H_8N^+$ (mz=130 a.m.u.) corresponding to the dominant mass-spectrum peak (Fig. 1) is produced due to the C_α–C_β (C10-C11) bond rupture in the initial molecule. The fraction of the complementary ion with the mass of mz=74 a.m.u. (i.e. $C_2H_4NO_2^+$) is about 1% of the main peak. Thus, the tryptophan molecule ionization proceeds mainly due to electron elimination by the methyleneindole group of the molecule at which the positive charge is predominantly localized.

In this mass-spectrum, we have found for the first time the production of the $C_8H_5N_2^+$, $C_9H_7N_2^+$ and $C_9H_9N_2^+$ doubly charged ions having the m/z=57.5, 64.5, 65.5, 79.5 a.m.u. masses. As our theoretical DFT approach-based analysis shows, some of them could be produced not only due to the direct initial molecule dissociation but also result from the secondary fragmentation of the 3-methyleneindole part of the parent tryptophan molecule.

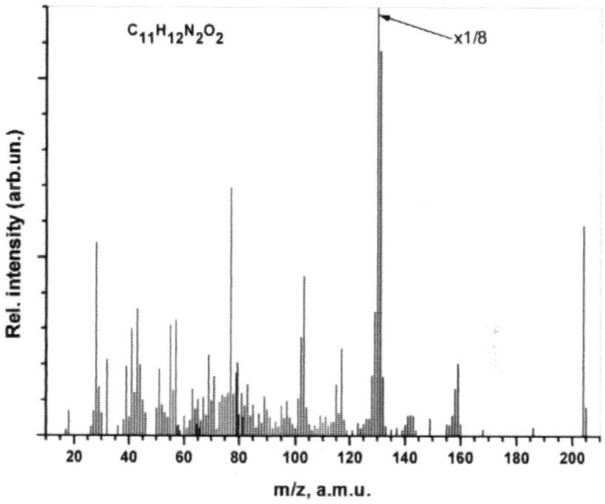

Figure 1. Tryptophan molecule mass-spectrum.

We have also determined the ionization threshold E_i for the parent tryptophan molecule. Its absolute value of 8.3±0.1 eV is slightly lower than the similar one obtained in our previous studies on the other aliphatic amino acids. The appearance potential E_{ap} for the $C_9H_8N^+$ ion (mz=130 a.m.u.) produced from the parent molecule was determined by us to be 9.1±0.1 eV and this value by 0.93 eV exceeds that obtained theoretically.

Thus, the analysis of the mass-spectrometric studies of the tryptophan molecule electron-impact fragmentation accompanied by theoretical calculations allowed the main dissociation mechanisms of the above molecule under the low-energy electron impact to be determined. Most of the peaks in the experimental mass spectrum were identified.

References

[1] V.S. Vukstich, A.I. Imre, L.G. Romanova, A.V. Snegursky. *J. Phys. B* **43,** 185208 (2010)
[2] *NIST Standard Reference Database*: *Chemistry Webbook*. http://webbook.nist gov.
[3] O. Plekan, V. Feyer, R. Richter, M. Coreno, K.C. Prince, *Mol. Phys.* **106**, 1143 (2008)

E-mail: snegrusky.alex@gmail.com

Content from this work may be used under the terms of the Creative Commons Attribution 3.0 licence. Any further distribution of this work must maintain attribution to the author(s) and the title of the work, journal citation and DOI.
Published under licence by IOP Publishing Ltd

Journal of Physics: Conference Series **635** (2015) 032049 doi:10.1088/1742-6596/635/3/032049

Experimental and theoretical study of the reaction of C_3N^- with acetylene and ethylene

Fredrik Linden*[1], Daniela Ascenzi†[2] Wolf Dietrich Geppert*[3]

*Department of Physics, University of Stockholm, Roslagstullsbacken 21, S-10691 Stockholm Sweden

†Dept. of Physics, University of Trento, via Sommarive 14, I-38100 Trento, Italy

Synopsis We investigated the reaction of C_3N^- with acetylene and ethylene both through experimental (Guided ion beam mass spectroscopy) as well as theoretical (ab initio calculations). The main experimental results as well as the potential energy surface illustrating possible reaction pathways for the formation of the observed reaction products are presented.

In 2004, the Cassini Plasma spectrometer (CAPS) on board the Cassini spacecraft [1] recorded a multitude of anions in Titan's ionosphere. Unfortunately, CAPS was designed for detection of electrons rather than anions, thus the resolution does not allow the identification of most anions, except CN^- and C_3N^- , which have been assigned with some confidence.

Our goal is to understand if the anion C_3N^- could, by reaction with unsaturated hydrocarbons, form larger negative species and thus serve as a first step in creating large anions in Titan's atmosphere.

Using the Guided Ion Beam Mass Spectrometer apparatus at the University of Trento, we identify the following main channels for the reaction with ethylene:

$$C_3N^- + C_2H_4 \rightarrow \begin{cases} CN^- + C_4H_4 & (1) \\ H_2C_4N^- + CH_2 & (2) \\ HC_4N^- + CH_3 & (3) \end{cases}$$

and the following reaction channels for acetylene:

$$C_3N^- + C_2H_2 \rightarrow \begin{cases} CN^- + C_4H_2 & (4) \\ HC_3N^- + C_2H^- & (5) \\ H_2 + C_4N^- & (6) \end{cases}$$

It is well known, mass spectrometric experiments do not yield any information about the presence of different isomeric structures in the reaction products. Hence theoretical modeling should be performed to assist in the interpretation of the experimental data and to determine the reaction pathways leading to the identified products. We have performed ab initio calculations using the Gaussian [2], $rMP2/6\text{-}311G{+}{+}^{**}$ level of theory. Minima and transition states of the reaction pathways were calculated and their identity confirmed by frequency calculations. Intrinsic reaction coordinate (IRC) calculations were carried out to certify that the transition states connect the correct minimas. The potential energy surface for the reaction with ethylene is shown in Fig. 1, while the one for acetylene will be presented at the conference.

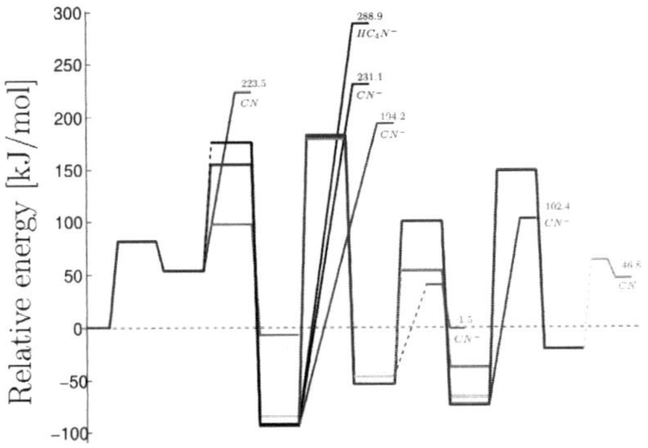

Figure 1. Potential energy surface for reaction C_3N^- with ethylene. Only ion products below 300 kJ/mol are indicated

Calculations indicate that some of the reaction pathways are either endothermic or hindered by high energy barriers, which make them quite unfeasible for Titan's atmosphere. However we see one exothermic product (C_4H_4) that could have great significance for formation of large anions (aerosols) on Titan.

Products from channel (2) is open for energies above 350 kJ/mol, H_2C_4N (anion product formed in channel 2), can contribute to formation of larger $C_{2n}N^-$ anions according [3].

References

[1] V. Vuitton *et al.* 2009 *Planetary and Space Science* **57** 1558

[2] M. J. Frisch *et al* 2009 *Gaussian 09, Revision* **A.02** (Gaussian Inc. Wallingford CT 2009)

[3] J. Zabka *at al* 2014 *International Journal of Mass Spectrometry* **367** 1

[1]E-mail: fredrik.linden@fysik.su.se
[2]E-mail: daniela.ascenzi@unitn.it
[3]E-mail: wgeppert@fysik.su.se

Content from this work may be used under the terms of the Creative Commons Attribution 3.0 licence. Any further distribution of this work must maintain attribution to the author(s) and the title of the work, journal citation and DOI.
Published under licence by IOP Publishing Ltd

Journal of Physics: Conference Series **635** (2015) 032050 doi:10.1088/1742-6596/635/3/032050

Cross sections for the ionization of tetrahydrofuran by light ions

Mingjie Wang*[1], Benedikt Rudek*, Ticia Buhr*, Daniel Bennett*, Gerhard Hilgers*, Hans Rabus*

* Physikalisch-Technische Bundesanstalt (PTB), Bundesallee 100, D-38116 Braunschweig, Germany

Synopsis Double differential cross sections (DDCS) for ionization of tetrahydrofuran by protons and helium ions were measured in the projectile energy range between 420 keV and 16.14 MeV at emission angles between 30° and 90° relative to the beam direction. The preliminary results for 420 keV protons indicate that the relative energy dependence of the experimental DDCS is in a better agreement with the modified Hansen-Kocbach-Stolterfoht (HKS) model [1] than with the recently reported calculation based on the dielectric response function (DRF) formalism [2].

Within the framework of the EMRP Joint Research Project "BioQuaRT" [3], a multi-scale track structure simulation program based on interaction cross sections of DNA constituents is under development [4]. As a prerequisite, ionization cross sections for model molecules representing DNA constituents are being determined.

Recently, double differential cross sections (DDCS) for ionization of tetrahydrofuran (THF) by proton and helium-ion impact were measured. The experiments were performed at the PTB ion accelerator facility [5] (Braunschweig, Germany) using monoenergetic protons with energies between 420 keV and 5.35 MeV, He^{+} ions of 1.66 MeV and 2.01 MeV and He^{2+} ions of 16.14 MeV.

The emitted electrons were detected with an electrostatic hemispherical electron spectrometer. The mean radius of the analyzer is 65 mm, its acceptance angle and the relative energy resolution are about 3° and 1%, respectively. The spectrometer can be rotated around the symmetry axis of the scattering chamber (i.e. the direction of the gas beam). The secondary electron spectra were measured at emission angles 30°, 45°, 60°, 75° and 90° relative to the direction of the incident particles. These spectra are being converted into absolute cross sections using the collected electron spectra from the collision of 400 eV electrons with THF by referring to previously published cross section data for ionization of THF by electrons [6].

Figure 1 shows the relative energy dependence of the experimental DDCS (red dots) at observation angle 30° for 420 keV proton impact on THF. The experimental values are compared with the results of the modified HKS model [1] (blue solid line) and of the recently reported DRF calculation [2] (green dashed line). The experimental and the theoretical data were normalized to the respective DDCS at 100 eV electron energy. The relative energy dependence of the experimental DDCS agrees better with the modified HKS calculation.

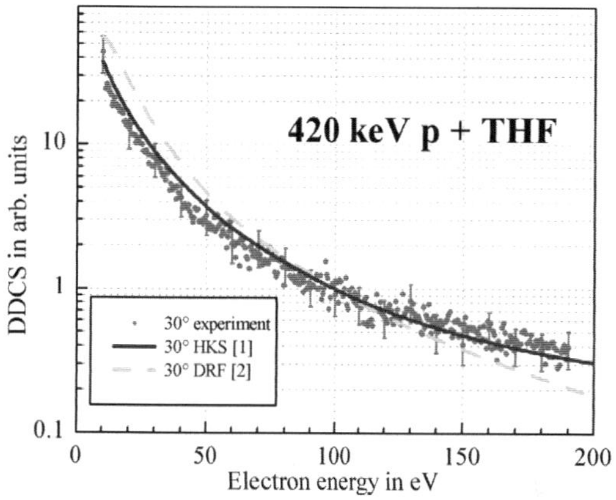

Figure 1. Comparison of the experimental (red dots) and the theoretical (blue solid [1] and green dashed [2] lines) DDCS as a function of the secondary electron energy at emission angle 30° for 420 keV proton impact on THF. The error bars indicate the estimated experimental uncertainties.

The authors thank Oliver Döhr and Manfred Hoffmann for operating the accelerators, and Andreas Pausewang and Wolfgang Helms for the technical support. This work was carried out within EMRP Joint Research Project SIB06 BioQuaRT. The EMRP is jointly funded by the EMRP participating countries within EURAMET and the European Union.

References

[1] M. A. Bernal *et al* 2006 *Nucl. Instrum. Meth. B* **251** 171
[2] P. de Vera *et al* 2015 *Phys. Rev. Lett.* **114** 018101
[3] Project website: www.ptb.de/emrp/bioquart.html
[4] H. Rabus *et al* 2014 *EPJ Web of Conf.* **77** 00021
[5] H. J. Brede *et al* 1980 *Nucl. Instr. and Meth.* **169** 349
[6] W. Y. Baek *et al* 2012 *Phys. Rev. A* **86** 032702

[1] E-mail: mingjie.wang@ptb.de

Content from this work may be used under the terms of the Creative Commons Attribution 3.0 licence. Any further distribution of this work must maintain attribution to the author(s) and the title of the work, journal citation and DOI.
Published under licence by IOP Publishing Ltd

Journal of Physics: Conference Series **635** (2015) 032051 doi:10.1088/1742-6596/635/3/032051

Dissociation rate, fluorescence and Infrared radiative cooling rates of Naphthalene studied in electrostatic storage Miniring.

C. Ortega*, N. Kono[α], M. Ji*, R. Brédy *, J. Bernard*, C. Joblin [†], A. Cassimi [£], Y. Ngono-ravache[£], L. Chen* and S. Martin* [1]

[*] Université de Lyon, F-69622, Lyon, France Université Lyon 1, Villeurbanne CNRS, UMR 5306, ILM
† Université de Toulouse, UPS-OMP, IRAP,Toulouse, France
[α] Department of Physics, Tokyo Metropolitan University 1-1 Minami-Osawa, Hchioji, Tokyo 192-0397, Japan
[£]CIMAP, GANIL Université de CAEN, Bd. H. Becquerel, Caen, France

Synopsis We report studies on fluorescence and Infra Red (IR) radiative cooling rates of naphtalene cation for a large time range (40ms). The dissociation rates versus internal energy have been estimated in a large energy range to reproduce the measured emitted neutral yield of the hot naphthalene.

Using action spectroscopy, we have studied the time evolution of internal energy distribution of naphthalene cations ($C_{10}H_8^+$) from 30µs to 40ms. The ions were produced in an ECR Nanogan ion source and stored in the small electrostatic storage ring call Miniring. The internal energy distribution of the population of the stored ion is broad and part of the hot molecules dissociate in the first three milliseconds. After 3 ms, the neutral emission is due to the collisions with the residual gas in the Miniring vacuum chamber (10^{-9}mbar). The fraction of hot molecules that has enough internal energy to dissociate has been estimated to be about 4% of the total number of the stored ions. This value is relatively constant for different experimental conditions of the ECR source. From 0.5 to 1.5 ms, 1% of store ions dissociated, whereas from 2.5 to 3.5 ms only 0.1% dissociated. In order to reproduce these neutral decay ratios from 30µs to 3ms and especially the ratio of 4%, we have simulated the evolution of the energy distribution assuming a Gaussian shape for the initial distribution. The cooling rates of fast fluorescence processes [1] have also been included in the calculations. The evolution of the internal energy distribution is very sensitive to the dissociation rates for a large range of internal energy. The dissociation rates were taken from the recent measurements of West et al[2]. In our simulations, for the time range up to 3ms, the extrapolation of their dissociation rates in the low energy range overestimates the neutral yield. We have modified the dissociation rates in the low energy range to reproduce our experimental decay. Therefore these results provide a limit for the dissociation rate at the corresponding low energy range.

Radiative cooling rates have been investigated for the naphthalene. We have studied the decay curves obtained after irradiation with the 532nm 1kHz Ekspla OPO LASER. Laser pulse was sent each ms. The figure 1 shows intensities of the first peaks of each decay curves versus the laser excitation time. The curve is nicely fitted using to exponential decays: a fast decay, attributed to fluorescence cooling process[1] and a slow decay attributed to the IR cooling process.

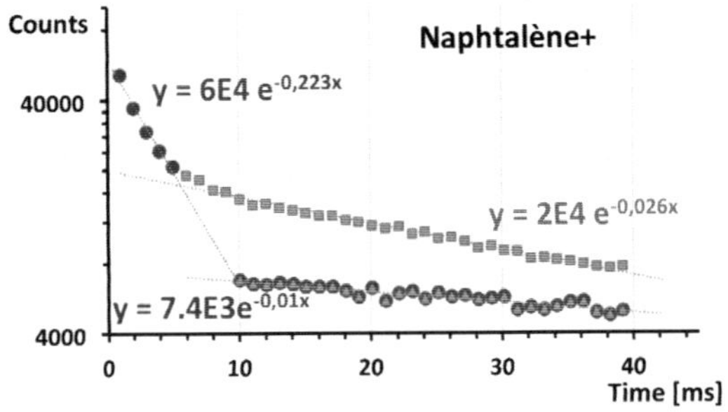

Figure 1. Fast (blue) fluorescence radiative cooling process and slow IR (red) cooling process of naphthalene. Background signal due to the collision with the residual gas (green).

References

[1] S. Martin *et al* 2013 *Phy. Rev. Lett.*. **110** 063003
[2] B. West *et al* 2012 *J. Phys. Chem.*. **116** 10999

[1] E-mail: smartin@univ-lyon1.fr

Content from this work may be used under the terms of the Creative Commons Attribution 3.0 licence. Any further distribution of this work must maintain attribution to the author(s) and the title of the work, journal citation and DOI.
Published under licence by IOP Publishing Ltd

Journal of Physics: Conference Series **635** (2015) 032026 doi:10.1088/1742-6596/635/3/032026

Influence of the projectile charge state on electron emission spectra from a Cu(111) surface

C D Archubi* [1]**, V M Silkin**†**, and M S Gravielle***

* Instituto de Astronomía y Física del Espacio (IAFE, CONICET-UBA), Buenos Aires, Argentina
† Departamento de Física de Materiales, Facultad de Ciencias Químicas, Universidad del País Vasco, Apdo. 1072, 20080 San Sebastian, Spain

Synopsis Double differential electron emission distributions produced by grazing impact of fast dressed ions on a Cu(111) surface are investigated focusing on the effects of the electronic band structure. The process is described within the Band-Structure-Based (BSB) approximation, which is a perturbative method that incorporates information of the electronic band structure of the solid. Differences in the behavior of the emission spectra for He^{+q}, Li^{+q}, Be^{+q} and C^{+q} projectiles with different charge states q are explained by the combined effect of the projectile trajectory and the projectile charge distribution.

In previous articles [1,2] we found that the electron emission distributions produced by grazing incidence of protons on different metallic surfaces present prominent signatures of the surface band structure, with pronounced shoulders due to the contribution of partially occupied surface electronic states (SESs). The aim of this work is to investigate to what extent such structures are affected by the charge state of the projectile.

To describe the electron emission process from the valence band we employ the BSB approximation, which is derived within the framework of the binary collisional formalism by including a precise representation of the surface interaction, given by the potential model of Ref [3]. This potential incorporates effects of the band structure of the metal, properly reproducing the projected energy gap and the energies of the surface and first image states. The interaction with the projectile is included by means of an effective projectile charge that depends on the transferred momentum, while the dynamic response of the medium to the incident projectile is obtained from the unperturbed electronic wave functions by using the linear response theory.

The BSB approach is applied to study double differential –energy and angle-resolved – electron spectra produced by He^{+q}, Li^{+q}, Be^{+q} and C^{+q} ions with different charge states q impinging on a Cu(111) surface. We focus on the contribution of partially occupied SESs to the electron emission process, analyzing separately the influence of the projectile path and the projectile charge distribution. Results show that these two factors - the trajectory and the projectile electron configuration - play different and complementary roles, being responsible for the behavior observed in the figure.

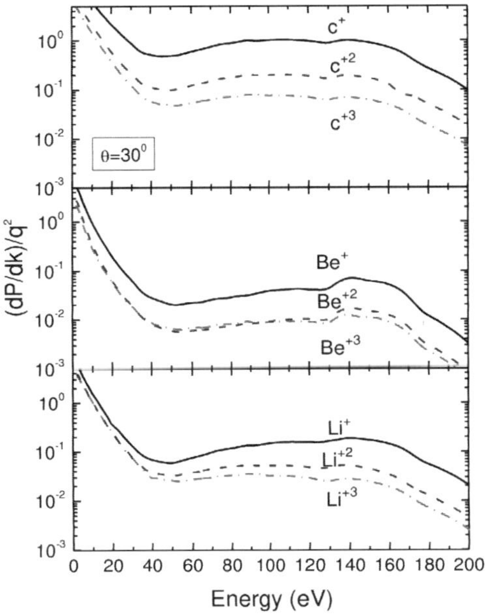

Figure 1. Differential probability of electron emission from the valence band, normalized by q^2, for the emission angle θ_e=30°, as a function of the electron energy. The following ions impinging on a Cu(111) surface, with incidence velocity v=2 a.u and angle α=0.1°, are considered: a) C^{+q}, b) Be^{+q}, c) Li^{+q}, with q=1,2,3.

References

[1] C. D. Archubi *et al* 2011 *Phys. Rev. A* **84** 012901
[2] C. D. Archubi *et al* 2014 *Phys. Rev. B* **89** 155421
[3] E. V. Chulkov *et al* 1997 *Surf. Sci.* **391** L1217

[1] E-mail: archubi@iafe.uba.ar

Content from this work may be used under the terms of the Creative Commons Attribution 3.0 licence. Any further distribution of this work must maintain attribution to the author(s) and the title of the work, journal citation and DOI.
Published under licence by IOP Publishing Ltd

Journal of Physics: Conference Series **635** (2015) 032031 doi:10.1088/1742-6596/635/3/032031

Guiding of 1 MeV proton microbeam through a tapered glass capillary

G.U.L. Nagy*, I. Rajta*, and K. Tőkési* [1]

* Institute for Nuclear Research, Hungarian Academy of Sciences (MTA Atomki), Debrecen, Hungary, EU

Synopsis The guiding of 1 MeV proton microbeam passing through a tapered-shape macroscopic insulator capillary was investigated. The intensity of the transmitted beam and the energy distribution of the transmitted particles were measured. We found that significant fraction of the beam can pass through the capillary without any significant energy loss indicating that the guiding electric field built up inside the capillary.

Charged particle beams are able to pass through insulator capillaries keeping their initial kinetic energy and initial charge state, even if the axis of the capillary is tilted with respect to the axis of the incident beam and thus the geometrical transmission would be impossible. The phenomenon is called "guiding" and caused by the self-organized charge-up of the insulator material due the interaction of the capillary inner wall and the incident ion beam.

The ion guiding effect is intensively studied, since it has many practical applications such as focusing and steering ion beams as passive elements (i.e. without external power supplies). The use of tapered capillaries allows the production of highly focused beams for microanalytical techniques and for single cell irradiations or beam extraction from vacuum to atmospheric environment without any dedicated and expensive apparatus [1,2,3].

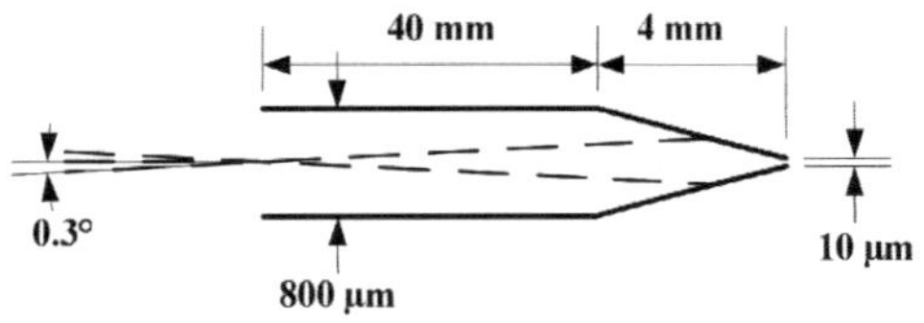

Figure 1. Schematic view of the capillary target and the focused proton beam with its divergence (not to scale).

In our experiments we studied the transmission of focused H^+ beam of 1 MeV energy through a tapered-shape insulator capillary made of borosilicate glass (see Figure 1). The intensity of the incident beam was measured by a beam chopper in the entrance of our target chamber. The intensity of the transmitted beam was measured by a Faraday-cup, the energy distribution of the transmitted particles was measured by a conventional SB type particle detector behind the capillary exit. The beam was point-like with respect to the target, the focus was on the capillary entrance, the divergence (half-opening angle) of the proton microbeam was 0.3°. Although the geometry of the target and the beam allowed only a few percent of direct transmission normalized to the intensity of the incident beam, we found that significant part (more than 30%) of the incoming beam passed through the capillary. In spite of this, the majority of the transmitted particles did not suffer considerable energy loss (see Figure 2), i.e. there was no inelastic collision between the projectile ions and the target atoms, which is the consequence that the guiding electric field built up inside the capillary.

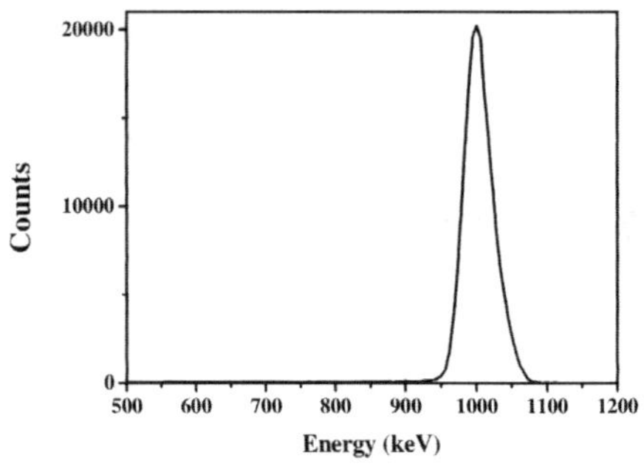

Figure 2. Energy distribution of the 1 MeV proton microbeam transmitted through the tapered glass capillary.

This work was supported by the Hungarian Scientific Research Fund OTKA Nos. NN 103279 and K103917 and by the COST Action CM1204 (XLIC).

References

[1] T. Ikeda *et al* 2011 *Surf. Coat. Tech.* **206** 859
[2] T. Nebiki *et al* 2006 *Nucl. Instr. Meth. Phys. Res. B* **249** 226
[3] J. Hasegawa *et al* 2011 *Nucl. Instr. Meth. Phys. Res. B* **269** 3087

[1] E-mail: tokesi@atomki.mta.hu

Content from this work may be used under the terms of the Creative Commons Attribution 3.0 licence. Any further distribution of this work must maintain attribution to the author(s) and the title of the work, journal citation and DOI.
Published under licence by IOP Publishing Ltd

Journal of Physics: Conference Series **635** (2015) 032032 doi:10.1088/1742-6596/635/3/032032

Protection and sensitization of normal and tumor cells to proton radiation by cerium oxide nanoparticles

J. L. Shinpaugh*[1], B. Swartz*, C. Putnam-Evans†, S. Das‡, and S. Seal‡

* Department of Physics, East Carolina University, Greenville, North Carolina, USA
† Department of Biology, East Carolina University, Greenville, North Carolina, USA
‡Department of Mechanical, Materials and Aerospace Engineering, Advanced Materials Processing and Analysis Center, and Nanoscience Technology Center, University of Central Florida, Orlando, Florida, USA

Synopsis Cerium oxide nanoparticles are investigated as radioprotectors and radiosensitizers for normal (healthy) and tumor cells for charged particle radiation. Results are presented for cell survival and DNA damage for *in-vitro* irradiation of normal breast epithelial cells and prostate tumor cells by 1.5 – 3.0 MeV protons.

Radiation therapy is widely used in the treatment of cancer. Methods to protect healthy tissue while enhancing damage to tumor cells are continually being sought. Recently, cerium oxide (CeO_2) nanoparticles have been shown to act as a protectant for normal cells while sensitizing tumor cells to x-rays [1, 2]. Use of radioprotectors and radiosensitizers could clearly increase the efficacy of the treatment.

We have initiated experiments to investigate the potential use of ceria nanoparticles as protectors/sensitizers for proton and ion beam therapies. Ceria nanoparticles (CNPs) prepared at the UCF Nanoscience Technology Center were transferred to ECU. Normal breast epithelial cells and prostate tumor cells cultured in the cell culture facility at ECU were treated with the CNPs and irradiated with 1.5 – 3.0 MeV protons for doses of 1 – 8 Gray in the ECU Accelerator Laboratory. Cell viability was measured with the MTT assay after 24 and 48 hours post irradiation.

DNA damage was assessed using the TUNEL assay; typical results for are shown in figure 1 of a confocal image of normal cells irradiated with 3 MeV protons for a dose of 2.8 Gy. Blue fluorescence indicates the location of the cell nuclei, with the DNA damage indicated by red. The top image (A) is for cells untreated with the CNPs, and the bottom image (B) is for cells treated with 100 nM concentration of the CNPs 24 hours prior to irradiation. The greater extent of DNA damage (ratio of red to blue) is clearly observed in the untreated cells, indicating protection provided by the CNPs.

Mechanisms of radioprotection and radiosensitization by the CNPs will be discussed.

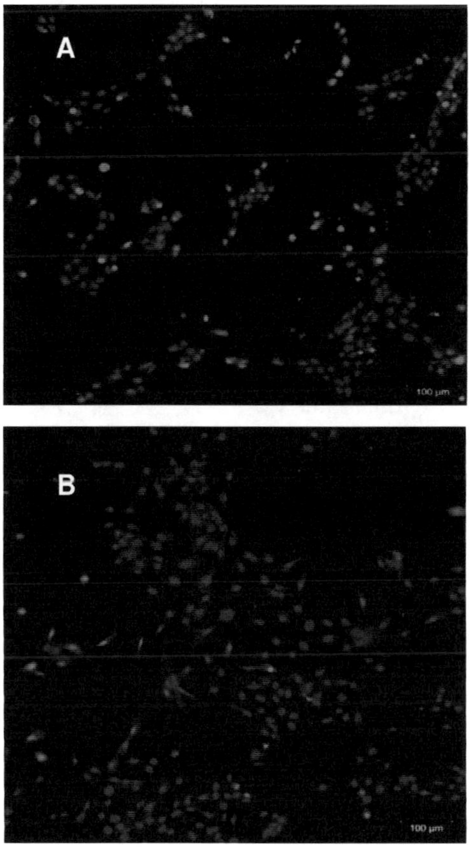

Figure 1. TUNEL assay showing DNA damage for untreated cells (A) and cells treated with ceria nanoparticles (B) following irradiation by 3 MeV protons. Cell nuclei are identified by blue fluorescence, with DNA damage labeled by a red fluorophore.

References

[1] R.W. Tarnuzzer *et al* 2005 *Nano Letters* **5**(12) 2573
[2] M.S. Wason *et al* 2013 *Nanomed.* **9** 558

[1] E-mail: shinpaughj@ecu.edu

Content from this work may be used under the terms of the Creative Commons Attribution 3.0 licence. Any further distribution of this work must maintain attribution to the author(s) and the title of the work, journal citation and DOI.
Published under licence by IOP Publishing Ltd

Dissociation of doubly charged Coronene as a function of internal energy

R. Brédy*[1], C. Ortéga*, M. Ji*, L. Chen*, J. Bernard* and S. Martin*

* Institut Lumière Matière, UMR5306 Université Lyon 1-CNRS, Université de Lyon 69622 Villeurbanne cedex, France

Synopsis We report on the dissociation of doubly charged Coronene (${C_{24}H_{12}}^{2+}$) as a function of internal energy. The mass spectra were obtained for internal energy measured up to 50 eV using the CIDEC method in collisions with F^+ and H^+ at 3 keV. The dehydrogenation processes of the molecule were observed over this large range of energy as well as the other weaker channels, loss of C_2H_n, $2C_2H_n$ and ionization. The branching ratios for the fragments ${C_{24}H_{12-n}}^{2+}$, ${C_{22}H_{12-n}}^{2+}$, ${C_{20}H_{12-n}}^{2+}$ and ${C_{24}H_{12-n}}^{3+}$ (n=0-6) were measured.

Polycyclic aromatic hydrocarbons (PAHs) are extensively studied since many decades due to their relevance e.g. in astrophysics in the attempt of explaining the observation of the unidentified infrared bands. The important role of PAHs as catalyst in the formation of H_2 in the photodissociation regions of the interstellar medium has also been suggested. The numerous photodissociation and collision experiments as well as theoretical investigations have shown that the main dissociation channels are the loss of H, H_2 and small hydrocarbon molecules such as C_2H_2. With a highly symmetric structure coronene ($C_{24}H_{12}$) is considered as a model PAH system. The emission of an even number of H atoms is in general favored but the related process whether the direct H_2 emission or the sequential H loss is still under debate [1].

In this work, the fragmentation of coronene induced by collision with F^+ and H^+ at 3 keV is investigated using the CIDEC method [2]. In this method the excitation energy of the doubly charged coronene ion is determined by measuring the kinetic energy loss of the negative scattered ion formed in the collision $F^+ + C_{24}H_{12} \longrightarrow F^- + {C_{24}H_{12}}^{2+}$. The measurements are performed in coincidence with time-of-flight mass spectrometry leading to mass spectra resolved in internal energy of the ${C_{24}H_{12}}^{2+}$ parent ion (Figure 1). As expected the dehydrogenation of coronene (loss of H, H_2) is the dominant process. Internal energy up to 50 eV have been measured. Over this wide range of energy the branching ratios for the fragments ${C_{24}H_{12-n}}^{2+}$, ${C_{22}H_{12-n}}^{2+}$ (C_2H_n loss), ${C_{20}H_{12-n}}^{2+}$ ($2C_2H_n$ loss) and ${C_{24}H_{12-n}}^{3+}$ (n=0-6) were measured to be about 90%, 4%, 1% and 5% respectively. These branching ratios are also presented in table 1 as a function of selected energy ranges. Stable ${C_{24}H_{12}}^{2+}$ parent ions are observed with internal energy of about 10 eV but it is noticeable that a small fraction of ${C_{22}H_{12-n}}^{2+}$ fragment is also observed at this energy. This may indicate a strong competition between the loss of H and C_2H_2 at low excitation energies while H/H_2 loss dominate at higher excitation energy as already observed for doubly charged anthracene ions [3].

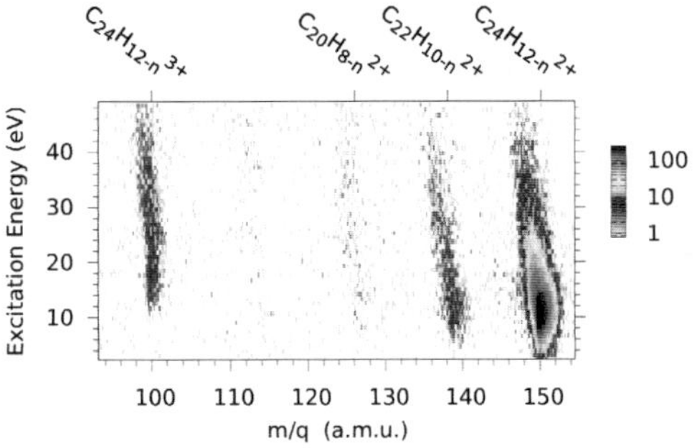

Figure 1. Excitation energy of ${C_{24}H_{12}}^{2+}$ as a function of the mass spectrum. ${C_{24}H_{12}}^{2+}$ are formed in 3 keV $F^+ + C_{24}H_{12} \longrightarrow F^- + {C_{24}H_{12}}^{2+}$ collisions.

Table 1. Branching ratios (%) as a function of the excitation energy of ${C_{24}H_{12}}^{2+}$ parent ion. Energy range in eV.

	1.5-10	10-19	19-27	27-49
${C_{24}H_{12-n}}^{2+}$	97.5	94.5	80	50
${C_{22}H_{12-n}}^{2+}$	2.5	3	7	14
${C_{20}H_{12-n}}^{2+}$		< 1	2	7
${C_{24}H_{12-n}}^{3+}$		2	11	29

References

[1] C. Paris *et al.* 2014 *J. Chem. Phys.* **140** 0204307

[2] L. Chen *et al.* 2007 *Phys. Rev. Lett.* **98** 193401

[3] S. Martin *et al.* 2012 *Phys. Rev.* **A85** 052715

[1]E-mail: richard.bredy@univ-lyon1.fr

Content from this work may be used under the terms of the Creative Commons Attribution 3.0 licence. Any further distribution of this work must maintain attribution to the author(s) and the title of the work, journal citation and DOI.
Published under licence by IOP Publishing Ltd

Interactions of amino acid enantiomers with chiral target in high and low energy collisions

O. Rebrov*[1], K. Kulyk*[2], M. Ryding[†], M. Stockett[‡], J. Alexander*, H. Zettergren*, H. Schmidt*, H. Cederquist*, R. D. Thomas*, E. Uggerud[†], M. Larsson*

[*] Stockholm University, Department of Physics, SE-106 91 Stockholm, Sweden
[†] Department of Chemistry, University of Oslo, N - 0315 Oslo, Norway
[‡] Department of Physics and Astronomy, Aarhus University, DK-8000 Aarhus, Denmark

Synopsis The study is focused on investigation of stereospecific interaction of amino acid enantiomers with chiral targets in collisions of 1 keV and ~1 eV energy using tandem mass spectrometry.

Investigation of stereospecific biomolecular interactions is an active area of research nowadays. Chiral forms of molecules can experience different interactions in chiral environment and thereby trigger different reaction mechanisms. Homochirality of amino acid and sugar enantiomers in living organisms results in stereospecificity of biomolecular processes. Therefore it is crucial in particular for pharmaceutics and drug discovery to be able to detect and separate optical isomers. The lack of theoretical basis able to describe processes that lead to enantioselectivity constitutes the need of fundamental research in this area.

Our study is focused on investigating interactions between singly protonated and neutral chiral molecules in the gas phase using tandem mass spectrometry. High and low energy collisional activation of L- and D-forms of amino acid cations with chiral and achiral target gases were studied. In order to make stereospecific interactions more feasible the amino acids with bulky side chain groups namely Phe, Trp and Met were utilized. The chiral (S)-(+)-2-butanol, racemic (+/-)-2-butanol and Argon were used as target gases. For collisions of 1 keV in a center-of-mass frame, the amount of transferred energy was sufficient to induce fragmentation via multiple channels. For all measured amino acids the most abundant products were produced due to the loss of NH_3 and combined loss of H_2O + CO. The fragmentation pathways and abundances of fragments revealed no difference for both amino acid enantiomers with all gases used. The collision energy was found to be too high to observe any kind of chiral effects. Thus the study of low energy collisions of around 1 eV in the center-of-mass energy with (S)-(+)-2-butanol and racemic (+/-)-2-butanol was performed. Additional to the main fragments that correspond to the loss of NH_3 and H_2O + CO, complexes of amino acids and 2-butanols were detected. The possible chiral impact in fragmentation and efficiency of adduct formation was elaborately investigated.

[1] E-mail: oleksii.rebrov@fysik.su.se

[2] E-mail: Kostiantyn.kulyk@fysik.su.se

Content from this work may be used under the terms of the Creative Commons Attribution 3.0 licence. Any further distribution of this work must maintain attribution to the author(s) and the title of the work, journal citation and DOI.
Published under licence by IOP Publishing Ltd

Slow ion interaction with N-methylglycine and N-acetylglycine

Jaroslav Kocisek[*,1], **Sergio Diaz Tendero**[†,2], **Dariusz Grzegorz Piekarski**[†], **Rudy Delaunay**[*], **Bernd A. Huber**[*], **Lamri Adoui**[*], **Manuel Alcami**[†], **Fernando Martín**[†,¶,‖], **Patrick Rousseau**[*], **Alicja Domaracka**[*] **and Janina Kopyra**[‡]

[*] CIMAP, UMR 6252 (CEA/CNRS/ENSICAEN/Universite de Caen Basse-Normandie), Bd. Henri Becquerel, BP 5133, 14070 Caen Cedex 5, France
[†] Departamento de Química, Modulo 13, Universidad Autónoma de Madrid, 28049 Madrid, Spain
[¶] Instituto Madrileno de Estudios Avanzados en Nanociencias (IMDEA-Nanociencia), Cantoblanco, 28049 Madrid, Spain
[‖] Condensed Matter Physics Center (IFIMAC), Universidad Autónoma de Madrid, 28049 Madrid, Spain
[‡] Chemistry Department, Siedlce University, 3 Maja 54, 08-110 Siedlce, Poland

Synopsis N-acetyl glycine and N-methyl glycine molecules in the gas phase are ionized by electron exchange with slow O^{6+} ions at an energy of 48 keV. After ionization, the methyl and acetyl substituted glycines dissociate into fragments analogous to that resulting from ionization and fragmentation of amino acids and peptides, respectively. N-acetylglycine which contains a peptide bond also effectively tautomerizes to the diol form. Such tautomerization is typical for amino acids, however, we show that the tautomerization mechanism of the N-acetylglycine is different.

We studied two N- substituted amino acids N-methylglycine and N-acetylglycine by means of the COLIMACON [1] experimental setup based at the ARIBE facility of the GANIL accelerator (Caen, France). The molecules were ionized in collisions with slow 48 keV O^{6+} ions. The ionization products were analyzed by coincidence mass spectrometry. The studied molecules together with the glycine molecule are sketched in Fig. 1. The figure demonstrates also the main differences observed in the present study.

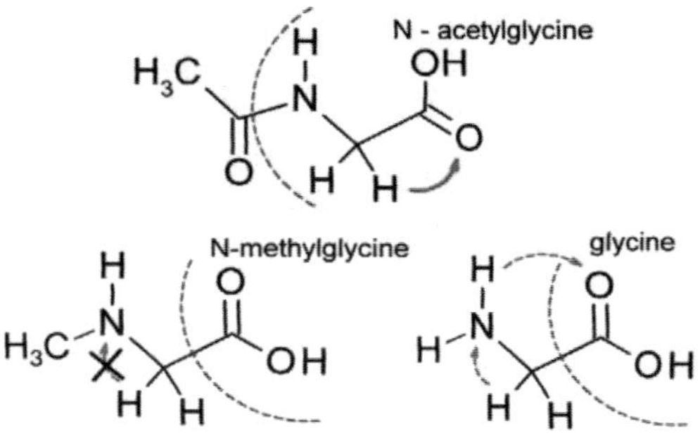

Figure 1. Fragmentation (blue) and tautomerization (red) of the studied N substituted glycine molecules in comparison with the glycine [2].

In the case of N-methylglycine, single ionization results in charge localization on the nitrogen atom and the molecule primarily decomposes by neutral carboxyl group loss in the process analogous to that of glycine molecule [2]. In the case of N-acetylglycine, charge delocalizes on the peptide bond. Therefore, fragmentation via COOH loss competes with the peptide bond dissociation and the acetyl cation dominates the mass specrum.

For N-acetylglycine molecule, we detect metastable parent ions which dissociate exclusively by neutral water release. Quantum chemical calculations indicate that the stabilization of the parent ion occurs via tautomerization to the diol form. Due to the charge delocalization within the molecule, the barrier towards direct hydrogen transfer between C_α and the carboxyl group lowers and direct hydrogen transfer becomes possible.

We show that the substitution which mimics the peptide bond significantly changes the behavior of the ion. The main difference lies in the charge distribution which is delocalized over the whole peptide bond in contrast to localization on the nitrogen atom which is typical for amino acids. We show also that the reactions typical for amino acids can be caused by different mechanism in peptide bonded systems.

Financial support from INCa-ITMO (Nr PC201307) within the Programme Plan Cancer 2009-2013 (Inserm) and from the XLIC COST action (CM1204) is gratefully acknowledged.

References

[1] M. Capron *at al* 2012 *Chem. Eur. J.* **18** 9321
[2] S. Maclot *at al* 2014 *Eur. Phys. J. D* **68** 149

[1]E-mail: jaroslav.kocisek@ganil.fr
[2]E-mail: sergio.diaztendero@uam.es

Content from this work may be used under the terms of the Creative Commons Attribution 3.0 licence. Any further distribution of this work must maintain attribution to the author(s) and the title of the work, journal citation and DOI.
Published under licence by IOP Publishing Ltd

XXIX International Conference on Photonic, Electronic, and Atomic Collisions (ICPEAC2015) IOP Publishing
Journal of Physics: Conference Series **635** (2015) 032055 doi:10.1088/1742-6596/635/3/032055

Charge transfer and formation of complexes in the He^+ collisions with the furan molecules

Tomasz J. Wasowicz*[1] and Boguslaw Pranszke†, ‡

* Department of Physics of Electronic Phenomena, Gdańsk University of Technology, 80-233 Gdańsk, Poland
† Institute of Experimental Physics, University of Gdańsk, 80-952 Gdańsk, Poland
‡ Present address: Gdynia Maritime University, 81-225 Gdynia, Poland

Synopsis Charge transfer and formation of the collision complexes have been studied experimentally in fragmentation of the furan molecules in collisions with He^+ cations. The excited atomic and diatomic fragments of furan have been identified using collision-induced luminescence spectroscopy. Charge transfer ionization of the furan molecules has been observed in production of helium atoms in the excited 1s4d $^1D_2$, $^3D_{1,2,3}$ states. The fragmentation yields of the excited fragments have been measured as a function of the projectile energies (velocities) in the range 5-1000 eV. Enhancement of the fragmentation yields occurring at lower velocities may indicate formation of the $[He\text{-}C_4H_4O]^+$ collision complexes prior to dissociation.

Damage induced in the living cells by impact of the charged particles, for example in hadrontherapy, is related to the modifications of the structural and chemical properties of their molecular constituents. These modifications originate from the interaction of the primary cation beams and the secondary particles generated along the track such as low energy electrons, radicals and ions [1]. Unlike electrons and radicals, interaction of ions (primary and secondary) with biological media are poorly understood despite their importance in various therapies. To determine the most sensitive part of the DNA or RNA molecular chains to the cation-induced bond rupture, ionization and fragmentation of 2-deoxy-d-ribose, uracil, thymine, adenine, thymidine and amino acids under ionic irradiation have been investigated in the gas phase and layers condensed on the surfaces (see e.g. [2-4]).

Furan, C_4H_4O, has a special significance for biology and chemistry, since its ring structure resembles that of the simple sugars, ribose and deoxyribose, in the backbone of the RNA and the DNA helix.

In this communication we present studies of charge transfer and the collision complexes formation observed in fragmentation of furan molecule in the collisions with the He^+ cations applying collision-induced luminescence spectroscopy [5]. In Figure 1 the emission spectrum obtained for the impact of the 1000 eV He^+ cations is presented. It reveals production of the excited hydrogen H(n=4,5) and carbon C($2p3s$ $^1P_1$) atoms and the diatomic CH($A^2\Delta$) and $C_2(d^3\Pi_g)$ fragments. It also shows formation of the excited He ($1s1d$ $^1D_2$) atoms in the electron transfer from the furan molecule. Figure 2 displays the velocity dependences of the fragmentation yields, σ_{rel}, for the excited CH($A^2\Delta$) and He ($1s4d$ $^3D_{1,2,3}$) fragments. The observed rapid increase of the CH($A^2\Delta$) fragmentation yield occurring at lower velocities may be regarded as an indication of the $[He\text{-}C_4H_4O]^+$ complex being formed prior to dissociation of furan molecules [5].

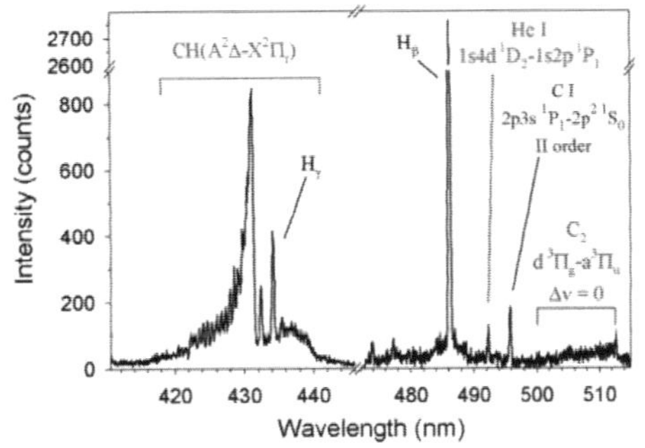

Figure 1. Emission spectrum measured at an energy of 1000 eV for He^++C_4H_4O collisions.

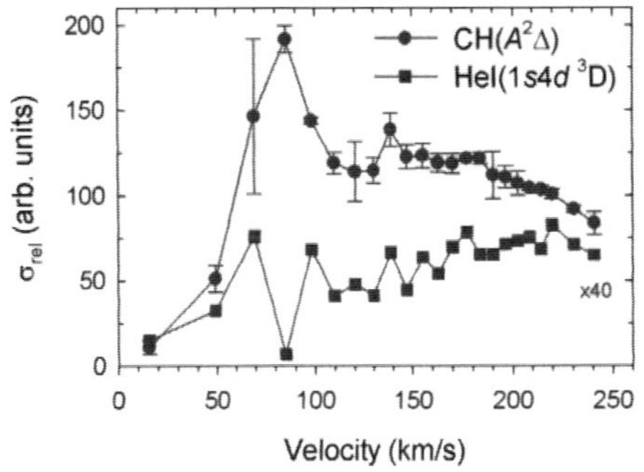

Figure 2. The CH($A^2\Delta$) and He I($1s4d$ ^3D) fragmentation yields obtained in the 15-250 km/s velocity range.

References

[1] M.C. Fuss *et al* 2010 *Chem. Phys. Lett.* **486** 110
[2] T. Schlatholter *et al* 2006 *Phys. Scr.* **73** C113
[3] Z. Deng *et al* 2005 *Phys. Rev. Lett.* **95** 153201
[4] S. Maclot *et al* 2011 *ChemPhysChem* **12** 930
[5] T. J. Wasowicz *et al.* 2015 *J. Phys. Chem. A* **119** 581

[1] E-mail: twasowicz@mif.pg.gda.pl

Content from this work may be used under the terms of the Creative Commons Attribution 3.0 licence. Any further distribution of this work must maintain attribution to the author(s) and the title of the work, journal citation and DOI.
Published under licence by IOP Publishing Ltd

XXIX International Conference on Photonic, Electronic, and Atomic Collisions (ICPEAC2015) IOP Publishing
Journal of Physics: Conference Series **635** (2015) 032056 doi:10.1088/1742-6596/635/3/032056

Water fragmentation by Li^{q+} and C^{q+} projectiles with MeV energies

W. Wolff[‡] [1], R. Schuch[†], D. Cariatore[+], S. Otranto[+], D. Fregenal*, G. Bernardi*, S. Suárez* [2]

[*] Centro Atómico Bariloche (CAB) Bustillo 9500, (8400) Bariloche – Argentina
[‡] Instituto de Física, Universidade Federal de Río de Janeiro (IF-UFRJ), Brazil
[+] Instituto de Física, Universidad Nacional del Sur, Bahía Blanca, Argentina.
[†] Department of Physics, Stockholm University, Alba Nova University Center, SE-106 91 Stockholm, Sweden

Synopsis We report on fragment distributions from water targets by impact of Li^{q+} and C^{q+} projectiles with MeV energies. We identify fragment species as function of emission energy by using an electrostatic spectrometer and as function of velocity by a time of flight mass spectrometer (pulsing the beam at the injector of the Pelletron accelerator). An improved Coulomb explosion model married to CTMC simulations allows us to calculate the energy of each fragment for different fragmentation channels.

In the last decades the irradiation of tumors by high energy protons or heavy ions has been considered as a standard treatment in cancer therapy. This fact has intensified the studies of water radiolysis as a simple way to understand the interaction of ionizing radiations with the human body.

The known high linear energy transfer within the Bragg-peak region, where projectiles are slowed down to less than MeV energies, lead the researchers to the investigation of heavy ion irradiation processes with water within that energy range. In particular, depending on the charge state of the transient water molecule prior the fragmentation, the ejected hydrogen ion(s) can acquire a broad range of energies.

In the present work, we report on fragmentation studies of water molecules by Li^{q+} and C^{q+} ions with MeV energies performed at the 1.7 MV Pelletron accelerator of the Centro Atómico Bariloche [1]. We used a cylindrical mirror electrostatic spectrometer [2] and a channeltron detector for the measurement of the positive charged fragments. We first measure the total fragment energy distributions for several angles of emission respect to the beam direction. Structures are observed in the energy spectra revealing the contribution of different fragmentation channels.

In order to identify the charge and mass of each measured fragment with a particular energy, we used the same spectrometer as a time of flight mass spectrometer. For this experiment the negative ion beam was pulsed at the injector of the Pelletron accelerator, just at the entrance of the acceleration tube. With the help of a standard coincidence circuit we collected TAC spectra that allowed us to discriminate the fragment species for fixed emission energies.

In Figure 1, we present the energy distribution of positive fragments from water by 3 MeV Li^{3+} impact at the detection angles of 10° and 90°. Fragments energies provided by a Coulomb Explosion (CE) model married to CTMC calculations are also shown. For example, H^+ (O^+) H means that around 14 eV the structure is formed by H^+ ions from the fragmentation channel $H^+ + O^+ + H$ and O^+ is the unobserved fragment ion. The CE model has been particularly modified to describe the H^+ + $(OH)^+$ fragmentation channel either originated in direct double electron removal or single ionization followed by autoionization. Present fragment ion spectra evidence contributions involving removal of up to 3 electrons.

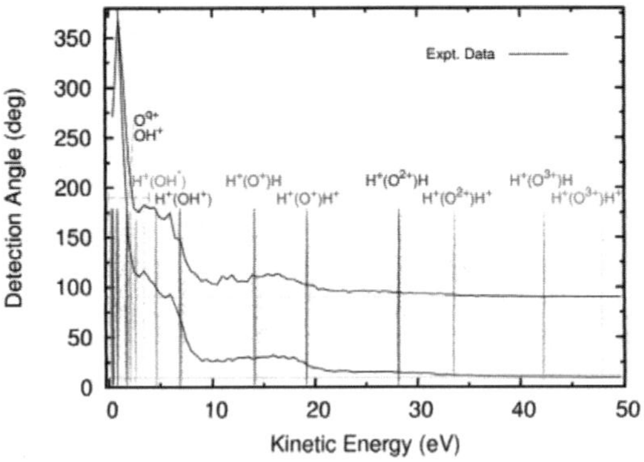

Figure 1. Experimental energy distributions of positive fragments from water by 3 MeV Li^{3+} impact at detection angles of 10° (lower spectra) and 90° (upper spectra). Vertical lines indicate calculated fragment energies labelled to the different fragmentation channels for detection angles from 0° - 180° (see text).

References

[1] S. Limandri, C. Olivares, L. Rodriguez, G. Bernardi and S. Suárez 2014 *Nucl. Instrum. And Meth. B* **318** 47-50
[2] S. Martínez, G. Bernardi, P. Focke, A. González and S. Suárez 2003 *J. Phys. B: At. Mol. Opt. Phys.* **36** 4813-4826

[1] E-mail: wania@ if.ufrj.br

[2] E-mail: suarez@ cab.cnea.gov.ar

Content from this work may be used under the terms of the Creative Commons Attribution 3.0 licence. Any further distribution of this work must maintain attribution to the author(s) and the title of the work, journal citation and DOI.
Published under licence by IOP Publishing Ltd

XXIX International Conference on Photonic, Electronic, and Atomic Collisions (ICPEAC2015) IOP Publishing
Journal of Physics: Conference Series **635** (2015) 032057 doi:10.1088/1742-6596/635/3/032057

Quantum dynamics study of the H_2 molecule confined in Single-Walled Carbon Nanotubes

Manel Mondelo-Martell[1], **Fermín Huarte-Larrañaga**[2],

Departament de Química Física & Institut de Química Teòrica i Computacional, Universitat de Barcelona, Barcelona 08028, Barcelona, Spain

Synopsis The quantum dynamics of the hydrogen molecule confined in carbon nanotubes is studied using the MCTDH algorithm. Novel insights on the coupling between the different degrees of freedom are found and discussed for this system.

The ever more important need to store low density gases, either potential fuels like H_2 or CH_4 or known pollutants such as NH_3 or other greenhouse gases, has been a driving force in the last few years for the study of novel materials with the capability to keep molecules adsorbed both efficiently and reversibly. Most of these new devices are based on nanostructured materials, such as zeolites, metal-organic frameworks and carbon nanotubes[1][2][3]. These solids present a huge surface-volume relation, and more importantly some substances have been seen to have a higher preference to be adsorbed in them than in regular surfaces. The study of this phenomenon has led to the discovery of important distortions in the energies and internal structure of the captured molecules. These effects are now known as *quantum confinement*. A deep understanding of the quantum confinement effects at a fundamental level would definitely help to the discovery and design of better storage devices. Moreover, it could help to develop other kind of devices, such as *quantum sievings*[4], or even open the way to the use of these materials to control chemical reactions, taking advantage of the distortions generated on the confined species[5].

In the present work we study the dynamics of the hydrogen molecule confined in the hollow cavity of a (8,0) Single-Walled Carbon nanotube. This is the narrowest carbon nanotube in which adsorption of hydrogen is energetically favourable, according to Molecular Dynamics studies[6]. Therefore, its cavity can be seen as being quasi one-dimensional and confinement effects are expected to be very important. The Multiconfiguration Time-dependent Hartree (MCTDH) method[7] has been used to perform the dynamic study in a rigorous and fully quantum approach. The study of the system allowed us to see that the confinement in nanometric cavities affects deeply to the dynamical structure of the trapped molecule, due to the tight potentials and the increase of the coupling between the different degrees of freedom.

[1]E-mail: manel.mondelo@ub.edu
[2]E-mail: fermin.huarte@ub.edu

References

[1] X. Ren *et al* 2011 *Chem. Eng. J.* **170** 10.1016/j.cej.2010.08.045

[2] L. Schlapbach, A. Züttel 2001 *Nature* **414** 10.1038/35104634

[3] R. Morris, P. Wheatley 2008 *Angew. Chemie Int. Ed. English* **47** 10.1002/anie.200703934

[4] Q. Wang *et al* 1999 *Phys. Rev. Lett.* **82** 10.1103/PhysRevLett.82.956

[5] T. Nielsen *et al* 2010 *ACS Nano* **4** 10.1021/nn1006946

[6] F. Huarte-Larrañaga, M. Albertí 2007 *Chem. Phys. Lett.* **445** 10.1016/j.cplett.2007.07.083

[7] H.-D. Meyer *et al* 1990 *Chem. Phys. Lett.* **165** 10.1016/0009-2614(90)87014-I

Content from this work may be used under the terms of the Creative Commons Attribution 3.0 licence. Any further distribution of this work must maintain attribution to the author(s) and the title of the work, journal citation and DOI.
Published under licence by IOP Publishing Ltd

Ionization and electron capture in 240 keV Kr^{8+} collisions with CO_2

Pragya Bhatt*[1], T. Sairam*, A. Kumar†, H. Kumar§ and C. P. Safvan*

*Inter University Accelerator Centre, Aruna Asaf Ali Marg, New Delhi-110067, India
†Department of Physics, Jamia Millia Islamia, New Delhi-110056, India
§Department of Physics and Astrophysics, University of Delhi, Delhi-110007, India

Synopsis The dissociation of CO_2^{q+} ions is studied in slow Kr^{8+} ion collisions with a CO_2 molecule using the technique of recoil ion momentum spectroscopy and multiple ion coincidence.

In collisions of slow highly charged ions with atoms/molecules, there is a finite probability that one or more electrons from the target atom/molecule are transferred to the projectile. The study of these electron transfer processes is important from fundamental point of view as it provides an insight into the ultra fast dynamics of such collision events and for practical applications as in astrophysics and fusion plasma [1]. Depending on the velocity of the incoming projectile, the processes of electron capture and/or ionization become dominant. Ionization and single electron capture processes are studied widely; multiple electron capture process is a debatable issue, however [2].

To study such fast processes, the techniques of Recoil Ion Momentum Spectroscopy and multiple ion detection are used in present work. The experimental setup described in [3] is modified to study the post collision projectile ions; a parallel plate electrostatic energy analyzer and a channel electron multiplier (CEM) are incorporated to disperse and detect these ions. The collisions of 240 keV Kr^{8+} ions generated from an ECR ion source with CO_2 are used to study the dynamical properties of the CO_2^{q+} (q=2-4) ions. The parent CO_2^{q+} ions and the fragment ions produced from their dissociation are detected by a dual micro channel plate coupled to a position sensitive delay line anode detector; the ejected electrons are detected by a CEM. The post collision Kr^{7+} ions (obtained from the capture of one electron from CO_2 by the incident Kr^{8+}) and the ejected electrons are used to mark the collision event one at a time using an OR gate; these are used to differentiate between the capture and ionization events. The multi-hit capability of the detection system makes possible the detection of all the fragment ions produced in the collision events. Typical time-of-flight (TOF) spectra of the ions produced due to ionization and electron capture events are shown in figure 1.

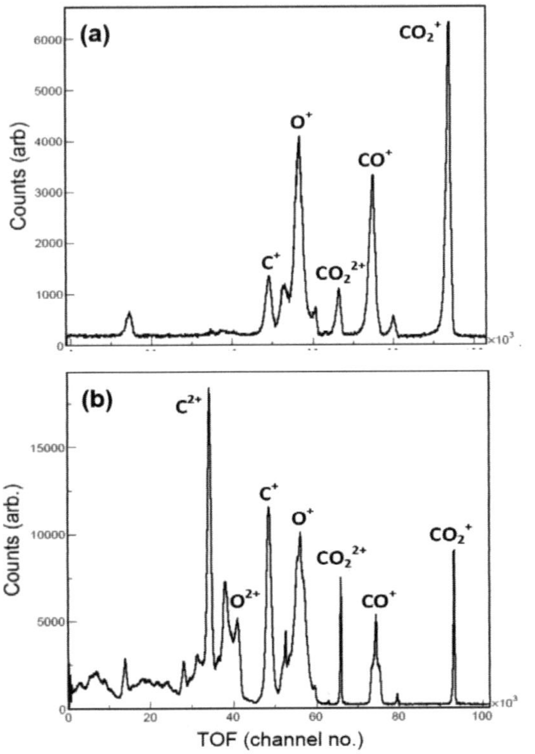

Figure 1. Time of flight spectra generated in 240 keV Kr^{8+} collisions with CO_2 using trigger by (a) Kr^{7+} (single electron capture) and (b) ejected electrons.

The effect of electron capture and ionization reactions on the angular correlation of the produced fragments and the kinetic energy release distribution of the CO_2^{q+} ions will be presented at the conference.

References

[1] R. K. Janev *et al* 1985 *Physics of highly charged ions* (Springer Series in Electrophysics, Vol. 13)
[2] F. Frémont *et al* 2002 *Int. J. Mol. Sci.* **3** 115
[3] A. Kumar *et al* 2014 *Int. J. Mass Spectrom.* **374** 44

[1] E-mail: pbpragya@gmail.com

Content from this work may be used under the terms of the Creative Commons Attribution 3.0 licence. Any further distribution of this work must maintain attribution to the author(s) and the title of the work, journal citation and DOI.
Published under licence by IOP Publishing Ltd

Fully differential cross sections for ionization of H_2 molecules by proton impact: single- and two-centre effects

M. F. Ciappina[*,1], **C. A. Tachino**[†], **R. D. Rivarola**[†], **S. Sharma**[‡], **M. Schulz**[‡]

[*] Max Planck Institute of Quantum Optics, Hans-Kopfermann Str. 1, D-85748 Garching, Germany
[†] Laboratorio de Colisiones Atómicas, Instituto de F´sica Rosario, CONICET y UNR, Argentina
[‡] Department of Physics and LAMOR, Missouri University of Science and Technology, Rolla, MO 65409, USA

Synopsis Fully differential cross sections for ionization of H_2 by 75 keV proton impact are compared with recent experimental data. Calculations were obtained by using a two-effective centre approximation within the CDW-EIS approach.

In this work, single ionization of hydrogen molecules by 75 keV proton impact is studied. Our goal is focused on the analysis of theoretical fully differential cross sections (FDCS), depending on the emitted electron momentum and on the transverse momentum transfer. To this end, different emission geometries and projectile kinematical conditions are considered. The present theory is compared with recent measurements made by Hasan *et al.* [1] .

Regarding the theory, a two-effective centre approximation (TEC) within the continuum distorted wave-eikonal initial state (CDW-EIS) approach, is employed to obtain FDCS. In this case, all the interactions present in the exit channel are considered on an equal footing. It should be mentioned that, as it happens for atomic targets, since the present differential cross sections depend on the momentum transfer, the interaction between the projectile and each molecular nuclei must be included in the formulation. Thus, it may play a relevant role in the corresponding calculations, depending on the energies of the emitted electron and momentum transfer values [2]. It is worth noting that the present theoretical method allows us then to compute two kind of FDCS, each one of them associated to different effects arising from the different interactions between the heavy particles during the reaction. In the so-called *two-centre FDCS*, interferences due to the presence of the two molecular nuclei is taken into account, whereas the *single-centre FDCS* are defined in such a way that only the effect related to the interaction between the projectile and each target nucleus is considered. This allows us to asses the contribution of each one of the mentioned interactions in the FDCS.

A reasonable good agreement between our theory and the measurements is achieved, in particular when a coplanar geometry where the electron emission plane coincides with the projectile dispersion one (the *scattering* plane) is analysed. Moreover, in some cases the agreement between experiment and the CDW-EIS–FDCS is better than for the case of more sophisticated schemes. However, substantial differences appear, for non-coplanar emission geometries and for large values of the momentum transfer. Also our approach appears to be perfectly suitable to tackle other *unconventional* electron emission geometries, for instance, those recently presented in [1], where the interference effects are clearly enhanced [3].

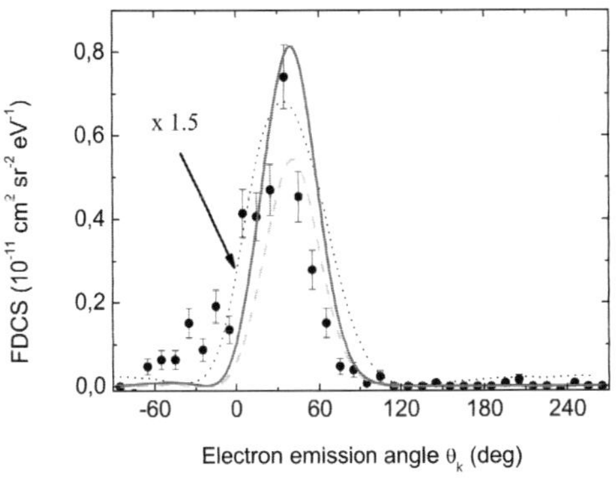

Figure 1. FDCS for electrons emitted with an energy of 14.6 eV into the scattering plane, as a function of the electron angle θ_k, for a momentum transfer $q = 0.9$ a.u. Dots: experiments [1], solid curves: two-centre FDCS, dashed curves: single-centre FDCS, dotted curves: M3DW results [1].

References

[1] A. Hasan *et al.* 2014 *J. Phys. B* **47** 215201

[2] M. Ciappina *et al.* 2008 *J. Phys. B* **41** 015203

[3] M. Ciappina *et al.* 2015, submitted for publication.

[1]E-mail: marcelo.ciappina@mpq.mpg.de

Content from this work may be used under the terms of the Creative Commons Attribution 3.0 licence. Any further distribution of this work must maintain attribution to the author(s) and the title of the work, journal citation and DOI.
Published under licence by IOP Publishing Ltd

Journal of Physics: Conference Series **635** (2015) 032060 doi:10.1088/1742-6596/635/3/032060

Toward laser spectroscopy of rotationally cooled CaH^+ ions trapped in a cryogenic linear Paul trap

Kunihiro Okada[1] and Masatoshi Kajita

[*] Department of Physics, Sophia University, 7-1 Kioicho, Chiyoda, Tokyo 102-8554, Japan
National Institute of Information and Communications Technology, Koganei, Tokyo 184-8795, Japan

Synopsis We have generated rotationally cooled CaH^+ ions with a cryogenic linear Paul trap toward laser spectroscopy. The direct observation of cold CaH^+ ions via laser Raman spectroscopy is now undergoing.

The study of possible time variations in fundamental constants plays an important role for the development of a ground unification theory (GUT) [1]. In addition to the fine structure constant α, the proton-to-electron mass ratio ($\beta = m_p/m_e$) is also one of important parameters for GUT, since β is related to $\dot{\alpha}/\alpha$ [1]. The ratio between a molecular vibrational transition frequency and an optical atomic transition yields the information of β. However, compared with the atomic transitions, precision measurements of the molecular transitions have not been progressed almost. In connection with this context, Kajita *et al.* proposed to detect possible time variation of β using a vibrational transition frequency of a single cold CaH^+ ion in a linear Paul trap (LPT) [2]. Then CaH^+ ions have been successfully produced by laser-induced reactions of $Ca^+(^2P_{1/2}) + H_2 \rightarrow CaH^+ + H$ in a LPT [3].

In this work we have developed a cryogenic LPT (Fig.1 (a)) in order to generate rotationally cooled CaH^+ ions. As shown in Fig.1 (b), the temperature at the LPT is cooled down to 13 K. Figure 1 (c) shows laser-induced fluorescence (LIF) images of Ca^+ Coulomb crystals in the cryogenic LPT before and after the laser-induced reactions. Since the mass of CaH^+ is a little larger than that of Ca^+, CaH^+ ions are distributed at the outside of the ion crystal. In this example, the number of CaH^+ ions is on the order of 10^3. In order to evaluate the rotational state distribution in $1^1\Sigma$ ($v = 0$), we have performed the rate-equation analysis including up to $J = 18$ rotational levels. The transition dipole moments and the rotational state constants were taken from the theoretical values [4] in order to include the transitions by black body radiations and spontaneous emissions. Figure 1 (d) shows plots of the time-dependence of the populations in $J = 0-3$ at 13 K, where we assume that the initial populations are the same as those at room temperature. The rotational state population of $1^1\Sigma$ ($v = 0$, $J = 0$) is evaluated to be 55% after 16 minutes.

We have been working on the laser Raman spectroscopy by $1^1\Sigma \rightarrow 2^1\Sigma$ ($\lambda = 405$ nm) excitations of rotationally cooled CaH^+ ions [4]. By inserting a half mirror in the detection optics, the Raman photons ranging from 430−750 nm [5] and the LIF from Ca^+ are simultaneously detected by a photomultiplier tube and a cooled CCD camera, respectively. At present we are searching for the resonance wavelengths of the rovibrational transitions in $1^1\Sigma \rightarrow 2^1\Sigma$, which are still unknown experimentally. The status of the experiment will be presented at the conference.

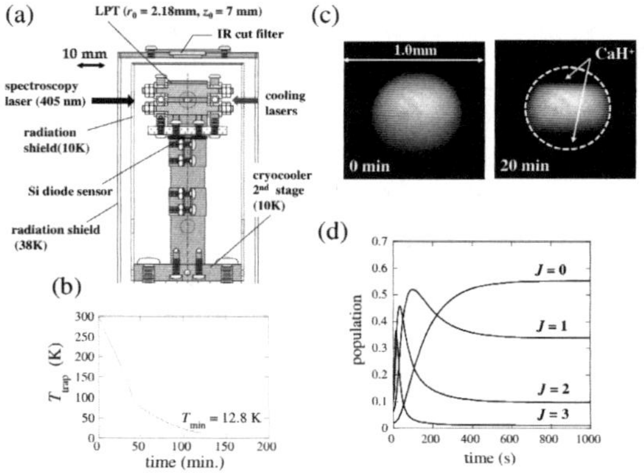

Figure 1 (a) Schematic drawing of the cryogenic LPT. (b) A plot of the wall temperature enclosing the LPT after turning on the cryocooler. (c) LIF images of Ca^+ ion crystals before and after the laser-induced reactions. (d) Simulation results of the rotational state populations of CaH^+ $1^1\Sigma$ ($v = 0$, J) as a function of time at 13 K.

References

[1] X. Calmet *et al.*, Eur. Phys. J. C **24**, 639
[2] M. Kajita *et al.*, J. Phys. B **42**, 154022
[3] N. Kimura *et al.*, Phys. Rev. A **83**, 033422
[4] M. Abe *et al.*, J. Phys. B **43**, 245102
[5] M. Abe *et al.*, CPL **521**, 31

E-mail: okada-k@sophia.ac.jp

Content from this work may be used under the terms of the Creative Commons Attribution 3.0 licence. Any further distribution of this work must maintain attribution to the author(s) and the title of the work, journal citation and DOI.
Published under licence by IOP Publishing Ltd

Journal of Physics: Conference Series **635** (2015) 032061 doi:10.1088/1742-6596/635/3/032061

Do linear molecules always dissociate along their axis? Intra-molecular scattering within Diiodoacetylene

Sankar De* [1], **H. Tezuka**[†], **P. Bhatt**[$], **G. Vesapidze**[§], **C. P. Safvan**[$], **J. Matsumoto**[†] **and H. Shiromaru**[† 2]

[*] Saha Institute of Nuclear Physics, 1/AF Bidhannagar, Kolkata, India
[†] Department of Chemistry, Tokyo Metropolitan University, Hachioji, Tokyo, Japan
[$] Inter-University Accelerator Centre, Aruna Asaf Ali Marg, New Delhi, India
[§] School of Natural Sciences and Engineering, Ilia State University, Tbilisi, Georgia

Synopsis We present here the coincidence study of multiply charged, dissociative diiodoacetylene. We find that instead of breaking up along the molecular axis as in the case of acetylene, the carbon ions of highly charged diiodoacetylene molecule tends to dissociate perpendicularly with respect to the iodine ones. This feature carries a strong signature of intra-molecular scattering between its breakup partners. The heavier iodines block the path of carbons and give rise to 'doughnut' – shaped momentum islands.

Highly charged ion (HCI) - molecule interaction opens up new channels to investigate the motional dynamics of nuclear configuration of multiply charged polyatomic species. Highly ionized molecules usually dissociate via Coulomb explosion where we intuitively predict for linear molecules that all the velocity vectors of the fragments will nearly align along the molecular axis. For example, C^{q+} and H^+ fragments from C_2H_2 are emitted in the same or opposite direction after collision with 30 keV/u Ar^{8+} projectiles [1] giving rise to 'butterfly'-shaped structures. However, this is not always the case.

Here we present a multi-coincidence time-of-flight (TOF) study of dissociative diiodoacetylene (C_2I_2) in collision with 3 keV/u Ar^{8+} projectiles from the ECR ion source at Tokyo Metropolitan University. From the TOF spectra obtained with (post collision) projectile charge state Ar^{7+} as trigger, we observed undissociated $C_2I_2^{q+}$ molecular ions with charge state 'q' as high as 3, thus proving the persistency of the molecular frame even at high charge states. The corresponding coincidence map shows highly charged symmetric ($C_2I_2^{4+} \rightarrow C_2I^{2+} + I^{2+}$) and asymmetric ($C_2I_2^{3+} \rightarrow C_2I^{2+} + I^+ \rightarrow C_2I^+ + I^{2+}$) two-body breakup channels.

With (post collision) projectile charge state Ar^{6+} as trigger, the TOF coincidence map in fig. 1 shows 'doughnut'-shaped islands for carbon and iodine ions. The shapes change with the charge states of the coincidence partners as the islands are correlated with the momentum and charge balance of the breakup fragments.

We analyzed the relative velocity vectors of any two coincidence partners for 'triple-hit' channels. On determining the complete momentum of coincidence partners, we observed that the carbon and iodine ions preferentially emit at 90° with respect to each other whereas C - C and I - I ions dissociate almost parallel to each other.

This confirms that while all the four atoms in highly charged C_2I_2 move apart from each other after Coulomb repulsion; the 'heavier' iodines at the ends stayed around their initial positions and obstructed the movement of the outgoing inner carbons. The 'doughnut'-shapes originate from the carbon ions emitted perpendicularly to the iodine ones. These unique shapes carry a signature of strong intra-molecular scattering between C^{m+} and I^{n+} ions. Monte Carlo numerical simulations are under way to help further understanding of the breakup process.

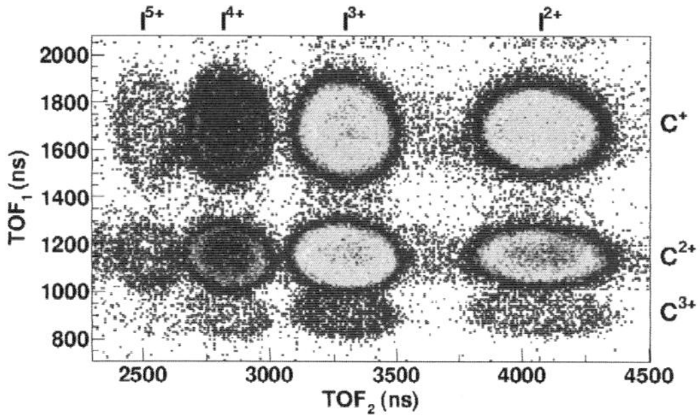

Figure 1. Coincidence map of carbon and iodine fragments from dissociative C_2I_2.

References

[1] S. De *et al* 2007 *J. Chem. Phys.* **127** 051101

[1] E-mail: sankar.de@saha.ac.in

[2] E-mail: shiromaru-haruo@tmu.ac.jp

Content from this work may be used under the terms of the Creative Commons Attribution 3.0 licence. Any further distribution of this work must maintain attribution to the author(s) and the title of the work, journal citation and DOI.
Published under licence by IOP Publishing Ltd

Journal of Physics: Conference Series **635** (2015) 032062 doi:10.1088/1742-6596/635/3/032062

Fragmentation of multiply ionized biomolecules induced by MeV heavy ions

S. O. Yoshida*, T. Majima*†1, T. Asai*, M. Matsubara*, H. Tsuchida*† and A. Itoh*†

* Department of Nuclear Engineering, Kyoto University, Kyoto 615-8540, Japan
†Quantum Science and Engineering Center, Kyoto University, Uji 611-0011, Japan

Synopsis Collision-induced ionization and fragmentation of gas-phase adenine molecules ($C_5N_5H_5$) have been investigated for single electron capture collisions of 1.2 MeV C^{2+} ions. Coincidence measurements between the number of ejected electrons and the time-of-flight of fragment ions allows us to determine charge states of intermediate parent ions r. Highly charged parent ions are observed up to $r \sim 6$ and the production of lighter fragment ions becomes predominant with increasing r. We found that kinetic energies of H^+ are significantly larger than those of other fragment ions and become nearly equivalent irrespective of r for $r > 1$.

Ion-induced fragmentation of DNA building blocks has gained increasing attention as fundamental processes in particle cancer therapy and radiation damage. In MeV heavy ion collisions, a large amount of energies are transferred from an incident ion to target electrons, leading to highly electronic excitation and multiple ionization of the target molecule. To understand the details of molecular fragmentation, we have extensively studied so far on the correlation between charged states of intermediate parent ions and fragmentation patterns for various polyatomic molecules [1, 2]. In this work, we extend measurements to an isolated biomolecular target of adenine ($C_5N_5H_5$).

The experimental procedure is given elsewhere [1]. Correlation data between charge states of intermediate parent ions and kinetic energies (KE) of fragment ions have been obtained under a single electron capture condition of 1.2 MeV C^{2+} ions. The reaction is written as

$$C^{2+} + C_5N_5H_5 \rightarrow C^{+} + C_5N_5H_5^{r+*} + n_e e^-,$$

where r represents the charge state of intermediate parent ions and is determined directly from the number of ejected electrons n_e: $r = n_e + 1$.

Figure 1 shows a result of total time-of-flight (TOF) mass spectrum of fragment ions. The main peaks are H^+ and CNH_2^+. Compared with slow ion collisions [3], relative yields of smaller fragment ions are high because of a large amount of energy transfers. It should be pointed out that fragment ions such as CNH_3^+, $CN_2H_3^+$ and $C_2N_3H_4^+$ are produced form proton migration in a molecule. Multiple ionization of up to $r \sim 6$ was observed by a direct electron counting technique [1-3]. By analyzing the coincidence data between the TOF and the number of ejected electrons, partial TOF spectra associated with a specific value of r were obtained. It is found that the intensities of lighter fragment ions increase with increasing r.

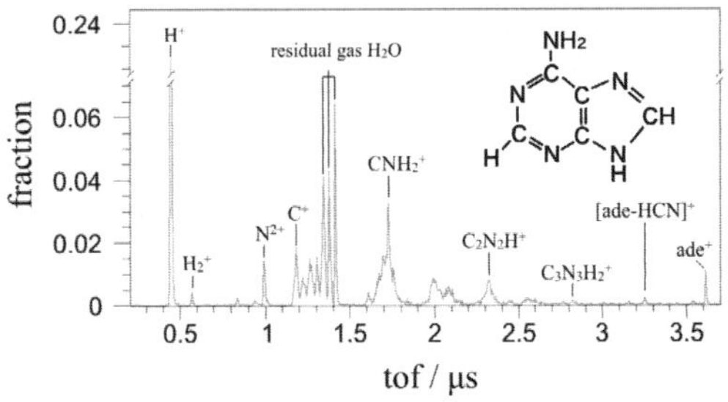

Figure 1. Total TOF spectrum of fragment ions obtained in collisions of 1.2 MeV C^{2+} with isolated adenine molecules.

Most of the fragment ions gain KE less than 10 eV, while H^+ gains significantly large KE extending to several tens of eV. It indicates a possibility of further fragmentation of DNA by these "high energy" protons [4]. The KE distribution of H^+ varies greatly for different values of r. As for $r = 1$, it peaks at 0 eV, while for $r > 1$ the KE peaks at about 6 eV with a tail extending to about 40 eV. It is pointed out that KE distributions are nearly independent of r for $r > 1$. While the reason is not clear at present, the simple Coulomb explosion model may not hold in the present collision system.

The work was supported by JSPS KAKENHI Grants No.24340096, No. 23760826 and Mizuho Foundation for the Promotion of Sciences.

References

[1] T. Majima *et al* 2014 *Phys. Rev. A* **90** 062711
[2] T. Majima *et al* 2004 *Phys. Rev. A* **69**, 031202(R)
[3] R. Brédy *et al* 2007 *Nucl. Instr. and Meth. B.* **261** 114
[4] Z. Deng *et al* 2006 *Phys. Rev. Lett.* **96** 243203

1 E-mail: majima@nucleng.kyoto-u.ac.jp

Content from this work may be used under the terms of the Creative Commons Attribution 3.0 licence. Any further distribution of this work must maintain attribution to the author(s) and the title of the work, journal citation and DOI.
Published under licence by IOP Publishing Ltd

Journal of Physics: Conference Series **635** (2015) 032063 doi:10.1088/1742-6596/635/3/032063

TILDA-V: A full-differential code for proton tracking in biological matter

M. A. Quinto* [1], **J. M. Monti**[† 2], **P. F. Weck**[# 3], **O. A. Fojón**[† 4], **J. Hanssen**[† 5], **R. D. Rivarola**[† 6], **and C. Champion*** [7]

[*]Université de Bordeaux, CNRS/IN2P3, Centre d'Etudes Nucléaires de Bordeaux Gradignan, Gradignan, France
[†]Instituto de Física Rosario, CONICET and Universidad Nacional de Rosario, Argentina
[#]Sandia National Laboratories, Albuquerque, NM 87185, USA

Synopsis Monte Carlo track structure codes are commonly used for predicting the radio-induced biological damages at the cellular scale. However most of them are based on semi-empirical cross sections as well as the use of water as surrogate of the biological matter. The aim of this work is to present our *home-made* Monte Carlo code, called TILDA-V, based on a quantum-mechanical cross section data base able to model all the proton- and secondary electron-induced collisions in both water and DNA.

Understanding the radiation-induced effects at the cellular level is of prime importance for predicting the future of irradiated biological organisms. Thus, whether it is in radiobiology to identify the DNA critical lesions or in medicine to adapt the radio-therapeutic protocols, an accurate knowledge of the numerous interactions induced by charged particles in living matter is required. To do that, Monte-Carlo track-structure codes represent the most suitable and powerful tools, in particular for modeling the full slowing-down of the ionizing particles in biological matter. However, it is worth mentioning that such numerical codes are reliable only if the input database used for modeling the charged particle induced interactions is precise and complete. In this context, the literature reports several numerical codes for proton and electron transport in water, the latter being commonly used as surrogate of the living medium.

The current work aims at going beyond this artifice with the development of an *event-by-event* Monte Carlo code - called ***TILDA-V*** - based on a complete set of multiple-differential and total cross sections for describing all the inelastic and elastic processes occurring throughout the slowing-down of protons in water and DNA [1]. ***TILDA-V*** (an acronym for **T**ransport d'**I**ons **L**ourds **D**ans **A**qua & **V**ivo) refers to an extension of the *TILDA* code previously developed by Champion *et al.* [2]. It is currently based on a complete set of quantum-mechanical cross sections for all the electron- and proton/hydrogen-induced interactions in water as well as in biological targets including the DNA nucleobases as well as the sugar-phosphate backbone [3-5].

Finally, a realistic description of the biological medium has been considered by modeling a typical nucleotide equivalent unit of hydrated DNA, namely, a nucleobase-pair plus a sugar phosphate group both surrounded by a hydration shell composed by 18 water molecules. Intra-comparisons between water and hydrated DNA in terms of proton range and stopping power will be presented in this work.

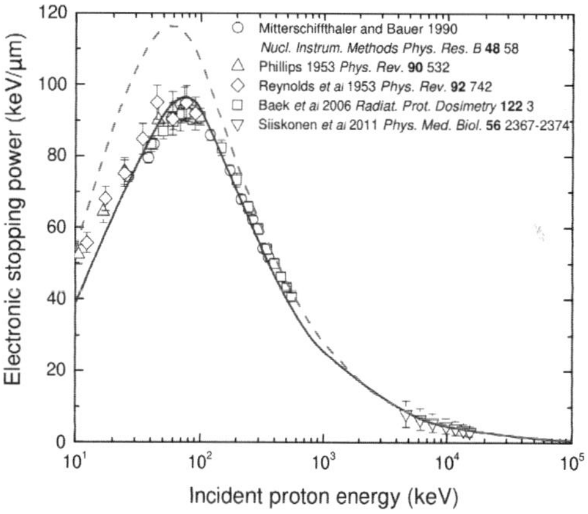

Figure 1. Electronic stopping power (induced by H^+/H^0) in water and hydrated DNA (blue and red line, respectively). Experimental data for water are reported for comparison.

References

[1] M. A. Quinto *et al* 2015 *J. Phys. Conf. Ser.* **583** 012049
[2] C. Champion *et al* 2005 *Radiat. Res.* **163** 222-231
[3] M. E. Galassi *et al* 2012 *Phys. Med. Biol.* **57** 2081-2099
[4] C. Champion *et al* 2012 *Phys. Med. Biol.* **57** 3039-3049
[5] J. M. Monti *et al* 2014 *Appl. Radiat. Isotopes* **83** 105-108

[1]E-mail: michele.arcangelo.quinto@gmail.com

Content from this work may be used under the terms of the Creative Commons Attribution 3.0 licence. Any further distribution of this work must maintain attribution to the author(s) and the title of the work, journal citation and DOI.
Published under licence by IOP Publishing Ltd

Journal of Physics: Conference Series **635** (2015) 032064 doi:10.1088/1742-6596/635/3/032064

Water versus DNA
A new deal for proton transport modeling in biological matter

C. Champion* [1], **M. A. Quinto**† [2], **J. M. Monti**† [3], **M. E. Galassi**† [4], **P. F. Weck**# [5], **O. A. Fojón**† [6], **J. Hanssen**† [7], **and R. D. Rivarola**† [8]

*Université de Bordeaux, CNRS/IN2P3, Centre d'Etudes Nucléaires de Bordeaux Gradignan, Gradignan, France
†Instituto de Física Rosario, CONICET and Universidad Nacional de Rosario, Argentina
#Sandia National Laboratories, Albuquerque, NM 87185, USA

Synopsis Water vapor is a common surrogate of DNA for modeling the proton-induced ionizing processes in living tissue exposed to radiations. The present study aims at scrutinizing the validity of this approximation and then revealing new insights into proton-induced energy transfers by a comparative analysis between water and realistic biological medium. In this context, self-consistent quantum mechanical modeling of the ionization and electron capture processes is reported within the continuum distorted wave-eikonal initial state framework for both isolated water molecules and DNA components impacted by proton beams.

In the context of biomolecular radiation damage, primary charged particles loose energy during interaction with biological tissues, potentially leading to the creation of secondary electrons in the medium. Modeling the radiobiological damages induced by such ionizing particles crossing the living matter requires a precise knowledge of the full radiation history, including the energy deposited during inelastic collisions as well as the kinetic energy transferred to secondary particles eventually emitted. Monte-Carlo methods are among the best-suited tools to achieve that goal since they provide an adequate description of the radio-induced energetic pattern at the finest scale, *i.e.* at the cellular or sub-cellular level. However, the reliability of such numerical methods heavily depends on the accuracy of the input data used in the simulations, namely, the interaction cross sections needed for describing the various collisional processes involved in the slowing-down of the charged particles in the medium of interest.

Thus, we recently proposed a series of quantum-mechanical models based on the first-Born approximation with correct boundary conditions and the continuum distorted wave-eikonal initial state methodology to describe the proton-induced ionization and electronic capture in water. Recently, both quantum-mechanical models were extended to handle more complex molecular targets and thoroughly tested in a series of studies dedicated to the description of the proton-induced ionization and electron capture on DNA components [1-3].

The current work aims at scrutinizing the appropriateness of water vapor as a surrogate for modeling the main ionizing proton-induced processes in human tissue and revealing new insights into proton-induced energy transfers in realistic biological medium.

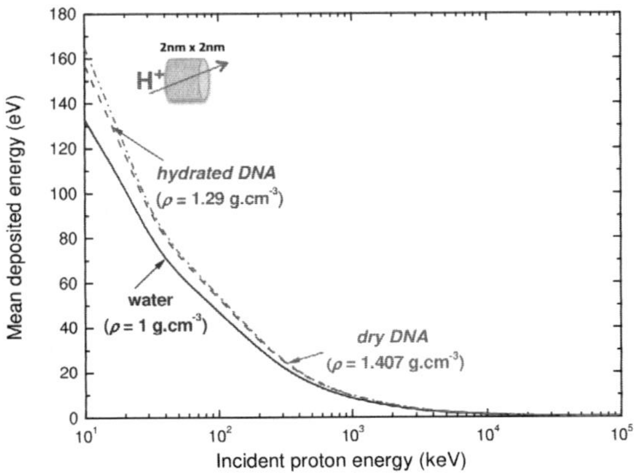

Figure 1. Mean deposited energy for proton-induced interactions in a DNA segment-size cylinder filled with water (blue line) or DNA (red line: dry DNA; green line: hydrated DNA).

References

[1] C. Champion *et al* 2010 J. *Phys. Med. Biol.* **55** 6053
[2] M. E. Galassi *et al* 2012 *Phys. Med. Biol.* **57** 2081
[3] C. Champion *et al* 2012 *Phys. Med. Biol.* **57** 3039

[1]E-mail: champion@cenbg.in2p3.fr
[2]E-mail: michele.arcangelo.quinto@gmail.com
[3]E-mail: monti@ifir-conicet.gov.ar
[4]E-mail: galassi@fceia.unr.edu.ar
[5]E-mail: pfweck@gmail.com
[6]E-mail: ofojon@fceia.unr.edu.ar
[7]E-mail: jocelyn.hanssen@univ-lorraine.fr
[8]E-mail: rivarola@fceia.unr.edu.ar

Content from this work may be used under the terms of the Creative Commons Attribution 3.0 licence. Any further distribution of this work must maintain attribution to the author(s) and the title of the work, journal citation and DOI.
Published under licence by IOP Publishing Ltd

Journal of Physics: Conference Series **635** (2015) 032065 doi:10.1088/1742-6596/635/3/032065

Hydrogen- *versus* proton-induced ionization in water vapor

C. Champion* [1], **J. M. Monti**† [2], **M. A. Quinto**† [3], **and R. D. Rivarola**† [4]

*Université de Bordeaux, CNRS/IN2P3, Centre d'Etudes Nucléaires de Bordeaux Gradignan, Gradignan, France
†Instituto de Física Rosario, CONICET and Universidad Nacional de Rosario, Argentina

Synopsis In the current work, neutral-hydrogen-induced ionization cross sections are calculated within both the continuum distorted wave-eikonal initial state and the 1st Born approximations. The results also obtained are intra-compared in terms of differential as well as total cross sections. Additionally, the rare available experimental data revealed fairly good with the theoretical predictions. Finally, we pointed out strong discrepancies with the proton's homologs.

In spite of its importance in many fields like astrophysics, plasma physics and radiobiology, hydrogen-induced interactions have been rarely experimentally investigated (see Refs. [1-2]). Moreover, among the existing experimental studies, only a few are focused on the *pure* ionization process. Indeed, during atom-target collisions, positive ions may be produced via many processes including target ionization, capture from the target as well as projectile ionization (electron loss). Thus, in the first two processes it is the target that is ionized whereas in the latter, the electron is extracted from the projectile. Under these conditions, it is clear that studying the *pure* hydrogen-induced ionization contribution requires the simultaneous detection of the charge state of both the projectile and the target.

In this context, collisions of hydrogen in water have received considerable attention during these last two decades, mainly due to the important role played by water, commonly used as surrogate of living matter, in particular in radiobiology for modeling the proton track structure in biological matter as well as in radiotherapy for improving the proton dosimetry in treatment planning. For more information, we refer any interested reader to Ref.[3] where we detailed our *home-made* TILDA-V Monte Carlo code based on a large set of quantum-mechanical cross sections able to describe the full proton slowing-down in a biological matter including both water and DNA components. Let us note that an accurate knowledge of all the proton- and hydrogen-induced interactions is all the more important that in the close vicinity of the Bragg peak (BP), namely, for proton energies around and below 100 keV, the electron capture becomes of the same order of magnitude, or even more important, than the ionization leading then to a substantial amount of atomic hydrogen in the projectile beam.

However, in spite of their importance, the cross section measurements for water ionization by neutral-hydrogen impact remain extremely rare. Besides, at least in part because of the absence of experimental data with which to compare, there has been little theoretical work done on neutral-impact ionizing processes. Many semi-empirical models were then proposed to overcome the lack of theoretical support.

The present work aims at providing a free parameter quantum mechanical model of the neutral-hydrogen-induced ionization of water molecules within both the *prior* CDW-EIS and the 1st Born approximation (FBA) [4].

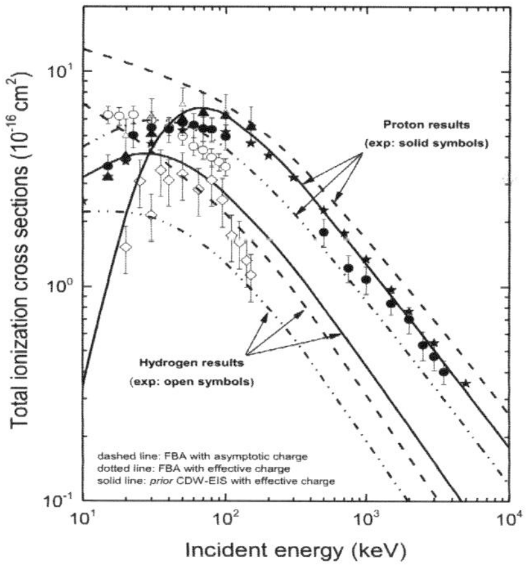

Figure 1. Total ionization cross sections of water molecule: neutral-hydrogen atoms *vs* protons.

References

[1] L. J. Puckett, G. O. Taylor, and D. W. Martin 1969 *Phys. Rev.* **178** 271
[2] R. J. McNeal, D. C. Clark, and R. A. Klinberg 1970 *Phys. Rev. A* **2** 131
[3] M. A. Quinto *et al* 2015 *J. Phys.: Conf. Ser.* **583** 012049
[4] C. Champion *et al* 2015 *Phys Rev A* submitted

[1] E-mail: champion@cenbg.in2p3.fr
[2] E-mail: monti@ifir-conicet.gov.ar
[3] E-mail: michele.arcangelo.quinto@gmail.com
[4] E-mail: rivarola@fceia.unr.edu.ar

Content from this work may be used under the terms of the Creative Commons Attribution 3.0 licence. Any further distribution of this work must maintain attribution to the author(s) and the title of the work, journal citation and DOI.
Published under licence by IOP Publishing Ltd

Journal of Physics: Conference Series **635** (2015) 032066 doi:10.1088/1742-6596/635/3/032066

Correlation between multiple ionization and fragmentation of small hydrocarbon molecules induced by fast heavy ions

T. Majima* [1], S.O. Yoshida†, T. Asai†, M. Matsubara†, H. Tsuchida*† and A. Itoh*†

* Quantum Science and Engineering Center, Kyoto University, Uji 611-0011, Japan
† Department of Nuclear Engineering, Kyoto University, Kyoto 615-8540, Japan

Synopsis We investigate correlations between multiple ionization and fragmentation processes of the acetylene, ethylene and ethane molecules in collisions with 0.58-1.2 MeV C^{+}, $^{2+}$ ions under specific charge changing conditions. The number of ionized electrons are observed directly by an electron counting technique. Fragmentation processes are discussed in terms of the charge state of parent ions prior to fragmentation. The correlation results reveal evidently roles of the charge state and the internal excitation in the molecular fragmentation processes.

Although a large number of studies of collision-induced dissociation of molecules have been performed so far, understanding of the mechanism is still challenging especially for large molecular targets. Most studies have tended to focus on specific fragmentation pathways which can be analyzed by coincidence measurements among the product ions. Understanding of the overall picture is still limited. One of the key parameters is the charge state of intermediate parent ions prior to fragmentation. To obtain the charge state, an electron counting method with a semiconductor detector is employed [1-3]. In this work, we apply this technique to collisions of 0.58–1.2 MeV $C^{+,\,2+}$ ions with the acetylene, ethylene, and ethane molecules (C_2H_{2n}, n = 1-3).

Experimental procedures are described elsewhere in detail [1]. In brief, product ions were analyzed by time-of-flight measurements with a position sensitive detector. Electrons emitted in collisions were detected with a semiconductor detector on a potential at +25 kV. Number distributions of the electrons are derived from pulse-height distributions of the signals. Charge states of outgoing projectile ions were analyzed to selectively study collision events in specific charge-changing conditions.

Figure 1 shows intensity fractions of product ions among the whole product ions from $C_2H_6^{r+}$* parent ions generated in single (1*e*)-capture and 1*e*-loss conditions of 0.56 MeV C^{+} ions. We see how distributions of product ions shift to smaller fragments with increasing the charge state *r*. Even at the same parent charge state, 1*e*-loss collisions exhibit higher degrees of H or H_2 loss than 1*e*-capture collisions, indicating higher internal energies accompanying in electron loss collisions. One can also see that H_3^{+} ions are generated only from doubly charged parent ions (r = 2). This is a good example of charge-specific reaction processes of multiply charged polyatomic molecular ions.

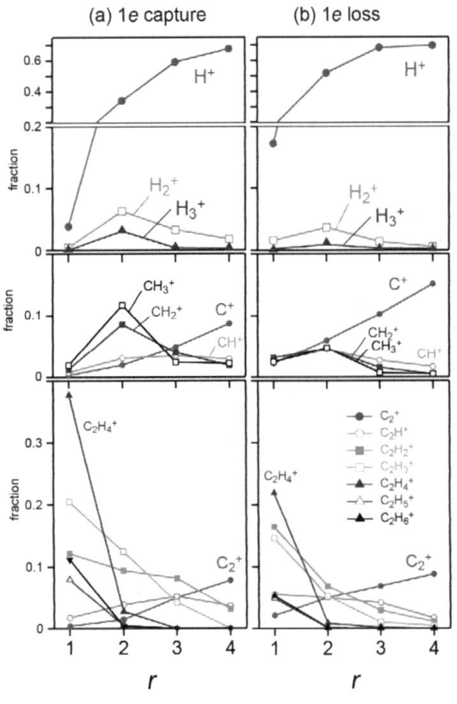

Figure 1. Fractions of product ions from $C_2H_6^{r+}$* parent ions generated in (a) 1*e*-capture and (b) 1*e*-loss conditions of 0.56 MeV C^{+} ions.

The work was supported by JSPS KAKENHI Grants No.24340096, No. 23760826 and Mizuho Foundation for the Promotion of Sciences.

References

[1] S. Martin *et al* 2008 *Phys. Rev. A* **77** 062513
[2] T. Majima *et al* 2004 *Phys. Rev. A* **69**, 031202(R)
[3] T. Majima *et al* 2014 *Phys. Rev. A* **90** 062711

[1] E-mail: majima@nucleng.kyoto-u.ac.jp

Content from this work may be used under the terms of the Creative Commons Attribution 3.0 licence. Any further distribution of this work must maintain attribution to the author(s) and the title of the work, journal citation and DOI.
Published under licence by IOP Publishing Ltd

Journal of Physics: Conference Series **635** (2015) 032067 doi:10.1088/1742-6596/635/3/032067

Measuring lifetimes of Polycyclic Aromatic Hydrocarbon fragments

M. Wolf[1], J. D. Alexander, M. H. Stockett, L. Giacomozzi, N. de Ruette, H. T. Schmidt, H. Zettergren, H. Cederquist

Department of Physics, Stockholm University, AlbaNova University Center, SE – 106 91 Stockholm, Sweden

Synopsis We investigate collisions of Polycyclic Aromatic Hydrocarbon cations with noble gases at center-of-mass energies close to 100 eV and below. Charged fragments are stored in a new version of the electrostatic ion beam trap ConeTrap. After storage for different well-defined times we extract the ions and use a reflectron based time-of-flight mass spectrometer to analyse the ion mass distributions. This will give information on secondary decay channels and the lifetime of the initially trapped fragment.

Here, we present a new experimental setup with which we aim to investigate long-term stabilities of fragments of Polycyclic Aromatic Hydrocarbon (PAH) molecules. These fragments may be produced in photo-absorption processes or by collisional induced dissociation (CID) in a gas cell. We will first focus on the latter processes as we have previously observed [1] that highly reactive PAH fragments are efficiently produced in the energy range around 100 eV and below by

$$PAH^+ + He \rightarrow [PAH - CH_x]^+ + CH_x + He$$

reactions. We will trap the $[PAH - CH_x]^+$ fragments and investigate how they cool down internally. This may occur through emission of radiation or by secondary fragmentation such as e.g.

$$[PAH - CH_x]^+ \rightarrow [PAH - C_3H_y]^+ + C_2H_{y-x}$$

or a combination of both. Due to the features of our ion trap the secondary fragments will in most cases also be trapped. After different storage times we will mass analyse the ions remaining in the trap in order to determine the intensity ratio between the primary and all secondary fragments. This will give the lifetime of the primary fragments.

One challenge in the design work has been to maintain the option to measure the fragment distribution directly from the collision cell [1]. To achieve this the trap had to be mounted without blocking the path between the gas cell and the fragment detector. This enables us to perform traditional CID experiments when all the trap electrodes are grounded. We thus developed a version of the electrostatic ion beam trap ConeTrap [2] with a large aperture (24 mm diameter). In the trap, the ions spend most of the time in regions with high potential. Therefore, the energy distribution of the secondary fragments will be similar to that of the primary fragments after extraction. The ion velocity will thus depend on the mass. On leaving the trap the ions pass through an Einzel lens which focuses them into a reflectron. This reflectron compensates for the energy spread and increases the time-of-flight separations between different fragments at an off-center MCP detector. A schematic is shown in figure 1. SIMION simulations indicate that we will be able to detect CH_x-loss for all PAH sizes up to that of coronene ($C_{24}H_{12}$) for 3 – 10 keV laboratory beam energies.

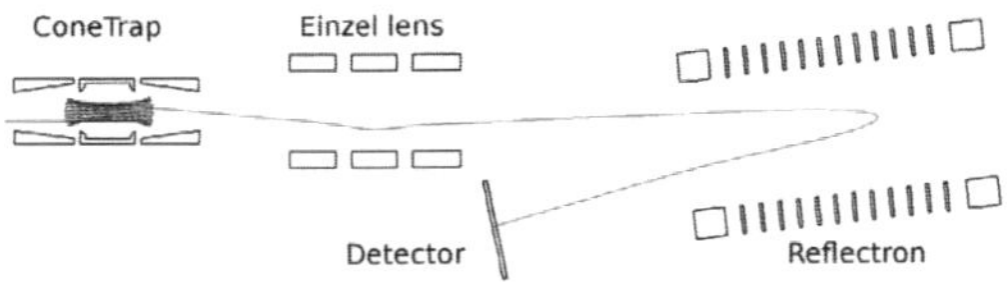

Figure 1. Schematic of the trap, lens, reflectron and MCP detector. One trajectory for an extracted fragment ion is shown. Both cone voltages are lowered for extraction while the central electrode is always grounded.

The current status is that trapping of PAH fragments has been demonstrated.

References

[1] M. H. Stockett *et al.* 2014 *Phys. Rev. A* **89** 032701

[2] H. T. Schmidt *et al.* 2001 *Nucl. Instr. Meth. Phys. Res. B* **173** 523

[1]E-mail: michael.wolf@fysik.su.se

Content from this work may be used under the terms of the Creative Commons Attribution 3.0 licence. Any further distribution of this work must maintain attribution to the author(s) and the title of the work, journal citation and DOI.
Published under licence by IOP Publishing Ltd

Journal of Physics: Conference Series **635** (2015) 032068 doi:10.1088/1742-6596/635/3/032068

Ion impact induced fragmentation of carbon dioxide.

Arnab Khan[1], **Lokesh C. Tribedi and Deepankar Misra**[2]

Department of Nuclear and Atomic Physics
Tata Institute of Fundamental Research, Homi Bhabha Road, Colaba, Mumbai 400005, India

Synopsis Momentum vectors of fragment ions produced in the Coulomb explosion of CO_2^{q+} (q = 3 − 5) in collision with highly charged ions are determined by the triple coincidence imaging technique. Kinematics as well as molecular structures of various explosion channels are reconstructed from the measured momentum vectors.

When a highly charged ion interacts with molecules one or many electrons can be knocked out from the molecule in various processes, namely, ionization, e-capture and e-capture-ionization etc. The probability that a process will be dominant mainly depends on the projectile velocity and the charge state. With the removal of electrons Coulomb interaction between various atomic centers increases and, as a result, the molecule may dissociate into the constituent atomic ions. The total charge acquired by the molecule gets divided into various daughter ions in the final state. Now the question is how the charge gets distributed inside a molecule? In case of diatomic molecules it is seen that symmetric charge distribution is more probable than asymmetric one. However, in case of poly-atomic molecules the situation is much more complicated. With the increase of the initial charge, the parent molecular ion gets fragmented in various combinations and with different dynamics. The sole aim of the recent study is to understand the decay dynamics of CO_2^{q+} upon the impact of 1 a.u. velocity Ar^{8+} ions. The velocity of the projectile is chosen in such a way that capture ionization is the most probable channel. In the past, several studies have been done in ionization dominant [1, 2] or capture dominant [3] regions, upon photon impact [4] and electron impact [5]. But the study of fragmentation dynamics of CO_2^{q+} in intermediate velocity regime is scarce. Most of these studies mainly concentrated on the decay dynamics of CO_2^{3+} or only to the kinetic energy release distribution studies.

Here we have used recoil ion momentum spectroscopy (RIMS) technique to study the fragmentation dynamics of CO_2^{q+} by employing a triple coincidence between all the three fragment ions. An effusive beam of CO_2 crosses perpendicularly with Ar^{8+} beam from an ECR based ion accelerator at TIFR, Mumbai. Electrons that are created in the interaction region are extracted and detected by a channel electron multiplier and serve as the start signal. Ions are projected onto a position and time sensitive detector. A representative Dalitz plot [6] of $CO_2^{3+} \rightarrow C^+ + O^+ + O^+$ fragmentation is shown in Figure 1. which shows a clear signature of concerted as well as step-wise fragmentation processes. It is also clear from the plot that the molecule bends significantly during fragmentation. We have not observed any asynchronous stretching [3] in the fragmentation of CO_2^{3+}.

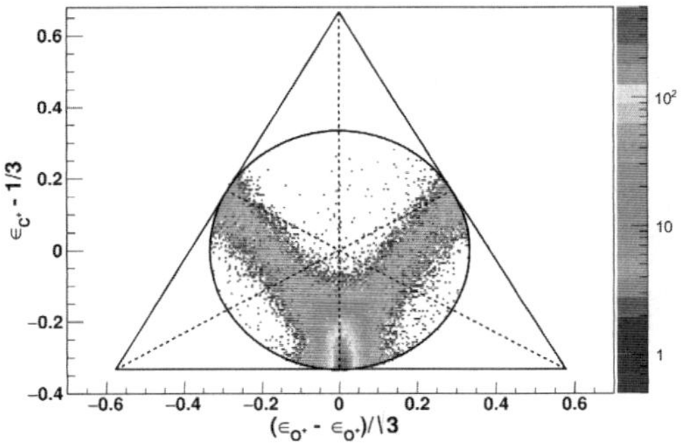

Figure 1. Dalitz plot for $CO_2^{3+} \rightarrow C^+ + O^+ + O^+$ fragmentation.

References

[1] M. R. Jana *et al* 2011 *Phys. Rev. A* **84** 062715

[2] B. Siegmann *et al* 2002 *J. Phys. B: At. Mol. Opt. Phys.* **35** 3755

[3] N. Neumann *et al* 2010 *Phys. Rev. Lett.* **104** 103201

[4] C. Wu *et al* 2013 *Phys. Rev. Lett.* **110** 103601

[5] C. Tian *et al* 1998 *Phys. Rev. A* **58** 3783

[6] R. H. Dalitz 1953 *Phil. Mag.* **44** 1068

[1]E-mail: arnab.khan@tifr.res.in
[2]E-mail: dmisra@tifr.res.in

Content from this work may be used under the terms of the Creative Commons Attribution 3.0 licence. Any further distribution of this work must maintain attribution to the author(s) and the title of the work, journal citation and DOI.
Published under licence by IOP Publishing Ltd

Journal of Physics: Conference Series **635** (2015) 032069 doi:10.1088/1742-6596/635/3/032069

Dissociation dynamics of nitrous oxide upon the impact of fast electrons and highly charged ions studied by a newly built 3-D focusing recoil ion momentum spectrometer

Arnab Khan [1], **Lokesh C. Tribedi and Deepankar Misra** [2]

Department of Nuclear and Atomic Physics

Tata Institute of Fundamental Research, Homi Bhabha Road, Colaba, Mumbai 400005, India

Synopsis Ionization and dissociation dynamics of N_2O has been studied upon the impact of fast electrons and highly charged ions using a two-stage 3-D focusing recoil ion momentum spectrometer. The kinematics of several fragmentation channels are studied by three-dimensional momentum imaging of fragments detected in coincidence. We observe the presence of both concerted and sequential breakup channels in the decay of N_2O^{3+} molecular ion.

Dissociation dynamics of nitrous oxide (N_2O) upon the impact of fast keV energy electrons and keV to MeV energy highly charged ions is studied. When one or more electrons are removed from a neutral molecule a molecular ion is formed. Depending on the amount of energy deposited several states of molecular ions can be populated. The decay products in a molecular breakup carry the information of transient molecular ions. The complete momenta and kinetic energy release (KER) distributions of the fragmentation of doubly and triply charged molecular ions have been deduced and compared with previous experimental as well as theoretical studies [1, 2, 3, 4]. Several studies are available on symmetrical molecule like CO_2 and asymmetric molecule like OCS. However, relatively less number of studies are available for the asymmetric, linear, tri-atomic molecule N_2O. Unlike OCS, where all three atoms are different N_2O has two N atoms which can be labeled as central (N_c) and terminal (N_t). Assignment of these two nitrogen atoms as central and terminal is difficult. However, under certain special conditions a partial assignment can be done.

The present experiments have been performed using a commercially available electron gun and an ECR based ion accelerator facility at TIFR, Mumbai. A newly build recoil ion momentum spectrometer [5] has been used to perform the experiments. It has been observed that N_2O^{2+} fragments mainly via two processes, namely, de-nitrogenation and de-oxygenation. It is also observed that de-nitrogenation is more probable than de-oxygenation. Figure 1. shows momentum distributions of fragmentation of N_2O^{2+} into O^+ and N_2^+. Along with this it is also seen that de-nitrogenation happens through the decay of a metastable state. Similarly we have also studied fragmentation N_2O^{3+}. It has been observed that the precursor ion decays via concerted and sequential processes.

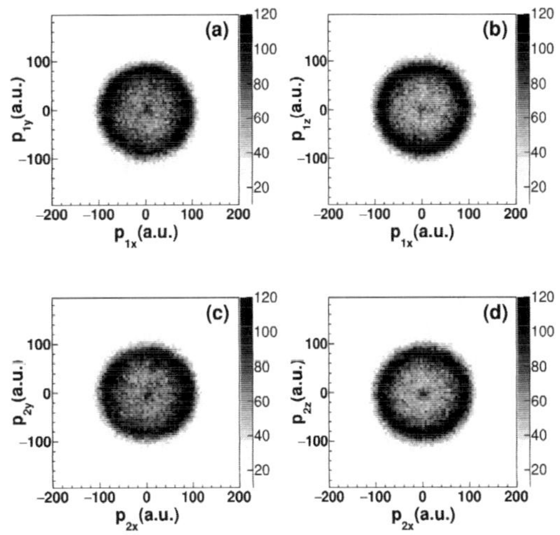

Figure 1. Momentum distributions for fragmentation of N_2O^{2+} into O^+ and N_2^+ (de-oxygenation). (a) p_x vs p_y, (b) p_x vs p_z for O^+ ions and (c) p_x vs p_y, (d) p_x vs p_z for N_2^+ ions.

References

[1] M. Ueyama *et al* 2005 *J. Chem. Phys.* **123** 154305

[2] S. Taylor *et al* 2006 *J. Chem. Phys.* **124** 204319

[3] P. Bhatt *et al* 2012 *Phys. Rev. A* **86** 052708

[4] R. Karimi *et al* 2013 *J. Chem. Phys.* **138** 204311

[5] A. Khan *et al* 2015 *Rev. Sci. Instrum.* **86** 043105

[1]E-mail: arnab.khan@tifr.res.in
[2]E-mail: dmisra@tifr.res.in

Content from this work may be used under the terms of the Creative Commons Attribution 3.0 licence. Any further distribution of this work must maintain attribution to the author(s) and the title of the work, journal citation and DOI.
Published under licence by IOP Publishing Ltd

Journal of Physics: Conference Series **635** (2015) 032070 doi:10.1088/1742-6596/635/3/032070

State-selective ion mobility measurement for NH_n^+ (n = 1 - 3) ions in helium gas at low temperature

K. Mori [1], H. Maki, A. Ninomiya, R. Isawa, and H. Tanuma [2]

Department of physics, Tokyo Metropolitan University, Minami-Ohsawa, Hachioji, Tokyo 192-0397, Japan

Synopsis We have measured ion mobility in cooled helium gas for NH^+, NH_2^+, and NH_3^+ at 77 and 4.3 K. In the measurement of NH^+ and NH_3^+, we have observed double peaks in the arrival time spectra, which correspond to different electronic states of molecular ions. These results can be regarded as the first observation of the metastable states in mobility measurement for these ions.

We have studied collision mechanism between ions and gases by using a low temperature ion-injected drift tube mass spectrometer cooled by liquid nitrogen and liquid helium.

Ion mobility is very sensitive to the interaction potential between a buffer gas and an ion. It is able to distinguish the electronic state of the ion by means of measurements of ion mobility, because the interaction potential depends on the electron state of the ion. We have already observed the separation of ground and metastable states for C^+, N^+, and O_2^+ ions produced by electron impact ionization in mobility measurements. Especially, the mobility of C^+ in He gas has been analyzed in detail by using of theoretical potential energy curves for both the ground and metastable states [1].

In this work, we have measured the mobility of NH_n^+ (n = 1 - 3) in He gas at 77 and 4.3 K. In Figure 1, we show typical arrival time spectra of NH_3^+ at low and high reduced electric fields at 4.3 K. At the high electric field, E/N = 65 Td, an arrival time spectrum has a single peak. On the other hand, at the low electric field, E/N = 2.6 Td, a double peak has observed in an arrival time spectrum. This result suggests that two peaks in the low electric field correspond to the ground and metastable states of the NH_3^+ ion, respectively.

According to theoretical calculations, the NH_3^+ ion in the ground state is planar (D_{3h}). On the other hand, that in the metastable electronic state has a pyramid structure (C_{3v}) [2]. It is reported that the molecular axis of ions in the drift tube is not random by collisions with buffer gases but orients to the electric field direction [3]. We suppose that a plane molecular ion flights like a Frisbee in the drift tube as the result of the collisional alignment. Such alignment gives the smaller collisional cross section between ions and gases and the shorter arrival time. Therefore, the fast peak in the low electric filed will correspond to the ground state of the NH_3^+ ion. On the other hand, the pyramid style ions have no alignment effect and larger cross sections than that of the planar molecular ions. Then the slower peak will be due to the metastable state of the NH_3^+ ion.

Also, in the measurement for NH^+, the separation of the peak has observed. However, the identification of the molecular states is quite difficult because of the unexpected behavior of the arrival time spectra.

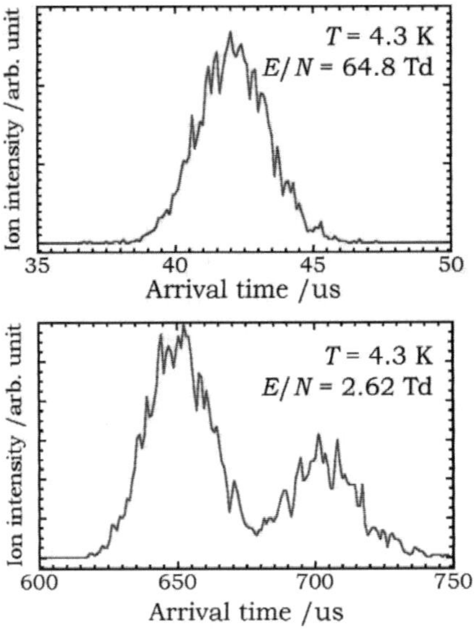

Figure 1. Arrival time spectra of NH_3^+ in He.

References
[1] S. Matoba *et al* 2008 *J. Phys. B* **41** 145205.
[2] A. R. Rossi and Ph. Avouris 1983 *J. Chem. Phys.* **79** 3413
[3] E. B. Anthony *et al* 1997 *J. Chem. Phys.* **106** 5413

E-mail: mori-kenta@ed.tmu.ac.jp

E-mail: tanuma-hajime@tmu.ac.jp

Content from this work may be used under the terms of the Creative Commons Attribution 3.0 licence. Any further distribution of this work must maintain attribution to the author(s) and the title of the work, journal citation and DOI.
Published under licence by IOP Publishing Ltd

Journal of Physics: Conference Series **635** (2015) 032071 doi:10.1088/1742-6596/635/3/032071

Double differential electron emission in ionization of Coronene by fast ions

Shubhadeep Biswas* [1], **C. Champion**† **and Lokesh C. Tribedi*** [2]

* Tata Institute of Fundamental Research, Mumbai 400005, India
† CNRS/IN2P3 Universite de Bordeaux, France

Synopsis We present the absolute differential electron emission cross section of coronene under fast bare O-ion impact. The results show some features which cannot be explained by simple ion-atom type collision result. The difference of the energy and angular distribution data with that for CH_4 and Ne may indirectly indicate the influence of collective excitation. We have performed the CB1 calculation for comparison.

Polycyclic aromatic hydrocarbon (PAHs) molecules are of great interest in recent times because of their astrophysical relevance. It has been realized that these molecules play an important role in physical evolution of inter stellar medium (ISM) [1]. According to the present understanding, different electronic excitations of these molecules are suitable candidates for explaining several unidentified astrophysical features. Therefore, to strengthen the claim, it has become relevant to understand the nature of physical as well as chemical changes of these molecules when these are exposed to different kinds of radiation in the ISM. Although, there are some photo-ionization and photo-fragmentation studies available, the interactions of ions with PAHs were not studied extensively. But it is extremely important because the energetic ions are inseparable components of the cosmic rays (CRs) present in the ISM. Only recently, the ion impact PAH studies have been performed in few cases concentrating on the fragmentation of gas phase monomers and clusters [2, 3].

The other important aspect regarding these molecules is the possibility of having low energy collective electronic response like plasmon excitation. These molecules comprise of a large delocalized π-electron cloud over the honycomb structured C skeleton. Because of the mutual electron-electron interaction, the whole electron cloud may respond collectively to the external perturbation. Although it was predicted long back [4], not much experimental studies exist in this respect. Keeping all these aspects in mind, in the present study, we have measured the absolute double differential cross section (DDCS) of electron emission in case of 3.75MeV/u O^{8+} ion impact on coronene ($C_{24}H_{12}$), which is one of the members of the PAH family. The measurements have been done in the energy range of 1 to 300 eV and in the angular range of 30° to 140°. An electrostatic hemispherical analyzer combined with channel electron multiplier has been used to detect the energy resolved electrons. As a theoretical investigation, we have carried out quantum mechanical calculations based on first-order Born approximation with correct boundary conditions (CB1).

A detailed comparison between the DDCS spectra of the coronene and the CH_4 indicates a clear difference in the low electron energy region. The analysis of the angular distribution of the DDCS also shows a strong forward backward asymmetry in that energy region, which cannot be explained in terms of pure ion-atom/small molecule collision models. We are in a process of understanding these unusual behaviors in terms of the collective excitation mechanism. The CB1 calculation, in general, provides a good qualitative description of the collision system, though there are discrepancies in the absolute values. Besides, this framework does not incorporate the collective excitation description [5].

[1] E-mail: shubhadeep@tifr.res.in

[2] E-mail: lokesh@tifr.res.in

References

[1] A. G. G. M. Tielens 2008 *Annu. Rev. Astron. Astrophys.* **46** 289

[2] A. I. S. Holm *et al* 2010 *Phys. Rev. Lett.* **105** 213401

[3] G. Reitsma *et al* 2013 *J. Phys. B: At. Mol. Opt. Phys.* **46** 245201

[4] H. W. Jochims *et al* 1997 *Int. J. Mass Spectrom. Ion Processes* **35** 167

[5] S. Biswas *et al* 2015 (under preparation)

Content from this work may be used under the terms of the Creative Commons Attribution 3.0 licence. Any further distribution of this work must maintain attribution to the author(s) and the title of the work, journal citation and DOI.
Published under licence by IOP Publishing Ltd

XXIX International Conference on Photonic, Electronic, and Atomic Collisions (ICPEAC2015) IOP Publishing
Journal of Physics: Conference Series **635** (2015) 032072 doi:10.1088/1742-6596/635/3/032072

Charge and energy flows in ionised thymidine

S Maclot[*,†,⊤ 1], **R Delaunay**[*,†], **D G Piekarski**[∥], **A Domaracka**[*], **B A Huber**[*], **L Adoui**[*,†], **F Martín**[∥,♯,‡], **M Alcamí**[∥,♯], **L Avaldi**[⊥], **P Bolognesi**[⊥], **S Díaz-Tendero**[∥], **P Rousseau**[*,† 2]

[*] CIMAP (UMR6552 CEA/CNRS/ENSICAEN/Unicaen), BP5133, 14070 Caen cedex 5, France
[†] Université de Caen Basse-Normandie, Esplanade de la Paix, CS 14032, 14032 Caen cedex 5, France
[⊤] Department of Physics, Lund University, PO Box 118, 22100 Lund, Sweden
[∥] Departamento de Química, Universidad Autónoma de Madrid, Cantoblanco 28049, Madrid, Spain
[♯] Instituto Madrileño de Estudios Avanzados en Nanociencia, Cantoblanco 28049, Madrid, Spain
[‡] Condensed Matter Physics Center (IFIMAC), Universidad Autónoma de Madrid, 28049 Madrid, Spain
[⊥] CNR-Instituto di Struttura della Materia, Area della Ricerca di Roma 1, Monterotondo Scalo, Italy

Synopsis We present a combined experimental and theoretical study of the ionisation and fragmentation of the nucleoside thymidine in the gas phase. Two sources of ionisation/excitation are used, namely UV photons and low-energy multiply charged ions, associated with coincidences measurements, respectively photoelectron/photofragment (PEPICO) and fragment/fragment. Coupling these experiments with quantum chemistry calculations, we obtain a complete picture of the fragmentation dynamics, in particular the charge and energy transfers within the molecular edifice.

The electronic and nuclear dynamics of complex molecular systems following sudden ionisation/excitation is a key of ultrafast chemical reactions. In particular, a charge migration due to hole formation could occur within the molecular structure at the fs timescale [1]. In the case of multiply charged systems, charges can be localised, e.g. due to the presence of "heavy" atoms, and the subsequent Coulomb explosion is associated with high kinetic energy fragments [2]. An intense and fast excitation is associated with intense fs laser electric field but also with collisions of multiply charged ions. Thus, a detailed knowledge of the response of excited complex molecular systems following photoionisation or ion collisions is of fundamental importance.

Moreover, with the development of radiotherapy including hadrontherapy, i.e. based on irradiation with protons or carbon ions, a better understanding of the radiation damage requires a multi-scale and -disciplinary approach. In this context, the molecular scale approach relies on studies, both experimentally and theoretically, of the ionisation/excitation of molecules of biological interest in the gas phase [3, 4, 5, 6]. Recent coupled experiment/theory works give a rather complete picture of the ion-induced fragmentation of amino acids [5, 6]. However a full comparison of experimental and theoretical results may suffer from the indeterminacy of the excitation energy transferred during the ion collision.

Here, we present a study of the ionisation/fragmentation of thymidine in the gas phase. We are coupling photoionisation and ion collisions experiments with quantum chemistry calculations in order to draw a complete picture of the charge and energy transfers.

Particularly, the comparison of the PEPICO mass spectrum with the calculated orbitals shows the strong dependence of the charge localisation with the excited orbitals. Using the PEPICO method, the fragmentation dynamics is followed as a function of the electronic excitation energy and the appearance energies associated with the opening of the main fragmentation channels are determined. The striking similarities between these energies and the calculated ones in the ground state show that the coupling between the electronic excitation and the vibrational one is highly efficient. Thus, we propose a method to evaluate the energy transferred into the system by ion collision using a fitting procedure of the ion-impact mass spectrum with the different PEPICO mass spectra. The obtained excitation energy distribution corresponds to the one expected.

References

[1] Calegari F *et al* 2014 *Science* **346** 336

[2] Erk B *et al* 2013 *Phys. Rev. Lett.* **110** 053003

[3] Bolognesi P *et al* 2009 *J. Phys. Chem.* A **113** 13593

[4] Castrovilli M C *et al* 2014 *J. Am. Soc. Mass Spectrom.* **25** 351

[5] Capron M *et al* 2012 *Chem. Eur. J.* **18** 9321

[6] Maclot S *et al* 2013 *J. Phys. Chem. Lett.* **4** 3903

[1]E-mail: sylvain.maclot@fysik.lth.se
[2]E-mail: patrick.rousseau@ganil.fr

Content from this work may be used under the terms of the Creative Commons Attribution 3.0 licence. Any further distribution of this work must maintain attribution to the author(s) and the title of the work, journal citation and DOI.
Published under licence by IOP Publishing Ltd

Journal of Physics: Conference Series **635** (2015) 032073 doi:10.1088/1742-6596/635/3/032073

An experimental apparatus for cold ion-atom collisions at the Cryogenic Storage Ring

Aodh P. O'Connor*, Manfred Grieser*, Florian Grussie*, Xavier Urbain†, Holger Kreckel* [1],

* Max-Planck-Institut für Kernphysik, Heidelberg, Germany
† Université Catholique de Louvain, Louvain-la-Neuve, Belgium

Synopsis Preparations for absolute rate coefficient measurements of ion-neutral reactions at the CSR

Gas-phase interstellar chemistry is dominated by two body reactions due to the low particle density of interstellar clouds. Neutral-neutral reactions are often endothermic and therefore potentially blocked at astrophysical temperatures. While anion-cation reactions are energetically favourable, the relatively low abundance of negative ions limit their influence on the chemical network. This leaves exothermic neutral-ion reactions to prevail as the foremost chemical processes in gas-phase interstellar chemistry.

With over 180 cosmic molecules detected to date, the molecular formation networks stem from a relatively small number of seemingly critical gateway reactions [1, 2]. Accurate measurements of these thermal reaction rate coefficients, combined with observations, are essential to develop astrochemical models and in turn our understanding of interstellar chemistry. Sensitivity studies have highlighted the most critical reaction rate uncertainties [3, 4]. Nevertheless, many of the astrochemical databases still rely heavily on semi-classical calculations, which are known to possess a high degree of uncertainty in the low temperature regime.

Despite the demand for accurate reaction rate coefficients, neutral-ion reaction studies remain a significant technical challenge with few measurements available. Newly emerging quantum calculations of reactions comprising three or more particles will still require accurate experimental data as a benchmark. Energy-resolved state-of-the-art measurements are represented by merged beams experiments [5, 6, 7] in which keV lab energy negative ions are photo-detached and the neutral products merged with atomic or molecular ions.

The development of merged beams apparatus surmounted many of the technical challenges hindering previous experimental measurements, namely the generation of a sufficiently intense neutral beam with well-defined spatial and momentum distribution and the possibility to precisely tune the collision energy. This enabled neutral-ion reaction measurements with relatively low kinetic temperatures, down to $\approx$ 70 K. However, difficulties arise in extending these measurements to astrophysically relevant temperatures as molecular ions extracted from conventional ion sources are often internally 'hot'. These 'hot' ions can potentially bring eVs of energy into the chemical process, influencing reaction rates.

In order to address these issues, we are developing an advanced ion-neutral collision apparatus for the Cryogenic Storage Ring (CSR) at the Max Planck Institute for Nuclear Physics in Heidelberg, Germany. In this experimental setup, molecular ions will be injected and stored in the CSR, where they can cool radiatively down to 10 K. Additional translational cooling via an electron cooler may be employed if required and desirable. Following this, a photo-detached beam of neutral atoms will be merged and co-propagated along a linear section of the CSR, with center of mass collisional energies in the meV range. By measuring the neutral and stored beam intensities and collecting reaction daughter products, accurate measurements of ion-neutral rate coefficients at truly astrophysical conditions will be performed.

References

[1] Eric Herbst 2001 *Chem. Soc. Rev.* **30** 168176

[2] V. Wakelam *et al* 2010 *Space Sci. Rev.* **156** 1-4 pp. 13-72

[3] V. Wakelam *et al* 2010 *A & A* **517** A21

[4] A. I. Vasyunin *et al* 2008 *ApJ* **672** 629

[5] V. M. Andrianarijaona *et al* 2009 *J. Phys: Conf. Ser:* **194** 012043

[6] A. P. O'Connor *et al* 2015 *ApJ* in press

[7] H. Kreckel *et al* 2010 *Science* **329** 69

[1]E-mail: holger.kreckel@mpi-hd.mpg.de

Content from this work may be used under the terms of the Creative Commons Attribution 3.0 licence. Any further distribution of this work must maintain attribution to the author(s) and the title of the work, journal citation and DOI.
Published under licence by IOP Publishing Ltd

Journal of Physics: Conference Series **635** (2015) 032074 doi:10.1088/1742-6596/635/3/032074

Double ionization of water molecule induced by swift H^+.

A. C. Tavares*, H. Luna†1, W. Wolff † and E C Montenegro†2

* Departamento de Física, Pontifícia Universidade Católica do Rio de Janeiro, 22451-900, Rio de Janeiro, RJ, Brazil
† Instituto de Física, Universidade Federal do Rio de Janeiro, 21941-972, Rio de Janeiro, RJ, Brazil

Synopsis Experimental cross sections for single and double ionization of H_2O by swift H^+ with energy ranging from 0.5 to 2.0 MeV are reported. In this energy range the ionization is the dominant collision process and charge transfer reactions can be disregarded. A multi-hit coincidence technique was used to measure the H^+ + OH^+ and H^+ + O^+ fragmentation channels. A model calculation allowed us to separate the post collisional contribution from the direct double ionization process, whenever an electron vacancy occurs in the 2a1 molecular orbital.

The ionization process leading to a multiple electron emission can be generated by various mechanisms, related to the strength of the collision dynamics, i.e. a hard or soft collision, or post-collisional ionization (PCI), i.e. Auger and Coster Kronig processes, electron shake-off, excitation followed by double Auger [1].

For the case of water molecule, the removal of one or more electrons from most of it molecular orbitals can be followed by fragmentation. The yields for the various possible molecular fragments depend in which one of the molecular orbitals the vacancies were produced. A key point in this whole process is the link between the population of vacancies and the branching ratios of the fragmentation products. These branching ratios are very difficult to calculate because the ejection of fragments occurs in conjunction with an equally fast rearrangement of the electrons in the orbitals.

For vacancies occurring in inner valence molecular orbital (eg. $2a_1$ in water), the Auger-like de-excitation channel can be open and competes with a prompt single-vacancy fragmentation. The relative yields of these two alternatives are unknown and difficult to be estimated. One of the reasons is because the Auger-like de-excitation results into two ionized species which can also be produced via dynamical two-step vacancies by the impinging projectile.

In this work double ionization cross sections leading to H^+ + OH^+ and H^+ + O^+ were measured using a multi-hit coincidence set-up. The experimental arrangement was previously described in reference [2] for the atomic target case. For a molecular target, opposite to the atomic ones, multiple ionization leads to the formation of multiple ionic fragments. In this case, the efficiencies related to ion collection (due to coulomb explosion) and double hit detection, have to be taken into account carefully. The procedures used in this work to obtain such efficiencies will be discussed in detail at the conference.

The comparison involving the measured ratios between double ionization cross sections and the single ionization channel (H_2O^+) with a model calculation based on first Born approximation allowed us to obtain the population of vacancies of the inner valence molecular orbitals. The contribution of an Auger like post collisional de-excitation from the $2a_1$ valence orbital to the double ionization was obtained. From the velocity dependence of the above ratios for proton and electron [3] impact it was possible to estimate the contributions from both the Auger-like and direct process (two-step ionization) to the total double ionization channel. It was found that Auger like process dominates for proton energies above 1.5 MeV while two-step process competes for lower energies.

References

[1] C.A. Tachino , M.E. Galassi and R.D. Rivarola *Phys. Rev.* A **80** 014701 (2009).
[2] W. Wolff, I. J. de Souza, A. C. Tavares, G. F. S. de Oliveira, and H. Luna, *Rev. Sci. Instrum* 83, 123107 (2012).
[3] S.W.J. Scully, J.A. Wyer, V. Senthil, M.B. Shah and E.C. Montenegro *Phys. Rev.* A **73**, 040701R (2006).

[1] E-mail: : hluna@if.ufrj.br
[2] E-mail: : montenegro@if.ufrj.br

Content from this work may be used under the terms of the Creative Commons Attribution 3.0 licence. Any further distribution of this work must maintain attribution to the author(s) and the title of the work, journal citation and DOI.
Published under licence by IOP Publishing Ltd

Journal of Physics: Conference Series **635** (2015) 032075 doi:10.1088/1742-6596/635/3/032075

Monte Carlo simulation for ion-molecule collisions at intermediate velocity

U R Kadhane* [1], **P M Mishra** [φ], **J Rajput** [Σ], **C P Safvan** [ϑ], **and S Vig** [λ]

[*] Department of Physics, Indian Institute of Space Science and Technology, Trivandrum 695547, India
[φ] Max-Planck Institut für Kernphysik, Heidelberg 69117, Germany
[Σ] Department of Physics and Astrophysics, University of Delhi, Delhi 110007, India
[ϑ] Inter-University Accelerator Centre, Aruna Asaf Ali Marg, New Delhi 110067, India
[λ] Department of Earth and Space Science, Indian Institute of Space Science and Technology, Trivandrum 695547, India

Synopsis Electronic energy loss distribution estimation is done under local density distribution using *Monte Carlo* simulations. These results are used to compare the experimental results of proton-polycyclic aromatic hydrocarbons (PAHs) and proton-nucleobase interactions at intermediate velocity collisions.

The ion-molecule collision study plays a major role in understanding fundamental atomic physics due to involvement of various processes such as direct (multiple) ionization, fragmentation, inner shell ionization preceding Auger effect etc. The fragmentation dynamics depends upon the columbic interaction between projectile and target, and hence depends on the energy transfer into the molecule. The analytical impact parameter treatment of energy loss is difficult for non spherical molecules. To probe this problem we have performed a *Monte Carlo* simulation for randomly generated trajectories passing through a non uniformly distributed electron density (obtained using density functional theory) representing the molecular volume. The electronic energy loss due to this interaction is estimated using the statistical energy loss deposition model [1] for PAHs [2] and nucleobases [3]. The trend in fragmentation yield of naphthalene upon proton (50-180 keV) and helium collision from our experimental data as a function of impact energy is reproduced with the help of simulated energy loss calculations (see Figure 1) [2].

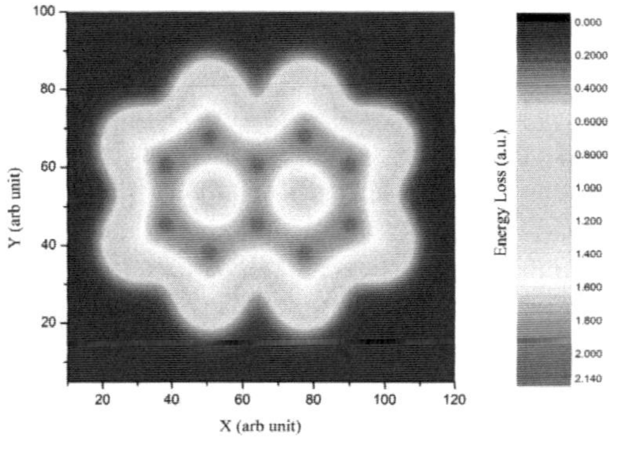

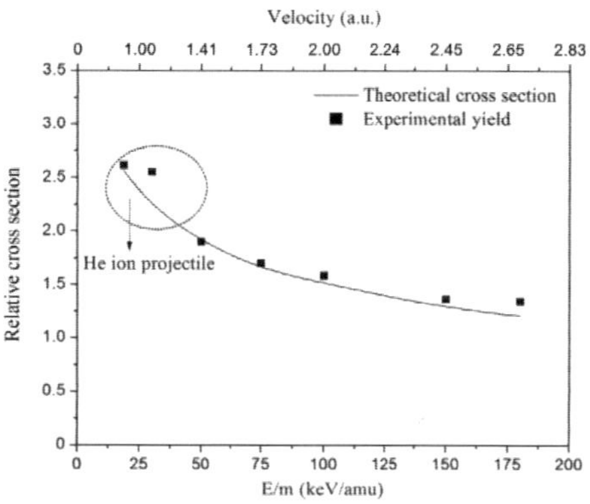

Figure 1. Electronic energy loss distribution in naphthalene obtained from *Monte Carlo* simulations (top) and Comparison of fragmentation yield with simulation results.

References

[1] N. M. Kabachnik *et al* 1998 *Phys. Rev. A* **57** 990
[2] P. M. Mishra *et al* 2014 *J. Phys. B: At. Mol. Opt. Phys.* **47** 085202
[3] P. M. Mishra *et al* 2014 *Nucl. Instrum. Meth B* **336** 12

[1] E-mail: umeshk@iist.ac.in

Content from this work may be used under the terms of the Creative Commons Attribution 3.0 licence. Any further distribution of this work must maintain attribution to the author(s) and the title of the work, journal citation and DOI.
Published under licence by IOP Publishing Ltd

Journal of Physics: Conference Series **635** (2015) 032076 doi:10.1088/1742-6596/635/3/032076

A geometric screening model for the calculation of ion-molecule electron removal cross sections from atomic contributions

H. J. Lüdde*[1], A. Achenbach*, T. Kalkbrenner*, H. C. Jankowiak*, M. Horbatsch†, and T. Kirchner†

* Institut für Theoretische Physik, Goethe-Universität, D-60438 Frankfurt, Germany
† Department of Physics and Astronomy, York University, Toronto, Ontario, Canada M3J 1P3

Synopsis We introduce a model that represents ion-molecule net capture and ionization cross sections as weighted sums of atomic cross sections. The weights are determined geometrically by analyzing the overlaps of the atomic cross sections drawn as circular disks around the nuclear centers of the molecule in a plane that is perpendicular to the projectile beam direction.

A simple way of calculating net cross sections for capture and ionization in ion-molecule collisions consists in adding up the respective ion-atom cross sections for all of the atoms that make up the molecule. This procedure is commonly referred to as Bragg's additivity rule (AR). It is based on the independent-atom model, in which the target is viewed as a collection of (undistorted) atoms centered at the equilibrium nuclear positions of the molecule. AR cross sections are independent of the orientation of the molecule relative to the projectile beam direction, and they do not differentiate between isomers, since the only structural information included is the molecular formula. Reasonably accurate results are usually obtained at collision energies above the Massey maximum, i.e., in regions in which the atomic cross sections are not too large. If they are, the AR tends to result in an overestimation of experimental data [1].

Similar observations have been made in the area of electron-molecule collisions, where AR total scattering cross sections agree well with measurements in fast collisions, but are too large below a certain impact energy. The range of applicability of the AR has been extended by introducing geometrical screening corrections, which reduce the weight of each atomic cross section that contributes to the molecular one [2].

In this contribution, we propose a similar model for heavy-particle collisions. The cross sections for net ionization and capture are represented by

$$\sigma_{\rm mol}^{{\rm net}\,x} = \sum_j^N s_j \sigma_j^{{\rm net}\,x},$$

where x stands for ionization or capture, the sum runs over the N atomic constituents of the target molecule, and $0 \leq s_j \leq 1$ are the screening coefficients, which we determine in the following way: For a given orientation of the molecule relative to the projectile beam (assumed to point towards the positive z-axis of a space-fixed coordinate system) we determine the positions of the nuclei in the perpendicular $(x-y)$ plane. Each of these points is the center of a circular disk with radius $r_j = \sqrt{\sigma_j^{{\rm net}\,x}/\pi}$ corresponding to the atomic net cross section $\sigma_j^{{\rm net}\,x}$, which we calculate by using the two-center basis generator method [3]. Depending on the orientation of the molecule and the size of the atomic cross sections the circular disks may overlap such that an atom in the background (i.e. at a larger z-value than others) is (partly) shielded and not completely 'visible' for the projectile (which is assumed to move on a straight-line path). The screening coefficient s_j is calculated as the ratio of the uncovered (visible) area σ_{pcm} to $\sigma_j^{{\rm net}\,x}$ with σ_{pcm} being calculated by a graphical pixel counting method (PCM). This procedure is repeated for a number of molecular orientations at each impact energy considered, and total cross sections to be compared with experimental data are obtained by averaging over all orientations.

At the conference we will show results of this procedure for proton collisions from a number of targets ranging from diatomic molecules such as H_2 and CO to intermediate-size hydrocarbons such as butane.

References

[1] A. Salehzadeh and T. Kirchner 2015 (to be published in *Physics Procedia*)

[2] F. Blanco *et al* 2010 *Phys. Lett. A* **374** 4420

[3] M. Zapukhlyak *et al* 2005 *J. Phys. B* **38** 2353

[1]E-mail: luedde@th.physik.uni-frankfurt.de

Content from this work may be used under the terms of the Creative Commons Attribution 3.0 licence. Any further distribution of this work must maintain attribution to the author(s) and the title of the work, journal citation and DOI.
Published under licence by IOP Publishing Ltd

Journal of Physics: Conference Series **635** (2015) 032077 doi:10.1088/1742-6596/635/3/032077

Proton-induced fragmentation of methane molecules

Arash Salehzadeh [1] **and Tom Kirchner** [2]

Department of Physics and Astronomy, York University, Toronto, Ontario, Canada M3J 1P3

Synopsis A theoretical study of the production of methane molecule fragments due to collisions with protons at impact energies ranging from 20 keV to a few MeV is presented.

We have investigated electron removal processes in the proton-methane collision system for an energy range of 20 keV to a few MeVs. Similar to a previous work [1], we have addressed the multi-centre nature of the problem by using a spectral representation of the molecular Hamltonian and a single-centre expansion of the initially populated molecular orbitals. The two-centre basis generator method is employed within the independent electron model to solve the single-electron Schrödinger equations. The solutions to the Schrödinger equations are then used to calculate properly orientation-averaged cross sections [2].

To obtain the molecule's fragmentation cross sections, the charge-state correlated cross sections are complemented with a collision dynamics independent dissociation model from [3]. The model assumes a variety of sequential decay-routes with fixed branching ratios for proton impact-induced CH_4 dissociations. According to the model, only single and double electron removal cross sections are taken into account and higher electron removal channels are neglected.

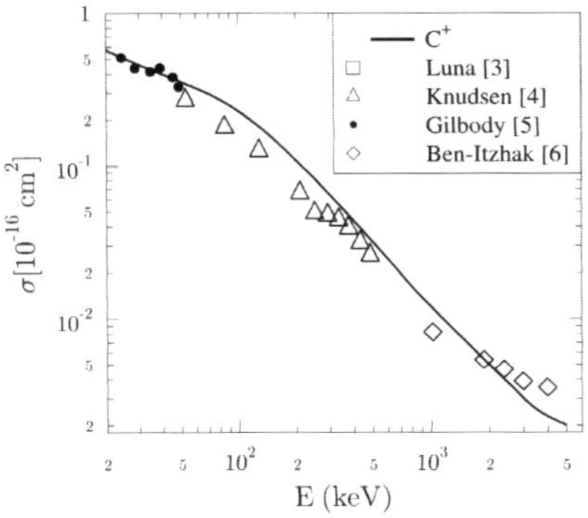

Figure 1. C^+ production cross section vs. impact energies for p-CH_4 collisions.

The fragmentation cross sections for the production of the CH_4 fragments (CH_4^+, CH_3^+, CH_2^+, CH^+ and C^+) have been calculated and compared against the measured cross sections. Overall, a good consistency is observed for all the fragments. At high energies, in particular, the agreement is strong. The cross sections for the production of fragments for which double electron removal plays an important role (CH_2^+, CH^+ and C^+) are very well predicted at low energies. The production cross section of C^+ is shown in Figure 1 while the CH_4^+ production cross section, where only single electron removal contributes, is displayed in Figure 2. The full set of results will be presented at the conference.

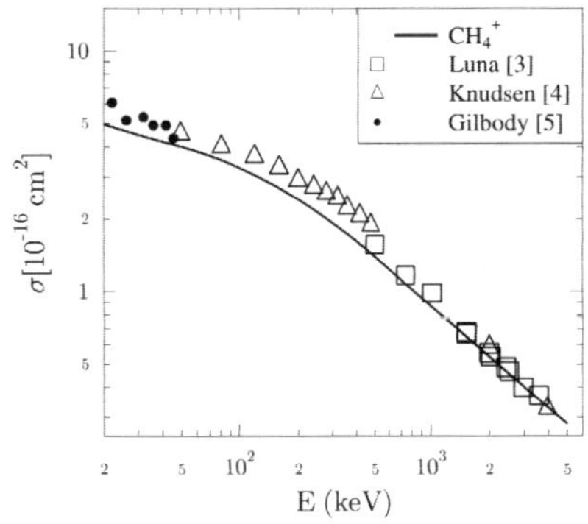

Figure 2. CH_4^+ production cross section vs. impact energies for p-CH_4 collisions.

References

[1] M. Murakami *et al* 2012 *Phys. Rev. A* **85** 052704

[2] A. Salehzadeh and T. Kirchner 2015 to be published in *Phys. Procedia*

[3] H. Luna *et al* 2003 *J. Phys. B* **36** 4717

[4] H. Knudsen *et al* 1995 *J. Phys. B* **28** 3569

[5] R. Browning and H. B. Gilbody 1968 *J. Phys. B* **1** 1149

[6] I. Ben-Itzhak *et al* 1994 *Phys. Rev. A* **49** 881

[1] E-mail: arash89@yorku.ca
[2] E-mail: tomk@yorku.ca

Content from this work may be used under the terms of the Creative Commons Attribution 3.0 licence. Any further distribution of this work must maintain attribution to the author(s) and the title of the work, journal citation and DOI.
Published under licence by IOP Publishing Ltd

Influence of capture to excited states of multiply charged ion beams colliding with small molecules

P. Montenegro[†1], **J. M. Monti**[†2], **O. A. Fojón**[†3], **J. Hanssen**[†4], **R. D. Rivarola**[†5]

[†] Instituto de Física Rosario, CONICET and Universidad Nacional de Rosario, Argentina

Synopsis Electron capture by multiply charged ions impacting on small molecules is theoretically investigated. Particular attention is paid to the case of biological targets. The interest is focused on the importance of the transition to excited final states which can play a dominant role on the total capture cross sections. Projectiles at intermediate and high collision energies are considered. Comparison with existing experimental data is shown.

The process of electron capture by multiply charged ion beams on atomic and molecular targets is a subject of main relevance in Medical Physics and Astrophysics. We focus here our research on small molecular targets, some of them of biological character, like water, which is usually employed to describe the living tissue, as for example in hadrontherapy. The interest of the present theoretical analysis resides in determining the role played by capture to excited states which is expected to dominate the total reaction. With such a goal in mind we employ the continuum-distorted-wave eikonal initial state (CDW-EIS) model which in its first-order has being formulated to give a better description of the reaction for atomic targets at intermediate impact energies, where its relative, the continuum distorted wave (CDW) model fails [1]. In order to achieve this more appropriate description a multiplicative eikonal phase distorting factor was introduced in the entrance channel to take into account the interaction between the projectile and the active electron to be captured and at same time to preserve the normalization of the initial state at all collision times. In the exit channel, a continuum distortion factor associated with the active electron-residual target interaction was chosen, in the same way as in CDW. Our previous calculations of total cross sections for small molecules were only given for capture to final excited states with small principal quantum number n for the case of proton impact [2-3]. A self-consistent field approximation is used to describe the initial molecular orbitals which are constructed as a linear combination of atomic wave functions corresponding to the different compounds of the target (MO-LCAO-SCF). The population analysis of the different orbitals is given by means of the neglect of differential overlap method (CNDO). The non-captured electrons are supposed to remain as frozen in their initial orbitals, considering that the collision time is smaller than the molecular vibrational and rotational ones. Effective continuum distorting factors are chosen in the exit channel, where effective coulombic charges are calculated from experimental orbital energies. Partial cross sections are determined for selective capture to each one of the final states with fixed quantum numbers n,l,m (where l and m are the angular and magnetic momentum ones, respectively), nl and n and total cross sections are computed by summing over all quantum numbers. Propensity rules for capture to different final states are also investigated. Calculations are compared with available experimental data.

References

[1] A. E. Martínez *et al* 1988 *Nucl. Instr. Meth. B* **34** 32
[2] M. E. Galassi *et al* 2010 *Phys. Rev. A* **81** 062713
[3] C. Champion *et al* 2012 *Phys. Med. Biol.* **57** 3039

[1] E-mail: montenegro@ifir-conicet.gov.ar
[2] E-mail: monti@ifir-conicet.gov.ar
[3] E-mail: fojon@ifir-conicet.gov.ar
[4] E-mail: jocelyn.hanssen@univ-lorraine.fr
[5] E-mail: rivarola@ifir-conicet.gov.ar

Content from this work may be used under the terms of the Creative Commons Attribution 3.0 licence. Any further distribution of this work must maintain attribution to the author(s) and the title of the work, journal citation and DOI.
Published under licence by IOP Publishing Ltd

XXIX International Conference on Photonic, Electronic, and Atomic Collisions (ICPEAC2015) IOP Publishing
Journal of Physics: Conference Series **635** (2015) 032079 doi:10.1088/1742-6596/635/3/032079

A time-resolved spectroscopic study of UV/visible emissions from the He^*/He_2^* (n=3) manifold states: formation channels for fast He_2^* production

R.J. Carman[1], **D.M. Kane**

Dept of Physics and Astronomy, MQ Photonics Research Centre, Macquarie University, Sydney, NSW 2109, Australia

We have investigated a fast formation channel yielding helium excited (bound) molecular states He_2^* from heavy-body conversion of n=3 manifold He^* atomic species, proposed by Frost *et-al* in 2001 (J.Phys.D, **34** 1569). Optical emission spectroscopy has been employed to measure the time-dependent intensity of ~20 atomic and molecular lines between λ=300-820nm from a short-pulse excited dielectric barrier discharge operating at 50-1000mbar. Correlations between atomic and molecular emission traces suggest that the singlet molecular state populations ($D^1\Sigma_u^+$, $E^1\Pi_g$, $F^1\Sigma_u^+$) can be attributed to conversion of the atomic $He^*(3^1S)$ state whilst the triplet molecular states are derived from the $He^*(3^3S)$ state.

Electronically excited (bound) He_2^* molecular states are known to play a major role in the kinetics of medium to high-pressure helium plasmas [1]. Key formation reactions include the three-body conversion of atomic $He^*(n=2)$ states and the di-electronic recombination of He_2^+ ions:

$$He^*(n=2) + 2He \Rightarrow He_2^*(^{1,3}\Sigma_{u,g}, {}^{1,3}\Pi_{u,g}) + He \quad (1)$$
$$He_2^+ + 2e \Rightarrow He_2^* + e \quad (2)$$

In 2001, Frost *et-al* [2] reported transient (τ<200ns) broad-band (molecular) emissions in the visible following pulsed laser state-selective pumping into the higher $He^*(n=3)^{1,3}P$ levels. They proposed the existence of a previously neglected fast formation channel of molecular states involving the creation of a short-lived (τ<5ns) intermediate activated He_2^* complex (denoted by $He_2^{*\ddagger}$) via two-body collisions (3). $He_2^{*\ddagger}$ subsequently undergoes fast radiative (4) and/or collisional (5) stabilization:

$$He^*(n=3, {}^{1,3}P) + He \Rightarrow He_2^{*\ddagger} \quad (3)$$
$$He_2^{*\ddagger} \Rightarrow He_2 + h\nu \quad (4)$$
$$He_2^{*\ddagger} + He \Rightarrow He_2^* + He \quad (5)$$
$$He_2^* \Rightarrow He_2 + h\nu \quad (6)$$

They also noted that the reactions (4)-(5) would compete with other known relaxation processes of $He_2^{*\ddagger}$ including associative ionisation and predissociation. To our knowledge, no follow-up studies of these observations have been reported, and the specific molecular states in (4) and (5) have yet to be identified. Moreover, due to weak LIF signals and low spectral resolution (Δλ~4nm), Frost *et-al* were unable to resolve the PQR branches within their emission bands to allow individual lines to be matched to known He_2^* transitions.

We have employed a short-pulse excited dielectric barrier discharge (DBD) to generate transient (~200ns) pulsed plasmas in helium at pressures of 50-1000mb. The experimental system and techniques are described in detail in [3]. DBD excitation is well suited to studies of heavy-body collisions in high pressure plasmas due to the intrinsic self-quenching of the discharge current pulse which limits the plasma electron density and subsequent electronic collisional processes e.g.(2) which may compete with heavy-body driven reactions.

The time-resolved intensities of ~20 selected He atomic lines and molecular bands between λ=300-820nm, originating from the n=3,4 manifolds, have been recorded using an Acton VM504 monochromator (Δλ~0.1nm resolution) and SRS430 multichannel analyzer (Δt~5ns resolution) with the detection system operating in photon counting mode. The principal results of our study show that band emission originating from both singlet and triplet He_2^* n=3 manifold molecular states ($D^1\Sigma_u^+$-$B^1\Pi_g$, $E^1\Pi_g$-$A^1\Sigma_u^+$, $F^1\Sigma_u^+$-$B^1\Pi_g$, $d^3\Sigma_u^+$-$b^3\Pi_g$, $e^3\Pi_g$-$a^3\Sigma_u^+$, $f^3\Sigma_u^+$-$b^3\Pi_g$) occurs concurrently with the atomic line emission from $He^*(n=3)$ states during the plasma excitation phase (when He_2^* formation via He_2^+/e recombination can be neglected). This suggests that a fast formation channel of molecular species is operative. Using a kinetic model based on conversion via an intermediate $He_2^{*\ddagger}$ state, the time-dependent emission curves from atomic/molecular transitions have been further analysed to determine possible correlations. Our results suggest that population transfer to the singlet molecular states ($D^1\Sigma_u^+$, $E^1\Pi_g$, $F^1\Sigma_u^+$) can be attributed to heavy-body conversion of the atomic $He^*(3^1S)$ state, thus linking $He_2^{*\ddagger}$ to high-v excited states of $F^1\Sigma_u^+(3d\sigma,3s)$ and $^1\Sigma_g^+(4f\sigma,3s)$. Similarly, the triplet molecular states are populated via $He^*(3^3S)$ conversion.

References

[1] F.Emmert *et-al* 1988 *J.Phys.D.* **21** 667
[2] M.J. Frost *et-al*, 2001 *J.Phys.B* **34** 1569
[3] R.J.Carman *et-al*, 2010 *J.Phys.D* **43** 025205

E-mail: robert.carman@mq.edu.au

Content from this work may be used under the terms of the Creative Commons Attribution 3.0 licence. Any further distribution of this work must maintain attribution to the author(s) and the title of the work, journal citation and DOI.
Published under licence by IOP Publishing Ltd

XXIX International Conference on Photonic, Electronic, and Atomic Collisions (ICPEAC2015) IOP Publishing
Journal of Physics: Conference Series **635** (2015) 032080 doi:10.1088/1742-6596/635/3/032080

Singlet and triplet states dissociation of doubly charged HDO, H_2O and D_2O as a function of internal energy

S. Martin* [1], **L. Chen***, **R. Brédy***, **J. Bernard*** **and A. Cassimi**†

* Institut Lumière Matière, UMR5306 Université Lyon 1-CNRS, Université de Lyon 69622 Villeurbanne cedex, France

† CIMAP, GANIL Université de Caen, Bd. H. Becquerel, Caen, France

Synopsis The fragmentation of HDO^{2+}, H_2O^{2+} and D_2O^{2+} is reported as a function of the excitation energy. The doubly charged ions were prepared selectively to triplet or singlet excited states in collisions with F^+ and H^+ at 3 keV, respectively. For HDO^{2+} the isotopic ratio, ratio of probabilities for the two dissociation channels $OD^+ - H^+$ over $OH^+ - D^+$, was estimated to be 4.5 ± 0.8 using H^+ and 3.8 ± 0.8 using F^+ projectiles. Results are compared with recent calculations.

Fragmentation of doubly charged water or heavy water ions has been studied with a large variety of excitation methods using photons, electrons, slow or fast ion impact. Two-body and/or three-body dissociation schemes have been predicted in theoretical calculations for each molecular energy state. For HDO^{2+} the isotopic ratio defined as the ratio of probabilities for the two dissociation channels $\frac{OD^+-H^+}{OH^+-D^+}$ was calculated to be 8.6 for the ground triplet state. For the first and second singlet states the ratios were found to be 3 and 15.7, respectively [1]. These calculations showed a strong dependence on the energy and the multiplicity of the HDO^{2+} molecular state.

In this work, we investigated the fragmentation of HDO^{2+}, H_2O^{2+} and D_2O^{2+} prepared selectively to singlet or triplet state due to the spin conservation rule using $H^+ + M \rightarrow H + M^{2+}$ (singlet states) or $F^+ + M \rightarrow F + M^{2+}$ (triplet states) collisions at 3 keV. M represents the mixture of HDO, H_2O and D_2O. Excitation energies of singlet and triplet states of the doubly charged parent ions with respect to the neutral ground state have been measured for different dissociation channels using the CIDEC method [2]. The doubly charged parent ions were identified by analyzing the correlation between the charged fragments. For HDO^{2+}, $OD^+ - H^+$ is the dominant dissociation channel for both multiplicities. For this channel the excitation energy of HDO^{2+} parent ion was measured to be 42.2 ± 0.5 eV and 40.7 ± 0.3 eV in H^+ and F^+ collisions. The isotopic ratio was estimated to be 4.5 ± 0.8 using H^+ and 3.8 ± 0.8 using F^+ projectiles. For F^+ projectiles the discrepancy between the measured and the theoretical isotopic value suggests that the initial vibrational and rotational states of neutral HDO play an important role in the dissociation dynamics of the doubly ionized molecules. The good agreement between the measured excitation energy and the calculated electronic energy levels indicates that the molecules are prepared to electronic excited states by electron capture directly from inner orbitals without vibrational energy deposition. This is also confirmed by our KER (kinetic energy release) measurements which are comparable to values obtained in photon ionization.

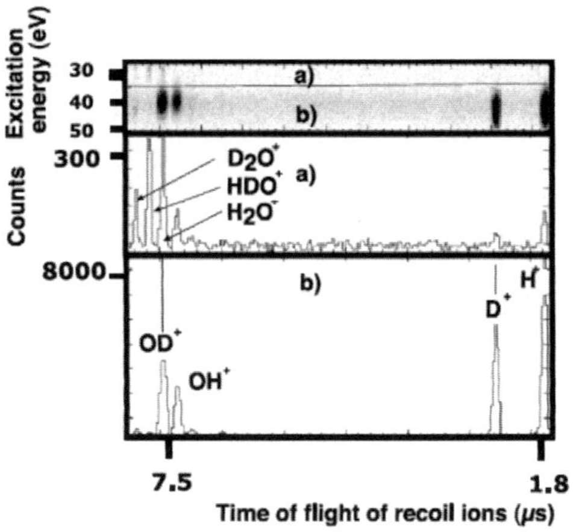

Figure 1. $H^+ + M \rightarrow H + M^{2+}$ (singlet states) collisions. Top panel: excitation energy (E_{exc}) versus time of flight of the recoil ions. (a) Mass spectrum for $E_{exc} < 34$ eV. (b) Mass spectrum for $E_{exc} > 34$ eV.

References

[1] B. Gervais *at al* 2009 *J. Chem. Phys.* **131** 024302

[2] S. Martin *at al* 2015 *J. Chem. Phys.* **142** 094306

[1]E-mail: smartin@univ-lyon1.fr

Content from this work may be used under the terms of the Creative Commons Attribution 3.0 licence. Any further distribution of this work must maintain attribution to the author(s) and the title of the work, journal citation and DOI.
Published under licence by IOP Publishing Ltd

Journal of Physics: Conference Series **635** (2015) 032081 doi:10.1088/1742-6596/635/3/032081

H_2 formation from Polycyclic Aromatic Hydrocarbon molecules

T. Chen[†1] M. Gatchell[†], M. H. Stockett[†], R. Delaunay[2,3], A. Domaracka[2], E. R. Micelotta[4], A. G. G. M. Tielens[5] P. Rousseau[2,3], L. Adoui[2,3], B. A. Huber[2], H. T. Schmidt[†], H. Cederquist[†], and H. Zettergren[†]

[†] Department of Physics, Stockholm University, SE-10691 Stockholm, Sweden

[2] CIMAP, UMR 6252, CEA/CNRS/ENSICAEN/Université de Caen Basse-Normandie, bd Henri Becquerel, BP 5133, F-14070 Caen cedex 05, France

[3] Université de Caen Basse-Normandie, Esplanade de la Paix, F-14032 Caen, France

[4] Université Paris Sud, Institut d'Astrophysique Spatiale, UMR 8617, 91405 Orsay, France

[5] Leiden Observatory, Leiden University, PO Box 9513, 2300 RA Leiden, The Netherlands

Synopsis In this work we study statistical fragmentation of Polycyclic Aromatic Hydrocarbon (PAH) molecules following collisions with keV ions. Dissociation and transition state energies for H- and H_2-emissions from PAHs have been calculated and a simple electronic stopping model has been used to calculate collision-induced internal PAH temperatures. We find that H_2 may be formed efficiently from pristine PAHs for internal ion temperatures above 2200 K.

Polycyclic Aromatic Hydrocarbons (PAHs) and molecular hydrogen (H_2) are present in many astrophysical environments [1, 2]. Broad emission features in the mid-infrared spectrum are observed in space and are in general attributed to infrared fluorescence of PAHs excited by ultraviolet radiation [3]. H_2 is the most abundant molecule in the universe and influences interstellar chemistry and star formation [1]. It is therefore of great interest to find the connection between PAHs and molecular hydrogen.

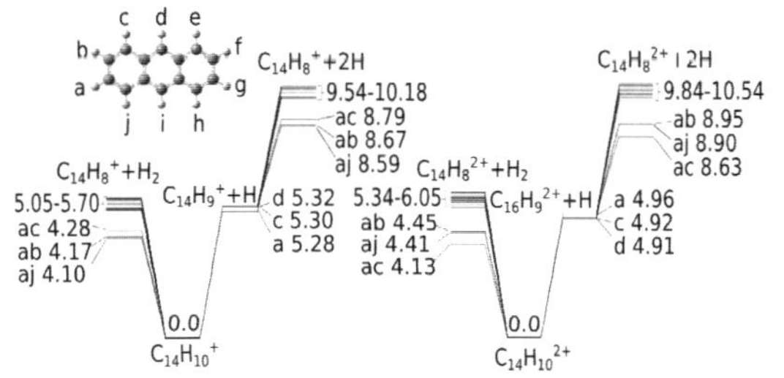

Figure 1. The adiabatic dissociation energies (in eV) for losses of H, H_2 and H+H from singly (left) and doubly (right) charged anthracene ($C_{14}H_{10}$) calculated with Density Functional Theory (B3LYP/6-311++G(2d,p)). The hydrogen atoms are labelled a-j.

We present experimental results for keV energy collisions between atomic ions and PAHs. We observe rich fragment mass spectra with peaks corresponding to the loss of e. g. H and 2H (H_2 or H+H) from singly charged PAHs. It is difficult to distinguish between H_2 and H+H contributions to $[PAH\text{-}2H]^+$ peaks by simple experimental techniques. In order to be able to discuss the origin of the $[PAH\text{-}2H]^+$ peaks, we have calculated dissociation energies and reaction barriers for H-, H_2- and H+H - emission from PAHs by means of quantum chemistry methods (B3LYP\6-311++G(2d,p)). Figure 1 shows the calculated adiabatic dissociation energies for H-, H_2- and H+H - emissions from singly and doubly charged anthracene ($C_{14}H_{10}$) [4].

We find that reaction barriers decrease the H_2-formation rates. As a consequence, this process is only effective for internal PAH temperatures exceeding, roughly, 2200 K, a temperature which is not reached for absorption of single photons below the Lyman limit (13.6 eV). In contrast, by using a simple model [5] for collision-induced electronic excitation, we find that PAHs may be strongly heated in collisions with ions and atoms such that H_2 molecules may be formed directly from pristine PAHs.

References

[1] A. G. G. M. Tielens 2005 *The Physics and Chemistry of the Interstellar Medium* **Cambridge University Press**

[2] A. G. G. M. Tielens 2013 *Reviews of Modern Physics* **85** 1021

[3] A. G. G. M. Tielens 2008 *Annu. Rev. Astron. Astrophys.* **46** 289

[4] T. Chen *et al* 2015 *Journal of Chemical Physics* **submitted**

[5] T. Chen *et al* 2014 *Journal of Chemical Physics* **140(22)** 224306

[1]E-mail: tao.chen@fysik.su.se

Content from this work may be used under the terms of the Creative Commons Attribution 3.0 licence. Any further distribution of this work must maintain attribution to the author(s) and the title of the work, journal citation and DOI.
Published under licence by IOP Publishing Ltd

XXIX International Conference on Photonic, Electronic, and Atomic Collisions (ICPEAC2015) IOP Publishing
Journal of Physics: Conference Series **635** (2015) 032082 doi:10.1088/1742-6596/635/3/032082

Structure, Ionization and Fragmentation of Neutral and Positively Charged Hydrogenated Carbon Clusters: $C_nH_m^{q+}$ $(n = 1 - 5, m = 1 - 4, q = 0 - 3)$

J.P.Sánchez* [1], **N.F. Aguirre*** [2], **S. Díaz-Tendero***, **M. Alcamí***, **F. Martín*** † ‡

* Departamento de Química, Módulo 13,Universidad Autónoma de Madrid, 28049, Madrid, Spain
† Instituto Madrileo de Estudios Avanzados en Nanociencias (IMDEA-Nanociencia), 28049 Madrid, Spain
‡ Condensed Matter Physics Center (IFIMAC), Universidad Autónoma de Madrid, 28049 Madrid, Spain.

Synopsis The $C_nH_m^{q+}$ family was studied at the DFT-B3LYP//CCSD(T) level of theory. Dissosiation energies, ionization potentials, geometries and frequencies were obtained. Different trends for these properteis have been observed and analyzed. The fragmentation behaviour has been studied within a combinatorial point of view.

Partially hydrogenated carbon clusters have been widely and heterogeneously studied. Theoretical modeling of their properties and their fragmentation behaviour provides valuable information to study the evolution of new molecules and ions detected in the Instersteallar Medium (ISM). It also helps to understand the composition and chemical processes taking place in the ISM[1]. Previously we have only considered bare carbon clusters [2], when hydrogen atoms are added, the complexity of the studied systems increases. In general, only a few works have considered this family of molecules systematically but only on small sequences[3]. This work intents to homogenize all the sparse data available in bibliography for the hydrogenated carbon clusters. From this study we obtain trends that they may be extrapolated to bigger and more complex molecules of the same family.

In this work the calculation of several properties for species of the $C_nH_m^{q+}$ familiy (for n=1-5; m=1-4; q=0-3) are carried out at DFT-B3LYP/6-311++G(3df,2dp)// CCSD(T)/6-311++G(3df,2dp) level of theory. The studied properties are: relative energies between isomers, ionization potentials, dissociation energies, vibrational frequencies and fragmentation reactions.

The results are interpreted in terms of both, energetic and entropic contributions. The energetic contributions are the values obtained for the relative energy of the isomers. For instance, Fig. 1 shows the distribution of all the considered molecules, note the scaling of the problem by increasing the system size on the top panel. The entropic contribution comes from the combinatorial choices of all possible reaction channels. Fig. 2 shows the combinatorial probability density for the dissociation processes leading to 1, 2, 3 and 4 fragments.

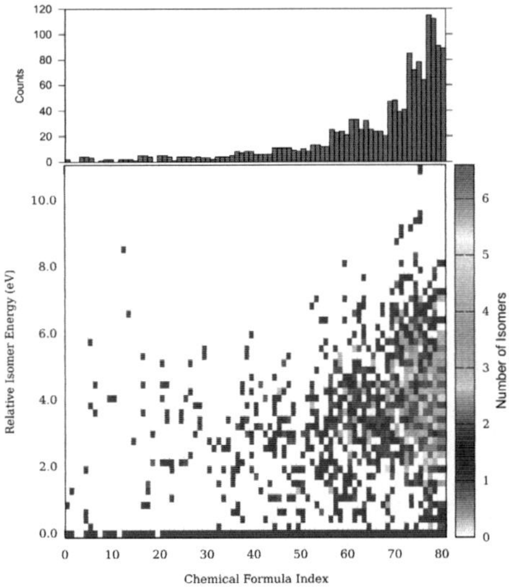

Figure 1. Bivariate histogram with the number of isomers as a function of the Chemical Formula Index $CFI[C_nH_m^{q+1}] = (q+1) + 4(m-1) + 4(n-1)$, and the relative isomer energy (from their ground state).

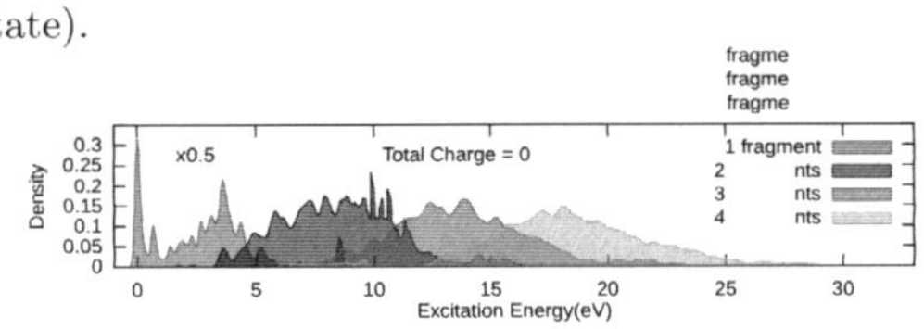

Figure 2. Combinatorial probability density to break down a $C_nH_m^{q+}$ molecule(or cation) into different number of fragments as a function of its internal energy

References

[1] M. Chabot *et al* 2010 *A&A* **A39** 524

[2] S. Díaz-Tendero *et al* 2005 *Phys. Rev. A* **71** 033202

[3] K.-C. Lau *et al* 2005 *J. Chem. Phys* **22** 122

[1]E-mail: juanp.sanchez@uam.es
[2]E-mail: nestor.aguirre@uam.es

Content from this work may be used under the terms of the Creative Commons Attribution 3.0 licence. Any further distribution of this work must maintain attribution to the author(s) and the title of the work, journal citation and DOI.
Published under licence by IOP Publishing Ltd

A numerical lattice approach for ionization and capture processes in ion-H_2O collisions.

L. Errea[1], **Clara Illescas**[2], **L. Méndez**[3], **I. Rabadán**[4], **J. Suárez**[5].

Departamento de Química, Módulo 13, Universidad Autónoma de Madrid, Cantoblanco 28049, Madrid, Spain

Synopsis We present a semi-classical numerical integration method for studying electron loss processes in ion collisions with water molecules. Capture and ionization cross section are calculated in the intermediate-high energy range, and the kinetic energy distribution of the ejected electrons is analyzed. The results are compared with those obtained with a classical Monte Carlo method.

Water collisions with ions are relevant in fields such as Astrophysics and Biomedicine. In particular, processes that involve the ejection of electrons play a key role in the biological damage that occurs in ion-beam cancer therapy. Since water is the main component of the cells, there is an increasing demand for detailed information on the ionization of water molecules by ion collisions.

We have carried out non-perturbative, classical and semi-classical, calculations of ionization cross sections in ion-water collisions at impact energies between 10 keV/u and 1 MeV/u, obtaining the kinetic energy distribution of the ejected electrons. The two methods are based on a one-electron approach, where the interaction of the active electron with the core is described by means of a three-center model potential. In both calculations, the projectile follows rectilinear trajectories and the target nuclei remain fixed during the collision. An orientation-averaged cross section is achieved by considering 12 different trajectory orientations. The classical Monte Carlo method has been described in refs. [1, 2].

In this work, we have implemented the direct numerical integration of the Time-Dependent Schrödinger equation in a Cartesian-coordinate lattice scheme. We have adapted the GridTDSE parallel code [3], which was strategically designed with a parallel computational architecture that allows for the massive memory allocations needed in the calculations. The wave function is discretized over a grid of points of the coordinate space, rendering a diagonal potential energy matrix, and the kinetic energy matrix operator is evaluated by applying a finite difference method.

We have implemented a Phillips-Kleinman operator [4] to prevent the electron from falling into an occupied orbital of the water molecule. A mask function is introduced at the edge of the grid in order to absorb the ionization and capture fluxes. The kinetic energy distribution of the ejected electrons is obtained from the probability density current at the border of the grid.

Figure 1 shows the probabilities of removing an electron from the $1b_1$ molecular orbital of the water molecule in collisions with protons at $E = 25$ keV and 100 keV.

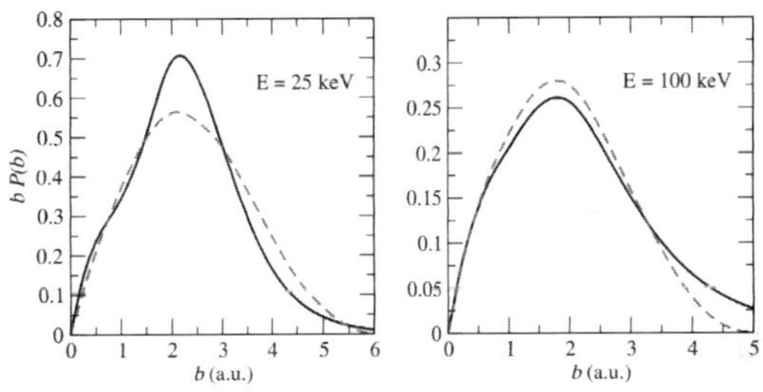

Figure 1. Probabilities for electron dettachment from the $1b_1$ molecular orbital in $H^+ + H_2O$ collisions multiplied by the impact parameter b. (—), numerical results; (- - -), classical results of ref. [1].

References

[1] C. Illescas *at al* 2011 *Phys. Rev. A* **83** 052704

[2] L. F. Errea *at al* 2011 *Phys. Rev. A* **87** 032709

[3] J. Suarez *at al* 2009 *Comp. Phys. Comm.* **180** 2025

[4] J. C. Phillips and L. Kleinman 1959 *Phys. Rev.* **116** 287

[1]E-mail: lf.errea@uam.es
[2]E-mail: clara.illescas@uam.es
[3]E-mail: l.mendez@uam.es
[4]E-mail: ismanuel.rabadan@uam.es
[5]E-mail: jaime.suarez@uam.es

Content from this work may be used under the terms of the Creative Commons Attribution 3.0 licence. Any further distribution of this work must maintain attribution to the author(s) and the title of the work, journal citation and DOI.
Published under licence by IOP Publishing Ltd

Electron capture and ionization processes in high velocity C_n^+, C-Ar and C_n^+, C-He collisions.

G.Labaigt[#], A.Jorge[¤], C.Illescas[¤], K. Béroff[*,1], A.Dubois[#,2], B.Pons[§,3], M.Chabot[†]

[#] Laboratoire de Chimie Physique-Matière et Rayonnement, Université Paris 06 and CNRS F-75231 Paris, France
[¤] Departamento de Qimica, Modulo 13, Universidad Autonoma de Madrid, 28049 Madrid, Spain
[*] Institut des Sciences Moléculaires d'Orsay, Université Paris Sud and CNRS F-91405 Orsay Cedex, France
[§] CELIA, Université de Bordeaux - CEA - CNRS, 33405 Talence, France
[†] Institut de Physique Nucléaire d'Orsay, Université Paris Sud and CNRS F-91406 Orsay Cedex, France

Synopsis: Single and double electron capture as well as projectile single and multiple ionization processes occurring in 125keV/u C_n^+-He, Ar collisions have been studied experimentally and theoretically for $1 \leq n \leq 5$. The Independent atom and electron (IAE) model has been used to describe the cluster-atom collision. The ion/atom-atom probabilities required for the IAE simulations have been determined by classical trajectory Monte Carlo (CTMC) and semiclassical atomic orbital close coupling (SCAOCC) calculations for the Ar and He targets respectively. In general the agreement between experiment and IAE simulations was good, with the exception of double electron capture leading to anionic C_n^- species.

Experiments have been performed at the Tandem accelerator in Orsay (France), using the AGAT setup [1], to study electron capture and ionization processes in carbon cluster –He, Ar collisions at high impact velocity. Single and double electron capture as well as projectile single and multiple ionization processes have been investigated in 125keV/u C_n^+-Ar, He collisions. Helium single and double ionization cross sections have also been measured in the 100-400 keV/u range (n=1,4). The present work extends and improves a recent study devoted to single and double electron capture in C_n^+-He collisions [1]. Indeed, in addition to the study of new processes and consideration of another target atom, we performed state of the art calculations of impact parameter probabilities for all the processes, in particular for electron capture by neutral C atoms that was fitted on the experiment previously. Also, the role of electronic correlations has been investigated within the SCAOCC approach by performing calculations with one, two or three active electrons [2].

We found a general good agreement between measured and IAE cross sections. For projectile ionization, IAE predictions are in very good agreement with experiment in C_n^+-He collisions, whereas IAE predicts right order of magnitude for the cross sections in C_n^+-Ar systems, in particular the large multiple-ionization cross sections. For neutralization of C_n^+, the agreement is better in the helium target case than in the argon target case. Due to the role of projectile ionization, IAE predicts cross sections smaller with argon than with helium, in accordance with the experiment. For He target ionization in C_n^+-He collisions the agreement is satisfactory which allows to extend the applicability of the IAE model, previously tested on projectile ionization and electron capture, to this process as well.

In contrast with the above cited processes, IAE predictions strongly overestimate double electron capture leading to anionic C_n^- species. This was already observed in the case of the helium target [1] and confirmed here in the argon target case [2]. We suspect the loss of the loosely bound electron to lie at the origin of the discrepancy. Indeed, for anions, electron emission is a probable relaxation channel, which is not measured in the present experiment and which deserves further investigation.

[1]E-mail: karine.beroff@u-psud.fr
[2]E-mail: alain.dubois@upmc.fr
[3]E-mail: pons@celia.u-bordeaux1.fr

References

[1] K.Béroff et al 2013 *J.Phys.B:At.Mol.Opt.Phys.* 46, 015201

[2] G.Labaigt et al 2015 *J.Phys.B:At.Mol.Opt.Phys.* 48, 075201

Content from this work may be used under the terms of the Creative Commons Attribution 3.0 licence. Any further distribution of this work must maintain attribution to the author(s) and the title of the work, journal citation and DOI.
Published under licence by IOP Publishing Ltd

Ion-pair dissociation of highly excited carbon clusters, size and charge effects

T.Launoy[§], M. Chabot[†], G. Martinet[†], T. Pino[*], A. Le Padellec[#], S.Bouneau[†], G. Féraud[¤], N.Do Thi[$], N.Vaeck[§], J.Liévin[§], J.Loreau[§], K. Béroff[*1]

[*] Institut des Sciences Moléculaires d'Orsay, Université Paris Sud and CNRS F-91405 Orsay Cedex, France
[†] Institut de Physique Nucléaire d'Orsay, Université Paris Sud and CNRS F-91406 Orsay Cedex, France
Institut de Recherche en Astronomie et Planétologie, Université Toulouse 3 and CNRS F-31028 Toulouse, France
[§] Chimie Quantique et Photophysique, Université libre de Bruxelles, Belgique

Synopsis: Ion-pair dissociation of a highly excited molecule is a relaxation process giving rise to emission of anionic and cationic fragments. We present first measurements of ion-pair dissociation of carbon clusters. We found that ion-pair relaxation is an ubiquitous, although very small, relaxation channel common to all sizes and charges of C_n^{q+} species produced in high velocity C_n^+-He collisions. Quantitative interpretation of measured branching ratios is conducted on the basis of a statistical approach i.e through listing of all possible final states.

Anion production in high velocity C_n^+-He collisions has been shown to proceed by three different mechanisms [1]. One of them is ion-pair dissociation in which a highly excited C_n^{q+} cluster, neutral (q=0) or positively charged (q≥1), relaxes by emission of one anionic and (q+1) cationic fragments. The AGAT set-up, situated nearby the Tandem accelerator in Orsay (France), is an ideal tool for studying this process. Indeed the set-up is based on a coincident recording of all fragments issued from the collision, identified in mass and charge (q=-1,0,1,2,3...).

In figure 1 are reported measured branching ratios (BR) for ion-pair dissociation of C_n^{q+} species as a function of the cluster charge q and cluster size n (symbols, see legend). The branching ratio is calculated as the ratio between measured cross sections for dissociation with anion and measured cross sections for dissociation without anion (v=2.25 a.u C_n^+ -He collision). As seen from figure 1 these BR are small (~few 10^{-4}) and little dependent on (n,q) with the exception of (n=2, q=1), and, in a lesser extent, (n=4, q=3).

Whereas ion-pair dissociation requires a large amount of energy (for instance dissociation of C_2^+ into C^{2+}/C^- requires more than 28 eV) the energetic criterion is not sufficient to interpret the results. Clearly the number of final states i.e the phase space open to ion-pair dissociation is playing the major role. We listed all possible final states for dissociation both with and without anions. Results for (q=1, n=2) and (q=1,n=3) will be presented at the conference and compared to experimental results.

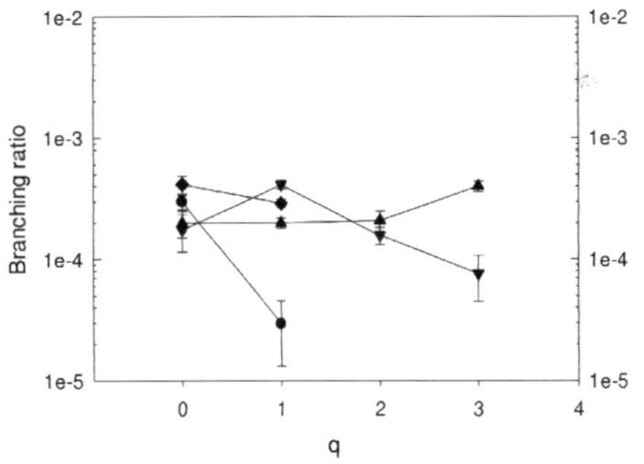

Figure 1. Measured ion-pair dissociation BR as a function of the cluster charge. Circles, diamonds, triangles down and triangles up refer to n=2,3,4,5 respectively; lines are to guide the eye.

References

[1] K.Béroff et al 2013 *J.Phys.B:At.Mol.Opt.Phys.* 46, 015201

[1]E-mail: karine.beroff@u-psud.fr
¤ present address : Departement f. Chemie and Biochemie, Universität Bern, 3012 Bern, Suisse
$ present address : Center for Computational Physics, Institute of Physics, VAST, Hanoi, Vietnam

Content from this work may be used under the terms of the Creative Commons Attribution 3.0 licence. Any further distribution of this work must maintain attribution to the author(s) and the title of the work, journal citation and DOI.
Published under licence by IOP Publishing Ltd

Journal of Physics: Conference Series **635** (2015) 032086 doi:10.1088/1742-6596/635/3/032086

Anion Chemistry on Titan: systematic studies of the growth and stability of large negative ions

Jan Žabka[*1], **Miroslav Polášek**[*], **Věra Křížová**[*], **Illia Zymak**[*], **Claire Romanzin**[†], **Christian Alcaraz**[†]

[*] J. Heyrovský Institute of Physical Chemistry of the ASCR, v. v. i., Dolejškova 3, Prague 8, 182 23, Czech Republic
[†]Laboratoire de Chimie Physique, Bât 350, Centre Universitaire Paris-Sud, 91405 Orsay, France

Synopsis The reactions of $C_{2n+1}N^-$ anions and their complexes with cyanoacetylene, HC_3N, and hydrogen cyanide, HCN, were studied experimentally using methods of tandem mass spectrometry. The reactions described in detail in this communication may represent a possible way of formation of large anions found in the ionosphere of Titan.

Cassini CAPS-ELS spectrometer revealed the presence of large negative ions in the ionosphere of Titan [1]. Recently, a mechanism has been proposed for the possible formation of these ions, in whith the cyanoacetylene, HC_3N, played a key role [2]. Chemical ionization technique (NCI and APCI-) were used successfully to prepare model complex $(HC_3N)_x.C_yN^-$ anions in the gas phase. The reaction itself and CID experiments was studied using a Waters Quattro Premier TM tandem quadruple mass spectrometer operating in negative ion mode and VG ZAB2-SEQ mass spectrometer. CID mass spectra of these anions, as well as their ion molecule reactions with HC_3N support the previously proposed reaction scheme [2,3]. Quantum chemistry calculations revealed details of the ion structures, energetics and reaction mechanisms. High-energy CID spectra of $(HC_3N)_x.C_yN^-$ anions showed a complexity of ionic and neutral products that can be expected to be formed by the high-energy ion precipitation observed at Titan [4].

Presented experiments show that in spite of its low abundance in Titan atmosphere [5], the cyanoacetylene is probably one of the most important species in the ionospheric chemistry of Titan.

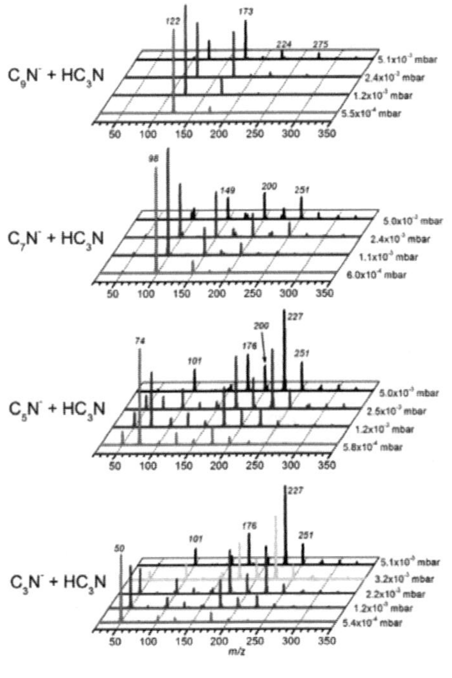

Figure 1. Pressure dependent mass spectra of reactions of C_xN^- anions (x =3,5,7,9) with HC_3N.

Acknowledgement

This work was supported by the Czech Science Foundation (grant No.14-19693S), the Ministry of Education Youth and Sports of Czech Republic (grant No. LD14024), COST Action CM1204 (XLIC), the RTRA "Triangle de la Physique (project GIN) and the French planetology program (PNP).

References

[1] A.J. Coates *et al* 2007 *Geophys. Res. Lett.* **34** L22103
[2] J. Žabka, M. Polášek, C. Romanzin, C. Alcaraz 2012 *Icarus* **219** 161
[3] J. Žabka, I. Ndiaye, C. Romanzin, C. Alcaraz, M. Polášek, 2014 *Int. J. Mass Spectrom.* **367** 1
[4] T.E. Cravens *et al* 2008 *Geophys. Res. Lett.* **35** L03103
[5] V. Vuitton *et al* 2009 *Planet. Space Sci.* **57** 15587

[1] E-mail: jan.zabka@jh-inst.cas.cz

Content from this work may be used under the terms of the Creative Commons Attribution 3.0 licence. Any further distribution of this work must maintain attribution to the author(s) and the title of the work, journal citation and DOI.
Published under licence by IOP Publishing Ltd

XXIX International Conference on Photonic, Electronic, and Atomic Collisions (ICPEAC2015) IOP Publishing
Journal of Physics: Conference Series **635** (2015) 032087 doi:10.1088/1742-6596/635/3/032087

Mobilities of Li^+ - (2-butanol)$_n$ (n = 1-2) ions in He gas

Kazunari Takaya[1], Meishi Nakahara, Yuya Hasegawa, Tetsuo Koizumi, Karin Takahashi[*] Shiro Matoba[**] and Hajime Tanuma[†]

Department of Physics, Rikkyo University, Tokyo 1718501, Japan
[*]Department of Physics, Toho University, Chiba 2748510, Japan
[**]Japan Atomic Energy Agency, Gunma 3701292, Japan
[†]Department of Physics, Tokyo Metropolitan University, Tokyo 1920397, Japan

Synopsis Mobilities of Li^+- (2-butanol)$_n$ (n = 1-2) ions in helium gas have been measured at 83 K and room temperature using a low temperature drift tube. The experimental mobilities are significantly smaller than the polarization limit values within the measured effective temperature range. Using the 12-6-4 potential, we have performed calculation of the mobilities in frame of the two-temperature momentum transfer theory. The agreement between the experimental and calculated mobilities is fairly good.

Mobility of an ion in a buffer gas is very sensitive to the shape of the interaction potential between the ion and the gas molecule. Generally, the long-range attractive interaction decides the cross section at low energy. In the drift tube method, the lower limit of collision energy between ions and gas molecules depends on the gas temperature, and the low collision energy can be achieved by cooling of the gas. In this work, we measured the mobilities of Li^+- (2-butanol)$_n$ (n = 1-2) ions in helium gas at 83 K and room temperature to investigate the influence of the long-range interaction on the mobility. The experimental results are shown in Figure 1 as a function of the effective temperature T_{eff}. The mobilities fall lower than the polarization limit K_{pol} at all temperature. In the small ions, the mobility has maximum around

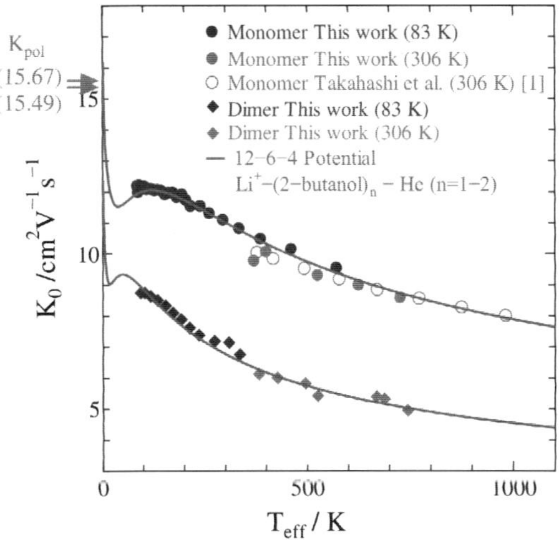

Figure 1. Mobilities of Li^+-(2-butanol)$_n$ (n = 1-2) in He at 83 K and 306 K. Lines indicate the calculated mobilites with 12-6-4 potential.

T_{eff} = 1000 K and tends to approach the polarization limit at the lower temperature. We have performed fitting of the experimental mobilities with 12-6-4 potential as a function of T_{eff}. Lines in Figure 1 indicate the calculated mobilites. The momentum transfer cross sections for Li^+-(2-butanol)$_n$ (n = 1-2) ions and He systems with 12-6-4 potential and the polarization (Orbiting) cross section are shown in Figure 2. The fact that momentum transfer cross sections are larger than the orbiting cross section, indicates that Li^+- (2-butanol)$_n$ (n = 1-2) ions in helium gas have the large dispersion interaction which is given by the intermediate-range attractive r^{-6} term in the interaction potential.

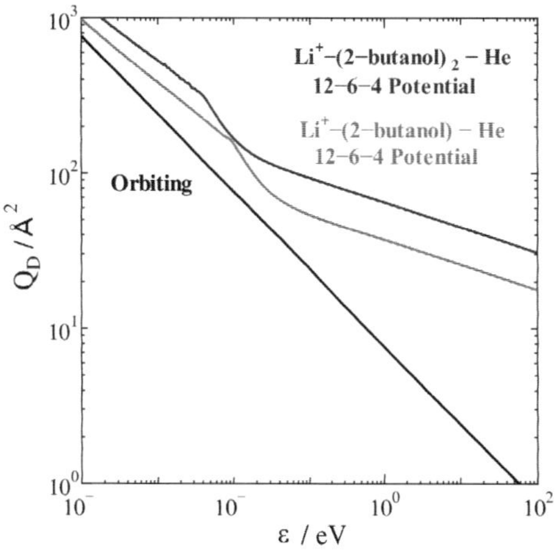

Figure 2. Momentum transfer cross sections with the classical calculation for Li^+-(2-butanol)$_n$ (n = 1-2).

References

[1] K. Takahashi *et al* 2013 *J. Chem. Phys.* **139** 084317

[1]E-mail: kazunari@rikkyo.ac.jp

Content from this work may be used under the terms of the Creative Commons Attribution 3.0 licence. Any further distribution of this work must maintain attribution to the author(s) and the title of the work, journal citation and DOI.
Published under licence by IOP Publishing Ltd

Induced ionization of small molecules: comparison of experiment with quantum and classical calculations

S. T. S. Kovács [1], P. Herczku, L. Sarkadi, L. Gulyás, Z. Juhász and B. Sulik

Institute for Nuclear Research, Hungarian Academy of Sciences (MTA Atomki), Bem tér 18/c, H-4026 Debrecen, Hungary

Synopsis The ion-impact induced ionization of small molecules was investigated both experimentally and theoretically. Electron energy spectra were taken by an energy dispersive electrostatic spectrometer at different observation angles. The obtained absolute electron emission cross sections were compared with the results of CDW-EIS and CTMC calculations. Both theories are in a reasonable agreement with the experimental data.

Ionization of atoms and molecules is of fundamental interest in atomic and molecular physics [1,2]. Besides that, ionization is important in many applications, for instance plasma physics, industrial irradiations and radiotherapy. It was discovered recently that free electrons are almost as effective as free radicals in causing DNA damage in living cells [3].

In the experiment H^+, He^+, C^+ and N^+ ions were provided by a 5 MV VdG accelerator in the 46-1000 keV/u energy range. The ion beams crossed gas jets of CH_4, H_2O and N_2. By applying a rotatable energy dispersive electrostatic spectrometer we determined double differential electron emission cross sections (DDCSs) in the 20°-160° observation angle range. In figure 1 we compare the energy spectrum of the electrons ejected in 1 MeV H^+ + CH_4 collisions at 90° observation angle with CTMC and CDW-EIS [4] calculations.

In the CTMC calculations we used a multi-center three-body approach. Our model is very similar to that of Illescas et al. [5] except that we considered the full three-body dynamics. A very good agreement was found between the experimentally observed DDCS and that obtained by the CTMC calculations in the entire electron energy range. This indicates that the classical trajectory method can be applied for calculation of the ionization properties of few atomic molecules by bare projectiles in this intermediate collision energy range (see Fig. 1).

Above ~12 eV electron energy the quantum mechanical CDW-EIS calculations also show a good agreement with the experiment, the two theories practically agree. Below 10 eV both theories overestimate the experimentally obtained DDCSs, but CTMC only slightly.

Similarly, good agreement was observed between the measurement and the theories for all studied observation angles. According to the calculations, single ionization is dominant for the studied collision system. For such small perturbation, both theories seem to be adequate for treating the electron emission process from small molecules.

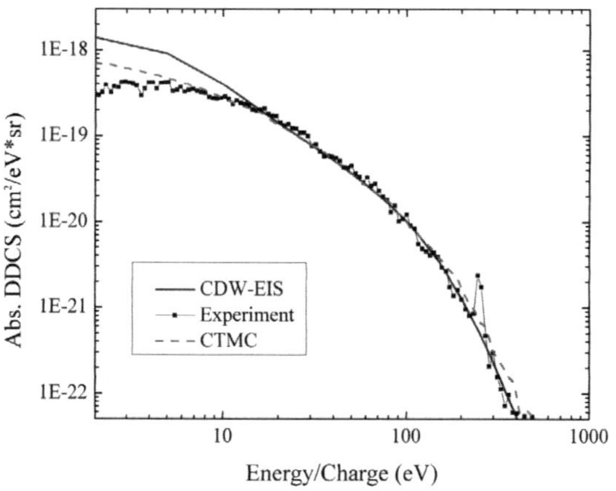

Figure 1. Comparison of the experimental electron spectrum with the results of CTMC and CDW-EIS calculations at 90° observation angle for 1 MeV H^+ + CH_4 collisions.

This work was supported by the Hungarian National Science Foundation OTKA (Grant No. K109440).

References

[1] W. E. Wilson *et al.* 1975 Phys. Rev. A **11** 1303
[2] G. Dujardin *et al.* 1985 Phys. Rev. A **31** 3027
[3] A.V. Solov'yov *et al.* 2009 Phys. Rev. E **79** 011909
[4] L. Gulyas *et al.* 2013 J. Phys. B **46** 075201
[5] C. Illescas *et al.* 2011 Phys. Rev. A **83** 052704

[1] E-mail: kovacs.sandor@atomki.mta.hu

Content from this work may be used under the terms of the Creative Commons Attribution 3.0 licence. Any further distribution of this work must maintain attribution to the author(s) and the title of the work, journal citation and DOI.
Published under licence by IOP Publishing Ltd

Dynamics of excited clusters of β–alanine in the gas phase

D.G. Piekarski*[1], M.J. Ramos§, S. Díaz-Tendero*[2]

* Departamento de Química, Módulo 13, Universidad Autónoma de Madrid, Cantoblanco 28049, Madrid Spain
§Departamento de Química e Bioquímica, Faculdade de Ciências, Universidade do Porto, 4169-007, Porto, Portugal

Synopsis We present a theoretical study of excited clusters of β–alanine molecules in the gas phase: (β–ala)$_n$, n=2–5. Classical molecular dynamics simulations performed for different internal excitation energies showed a thermal decomposition dependence with the cluster size. We also present an assessment study performed with different families of density functionals using the dimer, (β–ala)$_2$ as a benchmark system. M06-2X provides the best agreement for the relative energies of 20 isomers in comparison with the reference values computed with the MP2 method. The stability and reactivity of several cluster sizes have been investigated with this functional in combination with the 6-311++G(d,p) basis set.

Ionization and excitation of biomolecules and clusters of biomolecules, induced with highly charged ions, is a widely employed technique to study the stability of these species at a molecular level. In these studies, modeling has become a fundamental tool for understanding the fragmentation dynamics [1,2]. We recently studied doubly-positively charged and excited monomers of β-alanine in the gas phase [3]. In a further step, we here present our recent investigations concerning the stability of clusters of this molecule in the gas phase. This is a more complicated system because it presents additional type of interactions between the molecules such as hydrogen bonds, van der Waals forces, etc. When amino acids are embedded in a cluster of molecules, a protective effect of the environment against fragmentation has been observed [4]. Thus, we evaluate which are the main effects behind this behavior.

We have first performed classical molecular dynamics (MD) simulations using the general amber force field including dimer, trimer, tetramer and pentamer clusters of β–alanine. Several trajectories for internal energies in the range T=50-423K were carried out. With these simulations we studied the stability of the neutral clusters upon thermal excitations and we explore the potential energy surface, thus obtaining different structures.

In order to find a suitable computational tool for carrying out reliable electronic structure calculations for clusters of amino acids at a relatively low computational cost, an assessment based on density functional theory was performed. In our study we include different families of functionals, namely B3LYP, B97D, M06, M06-2X and MPWB1K in combination with 6-311++G(d,p) basis set. The MP2 method was taken as a reference for the chosen model system (β–ala)$_2$. The M06-2X method is shown to adequately reproduce the relative energies and geometries of 20 isomers of (β–ala)$_2$.

Finally, we have used this level of theory to evaluate the relative stability of numerous isomers of different cluster sizes (β–ala)$_n$, n=2-5. We have also studied ionization potentials and interaction energies. The possibility of inter- and intra-molecular reactions upon ionization and excitation of the clusters will be presented as well.

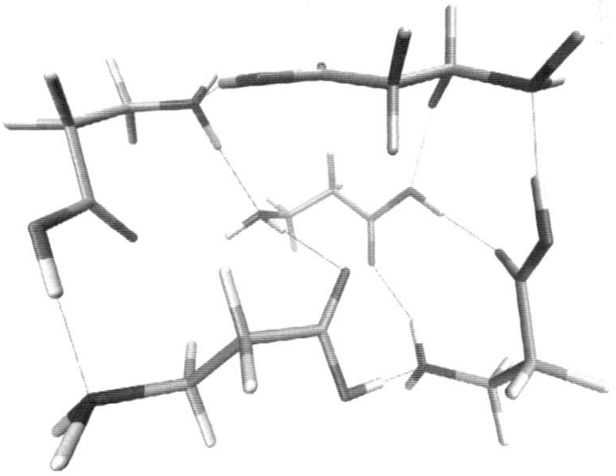

Figure 1. Stabilization of β–alanine clusters in the gas phase with intermolecular forces.

References

[1] S. Maclot *et al* 2013 *J. Phys. Chem. Lett.* **4** 3903
[2] M. Capron *et al* 2012 *Chem. Eur. J* **18** 9321
[3] D.G. Piekarski *et al* 2015 submitted
[4] S. Maclot *et al* 2011 *Chem. Phys. Chem.* **12** 930

[1] E-mail: dariusz.piekarski@uam.es

[2] E-mail: sergio.diaztendero@uam.es

Content from this work may be used under the terms of the Creative Commons Attribution 3.0 licence. Any further distribution of this work must maintain attribution to the author(s) and the title of the work, journal citation and DOI.
Published under licence by IOP Publishing Ltd

Journal of Physics: Conference Series **635** (2015) 032090 doi:10.1088/1742-6596/635/3/032090

"Water group" ion induced ionization and dissociation of major constituents of atmosphere of Titan

Miroslav Polášek[*,1], Věra Křížová[*], Ján Žabka[*], Illia Zymak[*]

[*] *J. Heyrovsky Institute of Physical Chemistry of the ASCR, v. v. i., Dolejškova 2155/3, Prague 8, 182 23 Czech Republic*

Synopsis Processes in high-energy collisions of positive and negative "water group" ions with thermal neutral constituents of atmosphere of Titan were studied. It was shown that internal energy deposited in nascent positive ions strongly depends on the nature of incident ions.

In the Saturnian magnetosphere, neutrals originating from the icy satellites and ring system are ionized producing plasma composed of H^+, H_2^+, and so-called "water group" ions O^+, HO^+, H_2O^+ [1]. The plasma is quite rapid, reaching the velocity of 170 km s^{-1} near Titan [1]. A precipitation of O^+ ions with keV energies was found at Titan by the Cassini CAPS spectrometer [2]. It has been speculated [3] that a few percent of O^+ flux is converted to O^- ions, which may play a role in Titan's negative ion chemistry.

The objective of this study was to measure the ions formed from the particular constituents of Titan's atmosphere (N_2, CH_4, C_2H_2, C_2H_4, HCN, HC_3N, etc.) when bombarded by the "water group" positive ions as well as negative O^- and HO^- ions at keV energies.

The experiments were done on a modified sector-type mass spectrometer having a small quadrupole mass analyzer attached to a collision/ionization chamber located between the magnetic and the electrostatic analyzer. The projectile ions were formed in an EI/CI ion source, accelerated to 8 keV and mass-selected by the magnetic sector. The slow secondary ions stemming from the neutral molecules admitted to the collision/ionization chamber were analyzed by the quadrupole mass analyzer.

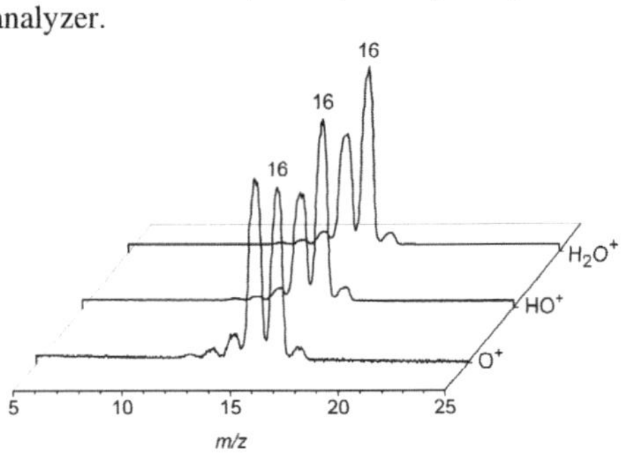

Figure 1. Mass spectra of methane obtained with ionization by 8 keV H_2O^+, HO^+ and O^+ ions.

In Figure 1, results are shown for positive "water group" ions colliding with gaseous methane molecules. It follows from these spectra that degree of fragmentation, i.e. content of internal energy, of ionized methane strongly depends on the nature of incident ion. For H_2O^+ the resulting mass spectrum closely resembles the EI spectrum, whereas for HO^+ and O^+ a larger extent of fragmentation was observed.

Mass spectra of methane obtained by collisions with O^- and HO^- ions (Figure 2) show even larger degree of fragmentation.

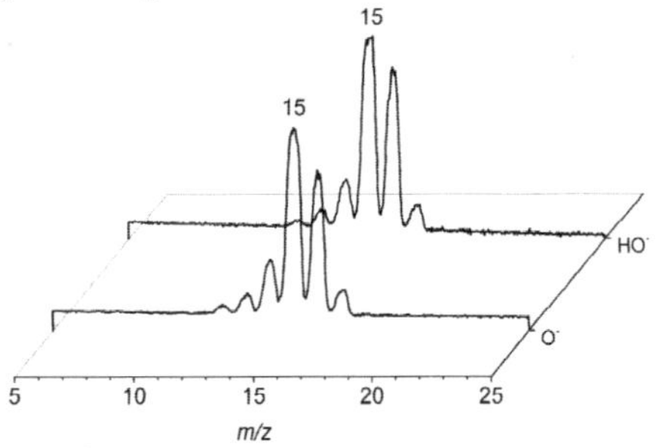

Figure 2. Mass spectra of methane obtained with ionization by 8 keV HO^- and O^- ions.

Analogous results were obtained for other constituents of the atmosphere of Titan showing that in spite of a large translational energy available in the collision system the internal energy of nascent secondary ions strongly depends also on the nature of incident ions.

This work was supported by the Czech Science Foundation (project 14-20915S).

References

[1] C.S. Arridge *et al* 2011 *Space. Sci. Rev.* **162** 25
[2] R.E. Hartle *et al* 2006 *Planet. Space Sci.* **54** 1211
[3] T. E. Cravens *et al* 2008 *Geophys. Res. Lett.* **35** L03103

[1] miroslav.polasek@jh-inst.cas.cz

Content from this work may be used under the terms of the Creative Commons Attribution 3.0 licence. Any further distribution of this work must maintain attribution to the author(s) and the title of the work, journal citation and DOI.
Published under licence by IOP Publishing Ltd

Journal of Physics: Conference Series **635** (2015) 032091 doi:10.1088/1742-6596/635/3/032091

Ion-induced ionization and fragmentation of organic molecules

Věra Křížová*[1], Ján Žabka*, Illia Zymak*, Miroslav Polášek*

J. Heyrovsky Institute of Physical Chemistry of the ASCR, v. v. i., Dolejškova 2155/3, Prague 8, 182 23 Czech Republic

Synopsis Processes in high-energy collisions of positive and negative ions with selected thermal neutral molecules were studied. It was shown that internal energy deposited in nascent positive ions strongly depends on the nature of incident ions.

The keV ion-induced processes in atoms and molecules have been a subject of numerous studies in the last few decades. These studies were motivated by an effort to understand molecular mechanisms of radiation damage of biological systems [1] as well as processes connected with star wind interactions with planetary atmospheres, interstellar clouds and other objects in space [2].

The range of ions used in these studies has been, however, rather limited. Usually, positive atomic ions with energies ranging from 3 to $n\times10^2$ keV were used as projectile ions. The large scale facilities such as cyclotrons have been used so far as the sources of these ions, which make these exciting experiments available to a limited number of scientists.

The objective of this study was to investigate the effects of mass, charge and chemical structure of the projectile ions on the amount of internal energy deposited in molecules systematically using the modified sector type mass spectrometer.

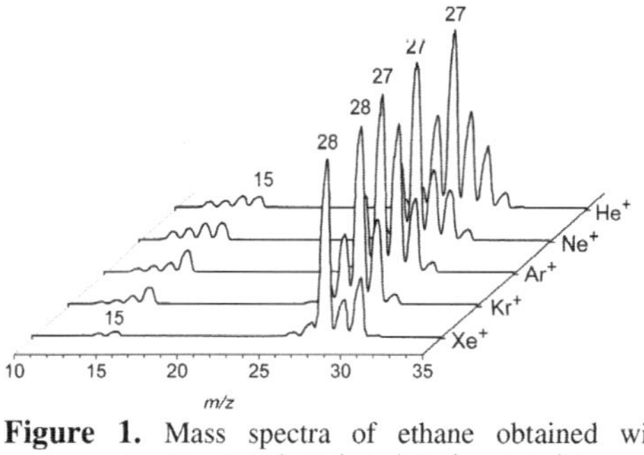

Figure 1. Mass spectra of ethane obtained with ionization by 8 keV He^+, Ne^+, Ar^+, Kr^+ and Xe^+ ions.

The experiments were performed on the sector-type mass spectrometer with a collision/ionization chamber placed in the region between magnetic and electric analyzer. The projectile ions were formed in an EI/CI ion source, accelerated to 8 keV and mass-selected by a magnetic sector. The secondary ions stemming from the collisions of these fast ions with neutral molecules were analyzed by a small quadrupole mass analyzer attached to the collision/ionization chamber perpendicularly to the incident ion beam.

In Figure 1, results are shown for singly charged noble gas ions colliding with ethane molecules. It follows from these spectra that degree of fragmentation, i.e. content of internal energy, of ionized ethane strongly depends on the nature of the incident ion. For the slightly endoergic electron transfer between Xe^+ and ethane, the resulting mass spectrum closely resembles its EI spectrum. Increasing the ionization energy and decreasing the size of incident ions leads to a significantly larger extent of fragmentation.

On the contrary, collisions of negative ions with methane (Figure 2) lead to a similar degree of fragmentation.

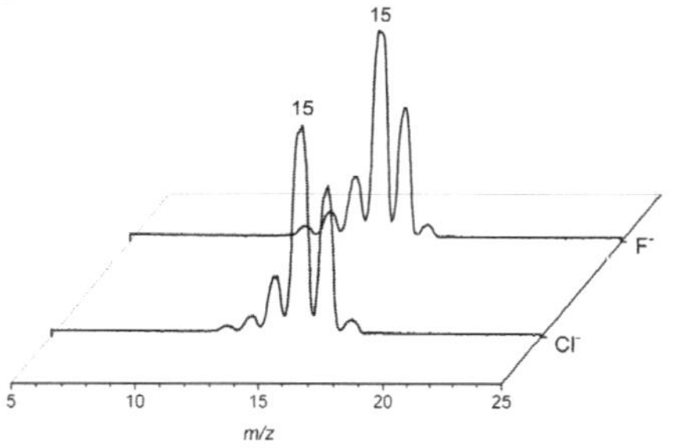

Figure 2. Mass spectra of methane obtained with ionization by 8 keV F^- and Cl^- ions.

Our results show that in spite of the large translational energy available in the collision system in the keV energy range, the internal energy of nascent secondary ions strongly depends also on the chemical nature of incident ions

This work was supported by the Ministry of Education Youth and Sports of the Czech Republic (grant No. LD14024).

Reference

[1] Gomez-Tejedor G G and Fuss M C 2012 *Radiation damage in biomolecular systems*(New York: Springer)
[2] Postma J *et al* 2010 ApJ **708** 435

[1] vera.krizova@jh-inst.cas.cz

Content from this work may be used under the terms of the Creative Commons Attribution 3.0 licence. Any further distribution of this work must maintain attribution to the author(s) and the title of the work, journal citation and DOI.
Published under licence by IOP Publishing Ltd

Modeling secondary particle tracks generated by high-energy protons in ater

F. Blanco[a], A. Muñoz[b], D. Almeida[c], F. Ferreira da Silva[c], P. Limão-Vieira[c], A. Verkhovtsev[d,e 1], L. Ellis-Gibbings[d], K. Krupa[d], A. Traore[d], and G. García[d,f]

[a] Departamento de Física Atómica, Molecular y Nuclear, Universidad Complutense de Madrid, 28040 Madrid, Spain

[b] Centro de Investigaciones, Energéticas Medioambientales y Tecnológicas (CIEMAT), Avenida Complutense 22, 28040 Madrid, Spain

[c] Faculdade de Ciências e Tecnologia, Universidade Nova de Lisboa, 2829-516 Caparica, Portugal

[d] CSIC, Instituto de Fisíca Fundamental, Serrano 113-bis, 28006 Madrid, Spain

[e] MBN Research Center, Altenhöferallee 3, 60438 Frankfurt am Main, Germany

[f] Centre for Medical Radiation Physics, University of Wollongong, Wollongong, NSW 2522, Australia

Synopsis We present interaction probability data of low-energy secondary electrons and positrons produced due to the proton impact. The probability distribution functions serve as input data for the Low Energy Particle Track Simulation (LEPTS) approach which allows one to include the effect of low-energy species in medical applications of radiation and in ion-beam cancer therapy, in particular.

The Low Energy Particle Track Simulation (LEPTS) code [1] is a powerful complementary tool to include the effect of low-energy electrons and positrons in medical applications of radiation. In particular, for ion-beam cancer treatments it provides a detailed description of the role of the secondary electrons abundantly generated in the vicinity of the Bragg peak, as well as the possibility of using transmuted positron emitters (^{11}C, ^{15}O) as a complement for ion-beam dosimetry.

In this study, we present interaction probability data, derived from experimental and theoretical studies, of low-energy secondary species produced due to the proton impact. The probability distribution functions, which serve as input data for the model, are also complemented with a comprehensive review of experimental [2] and theoretical [3, 4] cross sections available in the literature. We will also update the corrective factors for solid and liquid environments calculated by means of the IAM-SCAR method [5, 6]. Using these data, single electron and positron tracks in liquid water and pyrimidine have been simulated to provide information about energy deposition as well as the number and type of interactions taking place in any selected "nanovolume" of the irradiated area.

Acknowledgments: This work has been supported by the Spanish Ministry of Economía y Productividad (Project FIS2012-31320). We also acknowledge partial funding from the Portuguese Foundation for Science and Technology (FCT-MCTES), research grant PEst-OE/FIS/UI0068/2011,the EU/ESF COST Actions Nano-IBCT - MP1002 and CELINA CM-1301, and from the FP7 Multi-ITN Project "Advanced Radiotherapy, Generated by Exploiting Nanoprocesses and Technologies" (ARGENT) (Grant Agreement n°608163).

References

[1] F. Blanco *at al* 2013 *Eur. Phys. J. D* **67** 199

[2] M. E. Rudd, Y.-K. Kim, T. Märk, J. Schou, N. Stolterfoht, and L.H. Toburen, *Seco dary Electro S ectra from Charged Part cle I teract o s* (International Commission on Radiation Units and Measurements, Bethesda, MD, 1996) (ICRU 55)

[3] P. de Vera, R. Garcia-Molina, I. Abril, and A. V. Solov'yov 2013 *Phys. Rev. ett.* **110** 148104

[4] P. de Vera, R. Garcia-Molina, and I. Abril 2015 *Phys. Rev. ett.* **114** 018101

[5] F. Blanco and G. García 2003 *Phys. Rev. A* **67** 022701

[6] F. Blanco and G. García, 2004 *Phys. ett. A* **330** 230

[1]E-mail: verkhovtsev@iff.csic.es

Content from this work may be used under the terms of the Creative Commons Attribution 3.0 licence. Any further distribution of this work must maintain attribution to the author(s) and the title of the work, journal citation and DOI.
Published under licence by IOP Publishing Ltd

XXIX International Conference on Photonic, Electronic, and Atomic Collisions (ICPEAC2015) IOP Publishing
Journal of Physics: Conference Series **635** (2015) 032093 doi:10.1088/1742-6596/635/3/032093

Fusion reaction dynamics of fullerene molecules

Yang Wang[*,†,1], **Michael Gatchell**[§], **Henning Zettergren**[§], **Patrick Rousseau**[¶,||], **Tao Chen**[§], **Mark H. Stockett**[§], **Alicja Domaracka**[¶], **Lamri Adoui**[¶,||], **Bernd A. Huber**[¶], **Henrik Cederquist**[§], **Manuel Alcamí**[*], **and Fernando Martín**[*,†,‡]

[*] Departamento de Química, Módulo 13, Universidad Autónoma de Madrid, 28049 Madrid, Spain
[†] Instituto Madrileño de Estudios Avanzados en Nanociencia (IMDEA-Nanociencia), 28049 Madrid, Spain
[§] Department of Physics, Stockholm University, S-106 91 Stockholm, Sweden
[¶]CIMAP, UMR 6252, CEA/CNRS/ENSICAEN/Université de Caen Basse-Normandie, bd Henri Becquerel, BP 5133, F-14070 Caen cedex 05, France
[||]Université de Caen Basse-Normandie, Esplanade de la Paix, F-14032 Caen, France
[‡] Condensed Matter Physics Center (IFIMAC), Universidad Autónoma de Madrid, 28049 Madrid, Spain

Synopsis Systematic molecular dynamics simulations based on the self-consistent charge density functional tight-binding method have been performed for the molecular collisions inside clusters of fullerene molecules, after prompt atom knockouts by the keV ions. The thermodynamic and kinetic stabilities, as well as the bonding features, have been investigated for the most stable dumbbell dimers C_{118}^+, C_{119}^+ and C_{120}^+.

During the last decades, experiments have shown that fullerene fusion reactions can take place between C_{60} molecules and C_{58} or C_{59} radicals that are produced upon the fragmentation of C_{60} induced by laser irradiation, photoionization and heating, or ozone oxidation. In recent experiments [1, 2], covalently bound fullerene dimers C_{118}^+ and C_{119}^+ were observed in mass spectra resulting from collisions between clusters of C_{60} molecules and keV He^{2+}- or Ar^{2+}-ions. In this contribution, we present a systematic theoretical investigation [3] on the dynamics of C_n^+ + C_{60} ($n = 58, 59, 60$) collisions following prompt atom knockouts by the fast ions. This study will not only provide additional information to, and a deeper understanding of, the experimental results, but may shed some light on the general mechanisms of fullerene coalescence involving C_{58} or C_{59} fragments reacting with C_{60}. It might also be relevant to the formation and growth of fullerenes.

A large number of molecular dynamics (MD) simulations based on the self-consistent charge density functional tight-binding (SCC-DFTB) method have been carried out, with different initial orientations of fullerene cages and different center-of-mass collision energies. The DFTB method implemented here is based on the second-order expansion of the Kohn-Sham energy in terms of the charge density fluctuation. The resulting energy terms are calculated by applying the tight-binding approximation. The electronic and repulsive parts of the DFTB energy are computed using a set of empirical Slater-Koster interatomic parameters.

From the results of the simulations, we verify the feasibility of forming covalently bound fullerene dimers C_{118}^+ and C_{119}^+ under the experimental conditions, and confirm that the so-formed dimers are stable enough to survive on the experimental timescale. The statistics from the MD simulations indicate a much higher reactivity of C_{59}^+ and C_{58}^+ fragments compared to that of C_{60}^+. The formation of C_{119}^+ and C_{118}^+ dimers is almost barrierless and the MD simulations demonstrate that these dimers can be readily produced in less than 1 ps at collision energies below 1 eV. This is consistent with the experimental findings. Furthermore, we show that, the dumbbell dimer products C_{119}^+ and C_{118}^+ are both thermodynamically and kinetically stable so that they can survive to be detected on the experimental timescale of tens of microseconds.

References

[1] H. Zettergren *at al* 2013 *Phys. Rev. Lett.* **110** 185501

[2] F. Seitz *at al* 2013 , *J. Chem. Phys.* **139** 034309

[3] Y. Wang *at al* 2014 , *Phys. Rev. A* **89** 062708

[1]E-mail: yang.wang@uam.es

Content from this work may be used under the terms of the Creative Commons Attribution 3.0 licence. Any further distribution of this work must maintain attribution to the author(s) and the title of the work, journal citation and DOI.
Published under licence by IOP Publishing Ltd

Journal of Physics: Conference Series **635** (2015) 032094 doi:10.1088/1742-6596/635/3/032094

Electron interferences in the ionization of the dimers by s ift ion impact: Origin of high-order frequency

H. Agueny*,‡ [1], A. Makhoute†, A. Dubois*, J. P. Hansen‡

* Sorbonne Universités, UPMC Univ Paris 06, UMR 7614, Laboratoire de Chimie Physique-Matière et Rayonnement, F-75231 Paris Cedex 05, France

† UFR de Physique du Rayonnement et des Interactions Laser-Matière, Faculté des Sciences. Université Moulay Ismail, B.P. 11201, Zitoune, Meknès, Morocco

‡Department of Physics and Technology, Allegt. 55, University of Bergen, N-5007 Bergen, Norway

Synopsis We report evidence for interference phenomena beyond Young-type in electron emission spectrum from the dimers by fast ions. Prior measurements for H_2 have led to a wide discussion for which no definitive conclusions have been settled, whereas theoretical-side have failed to reproduce the phenomenon. The process under investigations is treated by using a nonperturbative semi-classical approach to solve the time-dependent Schrödinger equation. ur results are supported by an analytic model based on second- and third-order Born series.

Double-scattering effects were claimed to be observed and evoked to explain additional oscillatory structures in the measured spectra in fast ion-H_2 collisions. The first experimental observations of such additional structures were sparked off further experimental and theoretical efforts to enhance the understanding of the complete nature of the phenomenon [1]. In spite of the number of experiences performed in that context, no clear demonstration has been reached [2, 3, 4, 6]. In contrast, theoretical investigations have been unable to reproduce the phenomenon [7, 8]. This makes the numerical demonstration of the existence of such effects a real challenge.

During the conference we shall present our results stemming from a nonperturbative approach to solve the time-dependent Schrödinger equation in which all high-order mechanisms are taken into account in the dynamical stage. The purpose is to resolve this controversy and explain why calculations on H_2 do not get to any additional features related to high-order interference. This is achieved by investigating completely different energetic ion-molecule collisions, in which a clear footprint of high-order frequency in electron emission spectra is found. As is shown in Fig. 1 the spectrum shows up to four distinct harmonics from the full molecular treatment. In contrast, the corresponding results "2Rb" stemming from a pseudo-molecular model exhibit only first order frequency characteristic of Young-type interference. Our nonperturbative calculations are supported by a model derived from second- and third-order Born series. The latter reproduces the spectrum almost perfectly. We will also report an extension to a tri-atomic linear molecule which reveals additional and strong harmonics.

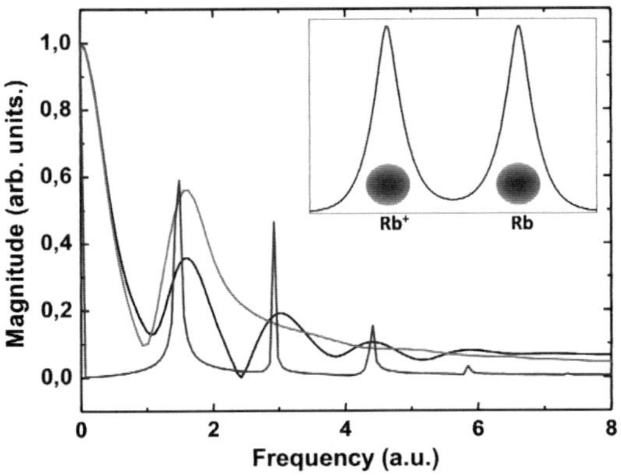

Figure 1. Frequency distributions from calculations for 63 Mev/u collisions of $H^+ - Rb_2^+$ (black curve) and $H^+ - 2Rb$ (red curve) compared with the Fourier spectrum from the third Born model (blue curve). The model $2Rb$ is resulting from a coherent sum of the ionization wave-function from two-independent monomers separated by a distance R_{ab}=9.2 a.u.. Inset: Initial Electronic probability density vs. spatial coordinate x ($|x|$<10 a.u.).

References

[1] N. Stolterfoht *et al* 2001 *Phys. Rev. ett.* **87** 023201

[2] N. Stolterfoht *et al* 2004 *Phys. Rev. A* **69** 012701

[3] S. Hossain *et al* 2005 *Phys. Rev. A* **72** 010701

[4] J. A. Tanis *et al* 2006 *Phys. Rev. A* **74** 022707

[5] J. L. Baran *et al* 2008 *Phys. Rev. A* **78** 012710

[6] D. Misra *et al* 2009 *Phys. Rev. A* **80** 062701

[7] N. Sisourat *et al* 2007 *Phys. Rev. A* **76** 012718

[8] L. Sælen *et al* 2010 *Phys. Rev. A* **81** 022718

[1]E-mail: h.agueny@gmail.com

Content from this work may be used under the terms of the Creative Commons Attribution 3.0 licence. Any further distribution of this work must maintain attribution to the author(s) and the title of the work, journal citation and DOI.
Published under licence by IOP Publishing Ltd

Journal of Physics: Conference Series **635** (2015) 032095 doi:10.1088/1742-6596/635/3/032095

Observation of recurrent fluorescence photons from small carbon cluster anions

Yuta Ebara*, Takeshi Furukawa* [1], Jun Matsumoto+, Hajime Tanuma*, Toshiyuki Azuma†, and Haruo Shiromaru+

*Department of Physics, Tokyo Metropolitan University, 1-1 Minami-Osawa, Hachioji, Tokyo 192-0397, Japan

+Department of Chemistry, Tokyo Metropolitan University, 1-1 Minami-Osawa, Hachioji, Tokyo 192-0397, Japan

†RIKEN, 2-1 Hirosawa, Wako, Saitama 351-0198, Japan

Synopsis We directly observed recurrent fluorescence from small carbon cluster anions C_6^- stored in an electrostatic ion storage ring TMU E-ring. This process, caused by the electronic transitions via low-lying electronic excited states after the inverse internal conversion (IIC), has been recently reported as fast cooling. In the present study, we succeeded in detection of emitted fluorescence. The observed lifetime was consistent with the theoretically estimated lifetime.

In an isolated environment, vibrationally excited molecules experience fragmentation, electron emission and IR radiation for the vibrational transitions. Recently, an additional rapid cooling process of emitting fluorescence has been experimentally revealed for the molecular ions stored in ion storage devices [1-3]. This emission called recurrent fluorescence is accompanied by the electronic transitions via low-lying electronic excited states after the inverse internal conversion (IIC) process, i.e., the conversion of internal energy from vibrationally excited states to electronically excited states (see Fig. 1).

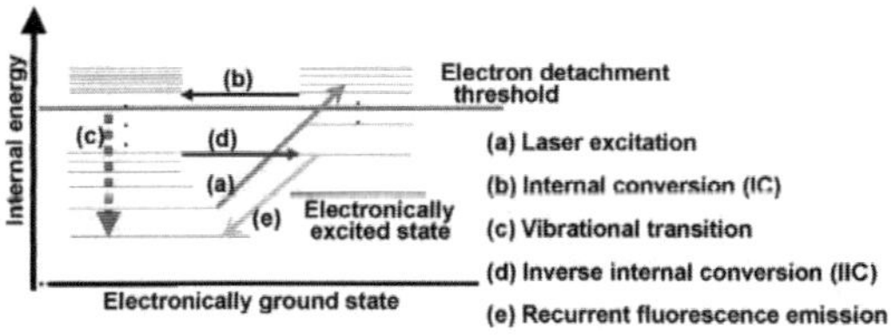

Figure 1. Jablonski diagram of the molecular anions with the low-lying electronic excited state.

It should be noted that the fast cooling rates observed in the studies reported so far [1-3] was derived from the yield of the neutral products through the dissociation or electron detachment. In this study, we succeeded in first observation of the recurrent fluorescence photons emitted from the electronically excited states $C^2\Pi_g^+$ (2.04 eV above the ground state) of C_6^- ions stored in an electrostatic ion storage ring, TMU E-ring.

Figure 2 shows the schematics of the experimental setup. 15 keV C_6^- ion bunches of typically less than 1 μs width, produced with a laser desorption ion source, were injected and stored in the TMU E-ring with a circulation period of 39.2 μs. After the storage time of typically 0.5 ms, pulsed laser light (a second harmonic light of the pulsed Nd:YAG laser, wavelength: 532 nm, energy: 9 mJ/pulse) was irradiated to excite the stored ions. We observed the delayed recurrent fluorescence from the stored C_6^- ions by a Pertier-cooled photomultiplier tube (PMT) for the single photon detection. In front of the PMT, four lenses, an interference filter (607 nm CWL, 80 nm FWHM), and an aperture for confining the field of view, were placed. All the optical devices and a PMT were covered with a dark box to reduce the background light.

The observed lifetime of the fluorescence was apparently longer than the intrinsic lifetime of the $C^2\Pi_g^+$ state, 730(50) fs [4]. The rapid decrease of the fluorescence was consistent with our theoretically estimated lifetime taking account of the IC and IIC processes.

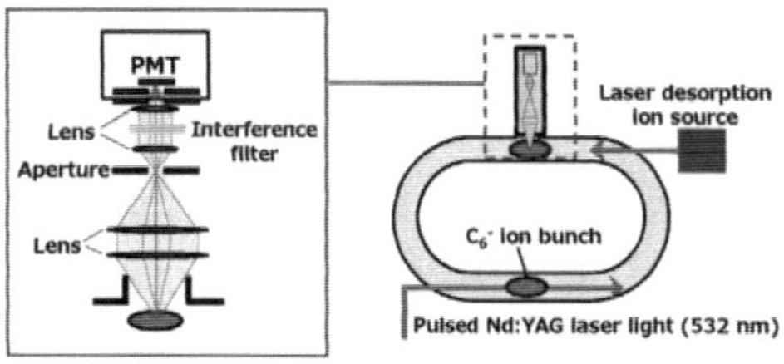

Figure 2. Schematic of the photon detection setup with the optical devices and a photomultiplier tube.

References

[1] S. Martin *et al* 2013 *Phys. Rev. Lett.* **110** 063003
[2] G. Ito *et al* 2014 *Phys. Rev. Lett.* **112** 183001
[3] V. Chandrasekaran *et al* 2014 *J. Phys. Chem. Lett.* **5** 4078
[4] C. Frischkorn *et al* 2001 *J. Chem. Phys.* 115 11185

[1] E-mail: takeshi@tmu.ac.jp

Content from this work may be used under the terms of the Creative Commons Attribution 3.0 licence. Any further distribution of this work must maintain attribution to the author(s) and the title of the work, journal citation and DOI.
Published under licence by IOP Publishing Ltd

Journal of Physics: Conference Series **635** (2015) 032096 doi:10.1088/1742-6596/635/3/032096

Effects of the environment on the uracil molecule ionization induced by $^{12}C^{4+}$ ion beam

P. Markush*[1], P. Bolognesi*[2], A. Cartoni[§], P. Rousseau[†], S. Maclot[†], R. Delaunay[†], A. Domaracka[†], J. Kocisek[†], M. Castrovilli[#], B. A. Huber[†] and L. Avaldi*

[*]CNR-ISM, Area della Ricerca di Roma 1, Monterotondo Scalo, Italy
[§]Dipartimento di Chimica, Università di Roma Sapienza, Roma, Italy
[†]CIMAP, CEA/CNRS/ENSICAEN/Unicaen, BP5133, 14070 Caen cedex 5, France
[#]CNR-IFN, Piazza Leonardo da Vinci 32, 20133 Milano, Italy

Synopsis In this study the fragmentation of isolated uracil molecules, uracil clusters and hydrated uracil clusters induced by $^{12}C^{4+}$ ions at 36 keV energy has been investigated. The mass spectra obtained by a TOF mass spectrometer are analyzed and compared to each other in order to see how the environment affects the fragmentation dynamics. The main differences between the mass spectra are highlighted and possible fragmentation pathways are proposed.

The studies of ionization induced by heavy charged particles are particularly important to understand the processes occurring during hadron therapy for cancer treatment. Gas phase studies allow to characterize the radiation damage of biomolecules, like DNA/RNA bases for example, at the molecular level. However, in a real biological system, the biomolecule is incorporated in a complex environment, which certainly affects its properties and behavior.

To unravel the role of the environment in the fragmentation of uracil, we have measured and compared the mass spectra of isolated uracil molecules and uracil clusters (Fig. 1). Even though the two spectra are very similar, differences can be observed in the intensity of different fragments and in the opening/closing of fragmentation channels.

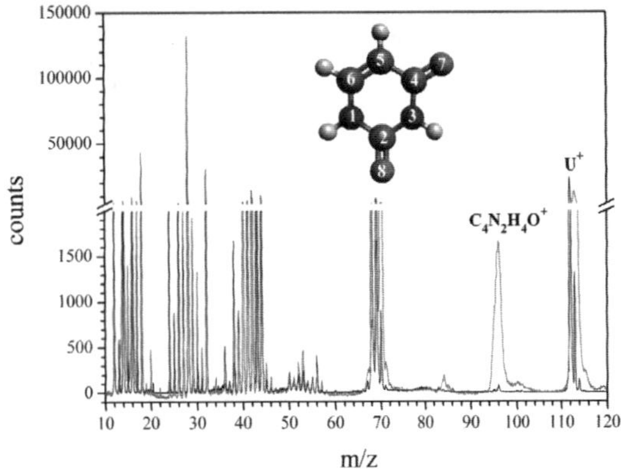

Figure 1. Comparison between isolated uracil molecules (black line) and uracil clusters (red line). The spectra are normalized to the same total area in the region up to the monomer.

For example, in the mass spectrum of uracil clusters a broad feature can be observed at m/z=96, attributed to the formation of $C_4N_2H_4O$ fragment ion due to the loss of an OH group by the parent ion. This fragment, which is almost negligible in the uracil molecules mass spectrum, indicates that O atoms may have weaker bonds within the clusters as a result of the hydrogen bridge bonds.

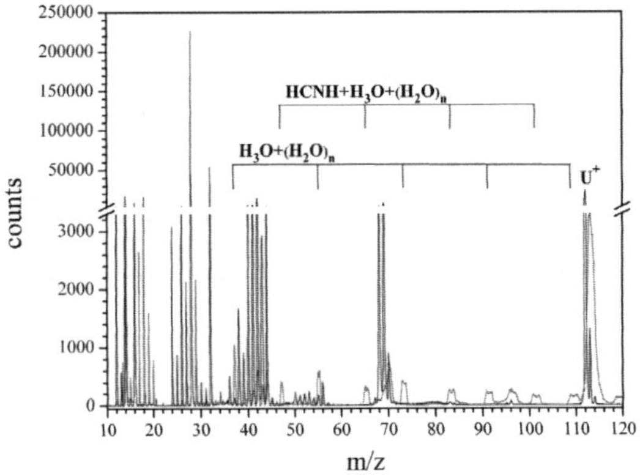

Figure 2. Comparison between isolated uracil molecules (black line) and hydrated uracil clusters (red line).

Many fragments observed in the mass spectra of the isolated uracil molecules and uracil clusters cannot be found in case of hydrated clusters (Fig. 2), confirming that the presence of the water molecule has a protecting effect on the nucleobases [1]. Nevertheless, in the case of the hydrated uracil clusters a series of hydrated fragments is observed.

References

[1] S. Maclot *et al* 2011 *ChemPhysChem.* 12 930

[1] E-mail: 1988.markus@gmail.com [2] E-mail: paola.bolognesi@cnr.it

Content from this work may be used under the terms of the Creative Commons Attribution 3.0 licence. Any further distribution of this work must maintain attribution to the author(s) and the title of the work, journal citation and DOI.
Published under licence by IOP Publishing Ltd

Journal of Physics: Conference Series **635** (2015) 032097 doi:10.1088/1742-6596/635/3/032097

Charge Transfer Cross Sections of Slow Singly Charged Lithium Ions in collisions with Hydrogen Isotope Molecules

T. Kusakabe* [1], **H. A. Sakaue**†, **and H. Tawara**†

* Department of Science, Kinki University, Kowakae, Higashi-osaka 577-8502, Japan
† National Institute for Fusion Science, Oroshi-cho, Toki 509-5292, Japan

Synopsis Charge transfer cross sections of $^7Li^+$ ions in collisions with H_2, HD, and D_2 molecules have been measured in the energy range from 1.4 to 4.0 keV. Significant differences in these results between collisions with H_2 and other hydrogen isotope molecules, the so-called target isotope effect, are observed below 2 keV. These observed cross sections for H_2 target are compared with available experimental and theoretical data.

Lithium are sometimes used for the purpose of plasma diagnostics in fusion devices. Therefore, cross-section data and any knowledge of collision dynamics for the charge transfer processes of lithium ions colliding with hydrogen molecules at low energies are known to be important the plasma modeling in controlled thermonuclear fusion research. However charge transfer cross section measurements of $Li^+ + H_2$ collisions are sparse, particularly at low collision energies. To the best of our knowledge, Ogurtsov *et al.* [1], and Cabrera-Trujillo *et al.* [2] reported the cross section measurements below 10 keV. Furthermore, few theoretical studies, e.g., Elizaga *et al.* [3] and Cabrera-Trujillo *et al.* [2], treated this charge transfer process.

We previously investigated the isotope effect on charge transfer by slow hydrogen ions (H^+, D^+) from hydrogen molecules (H_2, HD, D_2) [4]. The cross section ratios $\sigma(D_2)/\sigma(H_2)$ for the charge transfer decreased to a value smaller than unity below a collision energy of 1keV/u, and reached a value of about 0.6 at 0.18 keV/u. To establish the cross section data and obtain more general understanding of the isotope effect on the charge transfer in ion-molecule collisions of the hydrogen family, we have measured the charge transfer cross sections of $^7Li^+$ ions colliding with H_2, HD, and D_2 molecules in the energy range from 1.4 to 4.0 keV [5].

The $^7Li^+$ ion beam extracted from a surface ionization ion source was mass-analyzed with a Wien filter and was introduced into a collision cell in which target hydrogen gases of high purity were filled. The Li^+ ions and product Li atoms emerging from the cell after the collisions were charge-separated and detected with a position-sensitive micro-channel plate detector. Charge transfer cross sections were derived based on the growth rate method.

Figure 1 shows the present charge transfer cross sections of $^7Li^+$ ions colliding with H_2, HD, and D_2 molecules. Both the present charge transfer cross sections of HD and D_2 molecules are almost the same, but they become gradually smaller with a decrease in incident energy than those of H_2 molecules.

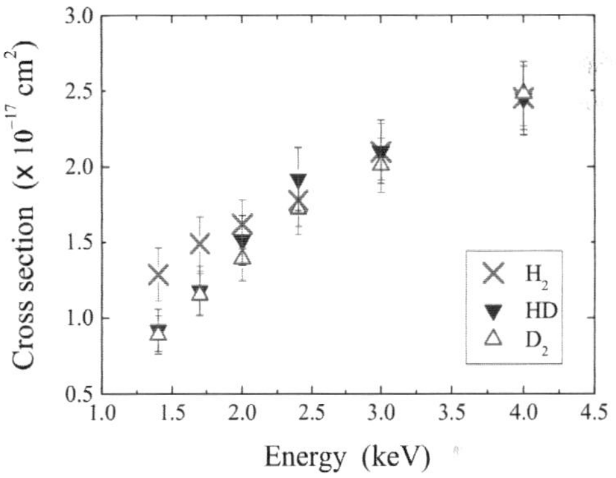

Figure 1. Charge transfer cross sections of $^7Li^+$ ions colliding with H_2, HD, and D_2 molecules [5].

The present charge transfer cross sections in collisions with H_2 molecules are found to reasonably agree with the theoretical results for the total charge transfer cross sections to Li ($1s^2\,2l$) reported by Elizaga *et al.* [3]. The experimental results measured by Ogurtsov *et al.* [1] are about 50% smaller than the present experimental data. Both the experimental and theoretical results of Cabrera-Trujillo *et al.* [2] are larger than the present data.

References

[1] G.N. Ogurtsov *et al* 1966 *Sov. Phys.-Tech. Phys.* **11** 362
[2] Cabrera-Trujillo *et al* 2010 *Phys. Rev. A* **82** 022718
[3] D. Elizaga *et al* 2000 *J. Phys. B: At. Mol. Opt. Phys.* **33** 2037
[4] T. Kusakabe *et al.* 2004 *Phys. Rev. A* **70**, 052710
[5] T. Kusakabe *et al* 2013 *Plasma Fusion Res.* **8** 2401100

[1] E-mail: kusakabe@phys.kindai.ac.jp

Content from this work may be used under the terms of the Creative Commons Attribution 3.0 licence. Any further distribution of this work must maintain attribution to the author(s) and the title of the work, journal citation and DOI.
Published under licence by IOP Publishing Ltd

ELASR- an ELectrostAtic Storage Ring in Riyadh for Atomic Collisions

Mohamed O. A. El Ghazaly[*†] [1]

[*] Astrophysics and Space Sciences Section, Jet Propulsion Laboratory/Caltech, Pasadena CA 91109, USA
[†] King Abdulaziz City for Sciences and Technology (KACST), Riyadh 1442, Saudi Arabia

Synopsis A new ELectrostAtic Storage Ring (ELASR) is being built at the King Abdulaziz City for Science and Technology (KACST), in Riyadh, Saudi Arabia. ELASR was designed to be the core of the new laboratory for atomic and molecular collisions at KACST. ELASR has been assembled, mounted and vacuum-tested. It is currently undergoing commissioning.

Electron-ion and photon-ion collisions are major reaction partners in ionized gases and plasmas both astrophysical and industrial. In naturally-occurring plasmas, such as interstellar clouds, stellar and planetary atmospheres, or comas of comets, electron–molecular-ion recombination, dissociation and ionization constitute key, fundamental molecular processes. Such reactions regulate energy balance and dominate most low- to high-temperature plasmas [1].

The advent of heavy-ion electrostatic storage rings has given strong impetus to the field of atomic collisions. The rings have enabled cutting-edge experiments to be performed on lifetime measurements and in the field of both electron collisions and laser interaction with molecular ions. In particular, electrostatic storage rings have opened up new applications to cluster and macromolecular ions of biological interest.

Following the first ELectrostatic Ion Storage ring in Aarhus (ELISA) [2], several ELISA-like storage rings have been built worldwide. Likewise, ELASR; the ELectrostAtic Storage Ring in Riyadh [3] follows the design of the pioneering storage ring ELISA. It is thereby intended to initiate and lead research programs that are similar to those that ELISA has been leading thus far [4]. Similar to ELISA, the ELASR storage ring features a racetrack, single-bend shaped ring [5]. It is designed to be the core of a new storage ring facility for atomic and molecular collisions at KACST [6]. Shown in Fig. 1 is a panoramic view of the ELASR as designed using Solidwork software.

Shown in Fig. 2 is a photograph of the ELASR ring. It has recently been assembled and vacuum-tested and is now being commissioned at KACST. Presented herein will be details of the design and construction of ELASR, as well as proposed first experiments.

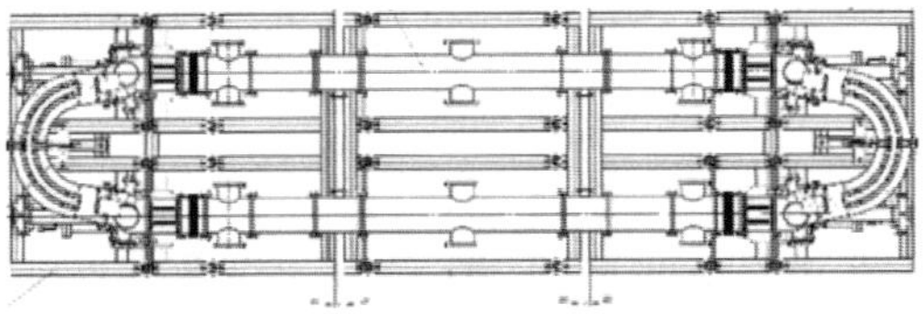

Figure 1. Top of view of ELASR as designed using Solidworks.

Figure 2. Photograph of the Riyadh ELASR electrostatic storage ring.

ELASR storage ring was developed and has been funded by the King Abdulaziz City for Sciences and Technology (KACST). The author acknowledges the support of the KACST, under the grant no.162-28/MOA_ElGhazaly.

References

[1] M. O. A El Ghazaly *et al* 2014 *J. Phys. Chem. A*, **118** (43) 10020–10027
[2] S. P. Møller 1997 *Nucl. Instrum. Meth. A* **394** 281-286
[3] M. O. A El Ghazaly 2015 *Results in Physics* **5** 60–61
[4] M. O. A. El Ghazaly *et al* 2013 *Nucl. Instrum. Meth. A* 709 76–84
[5] M. O. A El Ghazaly 2015 *Microscopy and Microanalysis* (*In press*).
[6] M. O. A El Ghazaly 2015 *EPJ Web of Conferences* **84** 05003

[† 1] E-mail: mohamed.el.ghazaly@jpl.nasa.gov

Content from this work may be used under the terms of the Creative Commons Attribution 3.0 licence. Any further distribution of this work must maintain attribution to the author(s) and the title of the work, journal citation and DOI.
Published under licence by IOP Publishing Ltd

Electrostatic Storage Ring for Physics and Space Studies

A. Chutjian* [1], M. O. A. El Ghazaly* [2], D. P. Mahapatra*† [3], A. Moradmand* [4]

* Astrophysics and Space Sciences Section, Jet Propulsion Laboratory/Caltech, Pasadena CA 91109, USA
† Department of Physics, Utkal University, Bhubaneswar-751004, India

Synopsis An electrostatic ion storage ring is being designed at JPL for application to present and future needs in solar/stellar physics, planet/exoplanet atmospheres, circumstellar clouds, and the interstellar medium.

Ongoing space experiments such as Hubble, Herschel/SPIRE, SOFIA, ALMA, Suzaku, and Newton – and the upcoming JWST, Solar Probe Plus, and ESM-ESA missions – place an ongoing need for lifetimes, spectroscopy, photon- and electron-ionization and dissociation cross sections of atomic and molecular ions [1]. To meet these needs in a flexible way, a program has been initiated at the Jet Propulsion Laboratory (JPL) to design and build an electrostatic storage ring (ESR). As is well known, the ESR is a mass-independent storage device [2] that makes use of purely electrostatic optics, and hence the storage of ions depends only on the kinetic energy of the species and not their mass [3]. One can store protons, as well as ionic clusters, amino acids, proteins, polyaromatic hydrocarbons (PAHs), tholins, and even dust grains. Other advantages of the ESR are that it provides a high-brightness ion beam, and is smaller and less-expensive than its magnetic-ring counterpart.

Since the successful introduction of the pioneering storage ring ELISA, a number of similar devices have been, or are being built. These include the KEK ring, the E-ring, the DESIREE double ring [3], and the recent ELectrostAtic Storage Ring in Riyadh (ELASR) [4]. FLSR is the first of existing comparable storage rings with a different optical lattice than that of ELISA [5]. The JPL ring will have a lattice layout similar to the FLSR, but with several features for simplifying the optics. It will have two long, ion-laser, ion-neutral or ion-electron beams sections to accommodate merged electron- and photon-dissociation and ionization experiments; to study dissociative recombination and its dynamics; as well as have the possibility of measuring lifetimes, vibration-rotation constants, and isotope ratios in mass-selected molecular and macro-molecular ions. The possibility of accommodating merged-beams experiments in the shorter, straight sections will also be explored, as well as formation of foci [5] for crossed-beams studies.

Shown in Fig. 1 is a conceptual design of the JPL ESR. It features a pseudo-racetrack, split-bend, symmetrically-shaped ring of eight beam inlet/outlet ports for beam injection or detection. The design features of the JPL ring based on SIMION [6] trajectory simulations and MAD [7] lattice calculation will be presented and discussed.

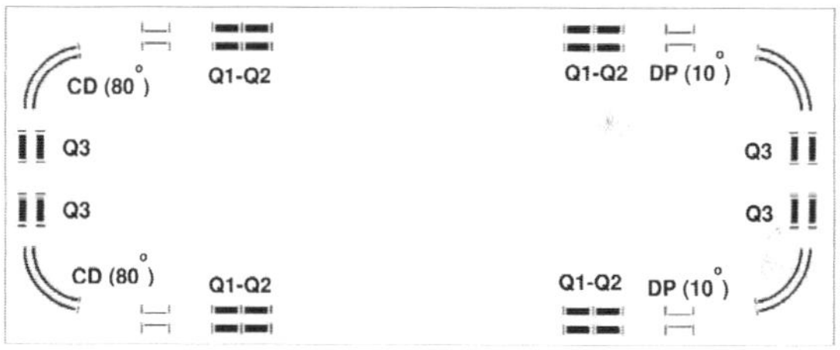

Figure 1. Schematic of the proposed JPL ESR with its double mirror-symmetry lattice and four super periods. The notation is: cylindrical deflectors (CD), deflector plates (DP), and quadrupole lenses (Q). The width × length at beam center is 1.4 × 3.7 m.

References

[1] A. Chutjian, *Uses and Construction of an ESR for Measurements in Planetary Physics and Astrophysics,* 2015 (JPL internal report, unpublished).
[2] S. P. Møller 1997 *Nucl. Instrum. Meth. A* **394** 281
[3] H. T. Schmidt *et al.* 2013 *Rev. Sci. Instrum.* **84** 055115
[4] M. O. A. El Ghazaly 2015 *Results in Physics* **5** 60
[5] K. E. Stiebing *et al.* 2010 Nucl. Instrum. Meth. A **614** 10
[6] D. Dahl 2000 *J. Mass Spectrom* **200** 3, and http://simion.com/
[7] MAD – Methodical Acceleration Design – at the CERN website, *http://mad.web.cern.ch/mad/*

[1] ara.chutjian@jpl.nasa.gov
[2] mohamed.el.ghazaly@jpl.nasa.gov
[3] durga.p.mahapatra@jpl.nasa.gov
[4] ali.moradmand@jpl.nasa.gov

Content from this work may be used under the terms of the Creative Commons Attribution 3.0 licence. Any further distribution of this work must maintain attribution to the author(s) and the title of the work, journal citation and DOI.
Published under licence by IOP Publishing Ltd

Origin of the sequential dissociation of $(CO_2)^{3+}$ by heavy ion impact

S. Yan*[1], X. L. Zhu*, W. T. Feng*, P. Zhang*, Y. Gao†, and X. Ma*[2]

* Insititute of Modern Physics, Chinese Academy of Sciences, Lanzhou 730000, China
† University of Chinese Academy of Sciences, Beijing 100049, China

Synopsis An experimental investigation of sequential dissociation of CO_2 by Ne^{4+} ion impact at incident energies 1.12 MeV was performed. Two initial states of $(CO_2)^{3+}$ ions were observed in the first fragmentation step. But for the second fragmentation step, our result suggests that both the metastable ions $(CO)^{2+}$ are populated in $^3\Sigma^+$ states.

If a triatomic molecule is multiply charged, the sequential and nonsequential dissociation processes will be invoked. Although these processes were studied extensively in the past [1, 2], the corresponding initial states, e.g. why these processes could take place, are not clearly so far.

We carried out the CO_2 fragmentation experiment in Reaction Microscope by Ne^{4+} ion impact at incident energies 1.12 MeV, aimed to investigate the origin of the sequential dissociation of $(CO_2)^{3+}$ ions. By coincidently detecting the $C^+/O^+/O^+$ ions, the Newton plot representing the momentum relation among the three ions, as well as the Kinetic Energy Release (KER) distribution of each step, are obtained.

As shown in figure 1, there is a circular ring in Newton plot, in which the momentum sum of the C^+ and the second O^+ ion is constant, this characteristic is consistent with the sequential fragmentation[2]. That is, in the first step the $(CO_2)^{3+}$ fragment into CO^{2+} and O^+, and in the second step another dissociation takes place inducing CO^{2+} fragment into C^+ and O^+. The corresponding events are chosen to reconstruct the total KER distribution. As displayed in figure 2, two narrow peaks (the corresponding peak values are 22.5 eV and 29 eV) are observed, which means that two decay channels can cause the sequential dissociation. The two peaks corresponding to lower and higher KER are labeled A and B. From the KER of the first and the second fragmentation steps obtained, we can investigate the fragmentation details of both channels mentioned above. It is found that, for channel A, the KER of peak value in the first step is 7.5 eV, while that in the second step is 15eV. But for channel B the corresponding values are 8 eV and 21 eV, respectively. This indicates that, for the first steps of channel A and B, the initial $(CO_2)^{3+}$ ions are populated in different states. But for the second steps, both the intermediate states ions $(CO)^{2+}$ are populated in $^3\Sigma^+$ states.

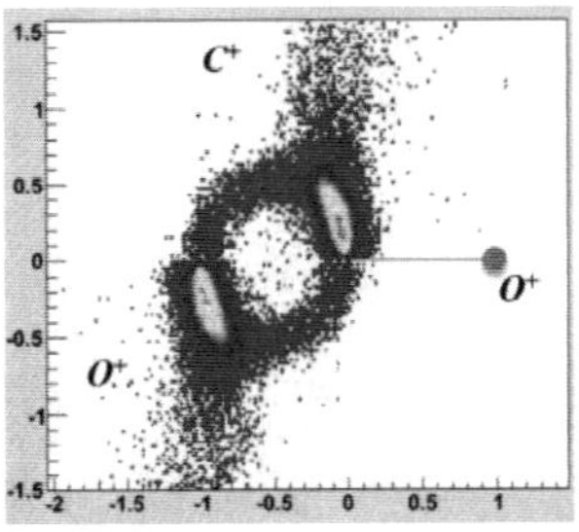

Figure 1. Newton plot of the $C^+/O^+/O^+$ ions from the fragmentation of $(CO_2)^{3+}$.

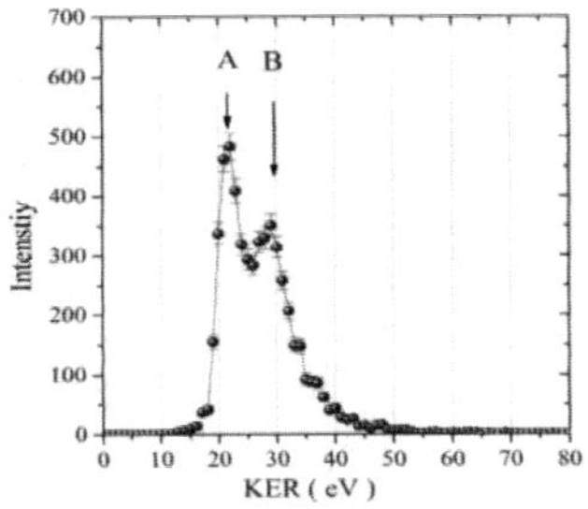

Figure 2. KER distribution of the $C^+/O^+/O^+$ ions from sequential dissociation of $(CO_2)^{3+}$.

This work is supported by NSFC under Grants No. 11304325, U1332128 and 11274317.

References

[1] N. Neumann *et al* 2010 *Phys. Rev. Lett.* 104 103201
[2] C. Wu, *et al* 2013 *Phys. Rev. Lett.* **110** 103601

[1] E-mail: yanshuncheng@impcas.ac.cn
[2] E-mail: x.ma@impcas.ac.cn

Content from this work may be used under the terms of the Creative Commons Attribution 3.0 licence. Any further distribution of this work must maintain attribution to the author(s) and the title of the work, journal citation and DOI.
Published under licence by IOP Publishing Ltd

Journal of Physics: Conference Series **635** (2015) 032101 doi:10.1088/1742-6596/635/3/032101

Screening effect during a collision in multiple ionization of rare gas dimers by highly charged ions

Tomoko Ohyama-Yamaguchi*[1] and Atsushi Ichimura†[2]

* Tokyo Metropolitan College of Industrial Technology, Shinagawa, Tokyo 140-0011, Japan

† Institute of Space and Astronautical Science, JAXA, Sagamihara, Kanagawa 252-5210, Japan

Synopsis Three-center, many electron dynamics is analyzed in multiple ionization of Ar_2 by highly charged ions. It is found that the dissociating ion-pair distribution is sensitive to a screening parameter during a collision.

Much attention has been called to the Coulomb explosion of molecules in collisions with slow (velocities of $v \ll 1$ atomic unit) highly charged ions. Recent progress of measurement techniques permits us to obtain the dissociating ion pair distribution produced in the collisions. Such findings are of great interest from the viewpoint of multi-center multi-electron dynamics. In contrast with covalent molecules, however, little effort has been devoted to rare gas dimers.

About ten years ago, we proposed a three-center Coulombic over-barrier model to describe sequential multiple ionization of rare gas dimer BC with highly charged ion A^{q+} [1]. In a recent work [2] we modified the model so as to incorporate the effect of partial screening for non-active target atomic site (B or C) in respective steps of electron removal during a collision. The measured result [3, 4] of ion pair distribution up to four electron removal in $Ar^{9+} + Ar_2$ collisions was best reproduced with the model by taking a screening parameter of $s = 0.4$ (see figure 1).

In the present work, we further consider the effect of screening in projectile site A^{q+} so as to make a consistent framework. The three charges in the three center Coulomb potential

$$U(\boldsymbol{r}) = -\frac{q_A}{|\boldsymbol{r}-\boldsymbol{R}_A|} - \frac{q_B}{|\boldsymbol{r}-\boldsymbol{R}_B|} - \frac{q_C}{|\boldsymbol{r}-\boldsymbol{R}_C|}$$

are effectively taken as

$$\begin{aligned} q_A &= q-(Q_B+Q_C)(1-s), \\ q_B &= Q_B+1, \\ q_C &= (1-s)Q_C, \end{aligned}$$

for the active electron at site B, and

$$\begin{aligned} q_A &= q-(Q_B+Q_C)(1-s), \\ q_B &= (1-s)Q_B, \\ q_C &= Q_C+1, \end{aligned}$$

for the active electron at site C.

The projectile screening is likely to weaken the charge asymmetry in the ion pair distribution. The experimental result is found to be reproduced by a parameter s smaller than 0.4.

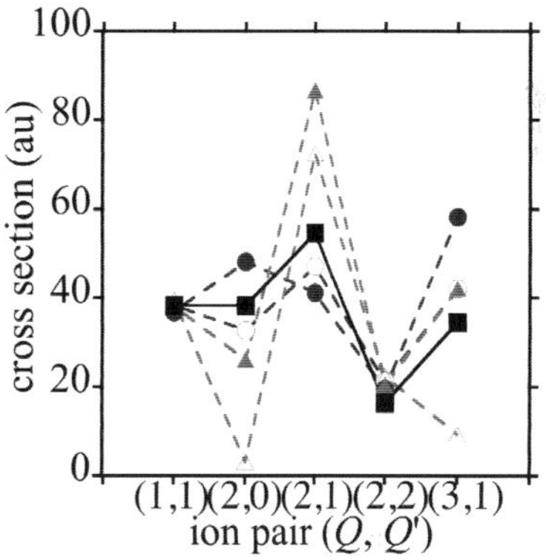

Figure 1. Ion pair (Q, Q') formation cross sections in $A^{9+} + Ar_2$ collisions. Calculation results with $s = 0.3$ and $s = 0.4$ are denoted by open (closed) triangles and open (closed) circles, respectively, where the screening effects only in the target (both in the target and projectile) are considered. The closed squares denote the experimental result [3].

References

[1] T. Ohyama-Yamaguchi and A. Ichimura, *Nucl. I strum. Methods* B **205**, 620 (2003).

[2] T. Ohyama-Yamaguchi and A. Ichimura, *Phys ca Scr ta* T **144**, 014028 (2011); *b d.* **T156**, 014043, (2013).

[3] J. Matsumoto *et al.*, *Phys. Rev. ett.* **105**, 263202 (2010).

[4] W.Iskandar *et al.*, *Phys. Rev. ett.* **113**, 143201 (2014).

[1]E-mail: yamaguti@s.metro-cit.ac.jp
[2]E-mail: ichimura@isas.jaxa.jp

Content from this work may be used under the terms of the Creative Commons Attribution 3.0 licence. Any further distribution of this work must maintain attribution to the author(s) and the title of the work, journal citation and DOI.
Published under licence by IOP Publishing Ltd

Journal of Physics: Conference Series **635** (2015) 032102 doi:10.1088/1742-6596/635/3/032102

The fragmentation of neon dimer induced by low energy O^{6+} ions

X. L. Zhu*[1], S. Yan*, W. T. Feng*, X. Ma*†[2], D. L. Guo*, Y. Gao*†, R. T. Zhang*, X. W. Cheng*† X. J. Zhang‡, D. M. Zhao*, S. Xu*, H. B. Wang*†, D. B. Qian*, Z. K. Huang*†

* Institute of Modern Physics, Chinese Academy of Sciences, Lanzhou, 730000, China
† University of Chinese Academy of Sciences, Beijing, 100049, China
‡ Henan University of Science and Technology, School of Physics and Engineering, Luoyang, 471023, China

Synopsis We have investigated the fragmentation channels Ne^+-Ne^+ and Ne^{2+}-Ne^+ from Ne_2 dimer induced by O^{6+} ions at impact energy of 15 keV/u. The Ne^+-Ne^+ fragmentation channel mainly results from three pathways: the interatomic coulombic decay of $Ne^{2+}(2p^4)$-Ne initial state, the direct columbic explosion and the radiation charge transfer. Nevertheless, the fragmentation channel Ne^{2+}-Ne^+ is predominant by the direct coulombic explosion.

Since Cederbaum et al. [1] predicted the Interatomic Coulombic Decay (ICD) in 1997, the rare gas atomic clusters have been widely investigated. However, the experimental investigation of dimers with the heavy ions and electron as the projectile is still scare [2-6]. Especially for neon dimer, only a few works were reported with heavy ions [3,5]. Utilizing the reaction microscope [7], we investigated the fragmentation of neon dimer induced by highly charged O^{6+} ion at the projectile energy of 15 keV/u.

For Ne^+-Ne^+ fragmentation channel, the kinetic energy release (KER) distribution shows two strong peaks in Figure 1(a). The left peak locates at 4.5 eV, indicating that the charge separation takes place at the internuclear distance of 3.2 Å, corresponding to the equilibrium inter-nuclear distance of groundstate Ne_2 dimer. According to the potential curve [8], the left peak results from the processes of direct columbic explosion (CE) and ICD. The maximum of right peak locates at 7.3 eV, corresponding to the internuclear distance of 2.0 Å. This distance is shorter than the equilibrium internuclear distance but consistent with the equilibrium internuclear distance of one-site dicationic states $Ne^{2+}(2p^4)$-Ne. So, the peak correspond to radiation charge transfer (RCT) from Auger decay of above-mentioned one-site states.

Figure 1(b) shows the KER distribution of Ne^{2+}-Ne^+ fragmentation channel. The large peak located at 9 eV, indicating that the charge separation takes place at the internuclear distance of 3.2 Å, corresponding to the equilibrium inter-nuclear distance of groundstate Ne_2 dimer. The outer-shell electron remove cross sections of neon is larger than the inner-shell, the contribution of ICD is small. Thus, the peak is mostly due to the CE.

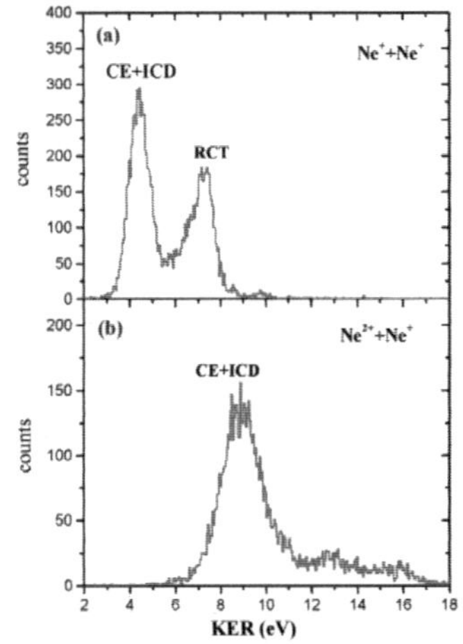

Figure 1. The the kinetic energy release distributions for (a) Ne^+-Ne^+ pair and (b) Ne^{2+}-Ne^+ pair induced by O^{6+} ion at impact energy 15 keV/u.

This work was partially supported by the 973 Program of China under Grants No. 2010CB832902 and by the NSFC of China under Grants Nos. 10979007, 10974207 and 11274317.

References

[1] L. S. Cederbaum et al *Phys. Rev. Lett.* **79** 4778
[2] J. Matsumoto et al *Phys. Rev. Lett.* **105** 263202
[3] W. Iskandar et al *Phys. Rev. Lett.* **113** 143201, *Phys. Rev. Lett.* **114** 033201
[4] J. Titze et al *Phys. Rev. Lett.* **106** 033201
[5] H.-K. Kim et al *Phys. Rev. A* **88** 042707, *Phys. Rev. A* **89** 022704
[6] S. Yan et al *Phys. Rev. A* **88** 042712, **89** 062707
[7] X Ma et al *Phys. Rev. A* **83** 052707
[8] S D Stoychev et al *J. Chem. Phys.* **129** 074307

[1] E-mail: zhuxiaolong@impcas.ac.cn
[2] E-mail: x.ma@impcas.ac.cn

Content from this work may be used under the terms of the Creative Commons Attribution 3.0 licence. Any further distribution of this work must maintain attribution to the author(s) and the title of the work, journal citation and DOI.
Published under licence by IOP Publishing Ltd

Journal of Physics: Conference Series **635** (2015) 032103 doi:10.1088/1742-6596/635/3/032103

Angular distributions of multiply charged fragments from dissociation of Nitrogen molecules by Xe^{9+} impact

Ajit Kumar*[1], T.Sairam†, Jyoti Rajput$, Lekha Nair* and C. P. Safvan†

* Department of Physics, Jamia Millia Islamia, Jamia Nagar, New Delhi, India
† Inter-University Accelerator Centre, Aruna Asaf Ali Marg, New Delhi, India
$Department of Physics and Astrophysics, University of Delhi, New Delhi India

Synopsis The angular distribution of the correlated molecular fragments with respect to the incoming beam direction for several of dissociation channels of multiply charged N_2 molecule on collision with Xe^{9+} was studied.

The angular distribution of the correlated molecular fragments with respect to the beam direction has been an active field of research both experimentally and by theoreticians due to the fundamental importance of these processes in many areas of science and technology. In the past two decades, significant experimental efforts have been devoted to studies of anisotropy in the angular distribution of fragments in fast highly charged ion-molecule collisions [1, 2]. The study of anisotropy in the angular distribution of fragments from highly charged molecular ions produced in slow highly charged ion collision is still mostly unexplored.

We present result on the angular distribution of fragmented nitrogen molecules emerging from collision between 450 keV Xe^{9+} projectile and neutral N_2. The measurements were made using a multihit, position sensitive, time of flight based recoil ion momentum spectrometer. The experiment was performed using the ECR ion source based low energy ion beam facility (LEIBF) at Inter-University Accelerator Centre (IUAC), New Delhi, India [3].

Several possible channels observed in the dissociation of multiply charged N_2 molecules formed by the impact of Xe^{9+} are studied. We have found that for the symmetric dissociation channels, the distribution of fragments is isotropic in nature. But in the case of asymmetric dissociation channels, the distribution of fragments are anisotropic. To confirm the nature of the distribution, we compared our result with results of Monte Carlo simulations for isotropic distributions of the fragments. We find that the simulated results match well with the angular distribution of fragments from symmetric channels. This shows that the angular distribution of fragments from symmetric channels is isotropic. But in the case of the angular distribution of fragments in asymmetric dissociation channels, the measured results do not match with the results of the simulation. We find that in any asymmetric dissociation channel the recoil ion fragment with the higher charge is emitted preferentially in the backward direction with respect to the beam direction. For example, in the dissociation channel $N_2^{3+} \rightarrow N^{2+} + N^+$, the N^{2+} fragment is preferentially observed in the backward direction as shown in figure 1 .

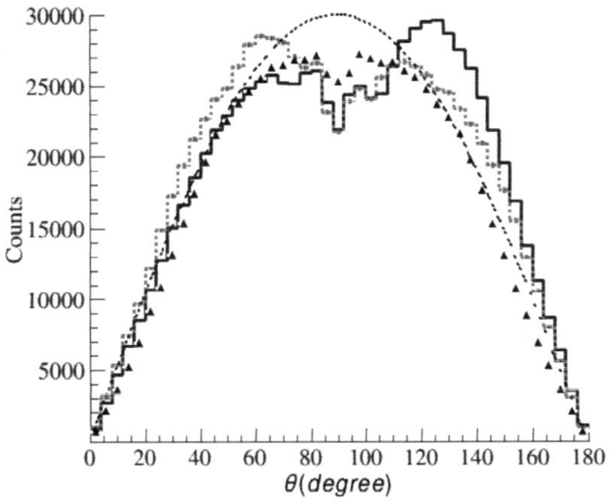

Figure 1. (color online) Angular distribution of fragments observed in the dissociation channel $N_2^{3+} \rightarrow N^{2+} + N^+$ with respect to projectile direction (θ). The blue (thick solid line) and the red (dashed line) curves represent the angular distribution of N^{2+} and N^+ fragments respectively. The smooth dotted curve is the sine theta distribution expected from an isotropic distribution while the black triangles show the expected distribution after taking into account the spectrometer acceptance.

References

[1] U Werner *et al* 1997 *Physical Review Letters*, **79**, 1662.
[2] V. Horvat *et al* 2013 *Physical Review A*,**88**, 022703
[3] A. Kumar *et al* 2014 *Int. J. Mass Spectrom.* **374** 44

[1] E-mail: ajit.gupta529@gmail.com

Content from this work may be used under the terms of the Creative Commons Attribution 3.0 licence. Any further distribution of this work must maintain attribution to the author(s) and the title of the work, journal citation and DOI.
Published under licence by IOP Publishing Ltd

XXIX International Conference on Photonic, Electronic, and Atomic Collisions (ICPEAC2015) IOP Publishing
Journal of Physics: Conference Series **635** (2015) 032104 doi:10.1088/1742-6596/635/3/032104

Electron-DDCS Measurements of Water by C and Si Ion beams

S Bhattacharjee*[1], S Biswas*, S. Kasthurirangan*, R. D. Rivarola†, J. M. Monti†, C. A. Tachino†, and L C Tribedi*[2]

* Tata Institute of Fundamental Research, Homi Bhabha Road, Colaba, Mumbai 400 005, India
† Institute de Fisica Rosario (CONICET-UNR), Universidad Nacional de Rosario, 2000 Rosario, Argentina

Synopsis Double differential distributions of electron emission from H_2O target under the impact of C^{6+} and Si^{13+} ions with energy 4 MeV/u have been measured. The state-of-the art CDW-EIS model provided a good agreement with the data for C-ions whereas a substantial deviation is found for the Si ions. The derived single differential cross sections and total cross sections were also compared with different models.

Interaction of highly charged ions with water molecules is of great importance in many areas including radiation therapy. Proton and carbon therapy are the most widely used techniques for treating cancer cells. Since water constitutes about 60% of the body it is important to understand the heavy ion collisions with this molecule in order to help modelling the radiation therapy. Most of the earlier investigations on ionization of water molecules use proton and He-ions [1-3]. Very few measurements have been reported so far regarding interaction of water [1-5] or other biomolecules [6-8]with other heavier ions

We have measured the double differential cross section (DDCS) of electron emission from a free water molecule under the impact of (4 MeV/u) C^{6+} and Si^{13+} ions at same velocity i.e. at energy 4 MeV/u. The ion beams were obtained using a 14 MV Pelletron tandem accelerator facility at TIFR, Mumbai. The angular and energy distributions of the e-DDCS were investigated over a wide energy region i.e. 1-600 eV to include the KLL Auger peak. The angular range was chosen to be between 20 and 160 degree. The single differential cross sections and total cross sections were also calculated. The measured cross sections were compared with the CDW-EIS (prior version) model. The calculated cross sections exhibit very good agreement with the theoretical model for C-ions [see Figure 1(a)]. A similar agreement was also reported by us in case of fast bare oxygen ions of energy 6 MeV/u [5] The deviation starts for the H-like Si ions [Fig 1(b)], which clearly tests the limit of validity of the model w.r.t atomic no. of projectiles as well as perturbation strength. A strong dependence of the projectile nuclear charge on the cross sections were seen by comparing the results of C^{6+} and Si^{13+} beams. The detail results will be presented along with earlier existing results.

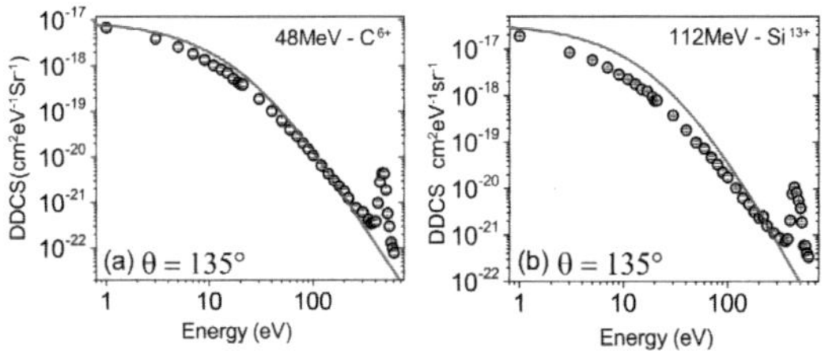

Figure 1: DDCS as a function of ejected electron energy for C^{6+} (a) and Si^{13+} (b) ions. In each plot, the solid red line corresponds to the CDW-EIS prior version

References

[1] L. H. Toburen et al. Radiat. Res. 82, 27 (1980).
[2] D. Ohsawa et al. Phys. Rev. A 72, 062710 (2005).
[3] C. Champion et al. Phys. Rev. A 75, 032724 (2007).
[4] C. Dal Capello et al. Nucl. Instrum. Methods Phys. Res.,Sect. B 267, 781 (2009).
[5] S. Nandi et al. Phys. Rev. A 87, 052710 (2013).
[6] A. N. Agnihotri et al. Phys. Rev. A 87, 032716 (2013).
[7] A. N. Agnihotri et al. Euro. Phys. Jour. D 66, 303 (2012).
[8] A. N. Agnihotri et al. Phys. Rev. A 85, 032711 (2012).

[1]E-mail: shamik127@gmail.com
[2]E-mail: lokesh@tifr.res.in

Content from this work may be used under the terms of the Creative Commons Attribution 3.0 licence. Any further distribution of this work must maintain attribution to the author(s) and the title of the work, journal citation and DOI.
Published under licence by IOP Publishing Ltd

Interatomic Coulombic Decay in Low Energy Ion-Dimer Collisions

W. Iskandar*[1], J. Matsumoto§, A. Leredde†, X. Fléchard‡, S. Guillous*, C. L. Zhou*, J. Rangama*, A. Méry*, B. Gervais*, D. Hennecart*, H. Shiromaru§ and A. Cassimi*

* CIMAP, CEA-CNRS-ENSICAEN-Université de Caen, BP 5133, F-14070 Caen cedex 5, France
§ Departement of Chemistry, Tokyo Metropolitan University, Hachiouji-shi, Tokyo 192-0397, Japan
† Physics Division, Argonne National Laboratory, Argonne, Illinois 60439, USA
‡ LPC Caen, ENSICAEN, Université de Caen, CNRS/IN2P3, 14050 Caen Cedex 04, France

Synopsis Low energy collisions between ions and neon dimers are investigated, focusing attention on the dissociation of the ionized molecular target. The results show a clear dependance of the processes with the projectile charge state and an unexpected appearance of inner-shell single electron capture from one site of the neon dimer leading to interatomic coulombic decay process and low energy electrons emission.

Interaction of slow multi-charged ions (MCIs) with dimers is of great interest in various scientific fields, not only to study the electron capture by the projectile, but also to have access to insightful information about the target and its relaxation processes [1, 2, 3]. We present here an experimental and theoretical study of collisions between MCIs and Ne dimers including the identification and analysis of the possible processes.

The experiment was performed at GANIL facility (Caen, France), on the low energy beam line. Three different projectiles O^{3+}, Ar^{9+} and Xe^{20+}, produced by an ECR ion source and accelerated to 15 qkeV, were used to collide with neon dimers provided by a supersonic jet. A COLTRIMS setup with multi-hit capability combined with a position-sensitive detector for scattered projectiles were employed to measure, in multi-coincidence, the projectile final charge state, the position and the time-of-flight of the ionic fragments. These measurements give access to the ionic fragments momentum and thus to the kinetic energy release (KER). The latter can be directly related to the internuclear distance prior dissociation (fig.1).

In the symmetric $Ne^+ + Ne^+$ fragmentation channel, we have identified three different competing processes [3] illustrated in fig.1. Beside the direct coulomb explosion (CE), wherein the projectile captures one electron from each center of the dimer which directly dissociates, two other electronic relaxation mechanisms were observed: the interatomic coulombic decay (ICD) and the radiative charge transfer (RCT). The ICD process results from the single capture of an electron from the 2s inner valence state by the projectile. Then, the excited ion decays to Ne^+ ($2p^{-1}$). The excess energy is transferred to the neighboring atom of the dimer by ejecting an outer valence electron from the 2p state, causing molecular ion fragmentation. For the RCT process, two electrons from the outer valence states of one center of the dimer are first captured by the projectile. Electron transfer from the doubly ionized center to the neutral one can then occur by photon emission leading to molecular dissociation into two Ne^+ ($2p^{-1}$) ions.

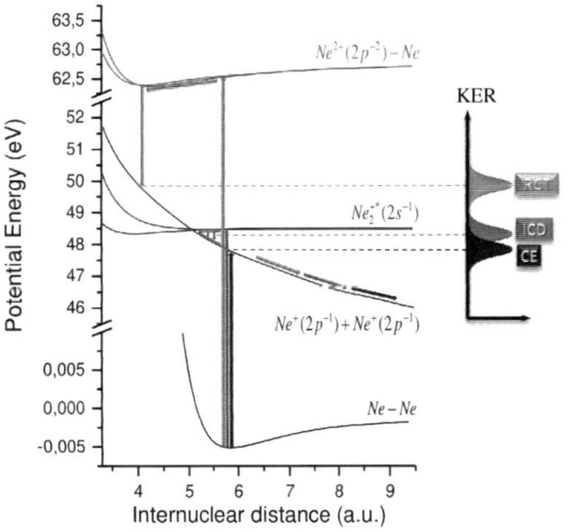

Figure 1. Illustration of the CE, RCT, and ICD processes in collisions between MCIs and Ne dimers.

For these three processes, the dependance on the initial and final projectile charge state [3] and, on the angular correlation between projectile scattering direction and dimer orientation has also been investigated, providing further insight into the dynamics of the collision.

References

[1] J. Matsumoto *et al.* 2010, Phys. Rev. Lett. **105** 263202
[2] W. Iskandar *et al.* 2014, Phys. Rev. Lett. **113** 143201
[3] W. Iskandar *et al.* 2015, Phys. Rev. Lett. **114** 033201

[1]E-mail: iskandar@ganil.fr

Content from this work may be used under the terms of the Creative Commons Attribution 3.0 licence. Any further distribution of this work must maintain attribution to the author(s) and the title of the work, journal citation and DOI.
Published under licence by IOP Publishing Ltd

Journal of Physics: Conference Series **635** (2015) 032106 doi:10.1088/1742-6596/635/3/032106

Atomic Site-Sensitive Processes in Slow Ar^{9+}-Ar_2 Collisions

W. Iskandar*[1], J. Matsumoto§, A. Leredde†, X. Fléchard‡[2], S. Guillous*, C. L. Zhou*, J. Rangama*, A. Méry*, B. Gervais*, D. Hennecart*, H. Shiromaru§ and A. Cassimi*

* CIMAP, CEA-CNRS-ENSICAEN-Université de Caen, BP 5133, F-14070 Caen cedex 5, France
§ Departement of Chemistry, Tokyo Metropolitan University, Hachiouji-shi, Tokyo 192-0397, Japan
† Physics Division, Argonne National Laboratory, Argonne, Illinois 60439, USA
‡ LPC Caen, ENSICAEN, Université de Caen, CNRS/IN2P3, 14050 Caen Cedex 04, France

Synopsis Electron capture processes for low energy Ar^{9+} ions colliding on argon dimer targets are investigated, focusing attention on charge sharing between the two argon atoms as a function of the molecular orientation and the impact parameter. Confrontation between experimental results and MC-COBM simulations have been investigated to provide access to atomic site sensitivity of the different fragmentation processes by measuring the angular correlation between the scattered projectile and the recoiling fragments.

Investigating an elementary reaction involving highly charged ions and molecules is a challenging subject: intramolecular charge redistribution after the collision may usually prevent the access to the primary process. Taking advantage of the low electron mobility in a dimer target, we present here a comparison between experimental and theoretical results giving access to the atomic site sensitivity for the collision system Ar^{9+}-Ar_2 at low energy regime.

The projectile scattering angle ϕ_{proj} is expected to be closely related to the impact parameter vector $\vec{b}$ in the molecular frame. For molecule orientations transverse to the projectile beam, the molecular frame is defined unambiguously by the angle $\phi_{Ar^{A+}}$, were Ar^{A+} is the most charged fragment (fig.1(d)). This angle is obtained by measuring in coincidence the position and time of flight of both ionic fragments. The transverse momentum transferred to the dimer center-of-mass during the collision is also inferred from these data sets, giving direct access to projectile scattering angle ϕ_{proj}. Then, the correlation between these two angles can be obtained by calculating the angular distributions in ϕ_{diff}=$\phi_{Ar^{A+}}$-ϕ_{proj}.

A preference for charge-asymmetric dissociation channels is observed [1], with a strong correlation between the projectile scattering angle and the molecular ion orientation as shown in the figure 1. This measurements provide clear evidence that projectiles distinguish each atom in the dimer and that electron capture from the closest atom to the projectile is favored. Monte Carlo simulations combined with Classical Over-the-Barrier Model (MC-COBM), by considering the Ar_2 dimer target as two independent Ar atoms fixed in space, are successfully compared to the data (fig.1). They give a new insight into the dynamics of the collision by providing, for the different fragmentation channels, the two-dimensional probability maps p($\vec{b}$) [2].

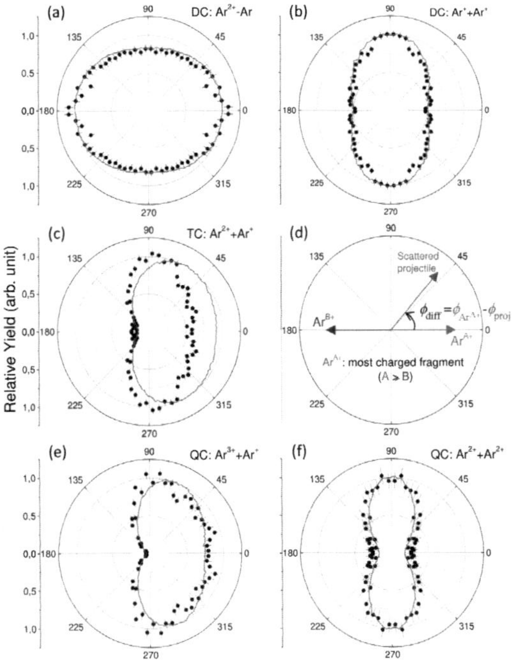

Figure 1. Angular distributions in ϕ_{diff} for the double, triple and quadruple electron capture and associated fragmentation channels. Experimental results (black dots) are compared to the MC-COBM calculations (red lines).

References

[1] J. Matsumoto *et al.* 2010, Phys. Rev. Lett. **105** 263202
[2] W. Iskandar *et al.* 2014, Phys. Rev. Lett. **113** 143201

[1]E-mail: iskandar@ganil.fr
[2]E-mail: flechard@lpccaen.in2p3.fr

Content from this work may be used under the terms of the Creative Commons Attribution 3.0 licence. Any further distribution of this work must maintain attribution to the author(s) and the title of the work, journal citation and DOI.
Published under licence by IOP Publishing Ltd

XXIX International Conference on Photonic, Electronic, and Atomic Collisions (ICPEAC2015) IOP Publishing
Journal of Physics: Conference Series **635** (2015) 032107 doi:10.1088/1742-6596/635/3/032107

Breakdown curves of carbon-based molecules for astrochemistry

M. Chabot†, K. Béroff*, P.Gratier#, A.Jallat†, V.Wakelam¤, J.P Sanchez§, N.Aguirre§, S.Diaz-Tendero§, M.Alcami§, F.Martin§, P.A Hervieux$

† Institut de Physique Nucléaire d'Orsay, Université Paris Sud and CNRS F-91406 Orsay Cedex, France
* Institut des Sciences Moléculaires d'Orsay, Université Paris Sud and CNRS F-91405 Orsay Cedex, France
# Institut de Radioastronomie Millimétrique, F-38406 Saint Martin d'Hères, France
¤Laboratoire d'Astrophysique de Bordeaux, Université de Bordeaux and CNRS, F-33270 Floirac, France
§ Universidad Autónoma de Madrid, Cantoblanco E-28049 Madrid, Spain
$ IPCMS, Université de Strasbourg, 67034 Strasbourg, France

Synopsis: Breakdown curves (BDC), which are energy dependent fragmentation branching ratios, constitute a kind of "identity card" of an excited molecule or cluster. We developed a method for constructing semi-empirical BDC, based on fragmentation measurements and structural known quantities of the considered species. Calculations of BDC have been performed within the statistical M3C theory. We will present a comparison of the two methods for some species and discuss application of these results to astrochemistry.

Carbon-based molecules are ubiquitous in astrophysical environments. The study of the composition and evolution of the molecular material is called astrochemistry. It involves the knowledge of a huge amount of data including fragmentation branching ratios (BR). The experimental group in Orsay (France) has developed a dedicated set-up (AGAT) for fragmentation recording of molecules and clusters of known charge and internal energy. Recently they showed how these data could be used to provide, within a statistical fragmentation context, BR for numerous physical and chemical processes of astrophysical interest [1]. The method relies on the construction of Breakdown curves (BDC, which are energy dependent BR) constrained by the experimental measurements.

On the other hand, BDC have been calculated within the Microcanonical Metropolis Monte Carlo method (M3C), recently improved [2]. Figure 1 shows, for the C_7 cluster, a comparison between the semi-empirical BDC and those obtained through M3C. The very good agreement validates so far the semi-empirical approach.

Whereas BDC of some hydrocarbon molecules were so constructed and used to predict BR for various processes [1] we aim at continuing those studies on $C_nH_m^{q+}$ species with larger m and n (n≤5, m≤4 q≤2) and, in a longer term, on nitrogenated species. BR of processes derived from the BDC will be included in the international astrochemical database KIDA (http://kida.obs.u-bordeaux1.fr) [3]. The impact of so-extracted BR in some astrophysical environments (dark clouds and photo-dissociation regions) will be presented and discussed.

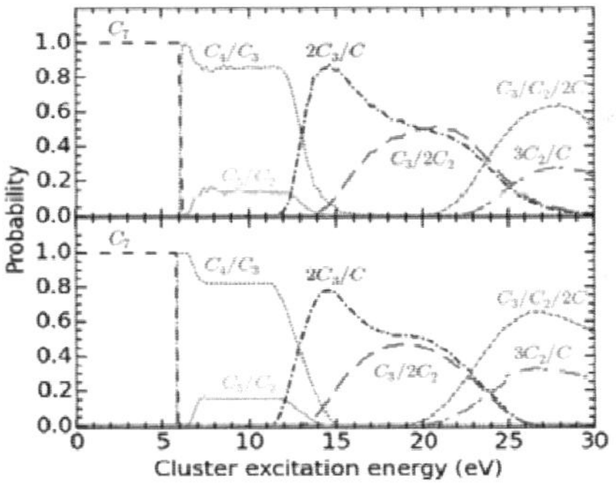

Figure 1 Breakdown curves for the C_7 cluster. Upper panel: theoretical M3C calculations. Lower panel: semi-empirical model.

References

[1] M.Chabot et al 2013 The Astrophysical Journal **771**, 90

[2] S.Diaz-Tendero et al 2005 Phy.ReV.A **71**, 033202 ; N.F.Aguirre et al, this conference

[3] V.Wakelam et al 2012 The Astrophysical Journal Suppl. Series **199**, 21; V.Wakelam et al 2015 The Astrophysical Journal Suppl. Series, in press.

[1]E-mail: chabot@ipno.in2p3.fr

Content from this work may be used under the terms of the Creative Commons Attribution 3.0 licence. Any further distribution of this work must maintain attribution to the author(s) and the title of the work, journal citation and DOI.
Published under licence by IOP Publishing Ltd

Journal of Physics: Conference Series **635** (2015) 032108 doi:10.1088/1742-6596/635/3/032108

Anion emission from H_2O molecules colliding with positive O^+ ions at keV energies: the role of dissociative excitation

J.-Y. Chesnel* [1], **Z. Juhász**†, **E. Lattouf***, **J. A. Tanis***,§, **B. A. Huber***, **E. Bene**†, **S. T. S. Kovács**†, **P. Herczku**†, **A. Méry***, **J.-C. Poully***, **J. Rangama***, **and B. Sulik**†

* Centre de Recherche sur les Ions, les Matériaux et la Photonique (CIMAP), 6 Bd Mar. Juin, F-14000 Caen, France
† Institute for Nuclear Research, Hungarian Academy of Sciences (MTA Atomki), H-4001 Debrecen, Hungary
§ Department of Physics, Western Michigan University, Kalamazoo, Michigan 49008 USA

Synopsis It is shown that negative ions can be ejected from gas-phase water molecules when bombarded with positive ions at keV energies. Emission of both H^- and heavier (O^- and OH^-) anions is observed. Electronic excitation and ionization processes are shown to play a decisive role in cation-induced anion formation.

We show that negative ions can be ejected from free water molecules by the impact of positive ions. The experiment was performed at the Grand Accélérateur National d'Ions Lourds (GANIL) in Caen, France. The ion beam crossed an effusive gas jet of H_2O. The collision-induced ionic fragments were selected according to their kinetic energy by means of an electrostatic analyzer. To avoid electron contamination of the anion spectra, the electrons and anions were magnetically separated so that only anions could reach the detector.

Figure 1 shows experimental double-differential cross sections (DDCS) for anion emission in 6.6-keV $^{16}O^+ + H_2O$ collisions. The main component of each spectrum is a broad, slowly decreasing structure. There, the relative contribution of H^- anions was derived from complementary time-of-flight (TOF) measurements. At forward angles (< 90°) pronounced peaks are observed at higher energies. These peaks are due to recoil anions formed in hard binary collisions occurring at small impact parameters [1-3], while the broad contribution results from soft many-body processes.

To further interpret our results, numerical simulations of the trajectories of the different fragments were performed. The method is similar to the one used in [4]. The calculated DDCS for H^- emission are shown in Fig. 1 (dotted curves). At low energies the calculated DDCS underestimate appreciably the experimental ones since in most of the simulated collisions at large impact parameters (> 1 a.u.) the kinetic energy transferred to the H atom is lower than the dissociation energy of an OH bond of H_2O in its ground state. However, electronic processes such as dissociative excitation and ionization are likely to lead to kinetic energy release (KER). By introducing into the model the same KER distribution as in [4], the simulated cross sections (open circles in Fig. 1) agree fairly well with the TOF data points (full circles) and reproduce satisfactorily the experimental H^- recoil peak. According to the model, without electronic excitation or ionization, the total cross section for H^- emission would be about 3 times smaller.

This work was supported by the Hungarian National Science Foundation OTKA (Grant No. K109440), the French-Hungarian Cooperation Program PHC Balaton (No. 27860ZL/TéT_11-2-2012-0028), and the French-Hungarian CNRS-MTA Cooperation (ANION project).

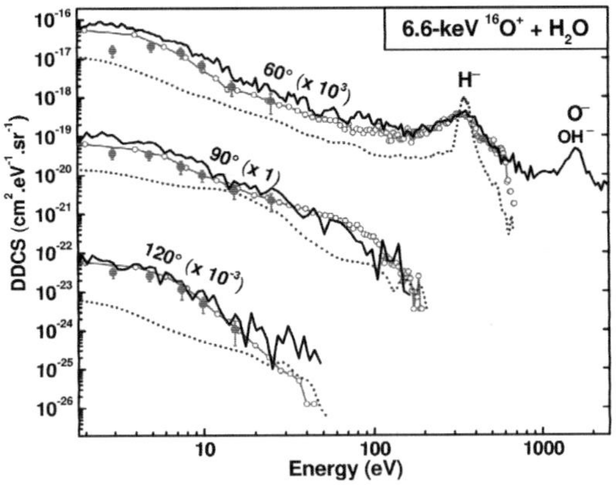

Figure 1. Full curve: Experimental DDCS for anion emission in 6.6-keV $^{16}O^+ + H_2O$ collisions; Full circles: H^- contribution from TOF measurements. Open circles: Simulated DDCS for H^- emission at different angles with KER; Dotted curve: Same without KER; Each spectrum is multiplied by the indicated factor.

References

[1] Z. Juhász *et al* 2012 *J. Phys. Conf. Ser.* **388** 102051
[2] Z. Juhász *et al* 2013 *Phys. Rev. A* **87** 032718
[3] Z. Juhász *et al* 2014 *J. Phys. Conf. Ser.* **488** 102024
[4] E. Lattouf *et al* 2014 *Phys. Rev. A* **89** 062721

[1] E-mail: jean-yves.chesnel@ensicaen.fr

Content from this work may be used under the terms of the Creative Commons Attribution 3.0 licence. Any further distribution of this work must maintain attribution to the author(s) and the title of the work, journal citation and DOI.
Published under licence by IOP Publishing Ltd

Effect of the Vibrational Excitation of CH_3^+ cations on their reactivity with CH_4

A. Lopes*, C. Romanzin*[1], B. Cunha de Miranda*, J. Zabka†, M. Polasek†, D. Ascenzi$, A. Cernuto$, and C. Alcaraz*€ [2]

* LCP, UMR 8000 CNRS-Université Paris Sud, Orsay, France
† J. Heyrovsky Institute of Physical Chemistry of the ASCR, Prague – Czech Republic
$ Atomic & Molecular Physics Lab., Università di Trento, Povo – Italy
€ Synchrotron SOLEIL, L'Orme des merisiers, St Aubin, France

Synopsis The VUV photoionisation of CH_3 radicals formed in a beam by pyrolysis of CH_3NO_2 precursors is used to produce CH_3^+ cations and control their vibrational excitation. The reactivity of CH_3^+ cations with methane is then studied on a Guided Ion Beam setup as a function of both collision energy and CH_3^+ internal energy.

The methyl carbocation is a reactive species which has been detected in a lot of gaseous environments fed by high energy sources such as the interstellar medium, planetary ionospheres as well as laboratory plasmas for methane conversion into higher hydrocarbons for instance. His reactivity with several hydrocarbons is well-known for the ground state whereas only few data about the reactivity of excited CH_3^+ ions are available in the literature although excited species are common in planetary atmospheres and plasmas.

Here, we report on experiments where CH_3^+ ions are generated via direct photoionization of $CH_3^\bullet$ methyl radicals, formed in a supersonic beam by pyrolysis of CH_3NO_2, thus allowing the production of CH_3^+ with a controlled degree of internal excitation when the photon energy is tuned between 9.8 and 12.5 eV on the DESIRS VUV beamline [1] at the French synchrotron SOLEIL.

In a first experiment, the CH_3^+ cation's excitation distribution is measured for each photon energy by the PEPICO (Photoelectron Photoion Coincidence) method using the DELICIOUS III spectrometer [2,3] of the SAPHIRS experiment. In this experiment, we collect all the photoelectrons and measure their kinetic energy in order to derive the distribution of internal energy of the parent ion CH_3^+ by the following equation: $h\nu - IP = E_c(e^-) + E_{int}(CH_3^+)$. We found that, regardless on the photon energy, a ground state population of CH_3^+ is produced and that, by increasing the photon energy, an additional population of very excited CH_3^+ ion is formed.

In the second experiment, $CH_3^\bullet$ radicals are formed by the same pyrolysis source which was recently adapted on the Guided Ion Beam setup, CERISES [4], devoted to the study of ion-molecule reactions. VUV photoionization of the $CH_3^\bullet$ radicals is used to produce the CH_3^+ cations which are then mass selected and guided to a reaction cell where they react with methane.

Absolute reaction cross-sections have been measured for the $CH_3^+ + CH_4$ and $CH_3^+ + CD_4$ systems as a function of both collision energy (0.1 – 4 eV) and photon energy. Reaction dynamics and its dependence on the CH_3^+ parent ion vibrational excitation will be discussed.

Acknowledgments

This work is supported by the RTRA "Triangle de la Physique" (project RADICAUX and NOSTADYNE) and the French Program of Planetology (PNP). A.C. thanks the COST XLIC Action for a STSM fellowship. D.A. acknowledges support from the Dept. Physics, University of Trento.

References

[1] Nahon L. *et al* 2012 *J. Synchrotron Radiat.* **19** 508-20
[2] G. A. Garcia et al 2013 *Rev. Sci. Instrum.* **84**, 053112 and 069902
[3] C. Alcaraz et al 2004 *J. Phys. Chem. A* **108** 9998-10009

Contact

[1] E-mail: claire romanzin@u-psud.fr
[2] E-mail: christian.alcaraz@u-psud.fr

Content from this work may be used under the terms of the Creative Commons Attribution 3.0 licence. Any further distribution of this work must maintain attribution to the author(s) and the title of the work, journal citation and DOI.
Published under licence by IOP Publishing Ltd

Journal of Physics: Conference Series **635** (2015) 032111 doi:10.1088/1742-6596/635/3/032111

Polyatomic ultralong-range Rydberg molecules in external fields

Javier Aguilera-Fernández [1] **and Rosario González-Férez** [2]

Instituto Carlos I de Física Teórica y Computacional and Departamento de Física Atómica, Molecular y Nuclear, Universidad de Granada, 18071 Granada, Spain

Synopsis We consider an ultralong-range polytonic Rydberg molecule formed by a Rydberg atom and two heteronuclear diatomic molecules. We focus here on the interaction of a Rb(n = 24, $l \geq 3$) Rydberg atom with two KRb diatomic polar molecules in their rotational ground state. The polar diatomic molecules are allowed to rotate in the electric fields generated by the Rydberg electron and core. We investigate the electronic structure of this ultralong-range polytomic Rydberg molecule, and rovibrational properties of the two polar dimers as an additional external field is applied.

Ultralong-range triatomic Rydberg molecules might be formed by a Rydberg atom and a heteronuclear diatomic molecule [1][2][3]. This species exhibits novel binding properties due to the interaction of the electric field induced by the Rydberg electron and core and the permanent dipole moment of the diatomic polar molecule. The coupling of the Rydberg electron and internal rotational states of the polar ground state dimer creates a series of undulating Born-Oppenheimer potential curves, which reflects the oscillating character of the Rydberg electron wave function. By applying additional electric fields of a few Vcm^{-1}, the electronic structure of these giant molecules could be controlled and manipulated [3][4].

These triatomic ultralong-range Rydberg molecules could be created in an ultracold mixture of atoms and molecules by using standard Rydberg excitation schemes. In this ultracold quantum gas of atoms and molecules, several ultracold heteronuclear diatomic molecules might interact with the electric field induced by the Rydberg electron and core. As a consequence, ultralong-range polyatomic molecule formed by a Rydberg atom and several ultracold polar diatomic molecules might be created. In this work, we consider a polyatomic molecule formed by a Rydberg rubidium atom and two ground state KRb polar molecules. We perform an extended and realistic treatment of the internal rotational motion of the diatomic molecules in the framework of the rigid rotor approximation [4]. The free rotation of the diatomic molecules in the polyatomic Rydberg molecule is now dressed due to the combined electric fields of the Rydberg electron and the core, and is characterized by a strong orientation and alignment along the field direction. We analyse the Born-Oppenheimer potentials of the Rb (n, $l \geq 3$)- 2(KRb) Rydberg polyatomic molecule as the internuclear separations between the two KRb dimers and the Rb$^+$ core vary and explore the effects of the electric field induced by the Rydberg atom on the rotational motion of the KRb-dimers, by studying the hybridization of their angular motion, orientation and alignment. Finally, we also investigate the impact of an additional external electric field on these adiabatic potential curves and their vibrational spectrum.

References

[1] S.T. Rittenhouse,and H.R. Sadeghpour, Phys. Rev. Lett. **104**, 243002 (2010).

[2] S.T. Rittenhouse, M. Mayle, P. Schmelcher and H.R. Sadeghpour, J. Phys. B **44**, 184005 (2011).

[3] M. Mayle, S. T. Rittenhouse, P. Schmelcher, and H. R. Sadeghpour Phys. Rev. A **85**, 052511 (2012).

[4] R. González-Férez, H. R. Sadeghpour and P. Schmelcher, New J. Physics **17**, 013021 (2015).

[1] E-mail: aguilerajavier@ugr.es
[2] E-mail: rogonzal@ugr.es

Content from this work may be used under the terms of the Creative Commons Attribution 3.0 licence. Any further distribution of this work must maintain attribution to the author(s) and the title of the work, journal citation and DOI.
Published under licence by IOP Publishing Ltd

Journal of Physics: Conference Series **635** (2015) 032112 doi:10.1088/1742-6596/635/3/032112

FT-ICR studies of anionic reactions for the chemistry of planetary ionospheres

C. Romanzin*[1], E. Louarn*, J. Lemaire*, J. Zabka†, M. Polasek†, J.-C. Guillemin$ and C. Alcaraz*[2]

* LCP, UMR 8000 CNRS-Université Paris Sud, Orsay, France
† J. Heyrovsky Institute of Physical Chemistry of the ASCR, Prague, Czech Republic
$ Ecole Nationale Supérieure de Chimie de Rennes, UMR 6226 CNRS, Rennes

Synopsis The rate constants of several ion-molecule reactions involving CN^- and C_3N^- anions and HC_3N molecules have been measured in a FT-ICR spectrometer. These measurements are of particular importance for the chemistry of Titan's ionosphere and interstellar medium.

Titan, Saturn's largest moon, exhibits a dense atmosphere characterized by a thick orange haze and mainly composed of molecular nitrogen and methane as well as numerous other organic compounds such as nitriles. The chemistry taking place in its atmosphere is complex and still not completely understood. Yet, results from the Cassini-Huygens mission have shown that ionospheric chemistry must play a more important role than previously thought. The discovery of CN^-, C_3N^- and C_5N^- together with a large amount of heavy cations and anions in the upper atmosphere [1-3] came indeed as a surprise and suggests that they contribute to the formation of aerosols particles. In this context, we have undertaken experimental investigations of ion-molecule reactions involving small anions such as CN^-, C_3N^- and C_5N^- in a common effort between several French and Czech groups [4-7].

Several results have already been obtained for the $CN^- + HC_3N$ reaction both in our group [4] and in the group of S. Le Picard *et al.* in Rennes [5,6]. In this contribution, we will present new measurements at 300 K [7] obtained in Orsay using a FT-ICR spectrometer - the MICRA setup [8] - for the following reactions:

$CN^- + HC_3N$
$C_3N^- + HC_3N$
$C_3N^- + HC_3{}^{15}N$
$C_3{}^{15}N^- + HC_3N$.

CN^- and C_3N^- parent ions have been produced by dissociative electron attachment on BrCN and BrC_3N respectively. After ejection of the electrons, pulses of increasing length of cyanoacetylene (HC_3N) target gas were injected to probe the kinetics of the reactions and derive their rate constant at 300 K. Several mechanisms, including fast proton transfer have been identified and will be discussed. Further studies are planned on the CERISES setup [9] at Orsay to characterize the reaction mechanism by measuring the absolute reaction cross sections and product velocity distributions as a function of collision energy.

Acknowledgments

RTRA - Triangle de la Physique (GIN, projet 2012-027T) ; Programme National de Planétologie (PNP) ; COST (Action CM0805 « The Chemical Cosmos ») ; France-Czech Republic Program CNRS-AVCR (n°20201) ; Czech Science Foundation (P208/11/0446), (Grant Nos. OC10046).

References

[1] A.J. Coates *et al*, 2007, Geophys. Res. Lett. 34
[2] A.J. Coates *et al*, 2010, Farad. Discuss. 147, 293-305
[3] V. Vuitton *et al*, 2009, Planet. Space Sci. 57, 1558-72
[4] J. Žabka *et al*, 2012, Icarus 219, 161-167
[5] S. Carles *et al*, 2011, Icarus, 211, 901-905
[6] L. Biennier *et al*, 2013, Icarus 227, 123
[7] C. Romanzin *et al*, (in preparation)
[8] G. Mauclaire *et al*, 2004, Eur. J. Mass Spectrom. 10, 155-162
[9] C. Alcaraz *et al*, 2004, J. Phys. Chem. A 108, 9998-10009

Contact

[1] E-mail: claire romanzin@u-psud.fr
[2] E-mail: christian.alcaraz@u-psud.fr

Content from this work may be used under the terms of the Creative Commons Attribution 3.0 licence. Any further distribution of this work must maintain attribution to the author(s) and the title of the work, journal citation and DOI.
Published under licence by IOP Publishing Ltd

Hydrogen migration observed in fragmentation of the pyridine molecules in collisions with the H^+, H_2^+, He^+ and He^{++} cations

Tomasz J. Wasowicz* [1]

* Department of Physics of Electronic Phenomena, Gdańsk University of Technology, 80-233 Gdańsk, Poland

Synopsis The hydrogen atom migration preceding fragmentation of the pyridine molecules by the H^+, H_2^+, He^+ and He^{++} collisions has been investigated in the 5-2000 eV energy range. The pyridine molecule structure is lacking of the NH group, thus formation of the NH radicals is an evidence of the H atom relocation prior to the cation-induced fragmentation. The NH($A^3\Pi$) fragmentation yields measured by detecting its $A^3\Pi \rightarrow X^3\Sigma^-$ fluorescence show that it depends on the selected projectile and the projectile velocity.

Many investigations have been devoted to understanding of the molecular reaction dynamics and chemical reactivity of molecular systems. In particular, selectively induced and controlled chemical reactions in molecular collisions have been studied by several different techniques, including molecular dynamics of photoexcited molecules [1,2], mode-selective excitation in ultrafast laser pulses [3], and more recently by site- and bond-selective dissociation in potassium collisions with pyrimidine bases of DNA [4].

The H atom migration observed in fragmentation of the pyridine (C_5H_5N) molecules induced by the H^+, H_2^+, He^+ and He^{++} collisions has been studied experimentally using collision-induced luminescence spectroscopy [5]. The collision energies of the cations were varied between 5 and 2000 eV and thus covered a velocity range from 30 to 440 km/s depending on the projectile mass.

In Figure 1 the emission spectra obtained for the He^+ and He^{++} cations are presented. The NH radicals have been identified by recording the $A^3\Pi \rightarrow X^3\Sigma^-$ emission. The ring of the pyridine molecule is built from five CH units and one N heteroatom [6]. Therefore to generate NH, it should undergo molecular rearrangements associated with migration of the hydrogen atom to form the NH group. Based on *ab initio* molecular calculations the plausible mechanisms for the NH formation will be proposed.

Figure 2 displays the velocity dependences of the NH($A^3\Pi$) fragmentation yields obtained in the H^+, H_2^+ and He^+ collisions. It is clearly seen that these functions have different shapes and in the case of H_2^+ and He^+ the NH($A^3\Pi$) production yields strongly depend on the cation velocity.

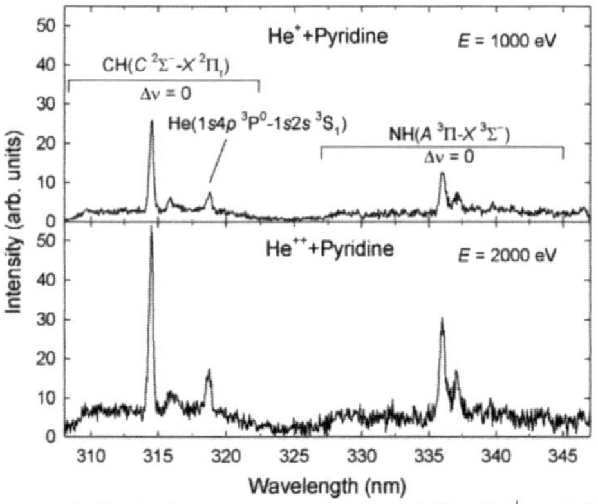

Figure 1. Emission spectra measured for He^+ and He^{++} collisions with the pyridine molecules.

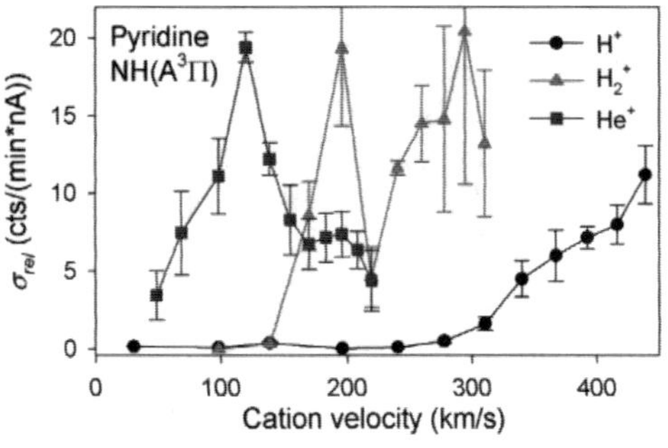

Figure 2. Fragmentation yields, σ_{rel}, of the excited NH($A^3\Pi$) fragments.

References

[1] H. Rabitz *et al* 2000 *Science* **288** 824
[2] M. Zubek *et al* 2014 *J. Chem. Phys.* **141** 064301
[3] E. D. Potter *et al* 1992 *Nature* **355** 66
[4] D. Almeida *et al* 2013 *Phys. Rev. Lett.* **110** 023201
[5] T. J. Wasowicz *et al.* 2015 *J. Phys. Chem. A* **119** 581
[6] T. J. Wasowicz *et al* 2014 *J. Phys. B* **47** 055103

[1] E-mail: twasowicz@mif.pg.gda.pl

Content from this work may be used under the terms of the Creative Commons Attribution 3.0 licence. Any further distribution of this work must maintain attribution to the author(s) and the title of the work, journal citation and DOI.
Published under licence by IOP Publishing Ltd

Journal of Physics: Conference Series **635** (2015) 032117 doi:10.1088/1742-6596/635/3/032117

Electron attachment properties of c-C_4F_8O: long-lived transient negative ion and plasma applications

Radmila Janečková*[1], Jaroslav Kočišek†, and Juraj Fedor†

* Department of Chemistry, University of Fribourg, Chemin du Musée 9, 1700 Fribourg, Switzerland
† J. Heyrovský Institute of Physical Chemistry, CAS, Dolejškova 3, 182 23 Prague, Czech Republic

Synopsis Dissociative and non-dissociative electron attachment to octafluorotetrahydrofuran (c-C_4F_8O) has been experimentally investigated. The partial absolute cross sections for all anionic fragments have been measured. An extraordinarily long-lived undissociated (parent) negative ion has been observed at non-thermal incident electron energy. The unusual properties of this anion explain attachment behavior of c-C_4F_8O in swarm experiments and the causes of its long lifetime are discussed.

The cyclic ether perfluorotetrahydrofuran (octafluorotetrahydrofuran, c-C_4F_8O) recently attracted considerable attention due to its electron attachment properties with many practical applications. This molecule promises to be a suitable alternative for plasma processing gasses. It has already been applied in Cherenkov radiation detectors [1], and it might effectively replace the high voltage insulating gasses (primarily SF_6) as it is environmentally more acceptable [2]. Since the fundamental electron driven processes of c-C_4F_8O are not well known, we have performed a detailed experimental study.

The experiment has been performed using a quantitative dissociative electron attachment (DEA) spectrometer with trochoidal electron monochromator in combination with ion time-of-flight (TOF) analyzer [3]. The measurements for absolute cross sections have been executed in a mixture of studied molecule c-C_4F_8O together with CO_2 and HCOOH used as calibration gasses with known cross sections and positions of resonances [4].

We present the partial absolute cross sections for negative-ion formation from electron attachment to c-C_4F_8O in the energy range 0-6 eV. We have observed rich fragmentation pattern from dissociative electron attachment at electron energies ~1.7 eV, ~2.5 eV and ~3.4 eV. We have recorded a long-lived parent anion (m/z = 216) formed at 0.9 eV.

We have explained the unusually long lifetime of the parent anion ($C_4F_8O^-$) as a consequence of very different geometries of the neutral molecule and the anion, and of high vibrational level density due to low frequencies of C–F vibrations. The long lifetime accounts for effective three-body attachment and it is thus crucial for the interpretation of high pressure electron transport data.

This work has been supported by SCIEX NMS-CH programme of CRUS, the project No. 12.224 - IDEA, and project No. PZ00P2_132357/1 of the Swiss National Science Foundation.

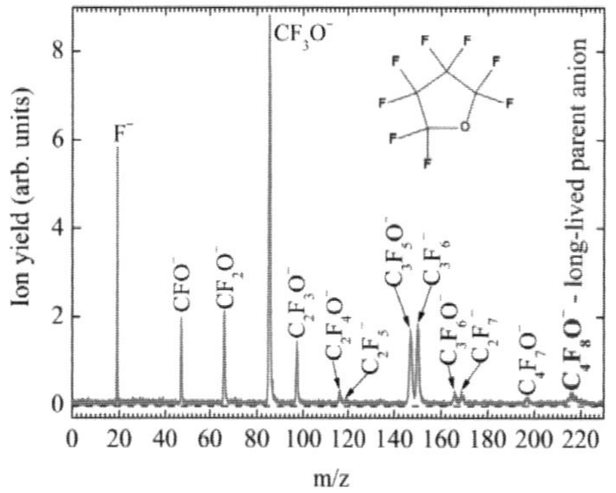

Figure 1. Mass spectrum at electron energy of 2 eV - rich fragmentation pattern of negative ion formation.

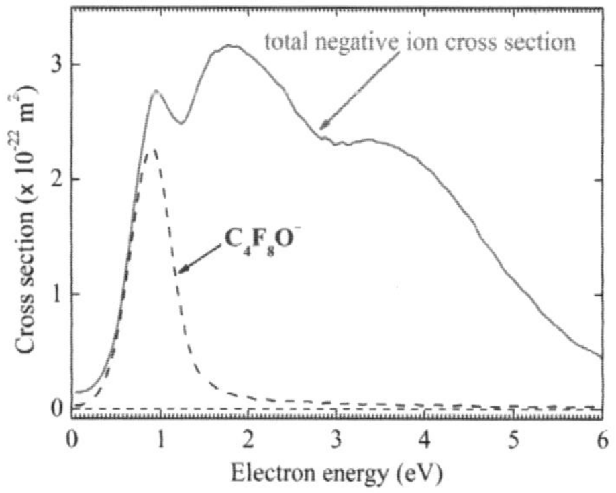

Figure 2. Full line - total negative ion cross section from electron attachment to c-C_4F_8O (sum of the partial cross sections for all anionic fragments). Dashed line – cross section for production of long-lived parent negative ion ($C_4F_8O^-$).

References

[1] M. Artuso *et al* 2006 *Nucl. Instrum & Meth. A* **558** 10.1016
[2] K. J. Kim *et al* 2004 *J. Vac. Sci. Technol. B* **22** 10.1116
[3] O. May *et al* 2009 *Phys. Rev. A* **80** 12706
[4] R. Janečková *et al* 2013 *Phys. Rev. Lett.* **111** 212301

[1] E-mail: radmila.janeckova@unifr.ch
Content from this work may be used under the terms of the Creative Commons Attribution 3.0 licence. Any further distribution of this work must maintain attribution to the author(s) and the title of the work, journal citation and DOI.
Published under licence by IOP Publishing Ltd

Journal of Physics: Conference Series **635** (2015) 032113 doi:10.1088/1742-6596/635/3/032113

Experiments at the Frankfurt low energy storage ring (FLSR)

F.King[1], K. E. Stiebing[2], R. Dörner, J. Müller, L. Ph. Schmidt, H. Schmidt Böcking

Institut für Kernphysik (IKF), Goethe Universität Frankfurt, Max von Laue Straße 1, D-60438 Frankfurt am Main, Germany

Synopsis FLSR has been designed for experiments on the dynamics of atomic and molecular collisions. After successful commissioning, an experiment on the dissociative recombination of $^4HeH^+$ has been started.

FLSR is an electrostatic storage ring for ions up to energies of 80 keV at IKF (see Fig.1). It has been designed for experiments on the dynamics of atomic and molecular collisions in complete kinematics, as well as for high precision, time resolved laser spectroscopy [1,2]. FLSR comprises four experimental/diagnostic sections with one region of enhanced ion density in each of it. First beam has successfully been stored in summer of 2013.

Figure 1. The **F**rankfurt **L**ow-Energy **S**torage **R**ing (**FLSR**) at IKF.

For first exploratory experiments on dissociative recombination of vibrational cool $^4HeH^+$, an electron gun has been installed in one of the experimental sections in "crossed beam" geometry with respect to the stored ion beam. For detection of the breakup of the neutralized fragments, a position resolving particle detector (MCP) has been installed at the related 0°-port.

In fig. 2 the distribution of the kinetic energy release (KER) from this experiment is shown (black dots). The KER has been calculated, selecting vibrational cool ($t_{storage} > 100$ ms), coincident ($\Delta t = 30$ns) and coplanar events ($\Delta\phi = 180° \pm 5°$ with respect to the center of gravity of the 2-D distribution of all events on the detector). Since in this experiment no trigger on events from the electron target is available, it

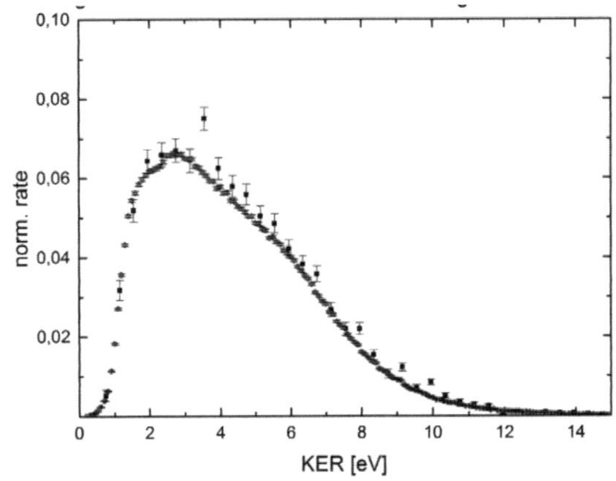

Figure 2. Preliminary results on the kinetic energy release in dissociative recombination of 20keV $^4HeH^+$-ions with electrons of 18eV kinetic energy (data points). Red dots: background from charge exchange with residual gas.

has been assumed that all fragments were created in the electron target. This is incorrect for the dominant background (red dots). While the method of background subtraction is unambiguous, the KER calibration by this method is only correct for those events that really originat in the electron target. The experiment was performed at a still moderate vacuum of about 2 $x10^{-10}$ mbar with an electron beam current of 2μA. The conditions are presently being improved by baking FLSR to better vacuum and enhancing the electron gun current. Furthermore, the installation of a gas target is discussed for experiments with trigger conditions.

References

[1] K.E. Stiebing et al. 2010, *Nucl. Instr. and Methods in Physics Research* **A614**, 10
[2] F. King et al. 2015 *Physica Scripta Topical Issues* to be published.

FLSR was made possible by a grant from "Innovationsfonds der Hessischen Landesregierung", Government of Hesse, Germany (2004–2007).

[1] E-mail: king@atom.uni-frankfurt.de
[2] E-mail: stiebing@em.uni-frankfurt.de

Content from this work may be used under the terms of the Creative Commons Attribution 3.0 licence. Any further distribution of this work must maintain attribution to the author(s) and the title of the work, journal citation and DOI.
Published under licence by IOP Publishing Ltd

XXIX International Conference on Photonic, Electronic, and Atomic Collisions (ICPEAC2015) IOP Publishing
Journal of Physics: Conference Series **635** (2015) 032115 doi:10.1088/1742-6596/635/3/032115

Fragmentation of H_2O molecules induced by singly charged projectiles

S. T. S. Kovács [1], P. Herczku, Z. Juhász, B. Sulik

Institute for nuclear Research, Hungarian Academy of Sciences (Atomki), Bem tér 18/c, H-4026 Debrecen, Hungary

Synopsis Ion impact induced fragmentation of H_2O molecules was experimentally studied. Energy and angular distribution of the charged fragments were measured by an energy dispersive electrostatic spectrometer. The measured spectra were converted into absolute double differential cross section (DDCS). The fragment ion spectra are different for H^+ and N^+ projectiles. From the fragmentation spectra we deduced the cross sections for the different ionization degrees of the target molecules.

By studying the ion impact induced fragmentation of small molecules different levels of information can be achieved. The fragmentation pattern is mainly determined by the perturbation strength of the projectile. According to the Bohr-Sommerfeld parameter ($k=q/v$), the interaction is governed by the charge over velocity ratio of the projectile: at small k values ($k<<1$) the target is only slightly perturbed, while at high k values ($k\geq1$) the projectile strongly interact with the target [1].

In this work we compare the fragmentation spectra of H_2O molecules induced by 1 MeV H^+ and 650 keV N^+ projectiles [2]. A jet of H_2O vapor was crossed by the ion beam supplied by the 5 MV VdG accelerator at Atomki, Debrecen. The energy and angular distribution of the charged fragments were measured by an energy dispersive electrostatic spectrometer in the angular range of 20°-160° relative to the incident ion beam. The measured spectra were converted to absolute DDCSs.

In figure 1 two different regions of the fragment ions can be identified: The peak below 2 eV contains dominantly heavy fragments (OH^+; O^{q+}). The more structured region above 2 eV is due to H^+ fragments. The energy of the H^+ ions is increasing towards higher ionization degrees of the H_2O^{q+} transient molecular ion (Coulomb explosion) [3]. Therefore H^+ fragments from the double, triple or four-fold ionized H_2O^{q+} molecules appear in the energy ranges of 4-15 eV, 16-28 eV and 30-38 eV respectively [4].

According to the energies of the H^+ fragments the maximal ionization degree was $q=2$ for H^+ projectiles, and $q=5$ for N^+ projectiles. The difference in the average degree of target ionization for the two, singly charged projectiles is originating from the difference in their impact velocities, and their effective charges in close collision events. While for N^+ the Bohr-Sommerfeld parameter is $k_{min}=0.74$ for the ionic, and $k_{max}=5.1$ for the nuclear charge, for H^+ it is as small as $k=0.16$.

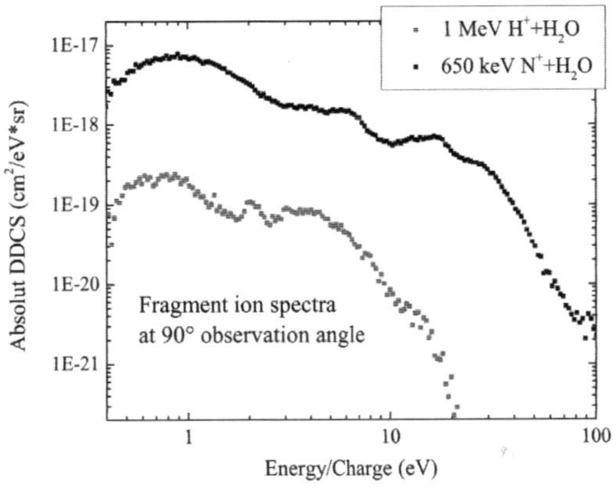

Figure 1. Fragment ion energy spectra for the two types of projectiles at 90° observation angle. The high energy protons for N^+ reveals to three, four- and five-fold ionization.

The theoretical work is under progress. Comparison of calculated partial multiple ionization cross sections (CTMC and CDW-EIS) to the extracted experimental values will be presented at the conference for a larger set of collision systems.

This work was supported by the Hungarian National Science Foundation OTKA (Grant No. K109440).

References

[1] B. Siegmann *et al* 2000 Phys. Rev. A **62**, 022718
[2] S. T. S. Kovacs *et al* 2013 Act. Phys. Deb. **XLVII** 87
[3] J. Rajput *et al* 2011 Phys. Rev. A **84** 052704
[4] H. Luna *et al* 2003 *J. Phys. B.* **36** 4717

[1] E-mail: kovacs.sandor@atomki.mta.hu

Content from this work may be used under the terms of the Creative Commons Attribution 3.0 licence. Any further distribution of this work must maintain attribution to the author(s) and the title of the work, journal citation and DOI.
Published under licence by IOP Publishing Ltd

Journal of Physics: Conference Series **635** (2015) 032118 doi:10.1088/1742-6596/635/3/032118

Two- and many-body effects in cation emission from H_2O molecules by O^+ impact at keV energies: Similarities between ionization of atoms and proton emission from molecules

B. Sulik*[1], Z. Juhász*, E. Lattouf†, J. A. Tanis†,§, B. A. Huber†, E. Bene*, S. T. S. Kovács*, P. Herczku*, A. Méry†, J.-C. Poully†, J. Rangama†, and J.-Y. Chesnel†

* Institute for Nuclear Research, Hungarian Academy of Sciences (MTA Atomki), H-4001 Debrecen, Hungary
† Centre de Recherche sur les Ions, les Matériaux et la Photonique (CIMAP), 6 Bd Mar. Juin, F-14000 Caen, France
§ Department of Physics, Western Michigan University, Kalamazoo, Michigan 49008 USA

Synopsis We demonstrate that energy and angle differential emission yields of fragment ions provide valuable information about the collisions of keV energy ions and molecules. For proton fragments, originating from larger molecules, there are striking similarities to the spectra of electrons emitted in fast ion-atom collisions.

The rich structure of the spectra of ejected electrons in fast ion-atom collision experiments provides deep insight into the details of the ionization process [1]. The key for clarity there is the large proton/electron mass ratio. Electrons move in the fields of much heavier centers.

Here we demonstrate that double differential emission yields of fragment ions provide similarly valuable information in ion-molecule collisions at keV energies. For protons emitted from heavier molecules, due to the large mass ratio, there are close similarities to ionization.

Figure 1 shows experimental double-differential cross sections (DDCS) for cation emission in 6.6-keV $^{16}O^+ + H_2O$ collisions. The experimental arrangement was identical with that of ref. [5]. The main component of each spectrum is a broad, slowly decreasing structure. There the relative contributions of H^+ and heavier fragments were derived from complementary time-of-flight measurements. The low energy part is mostly due to many-body processes with small momentum transfer, in analogy with the dipole-ionization dominated low energy electron emission in fast ion-atom collisions [1]. At forward angles (< 90°) pronounced peaks are observed at higher energies. These peaks are due to the different cations formed in hard binary collisions [2-5], similarly to electron emission. We note that the heavier oxygen ions are mostly recoils, appearing in the binary ridge, while proton emission is dominated by many-body effects, and it is almost isotropic at low energies. One should keep in mind that highly charged ion impact is somewhat different. There multiple charge-exchange processes, followed by Coulomb explosion, dominate the fragment ion spectra at low energies.

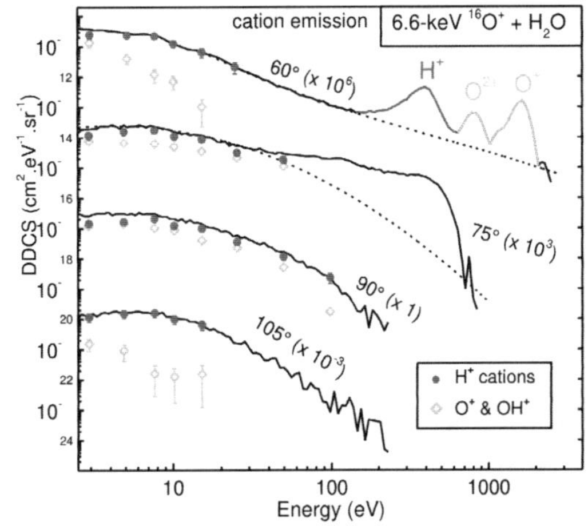

Figure 1. Full curves: Experimental DDCS for cation emission at various angles in 6.6-keV $^{16}O^+ + H_2O$ collisions; Full circles: H^+ contribution from TOF measurements. Open diamonds: O^+ and OH^+ contributions from TOF; Each spectrum is multiplied by the indicated factor.

This work was supported by the Hungarian National Science Foundation OTKA (Grant No. K109440), the French-Hungarian Cooperation Program PHC Balaton (No. 27860ZL/TéT_11-2-2012-0028), and the French-Hungarian CNRS-MTA Cooperation (ANION project).

References

[1] N. Stolterfoht, R.D. DuBois and R. D. Rivarola, *Electron emission in heavy Ion Collisions* (Springer, Berlin, 1997, pp 1-16)
[2] Z. Juhász *et al* 2012 *J. Phys. Conf. Ser.* **388** 102051
[3] Z. Juhász *et al* 2013 *Phys. Rev. A* **87** 032718
[4] Z. Juhász *et al* 2014 *J. Phys. Conf. Ser.* **488** 102024
[5] E. Lattouf *et al* 2014 *Phys. Rev. A* **89** 062721

[1] E-mail: sulik@atomki.mta.hu

Content from this work may be used under the terms of the Creative Commons Attribution 3.0 licence. Any further distribution of this work must maintain attribution to the author(s) and the title of the work, journal citation and DOI.
Published under licence by IOP Publishing Ltd

Positron-molecule interactions: theory and computation

M. T. do N. Varella[†] [1] **J. Charry**[⋆], **J. Romero**[⋆], **A. Reyes**[⋆]

[†] Instituto de Física, Universidade de São Paulo, São Paulo, SP, Brazil

[⋆] Department of Chemistry, Universidad Nacional de Colombia, Bogotá, Cundinamarca, Colombia

Synopsis We discuss the connection between the Feshbach Projection Operator approach to resonant positron annihilation on molecules and the Any-Particle Molecular Orbital Method to positronic molecules. This combination of formal and computational tools is a promising approach to the *ab initio* description of vibrationally-enhanced annihilation in molecular gases.

In the last decades, there has been an increasing interest on the physics and chemistry of positronic systems [1]. In low-energy positron beam experiments, the energy transfer to the vibrational degrees of freedom enhances the annihilation probability by forming positronic molecules (positron-molecule compounds) that undergo resonantly enhanced annihilation [1]. This process poses severe challenges to theory in view of the difficulties to describe electron-positron correlation and the vibration dynamics.

In this work, we report on the recent theoretical advances achieved for the description of positronic molecules and resonant annihilation. Our formulation is based on the Feshbach Projection Operator (FPO) approach to resonant annihilation [2, 3]. From the FPO expression for the annihilation parameter Z_{eff}, proportional to the annihilation rate in dilute gases, we construct a hierarchy of approximations that significantly facilitate the computations. The final expression relies on a local approximation to the potential energy surface (PES) of the positronic molecule, as schematically shown in Fig. 1: positron attachment gives rise to vibrational wave packets in the PES of the positronic system and hence to resonances in the stationary vibrational eigenvalues. This approximation is fairly simple from the theoretical standpoint, although it still challenges the available computational methods.

The key quantities of the theory are the positron binding energies, or, more specifically, the difference between the PES's of the positronic system and the isolated molecule. To compute this quantity, we employ perturbation techniques, namely second order Møller-Plesset (MP2) [4] and propagators [5], according to the Any Particle Molecular Orbital (APMO) method, implemented in the LOWDIN code [6]. We believe this combination of theory and computation is a promising strategy to pursue the *ab initio* description of resonant annihilation. We discuss the connection between the FPO formalism to resonant annihilation and the APMO method, also discussing some recently obtained results.

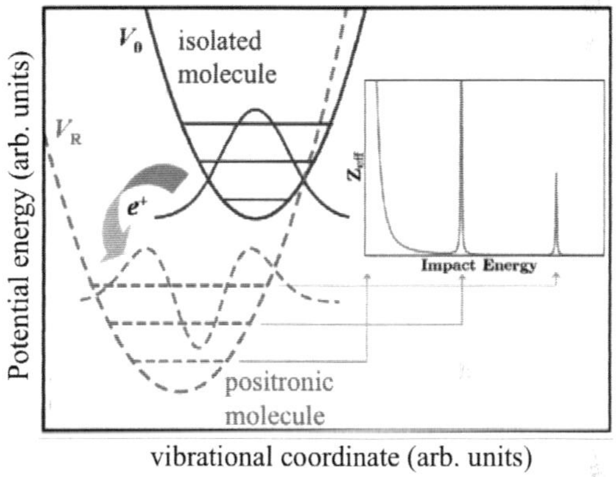

Figure 1. Schematic representation of resonant annihilation on molecules.

The authors acknowledge financial support from FAPESP and CNPq (Brazil), and also from UNAL and Colciencias (Colombia).

References

[1] G. F. Baking *et al* 2010 *Rev. Mod. Phys.* **82** 2557

[2] S. d'A. Sanchez *et al* 2009 *Phys. Rev. A* **80** 052710

[3] S. d'A. Sanchez *et al* 2011 *Phys. Rev. Lett.* **107** 103201

[4] J. Charry *et al* 2014 *Phys. Rev. A* **89** 052709

[5] J. Romero *et al* 2014 *J. Chem. Phys.* **141** 114103

[6] R. Flores-Moreno *et al* 2014 *Int. J. Quantum. Chem.* **114** 50

[1]E-mail: mvarella@if.usp.br

Content from this work may be used under the terms of the Creative Commons Attribution 3.0 licence. Any further distribution of this work must maintain attribution to the author(s) and the title of the work, journal citation and DOI.
Published under licence by IOP Publishing Ltd

Journal of Physics: Conference Series **635** (2015) 032120 doi:10.1088/1742-6596/635/3/032120

Ionization and Dissociation of PAH by Stellar Winds Application to H_2 Emission from Coronene in HD44179

J.-P. Champeaux*[1], P. Moretto-Capelle, P. Cafarelli, C. Deville, M. Sence, R. Casta

* Laboratoire Collisions Agrégats et Réactivité, CNRS-UMR 5589, Université Paul Sabatier Toulouse III, 118 Route de Narbonne, F-31062 Toulouse Cedex9, France

Synopsis - Recent results on dissociation of coronene by keV protons are presented and applied to the astrophysical context of HD44179 compact nebulae. We show that coronene dissociation by stellar wind can induce strong neutral molecular hydrogen emission and more generally, that stellar wind induced dissociation of carbonated molecules like PAHs inside compact nebulae (i.e before the astro-pause) can play an important role in the complex physico-chemistry of the nebulae media which could be competitive with other photodissociaiton or photochemistry mechanisms.

The physical interactions of polycyclic aromatic hydrocarbons (PAHs), detected by their IR signature in the interstellar media [1], with stellar wind particular radiation (protons, electrons) are key to understanding the life cycle of PAHs, their abundance and their role in the complex Astro-chemistry of the interstellar medium. In this context, we present experimental results on the ionization/fragmentation of an isolated coronene molecule by a keV proton, reproducing interactions between stellar winds and PAH molecules. Using multi-coincidence time of flight spectroscopy [2], we show that such ionization/fragmentation induces intense dehydrogenation processes in which the loss of an even numbers of hydrogen atoms and the detection of CH_2^+ cations as a possible H_2 precursor strongly suggest the formation of H_2 neutral molecules as show in Figure 1.

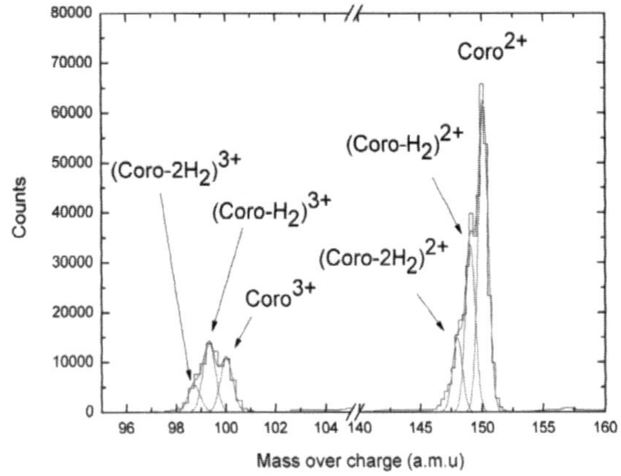

Figure 1. Zoom of coronene 2+ and 3+ fragmentation mass spectrum suggesting H_2 neutral emission.

Possible sceneries for H_2 production via $-CH_2$ precursors have been investigated by quantum chemical calculation.

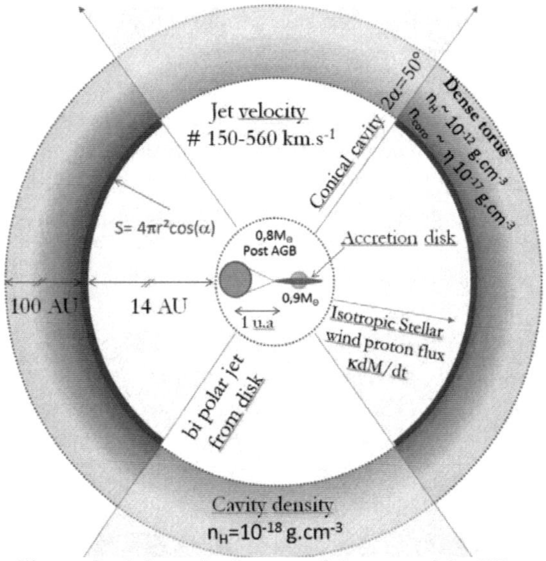

Figure 2. Schematic diagram of the core of the HD 44179 RR nebula.

We have evaluated the H_2 emission cross-section from the coronene/proton interaction at 100 and 1.6 keV to be 2.97×10^{-16} and 3.3×10^{-16} cm^2, respectively. A qualitative discussion off the formation rate of H_2 in the HD 44179 Red Rectangle (RR) nebula [3] (presented figure 2) leads to the conclusion that such processes could be very efficient, especially inside planetary nebulae rich in PAH molecules interacting with proton from high mass-loss rate stars (like asymptotic giant branch stars) or high velocity jets produced by an accretion disc [4].

References

[1] Leger A. et al. 1984 A&A 137 5
[2] J.-P. Champeaux et al. 2010 Phys. Chem. Chem. Phys. 12 5454
[3] J.D. Thomas et al. 2013 MNRAS 430 1230
[4] J.-P. Champeaux et al. 2014 MNRAS 441 1479

[1] E-mail: jean-philippe.champeaux@irsamc.ups-tlse.fr

Content from this work may be used under the terms of the Creative Commons Attribution 3.0 licence. Any further distribution of this work must maintain attribution to the author(s) and the title of the work, journal citation and DOI.
Published under licence by IOP Publishing Ltd

Hydrogen equation in spaces of arbitrary dimensions

M. Ya. Amusia*† 1

* The Racah Institute of Physics, the Hebrew University, Jerusalem 91904, Israel
† A. F. Ioffe Physical-Technical Institute, St. Petersburg 194021, Russian Federation

Synopsis We note that presenting Hydrogen atom Schrodinger equation in the case of arbitrary dimensions require simultaneous modification of the Coulomb potential that only in three dimensions has the form Z/r. This was not done in a number of relatively recent papers (see [1] and references therein). Therefore, some results obtained in [1] seem to be doubtful. Several required considerations in the area are mentioned.

1. The aim of the article [1] is to show how to treat using old Bohr's approach not only simple atoms but molecules. An important point in this development is reconciliation of quantum mechanics with Bohr's ideas. It is stated in [1] that in the infinite dimension quantum mechanics "morphs into classical mechanics".

However, the infinite-dimension system is not a clarifying model to describe physical or chemical objects. Far from being convincing are statements like "Hence the large-D limit, where 1/D→0 is closer to the real world (1/D=1/3) than is the often-used D=1 regime. Indeed, results obtained at large D usually resemble those for D=3". I must confess, it sounds too light weighted to be convincing.

2. But not the large-D limit per se is doubtful. Of great concern is the assumption that the radial part of the D-dimensional Schrödinger equation in Hartree units looks as follows:

$$\left\{-\frac{1}{2}\frac{\partial^2}{\partial r^2}+\frac{[l+(D-3)/2][l+(D-1)/2]}{2r^2}-\frac{Z}{r}\right\}\phi=E\phi, \quad (1)$$

where Z is the nuclear charge and l is the angular momentum. Obviously, this equation has no sense for $D=1$, since a finite angular momentum requires infinite speed of a rotating particle.

3. The equation (1) implies independence of the Coulomb potential upon D. However, to obtain the Coulomb law for a two-dimensional space, one has to consider Maxwell equations in a two-dimensional world. This equation for the considered case looks as

$$-\left(\frac{\partial^2}{\partial x^2}+\frac{\partial^2}{\partial y^2}\right)\varphi^{(2)}(r)=Z\delta^{(2)}(r), \quad (2)$$

where $\delta^{(2)}(r)$ is the two-dimensional delta-function and $r^2=x^2+y^2$. This leads to

$$\varphi^{(2)}(r)=-2Z\ln(r/r_0) \quad (3)$$

instead of Z/r r-dependence in the three dimensional case. Here r_0 is a cutoff length that depends upon the problem at hand.

4. If one takes into account the D-dependence of the Coulomb potential, derivations in [1] became meaningless. Indeed, by using an unrealistic equation (1), how one can believe that it illuminates quite realistic and physical postulates of Bohr?!

It is known since long ago a number of papers that consider two-dimensional Hydrogen atom (see e.g. [2]) that use for this or that reason the incorrect for the two-dimensional case potential Z/r instead of the correct one (3). It would be desirable to perform calculations with (3), instead of Z/r.

5. It would be of interest to find for completeness the shape of a "Coulomb" potential in the case of any arbitrary dimension D>3. This could be achieved by solving the equation

$$-\sum_{i=1}^{i=D}\frac{\partial^2}{\partial x_i^2}\varphi^{(2)}(r)=Z\delta^{(2)}(r)\cdot \quad (5)$$

Here $r^2=\sum_{i=1}^{i=D}x_i^2$.

This derivation would also permit to check whether indeed at $D\to\infty$ one arrives to a simple classical picture that I doubt.

Quite interesting would be an attempt to model experimentally the transition from Z/r to (3) by compressing an electron-proton-like system using analog of a flat conducting mirrors. Perhaps that could be achieved in cavities of rectangular shape.

7. There exists another case, when the Coulomb-like potential is of interest. I mean Newton's gravitation law. Its analogue for two-dimension world is definitely of at least theoretical interest. To do this, one has to use the equation of Einstein's general relativity and to apply them for a massive body in the field of point-like source of gravitational field.

References

[1] A. Svidzinsky, M. Scully, and D. Herschbach 2014, *Physics Today*, **1**, 33.
[2]]. G. Q. Hassoun 1981 *Am. J. Physics*, **49** (2), 143

1 E-mail: amusia@vms.huji.ac.il

Content from this work may be used under the terms of the Creative Commons Attribution 3.0 licence. Any further distribution of this work must maintain attribution to the author(s) and the title of the work, journal citation and DOI.
Published under licence by IOP Publishing Ltd

Radiative lifetime of the metastable $^1S_0$ level in Kr^{2+} measured using an electrostatic ion trap

Naoto Ojima*[1], Suguru Itoi*, Manabu Saito†[2], Yoichi Haruyama*

* Laboratory of Applied Physics, Kyoto Prefectural University, Kyoto 606-8522, Japan
† Department of Nuclear Engineering, Kyoto University, Kyoto 615-8540, Japan

Synopsis We have measured the radiative lifetime of the $^1S_0$ metastable level in Kr^{2+} using an electrostatic ion trap. The obtained lifetime was 17.4 ± 0.4 ms. Comparing with the previous results, our value is consistent with the theoretical predictions, but inconsistent with the experimental results even considering their errors.

The radiative lifetime of the $^1S_0$ metastable level in Kr^{2+} has been studied experimentally and theoretically by several groups [1]. The theoretical values reported by two groups (17.2 ms and 17.3 ms) were in good agreement with each other. On the other hand, the experimental values reported by two other groups (13.1 ± 0.6 ms and 14.8 ± 0.8 ms) were mutually inconsistent, and were smaller than the theoretical values. In this work, we measured the radiative lifetime of the $^1S_0$ metastable state of Kr^{2+} using an electrostatic ion trap of the ion-beam storage type [2, 3]. The trap was placed away from the ion source, and two differential evacuation systems were mounted between them, which enabled us to prevent the inflow of the ion-source gas into the trap. This minimized the influence of ion-beam losses in the trap due to collisions with residual gases.

Figure 1 shows the experimental arrangement. A Kr^{2+} beam of 2.4 keV was produced using an electron impact ion source. The beam was pulsed using an electrostatic chopper and was momentum analyzed in a 45° deflecting magnet. The beam pulse was then injected into the trap, and was multiply reflected between the entrance and exit-side electrodes, resulting in the storage of the ion-beam in the trap. The 350 nm photons emitted during the $^1S_0 \rightarrow {}^3P_1$ transition were measured with an optical system, which is composed of a view port, lenses, a bandpass filter (349 ± 5 nm), and a photomultiplier tube (PMT) as shown in figure 1. The count rate of photons as a function of time was recorded using a multichannel scaler (MCS). Since the count rate of the photons at a given time t is proportional to the number of Kr^{2+} ions in the $^1S_0$ state at the time t, we can obtain the lifetime of the $^1S_0$ state from the time spectrum recorded using the MCS. The trap was maintained at a pressure of 5×10^{-8} Pa during measurements.

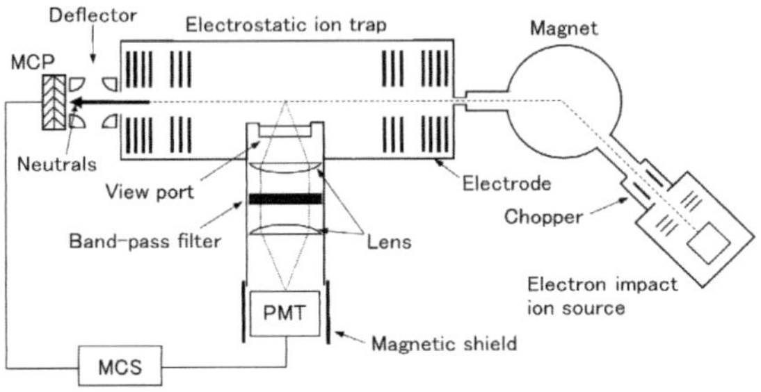

Figure 1. Experimental arrangement

The number of stored Kr^{2+} ions decreases with the trapping time due to electron capture or elastic collision with residual gases. Since the lifetime obtained from the photon time spectrum is composed of the radiative lifetime and the ion-storage lifetime, the accurate storage lifetime is necessary for the determination of the radiative lifetime. For that reason, the number of the stored Kr^{2+} ions as a function of time was monitored through the measurements of neutral Kr produced by double electron capture from residual gases in the trap.

To improve precision, three runs were done. About 100,000 beam pulses were injected into the trap per one run. The results yielded in the three runs agreed with each other within their errors. The lifetime of the $^1S_0$ metastable state of Kr^{2+} was determined to be 17.4 ± 0.4 ms from a weighted average of these results. Our value is in fair agreement with the previous theoretical values.

References

[1] E. Träbert 2012 *Phys. Scr.* **85** 048101, and references there in.
[2] D. Zajfman, O. Heber, L. Vejby-Christensen, I. Ben-Itzhak, M. Rappaport, R. Fishman, and M. Dahan 1997 *Phys. Rev.* A **55** R1577
[3] M. Saito, N. Ojima, S. Itoi, and Y. Haruyama 2015 *Phys. Rev.* A **91** 012508

[1] E-mail: n_ojima@mei.kpu.ac.jp
[2] E-mail: saito@nucleng.kyoto-u.ac.jp

Content from this work may be used under the terms of the Creative Commons Attribution 3.0 licence. Any further distribution of this work must maintain attribution to the author(s) and the title of the work, journal citation and DOI.
Published under licence by IOP Publishing Ltd

Journal of Physics: Conference Series **635** (2015) 042003 doi:10.1088/1742-6596/635/4/042003

Projectile coherence: The Van Cittert - Zernike theorem revisited

I. Fabre[* 1], **F. Navarrete**[* 2], **L. Sarkadi**[† 3] **and R. O. Barrachina**[* 4]

[*] Centro Atómico Bariloche and Instituto Balseiro, Comisión Nacional de Energía Atómica and Universidad Nacional de Cuyo, Av. Bustillo 9500, 8400 Bariloche, Argentina.

[†] Institute for Nuclear Research of the Hungarian Academy of Sciences (ATOMKI), H-4001 Debrecen, Pf. 51, Hungary

Synopsis We study the time evolution of an incoherent mixture of identical particles by solving the Liouville - von Neumann equation for the corresponding density matrix. We demonstrate the quantum mechanical equivalent of van Cittert - Zernike theorem, and apply these results to the quantitative analysis of the coherence of a beam of particles in atomic collisions.

Direct sunlight exhibits spatial coherence over a length of tens of μm at the Earth surface [1], even though it is produced by a conspicuously incoherent source, the Sun. This effect, where the light emitted from an incoherent source becomes approximately coherent at large distances, was demonstrated by Pieter H. van Cittert [2] in 1934 (see also [3]).

On the other hand, the study of the equivalent coherence properties of particle beams has largely relied on this and other results borrowed from light optics [4]. Even though this strategy is certainly sound for a qualitative analysis of particle coherence, it should be validated by means of a Quantum Mechanical approach in order to be applied to quantitative studies.

In this communication we analyze the time evolution of an incoherent mixture of particles of mass m, described by identical wave packets located at different positions upon a much larger region of dimension D. We evaluate the time evolution of corresponding density matrix $\langle \mathbf{r}|\rho|\mathbf{r}'\rangle$ by solving the Liouville - von Neumann equation. We exemplify these results by showing in figure 1 how the distribution of wavefronts for an incoherent mixture of particles traversing a slit smooths out at large distances.

Actually, it can be shown that a coherence length σ can be defined, such that for $|\mathbf{r}-\mathbf{r}'| < \sigma$ the system behaves coherently (i.e. ρ can be approximated by the product of pure states). In particular, σ coincides with the coherence length of each individual wave packet in the Fresnel limit, as expected; while in the Fraunhofer limit the standard expression of van Cittert - Zernike theorem is recovered, $\sigma \approx \gamma t/mD$. Here the factor γ depends on the shape of the incoherent distribution, and on how its width is defined. Finally, we show how these and other results of the present theoretical description can be of importance for the study of the coherence properties of particle beams in atomic collisions [5].

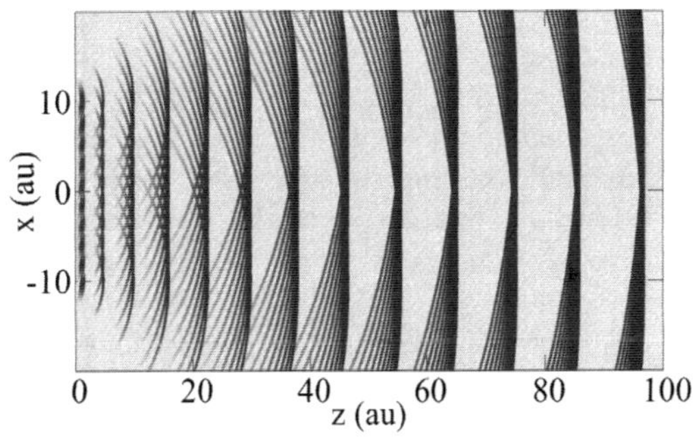

Figure 1. Wavefronts for an incoherent mixture of particles traversing a slit of width $D = 20$ au with velocity $v = 1$ au along the z direction. For the sake of simplicity, the individual wave packets are assumed to be of gaussian shape.

References

[1] H. Mashaal *et al* 2012 *Opt. Lett.* **37** 3516

[2] P. H. van Cittert 1934 *Physica* **1** 201

[3] F. Zernike 1938 *Physica* **5** 785

[4] A. D. Cronin *et al* 2009 *Rev. Mod. Phys.* **81** 1051

[5] K. N. Egodapitiya *et al* 2011 *Phys. Rev. Lett.* **106** 153202

[1]E-mail: ignacio.fabre@ib.edu.ar
[2]E-mail: navarrete@cab.cnea.gov.ar
[3]E-mail: sarkadil@atomki.mta.hu
[4]E-mail: barra@cab.cnea.gov.ar

Content from this work may be used under the terms of the Creative Commons Attribution 3.0 licence. Any further distribution of this work must maintain attribution to the author(s) and the title of the work, journal citation and DOI.
Published under licence by IOP Publishing Ltd

Partial cross sections for radiative recombination of bare ions : effects of relativity and radiation multipoles

L. Bettadj [1], M.K. Inal [2] and M.R. Boufatah [3]

Laboratoire de Physique Théorique, Département de Physique, Université de Tlemcen, 13000, Tlemcen, Algeria

Synopsis We derived an original expression for the partial cross sections for radiative recombination (RR) of highly charged bare ions in the nonrelativistic electric-dipole approach. Three sets of cross sections for RR to individual magnetic sublevels of the excited levels of various recombined H-like ions were computed: the first two in the electric-dipole approximation with the nonrelativistic and relativistic electron wavefunctions and the third one in the exact relativistic method including all multipole orders of the radiation field.

We present a systematic study of the relativistic and multipole effects, separately and combined, on the radiative recombination partial cross sections along an isoelectronic sequence and as a function of the incident-electron energy.

We have considered the case of radiative recombination of bare nuclei involving a single-electron system, with colliding electrons assumed to be unidirectional and monoenergetic. We have treated the radiative recombination process in the framework of the nonrelativistic electric-dipole approach and derived an original expression for the partial RR cross sections to a particular magnetic sublevel of H-like recombined ions [1].

The calculations have been performed for various recombined H-like ions with atomic numbers in the range $10 \leq Z \leq 92$. Several incident-electron kinetic energies ranging from 0.01 to 10 times the electron binding energy in the $1s_{1/2}$ ground state were considered and the radiative cascades following recombination into states with principal quantum number up to $n = 6$ were included. In order to stress the relative importance of relativistic and multipole effects as Z or the electron energy increases, three distinct calculations of the radiative recombination cross sections to individual magnetic sublevels of the recombined H-like ions were performed. The first two were made in the electric-dipole approximation with the nonrelativistic and relativistic Coulomb wavefunctions for both the continuum and bound states, and the third calculation was made using the exact relativistic treatment with all multipoles of the radiation field [2].

Comparing the results for the partial RR cross sections obtained in the three different treatments, we have shown how relativistic and multipole effects change with atomic number and with electron energy.

[1] E-mail: bettadj@gmail.com

[2] E-mail: m_inal@mail.univ-tlemcen.dz

[3] E-mail: boufatah@gmail.com

References

[1] L. Bettadj, M.K. Inal and M. Benmouna 2014 *J. Phys. B: At. Mol. Opt. Phys.* **47** 105205

[2] L. Bettadj, M.K. Inal, A. Surzhykov and S. Fritzsche 2010 *Nucl. Instrum. Methods Phys. Res.* B **268** 3509

Content from this work may be used under the terms of the Creative Commons Attribution 3.0 licence. Any further distribution of this work must maintain attribution to the author(s) and the title of the work, journal citation and DOI.
Published under licence by IOP Publishing Ltd

Time and space-resolved spectroscopic and imaging study of a laser-produced swine muscle tissue plasma

J. J. Camacho[1] **and L. Diaz** [2]

[1] Dpto. Química-Física Aplicada, Fac. Ciencias. Universidad Autónoma de Madrid, Cantoblanco 28049, Madrid Spain
[2]Instituto de Estructura de la Materia, CFMAC, CSIC, Serrano 121, 28006 Madrid, Spain

Synopsis We investigated the optical emission and imaging features of plasmas produced by a high-power transversely excited atmospheric CO_2 laser pulses on a swine muscle tissue sample in different vacuum conditions. The analyzed plasma shows electronically excited neutral Na, K, C, Mg, H, Ca, N, and O atoms, ionized C^+, C^{2+}, C^{3+}, Mg^+, Mg^{2+}, N^+, N^{2+}, Ca^+, O^+, and O^{2+} species and molecular band systems of CN, C_2, CH, NH, OH and CaOH. Time-resolved 2D emission spectroscopy is employed to study the expanded distribution of several species ejected during ablation. The expansion of the plume front was analyzed using Shock wave and Drag models.

Laser-produced plasmas (LPPs) are currently a topic of great interest in fundamental and applied areas such as fabrication of thin films by pulsed laser deposition, production of nanoparticles, spectrochemical analysis through laser-induced breakdown spectroscopy (LIBS) [1], ion source etc. Beyond traditional applications of LIBS, recent progresses lead to analysis of biological warfare agents and animal tissues.

In this work, we present a spatial and temporal analysis of the LPP plume generated on a biological tissue target. Time-resolved 2D emission spectroscopy is used to study the expanded distribution of different species ejected during ablation. Fig. 1 shows a schematic overview of the temporal history of the LPP sample. The temporal shape of the CO_2 laser pulse is also shown. Inset plots illustrate some space-resolved spectral images observed at different delays for a gate width time of 100 ns. The recorded spectral intensity is indicated by a pseudo-color. By tracking the maximum brightness displacement for different plasma species gives their average expansion velocities.

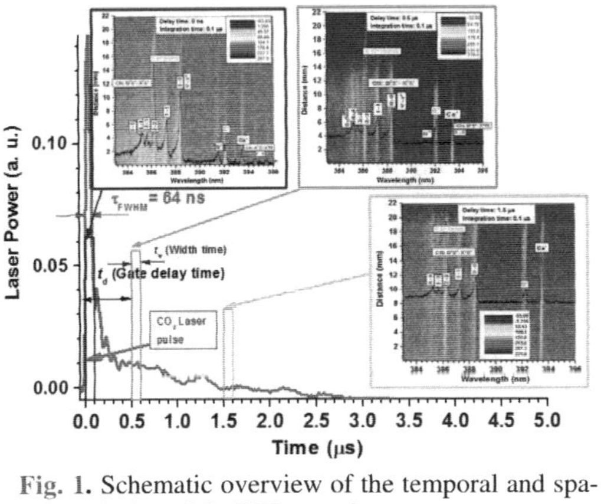

Fig. 1. Schematic overview of the temporal and spatial evolution of the LPP sample.

The expansion velocities of the ionized species towards the longitudinal direction are found to be increasing with degree of ionization. Plasma parameters such as electron density and temperature were measured from the spatial-temporal analysis of different specific species.

We used the imaging data to create position-time plots (Fig. 2) of the shockwave front at several background air pressures. The solid line represents the shockwave model fit. These images provide very useful information about the expansion and internal structure of the plasma-plume. Surface morphology of irradiated surface showed that increasing the pressure of the ambient gas, decreased the ablated mass.

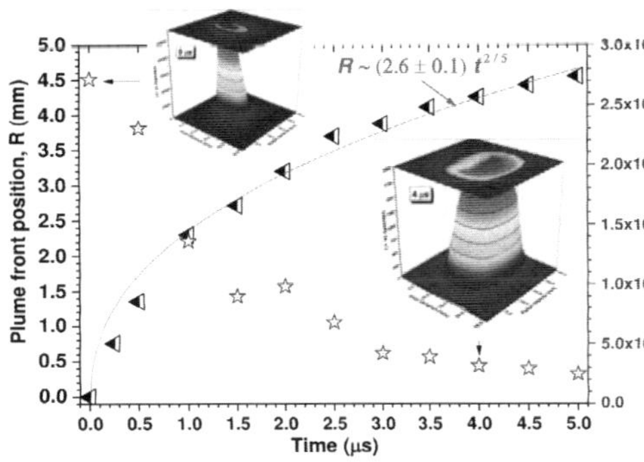

Fig. 2. *R-t* plot and velocities for the expanding plume front in vacuum (0.01 Pa). Insets show two images (0 and 4 μs) at 1.1 GW/cm^2 incident laser power density.

Acknowledgments We gratefully acknowledge the support received by the project: CTQ2013-43086.

References

[1] J. J. Camacho *et al* 2013 *Spectrochimica Acta Part B* **88** 203.

[1] E-mail: j.j.camacho@um.es

Content from this work may be used under the terms of the Creative Commons Attribution 3.0 licence. Any further distribution of this work must maintain attribution to the author(s) and the title of the work, journal citation and DOI.
Published under licence by IOP Publishing Ltd

Interaction of living cells ith a high dose proton microbeam

V. Mäckel*[1], N. Puttaraksa*, T. Kobayashi*, T. M. Kojima*, N. Imamoto† and Y. Yamazaki*

* Atomic Physics Laboratory, RIKEN, 2-1 Hirosawa, Wako, Saitama, 351-0198 Japan
† Cellular Dynamics Laboratory, RIKEN, 2-1 Hirosawa, Wako, Saitama, 351-0198 Japan

Synopsis We present current experiments studying the interaction of living cells to a high dose proton microbeam at the RIKE capillary cell irradiation setup.

The interaction of highly charged ions with single cells is an important topic of radiation biology. Especially focused ion beams down to few micrometer diameter, which enables the research of radiation effects on a cellular basis, has been proven to be extremely useful. This lead to an increasing number of microbeamlines devoted for biological research with the capability of delivering a defined number of ions down to sub-micrometer resolution into living cells.

We present current experiments studying the interaction of living cells to a high dose proton microbeam at the RIKEN capillary cell irradiation setup [1]. In a first set of experiments, we monitored the change of fluorescence signals of HeLa cells, after their nuclei have been irradiated. The cells have been genetically altered, such that they express green fluorescence protein (GFP) fused to histone H2B. Their nuclei were irradiated with 10^3 to 10^5 1 MeV protons during one second, focussed to a spot of few μm diameter. Directly after irradiation, the targeted area shows a reduction of the fluorescence signal (bleaching). Then the bleached area starts expanding while the total fluorescence signal in the irradiated area decreases, following an exponential decay with a half time of typically 30 s. Figure 1 shows an example of this bleaching taking place after irradiation of 25000 protons. This effect is different from usual photobleaching, since it only occurs in living cells, but not in dead or fixed cells. Also, it does not depend on the presence of the blue excitation light. The same bleaching effect is seen after staining the cells with Hoechst 33342, which attaches directly to the DNA.

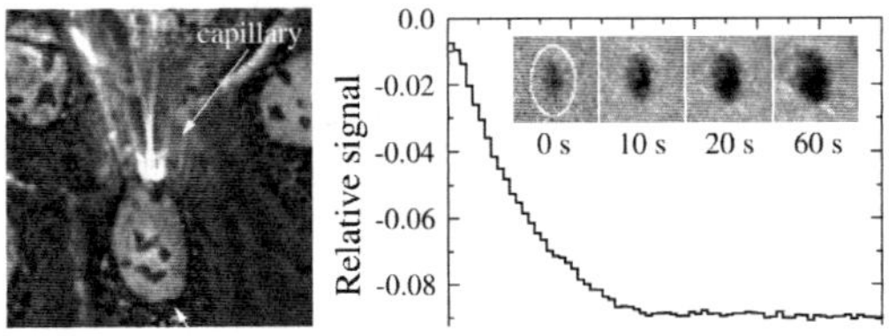

Figure 1. Figure 1 Left: HeLa cell expressing GFP in the nucleus before irradiation. The capillary is aiming at the nucleus from above. **Right:** Development of the GFP bleaching directly after irradiation. Shown is the relative mean signal of the area within the white oval (first picture at 0 s) in dependence of time after irradiation.

Secondly, we are studying the point of cell death within the cell cycle, following irradiation of the nucleus with high proton doses. Here we use fucci expressing HeLa cells, which change their fluorescence color during the cell cycle [2]. After irradiation of cell nuclei with various doses, we use time lapse microscopy for recording the phase in cell cycle by the fluorescence color and the cell death by the change in cell morphology [3].

References

[1] V. Mäckel *et al.*, 2014 *Rev. Sc . I strum.* **85** 014302

[2] A. Sakaue-Sawano *et al.*, 2008 *Cell* **132** 487

[3] N. Puttaraksa *et al.*, 2014 *Nuc. I st. Meth.* In Press, Corrected Proof

[1]E-mail: maeckel@riken.jp

Content from this work may be used under the terms of the Creative Commons Attribution 3.0 licence. Any further distribution of this work must maintain attribution to the author(s) and the title of the work, journal citation and DOI.
Published under licence by IOP Publishing Ltd

Ion extraction from a compact electron beam ion trap

Takayuki Nakajima, Tatsuhiko Kanehara, Yuki Yanaka, Junji Yatsurugi, and Nobuyuki Nakamura[1]

Institute for Laser Science, The University of Electro-Communications, Chofu, Tokyo 182-8585, JAPAN

Synopsis An ion extraction system has been installed at the compact electron beam ion trap CoBIT, which is a useful device for the spectroscopic studies of moderately charged ions. The extracted ions are used for definite charge state determination of the trapped ions observed in the spectroscopic studies.

An electron beam ion trap (EBIT) is a useful device for the spectroscopic studies of highly charged ions. It can produce highly charged few-electron ions through successive electron impact ionization of trapped ions. We constructed a compact EBIT [1] (CoBIT), several years ago for the spectroscopic studies of multiply and yet moderately charged ions relevant to astrophysical and fusion plasmas, such as Fe^{q+} with $q \sim 10$ [2] and W^{q+} with $q \sim 20$ [3]. In the spectroscopic studies, previously unreported lines have been observed and the charge state of the ions attributed for those lines has been identified by the appearance energy. For example, a line appearing when the electron energy exceeds the ionization energy of $A^{(q-1)+}$ is identified as a line from A^{q+}. However, the charge state identification based only on the appearance energy may have an error due to the following reasons: (1) the ionization energy (theoretical in most cases) is not accurate enough, (2) the electron energy is not accurate enough and has a width, and (3) recombination or charge exchange processes (or both) can reduce the charge state.

In this study, we have installed an ion extraction system to make definite charge state identification by the time of flight (TOF) method. The system consists of an einzel lens, an electrostatic bender, and a microchannel plate (MCP). The einzel lens is composed of three successive cylindrical electrodes installed inside the liquid nitrogen reservoir in the CoBIT chamber. The two electrodes at both ends are obliquely cut to be used as a deflector. The electrostatic bender installed just above the CoBIT chamber and just before the MCP is needed to separate the extracted ion beam from bremsstrahlung radiation arising at the electrodes in CoBIT, such as the electron collector.

The extraction system has been applied to identify the visible emission lines of Ho^{14+}, which has recently been suggested to be useful for the sensitive detection of the variation in the fine-structure constant [4]. During the spectroscopic observation, the trapped ions were periodically extracted by raising the potential at the middle drift tube. The extracted ion signal from the MCP was directly observed with an oscilloscope. The mass to charge ratio M/q of the extracted ions was determined from TOF calibrated with Ar ions. Consequently, definite identification of the visible emission lines of Ho^{14+} has been achieved. The application to the promethiumlike heavy ions [5] is also presented.

References

[1] N. Nakamura *et al.* 2008 *Rev. Sc . I strum.* **79** 063104
[2] N. Nakamura *et al* 2011 *APJ.* **739** 17
[3] A. Komatsu *et al.* 2012 *Plasma Fus o Res.* **7** 1201158
[4] V. A. Dzuba *et al.* 2015 *Phys. Rev. A* **91** 022119
[5] Y. Kobayashi *et al.* 2015 *oster rese ted th s co fere ce.*

[1]E-mail: n_nakamu@ils.uec.ac.jp

Content from this work may be used under the terms of the Creative Commons Attribution 3.0 licence. Any further distribution of this work must maintain attribution to the author(s) and the title of the work, journal citation and DOI.
Published under licence by IOP Publishing Ltd

XXIX International Conference on Photonic, Electronic, and Atomic Collisions (ICPEAC2015) IOP Publishing
Journal of Physics: Conference Series **635** (2015) 042008 doi:10.1088/1742-6596/635/4/042008

Towards precision β-decay measurements with laser cooled ^{35}Ar

R. D. Glover*[1], F. Lenaers*, Ph. Velten†, P. Finlay†, C. Couratin†, L. Hayen†, N. Severijns†, T. Bastin*

*Institut de Physique Nucléaire, Atomique et de Spectroscopie, Université de Liège, Liège 4000, Belgium
†Instituut voor Kern- en Stralingsfysica, KU Leuven, 3001 Leuven, Belgium

Synopsis We present modeling and preliminary experimental results towards the development of our magneto-optical trap (MOT) setup and polarisation scheme for ^{35}Ar. The atomic beam that is used for loading the MOT includes a novel transverse cooling arrangement based on a frequency broadened 'white-light molasses'. Our modeling shows that, under certain experimental conditions, such a setup can be beneficial in comparison to conventional transverse cooling techniques. We will report on experimental realization of such a setup.

Precision measurements in nuclear β-decay such as angular correlations and asymmetry measurements provide a sensitive means to test for non standard model (SM) physics such as deviations from maximal parity violation in the weak interaction. Recently, exciting new experiments have demonstrated that a magneto-optical trap (MOT) can be used as a tool to provide a novel source of short-lived radioactive isotopes, opening up new levels of precision [1]. At the same time, measurement of the β asymmetry parameter A_β in the β-decay of ^{35}Ar was identified as being a particularly sensitive candidate [2]. A single measurement of A_β in ^{35}Ar decay at the 0.5% level should enable determination of Cabibbo Kobayashi Maskawa (CKM) quark mixing matrix element $|V_{ud}|$ at a precision of 7×10^{-4} and with an improved ft-value a precision of 4×10^{-4} is expected. A measurement of this accuracy would clearly carry significant weight when compared to the world average of $|V_{ud}| = 0.97425(22)$ [3] which is at level of precision of 2×10^{-4}.

We are developing a magneto-optical trap for argon which will be used as a test apparatus for future experiments that will be conducted on-line at an accelerator facility with ^{35}Ar. Of critical importance are questions regarding the ultimate trapping efficiency that can be achieved and how to best generate a highly polarized sample which is fundamental for making a measurement of A_β.

In pursuit of generating a highly efficient system, we have conducted thorough modeling of the transverse cooling scheme for the atomic beam. We have investigated the collimation of our argon atomic beam using an optical molasses generated with a frequency modulated laser and compared this spectrally broadened 'white light molasses' with standard collimation techniques. This modeling shows that under certain conditions the white light molasses is preferable and can produce an atomic beam with higher phase space density as the resulting beam is shifted towards slower velocities [4]. Further, an optical molasses is geometrically relatively simple when compared to some standard techniques and therefore provides an attractive alternative. We report on the experimental realization of such a setup.

In parallel we are carrying out theoretical modeling of different possible polarization schemes that can be used to achieve a high polarization degree in a realistic MOT environment. We report on the progress of our theoretical modeling including the limitations of such schemes.

References

[1] D. Melconian *at al* 2007 *Phys. Lett. B* **649** 370-375

[2] N. Severijns and O. Naviliat-Cuncic 2013 *Physica Scripta* **T152** 014018

[3] J. Hardy and I. Towner 2009 *Phys. Rev. C* **79** 055502

[4] R. D. Glover and T. Bastin *J. Opt. Soc. Am. B* doc. ID 226250 (posted 16 January 2015, in press)

[1]E-mail: rohan.glover@ulg.ac.be

Content from this work may be used under the terms of the Creative Commons Attribution 3.0 licence. Any further distribution of this work must maintain attribution to the author(s) and the title of the work, journal citation and DOI.
Published under licence by IOP Publishing Ltd

Influence of oxygen content on the characteristics of DBD plasma actuator

R. S. Mangina[1], **C. L. Enloe**[2], **and G. I. Font**[3]

Department of Physics, United States Air Force Academy, CO 80840

The aerodynamic performance of an *AC* powered dielectric barrier discharge (DBD) plasma actuator immersed in pure oxygen and also in O_2 + noble gas mixtures is presented in terms of actuator's discharge structure and momentum coupling to the surrounding gas. At various gas pressures the impulse imparted to the freely-moving actuator body and its transient motion as a function of the applied voltage from 8-24 kV at f = 200 Hz is determined using laser interferometry. The present *E/N* results show that with increase of oxygen content the momentum coupling is better, especially in the negative species dominated forward discharge phase.

Understanding of plasma chemistry and collision dynamics of species in non-thermal plasmas produced in a DBD plasma actuator under the influence of time-varying applied voltages has been the focus in recent years for optimizing the actuator's performance at various geometries and operating regimes for aerodynamic flow-control [1]. The natural presence of electronegative O_2 in the ambient air forms negative ions, and also forms positive ions more easily than the N_2 in the discharge. We have investigated the actuator's performance by varying O_2 content in the mixture of O_2 + noble gas, and using 25 μm diameter wire exposed electrode (see Fig. 1) instead of conventional flat electrode [2]. As shown in Fig. 2, discharge is produced twice and quenched twice per AC cycle applied between asymmetrically placed wire electrode exposed to the surrounding gas and an encapsulated electrode.

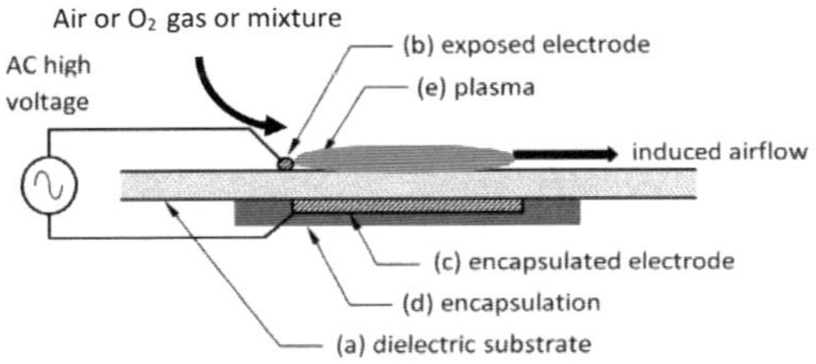

Figure 1. (Top) A schematic of a DBD plasma actuator with highly asymmetric electrodes set-up.

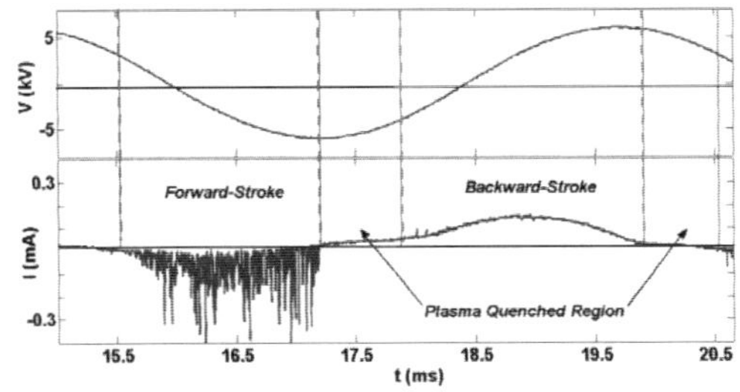

Figure 2. Voltage and current profiles of a DBD discharge with 25 μm diameter exposed-electrode at 6 kV of $V_{AC}(t)$ for the above geometry.

The use of wire exposed electrode shows improvement in the uniformity of discharge, especially during the positive going voltage half cycle as shown in Fig. 2, and in the momentum coupling compared to the conventional flat electrode [2]. We present the DBD plasma actuator body velocity (v), determined using laser interferometry, with respect to the discharge during each *AC*-half-cycle as a function of applied voltage. The plasma exerted force is followed by the negative force due to drag from surrounding gas on the actuator surface during plasma quenched phase are shown in Fig. 3. Most importantly the negative oxygen ions are a key element in producing plasma actuator's unidirectional net aerodynamic force even though the actuator is inherently AC device with no net charge transfer in either direction due to the presence of the dielectric barrier.

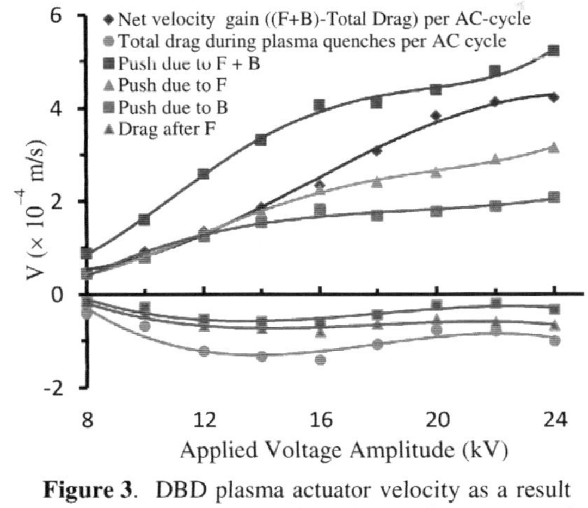

Figure 3. DBD plasma actuator velocity as a result of forward (F) and backward (B) discharges and drag during plasma-quenched phase in oxygen at atmospheric pressure as function of applied voltage.

References

[1] N. Benard *et al.*, 2014, *Expts. in Fluids*, **55**, 1846.
[2] C. L. Enloe *et al.*, 2004, *AIAA Journal*, **42**, 589-594.

[1] Email: Rao.Mangina@usafa.edu
[2] Email: Lon.Enloe@usafa.edu
[3] Email: Gabriel.Font@usafa.edu

Content from this work may be used under the terms of the Creative Commons Attribution 3.0 licence. Any further distribution of this work must maintain attribution to the author(s) and the title of the work, journal citation and DOI.
Published under licence by IOP Publishing Ltd

Discharge properties of macroscopic glass capillaries after irradiation by low intensity low energy single charged Ar ions

E. Giglio[1] , R. D. DuBois[2] , and K. Tőkési[3]

[1]Centre de Recherche sur les Ions, les Matériaux et la Photonique (CIMAP), F-14000, Caen, France, EU

[2]Missouri University of Science and Technology, Rolla MO 65409 USA

[3]Institute for Nuclear Research, Hungarian Academy of Sciences (ATOMKI), Hungary, EU

Synopsis The discharge properties of a macroscopic cylindrically shaped glass capillary, initially charged by a 1 keV Ar^+ beam tilted by 5° whith respect to the capillary axis is studied. Our experimental observations were compared to our Monte Carlo simulation, higlighting the importance of the surface conductivity.

While the guiding of ions through glass-capillaries due to charged patches is qualitatively understood, the complex nature of the electric conduction in such insulators makes quantitative predictions still a challenging task. Indeed, for a given ion beam, the guiding is entirely determined by the discharge dynamics of the charge patches at the surface, which in turn depend (i) on the electrical properties of insulators under ion beam irradiation and (ii) on the position and nature of grounded electrodes. In order to get some insights into the discharge dynamics of formally created charged patches, a combined experimental and theoretical study was performed. Experimentally, guiding of 1 keV Ar^+ ions through a macroscopic borosilicate glass capillary tilted by 5° with respect to the beam axis was studied. The capillary was inserted into an aluminum mount with the front surface covered with conductive paint. The current on the front surface and the transmitted current were measured, yielding absolute transmission rates. The aim was to measure the charging and discharging times for the capillary. This was done by centering the capillary on the entrance aperture and waiting until the transmitted intensity stabilized. Then the beam was blocked for varying times and the transmitted intensity was measured as a function of time. By plotting the ratio of the beam intensity immediately after the beam was re-injected with respect to the intensity just before the beam was blocked for the various blocking times, yields information about the capillary discharge time. This was done for several beam intensities ranging from 0.15 to 16 pA. Using a home-made numerical code [1], the same observables were simulated and compared to the experimental data. The theoretical model, on which the simulations are based, includes a realistic descriptions of the surface and bulk conductivity of glass capillaries. It was found that the charge relaxation of the former created patches depends strongly on the mobility of the charge carriers, giving boundary values for the effective mobility of the charge carriers. The capillary being grounded only at the entrance, surface conduction is privileged, and the simulations predict a non-linear discharge of the patches of the type $Q(t)=Q_0/(1+t/\tau)$, with τ=10s being the characteristic discharge time.

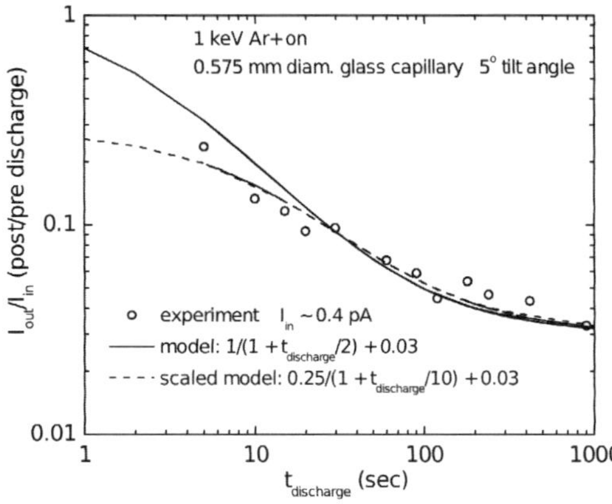

Figure 1. – Transmitted intensity as a function of the discharge time, normalized to the transmitted one, just before the beam was blocked, for a cylindrical capillary of 43 mm length, 0.575 mm inner diameter.

This work was supported by the TéT Grant No. TéT_12_FR-1-013-0007, by the Hungarian Scientific Research Fund OTKA No.NN 103279 and by the COST Action CM1204 (XLIC)

References:

[1] E.Giglio, R.D. Dubois, A.Cassimi, K.Tökési, NIMB B, doi:10.1016/j.nimb.2014.11.056

E-mail: giglio@ganil.fr

Content from this work may be used under the terms of the Creative Commons Attribution 3.0 licence. Any further distribution of this work must maintain attribution to the author(s) and the title of the work, journal citation and DOI.
Published under licence by IOP Publishing Ltd

Journal of Physics: Conference Series **635** (2015) 042011 doi:10.1088/1742-6596/635/4/042011

Combined theoretical and experimental study of the transmission of tilted ion beams through macroscopic conical glass capillaries

E. Giglio[1], H. Zhang[2], S. Guillous[1], A. Agnihotri[1], F. Ropars[1], W. Iskandar[1], A. Mery[1], H. Merabet[3], A. Cassimi[1]

[1]Centre de Recherche sur les Ions, les Matériaux et la Photonique (CIMAP), F-14000, Caen, France, EU

[2]School of Nuclear Science and Technology, Lanzhou University, Lanzhou, China

[3]Qatar University, P.O. Box 2713, Doha, Qatar

Synopsis The transmission as a function of the tilt angle of a 3 keV Ar^+ ion beam through a conical macroscopic glass capillary is studied theoretically and experimentally. It was found that the charge patches which are responsible for the ion guiding also compress the beam spatially in the direction orthogonal to the patches, resulting into an enhancement of the transmission with increasing tilt angle.

The transmitted intensity of a low intensity and low energy Ar^+ beam through a tilted insulating conical capillary is measured, for tilt angle between -1.5° and +1.5° with respect to the beam axis. The first 3rd of the borosilicate glass capillary has a cylindrical shape with 0.86 mm inner diameter, then becomes quickly conical with an outlet of 0.44 mm diameter and has a total length of 80 mm. Only the cylindrical part is covered with conductive paint. It was found that the transmitted intensity increases with the tilt angle up to 0.8°, then decreases until to vanish for larger angles (see figure 1), a behavior already shown by other authors [1].

Using a home-made numerical code [2], the same observables were simulated and compared to the experimental data. While the guiding of ions through glass-capillaries due to charged patches is qualitatively understood, the complex nature of the electric conduction in such insulators makes quantitative predictions still a challenging task. Indeed, for a given ion beam, the guiding is entirely determined by the discharge dynamics of the charge patches at the surface [3], which in turn depend (i) on the electrical properties of insulators under ion beam irradiation and (ii) on the position and nature of grounded electrodes. That is why the theoretical model, on which the simulations are based, include a realistic description of the surface and bulk conductivity of glass capillaries. It was found that the surface charge density of the patches increases with increasing tilt angle. The patches not only rotate/guide the beam but also compress it spatially in the direction orthogonal to the surface patch. As a result, the transmitted beam intensity increases with tilt angle. For zero tilt angle, this effect is absent, explaining the local minimum in the transmission curve in figure 1. For angle above 0.8°, the patches are not sufficiently intense and the guiding/rotating is not efficient enough, resulting into a lower and lower transmission.

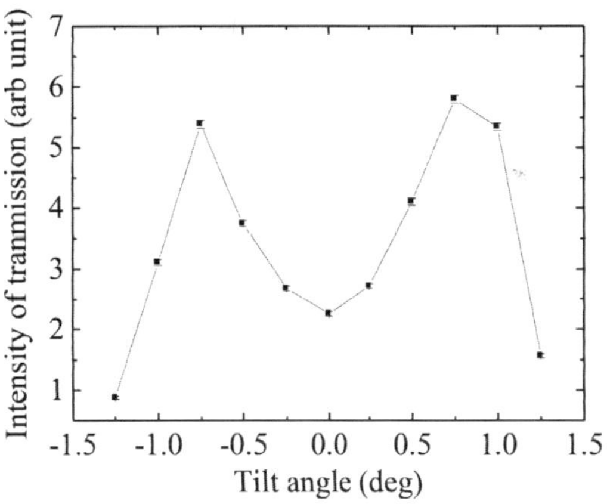

Figure 1. – Transmitted intensity of a 1 keV Ar^+ ion beam as a function of the tilt angle through a conical borosilicate glass capillary.

References:

[1] E. Gruber, N. Stolterfoht, P. Allinger, S. Wampl, Y. Wang, M.J. Simon, F. Aumayr, NIM B **340** (2014) 1

[2] E. Giglio, R.D. Dubois, A. Cassimi, K. Tőkési, NIMB B, doi:10.1016/j.nimb.2014.11.056

[3] N. Bundaleski, H. Khemliche, P. Rousseau, A. Cassimi, L. Maunoury, P. Roncin, J. Phys: Conf. Ser. **133** (2008) 012016

E-mail: giglio@ganil.fr

Content from this work may be used under the terms of the Creative Commons Attribution 3.0 licence. Any further distribution of this work must maintain attribution to the author(s) and the title of the work, journal citation and DOI.
Published under licence by IOP Publishing Ltd

Journal of Physics: Conference Series **635** (2015) 052001 doi:10.1088/1742-6596/635/5/052001

Subshell resolved inner shell ionization cross-sections of High Z elements by electron impact

H. V. Rahangdale*[1], D. Mitra†, M. Guerra♯, J. P. Santos♯, S. Saha*

* Saha Institute of Nuclear Physics, Kolkata, India

† Dept. of Physics, University of Kalyani, Kalyani, India

♯Centro de Física Atómica, CFA, Departamento de Física, Faculdade de Ciências e Tecnologia, FCT, Universidade Nova de Lisboa, 2829-516 Caparica, Portugal

Synopsis We have obtained the L sub-shell resolved electron impact ionization cross sections for high Z elements in the energy range 15-45 keV. The experimental results are compared with the DWBA and PWBA (Distorted/Plane Wave Born Approximation) based theoretical predictions and MRBEB (Modified Relativistic Binary Encounter Bethe) theory. Good agreement is found between DWBA based results and experiment for L3 sub-shell of all the elements studied.

Accurate cross-sections of inner shell ionization of atoms by electron impact are needed for many spectroscopic techniques like Electron Probe Micro Analysis(EPMA), Auger Electron Microscopy(AES), and also for theoretical studies of coupled electron-photon transport[1]. In spite of such need, enough experimental data are not available for inner shell ionization cross sections. Specifically the data for L and M shell are scarce, and also, mostly X-ray production cross sections are reported by researchers and less attention is paid to the extraction of ionization cross section data.

We have measured L X-ray production cross sections of Gold, Lead and Thorium by electron impact of energy 15−45 keV. Experimental setup consisted of electron gun, target ladder, Faraday cup, and Si PIN diode detector mounted inside a vacuum chamber as an energy dispersive spectrometer. Monoenergetic electrons of particular energy were bombarded on thin solid targets and the X-ray spectrum was recorded. Counts under L_α, L_β and L_γ peaks were obtained to get the production cross sections. Also the L_γ peak was de-convoluted into L_{γ_5}, L_{γ_1} and $L_{\gamma_{236}}$ lines. L_α, L_{γ_5}, L_{γ_1} and $L_{\gamma_{236}}$ cross sections were then converted into ionization cross sections as per the data analysis process described in detail in Ref. [2]. Figure 1 shows the L1, L2 and L3 cross sections of Lead along with DWBA [3] and MRBEB [4] theory results. Theoretical results of DWBA are obtained from analytical formulas provided by Bote el. al [5]. It is found that the agreement between theory and experiment is very good for L3 sub-shell for all the three elements studied. For L1 and L2, theory over-predicts the ionization cross section by about 30 − 40%. Difference between theory and experiment is more for L1 than for L2. The difference between experiment and theory can be attributed to some extent on the uncertainty in the relaxation parameters used. Also the discrepancy between theory and experiment is more for MRBEB as compared to the DWBA theory.

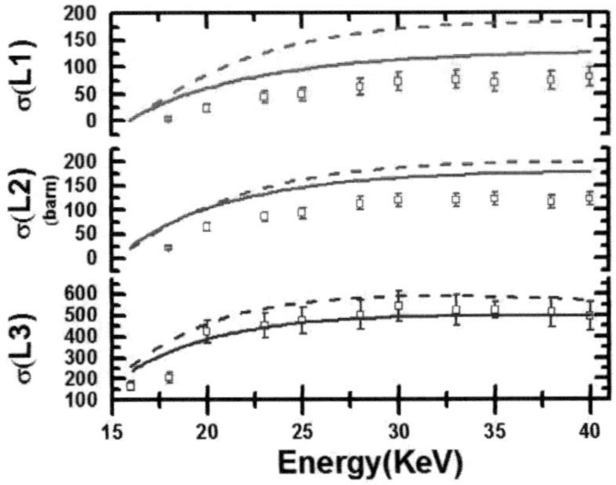

Figure 1. L sub-shell ionization cross sections of Pb. Solid and dashed lines represent DWBA and MRBEB theory results respectively.

References

[1] X. Llovet *et al* 2014 *J. Phys. Chem. Ref. Data*,**43**:013102

[2] H.V.Rahangdale *et al* 2014 *Phys. Rev. A*, **89**:052708.

[3] D. Bote *et al* 2008 *Phys. Rev. A*, **77**:042701

[4] M. Guerra *et al* 2012 *Int. J. Mass. Spectrom.*, **313**:1

[5] Bote *et al* 2009 *Atomic Data and Nuclear Data Tables*, 95(6):871 – 909.

[1]E-mail: hitesh.rahangdale@saha.ac.in

Content from this work may be used under the terms of the Creative Commons Attribution 3.0 licence. Any further distribution of this work must maintain attribution to the author(s) and the title of the work, journal citation and DOI.
Published under licence by IOP Publishing Ltd

Journal of Physics: Conference Series **635** (2015) 052002 doi:10.1088/1742-6596/635/5/052002

Simultaneous measurement of photorecombination and electron-impact ionization of Fe^{14+} ions

D. Bernhardt[1], A. Becker[2], M. Grieser[2], M. Hahn[3], C. Krantz[2], M. Lestinsky[4], O. Novotný[2,3], R. Repnow[2], K. Spruck[1,2], D. W. Savin[3], A. Wolf[2], A. Müller[1], and S. Schippers[1]

[1] Institut für Atom- und Molekülphysik, Justus-Liebig-Universität Gießen, 35392 Gießen, Germany
[2] Max-Planck-Institut für Kernphysik, 69117 Heidelberg, Germany
[3] Columbia Astrophysics Laboratory, Columbia University, New York, NY 10027, USA
[4] GSI Helmholtzzentrum für Schwerionenforschung, 64291 Darmstadt, Germany

Synopsis Cross sections for photorecombination (PR) and for electron-impact ionization (EII) of astrophysically relevant Mg-like Fe^{14+} ions have been measured by employing the electron-ion merged-beams technique at the Heidelberg heavy-ion storage ring TSR. Rate coefficients for PR and EII of Fe^{14+} ions in a plasma are derived from the measurements. Both agree with the most recent theoretical results.

Cross sections for photorecombination (PR) and for electron-impact ionization (EII) of Mg-like Fe^{14+} ions have been measured [1] (Fig. 1) by employing the electron-ion merged-beams technique at the Heidelberg heavy-ion storage ring TSR. The present findings together with our previous results for other Fe-M-shell and Fe-L-shell ions [2, 3] (and references therein) are relevant for the modelling of the charge balance in photoionized plasmas (PP) and collisionally ionized plasmas (CP). Moreover, the present results benchmark the most recent theoretical calculations for PR [4] and EII [5] of Fe^{14+}. In the relevant temperature ranges (PP and CP ranges for PR; CP range for EII) both theoretical results agree with the experimentally derived rate coefficients within the systematic experimental uncertainty.

The simultaneous measurement of PR and EII under the same experimental conditions provides a unique opportunity for a comparison of recombination and ionization resonances [6]. The present comparison between PR and EII (Fig. 1) shows that corresponding recombination and ionization resonances are rare. Only the $(2p^5\,3s^2\,3d\ ^3D_1)\ nl$ and $(2p^5\,3s^2\,3d\ ^1P_1)\ nl$ intermediate levels contribute to both PR and EII. Ionization resonances beyond the respective series limits at 794.95 and 807.29 eV are associated with higher $2p \to n'l'$ $(n' \geq 4)$ core excitations. The corresponding highly-excited intermediate levels strongly autoionize and therefore do not contribute significantly to PR.

References

[1] D. Bernhardt *et al.* 2014 *Phys. Rev. A* **90** 012702
[2] S. Schippers *et al.* 2010 *Int. Rev. At. Mol. Phys.* **1** 109
[3] M. Hahn 2014 *J. Phys.: Conf. Ser.* **488** 012050
[4] Z. Altun *et al.* 2007 *Astron. Astrophys.* **474** 1051
[5] K. P. Dere 2007 *Astron. Astrophys.* **466** 771
[6] J. Linkemann *et al.* 1995 *Phys. Rev. Lett.* **74** 4173

[1]E-mail: Stefan.Schippers@physik.uni-giessen.de

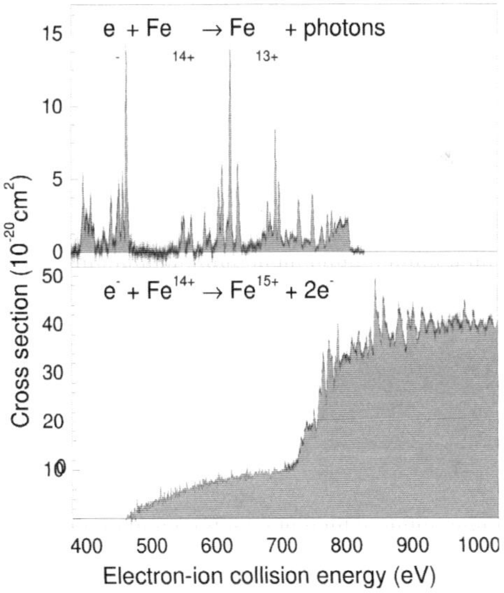

Figure 1. Measured cross sections [1] for PR (top) and EII (bottom) of $Fe^{14+}(3s^2)$. The resonances in both spectra are due to resonant dielectronic capture into $2p^5\,3s^2\,3d\,nl$ doubly excited intermediate levels. Subsequent decay of these levels via radiative transitions or via double Auger decay leads to net recombination or net ionization, respectively. The ionization resonances appear on top of a continuous background signal due to direct ionization and excitation-autoionization.

Content from this work may be used under the terms of the Creative Commons Attribution 3.0 licence. Any further distribution of this work must maintain attribution to the author(s) and the title of the work, journal citation and DOI.
Published under licence by IOP Publishing Ltd

Journal of Physics: Conference Series **635** (2015) 052003 doi:10.1088/1742-6596/635/5/052003

Electron-ion recombination of the open-4f-shell ion W^{19+}

N. R. Badnell[1], K. Spruck[2,3], C. Krantz[3], O. Novotný[3,4], A. Becker[3], D. Bernhardt[2], M. Grieser[3], M. Hahn[4], R. Repnow[3], D. W. Savin[4], A. Wolf[3], A. Müller[2], S. Schippers[2]

[1]Department of Physics, University of Strathclyde, Glasgow G4 0NG, United Kingdom
[2] Institut für Atom- und Molekülphysik, Justus-Liebig-Universität Gießen, 35392 Gießen, Germany
[3] Max-Planck-Institut für Kernphysik, 69117 Heidelberg, Germany
[4] Columbia Astrophysics Laboratory, Columbia University, New York, NY 10027, USA

Synopsis Electron-ion recombination of W^{19+} forming W^{18+} has been studied both experimentally and theoretically. An exceptionally large recombination rate coefficient was observed by the storage-ring experiment at zero electron-ion collision energy. It exceeds the expectation for nonresonant radiative recombination (RR) by more than two orders of magnitude. This follows the pattern seen previously for the neighboring charge states of tungsten. The results of our present theoretical calculations, which were carried-out within the Breit-Wigner partitioned framework for resonant electron-ion recombination of complex ions, agree with this observation.

Tungsten is used for the coating of parts of the inner walls of fusion devices such as ITER. Therefore, atomic collision processes with tungsten ions are of immediate interest for understanding the role of tungsten impurities in the fusion plasma. For example, the ionization balance of tungsten is determined by the interplay of electron-impact ionization and electron-ion recombination. Most of the required cross sections come from theoretical calculations which often bear large uncertainties and, thus, require benchmarking by experiment. To this end, we have focussed on tungsten ions with a particularly complex atomic structure, i.e, ions with a nearly half open $4f$ shell. The present study extends our previous work on electron-ion recombination of $W^{20+}([\mathrm{Kr}]\,4d^{10}\,4f^{8})$ [1, 2] and $W^{18+}([\mathrm{Kr}]\,4d^{10}\,4f^{10})$ [3] to $W^{19+}([\mathrm{Kr}]\,4d^{10}\,4f^{9})$.

The experiment was carried out by employing the electron-ion merged-beams technique at the heavy-ion storage ring TSR of the Max-Planck-Institute for Nuclear Physics in Heidelberg, Germany. As in the previous cases of W^{18+} and W^{20+}, an extremely large W^{19+} recombination rate coefficient has been observed at low electron-ion collision energies (Fig. 1). According to our present understanding, this is caused by resonant recombination involving many-electron processes which cannot fully be treated by the standard theory for electron-ion recombination. Nevertheless, they can be accounted for in an approximate manner by appealing to statistical theory [4]. Different implementations have already been applied successfully to the electron-ion recombination of W^{20+} [2, 5] and W^{18+} [3]. Our present calculations for W^{19+} are in good agreement with the experimental findings (Fig. 1). This puts much confidence in theory being able to provide reliable electron-ion recombination rate coefficients for other complex ions.

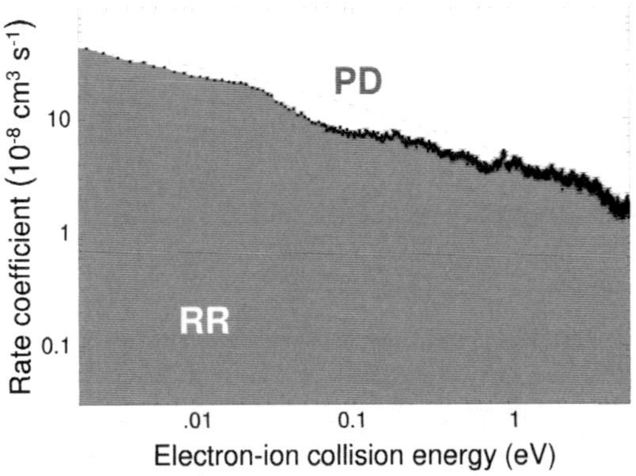

Figure 1. Measured rate coefficient (shaded curve) for the recombination of $W^{19+}([\mathrm{Kr}]\,4d^{10}\,4f^{9})$ ions with free electrons at low electron-ion collision energies. The dashed curve is the result of our "partitioned and damped" (PD) calculations. The dash-dotted curve is the theoretical rate coefficient for nonresonant radiative recombination (RR).

References

[1] S. Schippers *et al.* 2011 *Phys. Rev. A* **83** 012711
[2] N. R. Badnell *et al.* 2012 *Phys. Rev. A* **85** 052716
[3] K. Spruck *et al.* 2014 *Phys. Rev. A* **90** 032715
[4] V. Flambaum *et al.* 2002 *Phys. Rev. A* **66** 012713
[5] V. A. Dzuba *et al.* 2013 *Phys. Rev. A* **88** 062713

[1]E-mail: badnell@phys.strath.ac.uk

Content from this work may be used under the terms of the Creative Commons Attribution 3.0 licence. Any further distribution of this work must maintain attribution to the author(s) and the title of the work, journal citation and DOI.
Published under licence by IOP Publishing Ltd

The single ionization cross-section of Rb atoms by electron impact

V. Roman*[1], A. Kupliauskiene[†2], and A. Borovik*

*Department of Electron Processes, Institute of Electron Physics, Uzhgorod, 88017, Ukraine
†Institute of Theoretical Physics and Astronomy, Vilnius University, Vilnius, 01108, Lithuania

Synopsis The 5s+core single ionization cross-section of Rb atoms was calculated for the incident electron energies up to 600 eV by using the relativistic distorted-wave, Coulomb-Born and binary-encounter-dipole approximations. The accuracy of the data was analyzed by the comparison with experimental data. The best agreement was observed for the calculations performed in configuration interaction approximation.

The correct account for the contribution from indirect processes such as autoionization is very important problem of the theory of electron-impact ionization of atoms [1]. For instance, the single ionization of Rb atoms by electron impact was the subject of many theoretical studies (see e.g. [2] and references therein) but all of them failed to describe correctly the behavior of the experimental cross-section above the $4p^6$ excitation threshold. Meanwhile, there is a unique possibility to evaluate the accuracy of different theoretical techniques in accounting for indirect processes when along with the total ionization cross-section σ_{tot}, the autoionization cross-section σ_{aut} is also known experimentally [3].

In the present work we calculated the 5s+core single direct ionization cross-section of Rb atoms in relativistic distorted-wave (RDW), Coulomb-Born-Exchange (CBE) and binary-encounter-dipole (BED) approximations by using the FAC computer code [4]. In order to evaluate the role of correlation effects, the RDW and BED calculations were performed both in single configuration and configuration interaction (CI) approximations. In the case of CI calculations, the superposition of configurations $4p^6nl$ (n = 5, 6, 7, l = 0, 1, 2, 3), $4p^5nln'l'$ (nl = 5s, 6s, 5p, 4d, 5d, 4f, $n'l'$ = 5s, 6s, 5p, 6p, 4d, 5d, 4f, 5f, 5g) for the initial and $4p^6$, $4p^5nl$ (nl = 5s, 6s), $4p^4nln'l'$ (nl = 5s, 5p, $n'l'$ = 5s, 5p, 4d, 5d) for the final states was used to take into account the correlation effects. Figure 1 shows these data compared with the difference $\sigma_{tot}[5]$-$\sigma_{aut}[6]$ which represents the "true" 5s+core single direct ionization cross-section (see [3]).

As can be seen, significant discrepancies are observed in absolute values and energy behaviors of all calculated cross-sections around 10 eV and near the $4p^6$ ionization threshold. However, when compared with the "true" single direct ionization cross-section $\sigma_{tot}[5]$-$\sigma_{aut}[6]$ the RDW-CI and BED-CI data show the best agreement in the impact energy region just above the $4p^6$ ionization threshold. That points out the important role of correlation effects in electron impact ionization of inner shells in Rb atoms.

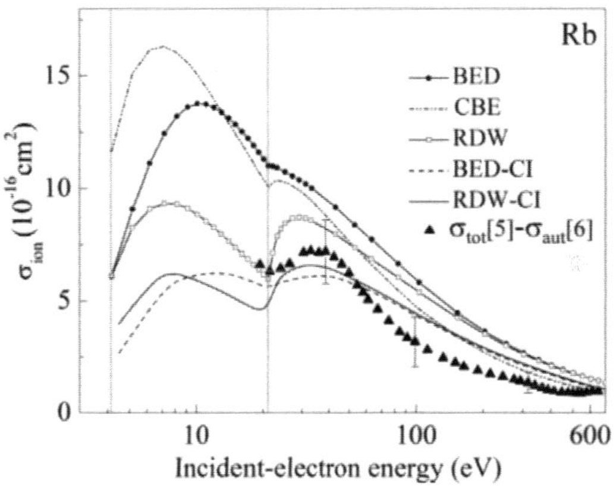

Figure 1. The single direct ionization cross sections of Rb atoms. Vertical dashed lines mark the 5s and $4p^6$ ionization thresholds at 4.18 and 20.71 eV [7], respectively.

This research was funded by the National Academy of Sciences of Ukraine (project 0112У002079).

References

[1]D.L. Moores, K.J. Reed 1994 *Adv. At. Mol. Opt. Phys.* **34** 301
[2]M. Łukomski *et al* 2006 *Phys. Rev. A.* **74** 032708
[3]A. Borovik *et al* 2012 *J. Phys. B.* **45** 045204
[4]M. F. Gu 2008 *Can. J. Phys.* **86** 675
[5] J. Tate, P. Smith 1934 *Phys. Rev.* **46** 773
[6]A. Borovik *et al* 2013 *J. Phys. B.* **45** 215201
[7]J. E. Sansonetti 2006 *J. Phys. Chem. Ref. Data* **35** 301

[1] E-mail: viktoriyaroman11@gmail.com
[2] E-mail: alicija.kupliauskiene@tfai.vu.lt

Content from this work may be used under the terms of the Creative Commons Attribution 3.0 licence. Any further distribution of this work must maintain attribution to the author(s) and the title of the work, journal citation and DOI.
Published under licence by IOP Publishing Ltd

Classification of the $5p^5 n_1l_1n_2l_2$ LSJ autoionizing states of Cs excited by 30 eV electron-impact

G. Kerevičius[1] **and A. Kupliauskienė**[2]

Institute of Theoretical Physics and Astronomy, Vilnius University, A. Goštauto 12, 01108 Vilnius, Lithuania

Synopsis Classification of the $5p^5n_1l_1(L'S')n_2l_2$ LSJ autoionizing states of Cs excited by 30 eV electron-impact was performed by using more accurate calculations of energy levels, excitation cross sections and autoionization probabilities obtained in the Dirac-Fock-Slater approximation. Calculated parameters were used for simulation of the ejected Auger electron intensity spectrum and identification of the experimental ejected-electron spectrum of Cs excited by 30 eV electrons taking into account the asymmetry of the angular distribution of Auger electrons.

Configuration interaction calculations of energy levels, 400 eV electron-impact excitation cross sections and autoionization probabilities of the $5p^5n_1l_1n_2l_2$ LSJ (n_1l_1 = 5d, 6s; n_2l_2 = 5d, 6(s – d), 7(s, p), 8s) autoionizing states of Cs were performed in [1] by using the jjJ coupling scheme. Calculated data were applied for an identification of 37 lines of the experimental ejected-electron spectrum [2] excited by 400 eV electrons, i. e. classification and identification of all levels of [2] was not performed. The main task of the present work is to perform a revised classification of calculated energy levels in the LSJ coupling scheme and apply it for a more precise and full identification of the experimental spectrum excited by 30 eV electrons and observed at 75° angle [2]. For this purpose, new calculations of energies, electron-impact excitation cross sections and autoionization probabilities more suitable for simulation of the intensities of ejected Auger electron spectrum were performed. The basis of mixed relativistic configurations was constructed by using the Flexible atomic code [3]. In order to optimize the local central potential with approximated exchange part, $5p^66s$ configuration was used. The following singly excited and 5p-core excited configurations were used to take into account the correlation effects: $5p^6n_0l_0$, n_0l_0 = (4, 5)f, (7 – 12)(s – d); $5p^5n_1l_1n_2l_2$, n_1l_1 = (4, 5)f, (5 – 7)d, (6 – 8)(s, p, f); n_2l_2 = (4 – 8)f, 5d, (6 – 11)(s – d).

Values of the level energies and simulated Auger electron emission spectra for 30 eV impacting electrons taking into account the asymmetry of the angular distribution of emitted Auger electrons were used for identification of the experimental spectrum measured at 75° with respect to the direction of impacting electrons [2]. The simulated spectrum is shown in figure 1 together with the measured spectrum [2] for comparison. Compared with the 400 eV case, largest intensities were obtained for the quartet and other dipole-forbidden lines as 30 eV energy of impacting electrons is close to the 5p-core excitation threshold. The latter lines blend with dipole-allowed lines which dominate in 400 eV case [1].

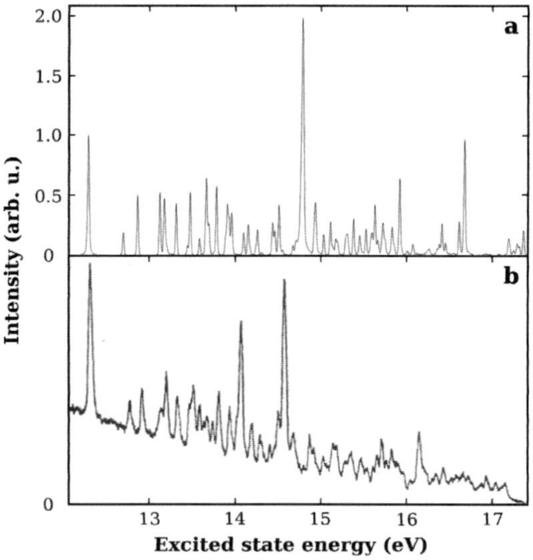

Figure 1. Simulated (a) and measured (b) [2] spectra of Auger electrons ejected from 5p-core excited states of Cs excited by 30 eV electron-impact.

This research was performed under the project VP1-3.1-ŠMM-07-K-02-013 by the European Social Fund under the Global Grant measure.

References

[1] A. Kupliauskienė 2011 *Phys. Scripta* **84** 045304

[2] V. Pejčev and K. J. Ross 1977 *J. Phys. B* **10** 2935

[3] M. F. Gu 2008 *Can. J. Phys.* **86** 675

[1]E-mail: gintaras.kerevicius@tfai.vu.lt
[2]E-mail: alicija.kupliauskiene@tfai.vu.lt

Content from this work may be used under the terms of the Creative Commons Attribution 3.0 licence. Any further distribution of this work must maintain attribution to the author(s) and the title of the work, journal citation and DOI.
Published under licence by IOP Publishing Ltd

Database of atomic parameters for plasma radiation modeling

A. Kupliauskienė[1], **P. Bogdanovich, R. Karpuškienė, R. Kisielius, and Š. Mikolaitis**
Institute of Theoretical Physics and Astronomy, Vilnius University, A.Goštauto 12, LT-01108 Vilnius, Lithuania

Synopsis The database ADAMANT of atomic parameters for the simulation of the electromagnetic radiation from plasma is established (www.adamant.tfai.vu.lt). The energy levels, radiative and autoionization transition probabilities, electron-impact excitation and ionization as well as dielectronic recombination rates for atoms and ions are constantly added to the database.

For the plasma simulation purposes, a newly-established database ADAMANT (**A**pplicable **DA**ta of **M**any-electron **A**tom e**N**ergies and **T**ransitions) containing various spectroscopic parameters of many-electron atoms and ions is being developed at Institute of Theoretical Physics and Astronomy, Vilnius University. This database contains parameters, involving free-electrons, such as autoionization probabilities, electron-impact excitation and ionization collision strengths, cross-sections and rates, dielectronic recombination rates as well as energy levels, radiative transition parameters (line strengths, oscillator strengths, transition probabilities) that are needed in modeling both high temperature (astrophysical, nuclear fusion) and low temperature plasma, such as planetary nebulae, working material of spectroscopic and medical devices.

Large amount of spectroscopic data for variuos atoms and ions has been produced so far (see, e.g., databases NIST [1], ADAS [2], or NIFS [3]). Unfortunately, it is rather complex task to apply these data for modeling purposes, because they are calculated in different approximations, with different computer code suites and atomic data accuracy. Therefore, it becomes difficult to match one set of data, e.g., energy levels, radiative transition probabilities, with another set of data involving free electrons, e.g. electron-impact excitation or ionization rates.

In ADAMANT, the atomic data sets for specific atom are produced by using the same computer code suites, applying the same approximation for inclusion of relativistic and correlation effects. They are generated using identical multireference wavefunction basis. Such an approach to produce data significantly reduces workload for data application, making possible an automatic data parsing in plasma modeling codes. The configuration interaction (CI) approximation by using numerical solutions of Hartree-Fock [4], relativistic Dirac-Fock-Slater [5] or quasirelativistic Hartree-Fock [6] equations is used for calculations of the same set of spectroscopic parameters.

ADAMANT database is constantly updated and extended with new results. At present it contains more than 2000 sets of parameters for some atoms and isoelectronic sequences of ions. Website for the database (www.adamant.tfai.vu.lt) provides general information about calculation methods, data formats, data accuracy assessment, it welcomes any comments on data or requests for new data production or accuracy improvement of existing data. We have an ambition to make this freely available ADAMANT database to serve not only as a source of highly-accurate data but also as a vehicle for other atomic data producers to disseminate their results.

This research was performed under the project VP1-3.1-ŠMM-07-K-02-013 by the European Social Fund under the Global Grant measure.

References

[1] http://www.nist.gov/pml/data/index.cfm

[2] http://open.adas.ac.uk/

[3] http://dbshino.nifs.ac.jp/

[4] Froese Fischer C 1987 *Comput. Phys. Commun.* **43** 355

[5] Gu M F 2008 *Can. J. Phys.* **86** 675

[6] Bogdanovich P and Rancova O 2008 *Phys. Scripta* **78** 045301

[1]E-mail: alicija.kupliauskiene@tfai.vu.lt

Content from this work may be used under the terms of the Creative Commons Attribution 3.0 licence. Any further distribution of this work must maintain attribution to the author(s) and the title of the work, journal citation and DOI.
Published under licence by IOP Publishing Ltd

Journal of Physics: Conference Series **635** (2015) 052007 doi:10.1088/1742-6596/635/5/052007

Ejected-electron spectra of Ba close to the $5p^6$ excitation threshold

V. Hrytsko[† 1], **A. Kupliauskienė**[‡ 2], **and A. Borovik**[† 3]

[†]Department of Electron Processes, Institute of Electron Physics, Uzhgorod, 88017, Ukraine
[‡]Institute of Theoretical Physics and Astronomy, Vilnius University, LT-01108 Vilnius, Lithuania

Synopsis The ejected-electron spectra arising from the decay of the $5p^5n_1l_1n_2l_2n_3l_3$ autoionizing states in Ba atoms were precisely measured for electron impact energies between 15 and 20 eV. The excitation threshold of the $5p^6$ subshell was evaluated at 15.63 ± 0.05 eV. The excitation thresholds and decay paths of the lowest states in $5p^55d6s^2$ and $5p^55d^26s$ configurations were determined.

The excitation of the $5p^6$ subshell in Ba atoms is affected by the strong mixing effects which result in formation of the $5p^5[6s^2 \times 5d6s \times 5d^2]nl$ configurations. The overlapping energy structure and the multichannel decay mode of corresponding autoionizing states are two main reasons for the situation when the lines in photoabsorption spectra [1] and in ejected-electron spectra [2], [3] are still unclassified. Meanwhile, it is well know that the accurate experimental data on excitation thresholds and decay channels of atomic levels is the first requirement for their accurate spectroscopic classification.

In order to determine the excitation thresholds and decay channels of the $5p^5n_1l_1n_2l_2n_3l_3$ lowest autoionizing states in Ba atoms we have measured the ejected-electron spectra in series, step-by-step for electron impact energy values between 15 and 20 eV with the increment step of 0.1 eV. The apparatus and measuring procedure were described elsewhere [4]. Spectra were recorded at the incident- and ejected-electron energy resolution of 0.2 eV and 0.06 eV, respectively. The incident-electron and ejected-electron energy scales were calibrated by using the photoabsorption data [5] for the autoionizing state at 15.80 eV. The uncertainties of energy scales were estimated to be ± 0.07 eV for incident and ± 0.05 eV for ejected electrons.

The excitation dynamics of ejected-electron spectra is shown in figure 1. As can be seen, in spectrum at 15.65 eV the only line 5 at 10.42 eV is observed. In accordance with our calculations based on the Dirac-Fock-Slater approach [6] this line can be assigned as $(5p^55d6s^2)^3P_0$ dipole-forbidden state at 15.63 eV. Thus, this is the lowest $5p^6$ -core excited atomic autoionizing state in Ba atoms with the single decay channel into the $5p^66s$ Ba^+ ground state.

The excitation thresholds and decay channels were determined for ten autoionizing states from $5p^55d6s^2$ and $5p^55d^26s$ configurations located between 15.63 and 16.70 eV. Most states possess the multichannel electron decay into the $5p^66s, 5d$ Ba^+ states.

This research was funded by the National Academy of Sciences of Ukraine (AB, VH).

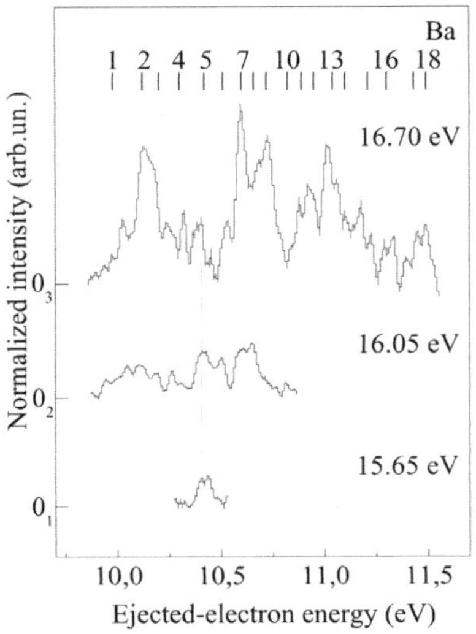

Figure 1. Ejected-electron spectra of Ba atoms for different electron impact energies. The polynomial background function was subtracted in all spectra.

References

[1] M.A. Baig *et al* 1984 *J. Phys. B* **17** 371

[2] D. Rassi and K.J. Ross 1980 *J. Phys. B* **13** 4683

[3] A. Borovik *et al* 1985 *Opt. Spektrosc.* **58** 988 (in Russian)

[4] A. Borovik *et al* 2005 *J. Phys. B* **38** 1081

[5] J.P. Connerade *et al* 1979 *Phil. Trans. R. Soc. A* **290** 327

[6] M.F. Gu 2008 *Can. J. Phys.* **86** 675-689

[1]E-mail: shargov11@gmail.com
[2]E-mail: alicija.kupliauskiene@tfai.vu.lt
[3]E-mail: sasha@aborovik.uzhgorod.ua

Content from this work may be used under the terms of the Creative Commons Attribution 3.0 licence. Any further distribution of this work must maintain attribution to the author(s) and the title of the work, journal citation and DOI.
Published under licence by IOP Publishing Ltd

Spectroscopic classification of the $5p^5n_1l_1n_2l_2n_3l_3$ autoionizing states in Ba atoms

V. Hrytsko[† 1], **A. Kupliauskienė**[‡ 2], **and A. Borovik**[† 3]

[†]Department of Electron Processes, Institute of Electron Physics, Uzhgorod, 88017, Ukraine
[‡]Institute of Theoretical Physics and Astronomy, Vilnius University, LT-01108 Vilnius, Lithuania

Synopsis The ejected-electron spectra of Ba atoms were measured in an electron impact energy range 15-100 eV. By comparative analysis of experimental and calculated data on excitation dynamics and excitation thresholds of lines the complete spectroscopic classification and decay channels for autoionizing states in $5p^55d6s^2$ and $5p^55d^26s$ configurations were determined.

The accurate spectroscopic classification of the $5p^5n_1l_1n_2l_2n_3l_3$ autoionizing states in Ba atoms faces the challenges of their multichannel decay to the $5p^6nl$ ionic states and of overlapping the lowest single- and double excited configurations (see e.g. [1], [2] and references therein). Consequently, the known spectroscopic classification of the atomic autoionizing states possesses substantially preliminary character [3].

In the present work the spectroscopic classification of lines in ejected-electron spectra of Ba atoms was carried out by comparative analysis of their excitation dynamics in a broad impact energy range and of calculated excitation thresholds, cross sections and decay rates of autoionizing states in $5p^55d6s^2$ and $5p^55d^26s$ configurations. The ejected-electron spectra were studied in an impact energy region from the excitation threshold of the $5p^6$ subshell up to 100 eV. The apparatus and measuring procedure were described in detail earlier [4]. The uncertainties of energy scales were estimated to be ±0.07 eV and ±0.05 eV for incident- and ejected electrons, respectively.

The calculations were performed by using singly $5p^66snl$ (n =6,...12; l =0,1,2 and nl =4f,5f) and 5p-core $5p^5nln'l'n''l''$ (nl =6s,5d; $n'l'$ =6s,7s,6p,5d,6d; $n''l''$ =7s,...,10s; 7p,...,10p; 5d,...,10d; 5f) excited configurations to take into account the correlation effects [5].

Of the 59 lines observed in spectra 17 were classified as corresponding to the multichannel electron decay of the $5p^55d6s^2, 5d^26s$ autoionizing states with excitation thresholds between 15.6 and 16.7 eV. The data for thirteen lowest states are presented in table 1.

The largest excitation efficiency possess the states from the $5p^55d^26s$ configuration. Most probable decay channel was transition into the $5p^66s\ ^2S_{1/2}$ Ba^+ state.

This research was funded by the National Academy of Sciences of Ukraine (BA, VH).

Table 1. Experimental (E_{exp}) and calculated (E_{calc}) energies, classification, decay channels and Auger yields (A^a) of the $5p^55d6s^2$, $5d^26s$ SJ states of Ba atoms.

E_{exp}	E_{calc}	State	Decay channel
15.63	15.60	$5d6s^2\ ^3P_0$	$6s^2S_{1/2}$
15.72	15.78	$5d^2(^3F)(^4D)6s^5D_0$	$6s^2S_{1/2}$
15.81	15.79	$5d6s^2\ ^3P_1$	$6s^2S_{1/2}$
			$5d^2D_{3/2}$
15.87	15.82	$5d^2(^3F)(^4D)6s^5D_1$	$6s^2S_{1/2}$
15.93	15.93	$5d^2(^3F)(^4D)6s^5D_3$	$6s^2S_{1/2}$
			$5d^2D_{3/2}$
16.01	15.86	$5d^2(^3F)(^4D)6s^5D_2$	$5d^2D_{3/2}$
16.10	16.09	$5d^2(^3F)(^4D)6s^5D_4$	$6s^2S_{1/2}$
			$5d^2D_{3/2}$
16.16	16.12	$5d6s^2\ ^3P_2$	$6s^2S_{1/2}$
16.25	16.27	$5d6s^2\ ^3F_4$	$6s^2S_{1/2}$
16.32	16.35	$5d6s^2\ ^3F_3$	$6s^2S_{1/2}$
16.42	16.48	$5d6s^2\ ^1D_2$	$6s^2S_{1/2}$
16.51	16.63	$5d^2(^3F)(^4D)6s^3D_1$	$6s^2S_{1/2}$
16.64	16.68	$5d^2(^3P)(^4D)6s^3D_3$	$6s^2S_{1/2}$
			$5d^2D_{3/2}$

References

[1] D. Rassi and K.J. Ross 1980 *J. Phys.* **13** 4683
[2] M.A. Baig *et al* 1984 *J. Phys.* **17** 371
[3] A.A. Borovik *et al* 1985 *O t. S ektrosc.* **58** 988
[4] A.A. Borovik *et al* 2005 *J. Phys.* **38** 1081
[5] M.F. Gu 2008 *Ca . J. Phys.* **86** 675-689

[1]E-mail: shargov11@gmail.com
[2]E-mail: alicija.kupliauskiene@tfai.vu.lt
[3]E-mail: sasha@aborovik.uzhgorod.ua

Content from this work may be used under the terms of the Creative Commons Attribution 3.0 licence. Any further distribution of this work must maintain attribution to the author(s) and the title of the work, journal citation and DOI.
Published under licence by IOP Publishing Ltd

Journal of Physics: Conference Series **635** (2015) 052009 doi:10.1088/1742-6596/635/5/052009

Positron and electron impact multiple-ionization of rare gases

C. C. Montanari* [1], **J. E. Miraglia*** [2]

* Instituto de Astronomía y Física del Espacio, Consejo Nacional de Investigaciones Científicas y Técnicas, Universidad de Buenos Aires, Ciudad Universitaria, C1428EGA, Buenos Aires, Argentina

Synopsis We present a theoretical treatment of the multiple-ionization by light projectiles, which is valid in an extended energy range (from near threshold to the keV region). It covers the direct ionization and the post-collisional electron emission, and takes into account charge and mass effects: projectile trajectory, finite momentum transferred, energy threshold. Our results for positron and electron-impact on Ne, Ar, Kr and Xe, are compared with a detailed compilation of experimental data of multiple and of total ionization cross sections. These values and previous ones for antiproton and proton-impact let us to present an extended picture of the particle-antiparticle multiple-ionization.

Electron and positron impact ionization is a very active field [1-3]. In this work we aim to describe the multiple ionization (MI) by light projectiles in an extended energy range by employing the continuum distorted-wave eikonal initial state approximation [4]. As this model is valid for heavy projectiles, we introduced corrections to account for the finite momentum transferred, the non-linear trajectory and the minimum energy [5]. The thresholds, calculated considering the mean kinetic energy of the outgoing electrons, describe reasonably well the experimental appearance energies from single to sextuple ionization [6]. Within the independent particle model, the MI is a multinomial combination of independent ionization probabilities of the different subshells [7-8]. We introduce the energy thresholds and the post-collisional branching ratios in the multinomial expression.

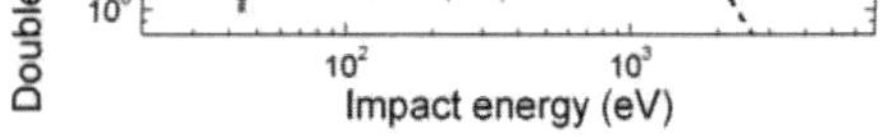

Figure 1. Double ionization cross section of Xe by impact of positron (thick-red line) and electron (thin black line). The dashed line indicates the direct ionization.

In figure 1 we display positron and electron results for the double-ionization of Xe, together with the experimental data available. The description of the experimental energy threshold is good. Note that for Xe, above 0.2 keV the Auger-type post-collisional ionization (PCI) is important. In figure 2 we display the total (or gross) ionization cross section of Xe by positron and electron-impact. Our positron results are above the electron ones around the maximum, in agreement with the data [9-10].

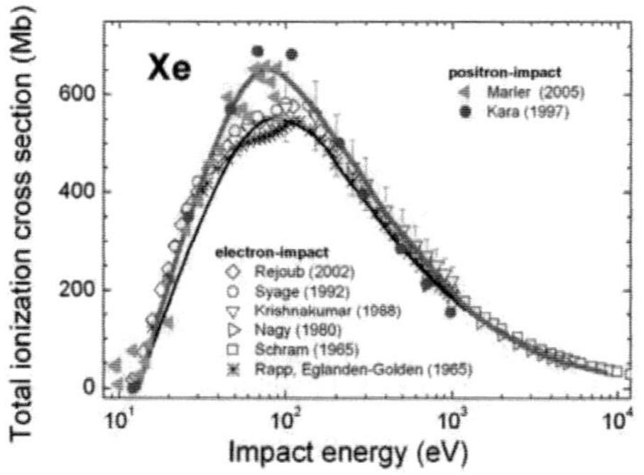

Figure 2. Total ionization cross sections by positron and electron impact. Notation, as in figure 1.

References

[1] Shevelko V and Tawara H 1998, *Atomic Multielectron Processes* (Berlin: Springer-Verlag)
[2] Laricchia G *et al* 2013, *Experimental Aspects of Ionization Studies by Positron and Positronium Impact* (Cambridge University Press)
[3] Chiari L and Zecca A 2014, Eur. Phys. J. D **68**, 297
[4] Fainstein P D *et al* 1988, J. Phys. B **21** 287
[5] Montanari C C *et al* 2014l, J. Phys. B **47** 105203
[6] Montanari C C and Miraglia J E 2015, Proceedings HCI 2014, J. Phys. Conf. Series, in press
[7] Cavalcanti E G *et al* 2002, J. Phys. B **35** 3937
[8] Spranger T and Kirchner T 2004, J. Phys. B **37** 4159
[9] Kara V *et al* 1997, J. Phys. B **30** 3933
[10] Marler J P *et al* 2005, Phys. Rev. A 71, 022701

[1]E-mail: mclaudia@iafe.uba.ar
[2]E-mail: miraglia@iafe.uba.ar

Content from this work may be used under the terms of the Creative Commons Attribution 3.0 licence. Any further distribution of this work must maintain attribution to the author(s) and the title of the work, journal citation and DOI.
Published under licence by IOP Publishing Ltd

Influence of Multiple Ionization on Charge State Distributions

D. W. Savin*[1] and M. Hahn*

* Astrophysics Laboratory, Columbia University, New York, NY 10027, USA

Synopsis We describe the influence of electron-impact multiple ionization (EIMI) on the ionization balance of collisionally ionized plasmas. Previous ionization balance calculations have largely neglected EIMI. Here, EIMI cross-section data are incorporated into calculations of both equilibrium and non-equilibrium charge-state distributions (CSDs). For equilibrium CSDs, we find that EIMI has only a small effect. However, for non-equilibrium plasmas the influence of EIMI can be important. In particular, we find that for plasmas in which the temperature oscillates there are significant differences in the CSD when including versus neglecting EIMI. These results have implications for modeling and spectroscopy of impulsively heated plasmas, such as nanoflare heating of the solar corona.

The spectrum emitted by a plasma depends on the CSD of the gas. For collisionally ionized plasmas, the CSD is determined by the corresponding rates for electron-impact ionization and recombination. In astrophysics, such plasmas are formed in stars, supernova remnants, galaxies, and galaxy clusters. Current CSD calculations generally do not account for EIMI, a process in which multiple electrons are ejected by a single electron-ion collision. We have estimated the EIMI cross sections for all charge states of iron using a combination of the available experimental data and semi-empirical formulae. We then modeled the CSD and observed the influence of EIMI compared to only including single ionization (cf., Figure 1).

One case of interest for astrophysics is nanoflare heating, which is a leading theory to explain the heating of the solar corona. In order to determine whether this theory can indeed explain coronal heating, spectroscopic measurements are being compared to modeled nanoflare spectra. Such models have attempted to predict the spectra of impulsively heated plasmas in which the CSD is time dependent. These nonequilbirium ionization calculations have so far ignored EIMI, but our findings suggest that EIMI can have a significant effect on the CSD of a nanoflare-heated plasma, changing the ion abundances by up to about 50%. A detailed discussion of this work has been published in [1].

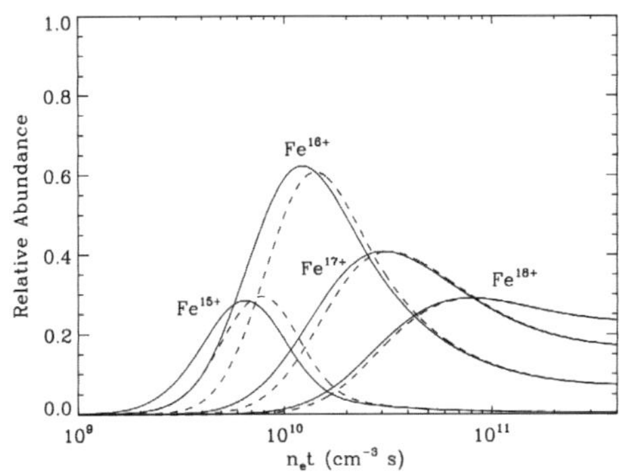

Figure 1. Ion abundances of Fe^{15+} to Fe^{18+} vs. scale $n_e t$, where n_e is electron density and t is time, following a sudden jump in temperature from 10^5 K to 10^7 K. The solid curves indicate the results with EIMI and the dashed curves are the calculations including only single ionization. It is clear that EIMI causes the charge states to evolve faster than if it is ignored.

References

[1] Hahn & Savin 2015, *Astrophys. J.* **800** 68

[1]E-mail: savin@astro.columbia.edu

Content from this work may be used under the terms of the Creative Commons Attribution 3.0 licence. Any further distribution of this work must maintain attribution to the author(s) and the title of the work, journal citation and DOI.
Published under licence by IOP Publishing Ltd

B-spline R-matrix with pseudo-states calculations for electron-impact excitation and ionization of beryllium

Oleg Zatsarinny [1] and Klaus Bartschat [2]

Department of Physics and Astronomy, Drake University, Des Moines, Iowa 50311, USA

Synopsis The B-spline R-matrix-with-pseudostates (BSR) method is employed to treat electron collisions with beryllium atoms. Predictions for elastic scattering, excitation, and ionization are presented for all transitions between the lowest 19 states of beryllium in the energy range from threshold to 150 eV. The sensitivity of the results is checked by comparing data obtained in different approximations with increasing number of pseudo-states.

Beryllium is used as a surface material in the JET project and for the plasma-facing walls in ITER. This calls for accurate electron–beryllium scattering results. Due to the lack of experimental data, researchers in plasma modelling currently have to rely entirely on theoretical predictions. For this reason, it is important to estimate the accuracy of the available theoretical datasets.

Extensive sets of theoretical calculations are indeed available. These data were obtained with advanced methods, such as R-matrix with pseudo-states (RMPS) [1,2], convergent close-coupling (CCC) [3], and time-dependent close-coupling (TDCC) [4]. All these calculations indicate a slow convergence of the close-coupling expansion and large effects from coupling to the target continuum.

In the present work, the B-spline R-matrix method [5] was used to study electron collisions with neutral beryllium over an energy range from threshold to 150 eV. The convergence of the results was checked by comparing data from different models with increasing number of target states, ranging from the distorted-wave approximation to the close-coupling method with 650 continuum pseudo-states. For most excitation cross sections we found very close agreement with the existing results [1-4]. This suggests an accuracy of those datasets within a few percent.

Two new points were revealed in the present calculations. For the first time we obtained converged results for the strong $2s^22p$ shape resonance at low energies (see Fig. 1). The position, 0.28 eV, and width, 0.37 eV, of this resonance differ considerably from numerous results obtained with methods based on model potentials [6] or complex rotation [7] .

Furthermore, while we obtained close (10%) agreement with RMPS and TDCC ionization cross sections [4] from the ground state, the predictions for ionizing the metastable $(2s2p)^3P$ state differ by a factor of 2 (c.f. Fig. 2). We also found large differences between ionization of the $(2s2p)^3P$ and ^1P states, due to a strong term-dependence of the valence orbitals.

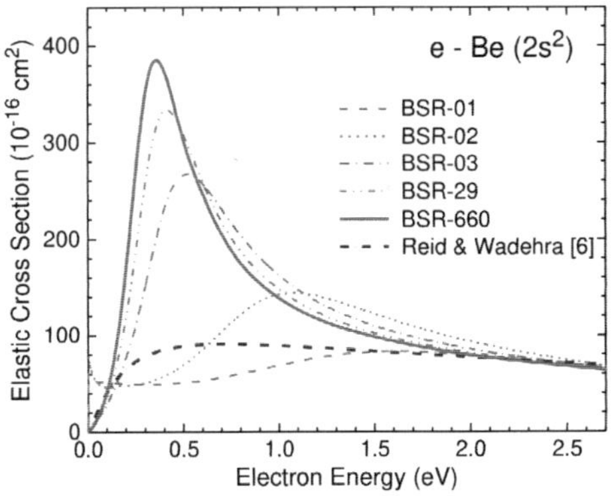

Figure 1. Model dependence of the cross sections obtained for elastic electron scattering from beryllium atoms in their ground state.

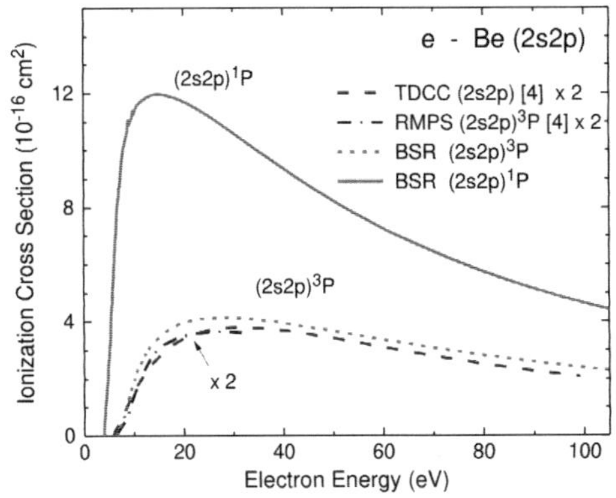

Figure 2. Electron-impact ionization cross sections for neutral Be from the excited 2s2p configuration.

This work was supported by the US National Science Foundation under grants PHY-1212450 and PHY-1430245, and the XSEDE allocation PHY-090031.

References

[1] K. Bartschat *et al* 1997 *J. Phys. B* **30** 5915
[2] C. Balance *et al* 2003 *Phys. Rev. A* **68** 062705
[3] D. Fursa and I. Bray 1997 *J. Phys. B* **30** 5915
[4] J. Colgan *et al* 2003 *Phys. Rev. A* **68** 032712
[5] O. Zatsarinny 2006 *Comp. Phys. Commun.* **174** 273
[6] D. Reid and J. Wadehra 2014 *J. Phys. B* **47** 225211
[7] T. Tsednee *et al* 2015 *Phys. Rev. A* **91** 022506

[1] E-mail: oleg.zatsarinny@drake.edu
[2] E-mail: klaus.bartschat@drake.edu

Content from this work may be used under the terms of the Creative Commons Attribution 3.0 licence. Any further distribution of this work must maintain attribution to the author(s) and the title of the work, journal citation and DOI.
Published under licence by IOP Publishing Ltd

Electron scattering from aluminum: B-spline R-matrix calculations

Oleg Zatsarinny*[1], Klaus Bartschat*,
Elizabeth Nagy†, Sergey Gedeon†, Viktor Gedeon†, and Vladimir Lazur†

* Department of Physics and Astronomy, Drake University, Des Moines, Iowa 50311, USA
† Department of Theoretical Physics, Uzhgorod State University, Uzhgorod, 88000, Ukraine

Synopsis A systematic study of angle-integrated cross sections for electron scattering from neutral aluminum is presented. The calculations cover ionization, elastic scattering, and excitation of the 14 states $(3s^2np)^2P^o(n=3, 4, 5, 6)$, $(3s^2ns)^2S$ $(n=4, 5, 6)$, $(3s^2nd)^2D$ $(n=3, 4)$, $(3s3p^2)^{4,2}P$, ^2D, ^2S, and $(3s^24f)^2F^o$ of aluminum. The sensitivity of the results is checked by comparing data obtained in different approximations, including a large-scale model with over 500 continuum pseudo-states.

Electron collisions with aluminum atoms have not been investigated thoroughly. The only cross-section measurements reported to date are for the 4s-3p and 3d-3p spectral transitions [1] and for the ionization process [2]. Theoretically, low-energy elastic cross sections have been considered by Ryabikh and Fabricant [3] in a two-state close-coupling approximation, whereas ionization has been thoroughly treated in [4] within advanced R-matrix with pseudo-states (RMPS) and time-dependent close-coupling (TDCC) methods.

In the present work, the B-spline R-matrix method [5] was used to study electron collisions with neutral aluminum over an energy range from threshold to 100 eV. The multi-configuration Hartree-Fock method with non-orthogonal orbitals in connection with B-spline expansions was employed for an accurate representation of the target wave functions. The convergence of the results was checked by comparing data from different models with increasing number of target states, ranging from the distorted-wave approximation to a close-coupling method with 550 continuum pseudo-states. Coupling to the target continuum was found to strongly affect (up to a factor of two) transitions from the ground state, and to a lesser extent the transitions between excited states.

The current predictions represent an extensive set of electron scattering data for neutral aluminum needed for plasma modeling. As seen in Fig. 1, our ionization cross sections are in excellent agreement with the measurements reported in [2] over a wide energy range.

The elastic cross section (c.f. Fig. 2), exhibits a complicated energy dependence due to different partial-wave and resonance contributions. Details will be presented at the conference. In particular, the shoulder at 0.34 eV is due to the $(3s^23p^2)^1S$ resonance, whereas the sharp peak at 3.7 eV is due to the $(3s3p^3)^3D^o$ resonance. Also shown in Figure 2 are the relative contributions from elastic scattering, excitation, and ionization processes to the total cross section.

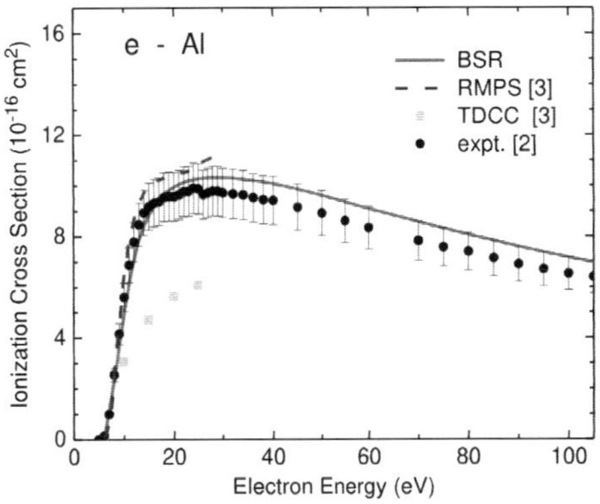

Figure 1. Cross sections for electron-impact ionization of atomic aluminum in its ground state.

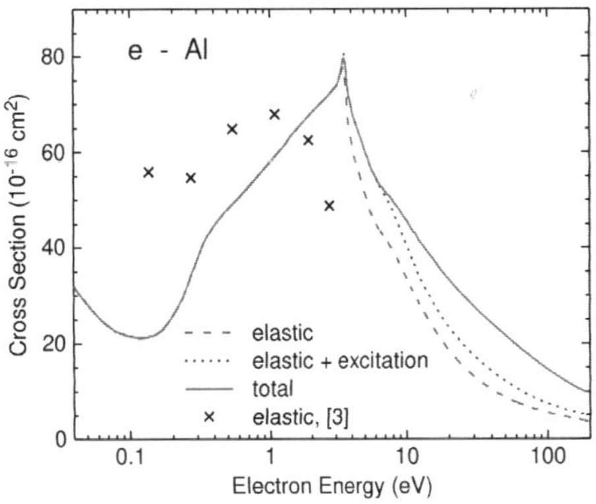

Figure 2. Elastic, elastic + excitation, and grand total cross sections for electron collisions with atomic aluminum in its ground state.

This work was supported, in part, by the US National Science Foundation under grants PHY-1212450 and PHY-1430245, and the XSEDE allocation PHY-090031.

References

[1] L. L. Shimon *et al* 1975 *Ukr. J. Phys.* **20** 229
[2] R. S. Freund *et al* 1990 *Phys. Rev. A* **41** 3575
[3] Ya. Ryabikh and I. Fabricant 1981 *J. Phys. B* **14** 349
[4] S. D. Loch *et al* 2012 *J. Phys. B* **45** 065201
[5] O. Zatsarinny 2006 *Comp. Phys. Commun.* **174** 273

[1] E-mail: oleg.zatsarinny@drake.edu

Content from this work may be used under the terms of the Creative Commons Attribution 3.0 licence. Any further distribution of this work must maintain attribution to the author(s) and the title of the work, journal citation and DOI.
Published under licence by IOP Publishing Ltd

XXIX International Conference on Photonic, Electronic, and Atomic Collisions (ICPEAC2015) IOP Publishing
Journal of Physics: Conference Series **635** (2015) 052013 doi:10.1088/1742-6596/635/5/052013

Atomic Zero Energy Resonance in Plasmas

C. F. Clauser [1] **and R. O. Barrachina** [2]

Centro Atómico Bariloche and Instituto Balseiro, Comisión Nacional de Energía Atómica and Universidad Nacional de Cuyo, Av. Bustillo 9500, 8400 Bariloche, Argentina.

Synopsis We analyze the emergence of zero-energy resonance effects in atomic processes occurring inside a plasma throughout a wide range of temperatures and densities. We demonstrate that in the case of dilute plasmas, like those used in magnetic confinement fusion, resonances tend to be blurred by plasma inhomogeneities, and the standard results corresponding to a pure Coulomb potential are recovered as an average effect. On the other hand, high density plasmas like those used in inertial confinement might be tuned to zero-energy resonances, leading to cross sections many times larger, at the threshold energy, than standard estimates.

A zero-energy resonance produces a cross section that at low energies might be many times larger than any reasonable estimate. This surprising effect, which was discovered in the late 1930s for the low-energy scattering of neutrons by ortho-hydrogen [1], might also occur in atomic processes near the threshold of opening reaction channels [2]. Actually, we recently uncovered their presence within a plasma for particular configurations of temperature and density [3]. Even though their effect on the reaction frequencies were mistakenly overestimated, they still might produce a sizable effect in the cross sections.

Let us consider an atomic transition at a threshold where the relative momentum k of a pair of charged partners, either in the initial or final configuration, vanishes. This is the case, for instance in photoionization or radiative recombination processes. According to the final-state interaction theory, the reaction cross section is affected by an enhancement factor $\mathcal{F}(k)$ which, for small values of k, is proportional to $\mathcal{F} \propto a^2/(1 + a^2k^2)$, where a is the s-wave scattering length for the corresponding two-body potential [2, 4]. The screening produced by the surrounding plasma leads to a dependence of a on the corresponding density ρ and temperature T. In order to illustrate this effect, in figure 1 we show the scattering length a as a function of ρ and T for an electron - proton interaction within a plasma (details of the calculation are provided in [3]). At each zero-energy resonance, the scattering length diverges, and so does the enhancement factor $\mathcal{F}$ at the $k = 0$ threshold.

Now, in a dilute plasma, even a small variation of ρ and T might encompass many zero-energy resonant conditions. Thus we evaluate how the cross section of a given atomic process is affected by these inhomogeneities. We demonstrate that in this case the averaged cross section tends to mimic that of a pure Coulomb potential [4], as if the Debye screening were producing no effect on the atomic process.

On the other hand, for the case of high density plasmas, the variations of ρ and T can be shown to be small in comparison with the separation between resonant conditions. This means that zero-resonances might occur within the plasma and should be taken into account.

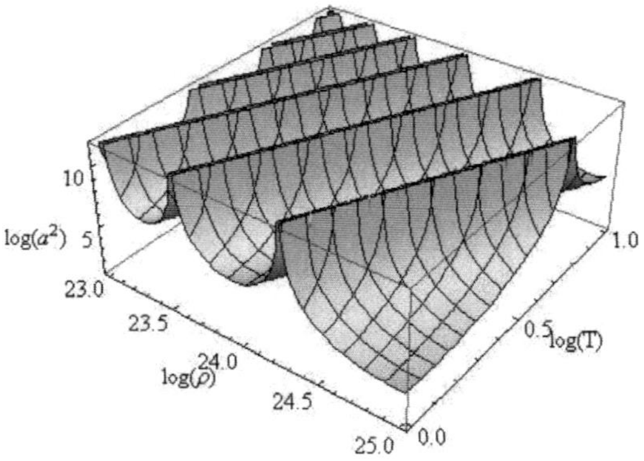

Figure 1. s-wave scattering length a [au], for a $e^- + H^+$ system inside a plasma, as a function of the corresponding density ρ [cm^{-3}] and temperature T [keV].

References

[1] J. Halpern *et al* 1937 *Phys. Rev.* **52** 142

[2] R. O. Barrachina 2005 *Nucl. Instr. and Meth. B* **233** 19

[3] C. F. Clauser and R. O. Barrachina 2015 *J. Phys. Conf. Ser.* **583** 012037

[4] C. R. Garibotti and R. O. Barrachina 1983 *Phys. Rev.* A **28** 2792

[1] E-mail: cesar.clauser@ib.edu.ar
[2] E-mail: barra@cab.cnea.gov.ar

Content from this work may be used under the terms of the Creative Commons Attribution 3.0 licence. Any further distribution of this work must maintain attribution to the author(s) and the title of the work, journal citation and DOI.
Published under licence by IOP Publishing Ltd

Journal of Physics: Conference Series **635** (2015) 052014 doi:10.1088/1742-6596/635/5/052014

Vortex lines and surfaces in positron-atom ionization collisions

F. Navarrete[*1] **and R. O. Barrachina**[*2]

Centro Atómico Bariloche and Instituto Balseiro, Comisión Nacional de Energía Atómica and Universidad Nacional de Cuyo, Av. Bustillo 9500, 8400 Bariloche, Argentina.

Synopsis We study the appearance of quantum vortices in the ionization of atoms by positron impact. We avoid the standard restrictions on the final-state variables, in order to study vortices as submanifolds of codimension 2 in the multidimensional space of definition of the T-matrix element. In particular we demonstrate that three isolated vortices previously observed under a collinear geometry, actually correspond to a two-dimensional cut on a single vortex line.

In the fall of 1926, some few months after the publication of Schrödinger's seminal papers on wave mechanics, Erwin Madelung proposed an alternative but mathematically equivalent formulation where "Schrödinger equation for one-electron problems can be transformed into the form of hydrodynamical equations" [1]. The corresponding velocity field is shown to be irrotational everywhere, except at vortices where the density vanishes. These quantum vortices can appear during the evolution of an atomic scattering process, and even survive up to macroscopic distances so as to leave their imprints on the corresponding transition matrix element T in the form of isolated zeros. In recent years, certain sharp minima previously found in the electron momentum distribution for the ionization of atoms by the impact of electrons and positrons were demonstrated to be quantum vortices in the velocity field associated with T. They have also been observed in the photoionization of atoms, and in ionization collisions by the impact of protons and antiprotons (see [2, 3] and references therein).

Usually, quantum vortices in ionization processes are studied as isolated zeroes on two-dimensionally constrained regions of the phase space of the multidimensional T-matrix element. These reductions are customarily achieved through restrictive geometries, as for instance the collinear arrangement, where the emitted electron and the projectile move along the same direction in the final state. However, vortices are submanifolds of codimension 2, and restrictive 2D geometries would provide only a glimpse of a much more complex structure. For instance, when tracked out of a collinear arrangement, a pair of vortices of opposite circulation previously observed in the ionization of Hydrogen (H) by positron impact [4], and a further vortex [2], which in a collinear geometry appears totally unrelated to them, turn out to correspond to a planar cut of a single vortex line, as shown in figure 1. Recently, a segment of a vortex line was also observed in a Coulomb-Born calculation of T for the electron-impact K-shell ionization of carbon [3]. Finally, in an effort to understand the topology of these quantum structures, we trace the vortices in the full multidimensional space of T for positron-atom ionization collisions, obtaining the first look of vortex surfaces, free of any restrictive geometry.

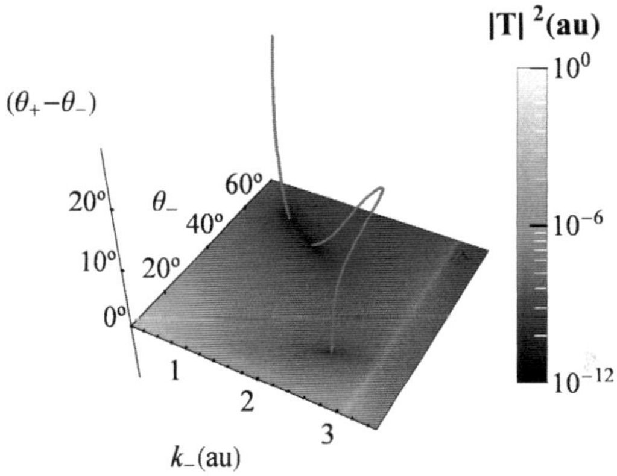

Figure 1. Vortex line (in red) of T for the 275eV ionization of H by positron impact. k_- and θ_- are the momentum and emision angle of the electron, and θ_+ is the emission angle of the positron. The density plot shows $|T|^2$ in the collinear geometry ($\theta_- = \theta_+$).

References

[1] E. Madelung 1926 *Naturwissenschaften* **14** 1004; 1927 *Z. Phys.* **40** 322

[2] F. Navarrete *et al* 2013 *J. Phys. B* **46** 115203

[3] S. J. Ward and J. H. Macek 2014 *Phys. Rev. A* **90** 062709

[4] F. Navarrete and R. O. Barrachina 2015 *J. Phys. B* **48** 055201

[1]E-mail: navarrete@cab.cnea.gov.ar
[2]E-mail: barra@cab.cnea.gov.ar

Content from this work may be used under the terms of the Creative Commons Attribution 3.0 licence. Any further distribution of this work must maintain attribution to the author(s) and the title of the work, journal citation and DOI.
Published under licence by IOP Publishing Ltd

Journal of Physics: Conference Series **635** (2015) 052015 doi:10.1088/1742-6596/635/5/052015

Critical nuclear charge and shape resonances for the two-electron Hamiltonian

Z.-C. Yan* [1], Y. K. Ho† [2]

* University of New Brunswick, Canada, and Wuhan Institute of Physics and Mathematics, China
† Institute of Atomic and Molecular Sciences, Academia Sinica, Taipei, Taiwan, Republic of China

Synopsis We report an improved calculation for the energy of the $2p^2\ ^3P^e$ state of H^- and a new determination of the critical nuclear charge Z_{cr} for the $2p^2\ ^3P^e$ state of the two-electron Hamiltonian, using the variational method in Hylleraas coordinates. The achieved precision for the energy and the critical nuclear charge is accurate, respectively, to 29 and 15 significant digits. We further investigate the resonance regime of $Z < Z_{cr}$ using the method of complex scaling.

For H^-, the energy of its ground state $1s^2\ ^1S^e$ has been accurately determined (see [1, 2, 3] and references therein). The critical value of Z, denoted by Z_{cr}, below which the two-electron ion becomes unbound, has also been determined by several authors with the latest entry in the literature giving 16-digit accuracy [3]. For the region when $Z < Z_{cr}$, the situation becomes more complicate, as the bound state now crosses the ionization threshold and turns into a shape resonance [4 - 6]. As a shape resonance lies in the scattering continuum, we cannot use the usual Rayleigh-Ritz bound principle to optimize its energy. In Refs. [6, 7] the complex resonance energies were determined using the method of complex scaling (see [8] and references therein). For the metastable $2p^2\ ^3P^e$ state in H^-, the critical value of Z has also been calculated but is only accurate to 3 digits: $Z_{cr}\approx 0.994$, see [9]. Furthermore, the resonance region where $Z < Z_{cr}$ has not been studied before, to the best of our knowledge. In this conference we will report an improved calculation for the bound state energy of $2p^2\ ^3P^e$ of H^- and an improved determination of Z_{cr} for the same configuration. The achieved precision for the energy and the critical nuclear charge is accurate to 29 and 15 significant digits, respectively.

The method of complex scaling is used to determine the complex energies of the resonance poles when the metastable bound state turns into a shape resonance when $Z < Z_{cr}$. We use basis set in Hylleraas coordinates to describe the correlations between the two electrons. For the shape resonances we also adopt CI-type bases to check consistence for the resonance energies and widths. Figures 1 and 2 show, respectively, the resonance energy and half-width as a function of $1/Z$.

ZCY is supported by NSERC of Canada and by CAS/SAFEA of China, and YKH by MOST of Taiwan, ROC.

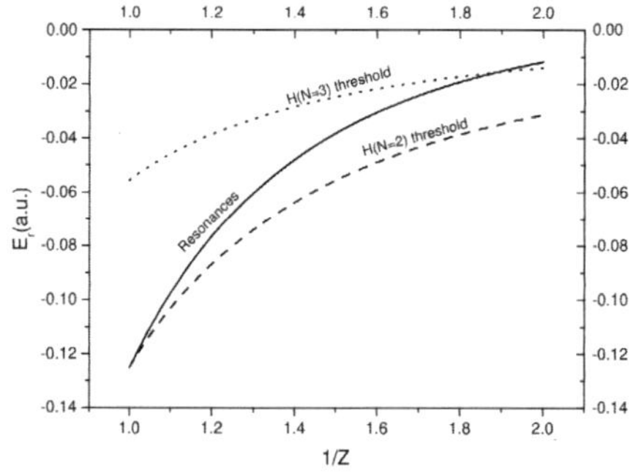

Figure 1. Resonance energy E vs $1/Z$.

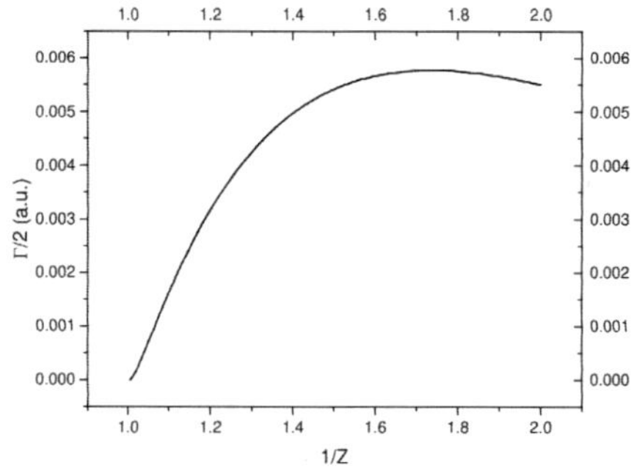

Figure 2. Resonance half-width $\Gamma/2$ vs $1/Z$.

References

[1] J. D. Baker *et al* 1990 *Phys. Rev. A* **41** 1247.
[2] I. A. Ivanov 1995 *Phys. Rev. A* **52** 1942.
[3] C. S. Estienne *et al* 2014 *Phys. Rev. Lett.* **112** 173001.
[4] F. H. Stillinger 1966 *J. Chem. Phys.* **45** 3623.
[5] W. P. Reinhardt 1977 *Phys. Rev.* A **15** 802.
[6] J. Dubau and I. A. Ivanov 1998 *J. Phys. B* **31** 3335.
[7] A. V. Sergeev and S. Kais 2001 *Int. J. Quantum Chem.* **82** 255.
[8] Y. K. Ho 1983 *Phys. Rep.* **99** 1.
[9] P. Serra *et al* 2001 *Phys. Rev. A* **64** 06250.

[1] E-mail: zyan@unb.ca
[2] E-mail: ykho@pub.iams.sinica.edu.tw

Content from this work may be used under the terms of the Creative Commons Attribution 3.0 licence. Any further distribution of this work must maintain attribution to the author(s) and the title of the work, journal citation and DOI.
Published under licence by IOP Publishing Ltd

XXIX International Conference on Photonic, Electronic, and Atomic Collisions (ICPEAC2015) IOP Publishing
Journal of Physics: Conference Series **635** (2015) 052016 doi:10.1088/1742-6596/635/5/052016

Hyperfine-changing transitions in ^{3}He II and other one-electron ions by electron scattering

Klaus Bartschat$^{*,\dagger\,1}$ **and H. R. Sadeghpour**$^{\dagger\,2}$

* Department of Physics and Astronomy, Drake University, Des Moines, Iowa, 50311, USA
† ITAMP, Harvard-Smithsonian Center for Astrophysics, Cambridge, MA 02138, USA

Synopsis We present short-range phase shifts and spin-exchange cross sections for elastic electron scattering from selected one-electron ions from He^+ to Fe^{25+}. The results are explained in a simple physical picture based on quantum defect theory.

The hyperfine-changing transition in $^3He^+$ at 3.46 cm (8.665 GHz) is used for precise determination of the ^{3}He/H abundance in the interstellar medium [1]. In models to obtain the ^{3}He/H ratio, non-LTE effects and line broadening due to electron collisions should be included [2]. Spin-exchange effects in highly charged ions are also of interest for many storage ring experiments.

Calculations using the Coulomb-Born and static exchange approximation, both in non-relativistic and relativistic form, were carried out by Augustin *et al* [3]. They focused on the angle-integrated spin-exchange cross section [4]

$$\sigma_{\rm SE} = \frac{\pi}{k^2}\sum_{\ell=0}^{\infty}(2\ell+1)\sin^2(\delta_\ell^t - \delta_\ell^s) \qquad (1)$$

for elastic collisions, where k and ℓ denote the electron's linear and angular momenta while the superscripts on the short-range potential phase shifts, δ_ℓ^t and δ_ℓ^s, refer to triplet (t) and singlet (s) scattering with total electron spin $S = 1$ or $S = 0$, respectively.

In a recent paper [5], we showed how highly accurate results for this problem can be obtained using Rydberg energy levels of neutral helium or those of the respective two-electron ions. The essential idea is to write the energy levels as

$$E_{nl} = -\frac{1}{2}\frac{Z^2}{n_\ell^*}, \qquad (2)$$

with an effective quantum number $n_\ell^* = n - \bar{\mu}_\ell$, and obtain the phaseshift from the relation

$$\lim_{E\to 0}\delta_\ell(E) = \pi\bar{\mu}_\ell. \qquad (3)$$

Figure 1 exhibits some examples. Spin-exchange cross sections for H and $^3He^+$ are shown in Fig. 2. Note the entirely different energy dependence, particularly at low energies.

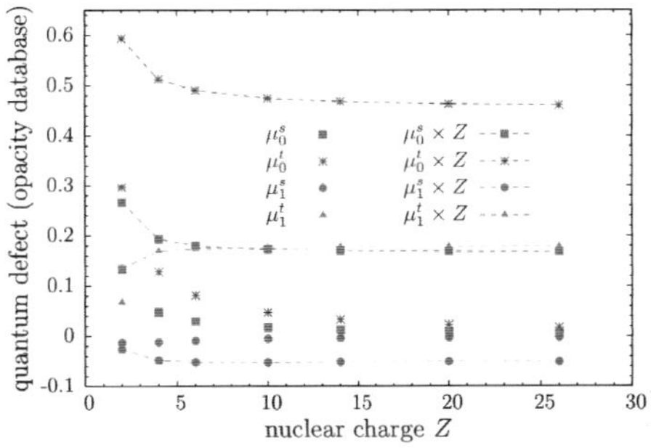

Figure 1. Quantum defects for two-electron systems as a function of the nuclear charge Z, based on data from the Opacity Project [6]. As expected, the quantum defects scale as $1/Z$.

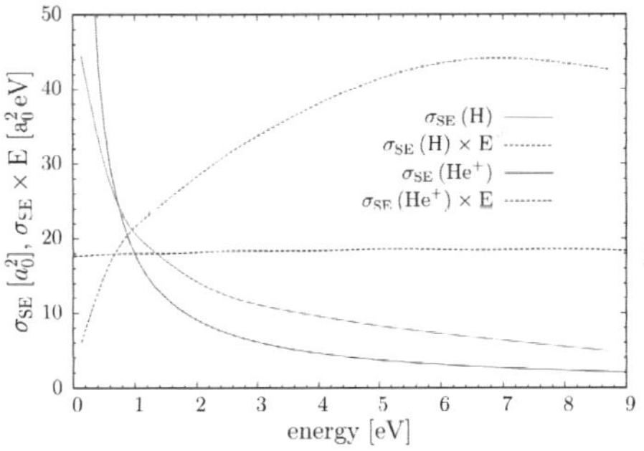

Figure 2. Spin-exchange cross sections for He^+ and atomic hydrogen [5].

This work was supported by the United States National Science Foundation through grants PHY-1068140, PHY-1430245, and ITAMP.

References

[1] T. M. Bania 1997 *ApJ Suppl. Ser.* **113** 353
[2] D. S. Balser *et al* 1995 *AJ* **100** 371
[3] J. Augustin *et al* 1989 *Z. Phys. D* **14** 312
[4] A. Dalgarno and M. R. H. Rudge 1965 *Proc. Roy. Soc. London A* **286** 519
[5] K. Bartschat and H. R. Sadeghpour 2014 *AJ* **788** 69
[6] N. Badnell *et al* 2014, The Opacity Project

[1]E-mail: klaus.bartschat@drake.edu
[2]E-mail: hrs@cfa.harvard.edu

Content from this work may be used under the terms of the Creative Commons Attribution 3.0 licence. Any further distribution of this work must maintain attribution to the author(s) and the title of the work, journal citation and DOI.
Published under licence by IOP Publishing Ltd

A study of the turn-up effect in the Electron Momentum Spectroscopy

F Menas [*,†1], C Dal Cappello[§2] and S Houamer[‡]

[*] Laboratoire de Physique et Chimie Quantique, Université Mouloud MAMMERI Tizi-Ouzou, BP 17,15000 Tizi-Ouzou, Algeria
[†]Ecole Nationale Préparatoire aux Etudes d'Ingéniorat, Route Nationale N°5, 16000 Rouiba, Algeria
[§] SRSMC, UMR CNRS 7565, University of Lorraine, BP 70239, 54506 Vandoeuvre–les-Nancy
[‡]Laboratoire de Physique Quantique et Systèmes Dynamiques, Département de physique, Faculté des sciences, Université Sétif1, 19000 Algeria

Synopsis We show that the PWIA model completely fails for any ionization of atomic hydrogen in its initial state or in an excited state. The results of our study are compared with those of the 3C or BBK model.

It is well known, since a long time, that the BBK model or 3C model [1] is able to reproduce all the data in an (e,2e) experiment when the incident energy is higher than 150 eV [2-3]. This model is considered as of infinitive order when treating the interaction between the two outgoing electrons. We decide to apply this 3C model in the case of the ionization of atomic hydrogen from n=1 to n=5 and to compare to the results given by the plane-wave impulse approximation (PWIA) which is commonly used in the Electron Momentum Spectroscopy (EMS) [4].These EMS experiments, which are a particular case of (e,2e) experiments, are actually performed in non-coplanar geometries: the two outgoing electrons have equal energies and are emitted at equal polar angle (45^0) while the initial momentum varies out of the scattering plane. Some authors [5-6] have shown that the PWIA only works for low values of the recoil momentum q. Recently it was proved that the DWIA (Distorted Wave Impulse Approximation) and the DWBA (Distorted Wave Born Approximation) models are necessary to be applied instead of the PWIA [7-9], even for low values of q. We also compare in this study our results to those given by the DWIA and DWBA models for any initial or excited state. For n=1 the 3C model gives a perfect agreement with the experiments of Lohmann et al [10-11] and the experiments of Weigold et al [12]. The agreement with the DWIA is less good, and the agreement with the PWIA is very bad, except for EMS experiments [10]. For the other n states there are no data up to now but we expect that the 3C model could give an excellent agreement too. For n=2 (and l=1) we observe no turn-up effect, the 3C and DWIA models have an amplitude which decreases when q decreases too. It is not the case for n=3 (and l=2): the TDCS goes to *a finite value different from 0* when q goes to 0 for the 3C, DWIA and DWBA models, contrary to the PWIA model. We also present results for the ionization of the 4f and 5g states.

References

[1] M. Brauner *et al* 1989 *J. Phys. B.* **22**,2265
[2] S. Jones and D. H. Madison 2000*Phys. Rev. A***62**,042701
[3] C. Dal Cappello, A. Haddadou, F. Menas, and A. C. Roy 2011*J. Phys. B* **44**,015204
[4] E. Weigold and I. E. McCarthy1976*Adv. At. Mol. Phys.***14**, 127
[5] C. E. Brion1986*Int. J. Quantum Chem.***29**, 1397
[6] S. Mazevet *et al* 1996 *J. Phys. B.***29**, L901
[7] M. J. Brunger *et al* 1994 *J. Phys. B.***27**, L597
[8] B. P. Hollebone *et al* 1995 *Chem. Phys.***196**, 13
[9] X. G. Ren *et al* 2005 *Phys. Rev. Lett.***94**,163201
[10] B. Lohmann and E. Weigold 1981 *Phys. Lett.* 86A, 139
[11] B. Lohmann *et al* 1984 *Phys. Rev.* A30, 758
[12] E. Weigold *et al* 1979 *J. Phys. B*12, 291

,† 1Email: menasferhat@ ummto.dz
§ 2Email:cappello@univ-metz.fr

Content from this work may be used under the terms of the Creative Commons Attribution 3.0 licence. Any further distribution of this work must maintain attribution to the author(s) and the title of the work, journal citation and DOI.
Published under licence by IOP Publishing Ltd

Charge modification of metal atoms: Catalysis of Water to Peroxide

Kelvin Suggs*‡, Filmon Kiros*, Aaron Tesfamichael‡, Zineb Felfli*[2] and Alfred Z. Msezane*[1]
*Center for Theoretical Studies of Physical Systems, Clark Atlanta University, Atlanta, Georgia 30314, USA
‡Department of Chemistry, Clark Atlanta University, Atlanta, Georgia 30314, USA

Synopsis Here we investigate the optimization of the catalytic behavior of Au, Ag, Pd, Rh, and Os atomic systems via charge modification and conclude that anionic atomic systems are optimal and therefore ideal for catalyzing the oxidation of water to peroxide, with anionic Os being the best candidate. This conclusion should also be applicable to the oxidation of methane to methanol without CO_2 emission, with the anionic Os being the best catalyst.

In [1] it was demonstrated that the fundamental mechanism of negative ion catalysis of water to peroxide using atomic Au^- is the formation of the anionic $Au^-(H_2O)_{1,2}$ molecular complex during the transition state, which breaks up the H-O bonds, thereby promot the formation of the H_2O_2 in the presence of O_2. The relatively large electron affinity (EA) of Au contributes to its anionic catalytic effectiveness. Our complex angular momentum (CAM) method [2] has been used to select the above metal atoms through the accurate determination of their large EAs for the investigations. Density Functional Theory (DFT) and dispersion corrected DFT approaches have been employed for the transition state calculations. Geometry optimization of the structural molecular conformation utilized the gradient-corrected Perdew-Burke-Ernzerhof parameterizations [3] of the exchange-correlation as implemented in DMol3 [4]. A tolerance of $1x10^{-3}$ Ha was used with a smearing value of 0.005Ha.

Transition state calculated results, given in Fig. 1 show the catalyzing effects of cationic, neutral, and anionic Ag, Au, Pd, Rh, and Os atoms when synthesizing water to peroxide. Clearly, the Au^+ is the least efficient of the given systems as shown by a barrier of ~1.3eV. Conversely, Os^- is energetically the most favored with a barrier of ~0.16eV as shown in Fig. 1(b).

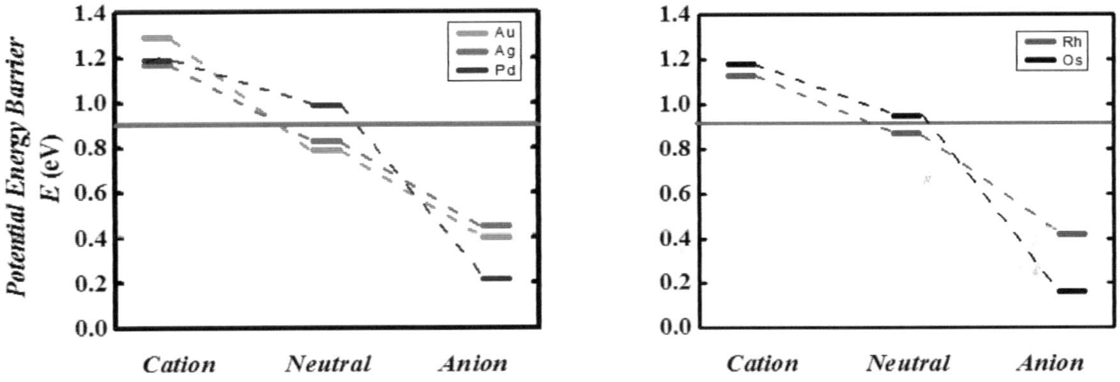

(a) (b)

Figure 1 (a) and (b): Potential energy transition state barrier calculations of cationic, neutral, and anionic Au, Ag, and Pd and Rh, and Os represented by orange, red, and blue and olive and black, respectively. The standard comparative reaction energy without catalysis is represented by the purple line.

The discovery that cationic systems increase the transition energy barrier in the synthesis of peroxide could render them important as inhibitors in controlling and regulating catalysis. These findings usher in a fundamental and comprehensive atomic theoretical framework for the generation of tunable catalytic systems. We conclude that anionic atomic systems are optimal for catalyzing the oxidation of water to peroxide, with Os^- being the preferred candidate among the atomic ions proposed in this work. Whether this is consistent across interdisciplinary systems, reaction types, and atomic cluster substrates is yet to be determined.

[1]E -mail: amsezane@cau.edu
[2]E -mail: zfelfli@cau.edu

The structure and dynamics of these atoms with relatively large EAs will facilitate the understanding of the recently synthesized Ag@Au@Pt nanosystem [5] with greatly enhanced plasmonic properties as well as lead to the creation of tunable supercatalysts.

References

[1] A. Tesfamichael *et al.* 2013 JNR 15 1333.
[2] D. Sokolovski *et al.* 2007 Phys. Rev. A **76** 012705
[3] A. Tkatchenko, and M. Scheffler 2009 Phys. Rev. Lett. **102** 73005
[4] DMol3 2011 Accelrys Software Inc.: San Diego, CA
[5] N. Deogratias *et al.* 2015 *Nano Res.* **8** (1) 271

Content from this work may be used under the terms of the Creative Commons Attribution 3.0 licence. Any further distribution of this work must maintain attribution to the author(s) and the title of the work, journal citation and DOI.
Published under licence by IOP Publishing Ltd

Atomic wavelength in a kappa-distribution plasma

Sabyasachi Kar*[1], Zishi Jiang†, Yongjun Cheng*, and Y. K. Ho#

*Center for Theoretical Atomic and Molecular Physics, Academy of Fundamental and Interdisciplinary Sciences, Harbin Institute of Technology, Harbin 150080, People's Republic of China
† College of Physical Science and Technology, Heilongjiang University, Harbin 150080, People's Republic of China
#Institute Atomic and Molecular Sciences, Academia Sinica, PO Box 23-166, Taipei, Taiwan 106, Republic of China

Synopsis We investigate the plasmas screening effect on the transition wavelength for helium atom using correlated exponential wave functions within the framework of Ritz variational principle. Transition wavelengths are presented as functions of plasma screening parameters.

It is well recognized that charged-particle velocity distributions in space plasmas environments are commonly non-Maxwellian. A family of distribution offering a good fit to measurements in some instances is the kappa distribution. A kappa distribution function is given by [1-6]

$$f_{\kappa}(v) = \left(\frac{m}{2\pi E_{\kappa}}\right)^{3/2} \frac{\Gamma(\kappa+1)}{\Gamma(\kappa-1/2)} \left(1+\frac{mv^2}{2\kappa E_{\kappa}}\right)^{-(\kappa+1)},$$

where m and v are the mass and velocity of the plasma electron, respectively, $E_{\kappa}(\cong(1-\frac{3}{2\kappa})E_T)$ is a characteristic energy in kappa (generalized Lorentzian) plasmas, and $\kappa(>3/2)$ is a spectral index, $E_T = \kappa_B T_e$, κ_B is the Boltzmann constant, T_e is electron temperature. We have incorporated the effective screening potential of the form $\varphi_{eff}(r)$ $=(Ze/r)\exp(-r/\lambda_{\kappa})$, where $\lambda_{\kappa}(=\mu_{\kappa}^{1/2}\lambda_D)$ is the effective screening length, with λ_D is the standard Debye length in Maxwellian plasmas and

$$\mu_{\kappa} = \mu_D\sqrt{\alpha^{\kappa}}, \quad \alpha^{\kappa} = \sqrt{(\kappa-1/2)/(\kappa-3/2)},$$

$\mu_D(\equiv\lambda_D^{-1})$ is Debye screening parameter.

In this work, we investigate the effects of kappa-distribution plasmas (or Lorentzian astrophysical plasmas) on the wavelength for the $2p^2\,{}^3P^e \rightarrow 2p3d\,{}^3D^o$ transition for helium atom using Ritz variational principle. The effect of Debye plasmas on Transition wavelengths of two electron systems are also reported in the literature [7-9]. The present calculations are based upon highly correlated exponential basis expansion [10] of the form : $\{x(\alpha_i r_1)x(\beta_i r_2)x(\gamma_i r_{12})\}$ with $x(\delta r)=\exp(-\delta r)$. The non-linear variational parameters α_i, β_i, and γ_i are generated in a pseudorandom way. The $2p^2\,{}^3P^e \rightarrow 2p3d\,{}^3D^o$ transition wavelength of helium as a function of spectral index κ and screening parameter μ_D is presented in figure 1. Due to the recent advancement in laser plasmas, experimental advancement in the $2p^2\,{}^3P^e \rightarrow 2p3d\,{}^3D^o$ transition wavelength, and importance of doubly excited states in astrophysical plasmas, we hope our findings will provide useful information to the research communities in astrophysics, plasma physics and atomic physics.

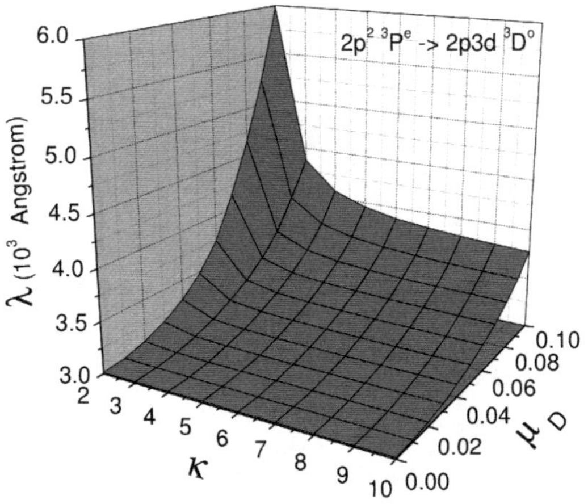

Figure 1. Wavelength for $2p^2\,{}^3P^e \rightarrow 2p3d\,{}^3D^o$ transition as a function of κ and μ_D.

Acknowledgments: ZJ & YC are supported by NNSF of China, and YKH by MOST of Taiwan.

References

[1] D. A. Bryant 1996 *J. Plasma Phys.* **56** 87
[2] N. Rubab, G. Murtaza 2006 *Phys. Scr.* **74** 145
[3] V. Pierrard, M. Lazar 2010 *Sol. Phys.* **267** 153
[4] G. Livadiotis et al 2013 *Space Sci. Rev.* **175** 183
[5] P. Jiang *et al* 2013 *Few-Body Syst.* **54** 1911
[6] P. Jiang *et al* 2013 *ICPEAC Book of Abstracts*
[7] S. Kar, Y.K. Ho 2007 *JQSRT* **107** 315
[8] Z. Jiang *et al* 2012 *Phys. Plasmas* **19** 033301
[9] S. Kar, Z. Jiang 2015 *Atomic Data and Nuclear Data Tables* **102** 42
[10] S. Kar *et al* 2014 *Phys. Plasmas* **21** 012105

[1] E-mail: skar@hit.edu.cn

Content from this work may be used under the terms of the Creative Commons Attribution 3.0 licence. Any further distribution of this work must maintain attribution to the author(s) and the title of the work, journal citation and DOI.
Published under licence by IOP Publishing Ltd

XXIX International Conference on Photonic, Electronic, and Atomic Collisions (ICPEAC2015) IOP Publishing
Journal of Physics: Conference Series **635** (2015) 052020 doi:10.1088/1742-6596/635/5/052020

Uncertainty Quantification of theoretical atomic and molecular collisional data

H. K. Chung[1] **and B. J. Braams**[2]

Atomic and Molecular Data Unit, Nuclear Data Section, International Atomic Energy Agency, A1400, Vienna, Austria

Synopsis Methods are discussed for uncertainty assessment for calculated atomic and molecular collisional data. Similarities and contrasts with Uncertainty Quantification for simulations of complex systems are emphasized.

The Atomic and Molecular (A+M) Data Unit at IAEA aims to provide evaluated and recommended databases in the area of A+M processes (and plasma-material interaction processes too) for nuclear fusion and other plasma applications. While available data are predominantly calculated data, there is not a good tradition of estimating uncertainties in calculated data and there is not a generally accepted formal procedure for evaluating and recommending such data. On the other hand, an editorial standard of Physical Review A requires that certain classes of atomic data must be published with uncertainty information and it specifies that advances in techniques for the evaluation of data and assessment of uncertainties qualify as "new physics" [1]. Therefore uncertainty assessment and evaluation and recommendation of theoretical A+M data is important for A+M database management and is also of independent scientific interest.

The science of uncertainty assessment for theoretical A+M collision data may be viewed from the perspective of Uncertainty Quantification (UQ) for simulation of complex systems [2], but there are important differences. The field of UQ has strongly emphasized complex systems such as the coupled atmosphere-ocean system or large engineering systems for which in many cases the basic equations are not well established, involve poorly known parameters and functional dependencies, include stochastic elements and give rise to chaotic behavior. A mathematical core of UQ, polynomial chaos, is concerned with uncertainty propagation for dynamical systems.

Atoms and small molecules may not be complex systems, but it is computationally hard to solve the many-body quantum mechanics governing the system. For these simple physical systems that are of high computational complexity a new science of uncertainty assessment needs to be developed, or at least a new branch of the developing science of UQ. We discuss possible approaches including the Unified Monte Carlo (UMC) method [3] to estimate uncertainties and their correlation structure, which is intensively used for nuclear scattering data.

The A+M Data Unit is encouraging work to develop guidelines for critically assessing theoretical atomic and molecular structure and collision data, taking into account the process and the quantity of interest as well as specific theoretical methods employed in calculations. A joint ITAMP-IAEA workshop was organized in July 2014 [4] to discuss sources of uncertainty in the physical models and computational methods employed for studying processes of our interest: electron collisions with atoms, ions, and molecules, heavy-particle collisions, and also electronic structure of atoms and molecules as input in a collision calculation. The progress towards guidelines for evaluation of theoretical A+M collision data will be reviewed here.

References

[1] Editorial in Phys. Rev. A 83, 040001 (2011).

[2] National Research Council Committee on Mathematical Foundations of Verification, Validation, and Uncertainty Quantification: "Assessing the reliability of complex models: mathematical and statistical foundations of verification, validation and uncertainty quantification." NAP Press, 2012.
http://www.nap.edu/openbook.php?record_id=13395.

[3] R. Capote and D. L. Smith. "An investigation of the performance of the Unified Monte Carlo method of neutron cross section data evaluation." Nuclear Data Sheets 109 (2008) 2768-2773.

[4] Joint IAEA-ITAMP Technical Meeting on Uncertainty Assessment for Theoretical Atomic and Molecular Scattering Data, ITAMP, Cambridge, Massachusetts, USA, 7-9 July 2014.
https://www-amdis.iaea.org/meetings/ITAMP/.

[1] E-mail: H.Chung@iaea.org

[2] E-mail: B.J.Braams@iaea.org

Content from this work may be used under the terms of the Creative Commons Attribution 3.0 licence. Any further distribution of this work must maintain attribution to the author(s) and the title of the work, journal citation and DOI.
Published under licence by IOP Publishing Ltd

Journal of Physics: Conference Series **635** (2015) 052021 doi:10.1088/1742-6596/635/5/052021

Low-energy ($E_0 = 65$ eV) electron-impact ionization of neon: Internormalized triple-differentical cross sections in 3D kinematics

XueGuang Ren*‡ [1], Sadek Amami§, Oleg Zatsarinny†, Thomas Pflüger*‡, Marvin Weyland*‡, Woon Yong Baek*, Hans Rabus*, Klaus Bartschat†, Don Madison§, Alexander Dorn‡

* Physikalisch-Technische Bundesanstalt, 38116 Braunschweig, Germany
‡ Max-Planck-Institute for Nuclear Physics, 69117 Heidelberg, Germany
§ Physics Department, Missouri University of Science and Technology, Rolla, Missouri 65409, USA
† Department of Physics and Astronomy, Drake University, Des Moines, Iowa 50311, USA

Synopsis We present a combined experimental and theoretical study on the low-energy ($E_0 = 65$ eV) electron-impact ionization of neon. The experimental data are compared to predictions from a hybrid second-order distorted-wave Born plus *R*-matrix approach (DWB2-RM), the distorted-wave Born approximation with inclusion of post-collision interaction (DWBA-PCI), a three-body distorted-wave approach (3DW), and a *B*-spline *R*-matrix (BSR) with pseudostates approach. Excellent agreement is found between experiment and the 3DW and BSR theories. The importance of PCI effects is clearly visible in this low-energy electron-impact ionization process.

The electron-impact ionization dynamics have now been well understood for simple systems such as atomic hydrogen and helium [1, 2, 3]. Recent studies for the ionization of neon by 100 eV electron-impact showed an unprecedented agreement between experiment and BSR predictions [4]. The physical effects of PCI as well as electron exchange and charge-cloud polarization in the projectile-target interaction are expected to become even more pronounced with decreasing projectile energy. Here, we investigate the low-energy ($E_0 = 65$ eV) electron-impact ionization of neon to thoroughly test state-of-the-art theoretical approaches.

The experimental data were measured using a reaction microscope [5, 6], which can cover nearly the entire 4π solid angle for the secondary electron emission. The measured cross sections are internormalized across all different scattering angles (θ_1) and ejected energies (E_2), which provide a thorough test ground for theory. As one example, the (e, 2e) triple-differential cross sections (TDCS) for $\theta_1 = -12.5°$ and $E_2 = 2$ eV are presented in Figure 1 for the scattering plane and the full-perpendicular plane. The experimental data are compared to various theoretical predictions from the 3DW, BSR, DWB2-RM, DWBA-PCI models. Excellent agreement is found between experiment and the 3DW and BSR theories. Significant discrepancies between DWB2-RM and experiment are observed near the projectile forward direction, while the DWBA-PCI model provides a clear improvement over the DWB2-RM calculations in this angular range. This indicates that PCI effects play a very important role in the low-energy ionization process studied here [7]. More results, including three-dimensional (3D) presentations of the TDCS, will be shown at the conference.

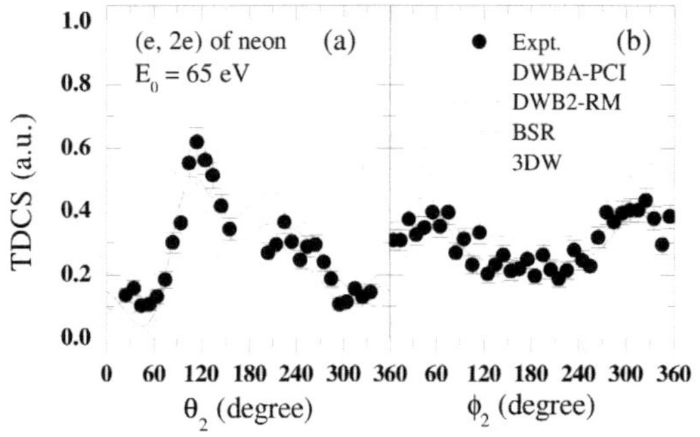

Figure 1. TDCS for the ionization of Ne ($2p$) presented as a function of the secondary electron (e_2) emission angle at $\theta_1 = -12.5°$ and $E_2 = 2$ eV. (a): TDCS in the scattering plane; (b): TDCS in the plane perpendicular to the incident beam direction.

References

[1] T. N. Resigno *et al* 1999 *Science* **286** 2474
[2] X. Ren *et al* 2011 *Phys. Rev. A* **83** 052711
[3] O. Zatsarinny *et al* 2011 *Phys. Rev. Lett.* **107** 023203
[4] T. Pflüger *et al* 2013 *Phys. Rev. Lett.* **110** 153202
[5] J. Ullrich *et al* 2003 *Rep. Prog. Phys.* **66** 1463
[6] X. Ren *et al* 2014 *J. Chem. Phys.* **141** 134314
[7] X. Ren *et al* 2015 *Phys. Rev. A* **91** 032707

[1] E-mail: xue.g.ren@ptb.de

Content from this work may be used under the terms of the Creative Commons Attribution 3.0 licence. Any further distribution of this work must maintain attribution to the author(s) and the title of the work, journal citation and DOI.
Published under licence by IOP Publishing Ltd

XXIX International Conference on Photonic, Electronic, and Atomic Collisions (ICPEAC2015) IOP Publishing
Journal of Physics: Conference Series **635** (2015) 052023 doi:10.1088/1742-6596/635/5/052023

B-spline R-matrix-with-pseudostates calculations for electron-impact excitation and ionization of oxygen

S. S. Tayal[*1] **and O. Zatsarinny**[†2]

[*] Department of Physics, Clark Atlanta University, Atlanta, GA 30314, USA
[†] Department of Physics and Astronomy, Drake University, Des Moines, Iowa 50311, USA

Synopsis A systematic study of angle-integrated cross sections for electron scattering from neutral oxygen is presented. The calculations cover ionization, elastic scattering and excitation of the first 21 states of oxygen with configurations $2s^22p^4$ and $2s^22p^3nl$ (n=3-5, l=0-3). The sensitivity of the results is checked by comparing data obtained in different approximations, including a large-scale model with over 1100 continuum pseudostates.

Low-energy collisions of electrons with oxygen atoms have been investigated thoroughly both experimentally and theoretically. We still, however, cannot claim that all excitation cross sections are known with high accuracy, especially at intermediate energies where coupling to the target continuum is expected to be strong.

In the present work, the B-spline R-matrix method (BSR) [1] is used to study electron collisions with neutral oxygen over an energy range from threshold to 150 eV. The multiconfiguration Hartree-Fock method with nonorthogonal orbitals in connection with B-spline expansions is employed for an accurate representation of the target wave functions. The convergence of the results was checked by comparing data from different models with increasing number of target states, including the extensive calculations with more than 1000 continuum pseudostates. We found that the excitation cross sections for transitions between all terms of the $2s^22p^4$ ground state configuration are well converged and can been established to an accuracy of a few percent, whereas the transitions to the $2s^22p^3nl$ valence states are strongly affected by coupling to the target continuum, with corrections reaching a factor of two for some dipole and quadrupole transitions.

As example, figure 1 represent the experimental and theoretical cross sections for the $2s^22p^4\ ^3P \rightarrow 2p^33s\ ^3S^o$ transition. Index in the BSR calculations indicates the number of target states into close-coupling expansion. We see considerable reduction of cross sections at low and intermediate energies with inclusion of continuum pseudostates, leading to overall better agreement with experimental data.

Figure 2 show our ionization cross-sections which are in excellent agreement with measurements in wide energy range. The present predictions represent an extensive and complete set of electron scattering data for neutral oxygen needed for plasma modeling

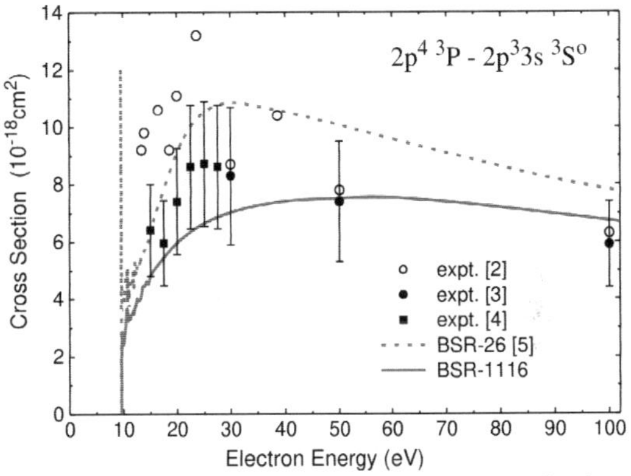

Figure 1. Electron-impact excitation of the $2p^33s\ ^3S^o$ state in oxygen.

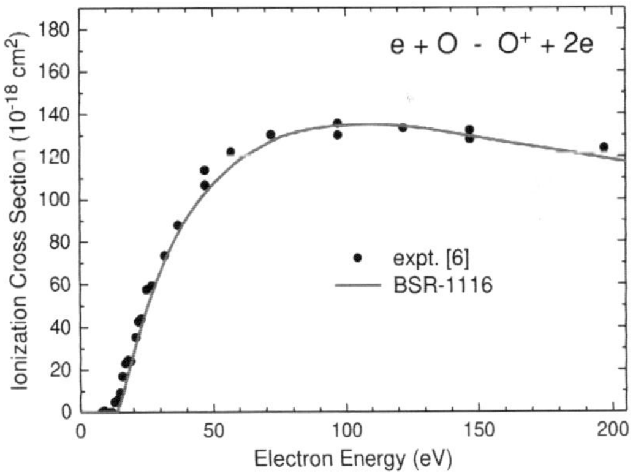

Figure 2. Cross sections for electron-impact ionization of oxygen from the ground state.

This work was supported by NASA under grant NNX11AB62G.

References

[1] O. Zatsarinny 2006 Comp. Phys. Commun. **174** 273
[2] E. E. Gulcicek and J. P. Doering 1988 J. Geophys. Res. **93** 5879
[3] I. Kanik *et al* 2001 J. Phys. B **34** 2647
[4] P. V. Johnson *et al* J. Phys. B **36** 4289
[5] O. Zatsarinny and S. Tayal 2002 J. Phys. B **35** 241
[6] E. Brook *et al* 1978 J. Phys. B **11** 3115

[1] E-mail: stayal@cau.ed
[2] E-mail: oleg.zatsarinny@drake.edu

Content from this work may be used under the terms of the Creative Commons Attribution 3.0 licence. Any further distribution of this work must maintain attribution to the author(s) and the title of the work, journal citation and DOI.
Published under licence by IOP Publishing Ltd

XXIX International Conference on Photonic, Electronic, and Atomic Collisions (ICPEAC2015) IOP Publishing
Journal of Physics: Conference Series **635** (2015) 052024 doi:10.1088/1742-6596/635/5/052024

Electron-impact excitation of Fe IX

S. S. Tayal*[1] and O. Zatsarinny†[2]

* Department of Physics, Clark Atlanta University, Atlanta, GA 30314, USA

† Department of Physics and Astronomy, Drake University, Des Moines, Iowa 50311, USA

Synopsis New extensive calculations are reported for electron collision strengths, rate coefficients and transitions probabilities for a wide range of transitions between the first 370 fine-structure levels of Fe IX in energy region up to $3p^55s$ states. It includes levels of the $3p^6$, $3p^53d$, $3p^54l$ *(l=0-3)*, $3p^55s$, $3s3p^63d$, $3s3p^64s$, $3s3p^64p$, $3p^43d^2$, $3s3p^53d^2$ configurations and some low-lying levels of the $3p^33d^3$ configuration.

Collisions of electrons with iron ions have been extensively investigated due to their importance in the astrophysical simulations, especially with the increased availability of high-resolution satellite observatories. Many strong emission lines have been observed and used for the temperature and density diagnostics of solar corona, but majority of the lines still remains unidentified.

The purpose of the present work is to perform elaborate and extensive calculations for the electron scattering from Fe IX by using highly accurate target wave-functions and by including fine-structure effects in the close-coupling expansions directly. The calculations were performed with the advanced BSR code [1] which employs the R-matrix method in the B-spline basis. To represent the target states, we use term-dependent non-orthogonal orbital sets, which allowed us to generate a more accurate description of the Fe IX target states than those previously employed.

As an example, figure 1 represents collision strengths and thermally averaged collision strengths for the $3s^23p^6\ ^1S_0$ - $3s^23p^54s\ ^1P_1$ transition of astrophysical importance. Good agreement in the background collision strengths is found with the distorted wave (DW) results [2] and the R-matrix (RM) plus Intermediate Coupling Frame Transformation (ICFT) calculations [3]. The resonance contributions to effective collision strength are significant at lower temperatures, as can be seen from the figure.

Overall comparison with the previous DW [2] and RM+ICFT calculations shows good agreement for transitions to the lower excited states belonging to the $3s^23p^53d$ and $3s3p^63d$ configurations. However, only 30% of transitions to the higher excited states are within the 50% tolerance. The worst agreement is found for excitations of the $3s^23p^43d^2$ levels, where our collision strengths are two to three order larger than the results from R-matrix calculations [3].

The present results considerably expand the existing data sets for Fe IX, allowing more detailed treatment of the available measured spectra from different space observatories. The present data can be also used for independent validation of the existing data sets.

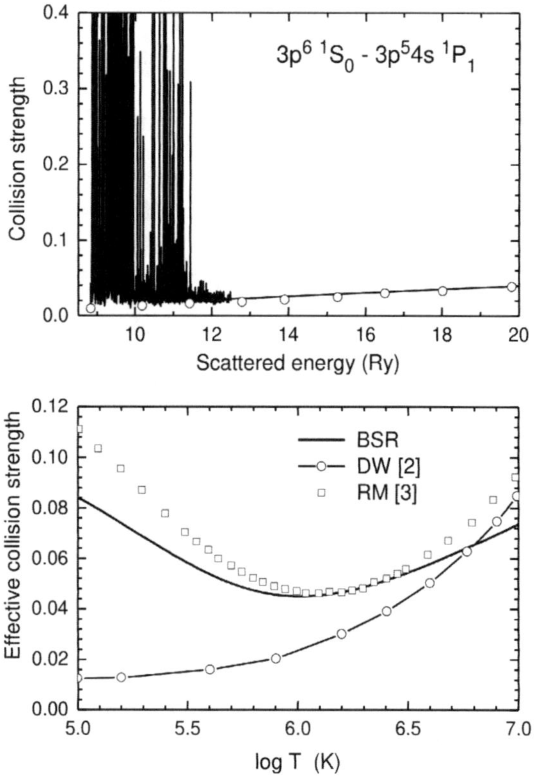

Figure 1. Collision strengths as a function of electron energy and thermally averaged collision strengths as a function of electron temperature for the $3s^23p^6\ ^1S_0$ - $3s^23p^54s\ ^1P_1$ transition in Fe IX.

This work was supported by NASA under grant NNX11AB62G.

References

[1] O. Zatsarinny 2006 Comp. Phys. Commun. **174** 273
[2] B. O'Dwyer *et al* 2012 A&A **537** A22
[3] P. J. Storey *et al* 2002 A&A **394** 753

[1] E-mail: stayal@cau.ed
[2] E-mail: oleg.zatsarinny@drake.edu

Content from this work may be used under the terms of the Creative Commons Attribution 3.0 licence. Any further distribution of this work must maintain attribution to the author(s) and the title of the work, journal citation and DOI.
Published under licence by IOP Publishing Ltd

XXIX International Conference on Photonic, Electronic, and Atomic Collisions (ICPEAC2015) IOP Publishing
Journal of Physics: Conference Series **635** (2015) 052025 doi:10.1088/1742-6596/635/5/052025

Resonant states and collision strengths of low energy electron scattering with hydrogen-like ions

Jun Li*[1], Bang Jiao Ye

* Department of Modern Physics, University of Science and Technology of China, Hefei 230026, Peoples Republic of China

Synopsis We calculate resonant parameters and collision strengths of low energy electron scattering with hydrogen-like ions and find the resonant width of the same resonant state becomes larger with the increasing of the nuclear charge Z.

The resonant states and collision strengths of low energy electron scattering with hydrogen-like ions have been studied with the R-matrix method [1]. Eighteen target states ($n <= 6$, $l <= 3$) have been employed to describe the hydrogen-like ions, which are optimized with the autostructure code [2]. Explicitly, we have calculated the $1s \rightarrow 1s$ elastic collision strengths below the $n = 2$ excitation threshold; the structures and peaks dominating the collision strengths are marked and identified. The resonant parameters are extracted by fitting the eigenphase sum [3]. Fig. 1 shows the elastic collision strengths and the eigenphase sum for different symmetries of Be^{3+} to show how to identify the peaks or resonances. We find the resonant width of the same resonant state becomes larger with the increasing of the nuclear charge Z, bearing in mind that generally the same state becomes more stable by increasing the nuclear charge, or the uncertainty of the lifetime becomes smaller, the change of the resonant width is consistent with the uncertainty principle.

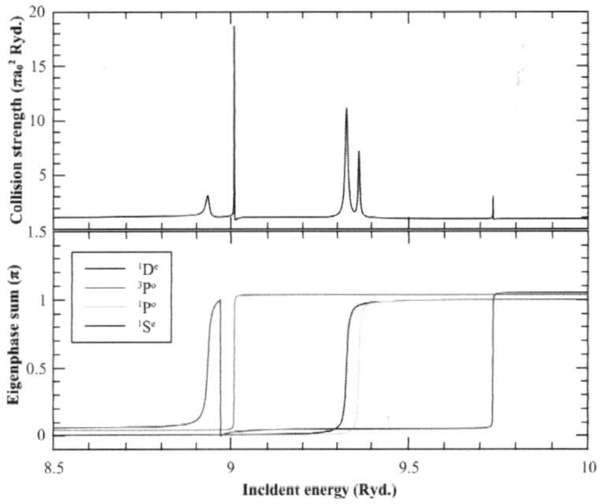

Figure 1. The elastic collision strengths (upper panel) and eigenphase sum (lower panel) of Be^{3+}.

[1]E-mail: lij1990@mail.ustc.edu.cn

References

[1] Keith A. Berrington *at al* 1995 *Comput. Phys. Commun.* **92** 290

[2] N.R. Badnell *at al* 2011 *Comput. Phys. Commun.* **182** 1528

[3] Song Bin Zhang *at al* 2010 *Phys. Rev. A* **81** 032707

Content from this work may be used under the terms of the Creative Commons Attribution 3.0 licence. Any further distribution of this work must maintain attribution to the author(s) and the title of the work, journal citation and DOI.
Published under licence by IOP Publishing Ltd

Exchange-assisted tunneling and enhancement of positron annihilation with inner-shell ellectrons

V. V. Flambaum[*] [1], **M. G. Kozlov** [†‡]

[*] School of Physics, The University of New South Wales, Sydney NSW 2052, Australia
[†]Petersburg Nuclear Physics Institute, Gatchina 188300, Russia
[‡]St. Petersburg Electrotechnical University "LETI", Prof. Popov Str. 5, 197376 St. Petersburg, Russia

Synopsis Exchange interaction and correlations radically change behaviour of of a quantum particle in a classically forbidden region. Exchange produces a power-law decay instead of the usual exponential decrease of the wave function. Exchange-assisted tunneling enhances annihilation of positron with inner-shell electrons and produces other observable effects.

Exchange interaction strongly influences the long-range behavior of localized electron orbitals and quantum tunneling amplitudes. It produces a power-law decay instead of the usual exponential decrease at large distances. For inner orbitals inside molecules decay is $1/r^2$, for macroscopic systems $cos(k_F r)/r^\nu$, where k_F is the Fermi momentum and $\nu = 3$ for one-dimensional, $\nu = 3.5$ for two-dimensional, and $\nu = 4$ for three-dimensional crystals [1]. Correlation corrections do not change these conclusions [1, 2, 3]. Slow decay increases the exchange interaction between localized spins and the underbarrier tunneling amplitude. The underbarrier transmission coefficients in solids (e.g., for point contacts) become temperature dependent.

Exchange-assisted tunneling enhances annihilation of positron with inner-shell electrons [4].

References

[1] V. V. Flambaum, Phys. Rev. A **79**, 042505 (2009).

[2] V. V. Flambaum Phys. Rev. A **80**, 055401 (2009).

[3] V.V. Flambaum, arXiv:0910.1155, EuroPhys. Lett. (EPL), **88**, 60009 (2009).

[4] M.G. Kozlov, V. V. Flambaum, Phys. Rev. A **87**, 042511 (2013).

[1]E-mail: v.flambaum@unsw.edu.au

Content from this work may be used under the terms of the Creative Commons Attribution 3.0 licence. Any further distribution of this work must maintain attribution to the author(s) and the title of the work, journal citation and DOI.
Published under licence by IOP Publishing Ltd

XXIX International Conference on Photonic, Electronic, and Atomic Collisions (ICPEAC2015) IOP Publishing
Journal of Physics: Conference Series **635** (2015) 052027 doi:10.1088/1742-6596/635/5/052027

Detecting Positron-Atom Bound States through Resonant Annihilation and Scattering

V. V. Flambaum[*1], **V. A. Dzuba**[*], **G. F. Gribakin**[*†2], **C. Harabati**[*]

[*] School of Physics, The University of New South Wales, Sydney NSW 2052, Australia

[†]School of Mathematics and Physics, Queen's University, Belfast BT7 1NN, Northern Ireland, UK

Synopsis A method is proposed for detecting positron-atom bound states by observing enhanced positron annihilation due to electronic Feshbach resonances. Relativistic coupled-cluster single-double approximation with third-order corrections is used to calculate positron binding energies and resonances for all atoms in Periodic Table.

In spite of wealth of predictions [1, 2, 3, 4], experimental verification of positron binding to neutral atoms is still lacking. We proposed a realistic experimental scheme to detect positron-atom bound states of open-shell atoms by measuring *resonant positron-atom annihilation* [5]. Therefore it is important to calculate positron binding energies to the ground state and excited configurations.

In resonant positron-atom annihilation process $A + e^+ \rightarrow A^* e^+ \rightarrow A^+ + 2\gamma$, the positron first loses energy by exciting the atom and becoming trapped in the bound state with the excited atom. It then annihilates with one of the electrons, and the resulting gamma quanta can be detected. The first step of the process is obviously reversible.

The observable effect may be obtained by averaging the normalised dimensionless annihilation rate $Z_{\rm eff} = \sigma_a k/(\pi r_0^2 c)$ over the energy distribution in the positron beam. Here σ_a is the resonant positron-atom annihilation cross section, k is the positron momentum, r_0 is the classical electron radius. The state of the positron bound to the excited atom, which lies above the atomic ground state, is a Feshbach resonance. Its total width Γ is the sum of the annihilation width Γ^a and the elastic width Γ^e. The latter gives the decay rate of this quasibound state in to $A + e^+$ the continuum, and also characterizes the probability of its formation in positron-atom collisions. Strong annihilation resonances require $\Gamma_\nu^e \gg \Gamma_\nu^a$.

The current study presents a first adaptation of the linear coupled-fluster single-double (SD) method with the third-order perturbation theory for the problem of positron-atom bound states. The calculations [6] for the positron binding energies to the closed-shell atoms (Be, Mg, Ca, Zn, Cd, and Hg) provide an estimate of an accuracy of our method ($\sim$ 100 meV) since various calculations for them are available in the literature [3]. In this approach, if the positron is found to be bound to an atom in a given configuration, it is assumed to be bound to all states (ground or excited) with the same configuration. In this way, we show that many atoms do bind a positron not only in the ground state, but also in excited states. The resonant and binding energies related to each other by $\varepsilon_{\rm r} = E_{\rm ex} - \varepsilon_{\rm b}$, where $E_{\rm ex}$ is the atomic excitation energy measured with respect to the ground. To observe a narrow resonance, one requires $\varepsilon_{\rm r} > 0$ ant that $\varepsilon_{\rm r}$ is smaller than positronium (PS) formation thereshold. According to these criteria, we have obtained many low lying resonances for many atoms [6].

References

[1] V. A. Dzuba, V. V. Flambaum, W. A. King, B.N. Miller, and O. P. Sushkov 1993 *Phys. Scr.* **T46**, 248.

[2] V. A. Dzuba, V. V. Flambaum, G. F. Gribakin, W. A. King, 1995 *Phys. Rev. A* **52**, 4541.

[3] J. Mitroy, M. W. J. Bromley, and G. G. Ryzhikh 2002 *J. Phys. B* **35** R81.

[4] V. A. Dzuba, V. V. Flambaum, G. F. Gribakin, and C. Harabati 2012 *Phys. Rev. A* **86**, 032503.

[5] V. A. Dzuba, V. V. Flambaum, and G. F. Gribakin 2010 *Phys. Rev. Lett.***105** 203401

[6] C. Harabati, V. A. Dzuba, and V. V. Flambaum 2014 *Phys. Rev. A***89** 022517

[1]E-mail: v.flambaum@unsw.edu.au
[2]E-mail: g.gribakin@qub.ac.uk

Content from this work may be used under the terms of the Creative Commons Attribution 3.0 licence. Any further distribution of this work must maintain attribution to the author(s) and the title of the work, journal citation and DOI.
Published under licence by IOP Publishing Ltd

Journal of Physics: Conference Series **635** (2015) 052028 doi:10.1088/1742-6596/635/5/052028

Periodic table of positronic atoms

V. V. Flambaum*[1], C. Harabati*, V. A. Dzuba*, G. F. Gribakin*†[2]

* School of Physics, The University of New South Wales, Sydney NSW 2052, Australia

†School of Mathematics and Physics, Queen's University, Belfast BT7 1NN, Northern Ireland, UK

Synopsis The calculation shows that more than half of the atoms in the Periodic Table bind the positron in their ground state. A number of atoms also bind positrons in excited states forming discrete spectrum and low-lying resonances.

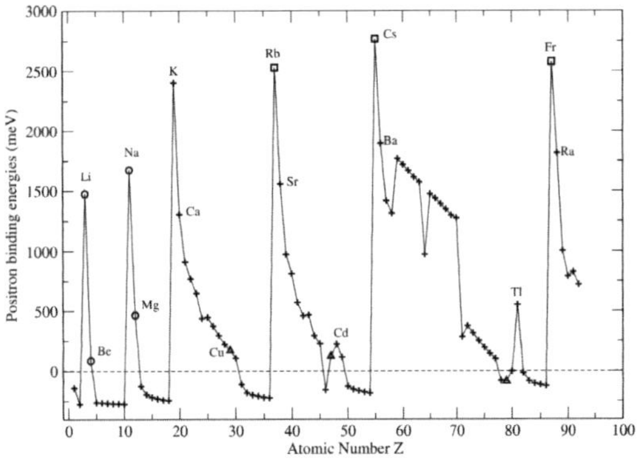

Figure 1. Recommended positron binding energies relative to the dissociation threshold e^++A. The results based on current study are shown with + sign. The ○ shows the results of the previous best calculations based on configuration interaction (CI) or stochastic variational methods (SVM). △ shows the previous result of the relativistic method MBPT+CI for Cu, Ag, and Au in our group. The binding energies of Rb, Cs, and Fr are obtained by linear extrapolation of the values of Li, Na, and K with respect to the ion(A^+) radius, which are marked by square □

Binding of positron to an atom with several valence electrons is a challenging problem. This is mainly due to the strong electron-positron correlation effects and virtual positronium (Ps) formation [1] . In Ref. [2] we presented calculations of the positron binding energies to all atoms in the Periodic Table. The most accurate previous calculations involving Li, Na, Ag, Cu, Au, Be, Mg, Ca, Zn, Sr and Cd atoms [3, 4] have been included and used to correct our values for the binding energies of positron to other atoms. We hope that the recommended positron binding energies to all atoms in the Periodic Table stimulate the experimentalist to detect the positron-atom bound states. The ordinary Periodic Table is constructed due to the similar chemical properties of atoms with the same number of valence electrons in external subshells. We have obtained similar periodicity in the positron-atom bound states, see the figure.

For atoms with ionization potential I such that $I > 6.80$ eV, the Ps-formation channel is closed. The closest decay channel will be e^++A. On the other hand, for $I < 6.80$ eV the lowest decay channel is Ps+A^+. On the figure, the positron binding energies ε_b have been presented relative to the decay channel e^++A. The binding energies relative to different channels are related by an equation

$$\varepsilon_b = \varepsilon_{Ps} - I + 6.80 \text{ eV}, \qquad (1)$$

where ε_{Ps} is the positron binding energy relative to the channel Ps+A^+, and 6.80 eV is the binding energy of positronium (Ps). We found that more than half of the atoms in the Periodic Table bind the positron in their ground state (i.e. stable relative to both channels). A number of atoms also bind positrons in excited states forming discrete spectrum and low-lying resonances. These resonances may be used for experimental detection of positron-atom bound states via resonant annihilation or scattering [5].

References

[1] V. A. Dzuba, V. V. Flambaum, W. A. King, B.N. Miller, and O. P. Sushkov 1993 *Phys. Scr.* **T46**.

[2] C. Harabati, V. A. Dzuba, and V. V. Flambaum 2014 *Phys. Rev. A***89** 022517

[3] J. Mitroy, M. W. J. Bromley, and G. G. Ryzhikh, in *New Directions in Antimatter Chemistry and Physics*, Eds. G. M. Surko and F. A. Gianturco, (Dordrecht: Kluwer, 2001).

[4] V. A. Dzuba, V. V. Flambaum, G. F. Gribakin, and C. Harabati, 2012 *Phys. Rev. A* **86**, 032503.

[5] V. A. Dzuba, V. V. Flambaum, and G. F. Gribakin 2010 *Phys. Rev. Lett.***105** 203401

[1]E-mail: v.flambaum@unsw.edu.au
[2]E-mail: g.gribakin@qub.ac.uk

Content from this work may be used under the terms of the Creative Commons Attribution 3.0 licence. Any further distribution of this work must maintain attribution to the author(s) and the title of the work, journal citation and DOI.
Published under licence by IOP Publishing Ltd

Electron recombination with multicharged ions via chaotic many-electron states

J. C. Berengut*[1], V. V. Flambaum*[2], V. A. Dzuba*, G. F. Gribakin*†[3], C. Harabati*, M. G. Kozlov‡♯

* School of Physics, University of New South Wales, Sydney NSW 2052, Australia
†School of Mathematics and Physics, Queen's University, Belfast BT7 1NN, Northern Ireland, UK
‡Petersburg Nuclear Physics Institute, Gatchina 188300, Russia
♯St. Petersburg Electrotechnical University "LETI", Prof. Popov Str. 5, 197376 St. Petersburg

Synopsis Electron recombination with multicharged ions that have open f shells (e.g. Au^{25+} and W^{20+}) is dominated by chaotic many-electron resonances. They cause the fluorescence yield to dominate entirely at low energies leading to an enhancement of the recombination rate up to 1000 times. At higher energies the opening of many autoionization channels leads to a rapid decrease of the fluorescence yield. We apply the statistical theory to express the recombination cross section as a sum over the doorway (dielectronic) resonances.

The electron recombination process at low energy is dominated by capture to a resonance state followed by fluorescence decay. The resonance state can either autoionize or radiate; when it radiatively decays to a level below the ionisation threshold, the capture is complete. Unlike in simple systems, the resonance state of a complex ion is a chaotic mix of many multiply-excited states. Each eigenstate $|\nu\rangle = \sum_k C_k^{(\nu)}|\varphi_k\rangle$ contains a large number N of *principal components* $|\varphi_k\rangle$ when the mixing is strong. Such eigenstates are called compound states. The only good quantum numbers are those exactly conserved due to the symmetry of the Hamiltonian of the ion: total angular momentum, its projection, and parity. Exact diagonalization of the many-electron Hamiltonian is impossible for the chaotic case. Fortunately, the statistical properties of the chaotic eigenstates enable one to avoid the determination of exact energy levels E_ν and compound state wave functions.

We have derived the recombination cross section of an electron with energy ε relative to threshold with a target ion (subscript i):

$$\sigma(\varepsilon) = \frac{\pi}{p^2}\frac{2J_\nu+1}{2(2J_i+1)}\omega_f \sum_k \frac{\Gamma_{ki}^{(a)}\Gamma_{\rm spr}}{(E_i+\varepsilon-E_k)^2+\frac{\Gamma_{\rm spr}^2}{4}}$$

where sum goes over the 'doorway' dielectronic states only. $\omega_f = \Gamma^{(r)}/(\Gamma^{(a)}+\Gamma^{(r)})$ is the fluorescence yield within the energy interval $\Delta\varepsilon << \varepsilon$ where the spectrum is dense. $\Gamma_{\rm spr}$ realises the spreading of the doorway state φ_k among the resonances. E_k can be approximated simply by the diagonal Hamiltonian matrix element $E_k = \langle\varphi_k|H|\varphi_k\rangle$. $\Gamma_{ki}^{(a)}$ is defined as an autoionization width of the doorway state. Our recent calculation based on the Slater determinants of the Hartree-Fock orbitals as doorway states is presented in the figure below.

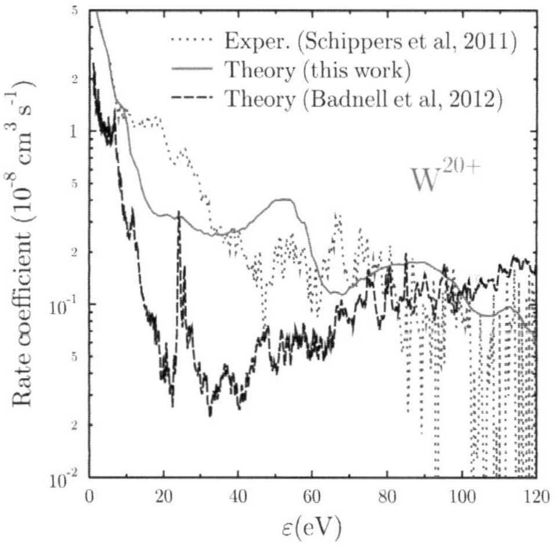

References

[1] V. A. Dzuba, V. V. Flambaum, G. F. Gribakin, C. Harabati, and M. G. Kozlov 2013 *Phys. Rev. A* **88** 062713.

[2] N. R. Badnell, C. P. Ballance, D. C. Griffin, and M. O'Mullane 2012 *Phys. Rev. A* **85** 052716

[3] S. Schippers, D. Bernhardt, A. Müller, C. Krantz, M. Grieser, R. Repnow, A. Wolf, M. Lestinsky, M. Hahn, O. Novotný, and D. W. Savin 2011 *Phys. Rev. A* **83** 012711

[4] V.V. Flambaum, M.G. Kozlov, G.F. Gribakin, arxiv:1404.4151

[1]E-mail: jcb@phys.unsw.edu.au
[2]E-mail: v.flambaum@unsw.edu.au
[3]E-mail: g.gribakin@qub.ac.uk

Content from this work may be used under the terms of the Creative Commons Attribution 3.0 licence. Any further distribution of this work must maintain attribution to the author(s) and the title of the work, journal citation and DOI.
Published under licence by IOP Publishing Ltd

Generalized oscillator strengths for the valence-shell excitations of xenon studied by the high-resolution fast-electron impact

Ya-Wei Liu[*†], **Long-Quan Xu**[*†], **Lin-Fan Zhu**[*†1]

[*] Hefei National Laboratory for Physical Sciences at Microscale and Department of Modern Physics, University of Science and Technology of China, Hefei, Anhui 230026, People's Republic of China

[†] Synergetic Innovation Center of Quantum Information and Quantum Physics, University of Science and Technology of China, Hefei, Anhui 230026, People's Republic of China

Synopsis Using the high-resolution fast-electron energy loss spectrometer, the generalized oscillator strengths (GOSs) of the valence-shell excitations of xenon were determined at a high resolution of 70 meV and a high incident electron energy of 1500 eV, and the momentum transfer dependence behaviors of GOSs of the $6s[3/2]_1$ and $6s'[1/2]_1$ excitations of xenon are discussed.

Xenon exists in the Earth's atmosphere to a small extent, and is widely applied in many areas such as lighting technique, laser and anesthetic technique. Xenon is the heaviest rare gas, and its GOSs can be used to test the theoretical methods severely. Furthermore, the differential cross sections (DCSs) of the valence-shell excitations of xenon vary strongly with the impact energy and scattering angle. So electron impact excitations of xenon can serve as a typical example for the study of electron collisions with heavy elements.

With such an importance, the high-energy electron scattering was used to determine GOSs of the $6s[3/2]_1$ and $6s'[1/2]_1$ excitations of xenon with the relative flow technique. The incident electron energy was set at 1500 eV and the energy resolution of 70 meV was used. The measured GOSs of the $6s[3/2]_1$ and $6s'[1/2]_1$ excitations of xenon are shown in Figures 1 and 2, and the experimental errors are also shown in the corresponding figures.

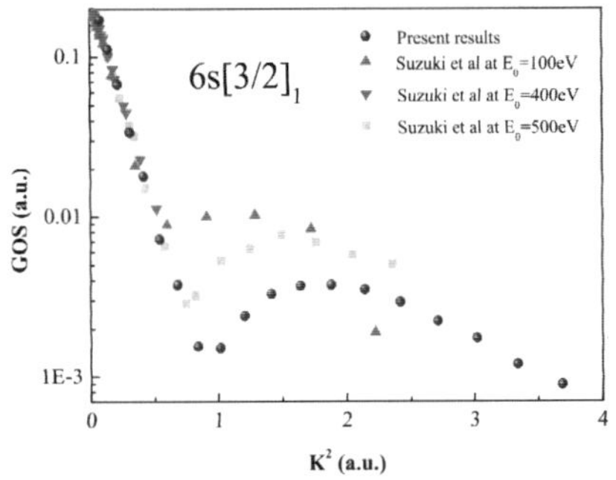

Figure 1. Absolute GOSs for $6s[3/2]_1$.

[1]E-mail: lfzhu@ustc.edu.cn

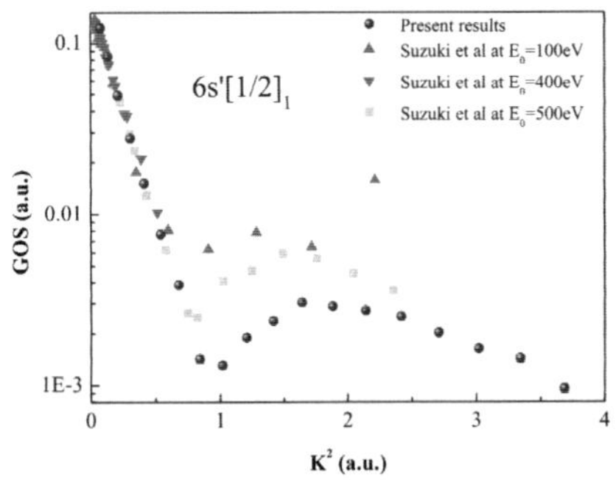

Figure 2. Absolute GOSs for $6s'[1/2]_1$.

It is obvious in Figures 1 and 2 that in the small momentum transfer region the present results at E_0=1500 eV are in agreement with the ones of Suzuki *et al* at E_0=100, 400 and 500 eV [1]. With the increasing of the momentum transfer, however, the discrepancies between the present results and the ones of Suzuki *et al* enlarge gradually. The squared momentum transfer K^2 of the minimum and maximum of the present results are larger than the ones of Suzuki *et al*.[1]. In addition, the values of the minimum and maximum of the present results are obvious lower than the ones of Suzuki *etal*. In a word, the behaviors of GOSs of the $6s[3/2]_1$ and $6s'[1/2]_1$ excitations of xenon with the increasing of incident electron energy is in agreement with the asymptotic behaviors for the inelastic electron scattering.

References

[1] T. Y. Suzuki *at al* 1991 *Phys. Rev. A.* **43** 5867

Content from this work may be used under the terms of the Creative Commons Attribution 3.0 licence. Any further distribution of this work must maintain attribution to the author(s) and the title of the work, journal citation and DOI.
Published under licence by IOP Publishing Ltd

Journal of Physics: Conference Series **635** (2015) 052031 doi:10.1088/1742-6596/635/5/052031

Collisional-radiative calculations for the $3p\,^1S_0 \rightarrow 3s\,^3P_1$ lasing line of neonlike germanium under anisotropic collision conditions

M. K. Inal[1] **and M. S. Bentotoche**[2]

Department of physics, Faculty of Sciences, University of Tlemcen, 13000 Tlemcen, Algeria

Synopsis Rate coefficients for collisional transitions between magnetic M-sublevels of 75 levels in neonlike germanium ions are calculated for both Maxwellian isotropic electron velocity distribution and monoenergetic electron beam. These are used in a steady-state collisional-radiative model to obtain differences between the $\Delta M = 0$ and $\Delta M = \pm 1$ gain components of the 19.6 nm $3p\,^1S_0 \rightarrow 3s\,^3P_1$ lasing line under anisotropic collision conditions in the amplifying plasma.

Most existing gain calculations for soft x-ray lasers pumped by electron collisions in laser-produced plasmas assume a Maxwellian isotropic electron velocity distribution. Fast electrons with a strongly anisotropic angular distribution could be present in plasmas when solid targets are irradiated by high-intensity ultrashort-duration laser pulses [1]. Such electrons can lead to different populations of magnetic M-sublevels after collisions with ions, affecting the polarization properties of laser lines.

In the present work, we examine differences between the $M = 0 \rightarrow M' = 0$ ($\Delta M = 0$) and $M = 0 \rightarrow M' = \pm 1$ ($\Delta M = \pm 1$) gain components of the 19.6 nm $3p\,^1S_0 \rightarrow 3s\,^3P_1$ lasing line of neonlike germanium under cylindrical symmetry conditions of the electron velocity distribution. We consider a velocity distribution model that includes both Maxwellian isotropic and monoenergetic beam electrons [2, 3]. We focus on the emission of the lasing line at the beginning of the plasma column far below saturation, where stimulated emissions can be disregarded. We use a steady-state collisional-radiative model to determine the populations of the upper $M = 0$ and lower $M' = 0, \pm 1$ sublevels of the lasing line as a function of electron density in the $10^{19} - 10^{22}\,\mathrm{cm}^{-3}$ range. This model includes collisional excitation and de-excitation, and spontaneous radiative decay among the 337 M-sublevels that arise from the 11 configurations $2p^6$, $2p^5 3l$, $2s2p^6 3l$ ($l = 0 - 2$) and $2p^5 4l$ ($l = 0 - 3$). The calculations were performed for various temperatures T_e of the Maxwellian component in the range $100 - 700$ eV and kinetic energies e_0 of the beam component from 1.5 to 20 keV at a beam fraction of 5%.

All basic atomic parameters needed for the calculations, including energy levels, spontaneous emission probabilities and collision strengths, were calculated with the flexible atomic code (FAC) [4]. However, some approximations were introduced to get the 4π-averaged collision strengths for optically allowed and forbidden transitions between M-sublevels belonging to levels with non-zero angular momenta, which cannot be provided by FAC. Those collision strengths are required to evaluate the Maxwellian-averaged rate coefficients. Other approximations used enabled us to have access to collision strengths for deexcitations by beam electrons as well, which also cannot be given by FAC.

Table 1 shows a typical example of the Maxwellian-averaged and beam rate coefficients for the transitions $3s\,^3P_1\,M_i = 0, 1 \rightarrow 3p\,^1D_2\,M_j = 0 - 2$, which are among the strongest ones.

Table 1. Maxwellian and beam rate coefficients ($10^{-10}\,\mathrm{cm^3 s^{-1}}$) for the $3s\,^3P_1\,M_i \rightarrow 3p\,^1D_2\,M_j$ excitations at various values of T_e and e_0, respectively.

	T_e (eV)			e_0 (keV)	
$M_i \rightarrow M_j$	200	400	700	2	10
$0 \rightarrow 0$	10.9	9.57	8.32	4.15	1.57
$0 \rightarrow 1$	8.19	7.18	6.24	5.27	3.67
$1 \rightarrow 0$	2.73	2.39	2.08	1.76	1.22
$1 \rightarrow 1$	8.19	7.18	6.24	3.12	1.18
$1 \rightarrow 2$	16.4	14.4	12.5	10.5	7.33

A maximum ratio of the $\Delta M = 0$ and $\Delta M = \pm 1$ gain components of the lasing line was found to be near 1.1 at $T_e = 100$ eV and $e_0 = 4$ keV.

References

[1] J. C. Kieffer *at al* 1993 *Phys. Rev.* E **48** 4648

[2] Z. Bedrane, M. K. Inal and S. Fritzsche 2009 *J. Phys. B: At. Mol. Opt. Phys.* **42** 055701

[3] A. K. Ferouani, M. K. Inal and G. Csanak 2013 *J. Phys. B: At. Mol. Opt. Phys.* **46** 085701

[4] M. F. Gu 2008 *Can. J. Phys.* **86** 675

[1]E-mail: m_inal@mail.univ-tlemcen.dz
[2]E-mail: m-sadek.bentotoche@mail.univ-tlemcen.dz

Content from this work may be used under the terms of the Creative Commons Attribution 3.0 licence. Any further distribution of this work must maintain attribution to the author(s) and the title of the work, journal citation and DOI.
Published under licence by IOP Publishing Ltd

Low-energy positronium scattering from noble-gas atoms

A. R. Swann[1], **J. A. Ludlow, and G. F. Gribakin**
Centre for Theoretical Atomic, Molecular, and Optical Physics,
School of Mathematics and Physics, Queen's University Belfast, Belfast BT7 1NN, United Kingdom

Synopsis van der Waals C_6 coefficients are calculated *ab initio* for positronium interactions with noble-gas atoms using the random-phase approximation with exchange (RPAE). These are scaled slightly to create our best predictions. Comparisons are made with existing configuration interaction and semi-empirical data.

Compared with electron-atom and positron-atom scattering, positronium (Ps)-atom scattering is difficult to treat theoretically, chiefly because both scattering objects have an internal structure [1].

Here we are concerned with low-energy Ps scattering from noble-gas atoms. We plan to study this problem by confining the electron-positron pair in a spherical cavity with a noble-gas atom at the centre. Finding the energy eigenvalues will enable the phase shifts (and hence all information about the scattering process) to be determined.

At low energies, the long-range interaction between the Ps atom and the noble-gas atom plays an important role in the scattering [2]. It is attractive and behaves asymptotically as

$$U(R) \simeq \frac{C_6}{R^6}, \tag{1}$$

where R is the distance between the two centres of mass and C_6 is the van der Waals coefficient for the pair of atoms. It is important to have accurate values of these C_6 coefficients because, e.g., they appear in the effective range expansion for the $l \geq 2$ phase shifts [3], and hence play a vital role in scattering cross sections.

As the first step toward including the long-range van der Waals interaction between the noble-gas and Ps atoms, we have performed *ab initio* calculations of the Ps–noble-gas C_6 coefficients in the random-phase approximation with exchange (RPAE) [4]. Mitroy and Bromley have previously calculated the C_6 coefficients for Ps–noble-gas interactions using the configuration interaction method for Ps-He and a semi-empirical method for the other pairs [5].

We began by calculating dynamic dipole polarizabilities α for both atoms over a range of imaginary frequencies. Then the C_6 coefficients were calculated by evaluating the integral

$$C_6 = \frac{3}{\pi} \int_0^\infty \alpha_{\mathrm{Ps}}(\mathrm{i}\omega)\alpha_X(\mathrm{i}\omega)\,\mathrm{d}\omega \tag{2}$$

numerically, where X represents the noble-gas atom.

Table 1 shows our best predictions of C_6 for the Ps–noble-gas pairs alongside the previous calculations by Mitroy and Bromley. Our best predictions were obtained by multiplying each *ab initio* value of C_6 by the ratio between the experimental and RPAE values of the static polarizability of the noble-gas atom. The maximum deviation from Mitroy and Bromley's values is 3%. We expect our values to be the most accurate data available for the C_6 coefficients for Ps–noble-gas scattering.

Table 1. C_6 values for Ps–noble gas pairs (in a.u.).

Pair	Present	Ref. [5]
Ps-He	13.45	13.37
Ps-Ne	26.69	26.74
Ps-Ar	99.00	98.50
Ps-Kr	147.4	144.1
Ps-Xe	228.2	221.6

We intend to use these data to develop an accurate theoretical description of low-energy Ps–noble-gas-atom collision processes.

References

[1] H. S. W. Massey and C. B. O. Mohr 1954 *Proc. Phys. Soc. A* **67** 695

[2] I. I. Fabrikant and G. F. Gribakin 2014 *Phys. Rev. A* **90**, 052717

[3] P. S. Ganas 1972 *Phys. Rev. A* **5** 1684

[4] M. Ya. Amusia and N. A. Cherepkov 1975 *Case Studies in Atomic Physics* **5** 47

[5] J. Mitroy and M. W. J. Bromley 2003 *Phys. Rev. A* **68** 035201

[1]E-mail: aswann02@qub.ac.uk

Content from this work may be used under the terms of the Creative Commons Attribution 3.0 licence. Any further distribution of this work must maintain attribution to the author(s) and the title of the work, journal citation and DOI.
Published under licence by IOP Publishing Ltd

Incident electron energy influence on energy positions and forms of the $3s3p^6np$ autoionising states of Ar

B. P. Marinković [1], J. J. Jureta, and A. R. Milosavljević

Institute of Physics, University of Belgrade, Pregrevica 118, 11080, Belgrade, Serbia

Synopsis Ejected electron spectra of Ar have been studied with high resolution at incident electron energies of 41.2 and 505 eV and at the constant ejection angle of 90°. The study was focused on the $3s3p^6np$ states and their behaviour at low and high incident electron energies.

The autoionising states in Ar have been studied in the past by photons, electrons and ions. The ejected electron energy spectra between 9 and 14 eV kinetic energies (24.76 to 29.76 of excitation energies) have been studied systematically in [1, 2 and 3]. This energy region is reach with autoionising features produced by excitation of the 3s electron from the ground $3s^23p^6$ configuration at low incident electron energy, while at higher incident electron energies the excitation of two 3p electrons or simultaneous excitation of 3p and 3s electrons can play important role in this energy region. The competition between these two excitation processes can produce energy shift and change the form of some states.

The apparatus used in the present measurement has been described in more details in [4]. It consists of a rotating non-monochromatic electron gun in angular range from 10° to 130° around the analyser axis, electrostatic lenses, high resolution hemispherical analyser operated at constant pass energy, and 7 channeltrons as detectors. Argon atom beam was provided into the interaction region by gas needle. The calibration point for ejected electron energy scale was taken from the mixture of helium and argon at 300 eV and 90° as an energy position of the $(2s2p)^1P$ state in helium at 60.13 eV [5].

In figure 1 we present high resolution ejected electron spectra of argon obtained at incident electron energies of 41.2 and 505 eV and at constant ejection angle of 90° with subtracted back-ground without normalization of data. The alignment of two spectra was done by using energy positions of features 3 and 3`what correspond to energy positions of the $3s3p^65s(^{3,1}S)$ and $3d(^{3,1}D)$ states, respectively. This alignment shows an energy shift of 0.070 eV between features 2-2` and a-a` respectively. The energy position of feature a` in the form of a deep corresponds to the energy position of the $3s3p^64p(^1P)$ state [1].

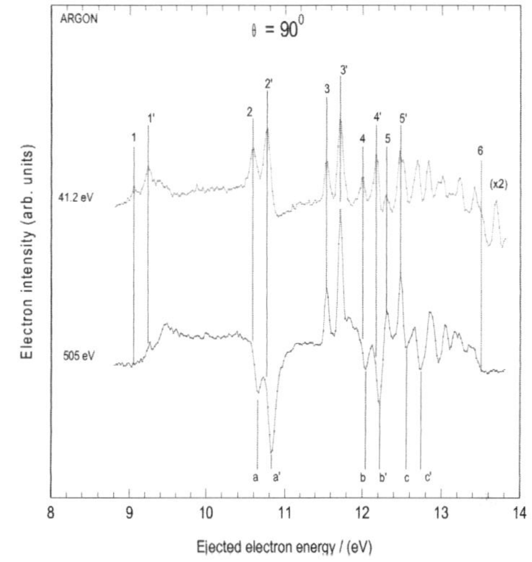

Figure 1. Ejected electron spectra of argon obtained at constant ejection angle of 90° and incident electron energies of 41.2 and 505 eV. The energy region of ejected electrons is from 9 to 14 eV (24.76 to 29.76 eV of excitation energy) with energy width per channel of 0.020 eV. Long vertical line between two spectra shows triplet and singlet states with energy separation of 0.18 eV. Vertical lines at the bottom labelled as small letters show resonances converging to the corresponding triplet and singlet states.

References

[1] R. P. Madden *et al* 1969 *Phys. Rev.* **177**, 136
[2] J. Fryar and J. W. McConkey 1976 *J. Phys. B: At. Mol. Phys.* **4**, 619
[3] J. Peresse *et al* 1972 *C. R. Acad. Sc. Paris* B-255
[4] J. J. Jureta *et al* 2014 *Int. J. Mass. Spectrom.* **365-366**, 114
[5] R. P. Madden and K. Codling 1963 *Phys. Rev. Lett.* **10**, 516.

[1] E-mail: bratislav.marinkovic@ipb.ac.rs

Content from this work may be used under the terms of the Creative Commons Attribution 3.0 licence. Any further distribution of this work must maintain attribution to the author(s) and the title of the work, journal citation and DOI.
Published under licence by IOP Publishing Ltd

XXIX International Conference on Photonic, Electronic, and Atomic Collisions (ICPEAC2015) IOP Publishing
Journal of Physics: Conference Series **635** (2015) 052034 doi:10.1088/1742-6596/635/5/052034

Completeness of the Configuration Interaction / Close Coupling expansion versus the relativistic formalism in R-matrix calculations

L. Fernández-Menchero[*1], **G. del Zanna**[†2] **and N. R. Badnell**[*3]
[*] Department of Physics, University of Strathclyde, Glasgow G4 0NG, United Kingdom
[†] Department of Applied Mathematics and Theoretical Physics, University of Cambridge, Cambridge CB3 0WA, United Kingdom

Synopsis R-matrix calculations have different sources of inaccuracy. We analyze the differences in the final results caused by differences in three aspects of the calculation: the atomic structure of the target, the completeness of the close coupling expansion, and the method used to include relativistic effects in the Hamiltonian. We present the electron-impact excitation of the ion Be-like Al^{9+} as a case study. We conclude that the completeness of the CI/CC expansion plays the most important role for the reliability of the results, mainly for the most highly-excited levels.

Reliable and accurate electron-impact excitation data are key to the successful spectroscopic diagnostic modeling of non-LTE plasmas.

We have compared and contrasted differences in such data for the benchmark Be-like Al^{9+} ion which we have calculated using the R-matrix method [1]. Such differences arise through: 1) differing degrees of approximation of relativistic effects, 2) uncertainties in atomic structure / configuration interaction (CI) expansion and 3) errors due to the lack of convergence of the close-coupling (CC) expansion. Error 3) is quantifiable and can be reduced systematically and reliably — to illustrate this we compare a new 98-level and previous 238-level CC Intermediate Coupling Frame Transformation (ICFT) R-matrix calculations [2]. We find that effective collision strengths to $n = 4$ levels are significantly enhanced over a wide range of temperatures by coupling to $n > 4$ levels. Uncertainty 2) is quantifiable but is more difficult to reduce and constrain as an error. To illustrate this, we compared 98-level and 238-level CI expansion calculations of line strengths and infinite energy plane-wave Born collision strengths. Again, transitions to $n = 4$ levels are most susceptible to lack of convergence but of the CI expansion this time. Differences 1) between different R-matrix treatments of relativistic effects are small, and negligible relative to 2) and 3). We illustrate this by a comparison of new 98-level CI/CC ICFT and Breit–Pauli (BP) R-matrix effective collision strengths which use the exact same atomic structure. This is to be expected for an element which lies below Zn.

Thus, we find to be false a recent conjecture that the ICFT approach may not be completely robust [3]. The conjecture was based upon a comparison of 98-level CI/CC Dirac Atomic R-matrix Code (DARC) effective collision strengths for Al^{9+} [4] with those from the 238-level CI/CC ICFT R-matrix calculations of [2]. The disagreement found recently is due to a lack of convergence of the CC expansion in the 98-level CI/CC DARC work, mainly for the $n = 4$ levels.

Similar considerations need to be made for other Be-like ions, and for other sequences.

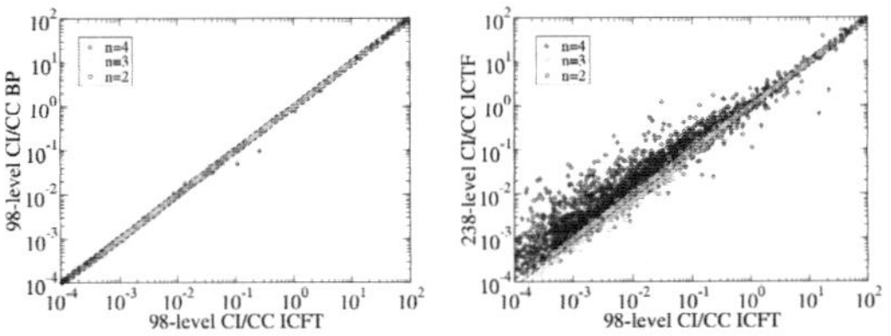

Figure 1. Comparison for the electron impact-excitation effective collision strengths for all transitions between the lowest 98 IC levels of the ion Al^{9+} at temperature $T = 10^6$ K. Left panel: BP vs ICFT, both for 98-level CI/CC expansion; right panel: 98- vs 238-level CI/CC expansion, both ICFT.

References

[1] L. Fernández-Menchero *et al* 2015 *Mon. Not. R. Astron. Soc. submitted.*

[2] L. Fernández-Menchero *et al* 2014 *Astron. and Astroph.* **566** A104

[3] K. M. Aggarwal and F. P. Keenan 2015, *Mon. Not. R. Astr. Soc.* **447** 3849

[4] K. M. Aggarwal and F. P. Keenan 2014 *Mon. Not. R. Astr. Soc.* **438** 1223

[1]E-mail: luis.fernandez-menchero@strath.ac.uk
[2]E-mail: gd232@cam.ac.uk
[3]E-mail: badnell@phys.strath.ac.uk

Content from this work may be used under the terms of the Creative Commons Attribution 3.0 licence. Any further distribution of this work must maintain attribution to the author(s) and the title of the work, journal citation and DOI.
Published under licence by IOP Publishing Ltd

Advances in atomic data for plasma diagnostics

L. Fernández-Menchero*[1], G. del Zanna†[2] and N. R. Badnell*[3]

* Department of Physics, University of Strathclyde. Glasgow G4 0NG, United Kingdom

† Department of Applied Mathematics and Theoretical Physics, University of Cambridge, Cambridge CB3 0WA, United Kingdom

Synopsis We present sets of new atomic data for astrophysical and fusion plasma diagnostics. We calculate electron-impact excitation collision strengths and effective collisions strengths for the Be- and Mg-like isoelectronic sequences up to Kr^{q+}. We use the R-Matrix Intermediate-Coupling Frame Transformation method. We also study the accuracy of calculations for highly-excited states and give extrapolation rules to estimate the effective collision strengths and their error at high n. We demonstrate the importance of a well converged CI/CC expansion in the R-matrix calculation to get good quality results, mainly for highly-excited states.

Atomic data requirements for modeling of fusion or astrophysical plasmas are becoming more and more demanding. Data bases, such as CHIANTI (www.chiantidatabase.org) or OPEN-ADAS (open.adas.ac.uk), require continual advancement as the demands of plasma modelers and observations grow.

Observations from several missions, such as SKYLAB and SOHO, detected spectral lines from ions in highly excited states ($n \geq 4$). To guarantee a good description of those levels, we require a good convergence of the Configuration Interaction (CI) basis set expansion of the target and the Close Coupling (CC) calculation.

We have carried-out Intermediate Coupling Frame Transformation (ICFT) R-Matrix calculations. The CI basis set for the Be-like sequence includes a total of 238 intermediate coupling levels from the configurations $1s^2\,\{2s^2, 2s2p, 2p^2\}$, and $1s^2\,2\{s,p\}\,nl$, with $n = 3-7$ and $l = s,p,d,f,g$ for $n = 4,5$ and $l = s,p,d$ for $n = 6,7$. For the Mg-like sequence, a total of 283 levels arising from the configurations $1s^2\,2s^2\,2p^6\,\{3s^2, 3s3p, 3s3d, 3p^2, 3p3d, 3d^2\}$ and $1s^2\,2s^2\,2p^6\,3\{s,p,d\}\,nl$, with $n = 4,5$ and $l = s,p,d,f,g$.

We present new data for the electron-impact excitation collision strengths Ω and Maxwell integrated effective collision strengths Υ for the Be-like isoelectronic sequence from B^+ to Kr^{32+} [1], and for the Mg-like isoelectronic sequence from Al^+ to Kr^{24+} [2]. We show data for non-Maxwellian velocity distributions too.

We also investigate extrapolation rules to estimate the electron-impact excitation effective collision strengths for highly excited states [3]. We focus in the H- and He-like isoelectronic sequences as a benchmark. Calculations for highly excited states are strongly limited by the quality of the atomic structure and the close coupling expansion for such terms/levels. Eventually, extrapolation from good-quality data for lower states gives a more accurate result than an unconverged calculation. We give an estimation of the error associated with the extrapolation too.

We look at differences between the R-matrix method and the distorted wave one for weak one-photon forbidden transitions ($J - J' = 0 - 0$). These differences are due to the transformation from the reactance matrix K to the transmission matrix T in an exact fashion, conserving the unitarity (AS-UDW), or in an approximate non unitary one (AS-DW) [4].

The completeness of the basis set in the CI/CC expansion of the target is the main source of error, mostly for the higher excited states. Other effects produce considerably smaller inaccuracies, like the method used to include relativistic effects in the Hamiltonian [5].

The present advances in atomic data for plasma modeling and diagnostics are available in CHIANTI and OPEN-ADAS data bases.

References

[1] L. Fernández-Menchero *et al* 2014 *Astron. and Astroph.* **566** A104

[2] L. Fernández-Menchero *et al* 2014 *Astron. and Astroph.* **572** A115

[3] L. Fernández-Menchero *et al* 2015 *Astron. and Astroph. in progress.*

[4] L. Fernández-Menchero *et al* 2015 *Astron. and Astroph. submitted.*

[5] L. Fernández-Menchero *et al* 2015 *Mon. Not. R. Astr. Soc. submitted.*

[1]E-mail: luis.fernandez-menchero@strath.ac.uk
[2]E-mail: gd232@cam.ac.uk
[3]E-mail: badnell@phys.strath.ac.uk

Content from this work may be used under the terms of the Creative Commons Attribution 3.0 licence. Any further distribution of this work must maintain attribution to the author(s) and the title of the work, journal citation and DOI.
Published under licence by IOP Publishing Ltd

Calculation of differential cross section for dielectronic recombination with one-electron uranium

K.N. Lyashchenko[*,1], **O.Yu. Andreev**[*]

[*]Faculty of Physics, St. Petersburg State University, ul. Ulyanovskaya 1, Petergof, St. Petersburg, 198504, Russia

Abstract

Calculation of the differential cross section for the dielectronic recombination with one-electron uranium within the framework of QED is presented. The contribution of the QED corrections and the interfernce of the photon multipoles is investigated.

Calculation of the dielectronic recombination with one-electron uranium within the framework of QED is presented [1]. We consider process where the initial state is presented by a one-electron ion of uranium being in the ground state and by an incident electron. The energy of the initial state is close to the energies of double-excited states ($(2s, 2s)$, $(2s, 2p)$, $(2p, 2p)$). The final state is given by a two-electron ion in one of the single-excited states ($(1s, 2s)$, $(1s, 2p)$) and by emitted photon. The process of dielectronic recombination is a resonant process. The resonances in the cross section correspond to the double excited states. In the resonant area the dielectronic recombination gives the main contribution to the cross section.

The calculation is performed with employment of the line-profile approach [2,3]. The one-photon exchange correction for the low-lying states is taken into account in all orders of the QED perturbation theory. The electron self-energy and vacuum polarization corrections are considered in the first order of the perturbation theory. We present results of the calculation of the total and differential cross section. The polarization properties are also considered and the Stokes parameters are presented. The contribution of the Breit interaction to the cross section and to the Stokes parameters is investigated. The results are compared with available experimental and theoretical data.

References

[1] K.N. Lyashchenko, O. Yu. Andreev 2015 Phys. Rev. A **91**, 012511

[2] O. Yu. Andreev, L. N. Labzowsky, G. Plunien, D. A. Solovyev 2008 Phys. Rep. **455**, 135

[3] O. Yu. Andreev, L. N. Labzowsky, A. V. Prigorovsky 2009 Phys. Rev. A **80**, 042514

[1]E-mail: laywer92@mail.ru

Content from this work may be used under the terms of the Creative Commons Attribution 3.0 licence. Any further distribution of this work must maintain attribution to the author(s) and the title of the work, journal citation and DOI.
Published under licence by IOP Publishing Ltd

Journal of Physics: Conference Series **635** (2015) 052037 doi:10.1088/1742-6596/635/5/052037

Line Profile Property of Autoionizing States

B. Tu, J. Xiao, K. Yao, Y. Shen, Y. Yang, D. Lu, Y. Q. Fu, B. Wei, C. Zheng, L. Y. Huang, R. Hutton, and Y. Zou[1]

Shanghai EBIT Laboratory, Institute of Modern Physics, Fudan University, and the Key Laboratory of Applied Ion Beam Physics, Chinese Ministry of Education, China

Synopsis This work studied the interference effect between dielectronic recombination and radiative recombination of highly charged tungsten ions. Our result shows a coexistence of Fano and Lorentzian line profile of autoionizing states.

Photon absorption spectroscopy is a powerful tool for uncovering the structure of atoms, molecules and solids. Symmetric Lorentzian and asymmetric Fano line shapes are fundamental spectroscopic signatures related to the structural and dynamical properties. Recently Ott *et al.* published a paper in Science[1], where they introduced a temporal-phase formalism, which enables mapping the Fano asymmetry parameter q to a phase φ of the time-dependent dipole response function, and presented experimental results of laser-transforming Fano profile into Lorentzian line shape after attosecond-pulsed excitation of autoionizing states in helium. This is a very important step forward of quantum phase control and would allow a general interpretation of effects such as electromagnetically induced transparency (EIT), lasing without inversion, and would have further application in amplification of extreme-ultraviolet light, or even x rays that resonantly interacting with atoms.

In this presentation, we show experimentally that an autoionizing state can have both Fano and Lorentzian behavior naturally, depending on the process involved. This study utilized the inverse process of photon absorption ionization, electron ion recombinations, including both resonant and non-resonant process. Dielectronic recombination (DR), as the main process of the resonant recombination, is considered as a two-step or indirect process. Whereas, the non-resonant recombination, also named as radiative recombination (RR), is considered as a direct process. DR and RR will interfere with each other when they have the same initial and final state, and the Fano profile is then expected, especially in highly charged high-Z elements, see work from LLNL[2], Heidelberg EBIT[3] and Tokyo EBIT[4].

In order to study the variation of the Fano profile (consequently the Fano parameter) of a given autoionizing state, we selected the DR resonances which go through an intermediate state with decay channels to both the final states with no excited electrons and the final states with two excited electrons. In the former case, both dielectronic and radiative recombination will occur, interference and hence the asymmetric line profile should be expected. In the latter case, only dielectronic recombination (so no interference) will take place, because radiative recombination of an electron with an ion in the ground state can not end up in a final state with more than one electron in excited orbit.

The experiment was done at the Shanghai electron beam ion trap (Shanghai EBIT)[5]. During the experiment, $W(CO)_6$ was continuously injected into the EBIT, and the electron beam energy was scanned through the resonant energies of KL_3L_3 DR of Be-like up to O-like W ions, while the X ray photons from the recombination were detected by a HPGe detector.

Our result shows that an autoionizing state can have dual Fano and Lorentzian line profile properties naturally, depending on the process involved. Our result implies that excitation of the state through different paths, for example one photon versus multi-photon excitation, or even one step versus multi-step excitation, can lead to different Fano profiles for the same resonant state.

References

[1] C. Ott *et al Science* **340**, 716 (2013)
[2] D. A. Knapp *et al Phys. Rev. Lett.* **74**, 54 (1995)
[3] A. J. G. Mart'ınez *et al Phys. Rev. Lett.* **94**, 203201 (2005)
[4] N. Nakamura *et al Phys. Rev. A.* **80**, 014503 (2009)
[5] D. Lu *et al Rev. Sci. Instrum.* **85**, 093301 (2014)

[1] E-mail: zouym@fudan.edu.cn

Content from this work may be used under the terms of the Creative Commons Attribution 3.0 licence. Any further distribution of this work must maintain attribution to the author(s) and the title of the work, journal citation and DOI.
Published under licence by IOP Publishing Ltd

Influence of Breit interaction on the cross section and circular polarization of radiation emitted by longitudinally-polarized electron impact excitation of highly charged Be-like ions

Zhan-Bin Chen, Chen-Zhong Dong [1], and Jun Jiang

Key Laboratory of Atomic and Molecular Physics and Functional Materials of Gansu Province, College of Physics and Electronic Engineering, Northwest Normal University, Lanzhou 730070, China

Synopsis: We investigated the contributions of the Breit interaction to the longitudinally-polarized electron impact excitation cross sections and circular polarizations of the $1s2s^22p_{3/2}(J = 2)\rightarrow 1s^22s^2(J = 0)$ line of highly charged Be-like ions using a fully relativistic distorted-wave method. It is found that the Breit interaction can modify the $M_f = -1$ and -2 sublevels cross sections by several orders of magnitude. These dramatic influences also lead to a remarkable decrease in the circular polarization of radiation, which characters becomes more evident as the incident energy and/or atomic number increases. And all these features are different from the conclusions for the linear polarization of radiation.

The polarization of x-ray radiation from highly charged ions undergoing collisions with an electron beam has been a topic of continuous fundamental interest for decades [1-3].

So far, however, most theoretical studies on the x-ray emission have dealt with the linear polarization. Early work on the excitation cross sections and circular polarization properties of the x-ray line emission was placed on relatively low-Z systems. Theoretical results are noticeably lacking for the cross sections and circular polarization properties in the high-Z domain. In recent years, some works have included the experimental verification of the importance of the Breit interaction for few-electron ions and the successful reinterpretation of the linear polarization of x-ray emission in high-Z ions [3]. Therefore, some similar effects of the Breit interaction should also be expected for the cross sections and circular polarizations of x-ray radiation.

In this study, longitudinally-polarized-electron impact excitation cross sections from the ground state to the individual magnetic sublevels of the excited state $1s2s^22p_{3/2}$ (J=2) of highly charged Be-like ions are investigated systematically [2]. The influences of Breit interaction on the excitation cross sections and the degree of circular polarizations of subsequent x-ray radiation are discussed in detail. It is found that the Breit interaction can modify the $M_f = -1$ and -2 sublevels cross sections by several orders of magnitude. These influences also lead to a remarkable decrease in the circular polarization of radiation, which character becomes more evident as the incident energy and/or atomic number increases. All these features are very different from the conclusions for the linear polarization of radiation.

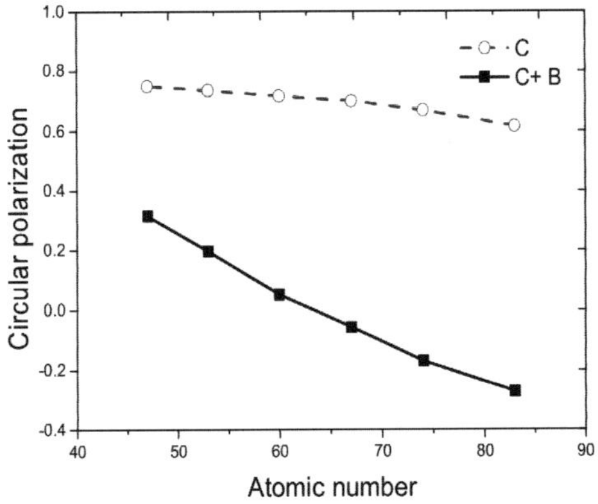

Figure 1. The degree of circular polarization of the transition line $1s2s^22p_{3/2}(J{=}2)\rightarrow 1s^22s^2(J = 0)$ for Be-like ions as functions of atomic number at incident polarized electron energy is four times the threshold energy. Here, C represents the values with inclusion of only the Coulomb interaction, and C+B represents the ones with the Breit interaction included.

This work was supported by the National Natural Science Foundation of China (Grant nos. 11274254 and U1332206).

References

[1] Z B Chen *et al* 2014 *Phys. Rev. A.* **90** 012703
[2] Z B Chen *et al* 2014 *Phys. Rev. A.* **90** 022715
[3] S Fritzsche *et al* 2009 *Phys. Rev. Lett.* **103** 113001

[1] E-mail: dongcz@nwnu.edu.cn

Content from this work may be used under the terms of the Creative Commons Attribution 3.0 licence. Any further distribution of this work must maintain attribution to the author(s) and the title of the work, journal citation and DOI.
Published under licence by IOP Publishing Ltd

Journal of Physics: Conference Series **635** (2015) 052039 doi:10.1088/1742-6596/635/5/052039

Cascades effects on the polarization properties of x-ray radiation emitted by electron-impact excitation of highly charged He-like Fe^{24+} ions

Z B Chen*†, C Z Dong *[1] and J Jiang*

*Key Laboratory of Atomic and Molecular Physics and Functional Materials of Gansu Province, College of Physics and Electronic Engineering, Northwest Normal University, Lanzhou 730070, China

†College of Science, National University of Defense Technology, Changsha, Hunan 410073, China

Synopsis: Detailed calculations using a fully relativistic distorted-wave method are carried out for the electron-impact excitation cross sections from the ground state $1s^2$ (J= 0) to the individual magnetic sublevels of the excited state $1s2s_{1/2}$ (J= 1) of highly charged He-like Fe^{24+} ions. The cascades effects to the cross sections and the linear polarizations of the $1s2s_{1/2}(J=1)\rightarrow 1s^2$ (J=0) line are investigated systematically. Our results show that the 1s2p, 1s3l, and 1s4l (l = s, p, d) excited states have significant effects and make the cross sections increase. These influences also lead to a remarkable decrease in the linear polarization of subsequent x-ray radiation. The present results agree with the previous theoretical results and the polarization measurements very well.

The polarization of He-like ions is of importance for it provides diagnostics of the physical conditions in terrestrial magnetic and laser produced fusion plasmas. Such polarization may potentially be applied to future studies of solar flares and other exotic plasmas [1]. Several studies therefore have been carried out in the past to better understand the role of angular and polarization properties of x-rays emitted from highly charged He-like ions. However, the accuracy of these works is often suspect, different calculations using similar methods/codes sometimes differ so much due to the influence of the higher order effects [2, 3].

In this study, the electron-impact excitation cross sections from the ground state $1s^2(J=0)$ to the individual magnetic sublevels of the excited state $1s2s_{1/2}(J=1)$ of He-like Fe^{24+} ions are calculated using a fully relativistic distorted wave (RDW) method. Special attention is paid to the cascade contributions from high-lying levels 1s2p, 1snl (n= 3, 4, 5, 6; l= s, p, d) to the cross sections and the polarization properties of the $1s2s_{1/2}(J=1)\rightarrow 1s^2(J=0)$ transition line [1]. Our results show that radiative cascades have large effects and make the polarization decrease. For the cross sections, the contributions of the 1s2p, 1s3l, 1s4l, 1s5l and 1s6l (l = s, p, d) levels are 140%, 300%, 70%, 40%, and 20% to the $M_f = 0$ sublevel cross sections, and are 106%, 112%, 52%, 40%, and 15% to the $M_f = \pm 1$ sublevel cross sections, respectively. For the polarizations, the cascades effects causes a change of the sign of the linear polarization and decrease linear polarization from 0.004 to −0.08 and 0.003 to −0.268 at 6.8 and 8 keV, respectively. The present results agree very well with the previous theoretical results and polarization measurements.

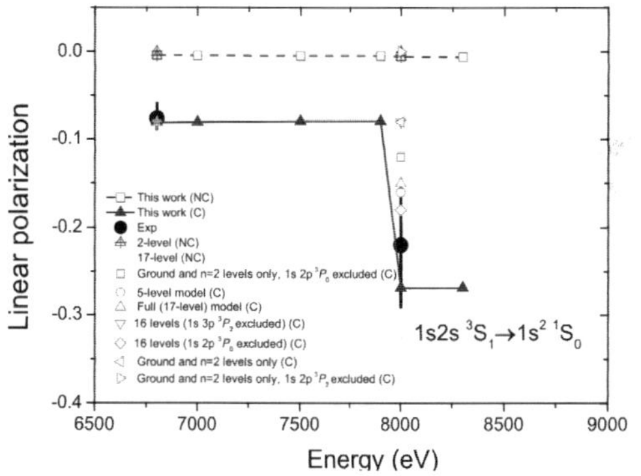

Figure 1. Calculated linear polarizations for the $1s2s_{1/2}(J=1)\rightarrow 1s^2(J=0)$ transition line of He-like Fe^{24+} ions, and comparison with the existing theoretical [5] and experimental values [4,5]. Here, NC represents the values without cascade contributions, and C represents the ones with cascade contributions included. The labeled '17-level' represent the results obtained from the full model [5] which included the 1s2, 1s2l (l = s, p), and 1s3l (l = s, p, d) configurations, the labeled '5-level' refer to separate calculations involving only the five configurations (1s2 $^1S_0$, 1s2s $^3S1$, 1s2p $^3P_0$, 1s2p $^3P_2$, and 1s3p $^3P_2$), and the labeled '2-level' refer to separate calculations involving only the two levels of each transition [5].

This work was supported by the National Natural Science Foundation of China (Grant nos. 11274254 and U1332206).

References

[1] Z B Chen *et al* 2015 *Phys.Scr* (In print)
[2] Z B Chen *et al* 2014 *Phys. Rev. A* **90** 012703
[3] Z B Chen *et al* 2014 *Phys. Rev. A* **90** 022715
[4] P Beiersdorfer *et al* 1996 *Phys. Rev. A* **53** 3974
[5] P Hakel *et al* 2007 *Phys. Rev. A* **76** 01271

[1]E-mail: dongcz@nwnu.edu.cn

Content from this work may be used under the terms of the Creative Commons Attribution 3.0 licence. Any further distribution of this work must maintain attribution to the author(s) and the title of the work, journal citation and DOI.
Published under licence by IOP Publishing Ltd

XXIX International Conference on Photonic, Electronic, and Atomic Collisions (ICPEAC2015) IOP Publishing
Journal of Physics: Conference Series **635** (2015) 052040 doi:10.1088/1742-6596/635/5/052040

Electron-impact ionization of tungsten ions

A. Borovik, Jr.[*1], **D. Schury**[*†], **B. Ebinger**[*], **K. Spruck**[*‡], **A. Becker**[*‡], **M. F. Gharaibeh**[*§], **J. Rausch**[*], **S. Schippers**[*], **A. Müller**[*]

[*] Institut für Atom- und Molekülphysik, Justus-Liebig Universität Giessen, Germany
[†] GSI Helmholtzzentrum für Schwerionenforschung GmbH, Darmstadt, Germany
[‡] Max Planck Institut für Kernphysik, Heidelberg, Germany
[§]Physics Department, Jordan University of Science and Technology, Irbid, Jordan

Synopsis Electron-impact ionization of tungsten ions has been measured in the energy range from threshold up to 1000 eV. Configuration-averaged distorted wave calculations were used to explain the experimental data.

In future Tokamaks, extreme conditions at the plasma edge require plasma-facing wall elements to possess high sustainability against sputtering. Due to its stability and robustness tungsten has been identified to be the most promising candidate material for the divertor plates in the upcoming largest thermonuclear device ITER. At the same time, huge loads of incident particles on the plate surfaces will make the presence of sputtered tungsten atoms (and, of course, ions) in the edge plasma unavoidable. Tungsten atoms and ions in a wide spectrum of charge states lead to significant energy losses which must be reliably estimated. In order to perform such estimations, reliable data on atomic processes of tungsten ions are of crucial importance [1].

We present electron-impact ionization cross sections (CS) of tungsten ions studied in the energy range from the corresponding signal onsets up to 1000 eV. Absolute CS values have been measured using the animated-beam technique implemented in a crossed-beams experiment. Additionally, a fine-step energy-scan technique has been employed to uncover fine structures in the studied CSs [2]. the ion charge states q=1, ..., 6, 8, 11, ..., 19 have been investigated. In order to explain the measured CSs, configuration-averaged distorted wave (CADW) calculations have been performed considering both direct- and indirect ionization mechanisms. Ions in long-lived excited states have been found to be present in the experiments and were also included in the calculations. Comparisons of the experimental and theoretical data shows good to very good agreement. In the energy ranges below CS maxima the dominant role of indirect ionization processes via excitation to autoionizing states is revealed. For high ionic charge states, the majority of such excited autoionizing states are supposed to decay via Auger processes, while, for lower charge states, radiative damping plays a significant role. The figure shows the CS for W^{19+} with comparison of the present experimental and theoretical data sets.

Together with the investigations on photoionization and dielectronic recombination [3, 4], present results were obtained within a comprehensive effort to provide accurate atomic data on photonic and electronic interactions of tungsten ions for further use in plasma modeling.

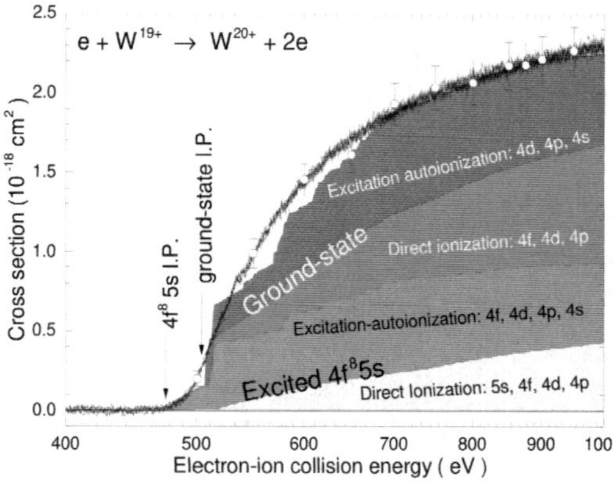

Figure 1. Ionization cross section of W^{19+}: experimental data (open dots, black line) compared to CADW calculations (individually shaded areas represent contributions of different ionization mechanisms).

References

[1] C. H. Skinner 2009 *Phys. Scr.* **T134** 014022

[2] J. Rausch *et al* 2011 *J. Phys. B* **44** 165202

[3] A. Müller *et al* 2011 *Phys. Scr.* **T144** 014052

[4] S. Schippers *et al* 2011 *Phys. Rev. A* **83** 012711

[1]E-mail: Alexander.Borovik@iamp.physik.uni-giessen.de

Content from this work may be used under the terms of the Creative Commons Attribution 3.0 licence. Any further distribution of this work must maintain attribution to the author(s) and the title of the work, journal citation and DOI.
Published under licence by IOP Publishing Ltd

Journal of Physics: Conference Series **635** (2015) 052043 doi:10.1088/1742-6596/635/5/052043

Parity violation in the resonance elastic electron scattering on He-like uranium

V A Zaytsev*[1], S Tashenov♯, A V Maiorova*, V M Shabaev*, and Th Stöhlker⋆∘•

* Department of Physics, St. Petersburg State University, Ulianovskaya 1, Petrodvorets, 198504 St. Petersburg, Russia

♯ Physikalisches Institut, Universität Heidelberg, D-69120 Heidelberg, Germany

⋆ GSI Helmholtzzentrum für Schwerionenforschung GmbH, D-64291 Darmstadt , Germany

∘ Helmholtz-Institut Jena, D-07743 Jena, Germany

• Institut für Optik und Quantenelektronik, Friedrich-Schiller-Universität Jena, D-07743 Jena, Germany

Synopsis Parity violation on the cross section of the resonance elastic electron scattering by He-like uranium ion is studied. It is assumed that the incident electron is polarized and tuned in resonance with one of the close-lying opposite-parity states.

Investigations of the parity nonconservaion (PNC) effects in atomic systems play a very important role in tests of the Standard Model at low-energy regime [1]. From the theoretical side, the heavy few-electron ions are the most appropriate candidates for these studies. This is due to a simple electronic structure that allows high-precision calculations of these systems. For the last years a vast number of different scenarios offering the detection of the PNC effects in heavy highly-charged ions were proposed (see, e.g., Refs. [2, 3, 4] and references therein). However, the breaking of the space symmetry in the process of electron scattering by heavy ions has not yet been discussed in the domain of atomic physics.

In the present contribution we study the PNC effect on the cross section of the elastic scattering of the polarized electrons by He-like uranium ion ($Z = 92$) being initially in the ground state. The energy of the incident electron is assumed to be tuned in resonance with one of the quasidegenerate $d_1 \equiv \left[\left(1s\ 2p_{1/2}\right)_0 6s\right]_{1/2}$ or $d_2 \equiv \left[(1s\ 2s)_0 6s\right]_{1/2}$ states [4], thus providing the enhancement of the parity violation. In order to investigate the PNC effect, we introduce the asymmetry parameter $\mathcal{A} = \left(\sigma_{1/2} - \sigma_{-1/2}\right) / \left(\sigma_{1/2} + \sigma_{-1/2}\right)$ where σ_{μ_i} is the differential cross section with μ_i designating the helicity of the incident electron. The deviation of this parameter from zero indicates the breaking of the spatial symmetry. In figure 1 it is depicted as a function of the scattering angle θ. From this figure one can see that the maximum value of $|\mathcal{A}|$ is expected for $\theta \sim 70^\circ$ at the energy of the incident electron being tuned in resonance with any of d_1 or d_2 state and amounts to about 8×10^{-8}. Though this asymmetry is too small to make possible performing the corresponding experiment it is worth to mention that one can expected the increase of the PNC effect by few orders of magnitude for the inelastic electron scattering.

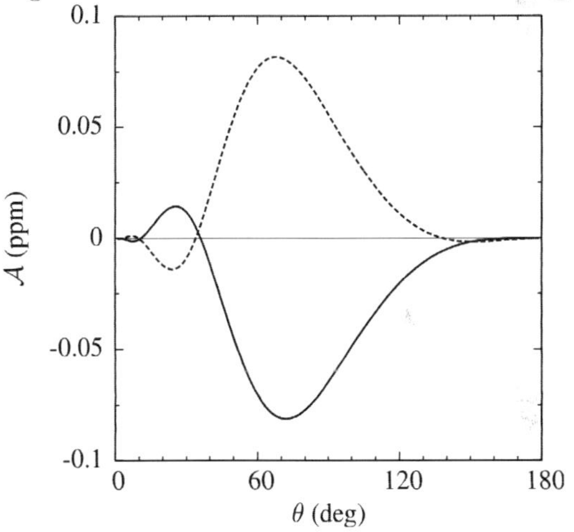

Figure 1. The solid and dashed lines are correspond to the energies of the incident electron being tuned in resonance with d_1 and d_2 states, respectively.

This work was supported by SPbSU (Grants No. 11.38.269.2014 and No. 11.38.237.2015).

References

[1] Khriplovich I B 2004 *Phys. Scr.* **T112** 52

[2] Shabaev V M *et al* 2010 *Phys. Rev. A* **81** 052102

[3] Gunst J *et al* 2013 *Phys. Rev. A* **87** 032714

[4] Zaytsev V A *et al* 2014 *Phys. Rev. A* **89** 032703

[1]E-mail: zaytsev@pcqnt1.phys.spbu.ru

Content from this work may be used under the terms of the Creative Commons Attribution 3.0 licence. Any further distribution of this work must maintain attribution to the author(s) and the title of the work, journal citation and DOI.
Published under licence by IOP Publishing Ltd

Antihydrogen produced in the charge transfer channel by the collision of antiprotons and positronium

R. Fathi[1], F. Shojaei[2] and M. A. Bolorizadeh[3]

[1&2]Faculty of Physics, Shahid Bahonar University of Kerman, Kerman, Iran
[3]Physics and Photonics Department, Graduate University of Advanced Technology, Kerman, Iran.

Synopsis: In the present study, the differential and total cross sections due to the collision of antiprotons and hydrogen atoms in the charge transfer channel is calculated to form antihydrogen. The state of the incoming particles was assumed to be plane wave or Coulomb distorted wave. The impact energies of 11.3, 13.3and 15.8 keV were assumed to be able to compare the results with the available experimental data for the total charge transfer channel.

The study of antiproton production is essential to understand CP violation and Lorentz invariance. Positronium being short lived, it is challenging to measure cross sections involving positronium atom. Here, it is intended to calculate the cross sections for the antihydrogen production in the reaction:

$$\bar{P} + Ps(1s) \rightarrow \bar{H}(nlm) + e^{+} \quad (1)$$

where $\bar{P}$ and $\bar{H}$ are antiproton and antihydrogen, respectively. The only available experimental work to measure cross sections for the reaction (1) is performed at three impact energies of 11.3, 13.3 and 15.8 keV.

In order to calculate the cross sections, correctly including all interaction terms in a quantum mechanical model, a three body model is assumed. The perturbation methods are usually valid at intermediate/high energies. Nonetheless, one could push the theory into lower energies, by assuming distorted Coulomb wave, it is possible to study the scattering at lower energies. However, 20 keV energy is very low for antiproton to have very good results. Nonetheless, the cross section for the charge transfer channel of reaction (1) is calculated both assuming the incoming and outgoing particles as plane wave and distorted Coulomb wave. The results are plotted in Figure 1 and compared with the available experimental work [1].

Three sets of data are plotted in figure 1. Present three-body model assuming plane wave for the initial antiproton and scattered electron (calling it set one) and assuming distorted Coulomb wave for the antiproton and electron (calling it set two) are compared with the measured cross sections [1].

At impact energy of 11.3 keV, the measured cross sections are higher than the first set, while it is lower than the second set. The error bars on

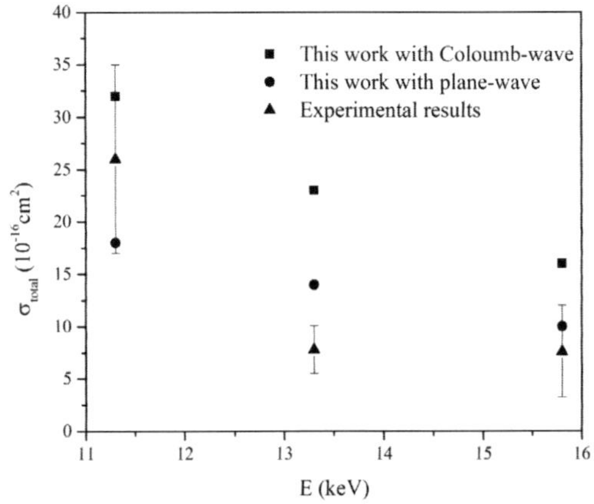

Figure 1. The calculated cross sections for the charge transfer channel, equation (1), compared with the experimental data of Morrison et al [1].

the experimental data could lead us to the conclusion that both theoretical results are justified by it. However, the other two experimental results are lower than the present two sets of calculated data, the first set being closer to the experimental one. This is not justified as we expect the distorted Coulomb wave representing the experiment better. The only conclusion that one can draw would be on the trend of change in all three data set. The second set of data is approaching the trend that the experimental measurement is presenting.

More experimental and theoretical data are needed to understand the reaction (1) in detail. This work will further be extended to higher energies (intermediate and high).

References:

[1] J. P. Merrison and et al, Hyperfine interactions 109 (1997) 313.

[1] E-mail: rfathi@ uk.ac.ir
[2] E-mail: fshojaei@uk.ac.ir
[3] E-mail: ma.bolorizadeh@kgut.ac.ir

Content from this work may be used under the terms of the Creative Commons Attribution 3.0 licence. Any further distribution of this work must maintain attribution to the author(s) and the title of the work, journal citation and DOI.
Published under licence by IOP Publishing Ltd

Journal of Physics: Conference Series **635** (2015) 052045 doi:10.1088/1742-6596/635/5/052045

Eletron-helium laser-assisted free-free scattering for incident energies from 30 - 200 eV: effects of polarization direction

B. A. deHarak* [1], **Benjamin Nosarzewski**†, **Mahsa Siavashpouri**‡, **N. L. S. Martin**‡

* Physics Department, Illinois Wesleyan University, P.O. Box 2900, Bloomington, IL 61702-2900, USA

† Department of Physics, Cornell University, 109 Clark Hall, Ithaca, New York 14853, USA

‡ Department of Physics and Astronomy, University of Kentucky, Lexington, KY 40506, USA

Synopsis We report on experiments that examine electron-helium scattering in the presence of an Nd:YAG laser field of 1.17 eV photons. At each incident electron energy (30, 60, and 200 eV), the laser polarization direction is varied within a plane perpendicular to the scattering plane. We compare our results with Kroll-Watson approximation calculations.

We are examining electron-helium elastic scattering in the presence of an Nd:YAG laser field ($h\nu = 1.17\,\text{eV}$). The first such laser-assisted free-free (LAFF) experiments were reported by Andrick and Langhans in 1976 [1], and Weingartshofer et al in 1977 [2]. Previously, we have reported results on such processes for incident energies from 50 – 350 eV for a fixed laser polarization direction [3].

Here, we report on experiments at three incident electron energies (30, 60, and 200 eV), where the laser polarization direction is varied within a plane perpendicular to the scattering plane. Specifically, we vary the polarization from being approximately parallel to the momentum transfer direction, to being perpendicular to it, in 30° increments. We compare our results with Kroll-Watson approximation (KWA)[4] calculations.

Of particular interest is the case where the polarization is perpendicular to the scattering plane, for which the KWA predicts vanishing cross section; other workers have found that the KWA tends to be inaccurate for those cases where it predicts small cross sections (e.g., [5, 6]). However, the KWA describes our results within the experimental uncertainties. A representive plot of our data and associated calculations is shown in figure 1.

This work has been supported by NSF Grant PHY-0855040 (NLSM), an IWU Artistic and Scholarly Development Grant and NSF Grant PHY-1402899 (BAd), and a Cornell Presidential Research Scholars Fellowship (BN).

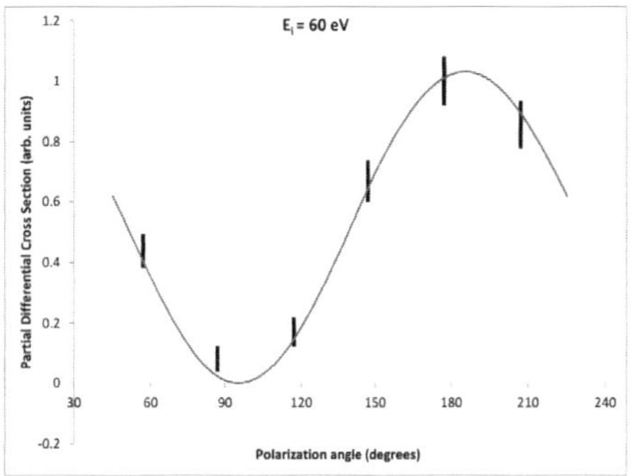

Figure 1. Angular distribution of electrons elastically scattered from helium that have emitted a single photon as a function of laser polarization direction (measured relative to the momentum transfer direction). Bars indicate experimental measurements (plus or minus one standard deviation). The solid line is a KWA calculation fitted to the data.

References

[1] D. Andrick and L. Langhans 1976 *J. Phys. B* **9** L1

[2] A. Weingartshofer, J. K. Holmes, G. Caudle, E. M. Clarke and H. Krger 1977 *Phys. Rev. Lett.* **39** 269

[3] B. A. deHarak, L. Ladino, K. B. MacAdam and N. L. S. Martin 2011 *Phys. Rev. A* **83** 022706

[4] N. M. Kroll and K. M. Watson 1973 *Phys. Rev. A* **8**, 804

[5] B. Wallbank and J.K. Holmes 2001 *Can. J. Phys.* **79** 1237

[6] M. O. Musa, A. MacDonald, L. Tidswell, J. Holmes, and B. Wallbank 2010 *J. Phys. B* **43** (17):175201

[1]E-mail: bdeharak@iwu.edu

Content from this work may be used under the terms of the Creative Commons Attribution 3.0 licence. Any further distribution of this work must maintain attribution to the author(s) and the title of the work, journal citation and DOI.
Published under licence by IOP Publishing Ltd

Atomic electron distribution and recoil effects follo ing beta decay of the halo nucleus helium-6

Gordon W.F. Drake [1], **Eva Schulhoff** [2]

Department of Physics, University of Windsor, Windsor, Ontario Canada N9B 3P4

Synopsis Following the β decay of a ^{6}He nucleus to form ^{6}Li, the atomic electrons adjust to the sudden change of nuclear charge from 2 to 3. The probabilities for electron shake-up and shake-off are calculated, including recoil effects, by the use of a Stieltjes imaging representation of the final states. The results will be applied to ongoing angular correlation studies in a search for new physics beyond the Standard Model.

There are currently several experiments in progress to search for new physics beyond the Standard Model by high precision studies of angular correlations in the β decay of the helium isotope $^6\text{He} \to {}^6\text{Li} + e^- + \bar{\nu}_e$ [1,2,3]. An essential part of the analysis is to understand the energy distribution and spectra of the recoil ions. After the β decay event, the atomic electrons suddenly find themselves in a $^6\text{Li}^+$ environment with nuclear charge $Z = 3$. The electrons redistribute themselves over all possible states of the $^6\text{Li}^+$ ion, including the continuum leading to $^6\text{Li}^{++}$ and $^6\text{Li}^{3+}$. Evidence for new physics beyond the Standard Model would reveal itself by an additional tensor coupling contribution to the weak interaction, in addition to the simple Gamow-Teller axial-vector mechanism [2,3].

We will present calculations employing Stieltjes imaging techniques in Hylleraas coordinates to study the probabilities for the shake-up and shake-off mechanisms, and especially the additional recoil accompanying the emission of the shake-off electrons. In the sudden approximation, one simply expands the initial ^{6}He atomic state in terms of the complete set of $^6\text{Li}^+$ states (including the ionization continuum) to obtain transition probabilities $P_{i\to f}$ in the form

$$P_{i\to f} = \left| \int \Psi_i^*(\text{He}) e^{-i\mathbf{K}\cdot\mathbf{r}} \Psi_f(\text{Li}^+)\, d\mathbf{r}_1 d\mathbf{r}_2 \right|^2 \quad (1)$$

where $\mathbf{r} = \mathbf{r}_1 + \mathbf{r}_2$, and $\mathbf{K}$ is the electronic recoil momentum for the daughter ion, and $K = |\mathbf{K}|$. Since $K_{\max} = 0.097$ a.u. for the β decay process, the factor of $e^{-i\mathbf{K}\cdot\mathbf{r}}$ can be expanded in powers of K, and terms up to order K^2 retained. Recoil terms of $O(K^2)$ come from both $S-S$ and $S-P$ transitions, but because of unitarity, they must sum to zero when summed over all shake-up and shake-off processes. The sum rule is illustrated by the results in Table 1 for the recoil coefficient $B_{i\to f}$, where $P_{i\to f}$ is expressed in the form

$$P_{i\to f} = A_{i\to f} + K^2 B_{i\to f} \quad (2)$$

and $149.2K^2 = E_{\text{rec}}$ is the ion recoil energy in keV. It is clear that the $S-P$ recoil term is canceled by the $S-S$ recoil contribution.

In summary, this work provides the first estimates of the recoil term for the ^{6}Li ions formed by the β decay of neutral ^{6}He atoms, and the validity of a sum rule is directly demonstrated. The total result $0.005(1)K^2$ for Li^{3+} is in reasonable accord with the value $0.0000412E_{\text{rec}} = 0.00615K^2$ obtained by Couratin *et al.* for the corresponding one-electron $^6\text{He}^+$ case [1]. Further results extended to higher accuracy for both the singlets and the triplets will be presented at the conference.

Table 1. Recoil contributions $K^2B_{i\to f}$ [see Eq. (2)] to probabilities for formation of Li^+, Li^{++}, and Li^{3+} following β decay of $^6\text{He}(1s^2\,{}^1S)$.

Ion	B_{S-S}	B_{S-P}	B_{Total}
Li^+	−0.665	0.573	−0.092
Li^{++}	−0.078	0.165	0.087
Li^{3+}	−0.009	0.014	0.005
Total	−0.752	0.752	0.000

References

[1] C. Couratin *et al.*, Phys. Rev. Lett. **108**, 243201 (2012).

[2] O. Naviliat-Cuncic and M. Gonzalez-Alonso, Ann. Phys. **525**, SI 600-619 (2013).

[3] G. Ban, D. Durand, X. Flechard *et al.*, Ann. Phys. **525**, SI 576 (2013).

[1] E-mail: gdrake@uwindsor.ca
[2] E-mail: schulh1@uwindsor.ca

Content from this work may be used under the terms of the Creative Commons Attribution 3.0 licence. Any further distribution of this work must maintain attribution to the author(s) and the title of the work, journal citation and DOI.
Published under licence by IOP Publishing Ltd

Muonic atom formation processes for carbon containing molecules

Go Yoshida* [1], **Kazuhiko Ninomiya***, **Makoto Inagaki***, **Wataru Higemoto**[†], **Takashi U. Ito**[†], **Naritoshi Kawamura**[§], **Kohichiro Shimomura**[§], **Yasuhiro Miyake**[§], **Taichi Miura**[‡], **Kenya M. Kubo**[⁏], **and Atsushi Shinohara***

[*] Graduate School of Science, Osaka University, Toyonaka, Osaka 560-0043, Japan
[†] Advanced Science Research Center, Japan Atomic Energy Agency, Tokai, Naka, Ibaraki 319-1192, Japan
[§] Muon Science Laboratory, High Energy Accelerator Research Organization (KEK), Tsukuba, Ibaraki 305-0801, Japan
[‡] Radiation Science Center, High Energy Accelerator Research Organization (KEK), Tsukuba, Ibaraki 305-0801, Japan
[⁏] Division of Natural Sciences, International Christian University, Mitaka, Tokyo 181-8585, Japan

Synopsis Muonic atom is an atomic system which has a negative muon substituted an electron. Muonic atom formation processes are strongly influenced by the chemical environment of muon capturing atom such as molecular structure, however, the initial processes of muon capture still have not been well investigated. In this study, we performed muon irradiation for gaseous carbon containing molecules to investigate the chemical environmental effect on muon capture processes.

A negative muon is one of the elementary particles (lepton) that has charge and approximately two hundred times heavier mass than that of an electron. When a negative muon is entered a substance, the muon gradually loses its kinetic energy by coulomb interaction, then the muon is captured by coulomb field of a nucleus in the substance and muonic atom is formed. When the muon is stopped in hydrogen containing substance, muon transfer process, becomes important. In this process, muonic hydrogen atom is firstly formed. Muonic hydrogen atom can diffuse in the substance easily like a neutron due to strong shielding effect of nuclear charge by muon. Thus, muonic hydrogen atom can approach to a nucleus of heavier atom and gives the muon to the deeper atomic levels in the heavier nucleus.

Captured muon has an individual quantum state like orbital electron, and the muon is highly excited state just after muonic atom formation. The captured muon immediately deexcites to muonic 1s state by emitting Auger electrons or characteristic X-rays (muonic X-rays). The energies of muonic X-rays are very high due to heavy mass of the muon. From muonic X-ray measurement, we can estimate initial quantum state of captured muon. Muon capture probabilities and initial quantum levels of captured muon are completely difference by the molecular structure of muon capturing atom and formation process of muonic atom (direct muon capture and muon transfer). However, this phenomenon still has not been well examined. In this study, to investigate muon capture process, low pressure simple carbon oxides; CO, CO_2 and mixtures with hydrogen; H_2+CO, H_2+CO_2 were used for muon irradiation samples. We measured muonic X-rays from these samples and estimated initial state of captured muon.

All experiments were performed at, J-PARC Japan. Sample gases below 100 kPa were put into a gas chamber and irradiated during 5 to 20 hour by muon beam with momentum of 18.8MeV/c. Muonic X-rays were measured by germanium semi-conductor detectors. Figure. 1 shows the muonic X-ray spectrum for H_2+CO_2 sample. From analyzing intensities of Lyman series muonic X-rays, we determined the muon capture probability of each atom and the initial quantum state of captured muon. In a case of a negative muon captured by carbon atom, a relative capture probability with oxygen atom and in an initial quantum state, are different between CO and CO_2 [1].

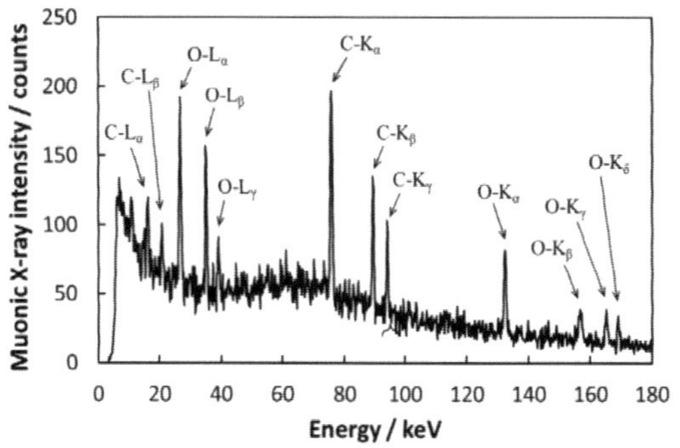

Figure 1. Muonic X-ray spectrum of H_2+CO

References

[1] G. Yoshida *et al* 2015 *J. Radioanal. Nucl. Chem.* **303** 1277

[1] E-mail: yoshidag12@chem.sci.osaka-u.ac.jp

Content from this work may be used under the terms of the Creative Commons Attribution 3.0 licence. Any further distribution of this work must maintain attribution to the author(s) and the title of the work, journal citation and DOI.
Published under licence by IOP Publishing Ltd

XXIX International Conference on Photonic, Electronic, and Atomic Collisions (ICPEAC2015) IOP Publishing
Journal of Physics: Conference Series **635** (2015) 052048 doi:10.1088/1742-6596/635/5/052048

Electron impact excitation of the low-lying $4s[3/2]_1$ and $4s'[1/2]_1$ levels in Ar atom

Y. Mochizuki*, H. Murai*, H. Kato*, M. Hoshino*,[1] and H. Tanaka*

*Department of Physics, Sophia University, Chiyoda-ku, Tokyo 102-8554, Japan

Synopsis We report measurements on differential cross sections (DCSs) for electron impact excitation of the low-lying levels, $4s[3/2]_1$ and $4s'[1/2]_1$, of the optically-allowed transitions in Ar atom in the electron energy range of 15 – 300 eV. Integral cross sections (ICSs) for these transitions obtained from the measured DCSs are compared with the BE*f*-scaled Born ICS.

We have been studying electron impact excitation of a number of atomic [1] and molecular [2] targets in order to extract absolute differential and integral cross sections (DCSs and ICSs), and more dedicated to obtain a comprehensive set of a measurements on optically-allowed transitions. These data have been compared with the BE*f*-scaled ICS, based on the Born approximation [3]. Such procedure has been carried out in order to verify the usefulness of the BE*f*-scaling law for some molecular targets. In the present work, we focus on Ar electronic excitation of the lowest-lying levels, $4s[3/2]_1$ and $4s'[1/2]_1$.

The present measurements were performed at Sophia University using a crossed-beam apparatus of a conventional high resolution electron spectrometer in incident electron energies of 15 – 300 eV at scattering angle range $15^\circ \leq \theta \leq 130^\circ$ with their energy resolution of about 45 meV (FWHM). The absolute scale of the DCS was obtained by the relative flow technique using the well-known elastic DCS of He [4]. ICSs for the excitation were obtained by extrapolating the measured DCSs. Errors in the DCS and ICS are estimated about 20% and 30% in this work, respectively. A typical energy loss spectrum of Ar atom is shown in Fig. 1, which has been recorded at an impact energy of 20 eV and scattering angle of 30°. As can clearly be seen in Fig. 1, that energy resolution was sufficient too large to resolve the four energy loss peaks of 4s transitions in the 11.3 – 12.0 eV energy-loss range.

The comparison of the present ICS results with the previous experimental ICS, generally speaking, shows a good agreement, although discrepancies are observed for the ICS in the low energy region.

The comparison the $3s[1/2]_1$ excitation in Ne of the present ICS results with the BE*f*-scaled ICS shows very good agreement over the wide energy region [1] in previous measurements. Furthermore, it is also found that the BE*f*-scaled ICSs for the transitions to both of levels in Ar reproduce well the experimental results, except for low energies below 30 eV.

In addition, we also compared our ICS data with the large-scale B-spline R-matrix calculations for Ar atom reported recently [5]. Further details will be discussed and presented at the poster session.

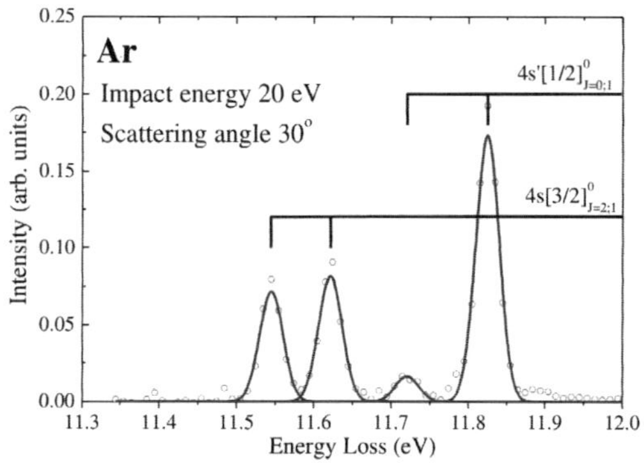

Figure 1. Typical energy loss spectrum of Ar atom.at an impact energy of 20 eV and scattering angle of 30°.

References

[1] M. Hoshino *et al* 2013 *J. Chem. Phys.* **139** 184301
[2] For example, K. Anzai *et al* 2012 *Eur. Phys. J. D* **66** 36
[3] Y.-K. Kim 2001 *Phys. Rev. A.* **64** 032713
[4] L. Boesten and H. Tanaka 1992 *At. Data Nucl. Data Tables.* **52** 25
[5] O. Zatsarinny *et al* 2014 *Phys. Rev. A* **89** 022706

[1]E-mail: masami-h@sophia.ac.jp

Content from this work may be used under the terms of the Creative Commons Attribution 3.0 licence. Any further distribution of this work must maintain attribution to the author(s) and the title of the work, journal citation and DOI.
Published under licence by IOP Publishing Ltd

Journal of Physics: Conference Series **635** (2015) 052049 doi:10.1088/1742-6596/635/5/052049

The theoretical studies on the electron-impact single ionization of Se^{3+}

D. H. Zhang * [1], **J. Q. Jing*, L. Y. Xie*, Y. Z. Qu**†, **J. Jiang* and C. Z. Dong ***

* Key Laboratory of Atomic and Molecular Physics & Functional Materials of Gansu Province, College of Physics and Electronic Engineering, Northwest Normal University, Lanzhou 730070, China
† College of Material Sciences and Optoelectronic Technology, University of the Chinese Academy of Sciences, Beijing 100049, China

Synopsis We have performed fully relatively level-to-level distorted-wave (LLDW) calculations for both direct ioonization (DI) and excitation autoionization (EA) of ground state Se^{3+} ion using the flexible atomic code (FAC). Large EA contributions come from the excitations 4p to *nl* and 3d to *nl*. Our calculated total DI+EA cross section have been compared with the recent experimental results. It suggests that the contributions of resonace-excitation double-autoionization process (REDA) to electron impact single ionization of Se^{3+} at the near threshold are more than 25%.

Electron-impact ionization is one of fundamental atomic process in hot dense plasmas occurring both in astrophysical and laboratory plasmas. Accurate atomic data including electron impact ionization (EII) cross section are of great importance to model and interpret the structure and dynamics of these plasmas, since one of the key steps in these studies is the knowledge of the charge state distribution.

Selenium is an important element in astrophysics. Recently, Alna'washi *et al.* [1] measured the absolute electron-impact single and multiple ionization cross-section using the dynamic-crossed-beams technique from threshold up to 1 kev and compared their experimental results with the Lotz semi-empirical formula prediction for direct ionization of the 4s and 4p subshells. Thus, no completed theoretical cross sections are available for comparison with the measurement.

In this work, we report fully relativistic lelel-to-level distorted-wave (LLDW) calculation for both direct ionization (DI) and excitation autoionization (EA) of Se^{3+} ion, by using the Flexible Atomic Code (FAC) [2]. In the EA calculation, the collisional excitation cross section is multiplied the explicit autoionization branching ratios [3].

For the single ionization threshold of the ground state for Se^{3+}, the present FAC result is 42.32 eV, NIST reference value of 42.95 eV, the experimental ionization onset is 42.2 ± 1.8 eV. The figure 1 shows the present calculated electron-impact single ionization (EISI) cross sections for Se^{3+} and the experimental results [1]. DI calculation included the contribution from direct removal of a 4p, 4s, 3d electron. $4s \rightarrow 4f, nl, (5 \le n \le 20, l \le 5)$ EA, and $3d \rightarrow nl, (4 \le n \le 20, l \le 5)$ EA were included into the calculation. The various important components to the total EISI cross section are plotted using colored areas. The gap between the calculation and the experiment from threshold up to 200 eV mainly come from the contributions of resonance-excitation double-autoionization (REDA) process. The REDA calculation are in progess.

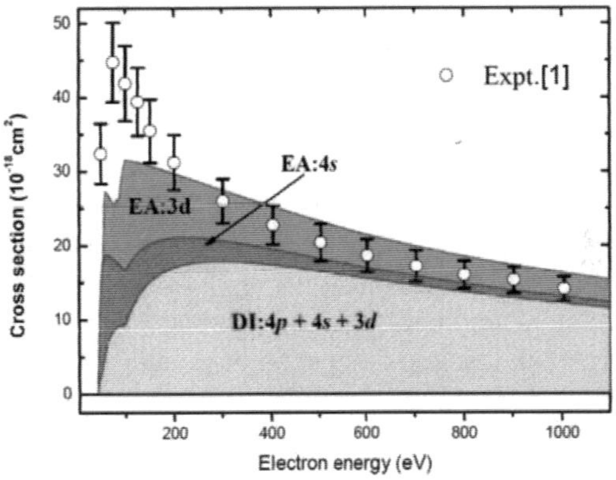

Figure 1. The calculated EISI cross sections for Se^{3+}. The total DI cross section and various EA cross section are shown by differently colored areas. The experimental results [1] are plotted using open circles.

This work was supported by the National Natural Science Foundation of China (grant nos 11464042, U1330117).

References

[1] G. A. Alna'washi *et al* 2014 *J. Phys.B* 47 105201
[2] M. F. Gu, 2008 *Can. J. Phys.*86 675
[3] D. H. Kwon *et al.* 2012 *Phys. Rev. A* 86 022701

[1] E-mail: zhangdh@nwnu.edu.cn

Content from this work may be used under the terms of the Creative Commons Attribution 3.0 licence. Any further distribution of this work must maintain attribution to the author(s) and the title of the work, journal citation and DOI.
Published under licence by IOP Publishing Ltd

XXIX International Conference on Photonic, Electronic, and Atomic Collisions (ICPEAC2015) IOP Publishing
Journal of Physics: Conference Series **635** (2015) 052050 doi:10.1088/1742-6596/635/5/052050

Observation of a resonance in the photodetachment of positronium negative ions

Koji Michishio*[1], Susumu Kuma[†], Tsuneto Kanai[†], Toshiyuki Azuma[†], Ken Wada[‡], Izumi Mochizuki[‡], Toshio Hyodo[‡], Akira Yagishita[‡] and Yasuyuki Nagashima*

[*] Department of Physics, Tokyo University of Science, Shinjuku, Tokyo 162-8601, Japan
[†] Atomic, Molecular and Optical Physics Laboratory, RIKEN, Wako, Saitama 351-0198, Japan
[‡] Institute of Materials Structure Science, High Energy Accelerator Research Organization (KEK), Tsukuba, Ibaraki 305-0801, Japan

Synopsis The positronium negative ion, $e^-e^+e^-$, is one of the simplest three body bound-states. As its constituents have a unique mass ratio, the ion is an attractive candidate for the investigation of the genuine three-body problem. In the present work, laser spectroscopy of this ion has been demonstrated in the UV region. A shape resonance near the positronium (n=2) formation threshold was clearly observed and the resonance energy was estimated.

The positronium negative ion (Ps^-), a bound-state of two electrons and a positron via Coulomb forces, is one of the simplest three body bound-states. As its constituents are only leptons with the same mass, Ps^- is an attractive candidate for the investigation of the quantum three-body problem.

Up to now, numerous theoretical studies have been conducted to explore its fundamental nature such as the bound-state properties and the annihilation rate. However experimental studies on Ps^- were limited to its observation [1] and the measurement of two-photon annihilation rate [2-4] due to weak Ps^- intensity.

In recent years, we have succeeded in generation of Ps^- efficiently by bombarding slow positrons onto alkali-metal coated surface [5]. This method has enabled us the first observation of the Ps^- photodetachment [6] and further optical studies. Here, we report on the laser spectroscopy of Ps^- in the UV region, where doubly-excited resonances and a shape resonance ($^1P^o$) have been predicted [7, 8].

The experiment was performed at KEK Slow Positron Facility. A slow positron beam from an electron LINAC was magnetically guided and injected onto a Na-coated polycrystalline tungsten surface. The Ps^- ions emitted from the surface were accelerated and then crossed with probe light in an electric-field free region. The light source was SHG (225 nm-231 nm) of a dye laser pumped by THG of a Nd:YAG laser. The fragment Ps atoms formed via the photodetachment were detected by a micro-channel plate.

The yields of the Ps atoms as a function of wavelength clearly showed a resonance structure, which was consistent with the shape resonance predicted theoretically [8]. Detail results will be presented at conference.

References

[1] A. P. Mills, Jr. 1981 *Phys. Rev. Lett.* **46** 717
[2] A. P. Mills, Jr. 1983 *Phys. Rev. Lett.* **50** 671
[3] F. Fleischer *et al.* 2006 *Phys. Rev. Lett.* **96** 063401
[4] H. Ceeh *et al.* 2011 *Phys. Rev. A* **84** 062508
[5] H. Terabe *et al.* 2012 *New J. Phys.* **14** 015003
[6] K. Michishio *et al.* 2011 *Phys. Rev. Lett.* 106 153401
[7] A. K. Bhatia and Y. K. Ho 1990 *Phys. Rev. A* 42 1119
[8] A. Igarashi *et al.* 2000 *New J. Phys.* **2** 17.1-17.4

[1] E-mail: michishio@rs.tus.ac.jp

Content from this work may be used under the terms of the Creative Commons Attribution 3.0 licence. Any further distribution of this work must maintain attribution to the author(s) and the title of the work, journal citation and DOI.
Published under licence by IOP Publishing Ltd

Journal of Physics: Conference Series **635** (2015) 052052 doi:10.1088/1742-6596/635/5/052052

Variational study of atomic hydrogen ionization by electrons impact

N.E.Houda Benmansour[3], Mohammed Sahlaoui[2], and Mammar Bouamoud[1]

[1] University Center of Naama, 45000 Naama, Algeria
[2] Ecole Préparatoire en Sciences et Techniques, 13000 Tlemcen, Algérie
[3] Theoretical Physics Laboratory, University AbouBekr Belkaid of Tlemcen, BP 119 Tlemcen 1300, Algeria

Triple differential cross section results of the electron impact ionization of hydrogen atom are performed in the so-called Schwinger-Born approximation of variational principle of Schwinger. In the theoretical calculations the first Born term was calculated analytically, and the second Born term was determined in the closure approximation using semi-analytical calculations. The theoretical results were corrected by the so-called Gamow factor in order to include post-collision effects, then compared with experiments performed with an incident electron energy of E_i = 250 eV and with those of other models. Better agreement was found between our results and experimental data even for high values of the ejected electrons energy (50 eV).

1. Introduction

We investigate in this work the ionization of hydrogen atom by electrons impact. The triple differential cross sections have been calculated in the framework of Schwinger-Born (SB) approximation of variational principle of Schwinger. The TDCS obtained in this work have been compared to the experimental results of Weigold *et al* [1] as well as to other theoretical ones [2],[3].

2. Theory

In the ionization of the atomic target by electron impact, the non-relativistic TDCS is given by

$$\frac{d^3\sigma}{dE_e d\Omega_e d\Omega_s} = \frac{1}{(2\pi)^2}\frac{k_e k_s}{k_i}|T_{\alpha\beta}|^2$$

$T_{\alpha\beta}$ is the matrix element describing the transition of the projectile target system from the initial state $|\alpha\rangle$ to the final state $|\beta\rangle$, which is the variational amplitude defined as

$$T_{\alpha\beta} = \sum_{i,j}\langle\beta|V|i\rangle\left(D^{-1}\right)_{ij}\langle j|V|\alpha\rangle$$

$$D_{ji} = \langle j|V - VG_T^{+}V|i\rangle$$

V is the target - projectile interaction and G_T^+ the target propagator.

In this study calculations were performed in the Schwinger-Born approximation, that is $|i\rangle = |\alpha\rangle$ and $|j\rangle = |\beta\rangle$ where $|\alpha\rangle = \varphi_{k_i}(r_0)\varphi_H(r_1)$ and $|\beta\rangle = \varphi_{k_s}(r_0)\psi_{k_e}^{(-)}(r_1)$. The incident and scattered particles are described by a plane wave. $\psi_{k_e}^{(-)}(r_1)$ is the continuum state of the ejected electron described by the Coulomb wave function which satisfies the incoming boundary condition.

To include the post-collision interaction (PCI) between the outgoing scattered and ejected electrons in the TDCS, the amplitude $T_{\alpha\beta}$ is multiplied by the Gamow factor [4].

3. Results and discussion

The figure shows the measurements of TDCS of Weigold *et al* [1] for electron-impact ionization of H for an incident energy E_i = 250 eV an ejected energy E_e = 50 eV and a scattered angle θ_s= 25^0 , our results of SB, first Born and second Born [5] approximations, and those of BBK model [2] and second Born approximation of Dal Cappello *et al* [3] including 294 intermediate states . SB results

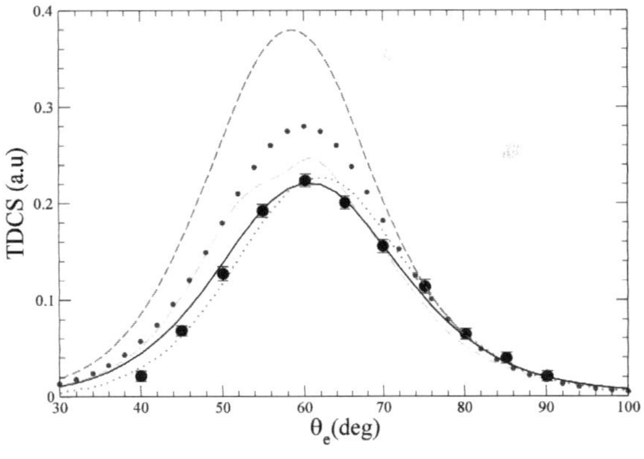

First Born:*small circles*. Second Born:*dashed line*. Dal Cappello Second Born:*dashed-dot*.BBK:*dot line*. SB:*fullline*. Experiments: *full circles with error bars*.

4. References

[1] E. Weigold *et al* 1979 *J. Phys. B: At. Mol. Phys.* **12** 291

[2] M. Brauner *et al* 1989 *J. Phys. B: At. Mol. Phys.* **22** 2265

[3] C. Dal Cappello *et al* 2013 *J. Phys. B: At. Mol. Opt. Phys.* **46** 145203

[4] A.S. Kheifets *et al* 2008 *J. Phys. B: At. Mol. Opt. Phys.* **41** 145201

[5] M. Sahlaoui and M. Bouamoud 2012 *J. Phys. B: At. Mol. Opt. Phys.* **45** 085201

Content from this work may be used under the terms of the Creative Commons Attribution 3.0 licence. Any further distribution of this work must maintain attribution to the author(s) and the title of the work, journal citation and DOI.
Published under licence by IOP Publishing Ltd

Ionization of atoms and molecules by electron impact: a three body wave treatment

S. Houamer[*,1], **C. Dal Cappello**[$] **and M. Chinoune**[*,£]

[*]LPQSD, Department of physics, Faculty of Science, University Sétif-1, 19000, Setif, Algeria
[$]SRSMC, CNRS 7565, Univ. of Lorraine, BP 70239, 54506 Vandoeuvre-les-Nancy, France

[£]USTHB, Faculty of Physics, Department of theoretical physics, Algiers 16111, Algeria

Synopsis Triple differential cross section for ionization of atomic and molecular targets by electron impact is calculated using approximate three body waves. The method is an alternative of the well known BBK model where the distortion of the ejected electron is taken into account. Results are compared with available data.

(e,2e) experiments constitute a privileged tool for studying the dynamics of electron impact ionization of atoms and molecules and more generally for contributing to the understanding of the N- body interaction problem. The final state consists of a three body continuum system which represents a challenge even for the simplest targets like hydrogen.

The ionization process at the multi differential level needs then sophisticated theoretical calculations [1]. We consider here an approach in which three different mechanisms are considered to investigate the ionization mechanism for some atoms and small molecules: (i)a first one where the ejection electron is represented by a Coulomb wave taking into account its interaction with the residual ion. (ii)a second one, where all the interactions in the final state are considered and supposed to be Pure Coulombic. (iii)and a third one where all the interactions are considered with a distortion of the ejected electron.

The three models are noted 1C, BBK-CW and BBK-dw1 for convenience. The method is applied to study the ionization process in the case of some atoms and molecules [2,3]. In figure 1 the TDCS is displayed in the case of Helium ionization at about 600 eV projectile energy. It is seen that the BBK-CW is able to exhibit an excellent agreement with experiments in shape and amplitude. We come to the conclusion that the interactions in the final state are to be considered especially in the recoil region. The distortion of the ejected electron is found to be important while investigating inner orbitals for the atomic and molecular targets.

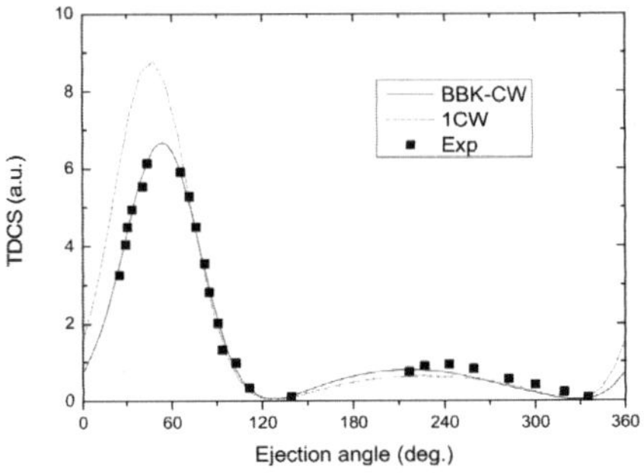

Figure 1. TDCS vs ejection angle for the ionization of helium by electron impact at fixed scattering angle $\theta_s = 6°$. Kinematical parameters are: Ei=598.6 eV, Ee=74 eV, Es=500 eV [3].

References

[1] M. Brauner, J. S. Briggs and H. Klar 1989 J. Phys. B **22,** 2265

[2] A. Lahmam-Bennani et al 2009 J. Phys. B42 5201 [3] A. Lahmam-Bennani et al 1984 Phys. Rev. A**30** 1511

[1]E-mail: s_houamer@univ-setif.dz

Content from this work may be used under the terms of the Creative Commons Attribution 3.0 licence. Any further distribution of this work must maintain attribution to the author(s) and the title of the work, journal citation and DOI.
Published under licence by IOP Publishing Ltd

Electron scattering by silver: excitation of the $4d^95s^2\ ^2D_{3/2}$ and $4d^{10}6s\ ^2S_{1/2}$ states

S. D. Tošić [1], V. Pejčev, D. Šević and B. P. Marinković

Institute of Physics, University of Belgrade, Pregrevica 118, Belgrade, Serbia

Synopsis Differential cross sections (DCSs) for electron impact excitation of the $4d^95s^2\ ^2D_{3/2}$ and $4d^{10}6s^2\ ^2S_{1/2}$ states of silver atom have been measured. Relative DCSs have been obtained by comparing intensities of these states relative to the resonant ^2P state at scattering angles from 10° to 150° and at electron impact energies from 10 eV to 60 eV.

The aim of our study is to provide reliable electron scattering data which are of great importance for a better understanding of electron interactions with silver atoms as well as for many scientific and practical applications [1]. The silver atom is very important in the field of laser cooling and trapping techniques.

Here we present experimentally obtained relative differential cross sections (DCSs) for electron impact excitation of the $4d^95s^2\ ^2D_{3/2}$ (4.304 eV) and $4d^{10}6s^2\ ^2S_{1/2}$ (5.276 eV) states of silver atom from the ground $4d^{10}5s\ ^2S$ state. The DCSs have been determined by using inelastic ($4d^95s^2\ ^2D_{3/2}$, $4d^{10}6s^2\ ^2S_{1/2}$)-to-inelastic ($4d^{10}5p\ ^2P_{1/2,3/2}$ resonance state) intensity ratios and relative DCSs for the resonance state. The results have been obtained at electron impact energies (E_0) of 10, 20, 40 and 60 eV and at scattering angles from 10° to 150° in steps of 10°.

Our scattering experiment is performed in crossed electron-atom beam arrangement in which an atomic beam is perpendicularly crossed by monoenergetic electron beam. The electron spectrometer ESMA [1] consists of a hemispherical electron energy selector with a hair-pin thermo-electron source and an energy analyzer of the same type with a channel electron multiplier as a detector. The analyzer can rotate from -30° up to 150° with respect to the zero angle defined by the axis of the electron beam.

The well collimated effusive Ag vapour beam has been produced by heating oven crucible containing silver atoms. All the elements of the system are placed inside a high vacuum chamber. The real zero position of the scattered signal has been determined from the symmetry at positive and negative scattering angles.

The inelastically scattered electron current intensities are detected as a function of scattering angle up to 150° and then converted to relative DCSs using the effective path length correction factors [2] determined for the present experimental conditions. Obtained relative inelastic DCSs for electron impact excitation of the $4d^95s^2\ ^2D_{3/2}$ and $4d^{10}6s^2\ ^2S_{1/2}$ states of silver at 40 eV electron impact energy are presented in figure 1.

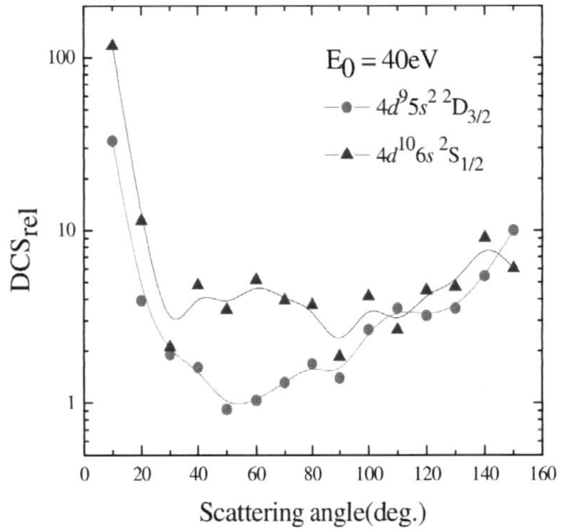

Figure 1. Relative differential cross sections for the excitation of the $^2D_{3/2}$ and $^2S_{1/2}$ states of silver atom at 40 eV electron impact energy as a function of scattering angle.

It is interesting and important to continue investigations of electron excitation processes in silver atoms in order to obtain a more profound knowledge and to obtain more complete sets of scattering data that could be used in theoretical modeling and numerical simulations.

References

[1] S. D. Tošić *et al* 2012 *NIMB* **279**, 53
[2] R. T. Brinkman and S. Trajmar 1981 *J. Phys. E: Sci. Instrum.* **14**, 245

[1]E-mail: seka@ipb.ac.rs

Content from this work may be used under the terms of the Creative Commons Attribution 3.0 licence. Any further distribution of this work must maintain attribution to the author(s) and the title of the work, journal citation and DOI.
Published under licence by IOP Publishing Ltd

XXIX International Conference on Photonic, Electronic, and Atomic Collisions (ICPEAC2015) IOP Publishing
Journal of Physics: Conference Series **635** (2015) 052055 doi:10.1088/1742-6596/635/5/052055

Origins of Very-Low and Zero-Energy Electron Structures in Strong-Field Ionization with Intense Mid-IR Pulses

B. Wolter*[1], C. Lemell†, M. Baudisch*, M. G. Pullen*, X.-M. Tong‡, M. Hemmer*, A. Senftleben•, M. Sclafani*, C. D. Schröter°, J. Ullrich°,+, R. Moshammer°, J. Burgdörfer†, J. Biegert*,⊕

* ICFO - Institut de Ciències Fotòniques, Mediterranean Technology Park, Castelldefels (Barcelona), Spain
† Institute for Theoretical Physics, Vienna University of Technology, Vienna, Austria
‡ Center for Computational Sciences, University of Tsukuba, Ibaraki, Japan
• Universität Kassel, Institut für Physik and CINSaT, Kassel, Germany
° Max - Planck - Institut für Kernphysik, Heidelberg, Germany
+ Physikalisch - Technische Bundesanstalt, Braunschweig, Germany
⊕ ICREA - Institució Catalana de Recerca i Estudis Avançats, Barcelona, Spain

Synopsis Intense long wavelength ($\lambda \geq 2\,\mu$m) laser pulses enable experiments in the tunneling ionization regime ($\gamma \ll 1$) and reveal surprising low electron energy features, which can not be described with the strong-field approximation (SFA). These features, universal for all target species, include the low-energy structure (LES), the very-low-energy structure (VLES) and the zero-energy structure (ZES). Using full 3D electron-ion coincidence detection in combination with our ultrafast 160 kHz mid-IR source, we reveal the entire 3D momentum spectrum well below 1 eV. Quantum and classical simulations allow for an interpretation of the LES, VLES and of the newly identified ZES.

In strong-field ionization for $\gamma \ll 1$, the post-tunneling behavior of the electron wave packet is expected to be dominated by the laser electric field rather than the Coulomb field of the parent ion. With the recent availability of intense mid-IR pulses ($\lambda \geq 2\,\mu m$) this simplification of tunneling-ionization has changed. Several unexpected low-electron-energy features such as the low-energy structure (LES) [1], the very-low energy structure (VLES) [2] and the zero-energy structure (ZES) [3] were found that could not be described within the strong-field approximation (SFA). To date classical calculations can explain the LES in terms of multiple recollisions of the tunnel-ionized electron at the Coulomb field of the residual ion. Yet the VLESs have not been quantitatively characterized and the origin of the ZES remained speculative.

Here we present a joint experimental and theoretical study of the 3D momentum space of photoelectrons which allows for a quantitative interpretation of all three structures [4]. Using a reaction microscope (ReMi) in combination with our ultrafast 160 kHz mid-IR source, we accurately resolve multiple orders of LES, the VLES and the ZES in momentum space. By comparing the experimental data with classical trajectory Monte Carlo (CTMC) and quantum simulations, we find agreement for all structures. Further analysis permits an identification of the VLES in terms of high-order two-dimensional Coulomb focusing. Moreover the ZES is conclusively attributed to electrons being recaptured into high-lying Rydberg states via frustrated tunnel ionization (FTI) [5] followed by subsequent field ionization by the static electric field of the ReMi.

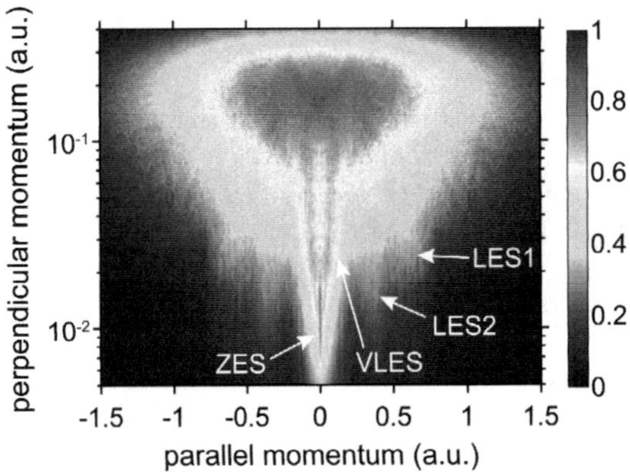

Figure 1. Experimental electron momentum distribution for argon interacting with 6-cycle mid-IR pulses with a peak intensity of $I_0 = 0.9 \times 10^{14}\,\mathrm{W/cm^2}$.

References

[1] C. Blaga *et al* 2009 *Nat. Phys.* **5** 335-338

[2] C. Y. Wu *et al* 2012 *Phys. Rev. Lett.* **109** 043001

[3] J. Dura *et al* 2013 *Sci. Rep.* **3** 2675

[4] B. Wolter *et al* 2014 *Phys. Rev. A* **90** 063424

[5] T. Nubbemeyer *et al* 2008 *Phys. Rev. Lett.* **101** 233001

[1]E-mail: benjamin.wolter@icfo.eu

Content from this work may be used under the terms of the Creative Commons Attribution 3.0 licence. Any further distribution of this work must maintain attribution to the author(s) and the title of the work, journal citation and DOI.
Published under licence by IOP Publishing Ltd

Journal of Physics: Conference Series **635** (2015) 052056 doi:10.1088/1742-6596/635/5/052056

Precision Relativistic eigenchannel R-matrix calculation of the spectroscopy and electron O^+ ion scattering in Oxygen system

De-Ling Zeng*‡, Xiang Gao‡ 1, Rui Jin†‡, Jia-Ming Li*†ǂ

* Key Laboratory for Laser Plasmas (Ministry of Education) and Department of Physics and Astronomy, Shanghai Jiao Tong University, Shanghai 200240, China

‡ Beijing Computational Science Research Center, Beijing 100193, China

† Department of Physics and Center for Atomic and Molecular Nanosciences, Tsinghua University, Beijing 100084, China

ǂ Collaborative Innovation Center of Quantum Matter, Beijing 100084, China

Synopsis The scattering matrices of Oxygen in various energy regions are calculated using our modified eigenchannel R-matrix method R-R-eigen. We can obtain abundant atomic dynamical data with accuracies determined based on analytical continuation properties of the scattering matrices. Our calculated discrete and autoionization spectrum can be readily examined with the experiment data. Our result agrees well with experiment. By comparing with available spectroscopic data using our newly proposed projected high dimensional quantum-defect graph (symmetrized), which is a generalization of Lu-Fano plots, we can readily determine the precision of scattering calculations in a systematical way.

Oxygen is one of the abundant elements in astronomy and astrophysics. The ratio between the two transitions $^2D^o_{5/2,3/2} \rightarrow {}^4S^o_{3/2}$ of O^+ ground state is used to diagnose the electron densities[1] in planetary nebulas. In the limit of high electron density, the line intensities are proportional to the radiative transition rates and the statistical weights of the ionic states. Our previous work has given a recommended ratio value $r(\infty) = 0.363 \pm 0.009$ [2], which is in the overlap range of two newly experimental measurements. In the limit of low electron density, the line intensities are proportional to the collision rate coefficients, which are related with electron collision cross sections. Over the past decades, a large amount of calculations have been carried out to investigate the collision cross sections. However the theoretical results between different calculations show distinct discrepancies.

Using our modified eigenchannel R-matrix method R-R-eigen [3], we can obtain abundant atomic dynamical data with accuracies determined based on analytical continuation properties of the scattering matrices. This method will help us determine calculation precisions of the collision cross sections, which in turn can make the precision of the related line ratio determined. A large number of partial waves would be included in the collision excitation calculation, here we choose 0^- partial wave as an example. The excitation energy from the O^+ ground state $[1s^2 2s^2 2p^3]^4S_{3/2}$ to $^2D_{5/2}$ is 0.244 Ryd, the energy of the colliding electron should be at least 1 Ryd to guarantee the rate integration. As the energy increases, more physical channels should be open. This is why the e+O^+ system is difficult to calculate. Using our modified R-matrix code R-R-Eigen, we calculated the eigenchannel quantum defects of the 0^- partial wave. All the quantum defects vary smoothly with the energy, and the same physical channels in different energy regions are smoothly connected. These show that we have properly considered the physical channels. With the physical parameters, we can obtain the discrete energy levels and autoionization resonant levels without missing any lines. By comparing with available spectroscopic data using our newly proposed projected high dimensional quantum-defect graph (symmetrized), which is a generalization of Lu-Fano plots, we can readily determine the precision of scattering calculations in a systematical way.

References

[1] M. J. Seaton and D. E. Osterbrock, 1957 Astrophys. J. 125, 66
[2] X. Y. Han, X. Gao, D. L. Zeng, J. Yan and J. M. Li, 2012 PHYSICAL REVIEW A 85, 062506
[3] X. Gao and J. M. Li, 2014 PHYSICAL REVIEW A **89**, 022710

‡1 E-mail: xgao@csrc.ac.cn

Content from this work may be used under the terms of the Creative Commons Attribution 3.0 licence. Any further distribution of this work must maintain attribution to the author(s) and the title of the work, journal citation and DOI.
Published under licence by IOP Publishing Ltd

Laser-assisted singly and doubly ionizing electron-helium collisions at high impact energy and large momentum transfer

Andrew A. Bulychev* [1], **Konstantin A. Kouzakov**† [2]

* Laboratory of Theoretical Physics, Joint Institute for Nuclear Research, Moscow Region, 141980 Dubna, Russia

† Department of Nuclear Physics and uantum Theory of Collisions, Faculty of Physics, Lomonosov Moscow State University, 119991 Moscow, Russia

Synopsis Electron-impact single and double ionization of a helium atom in the presence of laser radiation with low frequency and intensity is studied theoretically. The kinematical regime of high impact energy and large momentum transfer, with two fast electrons in the final channel, is considered. Fully differential cross sections corresponding to the laser-assisted $(e, 2e)$ and $(e, 3e)$ processes are calculated for different models of the target state. The laser-assisted cross sections are found to be more sensitive to the electron-electron correlations in helium than the field-free ones.

In the present work, we carry out a theoretical analysis of the laser-assisted $(e, 2e)$ and $(e, 3e)$ scattering in helium. A low-frequency laser field of relatively weak strength is considered, such that the Keldysh parameter is $\gamma \gg 1$. We focus on the kinematical regime of high impact energy and large momentum transfer, which is typical for electron momentum spectroscopy (EMS). This validates the use of the binary-encounter and first Born approximations for the S matrix of the process, thus strongly simplifying theoretical treatment of ionization dynamics, both in the absence and in the presence of a laser field. Another reason for inspecting the EMS kinematics is that in our recent theoretical study on laser-assisted EMS of helium with excitation of the residual He^+ ion [1] we found much higher sensitivity to electron-electron correlations in the target than in the field-free case. We also have shown that the laser-assisted $(e, 3e)$ EMS has even greater potential in this respect [2]. This potential can be realized most effectively by measuring angular distributions of a slow ejected electron for different multiphoton processes. Fig. 1 shows numerical results for different number of photons N when the laser electric field with frequency $\hbar\omega = 1.55\,\text{eV}$ and intensity $I = 1.55 \times 10^{11}\,\text{W/cm}^2$ is parallel to the incident electron momentum. The angular distributions appear to be strongly sensitive to the model of electron-electron correlations in helium (RHF, SPM, and BK), while in the absence of a laser field they are almost insensitive to this model. When the slow ejected electron is not measured, that is, in the $(e, 3-1e)$ case, we find the laser-assisted results to be practically indistinguishable from the field-free ones, thus indicating the applicability of the so-called Kroll-Watson sum rule.

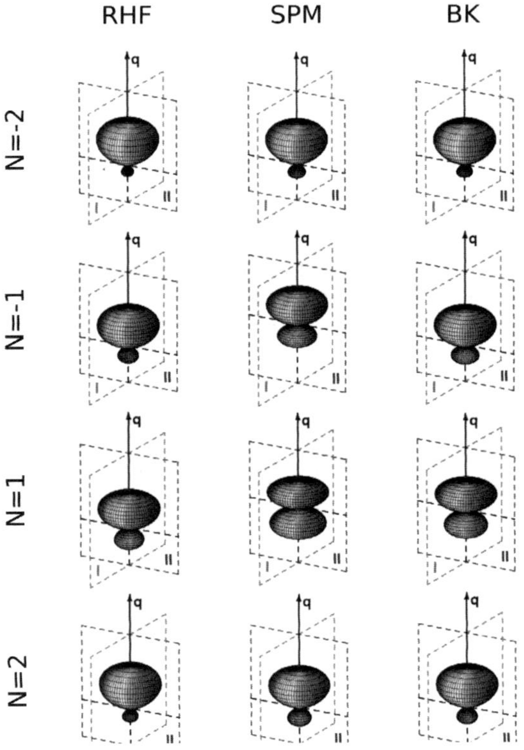

Figure 1. Angular distributions of the slow ejected electron in the laser-assisted $(e, 3e)$ EMS process in the case of symmetric kinematics at $E_0 = 6079\,\text{eV}$.

References

[1] A. A. Bulychev and K. A. Kouzakov 2014 *Eur. Phys. J. D* **68** 354

[2] A. A. Bulychev and K. A. Kouzakov 2015 *Phys. Rev. A* **91** 023413

[1]E-mail: bulychev@theor.jinr.ru
[2]E-mail: kouzakov@srd.sinp.msu.ru

Content from this work may be used under the terms of the Creative Commons Attribution 3.0 licence. Any further distribution of this work must maintain attribution to the author(s) and the title of the work, journal citation and DOI.
Published under licence by IOP Publishing Ltd

Journal of Physics: Conference Series **635** (2015) 052058 doi:10.1088/1742-6596/635/5/052058

Electron-impact ionization of W^{26+} ion

Valdas Jonauskas[1], **Aušra Kynienė**[2], **and Šarūnas Masys**[3]

Institute of Theoretical Physics and Astronomy, Vilnius University, A. Goštauto 12, LT-01108 Vilnius, Lithuania

Synopsis Electron-impact ionization for W^{26+} ion is investigated using Dirac-Fock-Slater approximation. The large scale calculations are performed for the excitations to over 800 configurations. Influence of radiative damping on the excitation-autoionization cross-sections is estimated.

Energy losses determined by radiation of impurities in various ionization stages have to be controlled for the successful ignition of a deuterium-tritium plasma in thermonuclear reactors. Ionization balance in the plasma is mainly established by electron-impact ionization and recombination processes.

The aim of the current work is to study influence of the excitations to the high-nl shells for electron-impact ionization of the W^{26+} ion which has two $4f$ electrons in the valence shell. The electron-impact excitations from the levels of the ground configuration to the levels of over 800 configurations of the W^{26+} ion are analyzed in order to find out contribution to the ionization cross-sections of the excitations to the high-nl shells ($n \leq 40$, $l \leq 6$). In addition, all electric dipole and Auger transition probabilities from the excited configurations are investigated to determine influence of radiative damping on the excitation-autoionization (EA) cross-sections. Besides to the electron-impact ionization cross-sections, the current study provides Maxwellian rate coefficients of the ionization process.

Our current results for contribution to the EA cross-sections of excitations to high-nl shells from the ground configuration are presented in Figure 1. The contribution of the EA channels from the high-nl shells ($n \geq 9$) is approximately equal to the contribution of the shells with $n \leq 8$.

Contribution of the excitations to shells with the orbital quantum number $l = 4$ is the largest for the excitations to the $n \leq 8$ and the higher ($9 \leq n \leq 40$) shells. The same tendency was observed for the W^{27+} ion the ground configuration of which has only one electron in the $4f$ valence shell [1].

The largest contribution to the EA process corresponds to the excitations from the $4p$ shell when investigation includes only the shells with $n \leq 8$. However, the contribution from the $4d$ shell dominates for the higher shells with $9 \leq n \leq 40$.

Current study shows that contribution to the total ionization cross-sections of excitations to high-nl shells amounts to about 15 % at the peak of the data. The total EA process contributes about 40 % to the ionization cross-sections. These results contradict conclusions about relatively small EA contribution to the cross-sections obtained from the semirelativistic calculations [2]. It can be explained by the fact that the configuration-average approach was used and important excitations to the shells with orbital quantum number $l = 4$ were omitted in the previous investigation.

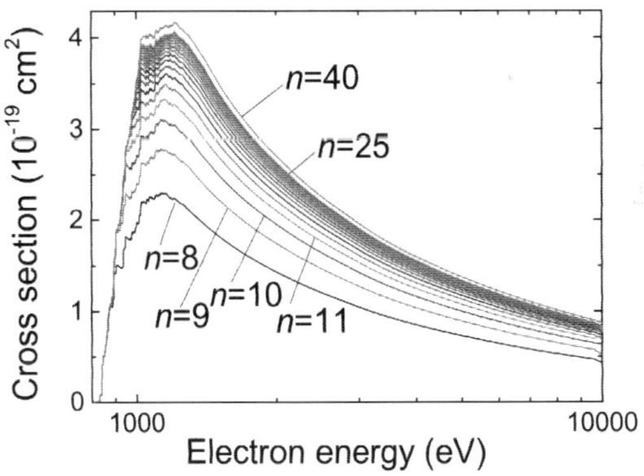

Figure 1. EA channels to the high-nl shells for the ground subconfiguration of the W^{26+} ion.

Acknowledgment: This research was funded by European Social Fund under the Global Grant Measure (No.: VP1-3.1-SMM-07-K-02-015)

References

[1] V. Jonauskas *et al* 2015 *Phys. Rev.* **A 91** 012715

[2] S. D. Loch *et al* 2005 *Phys. Rev.* **A 72** 052716

[1]E-mail: valdas.jonauskas@tfai.vu.lt
[2]E-mail: ausra.kyniene@tfai.vu.lt
[3]E-mail: sarunas.masys@tfai.vu.lt

Content from this work may be used under the terms of the Creative Commons Attribution 3.0 licence. Any further distribution of this work must maintain attribution to the author(s) and the title of the work, journal citation and DOI.
Published under licence by IOP Publishing Ltd

Journal of Physics: Conference Series **635** (2015) 052059 doi:10.1088/1742-6596/635/5/052059

Theoretical investigation of spectroscopic properties of W^{27+} ion

Valdas Jonauskas[1], **Aušra Kynienė, Pavel Rynkun, Laima Radžiūtė, Gediminas Gaigalas, Romas Kisielius, Sigitas Kučas, Šarūnas Masys, Gintaras Merkelis**

Institute of Theoretical Physics and Astronomy, Vilnius University, A. Goštauto 12, LT-01108 Vilnius, Lithuania

Synopsis Analysis of energy levels and radiative transition probabilities for the W^{27+} ion is performed using multiconfiguration Dirac-Fock method. Configuration interaction strength has been used to determine the influence of correlation effects.

Concentration of tungsten as an intrisic impurity in fusion devices has to be monitored in order to create and maintain the fusion reaction. Thus, reliable atomic data for various tungsten ions are needed for predicting possible plasma scenarios.

The aim of the current work is to investigate energy levels, radiative transition probabilities, and relativistic and correlation effects for W^{27+}. We focus our attention on the emission in spectral range $2-30$ nm where intensive radiation of tungsten ions has been theoretically and experimentally studied.

The investigation of energy levels and electric dipole, quadrupole, and octupole as well as magnetic dipole and quadrupole transitions is performed using multiconfiguration Dirac-Hartree-Fock method implemented in GRASP2K code [1]. The extended average level (EAL) approximation is employed in our study. Relativistic effects with quantum electrodynamic corrections are taken into account to provide accurate wavelengths and radiative transition probabilities.

The lowest configurations of the W^{27+} ion are shown in Figure 1. The levels having the largest lifetimes and levels to which radiative transition takes place from these levels are analyzed.

The ground configuration for the W^{27+} ion has one electron in the valence $4f$ shell. The first excited [Kr]$4d^{10}5s$ configuration can decay only through weak electric octupole transitions to the ground configuration. Previous study for the W^{25+} ion demonstrated that configuration mixing opens electric dipole transitions from the first excited [Kr]$4d^{10}4f^25s$ configuration to the ground one [2]. The same results were obtained for the [Kr]$4d^{10}4f5s$ configuration in the W^{26+} ion. However, the electric dipole transition from the [Kr]$4d^{10}5s$ configuration is forbidden for the W^{27+} ion by $\Delta J = 0, \pm 1$ selection rule.

[1]E-mail: valdas.jonauskas@tfai.vu.lt

Configuration interaction strengths T are calculated for the $4f$, $5l$ ($l = 0, 1, 2, 3, 4$), $4d^94f^2$, $4d^94f5s$ and $4d^94f5p$ configurations of the ion. The T value, divided by the statistical weight g_1 of the analyzed configuration has the meaning of the average weight of the admixed configuration in the expansion of the wave function for the initial configuration. The obtained data show that the core and core-core correlations play the major role compared to the valence and valence-valence correlations for all configurations among which transitions form the strongest lines in the spectrum of the W^{27+} ion. It has to be noted that the $4d^{-2}4f^2$ correlation is the strongest for these configurations.

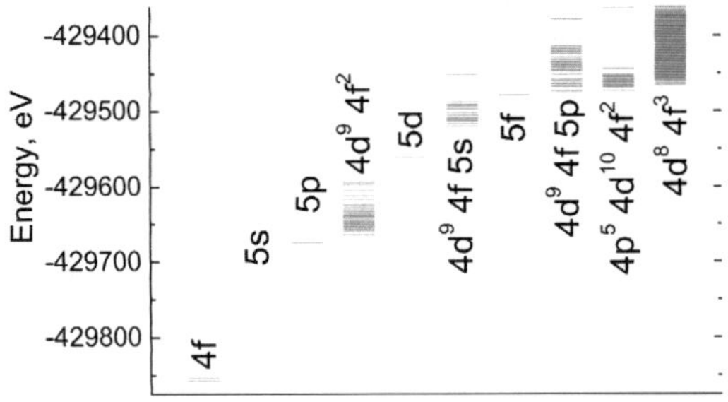

Figure 1. The lowest configurations of the W^{27+} ion.

Acknowledgment: This research was funded by European Social Fund under the Global Grant Measure (No.: VP1-3.1-SMM-07-K-02-015)

References

[1] P. Jönsson *et al* 2013 *Com ut Phys Commu* **184** 9

[2] A. Alkauskas *et al* 2014 *J Qua t S ect Rad at Tra sfer* **136** 108

Content from this work may be used under the terms of the Creative Commons Attribution 3.0 licence. Any further distribution of this work must maintain attribution to the author(s) and the title of the work, journal citation and DOI.
Published under licence by IOP Publishing Ltd

XXIX International Conference on Photonic, Electronic, and Atomic Collisions (ICPEAC2015) IOP Publishing
Journal of Physics: Conference Series **635** (2015) 052060 doi:10.1088/1742-6596/635/5/052060

Energy levels, transition rates and lifetimes for Li-like ions with $Z \leq 10$ in the $1s2s(^3S)3\ell$ states

J. P. Santos[†1], J. P. Marques[*], M. C. Martins[†], P. Indelicato[‡], E. P. Benis[⋆], T. J. M. Zouros[•], F. Parente[†]

[†] Laboratório de Instrumentação, Engenharia Biomédica e Física da Radiação (LIBPhys-UNL) e Departamento de Física da Faculdade de Ciências e Tecnologia da Universidade Nova de Lisboa, Monte da Caparica, 2892-516 Caparica, Portugal

[*] BioISI - Biosystems & Integrative Sciences Intitute, Faculdade de Ciências da Universidade de Lisboa, Campo Grande, C8, 1749-016 Lisboa, Portugal

[‡] Laboratoire Kastler Brossel, École Normale Supérieure, CNRS, Sorbonne Universités, UPMC Univ. Paris 06, Case 74; 4, place Jussieu, 75252 Paris CEDEX 05, France

[⋆] Department of Physics, University of Ioannina, GR 45110 Ioannina, Greece

[•] Department of Physics, University of Crete, P.O Box 2208, GR 71003 Heraklion, Greece

Synopsis Energy levels, transition rates and lifetimes for Li-like ions with $Z \leq 10$ in the $1s2s(^3S)3\ell$ states were calculated using the Dirac-Fock approach.

In the continuation of the study of the selective enhancement of $1s2sn\ell$ metastable states populated by cascades in single-electon transfer collisions of ions with He and H_2 targets [1], we calculated the energy levels, transition rates and lifetimes for Li-like ions with $Z \leq 10$ in the $1s2s(^3S)3\ell$ states states using the multi-configuration Dirac-Fock (MCDF) code of Desclaux and Indelicato [2, 3].

Table 1. Energy levels (EL), Auger and radiative transition rates (ATR, RTR, respectively) and lifetimes for Ne Li-like ions in the $1s2s(^3S)3s$ $^{2,4}S$ states

	$^2S_{1/2}$	$^4S_{3/2}$
EL (eV)	-1750.35	-1760.53
ATR (s^{-1})	3.23×10^{13}	3.02×10^{13}
RTR (s^{-1})	3.16×10^{10}	6.69×10^{10}
Lifetime (s)	3.10×10^{-14}	3.30×10^{-14}

The radiative and radiationless decay rates were calculated using the code in the single-configuration approach, with the Breit interaction and the vacuum polarization terms included in the self-consistent field calculation, and other QED effects, such as self-energy, included as perturbations [4]. Regarding the radiationless transitions, we have assumed a two-step process, in which the decay is independent from the ionization. Hence, the two electrons do not interact with each other and the core hole state interacts very weakly with the continuum electron, allowing for the transition rates to be calculated from perturbation theory. Initial-state wavefunctions were generated for configurations that contain one initial inner-shell vacancy while final state wavefunctions were generated for configurations that contain two higher shell vacancies. Continuum-state wavefunctions were obtained by solving the Dirac-Fock equations with the same atomic potential of the initial state, normalized to represent one ejected electron per unit energy. The preliminary results obtained for Ne Li-like ions in the $1s2s(^3S)3s$ $^{2,4}S$ states are listed in Table 1.

This work was supported by FCT Project No. PTDC/FIS/117606/2010, Portugal. We also thank the Allianz Program of the Helmholtz Association, contract No. EMMI HA-216 Extremes of Density and Temperature: Cosmic Matter in the Laboratory. BioISI is supported by the centre grant UID/MULTI/04046/2013 from FCT/MCTES/PIDDAC, Portugal. EPB and TJMZ are co-financed by the European Union (European Social FundESF) and Greek national funds through the Operational Program Education and Lifelong Learning of the National Strategic Reference Framework (NSRF)Research Funding Program: THALES. Investing in knowledge society through the European Social Fund (Grant No. MIS 377289).

References

[1] T. J. M. Zouros *at al* 2008 *Phys. Rev. A* **77** 050701

[2] J. Descalux 1975 *Comput. Phys. Commun.* **9** 31

[3] P. Indelicato, J. P. Desclaux 1990 *Phys. Rev. A* **42** 5139

[4] J. P. Santos *J. Phys. B: At. Mol. Opt. Phys.* 1999 **32** 2089

[1]E-mail: jps@fct.unl.pt

Content from this work may be used under the terms of the Creative Commons Attribution 3.0 licence. Any further distribution of this work must maintain attribution to the author(s) and the title of the work, journal citation and DOI.
Published under licence by IOP Publishing Ltd

Intensity ratio in EUV spectra of highly charged Fe ions: Experimental evaluation of astrophysical plasma model

Erina Shimizu*[1], Safdar Ali*, Hiroyuki A. Sakaue†, Daiji Kato†,‡, Izumi Murakami†,‡, Hirohisa Hara¶, Tetsuya Watanabe¶, and Nobuyuki Nakamura*[2]

* Institute for Laser Science, The University of Electro-Communications, Tokyo 182-8585, Japan
† National Institute for Fusion Science, Gifu 509-5292, Japan
‡ The Graduate University of Advanced Studies (SOKENDAI), Gifu 509-5292, Japan
¶ National Astronomical Observatory of Japan, Tokyo 181-8588, Japan

Synopsis We present spectra of highly charged Fe ions in the extreme ultraviolet region observed with a compact electron beam ion trap. The experimental intensity ratios obtained as a function of the effective electron density is used for evaluating plasma models used in the electron density diagnostics of the solar atmosphere.

Emission lines of highly charged Fe ions are important for astrophysical and laboratory plasma applications. In particular, emission in the extreme ultraviolet region are useful for the electron density diagnostics of the solar corona [1]. In such diagnostics, electron density is determined through the comparison of the observed emission line ratio with model calculation. For reliable diagnostics, it is thus important to test the model with spectra obtained with a well-defined laboratory plasma.

In this study, we use spectra obtained with a compact electron beam ion trap (EBIT) [2] to test the collisional radiative model calculation for the electron-density sensitive lines of Fe X-XV. In an EBIT, trapped ions with a narrow charge state distribution interact with a quasi-monoenergetic electron beam; thus EBIT spectra can provide a proper benchmark for testing theoretical models.

We have measured line ratios for density sensitive lines as a function of electron density and compared with a collisional-radiative model calculation. The electron density is experimentally determined from the simultaneously-measured electron beam radius and ion cloud distribution. The ion cloud in an EBIT generally spreads wider than the electron beam; therefore, the effective electron density should be obtained by taking the geometric overlap between the electron beam and the ion cloud into account. Liang et al.[3] suggested that the size of the ion cloud depends on ionic charge based on the comparison between the model calculation and the experimental line ratios obtained with the FLASH-EBIT. On the other hand, in this study, we have measured the dependence of the overlap on ionic charge by directly imaging both the electron beam and the ion cloud. This enables valid evaluation of the models describing density dependence.

Figure 1 shows a typical preliminary result obtained for the intensity ratio between the $3s3p\ ^3P_2 - 3s3d\ ^3D_3$ (233.9 Å) and $3s3p\ ^1P_1 - 3s3d\ ^1D_2$ (243.8 Å) transitions in Fe XV. As seen in the figure, discrepancy is found between the experiment and the model. In this study, the origin of the discrepancy is studied from several perspectives.

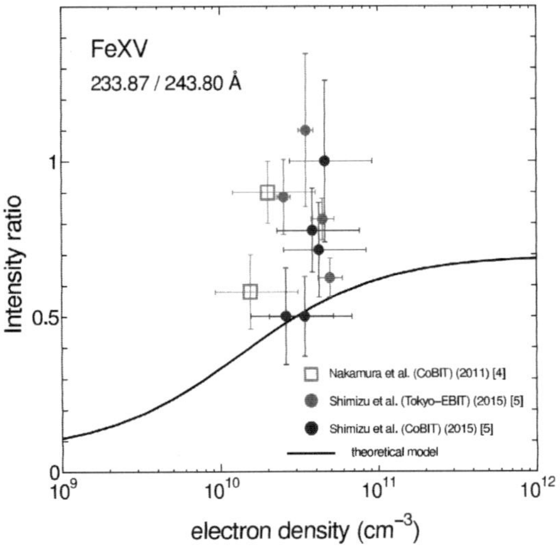

Figure 1. Intensity ratio between the lines at 233.9 Å and 243.8 Å in Fe XV.

References

[1] T. Watanabe *et al.* 2009 *A J* **692** 1294
[2] N. Nakamura *et al.* 2008 *Rev. Sc . I strum.* **79** 063104
[3] G. Y. Liang *et al.* 2009 *A J* **702** 838
[4] N. Nakamura *et al.* 2011 *A J* **739** 17
[5] E. Shimizu *et al.* 2015 *J. Phys.: Co f. Ser.* **583** 012019

[1]E-mail: e_ shimizu@ils.uec.ac.jp
[2]E-mail: n_ nakamu@ils.uec.ac.jp

Content from this work may be used under the terms of the Creative Commons Attribution 3.0 licence. Any further distribution of this work must maintain attribution to the author(s) and the title of the work, journal citation and DOI.
Published under licence by IOP Publishing Ltd

Use of Gas and Foil strippers for the production of He-like ionic beams in both pure ground state ($1s^2$) and mixed states ($1s^2$, 1s2s) for zero-degree Auger Projectile Electron Spectroscopy

A. Laoutaris*‡ [1], I. Madesis†‡ [2], A. Dimitriou†‡ [3], A. Lagoyannis‡, M. Axiotis‡, E. P. Benis¶, T.J.M. Zouros† ‡ [4]

*Dept. of Applied Physics, National Technical University of Athens, GR 15780, Athens, Greece
†Dept. of Physics, University of Crete, GR 71003, Heraklion, Greece
‡Tandem Accelerator Laboratory, INPP, NCSR Demokritos, GR 15310 Ag Paraskevi, Greece
¶Dept. of Physics, University of Ioannina, GR 45110 Ioannina, Greece

Synopsis We present the unique benefits of gas and foil pre- and post- stripping in the production of pure ground state ($1s^2$), as well as mixed ($1s^2$, 1s2s) state He-like beams. The availability of such beams in TANDEM accelerators is of vital importance in studies concerning the population mechanisms of the 1s2s2p ^4P metastable states produced in collisions of these He-like beams with gaseous targets.

The new research initiative APAPES funded by THALES* has so far established a new experimental station at the 5.5MV TANDEM of the NCSR "Demokritos" in Athens with a dedicated beamline for atomic collisions physics research [1].

The main apparatus, developed for zero-degree Auger projectile spectroscopy, has been put together to perform high resolution studies of electrons emitted at zero degrees with respect to the ionic beam in ion-atom collisions. This apparatus consists of a hemispherical deflector analyser equipped with a 4-element entry zoom lens and a 2-dimensional position sensitive detector in operation with a doubly-differentially pumped gas target. It is currently fully operational and used in performing systematic isoelectronic investigations of K-Auger spectra emitted from collisions of He-like ions with gas targets. Our goal is to provide a better understanding of the cascade feeding of the 1s2s2p ^4P metastable states produced in collisions of He-like and H-like ions with gaseous targets [2].

Our experimental results so far were limited to the case of capture measurements to a mixed state ($1s^2$, 1s2s^3S) of 12 MeV C^{4+} ion beam generated at the TANDEM terminal foil stripper. In order to perform measurements with pure as well as mixed state beams, it is necessary to incorporate: i) A gas stripper hosted inside the TANDEM terminal that will allow us to obtain pure ground state ionic beams [3], ii) A stripping point along the beam line of the accelerator hosting both foil and gas post strippers. This will allow us to obtain mixed state beams at different fractions vital for our research goals. Both options are under construction and should soon be fully operational.

The expected charge state intensities resulting from the stripping are calculated from a charge state analysis code using mainly the semi-empirical formulas of Nikolaev Dimitriev [4] and Sayer [5] along with the energy of the beam, its characteristics (Z, atomic mass) and incoming charge state.

Measurements of zero-degree Auger spectra following single state electron transfer to mixed state C^{4+} ($1s^2$, 1s2s ^3S) in collisions with various gaseous targets along with a thorough description of both gas and carbon foil strippers will be presented.

*Co-financed by the European Union and Greek national funds through OP: Education and Lifelong Learning, Research Program: THALES.

References

[1] I. Madesis *et al* 2015 *J. Phys. Conf. Ser.* 583 012014
[2] T.J.M. Zouros *et al* 2008 *Phys. Rev. A* 77 050701R
[3] E.P. Benis *et al* 2002 *Phys. Rev. A* 65 064701
[4] I. S. Dmitriev *et al* 1964 *J. Exptl. Theoret. Phys.* 47 615
[5] R.O. Sayer 1977 *Rev. Phys. Appl.* 12 1543

[1] E-mail: agglaou@gmail.com
[2] E-mail: adimitr@physics.uoc.gr
[3] E-mail: imadesis@physics.uoc.gr
[4] E-mail: tzouros@physics.uoc.gr

Content from this work may be used under the terms of the Creative Commons Attribution 3.0 licence. Any further distribution of this work must maintain attribution to the author(s) and the title of the work, journal citation and DOI.
Published under licence by IOP Publishing Ltd

Journal of Physics: Conference Series **635** (2015) 052063 doi:10.1088/1742-6596/635/5/052063

Investigation of the dependence of the energy resolution of a hemispherical deflection analyzer on the distance of the position sensitive detector from the focal plane

C. Nounis* , A. Laoutaris*, I. Madesis† ‡ , A. Dimitriou† ‡ 1, O. Sise§, E.P. Benis¶, T.J.M. Zouros† ‡

*Dept. of Applied Physics, National Technical University of Athens, GR 15780, Athens, Greece
†Dept. of Physics, University of Crete, GR 71003, Heraklion, Greece
‡Tandem Accelerator Laboratory, INPP, NCSR Demokritos, GR 15310 Ag Paraskevi, Greece
§Dept. of Science Education, Faculty of Education, Suleyman Demirel Univ., 32260, Isparta, Turkey
¶Dept. of Physics, University of Ioannina, GR 45110, Ioannina, Greece

Synopsis We investigate the energy resolution and line shape of a biased paracentric hemispherical deflector analyzer considering the distance between the exit focal plane and the detection plane. Equipped with a piezo-electrically controlled location of a 2D position sensitive detector we were able to accurately determine the optimum conditions for best energy resolution. The voltages of the injection lens of the spectrometer were also optimized for each location.

Most modern hemispherical deflector analyzers (HDAs) are equipped with an electrostatic injection zoom lens and a position sensitive detector (PSD). Practical geometrical constraints (fringing field correctors, grids, their mountings etc.) does not always allow the PSD placement at the optimal position, i.e. the first-order focal plane following 180° deflection. Page and Read [1] have shown in simulations that an optimal choice on the distance *h* between the focal plane and the detection plane might exist for which non-linearities in the energy of the electrons across the PSD become minimized.

Here, we present a study on the *h* dependence of the energy resolution and line shape in a biased paracentric HDA. The HDA is equipped with a 4-element entry zoom lens and a 2-dimensional PSD in operation with a doubly-differentially pumped gas target. It is a part of the experimental station at the 5.5MV TANDEM of the NCSR "Demokritos" in Athens developed in the new research initiative APAPES [2] funded by THALES. The setup is primarily dedicated to zero-degree Auger projectile spectroscopy, performing high resolution studies of electrons emitted in ion-atom collisions.

Recently, a piezo-electric motor on the shaft on which the PSD is supported has been installed (see Figure 1). This allows us to electronically control the distance *h* of the PSD with respect to the HDA exit focal plane thus enabling an experimental study of its effect on the measured line shape. Moreover, at different values of the distance *h* of the PSD, the injection lens voltages are optimized appropriately including this vital parameter in the study as well [3]. Cold cathode electron gun beams are mostly utilized while electrons resulting from ion-atom collisions at zero-degrees with respect to the ionic beam are complementing the study. In order to understand in detail our findings we also included simulations utilizing the SIMION 8.1 ion optics simulation package.

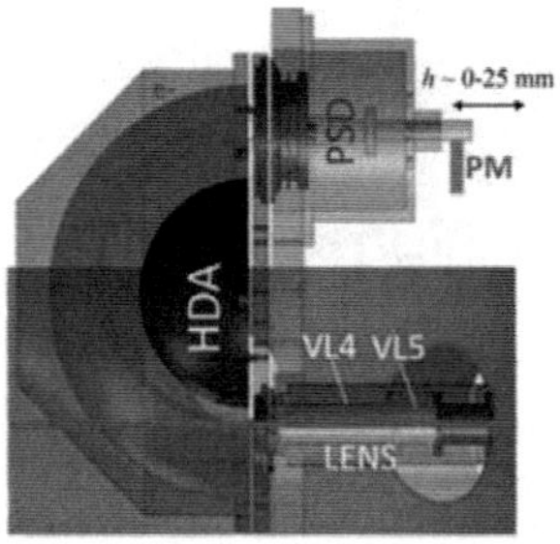

Figure 1. Schematic view of an HDA spectrometer. The PSD is placed a distance *h* from the focusing plane of the HDA.

This research has been co-financed by the European Union and Greek national funds through OP: Education and Lifelong Learning, Research Program: THALES.

References

[1] S.C. Page and F.H. Read 1995 *Nucl. Instrum. Meth. A* **363** 249
[2] I. Madesis *et al* 2015 *J. Phys. Conf. Ser.* 583 012014
[3] T.J.M. Zouros 2006 *J. Electron Spectrosc.* **152** 67

[1] E-mail: adimitr@physics.uoc.gr

Content from this work may be used under the terms of the Creative Commons Attribution 3.0 licence. Any further distribution of this work must maintain attribution to the author(s) and the title of the work, journal citation and DOI.
Published under licence by IOP Publishing Ltd

Rydberg-Stark states of Positronium for atom optics

A. M. Alonso*[1], B. S. Cooper, A. Deller, S. D. Hogan, T. E. Wall and D. B. Cassidy†[2]

Department of Physics and Astronomy, University College London, Gower Street, London, WC1E 6BT, UK

Synopsis Positronium atoms were produced in Rydberg states by means of a two-step optical excitation process (1s→2p→nd/ns). The $n = 11$ Rydberg-Stark manifold has been studied using different laser polarizations providing greater control over the electric dipole moment.

Positronium (the short-lived bound state of the electron and its antiparticle, the positron) has been produced in Rydberg states using a two-step optical excitation. This is feasible thanks to recent advances in positron trapping techniques [1] that allow intense pulses of positrons to be created, which are then directly implanted into an appropriate target to produce positronium (Ps). Since Ps is produced in nanosecond pulses, it is possible to address it with standard pulsed lasers. Ps was excited via similar methods to previous studies [2] using UV and IR lasers to excite the system to $n = 2$ and then to a high-n state, respectively. Fig. 1(a) shows the result of this process for values of $n \geq 9$.

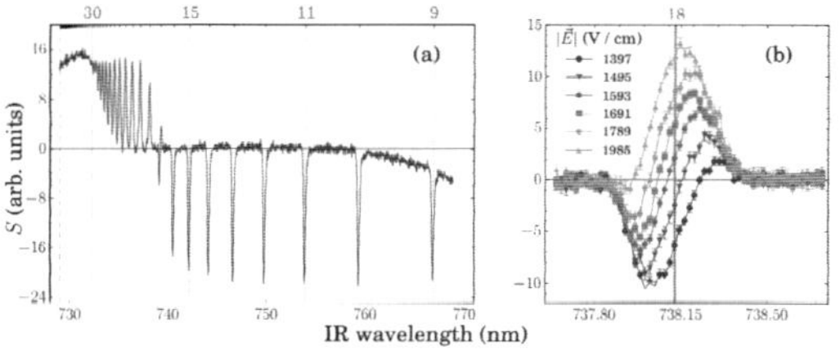

Figure 1. (a) Rydberg spectrum at zero electric field. (b) Stark-state dependent change in sign of the signal in $n = 18$ due to electric field ionisation.

Exciting Ps atoms to a Rydberg state prevents self-annihilation and provides the system with a large electric dipole moment that scales with n^2. This property has been exploited to control the motion of Rydberg atoms and molecules in recent years [3, 4].

Now that Rydberg Ps has been produced as reported in this study, it is possible to design a system to manipulate these atoms in such a way that would allow them to be focused, decelerated and trapped. However, this is only feasible if the dipole moment of the atom can be accurately selected, therefore Ps should be excited to a specific Stark state. This is shown in fig. 2. The outermost states in each manifold of Stark-states have the largest dipole moments and therefore allow for the most efficient control over the atomic motion. The difference in ionisation electric field for states on the long- and short-wavelength sides of each Stark manifold has been used to filter states with different dipole moments. This appears as a change in sign in the signal in fig. 1(b).

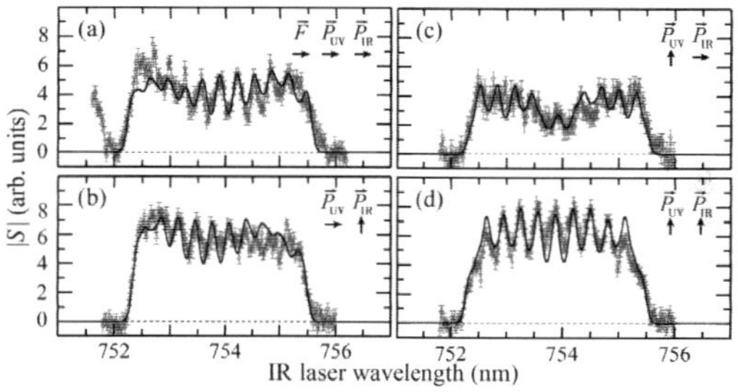

Figure 2. Ps Stark spectra encompassing all states in $n = 11$ for different laser polarizations with respect to the applied electric field $\vec{F}$.

These advancements will facilitate the design of Ps atom-optics which will be used for focusing, decelerating, and trapping. These techniques are ideally suited to the manipulation of Ps because although it moves at very high speeds, its kinetic energies are almost equal to those in beams of hydrogen and helium, for which deceleration and electrostatic trapping has previously been demonstrated. This will open the door to a wide range of new areas of research, ranging from spectroscopy to antimatter gravity measurements [5].

References

[1] C. M. Surko *et al* 2004 *Phys. Plasmas.* **11**, 2333

[2] D. B. Cassidy *et al* 2012 *Phys. Rev. Lett.* **108**, 043401

[3] T. Beeden and H. Metcalf 1981 *Phys. Rev. Lett.* **47**, 1726

[4] P. Lancuba and S. D. Hogan 2014 *Phys. Rev. A* **90**, 053420

[5] A. P. Mills and M. Leventhal 2002 *NIMB* **90**, 053420

[1]E-mail: Alberto.munoz.14@ucl.ac.uk
[2]E-mail: D.cassidy@ucl.ac.uk

Content from this work may be used under the terms of the Creative Commons Attribution 3.0 licence. Any further distribution of this work must maintain attribution to the author(s) and the title of the work, journal citation and DOI.
Published under licence by IOP Publishing Ltd

Recombination of H_2D^+ and HD_2^+ ions with electrons at 80 K.

Petr Dohnal[*] [1], Abel Kalosi[*], Artem Kovalenko[*], Radek Plašil[*], Juraj Glosík[*] [2]

[*] Charles University in Prague, Faculty of Mathematics and Physics, Department of Surface and Plasma Science, V Holešovičkách 2, 180 00, Prague 8, Czech Republic

Synopsis Stationary afterglow with Cavity Ring Down Spectrometer was utilized to probe the number densities of H_3^+, H_2D^+, HD_2^+ and D_3^+ ions in discharge and afterglow plasma. The recombination rate coefficients for recombination of H_2D^+ and HD_2^+ ions with electrons were obtained from the time decays of overall ion number density at different fractions of $[H_2]/[D_2]$.

The H_3^+ molecular ion plays an important role in chemical evolution of interstellar clouds and its recombination with electrons has been subject of extensive theoretical and experimental studies [1]. The deuterated isotopomers of H_3^+ (H_2D^+, HD_2^+ and D_3^+) are, according to recent chemical models [2], key elements in deuteration processes in interstellar medium but in contrast to H_3^+, studies on recombination of H_2D^+ or HD_2^+ with electrons (mainly in storage ion rings [3,4]) are rare and the agreement with theoretical predictions is not perfect [1].

This paper presents the first results of the study on recombination of H_2D^+ and HD_2^+ ions with electrons as obtained in Stationary Afterglow with Cavity Ring Down Spectrometer experiment [5].

The ions were produced in a pulsed microwave discharge in a mixture of He, Ar, H_2 and D_2. Typically [He] ~ 10^{17} cm^{-3}, [Ar] ~ 10^{14} cm^{-3} and $[H_2] + [D_2]$ ~ 10^{14} cm^{-3}. In the experiment the time evolutions of H_3^+, H_2D^+, HD_2^+ and D_3^+ number densities were probed in discharge and in the afterglow of the discharge using near infrared transitions listed in Table 1.

Table 1. Transitions used in present study.

ion	ν (cm^{-1})	lower state	upper state
H_2D^+	6466.53	$0\nu_2+0\nu_3(0,0,0)$	$2\nu_2+\nu_3(1,1,1)$
HD_2^+	6536.31	$0\nu_1+0\nu_3(0,0,0)$	$\nu_1+2\nu_3(1,1,1)$
H_3^+	6807.29	$0\nu_2^0(3,3)$	$3\nu_2^1(2,3)$
D_3^+	6848.50	$0\nu_2^0(2,2)$	$3\nu_2^1(3,2)$

The numbers in parentheses at each state in Table 1 are corresponding quantum numbers (J,K_a,K_b) for H_2D^+ and HD_2^+ and (J,G) for H_3^+ and D_3^+. For further details see refs. [5] and [6]. The measured dependence of relative populations of H_3^+, H_2D^+, HD_2^+ and D_3^+ ions on $[D_2]/([H_2] + [D_2])$ ratio is plotted in Figure 1.

The effective recombination rate coefficients were evaluated from the time decays of overall ion number density measured at different ratios of $[D_2]/([H_2] + [D_2])$. The first results show that at 80 K and [He] = 7.5×10^{17} cm^{-3}, the recombination rate coefficients for H_2D^+ and HD_2^+ are $(3.0 \pm 1.0)\times10^{-7}$ cm^{3}s^{-1} and $(1.4 \pm 0.8)\times10^{-7}$ cm^{3}s^{-1}, respectively and are in qualitative agreement with theoretical calculations [1]. We note that these reported values may be enhanced over pure binary recombination by third body recombination processes similar to those observed for H_3^+ and D_3^+ ions [7]. Further measurements and evaluations are in progress.

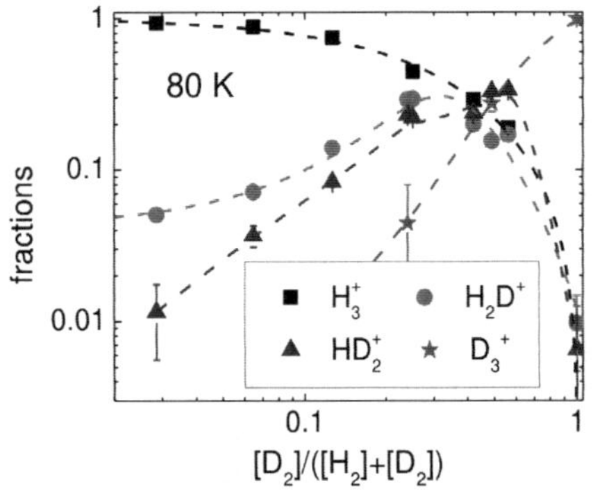

Figure 1. The dependence of relative populations of H_3^+, H_2D^+, HD_2^+ and D_3^+ on $[D_2]/([H_2] + [D_2])$ ratio as measured in SA-CRDS experiment. The dashed lines are for eye guiding only.

This work was partly supported by GACR (14-14649P and 15-15077S) and by GAUK 692214.

References

[1] L. Pagani *et al* 2009 *A&A* **494** 623
[2] C. D. Gay *et al* 2011 *Astrophys. J.* **737** 44
[3] S. Datz *et al* 1995 *Phys. Rev. A* **52** 4
[4] V. Zhaunerchyk *et al* 2008 *Phys. Rev.* A **77** 034701
[5] P. Dohnal *et al* 2012 *J. Chem. Phys.* **136** 244304
[6] P. Hlavenka *et al* 2006 *Int. J. Mass Spectrom.* **255** 170
[7] J. Glosik *et al* 2010 *Mol. Phys.* **108** 2253

[1] E-mail: petr.dohnal@mff.cuni.cz
[2] E-mail: juraj.glosik@mff.cuni.cz

Content from this work may be used under the terms of the Creative Commons Attribution 3.0 licence. Any further distribution of this work must maintain attribution to the author(s) and the title of the work, journal citation and DOI.
Published under licence by IOP Publishing Ltd

Journal of Physics: Conference Series **635** (2015) 052066 doi:10.1088/1742-6596/635/5/052066

Laser trapped ^{6}He as a probe of the weak interaction and a test of the sudden approximation

A. Leredde*[1], Y. Bagdasarova†, K. Bailey*, X. Fléchard‡ A. Knecht•, A. García†, R. Hong†, E. Liénard‡ P. Müller*, O. Naviliat-Cuncic⋆ T.P. O'Connor*, M. Sternberg†, D. Storm†, E. Swanson†, F. Wauters†, D. Zumwalt†

* Physics division, Argonne National Lab., Lemont, IL 60439 , USA
† University of Washington, Seattle, WA 98195, USA
‡ LPC-Caen, IN2P3-ENSICAEN-Université de Caen Basse-Normandie, Caen 14050 Cedex, France
⋆ Michigan State University, East Lansing, MI 48824, USA
• Paul Scherrer Institute, 5232 Viligent , Switzerland

Synopsis Correlations between β decay products are sensitive observables to look for exotic contributions to the weak interaction excluded by the Standard Model of particle physics. To precisely measure the $\beta-\nu$ angular correlation parameter, $a_{\beta\nu}$, in ^{6}He β decay, the radioactive atoms are trapped in a magneto-optical trap and the momentum of β particles and ^{6}Li recoil ions are measured in coincidence. The shake-off process, leading to higher charge states of the Li, plays an important role in this observation and must be studied carefully. Its study provides stringent tests of the sudden approximation for atomic calculations in this few electron system.

The study of $\beta-\nu$ correlations in the ^{6}He β decay is ideal to search for tensor-like contributions to the weak interaction. Its pure Gamow-Teller decay is free from branching and mixing ratios related uncertainties and its simple atomic and nuclear structures allow to perform ab initio calculations for recoil and radiative corrections.

Up to 2×10^{10} ^{6}He/s are produced via the ^{2}H(^{7}Li,^{3}He)^{6}He nuclear reaction by impinging a 15 μA deuteron beam, accelerated at 18 MeV with the tandem Van de Graaff at the University of Washington, onto a liquid lithium target. To be manipulated by lasers, noble gas atoms first have to be excited to their metastable state. An RF-discharge brings a small fraction of the atoms into the 2^3S_1 state which is then focused by means of a transverse cooling stage and injected in a zero crossing Zeeman slower. A 2-dimensional focusing setup sits at the zero-crossing, taking advantage of the quadrupolar magnetic field, and re-focus the slow atomic beam towards our first trap. The trapping efficiency is optimized by increasing the fluorescence signal from a large ^{4}He trap. For ^{6}He, an additional laser beam at 706 nm ($2^3P_2 \rightarrow 3^3S_1$) increases the detection efficiency and allows to detect less than 10 atoms in our trap.

The β decay is studied in a second trap isolated from the first by a low conductance tube and a shutter. The trapped atoms are periodically pushed out of the first MOT and then guided by a 2-dimensional blue-detuned optical dipole trap towards the second MOT. Surrounding the second trap are a β detector which triggers our acquisition system, an electrode array to collect the recoiling ions and a microchannel plate detector to detect them. Trap properties and dynamics can be studied by collecting Penning ions resulting from He(2s)+He(2s) and He(2s)+residual gas collisions.

Different charge states of the recoiling ^{6}Li ions, resulting from the β decay and potentially followed by single or double shake-off emission, can be distinguished by their time-of-flight distributions. However, due to their recoil energy distribution (up to 1.4 keV) and their angular correlation with the β particles, an accurate measure of $a_{\beta\nu}$ requires a precise knowledge of the shake-off probabilities. In particular, the dependence on recoil energy and electronic state population need to be studied. A similar study has been performed in the case of the ^{6}He$^+$ β decay [1] and found in very good agreement with theoretical calculations relying on the sudden approximation. Our experiment with a two-electrons system also allows to look for double shake-off emission and test the $e^- - e^-$ correlation.

Our dedicated setup will be presented at the conference along with early results for both $\beta-\nu$ correlations and shake-off emissions.

This material is based upon work supported by the U.S. Department of Energy, Office of Science, Office of Nuclear Physics, under contract numbers DE-AC02-06CH11357 and DE-FG02-97ER41020.

References

[1] C. Couratin *et al* 2012 *Phys. Rev. Lett.* **108** 243201

[1] E-mail: aleredde@anl.gov

Content from this work may be used under the terms of the Creative Commons Attribution 3.0 licence. Any further distribution of this work must maintain attribution to the author(s) and the title of the work, journal citation and DOI.
Published under licence by IOP Publishing Ltd

Journal of Physics: Conference Series **635** (2015) 052067 doi:10.1088/1742-6596/635/5/052067

Modified Binary-Encounter-Bethe Model for Electron Impact Ionization Cross Sections of Highly Charged Ions

M. Guerra*, **P. Amaro***, **J. Machado***, **J. P. Santos*** [1],

* Laboratório de Instrumentação, Engenharia Biomédica e Física da Radiação (LIBPhys-UNL), Departamento de Física, Faculdade de Ciências e Tecnologia, FCT, Universidade Nova de Lisboa, 2829-516 Caparica, Portugal

Synopsis The modified binary-encounter-Bethe model has been used in the calculation of direct ionization cross sections of ions with several charge states [1], going all the way to hydrogenlike Uranium [2]. In this work we present electron impact cross section values for hidrogenlike Mo, Dy, Au and Bi.

Electron impact ionzation cross sections (EICS) are used in a wide range of scientific and industrial applications, such as astrophysical plasmas, atmospheric science, X-ray lasers, magnetic fusion, radiation physics, accelerator physics and tumor therapy physics. The modified binary-encounter-Bethe (MBEB) model, and its relativistic counterpart (MRBEB) [3, 4], generates direct ionization cross sections for any neutral atom (or molecule), which are reliable in intensity (20%) and shape from the ionization threshold to thousands of keV [3].

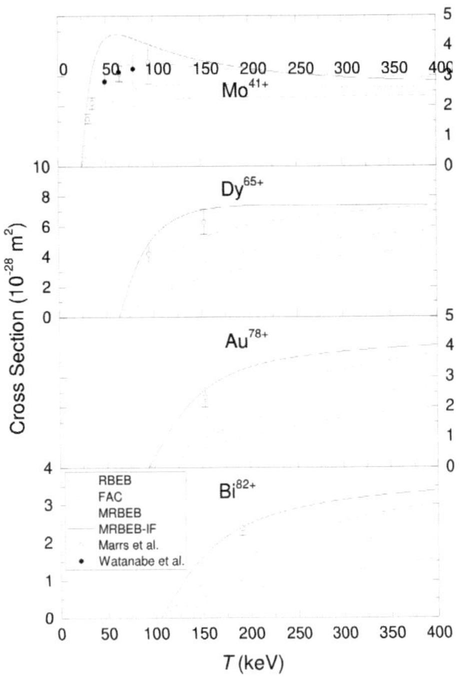

Figure 1. EICS for Mo^{41+}, Dy^{65+}, Au^{78+} and Bi^{82+}. Experimental results are from Marrs *et al.* [5] and Watanabe *et al.* [6]

Highly charged heavy ions have been given considerable attention as a way to investigate relativistic and quantum electrodynamic (QED) effects, as the strong Coulomb field, due to the heavy nucleus and low electron screening, enhances their contributions to the atomic energy levels. The Modified Binary-Encounter-Bethe (MRBEB) model, which rely only on two atomic parameters for EICS evaluation, the binding energy, B, and effective nuclear charge, Z_eff, has been used before to estimate the cross sections for ionization of highly charged Ar, Fe and Kr [1]. However, the influence of the autoionizing channels prevented us from assessing the reliability of the model as it only provides direct cross section values. Very recently, the same model, featuring a correction term called ionic factor (IF), to better describe the Z and incident electron's kinetic energy dependence on the cross section, was used for calculating cross sections for the ionization of U^{91+} to U^{88+} with very good results [2]. Nevertheless, the question about the reliability of the model for such targets remains, as the number of studied targets is scarce. In this work we have calculated EICS for hydrogenlike Mo, Dy, Au and Bi using the MRBEB-IF model, and the results are presented in Fig. 1.

References

[1] M. Guerra *at al* 2013 *Int. J. Mass Spectrom.* **348** 1-8

[2] M. Guerra *at al* 2015 *J. Phys. B* **In Press**

[3] M. Guerra *at al* 2012 *Int. J. Mass Spectrom.* **313** 1-7

[4] M. Guerra *at al* 2009 *J. Phys. Conf. Ser.* **194** 04204

[5] R. E. Marrs *at al* 1997 *Phys. Rev. A* **56** 1338

[6] H. Watanabe *at al* 2002 *J. Phys. B* **35** 5095

[1] E-mail: jps@fct.unl.pt

Content from this work may be used under the terms of the Creative Commons Attribution 3.0 licence. Any further distribution of this work must maintain attribution to the author(s) and the title of the work, journal citation and DOI.
Published under licence by IOP Publishing Ltd

Journal of Physics: Conference Series **635** (2015) 052068 doi:10.1088/1742-6596/635/5/052068

Triply Differential Measurements for Positron Impact Ionization of Argon

DW Mueller[†] [1], EJ Knudsen[†], R Boadle*, S Armitage[†], A Dorn[€], SJ Buckman*, and JP Sullivan*

[†]Department of Physics, University of North Texas, Denton, TX 76203, USA
*Research School of Physics and Engineering, Australian National University, Canberra, Australia
[€]Max-Plank-Institut für Kernphysik, Heidelberg, Germany

Synopsis: We report triply differential cross sections for positron impact ionization of argon. Among other features, we note evidence of post collision interactions.

Differential triple coincidence measurements for positron impact ionization can provide in-depth insight into antimatter interactions with matter. This information is critical for understanding the formation of antihydrogen and the subsequent fundamental studies of antimatter interactions. In addition, understanding these interactions is essential for modeling for tomographic imaging for medical purposes.

We report our first results of triply differential ionization cross section (TDCS) measurements for 190eV positrons on Argon. These triple coincidence measurements reflect the efficiency of the positron reaction microscope (PRM) technique in that all energies and nearly all angles are included in the measurements. An axial magnetic field prevents the loss of the electron and positron, and a weak axial electric field extracts the electron, positron and ion from the interaction region effectively resulting in 4π solid angle detection.

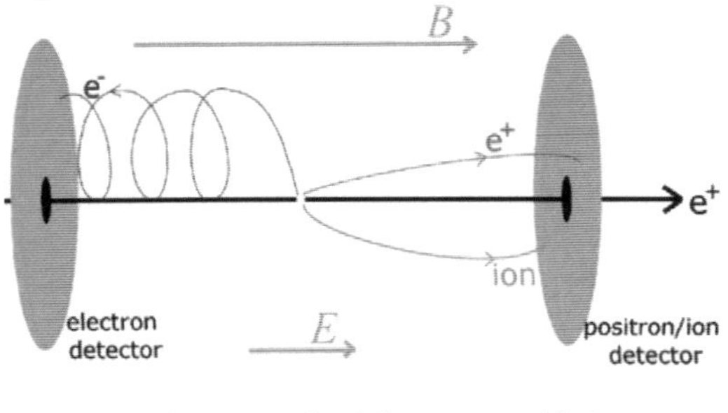

Figure 1 Positron Reaction Microscope with 4π detection

Measurement of positions where the electron and positron strike the detector and of the relative time of flight allows one to determine the kinematics of the ionization process.

A histogram of the ratio of measured longitudinal momenta shows a double peak structure. The smaller peak lies in the region where the ejected electron velocity is approximately ½ that of the scattered positron, corresponding to electron motion on the Wannier ridge. This graph includes the entire range of transverse momenta for each particle for each measured event.

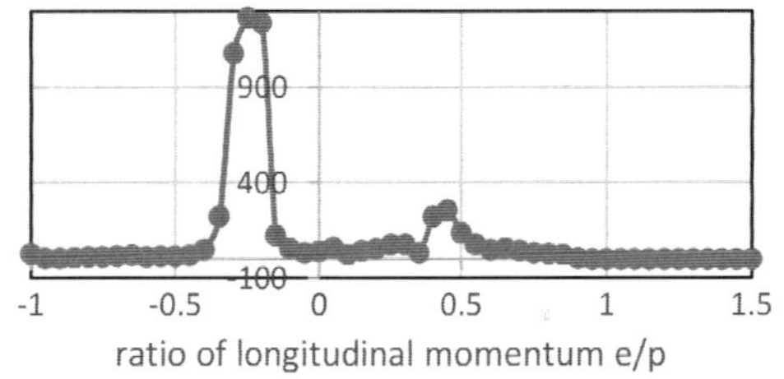

Figure 2 Histogram of the ratio of longitudinal momenta.

In addition, the positron scattering angle and the electron ejection angle for each event pair are determined although not presented here.

These measurements were performed using a room temperature atomic gas jet target with a thermal momentum distribution

It is worth noting that the data are not inconsistent with the measurements of Lucio *et al.* [1] which used a different technique and focused on comparison of electron and positron ionization of argon, but provides some additional insights into the processes being studied.

References

[1] Lucio *et al.* 2010 PRL **104** 163201

[1]**E-mail:** mueller@unt.edu

Content from this work may be used under the terms of the Creative Commons Attribution 3.0 licence. Any further distribution of this work must maintain attribution to the author(s) and the title of the work, journal citation and DOI.
Published under licence by IOP Publishing Ltd

Positronium beam production and scattering at low energies

S E Fayer[1], S J Brawley, M Shipman, L Sarkadi*, and G Laricchia

UCL Department of Physics and Astronomy, University College London, WC1E 6BT London, UK
* Institute for Nuclear Research of the Hungarian Academy of Sciences, H-4001 Debrecen, Pf. 51, Hungary

Synopsis We are now able to produce a positronium beam at energies in the range 1 − 400 eV, significantly lowering the previously achievable minimum of ∼ 7 eV and opening up the possibility of investigating subtle quantum mechanical effects such as those which give rise to low energy electron scattering phenomena (*e.g.* resonances and 'barrier transparency').

Positronium (Ps) is the lightest atom, comprised of a bound state of an electron and its antiparticle, a positron. Ps investigations can yield both tests of fundamental physics (*e.g.* QED due to its purely leptonic nature) and increase the understanding of atomic scattering. However, due to experimental and theoretical difficulties, knowledge on Ps scattering is scarce [1]. There remains, therefore, much to be learnt about Ps scattering, particularly at low energies and at UCL a monoenergetic Ps beam is being used to measure positronium total cross sections for noble gases and some simple molecules [2].

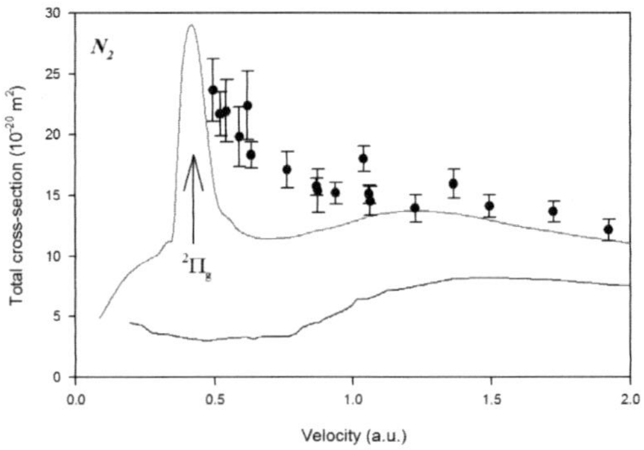

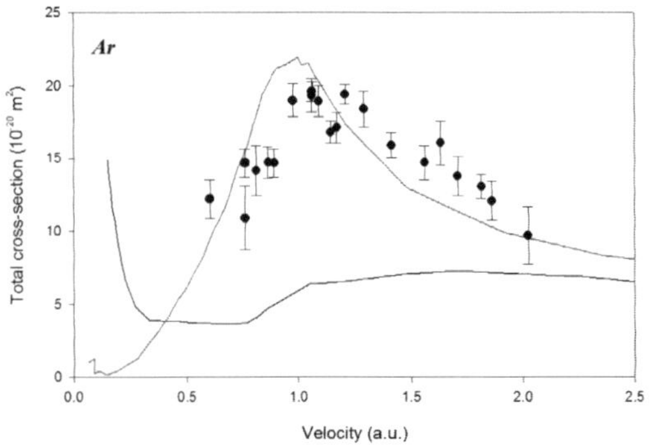

Figure 1. Total cross sections for N_2 and Ar for Ps [2], •; electron [3], [4] (—) and positron [5], [6] (—).

[1]E-mail: s.fayer@ucl.ac.uk

A description of the equipment used in this work may be found in the literature, *e.g.* [7]. Collimated Ps production is performed by passing a positron beam through a gaseous target [8]. Recently, we have achieved for the first time a beam at energies as low as 1 eV, equivalent to a velocity of 0.2 a.u., by using Ar as the production target [8]. Details of this work will be presented at the conference.

In recent years, a similarity between the total cross sections of Ps and electrons at the same velocity has been reported [2]. This similarity has been seen to extend to structures appearing in the electron total cross section [7]. Following the new developments described above, positronium total cross sections are now being measured for the first time at low energies for various targets including Ar, Xe (both possessing a Ramsauer-Townsend minimum in the electron cross section at the particle velocity of 0.15 and 0.23 a.u. respectively) and N_2 (which displays a pronounced $^2\Pi_g$ shape resonance near the velocity of 0.41 a.u.), as shown in figure 1. These results will be presented at the conference.

References

[1] G. Laricchia and H. R. J. Walters 2012 *Riv. Nuovo Cimento* **35** 305

[2] S. J. Brawley *et al* 2010 *Science* **330** 789

[3] Y. Itikawa 2006 *J. Phys. Chem. Ref. Data* **35** 31

[4] W. E. Kaupilla *et al* 1981 *Phys. Rev. A* **24** 725

[5] K. R. Hoffman *et al* 1982 *Phys. Rev. A* **25** 1393

[6] L. Chiari and A. Zecca 2014 *Eur. Phys. J. D* **68** 297

[7] S. J. Brawley *et al* 2010 *Phys. Rev. Lett.* **106** 263401

[8] M. Shipman *et al* 2014 *Eur. Phys. J. D* **68** 75

We are grateful to EPRSC for financial support under grant EP/J003980/1 and providing a studentship to SEF.

Content from this work may be used under the terms of the Creative Commons Attribution 3.0 licence. Any further distribution of this work must maintain attribution to the author(s) and the title of the work, journal citation and DOI.
Published under licence by IOP Publishing Ltd

An electrostatic positron beam for atomic and molecular collision experiments

S. E. Fayer*, A. Loreti*[1], Á. Kövér, S. L. Andersen†, and G. Laricchia***

* Department of Physics and Astronomy, University College London, Gower Street, London, WC1E 6BT
**Institute for Nuclear Research of Hungarian Academy of Science, Debrecen, P.O. Box 51, H-4001, Hungary
† Department of Physics and Astronomy, Aarhus University, DK-8000 Aarhus C, Denmark

Synopsis A new electrostatic positron beam at UCL is described. The beam incorporates a re-moderator to achieve brightness enhancement, time of flight facility for beam energy determination and improved signal to background ratio and a position sensitive detector for 2-D imaging of the beam spot. The apparatus is being employed for measuring the total cross section of positron scattering from krypton and helium in the energy range of (10 − 200) eV. Results will be presented at the conference.

Positrons are valuable tools in the understanding of the fundamental physics interactions between matter and antimatter, an endeavour which is of relevance also in material science, astrophysics and biological chemistry. Collision dynamics, in particular, may be revealed by performing energy- and angular-resolved measurements of the final-state particles following a scattering event. Such investigations are best performed in absence of external fields. The new electrostatic positron beam is being used to carry out total cross section, σ_T, measurements for positron scattering from He and Kr in a field-free region. This work has been stimulated by recent results [1, 2] of positron σ_T from helium and neon, also determined using an electrostatic beam. In these works, discrepancies in the range between $(3-20)$ eV have been found with respect to previous measurements on systems employing magnetic fields.

Table 1. Present angular discrimination, θ, in comparison with previous experiments

Experiment	θ
Present work	$\simeq 2^o$
Nagumo et al., [1, 2]	$\simeq 3^o$
Makochekanwa et al., [5]	$\simeq 18^o - 3^o$
Zecca et al., [6]	$\simeq 17^o - 5.4^o$
Dababneh et al., [7]	$\simeq 15^o - 20^o$
Beale et al., [8]	$\simeq 12^o - 6^o$

In these transmission experiments, the largest source of systematic errors comes from the accidental inclusion of forward scattered particles in the transmitted beam. This error leads to an artificially lower σ_T. In the current experiment, a Position Sensitive Detector (PSD) detects the transmitted positrons at the end of the beam line allowing for a significantly better angular discrimination against forward scattered particles when compared to others e.g., [5, 6, 7, 8] and comparable to that of [1, 2], as shown in table 1.

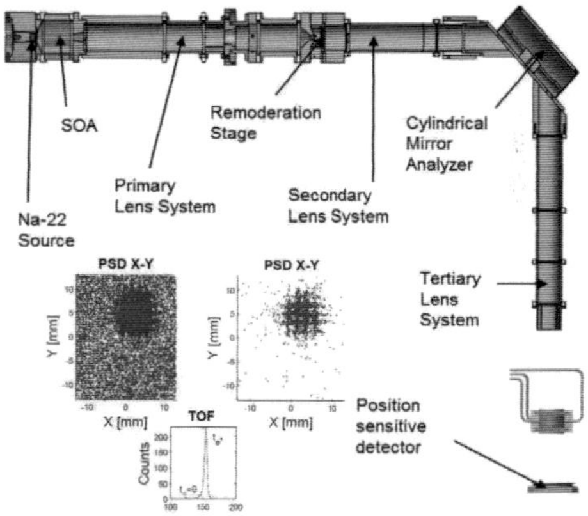

Figure 1. Schematic of the positron beam at UCL [3]. Refer to the text.

Secondary electrons at the re-moderator initiate a timing sequence terminated at the PSD, allowing for Time Of Flight (TOF) analysis. In figure 1, an overview of the positron beam is shown, together with a PSD image of the beam spot, before and after TOF selection. In the TOF plot, the γ-peak from positron annihilation against the gas cell walls (t=0), is easily separated from the incident positron peak.

References

[1] K. Nagumo *et al* 2011 *JPSJ* **80** 064301

[2] K. Nagumo *et al* 2012 *Euro Phys J.D* **66** 80

[3] Á. Kövér *et al* 2014 *Meas. Sci. Technol.* **25** 075013

[4] A. P. Mills 1980 *Appl. Phys. Lett.* **23** 189

[5] C. Makochekanwa *et al* 2011 *PRA* **83** 032721

[6] A. Zecca *et al* 2011 *Euro Phys J. D* **64** 317

[7] M. S. Dababneh *et al* 1980 *PRA* **22** 1872

[8] J. Beale *et al* 2006 *JPB: At. Mol. Opt. Phys.* **39** 1337-1339

[1]E-mail: andrea.loreti.13@ucl.ac.uk
We are grateful EPRSC for financial support (EP/E053521/1) and providing a studentship to SEF.

Content from this work may be used under the terms of the Creative Commons Attribution 3.0 licence. Any further distribution of this work must maintain attribution to the author(s) and the title of the work, journal citation and DOI.
Published under licence by IOP Publishing Ltd

XXIX International Conference on Photonic, Electronic, and Atomic Collisions (ICPEAC2015) IOP Publishing
Journal of Physics: Conference Series **635** (2015) 052072 doi:10.1088/1742-6596/635/5/052072

Electron correlation in the Quasi Sturmian approach for t o-electron continuum problems

A S Zaytsev°, C M Granados-Castro*, L U Ancarani*, G Gasaneo†,‡ and S A Zaytsev° [1]

° Pacific National University, Khabarovsk, 680035, Russia

*Equipe TMS, SRSMC, UMR CNRS 7565, Université de Lorraine, 57078 Metz, France

† Departamento de Física, Universidad Nacional del Sur, 8000 Bahía Blanca, Argentina

‡ Consejo Nacional de Investigaciones Científicas y Técnicas (CONICET), Argentina

Synopsis The driven equation describing an $(e, 3e)$ process on helium in the framework of the Temkin-Poet model is solved numerically using expansions of Convoluted Quasi Sturmian functions. As we observe first that the convergence rate of the solution is rather poor, we propose to significantly improve the expansion by introducing a phase factor corresponding the electron-electron interaction into the basis functions.

To describe two-electron continuum states we propose a basis set of Convoluted uasi Sturmian (C S) functions. For an energy E, these are expressed as a (complex energy plane) convolution integral of two one-particle uasi Sturmian [1] (noted $Q_{n_i}^{\ell_i(+)}(\sqrt{2E_i}; r_i)$, with energy E_i and distance r_i):

$$Q_{n_1 n_2}^{(\ell_1 \ell_2)(+)}(E; r_1, r_2) = \frac{1}{2\pi i} \int_{\mathcal{C}} d\mathcal{E} \; Q_{n_1}^{\ell_1(+)}(\sqrt{2\mathcal{E}}; r_1) Q_{n_2}^{\ell_2(+)}(\sqrt{2(E-\mathcal{E})}; r_2). \tag{1}$$

We wish to apply the C S set to build up the (outgoing (+)) scattering solution of the double ionization of He. Within the Temkin-Poet model, and to first order, the s-wave driven equation for the collision process takes the form [2]

$$\left[E + \frac{1}{2}\frac{\partial^2}{\partial r_1^2} + \frac{1}{2}\frac{\partial^2}{\partial r_2^2} + \frac{2}{r_1} + \frac{2}{r_2} - \frac{1}{r_>}\right] \chi(r_1, r_2) = -\frac{1}{(2\pi)^3}\frac{4\pi}{q^2}\left[2 - j_0(qr_1) - j_0(qr_2)\right] r_1 r_2 \times \frac{Z_e^3}{\pi} e^{-Z_e(r_1+r_2)}. \tag{2}$$

We set $E = 0.735$, $q = 0.24$ and $Z_e = 2 - 5/16$. To solve this equation we propose the expansion

$$\chi(r_1, r_2) = \sum_{n_1, n_2 = 0}^{N-1} C_{n_1, n_2} Q_{n_1, n_2}^{(+)}(r_1, r_2). \tag{3}$$

The C S functions asymptotic behavior in the Ω_0 region (where all three particles are well separated) is not correct since it misses out at least the phase factor

$$W_3(\mathbf{r}_1, \mathbf{r}_2) = -\frac{\rho}{\sqrt{2E}}\frac{1}{r_{12}} \ln\left(2\sqrt{2E}\rho\right) \tag{4}$$

corresponding to the Coulomb interelectronic r_{12} interaction [3] ($r_1 = \rho \cos\alpha, r_2 = \rho \sin\alpha$). Thus the feasibility of using the C S functions (1) for solving (2) is questionable. We nevertheless applied and explored the convergence of expansion (3) by increasing systematically N. The scattering solutions for different basis sizes presented in Figure 1 clearly show a divergent phase as a function of N.

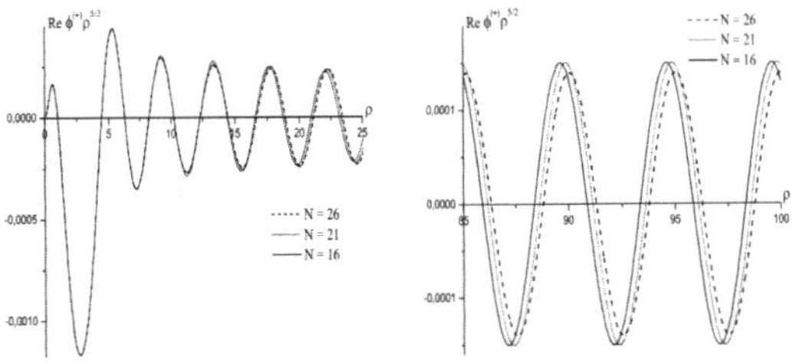

Figure 1. The real part of the solution $\phi^{(+)}\rho^{5/2} \equiv \chi(r_1, r_2)\rho^{1/2}2/\sin 2\alpha$ obtained with different N for $\alpha = \frac{\pi}{4}$, i.e., along the $r_1 = r_2 = \rho/\sqrt{2}$ diagonal.

The problem of slow convergence (or perhaps the lack of convergence) of the expansion can be removed by using modified C S functions

$$\widetilde{Q}_{n_1, n_2}^{(+)}(r_1, r_2) \equiv e^{i\mathcal{W}(r_1, r_2)} Q_{n_1, n_2}^{(+)}(r_1, r_2), \tag{5}$$

equipped with an *ad hoc* phase factor $\mathcal{W}(r_1, r_2)$ corresponding to the potential $\frac{1}{r_{12}}$. Moreover, such a modification of the basis set asymptotic behavior leads to appreciable changes in the magnitude of the solution.

Results will be presented at the conference.

References

[1] J. A. Del Punta, M. J. Ambrosio, G. Gasaneo, S. A. Zaytsev and L. U. Ancarani 2014 *J. Math. Phys.* **55** 052101

[2] G. Gasaneo, D. M. Mitnik, J. M. Randazzo, L. U. Ancarani and F. D. Colavecchia 2013 *Phys. Rev. A* **87** 042707

[3] M. R. H. Rudge 1968 *Rev. Mod. Phys.* **40** 564

[1] E-mail: zaytsevsa@pnu.edu.ru

Content from this work may be used under the terms of the Creative Commons Attribution 3.0 licence. Any further distribution of this work must maintain attribution to the author(s) and the title of the work, journal citation and DOI.
Published under licence by IOP Publishing Ltd

Ionization of Helium by electron and proton impact

M.J. Ambrosio* [1], **D.M. Mitnik***, **G. Gasaneo**†, **E.L. Gaggioli**† **and F.D. Colavecchia**°

* Instituto de Astronomía y Física del Espacio (IAFE, CONICET–UBA), Buenos Aires, Argentina.
† Departamento de Física, Universidad Nacional del Sur and Consejo Nacional de Investigaciones Científicas y Técnicas, 8000 Bahía Blanca, Buenos Aires, Argentina.
° División Física Atómica, Molecular y Óptica, Centro Atómico Bariloche and Consejo Nacional de Investigaciones Científicas y Técnicas, 8400 S. C. de Bariloche, Río Negro, Argentina.

Synopsis We apply the Generalized Sturmian Functions method to the double ionization of Helium by fast electron and proton impact, obtaining fully differentiated cross sections which allow for a better understanding of the few-particle dynamics.

Very recently the Generalized Sturmian Functions method [1] was applied to help advance the controversy regarding the $(e, 3e)$ process in the fast projectile and small momentum transfer regimes [2, 3], presenting a theoretical complete agreement between two theoretical approaches.

This led us to explore other kinematic conditions, still within the electronic projectile case. Our results in Fig. 1 are the theoretical counterparts to experimental measurements by Dorn *et al* [4]. There is agreement in shape and magnitude with the Convergent Close Coupling also in [4].

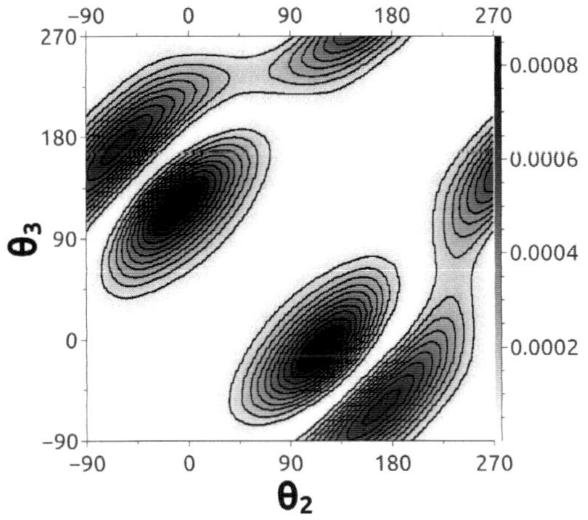

Figure 1. Fully differential cross section for the Helium double ionization by 2 keV electron impact. Momentum transfer is 0.6 a.u.. $E_2 = E_3 = 5$ eV.

Whereas there is an important ammount of experimental data sets for the electron-Helium collision, the same is not true for the proton impact case [5]. In a recent contribution, López *et al* [6] studied the ionization of Helium atoms by fast protons, with 700 keV and 6 MeV incident energies. In their distorted wave studies, they tried a variety of continuum descriptions, which lead to widely different cross section results. Motivated by their analysis, we applied the GSF method, which solves the three-body scattering equation exactly, to provide reliable cross sections (Fig. 2) for this collision problem.

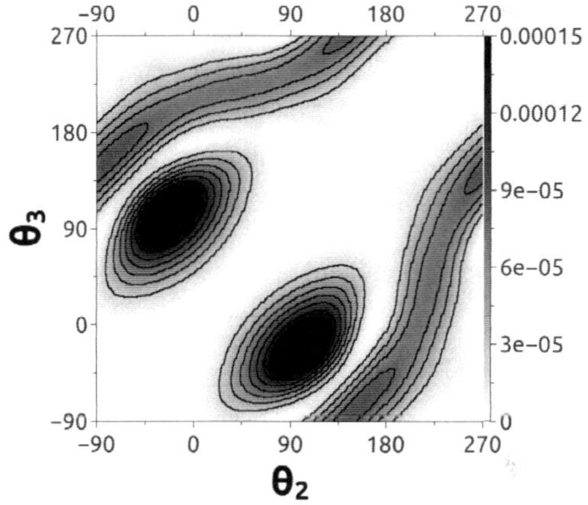

Figure 2. Fully differential cross section for the Helium double ionization by 700 keV proton impact. Momentum transfer is 0.9 a.u.. $E_2 = E_3 = 10$ eV.

With respect to Ref. [6], our FDCS (Fig. 2) present structures with a smoother angular dependence, and also show different shapes for the binary and recoil peaks.

References

[1] G. Gasaneo *et al* 2013 *Adv. Quant. Chem.* **67** 153-216
[2] M.J. Ambrosio *et al* 2015 *Phys. Rev. A* **91** 012704
[3] M.J. Ambrosio *et al* 2015 *J. Phys. B* **48** 055204
[4] A. Dorn *et al* 2001 *Phys. Rev. Lett.* **86** 3755-3758
[5] D. Fischer *et al* 2003 *Phys. Rev. Lett.* **90** 243201
[6] S.D. López *et al* 2011 *Phys. Rev. A.* **83** 062702

[1] E-mail: mja1984@gmail.com

Content from this work may be used under the terms of the Creative Commons Attribution 3.0 licence. Any further distribution of this work must maintain attribution to the author(s) and the title of the work, journal citation and DOI.
Published under licence by IOP Publishing Ltd

Journal of Physics: Conference Series **635** (2015) 052074 doi:10.1088/1742-6596/635/5/052074

Near-Threshold Dielectronic Recombination Studies of Si-Like Ions

J. Kaur,[*1] **T. W. Gorczyca,**[*2] **and N. R. Badnell**[†]

[*]Department of Physics, Western Michigan University, Kalamazoo, MI 49008-5252, USA
[†]Department of Physics, University of Strathclyde, Glasgow G4 0NG, United Kingdom

Synopsis Dielectronic recombination rate coefficients are computed for the Si-like isoelectronic sequence, focusing on the near-threshold resonances of S^{2+}.

We present dielectronic recombination (DR) calculations for Si-like ions, and the important S^{2+} case in particular. This work has been motivated by the astrophysical importance of the S^{2+} DR rate in determining the sulfur ionization balance in the Orion nebula, a photoionized plasma corresponding to low-energy electrons [1]. All the calculations are performed using the atomic structure and collision code AUTOSTRUCTURE [2], a multi-configuration Breit-Pauli (MCBP) method. Energy levels and transition rates are computed in LS- and IC- coupling approximations. Energy positions of near-threshold bound and resonance states are investigated in particular. Table 1 lists three prominent bound (N+1)-electron states and one strong resonance state, relative to the $S^{2+}(3s^2 3p^2\,{}^3P_0)$ threshold. On comparison with available NIST data, the theoretically predicted energies are overestimated by 0.25-0.45 Ryd. There are no NIST or other data available for the given resonance state [3].

Ionic state	Energy (Ryd)	
	Present	NIST
$S^+\ 3s3p^4(^2S_{1/2})$	-0.175	-0.623
$S^+\ 3s3p^4(^2P_{3/2})$	-0.142	-0.389
$S^+\ 3s3p^4(^2P_{1/2})$	-0.137	-0.386
$S^+\ 3s3p^3 3d(^4D_{7/2})$	+0.321	

We conclude from the overestimates observed in the bound energy values that the energy of the resonance is likewise overestimated. It could also mean that low-temperature (characteristic of photoionized plasmas) DR should be enhanced by the resonance position. Fig. 1 shows the computed DR cross-sections and rate coefficients for S^{2+}, including also the results corresponding to a global shift of low-lying, near-threshold resonances. The uncertainties in these resonances is seen to translate to a large uncertainty in the DR rate coefficient at lower-temperatures such as that of Orion nebula ($T \sim 10^4$ K). This work comprises a part of our program to assemble a general DR data base for all ions for modeling dynamic finite plasmas.

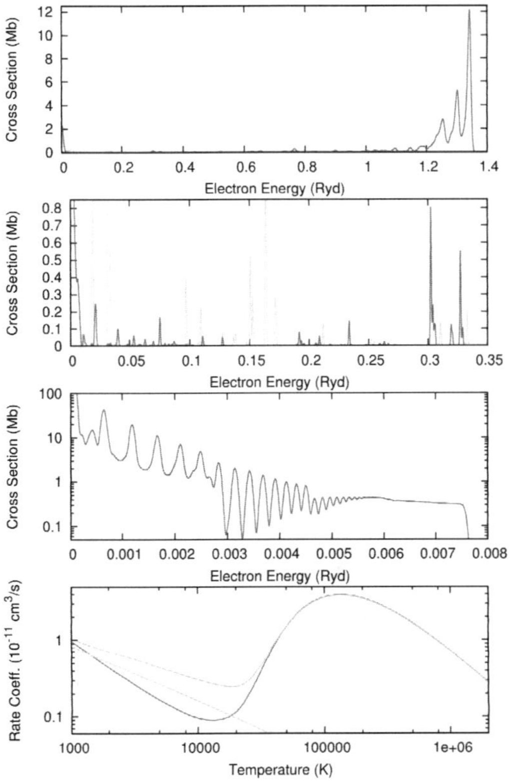

Figure 1. DR cross section (upper three panels, convolved with FWHM Gaussians convolutions of 0.01 and 0.001 Ryd in descending order) and Maxwellian-averaged rate coefficient (lower panel). *Ab initio* results are in red and those with all $(N+1)$-electron resonances shifted by -0.157 Ryd are in green. The computed RR rate coefficient is the cyan curve in the lower panel.

References

[1] Henry R B C *et al* 2012 *Astrophys. J* **749** 6

[2] N R Badnell 2011 *Comput. Phys. Commun.* **182** 1528

[3] Badnell N R, Ferland G J, Gorczyca T W, Nikolic D and Wagle G A *et al* 2015 *Astrophys. J* in press

[1]E-mail: fnu.jagjitkaur@wmich.edu
[2]E-mail: Thomas.Gorczyca@wmich.edu

Content from this work may be used under the terms of the Creative Commons Attribution 3.0 licence. Any further distribution of this work must maintain attribution to the author(s) and the title of the work, journal citation and DOI.
Published under licence by IOP Publishing Ltd

Journal of Physics: Conference Series **635** (2015) 052075 doi:10.1088/1742-6596/635/5/052075

Coplanar asymmetric angles and symmetric energy sharing triple differential cross sections for 200 eV electron-impact ionization of Ar (3p)

Zehra N. Ozer*[1], Sadek Amami†, Onur Varol*, Murat Yavuz*, Mevlut Dogan*[2], Don Madison†

* Department of Physics, e-COL Laboratory, Afyon Kocatepe University, 03200, Afyon, Turkey
† Department of Physics, Missouri University of Science and Technology, Rolla, Missouri, USA

Synopsis We have measured triple differential cross sections (TDCSs) for electron-impact ionization of the 3p shell of Ar at 200 eV incident electron energy. The experiments have been performed in coplanar asymmetric energy sharing geometry. The experimental results are compared with the theoretical models of three body distorted wave (3DW) and distorted wave Born approximation (DWBA).

Cross sections for electron-impact ionization have been measured and calculated since the early days of collision physics since understanding the kinetics and dynamics of collisions is relevant to many practical applications.

The agreement between theoretical predictions and experiment has been improving especially for light atomic targets. Recently, the challenge for the theory is to extend new accurate methods for many electron atoms [1]. It has been recently shown by theory that describing multiple interactions between projectile and target, i.e. higher order effects, are important for these targets at intermediate energies [2-3]. Very recently [4], very good agreement between experiment and both the perturbative 3DW and non-perturbative BSR (*B*-spline *R*-matrix with pseudostates) was found for 64 eV ionization of Ne [4] in full 3 dimensions.

In this study, we measured TDCSs for 200 eV incident electron energy and equal energy sharing for outgoing electrons with $E_{sc}=E_{ej}=92.12$ eV for ionization of the inner shell Ar (3p) orbital. The kinematical arrangement used in the experiment is termed the coplanar asymmetric geometry, in which the scattered electron is detected at a fixed forward angle θ_{sc} in coincidence with an ejected electron detected at varying angles θ_{ej}. (Obviously we do not know which electron is scattered and which one is ejected but we use this terminology only to distinguish between the two final state electrons.) The experimental results are compared with recent DWBA and 3DW calculations (Figure 1). Although the DWBA and 3DW results are very similar, the 3DW is in noticeably better agreement with experimental data. The results will be discussed at the conference in detail.

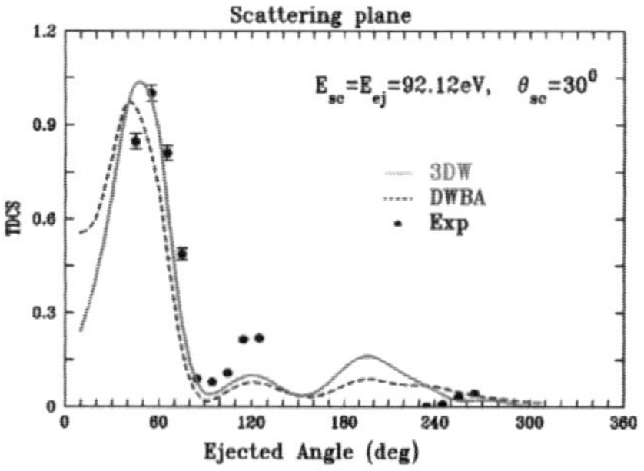

Figure 1. Experimental and theoretical TDCS results for Ar (3p) ionization at 200 eV electron impact.

Acknowledgments

This work was supported by TUBITAK through grant 109T738 and AKU, BAPK through grants 14.FENED.06 and 14.HIZ.DES.80. The theoretical work of SA and DM was supported by US National Science Foundation under Grant. No. PHY-1068237.

References

[1] M. A. Haynes and B. Lohmann 2001 *Physical Review A* 64 044701.
[2] O. Zatsarinny and K. Bartschat 2011 *Physical Review Letters* 107 023203.
[3] S. Amami *et al* 2014 *Physical Review A* 90(1) 012704.
[4] X. Ren *et al* 2015 *Physical Review A* 91 032707.

[1]E-mail: zehraerengil@aku.edu.tr

[2]E-mail: mdogan@aku.edu.tr

Content from this work may be used under the terms of the Creative Commons Attribution 3.0 licence. Any further distribution of this work must maintain attribution to the author(s) and the title of the work, journal citation and DOI.
Published under licence by IOP Publishing Ltd

Correlated transfer ionization in relativistic collisions of highly charged nuclei with light atoms

O.Yu. Andreev*[1], E.A. Mistonova*[2], A.B. Voitkiv†[3]

* Department of Physics, St. Petersburg State University, 198504 St. Petersburg, Russia
† Institute for Theoretical Physics I, Heinrich Heine University Düsseldorf, 40225 Düsseldorf, Germany

Synopsis We consider correlated transfer ionization in relativistic collisions between a highly charged ion and a light atom. We show that this process is strongly influenced by the generalized Breit interaction already at modest relativistic impact energies.

In a collision between an ion (projectile) and an atom with two or more electrons (target) one of atomic electrons can be captured into a bound state of the ion while another electron be emitted. Such a process is called transfer ionization.

There are few basic mechanisms contributing to transfer ionization in fast collisions. Depending on whether the electron-electron interaction (correlation) is crucial or not these mechanisms can be subdivided into correlated and uncorrelated ones. The uncorrelated mechanisms include so called independent capture–ionization and capture–shake-off whereas the correlated ones are electron-electron Thomas [1] and electron-electron Auger [2, 3, 4, 5].

A non-relativistic theory of transfer ionization in fast collisions with highly charged ions was presented in [2, 3, 4]. This theory can be applied when the typical velocities of the electrons in their initial bound states are much smaller than the collision velocity.

Here we present calculations of the double-differential cross section for correlated transfer ionization in relativistic collisions between bare nuclei and light atoms [5]. Nuclei with the nuclear charge $30 \leq Z \leq 92$ are considered. The calculations are performed within the relativistic QED theory.

In particular, it follows from the calculations that this process is profoundly influenced by the generalized Breit interaction already at modest relativistic impact energies (see Table 1 for illustration). This interaction qualitatively changes the shape of the emission pattern and strongly increases the emission. These effects can be verified experimentally by detecting high-energy electrons emitted in the laboratory frame under large angles ($\sim 120° - 180°$) with respect to the motion of the highly charged ion.

[1]E-mail: o.y.andreev@spbu.ru
[2]E-mail: jkfizfak@rambler.ru
[3]E-mail: voitkiv@tp1.uni-duesseldorf.de

Table 1. The total cross section for correlated transfer ionization in collisions with carbon. The first and second columns show the collision velocity and the colliding system, respectively, the third and fourth columns display the cross section calculated by ignoring (σ_{Coul}) and including (σ_{full}) the generalized Breit interaction, the last column gives the ratio $R = \sigma_{\text{full}}/\sigma_{\text{Coul}}$

v (a.u.)	Z	σ_{Coul} (mb)	σ_{full}(mb)	R
70	Zn^{30+}	2.74	4.93	1.8
	Zr^{40+}	5.61	9.51	1.7
	Nd^{60+}	8.91	17.74	2.0
90	Sn^{50+}	0.47	1.55	3.3
	Eu^{63+}	0.74	2.32	3.1
	Yb^{70+}	0.81	2.67	3.3
110	Yb^{70+}	0.04	0.39	8.3
	At^{85+}	0.06	0.49	7.7
	U^{92+}	0.07	0.92	13.1

References

[1] L. H. Thomas 1927 *Proc. R. Soc. Lond. A* **114** 561

[2] A. B. Voitkiv, B. Najjari, and J. Ullrich 2008 *Phys. Rev. Lett.* **101** 223201; A. B. Voitkiv 2008 *J. Phys. B* **41** 195201

[3] A. B. Voitkiv, X. Ma 2012 *Phys. Rev. Lett.* **86** 012709

[4] A. B. Voitkiv, 2013 *Phys. Rev. Lett.* **111** 043201

[5] O. Yu. Andreev, E. A. Mistonova, and A. B. Voitkiv 2014 *Phys. Rev. Lett.* **112** 103202

Content from this work may be used under the terms of the Creative Commons Attribution 3.0 licence. Any further distribution of this work must maintain attribution to the author(s) and the title of the work, journal citation and DOI.
Published under licence by IOP Publishing Ltd

(e,2e) coincidence study on the coherence of the $2s^2(^1S)$ and $2p^2(^1D)$autoionizing states of He

B. Paripás* [1], B. Palásthy* [2] and K. Tőkési† [3]

*Institute of Physics, University of Miskolc, 3515 Miskolc-Egyetemváros, Hungary
†Institute for Nuclear Research, Hungarian Academy of Sciences (MTA Atomki), Debrecen, Hungary

Synopsis The electron impact excitation of the autoionizing states of helium, their subsequent decay into the same He^+1s^{-1} final ionic state, and the interference of these processes have been studied. A constant ionic state (e,2e) experiment (CIS) is performed, isolating the final state by keeping the sum of the energies of the scattered-ejected electron pairs constant. The focus of this work is on the exchange interference of the $2s^2(^1S)$ and $2p^2(^1D)$ autoionizing states of helium. At around the critical energy fast variations were observed in the CIS spectra, which indicates the presence of exchange interference effects.

The coherence of autoionizing states of helium is studied via the measurement of interference of the given states. We concentrated the possible state-to-state interferences of the $2s^2(^1S)$ and $2p^2(^1D)$ autoionizing resonances at energies 57.8 eV and 59.9 eV (decaying to the same simple He^+1s^{-1} final state with E_F = 24.6 eV.), and their possible interference that was previously studied by others in a different way [1]. This state-to-state (exchange) interference (which is essentially differs from the Fano interference) occurs at a unique (critical) electron impact energy where the energy of the scattered electron from one reaction path equals the energy of the ejected electron released along the other path and vice versa: in that case the scattered-emitted electron pairs are indistinguishable.

Electron pairs are observed in coincidence by two cylindrical mirror analysers. Our experimental approach is based on the measurement of CIS (constant ionic state) spectra for the final state. These spectra were recorded in coincidence mode, by scanning simultaneously the transmission energies of the spectrometers in the opposite direction, so as to keep the sum of transmission energies constant.

Our "box" type cylindrical mirror analysers (CMA) have similar structure [2], but slightly different energy resolutions (A: 0,55%; B: 0,78% (FWHM)) and accepted scattering angular ranges (A: 43°-137°; B: 65°-115°).

Around the critical energy, a few tenths of eV modification in primary energy causes a significant change in the CIS spectra (Fig. 1.). Namely the overlapping peaks split into two in a resonance-like way, which can indicate the state-to-state interference effect [3].

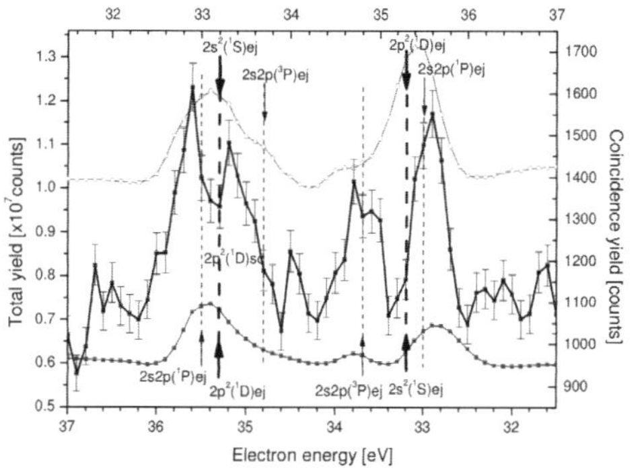

Figure 1. The CIS spectrum measured at 93.1 eV primary energy with E_0-E_F = E_A+E_B = 68.5 eV [3]. The energy spectrum measured by the A (open red circles, upper energy scale) and by the B (full blue squares, lower energy scale) spectrometers are also shown. The corresponding scattered-ejected electron peak-pairs are just opposite in the two spectra and shown by dashed lines. The spectrum in the middle is the coincidence CIS spectrum, in which the random coincidences (≈ 20%) are already subtracted. The error bars contain the statistical error of this background removal, too.

This research was (partially) carried out in the framework of the Centre of Excellence of Mechatronics and Logistics at the University of Miskolc. This work was supported by the Hungarian Scientific Research Fund OTKA Nos. NN 103279 and K103917 and by the COST Action CM1204 (XLIC).

References

[1] den Brink J P V, Nienhuis G, van Eck J and Heideman H M 1989 *J. Phys. B* **22** 3501.
[2] Paripás B, Palásthy B, Štuhec M and Žitnik M 2010 *Phys. Rev. A* **82** 032508-1–032508-1066.
[3] Paripás B, Palásthy B and Pszota G 2014 *Eur. Phys. J. D*, 69: 34

[1]E-mail: fizpari@uni-miskolc.hu [2]E-mail: fizpalb@uni-miskolc.hu [3]E-mail: tokesi@atomki.mta.hu

Content from this work may be used under the terms of the Creative Commons Attribution 3.0 licence. Any further distribution of this work must maintain attribution to the author(s) and the title of the work, journal citation and DOI.
Published under licence by IOP Publishing Ltd

Kinematically complete experiment for ionization-excitation of helium

S. Xu*[1], X. Ma*[2], Q. Zhao†, S. Yan*, P. Zhang*, X. L. Zhu*, J. Yang*, W. T. Feng*, D. M. Zhao*, S. F. Zhang*, D. C. Zhang*, and D. B. Qian*

*Insititute of Modern Physics, Chinese Academy of Sciences, Lanzhou 730000, China
†School of Nuclear Science and Technology, Lanzhou University, Lanzhou 730000, China

Synopsis We present a newly built reaction microscope for electron induced ionization-excitation of helium. Kinematically complete measurement with 2p state excitation is achieved by detecting the radiated photon, one electron and the He^+ ion in the final state in triple coincidence. Near threshold behavior of the ionization-excitation to 2p state is investigated.

Electron induced ionization-excitation (IE) of helium is a basic four-body Coulomb problem in which all the four charged particles are actively involved. It is much more challenging to both experiment and theory compared to direct ionization with the residual He^+ ion in the ground state. The 2s or 2p separated TDCS data, especially at low incident energy range where the high order effects are expected to play a significant role, would offer the most stringent test to theoretical models. However, only one experiment at high incident energy achieved the TDCSs for 2p state till now due to the small cross section and the low detection efficiency for multi-coincidence events [1].

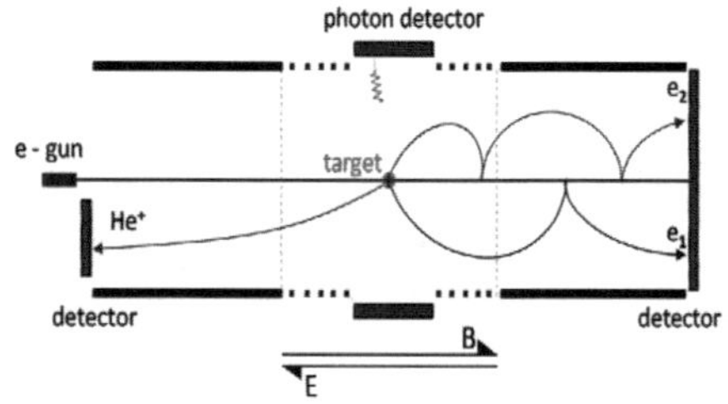

Figure 1. Schematic view of the setup.

Kinematically complete measurement for IE process of helium at low energy is achieved with our newly built reaction microscope. The (e, 2e) reaction microscope is equipped with two multi-channel plate photon detectors (Figure 1). The two electrons, the recoil ion and the photon in the radiative decay from np to 1s state in the final state are recorded either in the (2e+ion) or in the (γ+e+ion) triple coincidence mode. In the (γ+e+ion) case, IE to np state is well isolated from the ns state by the photon signal from decay of np state.

Figure 2 is the timing spectrum of the (γ+e+ion) triple coincidence events, where the IE events to np states could be well separated.

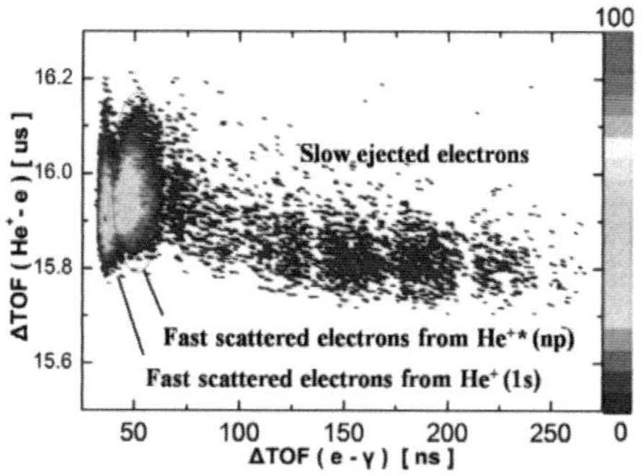

Figure 2. 2D plots for γ+e+ion triple coincidence events versus the time of flight differences ΔTOF(e-γ) and ΔTOF(He^+-e), the incident energy is 150 eV.

We also decreased the incident energy to several eV above the threshold to examine the near-threshold behaviors of IE to 2p process. As predicted by the famous Wannier model [2, 3], the back-to-back pattern of the two outgoing electrons is a key feature for ionization close to the threshold. It is interesting to examine whether the strong correlation from the bound-to-2p electron would disturb the emitting pattern of the two outgoing electrons in the IE process. Our preliminary result shows that there is a pronounced distribution beside the back-to-back pattern, which indicates that the strong enrollment of the 2p electron dramatically influences the behavior of the outgoing electrons.

Detailed discussion for the IE to 2p state in the threshold regime will be presented at the conference.

This work is supported by NSFC of China under Grants Nos 11204322 and 10979007.

References

[1] G. Sakhelashvili *et al 2005 Phys. Rev. Lett.* **95** 33201
[2] G. H. Wannier 1953 *Phys. Rev.* **90** 817
[3] G. H. Wannier 1955, *Phys. Rev.* **100** 1180

[1]E-mail:s.xu@impcas.ac.cn

[2]E-mail:x.ma@impcas.ac.cn

Content from this work may be used under the terms of the Creative Commons Attribution 3.0 licence. Any further distribution of this work must maintain attribution to the author(s) and the title of the work, journal citation and DOI.
Published under licence by IOP Publishing Ltd

Journal of Physics: Conference Series **635** (2015) 052080 doi:10.1088/1742-6596/635/5/052080

Electron impact *M*-subshell ionization of atoms at relativistic energies

A. K. F. Haque[1], M. Alfaz Uddin[1], M. A. R. Patoary[1], A. K. Basak[1] and B. C. Saha[2]

[1]Department of Physics, University of Rajshahi, Rajshahi-6205, Bangladesh
[2]Department of Physics, Florida A & M University, Tallahassee, Florida 32307, USA.

Synopsis We propose an extension of our recently modified simplified-improved-binary-encounter dipole (MUIBED) model [1] suitable even for relativistic energies. This new model (XMUBED) is applied to calculate the electron impact single *M*-subshell ionization cross-sections for atomic targets with various charges Z ($79 \leq Z \leq 92$). The parameters of the MUIBED model for the *M*-shell ionization calculations are modified to account well for fourteen atomic targets (between Au and U) for the inner *M*-subshell ionization cross sections.

Electron impact ionization cross section (EIICS) of inner subshells of atoms are required for different branches of physics, such as atomic physics, plasma physics, radiation physics, material analysis and astrophysics.

While for the *K*-shell, *L*-shell (total) and *M*-shell (total) there are few calculations in literatures, but for *L*-subshell and *M*-subshell targets, very few calculations are available, to the best of our knowledge. Moreover, the huge demand in applications for the ionization cross sections cannot be fulfilled by generating a database either by experiment or by quantum mechanical theories. This void in data and demand can be best fulfilled by using simple-to-use model that predict reasonable accurate results for inner *M*-subshell ionization of atoms. In this work, we report such a model that describes experimental results with reasonable accuracy for a wide range of targets and energy with only two subshell dependent generalized parameters.

In Figure 1. EIICS of U atom M-shell and its subshell (M1, M2, M3, M4 and M5) are shown as a function of incident electron energy. A comparison with experimental data [2, 3] and quantum results [4, 5] shows very good agreement with our calculated cross sections.

Details of our results will be presented at the conference.

E-mail: fhaque2001@yahoo.com

E-mail: uddinmda@yahoo.com

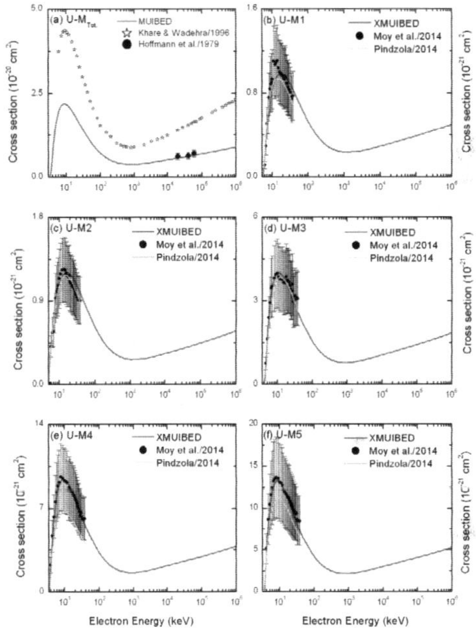

Figure 1

Figure 1. M-shell and M-subshell EIICS of U atom (a) U-M$_{Tot}$, (b) U-M1, (c) U-M2, (d) U-M3, (e) U-M4 and (f) U-M5.

References

[1] A. K. F. Haque *et al* 2011 *Advances in Quantum Chemistry* 61, 317.
[2] A Moy *et al* 2014 *J. Phys. B: At. Mol. Opt. Phys.* **47,** 055202.
[3] D. H. H. Hoffmann *et al* 1979 *Z. Phys. A* **293**, 187.
[4] M. S. Pindzola 2014 *Phys. Rev. A* **90,** 022708.
[5] S.P. Khare and J.M. Wadehra 1996 *Can. J. Phys.* **74**, 376.

Content from this work may be used under the terms of the Creative Commons Attribution 3.0 licence. Any further distribution of this work must maintain attribution to the author(s) and the title of the work, journal citation and DOI.
Published under licence by IOP Publishing Ltd

Journal of Physics: Conference Series **635** (2015) 052081 doi:10.1088/1742-6596/635/5/052081

Scattering of electrons with atomic Mo: free and metallic phases

Foram M Joshi*[1], K N Joshipura[†2]

*[1]G H Patel College of Engineering and Technology, Vallabh Vidyanagar - 388120, India
[†2]Sardar Patel University, Vallabh Vidyanagar - 388120, India

Synopsis The present paper reports comprehensive theoretical investigations on electron scattering with free atomic molybdenum and atomic Mo in solid. Spherical Complex potential formalism is used to derive total as well as inelastic cross-sections. The CSP-*ic* method, developed by us, is employed to extract ionization cross section from the inelastic cross section.

Electron impact ionization (EI) is one of the primary processes in plasmas and its cross section is of overriding importance to model fusion plasmas and astrophysical plasmas. In fusion plasmas it is well known that many kinds of impurity ions exist due to the sputtering of the wall materials. The understanding of the atomic processes of the elements with industrial importance such as Mo is also essential to control the impurity ions in plasmas.[1]

The calculations on atomic Molybdenum is also of interest in divertor studies in several experimental facilities, where a spectroscopic determination of impurity influx from the molybdenum surface is needed. One of the outstanding challenges in applied plasma physics is the development of non-mercury gas discharges for lighting. The discovery of graphene nearly a decade ago has stimulated intensive research efforts in atomically thin, two-dimensional (2D) crystals, such as transition-metal dichalcogenide (TMD) semiconductors and boron-nitride (BN).Compared to graphene, the presence of a bandgap in TMDs is more desirable for device applications. Recent experimental and theoretical work shows that these layered TMDs undergo a transition from indirect to direct bandgap when their thickness is reduced from bulk to monolayer, leading to a pronounced photoluminescence (PL). Among this family, molybdenum disulfide (MoS_2) is one of the most stable layered TMDs[2]. Atomically thin molybdenum disulfide (MoS_2) offers potential for advanced devices and an alternative to graphene due to its unique electronic and optical properties [3].

In this backdrop, the present paper reports comprehensive theoretical investigations on electron scattering with free atomic molybdenum and atomic Mo in solid. Presently we determine total ionization cross section Q_{ion}. We have employed the well-known spherical complex (optical) potential formalism i. e. (SCOP), which provides total elastic cross section Q_{el} and its inelastic counterpart Q_{inel} which includes Q_{ion}. We have already developed a method to extract ionization cross sections Q_{ion} from calculated inelastic cross sections Q_{inel}, by introducing a ratio function as in [4]. The calculated cross sections are examined as functions of incident electron energy along with available comparisons. Detailed analysis of the results will be presented in the Conference.

References

[1] H Watanabe *et al 2002 J. Phys. B: At. Mol. Opt. Phys.* **35** 5095
[2] N R Badnell *et al* 1996 *J. Phys. B: At. Mol. Opt. Phys.* **29** 3683
[3] R Yan *et al 2014 ACS nano* **8** 986
[4] S H Pandya *et al 2012 Int. J. of Mass Spectrometry* **323-324** 28

[1]E-mail: foram29@gmail.com
[2]E-mail: knjoshipura22@gmail.com

Content from this work may be used under the terms of the Creative Commons Attribution 3.0 licence. Any further distribution of this work must maintain attribution to the author(s) and the title of the work, journal citation and DOI.
Published under licence by IOP Publishing Ltd

Journal of Physics: Conference Series **635** (2015) 052082 doi:10.1088/1742-6596/635/5/052082

Separation and solid angle correction of the metastable 1s2s2p ^{4}P Auger yield produced in ion-atom collisions using the biased as cell technique: A tool for the determination of the population mechanisms

I. Madesis[†‡1], A. Laoutaris*[‡], S. Doukas[§], A. Dimitriou[†‡], E. P. Benis[¶], T. J. M. Zouros[†‡]

[†]Dept. of Physics, University of Crete, GR 71003, Heraklion, Greece
[‡]Tandem Accelerator Laboratory, INPP, NCSR Demokritos, GR 15310, Ag. Paraskevi, Greece
*Dept. of Applied Physics, National Technical University of Athens, GR 15780, Athens, Greece
[§]Dept. of Material Science and Engineering, University of Ioannina, GR 45110, Ioannina, Greece.
[¶]Dept. of Physics, University of Ioannina, GR 45110, Ioannina, Greece

Synopsis: We demonstrate an experimental approach to the solid angle correction needed for the metastable 1s2s2p ^{4}P state obtained in ion-atom collisions with gaseous targets. By applying a small bias voltage V_{GC} on the target gas cell, the ^{4}P contributions are separated into two peaks one arising from inside (which is shifted by e V_{GC}) and the other outside the gas cell which is unshifted. Using SIMION we compute the correction in each case and perform a consistency check on our approach.

A persisting problem in the analysis of Auger spectra obtained in collisions of He-like ions with gaseous targets using zero-degree Auger projectile spectroscopy is the determination of the correct contributions from the 1s2s2p $^4P_{1/2,3/2,5/2}$ metastable states to the measured total ^{4}P Auger electron yield. Their inherent long lifetime ($>10^{-9}$s) results in their decay all along the projectile path towards (and through) the spectrometer. Thus, the overall electron detection solid angle varies with the position of electron emission and therefore a considerable correction to the ^4P_J electron yields is needed. In the literature this correction has been treated either geometrically [1] or by SIMION simulations [2].

Another existing approach, which however has not been given sufficient attention, is adopted here. The separation of the ^4P_J yields produced inside and outside the target can be accomplished experimentally by applying a relatively small voltage (e.g. -20 Volts) to the gas cell [3]. Thus, the ^{4}P line now breaks in the spectrum into two separate components, i.e. one produced inside and the one produced outside the gas cell. The one produced inside the gas cell can be straightforwardly used with minimal correction in the determination of the ratio $R={}^4P/{}^2P$ of production cross sections. This ratio, has been of recent interest as large departures from the expected statistical value of R=2 have been reported lending themselves to various explanations as to the possible mechanisms involved [4-5].

Here we present results of the application of this method in collisions of 12 MeV C^{4+} with gaseous targets (H_2, He, N_2, Ne). The ratio R is then estimated, details of which will be provided. These results are part of the APAPES research initiative [6].

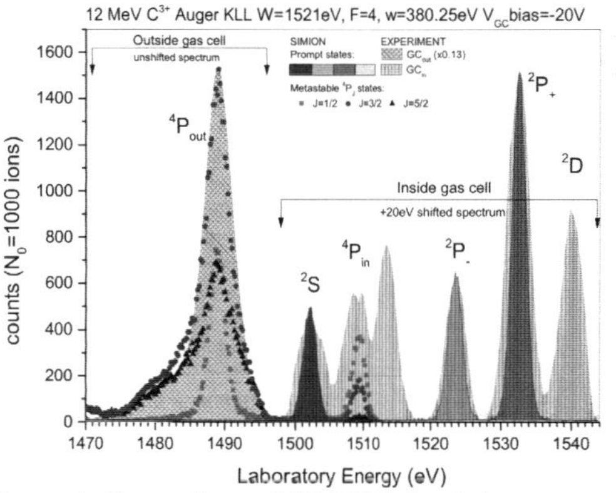

Figure 1. Comparison of SIMION simulations to experimental spectra. In both cases, the ^{4}P line is seen to be broken into two separate components, the unshifted line from electrons emitted outside the gas cell and the shifted line from electrons produced from inside the gas cell which is biased at V_{GC} = -20V. SIMION results (scaled to the experimental spectrum) are seen to nicely reproduce the line shapes of both the metastable ^{4}P state and the prompt doublet states.

*Co-financed by the European Union and Greek national funds through OP: Education and Lifelong Learning, Research Program: THALES.

References

[1] J. A. Tanis *et al* 2004 *Phys. Rev. Lett.* **92** 133201
[2] S. Doukas *et al* 2015 *Rev. Sci. Instrum.* **86** 043111
[3] D. H. Lee 1990 *Ph. D dissertation* KSU
[4] D. Röhrbein *et al* 2010 *Phys. Rev. A* **81** 042701
[5] T. J. M. Zouros *et al* 2008 *Phys. Rev. A* **77** 50701R
[6] I. Madesis *et al* 2015 *J. Phys. Conf. Ser.* **583** 012014

[1]E-mail: imadesis@physics.uoc.gr

Content from this work may be used under the terms of the Creative Commons Attribution 3.0 licence. Any further distribution of this work must maintain attribution to the author(s) and the title of the work, journal citation and DOI.
Published under licence by IOP Publishing Ltd

Journal of Physics: Conference Series **635** (2015) 052083 doi:10.1088/1742-6596/635/5/052083

Optimal operation of a 4-element injection lens in a hemispherical spectrograph: FDM/BEM simulation and experimental demonstration

G. Martínez*, I. Madesis[†‡], A. Dimitriou[†‡], A. Laoutaris[¶], O. Sise[§], and T. J. M. Zouros[†‡1]

[*] Dept. Física Aplicada III, Facultad de Física, UCM 28040-Madrid, Spain
[†] Dept. of Physics, Univ. of Crete, P.O Box 2208, GR 71003 Heraklion, Greece
[‡] Tandem Accelerator Laboratory, INPP, NCSR Demokritos, GR 15310 Ag Paraskevi, Greece
[¶] Dept. of Applied Physics, National Technical University of Athens, GR 15780, Athens, Greece
[§] Dept. of Science Education, Faculty of Education, Suleyman Demirel University, 32260 Isparta, Turkey

Synopsis We have investigated the voltage settings for the 4-element injection lens of a biased paracentric hemispherical deflection analyzer (HDA) with virtual entry aperture using numerical methods. The two lens electrode voltages were varied as free parameters while the electron optical properties of the analyzer were calculated from trajectories with test initial distributions for different pre-retardation factors in an effort to obtain improved energy resolution. The resulting lens voltages were then also tested on the existing HDA spectrometer at the Tandem Accelerator Laboratory, while simulated and measured line profile characteristics were compared, particularly at the best resolution working points.

The biased paracentric hemispherical deflection analyzers represent a novel class of HDAs, which use the lensing action of the strong fringing fields at its entry, to restore the first order focus characteristics of ideal HDAs in a controlled way [1,2]. Here, we extend previous investigations to the study of the electron optical properties of such an HDA as a function of lens voltages, and compare our results to experimental measurements taken with the HDA at the Tandem Accelerator Laboratory [3].

Figure 1(a) shows the schematic of the biased paracentric HDA equipped with a 4-element injection lens and position sensitive detector (PSD). The lens system plays a particularly important role in regulating the energy resolution of the analyzer since it defines the size of the virtual entry aperture. The two lens voltage combinations, V_{L4} and V_{L5}, were found to belong to the families of elliptical-like contours, as shown in Figure 1(b), and were investigated for different pre-retardation factors, $F = E_{s0}/qV_p$, where E_{s0} is the electron source energy, and V_p is the HDA plate voltage ($F = 1$ for no pre-retardation). The correlation between F and the energy resolution was evaluated by means of both experimental measurements and computer simulations of the entire trajectories including the injection lens system. Two numerical methods, the finite difference method (FDM) and boundary element method (BEM) [4], were employed in the solution of the electrostatic problem.

All four combinations of positive and negative voltages gave rise to energy resolution minima from which only the subset with the most practical values were further investigated in more detail.

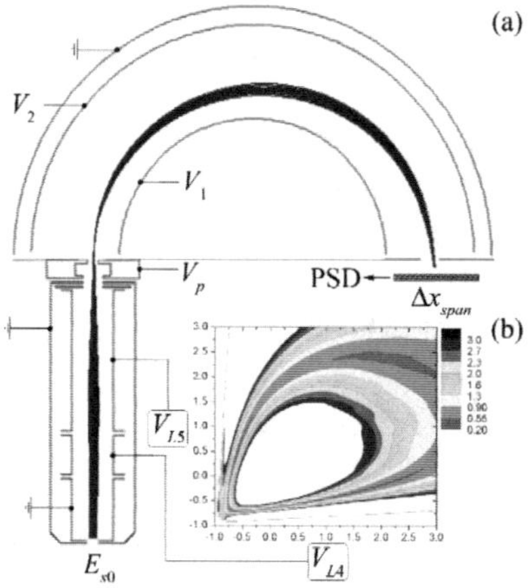

Figure 1. (a) Schematic view of the complete analyzer system. (b) 2-D contour map of the beam trace width Δx_{span} versus V_{L4} and V_{L5} (in kV) for $F = 1$.

This research has been co-financed by the European Union and Greek national funds through OP: Education and Lifelong Learning, Research Program: THALES.

References

[1] E.P. Benis *et al* 2008 *J. Electron Spectrosc.* **163** 28
[2] O. Sise *et al* 2010 *J. Electron Spectrosc.* **177** 42
[3] I. Madesis *et al* 2015 *J. Phys. Conf. Ser.* **583** 012014
[4] G. Martínez, M. Sancho, in: P. Hawkes (Ed.), *Advances in Electronics and Electron Physics* 81, Academic Press, New York, 1991, pp.1-41.

[1] E-mail: tzouros@physics.uoc.gr

Content from this work may be used under the terms of the Creative Commons Attribution 3.0 licence. Any further distribution of this work must maintain attribution to the author(s) and the title of the work, journal citation and DOI.
Published under licence by IOP Publishing Ltd

Journal of Physics: Conference Series **635** (2015) 052084 doi:10.1088/1742-6596/635/5/052084

Preliminary measurements of the Bremsstrahlung doubly differential cross section for electrons between 20 and 100 keV in Au

J. A. García-Alvarez[*,†], **N. L. Maidana**[* 1], **T. F. Silva**[*], **J. M. Fernández-Varea**[‡,*] **M. N. Martins**[*] **and V. R. Vanin**[*]

[*] Instituto de Física da Universidade de São Paulo, Travessa R 187, Cidade Universitária, CEP:05508:900, São Paulo, Brazil

[†] Instituto Superior de Tecnologías y Ciencias Aplicadas. Quinta de los Molinos, Ave. Salvador Allende, Esq. Luaces, CP:10600, Havana, Cuba

[‡] Facultat de Física (ECM and ICC), Universitat de Barcelona, Diagonal 645, E-08028, Barcelona, Spain

Synopsis Electron Bremsstrahlung cross sections were experimentally determined, for five kinetic energies, unfolding the measured photon spectra with an analytical detector response function.

The electron Bremsstrahlung doubly differential cross section (DDCS) in Au for kinetic energies $E = 22.4$, 42.0, 61.7, 80.8 and 100.8 keV was measured at the São Paulo Microtron.

Au targets were prepared by chemical vapor deposition over a C backing material, previously placed over a frame of amorphous C fiber, (30 mm × 15 mm × 0.3 mm) with a central hole of 10 mm in diameter. The target thickness $t_{\mathrm{Au}} = 32.6(18)\mu\mathrm{g/cm}^2$ was determined by Rutherford Backscattering Spectrometry at the LAMFI/USP tandem van de Graaff, employing a 2.2 MeV He^+ beam and two Si detectors at 120^0 and 170^0 with respect to the incident beam.

The Microtron delivers a continuous electron beam with dimensions of $\sim 2 \times 2\ \mathrm{mm}^2$. During the irradiations, the charge was collected both in the Faraday cup and the irradiation chamber that is electrically isolated from the beam line.

A planar HPGe detector positioned at $\theta = 48.8^0$ was used to acquire the Bremsstrahlung spectra. The solid angle Ω subtended by the detector with respect to the target was determined as done by Maidana et al [1], measuring the detector efficiency from the ^{57}Co γ-rays. Seltzer's efficiency model [2] was fitted to the experimental points with Ω as a linear adjustable parameter. The recorded pulse-height distributions were corrected for the background, pile-up and dead time effects.

The emitted photon spectra were obtained unfolding the net spectra with an analytical model that describes the detector response function (RF). Our RF considers the photon total absorption, the escape of Ge $K\alpha$ and $K\beta$ x-rays and the escape of photons after one Compton interaction. We adopt the relativistic impulse approximation to describe the incoherent scattering process in the Ge crystal. We also take into account the continuous spectrum produced by the escape of photo-electrons from the active volume. The included external interaction contributions to the RF were: the incoherent photon scattering in the detector Be window and the effect of photo-electrons created in the Ge dead layer that reach the active volume. We obtained the emitted Bremsstrahlung spectra with an energy bin $\Delta = 1$ keV, for photons between 14 keV (above the Au L x-ray region) and the tip region.

Our uncertainties are around 6–7% and its main contributor is the target thickness ($\sim 5\%$). The experimental DDCSs were compared with the theoretical data of Pratt et al [3] and Kissel et al [4] for Au. Experimental results are systematically above the theory, except for $E = 22.4$ and 42.0 keV in the tip region where both, theoretical and experimental data are in agreement.

References

[1] N. L. Maidana *et al* 2013 *Nucl. Instr. Meth. A.* **729** 371

[2] S. M. Seltzer 1981 *Nucl. Instr. Meth.* **188** 133

[3] R. H. Pratt *et al* 1977 *At. Data Nucl. Data Tables* **20** 175

[4] L. Kissel *et al* 1983 *At. Data Nucl. Data Tables* **28** 381

[1]E-mail: nmaidana@if.usp.br

Content from this work may be used under the terms of the Creative Commons Attribution 3.0 licence. Any further distribution of this work must maintain attribution to the author(s) and the title of the work, journal citation and DOI.
Published under licence by IOP Publishing Ltd

XXIX International Conference on Photonic, Electronic, and Atomic Collisions (ICPEAC2015) IOP Publishing
Journal of Physics: Conference Series **635** (2015) 052085 doi:10.1088/1742-6596/635/5/052085

L-shell ionization cross sections for silver by electron impact

Károly Tőkési[1] **and Takeshi Mukoyama**

Institute for Nuclear Research of the Hungarian Academy of Sciences (ATOMKI),
Bem tér 18/c, H-4026 Debrecen, Hungary

Synopsis The L-subshell ionization cross sections by electron impact have been calculated with the classical-trajectory Monte Carlo method. The results are compared with the quantum mechanical calculations and the cross sections by positron impact. It is demonstrated that the classical treatments are quite useful for electron-atom collisions.

The study on inner-shell ionization processes of atoms by electron impact is of fundamental importance in atomic collisions. Reliable ionization cross sections for atomic inner shells by low-energy electrons are essential to various application fields. Furthermore, it is interesting to compare the cross sections by electrons with those by positrons to elucidate the dependence of ionization cross sections on the charge of the projectile.

Recently we have shown that the classical-trajectory Monte Carlo (CTMC) method can well predict the L-shell ionization cross sections by positron impact [1]. In the present work we apply the similar approach to the L-shell ionization process by electron impact and calculate the L-shell ionization cross sections for Ag.

The CTMC is a kind of non-perturbative method based on the classical dynamics. We consider the target atom in the screened hydrogenic model. Then the system consists of three particles, i.e. the projectile electron, one of L-shell electron and the target nucleus. The classical equations of motion for three-body system are solved numerically for large number of trajectories. The initial conditions are the impact parameter, the position and the velocity of the target electron and they are selected randomly for each trajectory. The ionization cross section is determined as the ratio of the number for the ionization events to the total number of trajectories.

Figure 1 shows the L_1-shell ionization cross sections of Ag as a function of the incident energy of electron. The CTMC calculations are compared with the recent experimental values of Sepúlveda *et al.* [2] and the distorted-wave Born approximation (DWBA) [3]. The present results are slightly larger than the experimental values at low energies, but agree well with the experiment within the experimental error. The DWBA values are smaller than the CTMC values and the experimental data in low-energy region. However, all the values are in good agreement for high energies.

For comparison, the CTMC and the DWBA calculations by positron impact are also plotted in the figure. The cross sections by positron impact are smaller than those by electron in low-energy region due to the Coulomb repulsion for the incident positron, but becomes almost same in high-energy region.

From the present results it can be said that the CTMC method is useful to calculate inner-shell ionization cross sections by light projectiles.

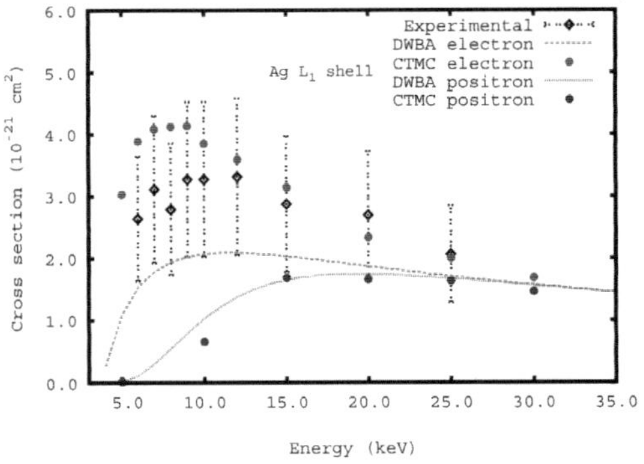

Figure 1. Comparison of the CTMC calculations of the L_1-shell ionization cross sections for Ag by electron impact with the experimental values and the DWBA calculations., together with the corresponding values by positron impact.

References

[1] T. Mukoyama *et al.* 2014 *Eur. Phys. J. D* **68** 342

[2] A. Sepúlveda *et al.* 2014 *J. Phys. B: At. Mol. Opt. Phys.* **47** 215006

[3] D. Bote *et al.* 2009 *At. Data Nucl. Data Tables* **95** 871

[1]E-mail: tokesi@atomki.mta.hu

Content from this work may be used under the terms of the Creative Commons Attribution 3.0 licence. Any further distribution of this work must maintain attribution to the author(s) and the title of the work, journal citation and DOI.
Published under licence by IOP Publishing Ltd

Relativistic effects in positronic alkali atoms

Takuma Yamashita[*1] **and Yasushi Kino**[*2]

[*] Department of Chemistry, Tohoku University, 980-8578, Sendai, Japan

Synopsis In order to investigate interactions between a positron/positronium and an atom/ion, we calculate positronic alkali atoms, a loosely bound state of an alkali ion and a positronium, as a three-body system composed of the positron, valence electron and residual ion core within a frozen core approximation. We optimize model potentials to reproduce energy levels and fine structure splittings of alkali atoms. The ion core polarization by a positron is determined more precisely than other previous studies. We found that the relativistic effects amazingly contribute to the binding energy of the positronic alkali atom. We also discuss the rearrangement effect on the relativistic corrections and roles of positronium-ion and atom-positron configurations.

A positron (e^+) is an anti-particle of an electron, having the same mass as the electron and equal but opposite charge. In matter, the positron is located far from the nuclei of atoms/molecules by the Coulombic repulsive interaction. A positron can be bound to an atom and form a positronium (Ps), the simplest hydrogen-like atom composed of a positron and an electron. The Ps formation is one of the most important reaction of the positron and atoms/molecules as both theoretical and experimental aspects. Recently, experimental techniques to control Ps atoms have been sophisticated rapidly. An energy tunable Ps beam [1] and a QED test for the ground state of Ps [2] have been accomplished. For the more quantitative estimation of experiments treating positrons/positroniums, high-precision theoretical study on interactions between a positron/positronium and an atom are indespensable.

A positronic alkali atom is one of the ideal systems to calculate atom-positron and ion-positronium interactions because we can regard this system as a simple three-body system consisting of the positron, valence electron and residual ion core within a frozen core approximation. When a positron attaches to an alkali atom (A), the valence electron transfers to the positron and forms a Ps atom because the ionization energies of alkali atoms (around 5 eV) are smaller than the binding energy of Ps (6.7eV). The residual ion core (A^+) polarizes the Ps atom and forms a loosely bound state of the positronic alkali atom (APs^+). Although a lot of theoretical studies have been done, calculated binding energies associated with a dissosiation threhold of A^++ Ps are scattered depending on the model potential between the positron/electron and ion core introduced along with the frozen core approximation [3, 4].

In our previous work [5], we pointed out that relaivistic effects can contribute to the binding mechanism of positronic alkali atoms by a simple approximation. Since the relativistic effects are much larger in a heavy atom than those in a light atom, many previous studies focus on atomic/molecular systems containing heavy atoms. In the case of positronic alkali atoms, only Li and Na atom can bind a positron. Relativistic corrections, therefore, have been considered to be small and not calculated precisely.

In this work, we examine the model potential and its role to the binding mechanism. In order to determine the polarization potential between the A^+ and e^+ precisely, we modify the electron exchange potential by introducing dependencies of the valence electron angular momentum and the inner electron kinetic energy. Then we calculate the relativistic corrections to the binding energy of $LiPs^+$ and $NaPs^+$ within the first-order perturbation theory, based on the Breit-Pauli interactions. We found that the relativistic effects amazingly contribute to the binding energy of APs^+, as shown in table 1. We also discuss how rearrangement, namely Ps + A^+ and A + e^+ configurations, affects the relativistic corrections in the loosely bound state.

Table 1. Relativistically corrected binding energy associated with a dissosiation threhold of A^++ Ps (singlet). Effects of relativistic corrections are shown in the bottom line. All energies are given in atomic units (a.u.).

	$NaPs^+$	Na (3s)
$-\varepsilon^{\rm nr}$	0.000 472 7	0.188 566
$\Delta\varepsilon^{\rm rel}$	−0.000 032 6	−0.000 292
$-\varepsilon^{\rm rel}$	0.000 487 8	0.188 878
$1-\varepsilon^{\rm nr}/\varepsilon^{\rm rel}(\%)$	3.09	0.155

References

[1] K. Michishio *et al.* 2012 *Appl. Phys. Lett.* **100** 254102

[2] A. Ishida *et al.* 2014 *Phys. Lett.* B **734** 338

[3] H. Han *et al.* 2008 *J. Chem. Phys.* **128** 244314

[4] J. Shertzer *et al.* 2010 *Phys. Rev.* A **81** 064505

[5] T. Yamashita and A. Irisawa and Y. Kino 2014 *Jpn. J. Appl. Phys. Conf. Proc.* **2** 011005

[1]E-mail: t.ymst@dc.tohou.ac.jp [2]E-mail: y.k@m.tohoku.ac.jp

Content from this work may be used under the terms of the Creative Commons Attribution 3.0 licence. Any further distribution of this work must maintain attribution to the author(s) and the title of the work, journal citation and DOI.
Published under licence by IOP Publishing Ltd

Journal of Physics: Conference Series **635** (2015) 052087 doi:10.1088/1742-6596/635/5/052087

Alignment of the excited states and angular distribution of subsequent decay following the radiative electron capture into lithium-like ions*

Y. L. Shi*[1], X. B. Liu*, D. H. Zhang†, L. Y. Xie†, C. Z. Dong†

* Department of Physics, Tianshui Normal University, Tianshui 741001, China
† Key Laboratory of Atomic and Molecular Physics & Functional Materials of Gansu Province, College of Physics and Electronic Engineering, Northwest Normal University, Lanzhou 730070, China

Synopsis The radiative electron capture into highly charged lithium-like U^{89+} ions is studied in the framework of the density matrix, based on Dirac's equation. Emphasis is placed on the population of $1s^22s2p_{3/2}$ J=1,2 excited states, as well as the angular distribution and polarization of subsequent characteristic photon emissions (included $E1$ and $M2$ transition).

Collisions of highly charged, heavy ions with atoms and free electrons have been studied intensively both at ion storage rings and at electron beam ion traps. In these collision studies, special interest has been devoted to the production of excited ionic states and on the measurement of their subsequent radiative decay. Indeed, the analysis of the bound-state transitions in high-Z ions plays a key role in our understanding of electron-electron and electron-photon interactions in the presence of strong fields, including important information about the relativistic and quantum electrodynamics effects in few-electron systems [1].

At high energies, the radiative electron capture (REC) is known as ones of the dominant processes for medium- and high-Z ions at storage rings. In this charge transfer process, a (quasi) free electron from either the (electron) cooler or some rest gas is captured into a bound state of the ion under the simultaneous emission of a photon. A great advantage of angle-resolved x-ray studies is that they are often much more sensitive to the magnetic and retardation effects than the analysis of the total (i.e., integrated over the angles) rates [2].

In this contribution, the REC into highly charged lithium-like U^{89+} ions is studied in the framework of the density matrix, based on Dirac's equation [3]. Detailed computations have first been carried out for the total and angle-differential cross sections (i.e. the angular distribution of the REC photon) of L-REC into the ground state $1s^22s^2$ J=0 and the low-lying excited state $1s^22s2p$ J=1,2 of final beryllium-like ion. Emphasis is placed on the population of $1s^22s2p_{3/2}$ J=1,2 excited states, as well as the angular distribution and polarization of subsequent characteristic photon emissions (included $E1$ and $M2$ transition). Strong angular anisotropy and linear polarization characteristics were found, and the possibility of observations from the experiment were also analyzed and discussed.

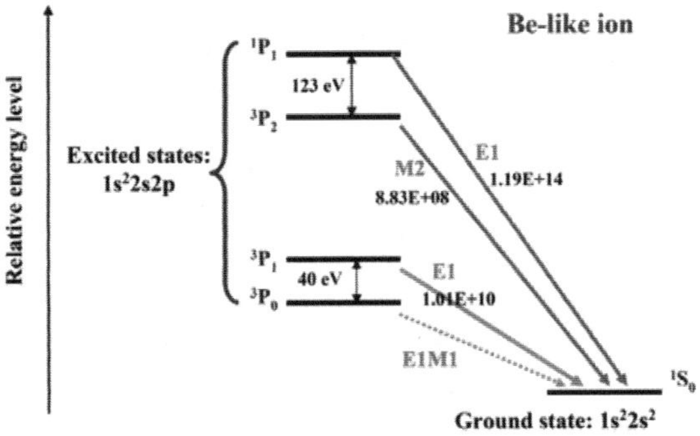

Figure 1. The relative energy level and main decay channel of beryllium-like U^{88+} ions.

References

[1] S. Fritzsche, P Indelicato, Th. Stöhlker, 2005 *J. Phys. B: At. Mol. Opt. Phys* **38** S707.
[2] J. Eichler and Th. Stöhlker, 2007 *Phys. Rep.* **439**, 1.
[3] Y. L. Shi et. al., 2015 *to be submitted.*

Acknowledgements

This work has been supported by the NNSF of China (No. 11464040, 11264033, 11274254) and the Scientific Research Foundation of the Higher Education Institutions of Gansu Province, China (No. 2014A-104).

[1] E-mail: shiyinglong331@163.com

Content from this work may be used under the terms of the Creative Commons Attribution 3.0 licence. Any further distribution of this work must maintain attribution to the author(s) and the title of the work, journal citation and DOI.
Published under licence by IOP Publishing Ltd

Ps slowing down process below the Ps break-up energy in Ar gas

Yosuke Sano* [1], **Yasushi Kino*** [2], **Toshitaka Oka*** [† 3], **Tsutomu Sekine**[† 4]

* Department of Chemistry, Tohoku University, Sendai, 980-8578 Japan

† Institute for Excellence in Higher Education, Tohoku University, Sendai, 980-8576 Japan

Synopsis We performed the positron annihilation age-momentum correlation measurement to investigate the Ps slowing down process in Ar gas. We determined kinetic energies of para-positronium (*p*-Ps) as a function of time. The initial kinetic energy of Ps was 3.7 eV which was somewhat higher than that predicted by the conventional Ps formation model. Momentum transfer cross sections were also determined below the Ps kinetic energy of 3.7 eV, and agreed well with other experimental values near thermal energy.

Positron Annihilation lifetime measurement in gas phase has been a powerful tool to investigate structures of atom/molecule. A positron injected into gas loses its kinetic energy and forms a positronium (Ps). The Ps consisting of a positron and an electron is the lightest atom, and can be another probe to investigate atom/molecule. The mechanism of the Ps formation in gas has been explained by Ore mechanisms, which proposes a suitable positron energy range for a Ps formation [1]. This energy range is called Ore gap. In the case of an Ar gas, the Ore gap ranges from 9 eV (Ps formation threshold) to 11.7 eV (1st electronic excitation threshold of Ar). Namely, the kinetic energy of Ps formed by the Ore mechanism is expected to range from 0 eV to 2.7 eV. The Ps formation cross-section in a positron-Ar collision is dominated in the positron energy range between 9 eV and 100 eV [2,3]. In a dense gas, energetic Ps having a kinetic energy of more than 6.8 eV would immediately break up into a positron and an electron by collisions with atoms [4], and other Ps having a kinetic energy of lower than 6.8 eV would survive and slow down until the annihilation takes place in the Ar gas. Since Ar is a monoatomic molecule and excitation energies are much higher than that of ro-vibrational excitation energy of polyatomic molecules, energy loss of Ps is very slow. Thus, Ps slowing down process can be measured in detail in Ar gas.

In this study, we investigated the time dependence of the Ps kinetic energy in Ar gas by the positron annihilation age-momentum correlation (AMOC) measurement in which positron annihilation lifetime spectroscopy and time dependence of Doppler broadening (DB) of annihilation photon are simultaneously obtained. The positron source (^{22}Na, 0.5 MPa) was placed in the center of a cylindrical high-pressure vessel with a diameter of 2.8 cm and a height of 5.0 cm. The PALS was performed with a fast-fast coincidence system by determining the time interval between detections of a 1.27 MeV gamma-ray from the β^+ decay of a ^{22}Na source and one of the 0.511 MeV annihilation photon by two BaF_2 scintillation detectors. From the other annihilation photon, the Doppler broadening was measured with an HPGe (high purity-Ge) detector, and center of mass energy of the annihilation pair was determined. Details of the experimental set-up was written in Ref. [5] where we measured a positron slowing down process. We used *p*-Ps annihilation gamma rays because the intrinsic lifetime of the *p*-Ps (0.125 ns) is much shorter than the slowing down process. Since the DB of the *p*-Ps is narrower than those from other annihilation processes, we could extract the *p*-Ps component from the energy spectrum of annihilation photons.

We obtained time dependence of Ps kinetic energy from that of the DB of *p*-Ps component. Comparing with a kinetic model of positronium slowing down [6], we determined momentum transfer cross sections below the positronium break-up threshold where no experimental studies have not been performed. The cross section agreed well with other experimental values [7,8] near thermal energy, and was close to the theoretical calculations [9,10]. The initial kinetic energy of Ps was 3.7 eV which was somewhat higher than the Ore gap.

References

[1] A. Ore 1949 *Univ. Bergen Arbok* No.9
[2] C. M. Surko *et al* 2005 *J. Phys.* B **38** R57
[3] G. Laricchia et al 2003 Radiat. Phys. Chem. 68 21
[4] S. Armitage et at 2002 Phys. Rev. Lett. 89 173402
[5] Y. Sano et al. 2014 Jap. J. App. Phys. Conf. Proc. 2 011004.
[6] Y. Nagashima et al 1995 Phys. Rev. A **52** 258
[7] Y. Nagashima et al *New Directions in Antimatter Chemistry and Physics* (Kluwer, Dordrecht, 2001) p. 291
[8] M. Skalsey et al 2003 Phys. Rev. A **67** 022504
[9] P.K. Biswas et al 2000 Chem. Phys. Lett. 317 129
[10] J. E. Blackwood et al 2002 J. Phys. B **35** 2661

[1]E-mail: y.sano@dc.tohoku.ac.jp [2]E-mail: y.k@m.tohoku.ac.jp

Content from this work may be used under the terms of the Creative Commons Attribution 3.0 licence. Any further distribution of this work must maintain attribution to the author(s) and the title of the work, journal citation and DOI.
Published under licence by IOP Publishing Ltd

Journal of Physics: Conference Series **635** (2015) 052089 doi:10.1088/1742-6596/635/5/052089

Strong effect of the electron spin on bremsstrahlung observed in the relativistic regime

Oleksiy Kovtun*[1], Valery Tioukine†, Andrey Surzhykov‡, Vladimir A. Yerokhin♮, Bo Cederwall♯ and Stanislav Tashenov*[2],

* Physikalisches Institut der Universität Heidelberg, 69120 Heidelberg, Germany
† Institut für Kernphysik Johannes Gutenberg-Universität Mainz, 55128 Mainz, Germany
‡ Helmholtz-Institut Jena, 07743 Jena, Germany
♮ Center for Advanced Studies, St. Petersburg State Polytechnical University, St. Petersburg 195251, Russia
♯Royal Institute of Technology, SE-10691 Stockholm, Sweden

Synopsis We have studied bremsstrahlung produced by polarized 2.1 MeV electrons colliding with a thin gold target. We have observed a strong correlation between the orientation of the electron spin and the angle of x-ray linear polarization at the x-ray high energy tip region. This indicates a dominance of the spin–orbit interaction in bremsstrahlung.

Atomic field bremsstrahlung is one of the most fundamental and dominant radiative processes in energetic collisions of electrons with atoms. Studies of this process included measurements of the total cross sections, angular distributions and, lately, polarization. Recently it was shown that in collisions of spin-polarized electrons with atoms the angle of x-ray linear polarization is tilted with respect to the reaction plane defined by the propagation directions of the incoming electron and the emitted x ray [1, 2, 3]. This effect was attributed to the spin–orbit interaction of the electron in the Coulomb field of the nucleus.

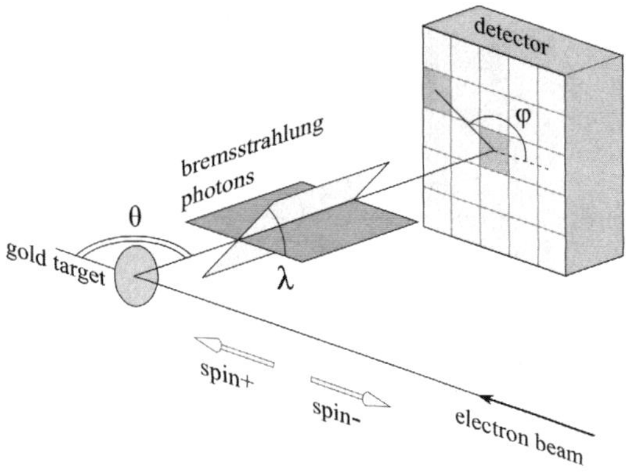

Figure 1. Scheme of the experiment.

We have studied bremsstrahlung produced in collisions of longitudinally polarized electrons of 2.1 MeV kinetic energy with gold atoms of a thin foil target. The x-ray linear polarization was measured with the Compton polarimetry technique applied to a segmented germanium detector, see Fig. 1. For this the Compton-scattered x rays and the recoiled electrons were detected in time-coincidences in separate detector pixels allowing to sample the complete azimuthal angular distributions of the scattered x rays, see Fig. 2. The tilt angles of x-ray linear polarization, determined from these distributions, were ±60° for opposite electron spin orientations. The strong correlation between the electron spin orientation and the tilt angle of x-ray polarization reveal a dominant role of the spin–orbit interaction in bremsstrahlung and Coulomb scattering at high collision energies.

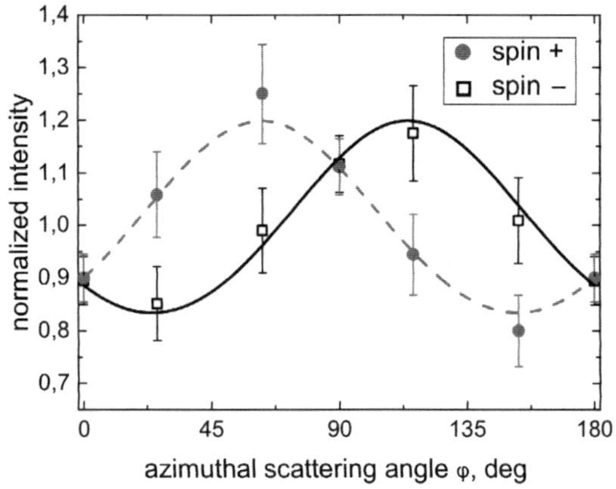

Figure 2. Azimuthal angular distributions of the Compton-scattered x rays produced by electrons polarized collinearly and anti-collinearly with respect to their propagation direction.

References

[1] S. Tashenov *at al* 2011 *PRL* **107** 173201

[2] R. Märtin *at al* 2012 *PRL* **108** 264801

[3] S. Tashenov *at al* 2013 *PRA* **87** 022707

[1]E-mail: kovtun@physi.uni-heidelberg.de
[2]E-mail: tashenov@physi.uni-heidelberg.de

Content from this work may be used under the terms of the Creative Commons Attribution 3.0 licence. Any further distribution of this work must maintain attribution to the author(s) and the title of the work, journal citation and DOI.
Published under licence by IOP Publishing Ltd

XXIX International Conference on Photonic, Electronic, and Atomic Collisions (ICPEAC2015) IOP Publishing
Journal of Physics: Conference Series **635** (2015) 052090 doi:10.1088/1742-6596/635/5/052090

Theoretical study of W^{27+} spectra in EBIT plasma

Valdas Jonauskas[1], **Aušra Kynienė, Šarūnas Masys, Romas Kisielius, Sigitas Kučas, Pavel Rynkun, Laima Radžiūtė, Gediminas Gaigalas, Gintaras Merkelis**

Institute of Theoretical Physics and Astronomy, Vilnius University, A. Goštauto 12, LT-01108 Vilnius, Lithuania

Synopsis Corona modeling of spectral lines for the W^{27+} ion is performed. Energy levels and radiative transition probabilities are calculated using GRASP2K and FAC codes. Electron-impact excitation rates are studied in the distorted wave approximation and are taken proportional to the electric dipole line strengths.

Emission of the tungsten ions that penetrate central regions of the fusion plasma leads to undesirable energy losses and has to be controlled for the successful performance of thermonuclear reactors. Transitions from many ions contribute to the line-of-sight measurements in the fusion plasma. The electron beam ion trap (EBIT) devices provide a unique opportunity to study emission mainly from the desirable ionization stage determined by the energy of the electron beam. Such plasma features low density of electrons that leads to the dominant population of the ground and long-lived levels.

The aim of the current work is to investigate the strongest lines of the W^{27+} ion in $2-30$ nm wavelength range using corona model. The excitations from all levels of the ground configuration with subsequent radiative cascades are investigated in this work. The total number of configurations included in the present study amounts to 52, they produce 20861 levels. Energy levels and electric dipole, quadrupole and octupole as well as magnetic dipole and quadrupole transitions are calculated using multiconfiguration Dirac-Hartree-Fock [1] and Dirac-Fock-Slater [2] methods. The electron-impact excitation rates from the levels of the ground configuration are calculated at the electron beam energy of 870 eV and the electron beam density of 10^{12} cm^{-3}. The Gaussian distribution function with a full width at half-maximum of 30 eV is used for the electron energy.

The investigations of the W^{29+} – W^{37+} ions in the EBIT plasma demonstrated that relative line intensities calculated by studying excitations from the corresponding ground levels are in quite good agreement with the data from the collisional-radiative modeling [3]. There the electron-impact excitation rates were considered as being proportional to the electric dipole transition probabilities because the plane-wave Born matrix element transforms to the matrix elements of the electric multipole transition operators with additional factors. Two approaches are used for modeling spectral lines of the W^{27+} ion in this work. In the first one, electron-impact excitation rates are obtained in the distorted wave (DW) approximation. In the second approach, the electric dipole line strengths are used instead of the electron-impact excitation rates because for the collision cross sections within the plane-wave Born approximation the leading term of the first order is proportional to the electric dipole (spin-allowed) transition probabilities divided by the third power of transition energy, i.e., is proportional to the transition line strength.

The intensive emission of the tungsten ions at 5 nm is determined by the $4d^9 4f^2 + 5d \to 4f$ transitions. Study of configuration interaction strengths reveals that the $4d^{-2}4f^2$ type correlation has the largest impact on these configurations. All corresponding configurations are included in corona modeling of spectral lines.

Large contribution to the spectrum is observed for the transitions from the $4d^9 4f5p$ and $4d^9 4f5f$ configurations to the ground one at the shorter wavelength side. The strongest lines in the $10-30$ nm wavelength range are formed by the $5d \to 5p$, $5g \to 5f$, and $4d^9 4f5p \to 4d^9 4f5s$ transitions.

Acknowledgment: This research was funded by European Social Fund under the Global Grant Measure (No.: VP1-3.1-SMM-07-K-02-015)

References

[1] P. Jönsson *et al* 2013 *Com ut Phys Commu* **184** 9

[2] M. F. Gu *et al* 2008 *Ca J Phys* **86** 675

[3] V. Jonauskas *et al* 2007 *J Phys* **40** 2179

[1]E-mail: valdas.jonauskas@tfai.vu.lt

Content from this work may be used under the terms of the Creative Commons Attribution 3.0 licence. Any further distribution of this work must maintain attribution to the author(s) and the title of the work, journal citation and DOI.
Published under licence by IOP Publishing Ltd

Linear polarization of x rays due to dielectronic recombination into highly charged ions

Chintan Shah[*1], **Holger Jörg**[*], **Zhimin Hu**[*], **Sven Bernitt**[†], **Hendrik Bekker**[†], **Michael A. Blessenohl**[†], **Daniel Hollain**[†], **Sebastian Weber**[*], **Stepan Dobrodey**[†], **Stephan Fritzsche**[‡], **Andrey Surzhykov**[♮], **José R. Crespo López-Urrutia**[†] **and Stanislav Tashenov**[*2]

[*] Physikalisches Institut der Universität Heidelberg, 69120 Heidelberg, Germany
[†] Max–Planck–Institut für Kernphysik, 69117 Heidelberg, Germany
[‡] Friedrich-Schiller-Universität Jena, 07743 Jena, Germany
[♮] Helmholtz-Institut, 07743 Jena, Germany

Synopsis The linear polarization of x rays produced by dielectronic recombination into highly charged ions was for the first time measured at an electron beam ion trap using a newly developed Compton polarimeter. The experimental results open a possibility for diagnostics of anisotropies of hot plasmas. We also demonstrate a high sensitivity of the x-ray polarization to the Breit interaction.

We report the first measurement of linear polarization of x rays emitted in the process of dielectronic recombination (DR) into highly charged ions. The krypton and xenon ions in the He- though O-like charge states were produced in an electron beam ion trap (EBIT) and the electron-ion collision energy was tuned into various KLL DR resonances. In these resonances an electron recombines into the L- shell of the ion and a bound electron is simultaneously excited from the K- to the L-shell. The polarization of the x rays emitted perpendicular to the electron beam propagation direction was analysed using the Compton polarimetry technique. For this the x rays were Compton-scattered in a block of beryllium or boron carbide. The scattered x rays were detected by an array of SiPIN diodes which sampled their azimuthal angular distribution, see Fig. 1. By fitting the Klein-Nishina formula to the measured scattering distributions we extracted the degrees of polarization of several DR transitions with a typical accuracy below 10%. The measured degrees of polarization range from -0.84 to 0.48 for krypton ions and from -0.43 to 0.53 for xenon ions. This measurement opens possibilities for polarization diagnostics of hot plasmas. Such diagnostics will be sensitive to the directionality of the electron-ion collisions revealing plasma anisotropies. Moreover, we have demonstrated that the polarization of the x rays emitted in the dielectronic recombination exciting the intermediate state $[1s2s^22p_{1/2}]_1$ is highly sensitive to the Breit interaction. The latter accounts for retardation and magnetic contributions to the Coulomb repulsion between the electrons. For both krypton and xenon ions, the experimental results for this resonance agree with the predictions that include the Breit interaction and they rule out by 2σ and 5σ, respectively, the calculations that treat the electron–electron interaction purely by the Coulomb force.

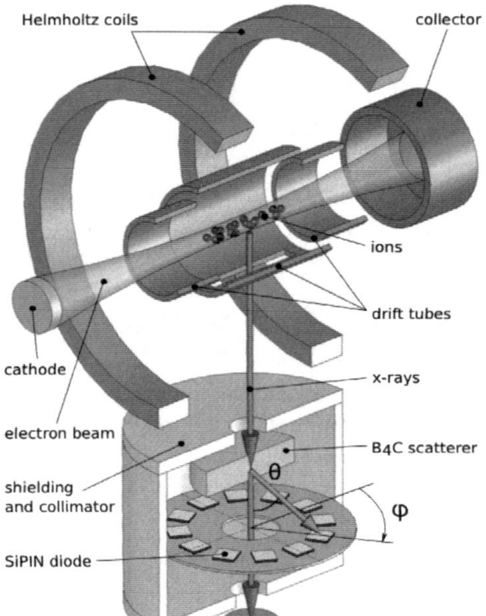

Figure 1. Scheme of the experiment: the ions are produced and trapped in an EBIT. The polarization of the x rays emitted in the electron-ion collisions is analysed using the Compton polarimetry technique.

[1]E-mail: chintan@physi.uni-heidelberg.de
[2]E-mail: tashenov@physi.uni-heidelberg.de

Content from this work may be used under the terms of the Creative Commons Attribution 3.0 licence. Any further distribution of this work must maintain attribution to the author(s) and the title of the work, journal citation and DOI.
Published under licence by IOP Publishing Ltd

X-ray spectroscopy of multicharged xenon ions at the EBIT plasma

D. Banaś* [1], Ł. Jabłoński*, P. Jagodziński†, A. Kubala-Kukuś*, D. Sobota*, M. Puchała*, M. Pajek*,

* Institute of Physics, Jan Kochanowski University, 25-406 Kielce, Poland

† Department of Physics, University of Technology, 25-314 Kielce, Poland

Synopsis X-ray spectra of multicharged xenon ions created in an electron-beam ion trap (EBIT) were measured both with a low-resolution silicon drift detector (SDD) and a high-resolution Johann/Johannson type crystal diffraction spectrometer. The measurement were performed in the electron-beam ion source (EBIS) using the transmission, leaky and pulsed modes of the ion source operation with trapped xenon ions excited by an electron beams of energies 9-15 keV. Moreover, the dynamics and equilibration of the EBIT plasma was studied by measuring the X-ray spectra for different electron energies and trapping times. The results will be discussed using available theoretical models describing the atomic processes in the EBIT plasma.

The measurements of X-rays emitted from electron-beam ion traps are important for various aspects of physics, ranging from atomic physics and astrophysics to diagnostics of fusion plasmas [1]. The low-resolution X-ray measurements deliver information important for understanding the dynamics and equilibration of the atomic processes occurring in the EBIT plasmas [2]. In turn, the high resolution studies of EUV/X-ray from the EBITs yield precise characteristics of a structure of few-electron ions and thus are important for testing the atomic structure calculations, in particular, using the relativistic multiconfiguration Dirac-Fock (MCDF) approach including ED corrections [3].

The X-ray measurements were performed using Dresden EBIS-A ion source at the Institute of Physics of Jan Kochanowski University in Kielce [4], which supplies a wide range of slow highly charged ions from bare ions of light elements to Ne-like and Ar-like ions of high-Z elements [5]. The maximum electron beam energy and current available for ionization of the trapped ions are equal to 25 keV and 200 mA, respectively. The ion source can work in three different operation modes, namely, in transmission, leaky and pulsed modes characterized by different distribution of ion charge states in the ion trap [5].

The low-resolution spectra of X-rays emitted from multicharged xenon ions, up to Xe^{40+}, were registered with a silicon drift detector (XFlash 5010, Bruker). This detector has energy resolution of about 127 eV for Mn Kα line. The XFlash has ability to measure X-ray emitted from EBIS-A synchronically with a time structure of a pulse, which allows to study time evolution of the ion charge states in the ion source as well as the electron energy dependence of the atomic processes in the EBIT plasma. The high-resolution X-ray spectra were measured with a compact multi-crystal diffraction spectrometer (INCAWave, Oxford Instruments) having energy resolution of about 1 eV for low-energy X-rays [6]. The spectrometer covers a wide range of photon energies from 70 eV to 15 keV, including thus both the extended ultraviolet (EUV) and X-ray regions. In the present configuration it is equipped with six diffraction crystals (LiF(220), LiF(200), PET, TAP, LSM060, LSM200), including two multilayers (LSM), which are installed in Johann or Johansson geometry. At the EBIS-A the spectrometer is operated in the "out-of-focus" geometry increasing a tolerance for positioning of the X-ray source.

Concluding, in this work the X-ray spectra emitted from interaction of xenon ions with electrons in the EBIS-A facility were measured for different electron excitation and ion trapping conditions by using both the energy dispersive X-ray SDD detector and wavelength-dispersive X-ray spectrometer. The results will be compared with predictions of multiconfiguration Dirac-Fock calculations and available models of atomic processes in the EBIT plasma.

References

[1] P. Beiersdorfer 2009 *Phys. Scr ta* **T134** 014010

[2] W. D. Chen *at al* 2007 *Phys. Plasmas* **14** 103302

[3] J. D. Gillaspy *at al* 2014 *Phys. Scr ta* **89** 114004

[4] D. Banaś *at al* 2010 *J. I strum.* **5** C09005

[5] G. Zschornack *at al* 2006 *Rev. Sc . I strum.* **77** 03A904

[6] Ł. Jabłoński *at al* 2015 *Nucl. I strum Meth.* in press

[1] E-mail: d.banas@ujk.edu.pl

Content from this work may be used under the terms of the Creative Commons Attribution 3.0 licence. Any further distribution of this work must maintain attribution to the author(s) and the title of the work, journal citation and DOI.
Published under licence by IOP Publishing Ltd

XXIX International Conference on Photonic, Electronic, and Atomic Collisions (ICPEAC2015) IOP Publishing
Journal of Physics: Conference Series **635** (2015) 052093 doi:10.1088/1742-6596/635/5/052093

Complete measurements of anisotropic x-ray emission following recombination of highly charged ions

Chintan Shah[*1], Pedro Amaro[*], René Steinbrügge[†], Christian Beilmann[*,†], Sven Bernitt[†], Stephan Fritzsche[‡], Andrey Surzhykov[♮], José R. Crespo López-Urrutia[†] and Stanislav Tashenov[*2]

[*] Physikalisches Institut der Universität Heidelberg, 69120 Heidelberg, Germany
[†] Max–Planck–Institut für Kernphysik, 69117 Heidelberg, Germany
[‡] Friedrich-Schiller-Universität Jena, 07743 Jena, Germany
[♮] Helmholtz-Institut, 07743 Jena, Germany

Synopsis We report the first systematic measurement of x-ray emission anisotropies following dielectronic, trielectronic and quadroelectronic recombination into He-like through O-like iron and krypton ions. Using the experimental data we calculated the polarization of the Kα x rays in an anisotropic plasma as a function of the plasma temperature. Unexpectedly, we found that the degree of polarization is dominated by previously neglected trielectronic and quadroelectronic recombination resonances.

X-ray emission asymmetries following resonant recombination into highly charged ions were studied using an electron beam ion trap (EBIT) of Max Planck Institute for Nuclear Physics in Heidelberg. Iron and krypton ions in the He-like through O-like charge states were populated in an EBIT and the electron-ion collision energy was scanned over the region of the K-shell recombination resonances. With the help of the excellent electron energy resolution of 6.5 eV for iron ions and 11.5 eV for krypton ions the dielectronic, trielectronic and quadroelectronic recombination transitions were systematically investigated. The x rays emitted in the decays of resonantly excited states were observed by two germanium detectors aligned along and perpendicular to the electron beam propagation direction and the corresponding intensities of the K-shell x-ray transition were recorded as a function of the electron collision energy. The x-ray emission asymmetries reveal alignment of the intermediate excited states and, thus, the polarization of the emitted x rays. Except for a few transitions the experimental results are in excellent agreement with the theoretical calculations done with FAC and RATIP computer codes.

This measurement allows for a systematic modeling of the polarization of the prominent Kα radiation emitted by hot anisotropic plasmas. The measurement of its polarization gives access to the anisotropy of the electron momentum distribution in the plasma. Using the experimental data we calculated the maximum polarization of the Kα x rays emitted by an anisotropic plasma as a function of the plasma temperature, see Fig. 1. Surprisingly, we found that the degree of the x-ray polarization is dominated by trielectronic and quadroelectronic recombination transitions which were previously neglected. These experimental results should play an important role in diagnostics of hot astrophysical plasmas of solar flares and active galactic nuclei and laboratory fusion plasmas of tokamaks and stellarators.

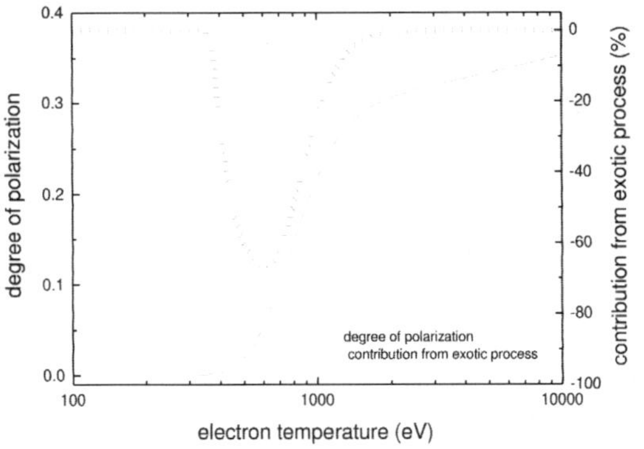

Figure 1. Maximum polarization of iron Kα x rays due to resonant recombination as a function of the plasma temperature. Exotic trielectronic and quadroelectronic recombination transitions dominate polarization in a broad range of temperatures.

[1]E-mail: chintan@physi.uni-heidelberg.de
[2]E-mail: tashenov@physi.uni-heidelberg.de

Content from this work may be used under the terms of the Creative Commons Attribution 3.0 licence. Any further distribution of this work must maintain attribution to the author(s) and the title of the work, journal citation and DOI.
Published under licence by IOP Publishing Ltd

Novel approaches to study low-energy electron-induced damage to DNA oligonucleotides

Jenny Rackwitz* [1], Miloš Lj. Ranković† [2], Aleksandar R. Milosavljević† [3], Ilko Bald*‡ [4]

* Department of Chemistry, University of Potsdam, Potsdam, Germany
† Institute of Physics, University of Belgrade, Belgrade, Serbia
‡ BAM Federal Institute of Materials Research and Testing, Berlin, Germany

Synopsis The novel approach of DNA origami structures as templates for precise quantification of various well-defined oligonucleotides provides the opportunity to determine the sensitivity of complex DNA sequences towards low-energy electrons.

Secondary low-energy electrons (LEEs) belong to the most important intermediates in DNA radiation damage. Strand breaks in the DNA backbone can be initiated by resonant electron attachment at different electron energies.[1] Recent experiments suggest that single strand breaks (SSBs) strongly depend on the nucleotide sequence due to efficient coupling of the electronic states of the nucleobases by stacking interactions.[2] A novel approach using DNA origami templates carrying different oligonucleotide target sequences provides access to efficient and systematic determination of electron induced DNA strand break cross sections (see Figure 1).[3]

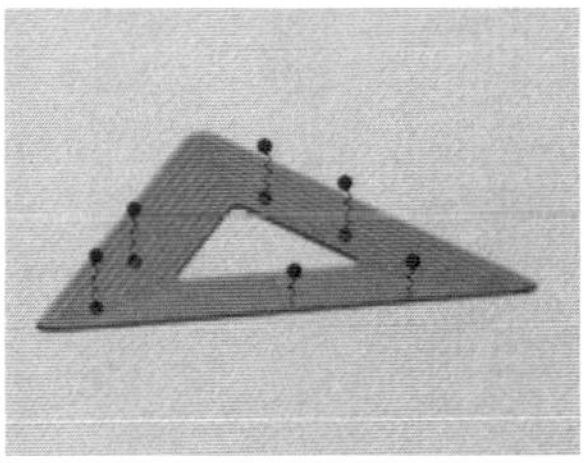

Figure 1. 3D model of a triangular DNA origami structure carrying six well-defined oligonucleotides and biotin at the 5' end.

We assembled a new instrument to irradiate specific oligonucleotide sequences on DNA origami templates with electrons of defined but variable energy from 2 to 50 eV (see Figure 2). First experiments are performed at 10eV, where a maximum of DNA single and double strand breaks was previously observed.[4]

The damaged oligonucleotides are analyzed with atomic force microscopy (AFM) and the well-defined shape of the DNA origami templates allows for a precise localization and thus quantification of DNA strand breaks at a single-molecule level. The influence of the DNA sequence, e.g the telomere sequence, shall be explored. Furthermore, potential radiosensitizers such as 2-Fluoroadenine are used to study their influence on DNA strand break cross sections.

Figure 2. 3D model of the instrument for oligonucleotide irradiation, equipped with vacuum pump, electron source, sample stage and manipulators.

Acknowledgement

We are grateful to a bilateral grant between Serbia and Germany through the German Academic Exchange Service (DAAD, project No. 57055766) supported by the German Federal Ministry of Education and Research.

References

[1] I. Baccarelli *et al* 2011 *Phys. Rep.* **508** 1.
[2] A. Keller *et al* 2014 *Sci. Rep.* **4** 7391.
[3] A. Keller *et al* 2012 *ACS Nano* **6** 4392.
[4] B. Boudaiffa *et al* 2000 *Science* **287** 1658.

[1] E-mail: rackwitz@uni-potsdam.de [2] E-mail: mrankovic@ipb.ac.rs [3] E-mail: vraz@ipb.ac.rs [4] E-mail: bald@uni-potsdam.de

Content from this work may be used under the terms of the Creative Commons Attribution 3.0 licence. Any further distribution of this work must maintain attribution to the author(s) and the title of the work, journal citation and DOI.
Published under licence by IOP Publishing Ltd

Journal of Physics: Conference Series **635** (2015) 062002 doi:10.1088/1742-6596/635/6/062002

Low Energy Electron Induced Reactions in Condensed Methanol

Sramana Kundu [1], Daly Davis [2], Vaibhav. S. Prabhudesai [3] and E. Krishnakumar [4]

Department of Nuclear & Atomic Physics, Tata Institute of Fundamental Research, Mumbai 400005, India

Synopsis Resonant formation of H^- and O^- ions and formation of ethylene glycol ($OHCH_2CH_2OH$) on electron impact is seen in the same energy range in condensed molecular films of methanol (CH_3OH). This could indicate a new reaction pathway involving bond breaking and bond formation at specific electron energies in condensed methanol.

Site selective reactions in molecules can be induced by low energy electrons (LEE) through dissociative electron attachment (DEA) [1]. Indications of selectivity in surface reactions could be seen in the electron induced desorption of negative ions from condensed molecular films [2]. Methanol (CH_3OH) is one of the simplest compounds with two different functional groups (CH_3 and OH). DEA to gas phase methanol showed the site-specific fragmentation of C-H and O-H bonds using incident electron energy as a control parameter [1]. A recent study of electron stimulated desorption (ESD) from condensed methanol has identified the formation of H^- and O^- by DEA using electrons of energy above 5 eV [3].

We have carried out ESD experiments from condensed methanol films of varying thickness using our newly built condensed phase experimental setup. We form cold molecular films by depositing them on a gold substrate cooled by a closed cycle helium cryostat. The ions desorbed from the surface following irradiation by a magnetically collimated and pulsed electron gun are analyzed using a time of flight mass spectrometer (ToFMS) mounted normal to the substrate. The composition of and any chemical changes taking place in the molecular film is probed in situ by using Fourier transform infrared (FTIR) spectrometry throughout the experiment.

The measurements on the desorbed anions show the resonant formation of O^- at 3 - 4 eV and H^- at 5 - 6 eV. FTIR measurements show signature of ethylene glycol ($OHCH_2CH_2OH$) in the same energy range (5 - 6 eV). Production of glycol in methanol ice indicates formation and recombination of (CH_2OH) radicals in the irradiated film. It is possible that (CH_2OH) and H^- are generated in the same channel in the 5 - 6 eV energy range. These observations suggest a new reaction path way involving bond breaking and bond formation at specific electron energies in condensed methanol.

References

[1] V.S. Prabhudesai *et al* 2005 *Phys. Rev. Lett.* **95** 143202

[2] L. Parenteau *et al* 1994 J. Chimi. Phys. **91** 1237

[3] Michael C. Boyer *et al* 2014 J. Phys. Chem. C **118 (39)** 22592

[1] E-mail: sramana@tifr.res.in

[2] E-mail: dalydavis@gmail.com

[3] E-mail: vaibhav@tifr.res.in

[4] E-mail: ekkumar@tifr.res.in

Content from this work may be used under the terms of the Creative Commons Attribution 3.0 licence. Any further distribution of this work must maintain attribution to the author(s) and the title of the work, journal citation and DOI.
Published under licence by IOP Publishing Ltd

Positive ion desorption from frozen N-heterocycles stimulated by high-energy electron impact

Fabio Ribeiro* [1], **Jessica Martins***, **Wania Wolff**† [2], **Heloisa Boechat-Roberty**‡ **and Maria Luiza Rocco***

* Instituto de Química, Universidade Federal do Rio de Janeiro, RJ, 21941-909, Rio de Janeiro, Brazil
† Instituto de Física, Universidade Federal do Rio de Janeiro, RJ, 21941-909, Rio de Janeiro, Brazil
‡ Observatório do Valongo, Universidade Federal do Rio de Janeiro, RJ, 20.080-090 Rio de Janeiro, Brazil

Synopsis Ion desorption measurements induced by the impact of electrons onto pyridine and pyrimidine ices will be presented. The data reported here is of relevance for chemical evolution models of astrophysical objects.

Nitrogen-Heterocyclic molecules (N-heterocycles) are one of most important biological class of molecules, since they make up the genetic material for all living organisms. Among them, pyridine (C_5H_5N) and pyrimidine ($C_4H_4N_2$) are prototypical aromatic molecules, in which the benzene ring is substituted by one and two nitrogen atoms, respectively. These molecules have also been the subject of intense radioastronomical surveys towards the interstellar-medium (ISM), although their detection was unsuccessful [1].

On the other hand, the presence of uracil traces in the organic extracts of the Murchison meteorite [2] may point to the incorporation of N-heterocycles into the solid component of the ISM, causing the molecular depletion observed in the gas-phase. Therefore, the desorption induced by photons and collisions with charge particles may be an important mechanism for leading neutral and ionic fragments into the gas-phase, acting as tracers for N-heterocycles.

In this work, we report the results of high-energy electron impact onto vapor deposited ices of C_5H_5N and $C_4H_4N_2$ by means of the electron stimulated ion desorption (ESID) technique. Desorbed cations from the icy surface were analyzed by time-of-flight mass spectrometry (TOF-MS). Both C_5H_5N and $C_4H_4N_2$ exhibit six distinct regions on their ESID spectra, containing ionic fragments bearing one to six members of the original heterocyclic ring. As reported in [3], five-membered fragment ions were observed for the first time in electron impact experiments for $C_4H_4N_2$.

Figure 1 shows several minor fragments from the ESID spectrum of C_5H_5N at 120 K, indicating strong fragmentation. Proton-transfer reactions also play an important role during desorption, as inferred by the intensity of the protonated parent ion ($C_5H_5NH^+$) at m/z = 80. Cluster ions have also been identified and their formation is discussed on the basis of ion-molecule interactions.

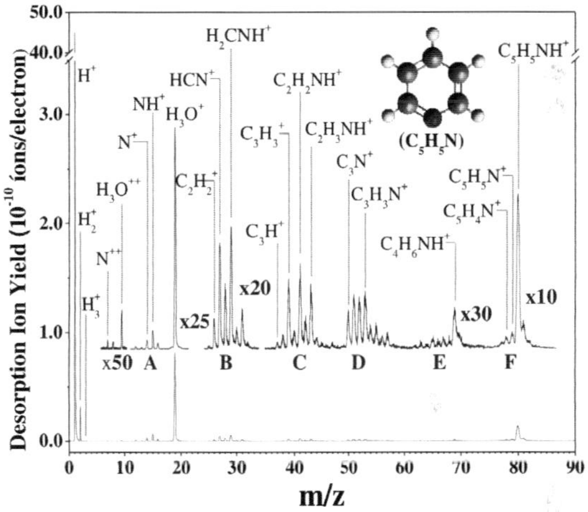

Figure 1. ESID spectrum of positive ions from pyridine ice (C_5H_5N) at 120 K due to 2.3 keV electron impact. Regions A to F are magnified in order to highlight the cracking pattern of solid pyridine.

The comparison between positive ion yields for C_5H_5N and $C_4H_4N_2$ will be presented. Several of the detected ions in this work are in agreement with cometary observations, and thus may act as an indicative of the presence of N-heterocycles on astrophysical objects.

References

[1] S. B. Charnley *et al.* 2005 *Adv. Space Res.* **36** 137
[2] Z. Martins *et al.* 2008 *Earth Planet. Sci. Lett.* **270** 130
[3] F. A. Ribeiro *et al.* 2014 *J. Phys. Chem. C* **118** 25978

[1] E-mail: fabioribeiro@iq.ufrj.br
[2] E-mail: wania@if.ufrj.br

Content from this work may be used under the terms of the Creative Commons Attribution 3.0 licence. Any further distribution of this work must maintain attribution to the author(s) and the title of the work, journal citation and DOI.
Published under licence by IOP Publishing Ltd

XXIX International Conference on Photonic, Electronic, and Atomic Collisions (ICPEAC2015) IOP Publishing
Journal of Physics: Conference Series **635** (2015) 062006 doi:10.1088/1742-6596/635/6/062006

Non-thermal desorption processes on nitrile-bearing astrophysical ice analogues studied by electron stimulated ion desorption

Fabio Ribeiro* [1], **Guilherme Almeida*, Lautaro Ramirez*, Jessica Martins*, Yunier Basabe*, Wania Wolff**[†], **Heloisa Boechat-Roberty**[‡] **and Maria Luiza Rocco*** [2]

[*] Instituto de Química, Universidade Federal do Rio de Janeiro, RJ, 21941-909, Rio de Janeiro, Brazil
[†] Instituto de Física, Universidade Federal do Rio de Janeiro, RJ, 21941-909, Rio de Janeiro, Brazil
[‡] Observatório do Valongo, Universidade Federal do Rio de Janeiro, RJ, 20.080-090 Rio de Janeiro, Brazil

Synopsis Methyl cyanide (CH_3CN) is the simplest organic nitrile and it is found in almost all astrophysical environments, including cometary comae, Titan's atmosphere and the interstellar medium (ISM), especially in star-forming regions and hot cores. In this work, the role of ion desorption caused by electron collisions onto a CH_3CN ice is discussed. New ionic species, including ion clusters are produced, pointing to the delivery of more complex species from ice surfaces to the gas-phase in the ISM.

Nitriles are very abundant in the gas-phase of a variety of astrophysical environments. Complex molecules, such as methyl cyanide (CH_3CN), ethyl cyanide (CH_3CH_2CN), vinyl cyanide (CH_2CHCN) and isopropyl cyanide ($(CH_3)_2CHCN$) have been detected towards the Sgr B2(N) star-forming region in the interstellar medium (ISM). Nitriles are also well known to be present in Solar System bodies, like cometary comae and on Titan's atmosphere. Despite their astrophysical relevance, little is known about their formation in the gas-phase.

The desorption process induced by photons, electrons or ions onto icy surfaces is an important mechanism for leading neutral or ionized molecular species to the gas-phase in the ISM. Non-thermal desorption processes, such as desorption induced by electronic transitions (DIET), may be responsible for delivering neutral and ionic fragments from simple nitrile-bearing ices to the gas-phase, contributing to the production of more complex molecules.

In this work, we have studied the fragmentation and desorption of cations from a CH_3CN ice at 120 K due to the impact of electrons with variable impact energy of up to 3 keV. The Electron Stimulated Ion Desorption (ESID) technique coupled to Time-of-Flight Mass Spectrometry (TOF-MS) was employed in order to analyze the positively charged ionic fragments which desorb from the surface at each impact energy. The experimental method is described in more detail elsewhere [1].

A sample ESID spectrum is presented in Figure 1. Several positive ions have been assigned, including the protonated parent molecule CH_3CNH^+, possible identification of $CH_3CH_2CN^+$ at m/z = 55 and the cluster ions $(CH_3CN)_2^+$ and $(CH_3CN)_2H^+$ (not shown here).

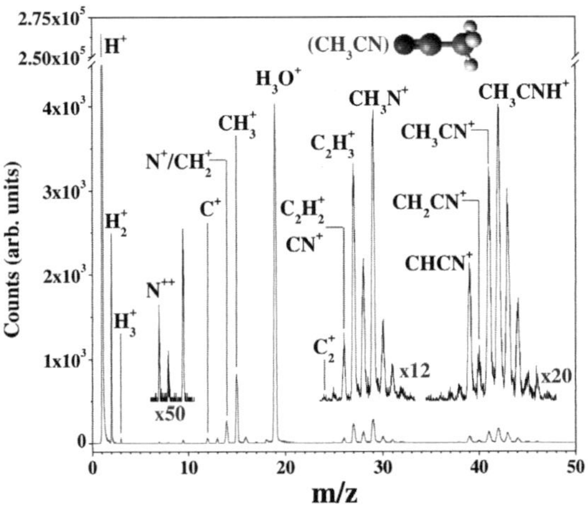

Figure 1. ESID spectrum of CH_3CN ice at 120 K due to 2.3 keV electron impact.

The energy dependence on the positive íon desorption yield will be presented. The desorption of molecular ionic fragments is believed to be caused by Coulomb explosion due to the creation of inner-shell holes followed by electronic relaxation via Auger decay, a process known as Auger stimulated ion desorption mechanism (ASID). The derived desorption yields per electron impact may contribute to chemical evolution models in different astrophysical objects, especially the Titan ionosphere and hot-cores in the ISM.

References

[1] F. de A. Ribeiro *et al* 2014 *J. Phys. Chem. C* **118** 25978

[1] E-mail: fabioribeiro@iq.ufrj.br
[2] E-mail: luiza@iq.ufrj.br

Content from this work may be used under the terms of the Creative Commons Attribution 3.0 licence. Any further distribution of this work must maintain attribution to the author(s) and the title of the work, journal citation and DOI.
Published under licence by IOP Publishing Ltd

XXIX International Conference on Photonic, Electronic, and Atomic Collisions (ICPEAC2015) IOP Publishing
Journal of Physics: Conference Series **635** (2015) 062007 doi:10.1088/1742-6596/635/6/062007

Channeling of Relativistic Electrons in Half-Wave Silicon Crystal and Corresponding Radiation

Y. Takabayashi* [1], **V.G. Bagrov**† [2], **O.V. Bogdanov**‡ [3], **Yu.L. Pivovarov**‡ [4], **T.A. Tukhfatullin**‡ [5]

*SAGA Light Source,8-7 Yayoigaoka, Tosu, Saga 841-0005, Japan
†National Research Tomsk State University, 634050 Tomsk, Russia
‡National Research Tomsk Polytechnic University, 634050 Tomsk, Russia

Synopsis The new experiments on channeling of 255 MeV in a 0.7 μm silicon half-wavelength crystal were performed at SAGA LS facility. Both experimental and simulated electron angular distribution after the crystal and corresponding radiation spectra reveal the number of peculiarities.

Suggestion that the straight half-wavelength crystal (HWC) may serve for particles deflection, due to effect of particle "mirroring" was made first in [1]. HWC means, that the channeled particle experiences only one collision with crystallographic plane. Recently, the authors of Ref. [2] experimentally demonstrated that planar channeled 2 MeV protons were successfully mirrored by a thin silicon HWC. In our works [3-4] we performed the detailed experimental investigation and computer simulations of channeling of 255 MeV electrons in the 1 μm thickness silicon HWC. This thickness is slightly greater than the thickness of a half wavelength crystal which is determined by the formula [1]

$$\lambda/2 = \pi \frac{d_p}{2\theta_c}, \quad \theta_C = \sqrt{\frac{2U_0}{E}},$$

where d_p is the distance between channeling planes, θ_c is the critical channeling angle, U_0 is the maximal value of potential energy of the particle in the field of continuous potential for planar channeling, E is the energy of the particle. Due to this fact the effect of mirroring [3-4] was not brilliant.

Recently, the new experiments on channeling of 255 MeV in a 0.7 μm silicon HWC were performed at SAGA LS facility. Here we present the results of the new experiments and comparison with computer simulation. Now, the thickness of the crystal (0.7 μm) was closer to the necessary for 255 MeV electrons value of HWC thickness, therefore the angular distribution of electrons after the HWC crystal revealed the number of peculiarities.

The specifics of trajectories as well as very small crystal thickness should effect on the channeling radiation spectra of relativistic electrons in a HWC. In this work we calculated radiation spectrum from electrons channeled in a HWC and performed the comparison with radiation spectrum of electron moving in an arc [4] (never studied experimentally).

Further perspectives for experimental and theoretical studies of channeling and radiation in a HWC are discussed, too.

References

[1] E. Tsyganov, A.Taratin *Nucl. Instr. & Meth. in Phys. Res.* 1995 **A363**, 511
[2] V. Guidi, *et al* 2012 *Phys. Rev. Lett.* **108** 014801
[3] Yu.L. Pivovarov and T.A. Tukhfatullin 2014 *J. Phys. Conf. Ser.* 517 012028
[4] Y. Takabayashi *et al. Nucl. Instr. & Meth. in Phys. Res.* 2015 (in Press doi:10.1016/j.nimb.2015.02.007).
[4] V.G. Bagrov, *et al J. Exper. and Theor. Phys.* 1982 **55** 835

[1] E-mail: takabayashi@saga-ls.jp
[2] E-mail: bagrov@tsu.ru
[3] E-mail: bov@tpu.ru
[4] E-mail: pivovarov@tpu.ru
[5] E-mail: tta@tpu.ru

Content from this work may be used under the terms of the Creative Commons Attribution 3.0 licence. Any further distribution of this work must maintain attribution to the author(s) and the title of the work, journal citation and DOI.
Published under licence by IOP Publishing Ltd

Journal of Physics: Conference Series **635** (2015) 062008 doi:10.1088/1742-6596/635/6/062008

Low-energy Electrons Interactions with Chemisorbed and Physisorbed Films of L-cysteine/Au(111)

Elahe Alizadeh* [1], **Sylvain Massey**† [2], **Léon Sanche**† [3] **and Paul A. Rowntree*** [4]

* Department of Chemistry, University of Guelph, Guelph, N1G 2W1, Canada

† Groupe en Sciences des Radiations, Faculté de Médecine et des Sciences de la Santé, Université de Sherbrooke, Sherbrooke, J1H 5N4, Canada

Synopsis We investigate electron attachment to chemisorbed and physisorbed L-cysteine/Au(111) films and its subsequent dissociation via electron stimulated desorption (ESD) measurements; More specifically, dissociative electron attachment (DEA) which in essentially, molecular dissociation into a neutral and anionic fragment following initial electron capture into short-lived molecular anion or *resonance* leads to the rupture of specific bonds, if the incident electron energies are within characteristic ranges (below 20 eV).

Studying self-assembled monolayers (SAMs) of protein building blocks is a promising approach to gain information on the complex interactions of proteins with metal surfaces. Thiol-containing amino acids cysteine ($C_3H_7NO_2S$) has been widely studied, becoming a standard for molecular self-assembly experiments, because of its mercapto group (-SH), which allows the formation of a self-assembled monolayers (SAMs) on metal surfaces [1].

Gold films with strongly Au(111) texture are prepared by evaporation on freshly cleaved mica sheets, and cleaned with sulfochromic acid in order to eliminate the organic contaminants. L-cysteine SAMs are prepared by immersion the cleaned Au in 1 mM solution of cysteine in methanol for overnight. Physisorbed samples are prepared as thin and uniform as possible onto gold by lyophilization (freeze-drying). The binding of L-cysteine to the Au surface and formation of SAMs have been analyzed using techniques such as X-ray photoelectron spectroscopy (XPS) and infrared reflection adsorption spectroscopy (IRRAS).

Dissociative electron attachment (DEA) onto chemisorbed and physisorbed L-cysteine films has been investigated utilizing a time-of-flight mass spectrometer set in negative ion mode of detection, coupled with a low-energy electron gun. Anion desorption yield functions are reported for more pronounced produced fragments from L-cysteine by measuring the anion yields as a function of the incident electron energy from 0 to 18 eV. These desorption curves exhibited purely resonant behavior with low-energy electrons (LEEs, $E_e < 20$ eV), indicating that the formation of transient anions is the predominant mechanism of anion fragments production from dissociation of L-cysteine in condensed-phase. Similar to the results observed from physisorbed films, ESD measurement of SAMs have shown that LEEs are able to efficiently decompose this amino acid via DEA and dipolar dissociation when the molecule is chemisorbed via the SH group to a gold surface.

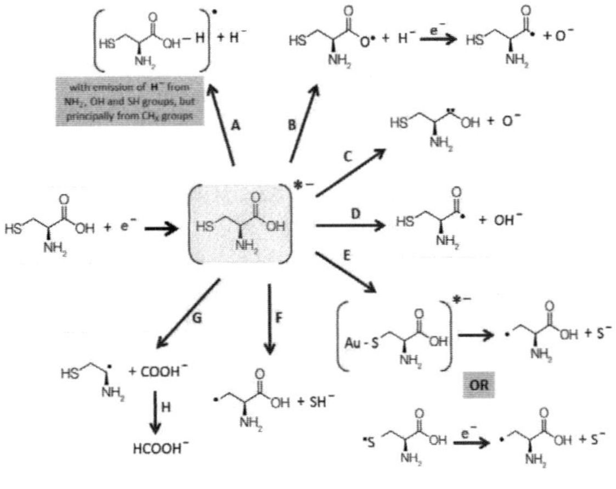

Figure 1. Possible dissociation pathways for formation of H^- (A), O^- (B and C), OH^- (D), S^- (E), SH^- (F) and $COOH^-$/$HCOOH^-$ (G) after LEE attachment to L-cysteine and producing a transient negative anion (cysteine)*$^-$.

References

[1] J. Kang and P.A. Rowntree, 2007 *Langmuir* 23 509

[2] E. Alizadeh and L. Sanche, 2012 *Chem. Rev.* **112** 5578

[1] E-mail: EAlizade@UoGuelph.ca

[2] E-mail: Sylvain.Massey@USherbrooke.ca

[3] E-mail: Leon.Sanche@USherbrooke.ca

[4] E-mail: Rowntree@UoGuelph.ca

Content from this work may be used under the terms of the Creative Commons Attribution 3.0 licence. Any further distribution of this work must maintain attribution to the author(s) and the title of the work, journal citation and DOI.
Published under licence by IOP Publishing Ltd

Journal of Physics: Conference Series **635** (2015) 062009 doi:10.1088/1742-6596/635/6/062009

Asymmetry in the excitation of plasmons by swift particles traversing a surface at oblique incidence

J. L. Gervasoni[*,†,‡ 1], **R. O. Barrachina**[*,†,‡ 2], **S. Segui**[*,‡ 3], **F. Navarrete**[*†‡ 4]

[*] Centro Atómico Bariloche (CNEA). Av. Bustillo 9500, 8400 S. C. de Bariloche, Argentina.
[†] Instituto Balseiro (CNEA - UNCuyo). Av. Bustillo 9500, 8400 S. C. de Bariloche, Argentina.
[‡] Consejo Nacional de Investigaciones Científicas y Técnicas (CONICET), Argentina.

Synopsis We analyze the asymmetry found in the production of plasmons by fast charged particles entering or leaving a semi-infinite material through a planar surface. In the framework of a semiclassical dielectric formulation, we study how this effect depends on the angle of incidence.

The study of plasmon excitation is of great relevance for the characterisation of surfaces, thin films, and nanoparticles using different charged-particles spectroscopic techniques.In this work we are interested in analyzing any difference that might appear in the excitation of plasmons when a projectile enters or leaves a semi-infinite material through a planar surface, considering a variety of experimental settings. We employ the specular reflection model and the extended pseudo-medium method [1] in order to assess the energy loss rate due to plasmon excitation along the projectile's trajectory.

We consider a particle of charge Z traversing the interface between vacuum and a medium of a given dielectric function $\epsilon(\mathbf{k}, \omega)$ located in the semispace $z < 0$ with a trajectory given by $\mathbf{v}t$ (being $\mathbf{v}$ the velocity). The energy loss rate dW/dt is calculated from the time derivative of the potential induced by the incident particle, evaluated along its trajectory. We obtain two contributions corresponding to the excitation of bulk and surface plasmons. The surface term, which acts on the external particle inside as well as outside the material, remains unchanged whether the particle enters or leaves the material. The bulk term, active only when the particle is inside the material, is made up of two contributions: one comes directly from the projectile, while the other comes from the image charge [2]. Our results (figure 1) show that the energy lost by a charged projectile traversing the planar surface of a semi-infinite material is essentially different for outgoing and incoming trajectories. While for outgoing trajectories the bulk plasmon production is almost identical to that in an infinite medium, an oscillatory structure is observed when the projectile penetrates the material. These oscillations are more pronounced for near-perpendicular trajectories and fade away with increasing angles of incidence. We also notice that, far from the surface, the energy loss rate tends to a constant value independent of the angle. This has been corrected from previous calculations, where a spurious dependence with the incidence angle persisted due to an approximation in the cutoff value of the wave vector $\mathbf{k}$ [3]. Here this approximation has been avoided, and the results are valid for any angle of incidence.

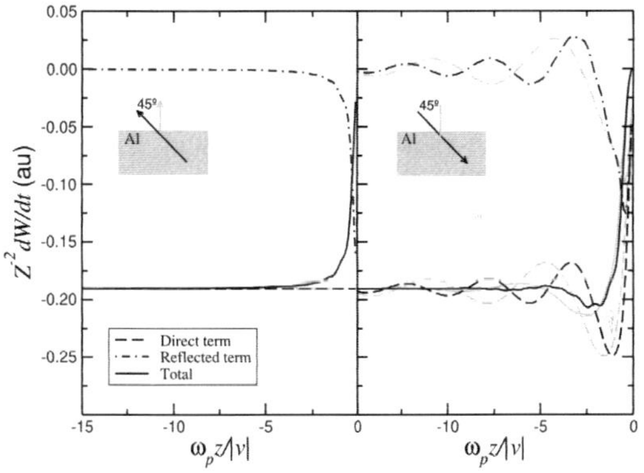

Figure 1. Bulk dW/dt for a charge leaving (left) and entering (right) a semi-infinite aluminum medium ($\omega_p = 0.55$ a.u.), in perpendicular (green) and oblique (black) trajectories. Contributions of the direct and reflected terms are included.

References

[1] J. L. Gervasoni, N. R. Arista 1992 *Surf. Sci.* **260** 329

[2] J. L. Gervasoni *et al* 2014 *Nucl. Instr. Meth. B* (in press)

[3] G. D. Mahan 1973 *Phys. Status Solidi B* **55** 703

[1]E-mail: gervason@cab.cnea.gov.ar
[2]E-mail: barrachina@cab.cnea.gov.ar
[3]E-mail: segui@cab.cnea.gov.ar
[4]E-mail: navarrete@cab.cnea.gov.ar

Content from this work may be used under the terms of the Creative Commons Attribution 3.0 licence. Any further distribution of this work must maintain attribution to the author(s) and the title of the work, journal citation and DOI.
Published under licence by IOP Publishing Ltd

XXIX International Conference on Photonic, Electronic, and Atomic Collisions (ICPEAC2015) IOP Publishing
Journal of Physics: Conference Series **635** (2015) 062010 doi:10.1088/1742-6596/635/6/062010

Plasmon decay in the context of the dielectric formalism

N. R. Arista[*†‡ 1], **J. L. Gervasoni**[*† 2], **S. Segui**[*‡ 3]

[*] Centro Atómico Bariloche (CNEA), Av. Bustillo 9500, 8400 S. C. de Bariloche. Río Negro, Argentina.
[†] Instituto Balseiro (CNEA and UNCuyo), Av Bustillo 9500, 8400 S. C. de Bariloche. Río Negro, Argentina.
[‡] Also member of Consejo Nacional de Investigaciones Científicas y Técnicas (CONICET), Argentina.

Synopsis We address the problem of quantifying the decay of plasmons excited in the electron gas of a condensed medium. Within the dielectric formalism, we use various methods to define and assess the damping constant γ.

The excitation of plasmons and their subsequent decay is of great interest in many branches of fundamental and applied physics. Charged-particles spectroscopic techniques (EELS, REELS, etc) show distinct features corresponding to the activation of plasmon modes by the interaction with the incident particle. The shape of the detected signals are determined to a large extent by way the plasmons interact with the medium and ultimately decay. Our aim in this work is to give an overview of different methods for calculating the parameters which give the measure of this decay.

Plasmons are the response of the electron gas in a solid to an external perturbation (e.g., a swift charged particle). In classical electrodynamic theory, this response is mediated by the dielectric function $\epsilon(\mathbf{k},\omega)$, which gives the relation between the induced and the external charged densities. The energy loss of the incident particle is given by the "Energy Loss Function" (ELF)

$$ELF(\mathbf{k},\omega) = \Im\left(-1/\epsilon(\mathbf{k},\omega)\right).$$

The zeroes of $\epsilon(\mathbf{k},\omega)$ yield the resonances identified as plasmon modes. The width of these resonances is determined by the way the plasmon energy is dissipated. A good measure of it is given by the *damping constant* γ in the well-known plasmon pole approximation for the dielectric function, $\epsilon(\mathbf{k},\omega) = 1 - \omega_P/(\omega(\omega + i\gamma))$. (here ω_P is the plasma frequency, determined by the electron gas density). Experimental determinations of γ [1], though scarce, show that it is almost negligible for low values of the momentum k and increases suddenly around a certain critical value k_c, beyond which collective excitations are damped by the individual electronic excitations. A more sophiticated method, based on a quantum formalism is that by Lindhard, which is free of empirical parameters and takes into account collective as well as individual excitations. Different methods can be used to calculate γ, from phenomenological to quantum approachs. In this work we make a thorough review of these methods, exploring the possibilities offered by each of them. As an example, figure 1 shows a map of the ELF calculated for aluminum using the Lindhard model for $\epsilon(k,\omega)$. The (bulk) plasmon resonance $\omega(k)$ distinctively shows up in the region $k < k_c$, $\omega < \omega_c$; at the critical point, the line widens as it enters the region where plasmons are heavily damped by the individual excitations. Using a combination of the Lindhard and the plasmon pole approximations, we obtain a fitting for $\gamma(k)$, which accurately reproduces the experimental results [1, 2, 3].

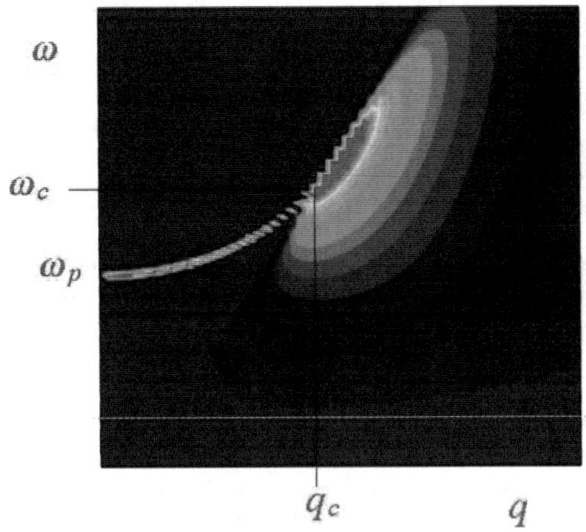

Figure 1. $ELF(k,\omega)$ for aluminum metal, using Lindhard model for the dielectric function.

References

[1] P. Zacharias 1974 *J. Phys. C: Sol. S. Phys.* **7** L26

[2] P. C. Gibbons *et al.* 1976 *Phys. Rev. B* **13** 2451

[3] C. H. Chen 1976 *J. Phys. C: Sol. S. Phys.* **9** L321

[1]E-mail: arista@cab.cnea.gov.ar
[2]E-mail: gervason@cab.cnea.gov.ar
[3]E-mail: segui@cab.cnea.gov.ar

Content from this work may be used under the terms of the Creative Commons Attribution 3.0 licence. Any further distribution of this work must maintain attribution to the author(s) and the title of the work, journal citation and DOI.
Published under licence by IOP Publishing Ltd

Study of electron transmission through a metallic capillary

A. R. Milosavljević[1], M. Lj. Ranković[1*], D. Borka[2], J. B. Maljković[1], R. J. Bereczky[3], B. P. Marinković[1] and K. Tőkési[3]

[1]Institute of Physics Belgrade, University of Belgrade, Pregrevica 118, 11080 Belgrade, Serbia
[2]Vinča Institute of Nuclear Sciences, University of Belgrade, P.O. Box 522, 11001 Belgrade, Serbia
[3]Institute for Nuclear Research, Hungarian Academy of Sciences, H-4001 Debrecen, Hungary

Synopsis In this work we study the transmission of charged particles through a single cylindrically shaped metallic capillary of microscopic dimensions with a large aspect ratio. We used electrons as projectiles. Our results suggest the existence of guiding of the electron beam by a metallic capillary.

In this work we investigate the guiding of electrons through straight and narrow metallic capillaries. The investigation of guiding of charged particles by metal capillaries is strongly motivated by the possibility of producing cheap alternatives to complex electron-optical tools.

We have performed both the measurements and the classical calculations of electrons guiding [1, 2, 3]. Since electrons cannot change their charge state, it is experimentally impossible to make a clear distinction between transmission of the primary and secondary electrons which correspond to elastic scattering. But using the classical simulation, we can identify and follow up all electron trajectories. We performed Monte Carlo simulation assuming that after each inelastic collision a secondary electron is generated with kinetic energy directly estimated from the energy transfer [2, 3]. The created secondary electron is treated as a primary electron and the trace of its path is followed in the successive simulation procedure.

Our experimental setup consisted of an electron gun, a straight cylindrical platinum capillary, an energy analyzer and a channeltron as a detector. Pressure inside the vacuum chamber was $7x10^{-7}$ mbar. An inner radius of the capillary was 3.3 mm, with 40.8 mm in length. Estimated electron beam width at the entrance of the capillary was 0.9 mm with 0.3° angle divergence. The initial electron energy was 200 eV. We found that a dominant fraction of electrons escaping from the capillary have energies smaller than the incident electron energy. Figure 1. shows the energy spectrum of electrons transmitted through a cylindrical platinum capillary.

E-mail: mrankovic@ipb.ac.rs

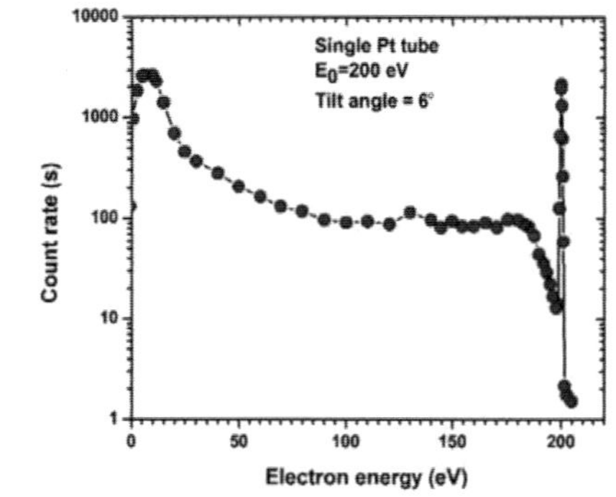

Figure 1. - Energy spectrum of electrons transmitted through platinum capillary [4].

This work was supported by the Ministry of education, science and technological develop ment of Republic of Serbia under the Grants OI 171020 and III 45005, by the Hungarian SRF OTKA No. NN 103279, and by the Bilateral Cooperation Program between the Hungarian and Serbian Academies. This work was conducted within the framework of the COST Action CM1301 (CELINA).

References

[1] A.R. Milosavljević *et al. 2007 Phys. Rev. A* **75** 030901(R).
[2] K. Schiessl, K. Tőkési, B. Solleder, C. Lemell, J. Burgdörfer *2009 Phys. Rev. Lett.* **102** 163201.
[3] B.S. Dassanayake, R.J. Bereczky, S. Das, A. Ayyad, K. Tőkési, J.A. Tanis *2011 Phys. Rev. A* **83** 012707.
[4] Milosavljevic et al, NIMB 2015

Content from this work may be used under the terms of the Creative Commons Attribution 3.0 licence. Any further distribution of this work must maintain attribution to the author(s) and the title of the work, journal citation and DOI.
Published under licence by IOP Publishing Ltd

Journal of Physics: Conference Series **635** (2015) 062012 doi:10.1088/1742-6596/635/6/062012

Electron induced degradation of condensed $Fe(CO)_5$ studied by electron stimulated desorption

Sylvain Massey* [1], Andrew D. Bass* [2], Elahe Alizadeh† [3] and Leon Sanche* [4]
* Département de médecine nucléaire et radiobiologie, Université de Sherbrooke, Sherbrooke, QC, J1H 5N4, Canada
†Department of Chemistry and Biochemistry, University of Guelph, Guelph, ON, N1G 2W1, Canada

Synopsis We have measured the electron stimulated desorption (ESD) yields of anions and cations from thin films of $Fe(CO)_5$ deposited onto both Xe or Pt surfaces. Measurements were made as functions of incident electron energy, incident current, and the thickness of both the $Fe(CO)_5$ and Xe layers. The desorption of C^- and O^- occur exclusively via dipolar dissociation (DD), while the desorption of $Fe(CO)x^-$ (x = 0–4) proceeds via DD and dissociative electron attachment (DEA). Threshold energies for the desorption of cationic fragments C^+, O^+, CO^+, and $Fe(CO)_x^+$ (x = 0–4) were measured. Only CO^+ appears to be produced by DD as opposed to dissociative ionization.

Focused electron beam induced deposition (FEBID) forms nano-scale structures by using highly focused electron beams (typical energy 1-30 keV) to dissociate surface-adsorbed "*precursor*" molecules. The non-volatile component then form a deposit on a substrate; the dimensions are determined by the electron beam, and by lower energy (i.e., < 50 eV) secondary electrons (SE) produced in the substrate. There is consequently much interest in identifying the mechanisms by which SE dissociate precursor molecules [1]. $Fe(CO)_5$ is a commonly employed precursor for conductive and magnetic structures. While the general mechanism of Fe deposition by FEBID (via the progressive elimination of CO) is well accepted, the precise role of SE remains unclear [2]. Our aim is to determine the role of LEEs (< 35 eV) in the dissociation of surface-adsorbed $Fe(CO)_5$ using ESD.

Experiments were performed using a previously described experimental system, consisting of a low energy electron gun and Time of Flight mass spectrometer within a UHV system [3]. $Fe(CO_5)$ samples were condensed directly onto a Pt substrate at 40 K or onto Xe spacer films to modulate the interaction of $Fe(CO)_5$ molecules with the metal. Anion and cation yields were recorded as of incident electron energy, incident current, and the thickness of both the $Fe(CO)_5$ and Xe layers. Representative data is presented in Figure 1.

The origin of fragments, via mechanisms such as DD and DEA was determined by reference to earlier work for simpler molecules (i.e., O_2) under the influence of image-charge effects [4]. The different behaviors of the yield functions depending of the deposition parameters permit a better understanding of the contribution of the dissociation mechanisms of $Fe(CO)_5$.

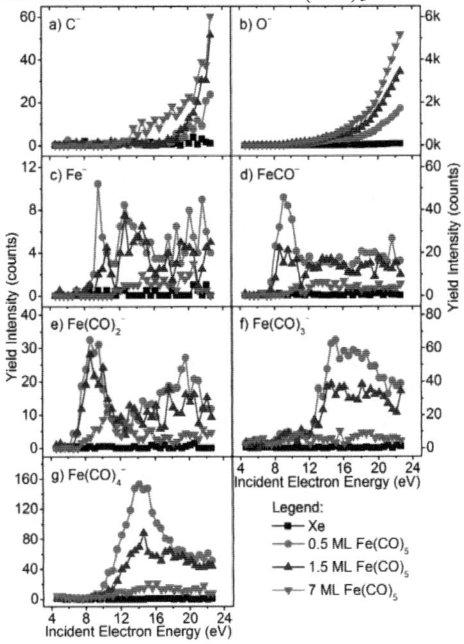

Figure 1. Anion yield functions for LEE impact (4–22 eV) on $Fe(CO)_5$ films of varying thickness condensed onto 2 ML of Xe at 40 K.

References

[1] W. F. van Dorp, C. W. Hagen. (2008) *J. Appl. Phys.* **104** 081301
[2] C. Hauchard, P. A. Rowntree. (2011) *Can. J. Chem.* **89** 1163
[3] M. Bazin, et al., (2009) *Phys. Chem. Chem. Phys.* **11** 1610
[4] H. Sambe, et al., (1987) Phys. Rev. Lett. **59** 236

[1] E-mail: Sylvain.Massey@Usherbrooke.ca
[2] E-mail: Andrew.Bass@Usherbrooke.ca
[3] E-mail: ealizade@uoguelph.ca
[4] E-mail: Leon.Sanche@Usherbrooke.ca

Content from this work may be used under the terms of the Creative Commons Attribution 3.0 licence. Any further distribution of this work must maintain attribution to the author(s) and the title of the work, journal citation and DOI.
Published under licence by IOP Publishing Ltd

Absolute vibrational cross sections for low energy electron (1-19 eV) scattering from condensed tetrahydrofuran (THF)

Vincent Lemelin*[1], Léon Sanche*[2], Elahe Alizadeh†[3] and Pierre Cloutier*[4]

* Département de médecine nucléaire et radiobiologie, Université de Sherbrooke, Sherbrooke, QC, Canada.

† Department of Chemistry and Biochemistry, University of Guelph, Guelph, Ontario, Canada.

Synopsis Absolute cross sections (CSs) for vibrational excitation were obtained for low energy electron impact with tetrahydrofuran (THF) molecules condensed on Ar at 19K. The absolute CSs were measured with a high resolution electron energy loss spectrometer for the 1 to 19 eV incident electron energy range. The values of the vibrational CSs are found to lie within the 10^{-17}cm^2 range and to show the effects of resonant excitation.

A principal goal in radiotherapy is to deliver an important dose to cancer cells without affecting the healthy cells. One way to achieve this goal is by targeted radionuclide therapy (TRT) in which an Auger electron emitter is delivered directly to cancer cells to provide a lethal dose [1]. Modelling such treatments requires multiple parameters to describe the transport and reactions of secondary species in biological matter. Thus, absolute cross sections for the interactions of low energy electrons (LEE) with DNA and/or its component sub-units *in the condensed phase* are important for accurate nanodosimetric calculations [2] applicable to cells. Here we measure absolute cross sections for vibrational excitation of tetrahydrofuran (THF*) condensed on solid Ar*, to serve as a convenient model for the sugar within the DNA backbone [3].

Vibrational electron energy loss spectra for one monolayer (ML) of THF deposited on 3 ML of Ar spacer on a Pt substrate at 19 K, were recorded for electron incident energies 'E_0' between 1 and 19 eV. As in earlier studies [4], spectra were taken off-specular with monochromatized electrons incident at an angle of 15° with respect to the substrate normal and the electron analyzer at 45° in the opposite azimuth. The vibrational excitations observed for condensed THF were consistent with earlier experiments [3].

As previously described [4], cross sections for particular vibrational losses were obtained by fitting each energy loss spectrum with multiple Gaussian functions each centered at the expected energy loss for each vibrational mode.

CS values for representative vibrational excitations of THF are shown as a function of E_0 in Fig.1. Each excitation exhibits a maximum near E_0=2.5 eV, though the energy varies slightly between modes. The same structure appears in other modes at higher energy loss. It is likely this structure is associated with formation of a transient negative ion or resonance. The largest CSs measured was ~2×10^{-17}cm^2. A second, wider, resonance was observed at E_0=6.5 eV for the ν_{10} and ν_8 modes at 146 meV and 163 meV respectively and in some higher modes (not shown here). This research is funded by CIHR.

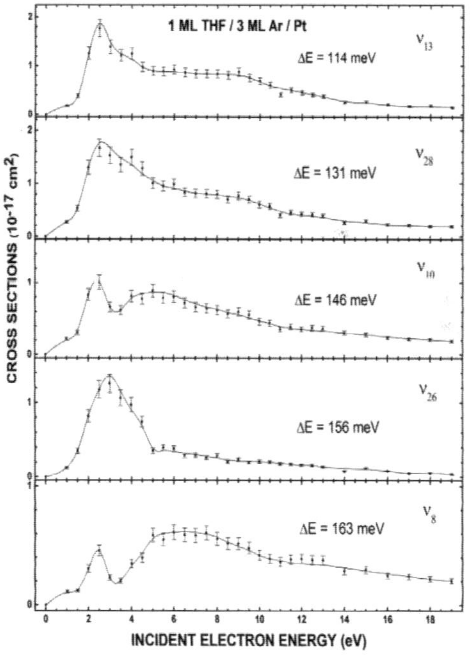

Figure 1. Representative absolute cross sections for electron impact excitation of the indicated vibrational modes of condensed THF molecules.

[1] E-mail: Vincent.Lemelin@USherbrooke.ca
[2] E-mail: Leon.Sanche@USherbrooke.ca
[3] E-mail: ealizade@uoguelph.ca
[4] E-mail: Pierre.Cloutier@USherbrooke.ca

References

[1] M. Michaud *et al* 2013 *Physical Review E* **87** 032701
[2] M. Rezaee *et al* 2014 *Medical Physics* **41** 072502
[3] M. Lepage *et al* 1998 *J. Chem. Phys.* **109** 5980
[4] M. Michaud *et al* 2009 *J. Chem. Phys* **137** 115103

Content from this work may be used under the terms of the Creative Commons Attribution 3.0 licence. Any further distribution of this work must maintain attribution to the author(s) and the title of the work, journal citation and DOI.
Published under licence by IOP Publishing Ltd

Interaction of low energy electrons with iron surface: Energy loss and penetration depths

D. Borka[1], and K. Tőkési[2]

[*] Institute for Nuclear Research, Hungarian Academy of Sciences (MTA Atomki), Debrecen, Hungary, EU
[†] Atomic Physics Laboratory (040), Vinča Institute of Nuclear Sciences, University of Belgrade, P.O. Box 522, 11001 Belgrade, Serbia

Synopsis: We present Monte Carlo simulation of low energy electrons backscattered from iron (Fe) surface. We take into account both elastic and inelastic collisions during the simulation. In our simulations the primary electron energy is 150 eV and the incidence angle of the electron beam with respect to the surface is varied between 1° and 90 °. The backscattered electron energy loss distributions for primary and as well for secondary electrons and the distribution of maximum electron penetration depths in the Fe sample were calculated using only the bulk and also the surface dielectric function.

The understanding of electron spectra backscattered from solid surfaces is important for many points of views, like it is important for a surface characterization and an elemental analysis using electrons either as excitation sources or as signals for electron spectroscopies. One of the well known way for the interpretation of the experimental data is the Monte Carlo simulation (MCS) of elastic and inelastic interactions of electrons in solids.

In this work we present Monte Carlo simulation of low energy electrons backscattered from iron (Fe) surface. In our theoretical model we take into account both elastic and inelastic collisions. To include the elastic scattering of electrons by Fe atoms we use the static field approximation with non-relativistic Schrödinger partial wave analysis [1]. To include the inelastic scattering we use the dielectric response formalism [2]. In our simulations the primary electron energy is 150 eV and the incidence angle of the electron beam with respect to the surface is φ = 2°. The results are obtained using the bulk and the surface dielectric function. We performed Monte Carlo simulation assuming that after each inelastic collision a secondary electron is generated with kinetic energy directly estimated from the energy transfer. After that, the created secondary electron is treated as a primary electron and the trace of its path is followed in the successive simulation procedure. In our theoretical results beside strong elastic peak we obtained significant inelastic contributions. We believe that our present calculations can be used to generate the input data for the theoretical investigation of collisions between electrons and a single and straight iron capillary.

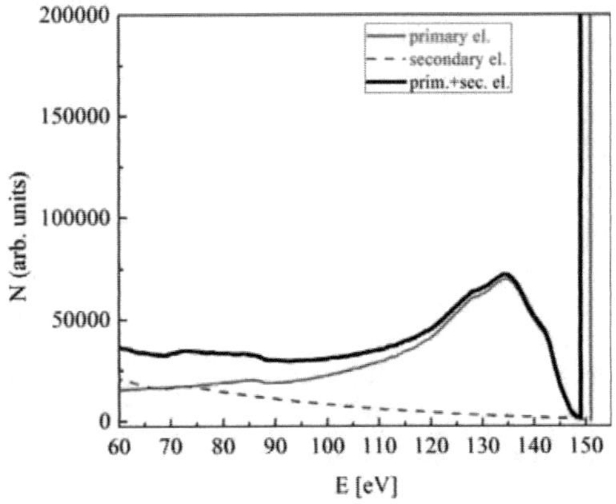

Figure 1. Energy loss distributions for primary (thin solid red line) and secondary electrons (dash blue line) using the surface energy loss function for Fe in the simulations. We also show the sum of energy loss distributions for primary and secondary electrons by thick solid black line. The initial electron energy is E = 150 eV and the incidence angle of the electron beam with respect to the surface is 2°.

References

[1] F. Salvat, R. Mayol, *1993 Comput. Phys. Commun. 74 358.*
[2] Z.-J. Ding, R. Shimizu, *1989 Surf. Sci. 222 313.*

[1]E-mail: dusborka@vinca.rs
[2]E-mail: tokesi@atomki.mta.hu

Content from this work may be used under the terms of the Creative Commons Attribution 3.0 licence. Any further distribution of this work must maintain attribution to the author(s) and the title of the work, journal citation and DOI.
Published under licence by IOP Publishing Ltd

Journal of Physics: Conference Series **635** (2015) 062016 doi:10.1088/1742-6596/635/6/062016

Extended reverse Monte Carlo method for extracting optical constants of thin Ni film from reflection electron energy-loss spectroscopy

H. Xu*, B. Da*, S.F. Mao‡, K. Tőkési†, and Z.J. Ding*[2]

*Hefei National Laboratory for Physical Sciences at Microscale and Department of Physics, University of Science and Technology of China, Hefei, Anhui 230026, P.R. China
‡ School of Nuclear Science and Technology, University of Science and Technology of China, Hefei, Anhui 230026, P.R. China
† Institute for Nuclear Research, Hungarian Academy of Sciences (ATOMKI), Debrecen, Hungary, EU

Synopsis An extended reverse Monte Carlo (RMC) method for obtaining optical constants of ultrathin films and solids is presented. Our results show that this method works very well in extracting some exquisite information from experimentally measured reflection electron energy loss spectroscopy (REELS) spectra. The energy loss function (ELF) of 100 nm nickel film in a relatively larger energy loss range of 0-200 eV has been obtained, which agrees well with other computational data. Employing this ELF, individual contributions to REELS intensity due to surface and bulk plasmon excitations are quantitatively separated for further study.

Precise information of optical properties for nanostructures and ultrathin films is becoming more important to material science. However, it is still difficult to accurately measure such optical data by optical methods for nanostructured materials. Fortunately, such information is essentially contained in and, therefore, may be extracted from a reflection electron energy loss spectroscopy (REELS) spectrum due to the shorter information depth of signal electrons compared with that of photons.

In this work we present an improved reverse Monte Carlo (RMC) method [1] so as to precisely describe complicated electron scattering process in REELS and thus to extract optical constants of a nano-structured nickel film. This method consists of an iterative process of global searching optical data modeling parameters by comparing Monte Carlo simulated REELS spectra with experimental measurement. In each iteration step, dielectric function $\varepsilon(\omega,q)$ and hence the energy loss function (ELF) is used as an input for calculation of electron inelastic cross section. The ELF parameters are modified after each iteration step according to the difference between simulated and measured REELS spectra, according to a Markov chain Monte Carlo (MCMC) sampling procedure.

It is then important to use an effective method for global searching of optimal ELF in hyperspace of parameters. Here we employ the Metropolis-Hastings algorithm in a simulated annealing method, and we adopted a tried-and-true semi-classical electron inelastic scattering cross-section model to quantitatively describe energy loss process of electron moving in surface region. Contributions from surface and volume plasmon excitations, interband transition, multi-scattering effect are taken into consideration in the Monte Carlo simulation.

Fig. 1 shows ELF of a nickel film extracted from REELS spectrum measured at 1 keV incident energy. Data from atomic photo-absorption experiments(blue triangle) and other literature(dark Cyan dots) are shown for comparison. Contributions of surface and volume excitations of REELS are further studied for several incident energies. The results show that the present data agrees well with other computational result and experimental data in the whole energy loss range.

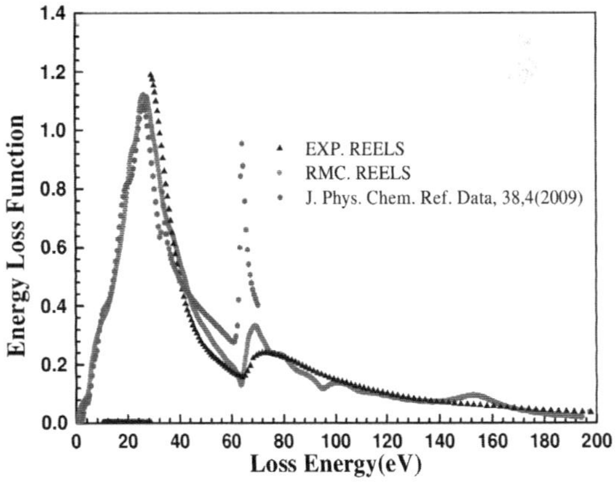

Figure 1. The energy loss function of Ni.

This work was supported by the Hungarian Scientific Research Fund OTKA Nos. NN 103279 and K103917 and by the COST Action CM1204 (XLIC).

References

[1] B. Da *et al* 2013 *J. Appl. Phys.* **113** 214303.

[2] E-mail: zjding@ustc.edu.cn

Content from this work may be used under the terms of the Creative Commons Attribution 3.0 licence. Any further distribution of this work must maintain attribution to the author(s) and the title of the work, journal citation and DOI.
Published under licence by IOP Publishing Ltd

Fragmentation mechanisms for Methane induced by 55 eV, 75 eV and 100 eV electron impact

B. Wei[1], X. Wang, Y. Zhang, R. Hutton and Y. Zou

Applied Ion Beam Physics Laboratory, Fudan University, Key Laboratory of the Ministry of Education Institute of Modern Physics, Fudan University, Shanghai 200433, China

Synopsis CH_4^{2+} break up into three or more fragments in one or two-step processes were studied, our result shows that there are three reaction channels to form CH_2^+, H^+, and H, one synchronous concerted reaction channel and two two-step reaction channels. For even more complicated fragmentation processes of CH_4^{2+} dications, the fragmentation mechanism can still be identified in the present measurements.

The fragmentation of CH_4^{2+} dications following 55 eV, 75 eV and 100 eV electron impact double ionization of methane was studied using a cold target recoil-ion momentum spectroscopy. From the measured momentum of each recoil ion, the momentum of the neutral particles has been deduced and the kinetic energy release distribution for the different fragmentation channels has been obtained. The doubly charged molecular ions break up into three or more fragments in one or two-step processes, resulting in different signatures in the data. We observed the fragmentation of CH_4^{2+} dications through different mechanisms according to the momentum of the neutral particles. For example, our result shows that there are three reaction channels to form CH_2^+, H^+ and H, one synchronous concerted reaction channel and two two-step reaction channels. For even more complicated fragmentation processes of CH_4^{2+} dications, the fragmentation mechanism can still be identified in the present measurements. The slopes of the peak in the ion-ion coincidence spectra were also estimated here, as they are also related to the fragmentation mechanism.

The neutral fragments produced in the two-step fragmentation shared the momentum with the charged fragments. However, in the synchronous concerted fragmentation, the neutral fragments did not take part in the Coulomb interaction, and thus there is no momentum sharing between the neutral fragments and the charged fragments. To identify the fragmentation mechanism of CH_4^{2+} ions into three bodies, Newton diagram was used in our analysis. Figure 1 show the Newton plot for the CH_4^{2+} diaction dissociating into CH_2^+, H^+ and H induced by 100 eV electrons. The momentum vector of H^+ ion defines the x axis, shown in Figure 1 as an arrow. The relative momentum vectors of CH_2^+ ions and neutral H are mapped in the upper and lower half respectively. The momentum of neutral H ranges from – 20 a.u. to + 20 a.u.. In the fourth quadrant, there are some neutral H atoms with a momentum in the same direction as H^+. While in the third quadrant, there are some neutral H atoms with a momentum in the same direction as CH_2^+. It indicates that neutral H shared momentum with H^+ and CH_2^+, respectively. Same measurement has been performed for CO_2. More detail can be found in Reference [1, 2].

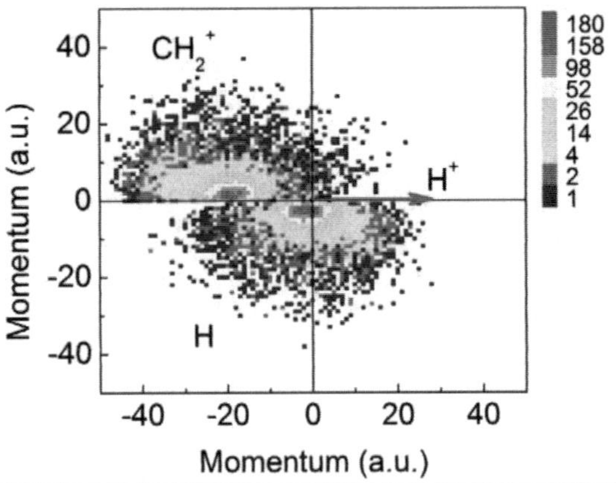

Figure 1. Newton diagram for the Coulomb explosion process of $CH_4^{2+} \rightarrow CH_2^+ + H^+ + H$, which induced by 100 eV electron impact.

References

[1] B. Wei, Y. Zhang, X. Wang, *et al*, *J. Chem. Phys.* **140**, 124303 (2014)
[2] X. Wang, Y. Zhang, D. Lu, *et al*, *Phys. Rev. A*, **90**, 062705 (2014)

[1] E-mail: brwei@fudan.edu.cn

Content from this work may be used under the terms of the Creative Commons Attribution 3.0 licence. Any further distribution of this work must maintain attribution to the author(s) and the title of the work, journal citation and DOI.
Published under licence by IOP Publishing Ltd

Use of heavy-Rydberg ion-pair states to probe dissociative electron attachment[††]

M. Kelley, S. Buathong, and F. B. Dunning[1]

Department of Physics and Astronomy, MS61, Rice University, Houston, Texas, USA

Synopsis Measurements of the velocity distributions of heavy Rydberg ion-pair states formed through electron transfer in collisions between Rydberg atoms and attaching targets are used to probe the dynamics of dissociative electron attachment.

Collisions between high-*n* Rydberg atoms and neutral targets can be discussed using the essentially-free electron model in which the excited Rydberg electron and core ion are treated as a pair of independent scatterers. Electron transfer in Rydberg collisions with molecules that attach free low energy electrons leads to creation of negative ions and such reactions have been widely used [1] to study electron capture at thermal, and subthermal, electron energies. As *n* decreases, however, the Rydberg electron cloud shrinks and the product positive-negative ion pairs must be formed in ever closer proximity. Their post-attachment electrostatic interactions therefore become important and, for thermal-energy collisions, an increasing fraction of them possess insufficient kinetic energy of relative motion to overcome their mutual Coulomb attraction and separate. They therefore remain bound forming so-called heavy-Rydberg ion-pair states [2] in which the positive-negative ion pair orbit each other at relatively large radii.

In the present work we exploit electron transfer reactions of the type

$$K(np)+ABC \rightarrow K^{+}+ABC^{-*} \rightarrow K^{+}..AB^{-}+C \quad (1)$$
$$\rightarrow K^{+}..A^{-}+BC$$

where $K^{+}..AB^{-}$ and $K^{+}..A^{-}$ denote ion-pair states, and ABC^{-*} a short-lived intermediate, to examine the properties (lifetimes, decay channels, energetics of dissociation) of excited intermediates.

The present experimental approach is illustrated in Fig. 1. Pulses of K(np) Rydberg atoms are photoexcited near the center of a gas cell where they interact with the target gas. A fraction of the bound ion pairs formed traveling in the plane defined by the atom and laser beams exit the gas cell through a narrow slit and enter an analysis region where they are detected, after a predetermined flight time, by dissociation induced by a pulsed field. The resulting ions are accelerated to a position-sensitive detector that records both their arrival times and positions.

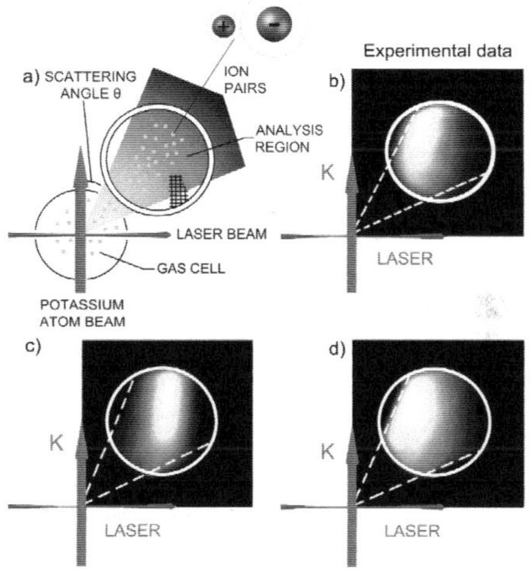

Figure 1. a) Schematic of experimental approach, b) measured spatial distribution of $K^{+}..\ CN^{-}$ ion pairs formed in K(14p)-BrCN collisions for a flight time of 40 μs, c) results of model simulations assuming all the excess energy of reaction appears in translation, and d) that some appears as rotational excitation of the CN^{-} fragment.

Figure 1 includes a typical position distribution at the PSD for $K^{+}..CN^{-}$ ion pairs formed in K(14p)-BrCN collisions together with the results of calculations using a Monte Carlo collision code that models the detailed reaction kinematics. The simulations demonstrate that the velocity distributions of the product ion pairs are sensitive to the assumed decay energetics. As a result comparisons between theory and experiment can provide new insights into the details of dissociative electron attachment.

[††]Research supported by the Robert A. Welch Foundation under Grant No. C-0734

References

[1] F. B. Dunning 1995 *J. Phys. B: At. Mol. Opt. Phys.* **28** 1645
[2] E. Reinhold and W. Ubachs 2005 *Mol. Phys.* **103** 1329

[1]E-mail: fbd@rice.edu

Content from this work may be used under the terms of the Creative Commons Attribution 3.0 licence. Any further distribution of this work must maintain attribution to the author(s) and the title of the work, journal citation and DOI.
Published under licence by IOP Publishing Ltd

Journal of Physics: Conference Series **635** (2015) 072003 doi:10.1088/1742-6596/635/7/072003

Integral electron impact ionization cross section of molecules through Coulomb crystallization of the product ions

Łukasz Kłosowski*[1], Mariusz Piwiński*, Michael Drewsen†

* Institute of Physics, Faculty of Physics, Astronomy and Informatics, Nicolaus Copernicus University in Toruń, Grudziadzka 5, 87-100 Torun, Poland

† Department of Physics and Astronomy, Aarhus University, DK-8000 Aarhus C, Denmark

Synopsis We present a new method for measuring electron impact ionization cross section of molecules through Coulomb crystallization of the product ions in a linear Paul trap.. The idea of electron impact ionization cross sections measurements is described, and the technique is analyzed in terms of its efficiency and some specific apparatus effects.

Trapped molecular ions have been subject of growing interest recently due to both studies of fundamental processes and various possible applications. Several methods of molecular ion production have been used: chemical reactions involving trapped atomic ions [1], laser ablation of a solid target [2], and photoionization [3]. Trapped molecular ions have been also observed as impurities (originating from residual gas in vacuum systems) of Coulomb crystals in experiments using high energy (~1 keV) electron impact ionization of atoms [4, 5, 6]. Moreover, electron collisional induced ions in Paul trap were applied for mass spectroscopy of residual gas in vaccum system [7].

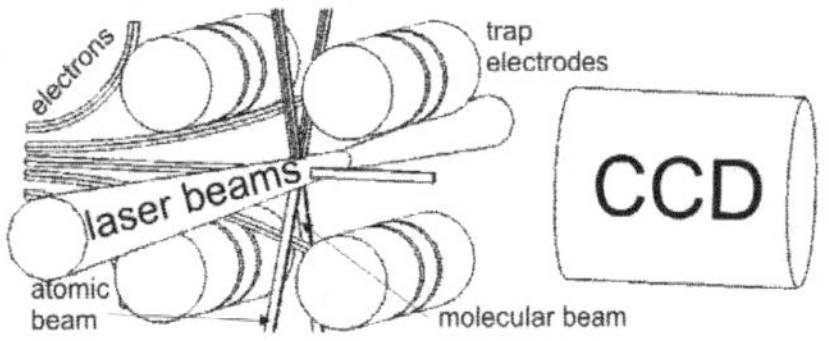

Figure 1. Schematic of the experimental geometry. All the beams are cross-fired in the central part of the trap.

In the proposed experiment (fig. 1) ionization of molecules (e.g. CO_2) with electrons of energies of the order of 10-100 eV will be studied. Molecules are ionized in both non-dissociative and dissociative channels. This way various ions are obtained from the same type of molecules. Additionally atomic ions (e.g. Ca^+) are prepared and laser-cooled to provide sympathetic cooling [1]. The ions form a multi-species Coulomb crystal. Example of numerically simulated ensemble of ions in equilibrium state is presented in fig. 2.

Composition of such crystal can be analyzed using images from CCD camera. The data derived for various electron impact energies can be used as a method to experimentally determine ionization integral cross sections (ICS). The accuracy of such ICS measurements is discussed with special attention to influence of the trap field on the geometry and energy of electron beam.

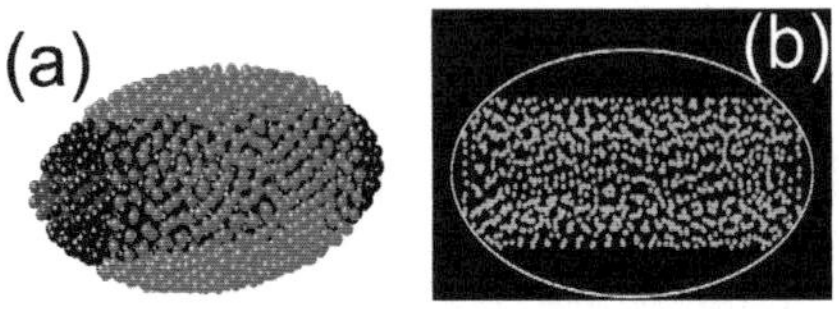

Figure 2. (a) Simulation of an ensemble of 3 species of ions (CO_2^+ – red, Ca^+ – black and CO^+ – blue) forming Coulomb crystal. (b) Expected image of such crystal - only the middle layer containing Ca^+ is visible. The loop denotes the crystal's edge.

In the future, experiments with higher electron energy resolution can be expected using digitally pulsed voltages on the trap electrodes [8] combined with gated electron beams coinciding with the zero voltage of the applied electrical pulses.

This work has been supported by the National Science Centre, Poland, project no. 2014/13/B/ST2/02684.

References

[1] K. Mølhave, M. Drewsen 2000 *Phys. Rev. A* **62** 401-411

[2] K. Chen *et al* 2011 *Phys. Rev. A* **83** 030501(R)

[3] X. Tong *et al* 2010 *Phys. Rev. Lett.* **105** 143001

[4] I. Waki *et al* 1992 *Phys. Rev. Lett.* **68** 2007-2010

[5] M. Raizen *et al* 1992 *J. Mod. Opt.* **39** 233-242

[6] P. Bowe *et al* 1999 *Phys. Rev. Lett.* **82** 2071-2074

[7] T. Baba, I. Waki 1996 *Jpn. J. Appl. Phys.* **35** L1134-L1137

[8] N. Kjærgaard, M. Drewsen *et al* (2003) *Phys. Rev. Lett.* **91** 095002

[1]E-mail: lklos@fizyka.umk.pl

Content from this work may be used under the terms of the Creative Commons Attribution 3.0 licence. Any further distribution of this work must maintain attribution to the author(s) and the title of the work, journal citation and DOI.
Published under licence by IOP Publishing Ltd

Journal of Physics: Conference Series **635** (2015) 072004 doi:10.1088/1742-6596/635/7/072004

On the additivity of scattering phases in collisions of electrons on endohedrals

M. Ya. Amusia*† [1] and L. V. Chernysheva† [2]

* The Racah Institute of Physics, the Hebrew University, Jerusalem 91904, Israel
† A. F. Ioffe Physical-Technical Institute, St. Petersburg 194021, Russian Federation

Synopsis It is demonstrated that an inner atom, either Ne or Ar, qualitatively affects the electron scattering phases upon an endohedral, in spite of the fact that the fullerene consists of 60 carbon atoms, while the atom staffed inside is only one. Calculations are performed in the one-electron Hartree-Fock (HF) and random phase approximation with exchange (RPAE) for the inner atom while the fullerenes shell is substituted by static potential without and with the polarization potential. The total endohedral scattering phase is a sum of atomic, Ne or Ar, and fullerenes C_{60} phases, contrary to the intuitive assumption that the total phases on C_{60} and Ne@C_{60} or Ar@C_{60} has to be the same.

1. We suggest after performing calculation of a concrete example that the total partial wave l phase $\delta_l^{A@C_N}$ of an electron scattered upon endohedral A@C_N is with good accuracy equal to the sum of scattering phases δ_l^A and $\delta_l^{C_N}$ of electrons upon atom A that is stuffed inside the fullerene C_N.

We generalize this result to any target that consists of loosely bound atoms. It is counter-intuitive, since it is reasonable to assume, at least in the case of the considered target A@C_N that the scattering is absolutely dominated by the very "big atom" so that the contribution of atom A can be neglected.

2. If the target consists of electrons and nuclei and exchange between incoming and target electrons is taken into account, one has for the place $\delta_l(0)-(n_l+q_l)\pi$, where q_l (n_l) is the number of bound electron states in the target itself (in $e+\text{target}$). Therefore, the behavior of phases as functions of E is qualitatively different in cases when we treat the target in Hartree or HF approximation. In these two cases the phases deviate from each other (in numbers of π) although the strength of the potential is almost the same.

3. The atom A was treated in HF and RPAE, while the fullerene was represented by square well potential $W_C(r)$ and polarization potential $V_C^{pol}(r)=-\alpha_C/[2(r^2+b^2)^2]$. The choice of potential parameters and some details of calculations are presented in [1].

4. Figure 1 illustrates the *property of additivity*, namely the fact that the following equation is valid with quite high accuracy

$$\delta_l^{A@C_{60}}(E)=\delta_l^A(E)+\delta_l^{C_{60}}(E). \quad (1)$$

Inclusion of $V_C^{pol}(r)$ modifies the phases, adding a number of bound states in $(e+C_{60})$. If the electronic structure of the target, namely exchange between the incoming and target electrons will be taken into account, the phases will increase, acquiring an addition (in units of π) to the phase, equal to the number of bound in the target electrons with the given angular momentum l.

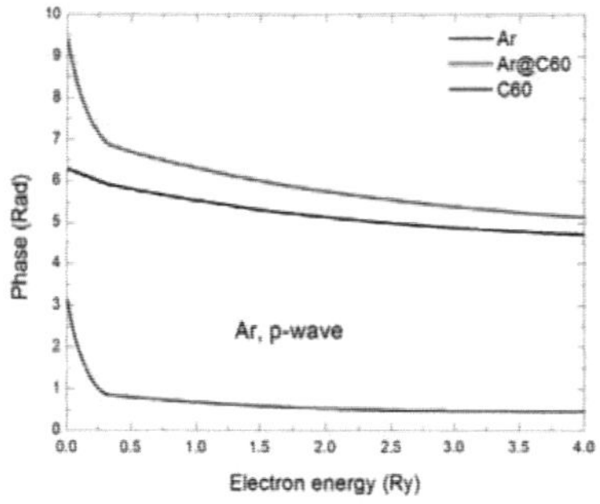

Figure 1. Elastic scattering p-phase of electrons upon atom Ar, fullerene C_{60}, and endohedral Ar@C_{60}.

5. Correlations between atomic electrons are essential, modifying the low-energy cross-section. The atomic cross-section at $E\to 0$ is much less than the cross-section upon C_{60}, but fullerene adds prominent structure. Phase additivity (1) does not lead to corresponding additivity in the cross-sections. The modification of the cross-section due to presence even of a single additional atom can affect not only elastic scattering but some other connected to it properties and characteristics of electron-target collisions.

References

[1] M. Ya. Amusia and L. V. Chernysheva 2015 *JETP Letters* **101**

[1] E-mail: amusia@vms.huji.ac.il

[2] E-mail: Larissa.Chernysheva@mail.ioffe.ru

Content from this work may be used under the terms of the Creative Commons Attribution 3.0 licence. Any further distribution of this work must maintain attribution to the author(s) and the title of the work, journal citation and DOI.
Published under licence by IOP Publishing Ltd

Journal of Physics: Conference Series **635** (2015) 072005 doi:10.1088/1742-6596/635/7/072005

Dissociation, dissociative excitation and ionization processes for N_2 and O_2 molecules by electron bombardment

M. Gochitashvili, R. Lomsadze, O.Taboridze, D. Kuparashvili

Physics Department, Tbilisi State University, Tbilisi, Chavchavadze 3, 0192, GEORGIA

Synopsis The results of measurements of N_2^+, O_2^+, N^+ and O^+ ions formation cross section, in the low electron energy range, from threshold to 400 eV are presented. For the e - O_2 , N_2 collision pare the excitation cross section of dissociative products - N , O and N^+, O^+ are determined in relatively high energy range 200 - 400eV. The intense oxygen ionic line OII (83.4 nm) is observed in vacuum ultraviolet region of emission spectrum. Excitation of this line is caused by formation of $2s\sigma_g$ vacancy of inner electronic shell of O_2 molecule. Was revealed, that the intensity of the formation of the dissociative products, in metastable states, strongly depends on the pressure condition into the ion source.

In this work the measurements are performed by mass-spectrometric (MS) device. The main MS assembly is composed with pumping unit and the electronics. In experiments we have used two different ion sources: electron impact ionization and high frequency discharge ion source.

Experimental results show, that the intensity of extracted atomic ions strongly depends on the pressure condition into ion source. This result can be explained by the production of metastable ions formed during collision [1]. In particular, at low pressure, the portion of extracted metastable ions is increases, while in high pressure condition the production of an ion in ground ^4S state is most favorable.

In the dissociative ionization processes the state – selected metastable $N^+(^1D,^1S,^5S)$ and $O^+(^2D,^2P)$ as well as ground $N^+(^3P)$, $O^+(^4S)$ ion beams are prepared and inelastic processes, employing low energy electron impact (from threshold to 400ev), are studied. This approach will be applied to measure cross sections of excitation processes in collision of metastable ions with molecules.

The Results of measurements of N^+ and O^+ ions formation cross section, in low electron energy range, from threshold to 120 eV, are presented in Fig.1. For the e-O_2, N_2 collision pare the excitation cross section of dissociative products - N, O and N^+, O^+ are determined in relatively high energy range, 200-400eV. The intense spectral oxygen ionic line OII (83.4 nm) are observed in vacuum ultraviolet range (50nm-120nm) of emission spectrum. Excitation of this line is caused by formation $2s\sigma_g$ vacancy of inner electronic shell of O_2 molecule.

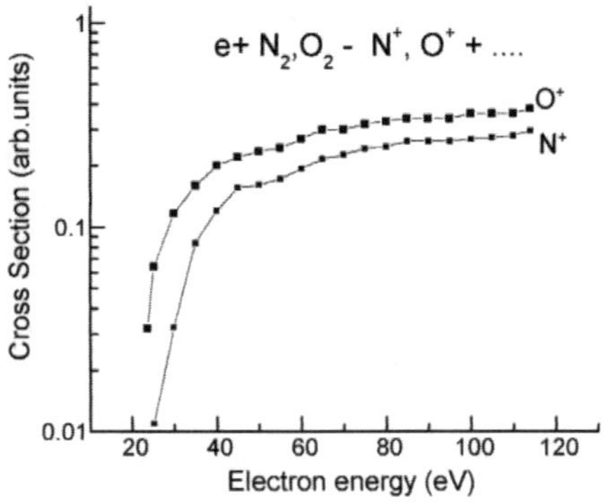

Figure 1. Cross section of formation of N^+ and O^+ ions in e – N_2, O_2 collisions.

References

[1] J.J.Corr *et al 2015 Int. J. Mass Spectrometry.* v376, 19.

E-mail: m.gochitashvili@hotmail.com

Content from this work may be used under the terms of the Creative Commons Attribution 3.0 licence. Any further distribution of this work must maintain attribution to the author(s) and the title of the work, journal citation and DOI.
Published under licence by IOP Publishing Ltd

Journal of Physics: Conference Series **635** (2015) 072006 doi:10.1088/1742-6596/635/7/072006

Dissociative Excitation of Thymine in the VUV.

C J Tiessen*, J D Hein*#, J A Trocchi*, W Kedzierski* and J W McConkey*

*Department of Physics, University of Windsor, Ontario, N9B 3P4, Canada
#Jet Propulsion Laboratory, California Institute of Technology, Pasadena, CA91109, USA.

Synopsis. Dissociative excitation of thymine into excited atomic fragments has been studied in the electron impact energy range from threshold to 435 eV. The main features in the spectrum are the H Lyman series lines.

Thymine ($C_5H_6N_2O_2$), as one of the four nucleobases of the double helical DNA structure, is a very important bio-molecule. Ever since it was demonstrated [1] that dissociative electron attachment (DEA) to such molecules at low energies provided a mechanism for double- and single-strand breaking within the DNA structure, there has been a need for quantitative electron scattering data to enable accurate modelling of damage produced by the low energy electron cascade which is released following entry of high-energy radiation into biological media. The present work presents data for electron-impact dissociative excitation of the molecule into individual atomic species.

A conventional crossed beam system [2] is used where a beam of thymine vapor from a stainless steel oven is crossed at right angles by an electron beam and the resultant radiation is detected in a direction perpendicular to both beams. A ½ metre Seya-Namioka VUV spectrometer coupled to a CsI coated channel electron multiplier is used to disperse and detect radiation in the 80-170 nm spectral region. Two types of experimental data are collected. In the first, photon emission spectra are recorded at a constant electron beam energy of 100 eV. In the second, a particular spectral feature is selected and the photon emission rate (relative cross section) is measured as the electron beam energy is scanned. Relative sensitivity of the system with wavelength is established using the well-known H_2 molecular spectrum excited by 100 eV electrons.

A sample of the results obtained is shown in Figure 1. Here the Lyman-α relative emission cross section is shown for electron energies from threshold to 435 eV. Emission spectra indicate that the Lyman series is excited prominently with no observable emissions from any other atomic species. This suggests that the probability of extracting and exciting C or N atoms from the ring is very small and dissociative excitation of O atoms is likewise minimal. Threshold excitation measurements show that simple elimination with excitation of the H atom from the molecule is unimportant. Break-up into molecular fragments dominates. A full discussion of the results obtained will be presented at the Conference.

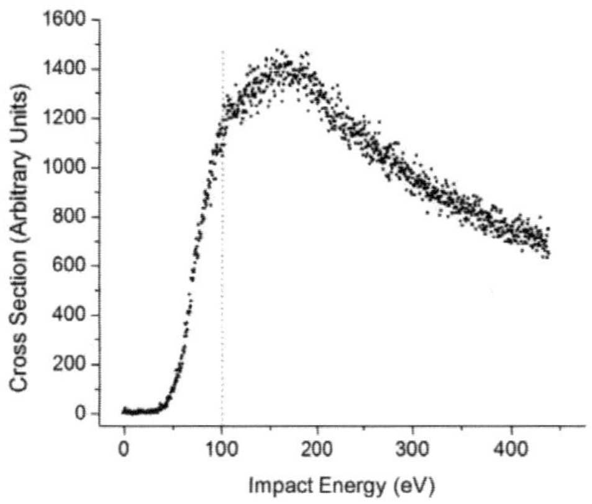

Figure 1. Relative emission cross section for the 121.6 nm H [2p-1s] transition.

The authors thank NSERC, Canada for financial support for this project. CJT and JAT acknowledge support from the University of Windsor "Outstanding Scholars" program. The mechanical and electrical workshop staff at the University of Windsor supplied expert technical help.

References.

[1] B. Boudaiffa et al, 2000, *Science*, **287**, 1658.
[2] J.D.Hein et al, 2013, *J Phys B*, **46,** 045202.

Content from this work may be used under the terms of the Creative Commons Attribution 3.0 licence. Any further distribution of this work must maintain attribution to the author(s) and the title of the work, journal citation and DOI.
Published under licence by IOP Publishing Ltd

Detection of N_2 metastable molecules using a solid N_2 matrix detector.

W Kedzierski*, C X Cerkauskas* and J W McConkey*

*Department of Physics, University of Windsor, Ontario, N9B 3P4, Canada

Synopsis. Excitation of metastable molecular N_2 by electron impact has been studied in the impact energy range from threshold to 300 eV using time-of-flight techniques and a novel solid matrix detector.

Metastable atoms and molecules are important species in a wide variety of applications ranging from low temperature electrical discharges to planetary atmospheres and astrophysical sources. Lifetimes of metastable states can range from tens or hundreds of microsecs to secs to hours or even longer. In our laboratory we have pioneered [1] the use of solid rare gas matrices at low temperatures to allow detection of metastable atomic species with an np^4 electron configuration, such as O (^{1}S) or (^{1}D), both prominent species in earth's upper atmosphere. In this technique the metastable particle impinges on the solid surface and forms an excimer which immediately radiates. The resultant photon (visible or near infra-red) is readily detectable. The effect of the excimer formation results in a lifetime reduction of the metastable species sometimes by as much as a factor of 10^7. In the present work we investigate the behaviour of a solid N_2 matrix as a metastable detector and use it to investigate the excitation of hitherto unknown molecular metastable species in N_2.

Metastable N_2 molecules produced in the interaction of electrons of carefully controlled energy with a thermal beam of N_2 in a crossed beam set-up have been studied in the energy range from threshold to 400 eV. The e-beam is pulsed and the metastables produced drift to a solid nitrogen target held at 10 K. Here they form excimers which immediately radiate. The resultant photons are detected using a photomultiplier-filter combination. Time-of-flight techniques are used to separate these photons from prompt photons produced in the initial electron-N_2 collision. The excimer emission is strongest in the green but still significant in the red spectral region. Quench plates in the path of the metastables to the detector allow a check for any Rydberg species which might be present. Excitation functions and threshold measurements help to identify the metastable states being observed and the excitation mechanisms which are responsible.

A sample of the results obtained is shown in Figure 1. Here the excitation probabilities are shown for metastables that produce excimers which decay with the emission of red and green photons respectively. Some significant points should be noted. First, the shapes of the two excitation functions in Figure 1 are clearly different from each other and from the shapes of known N_2 metastable excitation functions [2]. Second, the observed thresholds (~15 eV) are significantly higher than the thresholds of known N_2 metastable states such as a $^1\Pi_g$, $A^3\Sigma_u^+$ or E $^3\Sigma_g^+$. The data strongly suggest the existence of at least two hitherto unrecognised metastable states of N_2.

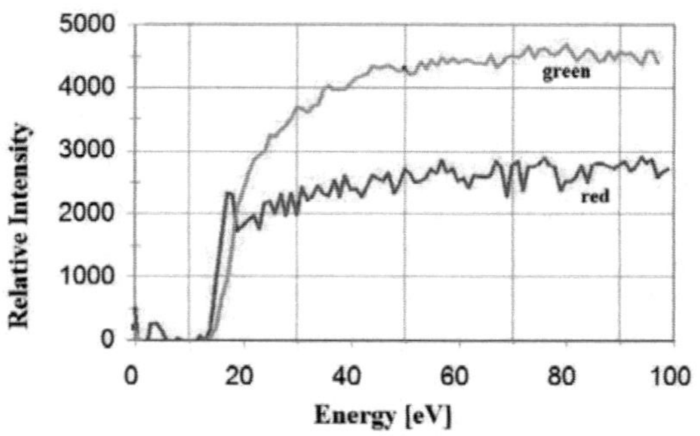

Figure 1. Relative excitation probabilities for production of N_2 metastables as a function of exciting electron energy. Excimer decay in the red and green spectral regions are indicated.

The authors thank NSERC and CFI, Canada, for financial support for this project. They also acknowledge the mechanical and electrical workshop staff at the University of Windsor for their expert technical help.

References.

[1] J W McConkey and W Kedzierski, 2014, Adv At Mol Opt Phys, **63,** 1.

[2] W L Borst and E C Zipf, 1971, Phys Rev A, **3**, 979.

Content from this work may be used under the terms of the Creative Commons Attribution 3.0 licence. Any further distribution of this work must maintain attribution to the author(s) and the title of the work, journal citation and DOI.
Published under licence by IOP Publishing Ltd

Journal of Physics: Conference Series **635** (2015) 072008 doi:10.1088/1742-6596/635/7/072008

High resolution (e, 2e + ion) study of low-energy electron-impact ionization and fragmentation of tetrahydrofuran

XueGuang Ren[*‡1], **Thomas Pflüger**[*‡], **Marvin Weyland**[*‡], **Woon Yong Baek**[*], **Hans Rabus**[*], **Joachim Ullrich**[*], **Alexander Dorn**[‡]

[*] Physikalisch-Technische Bundesanstalt, 38116 Braunschweig, Germany

[‡] Max-Planck-Institute for Nuclear Physics, 69117 Heidelberg, Germany

Synopsis We study the low-energy (E_0 = 26 eV) electron-impact induced ionization and fragmentation of tetrahydrofuran using an (e, 2e + ion) method. The momentum vectors and, consequently, the kinetic energies for all three final-state charged particles are determined. The ionic fragments are clearly identified in the experiment with a mass resolution of one atomic mass unit. The fragmentation pathways of tetrahydrofuran are investigated by measuring the ion kinetic energy spectra and the binding energy spectra where a binding energy resolution of 1.5 eV has been achieved.

The low-energy electrons produced by primary ionizing radiation penetrating biological tissue can effectively damage DNA [1]. Therefore, cosiderable attention has been paid on the study of slow electron collision with the buding blocks of DNA. Here, we investigate the processes of low-energy (E_0 = 26 eV) electron impact-induced ionization and ionic fragmentation of tetrahydrofuran (THF, C_4H_8O). This molecule is a surrogate of deoxyribose which forms the DNA backbone. The momentum vectors of all three final state charged particles, two outgoing electrons and one fragment ion, are detected in the so-called (e, 2e + ion) triple coincidence method [2, 3] using a reaction microscope and a pulsed photoelectron gun [3, 4]. Thus, the ion kinetic energy distributions (resolution about 1.3 meV) and the energy depositions or binding energy spectra (resolution ΔE = 1.5 eV) for the observed fragmentation processes are obtained.

The fragment products are identified with the measured high resolution mass spectrum where fragments differing by one atomic mass unit can be resolved. The measured mass spectrum of THF cations is presented in Figure 1. We can clearly identify all the cations. The fragmentation processes of tetrahydrofuran are studied by the measured ion kinetic energy spectra and the electron binding energy spectra [3].

In the present experiment, we also measured the triple and double differential (e, 2e) cross sections for individual molecular orbitals and product ions. From these, we can study the dynamics of electron scattering on biomolecules [5, 6]. Detailed results of these studies and a detailed analysis of the origins of the ionization and fragmentation processes of THF will be presented at the conference.

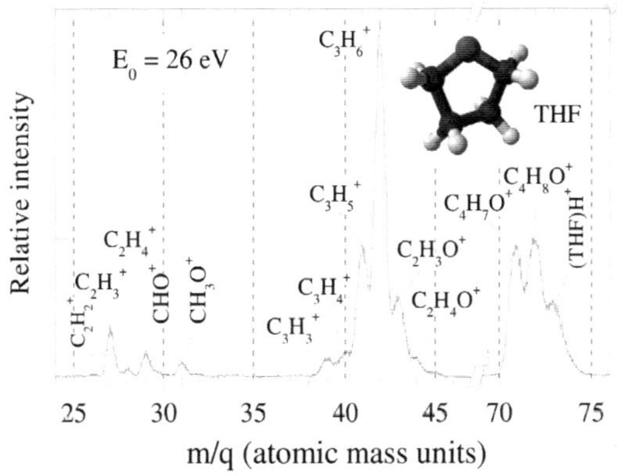

Figure 1. Fragment ion mass to charge (m/q) spectrum derived from the measured time-of-flight distribution upon low-energy electron-impact ionization and fragmentation of THF. Species detected are labeled in the figure.

References

[1] B. Boudaiffa et al 2000 *Science* **287** 1658

[2] S. Xu et al 2011 *Phys. Rev. A.* **83** 052702

[3] X. Ren et al 2014 *J. Chem. Phys.* **141** 134314

[4] J. Ullrich et al 2003 *Rep. Prog. Phys.* **66** 1463

[5] W.Y. Baek et al 2012 *Phys. Rev. A.* **86** 032702

[6] C.J. Colyer et al 2010 *J. Chem. Phys.* **133** 124302

[1]E-mail: xue.g.ren@ptb.de

Content from this work may be used under the terms of the Creative Commons Attribution 3.0 licence. Any further distribution of this work must maintain attribution to the author(s) and the title of the work, journal citation and DOI.
Published under licence by IOP Publishing Ltd

Variational and Path Integral Monte Carlo calculations on Helium Clusters Doped with Metastable Anions He^{*-} and He_2^{*-}

P. Villarreal* [1], **R. Rodríguez-Cantano***, **T. González-Lezana***, **R. Prosmiti***, **G. Delgado-Barrio***, **and F. A. Gianturco**†

* IFF-CSIC, Serrano 123, 28006 Madrid, SPAIN
† Institut für Ionenphysik und Angewandte Physic, Universität Innsbruck, Technikerstr. 25, A-6020 Innsbruck, AUSTRIA

Synopsis Variational calculations (T=0 K) on small He_N...He^{*-} and He_N...He_2^{*-} metastable clusters (N≤4), as well as Path Integral Monte Carlo (PIMC) simulations (T=0.4 K) on larger species are presented and discussed.

The calculations[1] have been carried out assuming additive pairwise-like potential surfaces. The underlying $He(^1S)$-$He^{*-}(^4P)$ potential curve and the $He(^1S)$-$He_2^{*-}(^4\Pi_g)$ anisotropic interaction have been recently estimated through accurate CCSD(T) calculations[2]. The He-He interaction is described by a semi-empirical potential[3].

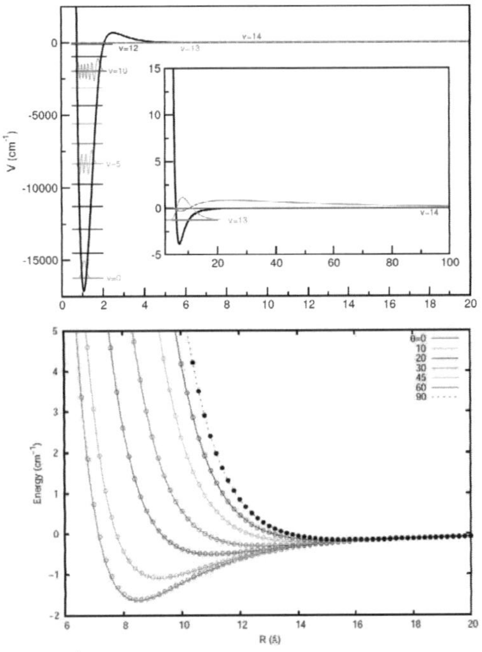

Figure 1. He-He^{*-} potential curve supporting 15 bound states, the last two ones being depicted in the inset (upper panel), and He-He_2^{*-} anisotropic interaction (lower panel).

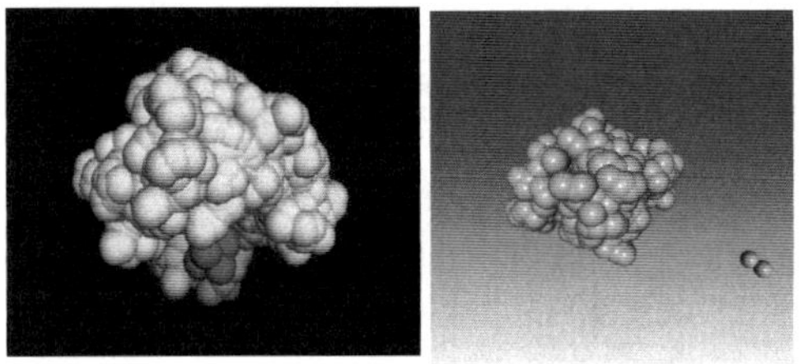

Figure 2. Left panel: snapshot from the PIMC simulation for the He_{32}-He^{*-} cluster at T = 0.4 K showing a bi-pyramid anionic structure surrounded by the rest of He atoms. Right panel: the simulation for the He_8-He_2^{*-} cluster shows instead the impurity far away the helium cluster.

For the atomic anion case, the interaction with helium presents a deep well near 1 Å followed by a small barrier and then a shallow minimum, see upper panel at **Figure 1**. Accordingly, as He atoms are added, a marked preference to form a bi-pyramid charged core He_7^-, with the rest of He atoms surrounding it, is obtained through PIMC simulations, see **Figure 2** (left panel).

In turn, the molecular anion He_2^{*-}, considered as a rigid rotor, tends to point towards a set of packed helium atoms which are placed at long distance from the anion, see **Figure 2** (right panel).

References

[1] R. Rodríguez-Cantano *et al* 2015 *J. Chem. Phys* **142**, 104303

[2] S. E. Huber and A. Mauracher 2014 *Mol. Phys.* **112**, 794

[3] R. A. Aziz and M. J. Slaman 1991 *J. Chem. Phys.* **94**, 8047

[1] E-mail: p.villarreal@csic.es

Content from this work may be used under the terms of the Creative Commons Attribution 3.0 licence. Any further distribution of this work must maintain attribution to the author(s) and the title of the work, journal citation and DOI.
Published under licence by IOP Publishing Ltd

Analytical and molecular dynamical investigations of the influence of molecular vibrations upon the (e,2e) electron momentum distributions of furan

F. Morini* [1], M. S. Deleuze*, N. Watanabe†, M. Takahashi†

* Center of Molecular and Materials Modelling, Hasselt University, Agoralaan Gebouw D, B-3590 Diepenbeek, Belgium

† Institute of Multidisciplinary Research for Advanced Materials, Tohoku University, Sendai 980- 8577, Japan

Synopsis The role of molecular vibrations has been theoretically investigated in the electronic ground state on the (e, 2e) valence orbital momentum profiles of furan by means of two complementary approaches. The first one relies upon the principles of Born-Oppenheimer Molecular Dynamics (BOMD), whereas the second one, referred as Harmonic Analytical Quantum Mechanical (HAQM) approach, includes on quantum-mechanical grounds the effect of vibrations in the structure factors.

This work [1] analyzes the influence of thermally induced nuclear dynamics (molecular vibrations) on the electron momentum profiles related to the valence orbitals of furan in its initial neutral electronic ground state. To this aim two approaches are used. The first one, referred as the Harmonic Analytical Quantum Mechanical (HAQM) approach [2], implies a decomposition of the structure factor into contributions for each vibrational eigenstates within the frame of the harmonic oscillator - rigid rotor approximation, including on quantum-mechanical grounds the effect of vibrations in the computed (e,2e) momentum distributions. The second one uses the principles of Born-Oppenheimer Molecular Dynamics (BOMD) in order to average (by virtue of the ergodic principle) over a large number of molecular structures the effect of temperature on (e,2e) momentum profiles [3]. In contrast with the HAQM approach, BOMD enables to cope with nonharmonic effects and couplings between vibrations and/or internal rotations as well as the chaotic nature of nuclear motions in polyatomic molecules. The two approaches enable consistent insights into the experimental momentum profiles obtained from newly performed Electron Momentum Spectroscopy measurements with improved energy and momentum resolutions. At 298K both approaches show that nuclear dynamics in the initial neutral electronic ground state has a strong influence on the $9a_1$ orbital momentum profile (Figure 1). The role of specific CH-stretching and bending vibrational modes of A_1 symmetry can be explained according to the symmetry characteristics of orbitals and their energy differences, in line with the Herzberg-Teller principle. This study shows how molecular vibrations and nuclear motions in the initial electronic ground state may affect electron momentum profiles, and gives new guidelines for discriminating these effects from other physical phenomena such as distorted wave effects or nuclear dynamics in the final state (comprising non adiabatic effects).

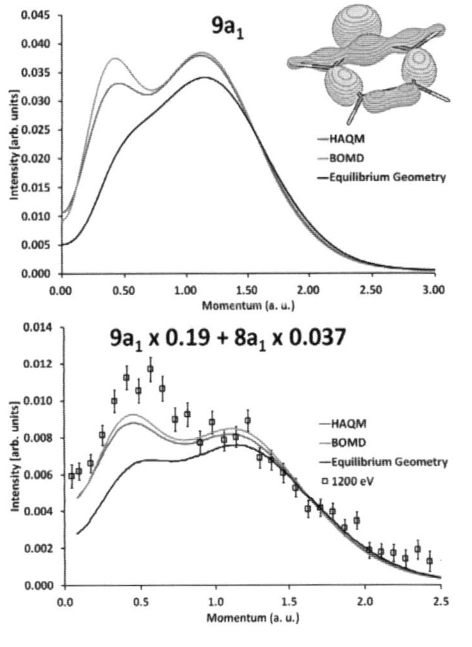

Figure 1. Theoretical momentum profiles for the $9a_1$ orbital (top), and comparison with the experiment (bottom).

References

[1] F. Morini *et al 2015 J. Chem. Phys.* **142** 094308
[2] N .Watanabe *et al 2012 J. Chem. Phys.* **137** 114301
[3] See e. g. S. H. R. Shojaei *et al 2013 J. Phys. Chem. A* **117** 1918 and references therein

[1]E-mail: filippo.morini@uhasselt.be

Content from this work may be used under the terms of the Creative Commons Attribution 3.0 licence. Any further distribution of this work must maintain attribution to the author(s) and the title of the work, journal citation and DOI.
Published under licence by IOP Publishing Ltd

Journal of Physics: Conference Series **635** (2015) 072011 doi:10.1088/1742-6596/635/7/072011

Signatures of Interatomic-Coulombic-Decay in electron-impact ionization of argon dimers

Thomas Pflüger[†,*,1], **Xueguang Ren**[†,*], **and Alexander Dorn**[*]

[†] Physikalisch-Technische Bundesanstalt, Bundesallee 100, 38116 Braunschweig, Germany
[*] Max Planck Institute for Nuclear Physics, Saupfercheckweg 1, 69117 Heidelberg, Germany

Synopsis We study electron-impact induced dissociation of small argon clusters at a projectile energy of 120 eV. Kinetic-energy-release (KER) spectra for the final charge states $2Ar^+$, $Ar^+ + Ar^{2+}$ and $Ar_2^+ + Ar^+$ of the Ar_2 and Ar_3 parent species and electron energies have been measured together with angular distributions of fragment ions. They are used to identify dissociation mechanisms such as Interatomic-Coulombic-Decay (ICD).

The investigation of small rare-gas clusters presents a unique opportunity in atomic physics. In a sense they can be thought of as bridging the gap between sole constituents and macroscopic matter, while still being simple enough to be understood on a particle level. For example, interesting processes were discovered in dimers such as Interatomic-Coulombic-Decay (ICD), which have been found to occur on very short time scales with a high efficiency compared to other possible pathways [1]. Experimentally, these processes have been found in neon dimers [2]. ICD is a known source of low energy electrons and has also been observed in water dimers [3]. The bulk of the research on these processes was carried out employing synchrotron radiation to create deep core holes.

With the help of a specially designed reaction microscope we have been able to carry out experiments on argon dimers using a low energy electrons beam of 120 eV as projectiles [4]. Briefly, in a reaction microscope, the imaging of the charged fragments onto time and position sensitive detectors is achieved by homogeneous electric and magnetic fields. These fragments are created by crossing a pulsed projectile electron beam, propagating along the spectrometer axis, with a cold target beam, created by supersonic expansion of the gas under investigation. From the time and position information, the 3D momentum vector of each charged particle can be retrieved.

We find signatures of ICD in the triply-charged final-state of Ar_2, as well as an interesting structure in the kinetic-energy-release spectrum of the doubly-charged final-state of Ar_2 pointing towards two separate mechanisms: a fast and a delayed one. Our results show general agreement with recently published results of Miteva et al. [5]. Furthermore, a significant amount of data for the dissociation of $Ar_3 \to Ar_2^+ + Ar^+$ could be retrieved.

We will present energy spectra for electrons and ions as well as their angular distributions.

References

[1] Cederbaum *et al.* 1997 *Phys. Rev. Lett.* **79** 4778

[2] Jahnke *et al.* 2004 *Phys. Rev. Lett.* **93** 163401

[3] Jahnke *et al.* 2010 *Nat. Phys.* **6** 193

[4] J. Ullrich *et al.* 2003 *Rep. Prog. Phys.* **66** 1463

[5] T. Miteva *et al.* 2014 *J. Chem. Phys.* **141** 064307

[1] E-mail: thomas.pfluegerptb.de

Content from this work may be used under the terms of the Creative Commons Attribution 3.0 licence. Any further distribution of this work must maintain attribution to the author(s) and the title of the work, journal citation and DOI.
Published under licence by IOP Publishing Ltd

XXIX International Conference on Photonic, Electronic, and Atomic Collisions (ICPEAC2015) IOP Publishing
Journal of Physics: Conference Series **635** (2015) 072012 doi:10.1088/1742-6596/635/7/072012

Towards electron-impact dissociation dynamics of biologically relevant molecules in a reaction microscope

Thomas Pflüger†,*1, Xueguang Ren†,*, Marvin Weyland†,*, Woonyong Baek†, Hans Rabus†, and Alexander Dorn*

† Physikalisch-Technische Bundesanstalt, Bundesallee 100, 38116 Braunschweig, Germany

* Max Planck Institute for Nuclear Physics, Saupfercheckweg 1, 69117 Heidelberg, Germany

Synopsis We present a new design of an advanced reaction microscope (REMI) for electron collisions with biologically relevant molecules. It will combine advancements which have been implemented for (e, 2e)-experiments, such as multi-hit detectors with a high detection efficiency, as well as new features for target creation and increased fragment acceptance.

Investigation of the ionization dynamics of simple atoms and molecules in electron collision experiments has provided significant background to enable the study of more complex targets, such as biologically relevant molecules(e.g. [1, 2, 3]). Our aim is to provide highly differential ionization cross sections and insight into fragmentation dynamics, especially of DNA constituents. Data of this nature – with special emphasis on low collision energies – has a high importance in the modeling of radiation damage to biological tissue [4] and in cancer treatment.

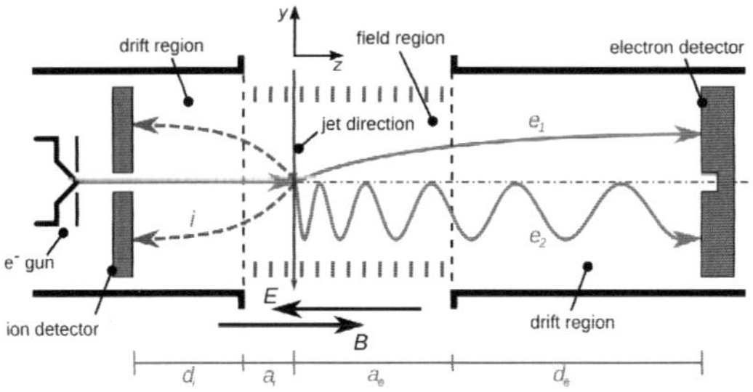

Figure 1. Schematic view of a REMI.

The experiment is based on an advanced reaction microscope (REMI) which was modified for electron impact studies [5]. Its general working principle can be seen in Figure 1. Briefly, a collimated molecular target beam is crossed with a monoenergetic, pulsed electron beam. All charged particles which emerge from a reaction are detected and their momentum vectors are reconstructed. Hence, highly detailed electronic information in form of e.g. triple-differential cross sections can be obtained. Furthermore, information on molecular fragmentation processes like kinetic energy release and molecular orientations are extracted by detecting the ionic fragments in coincidence with the electrons.

With respect to the conventional REMI technique several constrains had to be overcome. Since molecular dissociation can result in large fragment momenta, a short detector distance and/or a pulsed high-voltage extraction is required. Our studies show that the type of ramp applied makes a significant difference in the ion resolution and allows the operation of the spectrometer in two modes: a mass spectrometer mode and a momentum spectrometer mode. Additionally, the REMI measurement relies on the target species to be present in gaseous form. While some volatile DNA constituents or surrogates can be provided in seeded supersonic jet or by evaporation in an oven, less volatile molecules can be brought into vacuum by laser-induced acoustic desorption from a foil. Consequently, the new apparatus has the possibility to incorporate several methods of target creation.

Finally, since good energy resolution is paramount, projectile beam creation requires special attention. A specially designed electron gun with minimal dimensions was built, which can be inserted close to the interaction volume.

A detailed overview of the project and the employed methods and their performance will be presented.

References

[1] M. Dampc *et al.* 2011 *J. Phys. B* **44** 055206

[2] S. M. Bellm *et al.* 2012 *J. Chem. Phys.* **136** 244301

[3] W. Y. Baek *et al.* 2012 *Phys. Rev. A* **86** 032702

[4] M. U. Bug *et al.* 2012 *Int. J. Rad. Biol.* **88** 137

[5] J. Ullrich *et al.* 2003 *Rep. Prog. Phys.* **66** 1463

[1]E-mail: thomas.pflueger@ptb.de

Content from this work may be used under the terms of the Creative Commons Attribution 3.0 licence. Any further distribution of this work must maintain attribution to the author(s) and the title of the work, journal citation and DOI.
Published under licence by IOP Publishing Ltd

Momentum imaging of dissociative electron attachment in biologically relevant molecules

Marvin Weyland* †[1], **Xueguang Ren*** †, **Thomas Pflüger*** †, **Alexander Dorn**†, **Woon Yong Baek***, **Hans Rabus***

* Physikalisch-Technische Bundesanstalt, Bundesallee 100, 38116 Braunschweig, Germany

† Max-Planck-Institut für Kernphysik, Saupfercheckweg 1, 69117 Heidelberg, Germany

Synopsis We measured dissociative electron attachment in biologically relevant molecules using momentum imaging for negative ions in an apparatus that combines high resolutions of impact energy, fragment mass and fragment momentum. First investigations of the production of NH_2^--ions around the A_1 resonance at 5.7 eV impact energy show a clear dependence of the distribution of fragment dissociation angles on the projectile energy.

Since Pan et. al. [1] showed that dissociative electron attachment (DEA) an important process which can lead to the creation of strand breaks in DNA, much effort has been put into investigations of this process. Most experimental work however only identifies fragmentation channels while studies of the dissociation angles or kinetic energy transfer to the fragments are scarce. Our experimental setup is designed to perform those tasks. Based on the design of Adaniya et. al. [2], a pulsed electron beam from a photo-emission source with an energy spread of 200 meV (FWHM) is crossed with a supersonic gas jet. The created anion is pushed into the spectrometer by a pulsed extraction voltage applied after 1 μs. The potentials of spectrometer and extraction electrode as well as the extraction delay can be set to compress the time of flight of different momenta. This operation mode allows to significantly improve the mass resolution while still recording the momentum projection in the detector plane. Due to the axial symmetry of the problem, the complete information about the momentum distribution can be reconstructed.

An analysis of the distribution of dissociation angles in DEA to ammonia leading to production of NH_2^- ions shows a broad peak around 110° up to 5.7 eV electron impact energy, in agreement with measurements by Ram et. al. [3] who investigated H^--production at the same resonance. At higher energies, we observe a shift of the peak towards higher angles and a second peak in the forward direction that has not been described before.

The kinetic energy release of the reaction has its maximum at 0.95 eV and a FWHM of 0.5 eV. This distribution is constant over the investigated energy range from 4.3 eV to 7.3 eV, showing that excess energy has to be stored in internal degrees of freedom, i.e. a vibrationally excited state of the fragment.

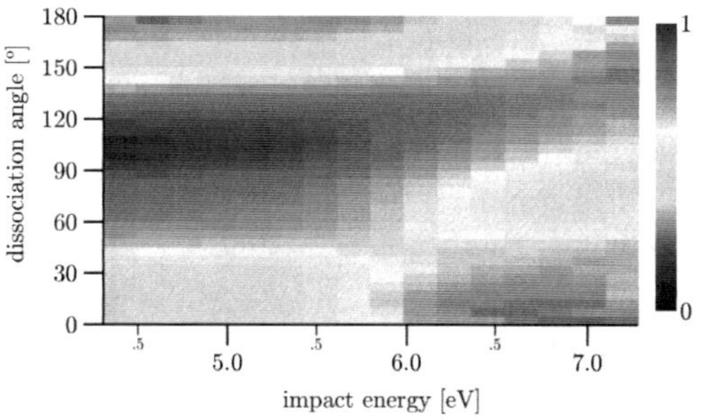

Figure 1. Distributions of fragment dissociation angles for the production of NH_2^- ions. Results have been normalized for every energy to compensate the varying total cross section.

The mass resolution of the setup has been demonstrated using the dissociation of furan around 6 eV. Here, production of $C_4H_3O^-$ (67 amu), C_4H^- (49 amu), C_2HO^-, (41 amu) and $C_3H_3^-$ (39 amu) can be observed [4]. Although the fragment with a mass of 39 amu is created with a significant amount of momentum, corresponding to a kinetic energy release up to 0.5 eV, it can be separated in time of flight from the fragment at 41 amu.

References

[1] X. Pan *et al* 2003 *Phys. Rev. Lett.* **90**(20) 208102

[2] H. Adaniya *et al* 2012 *Rev. Sci. Instrum.* **83** 023106

[3] N. B. Ram *et al* 2012 *J. Chem. Phys.* **136** 164308

[4] P. Sulzer *et al* 2006 *J. Chem. Phys.* **125** 44304

[1]E-mail: marvin.weyland@ptb.de

Content from this work may be used under the terms of the Creative Commons Attribution 3.0 licence. Any further distribution of this work must maintain attribution to the author(s) and the title of the work, journal citation and DOI.
Published under licence by IOP Publishing Ltd

XXIX International Conference on Photonic, Electronic, and Atomic Collisions (ICPEAC2015) IOP Publishing
Journal of Physics: Conference Series **635** (2015) 072014 doi:10.1088/1742-6596/635/7/072014

Cross sections for the CH_2^+ formation pathways via CH_4 fragmentation by electron impact

L Sigaud* [1] **and E C Montenegro**† [2]

* Instituto de Física, Universidade Federal Fluminense, 24210-346, Niterói, RJ, Brazil
† Instituto de Física, Universidade Federal do Rio de Janeiro, 21941-972, Rio de Janeiro, RJ, Brazil

Synopsis The mechanisms for CH_2^+ production, due to methane fragmentation by electron impact in the 22 to 400 eV energy range, were investigated using the DETOF technique. Absolute cross sections for the two possible channels, $CH_4^+ \rightarrow CH_2^+ + H_2$ and $CH_4^+ \rightarrow CH_2^+ + 2H$, and the corresponding kinetic energy distributions of CH_2^+ were obtained. Furthermore, the average kinetic energy release of H_2 and H neutrals for these two reactions were estimated as 0.04 eV and 2.1 eV, respectively. The dynamics governing these two pathways is discussed.

Methane is the major organic compound detected in many astrophysical environments, such as Titan's and the outer planets' atmospheres, and has also direct application in the physics of tokamak plasma devices.

For these reasons, its ionization and subsequent fragmentation by different projectiles have been the subject of many studies. Recently, the kinetic energy release (KER) of methane fragments produced in collision with electrons has been determined [1]. It was observed that two energy distributions contribute to the formation of CH_2^+, corresponding to different and well-known fragmentation pathways: (1) $CH_4^+ \rightarrow CH_2^+ + H_2$, and (2) $CH_4^+ \rightarrow CH_2^+ + 2H$.

In this work, these channels are thoroughly analyzed using the DETOF (Delayed Extraction Time-Of-Flight) technique, which can not only provide the disentanglement of the different energy distribution of slow fragments produced in a collisional process, identifying the different fragmentation pathways, but also measure the absolute cross section for each energy distribution of a fragment [2].

In short, the experimental setup consists of a pulsed electron gun operating, in this work, in the 22 to 400 eV energy range and a standard time-of-flight (TOF) spectrometer coupled with a gas cell. The DETOF technique consists in gradually and systematically increasing the delay time between the electron beam and the extraction field that guides the positively charged fragments into the TOF drift tube, in order to allow the faster fragments enough free flight time to leave the collection region. Thus, this delay time works as a velocity selector and the energy distributions can be obtained.

The absolute cross sections measured for the CH_2^+ fragment as a function of the electron impact energy can be seen in Figure 1, as well as the cross sections corresponding to fragmentation pathways (1) and (2).

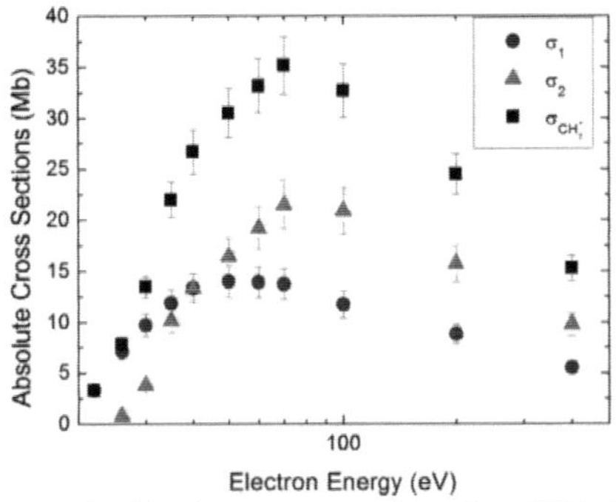

Figure 1. Absolute cross sections for CH_2^+ (black squares) and its two fragmentation pathways: σ_1 (blue circles) and σ_2 (red triangles), corresponding to channels (1) and (2), respectively.

The kinetic energy distribution associated to reaction (1), responsible for the release of a Hydrogen molecule, corresponds to very slow fragments, with a supra-thermal average energy of 0.04 eV, while the kinetic energy distribution associated to reaction (2), in which two Hydrogen atoms are produced, accounts for faster fragments, with average kinetic energy of 0.2 eV produced via pathway (2). It is noteworthy from Figure 1 that pathway (2) has a later onset than (1), but becomes more important as the impact energy increases. The kinetic energies of the neutral fragments were also obtained, being 0.3 eV for the hydrogen molecule in (1) and 2.1 eV for each hydrogen atom of (2).

References

[1] B Wei *et al* 2013 *J. Phys. B: At. Mol. Opt. Phys.* **46** 215205
[2] Natalia Ferreira, L Sigaud and E C Montenegro 2014 *J. Phys. Conf. Ser.* **488** 012042

[1] E-mail: lsigaud@if.uff.br
[2] E-mail: montenegro@if.ufrj.br

Content from this work may be used under the terms of the Creative Commons Attribution 3.0 licence. Any further distribution of this work must maintain attribution to the author(s) and the title of the work, journal citation and DOI.
Published under licence by IOP Publishing Ltd

XXIX International Conference on Photonic, Electronic, and Atomic Collisions (ICPEAC2015) IOP Publishing
Journal of Physics: Conference Series **635** (2015) 072015 doi:10.1088/1742-6596/635/7/072015

Predominance of single vacancy production in the formation of Oxygen dication

L Sigaud* [1], **Natalia Ferreira****, **and E C Montenegro**† [2]

* Instituto de Física, Universidade Federal Fluminense, 24210-346, Niterói, RJ, Brazil
** CEFET/RJ – Unidade Maracanã, 20271-110, Rio de Janeiro, RJ, Brazil
† Instituto de Física, Universidade Federal do Rio de Janeiro, 21941-972, Rio de Janeiro, RJ, Brazil

Synopsis Absolute cross sections for O_2^{++} production from O_2 fragmentation by 30-400 eV electron impact were measured using the DETOF technique. This data is compared to the molecular orbitals ionization cross sections of O_2, which were calculated using a simple and general procedure, and the fragmentation fractions arising from each molecular orbital were obtained. It was seen from the behaviour of the cross sections in this energy range that O_2^{++} results mainly from single vacancy production followed by post-collisional Auger-like deexcitation.

Although doubly charged molecular ions can be a key factor in the chemistry of energized media and planetary atmospheres, almost no absolute O_2^{++} cross section data are available in the literature [1,2], due to the intrinsic difficulty to separate fragments with same mass-to-charge ratio, O_2^{++} and O^+, in a time-of-flight spectrum.

In order to distinguish between these two moieties, the DETOF technique was employed. In short, the setup used was a pulsed electron gun operating in the 30 to 400 eV energy range and a standard time-of-flight (TOF) spectrometer coupled with a gas cell. The DETOF technique consists in a systematic increase of the delay time between the electron beam and the electrical field that guides the charged fragments into the TOF drift tube, so faster fragments have enough free flight time to leave the collection region. Since the O_2^{++} retains a Maxwell-Boltzmann energy distribution, it can be separated from the O^+, who gains kinetic energy due to the molecule breakup. Thus, absolute cross sections for homoisotopic O_2^{++} by electron impact was determined for the first time [3].

On the other hand, the ionization cross sections for each molecular orbital can be well scaled by a simple and general procedure that has provided good results for molecules even as complex as pyrimidine [4]. Afterwards, the fragmentation fractions of each molecular orbital can be inferred by matching the measured fragments cross sections with the calculated molecular orbital cross sections. In Figure 1, the calculated cross sections are shown below the respective molecular orbitals and the experimental cross sections are shown in the first column next to the corresponding fragment. The following column shows the fragment cross sections obtained with the fractions displayed in the matrix, using the calculated cross sections in order to match the experimental values.

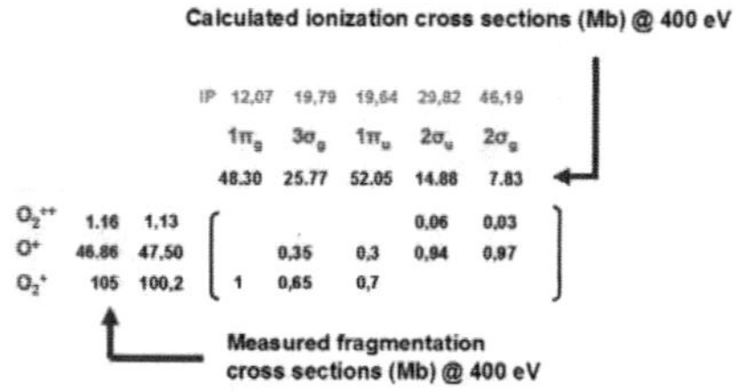

Figure 1. Oxygen fragmentation matrix (see text).

In Figure 2, the O_2^{++}/O_2^+ ratio for both experimental and calculated cross sections are shown, clearly reaching a constant value for higher electron energies. Since double ionization cross sections decrease much faster than single ionization ones with the increase of the projectile energy above ~100 eV, it can be concluded from the high impact energy data of Figure 2 that O_2^{++} formation results mainly from a post-collisional Auger-like deexcitation process after a single vacancy from either the $2\sigma_u$ or $2\sigma_g$ orbital is produced.

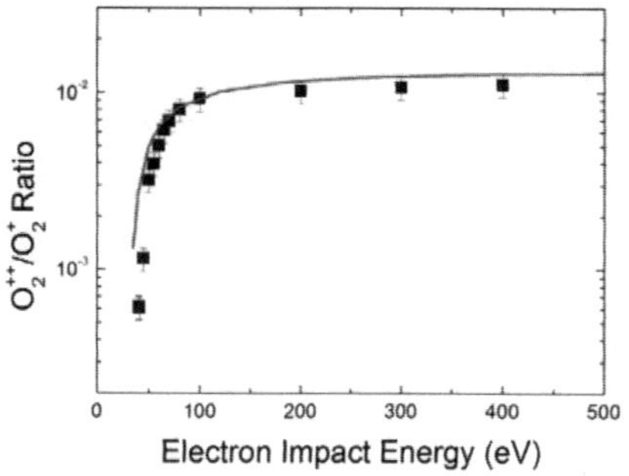

Figure 2. O_2^{++} to O_2^+ cross sections ratios for the experimental data (black squares) and the calculated cross sections (red line), with very good agreement.

References

[1] T.D. Märk 1975 *J. Chem. Phys.* **63** 3731
[2] J.N. Bull *et al.* 2015 *Phys. Rev. A* **91** 022704
[3] L. Sigaud *et al.* 2013 *J. Chem. Phys.* **139** 024302
[4] W. Wolff *et al.* 2014 *J. Chem. Phys.* **140** 064309

[1] E-mail: lsigaud@if.uff.br

[2] E-mail: montenegro@if.ufrj.br

Content from this work may be used under the terms of the Creative Commons Attribution 3.0 licence. Any further distribution of this work must maintain attribution to the author(s) and the title of the work, journal citation and DOI.
Published under licence by IOP Publishing Ltd

XXIX International Conference on Photonic, Electronic, and Atomic Collisions (ICPEAC2015) IOP Publishing
Journal of Physics: Conference Series **635** (2015) 072016 doi:10.1088/1742-6596/635/7/072016

Electron-DNA interaction in the presence of copper ion: comparison with iron-ion case

Hyung-Ah Noh*, Yeun-Soo Park†, and Hyuck Cho*[1]

* Department of Physics, Chungnam National University, Daejeon 305-764, South Korea
† Plasma Technology Research Center, National Fusion Research Institute, Gunsan 573-540, South Korea

Synopsis Combined effect of low-energy electron interaction and copper ion on DNA damage has been studied. And it has been compared with the previous study on iron ion.

For last a couple of decades, there have been the studies on the damages of biomolecules by low-energy electrons (LEEs) [1]. However, DNA in a real complex cell environment is surrounded by many other molecules like proteins, *metal ions*, and water molecules and interacts with these molecules. Therefore, it is necessary to consider the effects of the molecules surrounding DNA or DNA components when considering DNA damage by LEEs or by high energy radiations.

Metal ions, such as copper and iron, play an important role in reactive oxygen species generation from various carcinogens, leading to DNA damage. Many researchers have performed experiments with DNA-metal ion complex to investigate only the effects of metal ions on DNA damage; no electron irradiation.

Motivated by the possible *combined effects of LEE-irradiation and metal ions*, these authors, for the first time, had previously reported that DNA damages such as single-strand break or double strand break was *significantly enhanced* by LEE interactions in the presence of Fe ion [2].

In this presentation, we have studied the similar effects for the case of copper ion (Cu^{2+}) as a part of continuing studies. First, we have measured the DNA damage for the copper-ion concentrations at the same level as iron-ion case (Fig. 1). pBR322 plasmid DNA films with copper-ion concentrations of 1~7 mM were prepared by lyophilization technique and irradiated by electrons with 10 eV under ultra-high vacuum condition. DNA damage and the effects of copper ion on the damage by low energy electron collision were analyzed by 1 % agarose gel electrophoresis. The result is somewhat different from that of iron ion in Fig 2. In the Conference, we will present the more measurements at a few different electron energies and at the various concentrations of copper ion, and will attempt to analyze the differences in the mechanism of iron-ion and copper-ion in DNA damage.

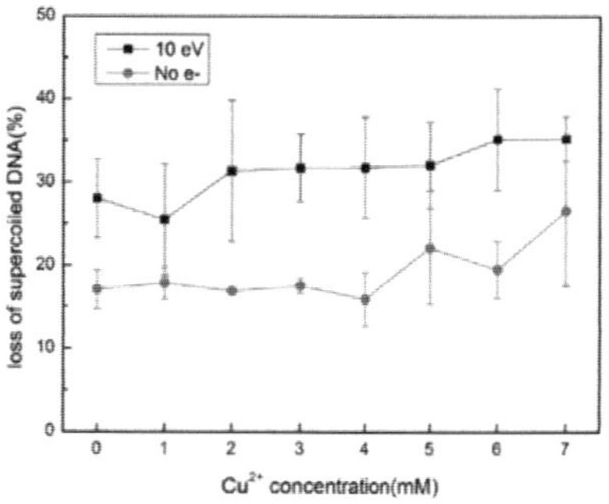

Figure 1. Loss of supercoiled DNA by Cu ion and LEE irradiation

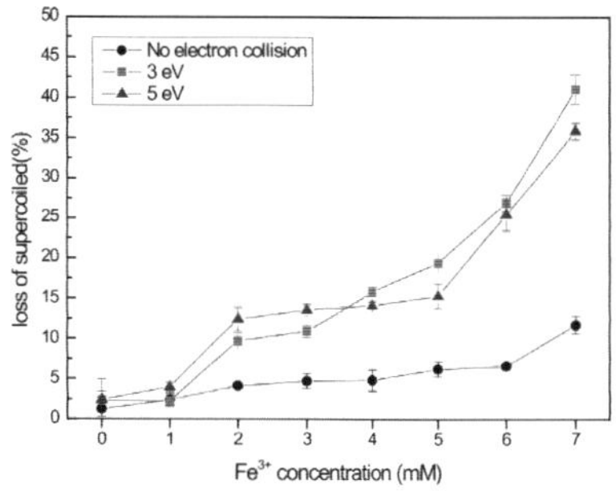

Figure 2. Loss of supercoiled DNA by the combined effects of Fe ion and LEE irradiation [2]

References

[1] B. Boudaiffa *et al 2000 Science* **287** 1658
[2] Y.S. Park, H. A. Noh and H. Cho *2012 Radiat. Res.* **177** 775

E-mail: hcho@cnu.ac.kr

Content from this work may be used under the terms of the Creative Commons Attribution 3.0 licence. Any further distribution of this work must maintain attribution to the author(s) and the title of the work, journal citation and DOI.
Published under licence by IOP Publishing Ltd

Journal of Physics: Conference Series **635** (2015) 072017 doi:10.1088/1742-6596/635/7/072017

Electron-ion collisional approach in the study of Rydberg and doubly-excited electronic states of H_2 and HeH^+

Mourad Telmini*[1], Soumaya Bezzaouia*[2], Houaida Oueslati*[3], Fatma Argoubi*[4], Islem Bouhali*[5], and Christian Jungen†[6],

* LSAMA, Department of Physics, Faculty of Science of Tunis, University of Tunis El Manar, Tunis 2092, Tunisia
† Laboratoire Aime Cotton du CNRS, Universite de Paris-Sud, F-91405 Orsay, France and Department of Physics and Astronomy, University College London, London WC1E 6BT, UK

Synopsis The halfium model combines the eigenchannel R-matrix method and the generalized multichannel quantum defect theory in prolate spheroidal coordinates to characterize Rydberg and doubly-excited states of two-electron molecular systems from first pricnciples. The background picture for the model is a collisional process between the outermost (Rydberg) electron on the ionic core. Various applications to molecular hydrogen and hydrohelium ion will be shown.

In this contribution, the case of (H_2^+ + e) reactions will be used as a prototype system to introduce electron-molecule collision computations and its extension to negative energies, allowing for investigation of bound states. Even for this simple case, there are several reaction channels to consider including elastic, inelastic and superelastic collision, dissociative recombination, dissociative excitation, dissociative ionization, etc. The processes proceed through H_2 intermediate doubly-excited states in the continuum and calculations require *ab initio* characterization of excited and doubly-excited molecular hydrogen on the same footing. Single channel quantum defect theory (QDT) considers the ion in a one specified state, whereas multichannel quantum defect theory (MQDT) considers the interaction of the Rydberg electron with multiple core states.

Matching of the electronic wavefunction between the asymptotic region and the core, described by eigenchannel R-matrix theory, yields to the computation of the short-range reaction matrix which contains in a compact shape all the information on the electronic correlations. These ideas, first developed in atomic Physics, have been extended to Rydberg molecular systems and electron-molecule collisions giving birth to the *halfium model.*

Indeed, the halfium model combines the variational eigenchannel R-matrix method and the generalized multichannel Quantum Defect Theory (*GMQDT*) in spheroidal coordinates. Moreover, the model uses a "*halfium particle*" as one-electron reference system to describe the molecular system in a simplified picture in the asymptotic region.

Several applications to H_2 Rydberg and doubly-excited states will be presented as well as more recent results of the generalization of the halfium model to the case of HeH^+ ion.

References

[1] M. Telmini and Ch. Jungen, *Phys. Rev. A 2003* **68**, 062704
[2] S. Bezzaouia *et al 2004 Phys. Rev. A* **70**, 012713
[3] S. Bezzaouia and M. Telmini, *AIP Conf. Proc 2007.* **935**, 183
[4] F. Argoubi *et al 2011 Phys. Rev. A* **83**, 052504
[5] H. Oueslati *et al 2014 Phys. Rev. A* **89**, 032501
[6] O. Motapon *et al 2014 Phys. Rev. A* **90**, 0012706
[7] F. Argoubi *et al 2015 EPJ Web of Conf.* **84** 04003
[8] I. Bouhali *et al 2015 EPJ Web of Conf.* **84** 04004

*1 E-mail: mourad.telmini@fst.rnu.tn
*2 E-mail: soumbez@yahoo.fr
*3 E-mail: houaida.oueslati@esac.rnu.tn
*4 E-mail: argoubi_fatma@yahoo.fr
*5 E-mail: islem.bouheli@yahoo.fr
†6 E-mail: christian.jungen@lac.u-psud.fr

Content from this work may be used under the ; Creative Commons Attribution 3.0 licence. Any further distribution of this work must maintain attribution to the a[…] the title of the work, journal citation and DOI.
Published under licence by IOP Publishing Ltd

The three-center parameterized continuum wave function and Dyson orbitals for the determination of the triply differential cross section of the simple ionization of CO_2 by electron impact

O. Alwan[1], **O. Chuluunbaatar**[2,3], **X. Assfeld**[4], **and B.B. Joulakian**[1]

[1] Universitè de Lorraine, SRSMC (UMR CNRS 7565), 1 bld Arago, bat. ICPM 57078 Metz Cedex 3, France
[2] Joint Institute for Nuclear Research, Dubna, Moscow Region 141980, Russia
[3] National University of Mongolia, Ulaanbaatar, Mongolia
[4] Universitè de Lorraine, SRSMC (UMR CNRS 7565), 54506 Vandoeuvre-lès-Nancy Cedex,France

Synopsis (e,2e) ionization of the linear three center CO_2 molecules by fast electron impact.

The triple differential cross section of the simple (e, 2e) ionization of CO2 is determined [1]. Its variation with the direction of the ejected electron is studied and compared to existing experimental [2] and a theoretical [3] results. The first-order perturbative Born procedure is applied where three-center Dyson type $1\pi_g$ orbital are employed for the bound electron. This orbital is obtained from coupled cluster results [4] by calculating the overlap between the N state of the target and the (N-1) state of the singly ionized ion [5] and constructed by linear combinations of Gaussian type orbitals. The ejected electron is described by a three-center continuum (ThCC) solution of the Schrödinger equation for a specific wave vector k satisfying the correct asymptotic boundary condition. This model given in the following form

$$\chi(\mathbf{k},\mathbf{r},\boldsymbol{\rho}) = \frac{\exp(\imath\mathbf{kr})}{(2\pi)^{3/2}} M_a\, {}_1F_1(\imath\alpha_a, 1, -\imath[kr_a + \mathbf{kr}_a]) \times M_b\, {}_1F_1(\imath\alpha_b, 1, -\imath[kr_b + \mathbf{kr}_b]) M_c\, {}_1F_1(\imath\alpha_c, 1, -\imath[kr_c + \mathbf{kr}_c]), \quad (1)$$

is an extension of the two-center continuum model developed in the past [6, 7], and lately applied to the ionization of CO_2 [3]. Here

$$M_j = \exp\left(-\pi\frac{\alpha_j}{2}\right)\Gamma(1-\imath\alpha_j), \quad j = a, c, b. \quad (2)$$

and $\alpha_j = -Z_j/k$ is the Sommerfeld parameter. Empirical values for the screening of the three nuclei of the target and for the Sommerfeld parameters of the three-center Coulomb continuum function are introduced.

[1]E-mail: osman.alwan@univ-lorraine.fr

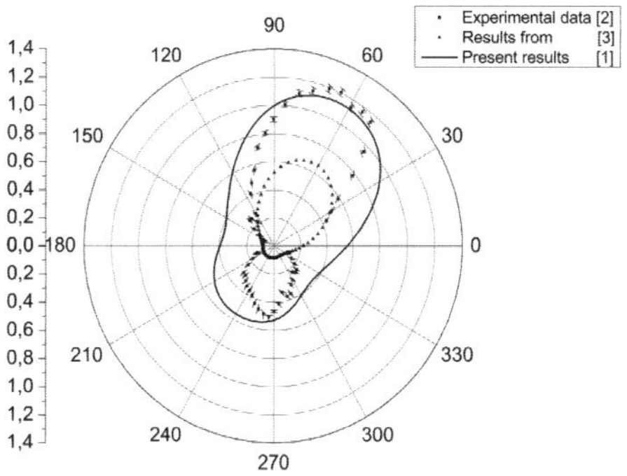

Figure 1. The variation, in polar representation, of the TDCS in terms of the ejection angle θ_e of the ionization of the $1\pi_g$ level of CO_2 for the empirical parameter $Z_a = 0.3$ for the final state. The energy of the scattered electron $E_s = 500eV$, detected at an angle $\theta_s = -6°$. The energy of the ejected electron $E_e = 37eV$.

References

[1] O Alwan, O chuluunbaatar, X. Assfeld, A Naja and B.B Joulakian, 2014 *J. Phys. : At. Mol. a d O t. Phys.* **47** 225201

[2] Lahmam-Bennani A, Staicu Casagrande E M and Naja A. 2009 *J. Phys. : At. Mol. a d O t. Phys.***42** 235205

[3] Chuluunbaatar O and Joulakian B B. 2010 *J. Phys. : At. Mol. a d O t. Phys.***43** 155201

[4] Oana C M and Krylov A I 2007 *J. Chem. Phys.***127** 234106

[5] Kaplan I G, Barbiellini B and Bansil A. 2003 *Phys. Rev.* .**68** 235104

[6] Joulakian B, Hanssen J, Rivarola R and Motassim A. 1996 *Phys. Rev. A*.**54** 1473

[7] Chuluunbaatar O, Joulakian B B, Tsookhuu Kh and Vinitsky S I. 2004 *J. Phys. : At. Mol. a d O t. Phys.***37** 2607

Content from this work may be used under the terms of the Creative Commons Attribution 3.0 licence. Any further distribution of this work must maintain attribution to the author(s) and the title of the work, journal citation and DOI.
Published under licence by IOP Publishing Ltd

XXIX International Conference on Photonic, Electronic, and Atomic Collisions (ICPEAC2015) IOP Publishing
Journal of Physics: Conference Series **635** (2015) 072019 doi:10.1088/1742-6596/635/7/072019

Effects of hydrogen bonding on pyridine resonances

Agnieszka Sieradzka[1], **Zdeněk Mašín and Jimena D. Gorfinkiel**

Department of Physical Sciences, The Open University, Walton Hall, Milton Keynes, MK7 6AA, United Kingdom

Synopsis R-matrix calculation for electron scattering from pyridine and the pyridine-water dimer have been carried out to investigate how the presence of water influences the position and lifetime of resonances. Pyridine has been chosen as a model of pyrimidinic DNA bases. Results show that the presence of 1 water molecule acting as the proton donor for the hydrogen bond leads to the shape resonances occurring in pyridine being shifted to lower scattering energies.

The collision of low-energy electrons ($E < 20$ eV, secondary products of the radiation interacting with biological material), with isolated DNA bases has been studied for more than a decade. A number of experimental works have been reported [1], focused on dissociative electron attachment (DEA). This process leads to the break-up of DNA via the formation of a temporary negative ion or resonance. From the theoretical point of view DEA is a highly complex process to describe, requiring use of sophisticated methods as well as large amount computational resources, even to treat the electronic part of the problem. Theoretical work has therefore concentrated on studying the formation of the temporary negative ions (except for a handful of small targets), and looked at model molecules as well as the bases themselves. DNA, however, is surrounded by water molecules. Understanding how solvation affects resonance formation (and the subsequent break-up) is therefore crucial.

We have performed calculations of electron collisions with pyridine (C_5NH_5) and the pyridine-water dimer using the R-matrix method [2]. Calculations for isolated pyridine at the SEP [3] and CC level revealed the presence of several shape and core excited resonances. Electronically inelastic cross sections as well as a detailed analysis of the resonances will be presented at the conference.

Similar calculations have been performed for pyridine-water. Results are consistent with earlier work on formic acid-water dimer [4] that shows a downshift of the shape resonance when the molecule acts as proton acceptor. The elastic cross section determined at SE level for both targets is shown in Figure 1. Two of the shape resonances are clearly visible in both cross sections.

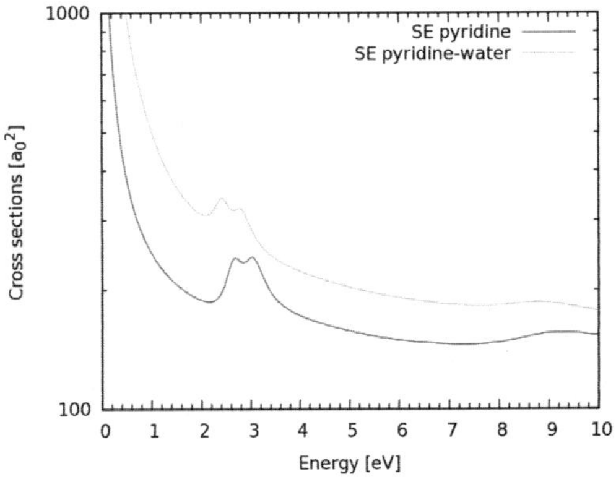

Figure 1. Low energy integral elastic cross section for electron scattering from pyridine and pyridine-water calculated using the R-matrix method.

References

[1] I. Baccarelli, I. Bald, F. A. Gianturco, E. Illenberger, and J. Kopyra, Phys. Rep. 508, 1 (2011).

[2] P. Burke, R-matrix theory of Atomic Collisions. Springer Series on Atomic, Optical, and Plasma Physics, 2011.

[3] A. Sieradzka, F. Blanco, M. C. Fuss, Z. Mašín, J. D. Gorfinkiel and G. García, JCPA, 2014, 118, 33, 6657-6663

[4] T. C. Freitas, K. Coutinho, M. T. do N. Varella, M. A. P. Lima, and S. Canuto, J. Chem. Phys. 138, 174307, 2013

[1]E-mail: agnieszka.sieradzka@open.ac.uk

Content from this work may be used under the terms of the Creative Commons Attribution 3.0 licence. Any further distribution of this work must maintain attribution to the author(s) and the title of the work, journal citation and DOI.
Published under licence by IOP Publishing Ltd

Potential electron scattering by the phosphorous systems P_n ($n = 1 \div 3$)

Sh.Sh. Demesh[1], V.I. Kelemen, E.Yu. Remeta[2]

Universitetska str. 21, Institute of Electron Physics of NAS of Ukraine, Uzhgorod 88017, Ukraine

Synopsis Independent atom model and the optical potential approach are used in order to calculate the differential and integral cross sections for the electron scattering by single-, two- and three-atom compounds of phosphorus. The potentials and atomic characteristics are determined in the local approach of the stationary and time-dependent density functional theory with account of some relativistic effects for the corresponding study of electron scattering process on the atoms from the molecules.

We have used the optical potential approach [1,2] for the $e + P_n$ ($n = 1 \div 3$) elastic scattering in the independent atom model (IAM) [3,4] for the cross section calculations. The static-exchange approximation in the SMC method were used with pseudopotentials in [5] for the calculation of integral (ICSs) and differential cross sections (DCSs) of electron scattering by X_2 and XH_3 molecules (X=N, P, As).

The DCSs of potential electron scattering by an *N*-atomic molecule in the IAM is given [3,4]

$$\frac{d\sigma_{el}}{d\Omega} = \sum_{m=1}^{N}\sum_{n=1}^{N}[f_m(\theta,k)f_n^*(\theta,k) + g_m(\theta,k)g_n^*(\theta,k)]\frac{\sin(sr_{nm})}{sr_{nm}}$$

Here , f_m, g_m are the scattering angle, the direct and spin-flip amplitudes due to the *m*-th atom of the molecule; θ ; while $k = \sqrt{2E}$, where *E* is the electron energy; r_{nm} is the distance between the *n*-th and *m*-th $s = 2k\sin(\theta/2)$ atoms (calculated by the GAMESS code). Elastic, momentum transfer and viscosity ICSs are calculated from the DCSs by integration along the scattering angle.

We have used the local relativistic parameter-free real optical potential approximation (RSEP LA). It contains static, local exchange, polarization (local correlation-polarization), scalar-relativistic and spin-orbit potentials [1,2].

Polarization potential at large distances has an asymptotic form $-\alpha_d(0)/2r^4$, where $\alpha_d(0)$ – dipole static polarizability of the *n*-th atom calculated in the local approximation of time-dependent density functional theory (for the P atom it is $30.65\,a_0^3$).

The cross sections are calculated in the 0.5÷30 eV energy region. On the Fig.1, as an example, we present the DCSs at 10 eV of impact energy. At all other impact energies *E* atomic and molecular DCSs have a similar special form. At small (<60°) and large (>150°) angles the DCSs [5] are close to our calculated atomic DCS while at intermediate angles they are between our DCSs for e+P and $e+P_2$. On the Fig.2 the elastic ICSs are presented. A similar behavior of the momentum transfer and viscosity ICSs was obtained.

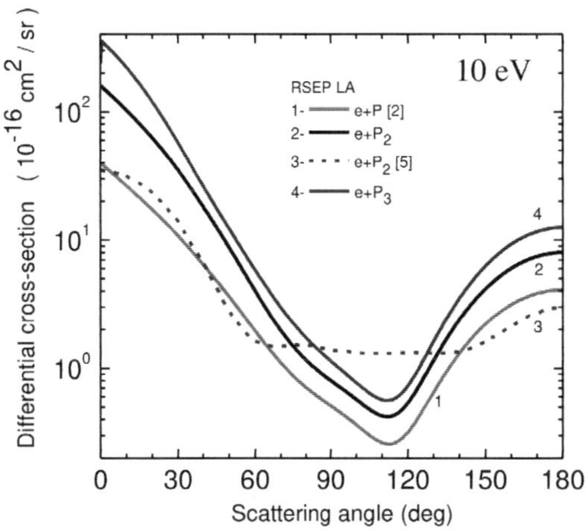

Figure 1. Differential cross sections at 10 eV.

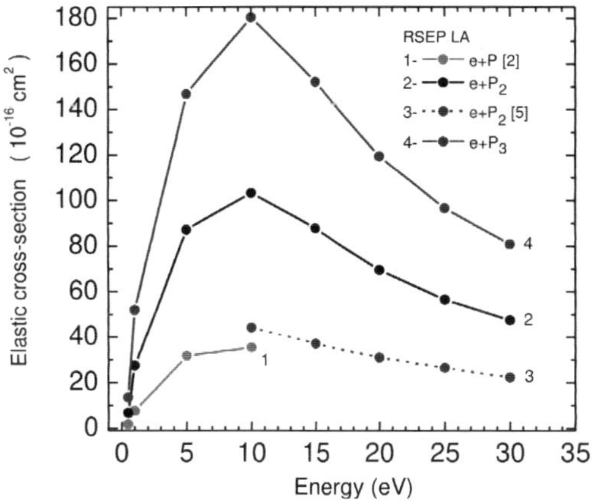

Figure 2. Elastic integral cross sections.

References

[1] Sh. Demesh *et al* 2014 *Contributed papers of the 6th CEPAS* (Bratislava, Slovakia) 65.
[2] V. Kelemen *et al* 2014 *Ukr. J. Phys.* **59** 569.
[3] D. Raj 1991 *Phys. Lett. A* **169** 571.
[4] P. Mozejko *et al* 2002 *Nucl. Instr. Meth. B.* **196** 245.
[5] M.H.F Bettega *et al* 1998 *J. Phys. B.* **31** 2091.

[1] E-mail: shandor.demesh@iep.org.ua
[2] E-mail: remetov@inbox.ru

Content from this work may be used under the terms of the Creative Commons Attribution 3.0 licence. Any further distribution of this work must maintain attribution to the author(s) and the title of the work, journal citation and DOI.
Published under licence by IOP Publishing Ltd

Time-resolved (e, 2e) electron momentum spectroscopy on the toluene S_1 excited state

Masakazu Yamazaki * [1], Yaguo Tang†, and Masahiko Takahashi* [2]
* Institute of Multidisciplinary Research for Advanced Materials, Tohoku University, Sendai 980-8577, Japan
† HFNL, University of Science and Technology of China, Hefei 230026, China

Synopsis We report a time-resolved (e, 2e) experiment on the toluene molecule in the $S_1(\pi, \pi^*)$ state. The toluene S_1 state was prepared by a 267 nm pump laser and probed with electron momentum spectroscopy using a 1.2 keV incident electron beam consisting of a train of 4 or 8 ultrashort electron pulses (~ 1 ps) with 0.3 or 0.7 ns intervals. In spite of the low data statistics, the experimental results clearly demonstrate that the present technique selectively probes ionization transition to two-hole-one-particle configurations that are hardly accessible from the ground electronic state.

Through many years of intensive studies, (e,2e) electron momentum spectroscopy (EMS) has been established as a means to look at molecular orbital patterns in momentum space [1,2]. Recently, we have developed time-resolved EMS (TREMS) [3] by combining EMS with the pump-probe technique using an ultrashort pulsed electron beam. The experimental feasibility of orbital imaging for molecular excited states has been successfully proved for the outermost orbital of the S_2 state of deuterated acetone with a lifetime of 13.5 ps [4]. Here, in order to further demonstrate the unique ability of TREMS, we have applied TREMS to a long-lived species, the first excited S_1 (π, π^*) state of toluene with a lifetime of ~80 ns.

Briefly, output from a 5-kHz Ti: sapphire femtosecond laser (<120 fs) was frequency tripled (267 nm) and was then split into a pump path and an electron-generation path. The pump laser (16 μJ, 2.5 kHz) was introduced into the vacuum chamber to excite toluene molecules. On the other hand, the 5-kHz electron-generation laser was directed toward a back-illuminated photocathode (40-nm, Ag film) to generate a train of 4 or 8 ultrashort electron pulses (~ 1 ps) with 0.3 or 0.7 ns intervals (100~200 pA) that induced EMS reactions of the excited molecules.

Fig. 1 shows a TREMS electron binding energy spectrum of the S_1 toluene. While the data statistics leaves much to be desired, one may see the ionization bands at around 4.5 eV and 17 eV, which can be attributed to the ionization from the S_1 state. The theoretical calculations using the symmetry-adapted cluster configuration interaction method support this observation, showing that the final ionic states are dominated by the two-hole-one-particle configurations in which the electron excitaion from the π to the π^* orbital is involved. This observation demonstrates the potential ability of TREMS to probe ionization of not only outermost orbitals but also of all other, more tightly bound orbitals in molecular excited states.

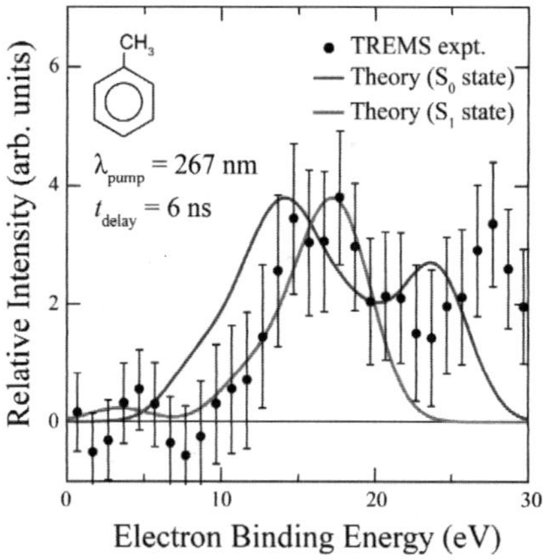

Figure 1. Comparison of TREMS binding energy spectra between experiment and theory for the S_1 excited state of toluene.

References

[1] E. Weigold and I. E. McCarthy 1999 *Electron Momentum Spectroscopy* (New York: Kluwer)
[2] M Takahashi 2009 *Bull. Chem. Soc. Jpn.* **82** 751
[3] M Yamazaki *et al* 2013 *Rev. Sci. Instrum.* **84** 063105
[4] M Yamazaki *et al* 2015 *Phys. Rev. Lett. in press*

[1] E-mail: masakazu@tagen.tohoku.ac.jp
[2] E-mail: masahiko@tagen.tohoku.ac.jp

Content from this work may be used under the terms of the Creative Commons Attribution 3.0 licence. Any further distribution of this work must maintain attribution to the author(s) and the title of the work, journal citation and DOI.
Published under licence by IOP Publishing Ltd

XXIX International Conference on Photonic, Electronic, and Atomic Collisions (ICPEAC2015) IOP Publishing
Journal of Physics: Conference Series **635** (2015) 072022 doi:10.1088/1742-6596/635/7/072022

An (e, 2e + ion) study of low-energy electron-impact ionization and fragmentation of carbon dioxide

Xing Wang ‡§ 1, **XueGuang Ren**‡* 2, **Alexander Dorn**‡ 3

‡ Max-Planck-Institute for Nuclear Physics, 69117 Heidelberg, Germany
§ School of science, Xi'an Jiaotong university, 710049 Xi'an, China
* Physikalisch-Technische Bundesanstalt, 38116 Braunschweig, Germany

Synopsis We study the ionization and fragmentation of carbon dioxide (CO_2) induced by low-energy ($E_0 = 66$ eV) electron-impact using a reaction microscope. The momentum vectors of all three charged final state particles, two outgoing electrons and one fragment ion, are detected with an (e, 2e + ion) triple coincidence method. The fragmentation pathways of carbon dioxide for formation of the CO_2^+, CO^+, O^+ and C^+ ions are investigated by measuring the ion kinetic energy spectra and the binding energy spectra where a binding energy resolution of 2.0 eV has been achieved in the experiment.

Electron-impact ionization of atoms and molecules is of fundamental importance in a wide range of scientific and practical applications. For molecular targets, the ionization process may populate dissociative states and result in neutral and positively charged fragments. As one of the fundamental constituents in planetary atmospheres, carbon dioxide (CO_2) has also been widely used in laser and gas discharge plasmas. Here, we investigate the processes of low-energy ($E_0 = 66$ eV) electron impact-induced ionization and ionic fragmentation of carbon dioxide (CO_2).

The observed reaction channels in the present low-energy electron-impact ionization and fragmentation of CO_2 are shown in Figure 1. The momentum vectors of all three final state charged particles, two outgoing electrons and one fragment ion, are detected in a (e, 2e + ion) triple coincidence method using a reaction microscope and a pulsed photoemission electron beam [1, 2]. The dissociated ions are usually created with higher kinetic energies than non-dissociated ions. Therefore, in order to cover a large solid angle for the detection of the fragment ions, the electric field is pulsed to about 15 V/cm for ion extraction. The electrons are measured by using lower electric field (1 V/cm) to obtain a better electron energy resolution. In the present work, the electron binding energy resolution of $\Delta E = 2.0$ eV has been obtained. Here, the binding energy is defined as the initial projectile energy minus the sum energy of the two final state electrons. The fragmentation pathways of CO_2 for formation of the CO_2^+, CO^+, O^+ and C^+ ions are investigated by the measured ion kinetic energy spectra and the electron binding energy spectra [2, 3]. In addition, we also measured the (e, 2e) differential cross sections for the individual molecular orbitals and the different fragmentation channels. A detailed analysis of the ionization and fragmentation processes of CO_2 will be presented at the conference.

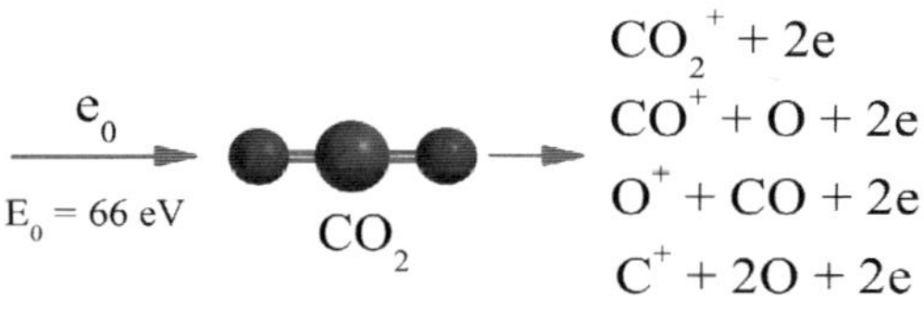

Figure 1. The reaction channels for the present low-energy electron-impact ionization and fragmentation of CO_2.

References

[1] J. Ullrich et al 2003 *Rep. Prog. Phys.* **66** 1463
[2] X. Ren et al 2014 *J. Chem. Phys.* **141** 134314
[3] S. Xu et al 2011 *Phys. Rev. A.* **83** 052702

[1]E-mail: xing.wang@mpi-hd.mpg.de
[2]E-mail: xue.g.ren@ptb.de
[3]E-mail: Alexander.Dorn@mpi-hd.mpg.de

Content from this work may be used under the terms of the Creative Commons Attribution 3.0 licence. Any further distribution of this work must maintain attribution to the author(s) and the title of the work, journal citation and DOI.
Published under licence by IOP Publishing Ltd

Journal of Physics: Conference Series **635** (2015) 072023 doi:10.1088/1742-6596/635/7/072023

Ab initio description of the fragmentation of $H_2O^+(\tilde{B}\ ^2B_2)$

Jaime Suárez [1], **L. Méndez** [2] **and I. Rabadán** [3]

Departamento de Química, Módulo 13, Universidad Autónoma de Madrid, 28049-Madrid, Spain

Synopsis A quantum-dynamical study of the fragmentation of $H_2O^+(\tilde{B}\ ^2B_2)$ is carried out by using wave packet propagations on *ab initio* potential energy surfaces connected by nonadiabatic couplings assuming a Franck-Condon initial wave packet from the ground state of the water molecule. The simulations indicate that a conical intersection between the $\tilde{B}\ ^2B_2$ and $\tilde{A}\ ^2A_1$ states of H_2O^+ allows the transfer of 80% of the initial wave packet within 30 fs, while the Renner-Teller coupling between the $\tilde{A}\ ^2A_1$ and $\tilde{X}\ ^2B_1$ states determines the fragmentation branching rations in the ps timescale.

The single ionization of water molecules by collisions with photons, electrons or ions leaves H_2O^+ in several possible electronic states: $\tilde{X}\ ^2B_1$, $\tilde{A}\ ^2A_1$ and $\tilde{B}\ ^2B_2$. The first two states are obtained after removal of one electron from the two highest occupied molecular orbitals of H_2O, and the corresponding cations do not fragment; however, the energy of the third state lies above the dissociation limits of the first two ones in a large fraction of geometry configurations, and it undergoes predissociation.

In a recent work [1], we have carried out a grid-based wave packet time-propagation in several potential energy surfaces (PESs), including nonadiabatic transitions between them. Both PESs and nonadiabatic couplings have been obtained *ab initio* in a set of $50 \times 76 \times 73$ grid points in Jacobi coordinates (R, r, θ). By considering four *ab initio* PES, and the corresponding nonadiabatic couplings, we have studied the fragmentation dynamics of water ions, assuming they are formed through Franck-Condon transitions from the ground state of the water molecule. Our simulations show that, within 10 fs, most of the initial wave packet in $\tilde{B}$ is transfered through a conical intersection to $\tilde{A}$. On the other hand, the inclusion of the Renner-Teller coupling between the $\tilde{A}\ ^2A_1$ and $\tilde{X}\ ^2B_1$ states at linear geometries produces fragmentation branching ratios that nicely reproduce the experimental values of [2].

We have also considered an alternative fragmentation mechanism that involves spin-orbit transitions from both $\tilde{B}$ and $\tilde{A}$ states to the repulsive $H_2O^+(\tilde{a}\ ^4B_1)$ state. This path leads to a small ion production due to the small fraction of the wave packet that explores the relevant crossing regions in the PESs, and the small coupling.

Figure 1(a) displays the time evolution of the populations of the $\tilde{X}$, $\tilde{A}$ and $\tilde{B}$ electronic states of H_2O^+ within the grid studied, for $K = 1$, with K the quantum number corresponding to the projection of the total angular momentum along the internuclear axis at linear geometry configurations. Figure 1(b) shows the probabilities of formation of the different ions and radicals.

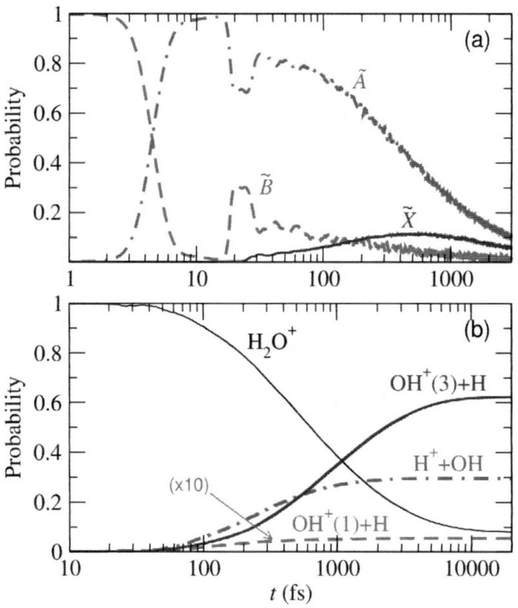

Figure 1. Time evolution of (a) the populations of $\tilde{X}\ ^2B_1$, $\tilde{A}\ ^2A_1$ and $\tilde{B}\ ^2B_2$ electronic states of H_2O^+ and (b) probability of production of the different fragments.

References

[1] J. Suárez *et al* 2015 *J. Phys. Chem. Lett.* **6** 72

[2] K. H. Tan *et al* 1978 *Chem. Phys. Lett.* **29** 299

[1]E-mail: jaime.suarez@uam.es
[2]E-mail: l.mendez@uam.es
[3]E-mail: ismanuel.rabadan@uam.es

Content from this work may be used under the terms of the Creative Commons Attribution 3.0 licence. Any further distribution of this work must maintain attribution to the author(s) and the title of the work, journal citation and DOI.
Published under licence by IOP Publishing Ltd

Journal of Physics: Conference Series **635** (2015) 072024 doi:10.1088/1742-6596/635/7/072024

Dissociative electron attachment to molecular chlorine

Krishnendu Gope* [1], **Vaibhav S Prabhudesai***, **Nigel Mason**† **and E. Krishnakumar*** [2]

* Dept. Nuclear and Atomic Physics, Tata Institute of Fundamental Research, Colaba, Mumbai 400005, India
† Dept. of Physical Sciences, The Open University, Walton Hall, Milton Keynes, MK7 6AA, UK

Synopsis We report the study of dissociative electron attachment to molecular chlorine in the energy range of 1 to 20 eV using velocity slice imaging technique. Based on the angular distributions at various electron energies, we identify the partial waves involved in the resonance formation that lead to DEA.

Velocity slice imaging (VSI) is a well-established technique to investigate the molecular dynamics involved in dissociative electron attachment (DEA) to simple molecules [1]. This technique enables one to measure the kinetic energy and angular distribution of the fragment ions produced in the DEA process. The kinetic energy distribution of the fragment ions points to the kinematics of the process leading to DEA. On the other hand the angular distribution of the fragment negative ions with respect to the incoming electron beam indicates the symmetry of the negative ion resonant states involved in the DEA process. This information in combination with the kinetic energy distribution unravels the dynamics of the DEA process.

Due to their high electron affinity, the halogen molecules form a class of interesting diatomic molecules. These molecules are known to show more than one resonances in the DEA channel with very little threshold for the formation of halide anion. This leads to the substantial kinetic energy release in the fragmentation process in DEA.

Particularly for chlorine, the DEA studies are important from the plasma processing point of view. Chlorine is known to affect the properties of electrical plasmas considerably and hence the efficiency of technologies utilizing the plasma, such as those generated in either pure chlorine or mixtures of chlorine with other gases. Chlorine being a highly oxidizing agent much attention has been paid to the investigation of its properties.

Dissociative electron attachment to chlorine has been extensively studied in the past with reports of measurements of absolute cross-section [2] and angular distribution in the limited angular range [3]. However the kinetic energy distribution and the angular distribution for the entire 0-360° angular range have not been reported so far. Here we report the study of DEA using velocity slice imaging at various resonances observed in the DEA to chlorine molecules [1]. The VSI spectrometer was modified with longer flight tube and larger (75mm diameter) phosphor screen based position sensitive detector [4].

In our measurements we measured the ion yield curve up to 90eV electron energy and we found that the chloride anion yield show three resonances at 0eV, 2.5eV and 5.6eV which is consistent with earlier measurements. The velocity slice images obtained at various electron energies across the resonances are shown in Fig. 1. In this poster we will describe the dynamics involved in DEA to chlorine molecule based on the kinetic energies and angular distributions obtained using these velocity slice images.

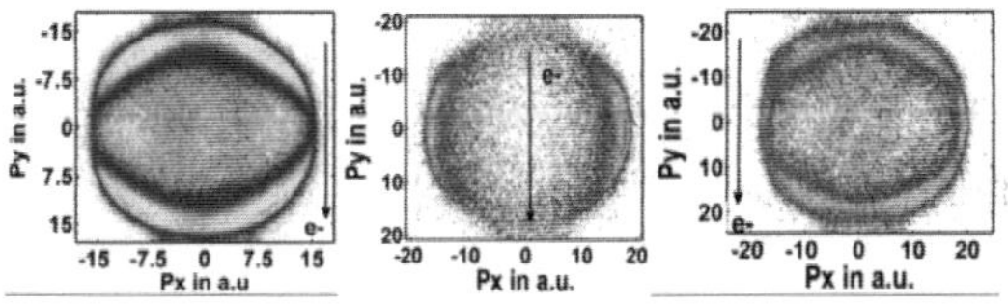

Figure 1. Image of Cl^- from Cl_2 at 2.5eV, 4eV and 6eV electron energy.

References

[1] Dhananjay Nandi, Vaibhav S. Prabhudesai, E. Krishnakumar and A. Chatterjee 2005 *Rev. Sci. Instrum.* **76** 053107
[2] M. V. Kurepa *et al.* 1978 *J. Phys. B: Atom. Mol. Phys.* **11** 3719
[3] R. Azria *et al.* 1982 *J. Phys. B: At. Mol. Phys.* **15** L569
[4] Ewelina Szymańska, Vaibhav S. Prabhudesai, Nigel J. Mason and E. Krishnakumar 2013 *Phys. Chem. Chem. Phys.* **15** 998

[1] E-mail: krishnendu@tifr.res.in
[2] E-mail: ekkumar@tifr.res.in

Content from this work may be used under the terms of the Creative Commons Attribution 3.0 licence. Any further distribution of this work must maintain attribution to the author(s) and the title of the work, journal citation and DOI.
Published under licence by IOP Publishing Ltd

Journal of Physics: Conference Series **635** (2015) 072025 doi:10.1088/1742-6596/635/7/072025

Electron ionization and low energy electron attachment to molecules of biological interest

M. A. Śmiałek*§ 1, K. Tanzer†, M. Neustetter†, S. Huber† , R.Schürmann‡, I. Bald‡ and S. Denifl† 2

* Gdansk University of Technology, Faculty of Ocean Engineering and Ship Technology, Department of Control and Energy Engineering, G. Narutowicza 11/12, 80-233 Gdansk, Poland
§ The Open University, Faculty of Science, Walton Hall, Milton Keynes MK7 6AA, United Kingdom
† Institute for Ion Physics and Applied Physics and Center of Molecular Biosciences Innsbruck, Leopold Franzens University of Innsbruck, Technikerstr. 25, 6020 Innsbruck, Austria ‡ Institut fr Chemie, Universitt Potsdam, Karl-Liebknecht-Str. 24-25, D-14476 Potsdam, Germany

Synopsis Ethylenediaminetetraacetic acid (EDTA) and 2-amino-2-(hydroxymethyl)-1,3-propanediol (TRIS) were investigated by electron impact ionization and low energy electron attachment. Both compounds are components of biological buffers and often are used as DNA stabilizers in irradiation studies. hus it is of a great importance to understand their potential interactions with radiation. Our results revealed that at least one of them, EDTA, may influence the experimental outcomes by O^- and OH^- production upon electron attachment.

Understanding the interactions of various types of radiation with the constituent cellular molecules (DNA in particular) underlay new DNA damage models formulation. Such models should be able to predict not only the patterns of ionizations but also the spectra of damage complexity that different types of radiation can induce in DNA.

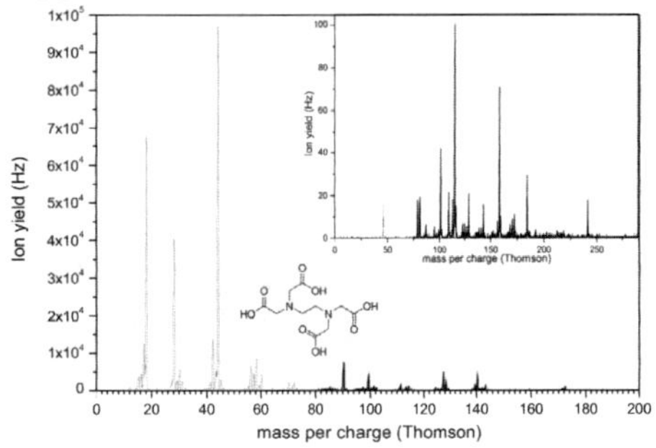

Figure 1. Positive ion mass spectrum of EDTA obtained at 70 eV incident electron energy and negative ions formed upon DEA at 0 eV electron energy (insert)

Hence knowledge of the relationship between the amount of energy deposited within a given region of the DNA helix and the type and severity of the damage that is produced is required [1]. In irradiation studies extracted plasmid DNA is being used that requires stabilisation, e.g. is suspended in EDTA solution. Therefore, the accurate determination of the influence of DNA stabilizers on the experimental outcomes is of great importance [2]. We have investigated by means of 70 eV electron impact and low energy (0-20 eV) electron attachment two most popular compounds used in DNA irradiation studies; EDTA often used as stabilizer [3] and TRIS used as OH radical scavenger [4], since it has already been noticed that EDTA may impact irradiation outcome [5].

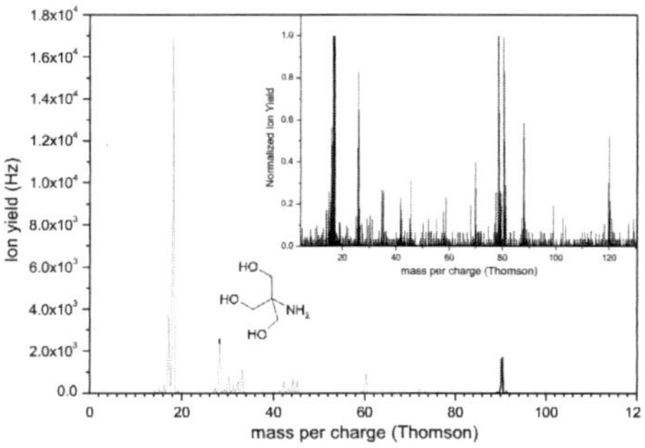

Figure 2. Positive ion mass spectrum obtained at 70 eV incident electron energy and negative ions formed upon DEA over the electron energy range 0-18 eV (insert) of TRIS

Our results indicate that upon LEE attachment as well as at 70 eV electron irradiations vast quantities of charged OH and O are being produced from both compounds. We also derived most probable dissociation channels throughout thermochemical calculations.

References

[1] M. A. Smialek *et al* 2009 *Radiat. Res.* **172** 529

[2] M. A. Smialek *et al* 2010 *Eur. Phys. J. D* **60** 31; 2011 *Eur. Phys. J. D* **62** 197; 2013 *Phys. Rev. E* **87** 060701(R)

[3] K. M. Prise *et al* 2000 *Int. J. Radiat. Biol.* **76** 881

[4] M. Folkard *et al* 1999 *J. Phys. B: At. Mol. Opt. Phys.* **32** 2753

[5] O. Boulanouar *et al* 2013 *J. Chem. Phys.* **139** 055101

[1]E-mail: smialek@pg.gda.pl
[2]E-mail: stephan.denifl@uibk.ac.at

Content from this work may be used under the terms of the Creative Commons Attribution 3.0 licence. Any further distribution of this work must maintain attribution to the author(s) and the title of the work, journal citation and DOI.
Published under licence by IOP Publishing Ltd

Journal of Physics: Conference Series **635** (2015) 072026 doi:10.1088/1742-6596/635/7/072026

Plasma screening effects in molecular hydrogen: modified energies and lifetimes of doubly excited states.

Andrés F. Ordóñez-Lasso[* 1], **José L. Sanz-Vicario**[† 2], **and Fernando Martín**[* 3]

[*] Departamento de Química,Módulo 13, Universidad Autónoma de Madrid, Cantoblanco 28049, Madrid, Spain
[†] Grupo de Física Atómica y Molecular, Instituto de Física, Universidad de Antioquia, Medellín, Colombia

Synopsis We perform a systematic study on the behaviour of energy positions and autoionization widths of metastable resonance states in the Hydrogen molecule subject to screened Coulomb interactions among all particles using an *ab initio* Feshbach configuration interaction method. We only focus on the $\mathcal{Q}_1$ series of doubly excited states lying between the first $H_2^+(1s\sigma_g)$ and the second $H_2^+(2p\sigma_u)$ ionization thresholds, for several spectroscopic molecular symmetries ${}^{1,3}\Sigma_{g\ u}$, ${}^{1,3}\Pi_{g,u}$ and ${}^{1,3}\Delta_{g,u}$. Special attention is given to the efficient and accurate method for the evaluation of screened Coulomb two-electron integrals in molecules when using configurations in terms of molecular orbitals described using a B-spline basis.

The electronic Hamiltonian describing two-electrons and two nuclei with charges Z_1 and Z_2 with pairwise screened Coulomb interactions is $\hat{H}=\hat{T}+\hat{V}$, where the kinetic term is (in a.u.) $\hat{T}=-\sum_{i=1}^{2}\nabla_i^2/2$ and the potential term is

$$\hat{V} = -\sum_{A=1}^{2}\sum_{i=1}^{2}\frac{Z_A e^{-\lambda r_{iA}}}{r_{iA}} + \frac{e^{-\lambda r_{12}}}{r_{12}} + \frac{Z_1 Z_2 e^{-\lambda R_{12}}}{R_{12}}$$

where λ is the screening parameter. In standard molecular structure calculations for bound states, the molecular orbitals are built as linear combinations of atomic orbitals expanded in terms of Gaussian basis functions (GTOs). In this particular case, all required molecular integrals involving screened Coulomb interactions and GTOs can be carried out analytically [1]. However, GTOs are not well suited basis functions to perform calculations to describe superexcited states (resonances) and electronic continuum states of H_2. Instead, screened molecular orbitals are obtained by diagonalizing the screened H_2^+ Hamiltonian using one-center expansions in terms of B-splines, i.e., $r^{-1}B_i^k(r)\mathcal{Y}_{lm}(\Omega)$ with the origin placed at the nuclear center of mass. The electron-nucleus and electron-electron molecular integrals are evaluated by expanding both screened interactions in partial waves

$$\frac{e^{-\lambda r_{mn}}}{r_{mn}} = -\lambda\sum_{\ell=0}^{\infty}[\ell]j_\ell(i\lambda r_<)h_\ell^{(1)}(i\lambda r_>)P_\ell(\cos\theta_{mn})$$

where $[\ell]=2\ell+1$, $P_\ell(x)$ is the Legendre polynomial, and $j_\ell(x)$ and $h_\ell^{(1)}(x)$ correspond to the spherical Bessel and Hankel functions, respectively.

Whereas the one-electron molecular integrals with screening can be readily computed with quadratures, the accurate computation of screened two-electron molecular integrals is more involved. For the latter, $\langle ab|e^{-\lambda r_{12}}/r_{12}|cd\rangle$, we first obtain and solve the corresponding screened Poisson equation, which is satisfied by the standard auxiliary function $Y^k(bd;r)$ within the electronic radial box $r\in[0,r_{max}]$ in which the B-splines are defined. We employ the Feshbach formalism to compute the resonance parameters, by first solving the $\mathcal{Q}$-projected eigenvalue problem $(\mathcal{Q}H\mathcal{Q}-\mathcal{E}_n)\phi_n=0$ for the autoionizing states and then widths are calculated using Fermi's golden rule, $\Gamma_n=2\pi\sum_\mu|\langle\phi_n|\mathcal{Q}H\mathcal{P}|\mathcal{P}\psi^{0-}_{\mu E=\mathcal{E}_n}\rangle|^2$, where $|\mathcal{P}\psi^{0-}_{\mu E=\mathcal{E}_n}\rangle$ is the non-resonant continuum state with appropriate incoming boundary conditions at the resonance energy [2]. The variation of energies and widths with the screening parameter λ has been successfully tested with our molecular code set at the internuclear distance $R\to 0$ (where $H_2 \to He$), for which accurate atomic values are available [3]. We show that the metastability properties of autoionizing states may change significantly when molecules are immersed in plasma environments.

References

[1] J. M. Ugalde and C. Sarasola 1997 *Int. J. Quantum Chem.* **62** 273

[2] I. Sánchez and F. Martín 1997 *J. Chem. Phys.* **106** 7720

[3] A. F. Ordóñez-Lasso, J. C. Cardona, and J. L. Sanz-Vicario 2013 *Phys. Rev. A* **88** 012702

[1]E-mail: andres.ordonez@estudiante.uam.es
[2]E-mail: jose.sanz@udea.edu.co
[3]E-mail: fernando.martin@uam.es

Content from this work may be used under the terms of the Creative Commons Attribution 3.0 licence. Any further distribution of this work must maintain attribution to the author(s) and the title of the work, journal citation and DOI.
Published under licence by IOP Publishing Ltd

Journal of Physics: Conference Series **635** (2015) 072028 doi:10.1088/1742-6596/635/7/072028

Ion-momentum imaging of dissociative-electron-attachment dynamics in CO_2, N_2O, HCCH and CF_4

M. Fogle*[1], A.L. Landers*, A. Moradmand*, J. Sartor*, D.S. Slaughter†, D.J. Haxton†, T.N. Rescigno†, C.W. McCurdy†⋆, Th. Weber†, A. Belkacem†, A.E. Orel⋆, S. Matsika‡

* Department of Physics, Auburn University, Auburn, Alabama 36849, USA
† Lawrence Berkeley National Laboratory, Chemical Sciences, Berkeley, California 94720, USA
⋆ Department of Chemistry and Applied Science, University of California, Davis, California 95616, USA
‡Department of Physics, Temple University, Philadelphia, Pennsylvania 19122, USA

Synopsis We present experimental results for dissociative electron attachment to CO_2, N_2O, HCCH and CF_4 from an ion-momentum imaging technique. From our measured ion-momentum results we extract fragment kinetic energies and angular distributions. We directly observe the dissociation dynamics associated with the formation of transitory negative ions. We compare the experimental results with *ab initio* electronic structure and fixed-nuclei scattering calculations and obtain good agreement.

In recent years, low energy dissociative electron attachment (DEA) interactions have been of interest to varying biological and technological applications. To study the dynamics resulting from DEA, we used an ion-momentum imaging apparatus based on the Cold Target Recoil Ion Momentum Spectroscopy (COLTRIMS) technique in which a molecular beam is crossed by a pulsed electron beam. The beam interaction takes place in a 4π pulsed electrostatic spectrometer that collects the anion fragments resulting from DEA. The molecular beam is formed by a supersonic expansion which results in a well-localized and cold target. An overview of the apparatus is given in Ref. [1].

Using this apparatus we have investigated the DEA dynamics for several small molecules: CO_2 at the 4 eV shape resonance [2] and the 8 eV Feshbach resonance [3]; N_2O at the 2.3 eV shape resonance [4]; HCCH at the 3 eV shape resonance [5]; and CF_4 near the 7 eV resonance.

The experimental ion-momentum results are compared to *ab initio* electronic structure and fixed-nuclei scattering calculations to gauged the resulting dynamics driven by DEA. In many cases, conical intersections play a pivotal role in driving the dynamics. Some of these systems exhibit non-axial recoil conditions indicative of a bending dynamics in the transitory negative ion state while others exhibit a direct axial recoil dissociation without any bending.

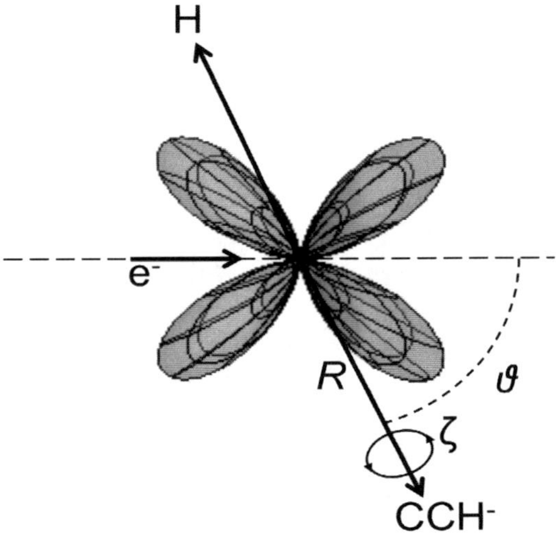

Figure 1. Modified axial recoil approximation for DEA to HCCH. The attachment amplitude lobes are shown along with the incoming electron and outgoing dissociation fragments in a rotated frame.

References

[1] A. Moradmand, J. Williams, A.L. Landers, M. Fogle 2013 *Rev. Sci. Instrum.* **84**, 033104
[2] A. Moradmand, D.S. Slaughter, A.L. Landers, M. Fogle 2013 *Phys. Rev. A* **88**, 022711
[3] A. Moradmand, D.S. Slaughter, D.J. Haxton, A.L. Landers, C.W. McCurdy, T.N. Rescigno, A. Belkacem, M. Fogle 2013 *Phys. Rev. A* **88**, 032703
[4] A. Moradmand, A.L. Landers, M. Fogle 2013 *Phys. Rev. A* **88**, 012713
[5] M. Fogle, D.J. Haxton, A.L. Landers, A.E. Orel, T.N. Rescigno 2014 *Phys. Rev. A* **90**, 042712

[1]E-mail: fogle@physics.auburn.edu

Content from this work may be used under the terms of the Creative Commons Attribution 3.0 licence. Any further distribution of this work must maintain attribution to the author(s) and the title of the work, journal citation and DOI.
Published under licence by IOP Publishing Ltd

Development of a method to measure molecular frame EELS cross sections

N. Watanabe[1], S. Yamada, T. Hirayama, A. Seki, D. Suzuki, and M. Takahashi

IMRAM, Tohoku University, Katahira 2-1-1 Aoba-ku Sendai 980-8577, Japan

Synopsis We have developed an apparatus for electron-ion coincidence experiments to examine high-energy electron scattering cross sections in molecular frame. By using the apparatus a preliminary experiment is performed for inner-shell excitations of N_2 at incident electron energy of 1.5 keV. The experimental result for the 1s $\rightarrow \pi^*$ excitation clearly shows that the electron-scattering cross section for the fixed-in-space N_2 molecule has successfully been obtained.

Electron-energy-loss spectroscopy (EELS) has been used as a powerful tool to investigate the electronic structures and excitation processes of atoms and molecules. A close examination of the momentum-transfer dependence of the electron-scattering cross sections allows one to infer the nature of excitation processes of interest as well as the symmetry and spatial extent of orbitals to which the target electron is excited [1]. It should be noted, however, that for molecules, what can be observed in EELS experiments is only the average value over all the orientations of the gaseous targets and that the spherically averaging results in substantial loss of information on electronic excitation dynamics and target electronic structure. To remove ambiguities in the analysis of the spherically averaged EELS cross sections, we have recently developed a method to perform molecular-frame EELS experiment by means of ion fragmentation of axial recoil. In this contribution we report the details of our apparatus developed and preliminary experimental results.

A schematic of our experimental setup is shown in Fig. 1. The spectrometer consists of an electron gun, a molecular beam source, a hemispherical electron analyzer, and an ion imaging spectrometer equipped with a position sensitive detector. A pulsed electron beam is crossed with a gas jet expanded from a nozzle. The electrons scattered at a particular angle of θ are energy analyzed using the hemispherical electron analyzer. Subsequently, ion extraction electric-field is applied to the scattering region to collect fragment ions produced by the electron impact. Within the axial recoil approximation, the direction of the ion recoil-momentum coincides with the molecular axis at the moment of the electron scattering.

By using the electron-ion coincidence spectrometer we have performed a preliminary molecular-frame EELS experiment for inner-shell excitations of N_2 at incident electron energy of 1.5 keV. The EELS cross section thus obtained for the 1s $\rightarrow \pi^*$ excitation at $\theta = 6°$ is shown in Fig. 2, where it is plotted as a function of the angle ϕ_K of momentum transfer vector $\boldsymbol{K}$ from the molecular axis. Under the experimental conditions employed, the inner-shell excitation can approximately be regarded as dipole-transition. In this case it is expected from scattering theory that the EELS cross section is proportional to $\sin^2\phi_K$. As can be seen in Fig. 2, the experiment shows good agreement with the expectation and this clearly indicates that the ϕ_K dependent EELS cross section has successfully been obtained.

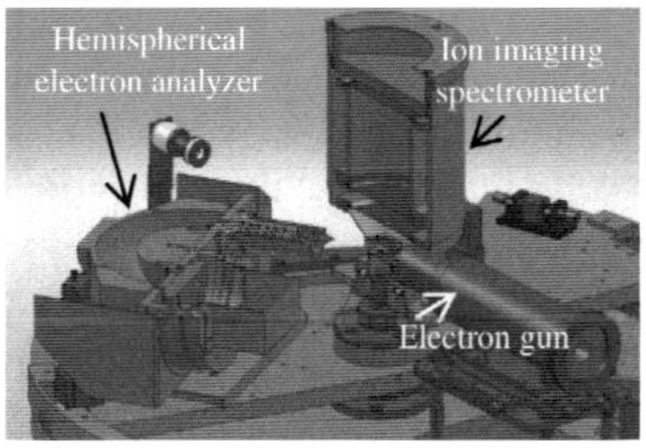

Figure 1. Schematic of our experimental setup.

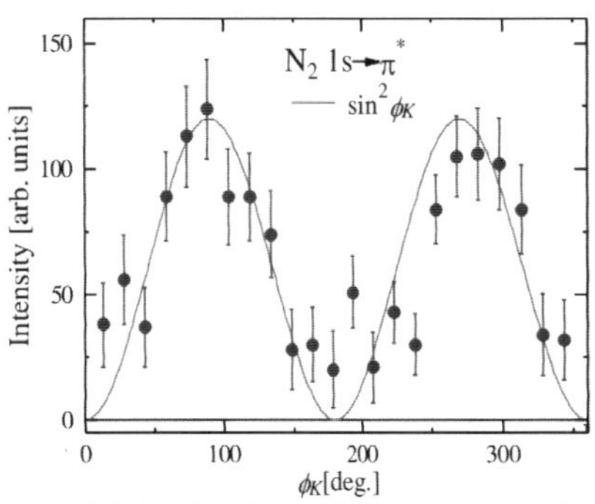

Figure 2. Molecular frame EELS cross section for the 1s $\rightarrow \pi^*$ excitation of N_2.

References

[1] M. Inokuti 1971 *Mod. Phys.* 43 297

[1] E-mail: noboru@tagen.tohoku.ac.jp

Content from this work may be used under the terms of the Creative Commons Attribution 3.0 licence. Any further distribution of this work must maintain attribution to the author(s) and the title of the work, journal citation and DOI.
Published under licence by IOP Publishing Ltd

EMS study of vibrational effects on electron momentum distributions of C_2H_4 and CH_2F_2

Noboru Watanabe[1], **Masakazu Yamazaki, and Masahiko Takahashi**[2]
IMRAM, Tohoku University, Katahira 2-1-1 Aoba-ku Sendai 980-8577, Japan

Synopsis We report an electron momentum spectroscopy study of vibrational effects on the electron momentum distributions for the outer valence orbitals of C_2H_4 and CH_2F_2. The symmetric noncoplanar (e,2e) experiment has been conducted at impact energy of 1.2 keV. Furthermore, a theoretical method of calculating electron momentum distributions has been developed with vibrational effects being involved. The experimental and theoretical results for these molecules strongly suggest that vibrational effects on electron momentum distributions tend to be appreciable for non-total symmetry molecular orbitals delocalized over some equivalent CH-bond sites.

The influence of molecular vibration on electronic wavefunction is of great importance for many physical and chemical processes. For instance, the appearance of dipole-forbidden lines in optical spectra has generally been explained by such influence; deformation of electronic wavefunctions owing to molecular vibration changes the dipole selection rule. Electron momentum spectroscopy (EMS), also known as binary (e,2e) spectroscopy, is a method that enables one to look at individual electron orbitals in momentum space, and thus may provide a powerful tool to investigate how molecular vibration affects molecular orbital patterns. In spite of their importance in EMS studies, however, investigations of vibrational effects on electron momentum profiles are scarce. In this contribution, we report our recent experimental and theoretical study of vibrational effects on electron momentum profiles for the outer-valence orbitals of C_2H_4 [1] and CH_2F_2 [2].

The symmetric noncoplanar (e,2e) experiment was carried out for the molecules at impact energy of E_0 = 1.2 keV. In the experiment, two fast outgoing electrons having equal energies (E_1=E_2) and making equal polar angles (θ_1=θ_2= 45°) with respect to the incident electron beam axis were detected in coincidence. In this way, (e,2e) data were obtained as a function of binding energy and the ion recoil momentum. We also performed plane-wave impulse approximation (PWIA) calculations with vibrational effects being involved.

Fig. 1 shows the experimental momentum profile for the $2b_1$ orbital of CH_2F_2. Also depicted in the figures are two kinds of theoretical calculations. One is theoretical momentum profile that has been calculated at the equilibrium geometry and the other is that involving vibrational effects. It can be seen from the figure that the equilibrium geometry calculation predicts typical 'p-type' momentum profile having a node at the momentum origin (p = 0). On the other hand the experiment clearly exhibits noticeable intensity near p = 0 and understanding of the observation thus requires consideration of additional effects that the equilibrium geometry calculation does not take into account. It is evident from the figure that the vibrational effects calculation gives substantial intensity near the momentum origin and reproduces the experimental result satisfactorily. The findings clearly indicate that molecular vibration plays an important role in the (e,2e) ionization process of the $2b_1$ orbital. By assessing contributions from individual normal modes, it has been revealed that the CH_2 asymmetric stretching and CH_2 rocking vibrational modes significantly affect the momentum distribution of the $2b_1$ orbital.

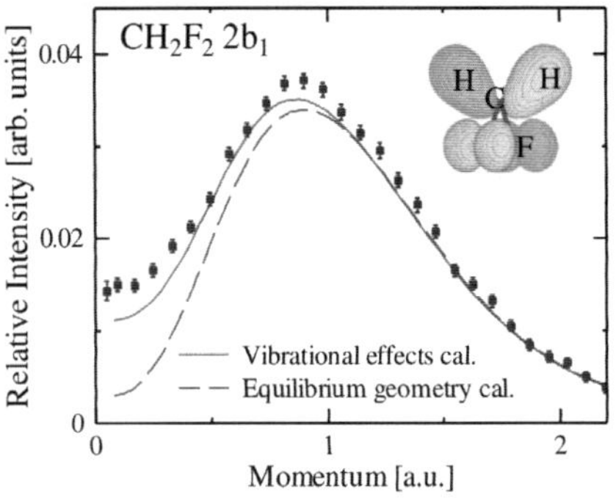

Figure 1. Momentum profiles for the $2b_1$ orbital of CH_2F_2.

References

[1] N. Watanabe, M. Yamazaki, and M. Takahashi, 2011 *J. Chem. Phys.* **137** 114301
[2] N. Watanabe, M. Yamazaki, and M. Takahashi, 2014 *J. Chem. Phys.* **141** 244314

[1]E-mail: noboru@tagen.tohoku.ac.jp
[2]E-mail: masahiko@tagen.tohoku.ac.jp

Content from this work may be used under the terms of the Creative Commons Attribution 3.0 licence. Any further distribution of this work must maintain attribution to the author(s) and the title of the work, journal citation and DOI.
Published under licence by IOP Publishing Ltd

Herzberg-Teller vibronic effects on electron-impact excitations of N_2O

Noboru Watanabe[1] **and Masahiko Takahashi**

IMRAM, Tohoku University, Katahira 2-1-1 Aoba-ku Sendai 980-8577, Japan

Synopsis We report a theoretical study on the electronic excitations of nitrous oxide (N_2O) induced by electron-impact. Momentum-transfer-dependent generalized oscillator strengths (GOSs) have been calculated for the low-lying electronic excitations using theoretical wave functions at the level of the equation-of-motion coupled-cluster singles and doubles treatment. In the calculation Herzberg-Teller vibronic effects are taken into account. The results thus obtained are found to be in overall agreement with experimental GOS profiles reported in the literature and reveal prominent roles of the bending vibration in the $B^1\Delta$ and $C^1\Pi$ transitions.

N_2O plays important roles in the chemistry of the upper atmosphere. Furthermore, N_2O is one of the simplest stable triatomic molecules and studies on the electronic excitations of this molecule are of interest from the viewpoint of molecular spectroscopy. For these practical and fundamental reasons, detailed information on electronic excitations of N_2O has been attracting broad interest from a diverse range of studies.

Angle-resolved electron-energy-loss spectroscopy (EELS) provides a powerful tool to investigate electronic excitations of molecules since the EELS cross section is directly related to the generalized oscillator strength (GOS). By examining the momentum-transfer dependence of GOS, it is possible to infer the nature of excitation processes of interest. Furthermore, GOS can be used as a sensitive probe of coupling between electronic states through molecular vibration, *i.e.* Herzberg-Teller vibronic effects [1]. Recently, Wang *et al* have reported the GOS profiles of the valence-shell electronic excitations of N_2O [2]. To give a detailed interpretation of their experimental results, comparisons with reliable theoretical calculations are desirable. The purpose of the present theoretical study [3] is to improve the knowledge of electron-impact excitations of N_2O. Emphasis is primarily put on Herzberg-Teller vibronic effects on electronic excitations of the molecule.

Fig. 1 shows two kinds of theoretical calculations for the $C^1\Pi$ transition of N_2O. One is theoretical GOS profile that has been calculated at the equilibrium geometry while vibronic effects were not taken into account. The other is the GOS profile that involves vibronic effects. Also depicted in the figure is an experimental GOS profile measured by Wang *et al* at $E_0 = 2.5$ keV [3]. It can be seen from the figure that the equilibrium geometry calculation substantially underestimates the experiment at small momentum transfer. The influence of molecular vibration is a possible source of the observed difference and indeed it gives a rational explanation of the experimental result. As can be seen from Fig. 1, the vibronic effects calculation gives much higher intensity at small momentum transfer than the equilibrium geometry calculation and reasonably reproduces the experimental result. By assessing contributions from individual normal modes, it has been revealed that the bending mode plays a leading role in the $C^1\Pi$ transition. Upon molecular bending the symmetry of N_2O is reduced from $C_{\infty v}$ to C_s. The $C^1\Pi$ state then splits into $3^1A'$ and $3^1A''$ states, whilst the adjacent $D^1\Sigma^+$ state correlates to the $4^1A'$ state in the bent geometry. The $3^1A'$ and $4^1A'$ states can interact with each other, and hence the intensity borrowing from the $D^1\Sigma^+$ transition takes place through the bending vibration.

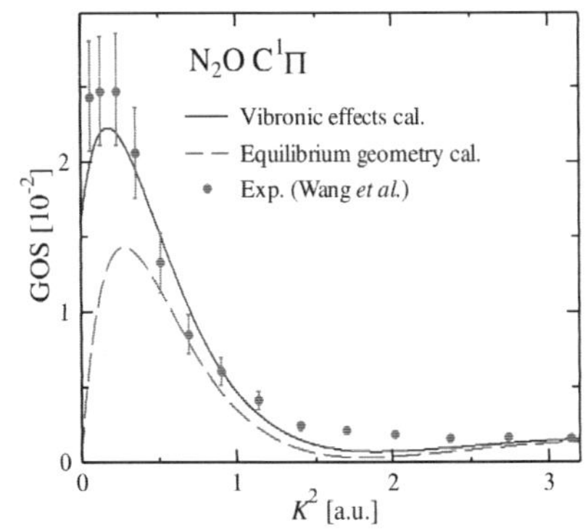

Figure 1. GOS profiles for the $C^1\Pi$ transition.

References

[1] N. Watanabe *et al* 2013 *J. Chem. Phys.* **138** 184311
[2] Y. Y. Wang, J. M. Sun, and L. F. Zhu 2010 *J. Chem. Phys.* **132** 124301
[3] N. Watanabe and M. Takahashi 2014 *J. Phys.B* **14732** 124301

[1]E-mail: noboru@tagen.tohoku.ac.jp

Content from this work may be used under the terms of the Creative Commons Attribution 3.0 licence. Any further distribution of this work must maintain attribution to the author(s) and the title of the work, journal citation and DOI.
Published under licence by IOP Publishing Ltd

Calculation of FDCS for the low and intermediate energy electron impact ionization of water molecules

G. Purohit*[1], P. Singh*[2], A. Dorn† and V. Patidar*

* Department of Physics, School of Engineering, Sir Padampat Singhania University, Bhatewar, Udaipur – 313 601, India

†Max-Planck-Institut für Kernphysik, Heidelberg, Germany

Synopsis Fully differential cross sections (FDCS) results for the electron impact ionization of water molecules taking place from $1b_1$, $1b_2$ and $2a_1$ orbitals are presented. The FDCS calculations have been performed in second Born approximation.

Fully differential cross sections (FDCS) for molecular ionization by electron impact, normally referred to as (e, 2e), represents a powerful tool to study the electronic structure of molecules as well as to examine the fundamental interaction between continuum electrons and molecules. Over the last few decades, there have been many theoretical and experimental studies performed for the (e, 2e) process with molecular targets. However most of these studies were for high incident-energy electron impact ionization of molecules [1].

Water molecule is the third most abundant molecule on the earth and is therefore of great importance in numerous multi-disciplinary research fields like radiobiology where it is commonly used as surrogate of the living matter [2, 3]. The knowledge of the collision dynamics of low-energy electrons with biological systems remain crucial so as to develop robust numerical models of charged particle tracking in biological matter.

Recently Singh et al. [4] reported fully differential cross section results for the electron and positron impact ionization of $3a_1$ orbital of the water molecule. We present in this communication the results of fully differential cross sections (FDCS) for the electron impact single ionization taking place from $1b_1$, $1b_2$ and $2a_1$ orbital of water molecule. We calculate FDCS for the ionization of water molecule in second order distorted wave Born approximation formalism in the low and intermediate energy ranges. We compare and discuss the salient features of FDCS observed for ionization taking place from different molecular orbitals of water.

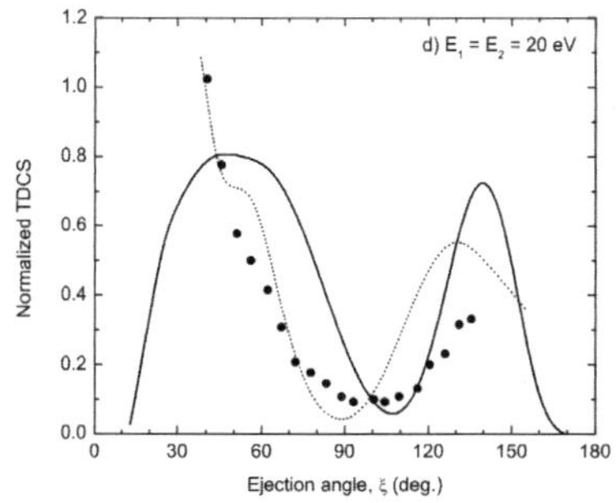

Figure 1: Triply differential cross sections for the electron impact ionization of the $3a_1$ molecular state of the water molecule in coplanar symmetric kinematics (*i.e.* $\phi_1 = 0$; $\phi_2 = 180^0$ and $\xi_1 = \xi_2 = \xi$) at incident electron energy 40 eV above the ionization potential (IP~15 eV). The solid line represents the second Born distorted wave Born approximation while the dotted line refers to the molecular distorted wave (MDW) approximation [5].

References

[1] E. Weigold and I. E. McCarthy, *Electron Momentum Spectroscopy* Kluwer Academic, Dordrecht/Plenum Publishers, New York, (1999).

[2] C. Champion, C. Le Loirec, and B. Stosic, *Int. J. Radiat. Biol.***88**, No.1-2, 62-65 (2012).

[3] B. Boudaiffa, P. Cloutier, D. Hunting, M. A. Huels, and L. Sanche, *Science* **287**,1658 (2000).

[4] P. Singh, G. Purohit, C. Champion and V. Patidar, Phys. Rev. A **89**, 032714 (2014).

[5] K. L Nixon, A. J. Murray, O. Al-Hagan, D. H. Madison, and C. Ning,*J. Phys. B: At. Mol. Opt. Phys.***43**, 035201 (2010).

[1] E-mail: ghanshyam.purohit@spsu.ac.in

*[2] E-mail: prithvi.singh@spsu.ac.in

Content from this work may be used under the terms of the Creative Commons Attribution 3.0 licence. Any further distribution of this work must maintain attribution to the author(s) and the title of the work, journal citation and DOI.
Published under licence by IOP Publishing Ltd

XXIX International Conference on Photonic, Electronic, and Atomic Collisions (ICPEAC2015) IOP Publishing
Journal of Physics: Conference Series **635** (2015) 072035 doi:10.1088/1742-6596/635/7/072035

Dissociative electron attachment to chlorine dioxide probed by velocity slice imaging

Krishnendu Gope[*1], **Vaibhav S Prabhudesai**[*], **Nigel Mason**[†] **and E. Krishnakumar**[*2]

[*] Dept. Nuclear and Atomic Physics, Tata Institute of Fundamental Research, Colaba, Mumbai 400005, India
[†] Dept. of Physical Sciences, The Open University, Walton Hall, Milton Keynes, MK7 6AA, UK

Synopsis We report the study of dissociative electron attachment (DEA) to chlorine dioxide (OClO) in the energy range of 1 to 10 eV using velocity slice imaging technique. DEA to OClO shows Cl^-, O^-, O_2^-, and OCl^- as fragment channels at various electron energies. Based on the momentum images of these ions we unravel the molecular dynamics leading to DEA.

Chlorine dioxide (OClO) is an important molecule in atmospheric chemistry as a major source of atomic chlorine in the stratosphere as it decomposes photo-chemically forming O_2 as the other product. Dissociative electron attachment (DEA) to OClO can also lead to atomic chlorine formation.

DEA to OClO has been studied in the past with reports of absolute cross-section measurements [1]. However, the dynamics that leads to DEA has not been reported so far. Here we report the study of DEA dynamics for OClO using velocity slice imaging technique.

We observed Cl^-, O^-, O_2^-, and OCl^- ions from the DEA measurements on OClO formed at various electron energies. The OCl^- being the most dominant channel was found to peak around 0.7 eV, whereas Cl^- and O_2^- signals peaked around 0.7 eV, 4 eV, 8 eV. Out of these the peak at 0.7 eV was found to be the dominant peak. The O^- ion yield was found to peak at 1.2 eV, 4 eV, 8 eV.

We carried out the momentum imaging of various fragment ions across different resonances using velocity slice imaging [2]. The VSI spectrometer was modified with longer flight tube and larger (75 mm diameter) phosphor screen based position sensitive detector [3]. For each of the ion fragments, we obtained the kinetic energy and angular distribution data. The velocity slice images obtained for Cl^- channel at 8 eV is shown in Fig. 1 as an example. For Cl^- channel, with very low energy threshold (-3.37 eV) [1], it was expected to carry good amount of kinetic energy from the 8 eV resonance. However, the low kinetic energy observed in Cl^- channel indicates an interesting many body fragmentation dynamics behind the DEA process. The angular distribution also shows a forward-backward asymmetry.

In this poster we will describe the dynamics involved in DEA to OClO based on the kinetic energies and angular distributions measured using velocity slice images for all the fragments observed at various resonances.

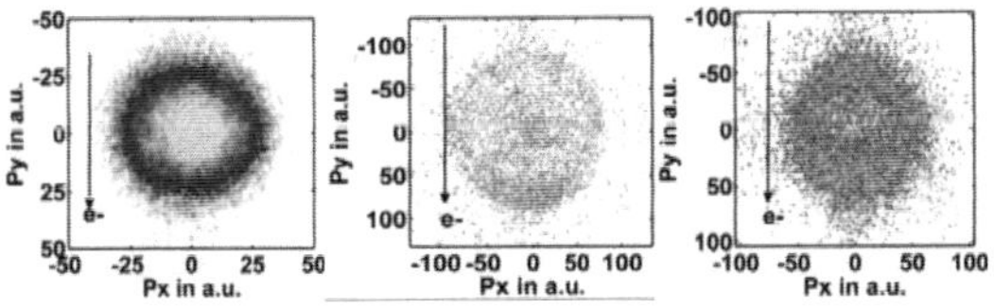

Figure 1. Momentum image of OCl^-, O^- and Cl^- from OClO at 0.7eV, 4eV and 8eV electron energy respectively.

References

[1] Gilbert Senn et al. 1999 *J. Phys. B At. Mol. Opt. Phys.* **32** 3615
[2] Dhananjay Nandi, Vaibhav S. Prabhudesai, E. Krishnakumar and A. Chatterjee 2005 *Rev. Sci. Instrum.* **76** 053107
[3] Ewelina Szyman´ska, Vaibhav S. Prabhudesai, Nigel J. Mason and E. Krishnakumar 2013 *Phys. Chem. Chem. Phys.* **15** 998

[1] E-mail: krishnendu@tifr.res.in
[2] E-mail: ekkumar@tifr.res.in

Content from this work may be used under the terms of the Creative Commons Attribution 3.0 licence. Any further distribution of this work must maintain attribution to the author(s) and the title of the work, journal citation and DOI.
Published under licence by IOP Publishing Ltd

Dissociative electron attachment dynamics of ozone using velocity slice imaging

Krishnendu Gope[* 1], **Vaibhav S Prabhudesai**[*], **Nigel Mason**[†] **and E. Krishnakumar**[* 2]

[*] Dept. Nuclear and Atomic Physics, Tata Institute of Fundamental Research, Colaba, Mumbai 400005, India
[†] Dept. of Physical Sciences, The Open University, Walton Hall, Milton Keynes, MK7 6AA, UK

Synopsis We report the study of dissociative electron attachment to ozone (O_3) in the energy range of 1 to 10 eV using velocity slice imaging technique. Based on the momentum images that we obtained for O^- and O_2^- channels at various electron energies we unravel the molecular dynamics leading to DEA.

Dissociative electron attachment (DEA) to ozone has been studied in the past including measurement of absolute cross-sections [1]. However, the dynamics that leads to DEA has not been reported so far. Here we report the study of DEA dynamics for ozone using velocity slice imaging technique.

Here we report the study of DEA dynamics of O_3 using velocity slice imaging [2]. The VSI spectrometer was modified with longer flight tube and larger (75 mm diameter) phosphor screen based position sensitive detector [3]. We observed O^-, and O_2^- ions from the DEA measurements on Ozone formed at various electron energies. The O^- being the most dominant channel was found to peak around 1.4 eV along with 3 eV and 7.5 eV whereas O_2^- signals peaked around 1.2 eV, 3.2 eV, 7.5 eV.

We carried out the momentum imaging of various fragment ions across different resonances using velocity slice imaging. Around 3 eV peak the angular distribution shows a forward-backward asymmetry in the O^- channel and that asymmetry is reversed in the O_2^- channel as shown in Fig. 1. Around 1.5 eV there is considerable kinetic energy release (~1.4 eV) in the O^- channel.

In this poster we shall describe the dynamics involved in DEA to ozone using velocity slice images for both the fragments observed at various resonances.

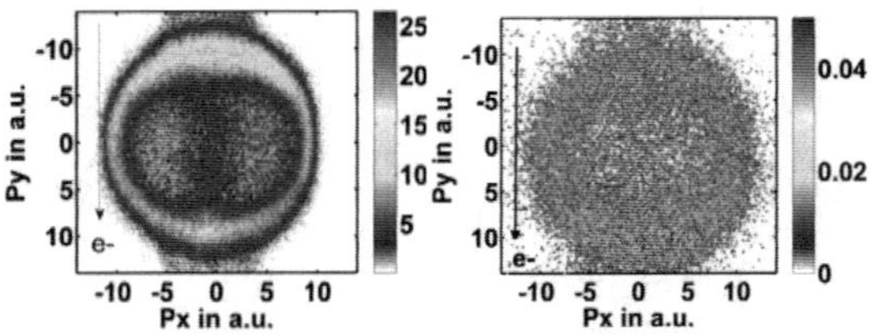

Figure 1. Image of O^- and $O_2^-$from O_3 at 2.5eV and 3.7eV electron energy.

References

[1] S A Rangwala et al. 1999 *J. Phys. B: At. Mol. Opt. Phys.* **32** 3795
[2] Dhananjay Nandi, Vaibhav S. Prabhudesai, E. Krishnakumar and A. Chatterjee 2005 *Rev. Sci. Instrum.* **76** 053107
[3] Ewelina Szyman´ska, Vaibhav S. Prabhudesai, Nigel J. Mason and E. Krishnakumar 2013 *Phys. Chem. Chem. Phys.* **15** 998

[1] E-mail: krishnendu@tifr.res.in
[2] E-mail: ekkumar@tifr.res.in

Content from this work may be used under the terms of the Creative Commons Attribution 3.0 licence. Any further distribution of this work must maintain attribution to the author(s) and the title of the work, journal citation and DOI.
Published under licence by IOP Publishing Ltd

Electron driven water formation from oxyhydrogen clusters in superfluid helium nanodroplets

Michael Renzler* [1], **Lorenz Kranabetter***, **Erik Barwa***, **Eugen Illenberger**†, **Paul Scheier*** [2]

* Institute for Ion Physics and Applied Physics, University of Innsbruck, Innsbruck, Austria
† Institute of Chemistry and Biochemistry, Freie Universität Berlin, Berlin, Germany

Synopsis Helium nanodroplets provide an enviroment that allow studies of chemical reactions at ultracold temperatures. We use these droplets as a matrix to study the formation of water upon electron bombardment of oxyhydrogen clusters

The production of water from the combution of oxyhydrogen is one of the most well known chemical reactions. It is highly exorthermic with an energy release of $572\,\frac{\mathrm{kJ}}{\mathrm{mol}}$ and in order to be ignited the oxyhydrogen mixture needs to be heated up to about 600 °C (at ambient pressure). In our study we employ a radically different method to observe a oxyhydrogen reaction. Superfluid helium nanodroplets (HND) offer an unique environment to study chemical reactions at ultracold temperatures and have been employed by our group to study a variety of different systems [1], [2].

HNDs were produced by supersonic expansion of helium (purity: 99.9999%) through a 5 µml nozzle at a temperature of 9.4 K and a stagnation pressure of 20 bar into vacuum, they are superfluid and have a temperature of 0.37 K. Under these conditions their size can be estimated to be $5 \cdot 10^5$ helium atoms per droplet. The HND then traversed through two differentially pumped pick-up cells where we introduced first O_2 and afterwards H_2. The dopants were able to move frictionless to the centre of the droplet were they formed oxyhydrogen clusters. Afterwards the doped HNDs were exposed to electron bombardment at various energies in a Nier-type ion source in order to form cations and anions respectively. The resulting ions were then mass-analyzed by a commercial time-of-flight instrument (Tofwerk AG, model HTOF) with a mass resolution of up to $R \approx 5000$.

The recorded mass spectra show series of O_n clusters with various hydrogen atoms attached to them. From the raw data we can extract cluster series of the different species to look for magic numbers with a specially developed software [3]. These magic numbers indicate ions with a higher binding energy compared to those of lower yield. Several findings can be gained from these cluster series:

1. There is a distinct odd-even oscillation in the ion yield of the oxygen clusters. When looking at the attachment of hydrogen to the oxygen clusters one has to differentiate between the cationic and the anionic case.

2. Even numbered oxygen cluster anions show a strong ion yield for the attachment of one hydrogen atom, whereas the signal drops considerably for the attachment of subsequent hydrogen atoms showing a distinct odd-even oscillation.

3. Odd numbered oxygen cluster anions on the other hand show a magic number for the attachment of H_2, thus indicating the formation of $O_{2n}H_2O^-$ complexes.

4. Even and odd numbered oxygen cluster-cations show magic numbers for the attachment of three hydrogen atoms, thus indicating the formation of $O_nH_3O^+$ complexes

As already mentioned magic numbers correspond to ions with an enhanced binding energy thus our findings indicate water formation on these oxygen clusters.

Acknowledgements: This work has been supported by the FWF Austrian Science Fund (FWF, P26635), Wien.

References

[1] M. Daxner *et al* 2014 *Angew. Chem. Int. Ed.* **53** 13528

[2] A. Mauracher *et al* 2014 *Angew. Chem. Int. Ed.* **53** 13794

[3] S. Ralser *et al* 2015 *Int. J. Mass Spectrom.* in press

[1] E-mail: michael.renzler@uibk.ac.at
[2] E-mail: paul.scheier@uibk.ac.at

Content from this work may be used under the terms of the Creative Commons Attribution 3.0 licence. Any further distribution of this work must maintain attribution to the author(s) and the title of the work, journal citation and DOI.
Published under licence by IOP Publishing Ltd

Journal of Physics: Conference Series **635** (2015) 072038 doi:10.1088/1742-6596/635/7/072038

Positron scattering from Biomolecules

J.R. Machacek[1*], W. Tattersall[†], R.A. Boadle[*], R.D. White[†], M.J. Brunger[‡], S.J. Buckman[*] and J.P. Sullivan[2*]

*Atomic and Molecular Physics Laboratories, Research School of Physics and Engineering, Australian National University, Canberra Australia 2601

†School of Engineering and Physical Sciences, James Cook University, Townsville Australia 4810

‡School of Chemical and Physical Sciences, Flinders University, Adelaide Australia 5001

Synopsis Experimental progress in positron scattering from biologically relevant molecules.

We will present our recent work on positron scattering from biologically relevant molecules. This program focuses on the nucleobases: Adenine, Guanine, Thymine, Cytosine and Uracil [1]. These molecules are solid at room temperature. This requires an oven to produce a vapour of the molecule of choice. The aim of this work is to provide absolute positron scattering information for the nucleobases. Currently, we are focusing on measurements of the grand total scattering cross section and the positronium formation cross section. Comparisons will be made between the available electron and positron scattering data including our work in a number of other targets of interest.

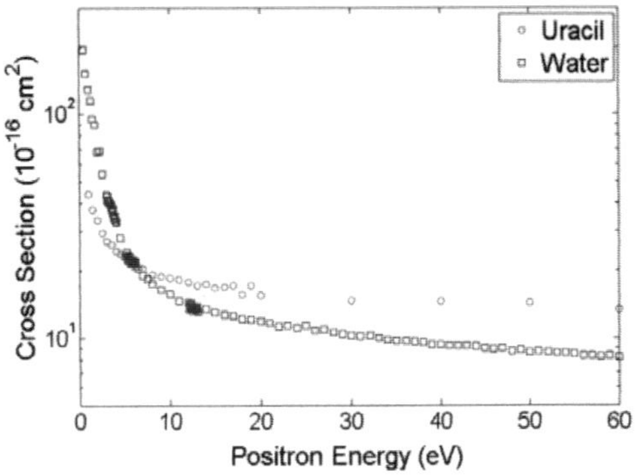

Figure 1. Total scattering cross section for uracil [1] and water [3]. The uncertainties have been suppressed.

The experimental measurements were carried out using the low-energy positron beam line at the Australian National University. The magnetically confined positron beam can be tuned from below 1 eV up to 200 eV with a variable energy resolution, which was typically between 60 and 100 meV. The oven temperature was monitored by a number of calibrated platinum resistive thermal devices (PRTDs) and a thermocouple, and controlled by varying a resistive heater (thermocoax). The number density is required for the normalization of the experimental results, and was determined via the vapour pressure curves for the target molecule combined with measurements of the oven temperature. Uncertainties in the temperature measurement and uncertainties in the vapour pressure curve constitute the largest systematic uncertainty. The uncertainty due to the statistical fluctuations in the measurement was typically less than 3 %. The positron energy-loss spectrum was used to determine the partial scattering cross sections. We will present details of the experimental procedure along with the design and operation of the oven based source. The results produced for the various targets will be presented.

References

[1] E. K. Anderson *et al* 2014 *J. Chem. Phys.* **141** 034306
[2] W. Tattersall *et al* 2014 *J. Chem. Phys.* **140** 044320
[3] C. Makochekanwa *et al* 2009 *New J. Phys.* **11** 103036

[1] E-mail: joshua.machacek@anu.edu.au

[2] E-mail: james.sullivan@anu.edu.au

Content from this work may be used under the terms of the Creative Commons Attribution 3.0 licence. Any further distribution of this work must maintain attribution to the author(s) and the title of the work, journal citation and DOI.
Published under licence by IOP Publishing Ltd

Journal of Physics: Conference Series **635** (2015) 072039 doi:10.1088/1742-6596/635/7/072039

Three-body fragmentation dynamics of CO_2^{q+} ($q = 3,\ 4$) induced by electron collision at 500 eV impact energy

Enliang Wang*,†, Xu Shan*,†, Zhenjie Shen*,†, Maomao Gong*,†, Yaguo Tang*,†, Xiangjun Chen*,†[1]

* Hefei National Laboratory for Physical Sciences at Microscale and Department of Modern Physics, University of Science and Technology of China, Hefei, Anhui, 230026, China

† Synergetic Innovation Center of Quantum Information and Quantum Physics, University of Science and Technology of China, Hefei, Anhui 230026, China

Synopsis Electron impact induced three-body fragmentation of CO_2^{q+} ($q = 3,\ 4$) is investigated. For CO_2^{3+}, non-sequential and sequential fragmentation processes are observed. For CO_2^{4+}, however, only non-sequential fragmentation processes are observed. The momentum correlations of each dissociative channel is analyzed by Newton diagram.

The fragmentation dynamics of molecule is one of the fundamental tasks in physics, chemistry as well as biology. The major challenges in this field are to understand the molecular bond breakage selectivity and to clarify which parameters control the bond cleavage. During the past decades, with the help of the rapid developing imaging techniques [1, 2], was it possible to reveal multi-body fragmentation dynamics for small molecules.

In this work, the three-body fragmentation dynamics of CO_2^{q+} ($q = 3,\ 4$) is studied:

$$CO_2^{3+} \rightarrow C^+ + O^+ + O^+, \quad (1)$$
$$CO_2^{4+} \rightarrow C^{2+} + O^+ + O^+, \quad (2)$$
$$\rightarrow C^+ + O^+ + O^{2+}, \quad (3)$$

where the molecular ions, CO_2^{q+} ($q = 3,\ 4$), are formed by electron collision at impact energy of 500 eV and the three final ions are detected in coincidence by a momentum imaging system.

The Newton diagrams of channel (1), (2) and (3) are shown in Figure 1 (a), (b) and (c), respectively. The relative momentum vector of O^+ is fixed on the x axis while C^+ and another O ion are plotted in the upper and the lower half of the plane. For channel (1), as shown in Figure 1 (a), the two dense islands correspond to linear and bending dissociation of CO_2^{3+}. In addition, as marked by the dashed circle, each of the islands drags a half circle tail, which is the evidence of the sequential fragmentation of CO_2^{3+} [3, 4]. In this situation, the two C-O bonds break stepwisely.

The structures in Newton diagrams of channel (2) and (3) are similar to the dense islands in that of channel (1) indicating that the two C-O bonds break simultaneously in channel (2) and (3). For channel (2), the two O^+ ions are indistinguishable and they share equivalent momentum. As a result, the island of O^+ in the lower half plane of Newton diagram, as shown in Figure 1 (b), located on a circle with unit radius. For channel (3), the momentum of O^{2+} is larger than that of O^+. In addition, the correlation angle of C^+-O^{2+} is also larger than that of C^+-O^+. This is because that the Coulomb repulsion between C^+ and O^{2+} is much larger than that between C^+ and O^+.

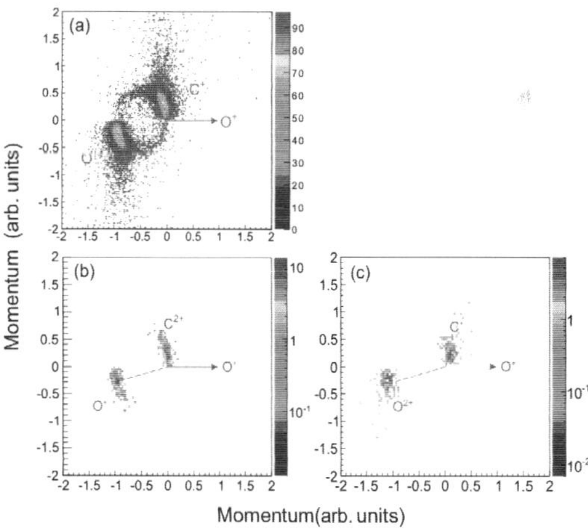

Figure 1. Newton diagrams for three-body fragmentation (a) for CO_2^{3+}, (b) and (c) for CO_2^{4+}.

References

[1] R. Dörner *et al* 2000 *Phys. Rep.* **330** 95

[2] J. Ullrich *et al* 2003 *Rep. Prog. Phys.* **66** 1463

[3] N. Neumann *et al* 2010 *Phys. Rev. Lett.* **104** 103201

[4] C. Wu *et al* 2013 *Phys. Rev. Lett.* **110** 103601

[1]E-mail: xjun@ustc.edu.cn

Content from this work may be used under the terms of the Creative Commons Attribution 3.0 licence. Any further distribution of this work must maintain attribution to the author(s) and the title of the work, journal citation and DOI.
Published under licence by IOP Publishing Ltd

Multi-center interference effect on (e, 2e) electron momentum spectroscopy

Enliang Wang*,†, Xu Shan*,†, Qiguo Tian*,†, Jing Yang*,†, Maomao Gong*,†, Yaguo Tang*,†, Shanshan Niu*,†, Xiangjun Chen*,† [1]

* Hefei National Laboratory for Physical Sciences at Microscale and Department of Modern Physics, University of Science and Technology of China, Hefei, Anhui, 230026, China

† Synergetic Innovation Center of Quantum Information and Quantum Physics, University of Science and Technology of China, Hefei, Anhui 230026, China

Synopsis The multi-center interference effect of three outermost molecular orbitals (MOs) of CF_4 is investigated using (e, 2e) electron momentum spectroscopy. We show that the observed oscillations are directly informative of the nature of molecular structure. By fitting the experimental results, the bond length of F-F of gas phase CF_4 is determined.

The wave-particle duality is one of the fundamental concepts of quantum mechanism. The wave like particle behaviors were mostly demonstrated through Young's double-slit experiment, in which the coherent addition of different outgoing amplitudes will present interference. The molecular double-slit interference, which was first proposed by Cohen and Fano [1] in 1966, provide a new way to investigate quantum interference effect in molecular scale. In 2003, Stia et al [2] suggested the interference in molecular (e, 2e) process in which the interference originate from the coherent addition of ionization amplitude from different atom centers, as shown in the inset of Figure 1.

For CF_4, the three outermost MOs (1e, $4t_2$, $1t_1$) are composed by $2p$ lone-pair electrons on fluorine atoms. For these MOs consisting of the F $2p$ atomic orbitals, the TDCS, $\sigma_{\mathrm{EMS}}(p)$, can be expressed as [3],

$$\sigma_{\mathrm{EMS}}(p) \propto \sigma_{\mathrm{A}}(p)[1+C_0 j_0(pR_{FF})+C_2 j_2(pR_{FF})], \quad (1)$$

where $\sigma_{\mathrm{A}}(p)$ is the TDCS of $2p$ orbital of F atom. j_i ($i = 0,\ 2$) are the spherical Bessel functions of order 0 and 2 and C_i ($i = 0,\ 2$) are the corresponding coefficients. R_{FF} is the internuclear distance between the F atoms. Obviously, the interference effect is presented in the last part of e.q. (1). To present the oscillation directly the theoretical TDCS of F atom is divided,

$$\sigma_{\mathrm{EMS}}(p)/\sigma_{\mathrm{A}}(p) \propto [1+C_0 j_0(pR_{FF})+C_2 j_2(pR_{FF})]. \quad (2)$$

In e.q. (2), $\sigma_{\mathrm{A}}(p)$ for the $2p$ orbital of F atom is calculated by distorted wave Born approximation employing B3LYP/aug-cc-pVTZ wave function. The interference factor for the three outermost MOs in large momentum range of CF_4 are shown in Figure 1. It is immediately clear that the experiments exhibit oscillatory structures. The interference factor, $h[1 + C_0 j_0(pR_{FF}) + C_2 j_2(pR_{FF})]$ is employed as a fitting curve for $4t_2$ to reproduce the experiment with R_{FF}, h, C_0, and C_2 being fitting parameters. The best fit to the experiment is presented in Figure 1 (b) by the dashed line. The value of R_{FF} is 3.9 Bohr which is in excellent agreement with 4.07 Bohr reported by electron diffraction[4].

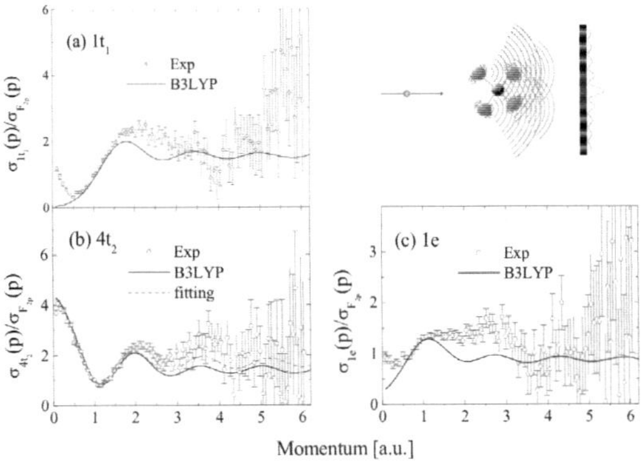

Figure 1. Experimental and theoretical interference effects. (a) $1t_1$, (b) $4t_2$, and (c) 1e molecular orbitals of CF_4.

References

[1] H. D. Cohen, U. Fano 1966 *Phys. Rev.* **150**, 30

[2] C. R. Stia *et al* 2003 *J. Phys. B* **36** L257

[3] N.Watanabe *et al* 2012 *Phys. Rev. Lett.* **108** 173201

[4] M. Fink *et al* 1979 *J. Chem. Phys.* **71** 5238

[1] E-mail: xjun@ustc.edu.cn

Content from this work may be used under the terms of the Creative Commons Attribution 3.0 licence. Any further distribution of this work must maintain attribution to the author(s) and the title of the work, journal citation and DOI.
Published under licence by IOP Publishing Ltd

Journal of Physics: Conference Series **635** (2015) 072041 doi:10.1088/1742-6596/635/7/072041

Electronic excitation to low-lying states of GeF_4 molecule by electron impact: A comparative study with CF_4 and SiF_4 molecules

S. Ohtomi*, M. Matsui*, Y. Mochizuki*, A. Suga*, H. Kato*, M. Hoshino*[1], D. Duflot†, P. Limão-Vieira‡ and H. Tanaka*

* Department of Physics, Sophia University, Tokyo 102-8554, Japan

†Laboratoire de Physique des Lasers, Atomes et Molécules (PhLAM), UMR CNRS 8523, Université de Lille, F-59655 Villeneuve d'Ascq Cedex, France

‡Laboratório de Colisões Atómicas e Moleculares, CEFITEC, Departamento de Física, FCT-Universidade Nova de Lisboa, Campus de Caparica, P-2829-516 Caparica, Portugal.

Synopsis We report on the measurements of the electron impact electronic excitation cross sections for XF_4 (X = C, Si and Ge) molecules at 100 eV, 5° scattering angle and 30 eV, 30° in the electron energy loss range 8.0 – 18 eV. For a target of GeF_4 molecule, the optically-forbidden behavior has been observed in the lower electron energy loss range.

CF_4 is one of the more commonly used etchant gases for dry-etching process, where the fluorine atoms created in CF_4/O_2 mixing plasma discharged on silicon surfaces, form SiF_4 molecule. GeF_4 molecule has been also used for the manufacturing of semiconductors. In spectroscopic studies, electron energy loss spectra (EELS) and photoabsorption cross sections of XF_4 (X = C, Si and Ge) molecules were obtained for understanding the optical diagnostics of chemical process in etching plasmas, e.g. refs [1, 2] and references therein. Recently, an angle-resolved EELS study in CF_4 has been reported [3]. Our group have also carried out a series of experimental measurements to probed the scattering dynamics of low energy electron collisions with tetrahedral (T_d) symmetry molecules such as CF_4 [4], SiF_4 [5] and GeF_4 [6]. In this work, we report on the measurements of the electron impact excitation cross sections for XF_4 molecules at 100 eV, 5° scattering angle and 30 eV, 30° by electron energy loss spectroscopy in order to observe the contribution of the optically-forbidden transitions.

The electron energy loss spectrometer used in this work has been described sufficiently in previous publication [7]. In brief, the spectrometer is a crossed-beam type with a hemispherical monochromator and analyzer pumped differentially and electron lens system controlled by computer-driven voltages. The setup was operated at impact energies from 1.5 to 300 eV and a scattering angle range from -6° to 130°, with typical energy resolution of 40 – 45 meV (FWHM). The electron energy scale was calibrated against the 19.37 eV Feshbach resonance in He atom and the angular scale were determined from the symmetry in the intensity profile of the He 2^1P excitation measured at the 0° nominal scattering angle. The absolute scale of the excitation cross sections has been obtained by the relative flow technique with He elastic differential cross sections as the reference species.

Comparison between electron energy loss spectra of XF_4 molecules at 100 eV incident energy, 5° scattering angle and at 30 eV, 30° have shown clear differences. In the case of the GeF_4 molecule, a shoulder structure is discernible and due to the optically-forbidden transition in the lower electron energy loss range than the lowest-lying optically-allowed transition, which have already assigned as $4\sigma^*$ transition.

Detailed discussion of the comparison among the electron energy loss spectra of XF_4 molecules and quantum chemical calculations, basically, EOM-CCSD method with aug-cc-pVQZ + Rydberg basis set using MOLPRO code will be presented at the conference.

References

[1] K. Kuroki *et al* 1992 *J. Chem. Phys.* **96** 6318
[2] K. Kuroki *et al* 1994 *J. Electron spectrosc. Relat. Phenom* **70** 151
[3] N. Watanabe *et al* 2011 *J. Chem. Phys.* **134** 064307
[4] L. Boesten *et al* 1992 *J. Phys. B* **25** 1607
[5] H. Kato *et al* 2012 *J. Phys. B* **45** 095204
[6] H. Kato *et al* 2012 *J. Chem. Phys.* **136** 134313
[7] H. Tanaka *et al* 1988 *J. Phys. B* **21** 1255

[1]E-mail: masami-h@sophia.ac.jp

Content from this work may be used under the terms of the Creative Commons Attribution 3.0 licence. Any further distribution of this work must maintain attribution to the author(s) and the title of the work, journal citation and DOI.
Published under licence by IOP Publishing Ltd

Cross sections for electron collisions with H_2O: elastic scattering and electronic excitation for the $\tilde{a}^3B_1$ and $\tilde{A}^1B_1$ states

M. Matsui*, H. Kato*, M. Hoshino*[1], P. Limão–Vieira[†], and H. Tanaka*

Department of Physics, Sophia University, Tokyo 102-8554, Japan.
Laboratório de Colisões Atómicas e Moleculares, CEFITEC, Departamento de Física, FCT-Universidade Nova de Lisboa, Campus de Caparica, P-2829-516 Caparica, Portugal.

Synopsis We report absolute differential cross sections (DCSs) and integral cross sections (ICSs) for the electronic excitation to the $\tilde{a}^3B_1$ and $\tilde{A}^1B_1$ states as well as the elastic scattering of H_2O molecules. The measurements are performed in the scattering angle range of 10 – 130°, and impact energies 15, 20, 30 eV for the electronic excitations and energy range 2.0 – 100 eV for the elastic scattering, respectively.

Electron impact excitation of water molecules is of particular relevance to radiation-driven chemistry in biological systems because several of the lowest-lying states are strongly dissociative, producing H atoms and OH radicals [1]. Itikawa and Mason noted in their review [2] that the angular differential cross sections (DCSs) and angular integral cross sections (ICSs) data for excitation of water by low-energy electron impact were poorly characterized. Therefore, in contrast to the elastic scattering, vibrational and rotational excitation, they were unable to recommend a cross sectional dataset for electronic excitation. Recently, the DCSs for electronic excitation to the $\tilde{a}^3B_1$ and $\tilde{A}^1B_1$ lowest-lying states, which correspond to the $1b_1$ to $4a_1$ transitions, have been reported by two different research groups [1, 3]. However, these results have shown serious discrepancies in the angular distributions of the electronic excitation cross sections of these lowest-lying states below 20 eV impact energy. In the present work, in order to fully address this discrepancy, we have performed comprehensive quantitative measurements of the electronic excitation cross sections at impact energy of 15, 20, 30 eV. In addition, we also report cross sections for elastic scattering from water molecules in the energy range of 2.0 – 100 eV.

The present measurements have been carried out using an original crossed-beam electron spectrometer described in detail many times before, see e.g. ref. [4]. Briefly, this spectrometer contains the hemispherical electrostatic monochromator and the analyzer pumped differentially, with virtual apertures, and a series of electron-lens systems to transport and focus the electron beam controlled by computer-driven voltages. The spectrometer was operated at fixed incident electron energy and over the scattered electron angular range of 15° to 130°. The energy resolution was typically 55 - 60 meV (FWHM). A transmission efficiency of the electrostatic analyzer has been decided most carefully and the absolute scale has been obtained by the standard relative flow technique using the well-known elastic DCSs of helium as the reference species. The H_2O sample was obtained from the vapour of ultra-pure water supplied by Wako Pure Chemical Industries, Ltd. The liquid sample was purified to remove dissolved gases by a repeated freeze-pump-thaw procedure. The molecular beam was produced effusively, kept at slightly elevated temperatures (~80° C) throughout the measurements.

In the present measurements, for the DCSs for electronic excitation to the $\tilde{a}^3B_1$ and $\tilde{A}^1B_1$ states, our present measurements and those of Ralphs *et al* [1] are generally consistent, but for the $\tilde{A}^1B_1$ state, we observed quantitative difference at the impact energy of 15 eV. The agreement between the present elastic DCSs and those of Nishimura and Danjo [5] is found to be good in all range of impact energies.

Further details of the results will be presented at the conference.

References

[1] K. Ralphs *et al* 2013 *J. Phys. B.* **46** 125201
[2] Y. Itikawa and N. Mason 2005 *J. Phys. Chem. Ref. Data* **34** 1
[3] P. A. Thorn *et al* 2007 *J. Phys. B.* **40**,697.
[4] H. Tanaka *et al* 1998 *Phys. Rev. A* **57** 1798
[5] H. Nishimura and A. Danjo, 1985 *J. Phys. Soc. Jpn* **54** 4

E-mail: masami-h@sophia.ac.jp

Content from this work may be used under the terms of the Creative Commons Attribution 3.0 licence. Any further distribution of this work must maintain attribution to the author(s) and the title of the work, journal citation and DOI.
Published under licence by IOP Publishing Ltd

Journal of Physics: Conference Series **635** (2015) 072043 doi:10.1088/1742-6596/635/7/072043

Electron-induced ionization of DNA components

C. Champion* [1] **and M. A. Quinto*** [2]

*Centre d'études nucléaires de Bordeaux Gradignan (CENBG), CNRS/IN2P3, Université de Bordeaux, France

Synopsis Ionization of biomolecules remains still today rarely investigated both experimentally and theoretically. In this context, the present work appears as one of the first quantum mechanical approaches providing a multi-differential description of the electron-induced ionization process of the main DNA components for impact energies ranging from the target ionization threshold up to about 10 keV. The cross section calculations are here performed within the 1st Born approximation framework in which the ejected electron is described by a Coulomb wave whereas the incident and the scattered electrons are both described by a plane wave.

Complex DNA lesions including clustered damages have been nowadays clearly identified as the most relevant precursors of elementary steps leading to critical biological lesions like chromosomal aberrations, genome mutagenesis as well as cellular death. Under these conditions and considering that the radio-induced damages in the genome of the living cells are - for about one third - due to the direct attack of the charged particles on the DNA components (the remaining being attributed to the highly reactive hydroxyl radicals produced by the absorption of the radiation in the surrounding aqueous medium), it clearly appears of prime importance to possess an accurate description of the direct ionization process of such complex biomolecules. However, although considerable progress has been made, the available studies on electron-induced ionization of biomolecules such as isolated DNA nucleobases as well as deoxyribose remain extremely rare in both the theoretical and the experimental approaches.

In this context, the current work appears as one of the first quantum mechanical approaches providing a multi-differential description of the electron-induced ionization process of the main DNA components for impact energies ranging from the target ionization threshold up to about 10 keV. The cross section calculations are here performed within the 1st Born approximation framework in which the ejected electron is described by a Coulomb wave whereas the incident and the scattered electrons are both described by a plane wave. The biological targets of interest, namely, the DNA nucleobases and the sugar-phosphate (SP) backbone, are here described by means of the Gaussian 09 system using the restricted Harthree-Fock method with geometry optimization. The theoretical predictions also obtained have shown a reasonable agreement with the experimental total ionization cross sections while huge discrepancies have been pointed out with existing theoretical models, mainly developed within a semi-classical framework [1-4].

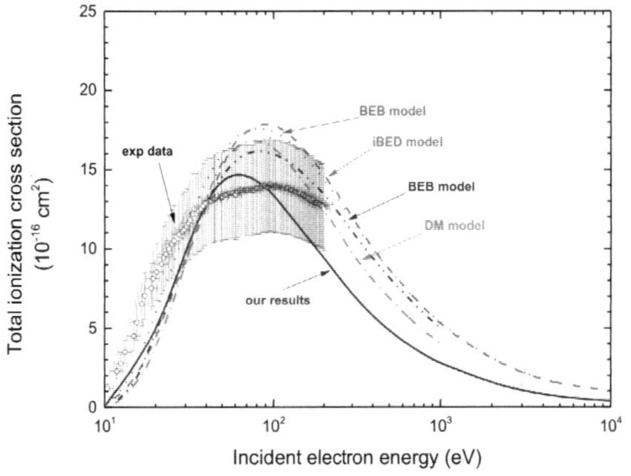

Figure 1. Electron impact total ionization cross section of thymine. Comparison between our results (blue line) [5], classical predictions taken from different sources [1-4] and experimental absolute total ionization cross sections [6].

References

[1] P. Możejko and L. Sanche 2003 *Radiat. Environ. Biophys.* **42** 201-211
[2] P. Możejko and L. Sanche 2005 *Radiat. Radiat. Phys. Chem.* **73** 77-84
[3] Y. K. Kim and M. E. Rudd 1994 *Phys. Rev. A* **50** 3954
[4] W. M. Huo, C. E. Dateo, and G. D. Fletcher 2006 *Radiat. Meas.* **41** 1202-1208
[5] C. Champion 2013 *J. Chem. Phys.* **138** 184306
[6] I. I. Shafranyosh *et al* 2008 *Technical Physics* **53(12)** 1536-1540

[1]E-mail: champion@cenbg.in2p3.fr
[2]E-mail: michele.arcangelo.quinto@gmail.com

Content from this work may be used under the terms of the Creative Commons Attribution 3.0 licence. Any further distribution of this work must maintain attribution to the author(s) and the title of the work, journal citation and DOI.
Published under licence by IOP Publishing Ltd

Journal of Physics: Conference Series **635** (2015) 072044 doi:10.1088/1742-6596/635/7/072044

Non-sequential and sequential fragmentation dynamics of OCS^{3+} in collision with 500 eV electron

ZhenJie Shen*†, MaoMao Gong*†, EnLiang Wang*†, Xu Shan*†, XiangJun Chen*†[1]

* Heifei National Laboratory for Physical Sciences at Microscale and Department of Modern Physics, University of Science and Technology of China, Hefei, Anhui, 230026, China

† Synergetic Innovation Center of Quantum Information and Quantum Physics, University of Science and Technology of China, Hefei, Anhui, 230026, China

Synopsis Non-sequential and sequential fragmentation dynamics of OCS^{3+} investigated by electron collision at impact energy of 500 eV is reported in this abstract. By using Dalitz plot together with momentum correlation spectrum various dissociative channels are distinguished.

The studies on fragmentation dynamics of simple molecules have been attracting continuous interests with the help of rapid developing momentum imaging techniques [1, 2]. In this work, the three-body fragmentation of triply ionized OCS molecule is reported. The triple ionized OCS molecular ion is created by electron collision,

$$e^- + OCS \to OCS^{3+} + 4e^-, \qquad (1)$$

where, the molecular ion, OCS^{3+}, will dissociate into three ions, O^+, C^+ and S^+ for which are detected in coincidence in experiment [3].

The fragmentation dynamics of OCS^{3+} is analyzed using Dalitz plot method, as shown in Figure 1 (a). The most intense areas represent direct and molecular bending fragmentation, where the two molecular bonds, O-C and C-S, break simultaneously in the fragmentation process. Besides the above two mechanisms, it is clear that there is "X" structure in the Dalitz plot indicating that there is some probability for OCS^{3+} dissociate through sequential fragmentation mechanism, for which the two molecular bond dissociate stepwisely. In this work, two different sequential fragmentation channels are distinguished,

$$OCS^{3+} \to O^+ + CS^{2+} \to O^+ + C^+ + S^+, \qquad (2)$$

$$OCS^{3+} \to OC^{2+} + S^+ \to O^+ + C^+ + S^+, \qquad (3)$$

where channel (2)/(3) corresponding to the right/left part of the "X" structure in the Dalitz plot. To separate the overlap of the sequential and molecular bending fragmentation, the momentum correlation of O^+ and S^+ are further analyzed. As shown in Figure 1 (b), the structures parallel to x or y axis originate from sequential fragmentation.

The momentum correlation of sequential fragmentation processes of channel (2) and (3) are shown in the Newton diagrams in Figure 1 (c) and (d), respectively. Both of the two channels exhibit similar circular structures in the Newton plot. Some differences, however, have also been clearly presented indicating that there maybe different mechanisms controlling the two sequential fragmentation channels.

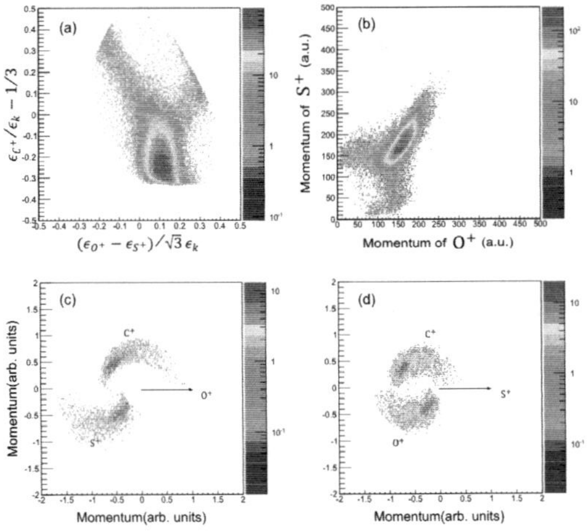

Figure 1. (a) Dalitz plot of OCS^{3+} dissociation. (b) Momentum correlation spectrum of O^+ and S^+. (c) Newton diagram of sequential fragmentation channel where the C-O bond breaks first. (d) Newton diagram of sequential fragmentation channel where the C-S bond breaks first.

References

[1] R. Dörner *at al* 2000 *Phys. Rep.* **330** 95

[2] J. Ullrich *at al* 2003 *Rep. Prog. Phys.* **66** 1463

[3] E. L. Wang *at al* 2013 *Rev. Sci. Instrum.* **84** 123110

[1] E-mail: xjun@ustc.edu.cn

Content from this work may be used under the terms of the Creative Commons Attribution 3.0 licence. Any further distribution of this work must maintain attribution to the author(s) and the title of the work, journal citation and DOI.
Published under licence by IOP Publishing Ltd

Electron-induced chemistry of cobalt tricarbonyl nitrosyl ($Co(CO)_3NO$) in liquid helium nanodroplets

Johannes Postler*[1], Michael Renzler*[2], Andrew M. Ellis†[3], Paul Scheier*[4],

* Institut für Ionenphysik und Angewandte Physik, Universität Innsbruck, Technikerstr. 25 / 3, A-6020 Innsbruck

† Department of Chemistry, University of Leicester, University Road, Leicester, LE1 7RH, UK

Synopsis Cobalt tricarbonyl nitrosyl ($Co(CO)_3NO$ - CTCN) is a widely used source for cobalt in focused electron-beam induced deposition (FEBID) processes. By doping helium nanodroplets at a temperature of 0.37 K with several $Co(CO)_3NO$ molecules we managed to create a multitude of molecular clusters and fragments. These clusters provide a similar environment to the materials used in FEBID. Dissociative electron attachment processes are hence studied in order to provide insight into factors limiting the achievable resolution in FEBID.

Focused electron-beam induced deposition (FEBID) is a promising technique to deposit metallic structures on substrates both in two and three dimensions. It consists of a jet of gaseous precursor molecules containing a metal compound and a focused electron beam. The electron beam is mostly produced by a scanning electron microscope and can be directed with sub-nanometer precision, hence allowing the printing of metallic films. However, due to the effects caused by secondary electrons, the actual smallest printable dot is two orders of magnitude larger than the precision of the electron beam position [1, 2]. Therefore, it is necessary to study effects of secondary electrons on the precursor molecules and investigate their potential dissociation channels upon DEA processes.

Here we focus on the potential precursor molecule cobalt tricarbonyl nitrosyl ($Co(CO)_3NO$). While a gas phase study is already available [3], a more realistic model for the actual FEBID environment might be given by clusters of the form $(Co(CO)_3NO)_n$ due to the involvement of thin layers of the precursor molecule in the actual process. We created these clusters by means of a helium nanodroplet source [4] and doping the pristine droplets with $Co(CO)_3NO$. Beside cluster formation, two more advantages over a pure gas phase study arise. One is the low temperature of 0.37 K, which quenches reaction channels possibly allowing the investigation of intermediate products. The other is the opportunity to add C_{60} as a model substrate to study the effects of a surface on the chemical processes involved.

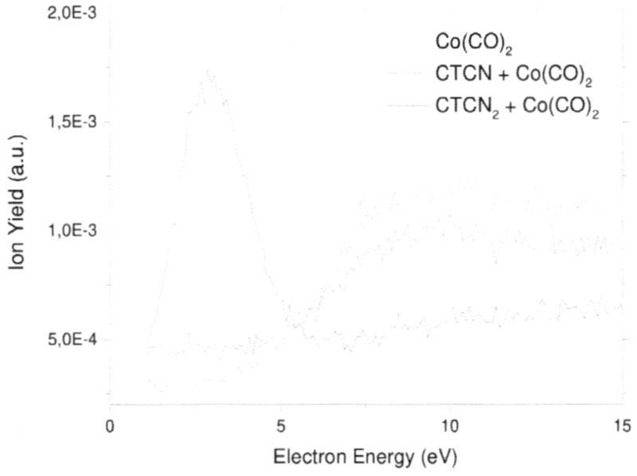

Figure 1. Anion yield profile for $Co(CO)_2$ in gas phase and attached to the $Co(CO)_3NO$ parent (CTCN).

We present anion yield profiles both for the gas phase and several clusters respectively their fragments. The cluster data show major differences when compared to the gas phase anion yields, as can be seen in figure 1. Preliminary data for $Co(CO)_3NO$ on a C_{60} surface will also be shown.

References

[1] R. Erni *et al* 2009 *Phys. Rev. Lett.* **102** 096101

[2] W. F. van Dorp *et al* 2005 *Nano Lett.* **5** 1303

[3] S. Engmann *et al* 2011 *Angew. Chem. Int. Ed.* **50** 9475

[4] L. An der Lan *et al.* 2011 *J. Chem. Phys.* **135** 044309

[1]E-mail: johannes.postler@uibk.ac.at
[2]E-mail: michael.renzler@uibk.ac.at
[3]E-mail: andrew.ellis@le.ac.uk
[4]E-mail: paul.scheier@uibk.ac.at

Content from this work may be used under the terms of the Creative Commons Attribution 3.0 licence. Any further distribution of this work must maintain attribution to the author(s) and the title of the work, journal citation and DOI.
Published under licence by IOP Publishing Ltd

XXIX International Conference on Photonic, Electronic, and Atomic Collisions (ICPEAC2015) IOP Publishing
Journal of Physics: Conference Series **635** (2015) 072046 doi:10.1088/1742-6596/635/7/072046

Positron and electron impact ionization of H_2O for coplanar and non-coplanar geometries

I. Tóth [1] **and L. Nagy** [2]

Faculty of Phyics, Babeş-Bolyai University, 400084 Cluj Napoca, Romania

Synopsis Calculated triple differential cross sections for the ionization of water by positron and electron impact are compared. Results are showed for coplanar and non-coplanar geometrical arrangements.

Ionization of water by different charged particles has attracted much interest in the last years [1–8]. Particularly interesting is to calculate the fully differential cross section, which provides the most details of this process.

Previously we have investigated the ionization of water by electron and positron impact in the scattering plane (defined by the momenta of the incident and scattered particles) for intermediate energies of the incident electron [7]. Recently we have extended our study on the ionization by electron impact to both coplanar and non-coplanar geometrical arrangements [8] and for low ejected electron energies. The obtained results showed good agreement with the experimental data, especially in the perpendicular plane (the plane perpendicular to the incident direction).

In our present work TDCSs (triple differential cross sections) are calculated for the ionization of water by positron impact and compared to their electronic counterpart in order to assess the effect of the projectile's charge on the cross section. The TDCSs are determined for both coplanar and non-coplanar geometries and for symmetric and asymmetric energy sharing of the outgoing particles.

At low energies the PCI (post collision interaction) effects play a significant role, therefore it is important to take them into account. These effects are incorporated into our models through the Gamow factor and the Coulomb distortion factor presented in [9], modified for positron projectiles too.

The calculations are performed within the distorted wave formalism, where the incident, scattered and ejected particles all are described by distorted waves. Multi-center Gaussian wavefunctions are employed in order to describe the initial state of the target. These functions are constructed as linear combinations of atomic-like Gaussian functions, which are contractions of Gaussian primitives. The distorted waves of the free particles are determined by employing two models, the TS (total screening) and TS*. In the first model the distorted wave associated with the faster outgoing particle is calculated in the spherically averaged potential of the target, while for the slower one the obtained distorted wave is determined in the averaged field of the residual ion. The TS* model describes both outgoing particles as moving in the spherically averaged potential field of the residual ion.

Unfortunately, no experimental data are available for positron impact ionization of water for the employed kinematical conditions. Nevertheless, the calculated electron cross sections are compared to the available experimental data and other theoretical TDCSs [1].

This work was supported by a grant of the Romanian National Authority for Scientific Research, CNCS UEFISCDI, project number PNII-ID-PCE-2011-3-0192

References

[1] K. L. Nixon *et al* 2010 *J. Phys. B: At. Mol. Opt. Phys.* **43** 035201

[2] L. Fernandez-Menchero *et al* 2014 *J. Phys. B: At. Mol. Opt. Phys.* **47** 035205

[3] D. H. Madison *et al* 2010 *J. At. Mol. Opt. Phys* **2010** 367180

[4] C. Y. Lin *et al* 2014 *Phys. Rev. A* **89** 012703

[5] M. Sahlaoui *et al* 2013 *J. Phys. B: At. Mol. Opt. Phys.* **46** 115206

[6] C. Champion *et al* 2006 *Phys. Rev. A* **73** 012717

[7] I. Tóth *et al* 2012 *Eur. Phys. J. D* **66** 21

[8] I. Tóth *et al* 2014 *Eur. Phys. J. D* **68** 369

[9] S. J. Ward *et al* 1994 *Phys. Rev. A* **49** 1049

[1]E-mail: istvan.toth@phys.ubbcluj.ro
[2]E-mail: lnagy@phys.ubbcluj.ro

Content from this work may be used under the terms of the Creative Commons Attribution 3.0 licence. Any further distribution of this work must maintain attribution to the author(s) and the title of the work, journal citation and DOI.
Published under licence by IOP Publishing Ltd

Journal of Physics: Conference Series **635** (2015) 072047 doi:10.1088/1742-6596/635/7/072047

Calculations of electron and positron scattering from vibrationally excited H^{2+} and H_2

Mark C. Zammit [1], Jeremy S. Savage, Dmitry V. Fursa, and Igor Bray

Department of Physics, Astronomy and Medical Radiation Sciences, Curtin University, GPO Box U1987, Perth, WA 6845, Australia

Synopsis Electron and positron scattering from the vibrationally excited H_2^+ and H_2 molecules were investigated using the adiabatic-nuclei convergent close-coupling method. Converged results are presented for a range of vibrationally excited states.

Collision data of electron and positron scattering from vibrationally excited molecules is particularly important in the modelling of fusion plasmas and in the calculation of gas and soft-condensed (biological) matter transport properties. Recent scattering studies indicate a large dependence on the initial vibrational state of the molecule [1, 2]. Utilising the *ab initio* adiabatic-nuclei convergent close-coupling (CCC) method, formulated in both the spherical [1] and spheroidal coordinate systems, we investigate electron scattering from vibrationally excited H_2^+ and positron scattering from vibrationally excited H_2. Converged results in both the multichannel and partial-wave expansions are presented over a broad energy range for elastic, vibrational excitation, electronic excitation, grand total and total ionisation cross sections.

In Fig. 1 we present dissociative excitation (DE) cross sections for electron scattering from H_2^+ in the initial vibrational state v_i. The DE cross sections have a major dependence on v_i. For example, at 20 eV the $v_i = 9$ DE cross section is 500% larger than the $v_i = 0$ state. As v_i increases the DE cross sections monotonically increase and the cross section peak shifts to lower energies.

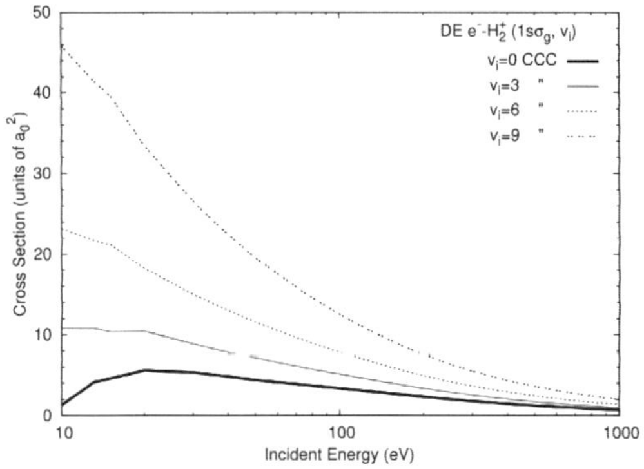

Figure 1. Dissociative excitation (DE) cross sections for electron scattering from H_2^+ in the vibrational state v_i.

The $0 \rightarrow 1$ vibrational excitation cross section for positron scattering from H_2 is presented in Fig. 2. Results are calculated directly from the adiabatic-nuclei approximation and are in excellent agreement with the measurements of Sullivan *et al.* [3].

Our preliminary results of positron scattering from H_2 in the initial vibrational state $v_i = 1$ indicate a major dependence on v_i in the low-energy region. At 3 eV the $v_i = 1$ grand total cross section is approximately 25% larger than the $v_i = 0$ cross section.

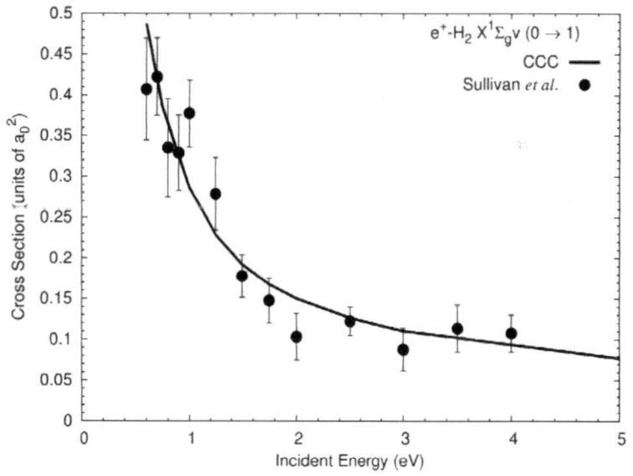

Figure 2. Vibrational $0 \rightarrow 1$ excitation cross section for positron scattering from H_2. Adiabatic-nuclei convergent close-coupling (CCC) results are compared with the measurements of Sullivan *et al.* [3].

References

[1] M. C. Zammit *et al.* 2014 *Phys. Rev. A.* **90** 022711

[2] M. C. Zammit *et al.* 2013 *Phys. Rev. A.* **88** 062709

[3] J. P. Sullivan *et al.* 2002 *Nucl. Instr. Meth. B.* **192** 3

[1] E-mail: mark.zammit@postgrad.curtin.edu.au

Content from this work may be used under the terms of the Creative Commons Attribution 3.0 licence. Any further distribution of this work must maintain attribution to the author(s) and the title of the work, journal citation and DOI.
Published under licence by IOP Publishing Ltd

Charge dependent adsorption of carbon dioxide on fullerenes

Stefan Ralser*[1], Alexander Kaiser*, Michael Probst*, Johannes Postler*, Michael Renzler*, Diethard K. Bohme†, Paul Scheier*[2]

* Institute for Ionphysics und Applied Physics, LFU Innsbruck, Technikerstraße 25, A-6020 Innsbruck
† Department of Chemistry, York University, Toronto, ON, Canada M3J 1P3

Synopsis A positive surface charge on a buckminsterfullerene leads to an enhanced first shell adsorption capacity of carbon dioxide in comparison to neutral or negatively charged species. The surface acts like a sponge that can adsorb and release carbon dioxide in a controllable manner.

Charge-dependent adsorption of CO_2 on cold (0.37 K) buckminsterfullerens (C_{60}) was investigated. It has been shown that the packing density in the first solvation shell around the fullerene is increased by a factor of ≈ 25% for positively charged C_{60} in comparison to negatively charged species. This was confirmed by both, mass spectrometric experiments, as well as theoretical investigations. We conclude, that the enhanced surface adsorption capacity origins from a reorientation of the CO_2s on the surface. A neutral or negatively charged C_{60} favors a flat orientation, whereas an orientation of the CO_2 axis perpendicular to the surface is preferred when a positive charge is present. This behavior can be explained by the electrostatic interaction between the surface charge and the negatively charged terminal O atoms of the CO_2. The "stand up" of the CO_2s decreases the intermolecular distances and therefore more molecules fit into the first shell. The experiment performed allows for a charge-selective, controllable adsorption capacity of CO_2 on buckminsterfullerens, however, the described mechanism should apply to other surfaces as well.

Experiment: The experiment has been described in detail elsewhere [1, 2]. In brief, superfluid helium nano droplets with a temperature of 0.37 K and an average size of 5×10^5 atoms per droplet are formed by supersonic expansion of helium into vacuum. After formation, the helium droplets are doped with C_{60} and CO_2, ionized and analyzed with the help of a time-of-flight mass spectrometer.

Figure 1 shows the abundances of the measured clusters. The structures between $40 < n < 50$ in the positive case and $30 < n < 40$ in the negative case are related to the closure of the first solvation shell. Second shell formation starts at a cluster size of $n = 49$ for cationic and $n = 39$ for anionic species, respectively. A comparison to our MD simulations indicates that a closed shell is possible over a certain range of n which shows in the experiment as a broad anomaly in the cluster size distribution rather than a sharp peak.

The abundances in figure 1 were obtained from the raw data by a fitting process with the help of our evaluation software IsotopeFit [3]

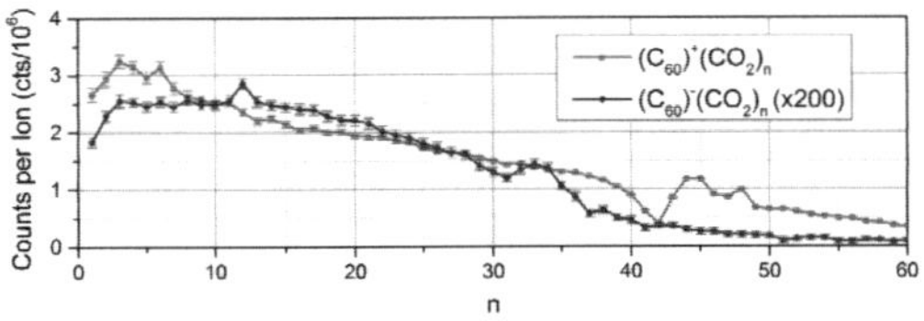

Figure 1. Ion stability measurements for $C_{60}^{+}(CO_2)_n$ and $C_{60}^{-}(CO_2)_n$ as a function of the number of adsorbed CO_2.

Calculations: MD Simulations were performed to allow insight in the dynamics of the system. We used force field parameters from references [4, 5] with a distributed charge of ± 1e on the fullerene. The simulations show a shell closure at $n = 33$ for anionic and $n = 49$ for cationic species, respectively. For low coverage numbers, the CO_2s form a "bald spot" on the fullerene whose dissipation with increasing coverage numbers is the origin for the broad stability anomalies measured in the experiment.

Support: Work supported by the FWF project P26635

References

[1] L. An der Lan *at al* 2011 *J. Chem. Phys.* **135** 044309

[2] H. Schöbel *at al* 2011 *J. Chem. Phys.* **63** 209-214

[3] S. Ralser *at al* 2015 *IJMS.* in press

[4] A. Martínez-Alonso *at al* 2000 *J. Phys. Chem.* **105** (1)135-139

[5] C.S. Murthy *at al* 1983 *Mol. Phys* **50** (3)531-541

[1]E-mail: stefan.ralser@uibk.ac.at
[2]E-mail: paul.scheier@uibk.ac.at

Content from this work may be used under the terms of the Creative Commons Attribution 3.0 licence. Any further distribution of this work must maintain attribution to the author(s) and the title of the work, journal citation and DOI.
Published under licence by IOP Publishing Ltd

XXIX International Conference on Photonic, Electronic, and Atomic Collisions (ICPEAC2015) IOP Publishing
Journal of Physics: Conference Series **635** (2015) 072051 doi:10.1088/1742-6596/635/7/072051

Polyatomic molecular structure retrieval using laser-induced electron diffraction

M. G. Pullen[1,*], B. Wolter[1], A.-T. Le[2], M. Baudisch[1], M. Hemmer[1], A. Senftleben[3], M. Sclafani[1], C. D. Schröter[4], J. Ullrich[4,5], R. Moshammer[4], C.-D. Lin[2], J. Biegert[1,6]

[1]ICFO-Institut de Ciencies Fotoniques, Mediterranean Technology Park, 08860 Castelldefels (Barcelona), Spain
[2]J. R. Macdonald Laboratory, Physics Department, Kansas State University, Manhattan, Kansas 66506-2604, USA
[3]Universität Kassel, Institut für Physik und CINSaT, Heinrich-Plett-Str. 40, 34132 Kassel, Germany
[4]Max-Planck-Institut für Kernphysik, Saupfercheckweg 1, 69117 Heidelberg, Germany
[5]Physikalisch-Technische Bundesanstalt (PTB), Bundesallee 100, 38116 Braunschweig, Germany
[6]ICREA-Institució Catalana de Recerca i Estudis Avançats, 08010 Barcelona, Spain

Synopsis Laser-induced electron diffraction is a developing dynamical imaging technique that is already able to probe molecular dynamics at few-femtosecond temporal resolutions and has the potential to reach the sub-femtosecond level. Here we provide the recipe for the extension of the technique to polyatomic molecules and we demonstrate the method by extracting the structure of aligned and anti-aligned acetylene (C_2H_2). We show that multiple bond lengths can be simultaneously imaged at high accuracy including elusive hydrogen containing bonds. Our results open the door to the investigation of larger complex molecules and the realization of a true molecular movie.

Conventional electron diffraction (CED) is able to determine internuclear distances at unsurpassed spatial accuracies, however, pushing the achievable temporal resolutions below a few-hundred femtoseconds has proved difficult. On the other hand, few-femtosecond structural dynamics of the O_2 molecule have already been reported using laser-induced electron diffraction (LIED) [1]. However, the real goal of LIED research is to probe polyatomic molecules that have many possible fragmentation channels. Disentangling information from these concurrent processes requires many capabilities that standard experimental apparatuses do not possess. Additionally, the low ionization potential of polyatomic molecules places strict limitations on the driving radiation. Therefore, up until now, LIED has been limited to diatomic molecules. Here, we provide the recipe to overcome these non-trivial hurdles and present the first simultaneous extraction of multiple bond lengths in aligned and anti-aligned acetylene molecules (C_2H_2). Moreover, we show how the electron recollisions induced by strong-field ionization can potentially be used to resolve sub-femtosecond dynamics.

Our experimental apparatus is composed of a combination of an intense 160 kHz, 3.1 μm source with a reaction microscope (ReMi) detection system. The mid-IR radiation generates high kinetic energy core penetrating re-scattering electrons and the ReMi is able to detected both electron and ion momentum distributions in full coincidence. Molecular structure information is hidden in the electron momentum distribution (Fig. 1) and can be retrieved following the quantitative re-scattering (QRS) theory. The obtained bond lengths agree well with the expected values for both alignments.

To our knowledge, our results show the first simultaneous extraction of multiple bond lengths from a polyatomic molecule using LIED [2]. Our unique experimental apparatus opens up the possibility of applying LIED to much larger and more complex molecules. Additionally, information about fragmentation processes is already available due to the coincidence capabilities of the ReMi detection system. Combining our methodology with the already presented few-femtosecond capabilities of LIED truly represents a realistic path to achieving the long sought after 'molecular movie'.

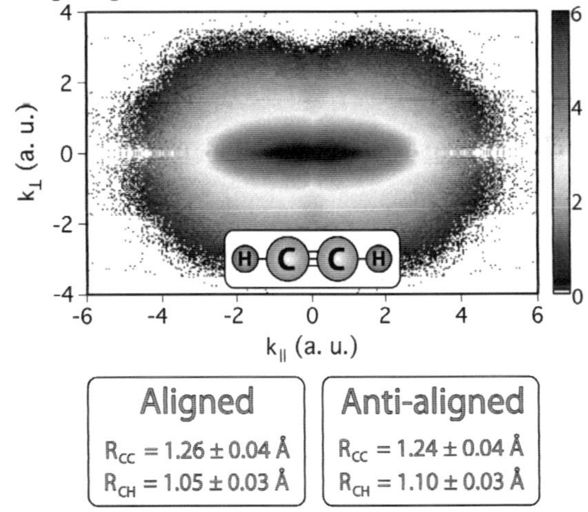

Figure 1. Electron momentum distribution (log scale) detected by the ReMi after the ionization of aligned C_2H_2 molecules by the high repetition rate 3.1 um source. Accurate C-C and C-H bond lengths can be simultaneously extracted from this diffraction pattern for both aligned and anti aligned molecules.

References

[1] C. Blaga *et al* 2012 *Nature* **483**, 194
[2] M. Pullen *et al* 2015 arxiv:1503.03294

* E-mail: michael.pullen@icfo.eu

Content from this work may be used under the terms of the Creative Commons Attribution 3.0 licence. Any further distribution of this work must maintain attribution to the author(s) and the title of the work, journal citation and DOI.
Published under licence by IOP Publishing Ltd

Electron attachment to molecular clusters

Michał Ryszka, Agnieszka Sieradzka[1], Rahul Pandey, Marcin Dampc, Zdeněk Mašín, Samuel Eden and Jimena D. Gorfinkiel[2]

Department of Physical Sciences, The Open University, Walton Hall, Milton Keynes, MK7 6AA, United Kingdom

Synopsis We study experimentally and theoretically the effect of hydrogen bonding in the formation of molecular temporary negative ions. A new experimental set up enables us, for the first time, to perform electron scattering experiments with selected clusters. This selectivity facilitates the comparison with theoretical results.

Electron attachment processes play a significant role in the radiation-induced damage of biological material. Experimental and theoretical work has, over the last 15 years, mainly focused on the study of electron attachment to gas phase (isolated) molecules. We aim to investigate the effects of hydrogen bonding and hydration in this process and how it affects molecular break-up.

A new experiment has been constructed at The Open University to irradiate biomolecular isomers and clusters with electrons from a high-resolution double-pass monochromator [1]. The resultant anions are analyzed by time-of-flight mass spectrometry. The experiment incorporates a Stark deflector developed by Küpper and coworkers [2] to deflect polar species in strong electric field gradients in order to enhance control over the target configurations. To date, comparisons with theory have been limited by the spread of biomolecular isomers and clusters in crossed-beam experiments.

As a complement to the Stark deflection approach, we are developing an additional experimental system to control neutral beams for electron attachment experiments. Stark deflection can provide target configuration specificity for relatively small molecules and complexes but is less practical for larger systems due to decreasing dipole moment/mass ratios and difficulty in producing rotationally cold beams. The second method will select only on the basis of mass but can be applied to large clusters. It involves the neutralization of mass-selected cluster anions by electron photo-detachment from weakly bound anionic states, with minimal change in stability and hence dissociation [3].

Concurrently, we are performing *ab initio* calculations to characterize the temporary negative ions (TNIs) of small molecular clusters. Recent software developments have enabled us to study electron-cluster scattering using the R-matrix method [4] as implemented in the UKRmol suite. In particular, by comparing these calculations with those for isolated molecules, we are able to investigate how the resonance energy and width are affected by clusterization. Linking TNIs to the anion yields detected experimentally will allow us to: (i) validate both experiments and calculations; (ii) shed light on the mechanism of anion formation.

References

[1] EPSRC project EP/J002577/1

[2] S.Trippel, Y. Chang, S. Stern, T. Mullins, L. Holmegaard and J. Kuepper, Spatial separation of state- and size-selected neutral clusters. *Phys. Rev. A*, 2012, 86, 033202.

[3] EPSRC project EP/L002191/1

[4] P. Burke, R-matrix theory of Atomic Collisions. *Springer Series on Atomic, Optical, and Plasma Physics*, 2011.

[1]E-mail: agnieszka.sieradzka@open.ac.uk
[2]E-mail: jimena.gorfinkiel@open.ac.uk

Content from this work may be used under the terms of the Creative Commons Attribution 3.0 licence. Any further distribution of this work must maintain attribution to the author(s) and the title of the work, journal citation and DOI.
Published under licence by IOP Publishing Ltd

Journal of Physics: Conference Series **635** (2015) 072053 doi:10.1088/1742-6596/635/7/072053

Energy distribution of slow fragments, OH^+ and O^+, in water dissociation by electron impact

Natalia Ferreira*[1], L Sigaud ** and E C Montenegro†[2]

*CEFET/RJ – Unidade Maracanã, 20271-110, Rio de Janeiro, RJ, Brazil

**Instituto de Física, Universidade Federal Fluminense, 24210-346, Niterói, RJ, Brazil

† Instituto de Física, Universidade Federal do Rio de Janeiro, 21941-972, Rio de Janeiro, RJ, Brazil

Synopsis: In this work, we present measurements of the energy distributions of OH^+ and O^+ produced in water molecule fragmentation by electron impact. We scanned the electron beam energy between 28 eV and 800 eV, examining the processes that occur from low energy transfer, below the double ionization threshold, up to processes occurring at the Oxygen K edge. The results show very different behaviors for the OH^+ and O^+ energy distributions for different electron energies.

Electron impact fragmentation of water has been widely studied for over 50 years [1]. The subject has great interest because it is a key process in many areas such as biology and medicine, due to the consequences for the human body submitted to ionizing radiation, or astrophysics, in connection to the study of planetary atmospheres irradiated by electrons from solar wind or secondary electrons from cosmic radiation.

The study of the kinetic energy distribution of ionic fragments of water has been performed by several methods, most of them measuring the more energetic fragments that dissociate via Coulomb explosion after multiple ionization [2, 3]. In the literature, the data available for the energy distribution of the slowest fragments, coming mainly from simple ionization, are very scarce.

The recently developed DETOF methodology (Delayed Extraction Time-Of-Flight Spectroscopy)[4] allows to obtain not only the kinetic energy distribution of the slow fragments produced by the interaction of electrons with water, for which this technique has higher sensitivity, but also the cross section of ions produced with a given average kinetic energy. We present here results for the kinetic energy distributions of the OH^+ and O^+ fragments, with the incident electron energy scanned between 28 eV and 800 eV.

For the OH^+ fragment, two distinct energy distributions could be assigned, one with an exponential shape, with average kinetic energy of 0.04 eV, and a Gaussian distribution, with average kinetic energy of 0.2 eV. Their relative contributions change with the electron energy. The exponential energy distribution is associated to OH^+ coming from dissociation of the single ionized molecule. The Gaussian distribution centered at 0.2 eV can be related with OH^+ coming from the double ionized molecule.

The cross section for the Gaussian distribution is very similar to the $OH^+ + H^+$ channel cross section [2] for high electron energies. For low impact energy, however, the cross section associated to the Gaussian distribution is higher than the double ionization one. This indicates that, for low electron energies, there is another mechanism generating fast OH^+ ions. This process begins to take place around 40 eV, which is near the threshold from the water molecule orbital 2a1, where an Auger emission is possible. This mechanism usually leads to a double ionized molecule, but if the electron energy is close to the ionization threshold, the molecule can be excited and it auto-ionizes, ejecting swift OH^+ fragments. After 70 eV, the Gaussian distribution cross section becomes very close to the double ionization ones. This behavior remains at least until 800 eV.

For the O^+ fragment, we identify three distinct kinetic energy distributions, with average energies of 0.03 eV, 0.2 eV and 1.2 eV. The slowest ions are the dominant group at low incident electron energy. As the electron energy rises and larger energy transfers are allowed, this group becomes less important. A detailed analysis of the origin and of the dynamic behavior of these three distributions will be presented at the conference.

References

[1] .W. McConkey *et al.* 2008 *Phys. Rep.* **466** 1
[2] Montenegro *et al.* 2007 *J. Electron. Spectrosc. Relat. Phenom.* **155** 81
[3] S. J. King *et al.* 2008 *Int. J. Mas. Spectrom.*, **277** 84
[4] Natalia Ferreira *et al.* 2014 *J. Phys.: Conference Series*, **488** 012042

[1] E-mail: nataferr@gmail.com

[2] E-mail: montenegro@if.ufrj.br

Content from this work may be used under the terms of the Creative Commons Attribution 3.0 licence. Any further distribution of this work must maintain attribution to the author(s) and the title of the work, journal citation and DOI.
Published under licence by IOP Publishing Ltd

Journal of Physics: Conference Series **635** (2015) 072054 doi:10.1088/1742-6596/635/7/072054

Elastic scattering of low-energy electrons with tetrahydropyran.

Alessandra Souza Barbosa [1], **Márcio H. F. Bettega** [2]

Departamento de Física, Universidade Federal do Paraná, Curitiba, Brazil

Synopsis In this work we present calculated cross sections for low-energy electron scattering by tetrahydropyran. We compare our cross sections with previous results for the close-related molecules cyclohexane [1] and 1,4-dioxane [2]. We also compare our calculated integral cross sections with experimental total cross sections available in the literature for this molecule [3].

In this work we report calculated cross sections for elastic scattering of low-energy electrons with tetrahydropyran ($C_5H_{10}O$) molecule. We employed the well-stablished Schwinger multichannel method (SMC) [4] implemented with pseudopotentials (SMCPP) [5] in the static-exchange (SE) and static-exchange-polarization (SEP) approximations for energies up to 30 eV. We compare our calculated cross sections with previous results obtained for cyclohexane [1] and 1,4-dioxane [2] molecules, since their geometrical structure are similar. In figure 1 we present the geometrical structures of the molecules. We also compare our calculated integral cross sections with experimental total cross sections available in the literature for this molecule [3].

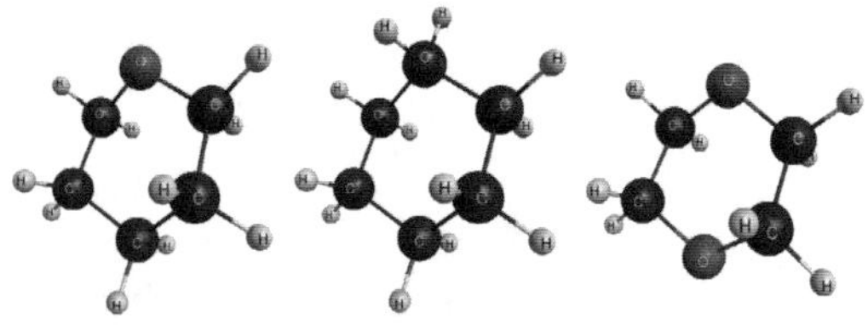

Figure 1. Geometrical structures of tetrahydropyran, cyclohexane and 1,4-dioxane molecules.

Recently we showed that calculated cross sections for elastic electron collisions with cyclohexane and 1,4-dioxane present some similarities among them [1]. These similarities were attributed to the geometrical structure of the molecules. In order to continue this study, we now compare tetrahydropyran cross sections with those obtained for the other close-related molecules. Tetrahydropyran and 1,4-dioxane can be seen as cyclohexane derivatives, with 1 ou 2 oxygen atoms replacing the CH_2 groups. Although the three molecules present similar geometrical structure in the ground state, one should note that while 1,4-dioxane and cyclohexane do not have permanent dipole moment, tetrahydropyran has.

In figure 2 we present a comparision between calculated differential cross sections (DCS) for elastic scattering of electrons by tatrahydropyran, cyclohexane [1] and 1,4-dioxane [2]. The results are presented in the SE approximation at energies between 15 and 30 eV. One can note that the DCSs are, in general, very similar in the oscilatory behavior and magnitude, presenting big discrepancies at lower angles where the dipole moment of the tetrahydropyran molecule increases the cross section abruptly.

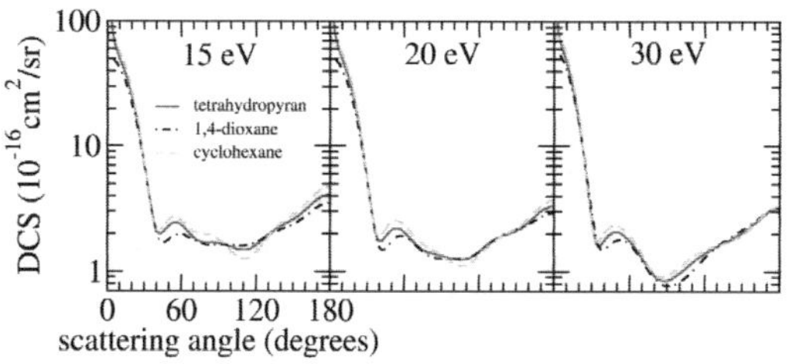

Figure 2. Calculated DCSs, in the SE approximation, for elastic scattering of electrons by tetrahydropyran, cyclohexane [1] and 1,4-dioxane [2] at energies 15 eV, 20 eV and 30 eV.

Acknowledgments: The authors acknowledge financial support from CNPq, CAPES and Finep (CT-Infra) and computational support Prof. Carlos de Carvalho at LFTC-DFis-UFPR and LCPAD-UFPR and from CENAPAD-SP.

References

[1] A. S. Barbosa and M. H. F. Bettega 2014 *J. Chem. Phys.* **141**, 244307.

[2] A. S. Barbosa and M. H. F. Bettega 2014 *J. Chem. Phys.* **140**, 184303.

[3] C. Szmytkowski, E. Ptasinska-Denga 2011 *J. Phys. B: At. Mol. Opt. Phys.* **44**, 015203.

[4] K. Takatsuka and V. McKoy 1981 *Phys. Rev. A* **24**, 2473; 1984 *Phys. Rev. A* **30**, 1734.

[5] M. H. F. Bettega, L. G. Ferreira, and M. A. P. Lima 1993 *Phys. Rev. A* **47**, 1111.

[1] E-mail: asb07@fisica.ufpr.br
[2] E-mail: bettega@fisica.ufpr.br

Content from this work may be used under the terms of the Creative Commons Attribution 3.0 licence. Any further distribution of this work must maintain attribution to the author(s) and the title of the work, journal citation and DOI.
Published under licence by IOP Publishing Ltd

Journal of Physics: Conference Series **635** (2015) 072055 doi:10.1088/1742-6596/635/7/072055

DEA to bare and water-solvated pyrimidine clusters

Michael Neustetter[*1], **Julia Aysina**[*2], **Filipe Ferreira da Silva**[†3], **Stephan Denifl**[*4],

[*] Institute for Ion and Applied Physics, University of Innsbruck, Technikerstr. 25, 6020 Innsbruck, Austria
[†] Laboratório de Colisões Atómicas e Moleculares, CEFITEC, Departamento de Física, Faculdade de Ciências e Tecnologia, Universidade Nova de Lisboa, Campus de Caparica, 2829-516 Caparica, Portugal

Synopsis In the present work we investigate the interaction of low energetic electrons with biologically relevant molecules. Our aim is to observe differences in the fragmentation pattern in the case of isolated molecules, bare clusters and water-solvated clusters. In this context we present the first results regarding electron attachment to nano-solvated biomolecule clusters.

For the understanding of radiation damage it is highly important to understand how the fragmentation process of biomolecules is modified by surrounding environment. For example, it was shown that dissociative electron attachment (DEA) to nucleobases embedded in helium droplets is strongly modified in terms of the molecular fragmentation pattern compared to the isolated molecule [1]. The studies with doped helium droplets represent a very first approach to investigate the effect of environment. Living cells consist of up to 70% water. Thus, to account for more realistic conditions in comparison to isolated biomolecules, we are using clusters of biomolecules solvated with several water molecules. Therefore, we built up a cluster source producing the mixed biomolecule/water clusters.

For the present study we chose pyrimidine which is a derivative of the nucleobases cytosine, thymine and uracil. Hence, pyrimidine ($C_4H_4N_2$) is often used as a model system for collision studies related to radiation damage. Compared to the mentioned nucleobases, pyrimidine has a much higher vapour pressure and thus it is experimentally easier to create the clusters of interest.

These clusters are produced via supersonic expansion into vacuum. To enhance the cluster efficiency we use argon as seeding gas. For the production of solvated pyrimidine clusters water is added to the expansion. The filament of a Nier type ion source provides the free electrons with a tunable kinetic energy in the range of interest. We are using a double focusing two sector field mass spectrometer to identify the created ions.

In this context we present the first results of electron attachment to pyrimidine/water clusters. The negative ion mass spectrum at the electron energy of ~0 eV is shown in figure 1, where the range between the pyrimidine dimer (160 Th) and trimer (240 Th) is plotted.

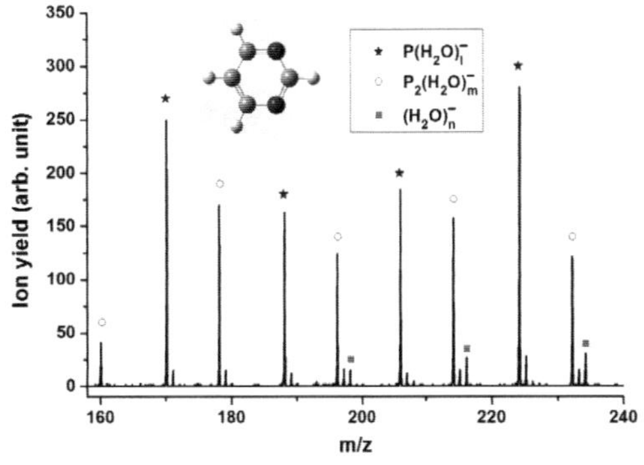

Figure 1. Detailed view of the mass spectrum obtained upon DEA to water-solvated pyrimidine clusters. One can easily distinguish between the solvated monomer (l = 5 − 8), the solvated pyrimidine dimer (m = 0 − 4) and the pure water clusters (n = 11 − 13).

In striking contrast to gas phase pyrimidine and water-solvated pyrimidine, where no anionic monomers are detectable, we are able to measure the parent anion $(C_4H_4N_2)^-$ as a fragmentation product from pure pyrimidine clusters. Furthermore we see evidence that water is preventing the ring fragmentation of the pyrimidine.

Acknowledgment: This work is partially supported by FWF, Vienna (P22665, P24443, I1015) and DFG (FOR1789). J. Aysina acknowledges a Lise Meitner grant from the FWF (M1445-N20).

References

[1] S. Denifl *et al.* 2006 *Phys. Rev. Lett.* **97** 043201

[1]E-mail: michael.neustetter@uibk.ac.at
[2]E-mail: julia.aysina@uibk.ac.at
[3]E-mail: f.ferreiradasilva@fct.unl.pt
[4]E-mail: stephan.denifl@uibk.ac.at

Content from this work may be used under the terms of the Creative Commons Attribution 3.0 licence. Any further distribution of this work must maintain attribution to the author(s) and the title of the work, journal citation and DOI.
Published under licence by IOP Publishing Ltd

Journal of Physics: Conference Series **635** (2015) 072057 doi:10.1088/1742-6596/635/7/072057

Partial ionization cross sections for positive fragments produced by electron impact on uracil

S. Diskin[*], **S. Eden**[†], **and P. J. M. van der Burgt**[*1]

[*] Department of Experimental Physics, National University of Ireland Maynooth, Maynooth, Co. Kildare, Ireland
[†] Department of Physical Sciences, The Open University, Walton Hall, Milton Keynes, MK7 6AA, United Kingdom

Synopsis Fragmentation of uracil in the gas phase has been studied using low-energy electron impact. Positive ions have been detected using a reflectron time-of-flight mass spectrometer. Mass spectra have been measured as a function of electron impact from 0 to 100 eV, and ion yield curves most of the positively charged fragments have been extracted.

We have measured mass spectra of positive ions for electron impact on uracil, with electron energies ranging from 0 to 100 eV in steps of 0.5 eV. Ions have been collected and detected using a reflectron time-of-flight mass spectrometer. Details about the experiment and the data analysis can be found in [1-3], which present results obtained for electron impact on cytosine, thymine and adenine.

A beam of uracil is generated by a resistively heated oven mounted in an expansion chamber, and the forward section of the beam effusing from a capillary in the oven passes through a skimmer into the collision chamber, where the beam is crossed by a pulsed electron beam. Positively charged fragments are mass resolved and detected using a reflectron time-of-flight mass spectrometer. LabView based data acquisition techniques are used to accumulate time-of-flight spectra as a function of electron impact energy.

Ion yield curves for most fragment ions of uracil have been obtained by fitting groups of adjacent peaks in the mass spectra with sequences of normalized Gaussians.

Tests of the electron beam current have shown that the current is constant down to 15 eV and decreases to about 60% at 8 eV. Because all ion yield curves are generated from a single dataset, and assuming that the detection efficiency of the reflectron mass spectrometer is mass independent, above 15 eV the yields of all fragments are on the same relative scale and are comparable. The ion yield curves have been normalized by comparing the sum of the ion yields to the average of calculated total ionization cross sections. Appearance energies have been determined for most ions.

Mass spectra, ion yield curves and appearance energies will be presented at the conference and will be compared with other research on uracil (see [4-6] and references therein).

The experiment also involves the implementation and development of a new source design using a 50 micron straight nozzle with argon gas acting as the buffer gas [4]. The new design will improve beam collimation and will enable formation of small nucleobase-water clusters when water vapour is introduced into the source. Other future plans include an investigation of electron impact on fluorouracil.

References

[1] P. J. M. van der Burgt 2014 *Eur. J. Phys. D* **68**, 135
[2] P. J. M. van der Burgt, F. Mahon, G. Barrett, and M. L. Gradziel 2014 *Eur. J. Phys. D* **68**, 151
[3] P. J. M. van der Burgt, S. Finnegan, and S. Eden 2015, submitted to *Eur. J. Phys. D*
[4] B. Barc, M. Ryszka, J. Spurrell, M. Damp, P. Limão-Vieira, R. Parajuli, N. J. Mason, and S. Eden 2013 *J. Chem. Phys.* **139**, 244311
[5] S. Denifl, B. Sonnweber, G. Hanel, P. Scheier, T. D. Märk 2004 *Int. J. Mass Spectrom.* **238**, 47
[6] I. I. Shafranyosh and M. I. Sukhoviya 2012 *J. Chem. Phys.* **137**, 184303

[1] E-mail: peter.vanderburgt@nuim.ie

Content from this work may be used under the terms of the Creative Commons Attribution 3.0 licence. Any further distribution of this work must maintain attribution to the author(s) and the title of the work, journal citation and DOI.
Published under licence by IOP Publishing Ltd

Journal of Physics: Conference Series **635** (2015) 072058 doi:10.1088/1742-6596/635/7/072058

Recent developments in R-matrix applications to molecular processes

Zdeněk Mašín*, **Alex Harvey**§, **Karel Houfek**‡, **Danilo S. Brambila**§, **Felipe Morales**§, **Jimena D. Gorfinkiel***, [1], **Jonathan Tennyson**†, **Olga Smirnova**§

* Department of Physical Sciences, The Open University, Walton Hall, Milton Keynes, MK7 6AA, United Kingdom
§ Max-Born Institute (MBI), Max-Born Strasse 2A, 12489 Berlin, Germany
‡ Institute of Theoretical Physics, Faculty of Mathematics and Physics, Charles University in Prague, Czech Republic
† Department of Physics and Astronomy, University College London, WC1E 6BT, United Kingdom

Synopsis We report on recent developments of the UKRmol suite, an implementation of the molecular R-matrix method and present examples of the calculations (e.g. electron scattering, photoionization, high harmonic generation, etc.) it has enabled.

Advances in experimental atomic and molecular physics (e.g. the advent of femto and attosecond laser pulses) and the increased need in technological areas for quantitative data involving the electronic continuum of molecules (e.g. electron-molecule scattering cross sections required to model industrial plasmas and the interaction of radiation with biological material) is driving the development of new methodology and software to treat correlated multielectron dynamics in molecules.

The UKRmol suite [1] is based on 30+ years of software development. Initially a set of codes that enabled the study of low energy electron scattering from molecules (and partly based on prior electron-atom scattering software) the suite has developed and is now used to study positron-molecule scattering, scattering from molecular clusters and to generate data necessary to perform calculations for photon induced processes (photoionization, high harmonic generation, etc) as well as other processes involving nuclear dynamics like associative detachment. In recent years, the suite has been significantly improved, both in terms of its efficiency (for example, the parallelization of various programs in the suite has allowed us to perform sophisticated calculations for electron scattering from uracil [2] as well as the bound and continuum electronic states of N_2 [3]) and the addition of new functionality (e.g. a program, CDENPROP [4], that determines transition moments between bound and (discretized) continuum states as well as Dyson orbitals has been developed). Many of these applications entail the calculation of scattering and other data (in particular, resonance energies and widths) for a large number of molecular geometries. A set of scripts has been developed that significantly simplify this task. The suite of codes and scripts are freely available for download [5].

We will present recent developments as well as some examples of the physics being investigated with the UKRMol suite.

References

[1] J.M. Carr, P.G. Galiatsatos, J.D. Gorfinkiel, A.G. Harvey, M.A. Lysaght, D. Madden, Z. Mašín, M. Plummer, J. Tennyson, H.N. Varambhia 2012 *Eur. Phys. J.* D **66** 58

[2] Z. Mašín and J. D. Gorfinkiel 2014 EPJD **68** 112

[3] D. A. Little and J. Tennyson 2014 *J. Phys. B: At. Mol. Opt. Phys.* **47** 105204

[4] Alex G. Harvey, Danilo S. Brambila, Felipe Morales, Olga Smirnova, arXiv:1401.0229 [physics.chem-ph]

[5] https://ccpforge.cse.rl.ac.uk/gf/project/ukrmol-in/, https://ccpforge.cse.rl.ac.uk/gf/project/ukrmol-out/, https://ccpforge.cse.rl.ac.uk/gf/project/ukrmol-scripts/

[1]E-mail: Jimena.Gorfinkiel@open.ac.uk

Content from this work may be used under the terms of the Creative Commons Attribution 3.0 licence. Any further distribution of this work must maintain attribution to the author(s) and the title of the work, journal citation and DOI.
Published under licence by IOP Publishing Ltd

XXIX International Conference on Photonic, Electronic, and Atomic Collisions (ICPEAC2015) IOP Publishing
Journal of Physics: Conference Series **635** (2015) 072059 doi:10.1088/1742-6596/635/7/072059

The cryogenic storage ring CSR for collision experiments with state-controlled and phase-space cooled molecular ion beams

A Becker,[*] K Blaum,[*] C Breitenfeldt,[†,*] F Fellenberger,[*] S George,[*] J Göck,[*] M Grieser,[*] F Grussie,[*] R von Hahn,[*] P Herwig,[*] J Karthein,[*] C Krantz,[*] H Kreckel,[*] S Kumar,[*] M Lange,[*] S Lohmann,[*] C Meyer,[*] P M Mishra,[*] O Novotný,[*,‡] A O'Connor,[*] R Repnow,[*] S Schippers,[¶] C D Schröter,[*] K Spruck,[¶,*] D Schwalm,[§,*] L Schweikard,[†] X Urbain,[||] S Vogel,[*] A Wolf[*1]

[*] Max Planck Institute for Nuclear Physics, D-69117 Heidelberg, Germany – [†] Institut für Physik, Ernst-Moritz-Arndt-Universität Greifswald, D-17487 Greifswald, Germany – [‡] Columbia Astrophysics Laboratory, Columbia University, New York, NY 10027, USA – [¶] Institut für Atom- und Molekülphysik, Justus-Liebig-Universität Gießen, D-35392 Gießen, Germany – [§] Weizmann Institute of Science, Rehovot 76100, Israel – [||] IMCN, Université Catholique de Louvain, B-1348 Louvain-la-Neuve, Belgium

Synopsis The cryogenic storage ring (CSR) for positive and negative atomic, molecular and cluster ions is starting operation at the Max Planck Institute for Nuclear Physics in Heidelberg.

Cryogenic electrostatic storage devices are being realized with two main motivations: (a) very long storage times of ion beams with multi-keV kinetic energies in extremely rarified background gas and (b) suppressing radiative (blackbody) excitation of the stored ions. First operation [1, 2] and results of cryogenic electrostatic ion beam storage devices were reported. Among these new devices, the cryogenic storage ring CSR [3] is unique by its ion beam energy range up to 300 keVq and a wide range of instruments for merged and crossed beams collision studies. Merged velocity-matched electron beams will serve for translational (phase-space) cooling of the stored ions and for electron-ion collision studies down to meV kinetic energies [4]. The electrostatic storage strongly extends the ion mass range compared to magnetic storage rings. In combination, a new tool emerges for studying the fragmentation of complex molecular ions following many types of collisional or radiative interaction. Many new options open up for state controlled reaction studies of systems relevant for molecular astrophysics, where low internal temperatures of the collision partners often are of particular importance.

Construction of the cryogenic storage ring has been completed. A 35 m long beam pipe encloses the ion beam round-trip orbit, containing all beam optics and diagnostics within a separate extremely-low-pressure vacuum system. It is cooled to 10 K, cryopumps being kept at 2 K, and itself encapsuled in a large toroidal cryostat. Ion-optics cryogenic cooling [3] and room-temperature ion beam storage were separately demonstrated in previous work, while a campaign to perform first long-time storage of molecular anion and cation beams is in progress. Implementation of the merged electron beam device [4] and a setup for velocity-matched neutral atomic beams [5] are upcoming.

First experiments will focus on laser photodetachment and photodissociation of stored molecular anions and cations with sensitivity to individual rotational levels. A wide range of the CSR experimental equipment has either been created [6] or will be added in the near future. This includes multicoincidence detectors for neutral and charged fragments from in-flight fragmentation of the stored ions. They will be complemented by mass-sensitive calorimetric detectors for neutral fragments with promising properties revealed in recent studies [7]. The CSR facility also includes a 300 keV accelerator platform suitable to interface with a wide range of ion-chemistry and plasma-type sources for cations and anions [8]. Hence, many types of collision and radiation induced fragmentation of complex molecular and cluster ions can be investigated with particular emphasis to internal state definition as well as to the detailed analysis of all products.

References

[1] M Lange *et al* 2010 *Rev Sci Instrum* **81** 055105
[2] H T Schmidt *et al* 2013 *Rev Sci Instrum* **84** 055115
[3] R von Hahn *et al* 2011 *Nucl Instrum Methods B* **269** 2871
[4] C Krantz *et al* 2011 *J Phys Conf Proc* **300** 122010
[5] A O'Connor *et al* *These Proceedings*
[6] K Spruck *et al* 2015 *Rev Sci Instrum* **86** 023303
[7] O Novotný *et al* *These Proceedings*
[8] P M Mishra *These Proceedings*

[1]E-mail: A.Wolf@mpi-hd.mpg.de

Content from this work may be used under the terms of the Creative Commons Attribution 3.0 licence. Any further distribution of this work must maintain attribution to the author(s) and the title of the work, journal citation and DOI.
Published under licence by IOP Publishing Ltd

Journal of Physics: Conference Series **635** (2015) 072060 doi:10.1088/1742-6596/635/7/072060

Theoretical Modeling of Mass Spectrometry

Néstor F. Aguirre*[1], Sergio Díaz-Tendero*, Paul-Antoine Hervieux⋆, Manuel Alcamí*,†,‡ and Fernando Martín*,†,‡

* Departamento de Química, Módulo 13,Universidad Autónoma de Madrid, 28049, Madrid, Spain
⋆ Institut de Physique et Chimie des Matériaux de Strasbourg, 23 rue du Loess, 67034 Strasbourg. France
† Instituto Madrileo de Estudios Avanzados en Nanociencias (IMDEA-Nanociencia), 28049 Madrid, Spain
‡ Condensed Matter Physics Center (IFIMAC), Universidad Autónoma de Madrid, 28049 Madrid, Spain.

Synopsis We present an implementation of the Microcanonical Metropolis Monte Carlo method based on statistical mechanics and electronic structure calculations. The method is designed to study any kind of fragmentation process. Here we show its capabilities to predict mass spectra of simple molecules.

When a highly energetic particle hits a neutral molecule, part of its energy is transferred to the molecule. If this energy exceeds the ionization energy, ejection of one electron occurs, which generates a molecular ion in an excited state. Then, the molecule breaks leading to a characteristic fragmentation pattern. As the excitation energy increases, the number of channels, abundance and variety of the ionized species also increase. The main hypothesis of our method is that the nature and extent of these reactions depend only on the ion's structure and its internal energy, irrespective of the ionization mechanism.

Under this hypothesis, the dynamical evolution of the many-body system is mainly guided by the accessible phase-space. Thus, statistical mechanics provides the appropriate theoretical framework to conduct this kind of simulations. In this context, we have developed the statistical Microcanonical Metropolis Monte Carlo (M_3C) method [1, 2], which is based on a specific random way to move in the phase-space until a region of maximum entropy is reached, where physical observables are computed by performing a statistical average. The minimum information that M_3C needs is the electronic energy, molecular geometry and vibrational frequencies for all possible molecules/fragments involved in the fragmentation process, and the main information provided is the breakup-curves (fragmentation probabilities as a function of the excitation energy). The mass spectrum is obtained by a weighted sum of these curves by using the internal energy distribution of the parent ion resulting from the process of electronic ionization.

In this contribution we will show some results for benchmark systems. In particular Fig. 1 shows the capabilities of M_3C to describe the fragmentation of the difluoromethane molecule (CH_2F_2) with their associated breakup-curves.

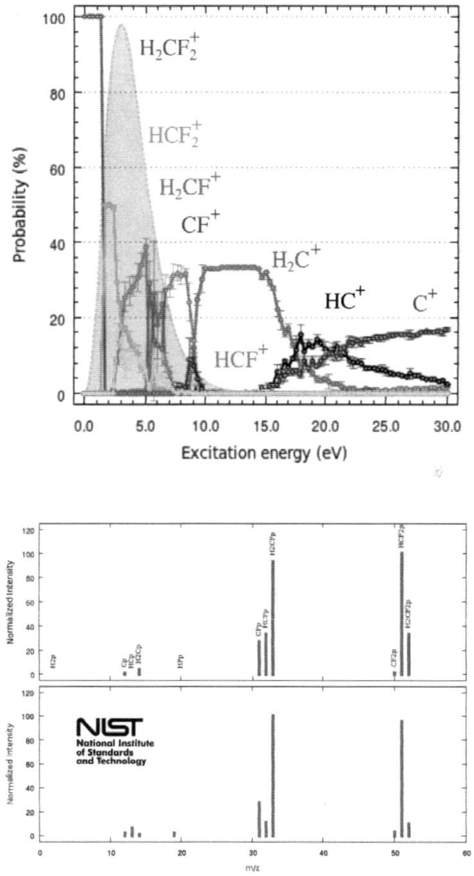

Figure 1. Fragmentation of CH_2F_2 molecule. Upper panel, breakup-curves. Middle/lower panel, theoretical/experimental [3] mass spectrum.

References

[1] S. Díaz-Tendero *et al* 2005 *Phys. Rev. A* **71** 033202

[2] N. F. Aguirre *et al* 2015. In preparation.

[3] P.J. Linstrom *et al* NIST Chemistry WebBook http://webbook.nist.gov (retrieved February 10, 2015)

[1]E-mail: nestor.aguirre@uam.es

Content from this work may be used under the terms of the Creative Commons Attribution 3.0 licence. Any further distribution of this work must maintain attribution to the author(s) and the title of the work, journal citation and DOI.
Published under licence by IOP Publishing Ltd

XXIX International Conference on Photonic, Electronic, and Atomic Collisions (ICPEAC2015) IOP Publishing
Journal of Physics: Conference Series **635** (2015) 072061 doi:10.1088/1742-6596/635/7/072061

Double differential cross sections for liquid water ionization by impact of fast electrons.

M.L. de Sanctis*[1], M.-F. Politis†, R. Vuilleumier††, C. Stia* and O. Fojón*

*Instituto de Física Rosario, UNR-CONICET, Pellegrini 250, (2000) Rosario, Argentina
†Laboratoire Analyse et Modélisation pour la Biologie et l'Environnement, UEVE, Bv. Mitterrand, 91025 Evry, France
††École Normale Supérieure, Dépt. de Chimie, UPMC, 24, rue Lhomond, 75005 Paris, France

Synopsis A theoretical study for the single ionization of water molecules in liquid phase by electron impact at high energies is presented. Through a first order model, we compute multiple differential cross sections considering an asymmetric coplanar geometry. The wavefunctions for a single liquid water molecule are obtained in a realistic way by using a Wannier orbital formalism. We compare our results with experiments for the vapor phase and previous theoretical results obtaining a good qualitative agreement.

We study theoretically the single ionization of liquid water by energetic electrons through a first-order model. This reaction is important in many domains such as plasma physics, fusion experiments, astrophysics, and even in the study of ionizing collisions on living matter. As the latter is composed mainly of liquid water, the knowledge of its ionization reaction is crucial to understand the damage provoked to the living tissue by the ionizing radiations. Moreover, the low energy electrons (products of the ionization) are of importance in the mechanisms that lead to cell alteration [1]. As in track structure analysis the liquid phase in cells is often modeled as the gas phase, a more realistic description for the liquid is required. However, to obtain an appropriate description for ionization of liquid water is a difficult task. So, several approximations are required to describe the reaction.

Multiple differential cross sections give the most detailed information on the mechanisms involved in charge transfer reactions. Particularly, angular distributions of secondary electrons allow the determination of the preferential directions or energies for electron ejection following the ionization of the target.

Therefore, we compute double differential cross sections (DDCS) averaged on the molecule orientations and scattering angles for electron projectiles in a coplanar geometry. These DDCS describe the angular distribution of the ejected electrons at fixed incident and ejection energies. The model (see our previous work [2] for more details) uses an independent electron approximation, neglecting the exchange effects as we consider asymmetric collisions. The initial state of the molecule in the liquid phase is obtained through a Wannier technique [3]. The projectile is described by a plane wave whereas the ejected electron is described by a Coulomb wave taking into account its interaction with the residual target.

We compare our DDCS with calculations for vapor [4,5] as well as with experiments also for this phase [6,7]. Fig. 1 shows DDCS for ionization of a single water molecule. As can be seen, a good qualitative agreement is found but significant differences between gas an liquid are observed at low ejection angles.

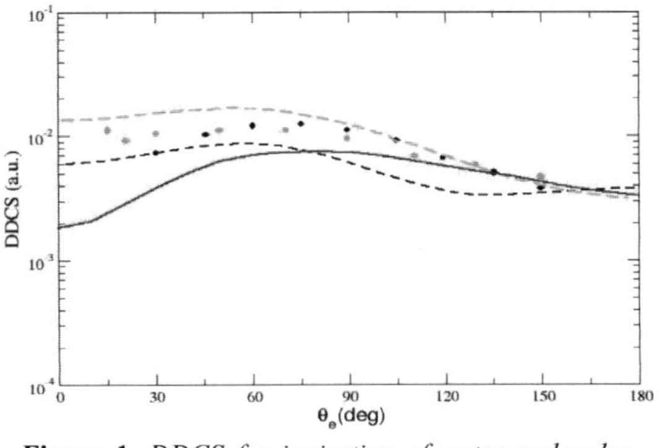

Figure 1. *DDCS for ionization of water molecules as a function of the ejection angle θ_e. Incident and ejection energies are E_i=500 eV and E_e=22 eV, respectively. For liquid, ▬, our results. For gas, ---, FBA calculations [4], ---, DWBA calculations [5], ●, experiments by Opal [6], ●, experiments by Bolarizadeh [7].*

References

[1] B. Boudaiffa *et al* 2000 *Science* **287** 1658
[2] M.L. de Sanctis *et al* 2012 *J. Phys. B* **45** 045206
[3] P. Hunt *et al* 2003 *Chem. Phys. Lett.* **376** 68
[4] C. Champion *et al* 2002 *J. Chem. Phys.* **117** 197
[5] S. Houamer *et al* 2007 *J. E. S. and R. P.* **161** 38
[6] M.A. Bolorizadeh *et al* 1985 *Phys. Rev. A* **33** 882
[7] C. B. Opal *et al* 1972 *Atomic Data* **4** 209

[1]E-mail: mldesanc@fceia.unr.edu.ar

Content from this work may be used under the terms of the Creative Commons Attribution 3.0 licence. Any further distribution of this work must maintain attribution to the author(s) and the title of the work, journal citation and DOI.
Published under licence by IOP Publishing Ltd

Experimental evidence for two decay channels in electron impact ionization and fragmentation of argon dimer

Elias Jabbour Al Maalouf* [1], **Xueguang Ren**†$ [2], **Alexander Dorn**† [2], **Stephan Denifl*** [1]

* Institute for Ion and Applied Physics, University of Innsbruck, Technikerstrasse 25, 6020-Innsbruck, Austria
† Max-Planck-Institut für Kernphysik, 69117 Heidelberg, Germany
$Physikalisch-Technische Bundesanstalt, 38116 Braunschweig, Germany

Synopsis We study the dissociative ionization of Ar dimer (Ar_2) by electron impact with the multiple particles coincidence (electrons and ions) method using a reaction microscope. From the measured kinetic energies of final state electrons and ions, we identify two decay channels which can contribute to the double peak structure in the observed kinetic energy release (KER) spectrum of argon Ar_2 dissociating into Ar^+- Ar^+.

In recent years, the interatomic Coulombic decay (ICD) has become known as an important decay process in the weakly bonded systems, e.g. in the Ar dimer studied here. The excess energy is transferred by exchange of a virtual photon to a neighboring atom, thereby, ionizing it by emitting an ICD electron [1]. Results show that ICD may be an abundant source of low energy electrons [2] which may further react in matter and cause the damage of biological tissue. In addition, the relaxation process of Radiative charge transfer (RCT) may occur in the Ar dimer in which one atom is doubly ionized, and an electron is transferred from a neighbor atom to the doubly charged Ar^{2+} ion. The system can relax by emitting a photon. Both ICD and RCT processes can result in two repulsive Ar^+ ions.

It has been observed upon ion- [3], electron- [4], and photon- [5] collisions with Ar dimer that the KER spectrum of the Ar^+- Ar^+ channel has a double peak structure. Calculations have shown that a contribution of this KER could be attributed to decay channel of ICD [6]. Other calculations suggest that RCT may also contribute in this KER [7].

In the present work, we study the ionization of Ar dimer by low energy electron impact using a reaction microscope [8] in which the kinetic energies of all final state particles are measure with the multiple particles coincidence method. The measured KER spectrum is presented in Figure 1, which shows two peaks: a peak at 3.8 eV and the other one at 5.2 eV. These two energies correspond respectively to the two internuclear distances of 3.8 angstrom and 2.8 angstrom in the Ar dimer [6] [7]. From the measured kinetic energies of 5 particles (two Ar^+ ions and three electrons), the binding energy or ionization energy for each channel are obtained. Thereby, we can identify the two decay channels (ICD vs. RCT) in Ar dimer. Detailed results will be presented at the conference.

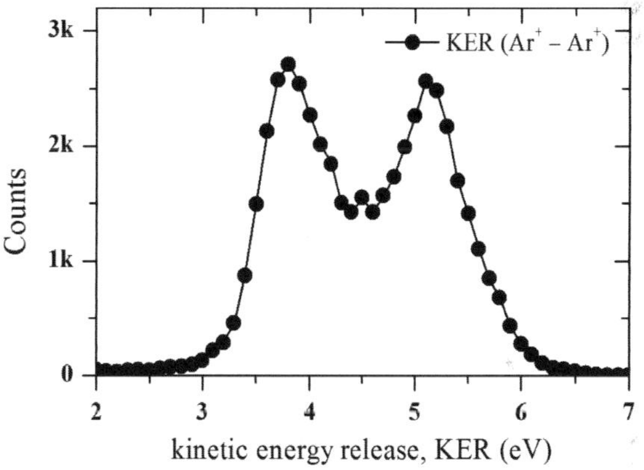

Figure 1. KER distribution for the Ar^+ - Ar^+ from the dissociation of the $(Ar_2)^{2+}$

This work was supported by DFG (FOR1789) and FWF (I1015).

References

[1] U. Hergenhahn 2011 *J. Elect Spec. Real. Phen.* **184** 78–90
[2] Hong-Keun Kim *et al* 2011 *PNAS* **108**. 29, 11821 - 11824
[3] H.-K. Kim *et al* 2013 *Phys. Rev. A* **88**, 042707
[4] S. Yan *et al* 2013 *Phys. Rev. A* **88**, 042712
[5] M. Kimura *et al* 2013 *Phys. Rev. A* **88**, 043414
[6] T. Miteva *et al* 2014, *J. Chem. Phys.* **141**, 064307
[7] Spas D. Stoychev and Alexander I. Kuleff 2008, *J. Chem. Phys.* **128**, 014307
[8] X. Ren *et al* 2014, *J. Chem. Phys.* **141**, 134314

[1] E-mail: elias.jabbour-al-maalouf@uibk.ac.at
$ [2] E-mail: ren@mpi-hd.mpg.de

Content from this work may be used under the terms of the Creative Commons Attribution 3.0 licence. Any further distribution of this work must maintain attribution to the author(s) and the title of the work, journal citation and DOI.
Published under licence by IOP Publishing Ltd

Journal of Physics: Conference Series **635** (2015) 072063 doi:10.1088/1742-6596/635/7/072063

Absolute Electron Ionization cross sections for Cytosine

M. A. Rahman and E Krishnakumar

Department of Nuclear and Atomic Physics, Tata Institute of Fundamental Research, Colaba, Mumbai 400 005, India

Synopsis Absolute partial and total electron ionization cross sections for the DNA base Cytosine are measured using Relative Flow Technique (RFT) up to 500 eV.

The interaction of ionizing radiation with living cells induces different types of damage to DNA and RNA [1]. These modifications of cellular DNA causes cytotoxic, mutagenic and carcinogenic lesions and are the results of variety of excitation and ionization processes producing large quantities of radicals, ions and secondary electrons within the track of primary radiation.

Monte Carlo track structure simulations [2] are frequently used in ion beam cancer therapy and in micro- and nanodosimetry to calculate the radiation transport in detail. These simulations require a complete set of cross sections for the biomolecules. In this context the data on DNA and RNA bases assume significance. No experimental measurement of the total and partial ionization cross sections of all these species has been reported, though theoretical results exist for total ionization cross sections for these bases [3-7].

The experimental setup is described in detail elsewhere [8]. Briefly, it consists of a magnetically collimated and pulsed electron gun, an effusive molecular beam formed by a capillary, a Time of Flight Mass Spectrometer (ToFMS) to mass select the ions, a pair of Micro-Channel Plates (MCP) in chevron configuration as detector, a Faraday cup to measure the incident electron current and the associated electronics to store the ion signal as a function of the mass and electron energy.

In the crossed beams experiment involving effusive beam from an oven where a solid sample is heated to 150 – 250 ^{o}C, the direct measurement of pressure is technically not possible due to the unavailability of appropriate manometers for high temperature applications. We overcome these difficulties by using the vapour pressure vs temperature data for the nucleobases available in the literature [9 - 11] as a measure of the pressure behind the capillary. The standard gas Ar is also flowed through the same oven and capillary for normalizing the cross section to absolute values. The molecule under investigation is prepared as a beam by heating the commercially available powder sample in the oven and allowing the sublimated molecules to effuse through a capillary directly into the interaction region. The partial cross section for prominent ions and total cross sections for Cytosine are shown in Figure 1.

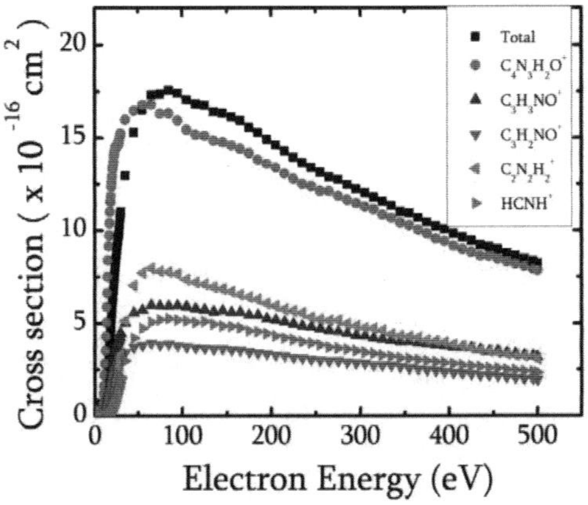

Figure 1. Partial and total ionization cross sections for Cytosine. (The partial cross sections for only dominant ions are given. All partial cross sections are multiplied by a factor of 4)

References

[1] B. Boudaiffa *et al.* 2000 *Science* 287 1658.

[2] H. Nikjoo *et al.* 1997 *Int J Radiat Biol* 71 467

[3] P. Bernhardt *et al.* 2003 *IJM S* 223 579

[4] P. Mozejko *et al.* 2003 *Rad Environ Biophys* 42 201

[5] M. Vinodkumar *et al.* 2013 IJMS 339 16

[6] W. M. Huo *et al.* 2006 *Rad. Measurements* 41 1202

[7] A. Peudon *et al.* 2006 *Rad. Prot. Dosi.* 122 128

[8] M. A. Rahman *et al.* 2012 *IJMS* 319 48

[9] C. J. Colyer 2011 *Thesis,* University of Adelaide.

[10] W. Zielenkiewicz, 2000 *J Chem Eng Data* 45 626.

[11] D. Ferro *et al* 1980 *Thermochim. Acta* 42 75

E-mail: rahman@tifr.res.in

E-mail: ekkumar@tifr.res.in

Content from this work may be used under the terms of the Creative Commons Attribution 3.0 licence. Any further distribution of this work must maintain attribution to the author(s) and the title of the work, journal citation and DOI.
Published under licence by IOP Publishing Ltd

Electron Ionization of Adenine

M. A. Rahman and E Krishnakumar

Department of Nuclear and Atomic Physics, Tata Institute of Fundamental Research, Colaba, Mumbai 400 005, India

Synopsis We present the measurements on absolute total and partial ionization cross sections and fragmentation studies for DNA base Adenine on electron impact using a newly designed experimental setup which uses Time of Flight Mass Spectrometer and the relative flow technique.

In order to model the radiation damage in biological tissues it is necessary to have absolute electron ionization cross section data on biological molecules. Theoretical calculations have been able to provide the total ionization cross sections for various molecules [1]. However, very limited data exist on partial ionization cross sections for molecules of biological importance. Considering this we have made the first-ever measurements of absolute partial ionization cross sections for all the nucleobases by electron impact.

The basic experimental setup and technique is described in detail elsewhere [2]. Briefly, it consists of a magnetically collimated and pulsed electron gun, an effusive molecular beam formed by a capillary, a Time of Flight Mass Spectrometer to mass select the ions, a pair of micro-channel plates (MCP) in chevron configuration as detector, a Faraday cup to measure the incident electron current and the associated electronics to store the ion signal as a function of the mass and electron energy. The cross sections are put on absolute scale using the relative flow technique with Ar as the standard. The pressures for the DNA bases are obtained from the known vapour pressure data as a function of temperature for the purpose of relative flow technique.

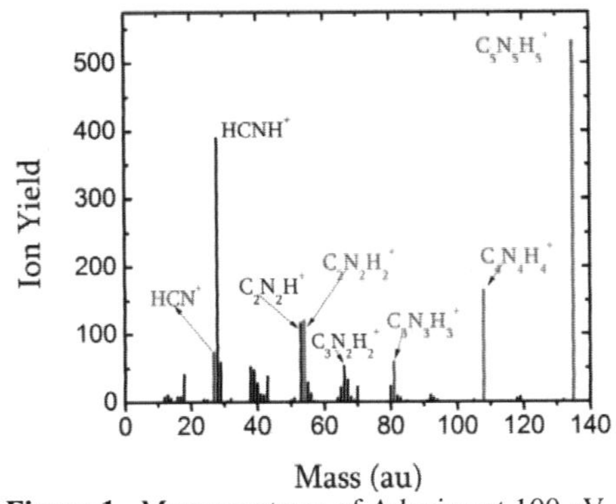

Figure 1. Mass spectrum of Adenine at 100 eV.

As shown in Fig.1, we observed prominent peaks corresponding to ions $C_nH_nN_n^+$ (n = 5 – 1) at mass 135 ($C_5N_5H_5^+$), 108 ($C_4N_4H_4^+$), 81 ($C_3N_3H_3^+$), 54 ($C_2N_2H_2^+$) and 27 (HCN^+) in the mass spectrum of adenine. Each of these ions is formed by the loss of one HCN unit in succession from the adenine molecule. This successive elimination of HCN units with such large intensities is exclusively observed only in the case of fragmentation of Adenine. If adenine is a pentamer of HCN formed by successive addition of HCN units then the probability of its fragmenting into $C_nH_nN_n^+$ (with n = 1 to 4) will be relatively higher and it has been observed in our measurements. Hence our data appear to support the theoretical prediction [3] that Adenine is formed in interstellar space by oligomerization of HCN.

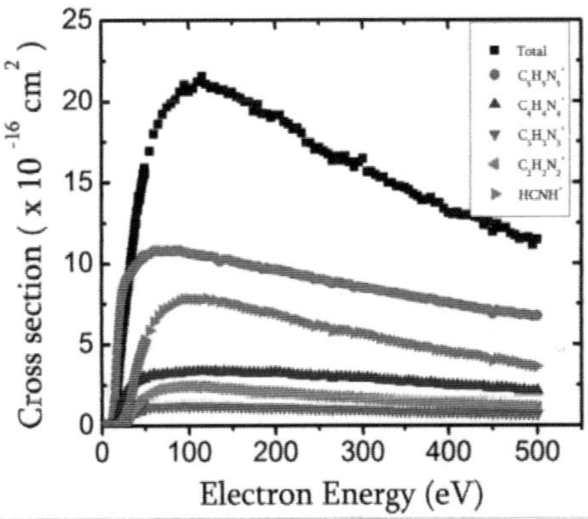

Figure 2. Absolute partial and total electron ionization cross sections for Adenine (The partial cross sections for only dominant ions are given. All partial cross sections are multiplied by a factor of 2).

The partial ion cross sections for the prominent ions are given in Fig. 2 along with the total ion cross sections.

References

[1] As listed by M. A. Rahman *et al* 2015 (This conference).

[2] M. A. Rahman *et al.* 2012 *IJMS* **319** 48.

[3] S. Chakraborti *et al.* 2000 *Astron. & Astroph.* **354** L6.

E-mail: rahman@tifr.res.in

E-mail: ekkumar@tifr.res.in

Content from this work may be used under the terms of the Creative Commons Attribution 3.0 licence. Any further distribution of this work must maintain attribution to the author(s) and the title of the work, journal citation and DOI.
Published under licence by IOP Publishing Ltd

Electron driven processes in sulphur containing compound: the case of dimethyl disulphide

Janina Kopyra [1], Jakub Władziński

Siedlce University, Faculty of Science, 3 Maja 54, 08-110 Siedlce, Poland

Synopsis Dissociative electron attachment to gas phase dimethyl disulphide (CH_3SSCH_3) has been studied by means of a crossed beams apparatus. Cleavage of the S–S bond is observed within a resonance in the energy range below 2 eV and visible preferentially via the appearance of the anionic fragment CH_2S^-.

Recent years have witnessed a remarkable increase of the interest in the studies of sulphur containing compounds [1,2]. This has been mainly motivated by the importance of these compounds in many bio-chemical reactions. As documented in a number of papers these compounds can be implicated in such processes as aging and/or oxidative stress [3]. On the other hand, when they are exposed to low energy electrons they can undergo reductive processes. Indeed, recent dissociative electron attachment experiments on cysteine [4], methionine [5] as well as thiolated nucleobases, 2-thiothymine [6] and 2-thiouracil [7], have shown that the presence of the S atom within the molecular structure strongly influences both the dissociation patterns as well as the attachment cross section.

In this contribution we will present the results from low energy electron driven processes in gas phase dimethyl disulphide CH_3SSCH_3. The results have been obtained by means of electron/molecule crossed beams apparatus. The negative ions formed in a collision region were extracted into a quadrupole mass analyser and detected by a single pulse counting technique.

The CH_3SSCH_3 molecule may serve as model system of sulphur containing peptides. It contains a disulphide bridge and hence can reflect the behaviour of the S–S bond between the two cysteine residues. The S–S bonds are responsible for the stabilization of the tertiary structure and play an important role in the folding of proteins. Therefore it is desirable to understand how the sulphur-containing compounds can be deteriorated by low energy electrons. The dissociation of CH_3SSCH_3 leads to the formation of the CH_2S^-, CH_3SS^-, HS^-, S^- and CH_3^- ions. The most dominant species is CH_2S^-, which appears at low energy range with a peak maximum at 0.88 eV and between 2-3 eV. In addition, we observe the appearance of a weak peak between 5-6 eV. Further fragments that are generated at high energy range are attributed to the S^- and CH_3^- anions.

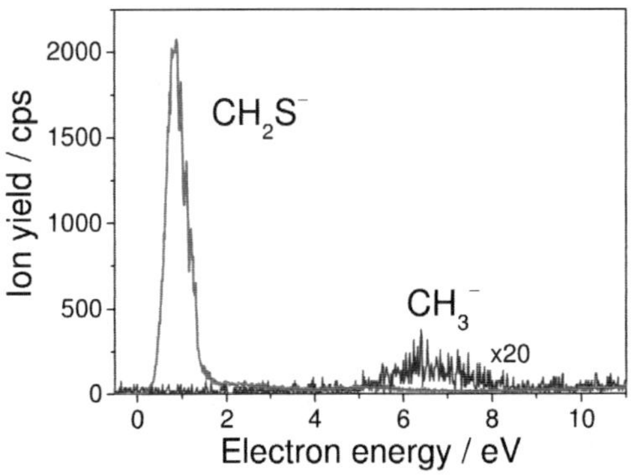

Figure 1. Ion yield curves for the dissociative electron attachment to CH_3SSCH_3 resulting in the formation of CH_2S^- and CH_3^-.

Acknowledgments

This work has been supported by the Polish Ministry of Science and Higher Education. This work was conducted within the framework of the EU/ESF COST Action MP1002 (Nanoscale insights into Ion Beam Cancer Therapy, Nano-IBCT).

References

[1] J.S. dos Santos *et al.* 2014 *Phys. Rev. A* **90** 052713
[2] A. Keller *et al.* 2012 *ACS NANO* **6** 4392
[3] B. Berlett *et al.* 1997 *J. Biol. Chem.* **272** 20313
[4] H. Abdoul-Carime *et al.* 2004 *Phys. Chem. Chem. Phys.* **6** 161
[5] J. Kopyra *et al.* 2012 *Phys. Chem. Chem. Phys.* **14** 8000
[6] J. Kopyra *et al.* 2014 *Phys. Chem. Chem. Phys.* **16** 5342
[7] J. Kopyra *et al.* 2014 *Phys. Chem. Chem. Phys.* **16** 25054

[1] E-mail: kopyra@uph.edu.pl

Content from this work may be used under the terms of the Creative Commons Attribution 3.0 licence. Any further distribution of this work must maintain attribution to the author(s) and the title of the work, journal citation and DOI.
Published under licence by IOP Publishing Ltd

Dissociative electron attachment to gas phase nucleobases: comparision of thymine and thiothymine

Janina Kopyra* [1], **Hassan Abdoul-Carime**† [2]

* Siedlce University, Faculty of Science, 3 Maja 54, 08-110 Siedlce, Poland

† Université de Lyon, F-69003, Lyon, France; Université Lyon 1, Villeurbanne; CNRS/IN2P3, UMR5822, Institut de Physique Nucléaire de Lyon

Synopsis Here we present the results of dissociative electron attachment to gas phase thymine and 2-thiothymine. A particular attention will be paid to the formation of the dehydrogenated molecular anions $(M - H)^-$, which preferentially occurs at low energy range below 2 eV. We will try to shed the light on the subtle differences visible within anionic yield curves that is the difference of the position and relative intensities of various overlapping peaks.

Studies of electron induced damage to biological systems have witnessed a remarkable growth in the last decade since it has been demonstrated that low energy electrons, copiously produced along the ionization tracks [1], can effectively cause strand breaks already at energy below the ionization threshold of DNA [2]. In the perspective of therapies combining the use of radiation and chemical agents such genotoxicity is highly desirable. Many efforts have been undertaken to understand the mechanism underlying the radiosensitizing effect induced by different chemical agents (*e.g.*, halouracils [3], cisplatin [4]). Thiolated nucleobases are known as anti-cancer drugs and can potentially develop radiosensitization properties.

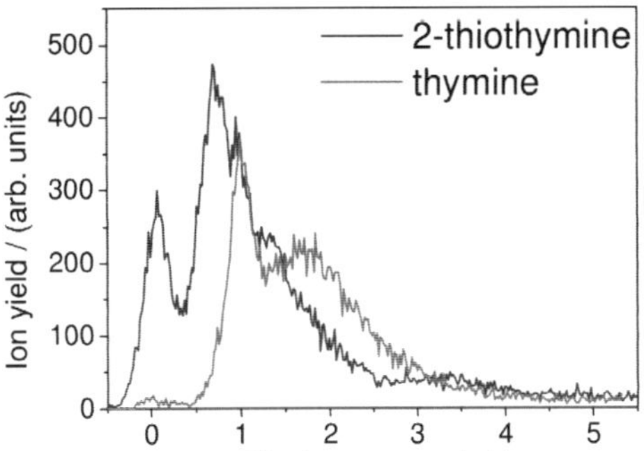

Figure 1. Ion yield curves for the dissociative electron attachment to thymine and 2-thiothymine resulting in the formation of $(M - H)^-$.

Here we present the results from dissociative electron attachment to gas phase thymine and 2-thiothymine. The results have been obtained utilizing the electron/molecule crossed beams apparatus. The yield of the negative ions, which are generated in the collision region is recorded as a function of the incident electron energy.

The fragmentation of both molecules under investigation is very rich and results in the formation of several fragment anions in a broad energy range 0-12 eV. However, the striking findings is that the fragmentation of the sulphur analogue, *i.e.*, thiothymine arise mainly at the sulphur site of the molecule [5]. This means that the presence of the S atom within the molecular structure facilitate the decomposition of the molecule initiated by the capture of low energy electrons. Further finding is the difference of the position and relative intensities of various overlapping peaks of $(M - H)^-$ in comparison to those recorded for thymine (Fig. 1). It appears that for thiothymine the dominant peak appears at substantially lower energy. In addition, a peak at the threshold energy is visible with a relatively high intensity.

Acknowledgments

This work has been supported by the Polish Ministry of Science and Higher Education, The University of Lyon 1 and the Centre National de Recherche Scientifique (CNRS/IN2P3) and the EU/ESF COST Action MP1002 (Nanoscale insights into Ion Beam Cancer Therapy, Nano-IBCT).

References

[1] T. Jahnke *et al.* 2010 *Nature Phys.* **6** 139.
[2] B. Boudaıffa *et al.* 2000 *Science* **287** 1658.
[3] H. Abdoul-Carime *et al.* 2001 *JACS* **123** 5354.
[4] J. Kopyra *et al.* 2009 *Angew. Chem. Int. Ed.* **48** 7904.
[5] J. Kopyra *et al.* 2014 *Phys. Chem. Chem. Phys.* **16** 5342.

[1] E-mail: kopyra@uph.edu.pl

[2] E-mail: hcarime@ipnl.in2p3.fr

Content from this work may be used under the terms of the Creative Commons Attribution 3.0 licence. Any further distribution of this work must maintain attribution to the author(s) and the title of the work, journal citation and DOI.
Published under licence by IOP Publishing Ltd

Dissociative recombination measurements of SH^+ with cold electrons

A Becker[a 1], **C Domesle**[a], **W D Geppert**[b], **M Grieser**[a], **C Krantz**[a], **R Repnow**[a], **D W Savin**[c], **D Schwalm**[a,d], **A Wolf**[a], **B Yang**[a,e], **O Novotný**[a,c 2]

[a]Max Planck Institute for Nuclear Physics, D-69117 Heidelberg, Germany
[b]Department of Physics, Stockholm University, AlbaNova, SE-10691, Stockholm, Sweden
[c]Columbia Astrophysics Laboratory, Columbia University, New York, NY 10027, USA
[d]Faculty of Physics, Weizmann Institute of Science, Rehovot 76100, Israel
[e]Institute of Modern Physics, Chinese Academy of Sciences, Lanzhou 730000, China

Synopsis We have experimentally investigated dissociative recombination of SH^+ with low-energy electrons. The merged-beams rate coefficient was measured at the TSR ion storage ring in Heidelberg, Germany. From these data the cross section was extracted and used to generate a plasma rate coefficient for plasma temperatures down to 10 K, which are relevant for astrochemical models. Furthermore, we have determined branching ratios of product excitation states utilizing a novel analysis method for imaging data.

Dissociative recombination (DR) of molecular ions with free electrons is a key neutralization process in plasma environments such as the cold interstellar medium (ISM). The DR rate coefficients for various molecular ions are needed to model ISM in order to deduce its physical properties and draw conclusions on dynamical processes. SH^+ is observed in various lines of sight. However, its high abundance cannot be reproduced by current astrochemical models, with deviations up to several orders of magnitudes [1]. This may be due to an underestimated DR rate coefficient for SH^+.

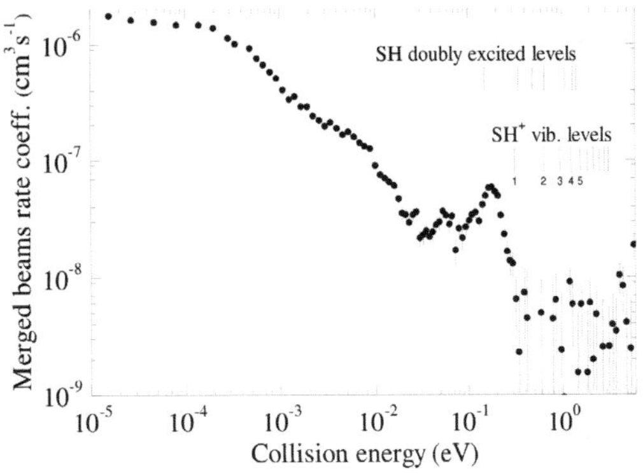

Figure 1. The measured merged beams DR rate coefficient for SH^+ as a function of mean center-of-mass collision energy. The vertical bars mark the known doubly excited electronic states of SH and the vibrational excitation energies of SH^+, all with respect to the $SH^+(X^3\Sigma^-)$ ground state. The rate coefficient enhancement at $\sim 0.04 - 0.2$ eV is attributed to an indirect DR process via neutral ro-vibrational Rydberg states in this energy range.

To this end we have experimentally investigated DR of SH^+ at the TSR ion storage ring of the Max Planck Institute for Nuclear Physics in Heidelberg, Germany. An SH^+ beam was merged with a cold electron beam produced by a photocathode. Utilizing an energy sensitive Si detector to identify neutral fragments, the recombination rate coefficient (see Fig. 1) was determined for collision energies ranging from sub-meV up to 6 eV, covering kinetic temperatures down to 10 K. The inferred DR plasma rate coefficient is not unusually low at the relevant temperatures, leading us to conclude that slow DR is not the reason for the high observed SH^+ abundances.

Moreover, we have determined branching ratios for various DR product excitation states by measuring fragment kinetic energy released on a position-sensitive microchannel plate detector. Here, most H-fragments missed the detector owing to the high mass ratio of the SH^+ constituents. With the low-emittance electron-cooled ion beam it was possible to deduce the product excitation states from the S-fragment distribution only. Our data show a preferred dissociation into a product state with sulfur excited to its $^1D_2$ level versus the ^3P ground level. This experimental method will be a valuable tool for analyzing imaging data of the heavier molecules to be stored at the new Cryogenic Storage Ring [2].

References

[1] Godard *et al* 2012 *Astron. Astrophys.* **540** 87

[2] R von Hahn *et al* 2011 *Nucl Instrum Meth B* **269** 2871

[1]E-mail: Arno.Becker@mpi-hd.mpg.de
[2]E-mail: Oldrich.Novotny@mpi-hd.mpg.de

Content from this work may be used under the terms of the Creative Commons Attribution 3.0 licence. Any further distribution of this work must maintain attribution to the author(s) and the title of the work, journal citation and DOI.
Published under licence by IOP Publishing Ltd

Journal of Physics: Conference Series **635** (2015) 072068 doi:10.1088/1742-6596/635/7/072068

Induced molecular dissociations as a radiation damage descriptor (Nanodosimetry)

A Traore[a], L. Ellis-Gibbings[a], A. Verkhovtsev[a], K. Krupa[a], A. Muñoz[b], F.Blanco[c], G. García[a,d]

[a] CSIC, Inst. Fis. Fundamental, Madrid 28006, Spain,

[b] Centre for Medical Radiation Physics, University of Wollongong, NSW 2522, Australia,

[c] Centro de Investigaciones, Energeticas Medioambientales y Tecnologicas (CIEMAT), Avenida Complutense, 28040 Madrid, Spain,

[d] Departamento de Física Atomica, Molecular y Nuclear, Universidad Complutense de Madrid, 28040 Madrid, Spain

Synopsis We present a nanodosimeter software based on Monte Carlo, termed: Low Energy Particle Track Simulation. For medical applications purposes, we propose bond breakings to describe the energy deposited taking water as an example. This tool aims to improve current treatment planning approach based on absorbed dose.

Understanding how the physical interaction of primary photons and electrons with biological media lead dose deposition is a fundamental issue in radiation therapy and radiation protection [1]. The main goal of this research area is to establish a clear relationship between the macroscopic radiation damage and the molecular dissociation occurring at the nanoscale level.

Conventional dosimetry correlates irradiation and biological damage through the amount of energy deposited (Joules) by unit mass (kilograms) or absorbed dose (in Grays) [2]. Although this absorbed dose is the main quantity of interest in treatment planning systems used in clinics, this parameter suffers from two limitations [3]: the absorbed dose is solely based on electronic ionization of the target neglecting therefore phenomena such as electronic excitation and vibration as well as dissociative electron attachment; this quantity is a macroscopic description of a nanoscopic processes and therefore cannot be complete even when considering only ionizations due to charge particle non equilibrium.

As an example, every 1 MeV photon produces 4 10^4 secondary particles and most of them lie below 70eV, hence their name Low Energy Electrons (LEEs). It has been demonstrated that those LEEs in dry and humid conditions, almost at their thermalization energy can induce single and double strand breaks [4]

Thus, these LEEs are highly relevant in order to improve the quality of dose planning and delivery in cancer patients. The main goal of radiation therapy is to spare the healthy sensitive tissues while irradiating the tumor exclusively.

In order to overcome these drawbacks, we come up with novel dosimeter software based Monte Carlo termed: Low Energy Particle Track Simulation. The LEPTS nanodosimeter provides deep insights of the dose deposition in the region of interest by considering all channels leading to bond breaking of biomolecules. Therefore, we aim to shift traditional concept of clinical dosimetry based on absorbed dose to a new descriptor, namely molecular dissociation nanophysics driven. We will present *in silico* experiment of dose prediction in water tissue equivalent media and hydrated prymidine, an analog of DNA constituent. The underlying physics, in the MC-LEPTS dosimeter prediction, driven by reliable atomic and molecular cross section data will be also presented.

[a,d] E-mail: g.garcia@csic.es

[a] E-mail: ali.traore@csic..iff.es

Acknowledgment: This work was supported by the Spanish Ministry of Econom ia y Productividad (Project FIS2012-31320) and by the People Programme (Marie Curie Actions) of the European Union's Seventh Framework Programme FP7/2007- 2013/ under REA – Grant Agreement n°608163

References

[1] *Radiation Damage in Biomolecular Systems* edited by G. Garcia, M. C. Fuss, (Springer London, 2012.

[2] H.-G. Menzel et al., *Dosimetry Systems for Use in Radiation Processing* (International Commission on Radiation Units and Measurements, Bethesda, MD, 2008) (ICRU 80)

[3] F. Blanco et al 2013 *Eur. Phys. J.* D **67,** 199

[4] L. Sanche, 2002 Mass. Spec. Rev. **21**, 1098

Content from this work may be used under the terms of the Creative Commons Attribution 3.0 licence. Any further distribution of this work must maintain attribution to the author(s) and the title of the work, journal citation and DOI.
Published under licence by IOP Publishing Ltd

Journal of Physics: Conference Series **635** (2015) 072069 doi:10.1088/1742-6596/635/7/072069

Electron induced fragmentation of sulphur containing biological prototypes: thiaproline and taurine

E. Lange* [1], **G. Meneses*** [2], **T. Cunha*** [3], **J. Krasuska†** [4], **J. Kopyra†** [5] **G. García‡** [6], **F. Ferreira da Silva*** [7], **P. Limão-Vieira*** [8]

* Laboratório de Colisões Atómicas e Moleculares, CEFITEC, Departamento de Física, Faculdade de Ciencias e Tecnologia, Universidade Nova de Lisboa, 2829-516 Caparica, Portugal

† Department of Chemistry, Siedlce University, 3 Maja 54, 08-110 Siedlce, Poland

‡ Instituto de Fsica Fundamental, Consejo Superior de Investigaciones Cientficas, Serrano 113-bis, 28006 Madrid, Spain

Synopsis Here we present negative ion formation driven by electron transfer in atom (potassium) molecule collision and dissociative electron attachment in sulphur containing compounds, probing the influence of sulphur atoms in the decomposition mechanism of two biological relevant molecules: thiaproline and taurine.

Sulphur containing nucleobases have been suggested as anticancer drugs.[1] Secondary electrons produced in the biological environment, as a result of the primary radiation interaction, may efficiently enhanced DNA/RNA loss of integrity. Such electron-induced decomposition mechanisms have been recognized on the basis of temporary molecular anions formation. However, there is still scarce information in the literature on the interaction of low-energy electrons with sulphur containing biologicaly relevant molecules. Notwithstanding, electron transfer has been acknowledge to be more attuned to the physiological environment processes in the decomposition of key molecular structures in several DNA/RNA constituents and other biological relevant molecules.[2, 3, 4]

Here we address electron capture and electron transfer induced decomposition of two sulphur containing biological prototypes, 4-thiaproline and taurine, the former a relevant amino acid with a key role in the human physiology and the latter suggested as a possible anti-cancer drug. [5]

In the electron transfer process by atom(potassium)-molecule collision experiments, a transient negative ion is formed, with the excess internal energy channelled into the different degrees of freedom resulting therefore in fragmentation. The resulting anionic fragments are time-of-flight (TOF) mass analysed. Moreover, dissociative electron attachment studies to sulphur containing molecules, such as taurine and 4-thiaproline are presented and compared with electron transfer studies upon potassium collisions

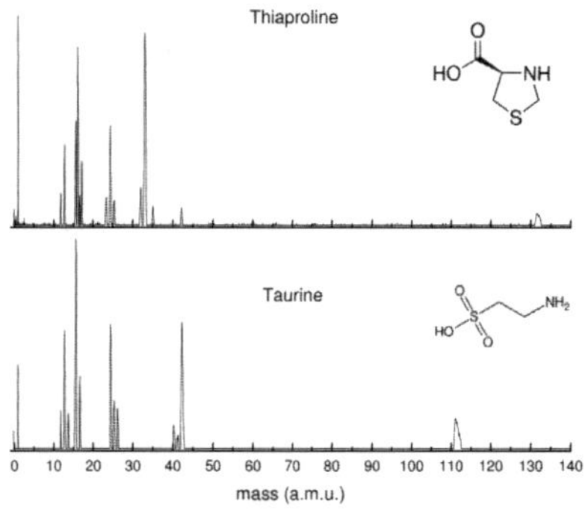

Figure 1. Anion mass spectra for 100 eV potassium impact on 4-thiaproline and taurine.

References

[1] G. B. Elion 1989 *Science* **244** 4900

[2] F Ferreira da Silva *at al* 2012 *Eur. Phys. J. D* **66** 78

[3] D. Almeida *at al* 2011 *PCCP* **13** 15657

[4] D. Almeida *at al* 2013 *PRL* **110** 023201

[5] M. Suo *at al* 2006 *Cancer letters* **237** 2

[1]E-mail: e.lange@campus.fct.unl.pt
[2]E-mail: g.meneses@campus.fct.unl.pt
[3]E-mail: t.cunha@campus.fct.unl.pt
[4]E-mail: jkrasuska06@gmail.com
[5]E-mail: kopyra@uph.edu.pl
[6]E-mail: g.garcia@csic.es
[7]E-mail: fr.silva@fct.unl.pt
[8]E-mail: plimaovieira@fct.unl.pt

Content from this work may be used under the terms of the Creative Commons Attribution 3.0 licence. Any further distribution of this work must maintain attribution to the author(s) and the title of the work, journal citation and DOI.
Published under licence by IOP Publishing Ltd

Theoretical and experimental study of (e,2e) ionization of the CO_2 ($1\pi_g$) molecule at 250 eV

Mevlut Dogan*[1], Zehra N. Ozer*, Murat Yavuz*, O. Alwan†‡, A. Naja†, B. B. Joulakian‡, Esam Ali+, C. Ning++, Don Madison+

* Department of Physics, e-COL Laboratory, Afyon Kocatepe University, 03200, Afyon, Turkey
† Lebanese University, LPM, EDST, Tripoli, Lebanon
‡ Universitè de Lorraine, SRSMC (UMR CNRS 7565), 1 bld Arago, bat. ICPM 57078 Metz Cedex 3, France
+ Missouri University of Science & Technology, Rolla, MO 65409, USA
++ Tsinghua University, Beijing 100084, People's Republic of China

Synopsis Triple differential cross sections (TDCSs) of the electron-impact ionization of carbon dioxide are measured in the coplanar asymmetric geometry, with incident electron energy value of 250eV, and ejected electron of 37eV. We will report the experimental results in comparison with the theoretical calculations of the M3DW and TCC (type 5) calculations.

Not only its importance in applied fields from astrophysics to plasma chemistry, but also being the main component in the atmospheres of Venus and Mars, Carbon dioxide is an important molecule and it is very desirable to have available sets of cross sections for this gas. On the other side, information on cross sections for ionization of such atmospheric molecules by electron impact is important both for understanding the basic processes of atomic collisions and controlling events in the ionosphere.

In recent years the two-center Coulomb continuum (TCC) model was extended to three-center targets like CO_2 for low equal energy sharing regime and also relatively higher incident energy asymmetric regime [1-4]. The molecular three body distorted wave (M3DW) model contains the post collision interaction (PCI) between scattered and ejected electrons to all orders of perturbation theory and the distorted wave Born approximations (DWBA) contains PCI only to first order. The M3DW and DWBA have yielded good with experiment for several different molecular targets. The aim of this work is to compare experimental and theoretical results for (e,2e) ionization of CO_2. For coplanar asymmetric kinematics, two outgoing electrons from the same ionization event are detected with specified kinetic energies and angles. The scattered electron was detected at three fixed angles (-10^0,-20^0,-30^0), while the ejected electron energy is fixed to 37 eV. The results are measured for ionization of the CO_2 ($1\pi_g$) orbital at an incident electron energy of 250 eV.

Figure 1 compares experimental results with the predictions of the DWBA, M3DW and TCC (type5) model. We will discuss the characteristics of the molecular results in combination with experiment and theory during the conference.

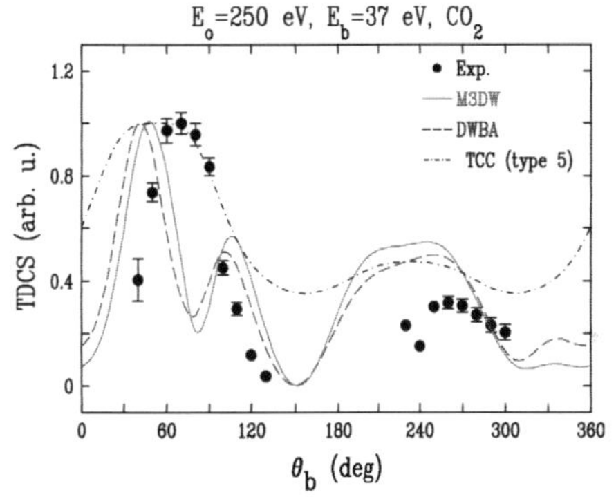

Figure 1. TDCS for CO_2 plotted as function of the ejection angle for the 37 eV ejected electron. The experimental results are compared to TCC, M3DW, and DWBA calculations.

Acknowledgments

This work was supported by TUBITAK through grant 109T738 and AKU, BAPK through grant 14.HIZ.DES.80. The theoretical work of EA and DM was supported by the US National Science Foundation under Grant. No. PHY-1068237.

References

[1] O. Chuluunbaatar *and* B. Joulakian 2010 *J. Phys. B: At. Mol. Opt. Phys.* 43 155201.
[2] M.J. Hussey *and* A.J. Murray 2005 *J. Phys. B: At. Mol. Phys.* 38 2965.
[3] A. Lahmam-Bennani *et al* 2009 *J. Phys. B: At. Mol. Opt. Phys.* 42 235205.
[4] O Alwan *et al* 2014 *J. Phys. B: At. Mol. Opt. Phys.* 47 225201

[1] E-mail: mdogan@aku.edu.tr

Content from this work may be used under the terms of the Creative Commons Attribution 3.0 licence. Any further distribution of this work must maintain attribution to the author(s) and the title of the work, journal citation and DOI.
Published under licence by IOP Publishing Ltd

Journal of Physics: Conference Series **635** (2015) 072071 doi:10.1088/1742-6596/635/7/072071

First Evidence of Interference effects in the ionization of N_2 molecule

Zehra N. Ozer*[1], Hari Chaluvadi†, Mevlut Dogan*, James Colgan‡, Don Madison†

* Department of Physics, e-COL Laboratory, Afyon Kocatepe University, 03200, Afyon, Turkey
† Department of Physics, Missouri University of Science and Technology, Rolla, Missouri, USA
‡Theoretical Division, Los Alamos National Laboratory, Los Alamos, New Mexico, USA

Synopsis We will present an experimental and theoretical investigation of triple differential cross sections for electron-impact ionization of nitrogen molecules at intermediate energies. A discussion of interference effects contained in the theoretical and experimental interference factors will be presented.

Double slit experiments with light depict the fundamentally probabilistic nature of quantum mechanical phenomena. The interference effects for quantum particles can be seen in triple differential cross sections (TDCSs) for electron impact ionization of diatomic molecules. Observation of such effects helps provide a detailed understanding of collision-induced reactions. The existence of Young-type interference patterns was recently observed in the ionization of diatomic molecules [1-2] and theoretically supported by calculations obtained using a molecular distorted-wave model. Evidence for double-slit interference effects was recently found for electron-H_2 scattering [3-5]. Although there have been several experimental attempts to find double-slit interference effects for electron-N_2 scattering, no experimental evidence had been found in prior work [6-8].

The I-factor, which is the ratio of molecular and atomic cross sections, represents a better test for interference than looking directly at the TDCS since it is not clear how interference effects will be manifested directly in the cross sections. In this work, we report a study looking for evidence of Young's double–slit interference in (e,2e) ionization of N_2 ($3\sigma_g$) by looking at the I-factor which, in this case, is the ratio of the molecular N_2 cross sections divided by the atomic nitrogen N cross sections [9]. Similar to the H_2 study, we do not have experimental data for ionization of atomic N. In this case, we use theoretical N cross sections calculated in the M3DW (molecular 3-body distorted wave) approximation as the denominator for both experiment and theory (as has been routinely done for heavy particle scattering [1,2]).

We compared experimental and theoretical (e,2e) cross sections and I-factors for intermediate energy electron-impact ionization of the N_2 molecule in the scattering plane. We found reasonably good agreement between the theoretical M3DW TDCS results and experiment. We will discuss the evidence for Young's double–slit interference in (e,2e) ionization of N_2 ($3\sigma_g$) at intermediate energies by looking at the I-factor during the conference.

Acknowledgments

This work was supported by the Scientific and Technological Research Council of Turkey (TUBITAK) through grant 109T738 and AKU, BAPK through grant 14.FENED.06 and 14.HIZ.DES.80. The theoretical work of HC and DM was supported by US National Science Foundation under Grant. No. PHY-1068237.

References

[1] D. Misra *et al* 2006 *Physical Review A*, 74(6), 060701.
[2] N. Stolterfoht *et al* 2001 *Physical Review Letters*, 87(2), 023201.
[3] D. Milne-Brownlie *et al* 2006 *Physical Review Letters*, 96(23), 233201.
[4] E. S. Casagrande *et al* 2008 *Journal of Physics B: Atomic, Molecular and Optical Physics*, 41(2), 025204.
[5] Z. N. Ozer *et al* 2013 *Physical Review A*, 87(4), 042704.
[6] J. Gao *et al* 2005 *Physical Review A*, 72(3), 032721.
[7] A. J. Murray *et al* 2006 *J. Phys. B: At. Mol. Opt. Phys*, 39, 3945-3956.
[8] L. R. Hargreaves *et al* 2009 *Physical Review A*, 80(6), 062704.
[9] H. Chaluvadi *et al* 2015 submitted to *J. Phys. B: At. Mol. Opt. Phys.*

[1]E-mail: zehraerengil@aku.edu.tr

[2]E-mail: mdogan@aku.edu.tr

Content from this work may be used under the terms of the Creative Commons Attribution 3.0 licence. Any further distribution of this work must maintain attribution to the author(s) and the title of the work, journal citation and DOI.
Published under licence by IOP Publishing Ltd

Double Ionization DNA bases by electron impact

A.Mansouri*[,1], M.F. Khelladi* and C. Dal Cappello[$]

*LPQSD, Department of physics, Faculty of Science, University Sétif-1, 19000, Setif, Algeria
[$]SRSMC, UMR CNRS 7565, University of Lorraine, BP 70239, 54506 Vandoeuvre-les-Nancy, France

Synopsis (e,3e) fivefold differential cross sections for the DNA bases are studied. The molecular wave functions are obtained by the multicenter wave functions from the Gaussain 03 program. To describe the correlations between ejected electron in the exit channel, we use the Ward-Macek method. The first order mechanisms involving the reaction are identified.

The ionization of biological molecules by electron impact is a fundamental step to understand the radiation damage in biological molecules and precisely in the DNA and RNA bases. It is well known that low energy electrons (less than 20 eV), resulting from the (e,3e) reaction are responsible for the main damage in biological systems[1].

We investigate here the double ionization of DNA molecules by electron impact in the experimental conditions of Casagrande et al. [2] where the scattered electron energy is fixed at 500 eV and the scattered angle θ_s = - 6°. The ejected electron energies are chosen to be (E_1,E_2) = (12eV,12eV), (37eV,12eV), (37eV,37eV). Fivefold cross sections(FDCS) are then calculated in the first Born approximation where the correlations between ejected electrons are described by the Ward-Macek method [3].

Firstly we calculate cross sections as a function of ejected energies and scattered angle for the DNA bases. We represent in (figure1) the thymine cases. the FDCS is maximum for small scattered angles and for 12 eV< (E1+E2) <17.5 eV. In fact this corresponds to the energy range of the electrons that cause most damages in the branches of the DNA[1]. In a second step, we calculate cross section by considering experimental conditions [2], as a function of ejected angles. The method is applied to study the FDCS for adenine, cytosine, guanine and thymine.

The results show that the two first order mechanisms TS1 and SO are identified in the structures of all four molecules sometimes with the same intensity, which means that TS1 mechanism is also present both with the SO. This is due to the softer description of the repulsion between ejected electrons during the reaction by the Ward-Macek method. As an example, we report in (figure2) the FDCS for the adenine molecule in the situation: (E_1,E_2) = (37eV,12eV).

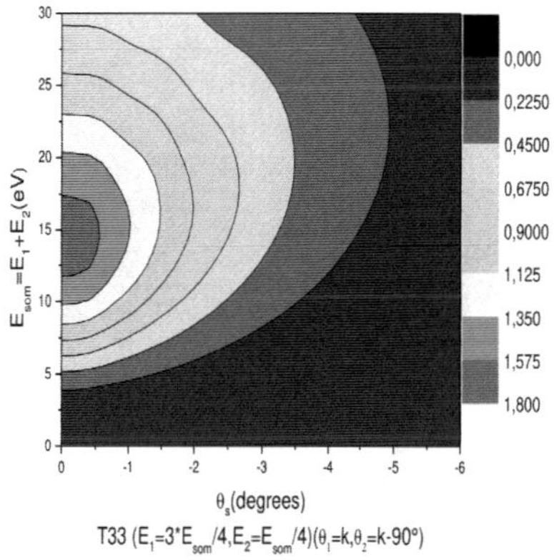

figure1: FDCS for thymine molecule .The scattered electron energy E_s = 500 eV and the scattering angle θ_s = - 6°.

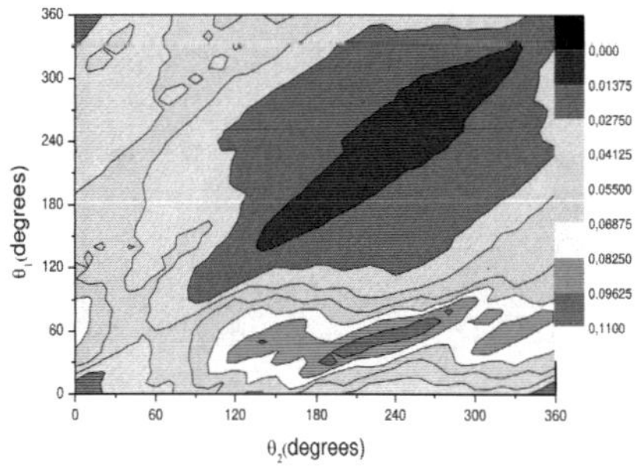

References

[1] B. Boudaiffa, P. Cloutier, D. Hunting, M.A. Huels, L. Sanche, *Science 287, 1658 (2000)*

[2] E M Staicu Casagrande C Li[1,2], A Lahmam-Bennani[1,2] and C Dal Cappello *J. Phys. B: At. Mol. Opt. Phys. 47 (2014) 115203*

[3] S. J. Ward and J. H. Macek, *Phys. Rev. A 49, 1049 (1994).*

[1]E-mail: mansouria@univ-setif.dz

Content from this work may be used under the terms of the Creative Commons Attribution 3.0 licence. Any further distribution of this work must maintain attribution to the author(s) and the title of the work, journal citation and DOI.
Published under licence by IOP Publishing Ltd

Electron Impact Ionization and Dissociation of N_3^+

P. Defrance[1], **M O. A. El Ghazaly** [1-2] **and J.J. Jureta**[1-3]

[1]IMCN, Catholic University of Louvain, Chemin du Cyclotron, 2, 1348 Louvain-la-Neuve, Belgium
[2] Astrophysics and Space Sciences Section, Jet Propulsion Laboratory/Caltech, Pasadena CA 91109, USA
[3] Institute of Physics, University of Belgrade, Belgrade 11081, Serbia

Synopsis Absolute cross sections for electron impact ionization and dissociation of N_3^+ ions have been measured in a crossed-beams apparatus. In this experiment, three product ions were collected separately: N_3^{2+}, N^+ and N_2^+. The single ionization process (formation of N_3^{2+}) shows a threshold around 22 eV with a maximum cross section of $(1.61 \pm 0.01)\times 10^{-17}$ cm^2 at 120 eV. The two dissociative reactions leading to N^+ and N_2^+ fragments are also analyzed in detail.

N_3^+ has been identified as an important ionic constituent of the lower atmosphere of Titan, and there has been a significant theoretical interest in determining its structure and reactivity. Its lowest energy states are a triplet open chain structure, while there is also a cyclic singlet form. These have been identified by means of photoionization experiments [1].

Molecular ions are found to be dominant constituents in many plasmas, whether astrophysical or industrial. They are subject to various excitation and ionization reactions, which, most of the time, result in dissociation of the molecular ion. Electron impact ionization and dissociation of single molecular ions is currently studied in a crossed beams experiment [2], and the measurement is a continuation of our studies of nitrogen containing species [3].

The set-up makes use of the animated crossed electron-ion beam method. N_3^+ molecular ions were formed in an ECR ion source [2], with pure ^{14}N, in order to avoid possible contamination. A relatively high nitrogen pressure (about $6.\times10^{-3}$ mbar) was required to obtain a significant N_3^+ ion current, which however never exceeded 1 nA, after acceleration at 8 keV kinetic energy. The mass selected ion beam crosses the electron beam at right angles. The electron energy can be tuned down to a few eV up to 2.5 keV, allowing observation of reactions from the energy threshold up to the asymptotic high energy limit. The ionic products resulting from of the e^- - N_3^+ interactions are separated from the primary beam by a 90° analyzing magnet and guided to the detector via a spherical electrostatic deflector.

Figure 1 shows the absolute cross section for electron impact ionization of N_3^+ leading to the formation of the doubly-charged N_3^{2+} ion, as function of the energy of the incoming electron. The cross section exhibits a maximum of $(1.61\pm 0.01) \times 10^{-17}$ cm^2 around 120 eV.

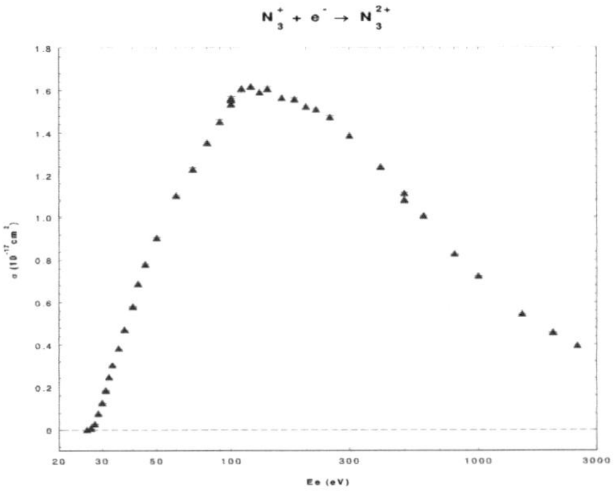

Figure 1. Absolute Cross sections for electron impact ionization of N_3^+.

Measurements were performed for the dissociative reactions leading to the formation of N^+ and N_2^+ fragments. These results are being analyzed, in order to accurately determine the cross section for each dissociative excitation and ionization reaction. Detailed results will be presented and discussed at the Conference.

References

[1] P.C. Samartzis et al. J. Chem. Phys. 123, 051101, 2005
[2] Lecointre *et al* 2006 J. Phys. B. 39, 3275-3297
[3] M. O A El Ghazaly, J.B.A. Mitchell, J.J. Jureta, P. Defrance, 2014 J. Phys. Chem.

E-mail: mohamed.el.ghazaly@jpl.nasa.gov

Content from this work may be used under the terms of the Creative Commons Attribution 3.0 licence. Any further distribution of this work must maintain attribution to the author(s) and the title of the work, journal citation and DOI.
Published under licence by IOP Publishing Ltd

Journal of Physics: Conference Series **635** (2015) 072075 doi:10.1088/1742-6596/635/7/072075

Double Differential Cross Section Measurements for Electron Impact Ionization of Atmospheric Gases

Melike Ulu*[1], Zehra Nur Ozer*, Murat Yavuz* and Mevlut Dogan*[2]

* Department of Physics, e-COL Laboratory, Afyon Kocatepe University, 03200 Afyonkarahisar, Turkey

Synopsis We present comprehensive experimental measurements of double differential cross sections (DDCS) for atmospheric gases, Ar, N_2, CO_2 and CH_4. DDCSs were obtained by measuring the energy and angular distributions of one of the two outgoing electrons which are indistinguishable by use traditional electron spectrometer with an electron analyzer.

Collision techniques has provided a description the structure of matter at the atomic or molecular level and explained natural laws. Applications of the results from collision physics are of most importance to atmospheric science, and meteorological phenomena. Recently, an intensive effort of experimental and theoretical work has been devoted to the study of ionization differential cross-sections of atoms and molecules by electron impact [1-4]. Doubly differential cross sections (DDCS) of single ionization, as a function of ejected energy, and the angle of the ejected electron, contain valuable information about both the collision dynamics and the internal structure of atomic or molecular systems.

The single ionizing collision between a projectile electron and a target ends up two of outgoing electrons called scattered and ejected electrons which are indistinguishable as feature of the ionization events. In the DDCS measurements, scattered or ejected electrons are detected by an electron energy analyzer rotating around the collision centre in a plane. Outgoing electrons from collision event have been focused at the entrance of the analyzer. The analyzer can be rotated around the target beam axis in the desired angle.

Ionization behavior of atmospheric gases are important especially ionosphere or plasma environment. We measured double differential cross-sections of Ar, N_2, CO_2 and CH_4 targets at intermediate electron impact energy (Figure 1). For all atom and molecules, we observed ionization peaks for lower ejected electron energies at $E_0 > 200$ eV as expected. We will present all data for other incident and ejected electron energies at conference.

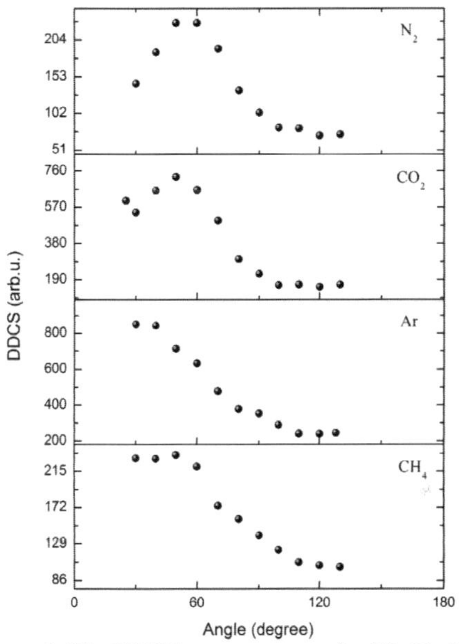

Figure 1. The DDCS measurements for 50 eV ejected electron energy of Ar, N_2, CO_2 and CH_4 at 250 eV incident electron energy

Acknowledgments

This work was supported by the Scientific and Technological Research Council of Turkey (TUBITAK) through grant 109T738 and AKU, BAPK through grant 14.FEN.EDE.05.

References

[1] M. Dogan et al 2013 J. Spect. 2013 192917
[2] M. Yavuz et al 2014 Act. Phys. Pol. A. 125 442
[3] M. Yavuz et al 2014 Can. J. Phys., 92(12) 1676-80.
[4]G. Bozkurt et al 2014, Act. Phys. Pol. A., 125,343.

[1] E-mail: mzekiler@aku.edu.tr

[1] E-mail: mdogan@aku.edu.tr

Content from this work may be used under the terms of the Creative Commons Attribution 3.0 licence. Any further distribution of this work must maintain attribution to the author(s) and the title of the work, journal citation and DOI.
Published under licence by IOP Publishing Ltd

Journal of Physics: Conference Series **635** (2015) 072076 doi:10.1088/1742-6596/635/7/072076

Measurements of the fragmentation cross sections of pyrimidine and tetrahydrofuran for electron energies of 50 eV, 100 eV and 300 eV

Alexander Arndt*[,1], Benedikt Rudek*, Woon Yong Baek* and Hans Rabus*

* Physikalisch-Technische Bundesanstalt (PTB), Braunschweig, Germany

Synopsis An experiment has been developed to study fragmentation cross sections of biomolecules by different radiation sources. In this study, cross sections for the fragmentation of pyrimidine and tetrahydrofuran molecules after impact ionization by incident electrons with energies of 50 eV, 100 eV and 300 eV have been measured.

As the production of strand breaks as physical damage to the deoxyribonucleic acid (DNA) molecule is difficult to observe directly, the induction of such lesions is commonly estimated from simulations of particle track structure. This involves simplifying assumptions, e.g. that each ionization has a fixed probability to lead to a break in the DNA backbone [1]. This probability is a free model parameter that is determined indirectly by radiobiological assays [2].

Within the framework of the EMRP Project "BioQuaRT" [3], we set up an experiment to determine the fragmentation cross sections of analogues of the DNA building blocks. These data will be used in an advanced track structure code which simulates radiation effects on DNA based on the radiation interaction cross sections of its constituents *without* any free parameters.

The experimental apparatus, based on a time-of-flight mass spectrometer (TOF), is designed for operation with different radiation sources. In this work, a pulsed electron gun was used (pulse length about 200 ns). Measurements were performed for electron energies of 50 eV, 100 eV and 300 eV. The target gases were pyrimidine and tetrahydrofuran which are analogues of two DNA bases and the DNA sugar, respectively.

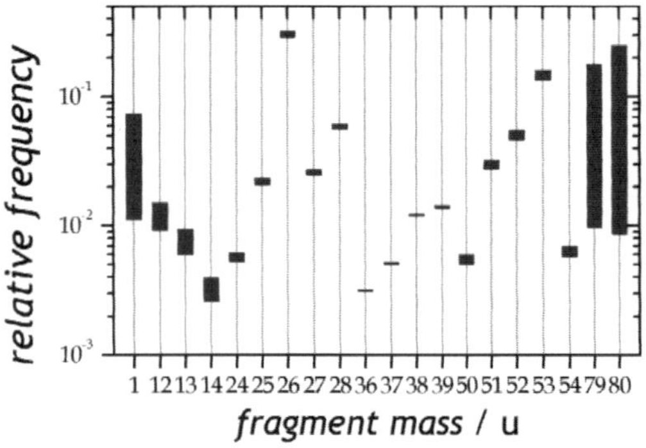

Figure 1. Relative abundance of fragment ions obtained after the ionization of pyrimidine by 100 eV electrons. The colored areas indicate the uncertainty due to the initial kinetic energy of the fragments.

To study the dependence of the spectrometer transmission on mass and initial kinetic energy of the fragment ions, different time gaps between ionization and extraction were used. This enables a more accurate determination of the initial distribution of fragments.

In addition, the kinetic energy release was estimated by comparing the peak shape within a TOF spectrum to results of SIMION simulations. The outcome indicates that it is possible to study the energy distribution of fragments carrying high kinetic energies such as hydrogen ions produced during a Coulomb explosion.

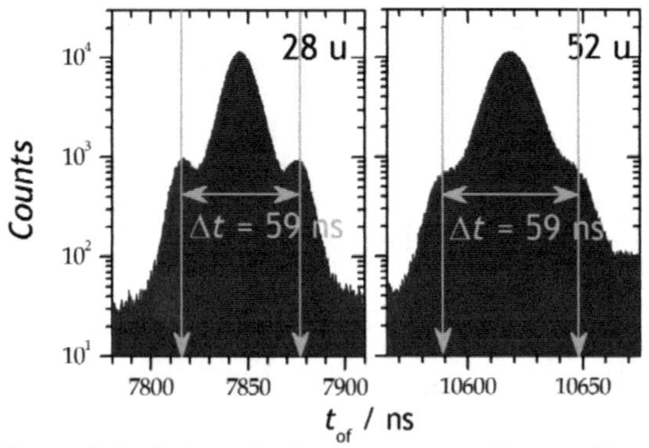

Figure 2. Peak shape for fragment masses 28 u and 52 u, observed after ionization of pyrimidine by 100 eV electrons From the measured time gap between the shoulders, a kinetic energy release of 2.6 eV can be estimated.

This work was supported within the EMRP Joint Research Project SIB06 BioQuaRT. The EMRP is jointly funded by the EMRP-participating countries within EURAMET and the European Union.

References

[1] G. Garty *et al* 2006 Radiat. Prot. Dosim. **122** 451
[2] G. Garty *et al* 2010 Phys. Med. Biol. **55** 761
[3] website: http://www.ptb.de/emrp/bioquart.html

[1]E-mail: alexander.arndt@ptb.de

Content from this work may be used under the terms of the Creative Commons Attribution 3.0 licence. Any further distribution of this work must maintain attribution to the author(s) and the title of the work, journal citation and DOI.
Published under licence by IOP Publishing Ltd

XXIX International Conference on Photonic, Electronic, and Atomic Collisions (ICPEAC2015) IOP Publishing
Journal of Physics: Conference Series **635** (2015) 072078 doi:10.1088/1742-6596/635/7/072078

High Resolution Electron Attachment to Water Clusters in Helium Droplets

Elias Jabbour Al Maalouf* [1], **Katrin Tanzer *** [1], **Paul Scheier*** [1], **Eugen Illenberger** † [2], **Stephan Denifl** [1]

* Institute for Ion and Applied Physics, University of Innsbruck, Technikerstrasse 25, 6020 Innsbruck, Austria
† Institut für Chemie, Physikalische und Theoretische Chemie Freie, Universität Berlin, Takustr. 3, 14195-Berlin, Germany

Synopsis: We have determined the resonance energies for free electron (up to 3eV) attachment to water clusters embedded in helium droplets. In comparison to pure water clusters, shifts and broadening of the resonance features were observed, which shows a pronounced dependence on the water cluster anion size.

Hydrated electrons are of importance in fusion reactions, water droplets in the atmosphere, and living tissue [1]. In a supersonic expansion experiment, it was shown that free electron attachment to water clusters forming $(H_2O)^-_n$ anions with $n \geq 11$ that are detectable by mass spectrometry were achieved [2]. In contrast, we previously studied electron attachment to $(H_2O)_n$ with $n \geq 2$ in helium droplets using a standard energy source with a resolution of around 1 eV [1]. The maximum of the resonance of low energy electron attachment to water clusters in this experiment was reported at 1.5 eV, which is shifted in comparison to that in the gas phase [1]. This is due to the fact that the projectile electron loses energy to penetrate the helium droplet. After penetration, the electron creates a bubble and moves freely in the superfluid helium droplet. In the case of pure helium droplets, the resonance of the electron capture by helium droplets has a maximum at 1.8-2.3 eV with a threshold at 1.26-1.32 eV, depending on the helium cluster size [3]. In the presented work, we studied electron attachment to water clusters in helium droplets using a high resolution electron beam (100 meV) formed by a hemispherical electron monochromator.

In the conducted experiment, precooled (9-11K) He gas at a pressure of 20 bars expands through a 5 μm nozzle into vacuum in a supersonic expansion. This further decreases the temperature of the He, creating helium nano droplets that are skimmed to a beam. The beam passes through a pick up cell containing water vapor. Water molecules are picked up and cooled down in the He droplet which leads to the formation H_2O^- clusters inside the He droplet. The electron beam crosses the doped helium beam. Formed $(H_2O)^-_n$ anions are selected using a quadrupole mass spectrometer and detected in a channel electron multiplier.

In contrast to the previous measurements, which have much less accuracy in terms of energy beam [1], we observe a sharp resonance for $(H_2O)^-_2$ with a threshold at 0.92 eV, a maximum at 1.14 eV and a full width at half maximum (FWHM) of 250 meV (see Figure 1). This results in a more precise value for the electron surface barrier of the helium droplet.

Measurements of electron attachment for different water cluster sizes in helium droplets were acquired. The resonance for $(H_2O)^-_3$ showed a maximum at 1.26 eV, shifted from that of $(H_2O)^-_2$. In addition, a structure appears at the high energy side broadening the peak. Shifting and broadening of the resonance continues as size of the $(H_2O)^-_n$ increases, for which $(H_2O)^-_{16}$ is shown as an example (see Figure 1).

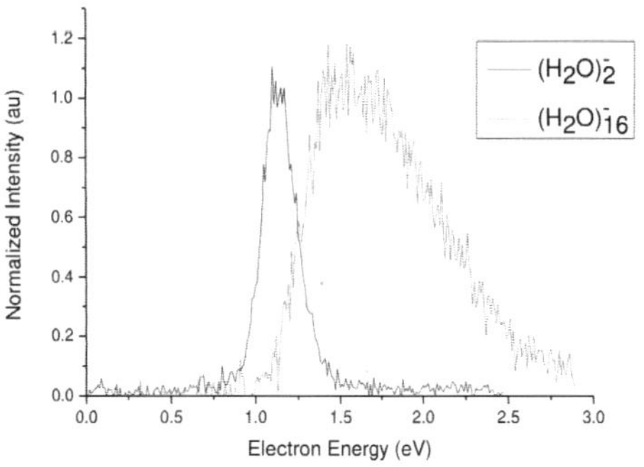

Figure 1. The ion yield of $(H_2O)^-_2$ and $(H_2O)^-_{16}$ clusters formed upon electron attachment to water clusters in helium droplets.

This work was supported by FWF (P24443).

References:

[1] Fabio Zappa et al 2008 J.A.C.S. 130, 5573–5578
[2] H. Haberland et al 1984 J. Chem. Phys. 81, 3742–3744
[3] Ulrich Henne and J. Peter Toennies 1998 J. Chem. Phys. 108, 9327

[1] E-mail: elias.jabbour-al-maalouf@uibk.ac.at

Content from this work may be used under the terms of the Creative Commons Attribution 3.0 licence. Any further distribution of this work must maintain attribution to the author(s) and the title of the work, journal citation and DOI.
Published under licence by IOP Publishing Ltd

Journal of Physics: Conference Series **635** (2015) 072079 doi:10.1088/1742-6596/635/7/072079

Electron impact cross-sections for biomolecules – completeness and self-consistency via swarm analysis

J. de Urquijo[1], M. Casey[2], D. Konvalov[2], M. J. Brunger[3], G. Garcia[4], Z. Petrovic[5] and R. D. White*[2]

[1] Universidad Nacional Autónoma de México, Cuernavaca, Mor., Mexico
[2] College of Science, Technology and Engineering, James Cook University, Townsville, Australia
[3] Flinders University, Adelaide, South Australia
[4] CSIC, Madrid, Spain
[5] Institute of Physics, Belgrade, Serbia

Synopsis The accuracy and completeness of electron impact cross-sections in water and tetrahydrofuran are assessed through comparison with experimental electron transport coefficients using the pulsed-Townsend experiment.

The determination of a comprehensive set of electron-biomolecule cross-sections is fundamental to understanding electron-induced processes arising in plasma medicine and radiation damage modelling. Formulation of complete sets is generally based on a critical assessment of available experimental 'beam' studies and theoretical calculations, and interpolations/extropolations. Issues of completeness and accuracy of cross-section sets arise.

Swarm experiments provide one of the key discriminating tests on the accuracy and completeness of cross-section sets. In swarm experiments, excess electrons under the action of an applied electric field are passed through a gas of known temperature and pressure. Measured currents are interpreted in terms of transport coefficients such as drift velocities, diffusion coefficients, and ionization/attachment rates. Various energy regimes of the cross-sections can be preferentially accessed by varying the electric field which modifies the velocity distribution of the electrons. Through comparison of measured transport coefficients with those calculated from transport theory/ simulation, one can assess the ability of the cross-section set to accurately treat particle, momentum, and energy balance. Cross-section degeneracy (i.e., different sets of cross-sections can produce the same field dependence of the transport coefficients) is an issue. Additional handles to assess the degeneracy, accuracy and completeness of the cross-section set can be through measurements in admixtures with well known cross-sections which can considerably modify the distribution function.

In this presentation we report on recent swarm measurements in the biomolecules of water and tetrahydrafuran (THF) using the pulsed-Townsend technique of the de Urquijo group. We present and assess the consistency of cross-section sets for water and THF (in various admixtures) through a comparison of calculated transport coefficients using a multi-term Boltzmann equation solution with the available experimental swarm measurements provides a discriminating test on consistency and accuracy of the cross-section sets. Sample results for electron transport in gaseous water/helium mixtures are presented in Figure 1 [1]. Further results will be presented for water/Ar mixtures and THF/N_2 mixtures.

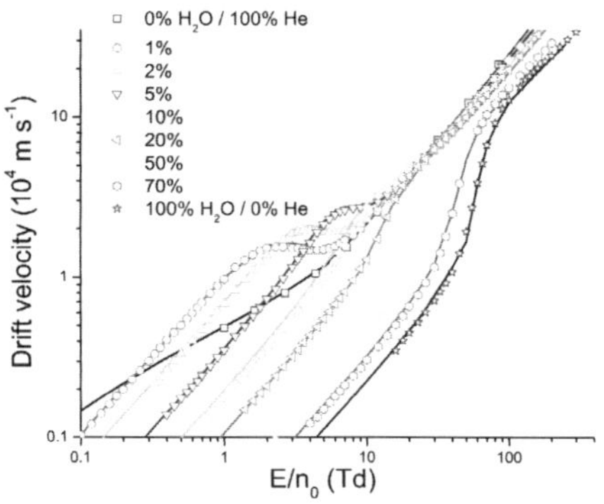

Figure 1. Comparison of experimental and calculated drift velocities for electrons in gaseous water/Ar mixtures.

References

[1] J. de Urquijo, E. Basurto, a M. Juárez, K. F. Ness, R. E. Robson, M. J. Brunger, and R. D. White (2014) *J. Chem. Phys.* **141**, 014308 (2014)

-

* E-mail: Ronald.White@jcu.edu.au

Content from this work may be used under the terms of the Creative Commons Attribution 3.0 licence. Any further distribution of this work must maintain attribution to the author(s) and the title of the work, journal citation and DOI.
Published under licence by IOP Publishing Ltd

Journal of Physics: Conference Series **635** (2015) 072080 doi:10.1088/1742-6596/635/7/072080

Molecular alignment resolved (e,2e) cross sections for H_2 at 38 eV impact energy

Khokon Hossen[* 1], **Alexander Dorn**[* 2], **and Xueguang Ren**[*† 3]

[*] Max Planck Institute for Nuclear Physics, 69117 Heidelberg, Germany
[†] Physikalisch-Technische Bundesanstalt, 38116 Braunschweig, Germany

Synopsis Fully differential cross sections for electron impact ionization of H_2 are determined experimentally using a reaction microscope. For dissociation of the final H_2^+ ion in its electronic ground state the measurement of the proton momentum allows to conclude on the molecular alignment during the collision. The cross sections obtained are compared to published theoretical data.

Kinematically complete studies of the electron impact ionization of small atomic and molecular systems give important information concerning dynamical correlation in few-particle reactions. Here we present a (e,2e) study for molecular hydrogen where the spatial alignment of the molecular axis at the instant of the collision is determined.

The molecular axis alignment with respect to the incoming projectile momentum should be relevant in multiple or higher-order scattering of both active electrons within the molecule. In addition for possible Young's double-slit type interferences due to electron scattering at two indistinguishable centers the molecular alignment determines the observed interference pattern.

In the past only a few (e,2e) experiments have resolved the alignment of the molecular axis. This was realized by studying reactions where the H_2^+ ion in its electronic ground state [1,2] or in excited states [3,4] dissociates. Then the emission direction of the proton allows to determine the molecular axis direction.

Using a reaction microscope we measured the (e,2e) cross section of H_2 molecules at the impact energy of $E_0 = 38$ eV where the outgoing electrons share 20 eV excess energy. For this case non-perturbative theoretical calculations demonstrated a strong alignment dependence of the cross section [5] which, so far, was not confirmed experimentally. Furthermore, strong discrepancies between theoretical results and experiments that do not resolve the molecular alignment angle were observed for the coplanar scattering geometry.

For the dissociative ionization of H_2, resonant autoionization can interfere with direct ionization into the dissociating ionic ground state at particular electron emission energies in the range of 5 - 10 eV. In order to separate pure direct ionization, we choose kinematics for $E_2 < 5$ eV at low impact energy where only few autoionizing states are populated. This is facilitated by significant improvements of the spectrometer's electron energy resolution and a new pulsed photoemission electron gun [6].

Results and a detailed discussion will be provided at the conference.

References

[1] A. Senftleben *et al* 2010 *J. Chem. Phys.* **133** 044302
[2] X. Ren *et al* 2012 *Phys. Rev. Lett.* **109** 123202
[3] M. Takahashi *et al* 2004 *J. Electron Spectrosc. Relat. Phenom.* **141** 83
[4] S. Bellm *et al* 2010 *Phys. Rev. Lett.* **104** 023202
[5] J. Colgan *et al* 2008 *Phys. Rev. Lett.* **101** 233201
[6] X. Ren *et al* 2014 *J. Chem. Phys.* **141** 134314

[1] E mail: Khokon.Hossen@mpi k.de
[2] E-mail: Alexander.Dorn@mpi-k.de
[3] E-mail: Xueguang.Ren@mpi-k.de

Content from this work may be used under the terms of the Creative Commons Attribution 3.0 licence. Any further distribution of this work must maintain attribution to the author(s) and the title of the work, journal citation and DOI.
Published under licence by IOP Publishing Ltd

Journal of Physics: Conference Series **635** (2015) 072081 doi:10.1088/1742-6596/635/7/072081

Formation and fragmentation of protonated molecules after ionization of molecular clusters with multiply charged ions

V. Vizcaino*[1], J.-C. Poully*, L. Schwob*, R. Delaunay*, J. Kočišek*, S. Eden†, J.-Y. Chesnel*, A. Méry*, J. Rangama*, L. Adoui* and B. Huber*

* CIMAP (UMR 6252) - CEA, CNRS, ENSICAEN, UCBN, Bd Henri Becquerel, BP 5133, 14070 Caen, France
† Dept. of Physical Sciences, The Open University, Walton Hall, Milton Keynes, MK7 6AA, United Kingdom.

Synopsis Collisions between neutral clusters of biomolecules (amino-acids and lactic acid) and keV ions in the gas-phase lead, in very large majority, to protonated species. We found that the target proton affinity and the nature of the projectile ion influence the protonation of monomers within these clusters. New fragment ion channels were also observed and have been assigned to the statistical dissociation of these protonated molecules formed upon ionization of the clusters. These new dissociation channels exhibit strong delayed fragmentation in the μs time scale.

Ionizing radiation-induced physical and chemical phenomena in biological matter are important, not only from a fundamental point of view, but also for a deeper understanding of radio- and hadron therapy mechanisms. Gas-phase investigations allow extracting the intrinsic properties of molecular systems, because all solvent effects are removed. In this contribution, we present our results on irradiation of gas phase amino acid and lactic acid clusters by ion beams at keV kinetic energy.

Experiments have been performed with a crossed-beam apparatus where the collimated molecular clusters beam formed in a gas aggregation chamber crossed the pulsed ion beam delivered by an Electron Cyclotron Resonance ion source. The cationic products were extracted from the interaction region orthogonally to both beams and mass-over-charge analyzed in a linear Wiley-McLaren time-of-flight mass spectrometer.

After ionization with O^{3+} at 30 keV, monomers and dimers are found protonated to a very large extent for all amino acids studied however, in the case of lactic acid, which has a lower proton affinity, the radical cation corresponds to around 12% of the protonated specie (Figure 1).

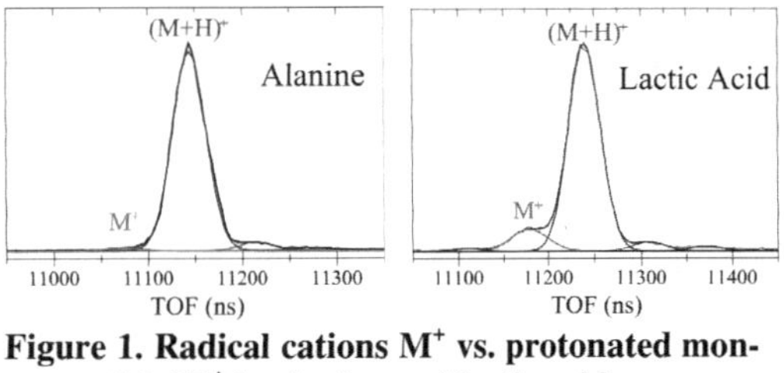

Figure 1. Radical cations M^+ vs. protonated monomers $(M+H)^+$ in alanine and lactic acid.

This ratio decreases to 8% after interaction with a beam of He^+ at 6 keV where the ionization occurs at smaller mean impact parameter and therefore higher excitation energy is transferred during the collision. This would indicate that proton transfer involves a potential energy barrier in ionized lactic acid clusters.

New fragments specific to the ionization/fragmentation of clusters have been found and are assigned to statistical fragmentation from the protonated molecule (Figure 2). Indeed, following multi-ionization, the clusters undergo a succession of evaporation and fission processes to accommodate the excess of charges and energy transferred in the collision. After these relaxation mechanisms, protonated monomers can still have a high internal energy and fragment in similar pathways (loss of CO) than in CID experiments.

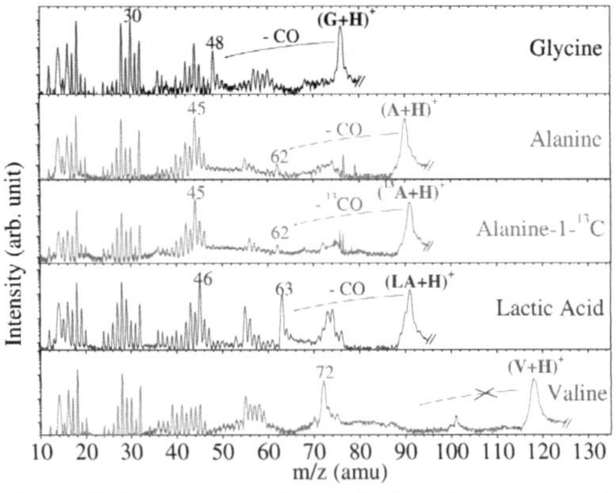

Figure 2. Cationic products (below the monomer) of the interaction between molecular clusters and O^{3+} ions at 30 keV.

V.V., J.K. and S.E. acknowledge, respectively, the financial support from the CRBN (12P04578), from INCa-ITMO (N° PC201307), and from the British EPSRC (EP/J002577/1 and EP/L002191/1).

[1] E-mail: vizcaino@ganil.fr

Content from this work may be used under the terms of the Creative Commons Attribution 3.0 licence. Any further distribution of this work must maintain attribution to the author(s) and the title of the work, journal citation and DOI.
Published under licence by IOP Publishing Ltd

Journal of Physics: Conference Series **635** (2015) 072082 doi:10.1088/1742-6596/635/7/072082

Cluster-Electron Interaction for Poly-Anion Production

F. Martinez* [1], **S. Bandelow**†, **G. Marx**†, **L. Schweikhard**†, **A. Vass**†

* Institute of Physics, University of Rostock, 18501 Rostock, Germany
† Institute of Physics, University of Greifswald, 17487 Greifswald, Germany

Synopsis By interaction with electrons in ion storage devices (ion-cyclotron-resonance and radio-frequency traps) negatively charged clusters of gold and aluminum have been produced up to the 6th and 10th charge state, respectively. The production of these poly-anions opens exciting new possibilities to measure their lifetimes, to monitor their relaxation schemes after laser radiation, as well as to probe their Coulomb barriers.

The properties of a nano-sized cluster, e.g. its geometrical shape, ionization potential, polarizability or dissociation energy, are determined not only by its size, but also by its charge state. In the present study, poly-anionic metal clusters are produced by electron attachment to cluster mono-anions stored in Penning and Paul traps [1].

In a Penning trap, charged particles are confined up to a trap-specific mass-to-charge limit, allowing simultaneous storage of both, cluster anions and electrons. In the so-called electron bath [2], attachment of electrons occurs if their energy is high enough to overcome the anions' Coulomb potential (Fig. 1). However, as the co-trapped electrons are produced along the axis of the trap potential, they exhibit a corresponding energy distribution and their actual energy during attachment remains somewhat uncertain.

In contrast, a technique for poly-anion production in a Paul trap at well-defined electron energies has been developed. By applying tailored radio-frequency trapping-voltage signals [3], an electron beam of defined energy can be guided through a linear Paul trap for electron attachment to stored cluster ions [1]. At appropriate storage conditions, the resulting poly-anions remain trapped and are available for further experiments.

Recent measurements on the cluster poly-anion abundance as a function of interaction time in the electron-bath of a cylindrical Penning trap confirm a sequential electron-attachment process. The electrons can be released from the trap while the poly-anionic clusters remain stored and are available for further investigation.

Among the topics of interest are lifetime measurements of meta stable poly-anionic clusters, i.e. systems too small to provide a positive binding energy for another surplus electron. However, the surplus electron is yet meta-stably bound to the cluster by means of the Coulomb potential, through which it tunnels and escapes at some point [4].

In addition, electron detachment may also be triggered by laser excitation of the cluster poly-anions [5]. It competes - as a function of both cluster size and charge state – with neutral atom evaporation.

Furthermore, by scanning the photon energy the electron detachment energy may be determined, thus probing the Coulomb-barrier heights. Respective results may then be compared with those from electron-attachment studies performed in Paul traps with well-defined electron energies.

The project was supported by a Collaborative Research Center of the DFG (SFB 652-TP A03).

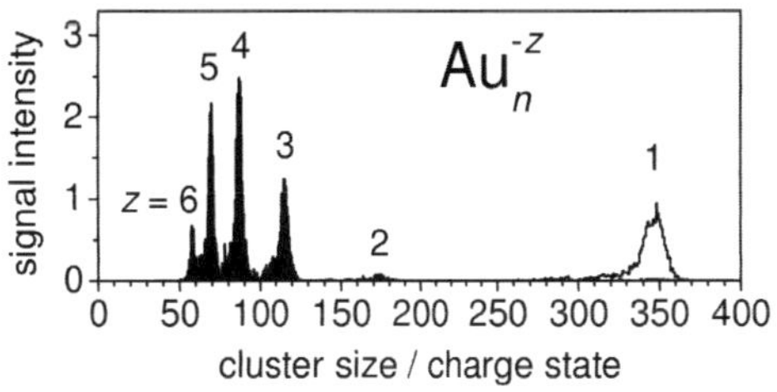

Figure 1. Gold cluster anions Au_n^{-z} before (open curve) and after application of an electron bath in a Penning trap (filled curve).

References

[1] F. Martinez *et al* 2013 *AIP Conf. Proc.* **1521** 230
[2] A. Herlert *et al* 1999 *Phys. Scr.* **T80** 200
[3] S. Bandelow *et al* 2013 *Int. J. Mass Spectrom.* **353** 4953
[4] F. Martinez *et al* 2015 *J. Phys. Chem. C* doi:10.1021/jp510947p
[5] A. Herlert *et al* 2012 *New J. Phys.* **14** 055015

[1] E-mail: franklin.martinez@uni-rostock.de

Content from this work may be used under the terms of the Creative Commons Attribution 3.0 licence. Any further distribution of this work must maintain attribution to the author(s) and the title of the work, journal citation and DOI.
Published under licence by IOP Publishing Ltd

Journal of Physics: Conference Series **635** (2015) 072083 doi:10.1088/1742-6596/635/7/072083

Theoretical investigation of the double differential cross section for electron impact ionization of methane molecule

Murat Yavuz*[1], Mohammed Sahlaoui†[2], Mammar Bouamoud, Boumedien Lasri‡, Mevlut Dogan***

* Department of Physics, e-COL Laboratory, Afyon Kocatepe University, Afyonkarahisar, Turkey
† Theoretical Physics Laboratory, Abou Bekr Belkaid University, Tlemcen, Algeria
‡ Dr Tahar Moulay University, Saïda, Algeria
** Department of Science and Technology, University Center of Naâma, Algeria.

Synopsis We present theoretical calculations of double differential cross sections (DDCS) for methane molecule. Angular distributions of secondary electrons have been measured after ionizing electron collisions with methane at primary energy of 350 eV using a conventional electron spectrometer. The theoretical model used is the first and second Born approximations with Gamow factor. It is found that these theoretical calculations are successful to describe the post-collision interaction effects due to the Coulomb long-range interaction between the outgoing electrons in the final state.

Over the past two decades there has been an increasing effort to fully understand the process of single ionization in electron-atom/molecule collisions from experimental and theoretical perspective. Much of this interest is focused on molecular structures upon the importance in a variety of applications as planetary, stellar and cometary atmospheres. Methane (CH_4) is one of the most rich and important hydrocarbon compounds. The electron impact cross sections of CH_4 have been studied by several experimental groups over a period of many years. However these studies limited to high impact energies for double differential cross section measurements. In this work, we report double differential cross sections in comparison with experimental measurements with two orders of Born approximations.

Experiments were performed at 350 eV incident electron energy. The apparatus originally developed for coincidence measurements of ejected and scattered electrons, i.e., triple differential cross sections. The same apparatus can also collect data on angular distribution of outgoing electrons simply disabling the coincident circuit [1, 2].

The double differential cross section is calculated by the integration of the triple differential cross section (TDCS) over the direction of one of the outgoing electrons (ejected or scattered). To compute the TDCS we use the formalism developed in the frame of the one Coulomb wave model given in [3, 4]. We multiply the TDCS by the Gamow factor to take into account the post-collision interaction (PCI) effects in the final continuum state. A considerably agreement is found between experimental and theoretical results (Figure 1).

This study presents 2nd Born approximation calculations of DDCS for electron impact ionization of methane molecule for the first time. The comparison results reported in this study will give a very good description of the ionization mechanisms for CH_4 by 350 eV electron impact.

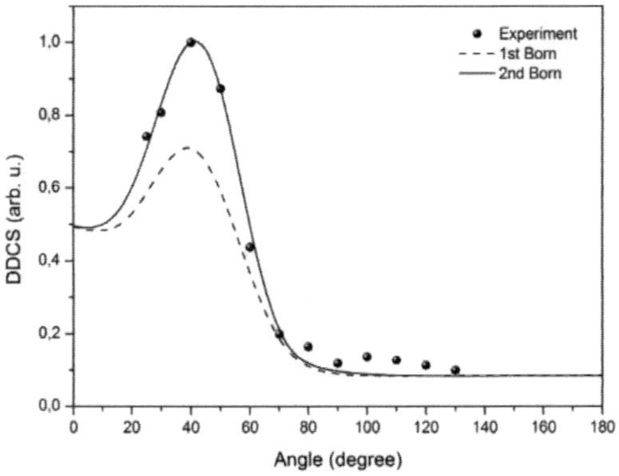

Figure 1. DDCS of methane molecule for incident electron energy 350 eV and secondary electron energy 150 eV.

Acknowledgments

This work was supported by the Scientific and Technological Research Council of Turkey (TUBITAK) through grant 109T738 and AKU, BAPK through grant 14.HIZ.DES.80.

References

[1] M. Yavuz *et al* 2014 *Act. Phys. Pol. A* 125 442
[2] M. Dogan *et al* 2013 *J. Spect.* 2013 192917
[3] M. Yavuz *et al* 2014 *Can. J. Phys.* 92(12) 1676
[4] M. Sahlaoui and M. Bouamoud *J. Phys. B* 45 085201

[1]E-mail: muratyavuz@live.com

[2]E-mail: mohammed.sahlaoui@gmail.com

Content from this work may be used under the terms of the Creative Commons Attribution 3.0 licence. Any further distribution of this work must maintain attribution to the author(s) and the title of the work, journal citation and DOI.
Published under licence by IOP Publishing Ltd

XXIX International Conference on Photonic, Electronic, and Atomic Collisions (ICPEAC2015) IOP Publishing
Journal of Physics: Conference Series **635** (2015) 072084 doi:10.1088/1742-6596/635/7/072084

Molecular structure of one-electron diatomic molecules subject to plasma screening and its effect on the dynamics.

Andrés F. Ordóñez-Lasso*[1], José L. Sanz-Vicario†[2] and Fernando Martín*

* Departamento de Química, Módulo 13,Universidad Autónoma de Madrid, Cantoblanco 28049, Madrid, Spain
† Grupo de Física Atómica y Molecular, Instituto de Física, Universidad de Antioquia, Medellín, Colombia

Synopsis The effect of plasma screening on the electronic and vibrational structure of one-electron diatomic molecules is analyzed by including screened Coulomb interactions among all their particles in the solution of the Schrödinger equation. The breakdown of the well-known separability of the Schrödinger equation in confocal elliptical coordinates, the lost of long-range Coulomb interaction and the non-degeneracy in both the united atom and separated atom limits bring new elements into the analysis of the modified binding energies (electronic and vibrational), adiabatic correlation rules, molecular Stark mixings due to polarization effects, unexpected presence of molecular shape resonances and changes in non-adiabatic and radiative couplings upon the variation of the screening strength.

The introduction of screening in the Coulomb interaction between charged particles appears in many fields of plasma, atomic, molecular and solid state physics. Screening effects may strongly modify the electronic and vibrational structure of molecules and their related properties and interactions. We analyze screening effects in the most simple one-electron molecules, H_2^+ and HeH^{2+}. The two-center (Z_1, e^-, Z_2) unscreened Coulomb problem is well known to be separable and exactly solved in confocal elliptical coordinates so that the electronic wavefunction can be written as the product $\Psi^m(\mathbf{r}; R) = \Lambda_{km}(\xi) M_{qm}(\eta) \Phi_m(\phi)$. The preservation of the number of nodes (k, q, m) from the united atom (UA) limit to the separated atom (SA) one fully determines the UA-SA correlation rules. In contrast, when screening is introduced this separability is lost and one is bound to solve the Schrödinger equation (even in the UA or SA limits), by diagonalizing the Hamiltonian that contains the screened potential $V(r_1, r_2, R) = -Z_1 e^{-\lambda r_1}/r_1 - Z_2 e^{-\lambda r_2}/r_2 + Z_1 Z_2 e^{-\lambda R}/R$. We use an expansion in the form $\Psi^m(\mathbf{r}; R) = \sum_{i\ell} c_{i\ell}^m (\xi^2 - 1)^{|m|/2} B_i(\xi) \Theta_\ell^m(\eta) e^{im\phi}$, in terms of B-splines $B_i(x)$ and associated Legendre polynomials $\Theta_\ell^m(x)$ and obtain very accurate energies and wavefunctions to construct the new distorted adiabatic correlation diagrams ($E_n[R]$ vs. R) for any arbitrary value of λ. The change against λ of the binding energy at the new equilibrium distances (R_e, $E_n[R_e]$) for those states that display a minimum ($1s\sigma_g$, $2p\pi_u$, $3d\sigma_g$, $4f\sigma_u$, etc) produces a varying number of supported vibrational states. For large enough λ, many molecular states of the Rydberg series become unbound. Real crossings among states in the unscreened case become now avoided ones in the presence of screening with an interchange of character that depends on the screening strength. In contrast to the unscreened case, where states do dissociate in the SA limit as Stark hybrids, the screened diatomic molecule dissociates into spherical atoms. However, very diffuse non adiabatic radial couplings among states dissociating to the same non-degenerate atomic shell in the SA limit show Stark mixing as R decreases due to polarization effects. Both radial and radiative dipolar couplings are computed and compared for $\lambda \in [0, 1.5]$ (including the unscreened exact case λ=0) to show that some couplings vanish at $R \to \infty$ when $\lambda > 0$. These effects indicate that screened one-electron molecules partially share properties pertaining more to the case of unscreened many-electron diatomic molecules dissociating into neutrals than to the unscreened one-electron molecule. Also, it is known that in the UA limit, the screened hydrogenic atoms like He^+ show a series of shape resonances in the continuum for $\ell > 0$ [1]. We show that these atomic shape resonances *survive* in the screened molecular case as $R > 0$. In addition, the screening conditions under which Borromean states appear are analyzed. All this structure analysis is required prior to perform any simulation for ion-atom collisions [2] or molecular photodynamics while under the influence of a Debye plasma.

References

[1] M. Bylicki, A. Stachów, J. Karwowski and P. K. Mukherjee 2007 *Chemical Physics* **331** 346

[2] Y. Zhang *et al* 2010 *Phys. Rev. A* **82** 052706

[1]E-mail: andres.ordonez@estudiante.uam.es
[2]E-mail: jose.sanz@udea.edu.co

Content from this work may be used under the terms of the Creative Commons Attribution 3.0 licence. Any further distribution of this work must maintain attribution to the author(s) and the title of the work, journal citation and DOI.
Published under licence by IOP Publishing Ltd

Theoretical analysis of positron-attachment to small molecules: Vibrational enhancement of positron affinities

Masanori Tachikawa*

*Quantum Chemistry Division, Graduate School of NanoBioScience, Yokohama City University, 22-2 Seto, Kanazawa, Yokohama 236-0027, Japan

Synopsis To theoretically demonstrate the binding of a positron to small polarized molecules, we have calculated the vibrational averaged positron affinity (PA) values for small molecules with multi-component quantum Monte Carlo and molecular orbital methods. Our most accurate prediction of the vibrational averaged PA values at the fundamental and overtone states for formaldehyde molecule are 31 and 36 meV, respectively.

The positron is now widely used in both scientific and technological areas. The detail mechanism of such processes, however, is still unclear in the molecular level. A positron affinity (PA) value, which is a binding energy of a positron to an atom or molecule, has now been experimentally measured by Surko and co-workers for many molecular species [1], based on the vibrational Feshbach resonance by incident low-energy positrons. Thus, in order to elucidate the mechanism of the positron binding to molecules, the theoretical analysis including the effect of molecular vibrations is indispensable.

In this study, we developed a new theoretical method for analyzing the effect of molecular vibrations on positron affinities. Our strategy is based on ab initio multi-component quantum Monte Carlo (QMC) [2] and molecular orbital (MCMO) [3, 4] methods for the electronic and positronic wave functions simultaneously, and the anharmonic vibrational state theory using quantum Monte Carlo (QMC) method [5, 6]. In the case of formaldehyde (CH_2O) molecule, the vertical PA value at the equilibrium position is predicted as +25(3) meV with QMC calculation. Applying the anharmonic vibrational analysis, the vibrational excitation of the C=O stretching mode drastically enhances the PA value, whereas the excitation of CH_2 rocking mode deenhances it. We confirmed that such PA variations arise from the change in both permanent dipole moment and dipole-polarizability at each vibrational excited state. Our most accurate prediction of the vibrational averaged PA values at the fundamental and overtone states are 31 and 36 meV, respectively, which strongly supports the conclusion that a positron can bind to formaldehyde [6].

E-mail: tachi@yokohama-cu.ac.jp

References

[1] G. F. Gribakin, J. A. Young, C. M. Surko, Rev. Mod. Phys. **82**, 2557 (2010), J. R. Danielson, J. J. Gosselin, C. M. Surko, Phys. Rev. Lett. **104**, 233201 (2010), J. R. Danielson, A. C. L. Jones, M. R. Natisin, C. M. Surko, Phys. Rev. Lett. **109**, 113201 (2012).
[2] Y. Kita, R. Maezono, MT, M. Towler, R. J. Needs, J. Chem. Phys. **131**, 134310 (2009), **135**, 054108 (2011).
[3] MT, R. J. Buenker, M. Kimura, J. Chem. Phys. **119**, 5005 (2003).
[4] MT, Y. Kita, R. J. Buenker, Phys. Chem. Chem. Phys. **13**, 2701 (2011), New J. Phys. **14**, 035004 (2012).
[5] K. Koyanagi, Y. Takeda, T. Oyamada, Y. Kita, MT, Phys. Chem. Chem. Phys. **15**, 16208 (2013), Y. Kita and MT, Eur. Phys. J. D **68**, 116 (2014).
[6] Y. Yamada, YK, MT, Phys. Rev. A **89**, 062711 (2014).

Content from this work may be used under the terms of the Creative Commons Attribution 3.0 licence. Any further distribution of this work must maintain attribution to the author(s) and the title of the work, journal citation and DOI.
Published under licence by IOP Publishing Ltd

XXIX International Conference on Photonic, Electronic, and Atomic Collisions (ICPEAC2015) IOP Publishing
Journal of Physics: Conference Series **635** (2015) 072086 doi:10.1088/1742-6596/635/7/072086

Potential energy surfaces for anion states of 5-chlorouracil

Fábris Kossoski, Márcio T. do N. Varella[1]
Instituto de Física, Universidade de São Paulo, São Paulo, SP, Brazil

Synopsis We present two-dimensional potential energy surfaces for the negative ions states π_1^*, π_2^* and $\sigma_{\rm CCl}^*$ of the potential radiosensetizer 5-chlorouracil. These were built along the C–Cl stretching mode (reactive coordinate) and an out-of-plane angular displacement of Cl atom (tuning coordinate), as obtained from scattering calculations performed with the Schwinger multichannel method. Analysis of the potential energy surfaces and quantum nuclear dynamics support that chloride elimination arises from an indirect mechanism around 1.5 eV, where the incoming electron is captured into a π^* orbital and is then transfered to the dissociative $\sigma_{\rm CCl}^*$ orbital.

Along the track of ionizing radiation large amounts of low-energy electrons are generated, which are very prone to be captured by biomolecules and trigger dissociative electron attachment (DEA) processes [1]. Halouracils, promising candidates for more efficient cancer therapies, are particularly susceptible to undergo DEA, which accounts for the radiosensitivity ability of these species [2]. However, the underlying molecular mechanisms for such process are still not well understood, and a more complete picture is required in order to fully explore the radiosensetizing activity.

When interacting with 5-chlorouracil, a low-energy electron may be imprisoned into delocalized π^* orbitals or into the localized $\sigma_{\rm CCl}^*$ orbital. At the neutral equilibrium geometry the character and energetics of the anion states were obtained by scattering calculations [3] and by electron transmission spectra measurements [4]. Here we present two-dimensional potential energy surfaces (PES) for the anion states of 5-chlorouracil, which allows for the evaluation of possible DEA mechanisms. The PES were built from a series of fixed-nuclei scattering calculations, performed with the Schwinger multichannel method implemented with pseudopotentials, in the static-exchange plus polarisation approximation.

Figure 1 shows the real component one-dimensional PES along the C–Cl stretching coordinate. Both π^* curves are nearly parallel to the neutral one, while the $\sigma_{\rm CCl}^*$ anion is clearly dissociative upon the bond stretching. The $\pi^*/\sigma_{\rm CCl}^*$ crossings suggest couplings between those states, which might explain the observed C–Cl bond breaking around 0 and 1.5 eV [2, 5, 6]. Since these couplings require a symmetry breaking vibration, an appropriate tuning coordinate must be taken into account, altogether with the reactive C–Cl coordinate. We considered the angular displacement of the chlorine atom away from the molecular plane as the second dimension for the PES, which showed to efficiently mix π^* and $\sigma_{\rm CCl}^*$ orbitals. Nuclear quantum dynamics performed on the PES provides estimates for the DEA cross sections.

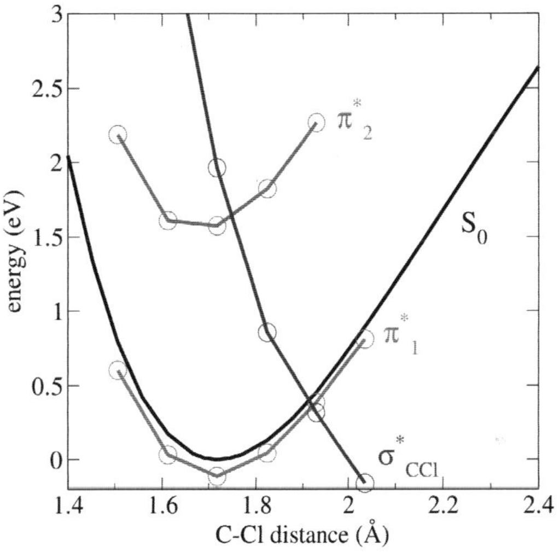

Figure 1. Potential energy curves for π^* and $\sigma_{\rm CCl}^*$ anion states and the neutral ground state of 5-chlorouracil, obtained along the C–Cl stretching coordinate.

The authors acknowledge financial support from Brazilian agencies FAPESP and CNPq.

References

[1] Boudaïffa *et al* 2000 *Science* **287** 1658

[2] H. Abdoul-Carime *et al* 2001 *J. Am. Chem. Soc.* **123** 5354

[3] F. Kossoski *et al* 2014 *J. Chem. Phys.* **140** 024317

[4] A. Scheer *et al* 2004 *Phys. Rev. Lett.* **92** 068102

[5] S. Denifl *et al* 2003 *J. Chem. Phys.* **118** 4107

[6] R. Abouaf and H. Dunet 2005 *Eur. Phys. J. D* **35** 405

[1]E-mail: mvarella@if.usp.br

Content from this work may be used under the terms of the Creative Commons Attribution 3.0 licence. Any further distribution of this work must maintain attribution to the author(s) and the title of the work, journal citation and DOI.
Published under licence by IOP Publishing Ltd

Potential Energy Curves and Generalized Oscillator Strength for Doubly Excited States of the Hydrogen Molecule

L. O. Santos[†1], **R. F. Nascimento**[*], **A. B. Rocha**[‡]. **N. V. de Castro Faria**[†], **Ginette Jalbert**[†]

[†]Instituto de Física, Universidade Federal do Rio de Janeiro, Cx. Postal 68528, Rio de Janeiro, 21941-972, RJ, Brazil

[*]Centro Federal de Educação Tecnológica Celso Suckow da Fonseca - 25620-003, Petrópolis, Brazil

[‡]Instituto de Qumica, Universidade Federal do Rio de Janeiro, Cx. Postal 68528, Rio de Janeiro, 21941-972, RJ, Brazil

Synopsis In this work we report calculations of potential energy curves in the $1.2a.u. \leq R \leq 100a.u.$ range at the Multireference Configuration Interaction (MRCI) level for doubly excited states of the H_2 molecule. The molecular state which dissociates into two H(2s) atoms is clearly identified. The Generalized Oscillator Strength as a function of transferred momentum for three doubly excited states is also presented.

Recently, our experimental group [1] observed the pair H(2s) + H(2s), measuring the neutral fragments in coincidence after the dissociation of H_2 induced by electron impact, confirming thus the existence of such dissociation channel. The purpose of the present work is to shed light on the description of the doubly excited states that may dissociate onto a pair H(2s) + H(2s). In order to do that, we have used an approach that mixes the standard CI method with MCSCF, known as MRCI [2] and implemented a Huzinaga basis-set (which consists of cartesian gaussian-type (12s, 6p, 3d, 1f)/[9s, 6p, 3d, 1f] and provides a total of 110 molecular orbitals: $29\sigma_g$, $29\sigma_u$, $22\pi_g$, $22\pi_u$, 4δ and 4ϕ) in the treatment of two-electron diatomic molecules allowing us to rise accurate potential curves at every internuclear distance. We have used an active space composed of 12 molecular orbitals, $n\sigma_g$ (n = 1-3), $n\sigma_u$ (n = 1-3), $n\pi_g$ (n = 1-2), $n\pi_u$ (n = 1-2) which comprise the main configurations of the electronic states from the Franck-Condon to the dissociative region. The main feature of our calculations is to provide for the first time an unambiguous characterization of the electronic state that goes, in the dissociation limit, to H(2s) + H(2s) fragments.

The Generalized Oscillator Strength (GOS) for some doubly excited states, in the vertical approximation, has been calculated as a function of the momentum transfer within Gauss-Hermite quadrature to the electronic coordinates integrals. The electronic states have been obtained as previously mentioned, i.e., with the MRCI approach. One should note that our technique takes into account highly correlated wavefunctions, allowing us to calculate transition matrices between ground and doubly excited states even in the first Born approximation approach.

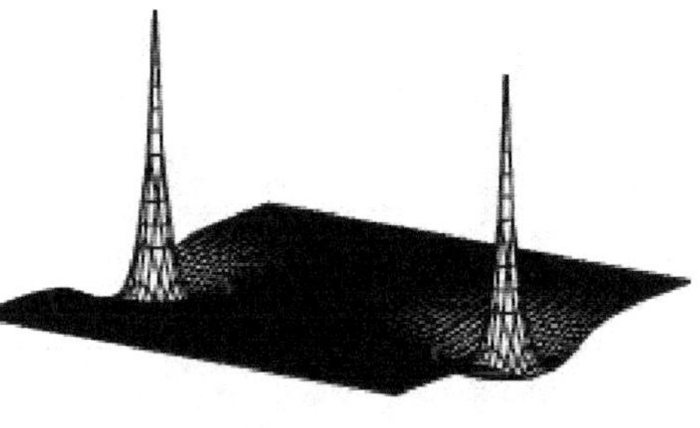

Figure 1. Electronic amplitude calculated at R=80 a.u. for the $^1\Sigma_g^+$ doubly excited state.

In Fig. 1, the electronic amplitude calculated for the doubly excited state $^1\Sigma_g^+$ at a internuclear distance R=80 a.u.. It is clear from this picture that, on the separated atoms regime, both atomic orbitals present one node which is characteristic of a 2s atomic orbital.

References

[1] J. Robert *et al* 2013 *Phys. Rev. Lett* **111**, 183203.

[2] B. H. Botch and T. H. Dunning, Jr. 1982 *J. Chem. Phys.* **76**, 6046.

[1]E-mail: leonardosantos@if.ufrj.br

Content from this work may be used under the terms of the Creative Commons Attribution 3.0 licence. Any further distribution of this work must maintain attribution to the author(s) and the title of the work, journal citation and DOI.
Published under licence by IOP Publishing Ltd

Focused Electron Beam Induced Deposition of 1,1-dichloro-1-silacyclohexane, silacyclohexane and 1,3,5-trisilacyclohexane; preliminary study on the role of low energy secondary electrons in the deposition process.

Oddur Ingólfsson* [1], **Ragesh Kumar TP***, **Cornelis W Hagen**[†], **Sangeetha Hari** [†].

[*] Science Institute and Department of Chemistry, University of Iceland, Dunhaga 3, 107 Reykjavik, Iceland
[†] Department of Imaging Physics, Delft University of Technology, Lorentzweg 1, 2628 CJ Delft, The Netherlands

Synopsis To study the potential role of low energy secondary electrons in focused electron beam induced deposition (FEBID) we have studied FEBID using 1,1-dichloro-1-silacyclohexane, silacyclohexane and 1,3,5-trisilacyclohexane. While the first of these compounds shows appreciable cross sections for dissociative electron attachment (DEA) in the gas phase, DEA is not observed for the latter two. Dissociative ionization, on the other hand is a fairly efficient decomposition path for all three compounds. The performance of these compounds in FEBID is compared with their decomposition through low energy electrons in the gas phase and discussed in context to the role of DEA and DI in FEBID.

Focused electron beam induced deposition (FEBID) is a direct writing technique suitable for the fabrication of three dimensional nanoscale structures. In this technique a tightly focused high-energy electron beam impinges on a surface under a constant stream of precursor molecules. Ideally these decompose through electron-induced processes leaving a well defined deposit, which spatial extension is confined to the focal with of the high-energy electron beam. However, through elastic and inelastic (ionizing) scattering of the high-energy electron beam, large amount of secondary electrons (SEs) are produced outside the focal point of the primary beam. These SEs are capable to dissociate precursor molecules through dissociative electron attachment (DEA), dissociative ionization (DI), neutral dissociation (ND) and dipolar dissociation (DD) - processes that have distinctly different characteristics with regards to the products formed and their energy dependency. Recent gas phase studies [1-3] showed that commonly used precursor molecules can have considerable DEA and DI cross sections and that these processes may thus play an important role in FEBID. Especially these processes are potentially responsible for broadening of the deposit beyond the focal width of the primary beam.

In the current contribution we put effort into comparing the relative role of DEA and DI in FEBID by comparing the performance of 1,1-dichloro-1-silacyclohexane (DCSCH), silacyclohexane (SCH) and 1,3,5-trisilacyclohexane (TSCH) in FEBID with gas phase, experiments where these molecules all decompose through DI while DEA is only observed for DCSCH [3].

For comparison we grew series of pillars using DCSCH and TSCH with increasing electron beam exposure time. The base diameter of the pillars grown with TSCH is saturating much sooner (at ~97nm) than those of DCSCH (at 160nm). Similarly the diameter of neighboring pillars, in close proximity to growing pillars, is found to be stronger influenced (broadened) in DCSCH. Figure 1 shows this effect for arrays of nanopillars from SCH and DCSC, deposited from top to bottom and left to right. While the first and last grown pillars for SCH shows more or less same diameter, there is a considerable difference for DCSCH.

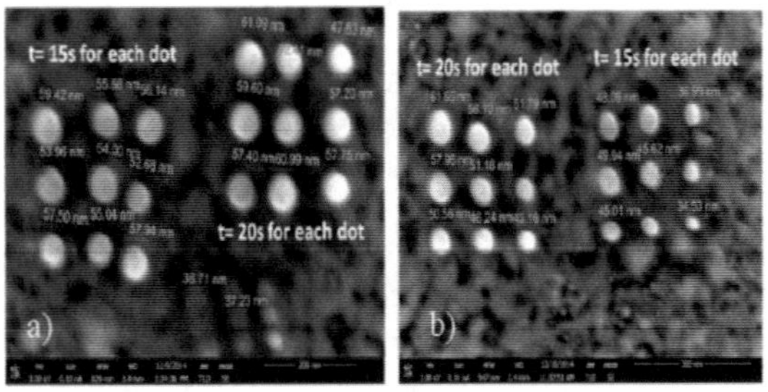

Figure 1. Array of nanopillars deposited with a) silacyclohexane and b) 1,1-dichloro-1-silacyclohexane.

In the current contribution we discuss these FEBID results in context to the efficiency of DI and DEA to these compounds in the gas phase.

Acknowledgement: The authors acknowledge support from the Icelandic Centre for Research and the University of Iceland Research Fund. RKTP acknowledges STSM financial support from the COST Actin CELINA; CM1301.

References

[1] S. Engmann *et al.* 2011 *Angew. Chem. Int. Ed.* **50** 9475
[2] S. Engmann, *et al.* 2012, *Phys. Chem. Chem. Phys.* **14** 14611
[3] E. H. Bjarnason, *et al.* 2011 *Int. J. Mass Spectrom.* **370** 39

[1] E-mail: odduring@hi.is

Content from this work may be used under the terms of the Creative Commons Attribution 3.0 licence. Any further distribution of this work must maintain attribution to the author(s) and the title of the work, journal citation and DOI.
Published under licence by IOP Publishing Ltd

Journal of Physics: Conference Series **635** (2015) 072089 doi:10.1088/1742-6596/635/7/072089

Three-body calculation of exotic molecules with negatively charged massive particle

Yasushi Kino* [1]

[*] Department of Chemistry, Tohoku University, Sendai 980-8578, Japan

Synopsis If the gravitino mass is in the region from a few GeV to a few 10's GeV, the scalar lepton X^- such as stau is most likely the next lightest supersymmetry particle. The negatively charged and long-lived X^- can form a Coulombic bound state, an exotic atom, with a nucleus. Energy levels and structures of exotic molecules consisting of two nuclei and X^- are precisely calculated. We discuss stabilities of the exotic molecule and intramolecular nuclear reactions.

A gravitino is the most important prediction of supergravity. The gravitino with a mass of O(1) MeV–O(10) GeV is very interesting in the sense that it is most likely the stable and lightest supersymmetric (SUSY) particle and can be a candidate of cold dark matter in the universe. Moreover, if the gravitino mass lies between a few GeV and a few 10s GeV and the next lightest SUSY particle is a charged scalar lepton X^- such as stau which is the supersymmetric partner of tau, the supergravity can be tested in high-energy collider experiments. Though the stau has never found becauseof its heavy mass, it should be created in the big-bang. The long-lived stau could survive during the big-bang nucleosynthesis, and they could form bound states together with positively charged nuclei as an exotic atom and molecule, which led to an enhancement of some nuclear reaction rates. The stau played as a catalyzer of nuclear reactions as in the muon catalyzed nuclear fusion process [1], and affects cosmological nuclear abundances [2-5]. Contribution of the reactions is expected to solve the calculated underproduction (~1000 times) of the primordial abundance of ^{6}Li and the overproduction (~3 times) of ^{7}Li [6] and, at the same time, to provide with information for constraining the lifetime and the primordial abundance of the elementary particle X^-. Since the number density of particles is high enough for three-body collision within the lifetime of the X^-, an exotic molecule with X^- might be created and also affect the big-bang nucleosynthesis. Though the probability of creating the X^- by a high energy accelerator, long-lived X^- may be able to form exotic atoms and molecules. Moreover, nuclear reaction induced by the exotic molecule may be used for the detection of X^-.

In this work, precise three-body calculations are performed for exotic molecules with X^-. Hereafter, it is not necessary to identify X^- with any particular particle such as the SUSY particle stau. The property of X^- we assume is that X^- has a charge $-e$, a mass $m_X > 100$ GeV and a lifetime $\tau_X > 10^3$ s and that X^- interacts with light nuclei (proton, deuteron, triton, helion, and alpha particle) via the Coulomb force only. It is essential for the calculation of the exotic molecules to employ appropriate nuclear interaction between two nuclei because the size of the molecule is close to that of nucleus. We assume Gaussian-shape charge distributions of the nuclei. The Gaussian size parameter is determined to reproduce observed root mean square of charge radii of the nuclei.

We examine mass dependence of X^- energy levels and structures of the exotic molecule. The result shows strong isotope effect. Stability of the exotic molecule is lost for the heteronuclear system.

References

[1] K. Ishida *et al* 2003 *J. Phys. G* **29** 2043
[2] M. Pospelov 2007 *Phys. Rev. Lett.* **98** 231301
[3] K. Hamaguchi *et al* 2007 *Phys. Lett. B* **650** 268
[4] M. Kamimura *et al* 2009 *Prog. Theo. Phys.* **650** 268
[5] M. Kusakabe *et al* 2014 *Astrophys. J. Suppl. Ser.* **214** 5
[6] M. Asplund *et al* 2006 *Astrophys. J.* **644** 229
[7] E. Hiyama *et al* 2003 *Prog. Part. Nucl. Phys.* **51** 223

[1] E-mail: y.k@m.tohoku.ac.jp

Content from this work may be used under the terms of the Creative Commons Attribution 3.0 licence. Any further distribution of this work must maintain attribution to the author(s) and the title of the work, journal citation and DOI.
Published under licence by IOP Publishing Ltd

Journal of Physics: Conference Series **635** (2015) 072090 doi:10.1088/1742-6596/635/7/072090

Spontaneous decay of small copper cluster anions Cu_N^-, $N = 3 - 6$

Mark H Stockett*† 1, Magdalena Kaminska†, Rodrigo F Nascimento†, Emma K Anderson†, Michael Gatchell†, Klavs Hansen*§, Henning Zettergen†, Henning T Schmidt†, and Henrik Cederquist†,

*Department of Physics and Astronomy, Aarhus University, Aarhus, Denmark
†Department of Physics, Stockholm University, AlbaNova University Center, SE 106 91, Stockholm, Sweden
§Department of Physics, University of Gothenburg, Gothenburg, Sweden

Synopsis We have measured the spontaneous neutral particle emission from beams of copper cluster anions (Cu_N^-, $N = 3 - 6$) stored in an electrostatic storage ring. These experiments have show that multiple structural isomers may be present in the beam.

We have measured spontaneous neutral particle emission rates for small copper cluster anions (Cu_N^-, $N = 3 - 6$) stored in one of the DESIREE (Double ElectroStatic Ion Ring ExpEriments) storage rings in Stockholm, Sweden [1]. The Cu_N^- clusters are produced in a cesium sputter ion source with a copper cathode. The ions are accelerated to 10 keV/q and a bending magnet is used to select the desired mass per charge before injection into the ring. Neutral particles - produced by electron detachment, unimolecular dissociation or background gas collisions - leave the ring and are collected by two detectors at different locations as shown schematically in the inset of Figure 1.

In this type of experiment, the ensemble of ions stored in the ring is produced with a broad distribution of internal excitation energies. The spontaneous neutral particle emission is then usually observed to follow a power law t^{-P} with $P \approx 1$ [2]. Radiative cooling depletes the population of hot ions without producing neutral particles, effectively quenching the power-law decay rate [3].

Like other cryogenic electrostatic storage devices [4], the excellent vacuum conditions in the DESIREE storage rings permit spontaneous decay to be observed over much longer timescales than in other experiments. We have taken advantage of this new capability to study surprising new phenomena in the decay of small metal clusters where multiple structural isomers may be present in the beam. This is most clearly seen in Cu_4^- (shown in Figure 1) and Cu_5^-, where two components are measured to decay with different characteristic lifetimes. We attribute the more slowly decaying components to linear isomers which lack a dipole-allowed radiative cooling transition. Cu_6^-, for which no stable linear isomer exists, shows a fast one-component decay.

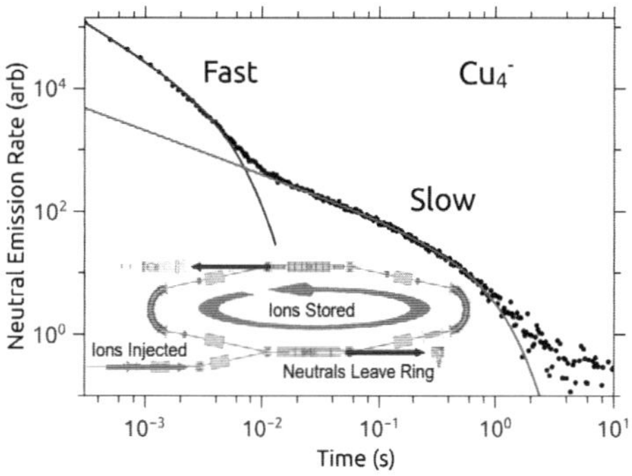

Figure 1. Neutral particle emission from stored Cu_4^- beam. In the inset we show a schematic representation of the experimental procedure.

A single-component decay is also observed for C_3^-, perhaps because only in this case does the *ground state* of the anion have a linear structure. Further, the power law exponent P for C_3^- is much less than unity, which may be due to the shape of the energy distribution. Planned experiments in which the stored ions will be excited by a laser will elucidate the details of this process.

References

[1] H. T. Schmidt *et al.* 2013 *Rev. Sci. Instrum.* **82** 056112

[2] K. Hansen *et al.* 2001 *Phys. Rev. Lett.* **87** 123401

[3] J. U. Andersen *et al.* 1996 *Phys. Rev. Lett.* **77** 3991

[4] M. W. Froese *et al.* 2011 *Phys. Rev. A* **83** 023202

[1]E-mail: stockett@phys.au.dk

Content from this work may be used under the terms of the Creative Commons Attribution 3.0 licence. Any further distribution of this work must maintain attribution to the author(s) and the title of the work, journal citation and DOI.
Published under licence by IOP Publishing Ltd

Journal of Physics: Conference Series **635** (2015) 072093 doi:10.1088/1742-6596/635/7/072093

Isotope effect in dissociative electron attachment to H_2 at 4 eV

J. Fedor*[1], J. Kočišek*, R. Janečková†

* J. Heyrovský Institute of Physical Chemistry, Czech Academy of Sciences, Dolejškova 3, 18223 Prague, Czech Republic

† Department of Chemistry, University of Fribourg, Chemin du Musée 9, CH-1700 Fribourg, Switzerland

Synopsis We report dissociative electron attachment cross sections for H_2 and D_2 at 4 eV and 14 eV resonances. The efficient signal-background separation, that has been enabled by utilizing time-and-position sensitive delay-line detector, reveals the isotope effect that is much stronger than previously believed.

The resonance created in e + H_2 collision at 4 eV can be viewed as the most fundamental resonance in electron-molecule scattering. At the same time, it is one of the broadest known resonances, which leads to very low cross section for production of H^- via dissociative electron attachment. Related to this is the strong isotope effect: due to seminal work of Schulz and Asundi [1] it is often quoted that cross sections of D^- production from D_2 is 200 times smaller than that of H^- from H_2. More recently, in an experiment with non-monochromatized electron beam, this ratio of cross sections has been determined to be 325 [2].

We have determined cross sections for dissociative electron attachment in H_2 and D_2 using a trochoidal electron monochromator in combination with an anion time-of-flight mass analyzer. The use of time-and-position sensitive delay-line detector for anion detection has enabled an efficient signal-background separation, which is crucial for measuring cross sections that are as low as in the present systems. We have measured cross sections at the 4 eV and 14 eV resonances, since in both energy ranges the DEA is a threshold process yielding slow fragments. The cross sections were calibrated against those of O^- from CO_2 and $HCOO^-$ from HCOOH [3].

Figure 1 shows cross sections for H^- from H_2 and D^- from D_2 at energy ranges of interest. For the 14 eV resonance the present energy-integrated cross sections (independent on electron energy resolution) are in very good agreement with recent measurements using the velocity-map imaging technique [4]. The ratio of integrated cross sections at the 4 eV resonance is 800. The critical re-examination of previous experiments shows, that the difference with present results is to a large degree caused by the insufficient separation of background signals in those experiments.

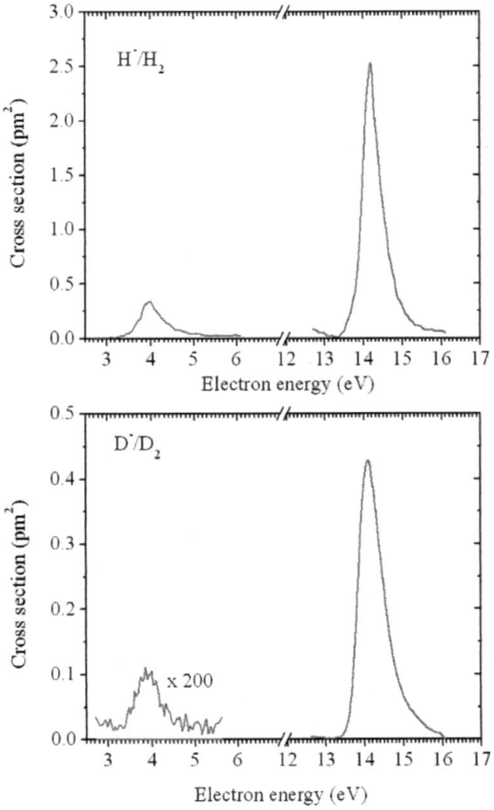

Figure 1. Cross sections for dissociative electron attachment to H_2 (top panel) and D_2 (bottom panel).

References

[1] G. J. Schulz and R. K. Asundi 1967 *Phys. Rev.* **158** 25

[2] I. Čadež *et al* 2008 *J. Phys. Conf. Ser* **133** 012029

[3] R. Janečková *et al* 2013 *Phys. Rev. Lett.* **111** 213201

[4] E. Krishnakumar *et al* 2011 *Phys. Rev. Lett.* **106** 243201

[1] E-mail: juraj.fedor@jh-inst.cas.cz

Content from this work may be used under the terms of the Creative Commons Attribution 3.0 licence. Any further distribution of this work must maintain attribution to the author(s) and the title of the work, journal citation and DOI.
Published under licence by IOP Publishing Ltd

XXIX International Conference on Photonic, Electronic, and Atomic Collisions (ICPEAC2015) IOP Publishing
Journal of Physics: Conference Series **635** (2015) 082001 doi:10.1088/1742-6596/635/8/082001

Resonances with natural and unnatural parities in positron-sodium scattering

Muhammad Umair* [1], **Svante Jonsell*** [2]

* Department of Physics, Stockholm University, SE-10691 Stockholm, Sweden

Synopsis Resonance energies and widths are reported and compare with other theoretical calculations.

We present an investigation of resonances with natural and unnatural parities in positron sodium scattering. The e^+Na system is formulated as a three-body system: a frozen core, a positron and an electron. In this way, the system resembles the hydrogen-positron system, but there are some qualitative differences. First, the interaction between the Na^+ core and the active electron is Coulombic only at very long range. At shorter distances the nuclear charge is only partially screened, which makes it necessary to employ some sort of model potential for the e^-Na^+ interaction. More generally this interaction should also include electron exchange, but this has not been explicitly included in our study. The e^+Na interaction is, like the e^+H interaction, at long distances dominated by the polarizability of the atom, giving a potential $\propto r^{-4}$. The polarizability of Na (162 atomic units) is however larger than that of H (4.5 atomic units), which allows the formation of a truly bound state of the e^+Na system [1]. Furthermore, the hydrogen atom has energy levels degenerate with respect to orbital angular momentum (neglecting fine structure) whereas the alkali atoms do not. Another characteristic of a positron-Na system is that the ionization energy of Na (0.19 a.u.) is smaller than the binding energy of Ps (0.25 a.u.), allowing for positronium formation even at zero impact energy.

S-, P-, and D-wave resonances are calculated for energies extending up to the positronium (Ps) $n = 2$ formation thresholds for natural parity. Resonance states for P^e and D^o are calculated for unnatural parity for energies extending up to the Ps, $n = 3$ formation threshold; which have not been previously reported. Below both positronium thresholds we have for each symmetry identified dipole series of resonances, with binding energies scaling in good agreement with expectations from an analytical calculation [2]. The presented results are compared with other theoretical calculations [3–6].

Explicitly correlated Gaussian trial functions expressed in Jacobi coordinates spanning all three rearrangement channels have been employed to include the correlation effects between the outer electron, the positron and the core. The complex scaling method has been used. Two different types of analytical model potentials are used to represent the interaction between the core and outer electron. Of the earlier works, our method is most similar to the stabilization method [4, 5], but allows more accurate description of resonances by going into the complex plane. It is also different in the coordinate system and functional forms it uses for representing the wave-function, allowing relatively large basis sets to be used.

[1]E-mail: muhammad.umair@fysik.su.se
[2]E-mail: jonsell@fysik.su.se

References

[1] J. Mitroy, M. W. J. Bromley and G. G. Ryzhikh 2002 *J. Phys. B* **35** R81

[2] A. Temkin and J. F. Walker 1965 *Phys. Rev. A* **140** A1520

[3] S. J. Ward, M. Horbatsch, R. P. McEachran and A. D. Stauffer 1989 *J. Phys. B* **22** 3763

[4] S. Kar and Y. K. Ho 2005 *Eur. Phys. J. D* **35** 453

[5] H. Han *et al* 2008 *Phys. Rev. A* **77** 012721

[6] L. Jiao, Y. Zhou, Y. Cheng and R. M. Yu 2012 *Eur. Phys. J. D* **66** 48

Content from this work may be used under the terms of the Creative Commons Attribution 3.0 licence. Any further distribution of this work must maintain attribution to the author(s) and the title of the work, journal citation and DOI.
Published under licence by IOP Publishing Ltd

Journal of Physics: Conference Series **635** (2015) 082003 doi:10.1088/1742-6596/635/8/082003

Development of an energy-tunable positronium beam apparatus using the photodetachment of the positronium negative ion

L. Chiari* [1], **K. Michishio*** [2], **N. Oshima**† [3], **Y. Nagashima*** [4]

* Department of Physics, Tokyo University of Science, 1-3 Kagurazaka, Shinjuku, Tokyo 162-8601, Japan
† National Institute of Advanced Industrial Science and Technology (AIST), Tsukuba, Ibaraki 305-8568, Japan

Synopsis We report on the development of an energy-tunable positronium beam apparatus at Tokyo University of Science. The positronium beam is generated using the photodetachment of the positronium negative ion, a bound state of one positron and two electrons. Positronium negative ions are efficiently emitted from a Na-coated tungsten surface by bombardment with a slow positron beam. The intermediate-energy positronium beam produced in this fashion can be used for fundamental studies of antimatter-matter interactions with solid surfaces.

Positronium is a bound state of a positron and an electron and is produced in gases, liquids and insulators. It may also form when positrons impinge on metallic surfaces and a fraction is re-emitted bound to a target electron [1]. However the maximum emission energy of positronium in that case is just a few eV and its neutral nature prevents the formation of a beam by electrostatic acceleration.

So far energy-tunable positronium beams have been produced using the charge-exchange reaction of energetic positrons with gas molecules [2], where the positronium energy is set by the energy of the primary positron beam. Here we report on the development of a new, monochromatic and energy-tunable positronium beam apparatus at Tokyo University of Science. The positronium beam is produced using the photodetachment of the positronium negative ion [3], a bound state of two electrons and one positron [4].

A slow positron beam is generated by a ^{22}Na isotope (~20 mCi) in conjunction with a solid Ne moderator. The positrons are guided by a strong axial magnetic field (600 G) into a buffer-gas trap containing a mixture of N_2 and CF_4 gases [5], where they thermalize and are accumulated to form the source for a pulsed beam (~100 Hz). The positron beam then passes through a buncher in order to reduce the pulse width.

The slow positron beam is subsequently accelerated to an energy of several keV before being focussed onto a Na-coated W (100) surface placed within a weaker magnetic field (~3 G). Efficient and durable emission of positronium negative ions from Na-coated W surfaces has been observed [6]. Those positronium negative ions are then accelerated using a static electric field and a positronium beam with a given kinetic energy is generated by photodetachment using a Q-switched Nd:YAG laser (1064 nm, 10 W) [7].

The incident energy of the positronium beam produced in this fashion will range from a few hundred eV to several keV. Hence, it will extend and complement the energy range that is and currently accessible with beams produced using the charge-exchange reaction with gas molecules [8]. The positronium beam at the target is anticipated to be around 1 mm in size with an intensity of the order of 100 positronium atoms per second.

Advantages of the present technique include the concurrent achievement of high-energy and ultra-high-vacuum compatibility. This will enable the investigation of solid surface structures, such as reflection high-energy positronium surface diffraction.

References

[1] P. G. Coleman 2000 *Positron beams and their Applications* (World Scientific, Singapore)
[2] N. Zafar *et al* 1996 *Phys. Rev. Lett.* **76** 1595
[3] K. Michishio *et al* 2011 *Phys. Rev. Lett.* **106** 153401
[4] Y. Nagashima *et al* 2014 *AIP Conf. Proc.* **1588** 27
[5] S. J. Gilbert *et al* 1997 *Appl. Phys. Lett.* **70** 1944
[6] H. Terabe *et al* 2012 *New J. Phys.* **14** 015003
[7] K. Michishio *et al* 2012 *Appl. Phys. Lett.* **100** 254102
[8] G. Laricchia and H. R. J. Walters 2012 *Riv. Nuovo Cimento* **35** 305

[1] E-mail: luca.chiari@rs.tus.ac.jp
[2] E-mail: michishio@rs.tus.ac.jp
[3] E-mail: nagayasu-oshima@aist.go.jp
[4] E-mail: ynaga@rs.kagu.tus.ac.jp

Content from this work may be used under the terms of the Creative Commons Attribution 3.0 licence. Any further distribution of this work must maintain attribution to the author(s) and the title of the work, journal citation and DOI.
Published under licence by IOP Publishing Ltd

Journal of Physics: Conference Series **635** (2015) 082004 doi:10.1088/1742-6596/635/8/082004

Theoretical electron impact ionization cross sections for atomic and ionic targets for E≤10 keV.

B. C. Saha[1], A. K. F. Haque[2], M. Atiqur R. Patoary[2], M. A. Uddin[2] and A. K. Basak[2]

Department of Physics,
[1]Florida A&M University, Tallahassee, FL-32307, USA,
[2]University of Rajshahi, Rajshahi-6205, Bangladesh.

Synopsis Electron impact ionization cross sections for E ≤10 keV are presented for O and its isoelectronic series (F^+, Ne^{2+} and Si^{6+} ions) and compared with experimental and other theoretical results.

The electron impact ionization cross sections (EIICS) are needed in various fields ranging from biological to chemical, plasma to astrophysics, and laser to medical physics. At energies above the breakup threshold the scattering in a three-body system remains a most challenging quantum mechanical problem. Although for simpler targets, like H and He, there are few accurate calculations in literatures, but for many electron targets the situation is rather very unfavorable. Moreover, the huge demand in applications for the ionization cross sections cannot be fulfilled easily by generating a database either by experimental or by quantal theoretical results. So resorts are made to fulfill this void by using *simple–to-use* models that predict accurate results. In this report a few such models that describe experimental results reasonably well for both atomic and ionic targets for a wider energy domain will be presented. To name a few, MBELL [1], GKLV [2] and NMBELL [3] models remain important in predicting EIICS for both neutral and ionic targets over a wide range of atomic numbers. These model includes the following effects explicitly: (i) correlation, (ii) ionic, (iii) relativistic and (iv) generalization of parameters including the asymptotic behavior. For almost all targets these effects restore the proper behavior of the EIICS as a function of energy. In Fig. 1 EIICS for neutral O and its isoelectronic series (F^+, Ne^{2+} and Si^{6+} ions) are shown as a function of E. A comparison with experimental and other theoretical results shows good agreement with our calculated cross sections.

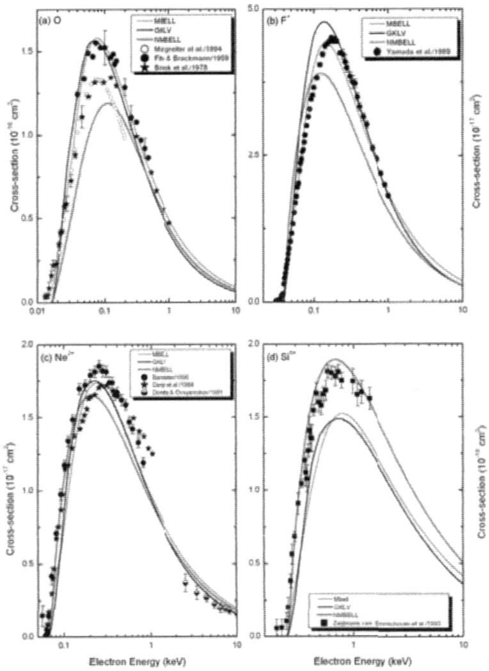

Figure 1. EIICS for O atoms and its isoelectronic, namely F^+, Ne^{2+} and Si^{6+} ions.

References

[1] A. K. F. Haque *et al* 2006 *Phys. Rev A* **73**, 012708.
[2] A. K. F. Haque *et al* 2007 *Eur. Phys. J. D.* **42** 203.
[3] M. A. R. Patoary *et. al.* 2011 *Int. J. of Quantum Chemistry* **111**, 923.

Content from this work may be used under the terms of the Creative Commons Attribution 3.0 licence. Any further distribution of this work must maintain attribution to the author(s) and the title of the work, journal citation and DOI.
Published under licence by IOP Publishing Ltd

Journal of Physics: Conference Series **635** (2015) 082005 doi:10.1088/1742-6596/635/8/082005

Quasi Sturmian basis functions in hyperspherical coordinates

J A Del Punta [*†‡ 1], **G Gasaneo** [†], **C M Granados-Castro** [*] **and L U Ancarani** [*]

[*] Equipe TMS, SRSMC, UMR CNRS 7565, Université de Lorraine, 57078, Metz, France
[†] Departamento de Física, Universidad Nacional del Sur - IFISUR, 8000 Bahía Blanca, Argentina
[‡] Departamento de Matemática, Universidad Nacional del Sur, 8000 Bahía Blanca, Argentina

Synopsis Quasi-Sturmian functions are an alternative set of basis functions useful to describe time-independent three body scattering problems. Their analytical closed form, for the Coulomb case, and some of their properties are presented here in hyperspherical coordinates.

An important theoretical issue when dealing with time-independent three-body scattering processes is how to impose the correct asymptotic behavior to the wave function. Many spectral methods use two-body basis functions that generally do not possess the correct behavior at large distances. One exception are the Generalized Sturmian Functions (GSF) [1], defined as to take into account the interactions of the problem, thus making them an efficient basis set.

In this work we propose a different set of functions, named Quasi Sturmian Functions (QSF). Like GSF, they posses the correct asymptotic behavior and, as an advantage, it is possible to express them with analytical representations. The definition, the properties and the efficiency of this alternative basis set (compared to GSF) have been presented for the two-body case [2].

For the three-body case, QSF in parabolic coordinates have been introduced [3]. We are interested here in using hyperspherical coordinates, consisting in one hyperradial (ρ) and five angular variables. Considering the latter as parameters, the hyperradial part of QSF is obtained by a generalization of the method presented in [2]. For a Temkin-Poet model, only the hyperangle $\alpha = \tan^{-1}(r_2/r_1)$ is retained, and we define the QSF to be the solutions of the non-homogeneous differential equation

$$\left[-\frac{1}{2\mu}\frac{d^2}{d\rho^2} + \frac{\lambda(\lambda+1)}{2\mu\rho^2} + \frac{C(\alpha)}{\rho} - E\right] Q_n^{(\pm)}(\alpha,\nu;\rho) = \frac{1}{\rho}\phi_{n,\lambda}(\nu;\rho)$$

where E is the total energy of the problem, λ and ν are fixed parameters and $C(\alpha)/\rho$ is the interaction potential. If $\phi_{n,\lambda}(\nu;\rho)$ is a Laguerre-type function, the QSF, and consequently its asymptotic behavior, have an analytical expression. For $\rho \to \infty$ we have:

$$\frac{Q_n^{(\pm)}(\nu,\alpha;\rho)}{\mathcal{Q}_n^{as}(\nu,\alpha)} \sim e^{\pm i\left(k\rho - \frac{\mu C(\alpha)}{k}\ln(2k\rho) + \sigma_\lambda - \frac{\lambda\pi}{2}\right)} \quad (1)$$

where σ_λ is the Coulomb phase shift and $\mathcal{Q}_n^{as}(\nu,\alpha)$ can be expressed in terms of a Gauss hypergeometric function. This is the expected Coulomb behavior for a α-dependent charge $C(\alpha)$, and it is illustrated in the figure.

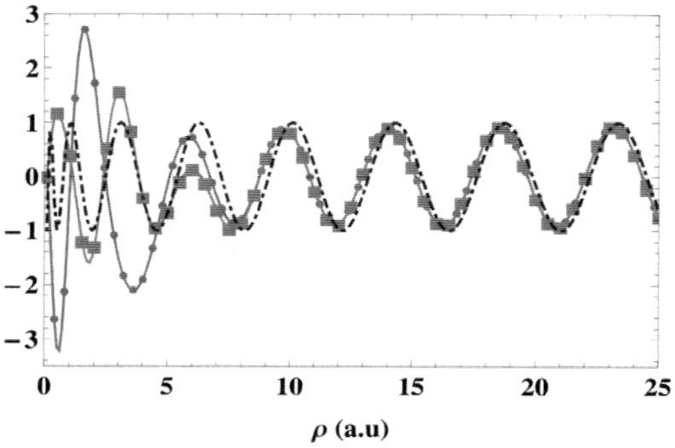

Figure 1. Real part of QSF (actually the ratio (1)), for $n = 4$ (line with solid dots) and $n = 7$ (line with solid squares), as function of ρ and for fixed $\alpha = \pi/4$. The dot-dashed line represents the real part of the asymptotic exponential function (1).

Taking advantage of their adequate asymptotic behavior, we plan to apply QSF basis sets to solve three-body scattering problems, such as double ionization processes, starting with a Temkin-Poet model. In particular, we will analyze the differences between standard extrapolation techniques and through QSF to extract the transition amplitude at infinite hyperradii.

References

[1] G. Gasaneo *et al* 2013 *Adv. Quantum Chem.* **67** 153.

[2] J. A. Del Punta *et al* 2014 *J. Math. Phys.* **55** 052101.

[3] S. A. Zaytsev and G. Gasaneo 2013 *J. At. Mol. Scs.* **4** 302.

[1] E-mail: jedelpunta@gmail.com

Content from this work may be used under the terms of the Creative Commons Attribution 3.0 licence. Any further distribution of this work must maintain attribution to the author(s) and the title of the work, journal citation and DOI.
Published under licence by IOP Publishing Ltd

Electrons transmitted through a plate capillary

Yingli Xue*[1], Deyang Yu*†[2], Junliang Liu*, Mingwu Zhang*, Bian Yang*, Yuezhao Zhang* and Xiaohong Cai*

* Institute of Modern Physics, Chinese Academy of Sciences, Lanzhou, 730000, China

Synopsis 1.5 keV electrons passed through a plated capillary were studied. We observed that the electrons were transmitted the capillary along its axis for smaller tilt angle, while for larger tilt angle, the electrons emerged from the capillary neither along the capillary axis nor along the beam incident direction, but kept their initial energy.

In 2002, N. Stolterfoht *et al* found the guiding effect of 3 keV Ne^{7+} ions passing through an insulating PET nanocapillaries [1]. Thereafter, considerable work [2] has been done to investigate slow highly charged ions (HCIs) interacting with inner surfaces of various insulating capillaries, showing that the guiding effect is prevalent in the transmission of slow HCIs.

Similar guiding phenomena were also observed for electrons transmitted through capillaries (see, e.g., [3-5]). However, the experiments showed that transmission efficiency of electrons was significantly lower than that of HCIs. Moreover, it revealed that portion of the transmitted electrons suffered significant energy-loss.

We investigated the transmission of 1.5 keV electrons through a plate capillary with a gap of 0.4 mm and a length of 20 mm. The capillary consists of a pair of high pure (99.999%) quartz glass plates of 120 mm long, 20 mm wide and 15 mm thick. The front and back ends of the capillary are grounded. The electron beam was produce by an electron gun with a cathode of LaB_6.

Figure 1 shows the dependence of the deflection angle on the capillary tilt angle. Contrast to previous results with slow HCIs [1, 2] or electrons at similar energy [3-5], we observed that the electrons were transmitted the capillary along its axis for smaller tilt angle. However, with the tilt angle increasing further, the deflection angle remained unchanged at first, then increased quickly and followed by a slightly slow increasing. We also observed that the transmitted electron fraction was higher than that of direct transmission. In addition, transmitted electrons kept their initial energy.

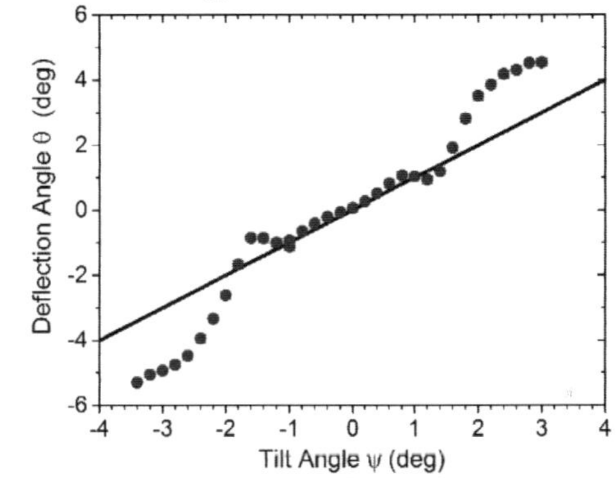

Figure 1. The dependence of the deflection angle on the capillary tilt angle for 1.5keV electrons passing through a plate capillary. The solid line represents $\psi=\theta$.

This work was partially supported by the NSFC Grants Nos. 11205224, 11275240 and U1332206.

References

[1] N. Stolterfoht *et al* 2002 *Phys. Rev. Lett.* **88** 133201
[2] C. Lemell *et al* 2013 *Prog. Surf. Sci.* **88** 237
[3] S. Das *et al* 2007 *Phys. Rev. A* **76** 042716
[4] A. Milosavljević *et al* 2007 *Phys. Rev. A* **75** 030901(R)
[5] B.S. Dassanayake *et al* 2010 *Phys. Rev. A* **81** 020701(R)

[1] E-mail: xueyl@impcas.ac.cn
[2] E-mail: d.yu@impcas.ac.cn

Content from this work may be used under the terms of the Creative Commons Attribution 3.0 licence. Any further distribution of this work must maintain attribution to the author(s) and the title of the work, journal citation and DOI.
Published under licence by IOP Publishing Ltd

Journal of Physics: Conference Series **635** (2015) 092002 doi:10.1088/1742-6596/635/9/092002

Stepwise contraction of the nf Rydberg shells in the 3d photoionization of multiply-charged xenon ions

S. Schippers[1], A. Borovik Jr.[1], T. Buhr[2], J. Hellhund[1], K. Holste[1], A. L. D. Kilcoyne[3], S. Klumpp[4], M. Martins[4], A. Müller[1], S. Ricz[5], S. Fritzsche[6,7]

[1] Institut für Atom- und Molekülphysik, Justus-Liebig-Universität Gießen, 35392 Gießen, Germany
[2] Physikalisch-Technische Bundesanstalt, 38116 Braunschweig, Germany
[3] Advanced Light Source, MS 7-100, Lawrence Berkeley National Laboratory, Berkeley, California 94720, USA
[4] Institut für Experimentalphysik, Universität Hamburg, 22761 Hamburg, Germany
[5] Institute for Nuclear Research, Hungarian Academy of Sciences, Debrecen, H-4001, Hungary
[6] Helmholtz-Institut Jena, 07743 Jena, Germany
[7] Theoretisch-Physikalisches Institut, Friedrich-Schiller-Universität Jena, 07743 Jena, Germany

Synopsis Photoionization of Xe^{3+}, Xe^{4+} and Xe^{5+} ions was measured in the energy range 670–750 eV, including the $3d$ ionization threshold. The cross sections exhibit a progressively larger number of sharp resonances as the ion charge state is increased. This clearly visualizes the re-ordering of the εf continuum into a regular series of (bound) Rydberg orbitals as the ionic core becomes more attractive.

Recently, first low-resolution results for 3d multiple photoionization of Xe^{q+} ($1 \leq q \leq 5$) from the photon-ion spectrometer PIPE at the Hamburg PETRA III synchrotron light source have been reported [1]. Here, new high-resolution measurements with an experimental energy spread of 160 meV are presented for $3 \leq q \leq 5$ [2] (Fig. 1). Pronounced resonance structures are observed for all ions but with an increasing number of resonances as the charge state of the ions is increased. In contrast to the usually rather complex cross sections for outer shell ionization of atoms and ions, the present inner-shell ionization cross sections can be interpreted straight-forwardly. The strongest resonances are associated with the photoexcitation of a $3d$ electron to an atomic nf subshell ($n = 4, 5, 6, ...$) and the subsequent multiple autoionization of the associated hole states. The $3d^{-1}_{3/2} - 3d^{-1}_{5/2}$ fine structure splitting ($\sim$ 13 eV) of the $3d$ hole leads to two distinct Rydberg series of resonances in each spectrum. For Xe^{3+}, for example, resonances with principal quantum numbers n from 4 to 7 can be clearly discerned. For Xe^{5+} the series of nf resonances could be observed up to $n = 9$. This progression of resonance structure with increasing charge state clearly visualizes the re-ordering of the εf continuum into a regular series of (bound) Rydberg orbitals as the ionic core becomes more attractive. The energies and strengths of the resonances are extracted from the experimental data and are further analyzed by relativistic atomic-structure calculations.

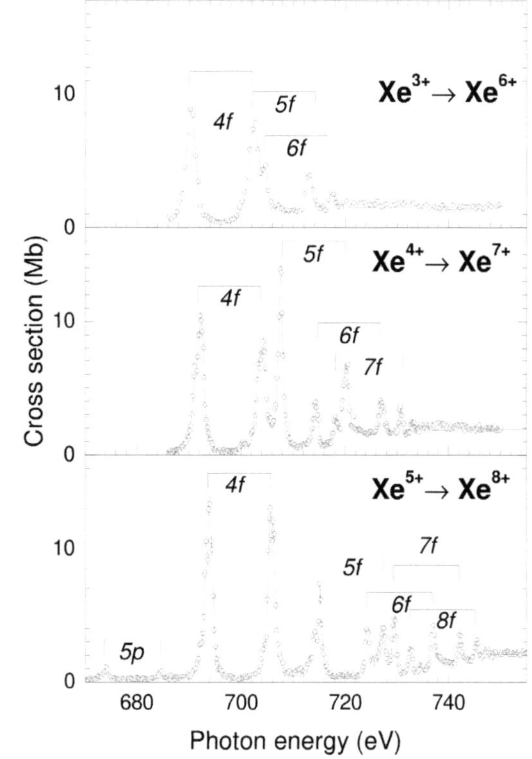

Figure 1. Measured (symbols) and fitted (full lines) absolute cross sections for triple ionization of Xe^{3+}, Xe^{4+}, and Xe^{5+} ions. Resonances are labeled by the $n\ell$ subshell to where the $3d$ electron is excited.

References

[1] S. Schippers *et al.* 2014 *J. Phys. B* **47** 115602

[2] S. Schippers *at al.* 2015 *J. Phys. B* in print; arXiv:1412.3592

[1]E-mail: Stefan.Schippers@physik.uni-giessen.de

Content from this work may be used under the terms of the Creative Commons Attribution 3.0 licence. Any further distribution of this work must maintain attribution to the author(s) and the title of the work, journal citation and DOI.
Published under licence by IOP Publishing Ltd

Journal of Physics: Conference Series **635** (2015) 092003 doi:10.1088/1742-6596/635/9/092003

Electron impact excitation of 4^1P_1 zinc state for 40 eV incidence energy

M. Piwiński [1], Ł. Kłosowski, D. Dziczek, S.Chwirot

Institute of Physics, Faculty of Physics, Astronomy and Informatics, Nicolaus Copernicus University in Toruń, Grudziądzka 5, 87-100 Toruń, Poland

Synopsis Coherence analysis version of the electron-photon coincidence technique was used to study the electron impact excitation of 4^1P_1 state of zinc atoms. The first experimental results of the Stokes and Electron Impact Coherence Parameters (EICPs) for 40 eV collision energy are presented.

The electron-photon coincidence studies deliver more complete information on the inelastic electron-atom collisions than any other experimental techniques. Therefore, such data represent a useful probe for both verification and improvement of theoretical models. The data available for Zn are scarce due to experimental problems associated with this atom, which recently has been proposed an interesting target for such studies [1,2,3]. Present work is a continuation of our systematic studies on electron impact excitation of atoms with two valence electrons, from the simplest case He [4,5] to more complicated, heavier elements with closed inner shells: Ca [6,7], Zn [3,8,9] and Cd [10,11,12].

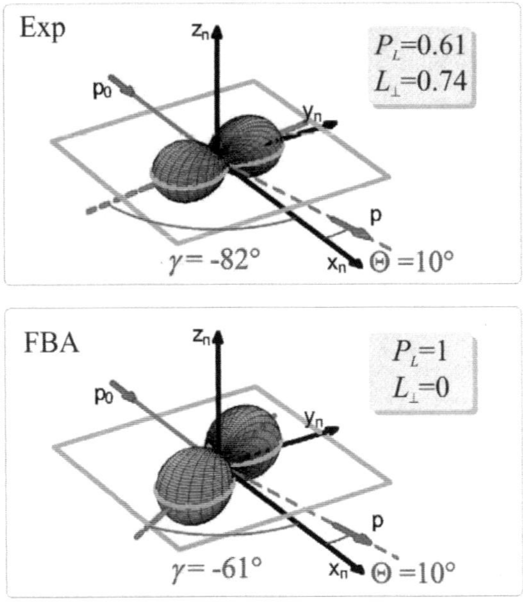

Figure 1. An example of graphical representations of angular electron charge cloud distribution corresponding to the EICP parameters for Zn 4^1P_1 excitation by 40 eV electrons for scattering angle $\Theta = 10°$. (Exp – experimental values, FBA – First Born Approximation predictions)

We present first results of Stokes parameters characterising electronic excitation of 4^1P_1 zinc state for 40 eV collision energy. The experiment was performed using electron-photon coincidence technique in the coherence analysis version. The apparatus and procedures were generally the same as in our previous studies [3].

The obtained raw values of the Stokes parameters were used to calculate the Electron Impact Coherence Parameters P_L, $L_\perp$, γ and P^+ (for details see review by Andersen *et al* [13]). The EICPs determine angular distributions of the electron charge cloud of the excited 4^1P_1 state of Zn. The results are shown in a graphical representation and compared with the First Born Approximation predictions (Figure 1).

The experimental values of Stokes and EICP parameters are analysed together with our previous results for Zn atoms for collision energy 100 eV [3], 80 eV [9] and 60 eV [8].

References

[1] M. Born and M. Strösser 2007 *J. Phys. D: Appl. Phys.* **40**, 3823–3828
[2] T. Das *et al* 2014 *Phys. Lett. A* **378**, 641–643
[3] M. Piwiński *et al* 2012 *Phys. Rev. A* **86**, 052706
[4] Ł. Kłosowski *et al* 2007 *Meas. Sci. Technol.* **18**, 3801–3810
[5] Ł. Kłosowski *et al* 2009 *Phys. Rev. A* **80**, 062709
[6] D. Dyl *et al* 1999 *J. Phys. B: At. Mol. Opt. Phys.* **32**, 837–844
[7] D. Dziczek *et al* 1998 *Acta Phys. Polon. A* **93**, 717–721
[8] M. Piwiński *et al* 2013 *Eur. Phys. J.-Spec. Top.* **222**, 2273–2277
[9] M. Piwiński *et al* 2015 *Phys. Rev. A* in press
[10] M. Piwiński *et al* 2002 *J. Phys. B: At. Mol. Opt. Phys.* **35**, 3821–3827
[11] D. Dziczek *et al* 2003 *Acta Phys. Polon. A* **103**, 3–11
[12] M. Piwiński *et al* 2006 *J. Phys. B: At. Mol. Opt. Phys.* **39**, 1945–1953
[13] N. Andersen *et al* 1988, *Phys. Rep.* **165**, 1–188

[1]E-mail: Mariusz.Piwinski@fizyka.umk.pl

Content from this work may be used under the terms of the Creative Commons Attribution 3.0 licence. Any further distribution of this work must maintain attribution to the author(s) and the title of the work, journal citation and DOI.
Published under licence by IOP Publishing Ltd

Journal of Physics: Conference Series **635** (2015) 092004 doi:10.1088/1742-6596/635/9/092004

Time delay in photoionization of half-filled shell atoms

V. K. Dolmatov[*,1], **A. S. Kheifets**[†], **S. T. Manson**[‡] **and P. C. Deshmukh**[§]

[*] Department of Physics and Earth Science, University of North Alabama, Florence, Alabama 35632, U.S.A.
[†] Research School of Physics and Engineering, Australian National University, Canberra ACT 0200, Australia
[‡] Department of Physics and Astronomy, Georgia State University, Atlanta, Georgia 30303, U.S.A.
[§] Department of Physics, Indian Institute of Technology Madras, Chennai-600036, India

Synopsis Time delay in photoionization of a half-filled shell atom is predicted to be strongly term-dependent and subject to the dramatic impact of a giant autoionization resonance. Photoionization of the Mn atom is chosen as a case study.

Time delay in photoelectron emission from atoms [1] has become a hot topic of modern studies. To date, however, neither the significance of autoionization resonances in photoionization time delay has been established, nor features in time delay brought about by the configuration of open-shell atoms have been uncovered.

In this work, we discuss the 3d- and 4s-photoionization of the Mn(...$3p^6 3d^5 4s^2$, ^{6}S) atom which has a $3d^5$ half-filled subshell in its configuration, as a convenient case study. In Mn, the resonance excitation from a $3p^6$ subshell to the half-filled $3d^5$ subshell occurs at $\hbar\omega_r \approx$ 50 eV, autoionizes primarily into 3d $\rightarrow$ f and 3d $\rightarrow$ p continua and has a big resonance width $\gamma \approx 2$ eV at a half-maximum of the 3d-photoionization cross section ($\sigma_{3d}^{\max} > 50$ Mb) [2, 3] (and references therein). Furthermore, the removal of a 4s-electron from Mn results in the ion-remainder in the state with a term Mn^+($4s^1$, ^{5}S) or Mn^+($4s^1$, ^{7}S).

In the present study, we use the "spin-polarized" random phase approximation with exchange (SPRPAE) [3] to calculate the τ_{3d}, $\tau_{4s}(^5S)$ and $\tau_{4s}(^7S)$ time delays for Mn 3d- and 4s-photoionization in the photon energy region which encompasses the 3p $\rightarrow$ 3d resonance. Calculated data are plotted in figure 1. Note how remarkably big all calculated time delays are, especially the τ_{3d} time delay, due to the impact of the 3p $\rightarrow$ 3d giant resonance. And τ_{3d}, obtained by accounting for both the 3d $\rightarrow$ f and 3d $\rightarrow$ p photoionization channels in the calculation (marked as the "3d, two-channel time delay" in figure 1), differs dramatically from τ_{3d} when only the dominant 3d $\rightarrow$ f continuum was accounted for [marked as the "3d, single-channel (3d $-$ f) time delay" in figure 1]. Note also that $\tau_{4s}(^5S)$ and $\tau_{4s}(^7S)$ differ strongly from each other, thereby revealing significant term-dependence in photoionization time delay in the half-filled shell Mn atom. These remarkable effects, elucidated above, should be general features of photoionization time delay in other half-filled shell atoms as well.

The support by NSF grant No. PHY-1305085 and DoE Grant No. DE-FG02-03ER15428 is acknowledged.

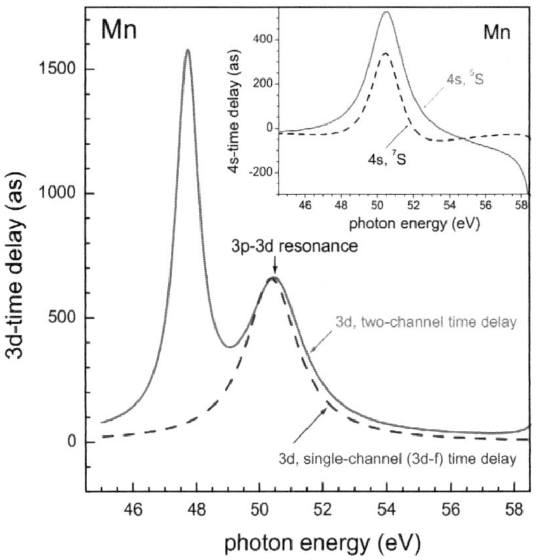

Figure 1. Calculated SPRPAE data for the τ_{3d}, $\tau_{4s}(^5S)$ and $\tau_{4s}(^7S)$ time delays upon the photoionization of Mn, as marked.

References

[1] M. Schultze *et al* 2010 *Sc e ce* **328** 1658

[2] B. Sonntag and P. Zimmermann 1992 *Re . Prog. Phys.* **55** 911

[3] M. Ya. Amus'ya, V. K. Dolmatov and V. K. Ivanov 1983 *Sov. Phys. - JETP* **58** 67

[1]E-mail: vkdolmatov@una.edu

Content from this work may be used under the terms of the Creative Commons Attribution 3.0 licence. Any further distribution of this work must maintain attribution to the author(s) and the title of the work, journal citation and DOI.
Published under licence by IOP Publishing Ltd

Journal of Physics: Conference Series **635** (2015) 092005 doi:10.1088/1742-6596/635/9/092005

Electron rescattering in strong-field photodetachment of F^-

O. Hassouneh*, S. Law*, S.F.C. Shearer*, A.C. Brown* and H.W. van der Hart* [1]

* Centre for Theoretical Atomic, Molecular and Optical Physics, School of Mathematics and Physics, Queen's University Belfast, Belfast BT7 1NN, United Kingdom

Synopsis The description of processes in mid-IR laser fields presents new challenges to theoretical atomic physics. We apply time-dependent R-matrix theory to study multiphoton detachment of F^-. Through comparison with the strong-field approximation, we observe signatures of rescattering in the analysis of the final wavefunction, demonstrating the accuracy of the time-dependent R-matrix approach.

The development and application of new laser technology continues to be at the forefront of science and technology. This ongoing development needs to be matched by the development of theoretical approaches capable of explaining the new phenomena seen in experiment. To this end, at Queen's University Belfast we have developed time-dependent R-matrix theory [2, 3]. This theory explicitly accounts for all the electrons in an atom, and is capable of studying their response to an external laser field.

When an atom is placed in an intense laser field, an escaping electron is driven back and forth by the oscillating field. If the escaping electron returns to the residual atom, a wide variety of highly non-linear processes can occur [1]. The maximum distance an electron can escape from the residual atom prior to return increases rapidly with increasing wavelength. At an intensity of 2×10^{14} W/cm^2, this maximum distance is 11 a_0 at 390 nm, 44 a_0 at 780 nm, and 234 a_0 at 1800 nm. Hence, an accurate theoretical description of atomic processes in the near-IR regime needs an accurate description of the moving electronic wavepacket at large distances.

Previously, we have demonstrated that time-dependent R-matrix theory can accurately describe the harmonic response of Kr and Xe atoms to strong light fields in the mid-IR regime, including harmonics up to order 275 [4]. However, the harmonic response is given by the Fourier transform of the time-dependent dipole moment, and, as such, it does not provide direct insight into the quality of the final-state wavefunction.

This quality can be assessed more accurately through its direct analysis. We have therefore carried out an investigation into multiphoton detachment of $F-$, and the influence of rescattering on this detachment. This process was previously investigated in [5, 6]. Through comparison with a strong-field approximation for multiphoton detachment of F^-, using the same pulse characteristics [7], we can clearly identify the signatures originating from the rescattering process.

We obtain an ejected-electron momentum distribution through analysis of the final wavefunction. The rescattered electrons can be clearly distinguished from the directly emitted electrons through the rings centered around a displaced origin on the z-axis. This displacement can be associated with the vector field at the moment of recollision. The overall characteristics of the momentum distribution are in agreement with the distributions obtained in [5, 6].

One of the advantages of time-dependent R-matrix theory is the capability to include the electron-repulsion interaction in detail. We have therefore carried out calculations using a Hartree-Fock representation of the residual F atom [8], and using a larger description, including double excitations out of the $2p^5$ shell [9]. When the basis set is increased, the momentum distributions have the same qualitative structure, but the overall magnitude of photodetachment is notably reduced.

References

[1] P. Corkum 1993, *Phys. Rev. Lett.* **71** 1994

[2] L.A.A. Nikolopoulos 2008, *Phys. Rev. A* **78** 063420

[3] L.R. Moore *et al* 2011, *J. Mod. Opt.* **58**, 1132

[4] O. Hassouneh *et al* 2014, *Phys. Rev. A* **90** 043418

[5] X.X. Zhou *et al* 2008, *Phys. Rev. A* **77** 053410

[6] B. Fetić *et al* 2011, *J. Mod. Opt.* **58**, 1149

[7] S.F.C. Shearer *et al* 2013, *Phys. Rev. A* **88** 033415

[8] E. Clementi *et al* 1974, At. Data Nucl. Data Tables **14** 177

[9] C. Niang and M. le Dourneuf, unpublished

[1] E-mail: h.vanderhart@qub.ac.uk

Content from this work may be used under the terms of the Creative Commons Attribution 3.0 licence. Any further distribution of this work must maintain attribution to the author(s) and the title of the work, journal citation and DOI.
Published under licence by IOP Publishing Ltd

XXIX International Conference on Photonic, Electronic, and Atomic Collisions (ICPEAC2015) IOP Publishing
Journal of Physics: Conference Series **635** (2015) 092006 doi:10.1088/1742-6596/635/9/092006

Transverse electron momentum distribution of spin orbit wave packets

A. S. Kheifets* [1] **and I. A. Ivanov**† [2]

* Research School of Physics and Engineering, Australian National University, Canberra ACT 0200, Australia

†Center for Relativistic Laser Science, Institute for Basic Science, Gwangju 500-712, Republic of Korea

Synopsis Transverse electron momentum distribution (TEMD) of the spin-orbit wave packets launched in a pump-probe sequential double ionization from the valence shell of Ar is calculated by solving the time-dependent Schrödinger equation (TDSE). Results are compared with a simple tunneling formula and recent measurements.

In a pump-probe sequential double ionization experiment on a close shell noble gas atom, dynamics of a coherent superposition of electronic states $^2P_{1/2,3/2}$ induced by the first ionization step can be probed and imaged using the time delayed second step (Fig. 1, [1]). Such a strong field tunneling ionization experiment on the valence $3p^6$ shell of Ar was reported recently [2].

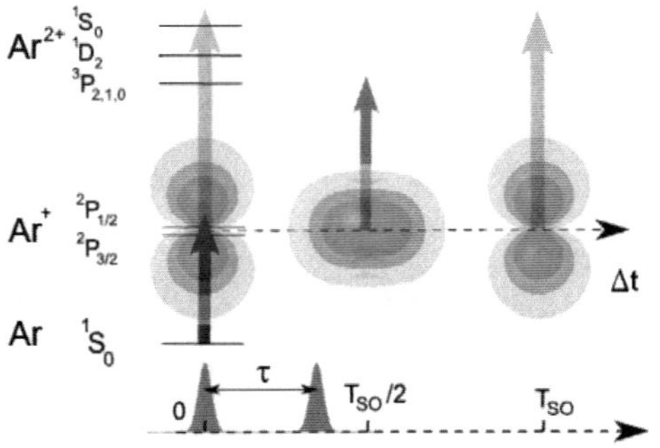

Figure 1. Schematic of the pump-probe sequential double ionization experiment (from Ref. [1])

Depending on the time delay τ between the pump and probe pulses, the populations of the $3p^5_m$ target states oscillate as

$$D_{m=0}(\tau) = 13/9 - 4/9\cos(2\pi\tau/T)$$
$$D_{|m|=1}(\tau) = 32/9 + 4/9\cos(2\pi\tau/T)$$

Here $2\pi/T = \Delta E$ is the spin-orbit energy splitting of the $P_{1/2,\,3/2}$ doublet. Accordingly, the TEMD is determined as

$$W(p_\perp,\tau) = \sum_{|m|=0,1} W_m(p_\perp)D_m(\tau) \qquad (1)$$

where the m-dependent momentum profiles are

$$W_m(p_\perp) = |\psi_{3p_m}(p_\perp)|^2 \exp\left(-\frac{\sqrt{2I_{3p}}}{E}p_\perp^2\right) \qquad (2)$$

Alternatively, the TEMD can be determined from solution of the TDSE in the single active electron approximation:

$$i\partial\Psi(\boldsymbol{r})/\partial t = \left(\hat{H}_{\rm atom} + \hat{H}_{\rm int}(t)\right)\Psi(\boldsymbol{r}) \,. \qquad (3)$$

Results of the calculations [3] for the time-differential TEMD

$$\Delta W(p_\perp,\tau) = \left[W(p_\perp,\tau) - \overline{W}(p_\perp)\right] p_\perp \qquad (4)$$

evaluated relative to the time average TEMD $\overline{W}(p_\perp)$ at $\tau = nT, nT/2$ are presented in Fig. 2.

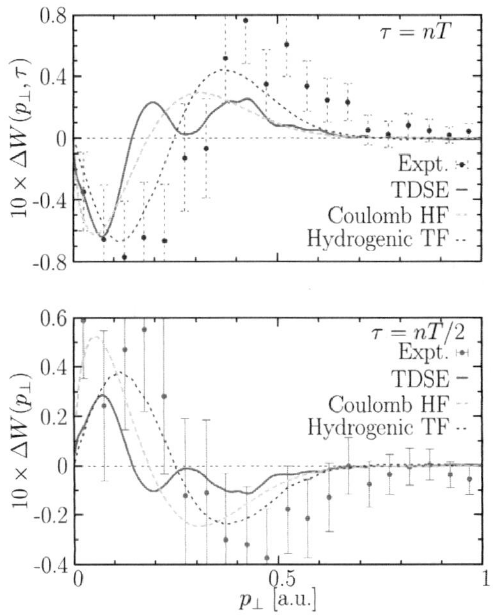

Figure 2. TEMD of Ar Eq. 4 as measured [2] and calculated using TEMD and the tunneling formula Eq. 2 with Hartree-Fock and hydrogenic orbitals

References

[1] H J Wörner and P B Corkum 2011 *J. Phys. B*, **44** 041001
[2] L. Fechner *et al.* 2014 *Phys. Rev. Lett.*, **112** 213001
[3] A. S. Kheifets and I. A. Ivanov 2014 *Phys. Rev. A*, **90** 033404

[1]E-mail: a.kheifets@anu.edu.au
[2]E-mail: Igor.Ivanov@anu.edu.au

Content from this work may be used under the terms of the Creative Commons Attribution 3.0 licence. Any further distribution of this work must maintain attribution to the author(s) and the title of the work, journal citation and DOI.
Published under licence by IOP Publishing Ltd

Relativistic calculation of the electron momentum shift in tunneling ionization

I.A.Ivanov [1]

Center for Relativistic Laser Science, Institute for Basic Science (IBS), Gwangju 500-712, Republic of Korea

Synopsis We describe a procedure for the solution of the time-dependent Dirac equation. The procedure is based on the relativistic generalization of the matrix iteration method. We use this procedure to study electron momentum distribution along the laser beam propagation direction for the process of the tunneling ionization of hydrogen atom. We found, in agreement with the experimental observations [1], that relativistic effects lead to appreciable deviation of the distribution from the strict left-right symmetry present in the non-relativistic case. The expectation value of the momentum along the laser beam propagation direction grows linearly with intensity and follows closely the behavior of the expectation value of the kinetic energy divided by the speed of light. These features agree with the experimental results [1].

Relativistic effects are generally believed to play important role for the ionization driven by electromagnetic fields of very high intensity and frequency. However, even for the infrared (IR) laser fields of moderate intensity relativistic effects can manifest themselves. For the processes with participation of a large number of photons the total momentum delivered to the atom can be non negligible. Work [1] demonstrated relativistic effects for the ionization process driven by pulses with wavelengths of 800 and 1400 nm under experimental conditions where a large number of photons is absorbed.

To study such phenomena we need to include relativistic effects into the theory describing evolution of the atomic system under the action of a laser pulse in time. We describe such an approach based on the numerical solution of the time dependent Dirac equation for an atom in presence of the external electromagnetic field:

$$\frac{\partial\Psi(\boldsymbol{r},t)}{\partial t} = \left(\hat{H}_{\text{atom}} + \hat{H}_{\text{int}}\right)\Psi(\boldsymbol{r},t) \qquad (1)$$

for the bispinor $\Psi(\boldsymbol{r},t)$ with:

$$\hat{H}_{\text{atom}} = c\boldsymbol{\alpha}\cdot\hat{\boldsymbol{p}}+c^2\beta+I\,V(r);\ \hat{H}_{\text{int}} = c\boldsymbol{\alpha}\cdot\boldsymbol{A}, \quad (2)$$

where $\boldsymbol{\alpha}$, β are Dirac matrices, $\boldsymbol{A}(\boldsymbol{r},t)$- vector potential of the laser pulse. To solve the time-dependent Dirac equation numerically we develop relativistic generalization of the matrix iteration method (MIM) [2]. The method has been often used in non-relativistic calculations [2], and is known to be reliable, accurate, and easily adaptable to complicated field geometries, such as fields of arbitrary ellipticity [3]. We performed calculations of the relativistic tunneling ionization of hydrogen atom driven by the linearly polarized pulse with the wavelength of 800 nm.

Figure I shows expectation value $\langle p_x\rangle$ of the momentum along the laser beam propagation direction as a function of the laser intensity. $\langle p_x\rangle$ grows linearly with intensity, and mimics very closely the curve showing dependence of the average kinetic energy divided by the speed of light, which is the behavior observed in the experiment [1].

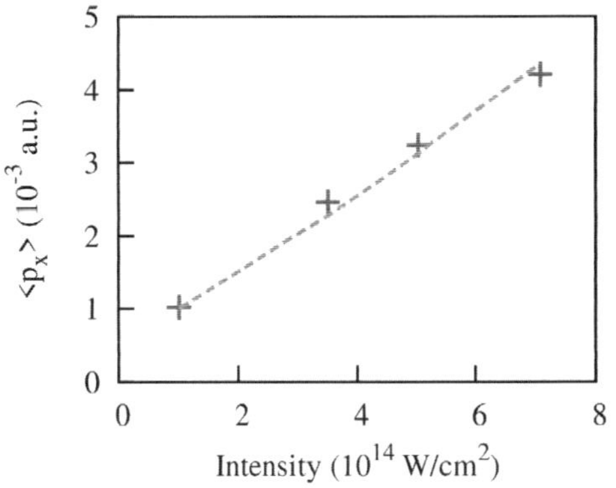

Figure 1. (Red) crosses: expectation value $\langle p_x\rangle$ as function of the field intensity. (Green) dash line: $\frac{1}{c}\langle E_{\text{kin}}\rangle$.

References

[1] C. T. L. Smeenk *at al* 2011 *Phys. Rev. Lett.* **106** 193002

[2] M. Nurhuda and F. H. M. Faisal 1999 *Phys. Rev. A.* **60** 3125

[3] I. A. Ivanov 2014 *Phys. Rev. A.* **90** 013418

[1]E-mail: igor.ivanov@anu.edu.au

Content from this work may be used under the terms of the Creative Commons Attribution 3.0 licence. Any further distribution of this work must maintain attribution to the author(s) and the title of the work, journal citation and DOI.
Published under licence by IOP Publishing Ltd

XXIX International Conference on Photonic, Electronic, and Atomic Collisions (ICPEAC2015) IOP Publishing
Journal of Physics: Conference Series **635** (2015) 092008 doi:10.1088/1742-6596/635/9/092008

Two-electron ions in strongly coupled plasma environment

S. Bhattacharyya* [1]

* Department of Physics, Acharya Prafulla Chandra College, Kolkata-700131, India

Synopsis Helium-like ions of carbon, aluminum and argon surrounded by strongly coupled plasma have been studied in the framework of ion-sphere potential. A methodology for evaluating two-electron correlated integrals in finite domain have been developed for this purpose.

Atoms placed in external plasma environment have received considerable attention from researchers due to their wide applications in various disciplines of science *e.g.* astrophysics, condensed matter physics, energy research etc. While dealing with plasma, it is customary to introduce a coupling parameter Γ which is defined as the ratio of the average Coulomb energy and the average kinetic energy of plasma particles. For weakly coupled plasma (WCP), $\Gamma < 1$ and for strongly coupled plasma (SCP), $\Gamma \geq 1$. Very recently, with the advent of Linac Coherent Light Sources it has been possible to create relatively long-lived high-density plasma of *Al* at homogeneous temperature and densities [1, 2]. The ionization potential depression (IPD) was observed [2] in the emitted spectra as a function of density by disappearance of spectral lines in hydrogen-like and helium-like *Al*. But accurate theoretical studies [3] to interpret these experimental results are extremely limited.

The effective potential in SCP is given by the well known Ion-sphere (IS) model [4]. The structure of the potential demands that the wave function is truncated at a finite distance .*e.* at the boundary of the ion-sphere. In the present work, we report precise non-relativistic energy values of $1sns$ $(^1S^e)$ $[n = 1-3]$ and $1sn'p$ $(^1P^o)$ $[n' = 2-4]$ states of helium-like *C*, *Al* and *Ar* within SCP environment. Explicitly correlated wavefunctions expanded in terms of Hylleraas basis have been used within a variational framework to calculate the energy eigenvalues. A methodology for evaluation of two-electron finite integrals is developed [5] that involves direct integration over the inter-electron coordinate.

Fig. 1 shows the energy values of Al^{11+} embedded in SCP. A detailed analysis of the present results show some novel phenomena [5] like *c de tal dege eracy* and *level-cross g* for the first time in case of two-electron ions embedded in strongly coupled plasma. Ionization potentials and spectral line positions for the ions embedded in intense plasma environments have been calculated [5] and compared with available experiments [1, 2, 6–8] and existing theoretical results [9]. The critical electron densities at which the spectral lines of helium-like *Al* cease to exist have also been calculated for comparison with experiments [2].

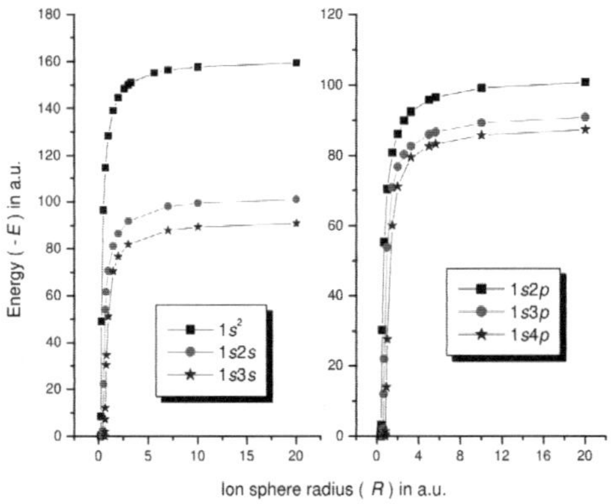

Figure 1. Energy values of $1sns$ $(^1S^e)$ $[n = 1-3]$ and $1sn'p$ $(^1P^o)$ $[n' = 2-4]$ states of Al^{11+}.

Due to spatial restriction imposed upon the wave function, the ion is subjected to an external pressure inside the sphere. The thermodynamic pressure experienced by the ions in their respective ground states are also reported.

References

[1] S. M. Vinko *et al.* 2012 *Nature* **482** 59

[2] D. J. Hoarty *et al.* 2013 *Phys. Rev. ett.* **110**, 265003 *a d refere ces there* .

[3] T. R. Preston *et al.* 2013 *H gh E ergy De s ty Phys cs* **9** 258

[4] S. Ichimaru 1982 *Rev. Mod. Phys.* **54** 1017

[5] S. Bhattacharyya *et al.* 2015 accepted in *Phys. Rev. A*

[6] M. Nantel *et al.* 1998 *Phys. Rev. ett.* **80** 4442

[7] A. Saemann *et al.* 1999 *Phys. Rev. ett.* **82** 4843

[8] N. C. Woolsey *et al.* 1998 *Phys. Rev. E* **57** 4650

[9] A. N. Sil *et al.* 2009 *Eur. Phys. J. D* **55**, 645

[1] E-mail: sukhamoy.b@gmail.com

Content from this work may be used under the terms of the Creative Commons Attribution 3.0 licence. Any further distribution of this work must maintain attribution to the author(s) and the title of the work, journal citation and DOI.
Published under licence by IOP Publishing Ltd

Journal of Physics: Conference Series **635** (2015) 092009 doi:10.1088/1742-6596/635/9/092009

Confinement Effects On Spin-Orbit Activated Interchannel Coupling

David A Keating* [1], **P. C. Deshmukh**† [2], **Steven T Manson*** [3]

* Department of Physics and Astronomy, Georgia State University, Atlanta, GA 30303
†Department of Physics Indian Institute of Technology Madras, Chennai 600036, India

Synopsis The spin-orbit activated interchannel coupling is a purely relativistic effect that alters the partial photoionization cross sections of spin-orbit doublet channels. Presented is an exploration of how the spin-orbit activated interchannel coupling is affected by confinement in a C_{60} cage.

At high enough Z relativistic effects become important contributors to even the qualitative nature of atomic properties. This is likely to be true for confined atoms as well.

One relativistic effect of interest is the spin-orbit activated interchannel coupling of a set of spin-orbit doublet channels. This interaction is possible thanks to the spin-orbit force breaking the degenerancy among the electrons of a subshell [1] allowing, in this case the $5d_{5/2}$ and $5d_{3/2}$ electrons of radon (Z=86), to interact. This phenomenon has been seen in the 3d cross sections of xenon, cesium, and barium [1].

To explore the effect confinement has on spin-orbit activated interchannel coupling, a theoretical study of the 5d subshell of radon confined in a C_{60} cage has been performed using the relativistic random phase approximation (RRPA) methodology [2]. The effects of the C_{60} potential modeled by a static spherical well which is reasonable in the energy region well above the C_{60} plasmons [3].

In order to determine which features in the photoionization cross section are solely due to confinement and which features are altered by the spin-orbit activated interchannel coupling, calculations are performed allowing coupling between the spin-orbit doublets and without coupling between them.

It is found that spin-orbit activated interchannel coupling significantly reduces the amplitudes of the partial cross sections of both the $5d_{5/2}$ and $5d_{3/2}$ of radon. There is also an extra structure in the $5d_{5/2}$ partial cross section, which is not present in the cross section without coupling allowed. In both partial cross sections without coupling allowed there is a double resonance feature, this is mirrored in the $5d_{5/2}$ partial cross section with coupling. But in the $5d_{3/2}$ these resonances trade relative amplitudes.

The confinement resonances become more defined in the $5d_{3/2}$ partial cross section with coupling than they were in the partial cross section without coupling. This is the opposite effect of the $5d_{5/2}$ partial cross section.

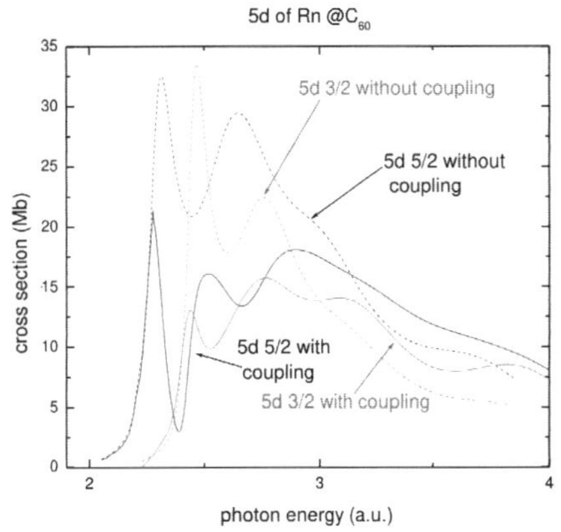

Figure 1. Partial photoionization cross sections of the $5d_{5/2}$ and $5d_{3/2}$ subshells of radon, the solid lines represent the calculations with coupling; the dashed lines represent the calculations without coupling allowed.

It is important to understand these alterations to the confined photoionization cross section of the 5d subshell of radon due to the spin-orbit activated interchannel coupling because of its effect on the other channels due to normal interchannel coupling.

References

[1] M. Ya. Amusia, L. V. Chernysheva, S. T. Manson, A. M. Msezane, and V. Radojevic, Phys. Rev. Lett. **88**, 9 (2002)
[2] W. R. Johnson and C. D. Lin, Phys. Rev A **20**, 964 (1979)
[3] V K Dolmatov, Adv. Quantum. Chem. **58**, 13 (2009)

[1] E-mail: dkeating2@student.gsu.edu
[2] E-mail: pcd@physics.iitm.ac.in
[3] E-mail: smanson@gsu.edu

Content from this work may be used under the terms of the Creative Commons Attribution 3.0 licence. Any further distribution of this work must maintain attribution to the author(s) and the title of the work, journal citation and DOI.
Published under licence by IOP Publishing Ltd

Journal of Physics: Conference Series **635** (2015) 092010 doi:10.1088/1742-6596/635/9/092010

Intra-cavity photodetachment microscopy and the electron affinity of germanium

Christophe Blondel[1], **David Bresteau**[2], **Cyril Drag**[3]

Laboratoire Aimé-Cotton, Centre national de la recherche scientifique, université Paris-sud, école normale supérieure de Cachan, bâtiment 505, F-91405 Orsay cedex, France

Synopsis Photodetachment microscopy is performed inside an optical cavity, on an ion beam and with an optical mode narrow enough not to blur out the output electron interferogram. This setting is used for an updated measurement of the electron affinity of germanium. Amplification of the photodetachment probability obtained in an optical cavity also opens the way to photodetachment microscopy with a p electron wave, even though photodetachment cross-sections are reduced considerably in this case, when compared to s-wave detachment, by the centrifugal barrier.

A linear optical cavity is set across a beam of Ge^- ions, itself produced by a cesium sputtering ion source. The optical cavity is injected with a single mode ring Ti:Sa laser, the wavenumber of which can be set either above the highest $^3P_2$ fine-structure excitation threshold, or just above the $^3P_1$, intermediate fine-structure threshold of the ^3P ground-term of Ge I. The freed photoelectron is accelerated towards an imaging detector by a uniform electric field of about 350 Vm^{-1}. When the initial kinetic energy is of the order of 50 m^{-1}, an electron interferogram can be observed, which semi-classically corresponds to the existence of a pair of trajectories bound to every detection point.

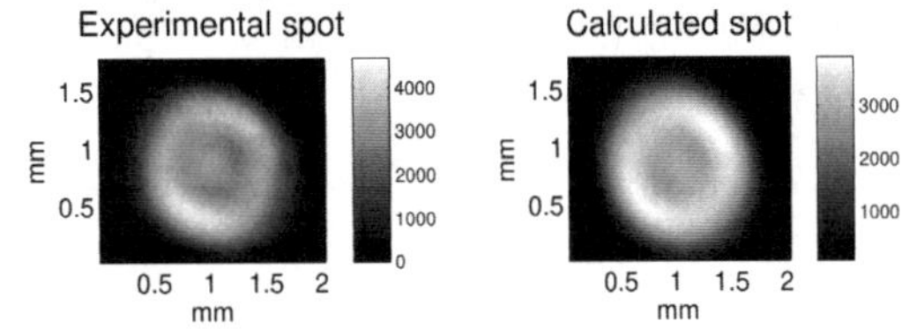

Figure 1. Intracavity-produced electron interferogram. In this example, photodetachment of Ge^- at wavenumber $\lambda^{-1} = 1135\,258.9(1)\,m^{-1}$, within an electric field of 357.6 Vm^{-1}, produces an electron interferogram of apparent energy $\epsilon = 44.2(15)\,m^{-1}$. Persistence of a ring pattern, with fringes not much more than 100 μm wide, shows that the ion beam diameter has been reduced to smaller values.

Fitting these interferograms with the expected form of the uniform acceleration Green function, one gets a measure of the electron's initial kinetic energy. The precision of the method can be three orders of magnitude larger than the precision of classical electron spectrometry. Provided that the two photoelectron energies produced by the two opposite wave vectors contained in the optical cavity are not too different, i.e. provided that the illumination angle remains close to 90 °, subtraction of the mean photoelectron energy from the photon energy directly provides a high accuracy measurement of the electron affinity.

A Doppler-free measurement of the photodetachment threshold can thus be produced, in an optical cavity, even more directly than with the usual double spot photodetachment microscopy method [1], at the only expense of a slight reduction of the contrast in the interferograms. The obtained value of the electron affinity of germanium is $994\,220.6\,(10)\ m^{-1}$ or $1.232\,6764\,(12)$ eV, one order of magnitude more precise and a little smaller than the last measured value $994\,249\,(12)\ m^{-1}$ or $1.232\,712\,(15)$ eV [2].

References

[1] C. Blondel, W. Chaibi, C. Delsart, C. Drag, F. Goldfarb & S. Kröger 2005 *Eur. Phys. J. D* **33** 335

[2] M. Scheer, R. C. Bilodeau, C. A. Brodie & H. K. Haugen 1998 *Phys. Rev. A* **58** 2844

[1]E-mail: christophe.blondel@u-psud.fr
[2]E-mail: david.bresteau@u-psud.fr
[3]E-mail: cyril.drag@u-psud.fr

Content from this work may be used under the terms of the Creative Commons Attribution 3.0 licence. Any further distribution of this work must maintain attribution to the author(s) and the title of the work, journal citation and DOI.
Published under licence by IOP Publishing Ltd

Journal of Physics: Conference Series **635** (2015) 092011 doi:10.1088/1742-6596/635/9/092011

Modulation of Attosecond Beating in Resonant Two-Photon Ionization

Á. Jiménez-Galán* [1], **L. Argenti***, **F. Martín***,†

* Departamento de Química, Módulo 13,Universidad Autónoma de Madrid, Cantoblanco 28049, Madrid, Spain

† Instituto Madrileño de Estudios Avanzados en Nanociencia (IMDEA-Nanociencia), Cantoblanco, 28049 Madrid, Spain

Synopsis We present a theoretical study of the photoelectron attosecond beating due to interference of two-photon transitions in the presence of autoionizing states. We show that, as a harmonic traverses a resonance, both the phase and frequency of the sideband beating significantly vary with photon energy. Furthermore, the beating between two resonant paths persists even when the pump and probe pulses do not overlap, thus permitting to reconstruct nonholographically coherent metastable wave packets. We characterize these phenomena with an analytical model that accounts for the effect of both intermediate and final resonances on two-photon processes.

Autoionizing states (AI) are hallmarks of electronic correlation, which shapes the reactivity of all many-body systems. Attosecond light sources provide a new means to investigate the role of such correlated states in atomic transitions, since one can map their evolution on a time scale smaller than their lifetime. The technique of reconstruction of attosecond beating in two-photon transitions (RABITT) [1], which has already been succesfully employed to study resonant processes [2], is particularly indicated to this task. So far, however, only transitions through either electronic bound states or autoionizing vibronic states without any appreciable contribution from the intermediate continuum have been considered. The phase of the complex transition amplitude was thus compatible with a change of π at the resonance energy.

In general, however, both the localized and continuum components of an intermediate AI contribute to the two-photon amplitude. We found that, when the two contributions are comparable, the phase exhibits a peaked excursion which returns to its original value, instead of a π jump. Furthermore, we predict that, for finite pulses, the frequency of the RABITT beating itself is altered. Models normally used to interpret these experiments do not contemplate the presence of resonances, while full *ab initio* calculations are either extremely demanding or outright unfeasible. For this reason, we developed a resonant analytical model which allows us to compute two-photon ionization spectra of attosecond pump-probe interferometric experiments, for arbitrary pulse sequences[3]. The model parameters can be retrieved either from experimental data or from selected *ab initio* simulations. We used this procedure to investigate, with RABITT, the two-photon ionization of helium in the region of the doubly excited states of the atom, obtaining excellent agreement with virtually exact *ab initio* simulations [4]. In particular, we show that both the frequency modulation and phase excursion of the transition through sp_2^+ and sp_3^+ ($^1P^o$) to $2p^2$ ($^1S^e$) are measurable.

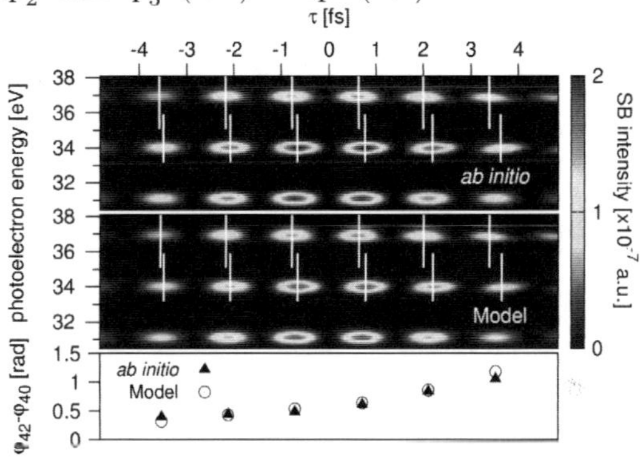

Figure 1. Photoelectron spectrum of He as a function of the pump-probe time delay. The presence of sp_2^+($\sim$ 35.5 eV) shifts the sidebands in opposite directions and modulates the beating frequency.

Finally, the beating between the overlapping sidebands of the two intermediate resonances persist even when the pump and probe pulses are separated. From this beating it is possible to reconstruct the metastable wavepacket generated by the pump pulse.

References

[1] P. M. Paul *et al.* 2001 *Science* **292**, 1689

[2] M.Swoboda *et al.* 2010 *Phys.Rev.Lett.* **104**, 103003

[3] Á. Jiménez-Galán, L. Argenti and F.Martín 2014 *Phys.Rev.Lett.* **113**, 263001

[4] L. Argenti and E. Lindroth 2010 *Phys. Rev. Lett.* **105**, 053002

[1]E-mail: alvaro.jimenez@uam.es

Content from this work may be used under the terms of the Creative Commons Attribution 3.0 licence. Any further distribution of this work must maintain attribution to the author(s) and the title of the work, journal citation and DOI.
Published under licence by IOP Publishing Ltd

Interelectron correlations in photoionization of outer shells near inner shell thresholds

M. Ya. Amusia*† [1], L. V. Chernysheva† [2], E. G. Drukarev**

* The Racah Institute of Physics, the Hebrew University, Jerusalem 91904, Israel
† A. F. Ioffe Physical-Technical Institute, St. Petersburg 194021, Russian Federation
** B. P. Konstantinov Petersburg Nuclear Physics Institute, Gatchina 188300, Russian Federation

Synopsis We have studied the role of virtual excitations of inner shells upon outer shell photoionization. The calculations were performed in the frames of the Random Phase Approximation with Exchange (RPAE) and its generalized version GRPAE that take into account variation of the atomic field due to electron elimination and the inner vacancies decay. We apply both analytic approximation and numeric computations. The results are presented for $3p$ electrons in Ar and for $4d$-electrons in Pd near inner shells thresholds. The effect considered proved to be quite noticeable.

1. The role of intershell interaction or electron correlations was disclosed theoretically long ago in the frame of RPAE. Theoretical predictions were confirmed experimentally very soon. At first, main attention was concentrated on studies of low photon energy photoionization, namely on elimination of electrons from outer, sub-valent and some intermediate shells.

2. Correlations are important also at high photon energy if interacting electrons belong to subshells with close values of energies, when the photon energy is high for both of them.

At the same time, if one considers photoionization of an outer shell close to the inner threshold, the photon energy is high for the outer shell, being low for the inner shell. There essential correlation effects are of importance due to near threshold jump of the inner shell photoionization cross-section. Qualitatively, it has been predicted long ago (see [1], Chap. 5). Here we employ the RPAE approximation and its generalized version GRPAE [1], both in analytic and numeric versions.

3. The analytic approach gives understanding of the physical mechanism of the inner shell action. It is connected to abrupt jump of the inner shell photoionization cross-section that leads to logarithmic divergence of the correction to the outer shell cross-section. It shows that correction can both enhance and decrease the resulting cross-section. However, it cannot predict the size of the correction with reasonable accuracy. To do this, numeric calculations in the frames of RPAE and GRPAE were performed.

4. The results obtained demonstrate the prominent role of inner shells in all considered cases. The contribution of continuous spectrum of inner shell excitations dominates over the discrete excitations, except close to the latter.

5. We found examples of both increase and decrease of the total outer shell photoionization cross section under the action of the inner electrons.

6. Figure 1 illustrates results of our RPAE calculations. It is seen strong enhancement of the 4d-subshell photoionization cross-section both above the ionization threshold and below the first excitation potential of the 3s subshell.

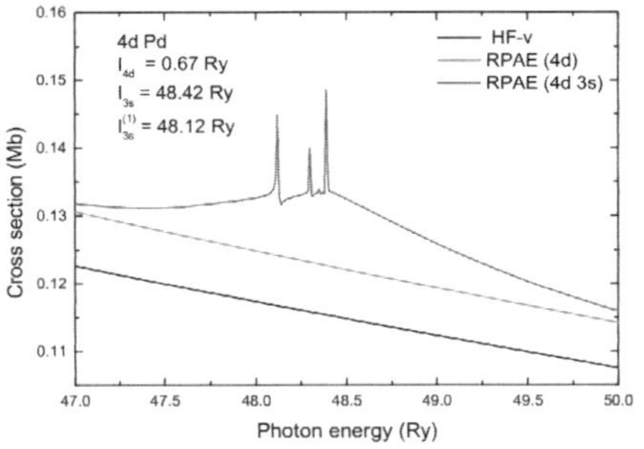

Figure 1. Photoionization cross-section of 4d electrons near 3s thresholds of Pd

6. The results present an area in photoionization studies that is far from being satisfactorily explored. In this area the intershell interaction is visible but relatively small, so it can be accounted for perturbatively, thus verifying the ability of the approaches employed to describe both, the photoionization amplitude far from its threshold and the intershell interaction.

References

[1] M. Ya. Amusia, *Atomic photoeffect*, Plenum Press, 1990

[1] E-mail: amusia@vms.huji.ac.il

[2] E-mail: Larissa.Chernysheva@mail.ioffe.ru

Content from this work may be used under the terms of the Creative Commons Attribution 3.0 licence. Any further distribution of this work must maintain attribution to the author(s) and the title of the work, journal citation and DOI.
Published under licence by IOP Publishing Ltd

XXIX International Conference on Photonic, Electronic, and Atomic Collisions (ICPEAC2015) IOP Publishing
Journal of Physics: Conference Series **635** (2015) 092013 doi:10.1088/1742-6596/635/9/092013

Merging quantum chemistry packages with B-splines for the multichannel scattering problem.

Carlos Marante* [1], **Jesús González***, **Inés Corral***, **Markus Klinker***, **Luca Argenti***, **Fernando Martín***,†

* Departamento de Química, Módulo 13,Universidad Autónoma de Madrid, Cantoblanco 28049, Madrid, Spain
† Instituto Madrileño de Estudios Avanzados en Nanociencia (IMDEA-Nanociencia), Cantoblanco 28049, Madrid, Spain

Synopsis To study the photoionization of polyelectronic systems by attosecond pulses, we take advantage of existing quantum chemistry packages for the description of correlated electronic states, and of hybrid Gaussian-B-splines basis for the representation of continuum orbitals. In our approach, a short-range region, which can host all the interacting electrons and is described by commercial packages, is matched to a long-range region which describes single-ionization states in terms of a close-coupling expansion. We validate this approach showing multichannel ionization results for He.

Recent breakthroughs in ultrafast-laser technology, and more specifically the generation of extreme ultraviolet (XUV) and x-ray attosecond pulses [1, 2], have made it possible to study electron dynamics at its natural time scale. Since such pulses can ionize the atom or molecule with which they interact, the theoretical description of the observables in attosecond pump-probe experiments requires a good representation of the system's ionization continua in a wide energy range. While this is possible for very small systems such as He and H_2 [3–6], in more complex systems this is still a challenge, due to the complicate short-range structure of polyelectronic functions. Indeed, whereas sophisticated methods to compute the electronic structure of bound molecular states are implemented in commercial quantum chemistry packages (QCP) (e.g., MOLCAS [7]), comparable tools for the continuum are not really available.

To tackle this problem, we are developing a merge between existing QCPs and current numerical scattering methods. To do so, we separate the electronic configuration space in a short-range region, which can host all the interacting electrons, and a long-range region in which one outer electron interacts with a finite number of correlated parent ions. This outer electron is expressed in terms of a hybrid Gaussian-B-spline basis (GABS) which combines short-range Gaussian functions, compatible with standard QCPs, with B-splines [8], which are appropriate to represent the continuum.

To illustrate the viability of our approach, here we present the results for the multichannel ionization of He, which is a necessary step towards the study of more complex systems. The eigenphases above the N=2 threshold and the photoionization cross section we obtain are in excellent agreement with state-of-the-art benchmarks [9]. The good results obtained together with the great flexibility of QCPs position our method as a strong candidate for the theoretical study of the ionization of poly-electronic systems.

[1]E-mail: carlos.marante@uam.es

References

[1] M. Chini *et al* 2014 *Nature Photonics* **8**, 178-186

[2] T. Popmintchev *et al* 2010 *Nature Photonics* **4**, 822-832

[3] A. Jiménez-Galán *et al* 2014 *Phys. Rev. Lett.* **113** 263001

[4] C. Ott *et al* 2014 *Nature* **516**, 374-378

[5] G. Sansone *et al* 2010 *Nature* **465**, 763-766

[6] D. Dowek *et al* 2010 *Phys. Rev. Lett* **104** 233003

[7] G. Karlstrm *et al* 2003 *Comput. Mater. Sci.* **28** 2, 222-239

[8] C. Marante *et al* 2014 *Phys. Rev. A* **90** 012506

[9] L. Argenti and R. Moccia 2006 *J. Phys. B* **39** 2773

Content from this work may be used under the terms of the Creative Commons Attribution 3.0 licence. Any further distribution of this work must maintain attribution to the author(s) and the title of the work, journal citation and DOI.
Published under licence by IOP Publishing Ltd

Slowly moving hydrogen-like ions in quantum plasma

T. K. Mukhopadhyay* [1]

* Department of Physics, Narula Institute of Technology, Kolkata 700109, India

Synopsis Precise non-relativistic energy eigenvalues of slowly moving hydrogen-like ions under quantum plasma environment have been estimated by using Ritz variational mathod. Present study shows that the energy levels gradually move towards the continuum as plasma electron density increases while level crossing phenomenon has been observed with the variation of ion velocity.

The studies on quantum plasma have got importance in several branches of physics such as nanoscience, laboratory plasma experiments and in astrophysical context. Quantum plasmas are generally made of electrons and ions or holes. It is well-known according to the Debye-Huckel theory [1, 2] that a static atom/ion feels screened Coulomb type potential while placed in a collision-less high temperature classical plasma. The screening parameter in this case is a function of electron number density (n_e)and temperature (T) of the plasma. In contrast, when the temperature of the plasma electrons approaches the Fermi-temperature, the equilibrium plasma electron distribution function changes from Maxwell-Boltzmann to Fermi-Dirac distribution. Under such condition, the quantum degeneracy effects start playing a significant role as the thermal de-Broglie wavelength for the plasma electrons becomes comparable or equal to that of average inter-electronic distance.

An arbitrary collision-less quantum plasma environment may be treated as a dielectric medium and the analytic form of dielectric function may be derived using Random Phase Approximation method [3]. This dielectric function can be used to express the effective potential felt by a slowly moving test charge as a sum of two components: *i)* near field potential originating from Debye-Hückel screening and *ii)* far-field wake potential [4]. In the present work, we have studied the influence of near field wake potential on the energy values of hydrogen-like carbon ion moving through Electron-Hole-Droplet (EHD) quantum plasma. The analytic form of the potential has been derived in the present work using Meijers G-function [5] and its identities. The variation method is then applied to obtain the binding energies of all states lying between $1s$ and $4f$ configuration of C^{5+} and the results are compared with other calculations [6] existing in literature.

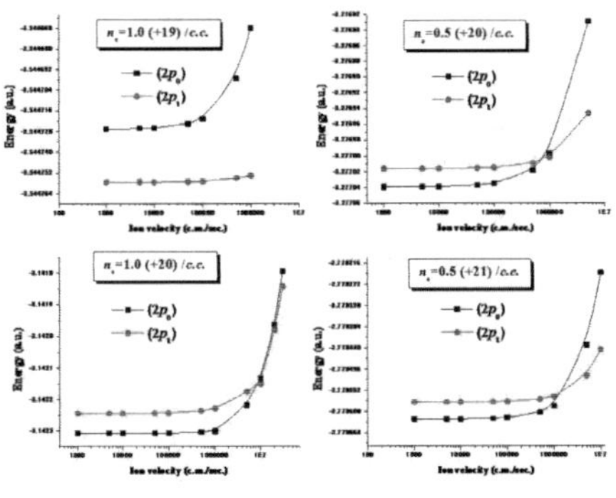

Figure 1. Energy values (a.u.) of $2p_0$ and $2p_1$ states of C^{5+} against ion velocity (cm/s) for different plasma electron densities (c.c.$^{-1}$).

Fig. 1 shows the variation of energy values of $2p_0$ and $2p_1$ states of C^{5+} against ion velocity for different plasma electron densities. *Level crossing* phenomenon and *incidental degeneracy* at some critical ion velocity have been observed for the first time in case of an ion moving in the quantum plasma environment.

References

[1] P. Debye and E. Huckel 1923 *Z. Phys.* **24** 185

[2] A.I. Akhiezer *et al.* 1975 *Plasma Electrodynamics, Linear Response Theory* Pergamon, Oxford **vol. 1**,

[3] D. Pines 1961 *Plasma Phys* **2** 5

[4] P. K. Shukla *et al.* 2006 *Phys. Lett. A* **359** 218

[5] L. S. Gradshteyn and I.M. Ryzhik 1994 *Table of Integrals, Series and Products, 7th ed.* Academic, New York

[6] H. Hu *et al.* 2012 *Phys. Rev. E* **85** 056406

[1]E-mail: drtapanmukherjee@gmail.com

Content from this work may be used under the terms of the Creative Commons Attribution 3.0 licence. Any further distribution of this work must maintain attribution to the author(s) and the title of the work, journal citation and DOI.
Published under licence by IOP Publishing Ltd

Journal of Physics: Conference Series **635** (2015) 092015 doi:10.1088/1742-6596/635/9/092015

Dressing Effects in the Attosecond Transient Absorption Spectra of Doubly-Excited States in Helium

L. Argenti*[1], Á. Jiménez-Galán*, C. Marante*, C. Ott†‡, T. Pfeifer†‡, F. Martín*¶§

* Departamento de Química, Módulo 13,Universidad Autónoma de Madrid, Cantoblanco 28049, Madrid, Spain
† Max-Planck Institut für Kernphysik, Saupfercheckweg 1, 69117 Heidelberg, Germany
‡ Center for Quantum Dynamics, Ruprecht-Karls-Universität Heidelberg, 69120 Heidelberg, Germany
§ Condensed Matter Physics Center (IFIMAC), Universidad Autónoma de Madrid, 28049 Madrid, Spain
¶ IMDEA-Nanociencia, Cantoblanco, 28049 Madrid, Spain

Synopsis Strong-field manipulation of autoionizing states is a crucial aspect of electronic quantum control. Recent measurements of the attosecond transient absorption spectrum (ATAS) of helium dressed by a few-cycle visible pulse [C. Ott *et al*, Nature **516**, 374 (2014)] provide evidence of the inversion of Fano profiles. With the support of accurate *ab-initio* calculations that reproduce the results of the latter experiment, here we investigate the new physics that arise from ATAS when the laser intensity is increased. In particular, we show that (i) previously unnoticed signatures of the dark $2p^2$ ^{1}S doubly excited state are observed in the experimental spectrum, (ii) inversion of Fano profiles is predicted to be periodic in the laser intensity, and (iii) the ac-Stark shift of the higher terms in the $sp^+_{2,n}$ autoionizing series exceeds the ponderomotive energy, which is the result of a genuine two-electron contribution to the polarization of the excited atom.

ATAS [1] has affirmed itself as a valid technique to monitor and control electronic wavepackets [1–3]. Control is achieved with a laser pulse that imparts an ac-Stark shift to the wavepacket components, thus altering their relative phases. The latter map onto the asymmetry of their resonant profiles in the ATAS spectrum [4]. This principle has been recently applied to a two-electron wave packet in the helium atom, formed by the $sp^+_{2,n}$ ^{1}P^o doubly excited states [3]. The experiment evidenced the Autler–Townes splitting of the $sp^+_{2,2}$ state, which is coupled to the $2p^2$ ^{1}S state, as well as inversion of the Fano profile in the higher terms of the $sp^+_{2,n}$ series as a function of both the time delay between the XUV and the VIS pulses, and the intensity of the VIS dressing field [4].

Here we show, by means of accurate *ab-initio* calculations [3, 5, 6] that reproduce the experimental results for the helium atom, that ATAS can be used to measure the non-perturbative response of autoionizing states to external dressing fields. In particular, we show that the Autler-Townes splitting of the dressed $2p^2$ ^{1}S "dark" state can be measured, and that the asymmetric resonant features in the spectrum undergo repeated inversions as the intensity of the driving laser is increased. The ac-Stark shift of the doubly-excited states, which are important parameters for the design of quantum-control protocols, increases non-linearly with the intensity of the dressing field; for the higher states of the autoionizing $sp^+_{2,n}$ series, it exceeds by more than twice the theoretical single-active-electron limit. Finally, we establish a relationship between the ATAS spectrum and the non-diagonal components of the XUV electrical susceptibility, thus extending multidimensional spectroscopy to the XUV domain (Fig. 1), which provides an alternative way to measure shifts of and coherences between states above the ionization threshold.

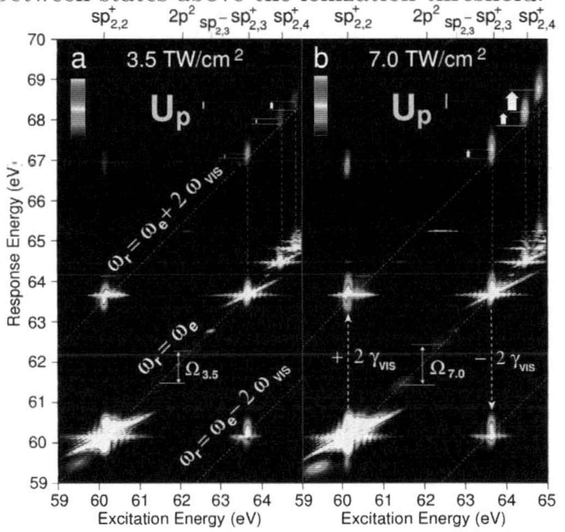

Figure 1. Reconstruction of the non-diagonal XUV electrical susceptibility $|\chi_{\rm nd}(\omega_{\rm r},\omega_{\rm e})|^2$ of dressed helium.

References

[1] E. Goulielmakis *et al.* 2010 *Nature* **466**, 739.
[2] A. R. Beck *et al.* 2014 *New J. Phys.* **16**, 113016.
[3] C. Ott *et al.* 2014 *Nature* **516**, 374.
[4] C. Ott *et al.* 2013 *Science* **516**, 374.
[5] L. Argenti and E. Lindroth 2010 *PRL* **105**, 053002.
[6] L. Argenti *et al.* 2015 *submitted*.

[1]E-mail: luca.argenti@uam.es

Content from this work may be used under the terms of the Creative Commons Attribution 3.0 licence. Any further distribution of this work must maintain attribution to the author(s) and the title of the work, journal citation and DOI.
Published under licence by IOP Publishing Ltd

Journal of Physics: Conference Series **635** (2015) 092016 doi:10.1088/1742-6596/635/9/092016

Rayleigh scattering of x–rays by many–electron ions

A. Surzhykov*[1], V. A. Yerokhin†, Th. Stöhlker*,‡ and S. Fritzsche*,‡

* Helmholtz–Institut Jena, D–07743 Jena, Germany

† Center for Advanced Studies, St. Petersburg State Polytechnical University, St. Petersburg 195251, Russia

‡ Friedrich–Schiller–Universität Jena, D–07743 Jena, Germany

Synopsis A theoretical investigation is presented for the elastic Rayleigh scattering of photons by many–electron ions. Special attention is paid to the question how the charge state and shell structure of the target ions affect the angular distribution and linear polarization of scattered light. Calculations are performed for helium–, neon– and argon–like heavy ions and for a wide range of the photon energies.

Investigations of the elastic Rayleigh scattering of photons by *bou d* atomic electrons have a long tradition going back to 1930's. During the last decades a number of experimental and theoretical Rayleigh studies have dealt with high–energetic incident radiation and heavy atomic targets, see Ref. [1] for further details. The analysis of the elastic scattering in such a relativistic regime provides valuable information about the structure of many–electron systems as well as their coupling to the electromagnetic radiation in the presence of very strong nuclear fields.

In order to better understand the interplay between the many–body, relativistic and non–dipole effects in the $\gamma + A \to \gamma + A$ process, we have recently explored the elastic scattering of x–rays by multiply–charged heavy ions [2, 3]. In our study, we considered the set–up of a typical synchrotron experiments, where the incident radiation is (completely) linearly polarized. For this case, we employed the second–order perturbation theory and the independent–particle approximation (IPA) in which the photon is scattered by a single (active) electron at a time, while the remaining electrons were kept "frozen" [3].

By making use of the IPA, detailed calculations have been performed for the angular distribution and the linear polarization of the scattered light. In Fig. 1, for example, we display the results for the scattering of 175 keV x–rays by helium–, neon– and argon–like gold ions. As seen from the figure, the angle–differential Rayleigh cross sections can largely vary with the charge state of the ion. The most pronounced effect can be observed at forward scattering angles where the cross section is enhanced by almost two orders of magnitude in going from helium–like Au^{77+} to argon–like Au^{61+} as a target. In contrast to the angular distribution, the linear polarization of the scattered Rayleigh photons is virtually not affected by the ionic shell structure even though it can be sensitive to the nuclear charge Z of the target and to the energy of the incident radiation.

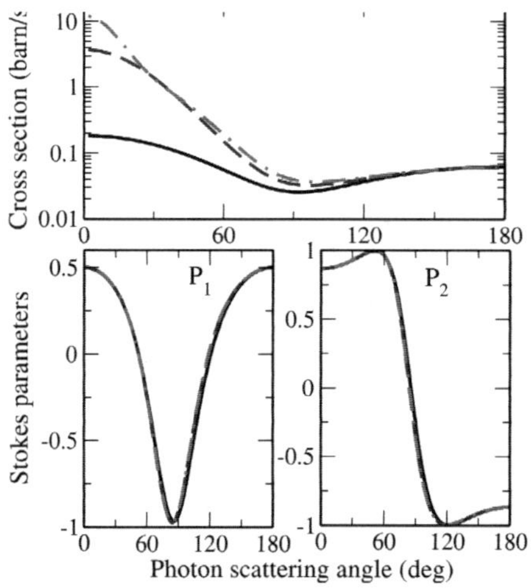

Figure 1. Angle–differential cross section (upper panel) and the linear–polarization Stokes parameters P_1 and P_2 (lower pannels) for the elastic Rayleigh scattering of linearly polarized x–rays by helium–like (solid line), neon–like (dashed line) and argon-like (dash-dotted line) gold ions in their ground state. Results are presented for the photon energy 175 keV and for the emission of the scattered light in the plane tilted by the angle $\phi = 30$ deg with respect to the polarization plane of the incident radiation.

References

[1] R. Pratt 2005 *Rad. Phys. Chem.* **74** 411

[2] A. Surzhykov *at al* 2013 *Phys. Rev. A* **88** 062515

[3] A. Surzhykov *at al J. Phys.* accepted

[1]E-mail: a.surzhykov@gsi.de

Content from this work may be used under the terms of the Creative Commons Attribution 3.0 licence. Any further distribution of this work must maintain attribution to the author(s) and the title of the work, journal citation and DOI.
Published under licence by IOP Publishing Ltd

XXIX International Conference on Photonic, Electronic, and Atomic Collisions (ICPEAC2015) IOP Publishing
Journal of Physics: Conference Series **635** (2015) 092018 doi:10.1088/1742-6596/635/9/092018

Non-perturbative B-spline R-matrix with pseudo-states calculations for electron-impact excitation-ionization of helium to the $n = 3$ states

Oleg Zatsarinny [1] **and Klaus Bartschat** [2]

Department of Physics and Astronomy, Drake University, Des Moines, Iowa 50311, USA

Synopsis We present fully-differential cross sections for electron-impact ionization plus simultaneous excitation of helium obtained from a non-perturbative close-coupling formalism with our B-spline R-matrix approach. Using a large number of pseudo-states we obtain excellent agreement with directly measured cross-section ratios for ionization leaving the residual He^+ ion in either the 1s ground state, the $n = 2$ (2s + 2p) excited states, or the $n = 3$ (3s + 3p + 3d) excited states.

Electron-impact ionization with simultaneous excitation is a highly challenging problem, both experimentally and theoretically. Taking e – He as the proto-type collision system, the challenges are due to the fact that all three electrons (the projectile and both target electrons) undergo significant changes in their respective quantum states. Hence the problem cannot be simplified further by effectively treating one of the target electrons as a "spectator" that remains in the 1s orbital.

In recent years, we have further developed our fully non-perturbative B-spline R-matrix (close-coupling) with pseudo-states (BSRMPS) method and applied it to both the direct ionization [1] and ionization with excitation problems [2,3]. Despite the success achieved in employing a projection approach to obtain the cross sections from excitation amplitudes of the positive-energy discrete (i.e., finite-range) pseudo-states, some skepticism remains about the validity of the approach. The method is indeed not exact [4], but it provides a practical interpolation scheme (currently the only one known) that should systematically improve when the density of pseudo-states is increased.

To check this hypothesis, we further optimized our parallelized BSR codes. This allowed us to increase the number of pseudo-states to 1,254 and thereby also consider ionization with excitation to the $n = 3$ states of He^+. Figures 1 and 2 show a comparison of our predictions with directly measured cross-section ratios [5] for ionization without excitation and ionization with excitation to He^+(3s+3p+3d). The agreement with the experimental data is excellent for both symmetric (Fig. 1) and asymmetric (Fig. 2) sharing of the excess energy between the two outgoing electrons. Normalizing to the well-known results for ionization without excitation will then allow for the determination of absolute cross sections for this process [6].

This work was supported by the United States National Science Foundation under grants PHY-1212450 and PHY-1430245, and the XSEDE allocation PHY-090031.

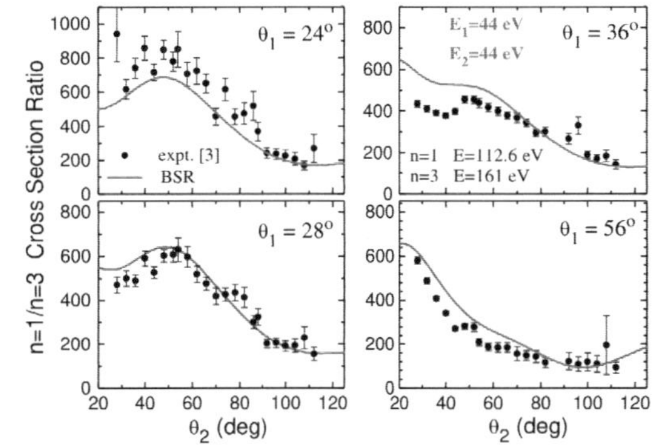

Figure 1. TDCS ratio for electron impact ionization of helium with the residual ion left in either the 1s ground state or the excited $n = 3$ states. The excess energy is shared equally between the two outgoing electrons.

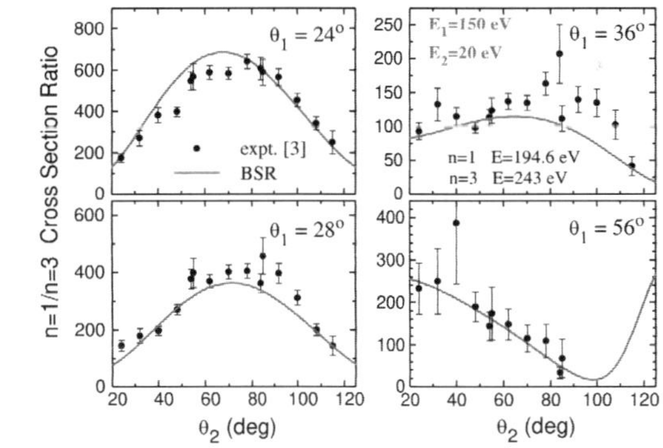

Figure 2. Similar to Fig. 1 for asymmetric sharing of the excess energy.

References

[1] O. Zatsarinny and K. Bartschat 2012 *Phys. Rev. A* **85** 062709
[2] O. Zatsarinny and K. Bartschat 2011 *Phys. Rev. Lett.* **107** 023203
[3] O. Zatsarinny and K. Bartschat 2014 *J. Phys. B* **47** 061001
[4] I. Bray *et al* 2014 *Phys. Rev. A* **90** 022710
[5] S. Bellm *et al* 2007 *Phys. Rev. A* **75** 042704
[6] S. Bellm *et al* 2008 *Phys. Rev. A* **78** 032710

[1] E-mail: oleg.zatsarinny@drake.edu
[2] E-mail: klaus.bartschat@drake.edu

Content from this work may be used under the terms of the Creative Commons Attribution 3.0 licence. Any further distribution of this work must maintain attribution to the author(s) and the title of the work, journal citation and DOI.
Published under licence by IOP Publishing Ltd

Journal of Physics: Conference Series **635** (2015) 092020 doi:10.1088/1742-6596/635/9/092020

Interference between one-photon and resonant two-photon ionization by intense pulses of a bichromatic VUV field

A. N. Grum-Grzhimailo*[1], E. V. Gryzlova*, E. I. Staroselskaya[†], J. Venzke[#], and K. Bartschat[#]
[*] Skobeltsyn Institute of Nuclear Physics, Lomonosov Moscow State University, 119991 Moscow, Russia
[†] Faculty of Physics, Lomonosov Moscow State University, 119991 Moscow, Russia
[#] Department of Physics and Astronomy, Drake University, IA 50311, Des Moines, USA

Synopsis Photoelectron angular distributions are theoretically described for coherent atomic ionization by short intense pulses of VUV radiation, containing both first and second harmonics, when the energy of the fundamental is in resonance with an intermediate state. The interference between different ionization pathways results in a 'left-right' asymmetry, which exhibits resonance profiles as function of the photon energy.

The high temporal coherency available from new-generation X-ray free-electron lasers (XFELs) pushes forward investigations of quantum coherent control of atomic and molecular processes in the VUV and X-ray range. We propose to manifest an interference of different pathways in atomic ionization by a bichromatic field containing first and second harmonics of a VUV laser with the electric field

$$E(t) = F(t)\left[\cos(\omega t + \varphi_1) + \eta\cos(2\omega t + \varphi_2)\right] .$$

Here $F(t)$ and ω are the field envelope and the frequency of the first harmonic, respectively, while η^2 ($\eta \geq 0$) is the ratio of first and second harmonic intensities. The radiation is linearly polarized along z axis. Two-photon ionization (2PI) by the first harmonic and one-photon ionization (1PI) by the second harmonic produce photoelectrons with equal energies, but different parities. The interference between the two ionization pathways manifests itself in an asymmetric photoelectron angular distribution (PAD)

$$W(\theta) = W_0\left(1 + \sum\nolimits_{n=1,2,...} \beta_n P_n(\cos\theta)\right)$$

with respect to the plane perpendicular to the electric field. This 'left-right' asymmetry is quantified as

$$A(\theta) = (W(\theta) - W(\pi-\theta))/(W(\theta) + W(\pi-\theta))$$

and formally originates from the parameters β_n with odd n. In the optical region, the 2PI-1PI interference has been predicted and observed for steady-state laser fields (e.g. [1-3]). Here we study this phenomenon theoretically for the case when ω is close to a resonant transition for pulses with finite time duration.

We consider ionization of the hydrogen atom with ω close to the energy of the 1s-2p transition. The PADs are calculated by numerically solving the time-dependent Schrödinger equation and also by employing the conventional first- and second-order perturbation theory for 1PI and 2PI, respectively.

Figure 1 exhibits a strong asymmetry, caused by interference of the two ionization pathways in the region of the 1s-2p excitation. This asymmetry depends strongly on the relative phase between the first and the second harmonics. It vanishes after averaging over the relative phase over the interval $(0,2\pi)$. Results for different pulse durations and intensities will be presented at the conference.

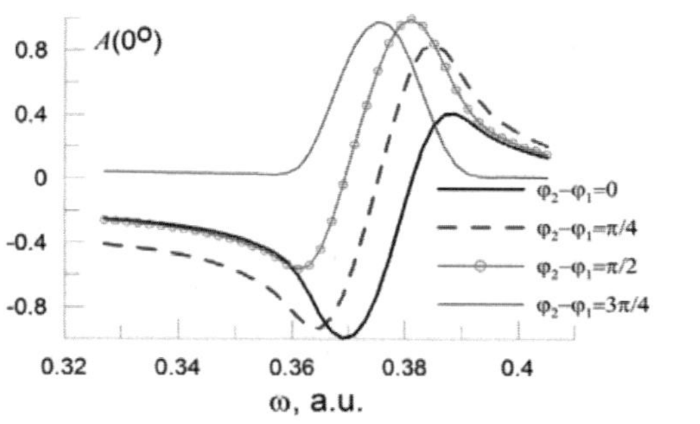

Figure 1. Left-right asymmetry $A(0^{\circ})$ as function of the photon energy for different relative phases of the first and second harmonics, as calculated within perturbation theory. A sine-squared pulse envelope $F(t)$ is assumed, with 40 optical cycles, a maximum intensity of 10^{12} W/cm^2, and $\eta^2 = 0.05$.

We thank G. Sansone, K. Prince, and K. Ueda for fruitful discussions. KB and JV acknowledge support from the United States National Science Foundation.

References

[1] N.A. Baranova *et al* 1992 *JETP Lett.* **55** 4391
[2] Y.-Y. Yin, C. Chen, D.S. Elliott, and A.V. Smith 1992 *Phys. Rev. Lett.* **69** 2353
[3] R. Yamazaki and D.S. Elliott 2007 *Phys. Rev. Lett.* **98** 053001

[1] E-mail: algrgr1492@yahoo.com

Content from this work may be used under the terms of the Creative Commons Attribution 3.0 licence. Any further distribution of this work must maintain attribution to the author(s) and the title of the work, journal citation and DOI.
Published under licence by IOP Publishing Ltd

Journal of Physics: Conference Series **635** (2015) 092021 doi:10.1088/1742-6596/635/9/092021

Pulse-shape effects in strong-field atomic ionization by an XUV pulse

Klaus Bartschat[*1] **Joel Venzke**[*], **John Emmons**[*], **Sean Buczek**[*],
Alexei N. Grum-Grzhimailo[†], **Elena V. Gryzlova**[†], **Igor A. Ivanov**[#&], **and Anatoli S. Kheifets**[&]

[*]Department of Physics and Astronomy, Drake University, Des Moines, Iowa 50311, USA
[†]Skobeltsyn Institute of Nuclear Physics, Lomonosov Moscow State University, Moscow 119991, Russia
[#]Center for Relativistic Laser Science, Institute for Basic Science, Gwangju 500-712, Republic of Korea
[&]Research School of Physics and Engineering, Australian National University, Canberra, ACT 0200, Australia

Synopsis Achieving a significant displacement, defined as a time integral of the vector potential taken over the pulse duration, in strong-field atomic ionization critically depends on the envelope function used for the electric field. Due to the sensitivity of theoretical predictions to the pulse details, an experimental realization of the effect appears to be a major challenge.

In a recent paper [1], strong-field ionization of atomic hydrogen as well as lithium driven by a short extreme ultraviolet (XUV) pulse was studied. The key results of the paper were some peculiar effects in the angular-momentum distribution of the ejected electron, provided the pulse alone caused a significant non-zero displacement of the electron without it leaving the laser focus. The displacement is defined as the time integral of the pulse vector potential taken over the pulse duration, i.e., essentially the second integral over time of the electric field. These effects should be visible in the photoelectron angular distribution [1].

Further examination [2] of the origin of the displacement, however, shows that its value critically depends on the assumption of a plateau in the envelope function of the electric field, and that the ramp-on is fine-tuned, via its length and/or the carrier envelope phase (CEP), in such a way that a drift velocity generated during the ramp-on phase can increase this displacement further. Seemingly minor variations in the electric field cause significant changes in the final results.

Specifically, we investigate n_1-n_2-n_3 pulses, where n_1, n_2, and n_3 denote the number of cycles in the sine-squared ramp-on, plateau, and sine-squared ramp-off phases of the pulse, respectively. Furthermore, we either set the envelope function of the electric field (**E**) or the vector potential (**A**) and then calculate the respective other field using the relationship $\mathbf{E}(t) = -d/dt\,(\mathbf{A}(t)/c)$.

Figure 1 shows the effects of two very similar pulses, which nevertheless differ slightly, since the envelope function of the electric field was set in the left panels while that for the vector potential was set on the right. Due to this variation, obtaining different results does not violate the principle of gauge invariance. For one of the cases, the displacement becomes significant and essentially grows proportional to the length of the plateau. In that case, indeed, there is a significant probability to find angular momenta of the ejected electron other than the expected p-character for a one-photon transition.

As usual, these theoretical predictions were obtained with idealized pulses, which do fulfill the basic requirements of Maxwell's equations for electromagnetic waves propagating in vacuum. In light of the sensitivity of the results, an experimental verification of the displacement effects appears to be very challenging.

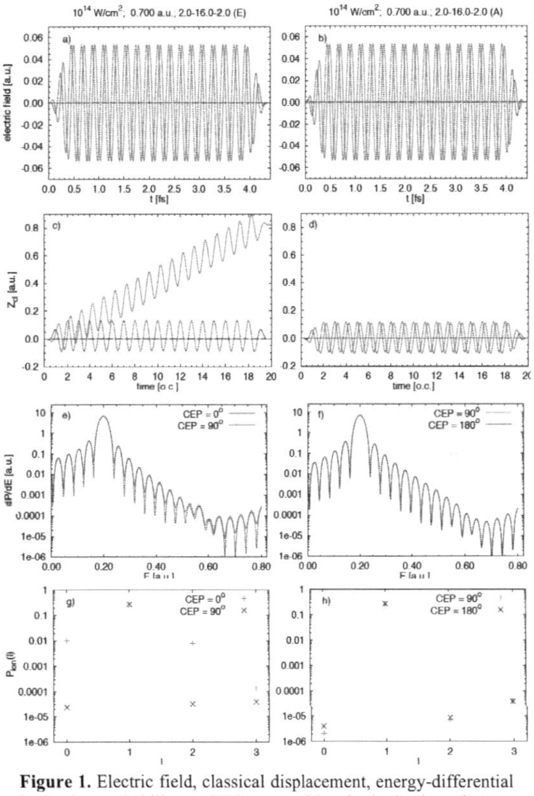

Figure 1. Electric field, classical displacement, energy-differential ionization probability, and l-decomposition for ionization of atomic hydrogen by a 2.0-16.0-2.0 pulse with a central frequency of 0.7 a.u. and a peak intensity of 10^{14} W/cm^2. In the left (right) panels, the envelope function of the electric field (vector potential) is set. The CEP used for panels a)-f) is indicated in panels e) and f).

This work was supported by the United States National Science Foundation under grants PHY-1068140, PHY-1430245, and the XSEDE allocation PHY-090031, and by the Australian Research Council under Grant No. DP120101805.

References

[1] I. A. Ivanov *et al* 2014 *Phys. Rev. A* **90** 043401
[2] K. Bartschat, J. Venzke, and A. N. Grum-Grzhimailo 2015, submitted to *Phys. Rev. A*

[1] E-mail: klaus.bartschat@drake.edu

Content from this work may be used under the terms of the Creative Commons Attribution 3.0 licence. Any further distribution of this work must maintain attribution to the author(s) and the title of the work, journal citation and DOI.
Published under licence by IOP Publishing Ltd

Journal of Physics: Conference Series **635** (2015) 092022 doi:10.1088/1742-6596/635/9/092022

Exploiting Rydberg blockade to probe strongly-coupled Rydberg atom pairs

X. Zhang* [1], **F. B. Dunning***, **S. Yoshida**†, **and J. Burgdörfer**†

* Department of Physics & Astronomy, MS61, Rice University, Houston, Texas, USA
† Institute for Theoretical Physics, Vienna University of Technology, Vienna, Austria, EU

Synopsis Rydberg blockade at very-high-*n*, *n*~300, is examined using strontium n^1F_3 Rydberg atoms excited in a small volume defined by two crossed tightly-focused laser beams. Strong but not complete blockade is observed and discussed with the aid of quantum calculations using a two-active-electron model. Nonetheless, the probability for creating one, and only one, Rydberg atom is large, >0.6, sufficient to enable study of strongly-coupled Rydberg atom pairs.

The goal of the present program is to produce pairs of high-*n*, *n*~300, Rydberg atoms with well-defined initial separations and then control and manipulate their interactions by exciting them to higher *n* states using techniques developed explicitly for high-*n* atoms that employ carefully-tailored sequences of one, or more, short electric field pulses [1]. If the atoms are initially well separated their interactions will be weak and they will evolve independently. Their interactions can be dramatically increased by transferring them to higher *n* levels using pulsed fields. Since both atoms are subject to the same pulses this enables creation of well-defined strongly-correlated two-electron wavepackets whose dynamics can be monitored by application of further probe fields. A necessary first step towards such studies is creation of single high-*n* Rydberg atoms at well-defined initial locations. We explore here the use of Rydberg blockade to accomplish this.

In the present experiments strontium atoms in a collimated beam are excited to n^1F_3 states by three-photon excitation via the 5s5p $^1P_1$ and 5s5d $^1D_2$ intermediate states using radiation at 461, 767, and 893 nm. The crossed 767 and 893 nm beams are strongly focused resulting in a localized excitation volume of $\sim 2\times10^{-7}$ cm^3. Measurements are conducted in a pulsed mode. The output of the 461 nm laser, which is not focused, is chopped into a series of ~100 ns-long pulses. Following excitation the number of Rydberg atoms created is determined by field ionization.

Figure 1 shows Rydberg atom number distributions typical of those obtained under optimal conditions. The distributions are strongly sub-Poissonian, pointing to sizeable blockade effects. Blockade, however, is not complete which, theory suggests, results because interactions between *n*F states are anisotropic. For certain relative atomic orientations interactions are strong, introducing pronounced blockade effects, for others interactions are weak and blockade effects minimal.

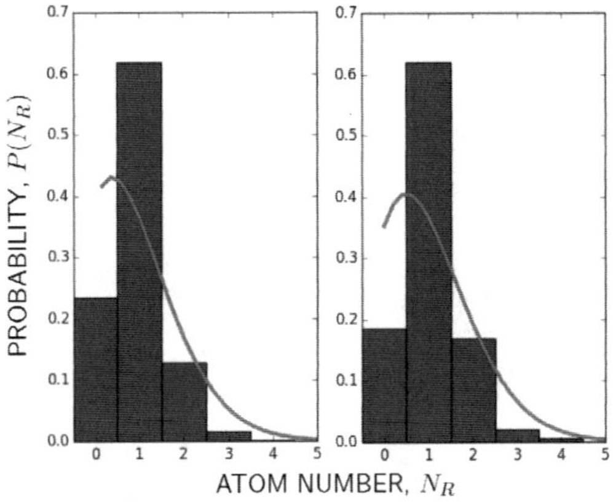

Figure 1. Measured Rydberg atom number distributions when the average number of Rydberg atoms created, $\bar{N}_R$, ~1. The curves show Poissonian distributions for the same $\bar{N}_R$.

Nonetheless, the probability for creating a single Rydberg atom is large, ~0.62. Thus, when using two excitation volumes the probability of producing a single Rydberg atom in each is sizeable, ~0.38, much larger than for forming two Rydberg atoms in a single volume, ~0.06. Since, with efficient Rydberg detection, it should be possible to discriminate against cases in which fewer than, or more than, two Rydberg atoms are created, study of Rydberg-Rydberg interactions under carefully controlled conditions becomes possible.

Research supported by the NSF, the Welch Foundation, and the FWF (Austria)

References

[1] F. B. Dunning *et al* 2009 *J. Phys. B: At. Mol. Opt. Phys.* **42** 022001

[1] E-mail: xz13@rice.edu

Content from this work may be used under the terms of the Creative Commons Attribution 3.0 licence. Any further distribution of this work must maintain attribution to the author(s) and the title of the work, journal citation and DOI.
Published under licence by IOP Publishing Ltd

Journal of Physics: Conference Series **635** (2015) 092023 doi:10.1088/1742-6596/635/9/092023

Production of ultra-long-range Rydberg molecules in a divalent atomic system

B. J. DeSalvo* [1]**, J. Aman*, F. B. Dunning*, T. C. Killian*, H. R. Sadeghpour**, S. Yoshida**[†]**, and J. Burgdörfer**[†]

[*] Department of Physics and Astronomy MS61, Rice University, Houston, Texas, USA

[**] ITAMP, Harvard-Smithsonian Center for Astrophysics, 60 Garden St, Cambridge, Massachusetts, USA

[†] Institute for Theoretical Physics, Vienna University of Technology, Vienna, Austria, EU

Synopsis Sr_2 Rydberg molecules are created through two-photon excitation in an ultracold strontium gas contained in an optical dipole trap and are detected through atom loss from the trap. First order perturbation theory employing the Fermi pseudopotential and effective s-wave and p-wave scattering lengths is used to evaluate the binding energies of the molecular levels.

Scattering of the excited electron in a Rydberg atom from a ground state atom introduces a novel chemical bond that can bind the two together to form an ultralong-range Rydberg molecule [1]. Their creation, however, requires ultracold temperatures so that the thermal energies of the atoms are lower than their binding energies, typically ~1-100 MHz, and high-density samples to ensure a high probability for finding two atoms with separations less than the radial extent of the Rydberg electronic wave function.

Whereas several studies have reported observation of Rydberg molecules using alkali Rydberg atoms, we report here their first observation in the alkaline earths using ^{84}Sr atoms held in an optical dipole trap (ODT). The molecules are created by two-photon excitation via the $5s5p\,^3P_1$ intermediate state using radiation at 689 and 319 nm. After excitation, atoms remaining in the trap are released and their number determined by absorption of 461 nm light tuned to the $^1S_0$-$^1P_1$ transition. Molecule creation is detected as trap loss and the molecular binding energies are determined by the shift of the molecular lines to the red of the atomic line.

Figure 1 shows the molecular binding energies observed using n^3S_1 Rydberg states with values of n from 29 to 36 together with the results of calculations employing first-order perturbation theory. The Rydberg electron-ground state atom interaction is described using a Fermi pseudopotential and optimized effective s-wave and p-wave scattering lengths of -13.2 and 8.4 a_0, respectively. The agreement between theory and experiment is excellent and the binding energies of the most bound levels obey a $1/(n\text{-}\delta)^6$ scaling law, where δ=3.371 is the n^3S_1 quantum defect. The same scaling law was seen earlier in the alkalis [2]. Not all the predicted lines are observed experimentally due to difficulties in resolving closely spaced lines and the reduced excitation strengths at the lower n.

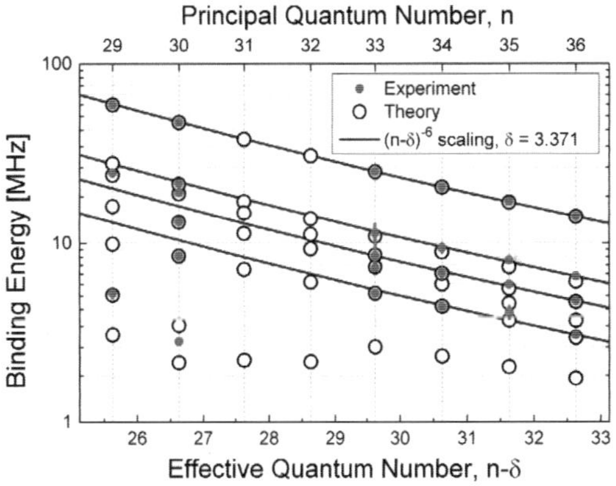

Figure 1. Dependence of the observed and calculated molecular binding energies on $n^*\equiv n\text{-}\delta$ for values of n from 29 to 36. The lines indicate $1/n^{*6}$ scaling. Only calculated levels with binding energies >1 MHz are included.

Research supported by the AFOSR, the NSF, the Welch Foundation, and the FWF (Austria)

References

[1] C. Greene *et al* 2000 *Phys. Rev.Lett.* **85** 2458
[2] V. Benkowsky *et al* 2009 *Nature* **458** 1005

[1] E-mail: jim.aman@rice,edu

Content from this work may be used under the terms of the Creative Commons Attribution 3.0 licence. Any further distribution of this work must maintain attribution to the author(s) and the title of the work, journal citation and DOI.
Published under licence by IOP Publishing Ltd

Trap losses due to near-resonant Rydberg dressing of cold atomic gases

B. J. DeSalvo*, J. A. Aman* [1], F. B. Dunning*, T. C. Killian*, S. Yoshida†, and J. Burgdörfer†

* Department of Physics & Astronomy MS61, Rice University, Houston, Texas, USA
† Institute for Theoretical Physics, Vienna University of Technology, Vienna, Austria, EU

Synopsis Studies of near-resonant dressing of cold strontium gases and BECs held in an optical dipole trap (ODT) with the 30^3S_1 Rydberg state reveal that, even when well detuned from resonance, such dressing can lead to a rapid decrease in the ground state atom population in the trap. This decrease is attributed to Rydberg atom excitation, which can trigger direct escape from the trap and population of long-lived metastable states.

Rydberg states can be used to tune the interaction between "ground state" atoms by introducing a small admixture of Rydberg character into their wavefunctions through dressing with optical radiation tuned near resonance with the transition to a Rydberg state. Here we examine the effects of introducing long-range interactions between strontium atoms contained in a dense cold thermal gas or BEC by Rydberg dressing.

Samples of cold ^{84}Sr atoms are prepared in an optical dipole trap (ODT) and dressed with the 30^3S_1 Rydberg state, which introduces nearly-isotropic repulsive interactions, utilizing the $^3P_1$ intermediate state and radiation at 689 and 319 nm. The ground state atom population in the trap is measured by releasing them from the trap. As it falls the atom cloud expands and is imaged by absorption of 461 nm light tuned to the $^1S_0 \rightarrow {}^1P_1$ transition. The kinetic energy acquired by a ground state atom upon excitation to a Rydberg state is such that many of them escape the ODT. In addition, even if remaining trapped, many decay to long-lived $^3P_0$ and $^3P_2$ states which also represents a loss of ground state atoms from the ODT.

Figure 1 shows trap losses observed using a cold thermal gas containing ~5×10^5 atoms at a density of ~4×10^{11} cm^{-3} as a function of overall two-photon detuning, Δ. Data are included for two-photon Rabi frequencies $\Omega_{eff}/2\pi$ in the range ~15-200 kHz and dressing times, t_D, ~80-800 μs. The product $\Omega_{eff}\times t_D$ is maintained approximately constant resulting in similar atom losses on resonance. As Ω_{eff} is increased the loss features broaden and sizeable trap losses are seen even when well detuned from resonance, an effect that is even more pronounced for (higher-density) BECs. The large trap losses can be attributed to a multistep process invoked previously [1,2] in which initial (off-resonant) excitation of Rydberg atoms enables the excitation of neighboring atoms facilitating the growth of localized groups of Rydberg atoms even when well detuned from resonance. Time-dependent studies of trap loss provided further support for this mechanism. While this mechanism will limit both the strength and duration of the dressing that can be applied to dense cold atom samples, it should be less important for atoms contained in an optical lattice whose separations can be better controlled.

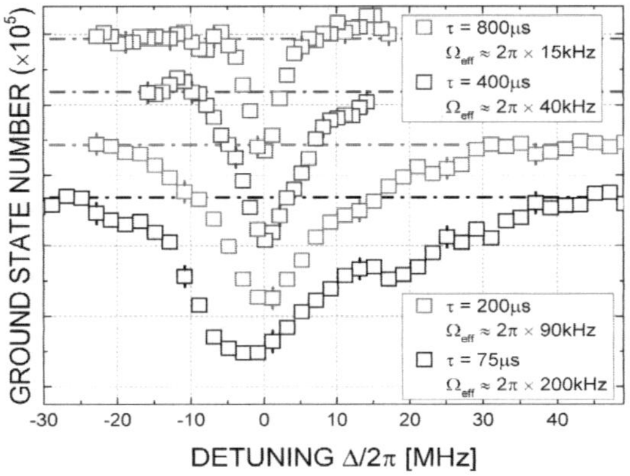

Figure 1. Measured decrease in the ground state population present in a cold thermal ^{84}Sr gas as a function of detuning Δ. Estimated Rabi frequencies, Ω_{eff}, and dressing times, t_D, chosen to keep the product $\Omega_{eff}\times t_D$ approximately constant, are as indicated

Research supported by the AFOSR, the NSF, the Welch Foundation, and the FWF (Austria)

References

[1] N. Malossi *et al* 2014 *J. Phys. Conf. Ser.* **497** 012031
[2] H. Schempp et al 2014 *Phys. Rev. Lett.* **112** 013002

[1]E-mail: jim.aman@rice.edu

Content from this work may be used under the terms of the Creative Commons Attribution 3.0 licence. Any further distribution of this work must maintain attribution to the author(s) and the title of the work, journal citation and DOI.
Published under licence by IOP Publishing Ltd

Journal of Physics: Conference Series **635** (2015) 092025 doi:10.1088/1742-6596/635/9/092025

Quantification of entanglement entropies for doubly excited resonance states in two-electron atomic systems

Yew Kam Ho* [1], **Chien-Hao Lin*** [2]

* Institute of Atomic and Molecular Sciences, Academia Sinica, Taipei, Taiwan

Synopsis In this work, we study the quantum entanglement for doubly excited resonance states in two-electron atomic systems such as the H^- and Ps^- ions and the He atom by using highly correlated Hylleraas type functions The resonance states are determined by calculation of density of resonance states with the stabilization method. The spatial (electron-electron orbital) entanglement entropies (linear and von Neumann) for the low-lying doubly excited states are quantified using the Schmidt-Slater decomposition method.

As quantum entanglement property plays a crucial role in areas such as quantum teleportation, quantum computation, and quantum cryptography [1], quantification of entanglement entropy in two interacting particles systems have attracted much attention (see [2-5], and references therein). In particular, considerable effort has been made on studies of entanglement for two-electron systems including model atoms, quantum dot systems, and natural two-electron atoms. In our recent works, we have studied the linear and von Neumann entropies of the ground and singly excited states in helium and the ground states of hydrogen negative ion and positronium ion [2-4], and have also quantified entanglement entropies for the doubly excited $2s^2$, $2s3s$ and $2p^2$ $^1S^e$ states in the helium atom [5]. Now we report results for the doubly excited $2s^2$ $^1S^e$ state in H^- and Ps^-.

For S-states we use Hylleraas-type wave functions to describe the system, with

$$\Psi_{kmn}(\mathbf{r}_1,\mathbf{r}_2)=\sum_{kmn}C_{kmn}\left\{\exp\left[-\alpha\left(r_1+r_2\right)\right]r_1^k r_2^m r_{12}^n+\left(1\leftrightarrow 2\right)\right\},$$

where 1 and 2 denote the electron 1 and electron 2, respectively, and r_{12} is the relative distance between the electron 1 and electron 2, and k, m, n are positive integers or zero. First we employ the stabilization method (see Figure 1) to determine the resonance energy and width (see Figure 2) for a particular resonance state. Once the wave function is obtained, the linear entropy and von Neumann entropy for such a state are quantified using the Schmidt-Slater decomposition method [3]. By systematically examining the convergence behaviors of calculated entropies by using different wave functions (up to 252 terms), we are able to establish benchmark values of entropies for doubly excited states in the two-electron atomic systems.

This work was supported by the Ministry of Science and Technology of Taiwan.

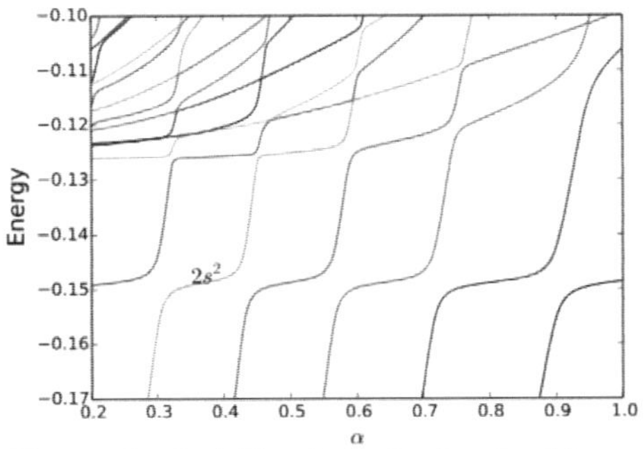

Figure 1. Stabilization plot for doubly excited states in H^-

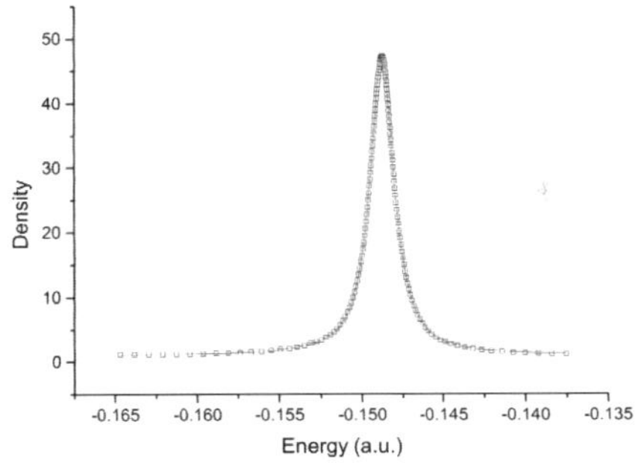

Figure 2. Fitting of density of resonance states for the $2s^2$ $^1S^e$ state in H^-

References

[1] M. A. Nielsen and I. L. Chuang, *Quantum Computation and Quantum Information*, Cambridge University press (2010)
[2] C.-H. Lin, Y.-C. Lin, Y. K. Ho, Y. K. 2013 *Few-Body Systems* **54** 2147
[3] C.-H. Lin and Y. K. Ho 2014 *Few-Body Syst.***55** 1141
[4] C.-H. Lin and Y. K. Ho 2014 *Phys. Lett. A* **378** 2861
[5] C.-H. Lin and Y. K. Ho 2015 to appear in *Few-Body Syst.*

[1] E-mail: ykho@pub.iams.sinica.edu.tw
[2] E-mail: b99202042@ntu.edu.tw

Content from this work may be used under the terms of the Creative Commons Attribution 3.0 licence. Any further distribution of this work must maintain attribution to the author(s) and the title of the work, journal citation and DOI.
Published under licence by IOP Publishing Ltd

Journal of Physics: Conference Series **635** (2015) 092026 doi:10.1088/1742-6596/635/9/092026

Laser-assisted ultrafast electron transfer through the continuum

P. Vindel-Zandbergen*[1], I.R. Sola*[2], C. Meier†[3]

* Departamento de Química Física, Universidad Complutense de Madrid, 28040, Madrid, Spain
† Laboratoire Collisions Agrégats, Réactivité IRSAMC, Université Paul Sabatier, 118 Route de Narbonne, 31062, Toulouse, France

Synopsis In this work we report control on the electron transfer between two separated nuclei in two different systems, H_2^+ and HeH^{2+}, assisted by strong attosecond laser pulses under the Local Control Theory frame

In recent years, advances in laser technology allowed the production of trains or isolated sub femtosecond pulses, with frequencies in the UV regime, paving the way to the development of a new scientific domain, called attosecond physics. These sub-femtosecond (hence attosecond) pulses are so short that electronic processes can be resolved in real time [1],[2]. Currently, attophysics techniques are mainly used to unravel ultrafast molecular (or condensed phase) processes.

The field of Quantum Control [3] deals with the control of the Hamiltonian dynamics by exerting an external operation on the system. The operation is necessarily unitary in the framework of coherent control. Thus one gets a modified, appropriately called controlled time evolution of the system dynamics which helps steer the system in the desired pathway towards some preset goal. From a theoretical point of view, different control schemes have been introduced.

In this work, the approach of Local Control Theory (LCT) [4] is followed. This latter approach has successfully been used to gain insight into the underlying dynamical processes leading to the desired control objective. Briefly, local control theory calculates the control field as a function of the instantaneous dynamics of the system at each time step.

The main purpose of this work is to combine quantum control techniques with attosecond processes. Within this scientific context the goal is the simulation of the dynamics of single electron processes in two systems, H_2^+ and HeH^{2+}, excited by intense ultrashort laser pulses in the femto- and attosecond scale and apply different formulations of Local Control theory (LCT) to obtain laser control pulses that lead us to population transfer and electron localization.

Considering separated fixed nuclei, the aim is to control the electron transfer from the Hydrogen to the proton, in the case of the H_2^+ system, and from the proton to an Helium atom or from an alpha particle to an Hydrogen atom, in the HeH^{2+} system, by means of intense and ultrashort laser pulses. First, the electron is excited and then the laser control field produces the reabsortion at the other nuclei when it travels throughtout the continuun. We also confirm the validity of the fixed nuclei results since laser control fields that emerge from 1D calculations work perfectly when nuclear motion is considered and we test how the initial kinetic energy affects the electron transfer.

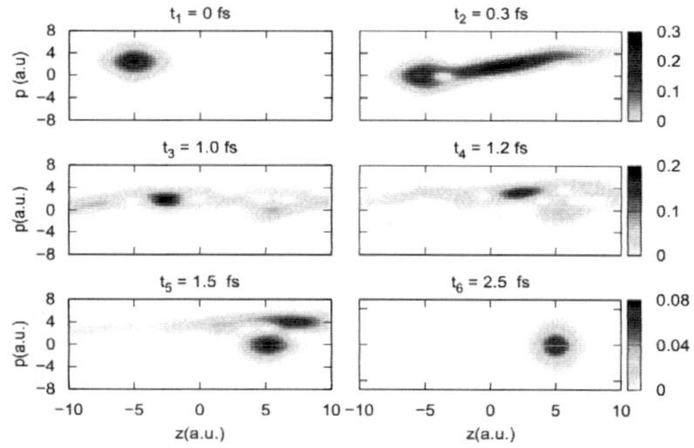

Figure 1. Husimi functions for different times showing the electron transfer from the left to the right proton

References

[1] M. F. Kling and M.J. Vrakking 2008 *Annu. Rev. Phys. Chem.* **59**463 -492

[2] F. Krausz and M. Ivanov 2009 *Rev.Mod.Phys.* **81** 163

[3] M. Shapiro and P. Brumer 2011 *Wiley-VCH Verlag GmbH & Co. KGaA.* **513-535**

[4] P. Peirce, M. A. Dahleh and H. Rabitz 1988 *Phys. Rev. A* **37** 4950-4964

[1]E-mail: pvzandbergen@ucm.es
[2]E-mail: isola@quim.ucm.es
[3]E-mail: chris@irsamc.ups-tlse.fr

Content from this work may be used under the terms of the Creative Commons Attribution 3.0 licence. Any further distribution of this work must maintain attribution to the author(s) and the title of the work, journal citation and DOI.
Published under licence by IOP Publishing Ltd

Journal of Physics: Conference Series **635** (2015) 092027 doi:10.1088/1742-6596/635/9/092027

Breit-Pauli electron-impact excitation collision strengths for transitions in O III

S. S. Tayal[*] [1] and O. Zatsarinny[†] [2]

[*] Department of Physics, Clark Atlanta University, Atlanta, GA 30314, USA

[†] Department of Physics and Astronomy, Drake University, Des Moines, Iowa 50311, USA

Synopsis Collision strengths for electron impact excitation of fine-structure transitions between the first 202 fine-structure levels of O III have been calculated using the Breit-Pauli B-spline R-matrix method. The close-coupling expansion includes levels of the $2s^22p^2$, $2s2p^3$, $2s^22p^33l$ *(l=0-2)*, $2s^22p^34l$ *(l=0-2)* and $2s2p^23l$ configurations. The present results considerably improve the existing oscillator strengths and collision strengths for O III.

Emission lines from oxygen ions have been observed in different astrophysical objects for many years, and the knowledge of the radiative and collision atomic parameters of oxygen ions in different ionization stages is crucial in modeling and understanding of processes occurring in the planetary nebulae and H II regions. The reliable theoretical study of low-energy electron scattering requires an accurate target description and a sufficient number of target states in the close-coupling expansion to achieve convergence and to account for cascade to lower levels. The purpose of the work reported here, therefore, is to perform elaborate and extensive calculations for the electron scattering from O III by using highly accurate target wave functions and by including fine-structure effects in the close-coupling expansion directly to make sure that relativistic effects are adequately accounted in the scattering calculation. The present calculations have been carried out with the B-spline Breit-Pauli R-matrix (BSR) code [1]. One of the distinct features of this code is the use of flexible term-dependent orbital sets in the description of both the target states and scattering system. This allows us to optimize the atomic wave functions for different states independently, resulting in a more accurate target description than those used in previous collision calculation.

Figure 1 represents collision strengths for the fine-structure $2p^2\ ^3P_0$-$^3P_1$ and $2p^2\ ^3P_1$-$^3P_2$ transitions in O III. A fine-energy grid has been used to resolve resonances in the near threshold energy region. The resonances make significant contributions to thermally averaged collision strengths, especially in the lower temperature region. Generally, all recent R-matrix calculations [2-4] agree to within 10% for the thermally averaged collision strengths for the forbidden transitions among the lowest five levels of the ground $2p^2$ configuration.

For highly excited states the agreement is more diverse. The comparison of effective collision strengths for a set of temperatures is presented in figure 2 where the ratios of our results with calculation [2] are shown. We have attempted to account for important physical effects. The Rydberg series of resonances converging to several excited levels are found to enhance collision strengths substantially.

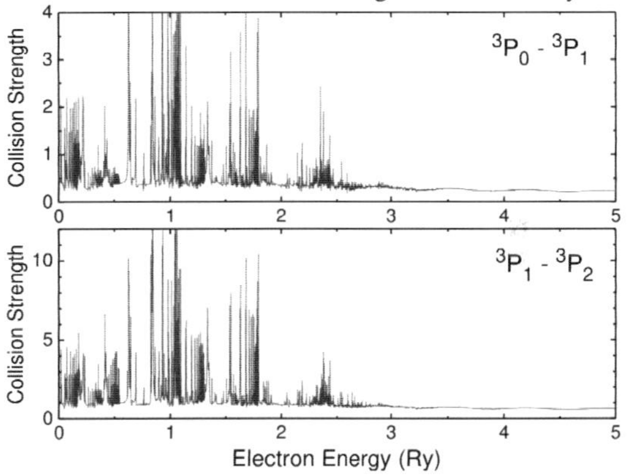

Figure 1. Collision strength for the fine-structure $2p^2\ ^3P_0$-$^3P_1$ and $2p^2\ ^3P_1$-$^3P_2$ transitions in O III.

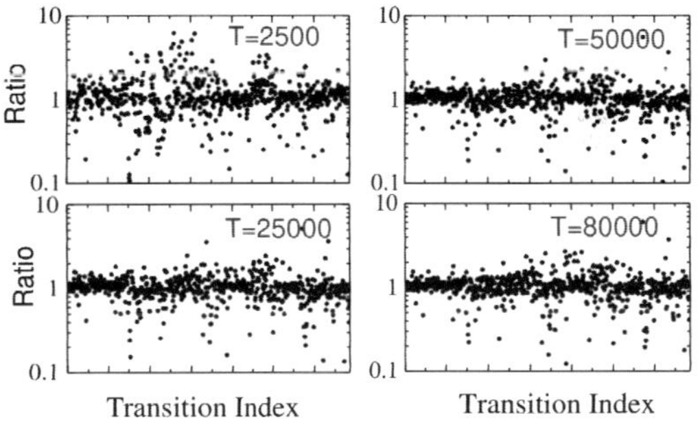

Figure 2. Comparison of our effective collision strengths with the R-matrix calculation [2].

This work was supported by NASA under grant NNX11AB62G from the Solar and Heliophysics program.

References

[1] O. Zatsarinny 2006 Comp. Phys. Commun. **174** 273
[2] K. Aggarwal and F. Keenan 1999 ApJS **123** 311
[3] E. Palay *et al* 2012 MNRAS **423** L35
[4] P. Storey *et al* 2014 MNRAS **441** 3028

[1] E-mail: stayal@cau.edu
[2] E-mail: oleg.zatsarinny@drake.edu

Content from this work may be used under the terms of the Creative Commons Attribution 3.0 licence. Any further distribution of this work must maintain attribution to the author(s) and the title of the work, journal citation and DOI.
Published under licence by IOP Publishing Ltd

Journal of Physics: Conference Series **635** (2015) 092028 doi:10.1088/1742-6596/635/9/092028

Steering the electron motion by two counter-rotating circularly polarized short intense laser pulses

Xiao-Min Tong[†,*1], **Nobuyuki Toshima**[†2]

[†] Graduate School of Pure and Applied Sciences, University of Tsukuba, Tsukuba 305-8571, Japan

[*] Center for Computational Sciences, University of Tsukuba, Tsukuba 305-8577, Japan

Synopsis We investigate atomic ionization in two counter-rotating circularly polarized short laser pulses by solving the three-dimensional time-dependent Schrödinger equation. Tuning the time-delay and intensity ratio between the two pulses provides us a way to steer the electron motion, in turn, the temporary profile of the atto-second pulse when the photoelectron radiatively recombines with the parent core.

Above threshold ionization (ATI) is an important process in atom intense-laser interactions. The process is closely related to the high order harmonic generation, which results in an atto-second pulse train or single atto-second pulse (SAP). Mashiko *et al* [1] has successfully generated a linearly polarized SAP by using one color two counter-rotating circularly polarized short laser pulses. To understand the mechanism and find a way to steer the electron motion, we develop a numerical method to solve the 3-dimensional time-dependent Schrödinger equation in spherical coordinates by using split-operator method in the energy representation under the single active electron approximation.

In the split-operator-method, we propagate the time-dependent wave function from time t to $t + \Delta t$ by first propagating the wave function in laser field free Hamiltonian H_0 for $\Delta t/2$, then, in the electron laser interaction $V(t)$ for Δt and finally in the H_0 for another $\Delta t/2$ as

$$\Psi(t+\Delta) = e^{-iH_0\Delta t/2}e^{-iV(t)\Delta t}e^{-iH_0\Delta t/2}\Psi(t).$$

Although there is no spherical or cylindrical symmetry in the laser-atom interactions, the present time propagator fully uses the spherical symmetry of H_0, and the local operator of $V(t)$ to simply and speed up the time-propagation of the above equation. Details about the numerical method can be found in Ref. [2]

Based on this numerical method, we investigate the momentum distribution of the ATI electron of hydrogen atom in two 800 nm, 5 fs FWHM (full width of half magnitude) counter-rotating laser fields as a function of the time-delay. Here we define that the lasers propagate along the z-direction and electric field is polarized on the xy-plane. Figure 1 shows a typical photoelectron momentum distribution on the xy-plane with a 3 fs of time-delay between the two pulses. The spectra look like a typical elliptical distribution with photoelectron concentrates along the major axis similarly to the spectra in the linearly polarized laser field and spinning around the center like a spectra in a circularly polarized laser field. We also study the spectra by tuning the relative laser intensities of the two-pulses and the results will be presented in the conference. The method can also be used to study the ATI process of atoms in a two-color circularly polarized laser field [3], which is of the current interest because the configuration can be used to generate circularly polarized harmonics.

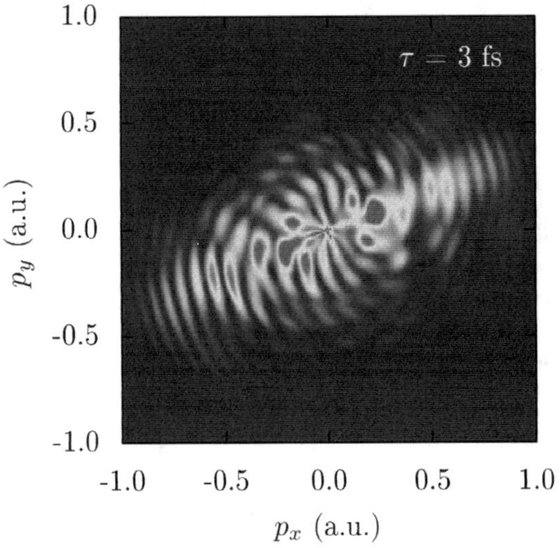

Figure 1. Electron momentum of hydrogen atoms in laser fields with intensity of 5×10^{13} W/cm^2.

References

[1] H. Mashiko *et al* 2008 *Phys. Rev. Lett* **100** 103906

[2] X. M. Tong *et al* 2000 *Phys. Rev. A* **61** 031401(R).

[3] C. Mancuso *et al* 2015 *Phys. Rev. A* (accepted)

[1]E-mail: tong@ims.tsukuba.ac.jp
[2]E-mail: toshima@bk.tsukuba.ac.jp

Content from this work may be used under the terms of the Creative Commons Attribution 3.0 licence. Any further distribution of this work must maintain attribution to the author(s) and the title of the work, journal citation and DOI.
Published under licence by IOP Publishing Ltd

XXIX International Conference on Photonic, Electronic, and Atomic Collisions (ICPEAC2015) IOP Publishing
Journal of Physics: Conference Series **635** (2015) 092029 doi:10.1088/1742-6596/635/9/092029

Photoexcitations of many-electron atoms and ions by twisted light

S. Fritzsche[†,♯,1], **D. Seipt**[†], **V. Serbo**[‡] **and A. Surzhykov**[†]

[†] Helmholtz-Institut Jena, Fröbelstieg 3, D-07743 Jena, Germany
[♯] Theoretisch-Physikalisches Institut, Friedrich-Schiller-Universität Jena, D-07743 Jena, Germany
[‡] Novosibirsk State University, 630090 Novosibirsk, Russia

Synopsis The photoexcitation of (multi-electron) atoms and ions by twisted light has been explored within the framework of the density matrix theory and Dirac's relativistic equation. It is shown that the population of the excited atoms and, hence, their fluorescence emission are sensitive with regard to the transverse linear momentum and the (projection of the) total angular momentum of the incident light.

Research on optical vortices, popularly known also as *twisted* light, has attracted much interest during the past two decades. These vortex beams, although freely propagating in space, can be understood as beams of photons (and similar for electrons), which individually carry a well defined orbital angular momentum (OAM) along their propagation direction. Indeed, the OAM of light is closely related to the spatial distribution of the electromagnetic field. — Today such optical vortex beams are found in various applications, from the manipulation of microparticles, to imaging and microscopy.

Although the spin and orbital angular momentum of twisted light and its interaction with matter has been explored in some detail, little attention has been paid so far to the atomic excitation and ionization processes in such light fields. A question of particular interest concerns the selection rules and how the bound-bound and bound-free transitions are modified by the OAM of the light. To address these questions, we have investigated the interaction of complex atoms with Bessel beams [1, 2]. Expressions were especially derived for the density matrix of the excited atomic states and the angular distribution as well as the polarization of the subsequent fluorescence radiation [3]. Apart from *head-on* collisions of atoms with twisted photons, we also considered the more realistic scenario in which a photon beam, prepared for instance as a single beam or the coherent superposition of two Bessel states with different projections m_1 and m_2 of the total angular momentum (TAM), collides with a macroscopic atomic target.

In Ref. [3], detailed calculations have been performed for the $3s\ ^2S_{1/2} \to 3p\ ^2P_{3/2}$ excitation of neutral sodium by a coherent superposition of two Bessel beams. For this photoexcitation, Fig. 1 displays the alignment parameters A_{2q}, $q = -2, ..2$ of the $^2P_{3/2}$ state, i.e. the relative population of the magnetic substates, as function of the *opening* angle of the beam, $\theta_k = \arctan(\kappa/k_z)$, and which defines the ratio of the transverse κ and longitudinal k_z linear momenta of twisted incident beams. As seen from this figure, the individual population of the substates occurs to be sensitive w.r.t. to both, the TAM of the beam and their opening angle.

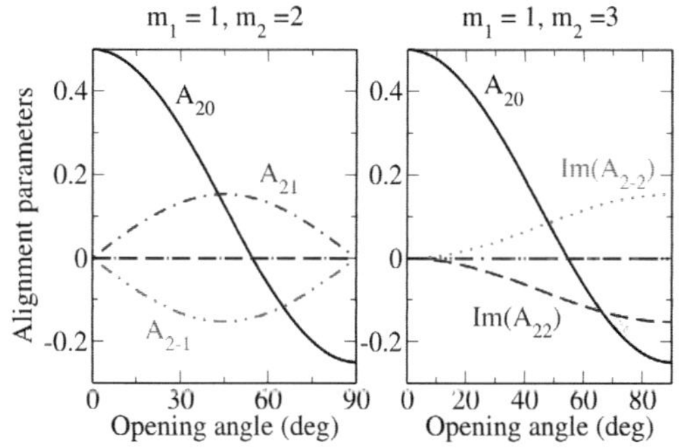

Figure 1. Alignment parameters A_{2q} of the $3p\ ^2P_{3/2}$ level of neutral sodium following the absorption of twisted light. The calculations were performed here for a coherent superposition of two Bessel beams with different projections m_1 and m_2.

We therefore hope that future measurements on the (angular and polarization) properties of the 'yellow' sodium line will provide further information about the interaction of the twisted light with atoms.

References

[1] O. Matula *et al* 2013 *J. Phys. B***46** 205002

[2] M. Scholz–Marggraf *et al* 2014 *Phys. Rev. A* **90** 013425

[3] A. Surzhykov *et al* 2015 *Phys. Rev. A* **91** 052515

[1] E-mail: s.fritzsche@gsi.de

Content from this work may be used under the terms of the Creative Commons Attribution 3.0 licence. Any further distribution of this work must maintain attribution to the author(s) and the title of the work, journal citation and DOI.
Published under licence by IOP Publishing Ltd

Journal of Physics: Conference Series **635** (2015) 092030 doi:10.1088/1742-6596/635/9/092030

Can angle-resolved x-ray measurements help determine small level splittings in highly charged ions?

Z. W. Wu[†,‡ 1], **N. M. Kabachnik**[§,♮], **A. Surzhykov**[†], **C. Z. Dong**[‡] and **S. Fritzsche**[†,♯]

[†] Helmholtz-Institut Jena, Fröbelstieg 3, D-07743 Jena, Germany

[‡] Key Laboratory of Atomic and Molecular Physics & Functional Materials of Gansu Province, P.R. China

[§] European XFEL, Albert-Einstein-Ring 19, D-22761 Hamburg, Germany

[♮] Skobeltsyn Institute of Nuclear Physics, Lomonosov Moscow State University, Moscow 119991, Russia

[♯] Theoretisch-Physikalisches Institut, Friedrich-Schiller-Universität Jena, D-07743 Jena, Germany

Synopsis The angular distribution and photon-photon angular correlation have been studied for x-ray emissions from two-step radiative cascades that proceed via two or more overlapping intermediate resonances. From this study, we suggest that accurate angle-resolved measurements of the x-ray emissions may serve as a tool for determining small level splittings in excited highly-charged ions.

During the last decades, the angular distribution and angle-resolved polarization of the characteristic x-ray emission from highly charged ions (HCI) have been studied extensively. Compared to the total decay rates of these ions, such angle-resolved properties were found to be more sensitive to subtle effects in the electron-photon and electron-electron interactions and, hence, helped provide new insight into the struture and dynamics of HCI.

Until the present, however, almost all experimental and theoretical investigations upon the angular distributions of HCI were performed for x-ray emissions from well-isolated fine-structure levels. In contrast, little attention was paid to radiative cascades that proceed via two or more overlapping intermediate resonances. When compared with well-isolated levels, such x-ray cascades are typically affected by spin-spin or spin-orbit interactions that may lead to a depolarization in the magnetic sublevel population of the intermediate levels. Such a depolarization can be seen in angle-resolved measurements of the associated x-ray emissions.

In order to explore these depolarizaton phenomena, expressions were derived for the angular distribution of 1st- and 2nd-step photons and the photon-photon angular correlation. In particular, we considered the radiative cascade

$$\begin{aligned} & 1s2p^2 \; J_i = 1/2, 3/2 \\ & \longrightarrow \; \gamma_1 + \left\{ \begin{array}{l} 1s2s2p \; J = 1/2 \\ 1s2s2p \; J' = 3/2 \end{array} \right\} \\ & \longrightarrow \; \gamma_1 + \gamma_2 + 1s^2 2s \; J_f = 1/2 \qquad (1) \end{aligned}$$

of lithium-like ions, which possess a simple level structure and are known to exhibit a level crossing of the two intermediate levels between $74 \le Z \le 79$ [1]. For W^{71+} ions, for example, Fig. 1 displays the angular distribution of the second-step photon γ_2 for four different intermediate level-splittings (in a.u.), if initially the $J_i = 3/2$ level has the alignment $\mathcal{A}_2 = 1.0$. As seen from this figure, a quite remarkable effect of the level-splitting is found for the angular distribution for small level splittings.

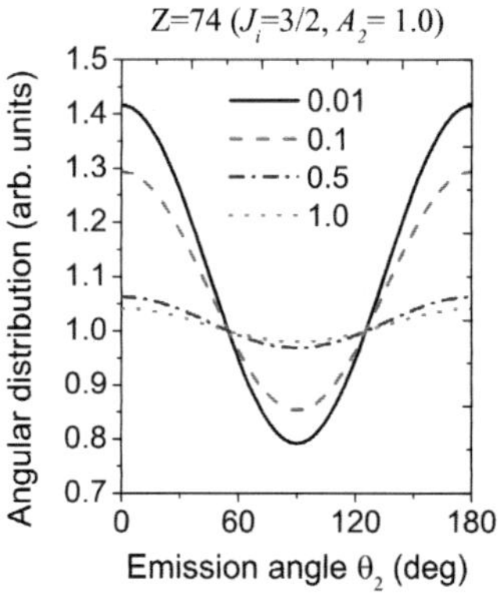

Figure 1. Angular distribution of the second-step photon emissions from the two-step radiative cascade (1). See text for further discussions.

Because of this strong dependence of the angular distribution upon the level-splitting of overlapping resonances, we suggest that accurate measurements of x-ray angular emissions may serve as a tool to determine small level splittings in highly charged ions [1]. Such measurements will be feasible with present-day x-ray detectors.

References

[1] Z. W. Wu *et al* 2014 *Phys. Rev. A* **90** 052515

[1]E-mail: zhongwen.wu@uni-jena.de

Content from this work may be used under the terms of the Creative Commons Attribution 3.0 licence. Any further distribution of this work must maintain attribution to the author(s) and the title of the work, journal citation and DOI.
Published under licence by IOP Publishing Ltd

XXIX International Conference on Photonic, Electronic, and Atomic Collisions (ICPEAC2015) IOP Publishing
Journal of Physics: Conference Series **635** (2015) 092031 doi:10.1088/1742-6596/635/9/092031

Animated-crossed-beam measurement of the photodetachment cross section of H^-

Matthieu Génévriez *[1], **Xavier Urbain***

* Institute of Condensed Matter and Nanosciences, Université catholique de Louvain, Louvain-la-Neuve B-1348, Belgium

Synopsis The photodetachment cross section of the hydrogen anion H^- was measured using the animated-crossed-beam technique in the range from 700 nm to 1064 nm. The values obtained are in excellent agreement with previous experiments and compelling theoretical works.

The photodetachment of the negative hydrogen ion H^- has attracted numerous studies due to the ion's prototypical character and the importance of electron correlations in that weakly bound system. It has become a commonly used benchmark from atomic theories and numerical methods, which have, over the decades, reached an overall good agreement. On the experimental side fewer studies exist, and only one [1] provides absolute cross sections at various photon energies. Recently, Vandrevraye *et al.* [2] have carried out an absolute measurement at the Nd:YAG laser wavelength (1064 nm, 1.1653 eV). Their value lies 1.5 σ above the commonly admitted theoretical value (see figure 1). This possible discrepancy, along with the absence of other measurements at the same wavelength, calls for further investigation to be carried out, which we performed in the range from 700 nm to 1064 nm.

In order to measure an absolute photodetachment cross section, it is necessary to disentangle the so-called volume effect. In our crossed beam setup, the finite size of the ion beam and of the laser beam defines an effective interaction volume, within which both the ion and the photon flux vary. The measured photodetachment signal is therefore an *average* over these local variations and the cross section is, in general, retrieved by assuming a certain distribution for both beams - uniform for the ions, Gaussian for the photons , thus causing inaccuracies.

The animated-crossed-beam method, originally developed for electron-ion collisions [3], eliminates the need for the knowledge of the beams distributions. Instead of fully characterizing the interaction volume, it aims at *removing* its influence on the measured signal by repeatedly swooping the laser beam across the ion beam, as realized by tilting a fused silica plate. After integration of the signal along the laser's vertical displacement, the cross section is expressed only in terms of easily measurable quantities, e.g., the laser power and the ion beam current.

The photodetachment cross section obtained in the range from 700 nm to 1064 nm [4] is in excellent agreement with previous experiments and compelling theoretical works, in particular the state-of-the-art calculation of Venuti and Decleva [5] shown in figure 1. We also confirm the commonly admitted value of $3.5 \times 10^{-21}\,m^2$ at 1064 nm (1.1653 eV).

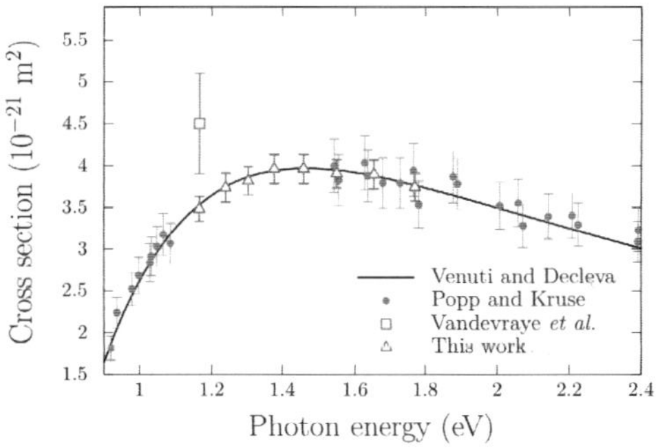

Figure 1. Photodetachment cross section as a function of the photon energy. Triangles: present work. Circles: Popp and Kruse [1]. Square: Vandevraye *et al.* [2]. Full line: calculation of Venuti and Decleva [5].

References

[1] H.-P. Popp, and S. Kruse 1976 *J. Quant. Spectrosc. Radiat. Transfer* **16** 683

[2] M. Vandevraye *et al.* 2014 *Phys. Rev. A* **90** 013411

[3] F. Brouillard and P. Defrance 1983 *Phys. Scr.* **1983** 68

[4] M. Génévriez and X. Urbain 2015 *Phys. Rev. A* **91** 033403

[5] M. Venuti and P. Decleva 1997 *J. Phys. B: At. Mol. Opt. Phys.* **30** 4839

[1]E-mail: matthieu.genevriez@uclouvain.be

Content from this work may be used under the terms of the Creative Commons Attribution 3.0 licence. Any further distribution of this work must maintain attribution to the author(s) and the title of the work, journal citation and DOI.
Published under licence by IOP Publishing Ltd

XXIX International Conference on Photonic, Electronic, and Atomic Collisions (ICPEAC2015) IOP Publishing
Journal of Physics: Conference Series **635** (2015) 092032 doi:10.1088/1742-6596/635/9/092032

Carrier- wave Rabi flopping signatures in high-order harmonic generation

M. F. Ciappina[a 1], **J. A. Pérez-Hernández**[b],**A. S. Landsman**[c], **T. Zimmermann**[d], **M. Lewenstein**[e,f], **L. Roso**[b] **and F. Krausz**[a,h]

[a] Max-Planck Institut für uantenoptik, Hans-Kopfermann-Str. 1, D-85748 Garching, Germany
[b] Centro de Láseres Pulsados (CLPU), Parque Científico, E-37008 Villamayor, Salamanca, Spain
[c] Max-Planck-Institut für Physik komplexer Systeme, Nöthnitzer Straße 38, D-01187 Dresden, Germany
[d] Physics Department, ETH Zurich, CH-8093 Zurich, Switzerland
[e] ICFO-Institut de Ciències Fotòniques, Mediterranean Technology Park, 08860 Castelldefels, Spain
[f] ICREA-Institució Catalana de Recerca i Estudis Avançats, Lluis Companys 23, 08010 Barcelona, Spain
[h] Department für Physik, LMU München, Am Coulombwall 1, D-85748 Garching, Germany

Synopsis We present theoretical investigations of carrier-wave Rabi flopping (CWRF) in real atoms by studying high-order harmonic generation in alkali species. We concentrate on the features of the third harmonic of a and K atoms. For pulse areas of 2π and a atoms, a characteristic unique peak appears, which, after analyzing the ground state population, we correlate it with the conventional Rabi flopping. n the other hand, for larger pulse areas and K atoms, CWRF occurs, and it is associated with a more complex structure in the third harmonic. These new characteristics indicate the breakdown of the area theorem, as was already demonstrated in narrow band gap semiconductors.

When the atom is modeled as a two-level system, the conventional Rabi flopping behavior and carrier-wave Rabi flopping (CWRF) phenomenon features are observed. However, it is well known that the two-level approximation breaks down when strong electric fields are applied, in particular in the CWRF regime. Experiments on narrow band gap semiconductors have shown a clear signature of CWRF, which manifested itself as a split in the third harmonic of the emitted light into the forward direction [1]. An important question emerges as to what extend CWRF could potentially be observed in real atoms [2]. One of the advantages of atoms (relative to semiconductors) is the possibility to employ longer laser pulses and, as a consequence, to explore a broader range of laser parameters. We compute high-order harmonic generation (HHG) spectra in K atoms in which the transition energy between the ground and first excited states is resonant with the driven light. We observe that HHG spectra of K atoms present a drastic change around the third harmonic (see figure 1) as pulse envelope area increases in contrast to the HHG spectra of Na (not resonant), where a single peak is seen in the third harmonic regardless of the envelope pulse area. We use accurate values for the atomic wavefunction of both ground and excited states as well as typical laser parameters, easily achievable with the current laser technology. As a consequence, the experimental confirmation of our results appears straightforward. Moreover, the CWRF phenomenon could emerge as a robust alternative for carrier-envelope phase (CEP) characterization for long pulses.

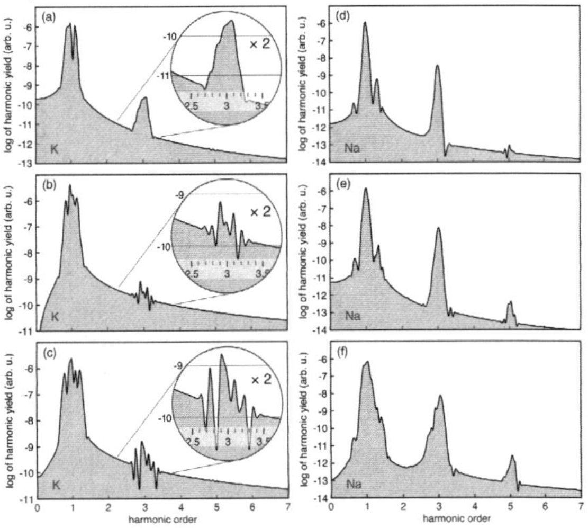

Figure 1. HHG spectra in K for the corresponding laser intensities $I = 3.158 \times 10^{11}$ W/cm^2 (panel a), $I = 5.614 \times 10^{11}$ W/cm^2 (panel b) and $I = 1.108 \times 10^{12}$ W/cm^2 (panel c) and an 18 fs (FWHM) driving pulse. Panels (d), (e) and (f) represent the HHG in Na for the same laser parameters. The insets of panels (a), (b) and (c) show a zoom of the third harmonic $\omega/\omega_0 = 3$.

References

[1] O. D. Mücke *et al.* 2001 *Phys. Rev. ett.* **87** 012058

[2] M. F. Ciappina *et al.* 2015 *Phys. Rev. ett.* **114** 143902

[1]E-mail: marcelo.ciappina@mpq.mpg.de

Content from this work may be used under the terms of the Creative Commons Attribution 3.0 licence. Any further distribution of this work must maintain attribution to the author(s) and the title of the work, journal citation and DOI.
Published under licence by IOP Publishing Ltd

Two-photon detachment of O^-: theory and experiment

Matthieu Génévriez * [1], **Xavier Urbain***, **Alain Cyr**†, **Kevin Dunseath** †, **Mariko Terao-Dunseath**† [2]

* Institute of Condensed Matter and Nanosciences, Université catholique de Louvain, B-1348 Louvain-la-Neuve, Belgium

† Institut de Physique de Rennes, UMR 6251 CNRS-Université de Rennes 1, Campus de Beaulieu, F-35042 Rennes cedex, France

Synopsis We report results from a new experiment and a new calculation of generalized cross sections for two-photon detachment of $O^-(^2P^o)$. Their mutual agreement is excellent and confirms earlier theoretical predictions.

The two-photon detachment of the $O^-(1s^22s^22p^5\ ^2P^o)$ state has not been widely studied, and the agreement between the few results available is not particularly good. The only existing experiment gives a generalized cross section of $4.2 \times 10^{-50}\,cm^4\,s$ at a wavelength of 1064 nm [1]. An early calculation based on perturbation theory and a model potential [2] however yields, after interpolation, a value of $1.8 \times 10^{-49}\,cm^4\,s$, more than four times larger. The results obtained using an adiabatic theory [3] are almost an order of magnitude larger than the experimental value. Clearly there is room for improvement. Here we report new experimental and theoretical results.

The two-photon detachment cross section has been measured, in the quadratic regime of detachment, with a 12 ns seeded Nd:YAG laser. The effects of the interaction volume were disentangled by repeatedly sweeping the laser beam across the ion beam [4]. The deconvolution of the detachment signal with the ion current density and its inverse Abel transformation were performed with Gaussian basis functions, and provided the instantaneous detachment probability under the assumption of a cylindrically symmetric laser beam. The absolute cross section is then readily expressed in terms of easily measurable quantities such as the pulse energy and the ion beam current.

Calculations were performed using the *R*-matrix Floquet (RMF) method. Wave functions for the three $1s^22s^22p^4$ states of the residual Oxygen atom, together with three pseudo-states, were built from Slater-type orbitals chosen to optimize the polarizability of the ground state [5]. The binding energy of O^- is -0.053800 Hartree, compared to the experimental value of -0.053695 Hartree. As the laser field breaks the spherical symmetry of space, the generalized cross sections for $O^-(|M_L| = 0, 1)$ are computed separately and subsequently averaged.

The results, in excellent mutual agreement, are presented in figure 1, together with previous experimental and theoretical data [1, 2].

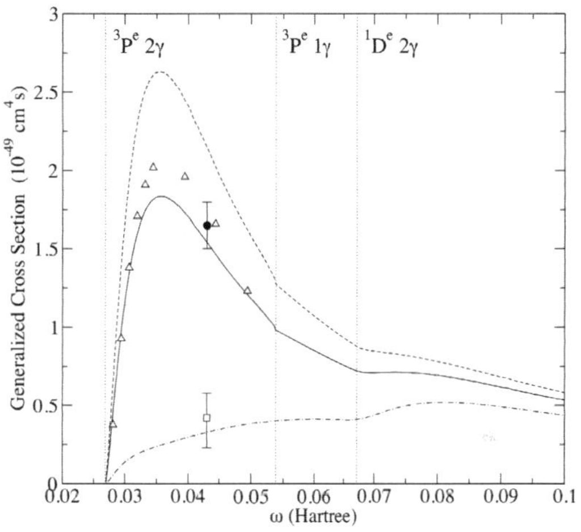

Figure 1. Generalized cross sections for 2-photon detachment of $O^-(1s^22s^22p^5\ ^2P^o)$ as a function of photon energy. Experiment: • current work, □ [1]. Theory: △ [2], RMF: - - - $|M_L| = 1$, - · - · $M_L = 0$, —— M_L-averaged cross section. Vertical lines indicate the position of thresholds.

References

[1] Stapelfeldt H, Brink C and Haugen H K 1991 *J. Phys. B: Atom. Mol. Opt. Phys.* **24** L437

[2] Robinson E J and Geltman S 1967 *Phys. Rev.* **153** 4

[3] Gribakin G F and Kuchiev M Y 1997 *J. Phys. B: Atom. Mol. Opt. Phys.* **30** L657

[4] Génévriez M and Urbain X 2015 *Phys. Rev. A* **91** 033403

[5] Cyr A 1998 *PhD thesis*, Université de Rennes 1, unpublished

[1] E-mail: matthieu.genevriez@uclouvain.be
[2] E-mail: mariko.dunseath-terao@univ-rennes1.fr

Content from this work may be used under the terms of the Creative Commons Attribution 3.0 licence. Any further distribution of this work must maintain attribution to the author(s) and the title of the work, journal citation and DOI.
Published under licence by IOP Publishing Ltd

The realization of the dipole(γ,γ) method and its application to determine the absolute optical oscillator strengths of helium

Lin-Fan Zhu[*†1], **Long-Quan Xu**[*†], **Ya-Wei Liu**[*†], **Xu Kang**[*†]

[*] Hefei National Laboratory for Physical Sciences at Microscale and Department of Modern Physics, University of Science and Technology of China, Hefei, Anhui 230026, People's Republic of China

[†] Synergetic Innovation Center of Quantum Information and Quantum Physics, University of Science and Technology of China, Hefei, Anhui 230026, People's Republic of China

Synopsis The dipole(γ,γ) method based on the inelastic x-ray scattering (IXS) technique is proposed and realized to measure the absolute optical oscillator strength (OOSs) for the valence-shell excitations of atoms and molecules. The OOSs of valence-shell excitations of helium are determined, and the good agreement with the previous results is observed.

Absolute optical oscillator strength of atom or molecule, which is a very important physical quantity, represents the transition probability between two quantum states. Therefore, exploring new experimental method to determine OOS is very useful and signaficant. Herein we proposed the dipole(γ,γ) method to determine the absolute OOSs of atoms and molecules with high accuracy for the first time. In the dipole(γ,γ) method, the high energy photons impact (~10 keV) is used and the dipole approximation condition of $q^2 \approx 0$ (i.e., at the small scattering angle) is used to simulate the photoabsorption process. The experimental arrangement is that a beam of monochromatic x-ray photon collide with gaseous atoms or molecules and the scattered photons are analysed by a spheric grating at the scattering angle of 2° to make sure that the squared momentum transfer satisfies the requirement of $q^2 < 10^{-2}$ a.u..

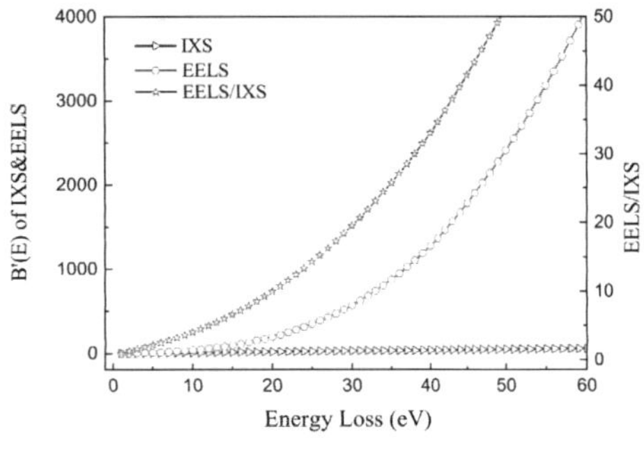

Figure 1. The Bethe-Born conversion factor of the dipole(γ,γ) method and EELS, which had been nomalized to a single point at the energy loss of 1eV.

Compared with the photoabsorption method based on the Beer-lambert law, the dipole(γ,γ) method is free from the line saturation effects that can seriously limit the accuracy of the measured photoabsorption cross-section for discrete transitions with narrow linewidth. Furthermore, the Bethe-Born factor of the dipole(γ,γ) method varies much more slowly with the excitation energy than that of the Electron Energy Loss Spectroscopy (EELS) method (see Figure 1). So more accurate absolute OOSs can be determined by the dipole(γ,γ) method. With the independent nomalization procedure, we measured the absolute OOSs of valence-shell excitations of atomic helium with the present dipole(γ,γ) method and compared them with the previous experimental and theoretical results, which are shown in Table 1.

Table 1. The OOSs of the $1s^{2\,1}S_0 \rightarrow 1snp\,^1P_1$ transitions in helium.

$1^1S \rightarrow$	3^1P	4^1P	5^1P	6^1P
This work	0.0740	0.0295	0.0156	0.0089
[1]	0.0739	0.0304	0.0154	0.0093
[2]	0.0741	0.0303	0.0152	0.0089
[3]	0.071			
[4]	0.0731	0.0299	0.0151	0.0086

It can be seen that the present high-resolution dipole(γ,γ) measurements are in excellent agreement with the results obtained by dipole(e, e) method and photoabsorption method as well as the ones of different theoretical calculations. Thus the dipole(γ,γ) method provides an reliable approach to obtain the benchmark data of OOSs of valence-shell excitations for gaseous atoms and molecules.

References

[1] Z. P. Zhong *et al* 1997 *J. Phys. B: At. Mol. Opt. Phys.* **30** 5305

[2] W. F. Chan *et al* 1991 *Phys. Rev. A* **44** 186

[3] S. Tsurubuchi *et al* 1989 *J. Phys. B: At. Mol. Opt. Phys.* **22** 2969

[4] W. L. Wiese *et al* 2009 *J. Phys. Chem. Ref. Data* **38** 565

[1]E-mail: lfzhu@ustc.edu.cn

Content from this work may be used under the terms of the Creative Commons Attribution 3.0 licence. Any further distribution of this work must maintain attribution to the author(s) and the title of the work, journal citation and DOI.
Published under licence by IOP Publishing Ltd

Journal of Physics: Conference Series **635** (2015) 092036 doi:10.1088/1742-6596/635/9/092036

Streaking temporal double slit interference by an orthogonal two-color laser field

Martin Richter*[1], Maksim Kunitski†, Markus Schöffler*, Till Jahnke*, Lothar P.H. Schmidt*, Min Li†, Yunquan Liu†,‡ and Reinhard Dörner*

* Institut für Kernphysik, Goethe-Universität Frankfurt, 60438 Frankfurt am Main, Germany
† State Key Laboratory for Mesoscopic Physics and Department of Physics, Peking University, Beijing 100871, China
‡ Collaborative Innovation Center of Quantum Matter, Beijing 100871, China

Synopsis We investigate electron momentum distributions from single ionization of Ar by two orthogonally polarized laser pulses of different color. The two-color scheme is used to experimentally control the interference between electron wave packets released at different times within one laser cycle.

The most prominent interference fringes emerging in strong-field ionization are the equidistant peaks in the electron energy distribution. These above threshold ionization (ATI) peaks [1] are spaced by the photon energy. They result from the interference of wave packets born periodically in time at subsequent laser cycles. Much less striking and only discovered in 2005 [2] are additional fringes resulting from the interference between wave packets born within one cycle at times where the vector potential is the same, but the direction of the electric field is opposite.

This intracycle interference pattern is typically hard to resolve in an experiment. In the present work we show that orthogonally polarized two-color pulses (OTC) can be used to turn the typically faint intracycle interference fringes into a dominating structure in the electron momentum distribution and at the same time they can be used as a tool to switch the regarding intracycle interferences on and off.

The orthogonal two-color pulses were focused into a supersonic gas jet of argon atoms and the three-dimensional (3D) momentum spectra of the resulting singly charged ions and corresponding electrons are measured in coincidence by means of COLd Target Recoil Ion Momentum Spectroscopy (COLTRIMS) [3]. The perpendicularly polarized second harmonic was generated by a 40fs, 780nm laser pulse in a collinear configuration. The intensity ratio of 780nm to 390nm was estimated to be 10:1.

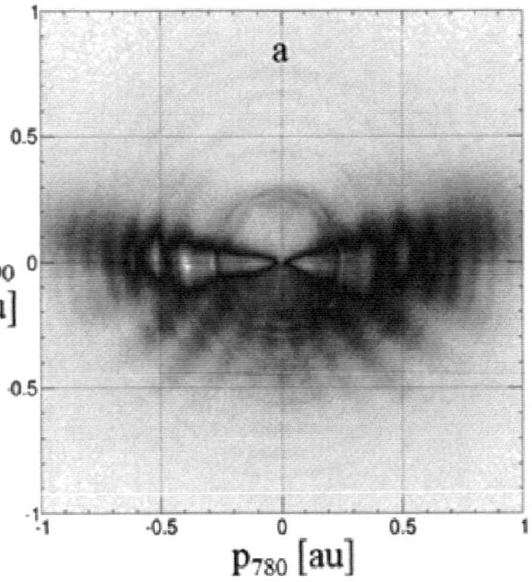

Figure 1. Electron momentum spectra in the polarization plane for the relative phase $\phi = \pi/2$ between the streaking field and ionizing field.

References

[1] P. Agostini *et al* 1979 *Phys. Rev. Lett.* **42** 1127

[2] F. Lindner *et al* 2005 *Phys. Rev. Lett.* **95** 040401

[3] R. Dörner *et al* 2000 *Phys. Rep.* **330** 95

[1] E-mail: richter@atom.uni-frankfurt.de

Content from this work may be used under the terms of the Creative Commons Attribution 3.0 licence. Any further distribution of this work must maintain attribution to the author(s) and the title of the work, journal citation and DOI.
Published under licence by IOP Publishing Ltd

Journal of Physics: Conference Series **635** (2015) 092037 doi:10.1088/1742-6596/635/9/092037

Spectroscopy for identification of plasma sources for lithography and water window imaging

Gerry O'Sullivan[*], **Padraig Dunne**[*], **Takeshi Higashiguchi**[†], **Bowen Li**[$], **Ragava Lokasani**[*], **Elaine Long**[*], **Hayato Ohashi**[#], **Fergal O'Reilly**[*], **Paul Sheridan**[*], **John Sheil**[*], **Emma Sokell**[*], **Chihiro Suzuki**[+], **and Tao Wu**[*].

[*]School of Physics, University College Dublin Belfield, Dublin 4, Ireland.
[†]Department of Advanced Interdisciplinary Science and Center for Optical Research (CORE), Utsunomiya University, Yoto 7-1-2, Utsunomiya, Tochigi 321-8585 Japan,
[$]School of Nuclear Science and Technology, Lanzhou University, Lanzhou, 730000, China
[#]Graduate School of Science and Engineering for Research, University of Toyama, Toyama, Toyama 930-8555, Japan
[+]National Institute for Fusion Science, 322-6 Oroshi-cho, Toki, Gifu 509-5292, Japan

Synopsis: We report on the result of an extensive study of the extreme ultraviolet (EUV) and soft x-ray spectra from laser produced plasmas of a number of medium and high Z metals undertaken to identify potential sources for use with multilayer mirrors as sources for lithography and water window imaging and cell tomography.

EUV lithography at 13.5 nm is currently being introduced. The choice of wavelength is based on the availability of Mo/Si multilayer mirrors (MLMs) with a reflectivity of ~70% at this wavelength and the optimum source has been identified as Sn, as lines from transitions of the type $4p^64d^n$ -$4p^54d^{n+1}$ + $4p^64d^{n-1}4f$ overlap in Sn^{10+} to Sn^{13+} overlap at this wavelength. The conversion efficiency (CE) of laser to EUV energy in a 2% bandwidth at 13.5 nm, corresponding to the reflectance bandwidth after ten reflections in a lithography tool and commercially available sources is now in excess of 4%. The theoretical maximum is significantly greater and recent experiments point towards a CE of ~7% being attainable [1].

Recently the development of LaB_4/C MLMs, with theoretical reflectance approaching 80% at this wavelength has led to proposals to introduce a future generation of lithography at this wavelength. The optimum sources were identified as laser produced plasmas of Gd and Tb, as n=4-n=4 transitions in their ions emit strongly near this wavelength. However to date, the highest CE obtained, for laser to EUV energy emitted within the 0.6% wavelength bandwidth of the mirror is only 0.8%, pointing to the need to identify other potential sources or selection of other wavelengths [2].

Sources for other applications are being developed. Conventional sources for soft x-ray microscopy use H-like line emission from liquid nitrogen or carbon containing liquid jets which can be focused using zone plates. Recently the possibility of using MLMs with n=4-n=4 emission from a highly charged Bi plasma was proposed [3] and subsequently the possibility of using Δn=1 transitions in 3rd row transition elements was identified [4]. All of these studies seek to identify spectral features that coincide with the reflectance characteristics of available MLMs, shown in the table below, determine the conditions under which they are optimized and establish the maximum conversion efficiencies obtainable. We have systematically studied the emission from laser produced plasmas of 4th and 5th row elements including the lanthanides.

Table 1. Peak reflectance wavelength and reflectivity of MLMs in the 'water window'.

MLM	Wavelength (nm)	Reflectivity (%)
Cr/V	2.42	9
Cr/Ti	2.73	17
TiO_2/ZnO	2.74	29
Cr/Sc	3.14	21
Cr/Sc B_4C	3.15	32.1
Cr/Sc	3.35	10

We have identified a number of potential candidates, especially 3d-4f arrays in the 4th row elements, 4d-5f transitions in the lanthanides and 4f-5g arrays in elements with Z>70.

References

[1] G O'Sullivan *et al* 2015*Phys. Scr.* (In press)
[2] K Yoshida *et al.* 2014 *Appl. Phys. Express* **7** 086202
[3] T Higashiguchi *et al.* 2012 *Appl. Phys. Lett.* **100**, 014103 (2012)
[4] B W Li *et al.* 2013 *Appl. Phys. Lett*, **102**, 041117

E-mail: gerry .osullivan@ucd.ie

Content from this work may be used under the terms of the Creative Commons Attribution 3.0 licence. Any further distribution of this work must maintain attribution to the author(s) and the title of the work, journal citation and DOI.
Published under licence by IOP Publishing Ltd

Attosecond time delays in the photoionization of noble gas atoms studied in TDLDA

Maia Magrakvelidze* [1], **Mohamed Madjet**†, **Himadri Chakraborty*** [2]

* Department of Natural Sciences, D.L. Hubbard Center for Innovation and Entrepreneurship, Northwest Missouri State University, Maryville, Missouri 64468, USA
† Qatar Environment and Energy Research Institute, Qatar Foundation, P.O. Box 5825, Doha, Qatar

Synopsis We perform time-dependent local density functional calculations of the quantum phase and time delays of valence photoionization of noble gas atoms. Results may be accessed by XUV-IR interferometric metrology.

In the standard RABITT (reconstruction of attosecond beating by interference of two-photon transitions) metrology, photoelectrons emitted by odd harmonics of an XUV pulse comb subsequently absorb or emit an IR photon, producing even-harmonic sidebands in the spectrogram. The ionization time delay is then determined by the ratio of the phase difference to the energy difference between these sidebands, directly connecting to the Wigner-Smith approach of the delay extraction from the emission quantum phase.

We compute these quantum phases and resulting Wigner-Smith time delays in the photoionization of valence electrons of various noble gas atoms, using a scheme of time-dependent local density approximation (TDLDA) [1] with Leeuwen and Baerends exchange-correlation functional. Analysis of the results with the goal to temporally access details of electron correlations has been carried out [2].

Numerical results for the $3p$ time delay of Ar (green curve) and the relative delay between $4s$ and $4p$ for Kr (red curve) are shown in Figure 1. For Ar, in addition, the calculations are compared with recent measurements [3] showing very good agreements. The minimum seen immediately above 45 eV is due to an accelerated ejection of $3p$ photoelectrons near the $3p$ Cooper minimum, which only exists in $3p \rightarrow kd$ ionization channel. However, the energy dependence of this delay becomes softer due to the participation of the $3p \rightarrow ks$ channel.

TDLDA $4s$–$4p$ relative delay for Kr exhibits a strong shape at lower energies centering 40 eV. This structure is due to the $4s$ Cooper minimum of Kr. In contrast, its $4p$ Cooper minimum at 67 eV generates a far weaker structure. This suggests the ability of the $3s$ minimum to affect the time behavior in a much stronger way than that of $3p$.

These results suggest significant differences in the details of correlations in the valence photoionization among various noble gas systems.

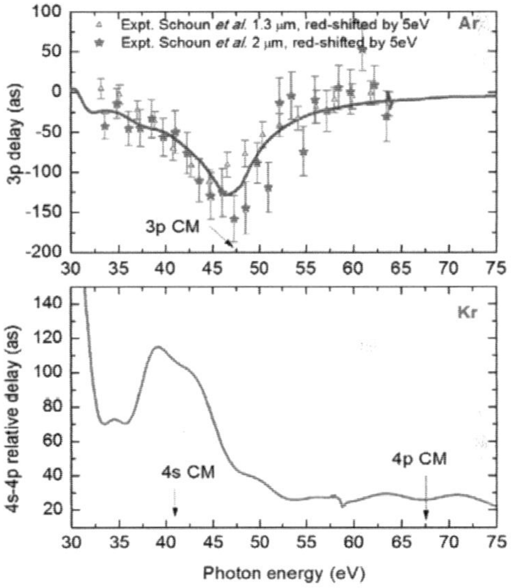

Figure 1. TDLDA Wigner-Smith time delay for Ar $3p$ and relative delay ($4s$–$4p$) for Kr valence electrons are shown. For Ar, the result is compared with the measured data [3].

This work was supported by the U.S. National Science Foundation.

References

[1] M. Madjet *et al* 2010 *Phys. Rev. A* **81** 013202
[2] M. Magrakvelidze *et al* 2015 *(Accepted)*
[3] S. B. Schoun et al. 2014 Phys. Rev. Lett. **112**, 153001.

[1] E-mail: maia@phys.ksu.edu
[2] E-mail: himadri@nwmissouri.edu

Content from this work may be used under the terms of the Creative Commons Attribution 3.0 licence. Any further distribution of this work must maintain attribution to the author(s) and the title of the work, journal citation and DOI.
Published under licence by IOP Publishing Ltd

Journal of Physics: Conference Series **635** (2015) 092039 doi:10.1088/1742-6596/635/9/092039

Ab-initio validation of a simple heuristic expression for the sequential-double-ionization contribution to the double ionization of helium by ultrashort XUV pulses

Aihua Liu [1] **and Uwe Thumm** [2]

Department of Physics, Kansas State University, Manhattan, Kansas 66506, USA

Synopsis We study two-photon double ionization of helium by short XUV pulses by numerically solving the time-dependent Schrödinger equation in full dimensionality within a finite-element discrete-variable-representation scheme. Based on the emission asymmetries in joint photoelectron angular distributions, we identify sequential and non-sequential contributions to two-photon double ionization for ultrashort pulses whose spectrum overlaps the sequential ($\hbar\omega > 54.4$ eV) and non-sequential (39.5 eV $< \hbar\omega <$ 54.4 eV) double-ionization regimes.

For sufficiently long extreme ultraviolet (XUV) pulses, sequential and non-sequential two-photon double ionization of helium atoms are distinguishable by the central XUV photon energy. If this energy is above the second ionization potential (54.4 eV), emission proceeds sequentially, otherwise non-sequential correlated double ionization occurs [1]. For ultrashort XUV pulses, this distinction breaks down if their spectrum overlaps the sequential and non-sequential regimes [2].

Solving the time-dependent Schrödinger equation (TDSE) *ab t o*, we investigate joint photoelectron angular distributions (JADs) for two-photon double ionization of helium atoms in the sequential and non-sequential regimes. We find that strong electronic correlation results in similar angular distributions over a large range of energy sharings between the two photoelectrons [3]. Based on the emission asymmetry in JADs (ranging from -1 for emission in opposite to 1 for emission into the same hemisphere), we distinguish "sequential" and "non-sequential" contributions to double ionization by linearly polarized ultrashort XUV pulses.

Figure 1 displays the "effective sequential double-ionization (SDI) yield" and double-ionization emission-asymmetry dependence on the XUV-pulse duration. We define the effective sequential yield as the product of the effective interaction time (pulse duration) and the integrated XUV-pulse-spectral intensity for $\hbar\omega >$ 54.4 eV (blue area in the inset of Fig. 1). For XUV pulses with a central photon energy of 50 eV, the SDI contribution reaches a maximum value at a pulse duration of 650 as, which we explain as being due to the competition between increasing temporal and decreasing spectral pulse widths.

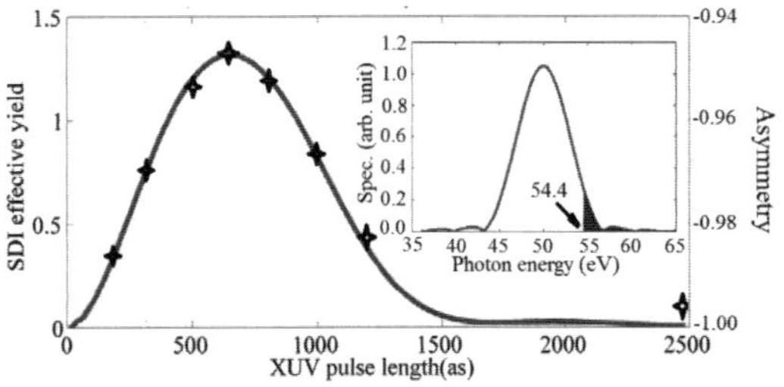

Figure 1. Effective SDI yield for helium in a 10^{14} W/cm^2 peak intensity XUV pulse with a central energy of 50 eV as a function of the XUV pulse duration. The stars are results of our *ab initio* TDSE calculations [1,3] for the double-emission asymmetry at extremely unequal energy sharing of the emitted electrons. The inset displays a typical XUV-spectral intensity, the blue area indicating the spectrum above the second ionization potential of helium.

This work is supported by the U.S. NSF and the Division of Chemical Sciences, Office of Basic Energy Sciences, Office of Energy Research, U.S. DoE.

References

[1] A. Liu and U. Thumm 2014 *Phys. Rev. A* **89** 063423.

[2] J. Feist *et al.* 2009 *Phys. Rev. ett.* **103** 063002.

[3] A. Liu and U. Thumm 2015, *Phys. Rev. A*, submitted for publication.

[1]E-mail: iwaa@phys.ksu.edu
[2]E-mail: thumm@phys.ksu.edu

Content from this work may be used under the terms of the Creative Commons Attribution 3.0 licence. Any further distribution of this work must maintain attribution to the author(s) and the title of the work, journal citation and DOI.
Published under licence by IOP Publishing Ltd

Journal of Physics: Conference Series **635** (2015) 092040 doi:10.1088/1742-6596/635/9/092040

Relativistic calculations of excitation and ionization probabilities in highly charged H-like ions exposed to intense laser fields

I. V. Ivanova*†, A. I. Bondarev*, I. A. Maltsev*†, D. A. Tumakov*, and V. M. Shabaev*[1]

* St. Petersburg State University, Ulianovskaya 1, Petrodvorets, St. Petersburg 198504, Russia

† SSC RF ITEP of NRC "Kurchatov Institute", Bolshaya Cheremushkinskaya 25, Moscow, 117218, Russia

Synopsis A method for the investigation of highly charged ions exposed to an intense laser field is developed and tested on various examples. The time-dependent Dirac equation for a hydrogenlike ion in presence of a strong laser field is solved numerically. The method is used for calculations of excitation and ionization probabilities in hydrogenlike ions. The interaction with the electromagnetic field is described within the dipole approximation in the length and velocity gauges.

Nowadays, laser technologies are rapidly developed and highly charged ions are ones of the most interesting objects that can be studied with intense laser fields [1, 2]. The time-dependent problems including an interaction with intense laser fields attract much attention during the past few years [3, 4].

In this work, a new method and code were designed for the investigation of highly charged ions exposed to an intense laser field. The theoretical consideration requires to solve the time-dependent Dirac equation. The method for solving the Dirac equation is proposed within the dipole approximation of a laser potential.

At the first stage the stationary problem in the Coulomb potential of the nucleus was solved within the dual kinetic balance (DKB) approach [5]. The DKB approach provides a solution of the stationary radial Dirac equation. A significant advantage of this approach is the absence of spurious states that makes our method more suitable for calculations of transition probabilities in comparison with others. At the second stage the time-dependent problem for a hydrogenlike ion in presence of a strong electromagnetic field was solved. The time-dependent Dirac wave function was expanded in terms of the stationary Dirac equation solutions. The expansion coefficients were found employing the Crank-Nicolson propagation scheme [6].

Based on the method developed, excitation and ionization probabilities were calculated for various ions. For instance, the multiphoton ionization process for the hydrogenlike tin ion (Z = 50) was investigated. The laser potential was chosen in the form of linearly polarized 20-cycle $\cos^2$-shaped pulse with a peak intensity of 5×10^{22} W/cm^2. The ionization probability was calculated in both length and velocity gauges (see Figure 1). The results obtained in two gauges coincide with each other. Moreover, our results are in a good agreement with the corresponding data from Refs. [3, 4]. This investigation is expected to be required for the future High-Intensity Laser Ion-Trap Experiment (HILITE).

This work was supported by SPbSU (Grant No. 11.38.269.2014), by RFBR (Grant No. 13-02-00630), by the Dynasty foundation, and by FAIR-Russia Research Centre.

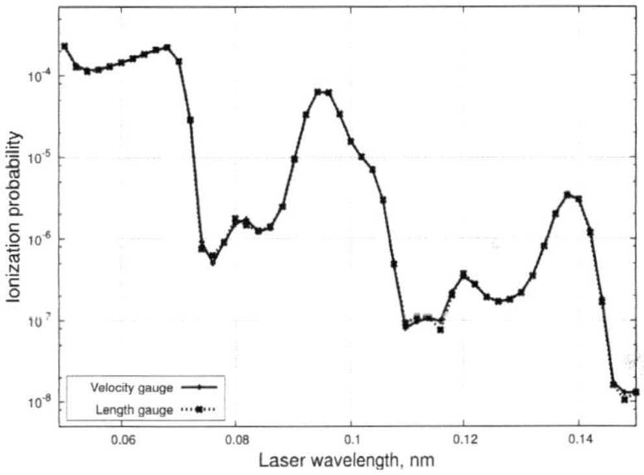

Figure 1. Ionization probability for the H-like tin ion as a function of the laser pulse wavelength.

References

[1] M. Richter *et al* 2009 *Phys. Rev. Lett.* **102** 163002

[2] M. Vogel *et al* 2012 *Nucl. Instrum. Methods Phys. Res. Sect. B* **285** 65

[3] Y. V. Vanne and A. Saenz 2012 *Phys. Rev. A* **85** 033411

[4] E. B. Rozenbaum *et al* 2014 *Phys. Rev. A* **89** 012514

[5] V. M. Shabaev *et al* 2004 *Phys. Rev. Lett.* **93** 130405

[6] J. Crank and P. Nicolson 1947 *Proc. Cambridge Philos. Soc.* **43** 50

[1]E-mail: shabaev@pcqnt1.phys.spbu.ru

Content from this work may be used under the terms of the Creative Commons Attribution 3.0 licence. Any further distribution of this work must maintain attribution to the author(s) and the title of the work, journal citation and DOI.
Published under licence by IOP Publishing Ltd

Effect of coulomb confinement resonances on time delay in Ne@C_{60}^{-5}

Ashish Kumar[* 1], H. R. Varma[* 2], P. C. Deshmukh[*,† 3], S. T. Manson[# 4], V. K. Dolmatov [$ 5] and A. S. Kheifets[! 6]

[*] School of Basic Science, Indian Institute of Technology Mandi, 175001, India
[†] Department of Physics, Indian Institute of Technology Madras, 600049, India
[#]Department of Physics and Astronomy, Georgia State University, Atlanta, GA 30303, USA
[$]Department of Physics and Earth Science, University of North Alabama, Florence, Alabama 35632, USA
[!]Research School of Physics and Engineering, Australian National University, Canberra, Australian Capital Territories 0200, Australia

Synopsis Wigner time delay for inner and outer shells of Ne@C_{60}^{-5} have been studied using relativistic random phase approximation (RRPA) [1]. Coulomb confinement resonances (CCR) due to the charged nature of fullerene cause prominent effects on the photoionization time delay of 1s subshell. These CCRs in 1s also results in significant modifications of the time delay in 2s and 2p ionization channels through interchannel coupling at energies far above their thresholds.

Sophisticated experimental techniques such as attosecond streaking and Reconstruction of Attosecond Bursts by Ionization of Two-photon Transitions (RABITT) [2] have opened up a new domain in photoionization experiments; it is now possible to measure the difference in delay time for the ejection of photoelectrons from different subshells of an atom with attosecond precision. A number of theoretical and experimental works have been reported in the recent past along this direction [3-6].

Dolmatov *et al.* [7] has predicted the re-emergence of confinement resonances at very high energies far away from the 2p threshold in Ne@C_{60}^{-5}. This revivification of confinement resonances was attributed to the presence of CCRs in the 1s subshell of Ne@C_{60}^{-5}. The present work reports the Wigner time delay in 2s ionization channels due to the presence of these CCRs in 1s subshell. The effect of these CCRs on the time delays of 2s and 2p subshells of Ne@C_{60}^{-5} are also studied in the region far above respective thresholds.

A model potential is used to simulate the effect of charged fullerene [7]. Wigner time delay is determined in the photon energy range of 880 to 920 eV within the frame work of relativistic random phase approximation (RRPA) which takes into account the important many electron correlations.

As an example, the time delay and phase shift for $2s \rightarrow \varepsilon p_{3/2}$ are shown in Fig. 1 ($2s \rightarrow \varepsilon p_{1/2}$ is essentially identical). In the region of CCRs (~ 882 -894 eV) the phase changes significantly. In each CCR region the phase shift initially increases and then decreases. As a result a flip in the sign of time delay profile is observed.

These rapid changes of phase shift in the region of CCRs are seen to enhance the time delay significantly.

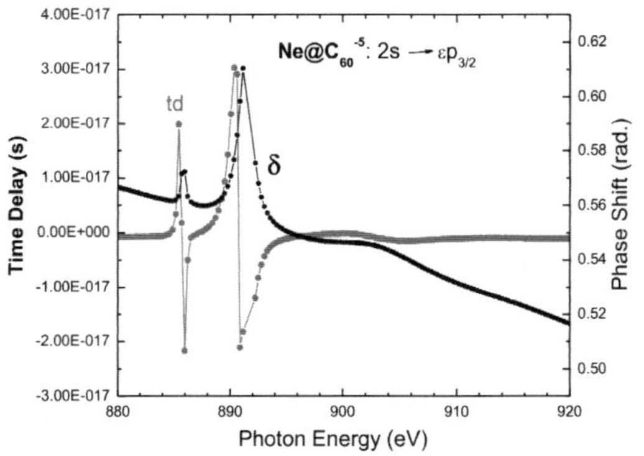

Figure 1. Wigner time delay (td) and phase shift (δ) for $2s \rightarrow \varepsilon p_{3/2}$ photoionization of Ne@C_{60}^{-5}.

The abrupt changes, which are observed above 895 eV, are absent in the region of ordinary confinement resonances (OCR).

In conclusion, CCRs play a major role in the time delay of electron from 1s subshell and also results in significant enhancement of time delay in 2s and 2p ionization channels.

References

[1] W R Johnson *et al* 1979 *Phys. Rev. A* **20** 978
[2] J M Dahlstrom *et al* 2012 *J. Phys. B.* **45** 183001
[3] M Schultze *et al* 2010 *Science* **328** 1658
[4] G Dixit *et al* 2013 *Phys. Rev. Lett.* **111** 203003
[5] P C Deshmukh *et al* 2014 *Phys. Rev. A* **89** 053424
[6] S Saha *et al* 2014 *Phy. Rev. A* **90** 053406
[7] V K Dolmatov *et al* 2009 *Phys. Rev. A* **80** 035401

[1]E-mail: ashishyadav002@gmail.com
[2] E-mail: hari@iitmandi.ac.in [3]E-mail: pcd@physics.iitm.ac.in [4]E-mail: smanson@gsu.edu
[5] E-mail: vkdolmatov@una.edu [6]E-mail: A.Kheifets@anu.edu.au

Content from this work may be used under the terms of the Creative Commons Attribution 3.0 licence. Any further distribution of this work must maintain attribution to the author(s) and the title of the work, journal citation and DOI.
Published under licence by IOP Publishing Ltd

Journal of Physics: Conference Series **635** (2015) 092043 doi:10.1088/1742-6596/635/9/092043

Electronic and nuclear dynamics of doubly core-ionized and excited states: Vibrationally resolved photoelectron and Auger spectra of small molecules

S. Oberli*[1], N. Sisourat*[2], P. Selles*[3], S. Carniato*[4],

* Université Pierre et Marie Curie, Paris Cedex 05, France

Synopsis In this work we want to assess the influence of nuclear dynamics on the features of the double core-hole photoelectron and Auger spectra of small isolated molecules by performing wave packet propagation on precomputed ab initio potential energy surfaces.

X-ray photoelectron spectroscopy (XPS), also known as ESCA, is a widely used method for chemical analysis. Since core electrons are highly localized around the nucleus, they encode element specific spectroscopic informations. Accordingly, a detailed analysis of the emitted electrons produced during the photoionization and the decay of the metastable states is a powerful mean to access electronic and structural properties of the ionized molecular states.

In 1986, Cederbaum and coworkers proposed to use double-core hole (DCH) spectroscopy [1], for which two inner-shell electrons are ejected, as an even more sensitive probe of the chemical composition of materials. In the last decade, the study of the spectroscopy of doubly core-ionized molecules received a strong impetus through the development of new generation synchrotron radiation and X ray free electron lasers (XFELs).

The two vacancies in the K shell can be observed either on a single (K^{-2}) or on different atomic sites ($K^{-1}K^{-1}$), the latter configuration allowing for instance to precisely probe the bond length [2]. The creation of core holes induces a rearrangement of the electronic distribution, which can lead to electronic excitation in valence orbitals and decay via electron emission (Auger process), see Figure 1.

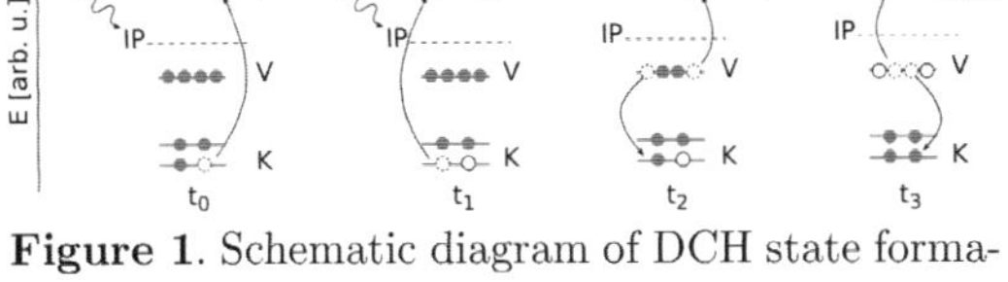

Figure 1. Schematic diagram of DCH state formation by sequential double ionization in the K shell, producing two photoelectrons (e_{ph}) in the continuum, followed by electronic relaxation in the valence (V) shell via Auger electron emission (e_{Auger}).

In previous theoretical studies, spectroscopic observables for DCH states were obtained in the frozen nuclei approximation [3], and molecular dynamics simulations were carried out for the Auger spectrum [4]. We will present our simulations on small isolated molecules interacting with (i) weak electromagnetic field such as synchrotron radiation, for which DCH states are created through single photon absorption; (ii) intense ultrashort pulse such as XFEL which induces sequential multiphoton absorption leading to DCH states. In the latter case, competing pathways complicate the theoretical investigations, as the sequential photoionization and the successive decays by electron emission channels are involved, depending on the lifetime of the core hole states.

Electronic structure simulations are carried out to get highly accurate ab initio potential energy surfaces for the various ionized/excited intermediate states and the final states reached via successive Auger decays. In this work we tackle to the challenging task to develop a method for computing vibrationally resolved photoelectron as well as Auger spectra of singly and doubly ionized small molecules, including the nuclear dynamics explicitly at a quantum level of theory.

References

[1] L. S. Cederbaum *at al* 1986 *J. Chem. Phys.* **85** 6513

[2] M. Nakano *at al* 2013 *Phys. Rev. Lett.* **110** 163001

[3] R. Santra *at al* 2009 *Phys. Rev. Lett.* **103** 013002

[4] L. Inhester *at al* 2012 *J. Chem. Phys* **136** 144304

[1]E-mail: solene.oberli@wanadoo.fr
[2]E-mail: nicolas.sisourat@upmc.fr
[3]E-mail: patricia.selles@upmc.fr
[4]E-mail: stephane.carniato@upmc.fr

Content from this work may be used under the terms of the Creative Commons Attribution 3.0 licence. Any further distribution of this work must maintain attribution to the author(s) and the title of the work, journal citation and DOI.
Published under licence by IOP Publishing Ltd

Journal of Physics: Conference Series **635** (2015) 092044 doi:10.1088/1742-6596/635/9/092044

Higher-order and unconventional low-energy structures

Ulf Saalmann[1], Elias Diesen & Jan-Michael Rost

Max Planck Institute for the Physics of Complex Systems
Nöthnitzer Straße 38 · 01187 Dresden · Germany

Synopsis Atoms or molecules exposed to intense near-infrared laser pulses produces a considerable amount of slow electrons forming peaks in measured photo-electron spectra. We show that these peaks, namely the so called low-energy structure (LES), higher-order variants and the very-low-energy structure (VLES), can be understood as different solutions of the same soft-recollision mechanism. Additionally we explain the formation of a "zero-energy structure" ("ZES") as a consequence of the release of Rydberg electrons by a static electric field.

Since the first observation [1] of the so-called low-energy structure — a strong peak at a few eV in the photo-electron spectrum of atoms or molecules exposed to linearly-polarized strong near-infrared laser pulses — a number of further measurements [2, 3, 4, 5] and theoretical investigations [6, 7, 8, 9] have been reported. The initially surprising observations has been reproduced by numerical propagation of the Schrödinger equation [1]. More importantly, in view of revealing the underlying mechanism, it became quickly clear that the observed structures can also be obtained by classical calculations [6].

By means of a proper analysis [7] of the underlying deflection functions, which directly determines the spectrum, it could be shown that *soft recollisions* (see the figure below for a sketch) are responsible for a bunching effect, which results in the formation of peaks. Apart from uncovering the mechanism this analysis revealed that there is a series of peaks, from which the 2nd one was seen in experiment recently [10]. Furthermore, there is another branch which induces a peak at even lower energies. Such a peak has also been found experimentally [4].

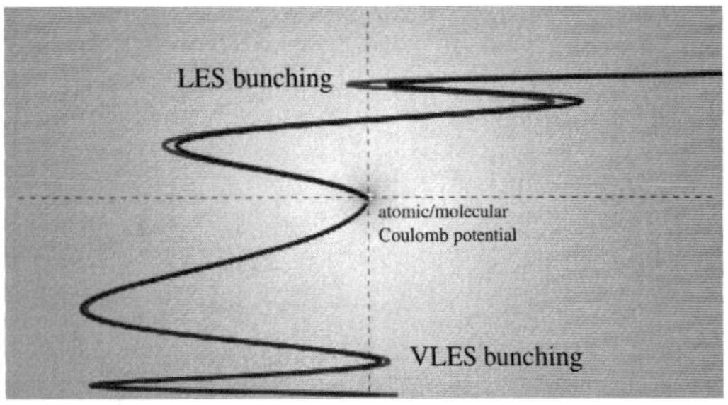

Figure: Characteristic trajectories forming the LES (upper half) and VLES (lower half). Bunching makes trajectories starting a *different* phases (blue vs. red) of the electric field to acquire *identical* final drift momenta.

To disentangle the underlying dynamics we study the peak formation in real time during the laser pulse recording the electron distribution in phase spaces. Neither the LES nor the VLES depend in the simplest non-trivial approximation on details of the potential rendering the respective peak positions universal and only dependent on the averaged vector potential of the laser field.

The recently observed "zero-energy structure" [5] differs from the before-mentioned features. It is shown that the ZES is due the extraction-field induced Stark dynamics of Rydberg electrons, i. e. the release of weakly bound electrons by a static field. The dynamical characteristics of this process are discussed.

References

[1] C. I. Blaga *et al.*, Nat. Phys. **5**, 335 (2009).
[2] F. Catoire *et al.*, Laser Phys. **19**, 1574 (2009).
[3] W. Quan *et al.*, Phys. Rev. Lett. **103**, 093001 (2009).
[4] C. Y. Wu *et al.*, Phys. Rev. Lett. **109**, 043001 (2012).
[5] J. Dura *et al.*, Sci. Rep. **3**, (2013).
[6] C. Liu and K. Z. Hatsagortsyan, Phys. Rev. Lett. **105**, 113003 (2010).
[7] A. Kästner, U. Saalmann, and J.-M. Rost, Phys. Rev. Lett. **108**, 033201 (2012); J. Phys. B **45**, 074011 (2012).
[8] C. Lemell *et al.*, Phys. Rev. A **85**, 011403 (2012).
[9] M. Möller *et al.*, Phys. Rev. A **90**, 023412 (2014).
[10] B. Wolter *et al.*, Phys. Rev. A **90**, 063424 (2014).

[1]E-mail: us@pks.mpg.de

Content from this work may be used under the terms of the Creative Commons Attribution 3.0 licence. Any further distribution of this work must maintain attribution to the author(s) and the title of the work, journal citation and DOI.
Published under licence by IOP Publishing Ltd

Journal of Physics: Conference Series **635** (2015) 092045 doi:10.1088/1742-6596/635/9/092045

Auger decay of the $4d^95s^25pnf$ excited states of Xe^{5+} ion

J.-M. Bizau*§ 1, D. Cubaynes*§, S. Guilbaud*, M.M. Al Shorman*, F. Penent†, P. Lablanquie†, L. Andric†, J. Palaudoux† and C. Blancard°

* ISMO, CNRS UMR 8214, Université Paris-Sud, Bât. 350, F-91405 Orsay cedex, France
§Synchrotron SOLEIL, L'Orme des Merisiers, St Aubin, BP48, F-91192 Gif-sur-Yvette, France
†LCPMR, CNRS & Univ. Paris 6, 11 rue Pierre et Marie Curie, F-75231 Paris Cedex 05, France
° CEA-DAM-DIF, Bruyère-le-Châtel, F-91297 Arpajon cedex, France

Synopsis The Auger decay of the 4d → nf (n = 4 to 6) photoexcitations in Xe^{5+} ion has been measured using electron spectrometry. MCDF calculations allow to interpret the recorded electron spectra.

Photoionization (PI) of free ionic species is a key process for plasmas modeling. Up to now, laboratory studies on multiply-charged ions were limited mainly to photoabsorption[1] and ionic spectrometry[2] experiments. Merged-beam setups in synchrotron radiation (SR) facilities provide absolute PI cross sections. Electron spectroscopy on ionic species is still a challenge due to the low target density. Its feasibility was demonstrated in the 90's on singly-charged atomic ions[3,4]. Since that time, only the high photon flux available at free electron lasers had allowed the observation of photoelectron spectra[5]. We present here the results of the first electron spectroscopy study performed on multiply-charged ions. The Auger spectra emitted in the decay of the $4d^95p5s^2nf$ resonances in Xe^{5+} ion, with n = 4 to 6, have been recorded.

The experiment was performed with the merged-beam setup of the PLEIADES beam line at SOLEIL French SR facility[6]. The Xe^{5+} ions were produced in an electron cyclotron resonance ion source (ECRIS). A cylindrical mirror electron analyzer (CMA), with its axis collinear to the ions and SR counter propagating beams, analyzed the kinetic energy of the electrons emitted in coincidence with the Xe^{6+} ions.

The Figure 1 shows an example of electron spectrum recorded at the photon energy of the $4d^95s^25p5f$ resonance (108.6 eV, upper panel). The Xe^{5+} ions are produced in the ECRIS in the 5p $^2P_{1/2}$ ground level and $^2P_{3/2}$ metastable level. At this photon energy, resonances from both levels can be excited. At least four lines are observed and can be identified with the help of multi-configuration Dirac Fock (MCDF) calculations. The calculated electron spectrum (lower panel) is reconstructed assuming a 45% $^2P_{1/2}$ and 55% $^2P_{3/2}$ population and is convoluted with a Gaussian profile simulating the experimental broadening of the electron lines. All the lines are issued from resonant PI processes of the type:

$$Xe^{5+}\ 5p\ ^2P_{1/2,3/2} + h\nu \rightarrow Xe^{5+}\ 4d^95s^25p5f$$
$$Xe^{5+}\ 4d^95s^25p5f \rightarrow Xe^{6+}\ 5s^2\ ^1S + e^-$$
$$Xe^{5+}\ 4d^95s^25p5f \rightarrow Xe^{6+}\ 5s5p\ ^{1,3}P + e^-$$

The good agreement between experimental and theoretical spectra shows that at this energy mainly Xe^{5+} ions in the $^2P_{3/2}$ metastable level are excited, and that the main decay channel leads with equal intensity to the $5s^2\ ^1S_0$ ground level (main line at 44 eV kinetic energy) and to the 5s5p $^3P_2$ excited level of Xe^{6+} ion (satellite line observed at 30 eV).

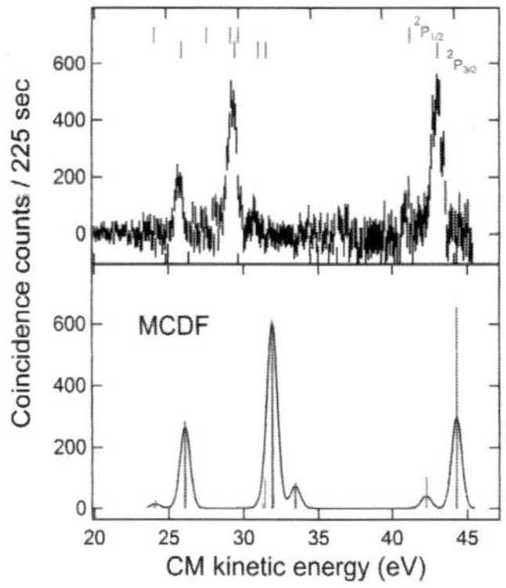

Figure 1. Electron spectrum recorded at 108.6 eV photon energy (top panel) compared to the reconstructed MCDF theoretical spectrum (bottom panel).

References

[1] Kennedy E T *et al* 1994 *Opt. Eng.* **33** 3984–92
[2] Kjeldsen H 2006 *J. Phys. B At. Mol. Opt. Phys.* **39** R325–77
[3] Bizau J M *et al* 1991 *Phys. Rev. Lett.* **67** 576–9
[4] Gottwald A *et al* 1999 *Phys. Rev. Lett.* **82** 2068–70
[5] Domesle C *et al* 2013 *Phys. Rev. A* **88** 043405
[6] Gharaibeh M F *et al* 2011 *J. Phys. B At. Mol. Opt. Phys.* **44** 175208

§ 1 E-mail: jean-marc.bizau@u-psud.fr

Content from this work may be used under the terms of the Creative Commons Attribution 3.0 licence. Any further distribution of this work must maintain attribution to the author(s) and the title of the work, journal citation and DOI.
Published under licence by IOP Publishing Ltd

Journal of Physics: Conference Series **635** (2015) 092046 doi:10.1088/1742-6596/635/9/092046

Theoretical characterization of the collective resonance states underlying the xenon giant dipole resonance

Yi-Jen Chen[*,† 1], **Stefan Pabst**[*], **Antonia Karamatskou**[*,†], **and Robin Santra**[*,† 2]

[*] Center for Free-Electron Laser Science, DESY, Notkestrasse 85, 22607 Hamburg, Germany

[†] Department of Physics, University of Hamburg, Jungiusstrasse 9, 20355 Hamburg, Germany

Synopsis We present an *ab initio* theoretical characterization of the two fundamental collective resonances underlying the xenon giant dipole resonance (GDR). This is achieved consistently by two complementary methods implemented within the many-body configuration-interaction singles (CIS) theory. We identify one resonance at an excitation energy of 74 eV with a lifetime of 27 as, and the second at 107 eV with a lifetime of 11 as. Insights on the nature of these resonances are provided, as well as information on previously unknown resonances in xenon.

The xenon GDR is not only a well-studied spectroscopic feature but also the key to understanding many start-of-the-art experiments on xenon exposed to strong midinfrared lasers or free-electron lasers. While the photoabsorption spectrum associated with the GDR has been characterized remarkably well, a question frequently overlooked is what exactly are the fundamental modes giving rise to the GDR.

We present a complete theoretical characterization of the resonance substructures underlying the xenon GDR. This is achieved consistently by two complementary methods implemented using CIS. The first method directly diagonalizes the many-electron Hamiltonian using the smooth exterior complex scaling technique. The second one involves a new application of the Gabor analysis to wave-packet dynamics.

We identify one resonance at an excitation energy of 74 eV with a lifetime of 27 as, and the second at 107 eV with a lifetime of 11 as. These excitation energies are expected to be more accurate than those given in the pioneering work of Wendin [1], since no approximate condition is invoked here to find the resonances. We demonstrate that many-body correlations result in qualitatively different substructures of the GDR: upon including the interchannel couplings between the $4d$ electrons, there are two far-separated resonances instead of three close-lying modes (see Fig. 1). As the wave functions of the two resonances are not dominated by any single particle-hole state, they correspond to plasma-like collective oscillations of the $4d$ shell as a whole.

Although the two resonances cannot be disentangled in the one-photon absorption cross section, a recent experiment at the free-electron laser FLASH, using an XUV nonlinear spectroscopy technique, has provided the first direct evidence for the two collective dipolar resonances [3]. Thus, it may be expected that experiments of this kind will offer an opportunity to test the results presented here.

In addition to the two resonances described above, we predict a new type of giant resonances in xenon. These new resonances have a total angular momentum larger than one, so they can only be accessed through multiphoton processes. This opens up possibilities to probe new quantum structures of matter with modern light sources.

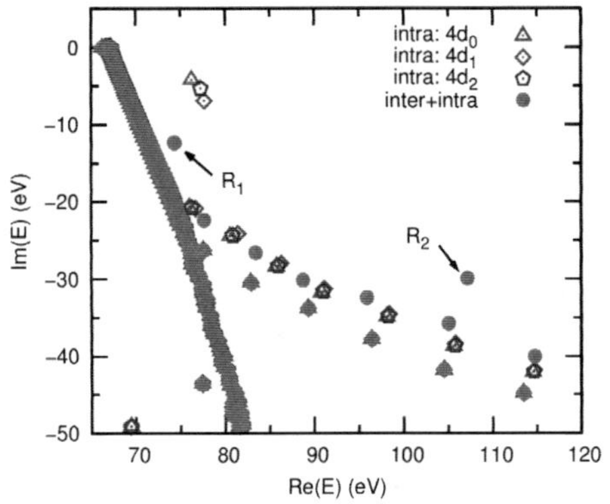

Figure 1. Complex energy spectra of the many-body Hamiltonian for the intrachannel ("intra") and full CIS ("inter+intra") models. From Ref. [2].

References

[1] G. Wendin, *J. Phys. B. At. Mol. Opt. Phys.* ***6*** *42 (1973)*.

[2] Y.-J. Chen *et al.*, *Phys. Rev. A* ***91*** *032503 (2015)*.

[3] T. Mazza and A. Karamatskou *et al.*, *Nat. Commun.*, accepted.

[1]E-mail: yi-jen.chen@desy.de
[2]E-mail: robin.santra@desy.de

Content from this work may be used under the terms of the Creative Commons Attribution 3.0 licence. Any further distribution of this work must maintain attribution to the author(s) and the title of the work, journal citation and DOI.
Published under licence by IOP Publishing Ltd

Journal of Physics: Conference Series **635** (2015) 092047 doi:10.1088/1742-6596/635/9/092047

Two-step semiclassical model for strong-field ionization with interference and multielectron polarization effects

N. I. Shvetsov-Shilovski[*1], **L. B. Madsen**[†], **E. Räsänen**[*], **J. Burgdörfer**[‡,○], **and K. Tőkési**[○]

[*] Department of Physics, Tampere University of Technology, Tampere FI-33101, Finland

[†] Department of Physics and Astronomy, Aarhus University, 8000 Aarhus C, Denmark

[‡] Institute for Theoretical Physics, Vienna University of Technology, A-1040, Vienna, Austria

[○] Institute of Nuclear Research of the Hungarian Academy of Sciences, Debrecen H-4001, Hungary

Synopsis We present a semiclassical model for above-threshold ionization with the inclusion of the Stark shift of the initial bound state, the Coulomb potential, and a polarization induced dipole potential capable to describe quantum interference. The model will be used to investigate the imprints of polarization effects in the interference structure of electron momentum distributions.

Semiclassical models applying a classical description of the electron after it has been promoted into the continuum by the laser field are widely used in strong-field physics. Semiclassical simulations can help to reveal the specific mechanism underlying the phenomena of interest and interpret them in terms of classical trajectories. Another appealing feature of semiclassical models is their numerical simplicity compared to the solution of the time-dependent Schrödinger equation.

Until recently, semiclassical models have not been able to describe quantum interference. A new approach developed in Ref. [1] allows to include interference effects into trajectory-based simulations of the above-threshold ionization (ATI). Within this approach each classical trajectory is associated with a certain phase, and the contributions of the trajectories ending up with a given final momentum are added coherently.

Here we develop a modified version of this semiclassical model for ATI with the inclusion of the Stark shift of the initial state, the Coulomb potential, and the polarization induced dipole potential capable to describe quantum interference. To this end, we combine the semiclassical model developed in Refs. [2, 3] with the approach of Ref. [1]. An example for momentum distribution of the ATI electrons calculated within our new model is shown in figure 1.

Our model will be used to investigate the role of the multielectron polarization effects in the formation of the interference structure in the photoelectron spectra and momentum distributions. Since the multielectron potential affects both the exit point at the tunnel and the electron dynamics in the continuum [3], a pronounced imprint of the polarization effects in the interference patterns is expected. Due to the rapid progress in experimental techniques, it is now possible to study interference structures of the photoelectron momentum distributions (see, e.g., Ref. [4]). Therefore, the expected results will provide a benchmark for the analysis of experimental data.

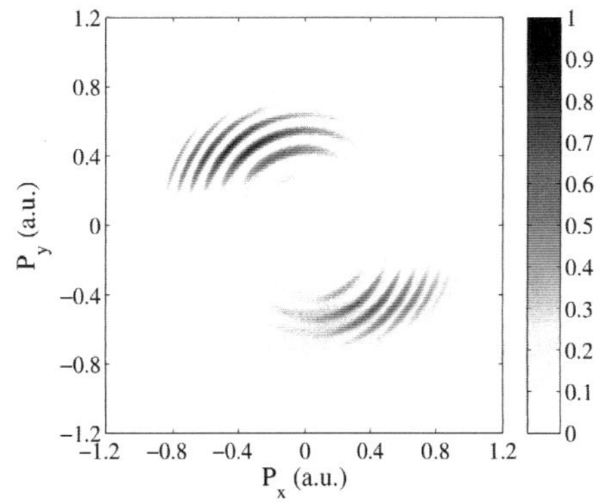

Figure 1. Photoelectron momentum distribution for the ionization of a hydrogen atom by a Ti:sapphire laser pulse (λ = 800 nm) with a duration of n_p = 6 cycles, peak intensity of 2.0×10^{14} W/cm^2, and ellipticity = 0.5 calculated within the semiclassical model.

References

[1] M. Li, J.-W. Geng, H. Liu *et al.* 2014 *Phys. Rev. Lett.* **112** 113002

[2] A. N. Pfeiffer, C. Cirelli, M. Smolarski, D. Dimitrovski, M. Abu-samha, L. B. Madsen, and U. Keller 2012 *Nat. Phys.* **8** 76

[3] N. I. Shvetsov-Shilovski, D. Dimitrovski, and L. B. Madsen 2012 *Phys. Rev. A* **85** 023428

[4] Ph. A. Korneev, S. V. Popruzhenko, S. P. Goreslavski *et al.* 2012 *Phys. Rev. Lett.* **108** 223601

[1]E-mail: n79@narod.ru

Content from this work may be used under the terms of the Creative Commons Attribution 3.0 licence. Any further distribution of this work must maintain attribution to the author(s) and the title of the work, journal citation and DOI.
Published under licence by IOP Publishing Ltd

Journal of Physics: Conference Series **635** (2015) 092048 doi:10.1088/1742-6596/635/9/092048

A dynamic ion-atom hybrid trap for high-resolution cold-collision studies

Alexander Dörfler*[1], Pascal Eberle*, Humberto da Silva Jr.†, Maurice Raoult†, Olivier Dulieu†, Stefan Willitsch*

* Department of Chemistry, University of Basel, Klingelberg-Strasse 80, 4056 Basel, Switzerland

†Laboratoire Aimé Cotton, CNRS, Université Paris-Sud XI, 91405 Orsay Cedex, France

Synopsis To push the frontier of studying ion-atom cold collisions, we present a new technique to precisely control collision energies in a hybrid trap setup.

The recent progress in the combined trapping of ions and atoms has paved the way to study collisions and chemical processes in a new physical regime at extremely low energies. Such experiments promise to elucidate the quantum character of collisions and enable to accurately characterise molecular interaction potentials and to study the details of chemical reaction mechanisms with high precision [1, 2, 3]. However, to observe quantum signatures in ion-atom collisions, an improved energy resolution compared to previous experiments is required [4]. Here, we introduce a new technique to precisely control collision energies in hybrid ion-atom experiments for this purpose.

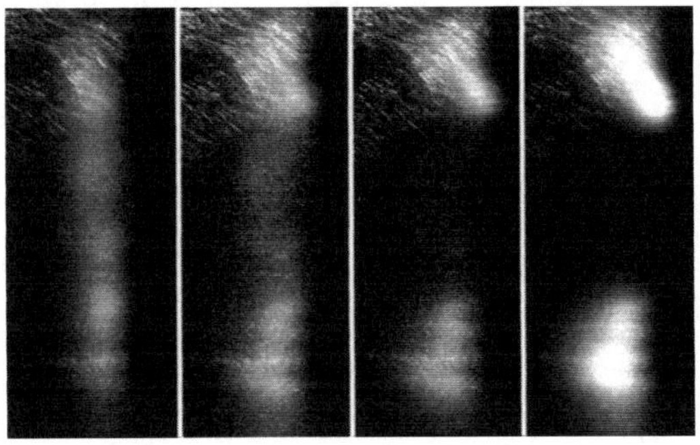

Figure 1. Images of an atom cloud during dynamic MOT operation (shuttling). This figure illustrates the effect of changing the hold time of an atom cloud in the off-center positions. From left to right: 5 ms, 10 ms, 15 ms and 25 ms hold time.

Previously an ultracold Rb atom cloud was stationary overlapping with a Coulomb crystal of Ca^+ ions in the trap center, which resulted in a large spread of collision energies. Inspired by previous experiments on moving optical molasses [5], we designed a scheme to create an ultracold atom cloud in an off-center position and use controlled radiation pressure to push it at a precise velocity through the interaction region with the cold ions. The atom cloud is then recaptured at a second off-center position on the other side of the trap center for re-use (Figure 1). This offers a new approach on controlled cold collision experiments yielding precise energies and a high energy resolution. Here we present a detailed characterization of the setup and first results on velocity-controlled collisions between Ca^+ ions and neutral Rb atoms in the cold regime.

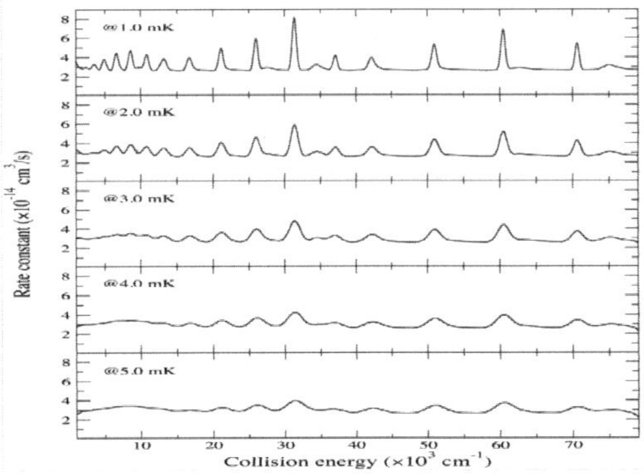

Figure 2. Theoretical rate constants for Ca^++ Rb radiative association and charge transfer obtained from quantum scattering calculations, convoluted for different collision energy resolutions. [6]

References

[1] S. Willitsch 2012 *Int. Rev. Phys. Chem.* **31**

[2] F. H. J. Hall *et al* 2011 *PRL* **107** 243202

[3] F. H. J. Hall, S. Willitsch 2012 *PRL* **109** 233202

[4] F. H. J. Hall *et al* 2013 *Mol. Phys.* **111** 2020

[5] W. Wohlleben *et al* 2001 *Eur. Phys. J. D* **15** 237

[6] H. Da Silva *et al* 2015 *ArXiv e-prints* **arXiv:** 1501.06385

[1]E-mail: alexander.doerfler@unibas.ch

Content from this work may be used under the terms of the Creative Commons Attribution 3.0 licence. Any further distribution of this work must maintain attribution to the author(s) and the title of the work, journal citation and DOI.
Published under licence by IOP Publishing Ltd

Journal of Physics: Conference Series **635** (2015) 092049 doi:10.1088/1742-6596/635/9/092049

Electron Spectroscopy of strontium in the vicinity of four-photon excitation of the $5p^2\ ^1S_0$ state

A. Dimitriou[† 1*], **S. Cohen**[† 2]

[†] Dept. of Physics, University of Ioannina, GR 45110 Ioannina, Greece

Synopsis We present experimental photoelectron energy spectra and angular distributions from multiphoton ionization of Sr by dye laser pulses of ~ns duration in the wavelength range of 715–737 nm and relatively low intensity ($\approx 4\times10^{11}$ W·cm^{-2}). The selected wavelength range comprises the highly correlated $5p^2\ ^1S_0$ autoionizing state, excited by four photons. Our results reveal the dominant (single and double) ionization pathways that were obscured in earlier experiments, based solely on the recording of ion and ionic-fluorescence yields.

We report on an electron energy analysis study conducted for elucidating the single and double ionization pathways when ground state strontium atoms interact with dye laser pulses of ≈5 ns duration and $\approx 4\times10^{11}$ W·cm^{-2} maximum intensity. Within the examined 715–737 nm wavelength range there are three-photon resonant, four-photon ionized bound states ($4d5p\ ^1P_1$ and $5s5f\ ^1F_3$) and the four-photon excited $5p^2\ ^1S_0$ highly correlated autoionizing state, located above the first ionization threshold of strontium. Electron spectra as a function of laser wavelength probed the accumulation of population in the excited $4d_j$ and $5p_j$ Sr^+ states. This signifies the absorption of at least two photons above the first ionization threshold ot the atom. However, contrary to what was anticipated in earlier work (performed using the same excitation and ionization scheme but based solely on the detection of ion and ionic-fluorescence yields [1]) the former states are found to be much more heavily populated than the latter ones. Thus, the main pathway to double ionization within this wavelength range is multiphoton ionization out of the $4d_{3/2,5/2}$ levels of Sr^+ and within the same laser pulse. Finally, the recording of photoelectron angular distributions from four- and five- photon ionization for selected laser wavelengths, reveales the dominant contributing partial waves at each ionization step. These latter results are compared to those obtained from relevant earlier studies on other Alkaline Earth atoms [2],[3].

This research has been co-financed by the European Union (European Social Fund – ESF) and Greek national funds through the Operational Program "Education and Lifelong Learning" of the National Strategic Reference Framework (NSRF) - Research Funding Program: THALES (ISEPUMA). Investing in knowledge society through the European Social Fund.

[*] **Present address:** Dept. of Physics, University of Crete, GR 71003, Heraklion, Greece

References

[1] I. Liontos 2008 *et. al* 2008 *J. Phys. B: At. Mol. Opt. Phys.* **41** 045601
[2] A. Dimitriou *et al* 2011 *J. Phys. B: At. Mol. Opt. Phys.* **44** 135001
[3] A. Dimitriou *et al* 2012 *J. Phys. B: At. Mol. Opt. Phys.* **45** 205003

[1] E-mail: adimitr@physics.uoc.gr
[2] E-mail: scohen@uoi.gr

Content from this work may be used under the terms of the Creative Commons Attribution 3.0 licence. Any further distribution of this work must maintain attribution to the author(s) and the title of the work, journal citation and DOI.
Published under licence by IOP Publishing Ltd

XXIX International Conference on Photonic, Electronic, and Atomic Collisions (ICPEAC2015) IOP Publishing
Journal of Physics: Conference Series **635** (2015) 092050 doi:10.1088/1742-6596/635/9/092050

Inner-shell ionization of carbon atoms using the R-matrix with time dependence (RMT) approach

Héctor F. Rey [1] and Hugo W. van der Hart [2]
Centre for Theoretical Atomic Molecular and Optical Physics,
Queen's University Belfast, Belfast, BT7 1NN, United Kingdom

Synopsis We use the R-matrix with time dependence approach to investigate inner-shell ionization of carbon atoms in ultra short high-frequency laser field with photon energies ranging from 170 to 245 eV. We present final-state populations for various subsets of photoemission channels at different peak laser intensities.

Over the last 15 years great strides have been made in the development of free-electron lasers (FELs) operating in the VUV - X-ray regime, such as the Linac Coherent Light Source (LCLS)[1], the Free-electron LASer in Hamburg (FLASH)[2] and the Spring-8 Angstrom Compact free-electron LAser (SACLA)[3]. In order to develop understanding of the interaction between such high-frequency laser light and matter, it is essential to consider the response of both inner and outer electrons.

Knowledge about the response of the carbon atom is of interest as it is the main constituent of organic molecules. To study its response to high frequency fields, we use the recently developed R-matrix with time dependence (RMT) approach [4]. This method combines the state to the art grid-based techniques of the HELIUM code [5] with the basis set method developed for the TDRM method [6]. The method maintains a full description of electron-electron correlation in an inner region surrounding the nucleus and joins this to an outer region wherein an ionized electron moves under the influence of the long range potential of the residual ion. It has already been proven successful to describe photoionization of systems with open outer shells, such as Ne^+[7] and C [8].

In the present study, we investigate the interaction between laser light with a photon energy between 170 and 245 eV and carbon atoms at various peak laser intensities. We obtain the emission yields for single-photon emission of a 2s or 2p electron, two-photon emission of a 1s electron, two-photon emission of a 2s or 2p electron and an approximation to three photon emission yield of a 1s electron at an intensity of 10^{17} W/cm^2. The probability that the residual C^+ ion is left with a hole in the 1s state is about 2%. 1s emission furthermore is the most likely two-photon emission process.

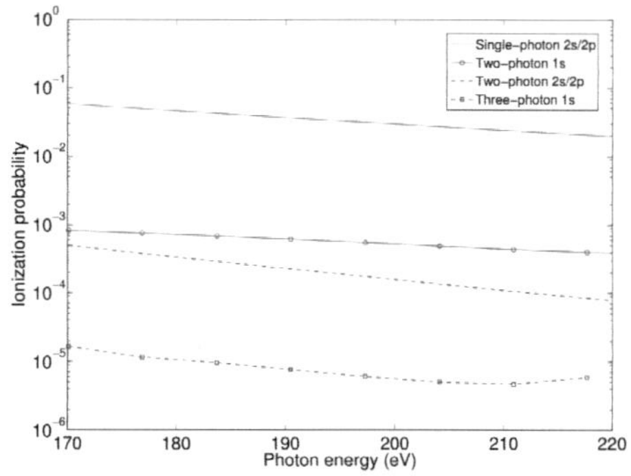

Figure 1. Photoionization yields for single- and two-photon emission of a 2ℓ electron and two- and three-photon emission of a 1s electron from C irradiated by a short laser pulse with a peak intensity of 10^{17} W/cm^2 as a function of photon energy.

References

[1] P. Emma et al., *Nature Photon.* **4**, 641647 (2010).

[2] W. Ackermann et al., *Nature Photonics* **1**, 336 (2009).

[3] M. Yabashi et al., *J. Phys. B: At. Mol. Opt. Phys.* **46**, 164001(2013).

[4] M. A. Lysaght, L. R. Moore, L. A. A. Nikolopoulos, J. S. Parker, H. W. van der Hart, and K. T. Taylor, in *Quantum Dynamic Imaging: Theoretical and Numerical Methods* Eds. A. D. Bandrauk and M. Ivanov (Springer, Berlin, 2011), pp. 107134.

[5] E.S. Smyth, J.S. Parker and K.T. Taylor, *Comput. Phys. Comm.* **114**, 1-14 (1998).

[6] H. W. van der Hart, M. A. Lysaght and P. G. Burke, *Phys. Rev. A* **76**, 043405 (2007).

[7] H. W. van der Hart and R. Morgan, *Phys. Rev. A* **90**, 013424 (2014).

[8] H. F. Rey and H. W. van der Hart, *Phys. Rev. A* **90**, 033402 (2014).

[1] E-mail: hreypereira01@qub.ac.uk
[2] E-mail: h.vanderhart@qub.ac.uk

Content from this work may be used under the terms of the Creative Commons Attribution 3.0 licence. Any further distribution of this work must maintain attribution to the author(s) and the title of the work, journal citation and DOI.
Published under licence by IOP Publishing Ltd

Application of R-Matrix with Time-dependence Theory to Double Ionisation Using a 2-electron Outer Region

Jack Wragg [1], **J. S. Parker** [2], **H. W. van der Hart** [3]

Centre for Theoretical Atomic, Molecular and Optical Physics, David Bates Building, The Queen's University of Belfast, Belfast BT7 1NN, Northern Ireland

Synopsis R-Matrix with Time-dependence (RMT) theory has been extended to cover double-ionisation processes. An application to photoionisation of He is demonstrated, with an emphasis on double-ionisation cross sections.

R-matrix with time-dependence (RMT) is an ab-initio method for solving the Schrödinger equation for many-electron atoms in the presence of short intense laser pulses [1, 2]. RMT makes use of the traditional R-Matrix division of configuration space into (i) a many-electron inner region and (ii) a region describing single-ionisation states, where the ionised electron is represented using finite-difference (FD) techniques. This work extends the RMT formalism to include a third region (iii) where double-ionisation states are represented through the inclusion of a two-dimensional finite difference grid. This double-ionisation grid is connected to single-ionised states (region (ii)) through standard RMT Bloch operator techniques.

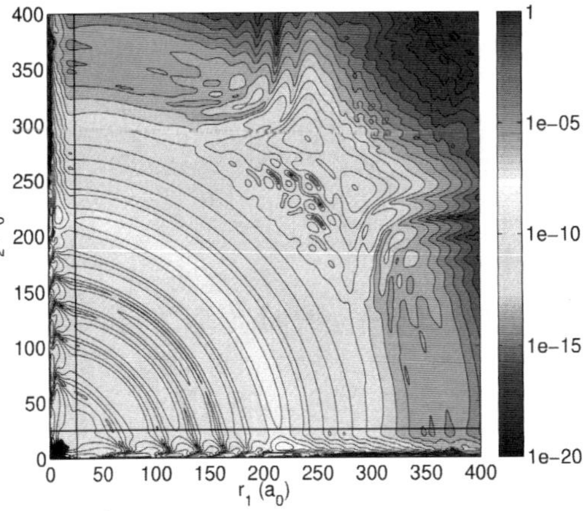

Figure 1. Two-electron He wavefunction density 2.335 fs after the start of a 15 cycle pulse of photon energy of 150 eV and a peak intensity of 10^{14} W cm^{-2}. r_1 and r_2 correspond to the radial distance of electron 1 and electron 2.

This extended RMT approach is applied to photoionisation of a He atom. The He wave function density, many cycles after the end of a short laser pulse, is shown in figure 1. High density (yellow) indicates the presence of significant double ionisation. Double ionisation yields for the non-sequential process are obtained from a momentum transform of the final He wavefunction. From these yields, cross-sections are obtained, and shown in figure 2 alongside experimental observations [3] and previously calculated theoretical results [4]. The RMT cross-sections agree well with experiment and established theory.

RMT can be extended to systems with more electrons than He. Work is ongoing to develop the inner-region R-matrix codes for the study of double ionisation processes in general atomic systems.

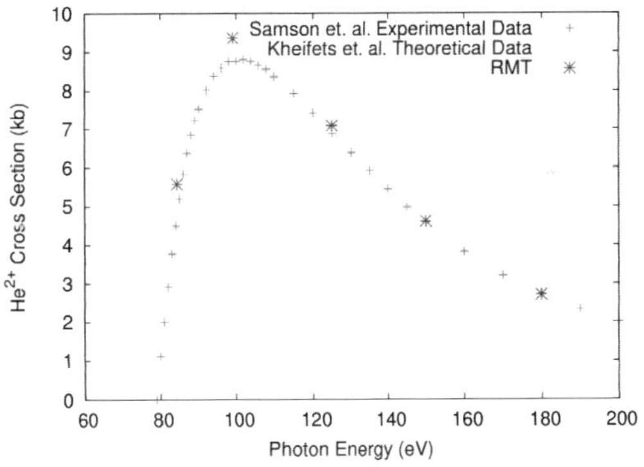

Figure 2. Single-photon double ionisation cross sections calculated using RMT alongside experimental data from [3] and theoretical data from [4].

References

[1] L. Nikolopoulos, J. Parker, and K. Taylor, Phys. Rev. A **78** (2008).

[2] L. Moore *et al.*, J. Mod. Opt. **58**, 1132 (2011).

[3] J. A. R. Samson *et al.*, Phys. Rev. A **57**, 1906 (1998).

[4] A. S. Kheifets and I. Bray, Phys. Rev. A **54**, R995 (1996).

[1] E-mail: jwragg01@qub.ac.uk
[2] E-mail: j.parker@qub.ac.uk
[3] E-mail: h.vanderhart@qub.ac.uk

Content from this work may be used under the terms of the Creative Commons Attribution 3.0 licence. Any further distribution of this work must maintain attribution to the author(s) and the title of the work, journal citation and DOI.
Published under licence by IOP Publishing Ltd

Journal of Physics: Conference Series **635** (2015) 092052 doi:10.1088/1742-6596/635/9/092052

Angular distributions of photoelectrons ejected from Neon in the presence of XUV and IR laser fields

R. Morgan*[1] and H. W. van der Hart†[2]

* Centre for Theoretical Atomic Molecular and Optical Physics
Queens University Belfast, Belfast, BT7 1NN, N. Ireland

Synopsis We use the R-matrix with time-dependence method to investigate the effects of two-colour laser fields on the angular dependence of photoelectron emission from Neon. Results are compared to both experimental and soft-photon approximation (SPA) results.

The rapid development of both high intensity and ultrashort light sources, e.g. free electron lasers (FELs) have led to some great breakthroughs in atomic science, unlocking the potential for experimental observation of electron dynamics with attosecond timescale resolution [1, 2, 3] and even control of these dynamics [4].

These advances however do come at a cost: the need for powerful theoretical methods which can describe and interpret experimental results on the dynamics of an atomic system that are simultaneously accurate and computationally feasible.

Theoretical models of varying complexity can be used to solve problems in these areas. These range from simple models which can cost very little in terms on computational power to complex models which describe the whole dynamics of an atomic system with high precision but are, in general, far more costly in terms of computing power. The latter contains ab initio calculations, such as the R-matrix with time-dependence (RMT) [5, 6, 7].

Recent experimental results of angular dependence of neon irradiated by intense two-colour fields have highlighted some interesting modulations in emission angle [8]. These experimental findings were compared to strong-field approximation calculations (SFA). However, the comparison showed some qualitative differences.

Using RMT we can investigate this system from first principles and compare outcomes to both experimental and theoretical model findings. Photoelectron angular distribution spectra are obtained for a range of intensities showing an increase in angular modulation and sideband formation as intensity increases. Our ab initio results are in very good agreement with experimental results for the central line; however for the second side band, our calculations compare better to the SFA model. We also compare RMT calculations with the soft-photon approximation [9].

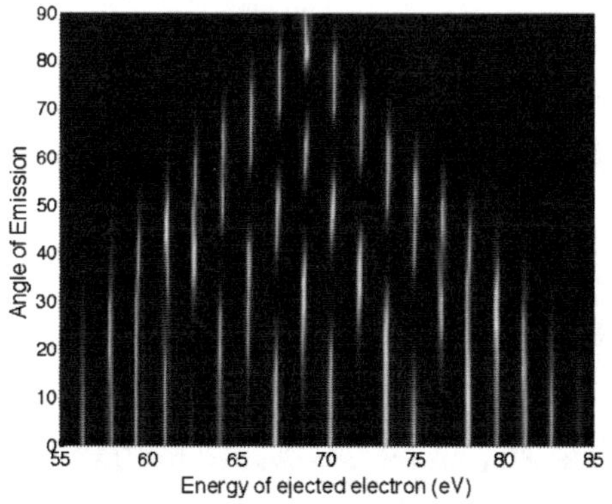

Figure 1. Photoelectron angular distribution spectra for the photoionisation of Neon irradiated by a combination of XUV laser at 90.5 eV photon energy and IR laser operating at 800 nm and intensity of 5×10^{12} W/cm^2.

References

[1] M. Hentschel *et al.*, Nature **414**, 509 (2001)

[2] M. Drescher *et al.*, Nature **419**, 803 (2002)

[3] M. Uiberacker *et al.*, Nature **446**, 627 (2007)

[4] A. Landsman *et al.*, Optica **1** 545 (2014)

[5] H. W. van der Hart, M. A. Lysaght and P.G. Burke Phys. Rev. A **76**, 043405 (2007)

[6] M. A. Lysaght, H. W. van der Hart and P. G. Burke, Phys. Rev. A **79**, 053411 (2009)

[7] M. A. Lysaght, *et al.*, Quantum Dynamic Imaging, Springer, 107 (2011)

[8] S. Düsterer *et al.*, J. Phys. B **46** 164026 (2014)

[9] A. Maquet *et al.*, Journal of Modern Optics **54** 1847 (2007)

[1]E-mail: rmorgan12@qub.ac.uk
[2]E-mail: h.vanderhart@qub.ac.uk

Content from this work may be used under the terms of the Creative Commons Attribution 3.0 licence. Any further distribution of this work must maintain attribution to the author(s) and the title of the work, journal citation and DOI.
Published under licence by IOP Publishing Ltd

XXIX International Conference on Photonic, Electronic, and Atomic Collisions (ICPEAC2015) IOP Publishing
Journal of Physics: Conference Series **635** (2015) 092053 doi:10.1088/1742-6596/635/9/092053

Photoionization of the neutral tungsten atom

B M McLaughlin* [1], **and C P Ballance**† [2]

*Centre for Theoretical Atomic and Molecular Physics (CTAMOP), School of Mathematics and Physics, Queens University of Belfast, Belfast BT7 1NN, UK

†Department of Physics, 206 Allison Laboratory, Auburn University, Auburn, AL 36849, USA

Synopsis Photoionization cross sections calculations for the neutral tungsten atom were obtained from large-scale close-coupling calculations within the Dirac-Coulomb R-matrix approximation (DARC) and compared with moderately resolved experimental measurements.

For the ITER tokamak, currently under construction at Cadarache, France, Tungsten (symbol W, atomic number 74) is the favoured material for the wall regions of highest particle and heat load in a fusion reactor vessel, with beryllium for regions of lower heat and particle load.

The lowest ground state level of neutral tungsten is $5p^6 5d^4 6s^2\ {}^5D_J$, with J=0, and requires only a single dipole matrix for photoionization. To make a meaningful comparison with existing experimental measurements, we statistically average the large-scale theoretical PI cross sections from the levels associated with the ground state $5p^6 5d^4 6s^2\ {}^5D_J[J = 0, 1, 2, 3, 4]$ levels and the $5d^5 6s\ {}^7S_3$ excited metastable level. As the experiments have a self-evident metastable component in their ground state measurement, averaging over the initial levels allows for a more consistent and realistic comparison to be made.

In the wider context, the absence of many detailed electron-impact excitation (EIE) experiments for tungsten and its multi-charged ion stages allows current photoionization measurements and theory to provide a road-map for future electron-impact excitation, ionization and di-electronic cross section calculations by identifying the dominant resonance structure and features across an energy range of hundreds of eV.

Results obtained from large-scale theoretical cross section calculations [1] for the total photoionization of the 4f, 5s, 5p and 6s orbitals of the neutral tungsten atom using the Dirac Coulomb R-matrix approximation (DARC: Dirac-Atomic R-matrix codes) will be presented and comparisons made with previous theoretical methods and prior experimental measurements.

Fig. 1, illustrates our DARC results compared with the available experimental work. Further details and a complete set of results will be presented at the meeting.

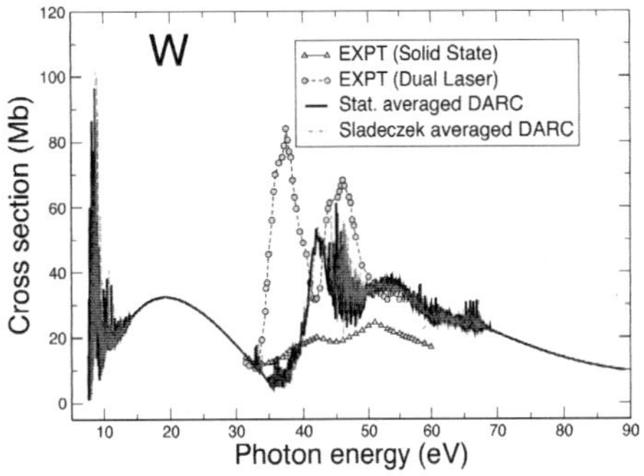

Figure 1. (Colour online)Single photoionization of neutral W over the photon energy range 8 eV - 100 eV, comparing weighted averaged theoretical calculations with several different experiments. The DARC results (645-levels approximation, solid black line), are the statistical average of the 5 levels associated the ${}^5D$ term ground state. The (dashed red-line) is the weighted DARC results of the lowest 6 levels, employing the mixing coefficients reported by Sladeczek and co-workers [2]. Solid circles with dashed line are the dual-laser experimental results of Costello and co-workers [3] and the solid triangles are the results from the experimental work of Haensel and co-workers [4]. The statistically averaged DARC PI cross sections were Gaussian convolved at a FWHM of 250 meV.

References

[1] C. P. Ballance and B. M. McLaughlin, J. Phys. B: At. Mol. & Opt. Phys., in press (2015).

[2] P. Sladeczek, H. Feist, M. Feldt, M. Martins and P. Zimmerman Phys. Rev. Lett. 70, 1483 (1995).

[3] J. T. Costello, E. T. Kennedy, B. F. Sonntag and C. L. Cromer, J. Phys. B 24, 5063 (1991).

[4] R. Haensel, K. Radler, B. Sonntag and C. Kunz, Solid State Commun. 7, 1495 (1969).

[1]E-mail: b.mclaughlin@qub.ac.uk
[2]E-mail: ballance@physics.auburn.edu

Content from this work may be used under the terms of the Creative Commons Attribution 3.0 licence. Any further distribution of this work must maintain attribution to the author(s) and the title of the work, journal citation and DOI.
Published under licence by IOP Publishing Ltd

Journal of Physics: Conference Series **635** (2015) 092054 doi:10.1088/1742-6596/635/9/092054

Photoionization of the 2p subshell in the Ar isonuclear sequence

Aarthi Ganesan* [1], **Sudha Deshmukh***, **Jobin Jose**‡, **Gagan B. Pradhan**#†
Vojislav Radojevic&, **Pranawa C. Deshmukh***† **and Steven T. Manson**ⱡ [2]

* Department of Physics, CPGS, Jain University, Bengaluru, India
‡ Manipal Centre for Natural Sciences, Manipal, Karnataka 576104
# Department of Physics, National Institute of Technology, Jamshedpur, 831014
& Department of Physics, Institute of Physics, Pregrevica, Beograd-Zemun, Serbia
† Department of Physics, IIT Madras, India
ⱡ Department of Physics and Astronomy, Georgia State University, Atlanta, USA

Synopsis Photoionization cross-sections and photoelectron angular distributions for Ar 2p subshell in Ar, Ar^{6+} and Ar^{8+} are calculated using the relativistic MultiConfiguration Tamm-Dancoff Approximation (MCTD) [1] that takes into account certain non-RPA correlations.

Extensive many-body calculations of the photoionization of the argon isonuclear sequence [2-5] using the Relativistic Random Phase Approximation (RRPA) [2, 6], RRPA *with* Relaxation (RRPA-R) [3, 4] and the relativistic Multi-Configuration Dirac Fock method [6] have drawn an important conclusion [8]: photoionization cross-sections of inner shells remain unaffected by the removal of outer-shell electrons. This conclusion was tested and re-established by Nasreen *et al* [3] using the RRPA which includes a significant portion of the electron correlation effects. Also calculations were done including relaxation effects in the RRPA formalism (RRPA-R) by Pradhan *et al* [5, 7]. The calculations showed that inner-shell cross-sections are affected by core-relaxation, but the angular distribution of the inner-shell photoelectrons is affected by the removal of electrons even in the absence of core-relaxation.

In the present work, the photoionization cross-section using RRPA-R and MCTD are reported. It is observed that the cross-section of the inner shell is altered when relaxation effect is included, so that the inner-shell ionic cross-sections differ significantly from the neutral. In the MCTD technique, the inclusion of electron correlations *via* configuration interaction in the initial state, and the interchannel coupling among ionization and excitation channels in the final state, indicate that, contrary to earlier belief, the threshold cross-section behavior is not unaffected by the removal of outer electrons, as seen in Fig 1. The photoelectron angular distribution (not shown) is also calculated.

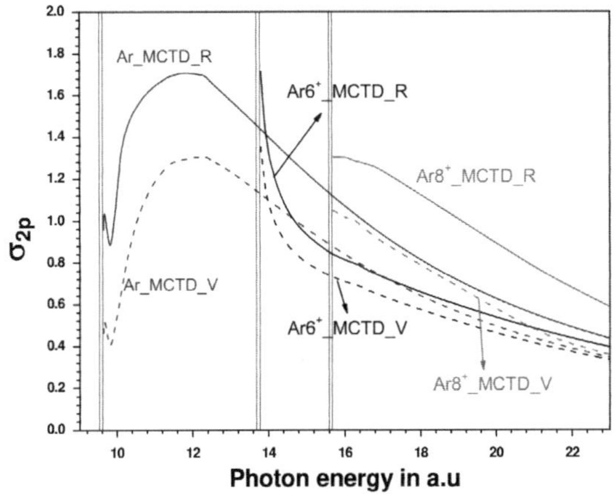

Figure 1. 2p subshell photoionization cross-section of Ar, Ar^{6+} and Ar^{8+} calculated using relativistic MCTD in length (R) and velocity (V) formulations. The vertical lines indicate the 2p thresholds at the various stages of ionization.

References

[1] V. Radojevic and W. R. Johnson 1985 *Phys. Rev. A* **31** 2991
[2] G. Nasreen *et al* 1989 *Phys. Rev. A* **40** 6091
[3] G. B. Pradhan *et al* 2010 *Phys. Rev. A* **81** 063401
[4] X. Liu and X. Ding 2013 *Eur. Phys. J. D* **67** 152
[5] G. B. Pradhan *et al* 2010 *Phys. Rev. A* **80** 053416
[6] W. R. Johnson and C. D. Lin 1979 *Phys. Rev. A* **20** 964
[7] M. Kutzner and V. Radojevic 1994 *Phys. Rev. A* **49** 2574
[8] S. T. Manson and J. W. Cooper 1968 *Phys. Rev.* **165** 126

[1] E-mail: aarthihari07@gmail.com
[2] E-mail: smanson@gsu.edu

Content from this work may be used under the terms of the Creative Commons Attribution 3.0 licence. Any further distribution of this work must maintain attribution to the author(s) and the title of the work, journal citation and DOI.
Published under licence by IOP Publishing Ltd

Inner-shell autoionization resonances along Mg isoelectronic sequence

K. Sindhu[*1], **H. R. Varma**[*2], **P. C. Deshmukh**[*†3], **and S. T. Manson**[#4]

[*]School of Basic Sciences, Indian Institute of Technology Mandi, Mandi 175005, India

[*†] Department of Physics Indian Institute of Technology Madras, Chennai 600036

[#]Department of Physics and Astronomy, Georgia State University, Atlanta, Georgia 30303, USA

Synopsis Studies of the 2s$\rightarrow$np autoionization resonances using a combination of Relativistic Random Phase Approximation (RRPA) and Relativistic Multichannel Quantum Defect Theory (RMQDT) of Mg isoelectronic sequence are reported.

Photoionization and photoexcitation studies of inner shell electrons in neutral, singly and multiply ionized atoms are required for understanding many astrophysical phenomena [1]. Several studies of photoionization of the Mg isoelectronic sequence have already been reported [2, 3]. However, to the best of our knowledge, no work has been reported on the inner shell 2s$\rightarrow$np autoionization resonances.

The Relativistic Random Phase Approximation (RRPA) [4] in conjunction with the Relativistic Multichannel Quantum Defect Theory (RMQDT) [5] is used in the present work to obtain detailed information about the autoionization resonance region. Such studies give information about the interference between ionization and excitation channels.

The present work reports autoionization studies of Mg, Al^+, Si^{2+}, P^{3+}, S^{4+}, Cl^{5+}, Fe^{14+}, Pd^{34+}, La^{45+}, Hg^{68+}, U^{80+} and Lr^{91+} resulting from the interference of bound to continuum (2p$\rightarrow$s,d) and bound to bound excitation channels (2s$\rightarrow$p). The total 2p ($2p_{1/2}$+$2p_{3/2}$) cross section in each case is periodic with respect to the effective quantum number (n*) and converges to the 2s threshold, as expected. The background value of σ_{2p} in the resonance region decreases as we go from low Z to high Z elements along the Mg isoelectronic sequence. In addition, the resonances become much more important compared to the background. The relativistic splitting of the resonances becomes pronounced at high Z.

In all the cases mentioned above, we have also studied the weighted average angular distribution asymmetry parameter (β_{2p}) as functions of n* and of the photon energy. In neutral Mg, the background value of β_{2p} is almost unity indicating the dominance of p$\rightarrow$d matrix elements over p$\rightarrow$s channel.

For Al^+, Si^{2+}, P^{3+}, S^{4+}, Cl^{5+}, Fe^{14+} background β_{2p} rises from unity which indicate that the dominance of p$\rightarrow$d channel decreases as Z increases. In the case of high Z elements the background value of β_{2p} is about ~1.5.

The quantum defect values of various members of the sequence at principal quantum number n=20 are calculated using the relativistic formula $\mu_{np} = n - n^* + \left(\kappa^2 - \beta^2\right)^{\frac{1}{2}} - |\kappa|$ where $\kappa = -(l+1)$ for $j = l + \frac{1}{2}$ and $\kappa = +l$ for $j = l - \frac{1}{2}$ [6], and are shown in Fig. 1. Both $p_{1/2}$ and $p_{3/2}$ quantum defects decrease as a function of Z, as expected. At low Z, the calculated quantum defects for $p_{1/2}$ and $p_{3/2}$ at each value of Z are essentially equal; but as Z increases, the splitting of $\mu_{p3/2}$ and $\mu_{p1/2}$ is evident, showing the importance of relativistic interactions at higher Z.

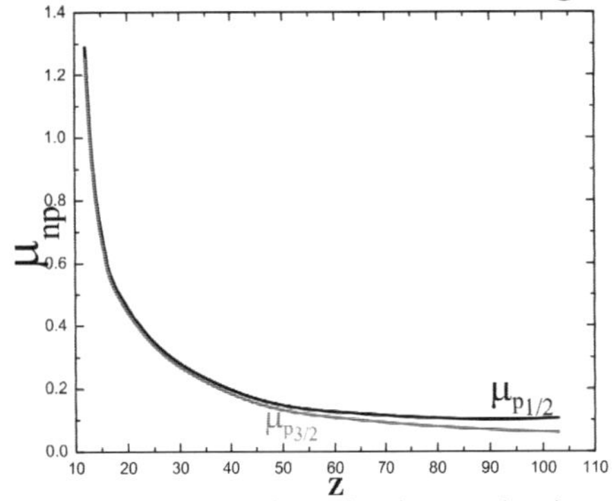

Figure 1. Quantum defects for the Mg isoelectronic sequence as functions of nuclear charge, Z.

[1] E-mail: sindhu_k@students.iitmandi.ac.in
[4] E-mail: smanson@gsu.edu

Reference

[1] J. P. Mosnier *et al* 1994 *Phys. Rev. A.* **49** 755
[2] K. Butler *et al* 1984 *Mon. Not. R. astr. Soc.* **209** 343
[3] J. A. Santana *et al* 2015 *Phys. Rev A.* **91** 022503
[4] W. R. Johnson *et al* 1979 *Phys. Rev. A* **20** 964
[5] C.M. Lee *et al* 1980 *Phys. Rev. A.* **22** 979
[6] M. J. Seaton 1983 *Rep. Prog. Phys.* **46** 167

Content from this work may be used under the terms of the Creative Commons Attribution 3.0 licence. Any further distribution of this work must maintain attribution to the author(s) and the title of the work, journal citation and DOI.
Published under licence by IOP Publishing Ltd

Journal of Physics: Conference Series **635** (2015) 092056 doi:10.1088/1742-6596/635/9/092056

Evaluation of photoexcitation and photoionization probabilities by trajectory method

A.B. Bichkov, V.V. Smirnov[1]

Saint Petersburg State University, Russia

Synopsis A new trajectory-based method of transition probability evaluation in quantum system is developed. Capabilities of the method are demonstrated on evaluating photo-excitation and photo-ionization probabilities in hydrogen atom under exposure of an ultra-short photopulse, and total photoionization probability in the helium atom.

Trajectory methods have many advantages, for example, they are among methods which account the electron-electron correlations. They are widely used when the direct solution of Schroedinger equation is impossible.

In our works [1, 2] we developed trajectory method of transition probability evaluation in quantum system. Here the new variant of this method, using Weyl symbols of initial and final states is represented. The transition probability w is represented in form

$$w = \frac{1}{\nu}\int W_f(q_l, p_l)\left|\exp(iS_l) - K\right|^2 W_i(q_0, p_0)dq_0dp_0$$

$$K = \frac{1}{\nu}\int \exp(iS_l)W_i(q_0, p_0)dq_0\,dp_0$$

where $S_l = -\int V\,dt$ – is integral of the interaction, W_i, W_f – Weyl symbols of initial and final states respectfully, q_0, p_0 – initial point of stationary path in phase space, q_l, p_l – final point, $\nu = (2\pi)^n$, n – is the coordinate space dimension.

The developed method was tested on the processes of photo-excitation and photo-ionization in the hydrogen atom by an ultrashort photo-pulse. Also we calculated total ionization crossection in helium atom.

For example, photoexcitation probability dependence on electric field carrier frequency ω for transition between the states 1s → 2p in hydrogen atom is shown in figure 1 for the field strength $E = 0.01$ and duration $\tau = 6$. Line 1 represents this method, line 2 – time-dependent perturbation theory. All values are in atomic unit system.

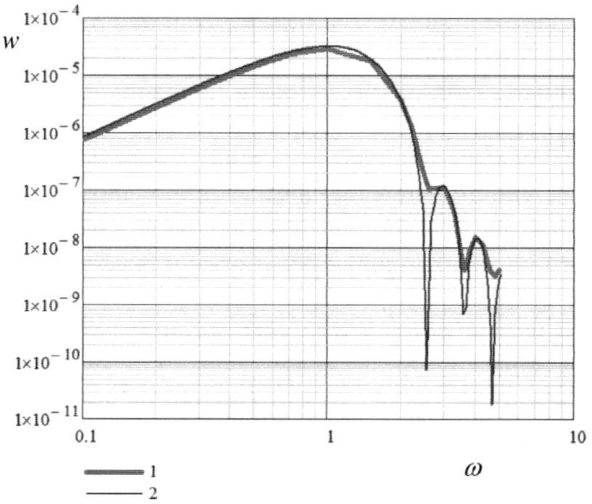

Figure 1. Photoexcitation probability dependence on field carrier frequency ω (atomic units).

The agreement with reliable theoretical results as well as with experimental results are rather good. The method works in wide range of interaction magnitudes. Method is especially relevant to solve the problems in ultraintense and ultrashort laser fields such as calculation of multiphoton multiple photoionization probabilities.

References

[1] Bichkov A.B., Smirnov V.V. 2013 *J. Phys. A: Math. and Theor.* **46** 015303

[2] Bichkov A.B., Smirnov V.V. 2013 *Laser Phys.* **23** 055302

[1]E-mail: valery_smirnov@mail.ru

Content from this work may be used under the terms of the Creative Commons Attribution 3.0 licence. Any further distribution of this work must maintain attribution to the author(s) and the title of the work, journal citation and DOI.
Published under licence by IOP Publishing Ltd

Evaluation of the probabilities of single and double photoionization of helium atom by trajectory method

A.B. Bichkov, A.S. Kozhina, V.V. Smirnov[1]

Saint Petersburg State University, Russia

Synopsis Capabilities of new trajectory-based method are demonstrated in calculation of single- and double-photoionization probabilities in helium atom. The method is suitable for wide range of field strengths and reproduces effects, attributed to electron-electron correlations.

The study of photoionization as a fundamental process is especially relevant in connection with development of ultrafast strong field radiation sources. Because of complexity of the theoretical description of the processes in atoms under the exposure of strong pulses, the study and the development of appropriate theoretical methods is very important. In spite of the progress in theory many interesting features of these processes are difficult to reproduce by means of the existing theoretical methods. It concerns, for example, the effects of electron-electron correlations. Rather popular in this connection are trajectory-based methods.

In previous works we developed the trajectory method for evaluating transition probability in quantum system [1]. Here we demonstrate capabilities of the new variant of our method using Weyl symbols of initial and final states in calculating probabilities of single- and double-photoionization of helium atom.

The calculation of the Weyl symbols for the groups of states, for example for the entire ionization continuum, is rather difficult even for the hydrogen atom. For this reason some simple models of final distributions were used. In spite of simplicity these models have given rather good results.

For example, photoionization probability dependence on electric field strength is shown in figure 1, where line 1 – total ionization probability, 2 – single ionization probability, 3 – double ionization probability. The lines 4, 5, 6 – the same values from [2]. Field parameters are the same as in work [2]. Our results correlate well with calculation of [2] and with our calculations based on previous variant of method [1]. Maximal discrepancy (about factor 2) we have for single-photoionization probability at very strong field, where saturation is observed.

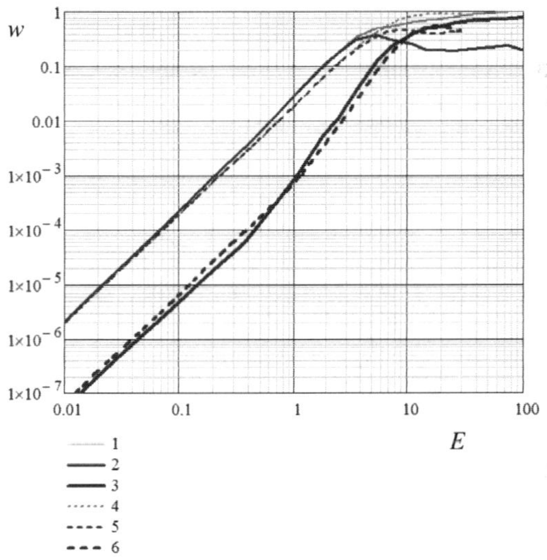

Figure 1. Photoionization probabilities dependence on electric field E (atomic units)

Gained results show that method works in wide range of the field strengths. It reproduces effects attributed to electron-electron correlations, such as stabilization of an atom in intense laser field [2].

References

[1] Bichkov A.B., Smirnov V.V. 2013 *Laser Phys.* **23** 055302

[2] Birkeland, *et al.* 2010 *Phys. Rev. Lett.* **104** 163002

[1]E-mail: valery_smirnov@mail.ru

Content from this work may be used under the terms of the Creative Commons Attribution 3.0 licence. Any further distribution of this work must maintain attribution to the author(s) and the title of the work, journal citation and DOI.
Published under licence by IOP Publishing Ltd

Journal of Physics: Conference Series **635** (2015) 092061 doi:10.1088/1742-6596/635/9/092061

Wigner time delay studies of photoionization of atomic zinc and cadmium

S. Saha* [1], **A. Mandal*** [2], **P. C. Deshmukh*** [3],
A. Kheifets[†] [4], **V. K. Dolmatov**[$] [5] **and S. T. Manson**[#] [6]

[*]Department of Physics, Indian Institute of Technology Madras, Chennai, 600036, India
[†]Research School of Physics and Engineering, The Australian National University, Canberra ACT 0200, Australia
[$]Department of Earth Science, University of North Alabama, Florence, AL 35632, USA
[#]Department of Physics and Astronomy, Georgia State University, Atlanta, 30303, USA

Synopsis Wigner time delay has been studied in the shape resonance region for the outer n$d\rightarrow\varepsilon f$ photoionization transitions in atomic Zn and Cd.

Photoionization of atomic Zn and Cd has been studied earlier within the framework of Relativistic Random Phase Approximation (RRPA) [1, 2]. In the present work, we report on the time delay in photoionization using the Wigner-Eisenbud time delay formalism [3-5]. Our main objective is to study the importance of the *f*-wave shape resonance, induced by the centrifugal barrier, in n$d\rightarrow\varepsilon f$ photoionization [6, 7]. In particular, RRPA calculations for the time delay for n$d\rightarrow\varepsilon f$ in atomic Zn and atomic Cd have been performed with all 3*s*, 3*p*, 3*d*, and 4*s* relativistic dipole channels coupled for Zn; and 3*d*, 4*s*, 4*p*, 4*d*, and 5*s* for Cd.

The results are shown in Figs. 1 and 2 for Zn and Cd respectively; both time delay and phase for the $3d_{5/2}\rightarrow\varepsilon f_{5/2}$ channel (Zn) and the $4d_{5/2}\rightarrow\varepsilon f_{5/2}$ channel (Cd) are presented. In both cases, significant time delay in the region of the rapidly increasing phase is evident. The other n$d\rightarrow\varepsilon f$ photoionization channels (not shown) exhibit exactly the same characteristics.

The *f*-wave dependence on energy, the rapid increase in the threshold region, is indicative of a resonance; a shape resonance in these cases. This well-known shape resonance comes about owing to the competition of the attractive Coulomb potential and the repulsive centrifugal potential which creates an effective double-well potential with a barrier between the two wells [8]. With increasing energy, the *f*-wave penetrates into the inner well creating the shape resonance. This increase in time delay should be a general phenomenon in the vicinity of shape resonances.

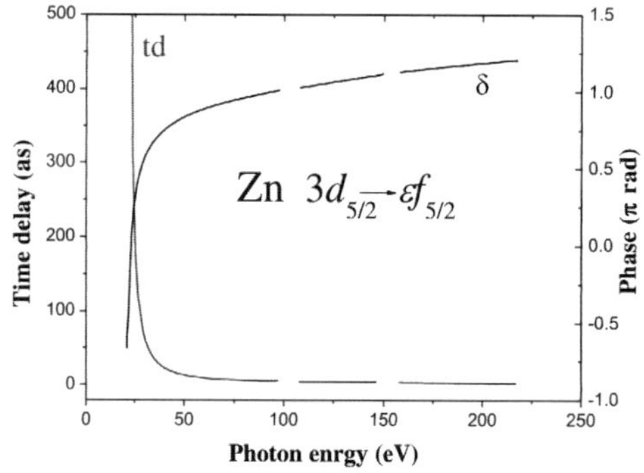

Figure 1. Wigner time delay (td) and phase (δ) for the $3d_{5/2}\rightarrow\varepsilon f_{5/2}$ photoionization channel in atomic Zn. The breaks in the curves represent regions of autoionizing resonances.

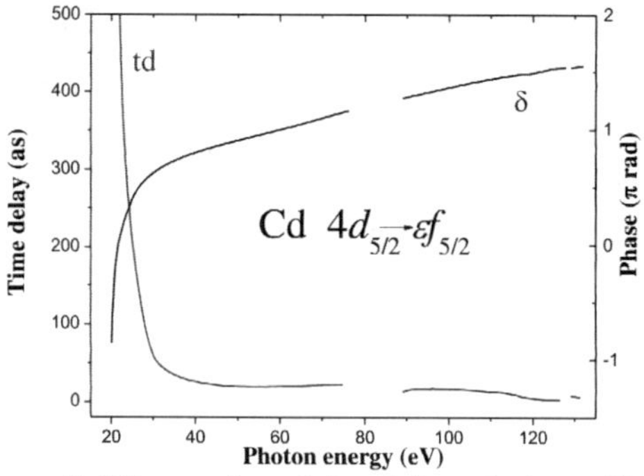

Figure 2. Wigner time delay (td) and phase (δ) for the $4d_{5/2}\rightarrow\varepsilon f_{5/2}$ photoionization channel in atomic Cd. The breaks in the curves represent regions of autoionizing resonances.

[1] E-mail: soumyajit@physics.iitm.ac.in
[2] E-mail: amankur@physics.iitm.ac.in
[3] E-mail: pcd@physics.iitm.ac.in
[4] E-mail: A.Kheifets@anu.edu.au
[5] E-mail: vkdolmatov@una.edu
[6] E-mail: smanson@gsu.edu

References

[1] W. R. Johnson *et al* 1982 *Phys. Rev. A* **25** 337
[2] W. R. Johnson and C.D. Lin 1979 *Phys. Rev. A* **20** 964
[3] M. Schultze *et al* 2010 *Science* **328** 1658
[4] L. E. Eisenbud 1948 *Ph. D. thesis*, Princeton Univ.
[5] E. P. Wigner 1955 *Phys. Rev.* **98** 145
[6] U. Fano and J. W. Cooper 1968 *Rev. Mod. Phys.* **40** 441
[7] A. Bohm 1986 *Quantum Mechanics* Springer: Berlin
[8] A. R. P. Rau and U. Fano 1968 *Phys. Rev.* **167** 7

Content from this work may be used under the terms of the Creative Commons Attribution 3.0 licence. Any further distribution of this work must maintain attribution to the author(s) and the title of the work, journal citation and DOI.
Published under licence by IOP Publishing Ltd

Journal of Physics: Conference Series **635** (2015) 092062 doi:10.1088/1742-6596/635/9/092062

Photodetachment of F^- by mid-infrared circularly polarized few-cycle pulses

S. M. K. Law* and S. F. C. Shearer*[1]

* Centre for Theoretical Atomic, Molecular and Optical Physics, Queen's University Belfast, Belfast BT7 1NN, U.K.

Synopsis Above threshold detachment (ATD) of the valence p-state electrons of F^- by strong circularly polarized few-cycle laser pulses is analysed within the framework of the adiabatic saddle point approach. We consider the emission spectra of the ejected photoelectrons and examine the qualitative effects arising from variation of the carrier-envelope phase (CEP), number of optical cycles and the orientation of the orbital angular momentum of the ground state.

The original Keldysh model [1] developed for infinitely long periodic laser pulses is extended to consider above threshold detachment (ATD) of negative ions with fully occupied outer p sub-shells by few-cycle laser pulses. In a previous study we have investigated the ATD process of F^- ions by linearly polarized pulses of femtosecond duration [2], and the aim of the present work is to adapt this study to consider the strong field photodetachment dynamics of F^- exposed to few-cycle circularly polarized laser pulses. The theory is applied to the low-energy part of the spectrum where the influence of the rescattering mechanism may be neglected.

This study focuses on numerical calculations for a $\sin^2$ N-cycle field with wavelength 1800 nm and peak intensity 1.3×10^{13} W/cm^2. The laser field is circularly polarized on the (x, y) plane with time-dependent vector potential $\mathbf{A}(t)$ given as

$$\mathbf{A}(t) = \frac{A_0}{\sqrt{2}} \sin^2\left(\frac{\omega t}{2N}\right) \begin{pmatrix} \cos(\omega t + \alpha) \\ \sin(\omega t + \alpha) \\ 0 \end{pmatrix}$$

where A_0 is its peak value, ω is the frequency and α is the carrier-envelope phase (CEP). The calculations involve summation over channels with different values of m = 0, ±1 simulated separately and with the statistical averaging of channels associated with the two final doublet atomic states $^2P_{1/2}$ and $^2P_{3/2}$ of F respectively.

A signature feature of the calculated electron momentum distribution probability maps is an angular anisotropy that is sharply reduced with an increase in the number of optical cycles within the pulse. A rich structure of well-pronounced interference patterns is observed in the photoelectron spectra for a high cycle number. These can be attributed to the prominence of $N + 1$ quantum wave packet interferences corresponding to the maximal electric field strengths. The profile of the electron spectra is found to be strongly influenced by the CEP indicating that it is a critical tool for controlling the preferential direction of electron emission in 3D momentum space. Additionally we report on the contribution of the individual magnetic state projections and show that the ratio of the $p_{\pm}$ orbitals is dependent on the rotational sense of the circular field.

Results for 2D photoelectron momentum and angular distributions parallel and perpendicular to the polarization plane will be presented at the conference. Finally our calculations are compared with other theoretical predictions [3, 4] and experimental data [5].

References

[1] L. V. Keldysh 1965 *Sov. Phys. JETP* **20** 1307

[2] S. F. C. Shearer and M. R. Monteith 2013 *Phys. Rev. A* **88** 033415

[3] I. Barth and O. Smirnova 2013 *Phys. Rev. A* **87** 013433

[4] S. Bivona, R. Burlon and C. Leone 2006 *Opt. Exp.* **14** 12576

[5] B. Bergues *et al* 2005 *Phys. Rev. A* **95** 263002

[1]E-mail: f.shearer@qub.ac.uk

Content from this work may be used under the terms of the Creative Commons Attribution 3.0 licence. Any further distribution of this work must maintain attribution to the author(s) and the title of the work, journal citation and DOI.
Published under licence by IOP Publishing Ltd

XXIX International Conference on Photonic, Electronic, and Atomic Collisions (ICPEAC2015) IOP Publishing
Journal of Physics: Conference Series **635** (2015) 092063 doi:10.1088/1742-6596/635/9/092063

Laser-assisted inelastic electron scattering by excited hydrogen atoms

Gabriela Buica[1]

Institute of Space Sciences, P.O. Box MG-23, Ro 77125, Bucharest-Măgurele, Romania

Synopsis We study the *inelastic* electron-H($2s$) scattering in a linearly polarized laser field in the domain of high scattering energies and moderate field intensities. We investigate the angular distributions and the resonance structure of the differential scattering cross sections.

The purpose of this work is to study the -*elast c* scattering of fast electrons by hydrogen atoms in the metastable $2s$ state in the presence of a linearly polarized laser field. For elastic scattering the initial and final states of the target are the same, while for inelastic scattering the final state differs from the initial one. We consider the excitation of metastable hydrogen to an arbitrary state accompanied by one-photon absorption. It is important to evaluate the contribution of the laser-assisted *elast c* electron-atom scattering to the total electron energy spectrum since in experimental studies it might be quite difficult to separate the signal of elastic and inelastic scattering channels [1].

Since the scattering process under investigation is a three-body problem, i.e., projectile, atomic target, and photon, the theoretical treatment becomes very complicated and we make several assumptions. We focus on moderate field intensities and fast projectile electrons in order to neglect the exchange scattering and the second Born approximation in the scattering potential. First, the interaction between the projectile electron and the laser field is described exactly by a Gordon-Volkov wave function. Second, the dressing of the hydrogen atom by the laser field is described within the first-order time-dependent perturbation theory in the field [2]. Finally, the interaction between the fast projectile electron and the hydrogen atom is treated in the first Born approximation [3]. Using the approach described in [4] we have obtained an *a alyt cal formula* for the differential cross section (DCS) in the laser-assisted inelastic e-H($2s$) scattering.

We analyze the angular distributions and the resonance structure of the DCS's for the atomic excitation of the $n = 4$ subshells. In Figure 1 we show the DCS's with respect of the photon energy for one-photon absorption corresponding to the $2s \to 4s$ (short dashed line), $2s \to 4p$ (dot-dashed line), $2s \to 4d$ (long dashed line), and $2s \to 4f$ (solid line) excitation processes of H($2s$), in the scattering geometry where the laser field is linearly polarized in the same direction along the momentum of the ingoing electron. The DCS's are normalized to the laser intensity and calculated at the incident projectile energy of 500 eV and the scattering angle $\theta = 5°$.

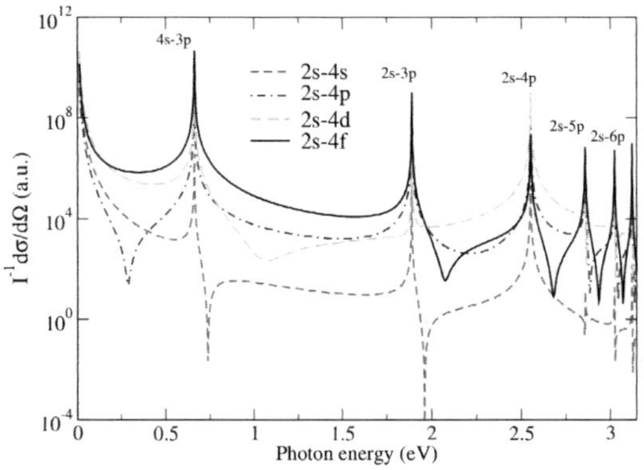

Figure 1. Differential cross sections for *elast c* laser assisted e(E_i)+H($2s$)→ e(E_f)+H($4l$) scattering process with the excitation of the $4l$ subshells as a function of the photon energy.

Our preliminary results indicate that the laser-dressing effects are significantly stronger than for the elastic $2s \to 2s$ collision [5] (static dipole polarizability α_{2s}=120 a.u), in particular at small scattering angles. For the studied range of photon energy our analysis shows important differences from the case of laser-assisted elastic scattering, that are mainly due to the intermediate resonances in the laser-atom interaction.

References

[1] C. J. Joachain *et al Atoms I te se aser F elds* (Cambridge 2012)

[2] V. Florescu *et al* 1986 *Phys. Rev. A* **34** 4641

[3] F. W. Byron Jr. *et al* 1984 *J. Phys.* **17** L295

[4] A. Cionga *et al* 1992 *Phys. Rev. A* **45** 5282

[5] A. Cionga *et al* 2001 *Phys. Rev. A* **64** 043401

[1]E-mail: buica@spacescience.ro

Content from this work may be used under the terms of the Creative Commons Attribution 3.0 licence. Any further distribution of this work must maintain attribution to the author(s) and the title of the work, journal citation and DOI.
Published under licence by IOP Publishing Ltd

Journal of Physics: Conference Series **635** (2015) 092065 doi:10.1088/1742-6596/635/9/092065

Microwave recombination and ionization

V. Carrat*, E. Magnuson*, A. Arakelyan*, and T. F. Gallagher*[1]

*Department of Physics, University of Virginia, Charlottesville, VA 22903, USA

Synopsis Laser excitation above the limit in the presence of a microwave field can lead to high lying bound states if it occurs at the correct phase of the microwave field, and phase dependent ionization is also observed. The excitation spectra of the high lying states in the presence of the field is a sequence of peaks straddling the limit, spaced by the microwave frequency.

When Rydberg atoms are excited in the presence of strong, 100 V/cm, 17, 38, and 79 GHz microwave field bound states within 50 GHz of the limit can be detected subsequent to exposure to the laser and microwave fields, even when the laser is tuned above the ionization limit. A bound state can result if the laser excitation above the limit occurs at a phase of the microwave field such that there is energy transfer from the electron to the microwave field as the electron leaves the ion. Using a laser which is amplitude modulated synchronously with a microwave field we have observed the expected phase dependent recombination by delaying the laser beam relative to the microwave field. At the opposite phase the field gives energy to the electron, and, if the laser is tuned below the limit we observe phase dependent ionization, as expected. The latter phase dependence is very similar to that observed in the ionization of He by intense IR pulses with synchronized attosecond pulse trains [1,2].

In addition, to the time domain measurements we have recorded frequency domain laser excitation spectra in the presence of the microwave field. When we detect the high lying states we observe a series of peaks, spaced by the microwave frequency, from as much as 300 GHz above the limit to 3000 GHz below the limit, with one peak just below the limit. The entire sequence exhibits the same ponderomotive shift as the limit and is shifted down in energy when the limit is depressed by a static field [3]. The stability of these atoms derives from the fact that in high lying states they rarely visit the ion core, where energy exchange with the microwave field can occur.

Both the recombination and the excitation spectra can be understood in terms of an extension of the Simpleman's model which takes into account the coulomb potential [4].

Finally, when Rydberg atoms initially in zero field are exposed to pulses of microwaves as strong as 100 V/cm we often, but not always, observe as much as 20% of the population left in high lying states which are easily ionized by a 40 mV/cm static field. This phenomenon was not previously observed due to the presence of stray electric fields, which only needed to be of order 10 mV/cm to destroy the high lying states. Population has also been left in Rydberg states in laser multiphoton ionization, for the same reason, the excited atoms do not return to the ion core during the laser pulse [5].

We believe the high lying states to be produced when the initial Rydberg state is brought into multiphoton resonance with the limit by the ponderomotive shift. This phenomenon was observed by Muller in simulations of multiphoton ionization of He [6]. He attributed several phenomena to it, including high harmonic generation, non sequential double ionization, and superponderomotive above threshold ionization.

References

[1] P. Johnsson *et al* 2007 *Phys. Rev. Lett.* **99** 233001
[2] P. Ranitovic et al 2010 *New J. Phys.* **12** 013008
[3] A. Arakelyan *et al* 2014 *Phys. Rev. A* **90** 013413
[4] E. S. Shuman *et al* 2008 *Phys. Rev. Lett.* **101** 263001
[5] R. R. Jones *et al* 1993 *Phys. Rev. A* **47** R49
[6] H. G. Muller 1999 *Phys. Rev. Lett.* **83** 3158

[1] E-mail: tfg@virginia.edu

Content from this work may be used under the terms of the Creative Commons Attribution 3.0 licence. Any further distribution of this work must maintain attribution to the author(s) and the title of the work, journal citation and DOI.
Published under licence by IOP Publishing Ltd

XXIX International Conference on Photonic, Electronic, and Atomic Collisions (ICPEAC2015) IOP Publishing
Journal of Physics: Conference Series **635** (2015) 092066 doi:10.1088/1742-6596/635/9/092066

The interaction of ultrashort laser pulses and exotic atoms

J.E. Calvert*, H. Xu*, A.J. Palmer*, X. Wang*, D. Kielpinksi*, I. Litvinyuk*, R.T. Sang*[1]

* Centre of Quantum Dynamics, Griffith University, Nathan, Queensland, Australia

Synopsis In this work we present results for the ionization of metastable neon with ultrashort laser pulses. In addition to improving on the accuracy of previous measurements of ion yield, we show comparisons to advanced TDSE based theory. This work presents preliminary experimental evidence of the effects of preparing the target atoms into particular angular momentum projection (m_j) states on the total ionization yield.

The interaction of ultrafast (<6fs) light pulses with matter is a topical field. This stems from the high peak electric field amplitude of such pulses, which is strong enough to cause ionization by tunneling or over-the-barrier ionization. This is the basis of several non-linear effects that are currently being studied [1]

In order to understand these processes it is necessary to understand the underlying principles, in this paper the ionization channels. Metastable ($^3P_2$) neon provides an excellent target for examining the effects of electron configuration the ionization yield, as it allows a clean comparison to ground state neon results. $^3P_2$ neon is also an interesting choice of atom as it has a low ionization energy of 5.1eV, allowing easy access to the above-the-barrier ionization regime.

The ionization yield is predicted by several models. The most basic involves treating the electric field as static and determining the DC tunneling rate - ADK theory [2]. This work also considers adjustments to ADK theory. More advanced treatments are provided by Klaus [3] and Tong [4], who apply approximations to solve the TDSE for $^3P_2$ neon in the laser field.

Experimentally, metastable neon is generated by supersonically expanding neon gas through a 250μm diameter nozzle, while passing a DC discharge through the nozzle. Electron collision promotes 0.01% of the neon atoms to the $^3P_2$ state.

Ultrafast laser light is produced by a commercially available laser system (Femtolasers Femtopower CE Phase Pro), which provides sub 6fs laser pulses centered around 760nm. Optics following the laser system allow the intensity of the laser light in the detection chamber to be varied between $1.5x10^{15}W/cm^2$ and $1x10^{13}W/cm^2$.

E-mail: R.Sang@griffith.edu.au

The detection system is a commercial cold target recoil ion momentum spectroscope (COLTRIMS) supplied by Roentdek. The optical system focuses the laser beam onto the atomic beam, causing ionization. Ions that are produced though the interaction are directed to the ion detector, which utilizes time of flight and position sensitive detection to record the ion yield and momentum of individual ions.

We have taken measurements comparing the ion yield of $^3P_2$ neon to the ground state of neon and compared them to the theoretical ion yield predictions. There is good agreement between these results and theory for low laser intensities. Discrepancies at higher powers are currently under investigation.

We also present preliminary results for the examination of the effects of angular momentum (m_j) dependence though the optical pumping of the target atoms on ion yield. Examination of the base ADK equation [2] indicates an ion yield dependence on the m_j state of the target atom. Relative changes in the atomic m_j states are achieved though optical pumping using the optical transition between the $^3P_2$ and $^3D_3$ states of neon. By pumping this transition with light of a varying polarization ellipticity, the population of the m_j states can be changed.

In conclusion, we will present data on the ionization yield of $^3P_2$ neon as a function of both laser intensity and investigate target atomic states to investigate the effect of the magnetic sublevel dependence on ionization.

References

[1] T. Brabec and F. Krausz 2000 *Rev. Mod. Phys.* **72** 545
[2] F.A. Ilkov *et al* 1993 *Laser Physics* **3** 298
[3] A.N. Grum-Grzhimailo *et al* 2010 *Phys. Rev. A.* **81** 043408
[4] X.M. Tong *et al* 2006 *Phys. Rev. A.* **74** 031405(R)

Content from this work may be used under the terms of the Creative Commons Attribution 3.0 licence. Any further distribution of this work must maintain attribution to the author(s) and the title of the work, journal citation and DOI.
Published under licence by IOP Publishing Ltd

Measurement of the spin polarization of electrons in a strong laser field

Alexander Hartung [1], Alina Laucke, Maksim Kunitski and Reinhard Dörner

Institut für Kernphysik, Goethe-Universität Frankfurt am Main, Max-von-Laue-Straße 1, D-60438 Frankfurt am Main

Synopsis We present first experimental data on the theoretically predicted effect [1] that photoelectrons created by ionization of noble gas atoms in an intense laser field are spin polarized.

As a fundamental property of the electron the spin plays an important role in the buildup of atoms, matter and its macroscopic properties, e.g. magnetism. Hence studying effects regarding the spin is highly interesting. Even 90 years after the famous Stern-Gerlach experiment was carried out new questions concerning the spin arise consistently. In 2013 I. Barth and O. Smirnova predicted theoretically that electrons, which were created by ionization of noble gas atoms by an intense infrared laser, are spin polarized [1]. They propose two effects to be responsible for the spin polarization. First, the probability for tunnel ionization by circular light depends on the sign of the magnetic quantum number m_l of the emitted electron [2]. This effect was partly confirmed experimentally by Herath et al. [3]. Second, the tunnel ionization probability is slightly different for different electron spin states. Due to spin-orbit-interaction electrons with spin parallel to m are less bound than those with spin antiparallel. The resulting spin polarization is presented in Fig 1.

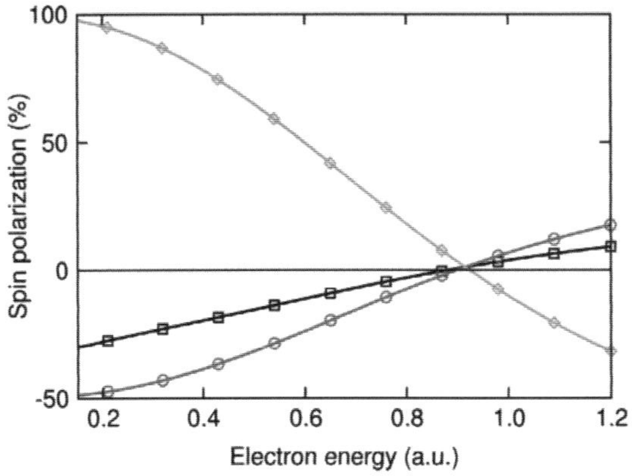

Figure 1. Spin polarization of photoelectrons emitted from krypton dependent on their kinetic energy. A circularly polarized 800nm laser with I = 1.8 · 10^{14} W/cm² was used for the simulation. Green curve with diamonds resolved on the $^2P_{1/2}$ state of the core; Red curve with circles resolved on $^2P_{3/2}$; Black curve with squares integrated over core states. From [1]

Here we present first experimental data on the effect, shown in Fig. 2. We ionized Xenon with an 800nm, 40fs, circularly polarized laser at 1.7 and 3.3 · 10^{14} W/cm². A Mott detector was used for the measurement of the spins. The results strongly indicate the existence of the theoretically predicted spin polarization. The measured strength of the effect as well as the dependences on the electron energy and laser intensity are in good agreement with theory.

Momentarily, further measurements with different gases and laser intensities are carried out to gain more precise comparison with theory.

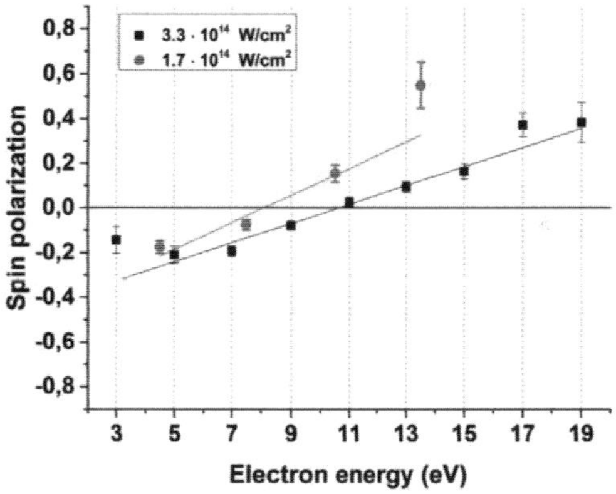

Figure 2. Measured spin polarization of photoelectrons emitted from Xenon at two intensities.

References

[1] I. Barth & O. Smirnova 2013 Phys. Rev. A **88**, 013401
[2] I. Barth & O. Smirnova 2011 Phys. Rev. A **84**, 063415
[3] T. Herath et al 2012 Phys. Rev. Lett. **109**, 043004

[1] E-mail: hartung@atom.uni-frankfurt.de

Content from this work may be used under the terms of the Creative Commons Attribution 3.0 licence. Any further distribution of this work must maintain attribution to the author(s) and the title of the work, journal citation and DOI.
Published under licence by IOP Publishing Ltd

XXIX International Conference on Photonic, Electronic, and Atomic Collisions (ICPEAC2015) IOP Publishing
Journal of Physics: Conference Series **635** (2015) 092070 doi:10.1088/1742-6596/635/9/092070

Design of mechanically compensated Penning trap for the study of ions in extreme laser field

Sugam Kumar* [1], **M. Vogel** [†], **W. Quint** [$], **S. Ringleb** [§], **Th. Stöhlker** [§] **and C.P. Safvan***

[*] Inter-University Accelerator Centre, Aruna Asaf Ali Marg, New Delhi, India.
[†] GSI Helmholtzzentrum für Schwerionenforschung, 64291 Darmstadt, Germany.
[$] Physikalisches Institut, Ruprecht Karls-Universität Heidelberg, 69120 Heidelberg, Germany
[§] Helmholtz-Institut Jena, 07743 Jena, Germany

Synopsis This paper describes the design of a mechanically compensated Penning trap as a versatile tool to confine atomic ions in static electric and magnetic fields for the study of non-linear processes such as multi-photon ionization from well-defined initial charge states interacting with a large range of photon energies by non-destructive observation.

We have designed and constructed a mechanically compensated Penning trap[1] as a tool to confine atomic ions in static electric and magnetic fields for the study of non-linear processes such as multi-photon ionization (MPI) from well-defined initial charged states and with a large range of photon energies by non-destructive observation. During the interaction of ions with extreme laser fields it is also possible to realize above-threshold ionization (ATI), tunneling ionization (TI) and above-the-barrier ionization (ABI) processes.

The experimental setup consists of a cylindrical Penning trap with 20mm inner diameter located at the magnetic field centre of a dry superconducting magnet which can produce homogeneous magnetic fields up to 6 T. In the region of ±10mm from the centre of magnet bore the homogeneity of magnetic flux density is better than $\pm 10^{-4}$ T [2]. A pulse-tube cooler is used to cool the trap electrodes, cryo-electronics and radiation shields. The trap will be cooled down to 4.2K and as a result the background pressure in the trap is expected to be 10^{-12} Torr.

We have chosen a 3-pole open-endcap (figure 1) cylindrical Penning trap with geometric compensation which consists of ring electrode and a pair of endcap electrodes with a separation distance of 17.4mm with an additional set of capture electrodes (length 20mm) on either side of endcap electrodes for dynamic capture of ions into the trap. These electrodes have a conical opening to accept laser beams up to an aperture of f/5. This geometry leads to a ring electrode with a large axial extension, such that ions in the trap centre are efficiently coupled to the ring. It is radially split 8-fold, for electronic excitation, rotating wall drive and detection of ion motions through Fourier-transform ion cyclotron resonance (FT-ICR). Two high quality factor resonators with resonant frequencies of 229kHz and 702kHz are employed to measure the axial frequencies whereas a helical resonator which amplifies multiples of fundamental frequencies is the used to measure cyclotron frequencies.

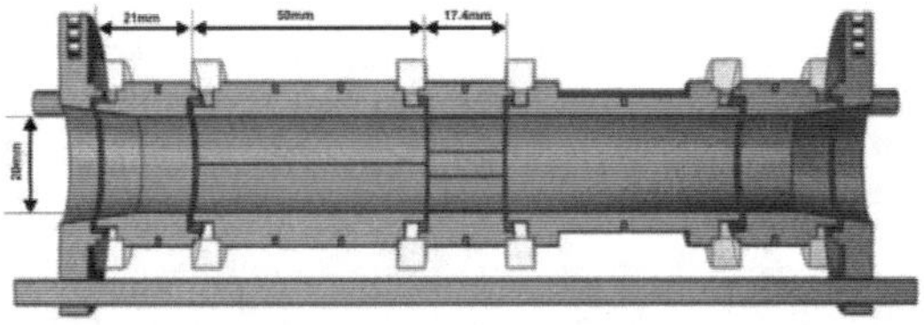

Figure 1. Cross section geometry of 3-pole mechanically compensated Penning trap with pair of capture electrodes with conical opening.

The FLASH Laser facility at DESY, Germany may be used as a source of high-intensity photons. It has the capability to produce photons with wavelengths down to 4.5 nm (275 eV) in pulses of a few femto-seconds duration resulting in a peak power of around 5GW with repetition rate of 10Hz. This allows photoionization studies well inside the non-linear regime. The plan also includes the use of the PHELIX laser with wavelength 1053 nm at GSI and POLARIS (1030 nm) and JETI (800 nm) at Jena, Germany with intensities up to order 10^{21}W/cm^2.

References

[1] G. Gabrielse *et al, Int. J. Mass Spectr. Ion Proc. 88 (1989) 319.*
[2] M.Vogel *et al NIMB, 285 (21020 65-71*

[1] E-mail: sugamkumar@gmail.com

Content from this work may be used under the terms of the Creative Commons Attribution 3.0 licence. Any further distribution of this work must maintain attribution to the author(s) and the title of the work, journal citation and DOI.
Published under licence by IOP Publishing Ltd

XXIX International Conference on Photonic, Electronic, and Atomic Collisions (ICPEAC2015) IOP Publishing
Journal of Physics: Conference Series **635** (2015) 092071 doi:10.1088/1742-6596/635/9/092071

Polarization transfer of photoionization in highly charged ions

K Ma, C Z Dong [1] and L Y Xie

Key Laboratory of Atomic and Molecular Physics & Functional Materials of Gansu Province, College of Physics and Electronic Engineering, Northwest Normal University, Lanzhou 730070, China

Synopsis The influence of polarization of incident photon and non-dipole terms which arises from the multipole expansion of electron-photon interaction on properties of subsequent x-ray radiation have been studied with the MCDF method and the density matrix theory. The results show that there is a linear relationship between the degree of linear polarization of the radiative decay and ones of the incident light, which can be used for diagnosing the polarization of light source. In addition, with increasing the photon energies, the non-dipole contribution to the degree of linear polarization of radiative decay following the inner-shell photoionization will also increase.

The information about the polarization transfer in the inner-shell photoionization (PI) is important for diagnosing laboratory and astrophysical plasmas as well as for understanding the PI experiment with polarized light source. In the polarization transfer of photoionization, the polarization properties of radiative decay following inner-shell PI are directly related to the magnetic cross section of PI. So far, theoretical studies of the PI mainly focus on the total cross sections, and there is only a few calculations for magnetic cross sections[1,2].

In this contribution, we apply the multi-configuration Dirac-Fock method[3] and the density matrix theory[4] to studying the magnetic sublevel cross section and related polarization of decay radiation for Mg-like ions.

In the practical calculations, we make use of the new component (PHOTO) of RATIP package[5] and extend it to calculate the present polarization transfer. The initial- and final-state wave functions of the target ions are generated with use of the atomic structure package GRASP [6] based on the MCDF method.

Figure 1 shows the relationship between the degree of linear polarization (DLP) of incident light and the DLP of transition from $2p^{-1}_{3/2}3s^2$ to $2p^6 3s$ (*Ll*) line following the inner-shell PI of Mg-like Fe^{14+}, Cd^{36+}, W^{62+} and U^{80+} ions. As seen from this figure, the DLP of incident light intensifies the DLP of the subsequent x-ray emission, it is so-called polarization transfer; On the contrary, the non-dipole terms weaken the DLP of the subsequent *Ll* x-ray, it is so-called depolarization effects. And then, the linear relationship between the DLP of the incident light and *Ll* x-ray emission following the inner-shell PI have been found in the present work, and this could be useful for diagnosing the DLP of a light source.

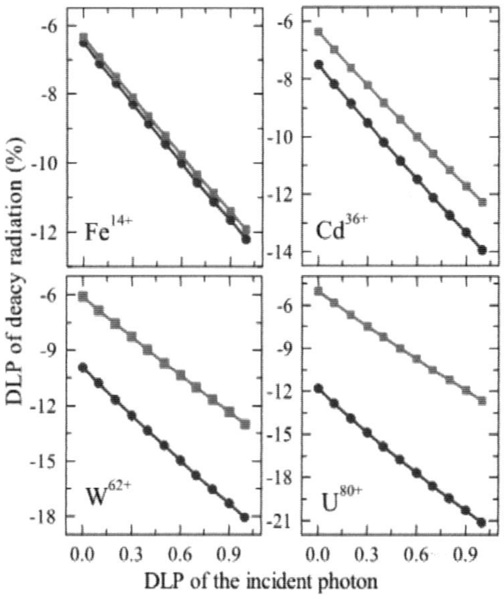

Figure 1. DLP of subsequent *Ll* line following PI as a function of the DLP of the incident photon. Calculations have been performed with (square) and without (circle) non-dipole contribution.

This work was supported by the National Natural Science Foundation of China (Grant nos.11274254, U1332206 and U1331122).

References

[1] K Ma et al 2014 *Chin. Phys. Lett.* **31** 053201
[2] K Ma et al 2014 *Chin. Phys. Lett.* **31** 103201
[3] I P Grant 2007 Relativistic Quantum Theory of Atoms and Molecules(Berlin: Springer)
[4] K Blum 1981 Density Matrix Theory and Applications (New York:Plenum Press)
[5] S Fritzsche 2012 *Comput. Phys. Commun.* **183** 1525
[6] F A Parpia et al 1996 *Comput. Phys. Commun.* **94** 2

[1] E-mail: dongcz@nwnu.edu.cn

Content from this work may be used under the terms of the Creative Commons Attribution 3.0 licence. Any further distribution of this work must maintain attribution to the author(s) and the title of the work, journal citation and DOI.
Published under licence by IOP Publishing Ltd

Journal of Physics: Conference Series **635** (2015) 092072 doi:10.1088/1742-6596/635/9/092072

Elastic scattering of linearly polarized hard x-rays

K.-H. Blumenhagen[a,b,c 1], **T. Gaßner**[a,b,c], **A. Gumberidze**[b,c], **R. Märtin**[a,b], **N. Schell**[e], **U. Spillmann**[b], **S. Trotsenko**[a,b], **G. Weber**[a,b] **and Th. Stöhlker**[a,b,c]

[a]Helmholtz-Institut Jena, 07743 Jena, Germany [b]GSI Helmholtzzentrum für Schwerionenforschung GmbH, 64291 Darmstadt, Germany [c]IOQ, Friedrich-Schiller-Universität Jena, 07743 Jena, Germany [d]ExtreMe Matter Institute EMMI, 64291 Darmstadt, Germany [e]Helmholtz-Zentrum Geesthacht GmbH, 21502 Geesthacht, Germany

Synopsis We investigated the elastic scattering of linearly polarized 175-keV photons by a neutral gold target. The angular-differential scattering cross section and the linear polarization of the scattered radiation were measured.

Quantum Electrodynamics (QED) effects in fundamental light-matter interactions are particularly pronounced in heavy (high-Z) atomic systems. Photons emitted by such systems provide experimental access to the detailed understanding of the underlying interaction. In the current contribution, we concentrate on elastic photon scattering (Rayleigh scattering) in the hard x-ray range namely in combination with a high-Z target. Although Rayleigh scattering (sometimes called virtual photoionization) is a very well-established process, experimental data for the hard x-ray range (beyond 100 keV) are rather rare, in particular with respect to the polarization properties of the scattered radiation [1]. Also with respect to theory this process is still a quite challenging task since it must be treated in second order perturbation theory [2]. Recently, photon polarization effects in elastic scattering were studied in a theoretical investigation and the assumed models can now be tested with our experimental data. Here, we report on an experiment where we investigated the elastic scattering of linearly polarized 175-keV x-rays by a thin gold foil. We note, for the first time, not only the angular distribution, but also the linear polarization of the scattered photons was measured for linearly polarized incident photons. The incident photon beam in our experiment was provided by the High Energy Material Science Beamline P07 [3] at the synchrotron PETRA III at DESY, Hamburg. Using a dedicated target chamber, the beam was scattered off a thin gold foil (1.036 μm) and the scattered photons were detected by standard germanium detectors as well as by a double-sided, position sensitive Si(Li) detector. The latter allows to study the linear polarization properties of the radiation by means of Compton polarimetry [4]. An overview of recent experiments where this technique was applied is given in [5]. Figure 1 shows two-dimensional position data from the polarimeter, see caption for details.

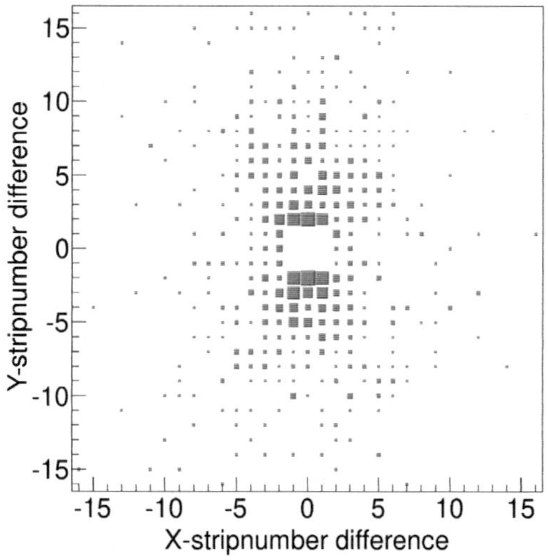

Figure 1. Preliminary data of the 2D position distribution of photons that were Compton-scattered inside the polarimeter crystal. The anisotropy indicates a strong linear polarization of the incident radiation, which consisted of photons that were elastically scattered in the gold target at a scattering angle of 120°.

References

[1] D.A. Bradley, O.D. Goncalves, P.P. Kane 1999 *Radiat. Phys. Chem.* **56** 125

[2] A. Surzhykov *et al* 2013 *Phys. Rev.* A **88** 062515

[3] N. Schell *et al* 2014 *Mater. Sci. Forum* **772** 57

[4] F. Lei *et al* 1997, *Space. Sci. Rev.* **82** 309

[5] G. Weber *et al* 2015 *J. Phys.: Conf. Ser.* **583** 012041

[1]E-mail: k.-h.blumenhagen@gsi.de

Content from this work may be used under the terms of the Creative Commons Attribution 3.0 licence. Any further distribution of this work must maintain attribution to the author(s) and the title of the work, journal citation and DOI.
Published under licence by IOP Publishing Ltd

Journal of Physics: Conference Series **635** (2015) 092073 doi:10.1088/1742-6596/635/9/092073

Transverse electron momentum distribution in tunneling and over the barrier ionization by strong-field laser pulses

J.E. Calvert* [1], S. Goodall*, X. Wang*, Han Xu*, A.J. Palmer*, I.A. Ivanov†#, A.S. Kheifets#, D. Kielpinski*, I.V. Litvinyuk* and R.T. Sang* [1]

* School of Natural Sciences and Centre for Quantum Dynamics, Griffith University, Nathan QLD 4111, Australia
† Centre for Relativistic Laser Science, Institute for Basic Science, Gwangju 500-712, Republic of Korea
# Research School of Physics and Engineering, The Australian National University, Canberra, ACT 0200, Australia

Synopsis We present the results of an investigation of the measurement of the transverse electron momentum (TEMD) distribution of atoms ionized though the interaction of strong-field light pulses of varying ellipticity. We show though experiment and theory that the TEMD for the tunneling regime and the over-the-barrier regime (OB) evolves in a qualitatively different way when polarization state of the ionizing laser changes ellipticity.

The interaction of strong field laser pulses with matter is highly non-linear with a large number of interesting interactions that can occur such as high harmonic generation, above threshold ionisation and double ionisation processes [1]. These effects are all the result of the ionisation of constituent components of the target matter. In the case of atomic gases, the strong laser field suppresses the atomic potential of the valence electrons, allowing the electron to tunnel into the continuum state, or in extreme cases of suppression, the electron is promoted 'over-the-barrier' (OB) into the continuum state. In order to understand these processes it is necessary to understand the underlying principles and in this paper we measure the variation of the TEMD resulting in a strong-field interaction as a function of the ellipticity of the driving laser light pulses.

The Keldeysh parameter is often used to determine the treatment of intense light-matter interactions as in the perturbative or the highly non-linear tunneling regime [1]. The transition from tunneling regime to the over-the-barrier (OB) regime is not obvious to determine. In this paper we will present both theoretical and experimental results that demonstrate that the measurement of the distribution of TEMD can be used to determine whether a strong-field light interaction with matter is in the OB or tunneling regime [2].

We will present the TEMD for the strong-field ionization of argon (Ar) using light pulses of varying ellipticity and compare this to TEMD obtained though the ionization of neon in the $^3P_2$ metastable state (Ne*). The ionization potentials (IP) of Ar and Ne* are 15.76ev and 5.07eV respectively. This difference in IP for the two target species facilitates the use of the same laser system to access the OB and tunneling regimes. The system is modeled using the time dependent Schroedinger equation [2].

We will demonstrate a good agreement between our theory and experiment with striking differences in the TEMD distributions for the tunneling and OB regimes. We will also show that the origin of the TEMD distributions is related to the change of transverse angular momentum exchanged in the interaction [3].

References

[1] T. Brabec and F. Krausz, Rev. Mod. Phys., **72**, 545 (2000)
[2] I.A. Ivanov, Phys. Rev. A 90, 013418 (2014).
[3] I.A. Ivanov, A.S. Kheifets, J.E. Calvert, S. Goodall, X. Wang, Han Xu, A.J. Palmer, D. Kielpinski, I.V. Litvinyuk and R.T. Sang in preparation (2015).

[1] E-mail R.Sang@griffith.edu.au

Content from this work may be used under the terms of the Creative Commons Attribution 3.0 licence. Any further distribution of this work must maintain attribution to the author(s) and the title of the work, journal citation and DOI.
Published under licence by IOP Publishing Ltd

Journal of Physics: Conference Series **635** (2015) 092074 doi:10.1088/1742-6596/635/9/092074

K-Shell Photoionization of the Oxygen Isonuclear Sequence

M F Gharaibeh[a,b 1], **J M Bizau**[c], **D Cubaynes**[c], **S Guilbaud**[c], **M M Al Shorman**[c], **I Q Ababneh**[b], **I Sakho**[d], **C Blancard**[e], **and B M McLaughlin**[f,g 2]

[a]Department of Mathematics, Statistics and Physics, P. O. Box 2713, Qatar University, Doha, Qatar
[b]Department of Physics, Jordan University of Sciences and Technology, Irbid 22110, Jordan
[c]Institut des Sciences Moléculaires dOrsay (ISMO), CNRS UMR 8214, Université Paris-Sud, Bât. 350, F-91405 Orsay cedex, France
[d]Department of Physics, UFR of Sciences and Technologies, University Assane Seck of Ziguinchor, Ziguinchor, Senegal
[e]CEA-DAM-DIF, Bruyéres-le-Châtel, F-91297 Arpajon Cedex, France
[f]Centre for Theoretical Atomic and Molecular Physics (CTAMOP), School of Mathematics and Physics, Queen's University of Belfast, Belfast BT7 1NN, UK
[g]Institute for Theoretical Atomic and Molecular Physics (ITAMP), Harvard Smithsonian Center for Astrophysics, MS-14, Cambridge, MA 02138, USA

Synopsis K-shell photoionization cross-section measurements for the oxygen isonuclear sequence from N-like to Li-like made at the SOLEIL Light Source are compared with SUNC, MCDF and R-matrix calculations.

Single and multiply ionization stages of C, N, O, Ne and Fe have been observed in the ionized outflow in the planetary nebulae NGC 4051, measured with the satellite XMM-Newton in the soft-x-ray region, the Seyfert galaxy NGC 3783, including UV imaging, x-ray and UV light curves, the 0.2 -10 keV x-ray continuum, the iron K-emission line, and high-resolution spectroscopy in the modelling of the soft x-ray warm absorber [1, 2].

Detailed measurements of the absolute K-shell single photoionization cross sections for N-like to Li-like oxygen ions obtained (for the first time) at the SOLEIL synchrotron radiation facility are compared with theoretical predictions made using the SCUNC, MCDF and RMPS methods. The theoretical work enables the identification and characterization of the very strong 1s $\rightarrow$ 2p transitions found in the spectra of these ions and other resonance features observed below the K-edge region in most of these ions.

The first measurements in the isonuclear sequence for O^{3+} were published recently [3], the experimental and theoretical data for the O^{+} and O^{2+} ions are in preparation. The O^{4+} and O^{5+} will be explored in May 2015 and all this data will be presented at the conference.

Fig. 1 illustrates the cross section results in the region of the 1s $\rightarrow$ 2p transition for B-like oxygen. Further details and a comprehensive set of results will be presented at the meeting.

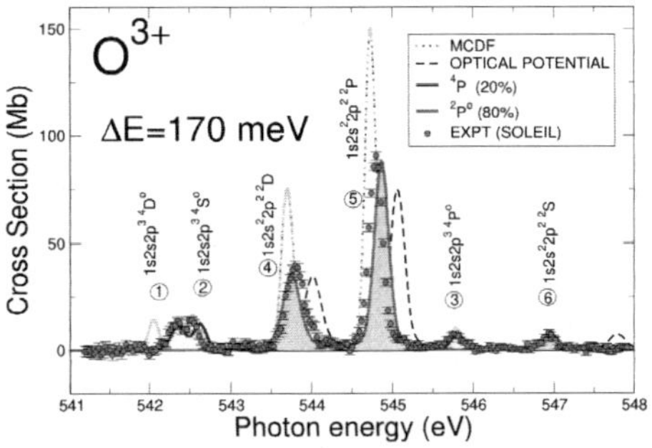

Figure 1. (Colour online) Photoionization cross sections for O^{3+} ions measured with a 170 meV band pass in the region of 1s $\rightarrow$ 2p photo-excitations. Solid points (magenta), experimental cross sections. The R-matrix (RMPS, solid red line, and blue line) results. Dotted line with solid green circles the MCDF calculations. The optical potential R-matrix results (dashed black) are from Garcia and co-workers [4]. Theoretical work was convoluted with a Gaussian profile of 170 meV FWHM and an approriate weighting of the ground and metastable states to simulate the measurements [3].

References

[1] P. M. Ogle *et al.*, Astrophys. J., 606, 151 (2004).
[2] A. J. Blustin *et al.*, Astron. Astrophys., 391, 453 (2002).
[3] B. M. McLaughlin *et al.*, J. Phys. B: At. Mol. & Opt. Phys, 47, 115201 (2014).
[4] J. Garcia *et al.*, Astrophys. J. Suppl. Ser., 158, 68 (2005).

[1]E-mail: mgharaibeh@qu.edu.qa
[2]E-mail: b.mclaughlin@qub.ac.uk

Content from this work may be used under the terms of the Creative Commons Attribution 3.0 licence. Any further distribution of this work must maintain attribution to the author(s) and the title of the work, journal citation and DOI.
Published under licence by IOP Publishing Ltd

Journal of Physics: Conference Series **635** (2015) 092075 doi:10.1088/1742-6596/635/9/092075

Quantum correlations in the radiative recombination into excited state of highly charged heavy ions

A. V. Maiorova[*,†,1], **S. Tashenov**[◇], **A. Surzhykov**[‡], **V. M. Shabaev**[†], **T. Stöhlker**[‡,⋆]

[*]Center for Advanced Studies, St.Petersburg State Polytechnical University, Polytekhnicheskaya 29, St.Petersburg 195251, Russia

[†] Department of Physics, St.Petersburg State University, Ulianovskaya 1, Petrodvorets, 198504 St.Petersburg, Russia

[◇] Physikalisches Institut, Universität Heidelberg, D-69120 Heidelberg, Germany

[‡] Helmholtz-Institut Jena, D-07743 Jena, Germany

[⋆] GSI Helmholtzzentrum für Schwerionenforschung GmbH, D-64291 Darmstadt , Germany

Synopsis We consider the radiative recombination of a free electron into the excited $2p_{3/2}$ state of heavy H-like ion with the subsequent decay into the ground $1s_{1/2}$ state. Quantum correlations between two photons emitted in the process are studied.

Radiative recombination (RR) is one of the basic processes that occurs in laboratory plasma, in collisions of heavy ions with electrons at ion storage rings and EBIT. During the last decades RR of highly charged heavy ions remains the subject of intense theoretical and experimental research (see Ref. [1] and references therein). Radiative recombination is a time-reversed photoionization, hence it allows to study this fundamental electron-photon interaction in large-energy and strong-field regime, which is not accessible in the direct channel. Moreover, the RR is very sensitive to the spin, relativistic and ED effects in the structure and dynamics of heavy atomic systems.

In the present work we investigate the radiative recombination of the relativistic electron with bare heavy ion into the $2p_{3/2}$ excited state of H-like ion and its subsequent Lyman-α_1 ($2p_{3/2} \to 1s_{1/2}$) decay. We consider the scenario where RR photon and subsequently emitted characteristic photon are detected in coincidence. In this case the axial symmetry of the RR-populated $2p_{3/2}$ state is removed and the magnetic sublevels of the highly charged ion are populated coherently [2]. We have employed the density matrix approach and relativistic Dirac's theory for description of such coherence between the magnetic substates [3].

We also apply our results to test Clauser-Horne-Shimony-Holt's (CHSH) inequality

$$|\Delta(\mathbf{a}, \mathbf{a}', \mathbf{b}, \mathbf{b}')| \leq 2, \quad (1)$$

which is one of a family of Bell inequalities. Here

$$\Delta(\mathbf{a}, \mathbf{a}', \mathbf{b}, \mathbf{b}') = E(\mathbf{a}, \mathbf{b}) - E(\mathbf{a}, \mathbf{b}') + E(\mathbf{a}', \mathbf{b}) + E(\mathbf{a}', \mathbf{b}'), \quad (2)$$

where $\mathbf{a}$, $\mathbf{a}'$ are the detector settings for the first photon and $\mathbf{b}$, $\mathbf{b}'$ for the second one, and $E(\mathbf{a}, \mathbf{b})$ is the correlation function for the photon pair.

[1]E-mail: maiorova@pcqnt1.phys.spbu.ru

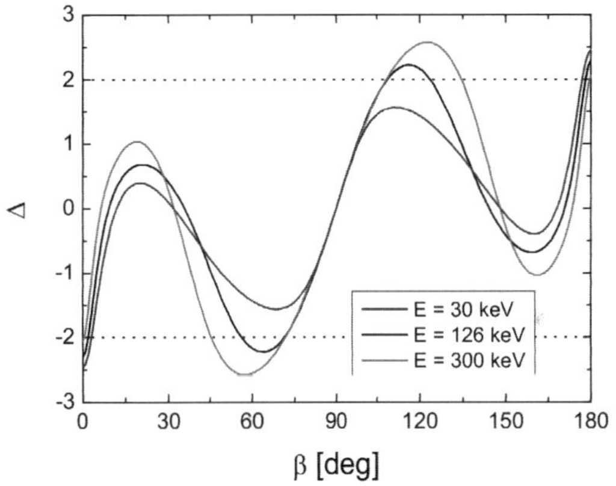

Figure 1. Test of CHSH inequality (2). β is angle between $\mathbf{a}$ and $\mathbf{b}$, $\mathbf{b}$ and $\mathbf{a}'$, and $\mathbf{a}'$ and $\mathbf{b}'$ [4].

This work was supported by SPbSU (Grants No. 11.38.269.2014 and No. 11.38.237.2015), by the Dynasty Foundation and by the grant of the President of the Russian Federation (Grant No. MK-1676.2014.2).

References

[1] J. Eichler and Th. Stöhlkler, Phys. Rep. **439**, 1 (2007).

[2] N. Andersen, J. Gallagher, and I.V. Hertel, Phys. Rep. **165**, 1 (1988).

[3] S. Tashenov *et al*, Phys. Rev. Lett. **113**, 113001 (2014).

[4] K. Blum, *De s ty Matr x Theory a d A l cat o s* (Springer, New York, 2012).

Content from this work may be used under the terms of the Creative Commons Attribution 3.0 licence. Any further distribution of this work must maintain attribution to the author(s) and the title of the work, journal citation and DOI.
Published under licence by IOP Publishing Ltd

Journal of Physics: Conference Series **635** (2015) 092076 doi:10.1088/1742-6596/635/9/092076

Precision Relativistic eigenchannel R-matrix calculation Of Ne^+ discrete energy region scattering matrix

Rui Jin*‡, Xiao-Ying Han†, Xiang Gao‡ 1, De-Ling Zeng+, Jia-Ming Li*+‡

* Department of Physics and Center for Atomic and Molecular Nanosciences, Tsinghua University, Beijing 100084, China

‡Beijing Computational Science Research Center, Beijing 100084, China

† Institute of Applied Physics and Computational Mathematics, Beijing 100094, China

+ Key Laboratory for Laser Plasmas (Ministry of Education) and Department of Physics and Astronomy, Shanghai Jiaotong University, Shanghai 200240, China

‡Collaborative Innovation Center of Quantum Matter, Beijing 100084, China

Synopsis With our relativistic eigenchannel R-matrix method R-R-Eigen, the physics parameters of Ne^+ in multi channel quantum defect theory (MQDT) are calculated directly, such as eigenchannel quantum defects, transformation matrix etc al. Various physical quantities can then be derived from a straightforward application of the MQDT procedure. Using our proposed projected high dimensional quantum-defect graph (symmetrized) - a generalization of Lu-Fano plots, we can readily determine the accuracies of the calculated MQDT physical parameters (ie. scattering matrix) systematically against the experimental energy levels. Based on the analytical continuation property of short-range scattering matrix below and above threshold, the origin of each resonant peak observed in the experiment can be designated. there is possibility to adjust calculated total photoionization spectra to fit the experiment.

Ne is the 10th abundant element in the universe, its energy structure and dynamics are important for astronomy and astrophysics. The photoionization of Ne^+ is carried out on Advanced Light Source with photon energy ranging from 40 to 71eV. The conventional R-Matrix calculation of this photoionization process is also reported [1]. However there exist discrepancies between the calculation and experiment, showing calculation is not convergent.

The weakness of conventional R-Matrix method is that when the basis set and target are not accurately convergent enough, we cannot tell which correlation is not well considered. Furthermore it's difficult to determine the calculation precisions.

Using our modified relativistic R-matrix code R-R-Eigen [2], we can directly calculate the short-range scattering matrices with good analytical properties corresponding to the physical parameters associated with the multichannel quantum defect theory (MQDT) for both the discrete and continuous energy regions of interest. Various physical quantities can then be derived from a straightforward application of the MQDT procedure. Through analytical continuation properties of short-range scattering matrices, the precision spectroscopic measurements of atomic energy levels can then be served as stringent tests of the accuracy of the short-range scattering matrices, i.e., related electron-ion collision data. We demonstrated that the precisions of scattering calculations can be determined readily in a systematical way by the Lu-Fano plot for Kr atom with only two ionization thresholds, i.e., Kr^+ ($2p^5\ ^2P_{3/2,1/2}$) [2]. However there are many cases of atomic spectra with multiple ionization thresholds. For example, the discrete energy levels of Ne^+ ions are related with five ionization thresholds, which are Ne^{2+} ($2p^4\ ^3P_{2,1,0}$, $^1D_2$ and $^1S_0$). Here we propose a projected high dimensional quantum-defect graph (symmetrized) as a generalization of Lu-Fano plots. In this way, we can readily determine the related scattering calculation precisions by using the spectroscopic data for general atoms with multiple ionization thresholds. We can also elucidate the intimate relations between the discrete energy levels and adjacent resonant autoionization spectra. For the case of Ne^+ photoionization, by using the precision spectroscopic data of the bound energy levels for benchmarks, the calculated total photoionization spectra are correctly designated and there is possibility to adjust it to fit the experiment.

References

[1] A. M. Covington *et al* 2002 PHYSICAL REVIEW A **66**, 062710

[2] X. Gao and J. M. Li 2014 PHYSICAL REVIEW A **89**, 022710

‡ 1 E-mail: xgao@csrc.ac.cn

Content from this work may be used under the terms of the Creative Commons Attribution 3.0 licence. Any further distribution of this work must maintain attribution to the author(s) and the title of the work, journal citation and DOI.
Published under licence by IOP Publishing Ltd

Transverse electron momentum distribution in strong field ionization: transition from tunneling to over the barrier ionization regimes.

I.A.Ivanov[*1], **A.S.Kheifets**[†2]

[*] Center for Relativistic Laser Science, Institute for Basic Science (IBS), Gwangju 500-712, Republic of Korea
[†] Research School of Physics and Engineering, The Australian National University

Synopsis We study transverse electron momentum distribution (TEMD) in strong field atomic ionization driven by laser pulses with varying ellipticity. We show, that the TEMD in the tunneling and over the barrier ionization regimes evolves in a qualitatively different way when the ellipticity parameter describing polarization state of the driving laser pulse increases.

The current understanding of strong field atomic ionization is based on the pioneering work by Keldysh which introduced the distinction between the multiphoton and tunneling ionization regimes. A finer distinction arises when one realizes that the Keldysh theory in its original form is not applicable for very high field strengths exceeding the over the barrier ionization (OBI) limit. Despite the fact that underlying physics is very different in the two regimes (a classically forbidden trajectory for tunneling and a classically allowed trajectory for OBI), the energy spectra and electron angular distributions as given by these two theories are not dissimilar. This makes it often difficult to distinguish the two regimes in the experiment [1].

We show, that the TEMD distributions are qualitatively different in the tunneling and the OBI regimes. In the tunneling regime, TEMD exhibits a cusp-like structure due to the Coulomb focusing effect at $p_\perp = 0$ for linear polarization and a Gaussian-like structure predicted by the Keldysh theory for circular polarization. We studied this transition from the cusp-like to the Gaussian structures in detail in the tunneling regime [2], and interpreted this transition as a gradual diminishing of the role of the Coulomb effects with a growing ellipticity of the laser pulse. Further study of the role of the Coulomb focusing effects has been reported in [3]. Behavior of the TEMD in the vicinity of the point $p_\perp = 0$ can described by means of an expansion:

$$\ln W(p_\perp) \approx B + A|p_\perp|^\alpha \ . \qquad (1)$$

For the TEMD $W(p_\perp = 0)$ to have a cusp $V(p_\perp = 0)$ should have an infinite derivative of some order, the gaussian behavior of the TEMD predicted by the Keldysh theory corresponds to $\alpha = 2$.

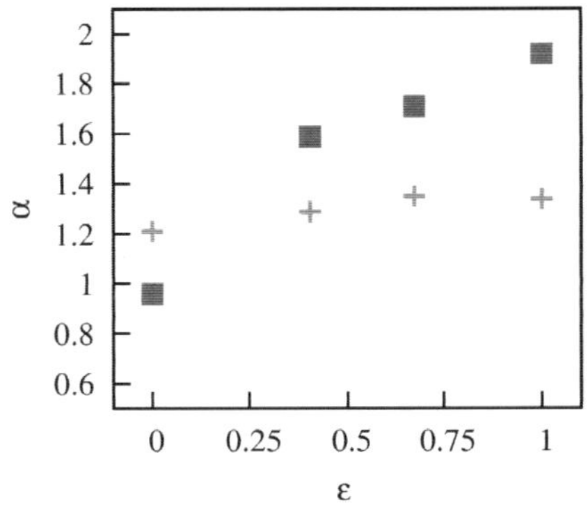

Figure 1. The fitting parameter α in Eq. (1) as a function of the ellipticity parameter ϵ for Ar (red) boxes and Ne* (green) crosses.

In Figure I we show behavior of α- parameter as a function of the the ellipticity parameter of the driving laser pulse for Ar (field intensity of 4.8×10^{14} W/cm^2, well below the OBI limit) and metastable Ne atom (field intensity of 2×10^{14}, which is above the OBI threshold).

For the Ar atom, the α parameter grows with ϵ reaching the value close to 2 for circular polarization, similar to the behavior we observed for the tunneling ionization of hydrogen [2]. For the metastable Ne* atom, the α parameter remains essentially flat, indicating that a cusp-like behavior is present for all ϵ in the range from linear to circular polarization.

References

[1] N. B. Delone and V. P. Krainov 1998 *Physics-Uspekhi.* **41** 469

[2] I. A. Ivanov 2014 *Phys. Rev. A.* **90** 013418

[3] A. S. Kheifets and I. A. Ivanov 2014 *Phys. Rev. A.* **90** 033404

[1]E-mail: igor.ivanov@anu.edu.au
[2]E-mail: a.kheifets@anu.edu.au

Content from this work may be used under the terms of the Creative Commons Attribution 3.0 licence. Any further distribution of this work must maintain attribution to the author(s) and the title of the work, journal citation and DOI.
Published under licence by IOP Publishing Ltd

Complex polarisabilities and hyperpolarisabilities of atoms

M. W. J. Bromley*[1], S. J. Grunefeld*, J. M. Rossi△, J. C. Berengut◇, L. Y. Tang♠, J. Jiang♣, Yongjun Cheng♡, J. Mitroy†

* School of Mathematics and Physics, The University of Queensland, Australia.
△ Computational Science Research Center, San Diego State University, U. S. A..
◇ School of Physics, University of New South Wales, Australia.
♠ Wuhan Institute of Physics and Mathematics, Chinese Academy of Sciences, Wuhan, P. R. China.
♣ Northwest Normal University, Langzhou, P. R. China.
♡ Harbin Institute of Technology, Harbin, P. R. China.
† School of Engineering, Charles Darwin University, Australia.

Synopsis The complex dipole polarisabilities of alkali atoms are computed for low-lying eigenstates. We present a unified method of calculation of the real part of the polarisability as well as two different imaginary contributions to the polarisability. One imaginary polarisability is related to the resonant damping of the cross-section both below and above threshold, whilst the other imaginary polarisability describes the photoionisation cross-section above threshold. We also present preliminary calculations of the complex hyperpolarisabilities of atomic clocks and how they are impacted by the linear and non-linear photon-atom interactions.

When an atom is exposed to an electric field the energies of its ground and excited states are altered and the electrons in the atom are seen to be polarized as per the Stark effect. Using time-dependent perturbation theory gives a straightforward method of computation of the real part of the polarisability which is related to the refractive index of an atomic gas [1].

The consequence of an imaginary part to a dipole polarisability is understood as being related to the scattering cross sections via the optical theorem. The first imaginary polarisability considered here describes the photon-atom absorption cross-section both below and above threshold, whilst the second describes the photon-atom ionisation cross-section above threshold. We are not aware of any previous calculation that presents both imaginary polarisabilities alongside the real part of the polarisability in a unified calculation. The total photon-atom scattering cross-section is a sum of such Rayleigh and ionisation processes. Furthermore, we use our pseudostate-based calculations to also compute the complex transition polarisabilities that are then related to the Raman and Compton photon-atom scattering cross-sections, giving a complete picture of the low-energy processes.

The complex polarisability of ground state lithium is shown in Fig. (1). This extends on from our calculations that previously were limited to the real part of the dipole polarisability of the Li($2s$) and Li($3s$) states [2], and only up to frequencies around the ionisation thresholds. We will present a survey of complex polarisability calculations that continues our series for various atoms and ions [3]. We will also present our initial calculations of the complex (second) hyperpolarisabilities of atoms and ions.

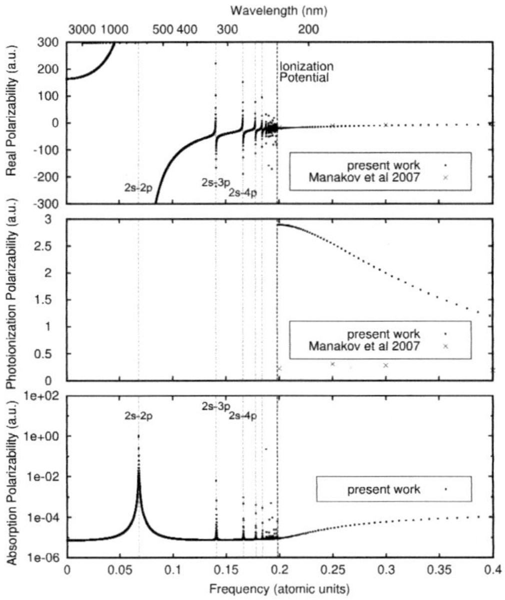

Figure 1. The dynamic complex polarisabilities of ground-state lithium ($2s$) as a function of frequency (in a.u.). (top) the real part of the polarisability, (middle) the ionisation polarisability, (bottom) the absorption polarisability.

(This work is dedicated to our colleague and mentor Jim Mitroy who recently passed away.)

References

[1] J. Mitroy *at al.* 2010 *J. Phys. B* **43** 202001

[2] L. Y. Tang *at al.* 2013 *Phys. Rev.A* **87** 032507

[3] J. Jiang *at al* 2015 *At. Data Nucl. Data Tables* **101** 158

[1]E-mail: brom@physics.uq.edu.au

Content from this work may be used under the terms of the Creative Commons Attribution 3.0 licence. Any further distribution of this work must maintain attribution to the author(s) and the title of the work, journal citation and DOI.
Published under licence by IOP Publishing Ltd

XXIX International Conference on Photonic, Electronic, and Atomic Collisions (ICPEAC2015) IOP Publishing
Journal of Physics: Conference Series **635** (2015) 092079 doi:10.1088/1742-6596/635/9/092079

Fano Resonances in the Time Domain

A. Kaldun[*1], **C. Ott**[*], **V. Stooß**[*], **A. Fischer**[*], **A. Blättermann**[*], **T. Ding**[*], **P. Raith**[*], **K. Meyer**[*], **M. Laux**[*], **J. Evers**[*], **C. H. Keitel**[*], **C. H. Greene**[†], **T. Pfeifer**[*2]

[*] Max-Planck-Institut für Kernphysik, Saupfercheckweg 1, 69117 Heidelberg, Germany
[†] Department of Physics, Purdue University, West Lafayette, IN 47907, USA

Synopsis The Fano phase formalism enables measurement and control of phase and amplitude of an emitting dipole. Here, we use this formalism to measure and understand the dynamics of bound atomic states in strong laser fields.

A universal phase formalism to understand laser control of absorption line shapes. We have recently shown and experimentally confirmed [1] that it is possible to map the Fano q-parameter onto the phase φ of the time dependent dipole response function. Using this concept, laser control of φ — and therewith the absorption line shape — can be established (see Fig. 1). We have experimentally realized this line-shape control by non-resonantly (ponderomotively) coupling the state to the continuum and thereby altering its phase using a laser pulse with a duration that is short compared to the lifetime of the excited atomic state. Furthermore, we control the sign of the absorption by varying only the phase φ. Thus, gain is achieved solely by modifying the phase of the polarization response after perturbative excitation in the absence of additional amplitude control or even inversion. We experimentally confirm this by applying the VIS laser field and thereby enhancing the extreme ultraviolet light intensity at the resonance positions of the singly excited states of He, which are normally absorbing.

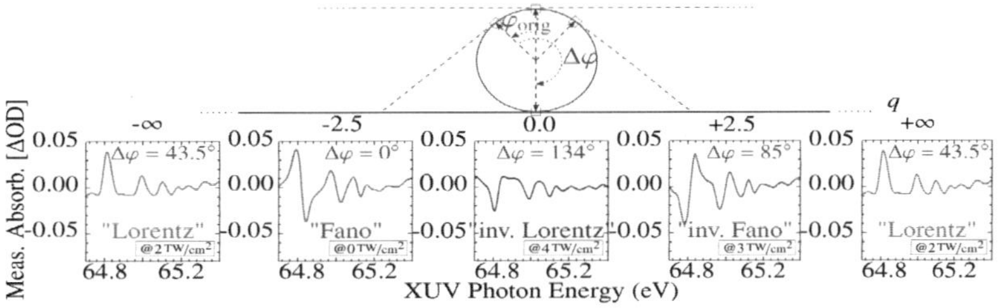

Figure 1: Mapping between the temporal response-function phase φ and Fano's q (line-shape asymmetry) parameter: $\varphi(q) = 2\arg(q - \imath) \quad \leftrightarrow \quad q(\varphi) = -\cot(\varphi/2)$. The experimental proof for this Fano-phase formalism is given by the measured line shapes of the doubly-excited $sp_{2,n+}$ series ($n \geq 5$) in He.

Generalization of the Fano-phase formalism for quantum-state holography. For the Fano-phase formalism as described above, we discussed non-resonant coupling of states. A generalization of the φ-to-q correspondence allows to observe *resonant* coupling of states by pulsed lasers [2], which is of substantial interest e.g. to fully understand the creation and probing of electronic wave packets [3].

By controlling the time delay τ and the intensity of an ultrashort near-visible (VIS) laser field with respect to an exciting attosecond-pulsed extreme-ultraviolet (XUV) field, it is possible to observe and separate the effects of different few-photon quantum paths that are involved in the coupling dynamics of the doubly excited states of He in a laser field. As a result, we explain the observed sub-cycle oscillations of the line shapes of doubly excited He states [2].

In essence, our generalized Fano-phase formalism provides a better understanding of laser coupling autoionizing states to other bound or continuum states by using precise measurements and the ensuing analysis of the exact variation of the different phase contributions with laser control. In addition, it allows for a systematic investigation of the transition from the perturbative to the strong-field regime of resonant & non-resonant laser coupling of electronic states.

References

[1] C. Ott, A. Kaldun, P. Raith, K. Meyer, M. Laux, J. Evers, C. H. Keitel, C. H. Greene, T. Pfeifer, in *Science* **340**,716-720 (2013).

[2] A. Kaldun, C. Ott, A. Blättermann, M. Laux, K. Meyer, T. Ding, A. Fischer, T. Pfeifer, in *Phys. Rev. Lett.* **112**, 103001 (2014).

[3] C. Ott, A. Kaldun, P. Raith, K. Meyer, M. Laux, Y. Zhang, A. Blättermann, S. Hagstotz, T. Ding, R. Heck, T. Pfeifer, in *Nature* **516**,374–378 (2014).

[1]E-mail: Andreas.Kaldun@mpi-hd.mpg.de
[2]E-mail: Thomas.Pfeifer@mpi-hd.mpg.de

Content from this work may be used under the terms of the Creative Commons Attribution 3.0 licence. Any further distribution of this work must maintain attribution to the author(s) and the title of the work, journal citation and DOI.
Published under licence by IOP Publishing Ltd

Journal of Physics: Conference Series **635** (2015) 092080 doi:10.1088/1742-6596/635/9/092080

Time-dependent multiconfiguration self-consistent-field methods for multielectron dynamics in intense laser fields

Takeshi Sato[*†1] **and Kenichi L Ishikawa**[*†2]

[*] Photon Science Center, School of Engineering, The University of Tokyo, Tokyo 113-8656, Japan

[†] Department of Nuclear Engineering and Management, School of Engineering, The University of Tokyo, Tokyo 113-8656, Japan

Synopsis Time-dependent multiconfiguration self-consistent-field methods with complete-active-space (TD-CASSCF) and occupation-restricted multiple-active-space (TD-ORMAS) models are developed. These methods enable highly flexible and systematic construction of the configuration interaction (CI) spaces, allowing a compact yet accurate description of multielectron dynamics in intense laser fields.

An important goal of high-field and ultrafast physics is the direct measurement and control of electron dynamics in atoms and molecules. To theoretically investigate multielectron dynamics in intense laser fields, the multiconfiguration time-dependent Hartree-Fock (MCTDHF) method has been developed [1, 2], in which the wavefunction $\Psi(t)$ is given in the CI expansion, $\Psi(t) = \sum_{\mathbf{I}} \Phi_{\mathbf{I}}(t)C_{\mathbf{I}}(t)$, with timely varying orbitals constituting Slater determinants $\Phi_{\mathbf{I}}(t)$. Though powerful, the MCTDHF method suffers from the exponential increase of the computational cost against the number of electrons.

To circumvent this difficulty, we have proposed the TD-CASSCF method [3], which classifies the occupied orbitals into *core* (doubly occupied at any time) and *active* (fully correlated) orbitals. We have also developed the TD-ORMAS method [4], which *further* subdivides the active orbitals into an arbitrary number of subgroups, and poses the *occupation restriction* in each subgroup. See Fig. 1. These methods enable a compact yet accurate description of multielectron dynamics in intense laser fields. This contribution describes the TD-CASSCF and TD-ORMAS methods and their applications (Fig. 2 as an example) to intense-field driven multielectron dynamics.

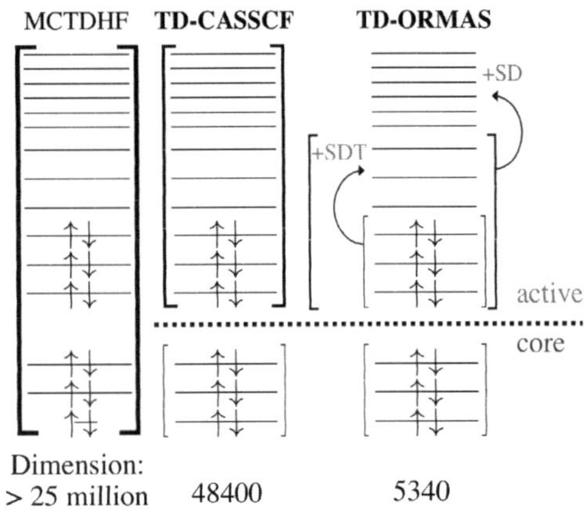

Figure 1. Examples of CI spaces in MCTDHF, TD-CASSCF, and TD-ORMAS methods for 12 electron systems with 15 occupied orbitals. Shown at the bottom are the number of determinants included in each method.

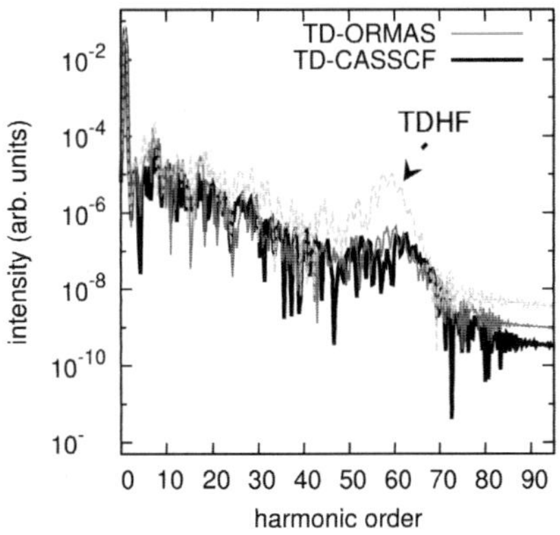

Figure 2. High harmonic generation spectrum of one-dimensional lithium hydride trimer model (12 electrons) exposed to a laser pulse with a wavelength of 750 nm and an intensity of 4×10^{14} W/cm^2, computed with various methods described in Fig. 1 and the conventional TDHF method. The MCTDHF calculation was not feasible due to the large CI dimension.

References

[1] J. Caillat *et al* 2005 *Phys. Rev. A* **71** 012712

[2] T. Kato and H. Kono 2004 *Chem. Phys. Lett.* **392** 533

[3] T. Sato and K. L. Ishikawa 2013 *Phys. Rev. A* **88** 023402

[4] T. Sato and K. L. Ishikawa 2015 *Phys. Rev. A* **91** 023417

[1]E-mail: sato@atto.t.u-tokyo.ac.jp
[2]E-mail: ishiken@atto.t.u-tokyo.ac.jp

Content from this work may be used under the terms of the Creative Commons Attribution 3.0 licence. Any further distribution of this work must maintain attribution to the author(s) and the title of the work, journal citation and DOI.
Published under licence by IOP Publishing Ltd

Journal of Physics: Conference Series **635** (2015) 092081 doi:10.1088/1742-6596/635/9/092081

Relativistic R-matrix calculation photoionization cross section of Xe and Xe@C_{60}

Bowen Li*,† 1, Gerry O'Sullivan† 2, Chenzhong Dong‡

* School of Nuclear Science and Technology, Lanzhou University, Lanzhou, 730000, China

† School of Physics, University College Dublin, Belfield, Dublin 4, Ireland

‡ College of Physics and Electronics Engineering, Northwest Normal University, Lanzhou, 730070, China

Synopsis The photoionization cross sections of Xe and Xe@C_{60} have been calculated using relativistic R-matrix method. Comparison between the present calculations and experimental results demonstrate that configuration interaction has an important effect.

The properties and behavior of endohedral fullerenes A@C_{60} have gained considerable interest in recent years [1-2]. An effective method to study such confined atoms is photoionization. The photoionization cross sections Xe (figure 1) and Xe@C_{60} (figure 2) [3-4]are calculated using the fully relativistic R-matrix method for both a spherical attaractive cavity and the newly proposed jellium potential [5].

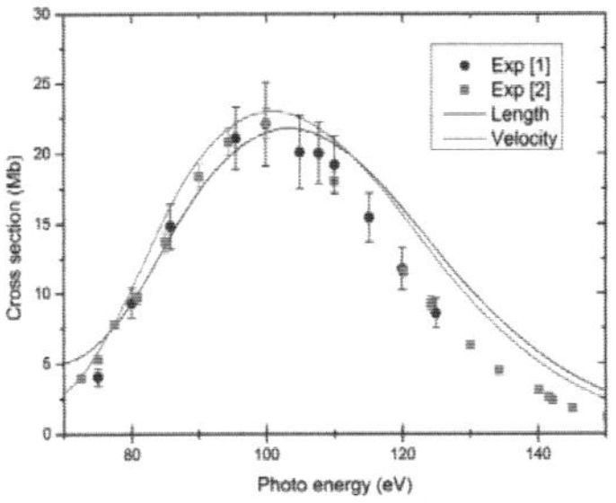

Figure 1. Photoionization cross section of Xe. Experimental measurements are also presented for comparison: Exp [1]: J. Phys. B 22 (1989) 841, Exp [2]: Phys. Rev. A 39 (1989) 3902.

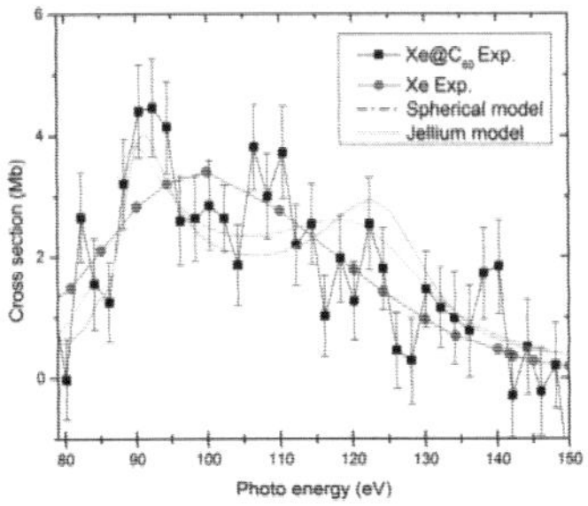

Figure 2. Comparison of the measured [6] and theoretical calculated Xe@C_{60} photoionization cross section.

Besides five main photoionization channels: 4d-εf,εp, 5p-εs,εd and 5s-εp, the most important configuration interaction based on $4d^2$-$4f^2$ promotions are also included in the final states. Comparison between the present calculations and other experimental and theoretical results demonstrate that configuration interaction plays an important role. It should be noted that for atoms like Xe, the 4d subshell does not overlap with C_{60} cage. It will be interesting to compare the influence of such confining effect on the photoionization cross sections of more hybridized atoms such as Ca, Y, Sc, etc (figure 3).

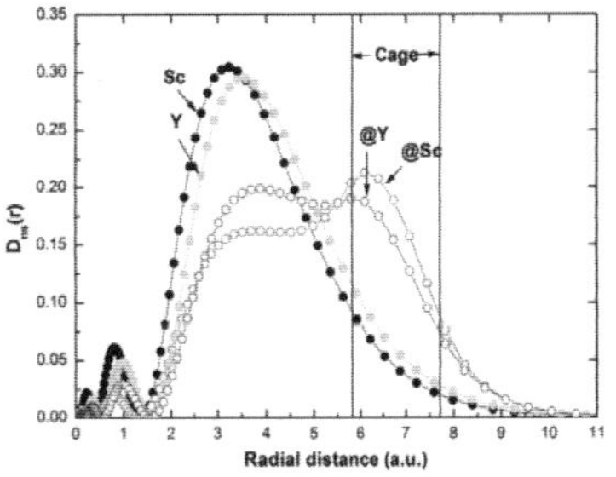

Figure 3. Valence *ns* electron density for *free* Sc and Y and confined @Sc and @Y. The location of model C_{60} cage potential is indicated by the vertical lines.

References

[1] W. Jaskólski 1996 *Phys. Rep.* 271 1

[2] V. K. Dalmatov 2009 *Adv. Quantum Chem.* 58 13

[3] B. Li *et al* 2012 *J. Phys.:Conf. Ser.* 388 022063

[4] B. Li *et al* 2013 *J. Phys. B* 46 155203

[5] A. V. Verkhovtsev *et al* 2012 *J. Phys. B* 45 215101

[6] A. L. D. Kilcoyne, *et al* 2009 *Phys. Rev. Lett.* 105 213001

† 1 E-mail: libw@lzu.edu.cn

2 E-mail: gerry.osullivan@ucd.ie

Content from this work may be used under the terms of the Creative Commons Attribution 3.0 licence. Any further distribution of this work must maintain attribution to the author(s) and the title of the work, journal citation and DOI.
Published under licence by IOP Publishing Ltd

Ultrafast two-photon Rabi oscillations in excited He driven by femtosecond intense laser fields

A. Hishikawa[1,*], M. Fushitani,[1] C.-N. Liu,[2] A. Matsuda,[1] T. Endo,[1] Y. Toida,[1] Y. Hikosaka,[3] T. Morishita[4]

[1] Department of Chemistry, Graduate School of Science, Nagoya University, Nagoya, Aichi 464-8602, Japan
[2] Department of Physics, Fu-Jen Catholic University, Taipei 24205, Taiwan
[3] Department of Environmental Science, Faculty of Science, Niigata University, Niigata, Niigata 950-2181, Japan
[4] Department of Engineering Science, The University of Electro-Communications, 1-5-1 Chofu, Tokyo 182-8585, Japan

Synopsis We demonstrate ultrafast multiphoton coherent excitation in near-infrared intense laser fields (≤10 TW/cm^2) using excited helium in the $1s2p$ (^{1}P) state as the target. Strong two-photon coupling between the $2p$ state and the $5f$ and $6f$ Rydberg states drives ultrafast Rabi oscillations in tens of femtosecond. Theoretical calculations solving the time-dependent Schrödinger equation were carried out to elucidate the strong field effects.

Coherent interaction between a quantum system and radiation fields plays a key role in chemical reaction control, information processing and ultrafast electronics. In a two-level system, for example, resonant interaction with a coherent radiation field results in a periodic oscillation of level populations, known as Rabi oscillation.

Rabi oscillation can also occur by coherent interactions with multiple photons. Combined with pulse shaping technique, multiphoton coherent interaction provides a unique means for an efficient population transfer to highly excited states in an ultrashort timescale.

Here we demonstrate femtosecond two-photon Rabi oscillations between $1s2p$ and $1snf$ (n = 5,6) Rydberg state of helium in ultrashort near-infrared (NIR) intense laser fields (795 nm, ≤10 TW/cm^2). The experiment was carried out at an extreme ultraviolet (EUV) free-electron laser (FEL) facility, SCSS Test Accelerator at RIKEN HARIMA institute [1]. The center wavelength was tuned to the 2^1P - 1^1S transition line (58.4 nm) to prepare He in the 2^1P excited state. A regenerative Ti:Sapphire laser amplifier system (795 nm, ~50 fs) synchronized with the FEL was used to generate NIR intense laser fields. The NIR laser beam was focused quasi-collinearly with the FEL beam by using a plano-convex lens with a long focal length (f = 1000 mm). Photoelectron spectroscopy of the He 2^1P state in the intense NIR laser fields was performed with a magnetic-bottle-type photoelectron spectrometer.

The observed photoelectron spectra are shown in Fig.1, where two peaks associated with resonances to the 1s5f and 1s6f states are identified. The 1s6f peak shows clear modulation in the peak intensity as a function of the NIR laser field intensity. To understand the experimental behavior, theoretical calculations based on time-dependent Schrödinger equation for He (2^1P) atoms in the presence of NIR laser fields is carried out. The obtained results (not shown) reproduced well the observed spectra, confirming that the modulation is due to the two-photon Rabi oscillations on a ten-femtosecond time scale.

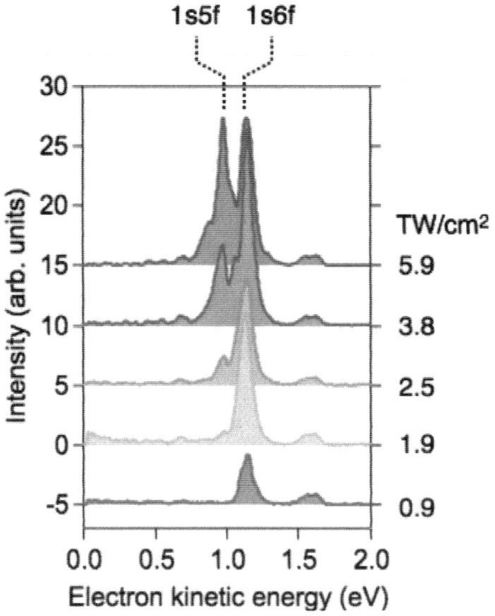

Figure 1. Photoelectron spectra of the excited He obtained for the NIR field intensities of 5.9, 3.8, 2.5, 1.9, and 0.9 TW/cm^2.

References

[1] M. Yabashi *et al* 2013 *J. Phys. B* **46** 164001.

* E-mail: hishi@chem.nagoya-u.ac.jp

Content from this work may be used under the terms of the Creative Commons Attribution 3.0 licence. Any further distribution of this work must maintain attribution to the author(s) and the title of the work, journal citation and DOI.
Published under licence by IOP Publishing Ltd

Journal of Physics: Conference Series **635** (2015) 092083 doi:10.1088/1742-6596/635/9/092083

Ionization of strontium atoms by ultrashort pulses of a Ti-Sapphire laser

A. Dimitriou[† 1*], **A. Marciniak**[‡ 2], **F.Lepine**[‡ 3], **C. Bordas**[‡ 4], **S. Cohen**[† 5]

[†] Dept. of Physics, University of Ioannina, GR 45110 Ioannina, Greece

[‡]Institute Lumière Matière, Université Lyon 1, UMR 5306, 10 rue Ada Byron 69622, Villeurbanne, France

Synopsis We present experimental photoelectron spectra obtained by ionizing two-valence electron Sr atoms with ≈25 fs Ti-Sapphire laser pulses and we show preliminary evidence that electron-electron correlation effects dominate for laser intensities $<0.4\times10^{13}$ W·cm^{-2}, while a single electron picture prevails when the laser intensity is increased further.

The two-electron Alkaline Earth atoms present dense manifolds of doubly excited autoionizing states above their first ionization threshold. Hence, their continua are structured and these states are dominated by electron-electron interaction effects. The interaction and subsequent ionization of Alkaline Earth atoms with intense laser fields reveals the absorption of photons within the structured continua. For laser pulses of long duration (~ns), there exist above-threshold-absorption pathways which are dominated by the presence of these doubly excited states and lead to the production of excited ions [1]. On the contrary, for pulses of short duration (~fs) energy level structure issues are of much less importance and one expects a single-active-electron picture to prevail [2,3]. In this case ions should be mostly produced to their ground state. This latter expectation is here experimentally examined by ionizing Sr atoms via ~25 fs, ~800 nm laser pulses, with intensities ranging from $I\approx2\times10^{12}$ W·cm^{-2} to $\approx5\times10^{13}$ W·cm^{-2}. A preliminary analysis of measured photoelectron spectra (recorded via a velocity map imaging (VMI) apparatus [4] at Institute Lumière Matière (ILM), Université Lyon 1) shows that for $I<0.4\times10^{13}$ W·cm^{2} electron-electron interaction dominates and population is created at excited states of Sr^{+}. As the laser intensity is further increased, however, the spectra are composed solely by a series of Above Threshold Ionization (ATI [3]) electron peaks (see Fig. 1) and consequently the single-active-electron picture appears to dominate. These findings suggest that the key factor controlling the dominance of one or the other mechanism is laser intensity, rather than pulse duration.

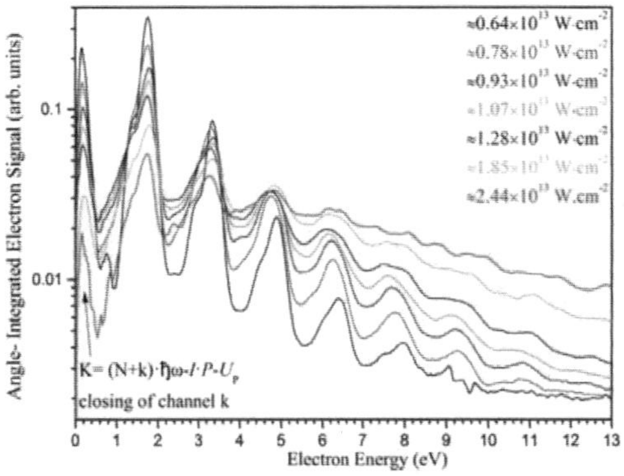

Figure 1. High laser intensity electron energy spectra of Sr ionized by 25 fs Ti-Saphire laser pulses.

This research has been co-financed by the European Union (European Social Fund – ESF) and Greek national funds through the Operational Program "Education and Lifelong Learning" of the National Strategic Reference Framework (NSRF) - Research Funding Program: THALES (ISEPUMA). Investing in knowledge society through the European Social Fund.

[*]**Present address:** Dept. of Physics, University of Crete, GR 71003, Heraklion, Greece

References

[1] I. Liontos *et al* 2008 *J. Phys. B: At. Mol. Opt. Phys.* **41** 045601
[2] M. J Nandor *et al* 1999 *Rhys. Rev. A.* **60** R1771
[3] P. Agostini *et al* 1979 *Phys. Rev. Lett.* **42** 1127
[4] A. T. J. B. Eppink and D. H. Parker 1997 *Rev. Sci. Instr.* **68** 3477

[1]E-mail: adimitr@physics.uoc.gr
[2]E-mail: alexandre.marciniak@univ-lyon1.fr
[3]E-mail: franck.lepine@univ-lyon1.fr
[4]E-mail: christian.bordas@univ-lyon1.fr
[5]E-mail: scohen@uoi.gr

Content from this work may be used under the terms of the Creative Commons Attribution 3.0 licence. Any further distribution of this work must maintain attribution to the author(s) and the title of the work, journal citation and DOI.
Published under licence by IOP Publishing Ltd

Journal of Physics: Conference Series **635** (2015) 092084 doi:10.1088/1742-6596/635/9/092084

High-Lying Rydberg States from Strong Field Interaction

Seyedreza Larimian*, Sonia Erattupuzha*, Raffael Maurer*, Christoph Lemell†, Stefan Nagele†, Shuhei Yoshida†, Joachim Burgdörfer†, Andrius Baltuška*, Markus Kitzler*, Xinhua Xie*[1]

* Photonics Institute, Vienna University of Technology, Gusshausstrasse 27, A-1040 Vienna, Austria

† Institute for Theoretical Physics, Vienna University of Technology, Wiedner Hauptstrasse 8-10, A-1040 Vienna, Austria

Synopsis We experimentally observed high-lying (n>238) Rydberg states which are formed through recapture of electrons after strong field interaction in atoms and small molecules using electron and ion coincidence detection. The high-lying Rydberg states are identified through weak DC field-induced ionization.

In general, high-lying Rydberg states of atoms and molecules have quite low ionization potentials which may lead to over-barrier ionization in weak electric fields. Recent theoretical and experimental results indicate that such states play an important role in the interaction of atoms and molecules with strong laser fields [1, 2].

In this submission, we report direct observation of post-pulse detachment of electrons from high-lying Rydberg states which have previously been recaptured by the ionic core after strong field ionization in atoms and small molecules. In the experiments, we measured electrons and ions from the strong field ionization of atoms and molecules in coincidence using the reaction microscope technique [3, 4]. Measurements were performed with 795 nm, 25 fs laser pulses with peak laser intensity on the order of 1×10^{14} W/cm^2 (peak electric field 3×10^8 V/cm). Weak DC fields (from 0.1 V/cm up to 30 V/cm) are applied not only to accelerate charged particles to the detectors but also to induce field ionization of high-lying Rydberg states. Three types of atomic gases (argon, helium, and neon) and four types of molecular gases (hydrogen, ethylene, acetylene and 1,3-butadiene) were measured in the experiments.

After tunneling ionization of the gases in the strong laser field, a small fraction of electrons can be recaptured to form Rydberg states when the laser pulse is past. Later, electrons from very high-lying Rydberg states will be re-ionized by the very weak DC field through over-barrier or tunneling ionization. In the measurements, we observed such DC field induced ionization of high-lying Rydberg states for all measured targets except for 1,3-butadiene. This refers to an instability of highly excited 1,3-butadiene resulting in molecular dissociation prior to re-ionization.

Measured photoelectron-photoion-coincidence (PEPICO) spectrum for argon is presented in Figure 1, in which we can clearly distinguish the signals originating from the strong field ionization and the weak DC field ionization.

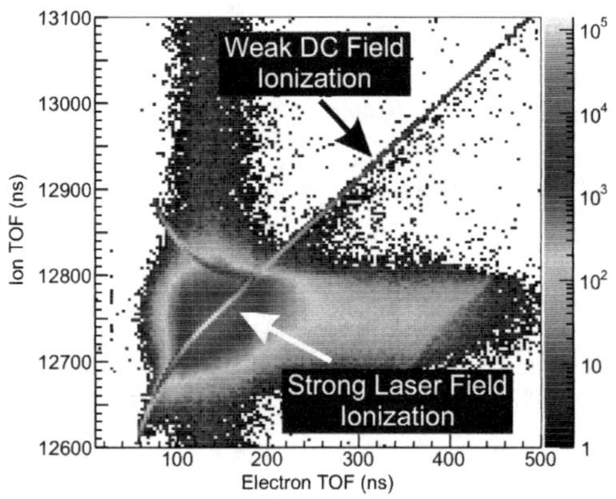

Figure 1. Measured PEPICO spectrum for argon.

We estimate the minimum primary quantum number n of the detected DC field-ionized Rydberg states from the ionization threshold $(-2\sqrt{F})$ for static field over-barrier ionization. For an electric field of 0.1 V/cm, we find that only Rydberg states with n>238 is depopulated by the weak DC field.

References

[1] T. Nubbemeyer *et al* 2008 *Phys. Rev. ett.* **101** 233001

[2] B. Wolter *et al* 2014 *Phys. Rev. A* **90** 063424

[3] R. Dörner *et al* 2000 *Phys cs Re orts* **330** 95

[4] X. Xie *et al* 2012 *Phys. Rev. ett.* **108** 193004

[1]E-mail: xinhua.xie@tuwien.ac.at

Content from this work may be used under the terms of the Creative Commons Attribution 3.0 licence. Any further distribution of this work must maintain attribution to the author(s) and the title of the work, journal citation and DOI.
Published under licence by IOP Publishing Ltd

Journal of Physics: Conference Series **635** (2015) 092085 doi:10.1088/1742-6596/635/9/092085

Ionization of noble gas atoms by few-cycle laser pulses: spatial interference effects

S. Borbély [*1], **A. Tóth**[†], **K. Tőkési**[‡] **and L. Nagy**[*]

[*] Faculty of Physics, Babeş-Bolyai University, Kogălniceanu Street No. 1, 400084 Cluj, Romania
[†] Department of Theoretical Physics, University of Debrecen, H-40410 Debrecen, PO Box 5, Hungary
[‡] Institute for Nuclear Research, Hungarian Academy of Sciences (ATOMKI), H-4026 Debrecen, P.O. Box 51, Hungary

Synopsis In the framework of the present work the ionization of atoms (H, He, Ne, Ar) by few-cycle XUV laser pulses is investigated by directly solving the time dependent Schrödinger equation. Our attention is focused on the investigation of the intra-pulse interference effects at the level of wave function dynamics, where we study in details how the resulting interference pattern is influenced by the shape and strength of the bounding Coulomb potential.

During the ionization of atoms by few-cycle laser pulses, beside the dominant ionization, secondary processes with significant impact on the final momentum distribution of the continuum electrons also occur. These are the result of interference between electronic wave packets following different spatial and temporal paths [1]. From the numerous possible interference scenarios [1] we focus our attention on the interference between electronic wave packets emitted at the same time (i.e., during the same quarter-pulse cycle), but following different spatial paths (spatial interference). The formed radial interference pattern, in a simplistic picture [2], is a result of the interference between the direct (i.e., unscattered) and the scattered wave packets, where the direct wave packet can be considered as a reference, while the scattered wave packet as a signal wave. In this picture, the interference fringe structure can be interpreted as the holographic mapping (HM) of the target atom's or molecule's state [2]. As a result of previous investigations [2, 3, 4] it is known that for a fixed atomic target the shape of the formed HM interference pattern is strongly influenced by the laser pulse through the z_0 parameter [2, 3]. This z_0 parameter measures the maximum distance reached by the electronic wave packets before the rescattering event leading to the formation of the HM interference pattern.

As a continuation of our previous works [3, 4] we have proposed to investigate how the atomic species of the target is influencing the shape of the HM pattern. In order to reach this goal we have performed calculations for the ionization of the noble gas targets by few-cycle laser pulses based on the direct numerical solution of the Schrödinger equation.

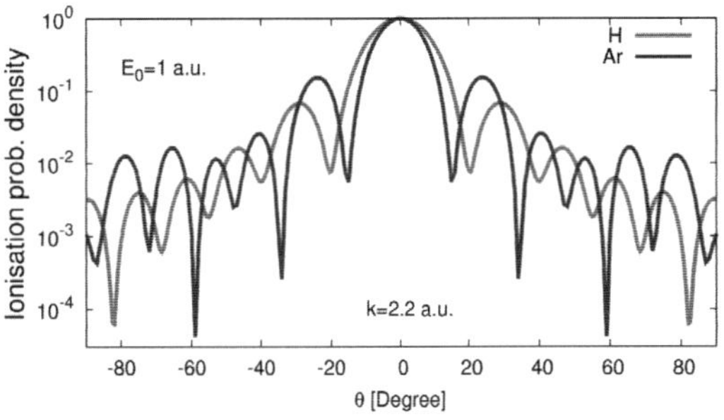

Figure 1. The angular distribution of photoelectrons ejected by the H and Ar targets with fixed $k = 2.2$ a.u. momentum. The parameters of the driving laser pulse are: $\omega = 0.4445$ a.u. (photon energy), $\tau = 28.27$ a.u. (pulse duration), $E_0 = 1$ a.u.(pulse strength).

In the calculated ionization spectra a strong dependence of the HM interference pattern on the atomic target was observed. This is illustrated on Figure 1, where the fixed momentum angular distribution of the continuum electrons is plotted for the H and Ar targets.

The present work was supported by a grant of the Romanian National Authority for Scientific Research (CNCS UEFIS-CDI) project number PN-II-ID-PCE-2011-3-0192, by the Hungarian Scientific Research Fund OTKA Nos. K103917 and NN 103279.

References

[1] X.-B Bian *et al* 2011 *Phys. Rev. A* **84** 043420.

[2] Y. Huismans *et al* 2011 *Nature* **331** 61.

[3] S. Borbély *et al* 2013 *Phys. Rev. A* **87** 013405.

[4] S. Borbély *et al* 2014 *Eur. Phys. J. D* **68** 339.

[1]E-mail: sandor.borbely@phys.ubbcluj.ro

Content from this work may be used under the terms of the Creative Commons Attribution 3.0 licence. Any further distribution of this work must maintain attribution to the author(s) and the title of the work, journal citation and DOI.
Published under licence by IOP Publishing Ltd

Post-collision interaction in Potassium 3*s* photoionization

S. Sheinerman[1], P. Lablanquie[2], L. Andric[2], K. Ito[2], J. Palaudoux[2], S-M. Huttula[3], M. Huttula[3], K. Jänkälä[3], J. Soronen[3], J-M. Bizau[4], S. Guilbaud[4], D. Cubaynes[4], and F. Penent[2]

[1] Department of Physics, St.Petersburg State Maritime Technical University, 198262 St.Petersburg, Russia
[2] LCPMR, CNRS & Université Paris 06, 11 rue Pierre et Marie Curie, 75231 Paris Cedex 05, France
[3] Department of Physics, MOMA-RC, P.O. Box 3000, 90014 University of Oulu, Finland
[4] ISMO, CNRS UMR 8214, Université Paris-Sud, Bâtiment 350, F-91405 Orsay, France

Synopsis The post-collision interaction (PCI) distortion of the Auger electron line shapes after K 3s photoionization is revealed in coincidence measurements and confirmed by calculations.

We present here a combined experimental and theoretical study of the influence of post-collision interaction (PCI) on the line shape of Auger photoelectrons emitted after 3*s* photoionization in Potassium. Emission of a 3*s* photoelectron leads to the creation of a 3*s* vacancy. The coupling of this vacancy with the outer 4*s* electron can form two different intermediate states of the K^{+*} ion: $3s^{-1}4s$ (^3S) or (^1S). These states differ energetically by 0.36 eV and have different widths, respectively 97 meV and 287 meV. Auger decays of these intermediate states lead to the final ionic states K^{2+} $3p^{-1}4s^{-1}$ ($^2P_{3/2}$) and ($^2P_{1/2}$) and to the emission of Auger electrons of ~5eV. Due to the small energy separation of the (^3S) and (^1S) intermediate states, the spectrum of the emitted Auger electrons shows overlapping contributions from each state.

From the other side it is known that the spectra of the low energy photoelectrons emitted from the inner shells and the Auger electrons are distorted by PCI [1] which is reduced in our case to the interaction between the photoelectron and Auger electron and to the interaction of the photoelectron with the ionic field which varies during the Auger decay. Hence both the PCI distortion and the contributions from two intermediate states affect the observed Auger electron energy distribution.

Experiments were done at the PLEIADES beam line of the SOLEIL synchrotron. A magnetic bottle time of flight spectrometer [2] was used to detect in coincidence the 3*s* photoelectron and the Auger electrons. This method allows us to select precisely the processes under study and to define the final ionic state. Our measurements cover the region of incident photon energies of ~ (1-7) eV above the threshold. We have calculated the spectra of the emitted Auger electrons taking into account the PCI effects within the semi classical [3] and quantum mechanical [4] models.

Both measurements and calculations show a significant distortion of the Auger electron spectrum. As an example we show in Figure 1 two spectra for the photon energy of 42.0 eV and the final ionic states ($^2P_{3/2}$) and ($^2P_{1/2}$). The line maxima are shifted and the contributions from the different intermediate states are distorted differently due to the large difference of their widths. The good agreement between measured and calculated energy distributions demonstrates the validity of our PCI approach.

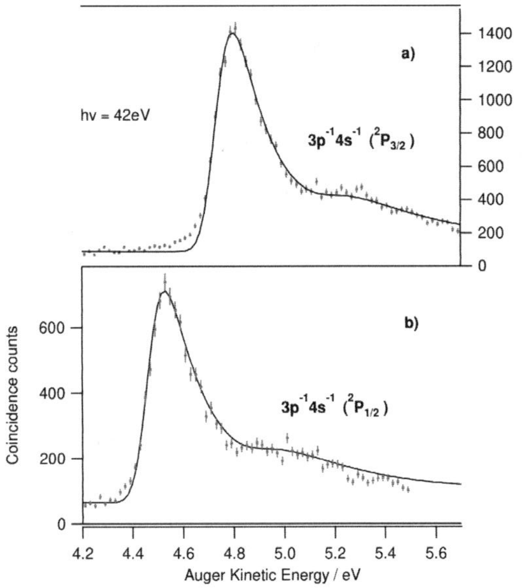

Figure 1. Auger spectra for the decay of the 3s hole, obtained at a photon energy 1.3eV above the $3s^{-1}$ (^3S) threshold. Red bars are experimental data, black lines are semi classical calculations. a) and b) select the different K^{2+} final states.

References

[1] M. Yu. Kuchiev and S. A. Sheinerman 1989 *Sov. Phys.- Usp.* **32** 569
[2] F. Penent *et al* 2005 *Phys. Rev. Lett.* **95** 083002
[3] P. van der Straten *et al* 1988 *Z. Phys. D* **8** 35
[4] M. Yu. Kuchiev and S. A. Sheinerman 1988 *J. Phys. B* **21** 2027

Content from this work may be used under the terms of the Creative Commons Attribution 3.0 licence. Any further distribution of this work must maintain attribution to the author(s) and the title of the work, journal citation and DOI.
Published under licence by IOP Publishing Ltd

Journal of Physics: Conference Series **635** (2015) 092087 doi:10.1088/1742-6596/635/9/092087

One and two-photon ionization of Argon by multicolor XUV radiation

Renata Della Picca* [1] **and Eva Lindroth**† [2]

*CONICET and Centro Atómico Bariloche, Av. Bustillo 9500, 8400 Bariloche, Argentina

†Department of Physics, Stockholm University, Alba Nova University Center, SE-106 91 Stockholm, Sweden

Synopsis We analyse theoretically the Argon photoelectron spectra produced by strong, extreme ultraviolet radiation of six colors: from the 11th to the 16th harmonics of ω_0 (800nm). The theoretical description shows good qualitative agreement with recent experimental results for two-extreme-ultraviolet-photon above-threshold ionization of Argon.

Techniques based on the process of High Harmonic Generation (HHG) have lead to the strong development of attosecond metrology that has taken place during the last decades [1]. The possibility for XUV-pump XUV-probe scenarios is a promising new development made possible through HHG from solid density plasmas [2, 3].

In a recent paper, Heissler *et al* [2] presented energy-resolved photo-electron (PE) spectra from non-resonant two-XUV-photon ionization of Argon. The XUV radiation used in Ref. [2] consists of higher-order harmonics (11th to 16th) of the original laser ω_0 ($\lambda = 800$ nm) generated by the so-called relativistic oscillating mirror (ROM) process. The PE spectra obtained show the above-threshold ionization (ATI) peaks separated by ω_0 corresponding to different combinations of two harmonic photons that lead the same excess energy.

In this work we present a model based on the Coulomb-Volkov (CV) approach and the Random Phase Approximation with Exchange (RPAE) to address this scenario theoretically. This theory combines an accurate target atom many-body model with a field-dressed description of the photo electron, accounting also for the interaction with the residual ion [4].

Within this context we analyse the Argon photo-electron spectra produced by the XUV radiation of six colors, from the 11th to the 16th harmonics of ω_0 as in the experiment of Heissler *et al* [2].

As an example, we show the PE spectrum in Figure 1 (a) at low energy, where absorption of only one photon take place, and in Figure 1 (b) at energies ranging from $22\omega_0 - IP$ to $32\omega_0 - IP$ (see upper horizontal scale), that are the energies expected from two-photon ATI caused by combination of the six harmonics. The experimental data is from a single shot PE spectrum presented in [2]. The theoretical spectrum is obtained by carefully choosing the laser pulse parameters to fit the experimental one.

To conclude, we point out that the relative intensities of the ATI peaks are very sensitive to the laser pulse parameter and target description. Comparison of our theoretical results with experimental data (including the average of several shots which is not show here) show good qualitative agreement.

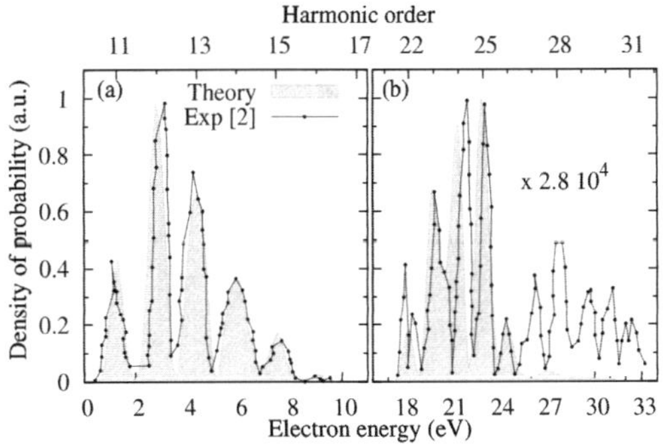

Figure 1. Argon photoelectron spectra. Line with dots (black): a single shot PE spectrum from [2]. Filled curve (green): current theoretical result. (a) Single photon and (b) two-photon PE spectrum region.

References

[1] F. Krausz and M. Ivanov 2009 *Rev. Mod. Phys.* **81** 163-234

[2] P. Heissler *et al* 2012 *New. J. Phys.* **14** 043025

[3] Y. Nomura *et al* 2009 *Nat. Phys.* **5** 124

[4] R. Della Picca and E. Lindroth 2015 *in preparation*

[1]E-mail: renata@cab.cnea.gov.ar

[2]E-mail: lindroth@fysik.su.se

Content from this work may be used under the terms of the Creative Commons Attribution 3.0 licence. Any further distribution of this work must maintain attribution to the author(s) and the title of the work, journal citation and DOI.
Published under licence by IOP Publishing Ltd

Coupling of frozen planet states of helium to its bound states: towards the preparation of two-electron nondispersive wave packets

Alejandro González [1], **Isabel Guzmán, and Javier Madroñero**

* Departamento de Física, Universidad del Valle, Cali, Colombia

Synopsis Frozen planet states of helium give rise to two-electron nondispersive wave packets. A study of the coupling of frozen planet states of helium to its bound states suggests a two photon process to populate such states. This is a first step in order to construct such wave packets.

The three body Coulomb problem (3BCP) is of paradigmatic importance since it bears dynamical and spectral features that establish a link between single and many-particle dynamics, with the absolute minimum of ingredients. Therefore, understanding the concerted motion of two electrons in the 3BCP is of fundamental importance for further understanding of more complex processes, e.g. non-equilibrium processes including chemical reactions.

The two-electron dynamics is in general irregular or chaotic, with only rather small domains of classical phase space occupied by regular, i.e., integrable motion. Regular regions in the phase space provide support for strongly correlated, but stable highly doubly excited states called frozen planet states (FPS). Under near-resonant periodic driving these states might transform into nondispersive wave packets (NDWP). However, there is no experimental evidence of the existence neither of FPS nor of these NDWP. From the theoretical point of view the characterization of NDWP posses a numerical challenge [1].

Here we investigate the coupling of FPS to bound states. For that purpose a two-dimensional model for helium is used [2]. By applying a special coordinate transformation the wave functions can be expanded in a basis where the Hamiltonian is a sparse band matrix of relatively small bandwidth and therefore numerically efficiently diagonalizable. The frozen planet states are not detected in the photoionization cross section from the ground state of the atom. However, they are rather strongly coupled to some excited bound states. This indicates that a two photon process might populate the so far non observed FPS (see figure 1) [3].

We go one step further and identify the relevant atomic states involved in the multiphoton process giving rise to the NDWP. This will help us to achieve an efficient numerical implementation of this problem beyond the planar approach.

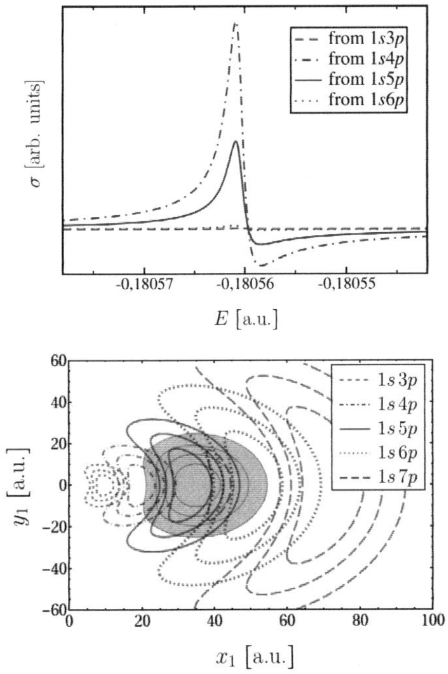

Figure 1. Top: Photoionization cross sections with different initial states $1s\,np$ near the energy of the ground S FPS of the series converging to the 4th ionization threshold (FPS_4^1). A significant coupling exists only for the initial states $1s\,4p$ and $1s\,5p$. Bottom: Projections of the wave functions for a fixed position of the inner electron of bound states $1s\,np$ and FPS_4^1. The FPS (shaded region) shows no overlap with $1s\,3p$ and $1s\,7p$, while $1s\,4p$ and $1s\,5p$ strongly overlap with it.

References

[1] J. Eiglsperger and J. Madroñero 2010 *Phys. Rev. A* **82**, 033422; J. Eiglsperger and J. Madroñero. In S. Wagner, *et al.*, editors, 2010 *High Performance Computing in Science and Engineering.* Pages 627-637 (Berlin: Springer)

[2] J. Madroñero and A. Buchleitner 2008, *Phys. Rev. A* **77**, 053402

[3] In preparation.

[1] E-mail: alejandro.gonzalez.melan@correounivalle.edu.co

Content from this work may be used under the terms of the Creative Commons Attribution 3.0 licence. Any further distribution of this work must maintain attribution to the author(s) and the title of the work, journal citation and DOI.
Published under licence by IOP Publishing Ltd

XXIX International Conference on Photonic, Electronic, and Atomic Collisions (ICPEAC2015) IOP Publishing
Journal of Physics: Conference Series **635** (2015) 092089 doi:10.1088/1742-6596/635/9/092089

Time delay anisotropy in photoelectron emission from isotropic helium

S. Heuser[1] †, Á. Jiménez-Galán[2], C. Cirelli[1], M. Sabbar[1], R. Boge[1], M. Lucchini[1], L. Gallmann[1,4], I. Ivanov[5,6], A. Kheifets[5], J. M. Dahlström[7,8,9], E. Lindroth[7], L. Argenti[2], F. Martín[2,3,10], U. Keller[1]

[1]Physics Department, ETH Zurich, 8093 Zurich, Switzerland, EU
[2]Departamento de Química, Módulo 13, Universidad Autónoma de Madrid, 28049 Madrid, Spain, EU
[3]Instituto Madrileño de Estudios Avanzados en Nanociencia, Cantoblanco, 28049 Madrid, Spain, EU
[4]Institute of Applied Physics, University of Bern, 3012 Bern, Switzerland, EU
[5]Research School of Physics and Engineering, The Australian National University, Canberra ACT 0200, Australia
[6]Center for Relativistic Laser Science, Institute for Basic Science (IBS), Gwangju 500-712, Republic of Korea, Asia
[7]Department of Physics, Stockholm University, AlbaNova University Center, SE-10691 Stockholm, Sweden, EU
[8]Max Planck Institute for the Physics of Complex Systems, Noethnitzerstr. 38, 01187 Dresden, Germany, EU
[9]Center for Free-Electron Laser Science, Luruper Chaussee 149, 22761 Hamburg, EU
[10]Condensed Matter Physics Center (IFIMAC), Universidad Autónoma de Madrid, 28049 Madrid, Spain, EU

Synopsis In contrast to expectations, we observe that the photoionization time delay from the $1s^2$ spherically symmetric ground state of He depend on the electron emission direction with respect to an external reference. We attribute the observed anisotropy to the interplay between different final quantum states, which become accessible once two photons are involved in the photoionization process. This is a universal effect, which needs to be taken into account for any study dealing with photoionization dynamics.

Attosecond electron dynamics have been studied with different pump-probe techniques each employing an extreme ultraviolet (XUV) pump-pulse to ionize the target system and an infrared (IR) probe-pulse, which subsequently probes the dynamics of the liberated electron. Thereby, relative timing information between electrons originating from different states within the same atom or different atoms can be extracted.

Alternatively, relative emission timing of electrons from the same initial state within the same target system but at different emission angles θ can also be studied. We define θ as the electron emission direction relative to the XUV-pump polarization axis.

Since photons carry a spin angular momentum of 1, the absorption of a single photon can excite electrons from an initial state into two different final states: $n_il_i \rightarrow E_f l \pm 1$, with n (l) representing the principle (orbital angular momentum) quantum number and E the photoelectron energy. As a consequence, if the ground state is spherically symmetric only one final state can be reached. However, as soon as two photons are involved in the ionization process, two different final states $1s \rightarrow E_ip \rightarrow E_fs/E_fd$ become accessible (Fig. 1a). Their interplay may give rise to anisotropic time delays.

Here, we present a rigorous experimental and theoretical investigation of the angle-dependent photoemission time delay of electrons removed from the spherically symmetric $^1S^e(1s^2)$ ground state of He.

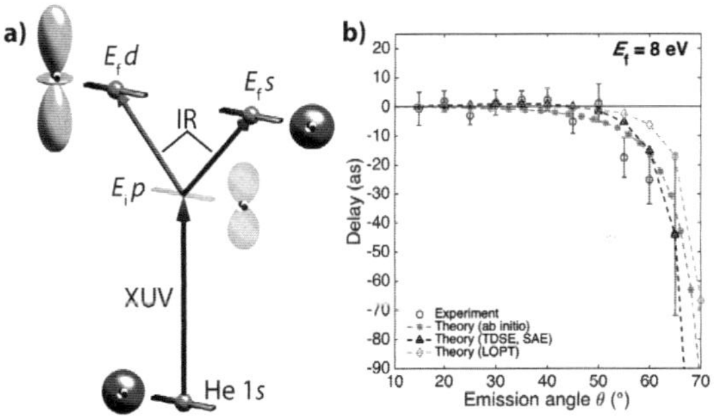

Figure 1. a) Accessible final states upon XUV-IR ionization: E_fs and E_fd. b) Angular dependence of experimental (blue points) and theoretical (dashed lines) data for a photoelectron energy of 8 eV.

We measure a significant angular variation of the photoionization time delay, which can be as large as 50 attoseconds (Fig. 1b). Our observation sheds light on a new general aspect for attosecond measurements resolving electron dynamics triggered by ultra-short pulses. Indeed, so far most measurement techniques that resolved attosecond electron dynamics have been based on a two-photon process: one photon to initiate and another one to probe the dynamics of the system. Therefore, the effect studied in this work is universal and has important implications on the electron dynamics in more complex systems such as e.g. molecules and condensed matter.

[1] E-mail: sheuser@phys.ethz.ch

Content from this work may be used under the terms of the Creative Commons Attribution 3.0 licence. Any further distribution of this work must maintain attribution to the author(s) and the title of the work, journal citation and DOI.
Published under licence by IOP Publishing Ltd

XXIX International Conference on Photonic, Electronic, and Atomic Collisions (ICPEAC2015) IOP Publishing
Journal of Physics: Conference Series **635** (2015) 092090 doi:10.1088/1742-6596/635/9/092090

Laser-Produced Plasma Spectroscopy of Medium to High-Z Elements in the 2 to 9 nm Spectral Region

Elaine Long[*1], Chihiro Suzuki[†], John Sheil[‡], Elgiva White[‡], Ragava Lokasani[*], Bowen Li[*], Paul Sheridan[*], Patrick Hayden[§], Padraig Dunne[*], Fergal O'Reilly[*] & Gerry O'Sullivan[*2]

[*] UCD School of Physics, University College Dublin, Belfield, Dublin 4, Ireland
[†] National Institute for Fusion Science, 322-6 Oroshi-cho, Toki 509-5292, Japan
[‡] Summer Intern at UCD School of Physics
[§] National Centre for Plasma Science & Technology and School of Physical Sciences, Dublin City University

Synopsis Much effort has gone into the development of plasmas as sources of radiation in the 2 to 9 nm region in recent times with particular interests in the "water window" region (2.3 to 4.4 nm) and the 6.X nm region. Plasmas were formed on a range of targets of medium to high atomic number (Z) elements using Nd:YAG lasers with various illumination parameters. Spectra were recorded using a 0.25 m grazing-incidence spectrograph and spectral analysis was carried out with the aid of Hartree-Fock with configuration interaction calculations using the Cowan suite of codes [1].

The region of the spectrum most commonly referred to as the "Water Window", between 2.3 and 4.4 nm, is of interest to those in the biomedical sciences due to its potential for high-contrast imaging based on the strong absorption by carbon and the low absorption by oxygen in this wavelength region [2].

The 6.X nm region is of interest to those in the semiconductor industry as a potential source for beyond extreme ultraviolet (BEUV) lithography [3] and metrology. One of the key advantages is that the short wavelength enables smaller lithographic feature sizes to be printed or imaged as dictated by Rayleigh's optical limit. The development of multilayer mirrors (e.g. LaN/B and LaN/B_4C) [4] which have relatively high throughputs in the 6.X nm region may enable BEUV lithography should an efficient source be developed.

The 0.25 m grazing-incidence spectrograph, which is equipped with two variable groove spacing gratings (1200 and 2400 grooves/mm), and a thermoelectrically cooled, back projection CCD camera with 2048 pixels in the spectral direction, is capable of recording spectra in the 1 to 20 nm region. Time-integrated spectra were recorded for several elemental targets using two Nd:YAG lasers with wavelengths of 1064 nm and pulse durations of 180 ps and 8 ns at a variety of laser power densities.

Figure 1 shows spectra for elements between Z = 12 and 68, produced using single, ~600 mJ, 8 ns pulses. The peak power density of the focused laser incident on the targets was approximately $4X10^{13}$ Wcm^{-2}. Hartree-Fock with configuration interaction calculations, performed using the Cowan suite of codes, were used as an aid to characterise the emission of the experimental spectra recorded.

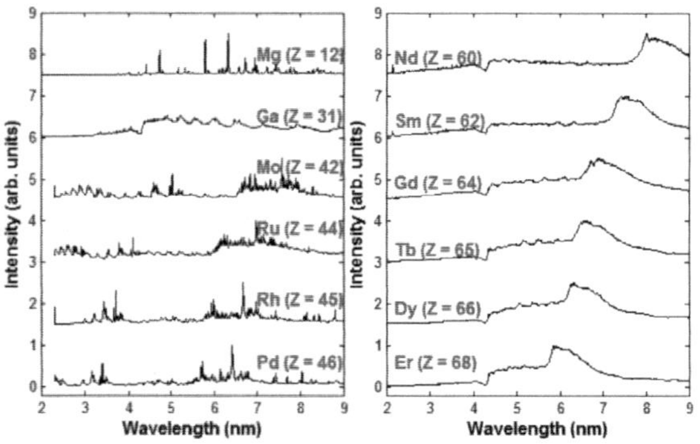

Figure 1. Laser-produced plasma spectra of medium to high-Z elements.

References

[1] R. D. Cowan, The Theory of Atomic Structure and Spectra (University of California Press, Berkeley, CA, 1981).

[2] P. A. C. Takman, H. Stollberg, G. A. Johansson, A. Holmberg, M. Lindblom, and H. M. Hertz, J. Microsc. **226**, 175 (2007).

[3] G. O'Sullivan, T. Cummins, P. Dunne, A. Endo, P. Hayden, T. Higashiguchi, D. Kilbane, B. Li, C. O'Gorman, T. Otsuka, E. Sokell and N. Yugami, Phys. Scr. **T156**,014105 (2013).

[4] I. A. Makhotkin, E. Zoethout, E. Louis, A. M. Yakunin, S. Müllender, and F. Bijkerk, J. Micro/Nanolith, MEMS and MOEMS, **11**, 040501 (2012).

[1] E-mail: elaine.long@ucdconnect.ie

[2] E-mail: gerry.osullivan@ucd.ie

Content from this work may be used under the terms of the Creative Commons Attribution 3.0 licence. Any further distribution of this work must maintain attribution to the author(s) and the title of the work, journal citation and DOI.
Published under licence by IOP Publishing Ltd

Journal of Physics: Conference Series **635** (2015) 092091 doi:10.1088/1742-6596/635/9/092091

Non-constant ponderomotive energy in above threshold ionization by intense few-cycle laser pulses

Renata Della Picca*[1], Ana A. Gramajo*, Diego G. Arbó†, Sebastián D. López† and Carlos R. Garibotti*

*CONICET and Centro Atómico Bariloche, Av. Bustillo 9500, 8400 Bariloche, Argentina
†Instituto de Astronomía y Física del Espacio IAFE (Conicet-UBA), Casilla de Correo 67 Suc. 28, C1428ZAA Ciudad Autónoma de Buenos Aires, Argentina

Synopsis We analyze the above threshold ionization of hydrogen atom by strong and short laser pulse with a simple envelope temporal dependence. We find additional peaks in the resulting ATI spectra originated from different values of the ponderomotive energy and the interference due to the electronic emission at different times.

Since the first observation in the eighties [1], Above Threshold Ionization (ATI) process has been extensively studied in the context of strong laser-matter interaction. In this process the target absorbs more photons than the minimum number required for ionization. As a result, the photo-electron (PE) spectra are characterized by peaks at energy values given by energy conservation [2]:

$$E = n\hbar\omega - I_p - U_p \quad (1)$$

where ω is the laser frequency and I_p is the ionization potential.

The ponderomotive energy U_p is the cycle averaged kinetic energy of an electron in an oscillating field. It is know that $U_p = (F_0/2\omega)^2$ when the electric field is described by a plane wave with amplitude F_0 [2]. However, the concept of U_p is questionable for few-cycle laser pulse. In this case, the field is commonly described as $F(t) = f(t)F_0 \cos(\omega t + \varphi)$ where the envelope $f(t)$ takes values starting from 0 and increasing to the maximum value 1, tipically at the middle of the pulse duration. Hence, different values of U_p are expected for each cycle and then, following Eq. (1), different energy position in the ionization spectra [3].

In this work, in order to analyze the envelope effect on the PE spectrum, we consider primarily a very simple and didactic envelope function with few steps. The calculations were done for ionization of the hydrogen atom from $1s$ within the semi classical Simple Man's Model (and also corroborated with the Coulomb-Volkov approach).

As we can see from the figure 1, each level of the field produces ATI peaks according to Eq. (1); and when another delayed step of same level is added, see figure 1 (b), the electrons emitted at different times generate a typical interference pattern in the PE spectrum [4].

During the conference we will show that this simple analysis can be easily extended to more realistic envelopes, as for example $f(t) = \sin^2(\pi t/\tau)$ case.

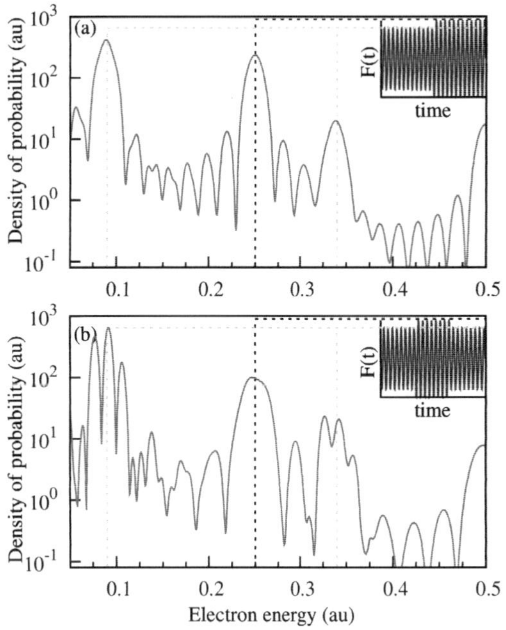

Figure 1. H(1s) ATI PE spectra with $F_0 = \omega = 0.25$ a.u. and 24 cycles. See insets for two envelopes: (a) one step and (b) symmetrical two steps.

References

[1] P. Agostini *et al* 1979 Phys. Rev. Let. **42** 1127

[2] J. Parker *et al* 1996 *J. Opt. Soc. Am.* B **13** 371

[3] M. Wickenhauser *et al* 2006 *Phys. Rev.* A **73** 011401R

[4] A. Gramajo *et al* 2015 *in preparation*

[1] E-mail: renata@cab.cnea.gov.ar

Content from this work may be used under the terms of the Creative Commons Attribution 3.0 licence. Any further distribution of this work must maintain attribution to the author(s) and the title of the work, journal citation and DOI.
Published under licence by IOP Publishing Ltd

Journal of Physics: Conference Series **635** (2015) 092092 doi:10.1088/1742-6596/635/9/092092

Upgrade of ESA-22D photoelectron spectrometer

Levente Ábrók*[1], Róbert Balog*†, Péter Herczku*†, Sándor Kovács*†, Dávid Hatvani*, Sándor Ricz*, Ákos Kövér*

* Institute for Nuclear Research, Hungarian Academy of Sciences (MTA Atomki), P.O. Box 51, H-4001 Debrecen, Hungary

† University of Debrecen, H-4032 Debrecen, Hungary

Synopsis Electron spectroscopy is a widely used method for observing the structure and fundamental processes in atomic systems. By employing energy- and angle-resolved photoelectron spectroscopy, valuable information can be obtained about ionization and excitation in photon-atom interactions.

The recently upgraded electron spectrometer, ESA-22D, was developed at the Division of Atomic Physics in Atomki, Debrecen. It was built for precise measurements of energy- and angular-distributions of electrons from the target area. As the spectrometer is cut into two halves we are able to measure two independent electron energy spectra simultaneously, allowing electron-electron coincidence measurements as well.

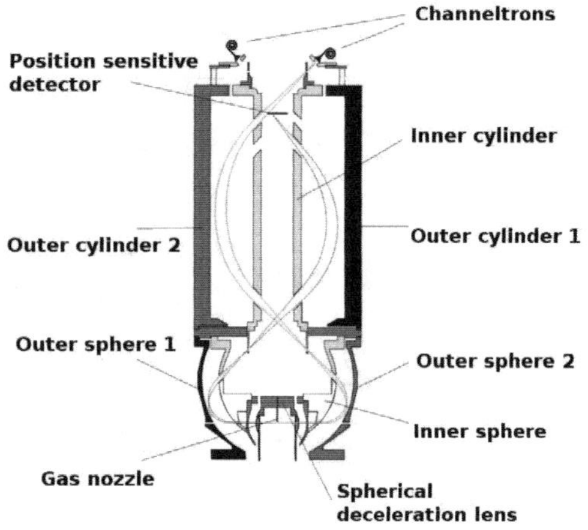

Figure 1. Basic design of ESA-22D electron spectrometer

The most significant improvements from our previous spectrometers [1] are the newly designed UHV vacuum chamber and detector holder, both being rotatable around different axes. Rotating the chamber around the beam axis gives us the opportunity to detect ejected photoelectrons not limited to the collision plane determined by the momentum and polarization vectors of the photons. As a result, our measurements can cover the 3-dimensional space, a unique feat not available before.

24 channel electron multipliers are placed in the holder at every 15°, allowing the simultaneous analysis of high resolution angular distributions. By rotating the holder around the spectrometer axis, the relative efficiencies of the individual detectors, as well as the angle dependent transmission function of the spectrometer can be determined. By summing the obtained intensities, measurement time is greatly reduced.

Furthermore, an electron gun was installed along the spectrometer axis, thus calibration can be performed at every electron energy using the elastically scattered electrons. A new data-gathering and -processing software was also developed using LabView, easing the control and evaluation of measurements.

With our measuring system we plan to study multi-photon single ionization of valence and inner shells of free atoms and molecules at the attosecond beamline being built in Szeged (ELI-ALPS) in order to investigate the channel interaction between the direct and indirect multi-photon ionization processes. We hope to resolve the disagreement between the theoretical end experimental data. We also plan to carry out measurements at the beamline of PETRAIII, Hamburg in order to measure the second order non-dipole contribution in a photon energy range where the dipole and first order non-dipole matrix elements have minimum.

This work was supported by the OTKA (Grant No. K104409).

References

[1] Ricz S., Kövér Á., Jurvansuu M., Varga D., Molnár J., and Aksela S., Phys. Rev **A65** (2002) 042707

[1]E-mail: abrok.levente@atomki.mta.hu

Content from this work may be used under the terms of the Creative Commons Attribution 3.0 licence. Any further distribution of this work must maintain attribution to the author(s) and the title of the work, journal citation and DOI.
Published under licence by IOP Publishing Ltd

Journal of Physics: Conference Series **635** (2015) 092093 doi:10.1088/1742-6596/635/9/092093

Ionization of atoms and molecules in a strong two-color field

Yonghao Mi [1], Nicolas Camus, Martin Laux, Lutz Fechner, Robert Moshammer and Thomas Pfeifer [2]

Max-Planck-Institute for Nuclear Physics, Saupfercheckweg 1, 69117 Heidelberg, Germany

Synopsis We experimentally investigated the asymmetry of photo-ion and photoelectron momentum spectra of argon and nitrogen by ionizing an argon-nitrogen gas mixture in a strong two-color (400 nm + 800 nm) laser field. By changing the time delay between the 400 nm laser pulse and the 800 nm pulse, the fan-like stripes in low-energy electron momentum shift from negative momentum to positive for both argon and nitrogen. A clear difference of asymmetric electron emission between Ar and N_2 is observed.

Two-color laser fields, which are synthesized by an 800 nm laser pulse and its second harmonic, have been demonstrated as an ideal tool to control the electronic wavepackets on attosecond time scales [1]. In order to understand the difference of tunnel ionization between different atoms and molecules, we perform an experiment by focusing the two-color laser field into an argon-nitrogen gas mixture in a Reaction Microscope [2].

The second harmonic laser pulse is generated by passing a 25 fs laser pulse at the central wavelength of 800 nm through a BBO crystal. Two color pulses with parallel polarization, overlapped temporally and spatially by using a two-color interferometer, are focused into the Reaction Microscope for ionizing the supersonic gas mixture.

With the above-mentioned experiment setup, we obtain the 3D photo-electron momentum from N_2 and Ar. The fan-like stripes in the low momentum region and the Freeman resonances [3] can be clearly observed for both nitrogen and argon electrons. By changing the time delay between the 400 nm laser pulse and the 800 nm laser pulse, the fan-like stripes shift from negative momentum to positive momentum for the two species. Figure 1 shows the asymmetric electron momenta when the relative phase delay between the two color pulses changes from 0.25 to 1.25 . The electron momentum from Ar is almost identical to the one from N_2, only a minor difference is obtained. As the directional electron emission is strongly modulated with the time-delay between the two-color pulses, we also investigate the asymmetry in the number of counts between electrons emitted with positive and negative longitudinal momentum for nitrogen and argon. A phase difference of asymmetric electron emission at a certain longitudinal momentum (the case of 0.5 a.u. and 0.2 a.u. is shown in Figure 2 (a) and (b), respectively), corresponding to tens of attosecond time delay, is observed between argon and nitrogen. The phase difference between atoms and molecules may help us to investigate the tunnel ionization process and its dependence on small details of the system potential or resonance structure.

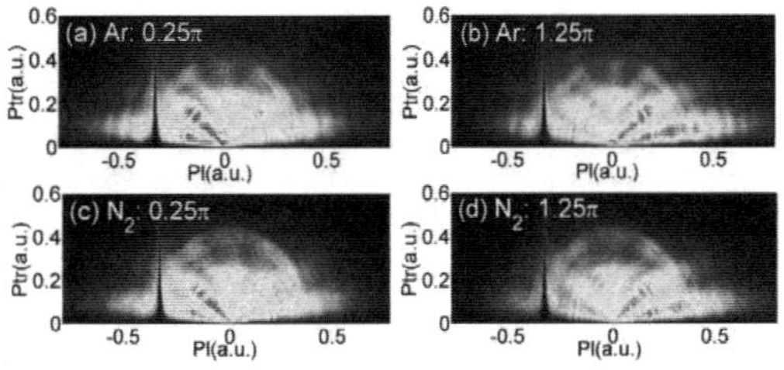

Figure 1. Asymmetric electron momentum distribution from N_2 and Ar for different relative phase between two-color fields

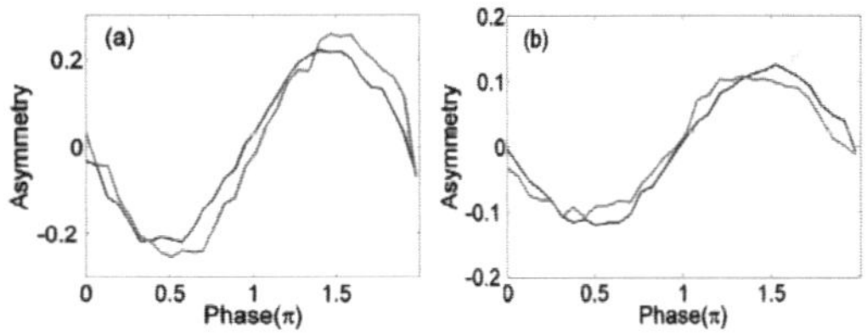

Figure 2. Asymmetry of electron emission between Ar (blue) and N_2 (red) at a certain longitudinal momentum ((a) 0.5 a.u. and (b) 0.2 a.u.) as a function of relative phase between 800 nm pulse and 400 nm pulse

References

[1] X. Xie *et al* 2012 *Phys. Rev. Lett.* **108** 193004
[2] J. Ullrich et al 2003 *Rep. Prog. Phys.* **66** 1463
[3] T.Morishita et al 2007 *Phys. Rev. A* **75** 023407

[1] E-mail: yonghao.mi@mpi-hd.mpg.de

[2] E-mail: thomas.pfeifer@mpi-hd.mpg.de

Content from this work may be used under the terms of the Creative Commons Attribution 3.0 licence. Any further distribution of this work must maintain attribution to the author(s) and the title of the work, journal citation and DOI.
Published under licence by IOP Publishing Ltd

K-shell width, fluorescence yield, and K_β/K_α intensity ratio calculation for Fe in the Dirac-Fock approach

J. P. Marques*[1], M. Guerra†, F. Parente†, J. P. Santos†, P. Indelicato‡, J. M. Sampaio*

* BioISI - Biosystems & Integrative Sciences Intitute, Faculdade de Ciências da Universidade de Lisboa, Campo Grande, C8, 1749-016 Lisboa, Portugal

† Laboratório de Instrumentação, Engenharia Biomédica e Física da Radiação (LIBPhys-UNL), Departamento de Física, Faculdade de Ciências e Tecnologia da Universidade Nova de Lisboa, Monte da Caparica, 2892-516 Caparica, Portugal

‡ Laboratoire Kastler Brossel, École Normale Supérieure, CNRS, Sorbonne Universités, UPMC Univ. Paris 06, Case 74; 4, place Jussieu, 75252 Paris CEDEX 05, France

Synopsis K-shell fluorescence yield, width, and K_β/K_α intensity ratio for Fe are calculated using the Dirac-Fock approach. Results are compared with available data.

The knowledge of accurate values of atomic decay rates and fluorescence yields is fundamental in quantitative analysis in X-ray spectroscopy, plasma physics, nuclear physics, dosimetry, radiation protection, among many other areas [1]. Despite the increasing number of theoretical and experimental work concerning the determination of fluorescence yields for several elements, the available data are still scarce. Iron is one of the most abundant elements in the universe and is of extreme importance in the analysis of stellar objects like stars and galaxies. Iron spectral lines are used in plasma diagnostics and help us to know physical properties such as temperature and electron density as well as for abundance studies [2].

In this work, we present a relativistic calculation of the K-shell width, fluorescence yield and K_β/K_α intensity ratio in Fe using the multi-configuration Dirac-Fock (MCDF) code of Desclaux and Indelicato [3, 4]. The code was used in a single-configuration approach, with Breit interaction and the vacuum polarization terms included in the self-consistent field calculations, and other QED effects included as perturbations.

Table 1. K-shell fluorescence yield (ω_K), width (eV), and K_β/K_α intensity ratio for iron

	ω_K	width	K_β/K_α
This work	0.378	1.14	0.134
McGuire [5]	0.386		
Krause [6]	0.34		
Campbell [7]		1.19	
Carreras *et al* [8]			0.14
Scofield [9]			0.139
Jankowsky*et al* [10]			0.134
Bé *et al* [11]			0.142

Since iron has an open outer shell ($3d^6$), electron coupling leads to 348105 possible radiative and radiationless transitions from the initial K hole configuration. Preliminary results for the K-shell fluorescence yield, width and K_β/K_α intensity ratio are consistent with the available data (Table 1).

This work was supported by FCT Project PEstOE/FIS/UI0303/2011, Portugal. We also thank the Allianz Program of the Helmholtz Association, contract No. EMMI HA-216 Extremes of Density and Temperature: Cosmic Matter in the Laboratory. BioISI is supported by the centre grant UID/MULTI/04046/2013 from FCT/MCTES/PIDDAC, Portugal.

References

[1] J. M. Sampaio *at al* 2014 *Phys. Rev. A* **89** 012512

[2] H. Hartman *at al* 2009 *J. Phys. Conf. Ser.* **130** 012010

[3] J. Descalux 1975 *Comput. Phys. Commun.* **9** 31

[4] P. Indelicato, J. P. Desclaux 1990 *Phys. Rev. A* **42** 5139

[5] E. J. McGuire 1970 *Phys. Rev. A* **2** 273

[6] M. O. Krause 1979 *J. Phys. Chem Ref. Data* **8** 307

[7] J. Campbell, T. Papp 2001 *At. Data Nucl. Data Tables* **77** 1

[8] A. Carreras *et al* 2003 *Rad. Phys. Chem* **67** 703

[9] J. Scofield 1974 *Phys. Rev. A* **9** 1041

[10] K. Jankowski, M. Polasik 1989 *J. Phys. B* **22** 2369

[11] M. -M Bé *et al* 1998 *Appl. Radiat. Isot.* **49** 1367

[1]E-mail: jmmarques@fc.ul.pt

Content from this work may be used under the terms of the Creative Commons Attribution 3.0 licence. Any further distribution of this work must maintain attribution to the author(s) and the title of the work, journal citation and DOI.
Published under licence by IOP Publishing Ltd

Relativistic calculations of atomic parameters in Ununoctium

J. M. Sampaio*[1], M. Guerra†, T. I. Madeira*, F. Parente†, P. Indelicato‡, J. P. Santos† and J. P. Marques*

* BioISI - Biosystems & Integrative Sciences Institute, Faculdade de Ciências da Universidade de Lisboa, Campo Grande, C8, 1749-016 Lisboa, Portugal

† Laboratório de Instrumentação, Engenharia Biomédica e Física da Radiação (LIBPhys-UNL), Faculdade de Ciências e Tecnologia da Universidade Nova de Lisboa, Monte da Caparica, 2892-516 Caparica, Portugal

‡ Laboratoire Kastler Brossel, École Normale Supérieure, CNRS, Sorbonne Universités, UPMC Univ. Paris 06, Case 74; 4, place Jussieu, 75252 Paris CEDEX 05, France

Synopsis K shell, and L and M subshells fluorescence and Coster-Kronig yields for the super-heavy element Uuo are calculated for the first time using the Dirac-Fock approach.

Ununoctium is the temporary IUPAC name for the transactinide element with $Z = 118$ to which corresponds the temporary element symbol Uuo. On the periodic table of the elements it is a group VIII element and the last one of the 7th period. Uuo has the highest atomic number and the highest atomic mass of all the elements discovered so far. The first decay of Uuo atoms was observed in 2002 at the JINR in Dubna. In 2006, researchers from JINR and LLNL announced that they had indirectly detected a total of three (possibly four) nuclei of Uuo-294 produced via collisions of Cf-249 atoms and Ca-48 ions [1].

Transition probabilities needed to derive the atomic yields shown in Table 1 were calculated with the relativistic general purpose MCDF code (MCDFGME) developed by Desclaux [2] and Indelicato. The MCDFGME code has been adapted since 1986 to work with super-heavy elements [3]. The nuclear charge distribution is described by the Fermi model with a nuclear mean-square radius obtained from fits to the experimental data of Angeli [4]. Initial and final state wavefunctions were calculated taking into account quantum electrodynamics (QED) radiative corrections, namely the self-energy and vacuum polarization [5].

Radiative transition rates were calculated making use of wavefunctions obtained in a complete self-consistent process including relaxation. We used the so-called optimized levels to obtain the wavefunctions and energies of all levels involved. No orbital relaxation was allowed between the initial and final bound state wavefunctions during the calculation of the radiationless transitions in order to ensure orthogonality. However, these transitions were calculated using the correct energies obtained in previous independent calculations of initial and final state wavefunctions and eigenvalues [6].

[1]E-mail: jmsampaio@fc.ul.pt

Table 1. K-, L-, and M-shell yields of Uuo.

Fluorescence yield		Coster-Kronig yield	
ω_K	9.73E-01	$f_{L_{12}}$	9.94E-03
ω_{L_1}	4.33E-01	$f_{L_{13}}$	4.31E-01
ω_{L_2}	6.86E-01	$f_{L_{23}}$	1.60E-01
ω_{L_3}	6.42E-01	$f_{M_{12}}$	4.74E-02
ω_{M_1}	3.39E-02	$f_{M_{13}}$	5.65E-01
ω_{M_2}	3.63E-02	$f_{M_{14}}$	1.09E-01
ω_{M_3}	5.17E-02	$f_{M_{15}}$	1.30E-01
ω_{M_4}	1.23E-01	$f_{M_{23}}$	9.09E-02
ω_{M_5}	1.10E-01	$f_{M_{24}}$	6.16E-01
		$f_{M_{25}}$	1.14E-01
		$f_{M_{34}}$	2.83E-02
		$f_{M_{35}}$	5.45E-01
		$f_{M_{45}}$	4.39E-02

This work was supported by FCT project PEstOE/FIS/UI0303/2011, Portugal and the Allianz Program of the Helmholtz Association, contract EMMI HA-216 Extremes of Density and Temperature. BioISI is supported by the grant UID/MULTI/04046/2013 from FCT/MCTES/PIDDAC, Portugal. M.G. and T.I.M. acknowledges the FCT fellowships SFRH/BPD/92455/2013 and SFRH/BPD/69627/2010, respectively.

References

[1] Yu. Ts. Oganessian *et al* 2006 *Phys. Rev.* C74 044602

[2] J. P. Desclaux 1995 *Comput. Phys. Commun.* 9 31

[3] P. Indelicato 1986 *J. Phys.* B 19 1719

[4] I. Angeli 2004 *At. Data Nucl. Data Tab.* 87 185

[5] P. Indelicato *et al* 2010 *Theor. Chem. Acc.* 129 495

[6] J. M. Sampaio *et al* 2013 *J. Phys.* B 46 065001

Content from this work may be used under the terms of the Creative Commons Attribution 3.0 licence. Any further distribution of this work must maintain attribution to the author(s) and the title of the work, journal citation and DOI.
Published under licence by IOP Publishing Ltd

XXIX International Conference on Photonic, Electronic, and Atomic Collisions (ICPEAC2015) IOP Publishing
Journal of Physics: Conference Series **635** (2015) 092096 doi:10.1088/1742-6596/635/9/092096

Creating coherent hole wave packets with strong-field pulses

Stefan Pabst[*,1] and Hans Jakob Wörner[†,2],

[*] Center for Free-Electron Laser Science, DESY, Notkestrasse 85, 22607 Hamburg, Germany
ITAMP, Harvard-Smithsonian Center For Astrophysics, 60 Garden Street, Cambridge, MA 02138, USA
[†] Laboratorium für Physikalische Chemie, ETH Zürich, Wolfgang-Pauli-Strasse 10, 8093 Zürich, Switzerland

Synopsis A new approach is identified that can generate highly coherent hole wave packets with multi-cycle strong-field pulses and stands in contrast to the usual idea of making the ionizing pulse as short as possible. The effect is demonstrated in atomic xenon with a spin-orbit wave packet involving the $5p_{3/2}^{-1}$ and $5p_{1/2}^{-1}$ ionic states. Tuning the driving frequency to an even fraction of the hole oscillation frequency ensures the creation of highly coherent holes.

Creating coherent electronic motion is atoms and molecules on ultrafast time scale is as exciting as challenging [1]. One specific type of electronic motion, which has attracted wide interest in the attosecond community, is hole migration [2]. By photoionization, an electronic hole motion can be triggered that moves within the parent system on a sub- to few-femtosecond timescale [2].

The hole generation process is often greatly simplified by exploiting the sudden approximation, which states that an electron is instantly removed from one specific state. Ionizing an electron with an attosecond or strong-field pulse can become a complex process for itself. Mechanisms such as interchannel coupling can very fast turn photoionization into a true many-body process. The coherence in the hole state that one wanted to create could be lost due to interchannel effects [3]. A coherent hole wave packet is, however, needed to see the subsequent hole migration dynamics.

Here, we proposed an alternative approach to create coherent hole wave packet that does not rely on attosecond XUV [3] or sub-cycle IR [4] pulses, which are very hard to generate. The main idea of this approach is that the system gets only ionized when the hole returns to its initial form. In this way, the hole creation is "in phase" with the hole oscillation. This ensures that the final hole state is fully coherent. In other words, the driving frequency ω of the pulse should be an even fraction of the hole oscillation frequency ΔE, i.e., $\Delta E = 2n\omega$. In Fig. 1 the hole coherence between $5p_{3/2}^{-1}$ and $5p_{1/2}^{-1}$ in xenon ($\Delta E = 1.3$ eV) is shown as a function of pulse duration and wavelength.

Besides the hole coherence, also the hole population is affected by ω. An enhanced hole population can be found when the system gets ionized "in phase" with the hole motion—exactly at the same frequencies as the maximum hole coherences are reached.

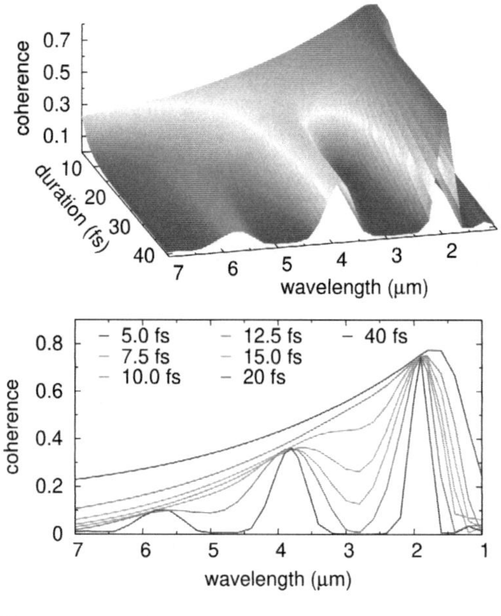

Figure 1. Hole coherence between the two ionic states $5p_{3/2}^{-1}$ and $5p_{1/2}^{-1}$ as function of pulse duration and wavelength.

References

[1] P. M. Kraus, *et al.*, PRL. **111**, 243005 (2013).

[2] S. Lünnemann, A. I. Kuleff, and L. S. Cederbaum, JCP **130**, 154305 (2009).

[3] S. Pabst, *et al.*, PRL **106**, 053003 (2011).

[4] A. Wirth, *et al.*, Science **334**, 195 (2011).

[1]E-mail: stefan.pabst@cfa.harvard.edu
[2]E-mail: woerner@phys.chem.ethz.ch

Content from this work may be used under the terms of the Creative Commons Attribution 3.0 licence. Any further distribution of this work must maintain attribution to the author(s) and the title of the work, journal citation and DOI.
Published under licence by IOP Publishing Ltd

Journal of Physics: Conference Series **635** (2015) 092097 doi:10.1088/1742-6596/635/9/092097

Wigner time delay in quadrupole photoionization channels of atomic Hg

Ankur Mandal*[1], Soumyajit Saha*, Tanima Banerjee*, Pranawa C. Deshmukh*[2], Anatoli Kheifets†, Valeriy K. Dolmatov$, Steven T. Manson**

*Department of Physics, Indian Institute of Technology Madras, Chennai, 600036, India

†Research School of Physics and Engineering, The Australian National University, Canberra ACT 0200, Australia

$Department of Physics and Earth Science, University of Northern Alabama, Florence, AL 35632, USA

**Department of Physics and Astronomy, Georgia State University, Atlanta, 30303, USA

Synopsis Time delay in quadrupole (E2) channels is studied in atomic Hg using the Relativistic Random Phase Approximation. Interchannel coupling is found to be extremely important and large time delays are found in the region of quadrupole (E2) Cooper minima.

Recent development of ultrafast sciences has enabled the access of quantum dynamics on the atomic time scale. Time delay in nonresonant photoionization has been measured to be a few tens of attoseconds [1]. It is thus important to study the process theoretically to understand time delay at a fundamental level.

Earlier studies have revealed the importance of non-dipole contributions to the photoionization parameters [2]. Though the cross-section for E2 transitions is usually four orders of magnitude less than that of dipole (E1) transitions, it has been found that we do have some energy regimes where E2 cross-sections dominate over E1, particularly near dipole Cooper minima [3] where it is necessary to incorporate E2 transition channels in the theory. Here we study the time delay in E2 channels. Since time delay in E1 channels is strongly affected by relativistic and correlation effects [4], it is important to include these effects in E2 channels too.

Hg has been chosen because some E1 and E2 studies already exist [5]. We have used the Eisenbud-Wigner formalism [6, 7] to calculate the time delay in which the time delay in a transition is given as $\tau = \hbar d\eta / dE$, where η is the phase of the transition matrix element. The Relativistic Random Phase Approximation (RRPA) [8] is employed in the calculations.

Time delay for the $6s \rightarrow \varepsilon d_{5/2}$, $\varepsilon d_{3/2}$ E2 channels are calculated, at various levels of truncation of the RRPA to gain some understanding of the importance of interchannel coupling.

The time delay for the $6s \rightarrow \varepsilon d_{5/2}$ E2 transition is shown in Fig. 1, along with the absolute value of transition matrix elements, for two levels of truncation of the RRPA calculation; 11 channels (all 6s and 5d transitions coupled) and 2 channels (only the 6s transitions coupled). Away from the Cooper minima, the time delay in this quadrupole channel is seen to be comparable to the dipole channels. As in the case of E1 photoionization (not shown), correlations due to interchannel coupling strongly affect the time delay in the E2 channels; interchannel coupling of the 6s with 5d subshell strongly influences the E2 time delay in the Cooper minimum region—the delay drops by about an order of magnitude owing to coupling with 5d. However, even at the 11-channel level, the time delay maximizes at about 50 as.

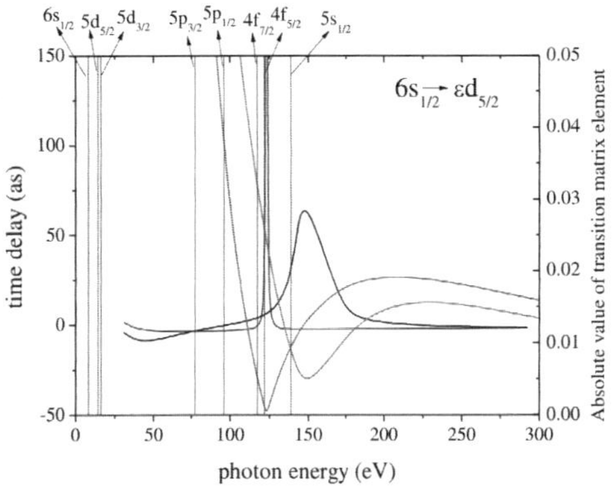

Figure 1. Time delay in the $6s \rightarrow \varepsilon d_{5/2}$ photoionization transition in Hg (left scale); 11 channel (black), 2 channel (blue). Also shown are the absolute values of the transition matrix elements (right scale); 11 channel (red), 2 channel (green). Vertical lines denote thresholds.

References

[1] M. Schultze *et al* 2010 *Science* **328** 1658
[2] G. B. Pradhan *et al* 2011 *J. Phys. B: At. Mol. Opt. Phys.* **44** 201001
[3] J. W. Cooper 1962 *Phys. Rev.* **128** 681
[4] S. Saha et al *Phys. Rev. A* **90** 053406
[5] T. Banerjee *et al* 2007 *Phys. Rev. A* **75** 042701
[6] E. P. Wigner 1955 *Phys. Rev.* **98** 145
[7] L. E. Eisenbud 1948 *Ph. D. thesis* Princeton Univ.
[8] W. R. Johnson and C D Lin 1979 *Phys. Rev. A* **20** 964

[1] E-mail: amankur@physics.iitm.ac.in

[2] E-mail: pcd@physics.iitm.ac.in

Content from this work may be used under the terms of the Creative Commons Attribution 3.0 licence. Any further distribution of this work must maintain attribution to the author(s) and the title of the work, journal citation and DOI.
Published under licence by IOP Publishing Ltd

XXIX International Conference on Photonic, Electronic, and Atomic Collisions (ICPEAC2015) IOP Publishing
Journal of Physics: Conference Series **635** (2015) 092098 doi:10.1088/1742-6596/635/9/092098

Relativistic dynamics for hydrogenlike systems

Tor Kjellsson*[1], Sølve Selstø†, Simen Bræck†, Eva Lindroth*

* Department of Physics, Stockholm University, AlbaNova University Center, SE-106 91 Stockholm, Sweden
† Faculty of Technology, Art and Design, Oslo and Akershus University College of Applied Sciences, N-0130 Oslo, Norway

Synopsis The interaction between super intense laser pulses and hydrogenlike systems is investigated. We aim to study cases where the field drives the electron up to velocities of $v \sim 0.5c$, where relativistic effects are believed to play a significant role. In order to investigate this, the time dependent Dirac equation (TDDE) has been solved numerically and comparisons are made to the predictions from solutions of the time dependent Schrödinger equation (TDSE).

The strong development in laser technology has lead to pulses with increasing intensities and higher photon energy. Given the extreme peak intensity expected in the future, for instance at the X-ray free electron laser facilities, it is of great importance to explore the role of relativistic effects in light-matter interaction predicted by the TDDE.

Doing so however poses quite some challenges. First, the spectrum of the time independent Dirac equation contains negative energy states (NES). Since these are connected to the description of positrons one might expect that they can be neglected as long as the field strength is far below the limit for electron-positron pair-production. This is however, in general, not the case since the NES are required to even get the correct non-relativistic limit [1, 2]. Together with the spin degree of freedom the NES make the basis set within each symmetry four times larger than in the non-relativistic case. Second, not only is the computational load heavy due to a larger basis set within each symmetry but the total number of symmetries required for convergence in a simulation seems to increase when the NES are included. Finally, for field strengths where a relativistic treatment is called for one also expects effects beyond the dipole approximation to be important. This increases the number of couplings significantly since the electron can now change m-quantum numbers even with linearly polarized light.

Here we will present various results obtained from a numerical solution of the full-dimensional TDDE beyond the dipole approximation. The spatial part of the vector field is represented through a power series expansion up to fourth order and the convergence of this expansion at different field strengths will be compared.

To handle the computational demand we employ complex scaling to minimize the basis set and a second order Magnus propagator for time propagation. The propagation is significantly speeded up by the use of a Krylov subspace technique and parallelized routines for matrix operations.

Figure 1 shows the ionization yield of hydrogen with the electron initially prepared in the ground state, exposed to a pulse of the form $\mathbf{A}(\omega t - kx) = \frac{E_0}{\omega}\sin^2\left(\frac{\omega t - kx}{2N}\right)\sin(\omega t - kx)\,\hat{\mathbf{e}}_z$. In this example the number of optical cycles is $N = 10$, the central frequency is $\omega = 3.5$ a.u. and the peak electric field strength E_0 is varied.

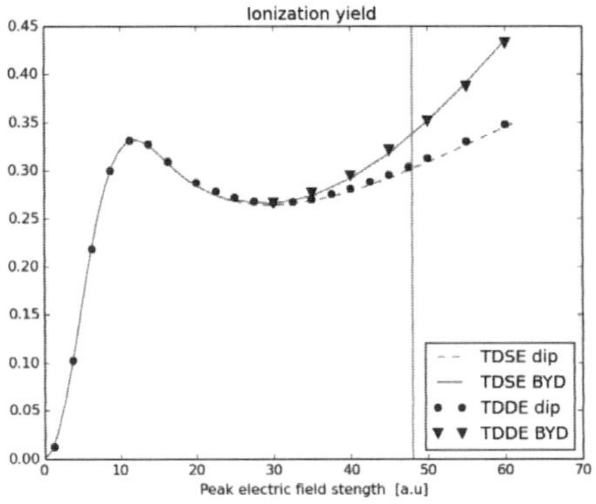

Figure 1. Ionization yield of hydrogen for the electron initially prepared in the ground state. The acronym BYD stands for beyond dipole approximation. The vertical line denotes the field strength where the electron obtains a velocity $v \sim 0.1c$.

References

[1] S. Selstø *et al* 2009 *Phys. Rev. A* **79** 043418

[2] Y. V. Vanne and A. Saenz 2012 *Phys. Rev. A* **85** 033411

[1]E-mail: tor.kjellsson@fysik.su.se

Content from this work may be used under the terms of the Creative Commons Attribution 3.0 licence. Any further distribution of this work must maintain attribution to the author(s) and the title of the work, journal citation and DOI.
Published under licence by IOP Publishing Ltd

XXIX International Conference on Photonic, Electronic, and Atomic Collisions (ICPEAC2015) IOP Publishing
Journal of Physics: Conference Series **635** (2015) 092099 doi:10.1088/1742-6596/635/9/092099

Coincidence photoelectron measurements following 2p photoionization in Mg

Emma Sokell*[1], P Bolognesi†, Stephan Safgren*, Elaine Long* and Lorenzo Avaldi†

• School of Physics, Science Centre North, University College Dublin, Belfield, Dublin 4, Ireland
† CNR-Istituto di Struttura della Materia , Area della Ricerca di Roma1 , 00015 Monterotondo Scalo, Italy

Synopsis Coincidence photoelectron spectroscopy has been used to test the two-step model of Auger decay following 2p photoionization in magnesium. Fitting of the data has yielded information on the dipole photoionization matrix elements as a function of photon energy.

Photoelectron-Auger electron coincident measurements have been used to investigate 2p photoionization in magnesium. Generally, the angular distribution of the Auger electron depends on both the alignment of the core hole state and the contributing partial waves of the Auger decays. In specific cases, such as Mg $2p_{3/2}$ photoionization followed by L_3-M_1M_1 Auger decay, only a single partial wave contributes to the Auger decay and a measurement of the angular distribution of the Auger electron yields information about the alignment of the core-hole state [1]. This particular case led to a complete experiment, in which the dipole matrix elements were determined at a photon energy of 80 eV, without measurement of the spin of the electrons involved [2]. Similar information about the dynamics of the photoionization process [1] can be achieved by coincidence measurements of the photo- and Auger electrons, by a proper choice of the photoelectron angle.

In the present work the $2p_{3/2}$ photoelectron and the L_3-M_1M_1 Auger electron of Mg have been measured in coincidence after angle and energy selection. The experiments were performed with the multi-coincidence end-station of the Gas Phase beamline at Elettra. This set-up contains ten hemispherical analyzers mounted on two independently rotatable frames. Three of the analyzers, in the plane of the linearly polarized incident light, direction ε, were fixed at 0, 30 and 60° with respect to ε, whilst the other seven, also mounted at 30° intervals in the same plane, were rotated in the plane perpendicular to the direction of propagation of the incident beam. This arrangement allows three triple differential cross-section (TDCS) data sets to be obtained simultaneously.

Sets of three TDCS measurements have been made at a range of photon energies. One made at 80 eV, allowed a direct comparison with previous measurements [3]. A simultaneous fit of the three TDCS indicates agreement with the predictions of the two-step model [1] and yields information on the dipole matrix elements at 80 eV. A similar fitting procedure carried out at other photon energies, shows that the two-step predictions become less valid as the photon energy is reduced and final state interactions, such as post-collision interactions, become significant.

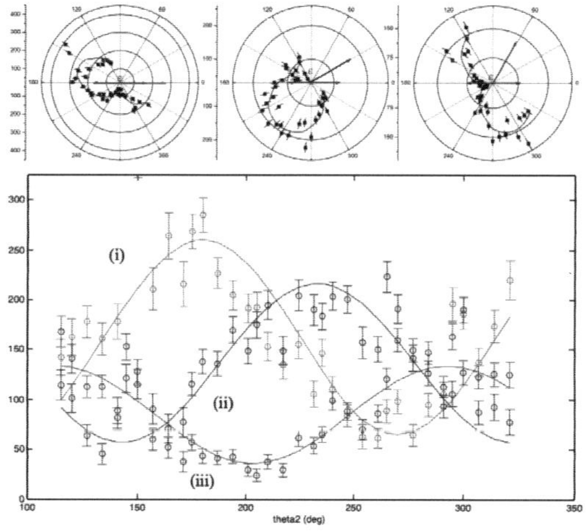

Figure 1. TDCS measured at 80 eV for three different photoelectron angles (indicated by the arrows) with respect to the polarization axis (ε), (i) 0°, (ii) 30°and (iii) 60°. The angle of emission of the Auger electron is indicated by the polar angle (theta2). The solid lines in the polar plots are smoothed representations of the data. The bottom panel shows a simultaneous fit to all three TDCS, using the two-step model [1].

This work was funded by Science Foundation Ireland, grant number SFI (08/RFP/PHY1117).

References

[1] V. Schmidt 1994 *Nucl. Instrum. Meth. in Phys. B.* 87 241
[2] A. Hausmann *et al* 1988 *Phys. Rev Lett. 61 2669*
[3] H. Lörch *et al. 1999 J. Phys. B.* 32 L371

[1] E-mail: Emma.Sokell@ucd.ie

Content from this work may be used under the terms of the Creative Commons Attribution 3.0 licence. Any further distribution of this work must maintain attribution to the author(s) and the title of the work, journal citation and DOI.
Published under licence by IOP Publishing Ltd

Journal of Physics: Conference Series **635** (2015) 092100 doi:10.1088/1742-6596/635/9/092100

Probing Xe electronic structure by two-color HHG

D. Faccialà[a], S. Pabst[b], B. D. Bruner[c], A. G. Ciriolo[a], S. De Silvestri[a], M. Devetta[a], N. Dudovich[c], M. Negro[a], H. Soifer[c], S. Stagira[a] and C. Vozzi[a 1]

[a] CNR–IFN & Politecnico di Milano (Italy)

[b] Center for Free-Electron Laser Science DESY (Germany) & ITAMP Harvard-Smithsonian Center for Astrophysics (USA)

[c] Department of Physics of Complex Systems, Weizmann Institute of Science (Israel)

Synopsis The aim of this study is probing the multi-electron behavior in xenon by two-color driven high harmonic generation. By changing the relative polarization of the two colors we were able to study different aspects of the multi-electron response.

We investigated the giant resonance in xenon [1, 2] by high-order harmonic generation (HHG) driven by a two-color field ($\lambda_1 = 1550\,\text{nm}$ and its second harmonic). We acquired HHG spectra as a function of the phase difference between the two colors, in two polarization configurations: parallel and orthogonal. The ratio between the second harmonic and the fundamental field was ≈ 0.4 for parallel polarization and ≈ 0.1 for perpendicular polarization. In parallel configuration, the presence of the second harmonic component produces two sets of trajectories with two different cutoffs: the upper branch corresponding to the more energetic trajectories and the lower branch, corresponding to the less energetic trajectories [4]. In the single active electron picture (SAE), the ionization probability of the upper branch is suppressed and only the lower branch is visible in the HHG spectrum. In our experimental results (fig. 1a) the upper branch is clearly visible, revealing an enhancement of the contribution of the more energetic trajectories with respect to SAE predictions. Theoretical calculations confirm this picture: the upper branch is not visible in the HHG spectrum in SAE (fig. 1b), while it appears when multi-electron effects are taken into account (fig. 1c). In orthogonal configuration the second harmonic acts as a displacement gate which selects specific continuum electron trajectories, depending on the relative phase between the two colors [3]. The HHG scan is reported in fig. 1d colormap along with the phase difference ϕ that maximizes harmonic emission for each harmonic order (red line). The theoretical expectation in the SAE approximation for ϕ is represented by the blue line in fig. 1d. A comparison between the two curves reveals two aspects: 1) the giant resonance increases the harmonic yield in the spectral region between 60 and 140 eV, allowing to observe the contribution of trajectories beyond the cutoff (70 eV); 2) the discrepancy between the experiment and the SAE model in the energy region corresponding to the giant resonance can be related to a change in the electron recombination time due to the collective electron correlation.

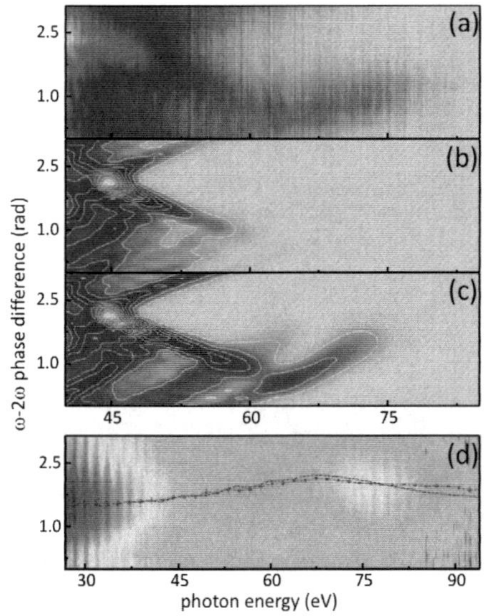

Figure 1. HHG scan in xenon as a function of ϕ. Parallel configuration: experiment (a) and Time-Dependent Configuration-Interaction Singles calculations without (b) and with (c) the contribution of electron correlation (linear scale). (d) Orthogonal configuration: harmonic spectra (colormap - log scale), experimental phase ϕ (red curve), SAE phase ϕ (blue curve).

References

[1] A. D. Shiner *et al* 2011 *Nat. Physics* **7** 464

[2] S. Pabst *et al* 2013 *Phys. Rev. Lett.* **111** 233005

[3] D. Shafir *et al* 2012 *Nature* **485** 343

[4] M. V. Frolov *et al* 2010 *Phys. Rev. A* **81** 063407

[1]E-mail: caterina.vozzi@polimi.it

Content from this work may be used under the terms of the Creative Commons Attribution 3.0 licence. Any further distribution of this work must maintain attribution to the author(s) and the title of the work, journal citation and DOI.
Published under licence by IOP Publishing Ltd

Observation of four-electron Auger processes

A. Müller© [1], A. Borovik Jr.©, T. Buhr©,†, J. Hellhund©, K. Holste©, A. L. D. Kilcoyne¶, S. Klumpp§, M. Martins§, S. Ricz©,‡, J. Viefhaus*, S. Schippers©

© Institut für Atom- und Molekülphysik, Justus-Liebig-Universität Giessen, 35392 Giessen, Germany
† Physikalisch-Technische Bundesanstalt, 38116 Braunschweig, Germany
¶ Advanced Light Source, Lawrence Berkeley National Laboratory, Berkeley, California 94720-8225, USA
§ Institut für Experimentalphysik, Universität Hamburg, 22761 Hamburg, Germany
‡ Institute for Nuclear Research of the Hungarian Academy of Sciences, 4001 Debrecen, Hungary
* FS-PE, DESY, 22607 Hamburg, Germany

Synopsis Absolute cross sections for triple photoionization of C^+ ions have been measured. Direct triple-Auger decay of K-shell excited states, i.e., an interaction process involving four electrons that results in the simultaneous emission of three electrons, could be unambiguously observed and quantified.

Multi-electron processes mediated by electron-electron interactions have attracted general interest ever since the early days of quantum mechanics. The free helium atom provides the minimum number of electrons necessary for electron correlation effects to occur which crucially determine the atomic structure of any He-like system and the decay probabilities of its excited states. The most important decay mechanism of doubly excited He or, more general, atomic systems with two or more electrons is the Auger process in which two electrons exchange energy so that one drops into a lower bound state while the second electron escapes to the continuum carrying away the transition-characteristic excess energy.

The next level of higher-order processes is reached when three electrons interact with one another. Carlson and Krause [1] identified double-Auger processes where one electron makes a transition to a deep atomic shell while the other two electrons are emitted simultaneously sharing the excess energy.

Recently, we have observed triple-Auger decay processes [2] (see figure 1) in which $C^+(1s2s^22p^2\,{}^2D,{}^2P)$ terms were selectively excited by irradiating singly charged carbon ions by VUV photons at the PETRA III synchrotron light source in Hamburg, Germany. Subsequent to the decay of the intermediate resonantly excited K-vacancy terms, product ions C^{2+}, C^{3+} and C^{4+} were recorded. Absolute cross sections for processes $\gamma + C^+(1s^22s^22p\,{}^2P) \rightarrow C^+(1s2s^22p^2\,{}^2D,{}^2P) \rightarrow C^{1+m+}(1s^22\ell^{4-m}) + me$ leading to net photoionization with $m = 1,2,3$ were determined. By making use of the outstanding brightness of PETRA III and the sensitivity of our experimental setup PIPE [3] the natural linewidths of the intermediate resonant states could be measured. By combining all this information, the absolute transition rates for radiative as well as single and multiple Auger decays were inferred. The triple-Auger decay rate was found to be 2.0×10^{10} s^{-1} for the $C^+(1s2s^22p^2\,{}^2D)$ and 9.7×10^{9} s^{-1} for the $C^+(1s2s^22p^2\,{}^2P)$ intermediate resonantly-excited terms investigated in this study. Both decay rates are estimated to have a 50% uncertainty. The ratios of single- to double- to triple-Auger rates of the ${}^2D$ and ${}^2P$ terms are about 100 : 2.7 : 0.013 and 100 : 3.3 : 0.013, respectively.

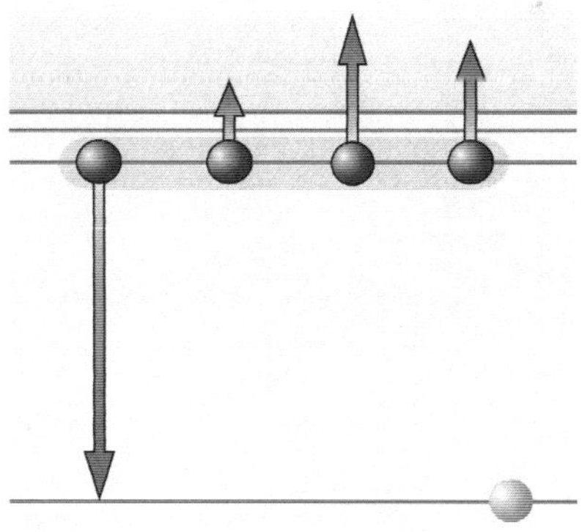

Figure 1. Three-electron-emission Auger process, for example in the decay of K-shell excited $C^+(1s2s^22p^2)$ ions.

References

[1] T. A. Carlson and M. O. Krause 1965 *Phys. Rev. Lett.* **14** 390

[2] A. Müller *et al* 2015 *Phys. Rev. Lett.* **114** 013002

[3] S. Schippers *et al* 2014 *J. Phys. B* **47** 115602

[1] E-mail: Alfred.Mueller@iamp.physik.uni-giessen.de

Content from this work may be used under the terms of the Creative Commons Attribution 3.0 licence. Any further distribution of this work must maintain attribution to the author(s) and the title of the work, journal citation and DOI.
Published under licence by IOP Publishing Ltd

e-Zn inelastic scattering at 80 eV

M. Piwiński* [1], Ł. Kłosowski*, D. Dziczek*, S.Chwirot*, D. V. Fursa† and I. Bray†

†*Institute of Physics, Faculty of Physics, Astronomy and Informatics, Nicolaus Copernicus University in Toruń, Grudziądzka 5, 87-100 Toruń, Poland

†Department of Physics, Astronomy and Medical Radiation Sciences, Curtin University, GPO Box U1987, Perth, WA 6845, Australia

Synopsis Stokes and Electron Impact Coherence Parameters (EICPs) for electronic excitation of 4^1P_1 state of zinc atoms have been measured for incident electron energy of 80 eV. The experimental data are presented together with Convergent Close Coupling (CCC) theoretical predictions. Our results are compared with recently published Relativistic Distorted–Wave Approximation (RDWA) calculations.

There has been a growing interest in studies of impact excitation of Zn atoms due to possible application of Zn in novel light sources and to a serious discrepancy between theory and experiment for the linear polarization Stokes parameter P_2 determined using spin-polarised electron impact excitation of the $(3d^{10}4s5s)^3S_1$ Zn state [1-4]. Therefore, it is important, to build up a body of knowledge relevant to electron impact on zinc with aim to test various aspects of the existing theoretical methods.

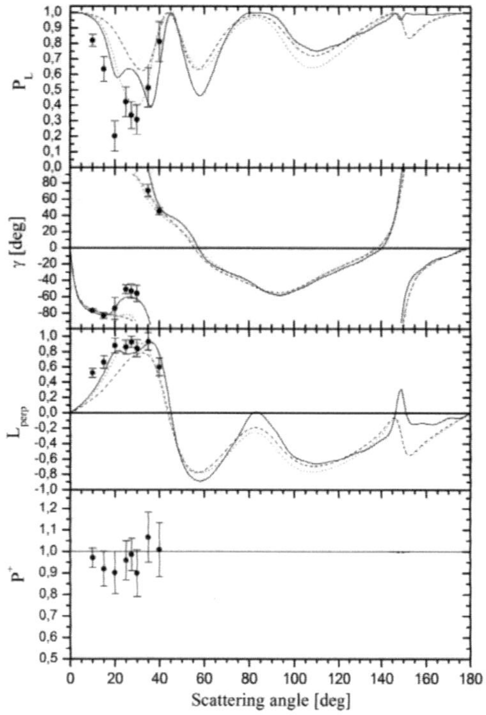

Figure 1. EICP parameters (P_L, γ, L_{perp}, P^+) for electronic excitation (80 eV) of 4^1P_1 Zn state. Experimental data (•) are presented together with, (—) CCC, (– –) SCRDWA and (· · ·) MC RDWA theoretical predictions [17].

This work is a continuation of our study on electron–zinc atom collisions [5-7], which is a part of our broader research on He-like atoms: Ca [8, 9], Cd [10–12] and He [13–14]. We present the set of experimental values of Stokes and Electron Impact Coherence Parameters (defined by Andersen *et al.* [15]) for 80 eV incident electron-energy, together with Convergent Close Coupling (CCC) theoretical predictions (for details see Napier *et al.* [16]). Our results are compared with the recently published results of RWDA modelling [17] of the same excitation process (Figure 1).

References

[1] L. Pravica *et al* 2011 *Phys. Rev. A* **83**, 040701
[2] J. F.Williams *et al* 2012 *Phys. Rev. A* **85**, 022701
[3] K. Bartschat and O. Zatsarinny 2013 *Phys. Rev. A* **87**, 016702
[4] C. J. Bostock *et al* 2013 *Phys. Rev. A* **87**, 016701
[5] M. Piwiński *et al* 2012 *Phys. Rev. A* **86**, 052706
[6] M. Piwiński *et al* 2013 *Eur. Phys. J.-Spec. Top.* **222**, 2273–2277
[7] M. Piwiński *et al* 2015 *Phys. Rev. A* in press
[8] D. Dyl *et al* 1999 *J. Phys. B: At. Mol. Opt. Phys.* **32**, 837–844
[9] D. Dziczek *et al* 1998 *Acta Phys. Polon. A* **93**, 717–721
[10] M. Piwiński *et al* 2002 *J. Phys. B: At. Mol. Opt. Phys.* **35**, 3821–3827
[11] D. Dziczek *et al* 2003 *Acta Phys. Polon. A* **103**, 3–11
[12] M. Piwiński *et al* 2006 *J. Phys. B: At. Mol. Opt. Phys.* **39**, 1945–1953
[13] Ł. Kłosowski *et al* 2007 *Meas. Sci. Technol.* **18**, 3801–3810
[14] Ł. Kłosowski *et al* 2009 *Phys. Rev. A* **80**, 062709
[15] N. Andersen *et al* 1988 *Phys. Rep.* **165**, 1–188
[16] S. A. Napier *et al* 2009 *Phys. Rev. A* **79**, 042702
[17] T. Das *et al* 2014 *Phys. Lett. A* **378**, 641–643

[1]E-mail: Mariusz.Piwinski@fizyka.umk.pl

Content from this work may be used under the terms of the Creative Commons Attribution 3.0 licence. Any further distribution of this work must maintain attribution to the author(s) and the title of the work, journal citation and DOI.
Published under licence by IOP Publishing Ltd

Ion yields and electron spectra of core-ionized Ar with Free-Electron Lasers

H. Banks*[1], A. O. G. Wallis*[2], A. Emmanouilidou*[3]

* Department of Physics and Astronomy, University College London, Gower Street, London WC1E 6BT, United Kingdom

Synopsis Hollow states, containing inner-shell holes, are created by exposure to a Free-Electron Laser (FEL). We compute ion yields and electron spectra, which are experimental observables.

At X-ray energies, the photoionization of core electrons is more probable than the photoionization of valence electrons. This allows for the production of hollow states [1, 2], which can be detected experimentally by the resultant ion yields and the electron spectra. These states offer a potential basis for two-site double core hole (tsDCH) spectroscopy [3, 4].

We have considered a variety of photon energies for different core hole states in argon; 200, 260, 315 and 360 eV for zero, one, two and three core holes, respectively. At low energy, the FEL pulse results in the ionization of 2p electrons, which is then followed by Auger decay. As each ionization of a core electron eventually leads to an Auger transition, this two-step sequence predominately populates even-charged ionic states. As the photon energy is increased, 2s electrons can also be ionized, introducing Coster-Kronig[5] transitions. Because these transitions conserve the number of core holes, they will be followed by additional Auger decay, favouring populating odd-charged ionic states.

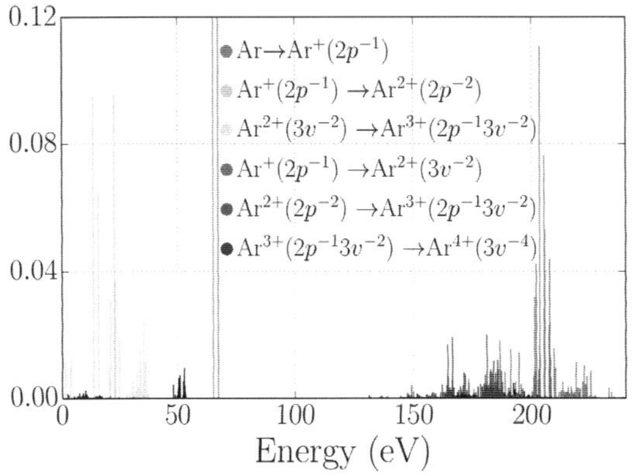

Figure 1. The one-electron spectra for argon, following a pulse of 5×10^{15}Wcm^{-2} intensity, with 315 eV photons and 10 fs duration.

We calculated the ion yields and electron spectra resulting from the above processes. The ion yields are not sufficient to detect the production of individual double core hole states. Therefore we use the one electron spectra for an unambiguous detection of a double core hole state, see figure 1.

We obtain the Auger spectrum of the double core hole state, by considering two electrons in coincidence. In particular the spectra resulting from considering an Auger and a photo-electron in coincidence is shown in figure 2. The Auger spectrum provides another experimental observable of hollow state formation.

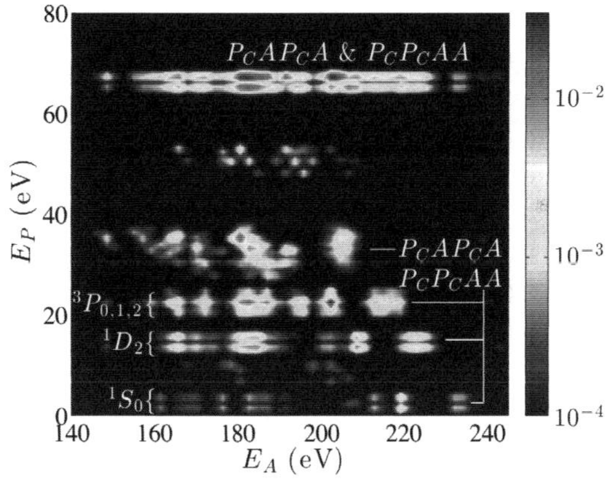

Figure 2. The Auger and photo-electron coincidence spectra for argon, following a pulse of 5×10^{15}Wcm^{-2} intensity, with 315 eV photons and 10 fs duration.

References

[1] A. Wallis *et al* 2014 *Phys. Rev. A* **89** 063417

[2] A. Wallis *et al* 2015 *arXiv* **1502** 00907v1

[3] P. Salen *et al* 2012 *Phys. Rev. Lett.* **108** 153003

[4] N. Berrah *et al* 2011 *PNAS* **108** 16912

[5] D. Coster and R. Kronig 2012 *Physica* **2** 13

[1]E-mail: h.banks@ucl.ac.uk
[2]E-mail: alisdair.wallis@googlemail.com
[3]E-mail: a.emmanouilidou@ucl.ac.uk

Content from this work may be used under the terms of the Creative Commons Attribution 3.0 licence. Any further distribution of this work must maintain attribution to the author(s) and the title of the work, journal citation and DOI.
Published under licence by IOP Publishing Ltd

XXIX International Conference on Photonic, Electronic, and Atomic Collisions (ICPEAC2015) IOP Publishing
Journal of Physics: Conference Series **635** (2015) 092104 doi:10.1088/1742-6596/635/9/092104

Many-body calculations on attosecond delays in photoionization

Eva Lindroth [1], **Marcus Dahlström** [2] **and Jimmy Vinbladh**

Department of Physics, Stockholm University, Alba Nova University Center, SE-106 91 Stockholm, Sweden

Synopsis We discuss how attosecond delays in laser-assisted photoionization can be computed using perturbation theory based on two-photon matrix elements.

In time-delay experiments attosecond XUV pulses are used to photoionize target atoms at well-defined times, followed by a probing process in real time by a phase-locked, infrared laser field. In this way, the laser field serves as a "clock" to monitor the ionization event. However, the observable delays do not correspond directly to the delay associated with single-photon ionization. Instead, a significant part of the observed delay originates from a measurement induced process, which obscures the single-photon ionization dynamics. To understand the experimental results it is thus necessary to describe the interaction with both light fields.

We will discuss in detail how attosecond delays in laser-assisted photoionization can be computed using perturbation theory based on two-photon matrix elements. Special emphasis is laid on above-threshold ionization, where the electron interacts with an infrared field after photoionization by the extreme ultraviolet field. Correlation effects are introduced using diagrammatic many-body theory to the level of the random-phase approximation with exchange (RPAE) [1, 2]. Strategies to go beyond that will also be discussed. Our aim is to provide an ab initio route to correlated multi-photon processes that are required for an accurate description of experiments on the attosecond time scale. Results will be shown for photoionization of the M-shell of argon atoms, where experiments have been carried out using the technique *Resolution of Attosecond Beating by Interference of Two-photon Transitions* (RABITT). An influence of autoionizing resonances in attosecond delay measurements is observed. Further, it is shown that the delay depends on both detection angle of the photoelectron and energy of the probe photon.

We will also discuss attosecond delays in double ioniztion [3] and show preliminary results for calculations on heavier systems where a relativistic treatment is called on.

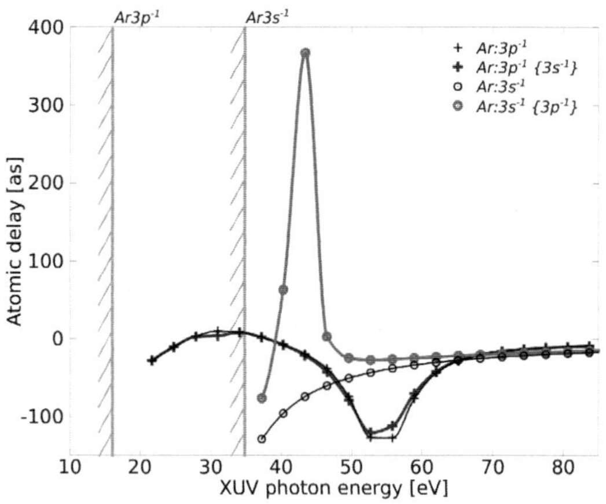

Figure 1. Atomic delay in argon observed along the polarization axis of the fields for ionization from the $3s$ and $3p$ orbitals in Argon The correlated orbitals in the RPAE-model are given in the curly brackets. The IR photon energy is 1.55 eV.

References

[1] J. M. Dahlström. T. Carette and E. Lindroth 2012 *Phys. Rev. A* **86** 061402(R)

[2] J. M. Dahlström and E. Lindroth 2014 *J. Phys. B* **47** 124012

[3] E. P. Månsson *et al* 2014 *Nat. Phys.* **10** 207

[1]E-mail: Eva.Lindroth@fysik.su.se
[2]E-mail: Marcus.Dahlstrom@fysik.su.se

Content from this work may be used under the terms of the Creative Commons Attribution 3.0 licence. Any further distribution of this work must maintain attribution to the author(s) and the title of the work, journal citation and DOI.
Published under licence by IOP Publishing Ltd

Journal of Physics: Conference Series **635** (2015) 092105 doi:10.1088/1742-6596/635/9/092105

Generalized Sturmian treatment of atomic double photoionization by n-photon absorption at low intensity laser fields

J M Randazzo*,†1, L U Ancarani$, D Mitnik§,†, C M Granados-Castro$, G Gasaneo‡,†, F D Colavecchia*,†

* División Física Atómica, Molecular y Óptica, Centro Atómico Bariloche, 8400 S. C. de Bariloche, Río Negro, Argentina.

§ Instituto de Astronomía y Física del Espacio, FCEyN Universidad de Buenos Aires, Argentina.

$ Equipe TMS, SRSMC, UMR CNRS 7565, Université de Lorraine, 57078 Metz, France.

‡ Departamento de Física, Universidad Nacional del Sur, 8000 Bahía Blanca, Buenos Aires, Argentina.

† Consejo Nacional de Investigaciones Científicas y Técnicas, CONICET, Argentina.

Synopsis Perturbative driven equation series for single and double ionization by n-photon absorption of atoms are solved in a stationary scattering framework with Generalized Sturmian Function expansions.

We apply the Generalized Sturmian approach for the study of single and double photoionization (DPI) of two electron atoms by n-photon absorption at low intensity laser fields. The initial unperturbed state corresponds to a bound state of the system, which we find by solving the Schrödinger equation:

$$[H_0 - E_0]\,\Psi_0 = 0. \tag{1}$$

The scattering state describing the dynamics of the system after absorption of one photon of energy ω satisfies the driven equation:

$$[H_0 - E_1]\,\Psi^+_{sc,1} = -W\Psi_0, \tag{2}$$

where $E_1 = \omega - E_0$ and W is the field operator in either the length (L) or velocity (V) gauges.

The three–body wave functions Ψ_0 and $\Psi^+_{sc,1}$ are both expanded with spherical Generalized Sturmian Functions (GSF) [1]:

$$\Psi = \sum_{\nu} a_\nu \mathcal{A} \frac{S_{n_1}(r_1)}{r_1}\frac{S_{n_2}(r_2)}{r_2}\mathcal{Y}^{L,M}_{l_1 l_2}(\hat{\mathbf{r}}_1,\hat{\mathbf{r}}_2) \tag{3}$$

where $\mathcal{A}$ is the exchange operator and $\mathcal{Y}^{L,M}_{l_1 l_2}$ are spherical biharmonics. The expansion coefficients a_ν are obtained by solving the linear system of equations which arises after projection by the entire basis. The GSF representation of both bound and continuum three–body wave functions leads to very accurate triple differential cross sections for single photon double ionization as illustrated in figure 1.

If two or more photons of energy ω are absorbed, we have to consider the corresponding wave functions which describe the dynamics at each stage:

$$[H_0 - E_n]\,\Psi^+_{sc,n} = -W\Psi^+_{sc,n-1} \tag{4}$$

where $E_n = \omega n - E_0$. Solving equations (2) and (4) with GSF involve continuum to continuum type matrix elements; in the first one, on the left hand side, and in the second one, on both sides. The structure and convergence of these integrals have been discussed for other ionization processes [4] and will be investigated here for $n > 1$ in both velocity and length gauges.

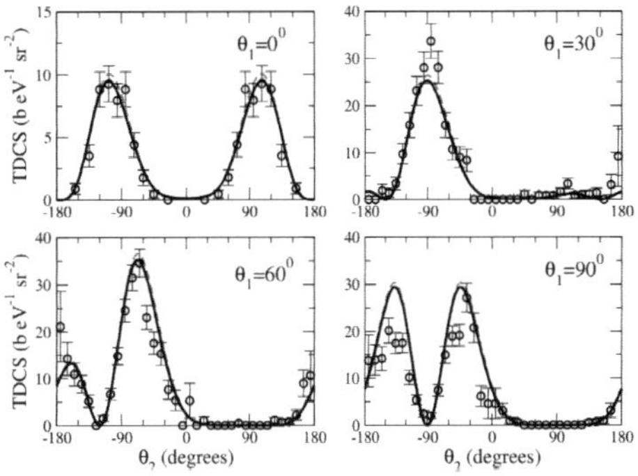

Figure 1. Triple differential cross section (TDCS) for DPI of He at 20 eV and equal energy sharing for different geometries. White circles: experimental data [2], black line: exterior complex scaling (ECS) results [3]. Red line: GSF results [5].

We will present single photon double ionization of helium for 20 and 40 eV, and then examine the solutions of equation (4) for $n > 1$ analyzing the applicability of the iterative series with a focus on the computational requirements for each n.

References

[1] G. Gasaneo *et al* 2013 *Adv. Quantum Chem.* **87** 153

[2] H. Bräuning *et al* 1998 *J. Phys. B* **31** 5149

[3] C. W. McCurdy *et al* 2004 *Phys. Rev. A* **69** 032707

[4] J. M. Randazzo *et al* 2011 *Phys. Rev. A* **84** 052715

[5] J. M. Randazzo *et al* submitted to *Eur. Phys J. D*

[1]E-mail: randazzo@cab.cnea.gov.ar

Content from this work may be used under the terms of the Creative Commons Attribution 3.0 licence. Any further distribution of this work must maintain attribution to the author(s) and the title of the work, journal citation and DOI.
Published under licence by IOP Publishing Ltd

Journal of Physics: Conference Series **635** (2015) 092106 doi:10.1088/1742-6596/635/9/092106

High-order harmonic generation driven by plasmonic fields: the velocity gauge

A. Chacón[a], M. Lewenstein[a,b] and M. F. Ciappina[c 1]

[a] ICFO-Institut de Ciències Fotòniques, Mediterranean Technology Park, 08860 Castelldefels, Spain
[b] ICREA-Institució Catalana de Recerca i Estudis Avançats, Lluis Companys 23, 08010 Barcelona, Spain
[c] Max-Planck Institut für uantenoptik, Hans-Kopfermann-Str. 1, D-85748 Garching, Germany

Synopsis We present theoretical investigations of high-order harmonic generation (HHG) driven by plasmonic fields in the velocity gauge. Field-enhanced HHG using plasmonics, generated starting from engineered nanostructures, requires no extra cavities or laser pumping to amplify the pulse power. ne of the main characteristics of these fields is the spatial variation in a nanometric scale and, as a consequence, strong modifications in the electron dynamics are observed. Up to now all the developed theoretical tools and numerical schemes have employed the length gauge and no attempts for implementation of approaches based on the velocity gauge have been made.

The fundamental phenomenology of HHG in atoms based on plasmonics fields can be briefly summarized as follows: a femtosecond low-intensity pulse is coupled to the plasmon mode inducing a collective oscillation of free charges within the metal. These free charges redistribute the electric field around each of the metallic nanostructures, thereby forming a *hot s ot* - a tiny region of highly enhanced electric field. The enhanced field exceeds the threshold of HHG, thus by injection of a noble gas onto the spot of the enhanced field, HHG are generated [1].

All the numerical and semiclassical approaches to study HHG in atoms and molecules are largely based on the assumption that the laser electric field is spatially homogeneous in the region where the electron dynamics takes place. Due to the strong confinement of plasmonic hot spots, the laser electric field is clearly no longer spatially constant in the region where the electron dynamics takes place, and consequently important changes in the laser-matter processes would occur.

Up to now numerous studies of HHG driven by plasmonic fields have been made (see e.g. [2, 3]), but in all of them the length gauge (LG) of the theoretical formulations was used. This is mainly due the fact that it is the laser electric field the quantity obtained from finite-difference time-domain (FDTD) simulations [3]. From a conceptual viewpoint, however, there is no a preferred gauge and the numerical outcomes should to be independent of this choice. In this work we present simulations of HHG in atoms driven by plasmonic fields employing the velocity gauge (VG). From the comparison between LG and VG we validate our approach and in addition we discuss the advantages and drawbacks to use the VG from a numerical and computational viewpoints.

[1]E-mail: marcelo.ciappina@mpq.mpg.de

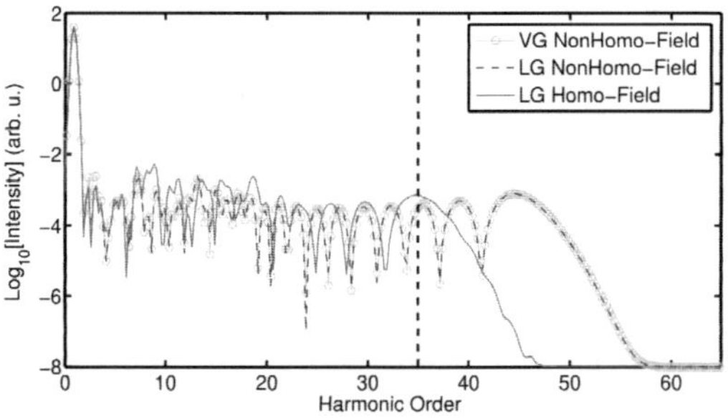

Figure 1. HHG spectra for a 1D-model He^+ ion. The laser intensity and wavelength are $I = 2 \times 10^{14}$ W/cm^2 and $\lambda = 800$ nm, respectively. The laser pulse has a $\sin^2$ shape with two cycles (5.4 fs) of total time duration. The spatial non-homogeneous field has a linear dependence and we have employed an $\varepsilon = 0.0075$ (see [2] for more details). The dotted line indicates the semiclassical cutoff [4].

References

[1] S. Kim *et al* 2008 *Nature (o do)* **453** 757

[2] M. F. Ciappina *et al* 2012 *Phys. Rev. A* **85** 033828

[3] M. F. Ciappina *et al* 2012 *O t. Ex .* **20** 26261

[4] M. Lewenstein *et al* 1994 *Phys. Rev. A* **49** 2117

Content from this work may be used under the terms of the Creative Commons Attribution 3.0 licence. Any further distribution of this work must maintain attribution to the author(s) and the title of the work, journal citation and DOI.
Published under licence by IOP Publishing Ltd

Journal of Physics: Conference Series **635** (2015) 092107 doi:10.1088/1742-6596/635/9/092107

Radiative double electron capture by bare ions in collisions with gas targets

P. N. S. Kumara*[1], A. Simon†, D. S. La Mantia*, A. Kayani*, J. A. Tanis*

* Department of Physics, Western Michigan University, Kalamazoo, MI 49008 USA
† Department of Physics, University of Notre Dame, Notre Dame, IN 46556 USA

Radiative double electron capture (RDEC) is a fundamental atomic process in ion-atom collisions, in which two electrons are captured to bound states of the projectile with the simultaneous emission of a single photon. To date, RDEC has been successfully measured only in collisions of bare O^{8+} and F^{9+} ions with thin-foil carbon targets. In this work we report progress on investigating RDEC in collisions with gas targets.

Radiative double electron capture (RDEC) occurs when a highly-stripped ion captures two electrons accompanied the simultaneous emission of a photon. RDEC by fully-stripped ions can be treated as the time reversed process of double photoionization of the ions. Although RDEC has very small cross sections compared to single radiative electron capture (REC), it provides a clean way to study electron correlation under ideal conditions when only two electrons are involved in the collision.

To date, five experiments [1,2,3,4,5] have been done in the search for RDEC. The first positive results were reported in 2010 [3] for collisions of 38 MeV O^{8+} on a thin carbon foil. Three experiments [1,2,5] were done using highly-charged projectile ions incident on both solid and gas targets at the GSI facility in Darmstadt, but evidence of RDEC was not observed. However, from these studies it was possible to estimate the upper limit for the RDEC cross section using the principle of detailed balance. Figure 1 shows the estimated total cross sections from these five experiments compared with the recent theoretical calculations [6] done by Mistonova and Andreev.

New measurements are underway using the tandem Van de Graaff accelerator facility at Western Michigan University for bare fluorine ions with energies of ~2.5 MeV/u incident on gas targets. These gas target studies have the advantage of not introducing contaminants into the measured results as was the case for fluorine on the C target [4]. The new measurements will be done for targets of He, Ne and N_2 and the results will be compared with the previous experiments and with theoretical calculations [6].

At present a fully-stripped O^{8+} beam is being used with an Ar gas target to calibrate the data acquisition setup. The Ar K x rays at ~3 keV come in the x-ray energy spectrum just at the low point of the RDEC x rays that are expected.

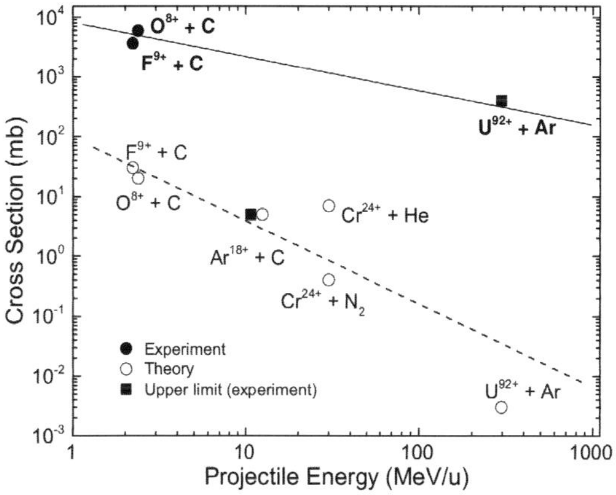

Figure 1. Comparison of experimentally estimated and theoretically calculated total RDEC cross sections.

The new measurements are expected to take fairly long times (about 1 to 2 months) to observe RDEC because of the comparatively small total cross sections and also because of the use of gas targets. The results are expected to shed new light on RDEC and the electron correlation effects involved in the process.

This project is funded in part by the NSF.

References

[1] A. Warczak et al., Nucl. Instrum. Methods Phys. Res. B **98**, 303 (1995).
[2] G. Bednarz et al., Nucl. Instrum. Methods Phys. Res. B **205**, 573 (2003).
[3] A. Simon et al., Phys. Rev. Lett. **104**, 123001 (2010).
[4] T. Elkafrawy et al., Phys. Scr. T **156**, 014047 (2013).
[5] N. Winters et al., Phys. Scr. T **156**, 014048 (2013).
[6] E. A. Mistonova and O. Yu. Andreev, Phys. Rev. A **87** 034702 (2013).

E-mail: pathirannehelag.kumara@wmich.edu

Content from this work may be used under the terms of the Creative Commons Attribution 3.0 licence. Any further distribution of this work must maintain attribution to the author(s) and the title of the work, journal citation and DOI.
Published under licence by IOP Publishing Ltd

Journal of Physics: Conference Series **635** (2015) 092109 doi:10.1088/1742-6596/635/9/092109

CRDS investigation of line shapes of the nitrogen-broadened oxygen *B*-band transition

S. Wójtewicz, P. Masłowski, A. Cygan, P. Wcisło, M. Zaborowski, M. Piwiński*[1], R. Ciuryło, D. Lisak

Institute of Physics, Faculty of Physics, Astronomy and Informatics, Nicolaus Copernicus University in Toruń, Grudziadzka 5, 87-100 Torun, Poland

Synopsis We present precise line-shape measurements of the nitrogen-broadened P9 P9 oxygen *B*-band transition occurring near 689 nm. Data were obtained using the optical frequency comb-assisted Pound-Drever-Hall-locked frequency-stabilized cavity ring-down spectrometer. Several line-shape models describing physical effects such as Dicke narrowing, the speed dependence of collisional broadening and shifting, and the correlation between velocity- and phase-changing collisions were used in the analysis. The multispectrum fitting technique is used to minimize correlation between line-shape parameters. Observed line narrowing is mostly determined by the speed dependence of the collisional broadening.

We present here precise line-shape study of the nitrogen-broadened P9 P9 $^{16}O_2$ *B*-band transition. The spectra were measured in a O_2–N_2 mixture containing 3.00(1)% of O_2, in wide pressure range (10–400 Torr). Experimental line shapes of oxygen were measured with the optical frequency comb-assisted Pound–Drever–Hall-locked frequency-stabilized cavity ring-down spectrometer (OFC-assisted PDH-locked FS-CRDS) [1]. This setup enables us to achieve high spectral resolution (at kHz level) and high signal-to-noise ratio (exceeding 10^5) spectra of weak transitions.

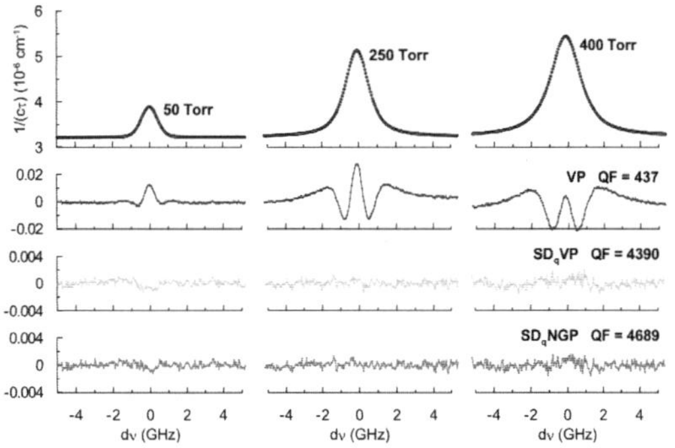

Figure 1. Measured spectra of the N_2-broadended O_2 P9 P9 line and residuals from the multispectrum fits of the VP, SDVP, and SDNGP. The value of the quality of the fit (QF) is given for each profile.

In the line-shape analysis we used several models taking into account such physical effects as Dicke narrowing [2], the speed dependence of collisional broadening and shifting [3], and the correlation between the velocity- and phase-changing collisions [4]. The multispectrum fitting technique was used to minimize correlation between line-shape parameters. In Fig. 1 selected measured spectra of the P9 P9 transition are shown with residuals from fits of the Voigt profile (VP), the speed-dependent Voigt profile (SDVP), and the speed-dependent Nelkin-Ghatak profile (SDNGP). It can be seen that commonly used Voigt profile is not sufficient to describe the measured line shapes. Incorporation of the speed-dependent effects to the model improves the QF by an order of magnitude. Taking both the speed dependence and velocity-changing collisions into account in the SDNGP further improves the QF, but only by 7% comparing to the SDVP. All of the speed-dependent profiles give consistent values of the collisional broadening coefficients γ_L/N to within 0.4%. The results for models including only Dicke narrowing differs by up to 1% depending on the velocity-changing collision model. The VP leads to systematic error of about 5%. The analysis revealed that the observed line narrowing is mainly described by the speed dependence of collisional broadening and the influence of Dicke narrowing is not very significant in the investigated case.

References

[1] A. Cygan *at al* 2013 *Eur. Phys. J.-Spec. Top.* **222** 2119

[2] R. H. Dicke 1953 *Phys. Rev.* **89** 472

[3] P. R. Berman 1972 *J. Quant. Spectrosc. Radiat. Transfer* **12** 1331

[4] S. G. Rautian and I. I. Sobelman 1967 *Sov. Phys. Usp.* **9** 701

[1]E-mail: piwek@fizyka.umk.pl

Content from this work may be used under the terms of the Creative Commons Attribution 3.0 licence. Any further distribution of this work must maintain attribution to the author(s) and the title of the work, journal citation and DOI.
Published under licence by IOP Publishing Ltd

Journal of Physics: Conference Series **635** (2015) 092110 doi:10.1088/1742-6596/635/9/092110

Electron-Impact Ionization of Se^{16+} and Se^{17+} Ions

Erdi A. Bleda* [1], Yavuz Tıkman*, Zikri Altun* [2]

* Department of Physics, Marmara University, 34722, Istanbul Turkey

Synopsis Electron-impact ionization cross sections for Se^{16+} and $\mathbf{Se^{17+}}$ ions have been calculated using semirelativistic configuration-average distorted-wave (CADW) method for both the direct and indirect ionization contributions. The excitation-autoionization contributions originating from the excitations of 2s and 2p subshels were calculated using a semirelativistic level to level distorted-wave (LLDW) method. Direct ionization cross section calculations involving 2s,2p,3s, and 3p subshells were also calculated using the LANL collisional atomic code and comparisons are made.

The ionization of atoms by electron impact is one of the most fundamental electron collision processes and plays an important role in many areas of research. The modeling and diagnostic of astrophysical and laboratory plasmas need atomic data characterizing the various collision processes that might occur in them. In this study, we investigate the electron-impact ionization of Se^{16+} and Se^{17+} ions as function of incident electron energy.

The direct ionization cross sections of 2s, 2p, 3s, and 3p electrons have been calculated within the semirelativistic CADW method[1]. The excitation-autoionization contributions originating from the inner shell excitations of the type

$$2s^2 2p^6 3s^2 3p^6 \rightarrow 2s^2 2p^5 3s^2 3p^6 nl,\ 2s^1 2p^6 3s^2 3p^6\ nl$$

for Se^{16+} (n=4–8, l=0–3) and

$$2s^2 2p^6 3s^2 3p^5 \rightarrow 2s^2 2p^5 3s^2 3p^5 nl,\ 2s^1 2p^6 3s^2 3p^5 nl$$

for Se^{17+} (n=3–8, l=0–3) have been calculated using both the CADW and LLDW methods[2] including autoionization branching ratios. The energies and bound orbitals needed for the calculations were obtained from the Hartree-Fock Relativistic approximations[2] including the mass velocity and Darwin corrections.

The results for Se^{16+} and Se^{17+} are shown in Figure 1a and 1b, respectively. The curves labeled as Los Alamos in these figures represent total configuration-average direct electron-impact ionization cross sections within distorted-wave approach calculated using the codes developed at the Los Alamos National Laboratory[3]. The agreement between our total direct CADW results labeled as "CADW direct" and the Los Alamos curve is quite good.

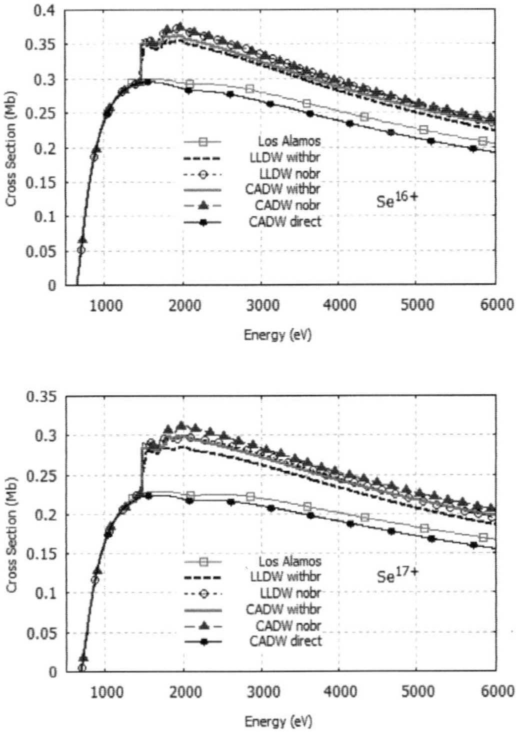

Figure 1. Electron-impact ionization cross sections for Se^{16+} and Se^{17+} The total CADW direct ionization cross sections are labeled as CADW direct in these figures. CADW direct plus LLDW and CADW excitation-autoionization contributions with and without branching ratios are represented by withbr and nobr extensions Labels "withbr" and "nobr" in the curves represent the results with and without branching ratios.

Acknowledgement

We thank BABKO of Marmara University and National Energy Research Scientific Computing Center (NERSC) in Oakland, California for computational support.

References

[1] M. S. Pindzola, D. C. Griffin, and C. Bottcher, 1986, Atomic Processes in Electron-Ion and Ion-Ion Collisions, ed. F. Brouillard, NATO ASI B 145, 75
[2] Cowan R.D., 1981, The Theory of Atomic Structure and Spectra. Berkeley CA: University of California Press
[3] Los Alamos National Laboratory, USA, http://aphysics2.lanl.gov/tempweb/lanl/

[1] E-mail: ebleda@marmara.edu.tr

[2] E-mail: zikalt@marmara.edu.tr

Content from this work may be used under the terms of the Creative Commons Attribution 3.0 licence. Any further distribution of this work must maintain attribution to the author(s) and the title of the work, journal citation and DOI.
Published under licence by IOP Publishing Ltd

Journal of Physics: Conference Series **635** (2015) 092111 doi:10.1088/1742-6596/635/9/092111

Testing a method for solving TDSE in momentum space with a one-dimensional model

Vladimir D. Rodríguez* [1]

* Departamento de Física, FCEyN UBA and Instituto de Física de Buenos Aires, CONICET
Pabellón 1, Ciudad Universitaria, 1428 Buenos Aires, Argentina

Synopsis One-dimensional models may be used to get insight into a recently proposed method to solve the TDSE in momentum space. At first a delta-function potential giving rise to a separable potential in momentum space, is employed. In this case, the method gives exact results remarkably. Poschl-Teller potentials supporting more than one bound-states are proposed as a further test.

Recently, a procedure to solve the time dependent Schrödinger equation (TDSE) in momentum space for studying strong field Hydrogen ionization has been proposed [1]. This method simplifies the problem by using a model potential that is a sum of separable (non-local) potentials. The model potential accounts for one or more bound states of the real Hydrogen atom potential. The procedure has a number of good features. It is gauge independent since the model is exactly solved. It accounts for continuum-continuum coupling and, therefore, multiphoton ionization is well represented. Like any exact solution, it is unitary. It may also be improved by increasing the number of bound states considered. For instance, in the resonant case, it is able to show the Rabi oscillations as well as the splitting in the above threshold ionization peaks (Autler-Townes doublets) [1]. Nevertheless, the comparison with exact numerical TDSE solutions is good at a qualitative level but some differences still remain at a quantitative level.

In this work, the theory applied in a simplified model such as a particle bounded to a one-dimensional delta-function potential is examined. The Fourier Transform of such a potential is constant, thus trivially separable in momentum space. The exact wave function satisfies:

$$i\frac{\partial}{\partial t}\tilde{\psi}(p,t)=\frac{1}{2}(p+A(t))^2\tilde{\psi}(p,t)-\frac{Z}{\sqrt{2\pi}}\psi(0,t)$$

The integral equation for $\psi(0,t)$ is

$$\psi(0,t)=\phi(0,t)+iZ\int_0^t dt'\psi(0,t')G(t',t)$$

where

$$\phi(0,t)=\frac{1}{\sqrt{2\pi}}\int_{-\infty}^{\infty}dp\, e^{-i\int_0^t dt'\frac{1}{2}(\mathbf{p}+\mathbf{A}(t'))^2}\tilde{\varphi}_i(p)$$

and

$$G(t',t)=\frac{1}{\sqrt{2\pi i}}\frac{e^{-i\int_{t'}^{t}dt''\frac{\mathbf{A}(t'')^2}{2}}\, e^{i\frac{\left(\int_{t'}^{t}dt''\mathbf{A}(t'')\right)^2}{2(t-t')}}}{\sqrt{t-t'}}$$

is the propagator playing the role of kernel in the integral equation.

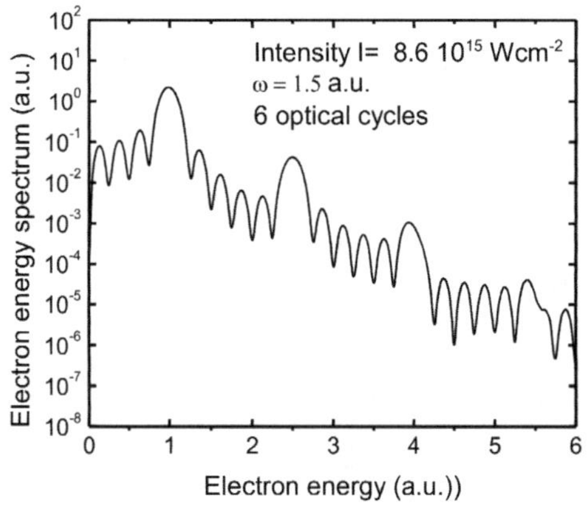

Figure 1. Electron energy spectrum by interaction of the model atom with a six cycles sine square pulse. The peak intensity is equal to 10^{15} Wcm^{-2} an the photon energy to 0.5 a.u.

The figure shows the spectrum obtained by the interaction between our one-dimensional model and a six-cycle laser pulse of 10^{15} Wcm^{-2} intensity peak. The spectrum exhibits several ATI peaks separated by the photon energy. The minor peaks arise from the pulse envelope Fourier Transform.

One-dimensional Poschl-Teller potential supporting more bound states will be used for further testing of the theoretical approach in [1].

References

[1] H. M. T. Nganso *et al* 2013 *Phys. Rev. A. Phys.* **87** 013420

[1] E-mail: vladimir@df.uba.ar

Content from this work may be used under the terms of the Creative Commons Attribution 3.0 licence. Any further distribution of this work must maintain attribution to the author(s) and the title of the work, journal citation and DOI.
Published under licence by IOP Publishing Ltd

A semi-analytical model for strong field atomic ionization

Vladimir D. Rodríguez* [1]

* Departamento de Física, FCEyN UBA and Instituto de Física de Buenos Aires, CONICET
Pabellón 1, Ciudad Universitaria, 1428 Buenos Aires, Argentina

Synopsis A semi-analytical model is proposed to solve the TDSE. The method leads to an integro-differential equation for the initial state amplitude to be numerically solved. Closed expressions for the ionization amplitudes in term of the survival amplitude are obtained. The method is applied to Hydrogen ionization by a laser pulse.

The highly non-linear interaction of atoms with intense laser pulses has demanded increasing effort for either solve the time dependent Schrödinger equation (TDSE), or proposing semi-analytical models keeping simple interpretation while still describing the essential phenomena. Among the first ones an interesting numerical method for solving TDSE in momentum space has been recently proposed [1]. In the second group, most of the best known and successful semi-analytical methods are related to the Strong Field Approximation (SFA) [2].

In [1], the method is able to be improved by adding a number of bound states supported by the model potential. A set of coupled Volterra type integral equations has to be numerically solved. Following this line, a theoretical method accounting for a number of bound states coupled among them and with the continuum is proposed in this study. However, a simplified treatment for the continuum-continuum coupling is made. This simplification is related to that made in the derivation of the SFA [2].

As in [1], the velocity gauge is used. A sine-square envelope laser pulse applied to Hydrogen atom is considered. The Hamiltonian (a.u. are used) is:

$$H = H_{\text{at}} + \hat{\mathbf{p}}.\mathbf{A}(t) = \frac{\hat{\mathbf{p}}^2}{2m_e} + V_{\text{T}}(\mathbf{r}) + \hat{\mathbf{p}}.\mathbf{A}(t)$$

The TDSE for the state $|\psi(t)\rangle$ with the initial condition $|\psi(0)\rangle = |i\rangle$ has to be solved. $|\psi(t)\rangle$ is expanded in a complete set of H_{at} eigenstates $|\psi(t)\rangle = a_i(t)e^{-i\varepsilon_i t}|i\rangle + \sum_{n\neq i} a_n(t)e^{-i\varepsilon_n t}|n\rangle + \int d\mathbf{k}\, a_{\mathbf{k}}(t)e^{-i\varepsilon_K t}|\mathbf{k}\rangle$.

Using the above expansion in the TDSE,

$$\dot{a}_i(t) = -\int_0^t dt' K(t,t') a_i(t)$$

is obtained for the initial-state amplitude.

The integro-differential equation kernel is

$$K(t,t') = \sum_{n\neq i} [\hat{\mathbf{p}}_{\text{in}} \cdot \mathbf{A}(t)][\mathbf{p}_{\text{ni}} \cdot \mathbf{A}(t')] e^{-i\varepsilon_n (t-t')}$$
$$+ \int d\mathbf{k} [\hat{\mathbf{p}}_{\text{ik}} \cdot \mathbf{A}(t)][\hat{\mathbf{p}}_{\text{ki}} \cdot \mathbf{A}(t')] e^{-i\varepsilon_k (t-t')} e^{i\mathbf{k}\cdot\boldsymbol{\alpha}(t,t')}$$

Two main approximations are made to obtain this equation. Coupling between the initial-state and all the other states are kept. The other couplings are neglected excepting the continuum-continuum ones which are approximated by $\hat{\mathbf{p}}_{\text{kk'}} \approx \mathbf{k}\,\delta(\mathbf{k}-\mathbf{k}')$. In spite of these approximations, the initial-state decay and the ATI peaks are well described.

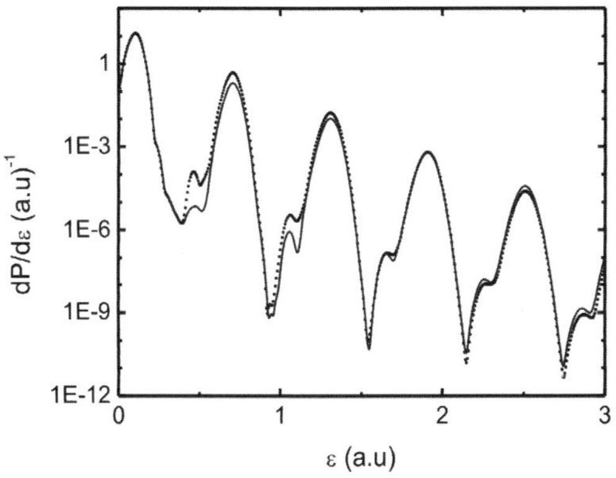

Figure 1. Electron energy spectrum by interaction of the Hydrogen atom with a twenty-cycle sine-square pulse. The electric field peak amplitude is equal to 0.03 a.u., and the photon energy is 0.6 a.u.

The electron spectrum obtained by Hydrogen interacting with a twenty-cycle, 3.1×10^{15} Wcm^{-2} peak-intensity laser pulse is shown in figure 1. The spectrum exhibits several ATI peaks separated by the photon energy. The present results along with those obtained by using the QPROP code for solving the TDSE [3] are displayed.

References

[1] H. M. T. Nganso *et al* 2013 *Phys. Rev. A. Phys.* 87 013420
[2] F. Kraus and M. Ivanov 2009 *Rev. Mod. Phys.* 81 163
[3] D. Bauer and P. Koval, 2006 *Comput. Phys. Commun.* 174 296

[1] E-mail: vladimir@df.uba.ar

Content from this work may be used under the terms of the Creative Commons Attribution 3.0 licence. Any further distribution of this work must maintain attribution to the author(s) and the title of the work, journal citation and DOI.
Published under licence by IOP Publishing Ltd

Journal of Physics: Conference Series **635** (2015) 092113 doi:10.1088/1742-6596/635/9/092113

Improved Weisskopf-Wigner approximation for atomic ionization by short laser pulses

M. G. Bustamante†, **V. D. Rodríguez**† [1]

†Departamento de Física and Instituto de Física de Buenos Aires, Argentina

Synopsis An improved Weisskopf-Wigner approximation allows to obtain a closed form for the survival amplitude and the ionization amplitudes of atoms subject to laser pulses. The improvement avoids the numerical solution of a Volterra integro-differential equation obtained by modeling full initial state coupling to the continuum as well as transitions between initial state and other discrete states. The calculations are made in the velocity gauge. The improved approximation compares well with a full TDSE numerical solution for Hydrogen.

In a recent work we have introduced a model to deal with atomic ionization by short laser pulses [1]. The model accounts for full initial state coupling to the continuum and discrete spectrum, while discrete-discrete, discrete-continuum and continuum-continuum transitions are neglected. We have obtained a single Volterra integro-differential equation for the initial state amplitude. This equation is easily solved using the Goldfine prescription [2]. We have analytically shown that the sum of all transition probabilities is unity, i.e., the model is unitary.

In this communication we consider the model in the velocity gauge and report two analytical approximations (closed-forms) to the integro-differential equation: one, closely related to the Weisskopf-Wigner approximation, and the second one, an improved version of it.

To test the model and its approximations we show in figure 1 the ionization spectra of Hydrogen under a 20 cycles XUV laser pulse ($\omega = 0.6$ a.u.). Two laser intensities are examined, $E_0 = 0.2$ a.u. (1.4×10^{14} W/cm^2) and $E_0 = 0.4$ a.u. (5.6×10^{14} W/cm^2). The results obtained with the model and its approximations for the electron spectrum, are compared with full time dependent Schrödinger equation simulation performed with the prop [3]. The comparison is rather good for the entire electron energy interval, at the smaller intensity. The full model and both Weisskopf-Wigner approximations are close each other and follow the TDSE results quite well. For the larger electric field amplitude, the improved Weisskopf-Wigner approximation remains close to the full model calculation, which in turn keeps near prop simulations within 15 percent. The simplest Weisskopf-Wigner approach depart considerably from TDSE results.

[1] E-mail: vladimir@df.uba.ar

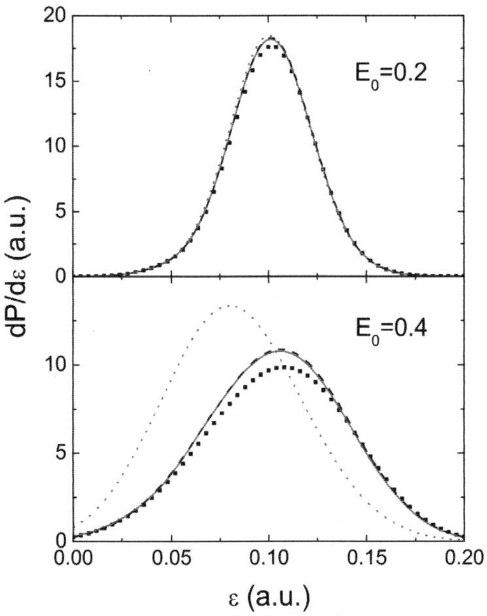

Figure 1. Hydrogen ionization spectra (first peak) for a 20 cycles laser pulse with $\omega = 0.6$ a.u. The laser field amplitudes are $E_0 = 0.2$ a.u. and $E_0 = 0.4$ a.u. Colors online. Full line (red), unitary model; dashed line (blue), improved Weisskopf-Wigner approximation; dotted line (green), Weisskopf-Wigner approximation; squares: exact calculations with prop code [3].

References

[1] M. G. Bustamante and V. D. Rodríguez, unpublished.

[2] A. Goldfine 1977 *Mathemat cs of Com utat o* , Vol. 31, No. 139, 691-707

[3] D. Bauer and P. Koval 2006 *Com ut. Phys. Comm.*, **174** 396-421

Content from this work may be used under the terms of the Creative Commons Attribution 3.0 licence. Any further distribution of this work must maintain attribution to the author(s) and the title of the work, journal citation and DOI.
Published under licence by IOP Publishing Ltd

Journal of Physics: Conference Series **635** (2015) 092114 doi:10.1088/1742-6596/635/9/092114

Photoionization of the ground $3s^2\ ^1S_0$ and excited $3s3p\ ^3P_{2,1,0}$ states of Fe XV

L Y Xie * [1], C Ma*, C Z Dong*, J G Wang† and Y Z Qu††

* Key Laboratory of Atomic and Molecular Physics & Functional Materials of Gansu Province, College of Physics and Electronic Engineering, Northwest Normal University, Lanzhou 730070, China

† Key Laboratory of Computational Physics, Institute of Applied Physics and Computational Mathematics, P.O. Box 8009-26, Beijing 100088, China

†† College of Material Sciences and Opto-electronic Technology, University of the Chinese Academy of Sciences, Beijing 100049

Synopsis The near-threshold photoionization cross sections of the ground $3s^2\ ^1S_0$ and excited $3s3p\ ^3P_{2,1,0}$ states of Fe XV ion are investigated using the fully relativistic *R*-matrix method. The dominant resonances are analyzed and identified. The present results are compared with early theoretical values available in the literature.

Photoionization (PI) of highly charged iron ions are particular interested owing to their importance in the modeling of astrophysical plasmas and in laboratory sources such as inertial and magnetic confinement fusion devices [1]. Over the past two decades, a number of theoretical efforts have been devoted to the study of PI cross sections of Fe XV. Buttler et al [2] carried out the calculations for the $3s^2(^1S_0)$ ground stat near the ionization threshold in the close-coupling approximation using the R-matrix method under the Opacity Project (OP). Haque and Pradhan [3] obtained cross sections ranges extend to the *L*-shell threshold using a combination of Breit-Pauli R-matrix (BPRM) and relativistic random-phase approximations plus multichannel quantum-defect theories. Bautista [4] calculated cross sections from the first ionization threshold to the *K*-shell threshold using the close-coupling R-matrix method including 99 LS target terms. Recently, Simon et al [5] measured cross sections in the range from 450 to 1100 eV using an electron beam ion trap. Hasoglu et al [6] investigated Radiation damping in the photoionization of Fe XV. Although, all of those studies are carried out for the ground state PI cross sections of Fe XV, the obvious differences still exist between different calculation results, even for the near-threshold photoionization [2-4].

In this study, we performed comprehensive fully relativistic R-matrix calculations for both PI cross sections of the ground $3s^2\ ^1S_0$ and excited $3s3p\ ^3P_{2,1,0}$ of Fe XV in the near-threshold region by considering elaborate configuration interaction (CI) using the DARC code [8]. To analyze most of the dominant resonances, the detailed calculations are also carried out for resonant energies, and resonant strengths using the Flexible Atomic Code (FAC). The present results are compared with early theoretical calculations.

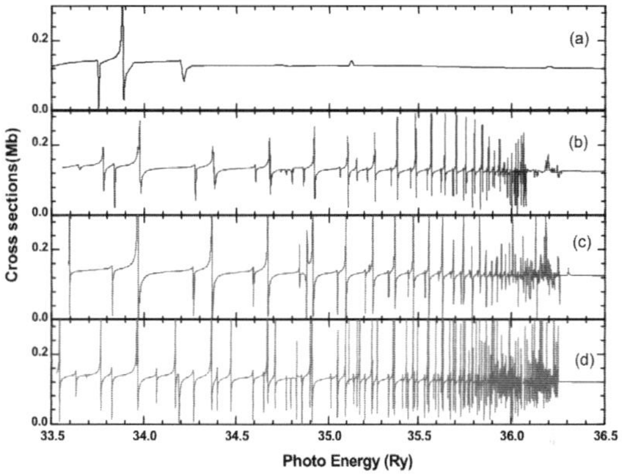

Figure 1. Photoionization cross section of the Fe XV $3s^2\ (^1S_0)$ ground state (a) the OP cross section [2]; (b) the BPRM results[3] (c) the close-coupling R-matrix results [4]. (d) the present results (in length gauge).

The work was supported by the National Natural Science Foundation of China (Grant Nos. 11274254, U1331122, and U1330117, 2013CB922200).

References

[1] T.W.Gorczyca *et al* 2000 *Phys.Rev.A* 63 010702(R)
[2] K.Buttler *et al* 1993 *J. Phys. B: At. Mol. Opt. Phys.* 26 4409
[3] N. Haque *et al* 1999 *Phys. Rev. A* 60 R4221
[4]M. A. Bautista 2000 *J. Phys. B: At. Mol. Opt. Phys* 33 L419
[5] M. C. Simon *et al* 2010 *Phys. Rev. Lett.* 105 183001
[6] M. F. Hasoglu *et al* 2012 *Phys. Rev. A* 85 040701(R)

[1] E-mail: xiely@nwnu.edu.cn

Content from this work may be used under the terms of the Creative Commons Attribution 3.0 licence. Any further distribution of this work must maintain attribution to the author(s) and the title of the work, journal citation and DOI.
Published under licence by IOP Publishing Ltd

Photoionization of Cl–like Argon using the Breit–Pauli R–matrix method

N. B. Tyndall*[1], C. A. Ramsbottom*[2], A. Hibbert*[3],

* School of Mathematics & Physics, The Queen's University of Belfast, Belfast BT7 1NN, Northern Ireland

Synopsis Here we present the photoionization cross sections for the ground and metastable states of Cl–like Argon by exploiting the fully relativistic Breit–Pauli **R**–matrix computer codes to determine these transitions of interest. We compare our work with previous theoretical and experimental results and present a detailed investigation into the model of Ar III, the resonant structure and identification process.

Investigation into the structure of ionic targets is essential for the generation of accurate radiative transitions, of which the bound–free photoionisation (PI) process is to be discussed. Such data is useful in modelling various synthetic spectra through packages such as Cloudy [1]. We focus solely on the PI of Ar II governed by,

$$\begin{array}{c} h\nu + \text{Ar II} \rightarrow \text{Ar III} + \text{e}^- \\ \searrow \quad \nearrow \\ (\text{Ar II})^* \end{array} \qquad (1)$$

in which a direct PI pathway leaves the target Ar III plus the ejected photoelectron, but also indirectly via autoionizing bound states, where the photoelectron is firstly excited into a more energetic state prior to ejection.

Results are obtained through the implementation of large scale Breit–Pauli **R**–matrix computer codes, to include the one body relativistic operators [2] by availing of the well established R–matrix theory [3]. A delineation of configuration space by the **R**–matrix boundary ($r = 16.6$ a.u. in this study) is chosen so that all electron correlation effects are contained within the internal region. Accurate wavefunctions are calculated and then matched at the boundary r, for all initial and final state transitions of interest.

We have included the HF orbitals of Clementi and Roetti [4] for all $n = 1, 2, 3$ orbitals within this work. For a more accurate target wavefunction representation, we have also optimised the $n = 4$ orbitals on the lowest quintet states observed by NIST through the computer package CIV3 developed by A. Hibbert [5]. Throughout the calculation, a closed $1s^2 2s^2 2p^6$ core is maintained which results in a total of (64 LSπ) 124 Jπ levels arising from a 7 configuration close–coupling expansion of the wavefunction. Additional configuration interaction terms are also required to allow for electron correlation effects.

Once an accurate model for Ar III is established, we then compute all PI cross sections in the length gauge defined by,

$$\sigma = \frac{4\pi^2 a_0^2 \alpha\omega}{3g_i} \sum_{L,lf} |(\Psi_f^- || \mathbf{D} || \Psi_i)| \qquad (2)$$

where the wavefunctions of the initial bound state, Ψ_i and final free state of Ψ_f^- are computed over a very fine energy mesh to allow for sharp resolution of the spectra. $a_0 \equiv$ bohr radius, $\alpha \equiv$ fine structure constant, $\omega \equiv$ photon energy and $g_i \equiv$ statistical weight of initial state.

We compute and present transitions within the intermediate coupling frame for the ground and metastable initial odd states of $J = 3/2$ and $J = 1/2$ to accessible, dipole allowed final states through Equation (2). A direct comparison with M. Covington et al [6] can be achieved by performing both a 10meV gaussian convolution of the spectra and a weighting comprising of the two initial states. We also conduct a resonance identification procedure using the **QB** technique [7] which exploits the analytical properties of the **R**–matrix to identify autoionizing bound states of the indirect path in Equation (1).

References

[1] Ferland et al, *PA Soc Pac* ***110*** *761 – 778, 1998*

[2] Scott et al, *J Phys B* ***13*** *4299 – 4314, 1980*

[3] Burke et al, *J Phys B* ***8*** *16 2620 – 2639, 1975*

[4] Clementi et al, *ADNDT* ***14*** *177 – 478, 1974*

[5] Hibbert, *CPC* ***9*** *141 – 172, 1975*

[6] Covington et al, *PRA* ***84****, 2011*

[7] Quigley et al, *CPC* ***114*** *225 – 235, 1998*

[1]E-mail: ntyndall01@qub.ac.uk
[2]E-mail: c.ramsbottom@qub.ac.uk
[3]E-mail: a.hibbert@qub.ac.uk

Content from this work may be used under the terms of the Creative Commons Attribution 3.0 licence. Any further distribution of this work must maintain attribution to the author(s) and the title of the work, journal citation and DOI.
Published under licence by IOP Publishing Ltd

Journal of Physics: Conference Series **635** (2015) 092116 doi:10.1088/1742-6596/635/9/092116

Adiabatic approximations for electrons interacting with ultrashort high-frequency laser pulses

Koudai Toyota[1], **Ulf Saalmann**[2], **Jan M. Rost**[3]
Max Planck Institute for the Physics of Complex Systems, Nöthnitzer Str. 38, Dresden 01187, Sachsen, Germany

Synopsis Adiabatic approximations in high-frequency regime is developed, which is capable of treating both photo absorption and non-adiabatic transitions in equal footing.

The latest advanced experimental techniques can generate coherent light in X-ray range with pulse duration less than 100 attoseconds. Such noble light sources has opened up the new research area so-called high-frequency regime [1]. On the other hand, high-frequency regime has been studied in theory more than three decades in terms of the high-frequency Floquet theory (HFFT) [2]. However, it can only be applied for monochromatic laser fields i.e. *infinite* pulse length. It is thus highly desirable developing new theory which can adequately treat *finite* pulse length to study ionization dynamics with temporal resolutions comparable with electron's orbiting time. In [3], we invented the envelope Hamiltonian to achieve this goal, which is written in the form of

$$H_{\rm env}(t) = H_0(t) + \sum_{|n|=1,2} V_n(\vec{r};t)e^{-in\omega t}, \quad (1)$$

$$H_0(t) = -\frac{1}{2}\nabla^2 + V_0(\vec{r};t). \quad (2)$$

It is similar to usual Floquet expansion in the Kramers-Henneberger frame, but effects of an envelope function of a pulse is remained through the time dependency of the coefficients V_n by a certain simple devise. The envelope Hamiltonian decompose the full Hamiltonian into the effective Hamiltonian of the electron $H_0(t)$ of *slow* time scale and the photo absorption/emission channels $V_n(t)e^{-in\omega t}$ of *fast* time scale. This feature enables us to develop the adiabatic approximations for photo ionization. Let $|0(t)\rangle$ and $|\vec{k},t\rangle$ be the ground state and scattering state of momentum k for $H_0(t)$. Then up to the first order, it is found that the photo electron spectrum can be written by $P(\vec{k}) = \sum_{|n|=0,1,2} |C_{n\omega}(\vec{k})|^2$, where $C_{n\omega}(\vec{k}) = \int_{-\infty}^{\infty} dt M^{(n)}(\vec{k},t)e^{-i\Phi_n(t)}$. The matrix element $M^{(n)}(\vec{k},t)$ is given by $M^{(0)}(\vec{k},t) = -i\langle\vec{k},t|\frac{\partial}{\partial t}|0(t)\rangle$ for $n=0$, and $M^{(n\neq 0)}(\vec{k},t) = \langle\vec{k},t|V_n(\vec{r},t)|0(t)\rangle$, for $n \neq 0$. The phase $\Phi_n(t)$ is given by $\Phi_n(t) = \int_{-\infty}^{t} E_0(t')dt' + n\omega t - Et$. The quantities $C_{0\omega}$ and $C_{n\omega}$ ($n \neq 0$) represents the non-adiabatic transitions and $n$photon absorption (emission) for $n > 0$ ($n < 0$), respectively. Fig. 1 shows the ionization yield for a model one-electron system as a function of pulse duration. We find that the envelope Hamiltonian $H_{\rm env}(t)$ perfectly reproduces the results of the full time-dependent Schrödinger equation. For long pulse limit, single photon ionization (red) dominate the dynamics. It is found as the pulse duration becomes shorter that the non-adiabatic transitions (blue) plays a significant role.

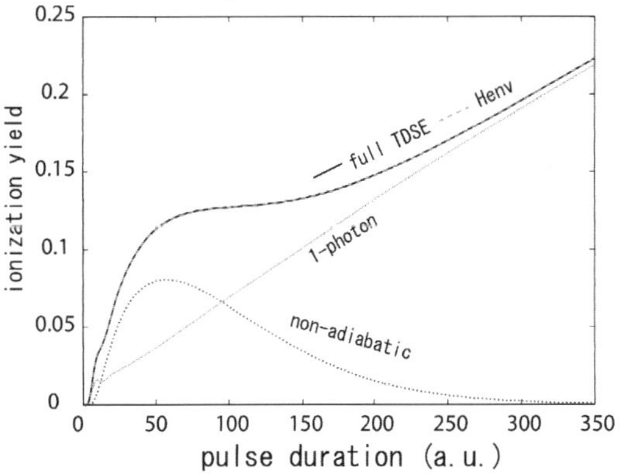

Figure 1. Ionization yield for a model one electron system in one-dimension as a function of pulse duration.

References

[1] M. Chini, et. al. 2014 *Nat. Photo.* **8** 178

[2] M. Gavrila and J. Z. Kaminski 1984 *Phys. Rev. Lett.* **52** 613

[3] K. Toyota, et. al. 2014 *arXiv* **1408** 4541

[1]E-mail: koudai.toyota@gmail.com
[2]E-mail: us@pks.mpg.de
[3]E-mail: Rost@pks.mpg.de

Content from this work may be used under the terms of the Creative Commons Attribution 3.0 licence. Any further distribution of this work must maintain attribution to the author(s) and the title of the work, journal citation and DOI.
Published under licence by IOP Publishing Ltd

XXIX International Conference on Photonic, Electronic, and Atomic Collisions (ICPEAC2015) IOP Publishing
Journal of Physics: Conference Series **635** (2015) 092117 doi:10.1088/1742-6596/635/9/092117

Laser cooling and trapping of thulium atoms for further investigation of collisional properties

E. Kalganova[*†‡ 1], **G. Vishnyakova**[*†‡ 2], **A. Golovisin**[*†‡], **D. Tregubov**[*†‡], **D. Sukachev**[*‡], **S. Fedorov**[*†‡], **K. Khabarova**[*‡], **A. Akimov**[*†‡], **N. Kolachevsky**[*†‡], **, and V. Sorokin**[*‡]

[*] P. N. Lebedev Physical Institute of RAS, 53 Leninsky prosp., Moscow 119991, Russia

[†] Moscow Institute of Physics and Technology,9 Institutsky per., Dolgoprudny, Moscow region, 141700, Russia

[‡] Russian Quantum Center, 100 Novaya St., Skolkovo, Moscow 143025, Russia

Synopsis We demonstrated deep laser cooling of thulium atoms on a weak closed transition in magneto-optical trap and reloading of an atomic ensemble into an optical lattice. 10^5 atoms with a temperature 10 μK were trapped.

Thulium is a rare-earth element with a high ground state magnetic moment ($4\mu_B$). Therefore, it is a promising object for study dipole-dipole interactions and for quantum simulations. Furthermore, a lot of low-field Fano-Feshbach resonances are expected in an ultracold thulium gas. Our group deals with an investigation of collisional properties of ultracold thulium atoms.

The first step towards research of cold collisions is an obtaining of an ensemble of cold thulium atoms. For this purpose we perform a laser cooling of thulium in a magneto-optical trap (MOT) [1]. A loading of MOT is fulfilled from an atomic beam decelerated by Zeeman slower. A cooling process consists of two stages.

For the first-stage cooling and Zeeman slowing we use a strong transition $4f^{13}(^2F^o)6s^2, J_g = 7/2, F_g = 4 \rightarrow 4f^{12}(^3H_5)5d_{3/2}6s^2, J_e = 9/2, F_e = 5$ at wavelength 410.6 mn (Tabel 1). It is worth to note that Lande g-factors of lower and the upper states for this transition are very close to each other. Because of this Sub-Doppler cooling mechanism works effectively directly in MOT. That allows to obtain a temperature about 25 μK during the first-stage cooling by working with a large detuning of cooling light [2]. However, the number of atoms in this case significantly reduces.

Table 1. Properties of cooling transitions in Tm.

	First stage	Second stage
Wavelength	410.6 nm	530.7 nm
Linewidth	10 MHz	350 kHz
Doppler limit	240 μK	9 μ

To achieve lower temperature without decreasing of number of atoms the second-stage cooling at weak closed transition $4f^{13}(^2F^o)6s^2, J_g = 7/2, F_g = 4 \rightarrow 4f^{12}(^3H_6)5d_{5/2}6s^2, J_e = 9/2, F_e = 5$ at wavelength 530.7 mn is used. A Sub-Doppler cooling mechanism also works in MOT for this transition. The dependence of a temperature after second-stage cooling on a detuning is shown on the figure 1.

After cooling and trapping in MOT, thulium atoms are recaptured in an optical lattice. Its confinement potential is formed by a laser beam at wavelength 532 nm. The beam with a power up to 4 W is focused into the atomic cloud, forming an optical trap with an estimated depth of 100 μK. An efficiency or reloading is about 60%.

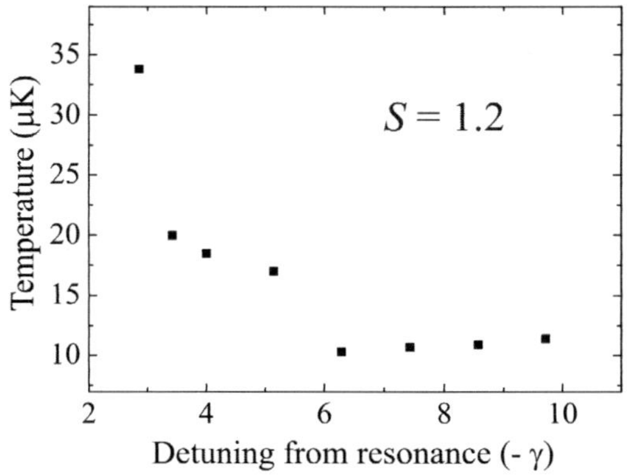

Figure 1. Temperature of atoms versus the detuning of laser beams, a saturation parameter $S = 1.2$.

Thus, an ensemble of 10^5 thulium atoms with a temperature about 10 μK was loaded an optical lattice. Now we are working on enhancement of a signal to noise ratio for a further study of thulium collisional properties.

References

[1] D. Sukachev *at al* 2010 *Phys. Rev. A.* **82** 011405

[2] D. Sukachev *at al* 2010 *JETP Letters* **92(10)** 703

[1]E-mail: kalganova.elena@gmail.com
[2]E-mail: gulnarav7@gmail.com

Content from this work may be used under the terms of the Creative Commons Attribution 3.0 licence. Any further distribution of this work must maintain attribution to the author(s) and the title of the work, journal citation and DOI.
Published under licence by IOP Publishing Ltd

Alkali-atom photoionization and negative-ion photodetachement processes in the presence of external fields

P. Giannakeas[*1], **F. Robicheaux**[*2], **and C. H. Greene**[*3]

[*]Department of Physics and Astronomy, Purdue University, West Lafayette, Indiana 47907, USA

Synopsis Processes of alkali-atom photoionization and negative ion photodetachement in the presence of external fields are investigated. The corresponding photoabsorption spectra are interpreted by employing the theoretical framework of Schwinger variational principle and the local frame transformation theory.

Two-photon ionization of an alkali-metal atom in the presence of a uniform electric field is investigated using a standardized form of local frame transformation (LFT) and generalized quantum defect theory [1]. The relevant long-range quantum defect parameters in the combined Coulombic plus Stark potential is calculated with eigenchannel R-matrix theory applied in the downstream parabolic coordinate η. The present formulation permits us to express the corresponding microscopy observables in terms of the local frame transformation, and it gives a critical test of the accuracy of the Harmin-Fano theory[2, 3] .

The negative-ion photodetachement in a uniform magnetic field is also investigated where polarization of the photon is chosen to be perpendicular to the field. The LFT theory and a new non-perturbative method in terms of the Schwinger variational principle are employed interpreting the corresponding photoabsorption spectra. In particular, the physics of *all* the closed channels due to the magnetic field is included and compared with previous studies [4]. Indeed, the inclusion of the closed channel manifold induces non-trivial effects, particularly in the regime of large atom-electron scattering lengths. Fig.1 illustrates the differences between the photoabsorption probabilities with (solid line) and without (dashed line) the physics of the closed channels. These differences emerge due to a bound state which is collectively supported by all the closed channels.

Furthermore, the concept of collective bound states is generalized and studied in negative-ion photodetachement in the presence of parallel and crossed electric and magnetic fields. We observe that such bound states yield intriguing features in the collisional aspects of atom-electron scattering which can be manipulated by adjusting the strength of the electric and magnetic fields.

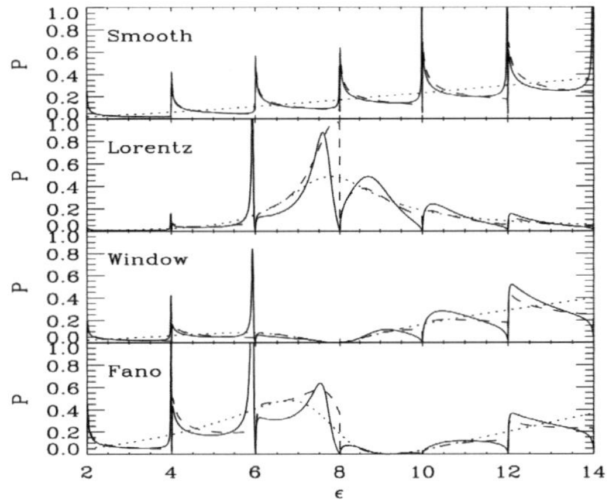

Figure 1. The p-wave photoabsorption probability for different cases of free-field resonances. In all panels, the dotted line is for photoabsorption spectrum without the external field, the dashed line is the approximation that includes the external field but does not include the effect of closed channels, and the solid line is the full calculation including the magnetic field. The "Smooth" case does not have a resonance. The "Lorentz", "Window", and "Fano" cases only differ in the energy dependence of the dipole matrix element when no field is applied.

This work was supported by the U.S. Department of Energy, Office of Science, Basic Energy Sciences, under Award numbers DE-SC0010545 (for PG and CHG) and DE-SC0012193 (for FR).

References

[1] P. Giannakeas, F. Robicheaux, and C. H. Greene, arXiv:1410.5360.

[2] U. Fano, Phys. Rev. A **24**, 619 (1981).

[3] D. A. Harmin, Phys. Rev. A **26**, 2656 (1982)

[4] C. H. Greene, Phys Rev. A **36**, 4236 (1987).

[1]E-mail: pgiannak@purdue.edu
[2]E-mail: robichf@purdue.edu
[3]E-mail: chgreene@purdue.edu

Content from this work may be used under the terms of the Creative Commons Attribution 3.0 licence. Any further distribution of this work must maintain attribution to the author(s) and the title of the work, journal citation and DOI.
Published under licence by IOP Publishing Ltd

Journal of Physics: Conference Series **635** (2015) 092119 doi:10.1088/1742-6596/635/9/092119

Autoionizing states of Spatially Confined Helium Atom

Jayanta K. Saha [1]

Indian Association for the Cultivation of Science, Jadavpur, Kolkata-700032, India

Synopsis The effect of spatial confinement on the resonance parameters of the helium atom have been studied. The transformation to the free atom regime and the harmonic confinement regime are discussed in detail in dependence of the tuning parameters of the confined system.

Structural properties of atomic systems placed under different confinements change appreciably compared with those of free atoms. It is now possible to do control experiments with confined quantum systems. Thus in the last few years, bulk of investigations on the properties of spatially confined systems are being done with a view of their possible applications to model atoms within cavities, atoms under pressure, impurities in quantum dots or nano crystals, nanopores, fullerenes and foreign atoms in liquid helium environment. A detail discussion on these applications can be found in [1]. Although the confined hydrogen and helium atom has been widely studied by several techniques [1], very few attempts have been made towards the quasi bound or resonance states of one electron impurity atom in an isolated quantum dot as well as in case of confined two electron systems. Here the helium atom is confined by spherically symmetric finite oscillator potential [2] of the type

$$V_c(r) = -V_0(1 + c_w r)e^{-c_w r} \qquad (1)$$

where, V_0 is the depth of the potential well fixed at 0.2 *a.u.* and cavity constant c_w is defined as $c_w = \frac{1}{\Delta\sqrt{V_0}}$; Δ being the width of the potential well. The width (Δ) of the potential well ranges from 0.001 *a.u.* to 1000 *a.u.* It has been shown that the potential given by equation (1) breaks the orbital angular momentum (l) degeneracy in Coulomb field for the energy levels of hydrogen-like atoms [figure 1]. Subsequently, the resonance parameters of $^1S^e$ states due to *2sns* [n = 2-5] and *2pnp* [n = 2-5] configuration of confined helium below N = 2 ionization threshold of He^+ in the framework of stabilization method [3] has been studied by using explicitly correlated Hylleraas type basis set [3]. It is evident from figure 2 that for a fixed cavity depth (V_0), width of the corresponding resonance states shows oscillatory behavior *w.r.t.* the width (Δ) of the quantum cavity which is consistent with few available theoretical estimates [2]. Moreover for higher excited states this oscillation is more pronounced. More detailed reasoning of this phenomenon will be discussed in the meeting.

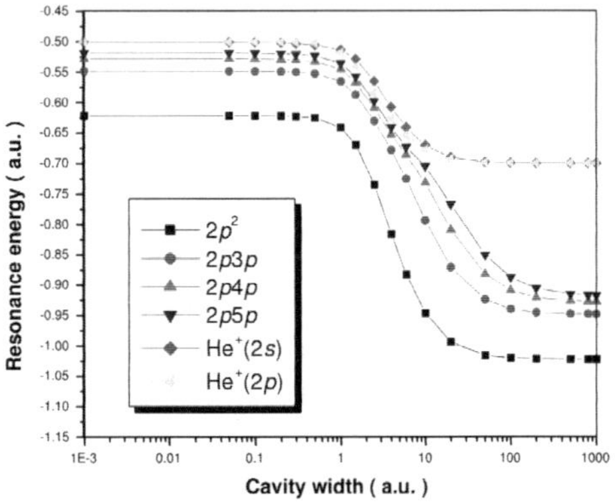

Figure 1. Resonance position of *2pnp* $^1S^e$ states of confined helium vs. width of the cavity

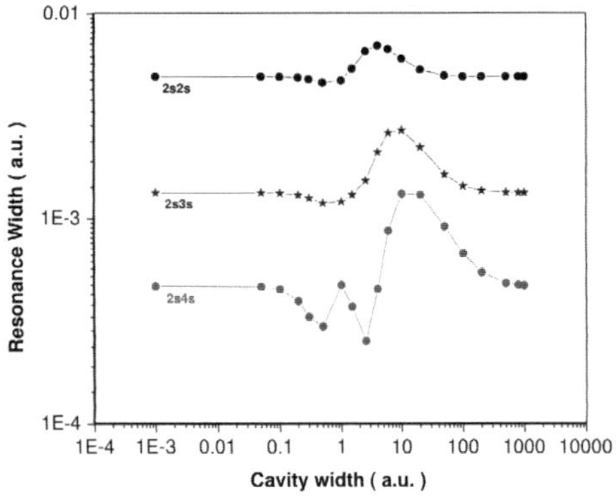

Figure 2. Resonance width of *2sns* $^1S^e$ states of confined helium vs. width of the cavity

References

[1] J. Sabin and E. Brandas (Ed.) 2009 *Adv. Quantum Chem.* **58**

[2] S. Chakraborty and Y. K. Ho 2011 *Phys. Rev. A* **84** 032515

[3] J. K. Saha and T. K. Mukherjee 2009 *Phys. Rev. A* **80** 022513

[1]E-mail: spjks@iacs.res.in

Content from this work may be used under the terms of the Creative Commons Attribution 3.0 licence. Any further distribution of this work must maintain attribution to the author(s) and the title of the work, journal citation and DOI.
Published under licence by IOP Publishing Ltd

Doublet and Quartet states of Li-like ions under hot plasma

Jayanta K. Saha[1]

Indian Association for the Cultivation of Science, Jadavpur, Kolkata-700032, India

Synopsis Precise non-relativistic energy eigenvalues and other structural properties of plasma embedded Li-like ions are estimated with explicitly correlated Hylleraas type basis set.

Structural properties and spectroscopic behavior of quantum mechanical systems under various external confinements have become an active area of research for last few decades. The modification of energy levels and other spectral properties of these systems due to the presence of the confining potential provide very important information about the external environment surrounding the system. It is well-known that Screened Coulomb potential (SCP) [1] of the type

$$V(r) = -Z\frac{e^{-\lambda r}}{r} \tag{1}$$

can mimic a weakly coupled hot plasma background where the screening parameter (λ) is a function of temperature (T) and number density (n_e) of it [1].

In the present work, modification of the structural properties of plasma embedded H-like, He-like and Li-like ions are investigated in the framework of Rayleigh-Ritz variational technique. The wave functions for both He-like and Li-like ions are expanded in explicitly correlated Hylleraas type basis set which fully takes care of electron-electron correlation effect. For He-like systems, the necessary correlated basis integrals arising for SCP are well known. In case of Li-like systems, correlated basis integrals are evaluated following Drake and Yan [2]. The two particle screened Coulomb part of the potential is tackled by expanding it in two ways: Taylor's expansion and Gegenbauer expansion. The first expansion works only when the screening parameter is low *i.e.* the corresponding plasma electron number density (n_e) is less. But the later is useful for a large range of screening parameter. The doublet states arising from $1s^2ns$ [$n \geq 2$] and quartet states due to $1snsn's$ [$n \geq 2$; $n' \geq 3$] configuration of Li-like atoms are considered. In general, it has been noted that the three-electron energy levels approaches towards the two-electron threshold thereby reduces the ionization potential and similarly the two-electron energy levels gradually merges to the one-electron continuum. Different expectation values *e.g.* the one particle moments $\langle r^n \rangle$ [$n = 1-3$], the two-particle moments $\langle r_{ij}^n \rangle$ [$n - 1 - 2$], the electron density at the nucleus, Specific mass shift, Nuclear magnetic shielding constant, Diamagnetic susceptibility etc. are estimated as a function of plasma screening parameter.

References

[1] J. K. Saha and S. Bhattacharyya, T. K. Mukherjee and P. K. Mukherjee 2009 *J. Phys. B* **58** 245701

[2] G. W. F. Drake and Z.-C. Yan 1995 *Phys. Rev. A* **52** 3681

[1]E-mail: spjks@iacs.res.in

Content from this work may be used under the terms of the Creative Commons Attribution 3.0 licence. Any further distribution of this work must maintain attribution to the author(s) and the title of the work, journal citation and DOI.
Published under licence by IOP Publishing Ltd

Measurement of the 5D level polarizabilities in laser cooled Rb atoms

D. Tregubov*,†,•[1], S. Snigirev*,†[2], A. Golovizin*,†,•, S. Pyatchenkov*,†, D. Sukachev*,†, A. Akimov*,†, V. Sorokin*,†, N. Kolachevskyv*,†,•

* P.N. Lebedev Physical Institute, Leninsky Prospekt 53, 119991 Moscow, Russia
† Russian Quantum Center, 100 Novaya St., Skolkovo, Moscow 143025, Russia
• Moscow Institute of Physics and Technology, 141704 Dolgoprudny, Moscow region, Russia

Synopsis We report on accurate measurements of the scalar α_S and tensor α_T polarizabilities of the $5D$ fine structure levels $5D_{3/2}$ and $5D_{5/2}$ in Rb.

Study of atomic and molecular polarizabilities remains an important task in atomic physics. Accurate measurements of polarizability facilitate progress in sophisticated atomic structure calculations and the theory of heavy atoms, which results in more precise predictions for other important atomic parameters (see, e.g., [1]). Measurements of polarizabilities become even more crucial in applications for modern optical atomic clocks. Predictions of the magic wavelength in optical lattice clocks [2] and accurate estimation of the blackbody radiation shift require precise knowledge of static and dynamic polarizabilities [3].

Polarizabilities of ground and highly excited states of alkali atoms are known quite accurately. In case of Rb, the polarizabilities of $5S$ and $5P$ levels have been measured long ago with high precision while being in agreement with theory [4]. However, considering intermediate states there are some difficulties both theoretically and experimentally. For example, $5D$ level in Rb. There are several theoretical predictions which differ up to 20% [5, 6].

We report on accurate measurements of the scalar α_S and tensor α_T polarizabilities of the 5D fine structure levels $5D_{3/2}$ and $5D_{5/2}$ in Rb-87 [7]. α_S and α_T describe the dependance of α on total angular momentum F and its projection m_F:

$$\alpha = \alpha_S + \alpha_T P(F, m_F) \tag{1}$$

Rb atoms were laser cooled in a regular six-beam magneto-optical trap, forming a cloud at the center of a plane capacitor. Excitation to $5D$ level was performed in a cascade mode, the population was measured by spontaneous emission.

Optical pumping of atoms to a specific magnetic sublevel allowed us to measure the tensor polarizability component with relatively high precision. In order to perform optical pumping we applied a circularly σ^+-polarized pump pulse to transfer atoms to the $5P_{3/2}(F = 3, m_F = +3)$ magnetic sublevel. Using the probe $5P \rightarrow 5D$ pulse with certain polarization (σ^+ or σ^-) we defined both scalar and tensor polarizabilities.

The measured values (in atomic units) $\alpha_S(5D_{3/2}) = 18400(75)$, $\alpha_T(5D_{3/2}) = 750(30)$, $\alpha_S(5D_{5/2}) = 18600(76)$ and $\alpha_T(5D_{5/2}) = 1440(60)$ show reasonable correspondence to previously published theoretical predictions, but are more accurate. We demonstrated a relative uncertainty of 4% for the tensor polarizability and 0.4% for the scalar polarizability, which is comparable to accurate measurements in ground-state alkali atoms. Our result is close to the theoretical prediction [5].

References

[1] Mitroy J, Safronova M S and Charles W Clark 2010 *J. Phys. B* **43**, 202001

[2] Katori H, Takamoto M, Palchikov V G and Ovsiannikov V D 2003 *Phys. Rev. Lett.* **91**, 173005

[3] Beloy K, Safronova U I and Derevianko A 2006 *Phys. Rev. Lett.* **97**, 040801

[4] Daniel A. Steck, Rubidium 87 D Line Data, available online at http://steck.us/alkalidata

[5] Kamenski A A and Ovsiannikov V D 2006 *J. Phys. B: At. Mol. Opt.Phys.* **39**, 2247

[6] Kondrat'ev D A, Beigman I L and Vainshtein L A 2008 *Bull. Lebedev Phys. Inst.* **35**, 355

[7] Snigirev S, Golovizin A, Tregubov D, Pyatchenkov S, Sukachev D, Akimov A, Sorokin V and Kolachevsky N 2014 *Phys. Rev. A* **89**, 01251

[1] E-mail: treg.dim@gmail.com
[2] E-mail: ss@rqc.ru

Content from this work may be used under the terms of the Creative Commons Attribution 3.0 licence. Any further distribution of this work must maintain attribution to the author(s) and the title of the work, journal citation and DOI.
Published under licence by IOP Publishing Ltd

Journal of Physics: Conference Series **635** (2015) 092122 doi:10.1088/1742-6596/635/9/092122

Control of strong-field ionization with two-color laser pulses

N. Camus* [1], **L. Fechner***, **D. G. Arbó****, **C. Lemell**×, **S. Nagele**×, **J. Ullrich***†, **T. Pfeifer***, **S. D. López****, **J. Burgdörfer**×, **R. Moshammer*** [2]

* Max-Planck-Institut für Kernphysik, Heidelberg, 69117, Germany
** Institute for Astronomy and Space Physics, IAFE (Conicet-UBA), Buenos Aires, Argentina
× Institute for Theoretical Physics, Vienna University of Technology, Austria
† Physikalisch-Technische Bundesanstalt, 38116 Braunschweig, Germany

Synopsis Experimental results and theoretical analysis of the ionization process of argon atoms interacting with linearly polarized two-color fields ($\lambda_1 = 800\,nm$, $\lambda_2 = 400\,nm$) are presented. We observe complex asymmetry patterns in the measured three-dimensional momentum distributions.

Strong-field ionization of atoms with near infra-red laser pulses reveals a rich structure in the momentum distribution of the liberated electron. The patterns observed (Fig. 1) arise from the interferences of multiple pathways leading to the same final momentum state. They retain precise information about the creation of electron wave-packets in the laser field and the phase they accumulate during the propagation in the combined laser and ionic Coulomb field [1].

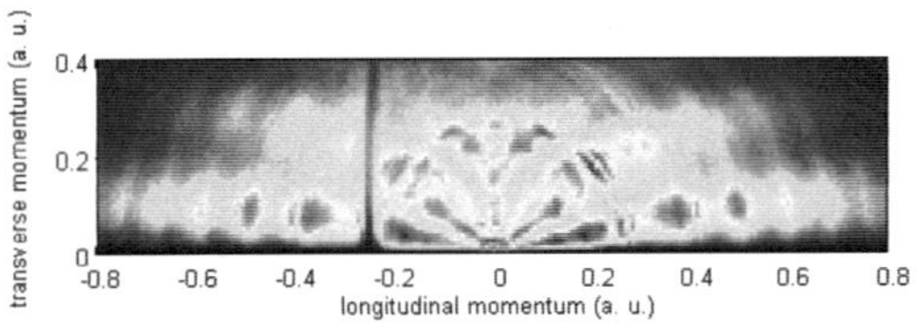

Figure 1. Electron momentum distribution for ionization of Ar by 800 nm, 25 fs pulses (intensity $2 \times 10^{14}\,\mathrm{W/cm^2}$).

We study the ionization of argon with shaped two-color laser fields. They are formed by the combination of 800 nm femtoseconds laser pulses and their frequency-doubled component (400 nm).

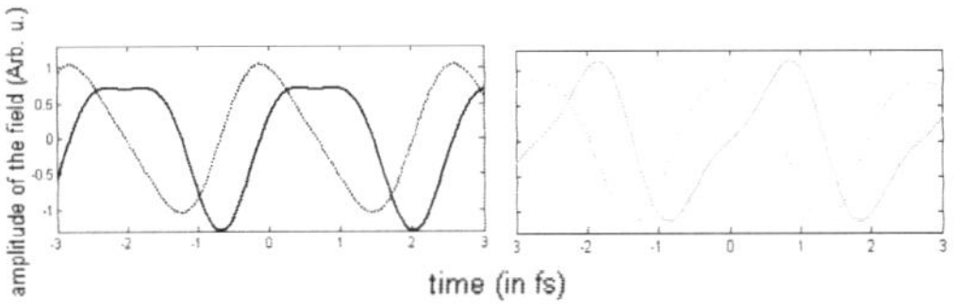

Figure 2. Shapes of the laser electric field (solid line) and its vector potential (dotted line) for different phases between red and blue carriers: $0.25\,\pi$ (left) and $0.5\,\pi$ (right).

By controlling the relative phase between the red and blue carrier waves (Fig. 2), we can control the ionization rate via the temporally modified electric field strength. More importantly this shaped field controls also the trajectories of the ionized electron and the accumulated phase leading to modified interference structures in the 3-dimensional momentum distributions. We precisely monitor the modifications of the interference structures (Fig. 3).

Electron momentum distributions are recorded with a reaction microscope and compared to simulated spectra.

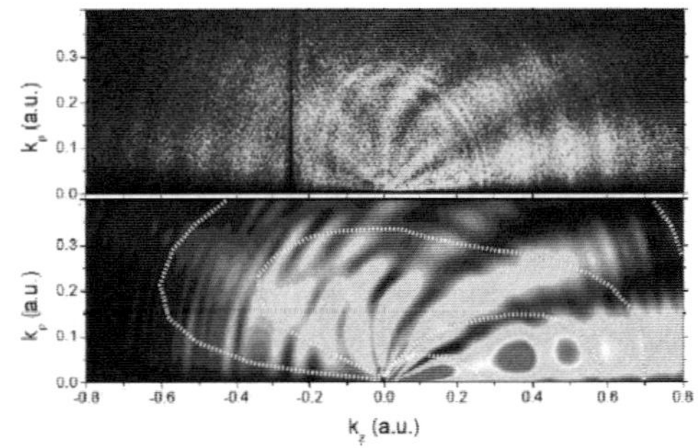

Figure 3. Measured (upper panel) and TDSE (lower panel) momentum distributions of photoelectrons for relative phase $\varphi = 0$. The contour lines on the lower panel mark isolines of the CTMC simulation.

We also perform a study of classical-quantum correspondence in the simulation of laser-atom interactions. We find a surprisingly good agreement between CTMC and TDSE simulations which might open the pathway for an improved understanding of previously unexplained features observed in experimental spectra.

References

[1] D. G. Arbó *et al* 2006 *Phys. Rev. Lett.* **96** 143003

[1] E-mail: nicolas.camus@mpi-hd.mpg.de
[2] E-mail: robert.moshammer@mpi-hd.mpg.de

Content from this work may be used under the terms of the Creative Commons Attribution 3.0 licence. Any further distribution of this work must maintain attribution to the author(s) and the title of the work, journal citation and DOI.
Published under licence by IOP Publishing Ltd

XXIX International Conference on Photonic, Electronic, and Atomic Collisions (ICPEAC2015) IOP Publishing
Journal of Physics: Conference Series **635** (2015) 092123 doi:10.1088/1742-6596/635/9/092123

Generalized Sturmian approach to extracting transition amplitudes for two-photon ionization of atoms by electromagnetic pulses

A.I. Gómez[†1], **G. Gasaneo**[†] **and D.M. Mitnik**[‡]

[†] Departamento de Física, Universidad Nacional del Sur - IFISUR, Bahía Blanca, C.P. 8000, Buenos Aires, Argentina

[‡] Instituto de Astronomía y Física del Espacio, CONICET-UBA, C.C. 67, Suc. 28 (C1428ZAA), Ciudad Autónoma de Buenos Aires, Buenos Aires, Argentina

Synopsis In this work we present the application of the Generalized Sturmian basis to the process of photonionization by an electromagnetic pulse in the framework of a perturbation theory. The Generalized Sturmian basis have the proper asymptotic behavior allowing us to extract the transition amplitudes directly from the coefficients of expansion.

In the last years there have been impressive progress in attosecond physics [1, 2] mainly because of the fast experimental development of ultrafast lasers, which enabled the temporal study of dynamic processes at the natural atomic time scale. In turn, this led to the development of theoretical and computational methods to study ultrafast phenomena with high resolution in the time domain. Our aim is to describe the interaction of atoms with trains of laser pulses. For that reason we start analyzing the details on the structure corresponding to atoms. In particular we will consider the case where up to two photons are absorbed.

Our starting point is the time-dependent Schrödinger equation

$$\left[i\frac{\partial}{\partial t} - H_a - \lambda W\right]\Psi(t,\mathbf{r}) = 0 \qquad (1)$$

where H_a is the non-relativistic atomic Hamiltonian, W is the perturbation that contains the interaction with the electromagnetic field and λ defines the field intensity. In the perturbative scheme λ is a small parameter. We propose an expansion in powers of the field intensity λ, $\Psi(t,\mathbf{r}) = \sum_n \lambda^n \psi^{(n)}(t,\mathbf{r})$, which yields a coupled system of differential equations for successive orders $\psi^{(n)}(t,\mathbf{r})$. To represent the time evolution of the system we use Fourier representation for both wave function and the electromagnetic field.

We make use of Generalized Sturmian Functions (GSF) [3, 4] to solve the time-independent equations obtained. The ionization amplitude is extracted from the asymptotic limit of the scattering functions $\widetilde{\psi}^{(n)}_{sc}$ to the order solved , simply, as the sum of expansion coefficients.

For the case in which the system is a hydrogen atom interacting with a laser pulse, the first-order equation has an analytic solution. This allowed us to test the method and verify its efficiency (see figure 1).

In this work we extend this methodology concerning the interaction of atoms with up to two photons.

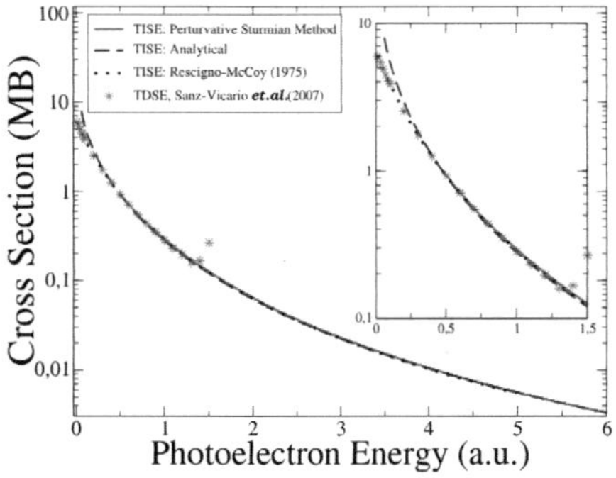

Figure 1. Cross section of H atom in mega-barns versus photoelectron energy in atomic units. Our results after solving the perturbative TISE with Sturmian method (red, solid line), are compared with the exact analytical result (blue, dashed line) and with the from [5](black, dotted line), and with the calculated one after solving the TDSE (green, star points) by Sanz-Vicario [6].

References

[1] F. Krausz y M. Ivanov 2009 *Rev. Mod. Phys* **81** 163

[2] M. Drescher *et al.* 2002 *Nature* **419** 803

[3] G. Gasaneo *et al.* 2013 *Adv. Quantum Chem.* **67** 153.

[4] D.M. Mitnik *et al.* 2011 *Comp. Phys. Comm.* **45** 1145

[5] T.N. Rescigno and V. Mckoy 1975 *Phys. Rev. A* **12** 522

[6] J.L. Sanz-Vicario *et al.* 2007 *J. Electron Spectrosc. Relat. Phenom.* **161** 182

[1]E-mail: unilo.2099@gmail.com

Content from this work may be used under the terms of the Creative Commons Attribution 3.0 licence. Any further distribution of this work must maintain attribution to the author(s) and the title of the work, journal citation and DOI.
Published under licence by IOP Publishing Ltd

Journal of Physics: Conference Series **635** (2015) 092124 doi:10.1088/1742-6596/635/9/092124

HILITE - Ion Trap for Studies with Intense Laser Pulses

S. Ringleb[A 1], M. Vogel[B 2], S. Kumar[C 3], W. Quint[B,D 4], G.G. Paulus[A,E 5], Th. Stöhlker[A,B,E 6]

[A] Institut für Optik und Quantenelektronik, Friedrich-Schiller-Universität, 07743 Jena, Germany
[B] GSI Helmholtzzentrum für Schwerionenforschung GmbH, 64291 Darmstadt, Germany
[C] Inter-University Accelerator Centre, 110067 New Delhi, India
[D] Physikalisches Institut, Ruprecht Karls-Universität Heidelberg, 69120 Heidelberg, Germany
[E] Helmholtz-Institut Jena, Fröbelstieg 3, 07743 Jena, Germany

Synopsis We present our Penning Trap setup which is designed for capture, confinement and preparation of well-defined ion clouds for use in experiments with high-intensity lasers. We explain the experimental setup and the techniques used to capture, confine, manipulate and detect the ions inside the Penning trap. We give an overview of the status of the project and the planned procedures to measure the laser-focus shape in situ.

We are currently devising an experimental setup which allows the preparation of well-defined ion targets for the interaction with high-intensity and high photon-energy lasers. The ion-target preparation is supported by an open-endcap cylindrical Penning trap, which can store atomic ions of an arbitrary charge state up to bare uranium U^{92+} [1].

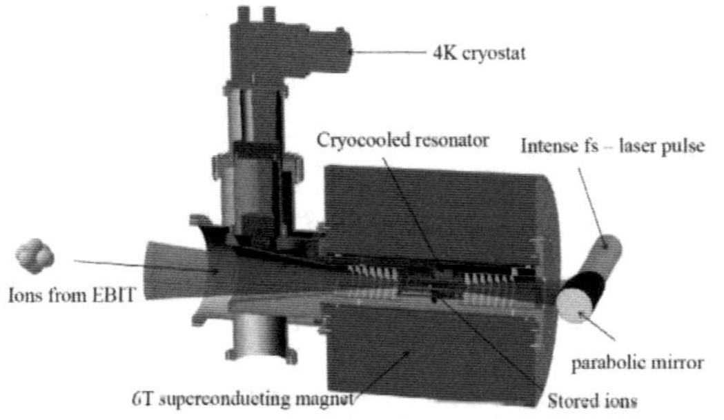

Figure 1. Experimental setup consisting of a superconductive magnet with a warm bore and an open-endcap Penning trap with three trap electrodes and two conical capture electrodes. The laser focus is in the trap center.

We have established a superconducting magnet with a horizontal warm bore and a magnetic field strength of up to 6 T. The Penning trap is a mechanically compensated trap consisting of three trap electrodes and two conical electrodes for dynamic capture of externally produced ions. The trap is optimized to allow the non-destructive detection also of small particle numbers over a large range of species using the FT-ICR (Fourier Transform Ion cyclotron resonance) technique. We will apply the SWIFT (Storage Waveform Inverse Fourier Transform) technique to select one or more defined ion species of an arbitrary mass and charge state by resonant ejection of the non-wanted ions. Application of the rotating wall technique will allow to define the ion number density up to values of about 10^9 cm^{-3}. It also allows good control over the ion-cloud shape by appropriate adjustment of the trapping parameters. The ions can be resistively cooled to kinetic energies of around 0.1 meV [2], which allows a well-defined nearly step-like frontier of the ion cloud. The ion-cloud position with respect to the laser focus may be controlled by an induced asymmetry of the trap voltages providing an accuracy of positioning in the micrometer scale.

With such a well-defined ion target, we intend to probe the laser-focus geometry in situ, yielding the intensity distribution inside the laser focus without the need for laser attenuation or outcoupling. With this configuration we also intend to investigate the ionization behaviour of ions under the influence of extreme laser fields up to an intensity of 10^{20} W/cm². To this end, we optimize the focusing parameters as well as our target ion cloud to maximize the number of particles which are exposed to an intensity above the ionization threshold. Since the whole setup is designed to be moved readily we are able to perform studies with a variety of high-intensity lasers, e.g. PHELIX at GSI, JETI and POLARIS in Jena or FLASH at DESY.

We present the experimental outline, the devised setup with its current state and the intended experiments with their possible benefit.

References

[1] M. Vogel *et al* 2009 *Nucl. Instr. Meth. Phys. Res. B* 285 65 - 71
[2] M. Vogel *et al* 2014 Phys. Rev. A 90 043412

[1]E-mail: stefan.ringleb@uni-jena.de
[2]E-mail: m.vogel@gsi.de
[3]E-mail: sugamkumar@gmail.com
[4]E-mail: w.quint@gsi.de
[5]E-mail: gerhard.paulus@uni-jena.de
[6]E-mail: t.stoehlker@gsi.de

Content from this work may be used under the terms of the Creative Commons Attribution 3.0 licence. Any further distribution of this work must maintain attribution to the author(s) and the title of the work, journal citation and DOI.
Published under licence by IOP Publishing Ltd

Journal of Physics: Conference Series **635** (2015) 092125 doi:10.1088/1742-6596/635/9/092125

High-order harmonic generation from confined Rydberg atoms

Kudiyar Orazymbetov*, Erdi A. Bleda*[1], Zikri Altun*[2], Turker Topcu†[3]

* Department of Physics, Marmara University, 34722, Istanbul Turkey

† Department of Physics, University of Nevada, Reno, Nevada 89557, USA

Synopsis We report results from our simulations of High Harmonic Generation (HHG) from a confined atom in a Rydberg state. We find that for the $n = 2$ excited state of H the cut-off of the harmonic spectrum is substantially extended compared to that for a free atom at the expense of the harmonic yield. This effect is dependent on the radius of the confining shell for a given n. We also observe that the confined spectrum exhibits cusps similar to those seen in the HHG spectra from ground state atoms in the presence of Cooper minima.

Dynamically rich nature of the high-order harmonic generation (HHG) process lends itself to a variety of ways to extend the harmonic cut-off frequency. In this work, we investigate HHG from a Rydberg atom confined in an attractive shell as HHG from Rydberg states have already shown interesting physics and potential for extending the harmonic cut-off frequency [1, 2].

We solve the time-dependent Schrödinger equation within a one-dimensional s-wave model, where we use a $800n^3$ nm laser pulse with 4 cycles at FWHM. The laser intensity is $(3.5/n^8) \times 10^{14}$ W/cm^2. The cage is modeled as a -8.22 eV deep spherical shell potential with 5.8 $\mathring{A}$ inner radius and 7.69 $\mathring{A}$ outer radius. These parameters correspond to a C_{60} cage [3].

We compare the spectra for a confined atom with that for a free H atom to see how the modified properties of the caged atom affect its HHG spectrum. The results are shown in Fig. 1. We perform three calculations: first we calculate the HHG spectra for free (pink) and confined (red) atoms in the $n = 2$ state using the same laser frequency ω_0 and intensity I. Because part of the 2s state lies inside the attractive cage potential, its ionization potential I_p is increased relative to the free atom. This results in a decreased tunneling rate which drastically reduces the HHG yield. Surprisingly, although the cut-off frequency at $3.17U_p$ does not depend on I_p, it is also shifted by almost 30 harmonic orders. Here U_p is the ponderomotive potential. In order to compare the caged spectrum with that for a free atom with the same level of yield, we perform a third calculation for a free H atom where we now fix the Keldysh parameter γ at 0.57 (blue dotted). In this case, we fix I and the increase ω_0 to match the γ for the caged spectrum (red). Increased ω_0 results in decreased U_p, which reduces the cut-off by ~150 harmonic orders. We estimate that ~120 out of these 150 orders are associated with the increase in ω_0.

We also performed the same set of calculations for the 1s and the 4s states. However, we did not see any dramatic shift in the cut-off frequency. For $n = 2$, the last peak of the bound state wave function lies inside the cage, suggesting that the cage radius relative to the spatial extent of the initial state wave function is important. The cusps seen in the spectrum of the caged H in Fig. 1 (red) resemble those seen in HHG spectra from ground state atoms due to Cooper minima when photoionization is present in addition to tunneling. We investigate the mechanism behind these cusps through extensive calculations of electron flux through the C_{60} surface.

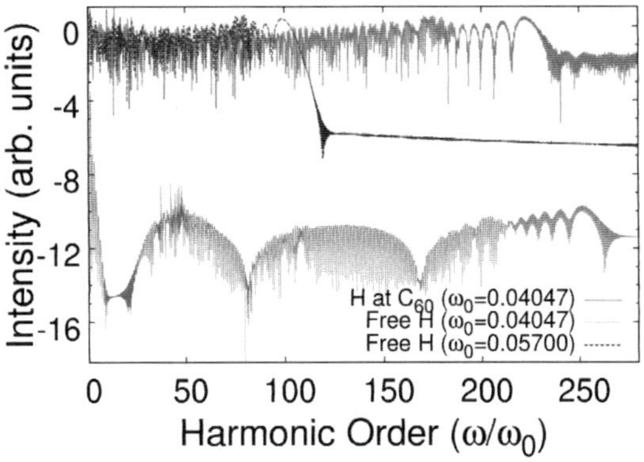

Figure 1. HHG spectra for free and confined H.

References

[1] E. A. Bleda, I. Yavuz, Z. Altun, and T. Topcu, Physical Review A **88**, 043417 (2013)

[2] Z. Zhai, Q. Zhu, J. Chen, Z. C. Yan, P. Fu, and B. Wang, Phys. Rev. A **83**, 043409 (2011).

[3] V. K. Dolmatov, G. T. Craven, E. Guler, and D. Keating, Physical Review A **80**, 035401 (2009)

[1]E-mail: ebleda@marmara.edu.tr
[2]E-mail: zikalt@marmara.edu.tr
[3]E-mail: ttopcu@unr.edu

Content from this work may be used under the terms of the Creative Commons Attribution 3.0 licence. Any further distribution of this work must maintain attribution to the author(s) and the title of the work, journal citation and DOI.
Published under licence by IOP Publishing Ltd

Journal of Physics: Conference Series **635** (2015) 092126 doi:10.1088/1742-6596/635/9/092126

Effect of the fullerene confining potential in the $1s \to 2p$ resonant transition of H

A L Frapiccini* [1], **G Gasaneo***, **D Mitnik**† **and F D Colavecchia**‡

* IFISUR-Dpto. de Física, Universidad Nacional del Sur, 8000 Bahía Blanca, Argentina, and CONICET
† Instituto de Astronomía y Física del Espacio (IAFE) and Departamento de Física, Universidad de Buenos Aires, C1428EGA Buenos Aires, Argentina and CONICET
‡ División Física Atómica, Molecular y Optica, Centro Atómico Bariloche, 8400 S. C. de Bariloche, Río Negro, Argentina and CONICET.

Synopsis In this work we study the effect of the confinement of H atom in a fullerene cage for the $1s \to 2p$ resonant transition of the atom interacting with a laser pulse. We present a novell implementation of the Generalized Sturmian Functions in the time-dependent frame to numerically solve this problem calculating the population of bound states and spectral density.

Atoms trapped inside a hollow spacial carbon cage, A@C_N, also called endohedral or doped fullerenes, are of interdisciplinary significance and attractive nano-objects for investigation [1]. Endohredral atoms are the subject of intense interest due to the ability of the enclosing fullerene cage to shield the atom from the surrounding environment. This gives rise to many proposed applications, such as qubits for quantum computing [2], the dating of meteorite and comet impacts [3], and as diagnostic agents in medicine [4]. Endohedral atoms can be confined by various fullerenes such as C_{60},C_{70}, and C_{82}. Due to the high symmetry of C_{60}, this was the favored fullerene for theoretical studies.

Photoionization and photoelectron spectroscopy is an effective method to investigate structure and properties of these bodies.

In this work we present an implementation of the Generalized Sturmian Functions [5] (GSF) within the solution of the time-dependent Schrödinger equation (TDSE), to study a simple case of the photoionization of the bare H atom and the caged H@C_{36} and H@C_{60} interacting with an electromagnetic pulse of finite duration.

We write the time-dependent Schrödinger equation (TDSE) for a two particle system interacting with an external field in the form

$$\imath\frac{\partial}{\partial t}\Psi(\mathbf{r},t) = H(\mathbf{r},t)\Psi(\mathbf{r},t) \tag{1}$$

where the hamiltonian can be written as

$$H(\mathbf{r},t) = H_0 + H_{int} \tag{2}$$

with H_0 the unperturbed hamiltonian, and H_{int} the interaction with the field.

Following the approach given by Palacios et al [6], we consider the evolution of the wave packet at $t > t_{final}$ under the atomic, time-independent hamiltonian H_0, thus the propagation of the wave packet from $t = t_{final}$ to infinity is equivalent to solve the time-independent Schrödinger equation given by

$$(E - H_0)\Psi_{sc}(\mathbf{r}) = \Psi(\mathbf{r}, t_{final}) \tag{3}$$

where Ψ_{sc} has outgoing boundary conditions, and $\Psi(\mathbf{r}, t_{final})$ is the wave packet at the end of the pulse. The ionization amplitude can be extracted from the scattering function at sufficiently large values of the radius r.

We compare the bare and confined H atom in the case of a (central) photon with energy 0.375 a.u., which corresponds to the $1s \to 2p$ resonant transition of H. The parameters of the model potential represents the C_{36} and C_{60} shells, and we refer to these two cases as endofullerenes H@C_{36} and H@C_{60}. The results for the $1s$ and $2p$ bound states population during the interaction with the pulse are presented, as well as the spectral density.

References

[1] Dolmatov V K 2009 *Adv. Qua tum Chem.* **58** 13.

[2] Ju C, Suter D and Du J 2007 *Phys. Rev. A* **75** 012318.

[3] Becker L, Poreda R J and Bunch T E 2000 *Proc. Natl. Acad. Sc . U.S.A.* **97** 2979.

[4] Bakry R et al. 2007 *I t. J. Na omed c e* **2** 639.

[5] Gasaneo G et al 2013 *Adv. Qua tum Chem.* **67** 153.

[6] Palacios A, McCurdy C W and Rescigno T N 2007 *Phys. Rev. A* **76** 043420.

[1] E-mail: afrapic@uns.edu.ar

Content from this work may be used under the terms of the Creative Commons Attribution 3.0 licence. Any further distribution of this work must maintain attribution to the author(s) and the title of the work, journal citation and DOI.
Published under licence by IOP Publishing Ltd

COBRA: a zero-deadtime camera based multiparticle detector

X Urbain* [1], D Bech* , J-P Van Roy†, M Géléoc‡, S J Weber‡, A Huetz§ and Y Picard§ [2]

* Institute of Condensed Matter and Nanosciences, Université catholique de Louvain, B-1348 Louvain-la-Neuve, Belgium

†Physics Department, University of Namur, B-5000 Namur, Belgium

‡ Laser, Interactions and Dynamics Laboratory (LIDyL), DSM/IRAMIS, CEA-Saclay, F-91191 Gif-sur-Yvette,France

§ Institut des Sciences Moléculaires d'Orsay, UMR 8214, Université Paris-Sud, Orsay Cedex, France

Synopsis A new time and position sensitive detector is described, which combines an MCP stack with a phosphor screen, a CMOS camera and a waveform digitizer. Particle identification relies on the correlation between brightness of phosphor screen spots and amplitude of electrical signals (COBRA).

Two main families of position sensitive detectors are in use, depending on the type of anode backing the MCP, either pixelated or delay-line anodes (DLA) with fast electronic readout [1], or phosphor screens with optical imaging. The latter technique lacks temporal information, unless combined with a second camera with fast optical shutter [2], a segmented photomultiplier [3], or a low resolution delay line [4].

Combining the best of both techniques, the COBRA method [5] correlates the brightness of the phosphor screen spots to the amplitude of the fast MCP signal. The latter is directly proportional to the amount of electrons generated at the MCPs which hit the phosphor, providing a marker to label each particle. Pulse amplitudes and spot brightnesses may thus be sorted in order to retrieve time and position coordinates of individual hits.

The detector (figure 1) consists in a Z-stack of imaging grade microchannel plates, a P46 phosphor screen, and a kHz CMOS camera controlled by a FPGA. The amplified MCP signal is fed into a waveform digitizer for further processing.

The detector was first tested on the two- and three-particle breakup of hot and cold H_3^+ ions upon electron transfer from potassium, producing kinetic energy release spectra in agreement with the literature [6].

The ability of our COBRA detector to collect and to analyze electron pairs was tested on a COLTRIMS apparatus, to study the photo-double-ionization of Xe with XUV photons produced by high-harmonic generation at the PLFA laser beamline (CEA-Saclay, France). Electron energy correlation maps and distributions of mutual angle of ejection were obtained.

The COBRA detection system may be implemented wherever MCPs are used in combination with a phosphor screen. The rapid progress in high frame-rate CMOS sensors makes this approach perfectly suited to ultrafast laser experiments.

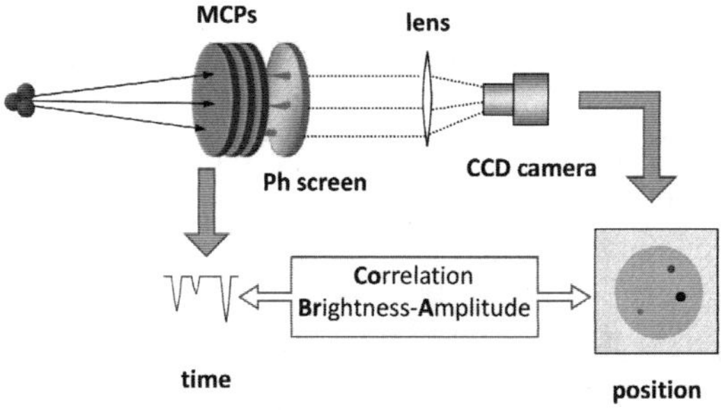

Figure 1. Schematics of the COBRA detector. Light spots and electrical pulses are sorted according to their brightness and amplitude, respectively.

References

[1] Jagutzki O *et al* 2002 *IEEE Trans. Nucl. Sci.* **49 2477**

[2] Strasser D *et al* 2000 *Rev. Sci. Instrum.* **71 3092**

[3] Amitay Z and Zajfman D 1997 *Rev. Sci. Instrum.* **68 1387**

[4] Ismail I *et al* 2005 *Rev. Sci. Instrum.***76 043304**

[5] Urbain X *et al* 2015 *Rev. Sci. Instrum.* **86 023305**

[6] C. Laperle *et al* 2004 *Phys. Rev. Lett.* **93 153202**

[1]E-mail: xavier.urbain@uclouvain.be
[2]E-mail: yan.picard@u-psud.fr

Content from this work may be used under the terms of the Creative Commons Attribution 3.0 licence. Any further distribution of this work must maintain attribution to the author(s) and the title of the work, journal citation and DOI.
Published under licence by IOP Publishing Ltd

Journal of Physics: Conference Series **635** (2015) 092128 doi:10.1088/1742-6596/635/9/092128

Multiple Cooper minima in ground state E2 photoionization of high Z atoms

Tanima Banerjee*[1], Ankur Mandal*, Pranawa C. Deshmukh*[2] and Steven T. Manson[†]

[*] Department of Physics, Indian Institute of Technology Madras, Chennai 600036, India
[†] Department of Physics and Astronomy, Georgia State University, Atlanta, 30303, USA

Synopsis Quadrupole photoionization from valence and intermediate subshells of atomic mercury, radon and radium have been studied in the present work. Multiple Cooper minima in the E2 channels have been found even when only *intra-shell inter-channel coupling* is included, in contrast with the dipole case in which multiple Cooper minima were found only when *inter-shell interchannel coupling* was included.

With advanced sources and detectors, present day photon matter interaction studies require theoretical tools often beyond the conventional theoretical models. The ionization probabilities are computed from the transition matrix elements. These matrix elements at certain photon energies go through a minimum, namely Cooper minimum [1] which is manifest in the photoionization cross-section. In the well known 'dipole approximation', in some cases, even multiple Cooper minima have been found. However they were found either in ionization from the excited states [2, 3] or were found to be induced by inter-shell interchannel coupling [4, 5].

In the present we investigate the quadrupole (E2) matrix elements, which have been found to be of importance even at low energies due to certain dynamical factors. A many body theory, namely the RRPA [6] is used which includes most of the important electron correlations, and also the relativistic terms. Calculations were done at various levels of truncation to identify the electron correlation effects clearly. It was found that in moderately high Z atoms, namely krypton and xenon, intrashell interchannel coupling between quadrupole photoionization channels already led to the occurrence of multiple Cooper minimum [7]. In the present work, some more studies have been carried out to test the generality of this phenomenon. Toward that goal, we studied the E2 transitions in atomic mercury. As this is a high Z atom, the relativistic effects are expected to be strong. E2 photoionization from the 5d subshell was studied at different levels of truncation, starting from coupling channels only from 5d subshell itself (intrashell coupling), up to the inclusion of all the E2 channels from all subshells with principal quantum number n = 6, 5 and 4. It was found that E2 transition matrix elements for transitions from the 5d subshell have two Cooper minima, one at photon energy of ~300 eV of photon energy, and the second at ~1750 eV of photon energy. The multiple Cooper minima were observed even at the *intrashell* coupling level, as in the case of E2 photoionization of Kr and Xe, as shown in the Fig.1.

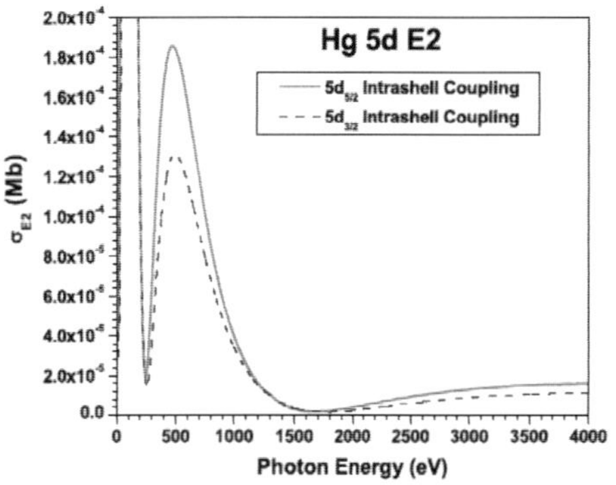

Figure 1. Hg 5d E2 photoionization cross-section using RRPA coupling channels only from 5d shells. Solid line shows cross-section for ionization from $5d_{5/2}$ subshell and the dashed line indicated photoionization cross-section from $5d_{3/2}$ subshell

Multiple Cooper minima in the E2 channels thus seem to occur even when only *intra-shell inter-channel coupling* is included, in contrast with the dipole case in which multiple Cooper minima were found only when *inter-shell interchannel coupling* was included.

References

[1] J. W. Cooper 1962 *Phys. Rev.* **128** 681
[2] J. Lahiri *et al* 1982 *Phys. Rev. Lett.* **48** 614
[3] A. Z. Msezane *et al* 1982 *Phys. Rev. Lett.* **48** 473
[4] P. C. Deshmukh *et al* 1986 *Phys. Rev. A.* **34** 5162
[5] T. Banerjee *et al* 2007 *Phys. Rev.A .* **75** 042701
[6] W. R. Johnson *et al* 1979 *Phys. Rev. A.* **20** 964
[7] T. Banerjee *et al* 2007 *J. Phys.: Conf. Ser.* **80** 012001

[1]E-mail: tanimaban@gmail.com
[2] E-mail: pcd@physics.iitm.ac.in

Content from this work may be used under the terms of the Creative Commons Attribution 3.0 licence. Any further distribution of this work must maintain attribution to the author(s) and the title of the work, journal citation and DOI.
Published under licence by IOP Publishing Ltd

Time delays in above-threshold ionization

S. Nagele*[1], S. D. López†[2], D. Arbó†, and J. Burgdörfer*

* Institute for Theoretical Physics, Vienna University of Technology, 1040 Vienna, Austria, EU

† Institute for Astronomy and Space Physics, IAFE (CONICET-UBA), 1428 Buenos Aires, Argentina

Synopsis We theoretically analyze time-resolved above-threshold ionization as probed by an interferometric measurement technique. We investigate the time delays in the ATI process as a function of the number of involved photons and the intensity of the ionizing UV field and identify IR-probe field induced contributions to the time shifts.

Progress in the creation of ultrashort light pulses during the last decade has enabled the time-resolved study of photoemission with attosecond precision (1as $= 10^{-18}$s). First pioneering experiments, using either attosecond streaking or the complementary interferometric RABBIT technique, have been performed for photoemission from tungsten surfaces [1] and photoionization of rare gas atoms [2, 3, 4]. The experimental progress has triggered considerable theoretical efforts to understand photoionization and the formation and emission of electronic wavepackets from a time-dependent perspective. Substantial progress has been made for time-resolved *single-photon* and *two-photon* ionization.

Very recently, experimental results for time delays in above-threshold ionization (ATI) by UV radiation have been reported [5] which have been obtained using an interferometric probing technique employing an additional IR field, analogously to the well-known RABBIT approach [3, 4]. A theoretical analysis of the temporal structure of *multi-photon* ATI has not yet been put forward.

In this contribution we theoretically analyze the ATI process for argon in the time domain by solving the time-dependent Schrödinger equation (TDSE) in the single-active electron approximation using a suitable model potential. We aim at understanding the timing of ATI of Ar by a strong 400 nm UV pulse interferometrically probed by a weaker 800 nm IR field. We compare our TDSE results (Figure 1) with approximate calculations employing the strong-field approximation (SFA) and the Coulomb-Volkov approximation (CVA) to identify the relevant contributions to the ATI time shifts. At the experimental intensities ($> 10^{13}$ W/cm^2) the ATI spectrum results from the interference of multiple ionization pathways featuring the absorption and emission of several UV photons of which all contribute to the individual ATI peaks. We investigate the complex strong-field multi-photon process as a function of the intensity and wavelength of the ionizing UV field.

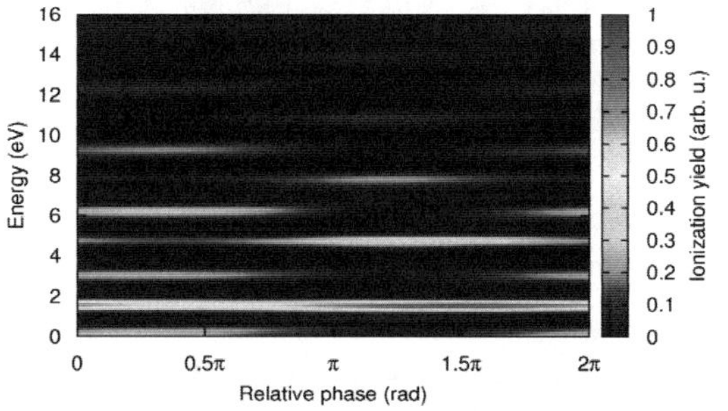

Figure 1. Simulated ATI spectrum for electron emission from Ar along the direction of the laser polarization axis as a function of the relative phase between the ionizing UV pulse ($I=8\cdot10^{13}$ W/cm^2) and the probing IR field ($I=4\cdot10^{11}$ W/cm^2).

In order to identify the influence of the long-range Coulomb tail and the presence of Rydberg series of excited states below the ionization threshold, we compare our results for Ar with time-resolved multi-photon above-threshold ionization from short-ranged Yukawa-type potentials.

This work was supported by the FWF-Austria (SFB NEXTLITE, SFB VICOM), by CONICET (PIP100386), by the OeAD (WTZ AR 03/2013), by the NSF through XSEDE resources (TG-PHY090031), and through computational resources at the Vienna Scientific Cluster (VSC).

References

[1] A. L. Cavalieri *et al* 2007 *Nature* **449** 1029
[2] M. Schultze *et al* 2010 *Science* **328** 1658
[3] K. Klünder *et al* 2011 *Phys. Rev. Lett.* **106** 143002
[4] C. Palatchi *et al* 2014 *J. Phys. B* **47** 245003
[5] L. Zipp *et al* 2014 *Optica* **1** 361

[1] E-mail: stefan.nagele@tuwien.ac.at [2] E-mail: sebastlop@iafe.uba.ar

Content from this work may be used under the terms of the Creative Commons Attribution 3.0 licence. Any further distribution of this work must maintain attribution to the author(s) and the title of the work, journal citation and DOI.
Published under licence by IOP Publishing Ltd

Journal of Physics: Conference Series **635** (2015) 092132 doi:10.1088/1742-6596/635/9/092132

K-shell Photoionization of Atomic Cl

Felfli Z.[†1], **Manson S. T.**[*2] **and Msezane A. Z.**[†3]

[†]Center for Theoretical Studies of Physical Systems, and Department of Physics, Clark Atlanta University, Atlanta, Georgia 30314, USA

[*] Department of Physics and Astronomy, Georgia State University, Atlanta, Georgia 30303, USA

Synopsis Recent measurements of the photoionization of atomic Cl in the vicinity of the 1s thresholds have motivated the present R-matrix calculation which takes into account relativistic effects *via* the Breit-Pauli operator. The computer code CIV3 of Hibbert and Glass and Hibbert, which also includes relativistic effects, is used to obtain the discrete wavefunctions. These are constructed with orbitals generated from a carefully-chosen large scale configuration interaction expansion. The open-shell nature of the Cl atom translates into the existence of actually four 1s thresholds, $^3P^o_{0,1,2}$ and $^1P^o_1$. The results are analyzed with particular focus on the resonances leading up to the four thresholds, and the various effects that dominate the cross sections in this energy range are unraveled.

When energetic x-rays interact with atoms (molecules), many processes take place simultaneously and must all be taken into account. This is a complex problem of multi-electron correlations that manifest *via* various types of measurable relaxation effects, such as x-ray emission (fluorescence) and Auger decay; to replicate measurement, all these processes must be included at the same level of approximation.

Prompted by the recent high-resolution measurement [1], photoionization of Cl in the vicinity of the $1s^{-1}$ thresholds is studied using the R-matrix method [2].

The process of interest is:

$h\nu + 1s^2....3p^5 \rightarrow 1s....3p^5np$

An intermediate resonance state can either decay *via* autoionization, leading to ionization of Cl into the Cl^+ ion plus a free electron, or it can radiatively stabilize, leading to no charge change of the atom; there are participator Auger and radiative decays to the main lines, and there is the more important spectator Auger and radiative decays to the satellite lines. The decay pathways of radiative stabilization redirect some of the initial photoabsorption amplitude, giving a damped photoionization cross section relative to the photoabsorption cross section.

These effects are taken into account in our calculation. For the construction of the target wave functions of the atom, the radial orbitals are optimized on adequately chosen energy levels of states that are crucial to the collision problem, such as those of the relevant core excited states of the atomic system. Our orbital basis is augmented by the inclusion of a number of correlation functions chosen to improve the energies of the ionic thresholds and the ground state of the ionic target and account for orbital relaxation following inner shell photoionization. The photoabsorption and photoionization cross sections are calculated within the framework of the R-matrix wherein participator decay is taken into account as the wavefunctions involved in the autoionization transition matrix elements are included in the close-coupling expansion of the total wavefunction of Cl. To account for spectator Auger decay to the infinite number of continua, the R matrix is modified by using an optical potential [3], wherein the energies of the core-excited states of the target are analytically continued into the complex plane via, $E \rightarrow E - i\Gamma/2$, where Γ is the spectator Auger decay width and is computed in lowest-order perturbation theory using the code AUTOSTRUCTURE [4]. Results will be contrasted with the measurement [1] at the meeting and details of the calculation will be presented and analyzed.

Acknowledgments

Research was supported by U.S. DOE Office of Basic Energy Sciences and the CAU CFNM, NSF-CREST award #HRD-1137751. The NERSC computing facilities are greatly appreciated

References

[1] W. Stolte *et al* 2013, Phys Rev A **88**, 053425
[2] A. Hibbert 1975, *Comput. Phys. Commun.* **9**, 141
[3] T. W. Gorczyca and F. Robicheaux 1999, Phys. Rev. A **60**, 1216
[4] N. R. Badnell 1986, J. Phys. B **19**, 3827

[1]zfelfli@cau.edu
[2]smanson@gsu.edu
[3]amsezane@cau.edu

Content from this work may be used under the terms of the Creative Commons Attribution 3.0 licence. Any further distribution of this work must maintain attribution to the author(s) and the title of the work, journal citation and DOI.
Published under licence by IOP Publishing Ltd

Excited state distribution and spin-effects in strong-field excitation of neutral Helium

Henri Zimmermann*[1], Sebastian Eilzer* and Ulli Eichmann*,†[2],

* Max-Born-Institute Berlin, Max-Born-Straße 2A, 12489 Berlin, Germany

† Institut für Optik und Atomare Physik, Technische Universität Berlin, 10623 Berlin, Germany

Synopsis We investigated the principal quantum number n distribution of excited states resulting from the interaction of Helium with strong, short laser pulses. We find excellent agreement with predictions of the semiclassical frustrated tunneling ionization (FTI) model [1] as well as fully quantum mechanical calculations. Furthermore, the excitation process directly populates triplet excited states due to the breakdown of the Russel-Saunders coupling scheme for high orbital angular momentum l states of Helium, which are predominantly populated in the strong laser field.

Excitation of neutral atoms via frustrated tunneling ionization [1] is a recently established, important exit channel of atoms subjected to a strong short-pulse laser field. While processes such as the acceleration of the excited neutral atoms in strong intensity gradients ([2],[3]) and Coulomb explosion without double ionization in strong laser field fragmentation of H_2 [4] are convincingly explained within the FTI model, the experimental confirmation of the important prediction on the excited state distribution has been missing so far.

In our experiments we excite neutral Helium atoms under conditions characterized by the tunneling regime of strong-field physics using femtosecond pulses from a Ti:Sapphire laser and employ the method of pulsed field ionization in conjunction with direct measurement of the excited neutral atoms to analyze the distribution of excited states. This approach enables us to obtain a measurement signal proportional to the amount of initially excited atoms with a detection efficiency of approximately 30 %.

Our measurements confirm the predictions [5] of the FTI model as well as quantum mechanical calculations [6] using the single-active electron (SAE) approximation (see figure 1) showing that one can model excitation in the tunneling regime by an almost purely classical process.

Moreover, we find that the excitation process also results in an efficient population of triplet excited states of Helium [5], which results from the fact, that states with high orbital angular momentum $l > 3$ are accessible under our experimental conditions. For these high l states the strict separation into singlets and triplets is removed due to the breakdown of the Russel-Saunders coupling scheme. The direct excitation of triplet states is then possible via the admixture of their singlet component.

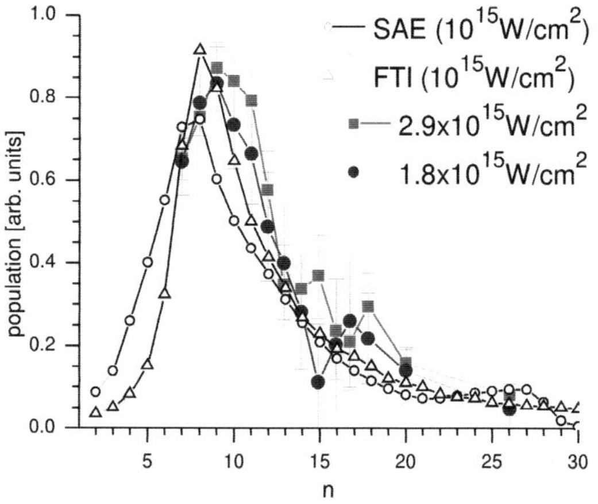

Figure 1. Comparison of measured (filled red squares and blue circles) and calculated (open triangles and circles) n-distributions of Helium.

References

[1] Nubbemeyer T, Gorling K, Saenz A, Eichmann U and Sandner W 2008 *Phys. Rev. Lett.* **101** 233001

[2] Eichmann U, Nubbemeyer T, Rottke H and Sandner W 2009 *Nature* **461** 1261

[3] Eilzer S, Zimmermann H and Eichmann U 2014 *Phys. Rev. Lett* **112** 113001

[4] Manschwetus B, Nubbemeyer T, Gorling K, Steinmeyer G, Eichmann U, Rottke H and Sandner W 2009 *Phys. Rev. Lett.* **102** 113002

[5] Zimmermann H, Buller J, Eilzer S and Eichmann U 2015 *Phys. Rev. Lett.* **114** 123003

[6] Khosravi E and Saenz A (unpublished)

[1]E-mail: Henri.Zimmermann@mbi-berlin.de
[2]E-mail: Ulli.Eichmann@mbi-berlin.de

Content from this work may be used under the terms of the Creative Commons Attribution 3.0 licence. Any further distribution of this work must maintain attribution to the author(s) and the title of the work, journal citation and DOI.
Published under licence by IOP Publishing Ltd

Relativistic photoionization delays and the role of auto-ionizing resonances

M. Huppert*[1], I. Jordan*, S. Pabst†,‡, H.J. Wörner*[2]

* Laboratory for physical chemistry, ETH Zurich, Vladimir-Prelog-Weg 2, CH-8093 Zurich, Switzerland
† Center for Free-Electron Laser Science, DESY, Notkestrasse 85, 22607 Hamburg, Germany
‡ ITAMP, Harvard-Smithsonian Center for Astrophysics, 60 Garden Street, Cambridge, MA 02138, USA

Synopsis We measured the relative ionization delays into the two spin-orbit components of the electronic ground states of Xe^+ and Kr^+ under the influence of auto-ionizing intermediate and final states. The results are in good agreement with state-of-the-art calculations based on the time-dependent configuration-interaction singles (TDCIS) approach.

Attosecond time-resolved photoelectron experiments have been performed by combining an actively-stabilized pump-probe setup with a magnetic-bottle time-of-flight spectrometer. We have measured the relative ionization delays between photoelectrons associated with the two spin-orbit components ($^2P_{3/2}$, $^2P_{1/2}$) of the electronic ground states of Xe^+ and Kr^+ by using the RABBIT technique [1]. A Ti:Sa laser (800 nm, 30 fs, 1.5 mJ, 1 kHz) is used to generate an XUV pulse train in a semi-infite gas cell filled with argon.

An inherent difficulty of measuring photoionization delays is finding a reference time and subtracting the group delay of the attosecond pulses [3]. Measuring relative delays between two spin-orbit states allows us to overcome both, and determine ionization delays which are independent of XUV and IR spectral phases.

The delays in the observed energy range (17 eV - 31 eV) for xenon are up to 50 as (Fig. 1) and significantly less for krypton. In xenon we observe a clear dependence of the measured photoionization delay around 21.7 eV (sideband 14) on the exact center wavelength of the laser pulses driving the high-harmonic generation.

This observation is consistent with the effect of auto-ionizing states ($5s5p^65d$ and $5s5p^67s$) of the neutral, which are accessible by a resonant two-photon transition (XUV + IR). A strong phase variation over the resonances leads to a significant influence onto the photoionization delay in the region of auto-ionizing states.

The relative delays and their modification by the auto-ionizing states are in good agreement with calculations based on the recently-developed time-dependent configuration-interaction singles (TDCIS) approach [2].

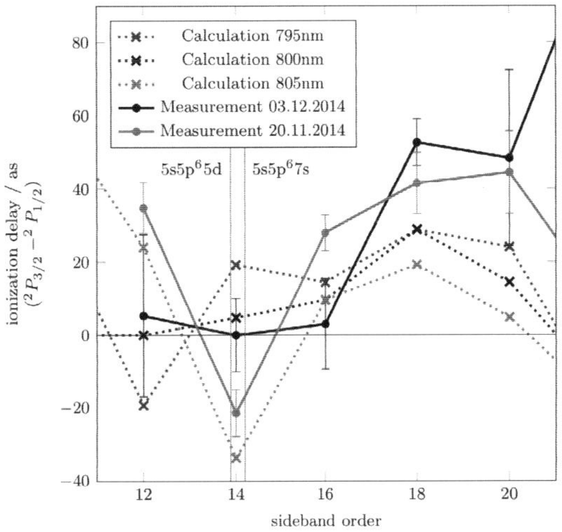

Figure 1. Measured relative ionization delays ($^2P_{3/2}$ - $^2P_{1/2}$) in xenon compared to ab-initio calculations (dotted lines). The most relevant accessible auto-ionizing states of the neutral atom are marked.

In krypton we investigate the influence on the photoionization delay around 26 eV (harmonic 17) due to an auto-ionizing state ($4s4p^66p$) which is accessible by a single photon XUV transition. Effects similar to those predicted in [4] are expected.

References

[1] P.M. Paul *et al.* 2001 *Science* **292** 1689

[2] S. Pabst *et al.* 2013 *Eur.Phys.J.Spec.Top.* **221** 1

[3] K. Klünder *et al.* 2011 *Phys. Rev. Lett.* **106** 169904

[4] Á. Jiménez-Galán *et al.* 2014 *Phys. Rev. Lett.* **113** 263001

[1]E-mail: martin.huppert@phys.chem.ethz.ch
[2]E-mail: hansjakob.woerner@phys.chem.ethz.ch

Content from this work may be used under the terms of the Creative Commons Attribution 3.0 licence. Any further distribution of this work must maintain attribution to the author(s) and the title of the work, journal citation and DOI.
Published under licence by IOP Publishing Ltd

A Systematic Study of Interference Effects in Non-Sequential Double Ionisation

Andrew S Maxwell [1], **Carla Figueira de Morisson Faria** [2],

University College London Department of Physics & Astronomy
Gower Street London WC1E 6BT, United Kingdom

Synopsis We perform a detailed study of interference in time-delayed NSDI showing significant effects on the momentum dependent transition amplitude. We also provide analytical arguments to back up our findings.

The importance of quantum interference in strong-field Non-Sequential Double Ionisation (NSDI) has become an open question. Up to the past year, it was widely believed that it did not play a significant role. This may be attributed to the success of classical NSDI models [1, 2] and to studies assessing the classical-quantum correspondence in the context of the electron impact ionization mechanism, in which the first electron, by re-colliding with the core, immediately releases a second electron [4]. In particular it has been shown that interference effects present in the momentum dependent probability distribution will not survive integration over momentum components perpendicular to the laser field.

Recently however it has been argued, in the context of the re-collision with subsequent ionization (RESI) mechanism, that quantum interference is of major importance [3]. It has been found that the interference between the transition amplitudes associated with different excitation channels may lead to breaking of a fourfold symmetry in the momentum distribution, resulting in distributions to be concentrated in the second and fourth quadrants of the plane spanned by parallel momentum components, agreeing with experimental findings [5].

Using the strong field approximation (SFA), we present an analysis of the types of interference that can take place in NSDI, which may occur (a) between events which are displaced by a half cycle and those present due to the symmetry of indistinguishable electrons, (b) between different channels of excitation for the second electron and (c) between different electron orbits.

Our results show significant effects in the electron momentum distribution. The most notable effect arising from the interference of events shown in figure 1. This coherent superposition of events gives rise to diagonal and hyperbolic features in the probability distribution that can be partially recovered by a fully analytical analysis of the symmetry transformed actions. The results also show symmetry breaking of the momentum probability distribution with the original distribution being distorted towards the 1st and 3rd quadrants, which is more consistent with experimental results [5]. Further work investigates the effect of the spatial geometry of the bound states to which excitation occurs, which is contained in momentum-dependent prefactors.

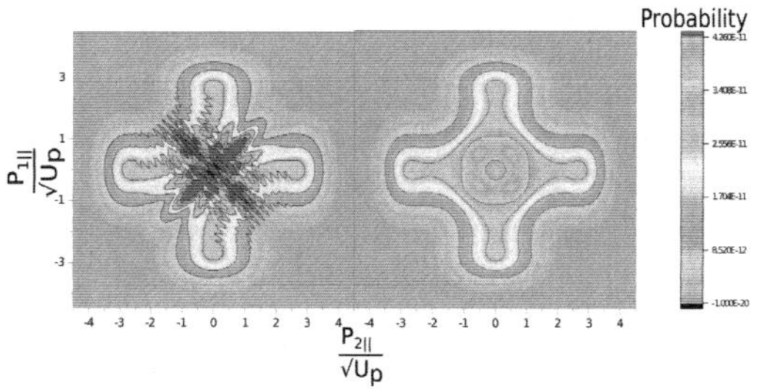

Figure 1. Shows the probability distribution for Argon from a coherent sum of symmetrised events (left) vs an incoherent sum of events (right). The orbital-dependent pre-factors have been neglected.

These results show clear evidence of quantum interference in RESI, which demonstrates that classical models leave out important effects.

References

[1] C. Figueira de Morisson Faria and X. Liu 2011 *J. Mod. Opt.* **58, 1076**

[2] W. Becker, XiaoJun Liu, Phay Jo Ho, and J. H. Eberly 2012 Rev. Mod. Phys. **84, 1011**

[3] Hao, X., Jing Chen, W. Becker *et al* 2014. *Phys. Rev. Lett.* **112(7)**

[4] C. Figueira de Morisson Faria, X. Liu, and W. Becker 2004*Phys. Rev. A* **69, 021402**

[5] Y. Q. Liu, *et al* 2008 *Phys. Rev. Lett.* **101, 053001**

[1]E-mail: ucapsam@ucl.ac.uk
[2]E-mail: c.faria@ucl.ac.uk

Content from this work may be used under the terms of the Creative Commons Attribution 3.0 licence. Any further distribution of this work must maintain attribution to the author(s) and the title of the work, journal citation and DOI.
Published under licence by IOP Publishing Ltd

XXIX International Conference on Photonic, Electronic, and Atomic Collisions (ICPEAC2015) IOP Publishing
Journal of Physics: Conference Series **635** (2015) 092137 doi:10.1088/1742-6596/635/9/092137

Phase Measurement of a Fano Resonance Using Tunable Attosecond Pulses

A. Jiménez-Galán[‡], M. Kotur[*1], D. Guénot[*], D. Kroon[*], E. W. Larsen[*], M. Louisy[*], S. Bengtsson[*], M. Miranda[*], J. Mauritsson[*], C. L. Arnold[*], S. E. Canton[⋆], M. Gisselbrecht[*], T. Carette[†], J. M. Dahlstrom[†], E. Lindroth[†], A. Maquet[¶], L. Argenti[‡], F. Martín[‡◇§], A. L'Huillier[*]

[‡]Dep. Química, Módulo 13,Universidad Autónoma de Madrid, Cantoblanco 28049, Madrid, Spain
[*]Dep. Physics, Lund University, P. O. Box 118, SE-22100 Lund, Sweden
[⋆]Dep. Synchrotron Radiation Instrumentation, Lund University, P.O. Box 118, SE-22100 Lund, Sweden
[†]Dep. of Physics, Stockholm University, AlbaNova University Center, SE-10691 Stockholm, Sweden
[¶]Laboratoire de Chimie Physique-Matière et Rayonnement, UPMC, 75231 Paris, France
[◇]Condensed Matter Physics Center (IFIMAC), Universidad Autónoma de Madrid, 28049 Madrid, Spain
[§]IMDEA-Nanociencia, Cantoblanco, 28049 Madrid, Spain

Synopsis We study photoionization of argon atoms close to the $3s^23p^6 \rightarrow 3s^13p^64p$ Fano resonance using an attosecond pulse train and a weak infrared probe field. An interferometric technique combined with tunable attosecond pulses allows us to determine the phase of the photoionization amplitude as a function of photon energy. We interpret the experimental results using an analytical two-photon model based on the Fano formalism and obtain quantitative agreement.

Correlated electron dynamics induced by photoabsorption is a process of fundamental importance in nature. The development of tabletop attosecond sources in the extreme ultraviolet (XUV) has opened up possibilities of accessing this dynamics directly in the time domain. In particular, correlation is responsible for the Auger decay, a resonant process that dramatically alters atomic photoionization spectra.

When a continuum state belonging to an open ionization channel interacts with a bound state belonging to a closed channel, autoionization may occur. Interference between the different ionization pathways leads to characteristic asymmetric Beutler-Fano profiles in the photoionization cross-section [1]. Studying this interaction in the time domain has been a major goal of attosecond science since the early days [2]. A method to do so is the so-called reconstruction of attosecond beating by interference of two-photon transitions (RABITT) technique [3], in which a train of XUV attosecond pump pulses is combined with a weak infrared (IR) probe with a variable time delay. The train, consisting of odd harmonics of the fundamental IR frequency, initiates the photoionization process. Further absorption or emission of an IR photon gives rise to sidebands. Due to the interference between two quantum paths, the sideband intensities oscillate as a function of the pump-probe delay. From the phase of these oscillations it is possible to obtain the phases of two-photon transition-amplitude and, in turn, of photoemission timing itself.

The RABITT technique has already been used to measure the effect of intermediate bound states on atomic two-photon transitions [4]. In this work, we present a study of the photoionization of argon using an interferometric method based on the RABITT technique, in the proximity of the $3s^23p^6 \rightarrow 3s^13p^64p$ autoionizing resonance, in which a tunable attosecond pulse train is used to initiate the photoionization process. As expected, changing the detuning between the resonance and the harmonic that is closest to it strongly affects the phases of the two adjacent sidebands. However, the phase profile clearly differs from the π jump observed for purely bound states. Our measurements are in quantitative agreement with the predictions of the analytical two-photon model in [5], which we extended to the multichannel case and parametrized with *ab initio* one-photon transition matrix elements obtained with a multi-configuration Hartree-Fock approach [6].

References

[1] U. Fano 1961 *Phys. Rev.* **124** 1866.
[2] M. Drescher *et al.* 2002 *Nature* **419** 803.
[3] P.M. Paul *et al.* 2001 *Science* **292** 1689.
[4] M. Swoboda *et al.* 2001 *Phys. Rev. Lett.* **104** 103003.
[5] A. Jiménez-Galán *et al.* 2014 *Phys. Rev. Lett.* **113** 263001.
[6] T. Carette *et al.* 2013 *Phys. Rev. A* **87** 023420.

[1]E-mail: marija.kotur@fysik.lth.se

Content from this work may be used under the terms of the Creative Commons Attribution 3.0 licence. Any further distribution of this work must maintain attribution to the author(s) and the title of the work, journal citation and DOI.
Published under licence by IOP Publishing Ltd

Tunneling time in strong field ionisation

A.S. Landsman *,†[1], **T. Zimmermann** *,†, **and S. Mishra** †

* Max Planck Institute for the Physics of Complex Systems, Dresden, Germany

† Seminar of Applied Math, ETH Zurich, Zurich, Switzerland

Synopsis Tunneling time definitions are calculated in the adiabatic limit of strong field tunnel ionization by the exact solution of the Schrodinger equation in a static electric field. We find that all definitions give finite tunneling times. In addition, the short-range potential approximation agrees well with the exact solution as long as one stays sufficiently far from over-the-barrier ionization regime.

The question of how long it takes a particle to tunnel through a potential barrier has remained unresolved for many decades. Among the best-known theoretical predictions are the *Buttiker-Landauer, Pollack-Miller, Eisenbud-Wigner*, and *Larmor times*, which were derived based on very different physical considerations. Recently, the interest in tunneling time has renewed due to the introduction of the attoclock, which identifies tunneling time with the delay (relative to the peak of the ionizing ultra-short laser pulse) in the appearance of the most probable electron trajectory at the tunnel exit [1, 2]. Most recent attoclock measurements are presented in [3] and show a small (less than 100 as), but nonvanishing tunneling delay time. This nonvanishing tunneling time remains even if a non-adiabatic theory based on PPT (rather than the TIPIS model), which includes a non-zero velocity at tunnel exit, is used to reconstruct the time that the electron appears in the continuum [4].

In [3], the attoclock measurements were compared to analytic calculations done under the assumption of a short-range adiabatic potential, resulting in a static triangular barrier. Here, we present calculations of the four well-known definitions of tunneling time obtained by the exact solution of the Schrodinger equation in the static electric field [5]. These results show very good agreement with the short-range potential approximation in Helium over the intensity range corresponding to the recent attoclock experiments. For Hydrogen, the agreement between the exact solution and the short-range potential approximation breaks down at higher peak electric field strengths, above $F > 0.1$ au.

In general, we find very good agreement for the four definitions of tunneling time between the exact solution and the short-range potential approximation, as long as the barrier width obtained within the short-range potential approximation is similar to the barrier width given by the exact potential of the combined Coulomb-laser field (this is true sufficiently far from over-the-barrier ionisation regime.) In particular, although ionization probabilities are exponentially dependent on barrier-shape, all four tunneling time definitions seem to have a much weaker linear dependence. Hence, small changes in barrier shape result in correspondingly small changes in tunneling time.

All four definitions of tunneling time predict that the delay inside the barrier is finite, at least in the static field limit, corresponding to $\gamma \to 0$

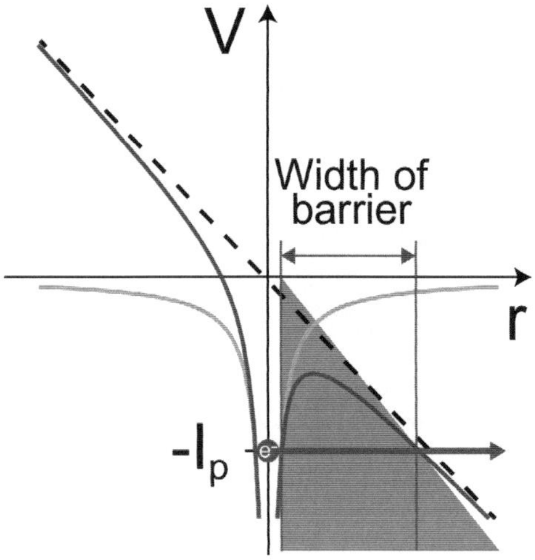

Figure 1. Tunneling barrier in strong field ionisation.

References

[1] P. Eckle *at al* 2008 *Science* **322** 1525

[2] A.S. Landsman *at al* 2015 *Phys. Rep.* **547** 1024

[3] A.S. Landsman *at al* 2014 *Optica* **1** 343

[4] A.S. Landsman *at al* 2014 *J. Phys. B* **47** 20

[5] T. Zimmermann, S. Mishra, D. Gordon, and A. S. Landsman, preprint

[1]E-mail: landsman@pks.mpg.de

Content from this work may be used under the terms of the Creative Commons Attribution 3.0 licence. Any further distribution of this work must maintain attribution to the author(s) and the title of the work, journal citation and DOI.
Published under licence by IOP Publishing Ltd

XXIX International Conference on Photonic, Electronic, and Atomic Collisions (ICPEAC2015) IOP Publishing
Journal of Physics: Conference Series **635** (2015) 092139 doi:10.1088/1742-6596/635/9/092139

Wave packet retrieval by multi-photon quantum beat spectroscopy in helium

R. Pazourek*[1], M. Reduzzi†‡, P. A. Carpeggiani†, G. Sansone†‡, M. Gaarde*, and K. Schafer*

* Department of Physics and Astronomy, Louisiana State University, Baton Rouge, LA 70803, USA
† Dipartimento di Fisica Politecnico Milano, Piazza Leonardo da Vinci 32, 20133 Milano, Italy, EU
‡ CNR-IFN Piazza Leonardo da Vinci 32 20133 Milano, Italy, EU

Synopsis We show that we can probe the components of an attosecond bound electron wave packet by mapping the quantum beat signal produced by a synchronized delayed few-cycle infrared pulse into the continuum. In addition, spectrally overlapping peaks that result from one-, two- or three-photon processes from more or less deeply bound states can in principle be interferometrically resolved with high resolution.

Attosecond pulses interacting with atoms or molecules create a broad band electron wave packet (EWP). In order to capture and ultimately control the localized electron motion, it is necessary to be able to reconstruct the excited bound states and their initial coherence. A delayed synchronized few-cycle infrared (IR) probe pulse that ionizes the excited bound states can be used as either a which-way interferometer or a quantum beat spectrometer. Such an interferometric pump-probe technique has been shown to be able to determine the bound wave packet components excited by an attosecond pulse around threshold for helium [1].

In this contribution, we propose a similar setup for measuring quantum beats separately from which-way interferences by initially exciting only a purely bound wave packet of $1snp$-states in helium by using a shaped attosecond pulse. In Fig. 1 we show a simulated spectrogram as a function of delay between attosecond and IR pulses (left) and the spectrally resolved beating thereof (right). The spectral analysis shows with high resolution the energy differences between different bound states as long as the temporal beating signal can be resolved by the delay steps. The vertical width is determined by the spectral overlap of different contributions, while the horizontal width results from the length of the time series. The high frequency beatings result from the interference between $1s2p + 3\gamma_{\rm IR}$ and $1snp + \gamma_{\rm IR}$ $(n > 3)$. Processes involving a different number of photons can be parity selected by either looking along a certain direction or at angle-integrated spectra.

While the beat frequencies in the Fourier analyzed signal are unambiguously given by the energy difference of the involved bound states no matter how many photons were involved in the ionization process, phase information of the initial wave packet is encoded in the absolute phase of the Fourier signal [2]. Analyzing the phase of the beat signals allows us in principle to distinguish the phase difference between one, two, and three photon ionization processes. In addition to the initial phases of the bound states also a possible phase difference of the ionization step is included. The latter is small for contributions from a one-photon process, however, two- or three-photon processes acquire a more complicated phase depending on the length and intensity of the IR pulse.

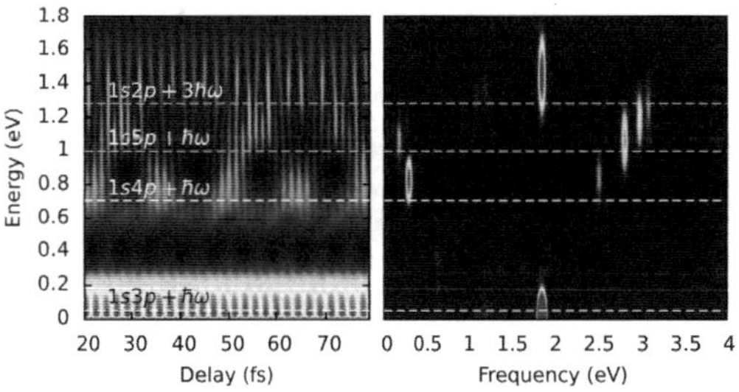

Figure 1. Simulated spectrogram as a function of XUV and IR delay along the polarization axis for collinear pulses (left) and Fourier transform of the time signal at each energy (right).

This project has received funding from the NSF under grant PHY-1307083, from the European Unions Horizon 2020 research and innovation programme under the Marie Sklodowska-Curie grant agreement No 641789 and from the Italian Ministry of Research (Project FIRB No. RBID08CRXK).

References

[1] J. Mauritsson *et al* 2010 *Phys. Rev. Lett.* **105** 053001

[2] K. Klünder *et al* 2013 *Phys. Rev. A* **88** 033404

[1]E-mail: rpazourek@lsu.edu

Content from this work may be used under the terms of the Creative Commons Attribution 3.0 licence. Any further distribution of this work must maintain attribution to the author(s) and the title of the work, journal citation and DOI.
Published under licence by IOP Publishing Ltd

Journal of Physics: Conference Series **635** (2015) 092140 doi:10.1088/1742-6596/635/9/092140

Mass scaling in photoassociation of spin-singlet atoms

Mateusz Borkowski[1], Piotr Morzyński[1], Roman Ciuryło[1], Piotr S. Żuchowski[1], Paul S. Julienne[2], Axel Görlitz[3], T. C. Killian[4], Yoshiro Takahashi[5]

[1]Institute of Physics, Faculty of Physics, Astronomy and Informatics, Nicolaus Copernicus University, Grudziadzka 5, 87-100 Torun, Poland
[2]Joint Quantum Institute, University of Maryland and National Institute of Standards and Technology, College Park, Maryland 20742, USA
[3]Institut für Experimentalphysik, Heinrich-Heine-Universität Düsseldorf, Universitätstraße 1, 40225 Düsseldorf, Germany
[4]Rice University, Department of Physics and Astronomy, Houston, Texas, 77251, USA
[5]Department of Physics, Graduate School of Science, Kyoto University, Kyoto 606-8502, Japan

Synopsis Photoassociation spectroscopy, based on forming of molecules from colliding atoms in the presence of light, is a priceless tool for the study of atomic interactions. It enables direct measurements of bound state energies, both in excited and ground state molecules. Applications include determinations of *s*-wave scattering lengths, as well as atomic state lifetimes. In this work we present the results of our research on the mass-scaling behaviour of molecular bound state energies and collisional properties in spin-singlet atoms. We will concentrate on two such species: strontium and ytterbium, both of which offer several stable isotopes, enabling mass tuning of the system's properties.

Two-valence-electron atoms, particularly strontium and ytterbium, have special features that make them particularly attractive from the standpoint of ultracold physics. Their $^1S_0$ ground state makes them insensitive to external magnetic fields. Narrow intercombination lines present in these systems (7.5 kHz in Sr and 181 kHz in Yb) enable magneto-optical trapping at microkelvin temperatures. A doubly-forbidden $^1S_0 \to {}^3P_0$ line in both species is used as atomic references in optical clocks. Finally, both species have an array of stable isotopes (five bosons and two fermions in ytterbium, three bosons and a fermion in strontium), which enables mass tuning of collisional properties of cold gases of these atoms simply by switching the isotope. The increased interest in ultracold gases of divalent atoms has resulted in spectacular achievements, including degenerate quantum gases in almost all of the isotopes of both species. Photoassociation spectroscopy [1], which enables direct measurement of molecular bound state energies of molecules formed by colliding cold atoms in the presence of light, has also been extensively performed with divalent atoms. In this work we concentrate on the mass-scaling behaviour of bound state energies of molecules formed in photoassociation spectroscopy, as well as the properties of collisions of divalent atoms.

The first presented system is the case of two-color photoassociation of ytterbium with rubidium atoms [2]. This system is of particular importance as it bridges the gap between alkali and divalent species. Based on data from photoassociation spectroscopy we have fitted an *ab initio* based single-channel model of the ground state Rb-Yb interaction. Using this model we have determined the *s*-wave scattering lengths for all isotopologues of this system.

In the case of the RbYb system, a mass scaled model was obtained using just one potential curve. A different case is seen the Sr_2 dimer near the $^1S_0 + {}^3P_1$ asymptote [3]. Here, not only the nonadiabatic mixing between two potential curves (0_u^+ and 1_u) has to be considered to even obtain a good fit to photoassociation data for one isotope, but also an additional 0_u^+ curve perturbs the mass-scaling behaviour of the bound state energies in this molecule.

The last presented system is the $^1S_0 + {}^3P_1$ 0_u^+ curve in the ytterbium dimer [4], for which we have constructed a mass scaled model of the atomic interactions, again by fitting to photoassociation data. We will show that the rotational structure of photoassociation spectra near the intercombination line of ytterbium depends heavily on the properties of the ground state collision due to the emergence of shape resonances.

References

[1] K. M. Jones *et al.* 2006 *Rev. Mod. Phys.* **90** 032713

[2] M. Borkowski *et al.* 2013 *Phys. Rev. A* **78** 483

[3] M. Borkowski *et al.* 2014 *Phys. Rev. A* **88** 052708

[4] M. Borkowski *et al.* 2009 *Phys. Rev. A* **80** 012715

Content from this work may be used under the terms of the Creative Commons Attribution 3.0 licence. Any further distribution of this work must maintain attribution to the author(s) and the title of the work, journal citation and DOI.
Published under licence by IOP Publishing Ltd

Journal of Physics: Conference Series **635** (2015) 092141 doi:10.1088/1742-6596/635/9/092141

Theoretical analysis of attosecond quantum beat spectroscopy of helium excited states

T. Zimmermann*,‡ 1, A. S. Landsman*,‡, M. Lucchini†, A. Ludwig†, L. Kasmi†, J. Herrmann†, L. Gallmann†, A. Scrinzi§, U. Keller†

* Seminar for Applied Mathematics, ETH Zürich, CH-8093 Zürich, Switzerland
† Department of Physics, ETH Zürich, CH-8093 Zürich, Switzerland
‡ Max Planck Institute for the Physics of Complex Systems, Dresden, Germany
§ Fakultät für Physik, Ludwig-Maximilians-Universität München, 80333 München, Germany

Synopsis Photoelectron spectra measured in the attosecond quantum beat spectroscopy experiment on helium were computed solving the single-electron time-dependent Schrdinger equation. Final momenta of photoelectrons were obtained from the probability flux at the boundary of the simulation region using the time-dependent surface flux (t-SURFF) method [1]. Reaching a very good agreement with the experimental results, simulations allow for deeper understanding of processes underlying the quantum beating, which-way interference and polarization of the excited state.

The quantum-beat spectroscopy with an attosecond resolution is now possible using tailored short attosecond pulse trains which populate several excited states. The excited electron wavepacket is subsequently ionized by few-cycle IR pulses leading to emission lines, which beat with a frequency proportional to the energy gap between states involved in the two-color photoionization process.

In a recent experiment [3], an excited wavepacket was created in helium while avoiding a direct excitation to continuum states. The suppression of the "which-way interference" with continuum states allows for clear observation of quantum-beats as well as of the effects of the IR field on the excited states when the attosecond pulse train and IR pulse overlap.

As can be seen in Fig. 1, resulting spectra are relatively complicated and their proper analysis is best done by solving the time-dependent Schrödinger equation (TDSE). Although the full solution of the two-electron TDSE is very costly, in this case the experimental setup and electronic structure of helium allow for using the single-electron model. We have employed this model with the pseudo-potential V (r) = (1 + expr)/r, where = 2.1325. The solution was obtained in the velocity gauge and the photoelectron spectra were calculated by the time-dependent surface flux (tSURFF) method [1, 2].

An excellent agreement between the experimental and calculated spectra was reached. Using an intuitive population analysis and restricted space dynamics, the TDSE analysis provided an interpretation of spectral features observed including a directional signal due to the polarization of the excited state. In addition, simple explanation of the CEP dependence was provided in terms of parity of electrons originating from different p states.

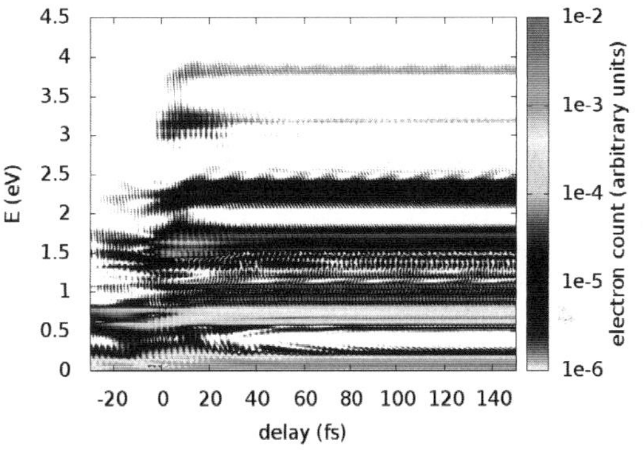

Figure 1. Simulated photoelectron spectrum of He as a function of the delay between the IR femtosecond pulse and the XUV attosecond pulse train (positive delays correspond to XUV arriving first). Oscillations of the emission lines clearly appear in the region of pump-probe overlap and extend to positive delays. These spectral features can be attributed mainly to two-color ionization involving one XUV photon and one to three IR photons.

References

[1] L. Tao and A. Scrinzi 2012 *New J. Phys.* **14**, 013021

[2] V. Majety, A. Zielinski, A. Swoboda, and A Scrinzi, The tRecX PDE solver (unpublished)

[3] M. Lucchini, A. Ludwig, T. Zimmermann, L. Kasmi, J. Herrmann, A. Scrinzi, A. S. Landsman, L. Gallmann, and U. Keller, submitted

[1]E-mail: tomaszi@sam.math.ethz.ch

Content from this work may be used under the terms of the Creative Commons Attribution 3.0 licence. Any further distribution of this work must maintain attribution to the author(s) and the title of the work, journal citation and DOI.
Published under licence by IOP Publishing Ltd

Journal of Physics: Conference Series **635** (2015) 092142 doi:10.1088/1742-6596/635/9/092142

Measuring the $^2D_{3/2}$ Ni$^-$ excited state lifetime in DESIREE

KC Chartkunchand† [1], O. M. Hole*, E. Bäckström*, R. F. Nascimento*, M. Kaminska*, V. T. Davis†, P. A. Neill†, R. D. Thomas*, S. Mannervik*, J. S. Thompson†, D. Hanstorp‡ [2], H. T. Schmidt*, and H. Cederquist*

† Department of Physics, University of Nevada, Reno, Nevada 89557, United States

* Department of Physics, Stockholm University, SE-106 91 Stockholm, Sweden

‡ Department of Physics, University of Gothenburg, SE-412 96 Gothenburg, Sweden

Synopsis Experimental studies of the radiative lifetime of the sole excited state in the nickel anion are presented. Beams of Ni$^-$ stored in DESIREE at cryogenic temperatures were subject to photodetachment at different photon energies and the resulting neutrals collected. Preliminary analysis of the decay in neutral production yields a radiative lifetime of approximately 15 seconds for the $3d^94s^2$ $^2D_{3/2}$ excited state.

Negative ions are challenging systems to study both experimentally and theoretically. Their tenuous nature is governed by electron correlation effects which are necessary for these systems to exist. Measurements of properties such as lifetimes of negative ion excited states thus provide stringent tests of theories that attempt to accurately describe these interactions [1].

We present studies of Ni$^-$ conducted at DESIREE, a unique electrostatic ion storage ring with two cryogenically-cooled rings sharing a common straight section to allow for merged-beam collisions experiments [2, 3]. By cooling the rings to 13 K, pressures on the order of 10^{-14} mbar are achieved, greatly reducing ion losses due to collisions with residual gases. DESIREE's capabilities were recently demonstrated by the radiative lifetime measurement of the $3p^5$ $^2P^o_{1/2}$ excited state of the sulfur negative ion [4].

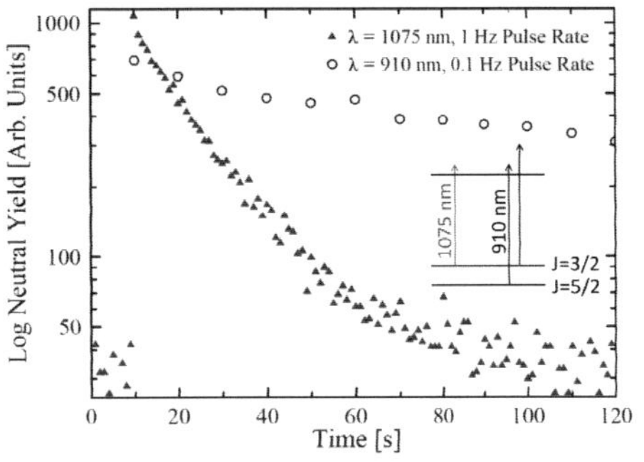

Figure 1. Neutral yield for Ni$^-$ photodetachment at 1075 nm (▲) where only ions in the upper fine structure level can be photodetached and at 910 nm (○) where the photon energy is sufficient for detachment of ions occupying both fine structure levels. The shorter decay time at 1075 nm compared to 910 nm is ascribed to the spontaneous decay.

For the present experiment Ni$^-$ was stored in one of DESIREE's rings, with storage lifetimes of ~250 s. Laser light at 1075 nm and 910 nm was used to photodetach the circulating Ni$^-$ beam. The different wavelengths allow us to selectively detach from a given state in Ni$^-$, with the 1075 nm wavelength detaching only the $^2D_{3/2}$ excited state and the 910 nm wavelength detaching both the excited state and the $^2D_{5/2}$ ground state. By pulsing the laser light at fixed time intervals and measuring the resulting yields of neutral products the ground and/or excited state population(s) of Ni$^-$ in the beam was measured as a function of time.

Neutral yields as a function of time after photodetachment are shown in Figure 1. While the signal at 910 nm decays slowly, with contributions to the lifetime dominated by the Ni$^-$ ground state, the signal at 1075 nm decays much more rapidly. From our choice of wavelength this must be purely due to decay of the $^2D_{3/2}$ excited state population. Preliminary data analysis, including those taken at slightly higher pressures for use in extrapolating back to zero residual gas pressure, indicate a radiative lifetime of ~15 s for the $^2D_{3/2}$ excited state.

References

[1] D. J. Pegg 2004 *Reports on Progress in Physics* **67** 857

[2] R. D. Thomas *et al.* 2011 *Review of Scientific Instruments* **82** 065112

[3] H. T. Schmidt *et al.* 2013 *Review of Scientific Instruments* **84** 055115

[4] E. Bäckström *et al.* 2015 *Physical Review Letters* **In Press**

[1]E-mail: kiattichartc@gmail.com
[2]E-mail: dag.hanstorp@physics.gu.se

Content from this work may be used under the terms of the Creative Commons Attribution 3.0 licence. Any further distribution of this work must maintain attribution to the author(s) and the title of the work, journal citation and DOI.
Published under licence by IOP Publishing Ltd

XXIX International Conference on Photonic, Electronic, and Atomic Collisions (ICPEAC2015) IOP Publishing
Journal of Physics: Conference Series **635** (2015) 092144 doi:10.1088/1742-6596/635/9/092144

X- ray wavelengths and Auger transition energies of $1s2p^4$ (^2S, $^{2,4}P$, ^2D) resonances in B-like (Z = 6-18) ions

I Sakho [*1], **M Sow** [†2] **and A Wagué** [††3]

[*] Department of Physics, UFR of Sciences and Technologies, University Assane Seck of Ziguinchor, Ziguinchor, Senegal

[†] Department of Physics, Atoms Laser Laboratory, Faculty of Sciences and Technologies, University Cheikh Anta Diop, Dakar, Senegal

[††] Department of Physics, Atoms Laser Laboratory, Faculty of Sciences and Technologies, University Cheikh Anta Diop, Dakar, Senegal

Synopsis Please Auger energies and X-ray wavelengths of the 1s $2p^4$ (^2S, $^{2,4}P$, ^2D) resonances in B-like ions (Z = 6-18) are reported. New Auger energies and wavelengths for B-like N^{2+}, F^{4+}, Na^{6+}, Al^{8+}, P^{10+}, S^{11+}, Cl^{12+}, and Ar^{13+} ions are tabulated as benchmarked values for future experimental and theoretical studies.

Studies of core-excited doublet and quartet many electron systems are very challenging because of their importance for the diagnostic of astrophysical and laboratory plasma. For boron isoelectronic sequence, Auger energies of the 1s $2p^4$ (^2S, $^{2,4}P$, ^2D) resonances have been investigated in the past on the both side of experiment and theory. The most recent saddle-point variation and saddle-point complex-rotation (SPCR) calculations of Sun *et al.* [1] has been used to solve earlier theoretical discrepancies and to identify some former unknown experimental lines from $1s2p^4$ resonances in B-Like C^+, O^{3+}, and Ne^{5+} ions. However, the high-relativistic calculations reported by Sun *et al.* [1] have been limited to the 1s $2p^4$ (^2S, $^{2,4}P$, ^2D) resonances in B-like C^+, O^{3+}, Ne^{5+}, Mg^{7+}, and Si^{9+} ions.

We present theoretical Auger energies and X-rays wavelengths of the $1s2p^4$ levels in the entire boron isoelectronic C^+ to Ar^{13+} using the Screening constant by unit nuclear charge (SCUNC) method compared with existing literature data. Overall, the present results published [2] for B-like C^+, O^{3+}, Ne^{5+}, Mg^{7+}, and Si^{9+} ions agree well with various theoretical and experimental data. The tabulated new Auger energies and X-ray wavelengths for the 1s $2p^4$ resonances in B-like N^{2+}, F^{4+}, Na^{6+}, Al^{8+}, P^{10+}, S^{11+}, Cl^{12+}, and Ar^{13+} ions may be benchmarked values for future experimental and theoretical studies.

[1] E-mail: aminafatima_sakho@yahoo.fr

[2] E-mail: malickitna@yahoo.fr

[3] E-mail: ahmadou.wague@ucad.edu.sn

Tables 1 and 2 illustrate the agreement between theories in the particular case of B-like Ne^{5+} ion. Results for other B-like ions compared with theory and experiment will be presented in the meeting.

Table 1. Auger energy (in eV) for some transitions in Ne^{5+} ion from the 1s $2p^4$ ^2S initial state.

Levels	SCUNC	SPCR	MDFC
$1s^2 2p^2$ ^2S	1042.48	1043.28	1045.23
$1s^2 2p^2$ ^2D	1054.21	1054.94	1057.50
$1s^2 2s 2p$ ^1P	1070.75	1071.48	1073.72
$1s^2 2s 2p$ ^3P	1086.82	1087.53	1091.25

SCUNC: screening constant by unit nuclear charge data, present results [2].
SPCR: saddle-point variation and saddle-point complex-rotation results of Sun *et al.* [1].
MCDF: multiconfiguration Dirac-Fork values of Chen and Crasemann [3].

Table 2. Theoretical wavelength (in Å) for Ne^{5+} ion.

Transition	SCUNC	SPCR	MDFC
$1s2p^4\ ^4P \to 1s^22p^3\ ^4S^\circ$	14.064	14.069	14.062
$1s2p^4\ ^2S \to 1s^22p^3\ ^2P^\circ$	13.998	14.006	13.987
$1s2p^4\ ^2P \to 1s^22p^3\ ^2P^\circ$	14.121	14.123	14.126
$1s2p^4\ ^2P \to 1s^2 2p^3\ ^2D^\circ$	14.017	14.029	14.038
$1s2p^4\ ^2D \to 1s^2 2p^3\ ^2P^\circ$	14.152	14.154	14.122
$1s2p^4\ ^2D \to 1s^22p^3\ ^2D^\circ$	14.053	14.060	14.048

SCUNC [2]; SPCR [1]; MCDF [3].

References

[1] Y. Sun, B. C. Gou, and C. Chen, Phys.Rev. A **87**, 032509 (2013).
[2] I. Sakho, M. Sow and A. Wagué, Phys. Scr. **90**, 045401 (2015).
[3] M. H. Chen and B. Crasemann, At. Data Nucl. Data Tables **38**, 381 (1988).

Content from this work may be used under the terms of the Creative Commons Attribution 3.0 licence. Any further distribution of this work must maintain attribution to the author(s) and the title of the work, journal citation and DOI.
Published under licence by IOP Publishing Ltd

XXIX International Conference on Photonic, Electronic, and Atomic Collisions (ICPEAC2015) IOP Publishing
Journal of Physics: Conference Series **635** (2015) 092145 doi:10.1088/1742-6596/635/9/092145

Optical transitions at the *5s-4f* level crossing with high sensitivity to a time variation of the fine structure constant α

A. Windberger[* 1], **P. O. Schmidt**[†], **J. R. Crespo López-Urrutia**[*]

[*] Max-Planck-Institut für Kernphysik, Saupfercheckweg 1, 69117 Heidelberg, Germany
[†] Physikalisch-Technische Bundesanstalt, Bundesallee 100, 38116 Braunschweig, Germany

Synopsis The optical spectra of Nd-like W, Re, Os, Ir, and Pt ions were studied in charge states close to the predicted 5s-4f level crossing. After identifying the lines, we infer possible values for optical transitions in Ir^{17+} with very high sensitivity to a variation of the fine-structure constant α.

We study the Nd-like ions W^{14+}, Re^{15+}, Os^{16+}, Ir^{17+}, and Pt^{18+} around the predicted 5s-4f level crossing. A systematic investigation of the hitherto unknown optical spectra of Ir ions in various charge states was needed (see Fig. 1) for charge state assignment. By using the smooth scaling of the intra-configuration transition energies as a function of the atomic number Z , we assign the observed spectral lines in the spectra of all Nd-like ions under study [1].

These results agree better with calculations based of Fock-space coupled cluster methods than with configuration interaction and relativistic Hartree-Fock predictions. Measurements of the strongest lines in Ir^{17+} showing their Zeeman splitting confirmed these identifications. From a total of 85 spectral lines, 47 were assigned to M1 transitions. Further M1 identifications and a few E1 identifications in Ir^{17+} based on the Ritz combination principle allowed us to establish the energy separation of the $4f^{12}5s^2$ and $4f^{13}5s$ configurations, confirming the level crossing at $Z = 77$.

Furthermore, we infer with high probability the value for the transition energy of a highly forbidden optical M2/E3 transition in the Ir^{17+} ion, which according to recent predictions should have the highest sensitivity to an α variation [2], and also determine the transition energies of proposed optical clocks lines in Hf^{12+} and W^{14+}. These latter ones are particularly insensitive to blackbody radiation, laser light shifts, and external perturbations. Having experimental values for such candidate transitions is crucial for future laser spectroscopic studies aiming at the development of very accurate frequency standards, and for searches of a possible cosmological variation of α. Sympathetic cooling, a technique recently demonstrated for highly charged ions using Ar^{13+} in [3], will enable the application of frequency metrology to these large class of ions.

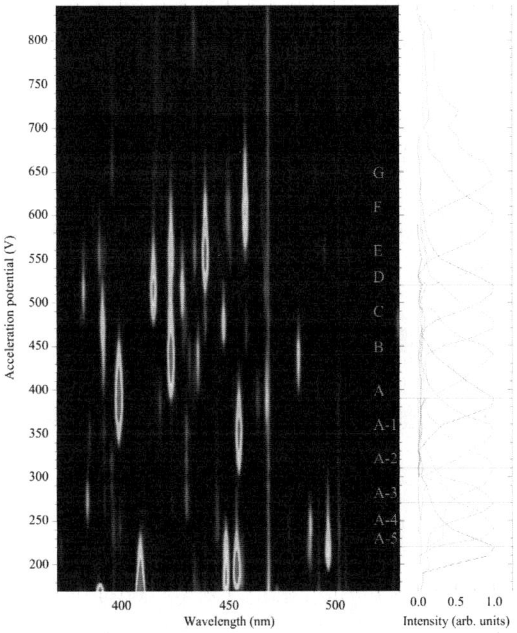

Figure 1. Low resolution composite spectral map of Ir ions in growing charge states as a function of the uncorrected acceleration potential. By a combination of several methods, the spectrum marked with B was identified as that corresponding to the Ir^{17+} ion [1]. Right: Acceleration-potential dependent intensity curves used for charge state identification.

References

[1] A. Windberger *et al.*, 2015 *Phys. Rev. Lett.* **114**, 150801

[2] J. C. Berengut, V. A. Dzuba, and V. V. Flambaum, 2010 *Phys. Rev. Lett.* **105**, 120801

[3] L. Schmöger *et al.*, 2015 *Science* **347**, 1233

[1]E-mail: a.windberger@mpi-hd.mpg.de

Content from this work may be used under the terms of the Creative Commons Attribution 3.0 licence. Any further distribution of this work must maintain attribution to the author(s) and the title of the work, journal citation and DOI.
Published under licence by IOP Publishing Ltd

Journal of Physics: Conference Series **635** (2015) 092146 doi:10.1088/1742-6596/635/9/092146

Absolute radiative and Auger transition rates of K-shell excited few-electron iron ions

R. Steinbrügge[1]**, S. Bernitt, J. K. Rudolph, J. R. Crespo López-Urrutia**

Max-Planck-Institut für Kernphysik, Saupfercheckweg 1, 69117 Heidelberg, Germany

Synopsis The iron Kα lines are among the most prominent features in the x-ray spectra of many celestial sources. However, transition rates used to model these spectra were only known by theoretical calculations. Here we present an absolute measurement of iron K-shell transition rates for Li-like to C-like iron ions. These ions were created with an electron beam ion trap and excited with x rays from the PETRA III synchrotron source. Measuring x-ray fluorescence and Auger decay simultaneously allows to determine absolute decay rates independent of most experimental parameters.

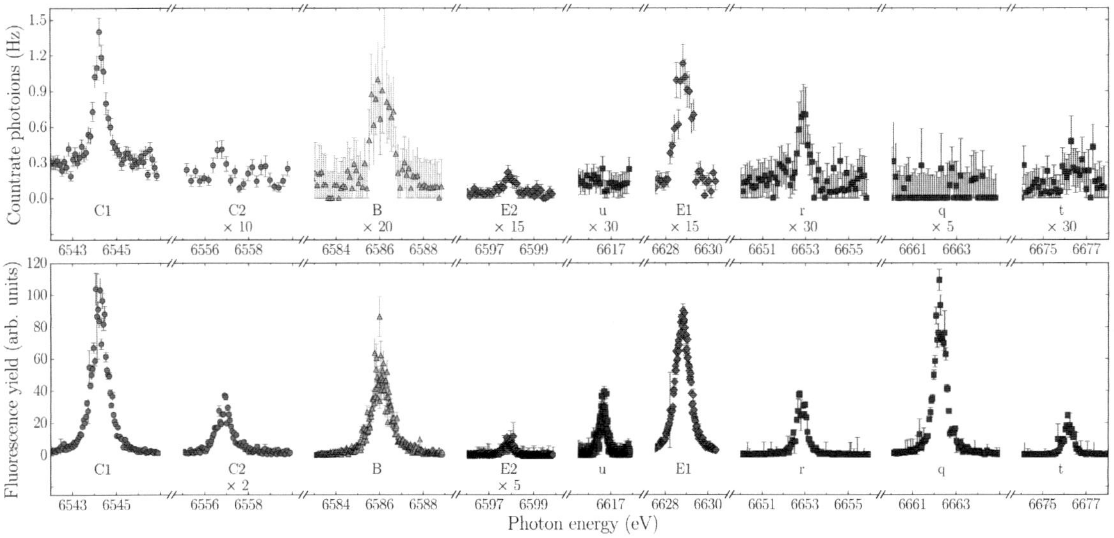

Figure 1. Measured photoionization (top) and fluorescence spectra (bottom) of highly charged iron ions [1]. Shown are resonances in C-like (red circles), B-like (yellow triangles), Be-like (green diamonds), and Li-like (blue squares) charge states. Part of the data is scaled up by the factors denoted below the labels.

The spectrum of highly charged iron ions provides rich information about the dynamics in X-ray binary stars and active galactic nuclei. To model measured spectra, a precise knowledge of the transition rates is needed. While transition energies have been accurately measured before, only theoretical values for transition rates were available. We present a measurement of radiative and Auger rates for K-shell transitions in Li-like to C-like iron ions. These were produced and trapped in the transportable electron beam ion trap FLASH-EBIT and resonantly excited with X-ray photons at PETRA III. The subsequent radiative decay was measured with semiconductor photon detectors, while the Auger decay was observed by detecting the change of the ion charge state. By taking ratios of the photoionization yield and the simultaneous recorded fluorescence, we suppress setup-dependent uncertainties. Together with natural line widths [2], this allows us to determine absolute values for the radiative and Auger transition rates. Furthermore, by analyzing the angular distribution of the fluorescence photons, we resolved different radiative decay channels in the Be-like and C-like system. Although only short measurement time was available, a better-than-10% accuracy could be achieved in the Li-like system [1]. The used method can easily be transferred to other charge states and elements.

References

[1] R. Steinbrügge *et al* 2015 *Phys. Rev.* A **91** 032502

[2] J. K. Rudolph *et al* 2013 *Phys. Rev. Lett.* **111** 103002

[1]E-mail: rene.steinbruegge@mpi-hd.mpg.de

Content from this work may be used under the terms of the Creative Commons Attribution 3.0 licence. Any further distribution of this work must maintain attribution to the author(s) and the title of the work, journal citation and DOI.
Published under licence by IOP Publishing Ltd

XXIX International Conference on Photonic, Electronic, and Atomic Collisions (ICPEAC2015) IOP Publishing
Journal of Physics: Conference Series **635** (2015) 092147 doi:10.1088/1742-6596/635/9/092147

Two-photon absorption using off-resonant excitation with ultrashort X-ray pulses

J. Szlachetko[* †1], J. Hoszowska[§], J.-Cl. Dousse[§], B. Patterson[*], R. Abela[*], M. Pajek[†], C.J. Milne[*2]

[*]Paul Scherrer Institute, Villigen-PSI, Switzerland

[†]Insitute of Physics, Jan Kochanowski University, Kielce, Poland

[§] Department of Physics, University of Fribourg, Fribourg, Switzerland

Synopsis High energy resolution off-resonant X-ray emission spectroscopy (HEROS) was applied to study one-photon (OPA) and two-photon absorption (TPA) processes in metallic copper generated by ultra-short hard X-ray FEL pulses from the Linac Coherent Light Source (LCLS). The OPA and TPA signals were measured using a multiple-crystal von Hamos diffraction spectrometer allowing observation of both the intermediate and final states in the TPA process. The universal scaling of TPA/OPA cross sections is discussed.

Until recently, the nonlinear interaction of Ångstrom X-ray wavelengths with matter has remained an unexplored regime. It is only within the last few years that the development of X-ray free electron laser sources (XFELs) has permitted the field of nonlinear optical research to be extended into the hard X-ray regime [1]. This has allowed the first observations of processes with very low cross-sections, including double-core hole formation [2], two-photon absorption [3], amplified spontaneous x-ray emission [4], plasma creation [5] and X-ray-optical wave mixing [6].

Among the different mechanisms for nonlinear X-ray interaction with matter, the two-photon absorption (TPA) process exhibits third-order photon-atom interactions. In contrast to one-photon absorption processes, which are determined by dipole-allowed transitions, the TPA mechanism requires excitation between states of the same parity. We have investigated TPA processes for electronic core-levels using off-resonant X-ray emission spectroscopy at an XFEL. The results indicate a strong correlation between one-photon and two-photon mechanisms, with the TPA process being purely sequential and mediated by excitation into a virtual electronic intermediate state.

In the present work, we employed high energy resolution off-resonant X-ray emission spectroscopy [7] (HEROS) to measure one- and two-photon absorption signals (OPA and TPA respectively) in metallic copper generated by ultra-short hard X-ray pulses. The experiment was performed at the Coherent X-ray Imaging instrument at the Linac Coherent Light Source (Menlo Park, USA) XFEL. For x-ray emission detection at around 8040eV ($K\alpha_{1,2}$ x-ray lines, $2p_{3/2}$->1s and $2p_{1/2}$->1s transitions) we employed a spectrometer operated in a multiple-crystal dispersive von Hamos geometry.

We will report on extracted TPA cross-sections and examine the X-ray fluence thresholds at which this process becomes dominant. We find a strong relationship between one-photon and two-photon absorption cross-sections as a function of incoming X-ray energy. Experimental data suggests a universal scaling of TPA/OPA cross-sections, which explains the saturation effects of TPA signal at highest fluences. Thanks to our experimental approach, we track the yields of both the intermediate and final states during the course of the TPA process. The measured cross-section for the second absorption step in TPA has been found to be four orders of magnitude higher than the first absorption event. With the proliferation of experiments using ultrashort, intense X-ray pulses for a variety of measurements at XFELs, measuring TPA cross-sections is critically important for establishing the regimes over which these processes can either be avoided or enhanced. This is especially relevant for X-ray photon energies close to ionization thresholds, where TPA cross-sections can be significant.

We would like to acknowledge M. Messerschmidt, S. Boutet, and G. Williams from the CXI instrument of LCLS and W. Błachucki, Y. Kayser, J. Sà, C. David, G. Smolentsev, J. A. van Bokhoven, M. Nachtegaal, T. Penfold, and G. Knopp for help with experimental support and results discussions.

References

[1] P. Emma *et al* 2010 *Nature Photon.* **4** 641, J. Amann, *et al* 2012 *Nature Photon.* **6** 693
[2] L. Young *et al* 2010 *Nature* **466** 56, K. Tamasaku *et al* 2013 *Phys. Rev. Lett.* **111** 043001
[3] K. Tamasaku *et al* 2014 *Nature Photon.* **8** 313
[4] N. Rohringer *et al* 2012 *Nature* **481** 488, M. Beye *et al* 2013 *Nature* **501** 191
[5] S.M. Vinko *et al* 2012 *Nature* **482** 59
[6] T.E. Glover *et al* 2012 *Nature* **488** 603
[7] J. Szlachetko *et al* 2012 *Chem. Comm.* **48** 10898, M. Kavčič *et al* 2013 *Phys. Rev. B* **87** 075106, C.J. Milne, T.J. Penfold, M. Chergui 2014 *Coord. Chem. Rev.* **277** 44, J. Szlachetko *et al* 2014 *Struct. Dyn.* **1** 021101

[1] E-mail: jakub.szlachetko@psi.ch
[2] E-mail: chris.milne@psi.ch

Content from this work may be used under the terms of the Creative Commons Attribution 3.0 licence. Any further distribution of this work must maintain attribution to the author(s) and the title of the work, journal citation and DOI.
Published under licence by IOP Publishing Ltd

Journal of Physics: Conference Series **635** (2015) 092148 doi:10.1088/1742-6596/635/9/092148

Photoionization dynamics: Transition and scattering delays

J. Caillat[a 1], M. Vacher[b], R. Gaillac[a], A. Maquet[a] and R. Taïeb[a]

[a] UPMC, CNRS, UMR 7614, Laboratoire de Chimie Physique – Matière et Rayonnement, Paris, France

[b] Department of Chemistry, Imperial College London, London, SW7 2AZ, UK

Synopsis The advent of attosecond science, with main objectives to probe and to control ultrafast dynamics in quantum systems, brought in the past few years a fresh look on photoemission. This very fundamental process was indeed recently revisited in the time domain, in a series of pioneering experiments [1] that were soon accompanied by an intense theoretical activity devoted to the interpretation of the reported "photoemission delays". At the conference, we will present some contributions of our group to this topic [2], with a focus on the characterization of photoemission dynamics in terms of "scattering delay" and "transition delay" as evidenced in numerical experiments performed on simple model systems.

Resolving electron motion in atoms and molecules on its natural attosecond (as) scale is way beyond the temporal resolution of available detection devices. The techniques developed to achieve such sub-fs resolution thus rely on interferometric setups, and the reported times are actually group delays derived from phase measurements, involving coherent photoemission processes. Therefore, the analysis of the experimental data and the related theoretical development ask for rigorous and unambiguous definitions and interpretations of these phases, and of the inferred group delays.

It is now accepted that a "scattering delay" [4] affects the dynamics of any photoemission process. However, the simplicity of the underlying physics is not fully recognized yet. Formally, such delays are imprinted in the phase shifts of the photoelectron wave-functions, which are commonly expressed on the basis of incoming waves. In this framework, the "scattering phase" associated to photoemission appears as the argument of the transition amplitude, thus obscuring the significance of the delay – which may be misinterpreted for example as a transition duration. Here, we will present the benefits of working with the continuum wave-functions selected by the transitions (SCWF) [3], which (i) carry all the information related to the continuum reached by photoabsorption, (ii) are defined independently of the arbitrary basis one chooses to work with and (iii) are real valued for single-photon transitions. They provide a clear-cut interpretation of the scattering delays.

In higher order processes, the SCWF comes with an additional complex phase, as soon as the transition is resonant. We will show that the group delay associated with that phase can now be interpreted as a "transition delay", and how it can be accessed experimentally in a straightforward reinterpretation of the RABBIT interferometric technique, initially designed for the characterization of coherent XUV pulses [5].

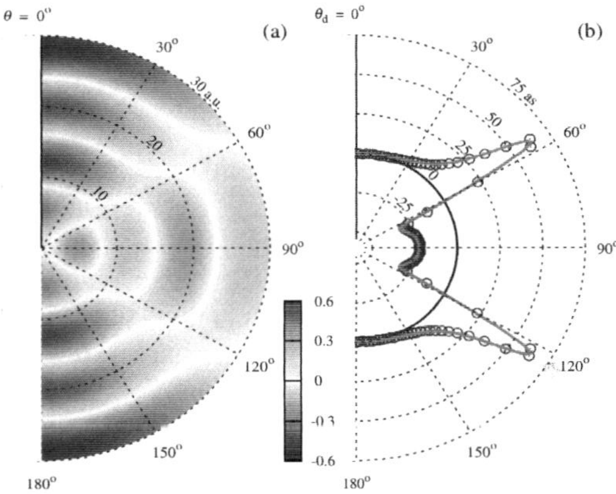

Figure 1. $He^+(2p_0)$ @ 17.06 eV. (a) Real valued SCWF . (b) Ionization delays against the detection direction θ_d: time-dependent simulations (circles); scattering group delays (full line). The delays are defined and computed with $\theta_d = 0$ as reference.

[1]E-mail: jeremie.caillat@upmc.fr

References

[1] S. Haessler *et al* 2009 *Phys. Rev. A* **80** 011404; M. Schultze *et al* 2010 *Science* **25** 1658 ; K. Klunder *et al* 2011 *Phys. Rev. Lett.* **106** 143002

[2] J. Caillat *et al* 2011 *Phys. Rev. Lett.* **106** 093002; R. Gaillac *et al* 2015 *submitted*; M. Vacher *et al* 2015 *in preparation*

[3] H. Park *et al* 1996 *J. Chem. Phys* **104** 4554

[4] E. P. Wigner 1955 *Phys. Rev.* **98 145**

[5] P. M. Paul *et al* 2001 *Science* **292 1689**

Content from this work may be used under the terms of the Creative Commons Attribution 3.0 licence. Any further distribution of this work must maintain attribution to the author(s) and the title of the work, journal citation and DOI.
Published under licence by IOP Publishing Ltd

Journal of Physics: Conference Series **635** (2015) 102001 doi:10.1088/1742-6596/635/10/102001

Surface Plasmon stimulated photoemission process at metal gratings : Theory and Experiments

A. Sadou* , M. Raynaud †, G. Duchateau* , G. Geoffroy* , P. Martin* and H. Jouin*

* Univ. Bordeaux, CNRS, CEA, CELIA, UMR 5107, F-33400 Talence, France
† CEA/DSM/IRAMIS/LSI, CNRS, Ecole Polytechnique, 91128 Palaiseau, France

Synopsis : Energy spectra related to the Surface Plasmon enhanced electron emission process at gold surfaces have been measured and calculated. Comparisons between theoretical and experimental spectra allow to obtain the amplitude of the Surface Plasmon field.

Surface Plasmon (SP) excitation by femto-second laser pulses at metal gratings or at metallic nanoparticles has attracted considerable interest over the last decade. As SP excitation allows to concentrate and channel the incident light and also to induce a strong absorption of the laser energy, the number and the kinetic energy of the ejected electrons is drastically increased with respect to the case of a standard photoelectric effect even for modest laser intensities in the range 10^9 – 10^{11} W/cm^2 [1]. This process can lead to the generation of intense and short electron sources since the SP lifetime is of the same order as the laser pulse duration.

We have developed a model [2] in which the SP stimulated photoemission process is treated as a two steps mechanism. The first one which considers electronic transitions from the metallic conduction band towards the solid continuum due to the SP field has been initially treated by means of the perturbative quantum-mechanical approach called Jellium-Volkov (JV) [3]. This description has been recently improved through the numerical resolution of the Time Dependent Schrödinger Equation (TDSE) in one spatial dimension (the coordinate perpendicular to the surface) [3-4]. In the second step, we describe classically the motion of the freed electrons in the two components of the inhomogeneous oscillating SP field outside the metal in which they gain more or less energy depending on their initial conditions provided by the quantum calculations of the first step in terms of ejection energy, emission angle and instants of release. The enhancement factor ($\eta = E_{SP} / E_0$, with E_{SP} and E_0 the amplitudes of the SP field and of the laser one, respectively) which is the only free parameter of the model is obtained by means of comparisons between theoretical and experimental energy spectra. Experiments have been performed by focusing the laser beam (800 nm, 58 fs and intensities in the range 10^9 – 10^{10} W/cm^2) with p-polarization on a sinusoidal gold grating with 150 grooves per millimeter at an angle of incidence of 65° required for SP excitation in these conditions.

When the JV approach is used to describe the first step, calculations lead to a good agreement with experimental spectra for η values around 120 in the intensity range studied here [2] (η=135 in the case presented in Fig.1). However, this value is overestimated due to the somewhat inaccurate primary energy spectra provided by the JV method. Preliminary calculations using the TDSE resolution seem to indicate that the η factor is of the order of those calculated in [5] for continuous SP excitation (some tens).

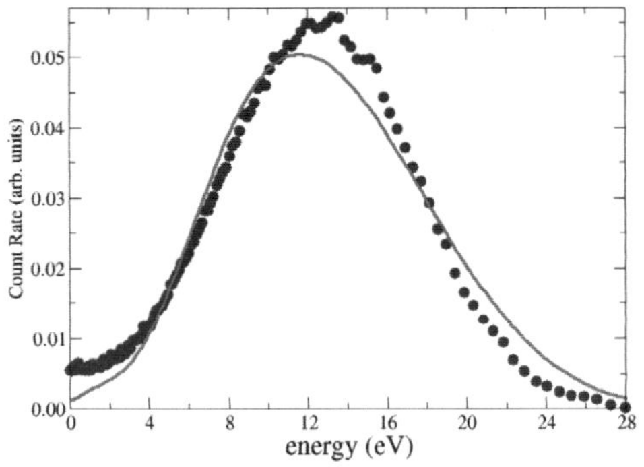

Figure 1. SP enhanced emission spectra of a gold grating as a function of electron energy for a 800 nm, 58 fs FWHM laser pulse of intensity 8.1 GW/cm^2. ● : Experiment , full line : calculation with η=135.

References

[1] J. Kupersztych *et al* 2001 *Phys. Rev. Lett.* **86** 5180
[2] H. Jouin *et al* 2014 *Phys. Rev. B* **89** 195136
[3] M.N. Faraggi *et al* 2007 *Phys. Rev. A* **76** 012903
[4] S.V. Yalunin *et al* 2011 *Phys. Rev. B* **84** 195426
[5] W.H. Weber and G.W. Ford 1981 *Opt. Lett.* **6** 122

Content from this work may be used under the terms of the Creative Commons Attribution 3.0 licence. Any further distribution of this work must maintain attribution to the author(s) and the title of the work, journal citation and DOI.
Published under licence by IOP Publishing Ltd

Quantum-mechanical simulation of attosecond streaked photoemission from Mg-covered W(110) surfaces

Qing Liao and Uwe Thumm[1]

Department of Physics, Kansas State University, Manhattan, Kansas 66506, USA

Synopsis We apply a quantum-mechanical model to simulate infrared-streaked photoelectron emission by an ultrashort extreme ultraviolet pulse from adsorbate-covered metal surfaces. Incorporating effects of energy-dependent electron mean-free paths, the properties of initial states, photoelectron energy dispersion, and the screening of the streaking field, this model is able to reproduce recently measured photoelectron spectrograms and adsorbate-thickness-dependent photoemission time delays.

Attosecond streaking, using an isolated attosecond extreme ultraviolet (XUV) pump pulse and a delayed few-cycle infrared (IR) streaking pulse, is able to time-resolve electron photoemission from atoms and condensed matter systems on an attosecond time scale [1]. A recent streaked photoemission experiment with ultrathin Mg adsorbate films on a W(110) substrate [2] reveals a monotonic dependence of the relative photoemission time delay $\Delta\tau_{4f-2p}$ between W(4f) and Mg(2p) core-level (CL) photoelectrons and a non-monotonic dependence of the relative photoemission time delay $\Delta\tau_{CB-2p}$ between conduction band (CB) and Mg(2p) CL photoelectrons on the adsorbate thickness.

In order to interpret the measured relative photoemission time delays from the Mg/W(110) systems, we apply a quantum-mechanical model [3, 4]. By taking the energy-dependent electron mean-free path (MFP), the properties of initial states (including band-structure and wave-function-localization effects), energy dispersion, and the screening of IR streaking field at the surface into account, our numerical results [5] reproduce measured [2] streaked photoemission spectra and relative photoemission time delays as a function of the Mg-film thickness (Fig. 1).

As input for our numerical simulation, we only employ measured and calculated MFPs and an IR skin depth of 2 Å [2, 3]. In addition, we describe the dispersion of released CB electrons inside the W substrate based on an adjusted effective electron mass (0.86 a.u.). We assume free-electron propagation of CB electrons in the Mg adsorbate film and for electrons that are released from Mg(2p) and W(4f) CLs, to match the relative photoemission delay between CB and W(4f) photoelectrons for the adsorbate-free W(110) surface. Our quantum-mechanical results also agree with classical free-electron calculations [2] for $\Delta\tau_{4f-2p}$.

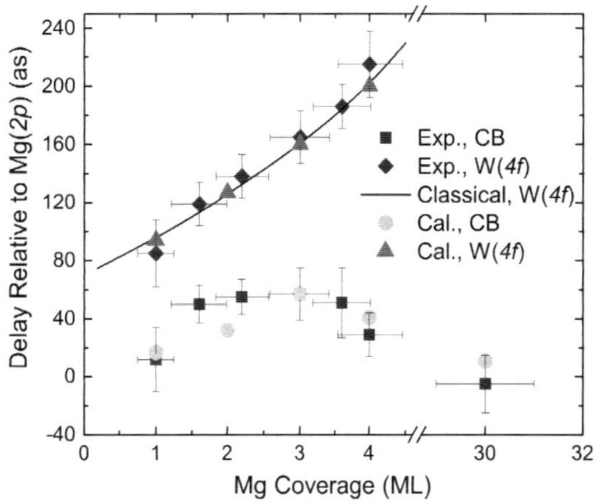

Figure 1. Measured [2] and classically [2] and quantum-mechanically [5] calculated streaking time delays relative to Mg(2p) emission for CB and W(4f) photoelectrons as a function of the Mg coverage on a W(110) substrate. The central XUV photon energy is 118 eV.

This work was supported by NSF Grant PHY 1068752 and the Chemical Sciences, Geosciences, and Biosciences Division, Office of Basic Energy Sciences, Office of Science, U.S. DOE.

References

[1] U. Thumm *et al.*, Handbook of Photonics **1**: "Attosecond physics", Chapter XIII: "Attosecond streaking spectroscopy of atoms and solids" (Wiley, 2015).

[2] S. Neppl *et al.*, Nature **517**, 342 (2015).

[3] . Liao and U. Thumm, Phys. Rev. Lett. **112**, 023602 (2014).

[4] . Liao and U. Thumm, Phys. Rev. A **89**, 033849 (2014).

[5] . Liao and U. Thumm, Phys. Rev. Lett., submitted for publication.

[1]E-mail: thumm@phys.ksu.edu

Content from this work may be used under the terms of the Creative Commons Attribution 3.0 licence. Any further distribution of this work must maintain attribution to the author(s) and the title of the work, journal citation and DOI.
Published under licence by IOP Publishing Ltd

Attosecond plasmonic streaking from gold nanospheres

Jianxiong Li[1] **and Uwe Thumm**[2]

James R. Macdonald Laboratory, Kansas State University, Manhattan, Kansas 66506, USA

Synopsis To study time-resolved photoemission from gold nanospheres, we introduce a semi-analytical quantum-mechanical model, including the plasmonic near-field-enhancement of the streaking field at the surface of the nanosphere. For an 800 nm incident near infrared (NIR) pulse and 10 nm diameter gold nanosphere, we find that the presence of the nanoplasmonic field enhances the oscillation amplitude by a factor of 3.

In this work, we present a semi-analytical quantum-mechanical model to study attosecond streaking from gold nanpspheres, including the plasmonic near-field-enhancement of streaking field at the surface.

We use Mie theory [1] to calculate the plasmonic field near 10 to 200 nm gold nanospheres, driven by incident NIR or visible light pulses, based on a complex-valued frequency dependent dielectric function that was adjusted to experimental data [2]. Figure 1 shows the total field intensity distribution (incident field + plasmonic response), normalized to the incident field intensity, with a maximum enhancement of about 14.

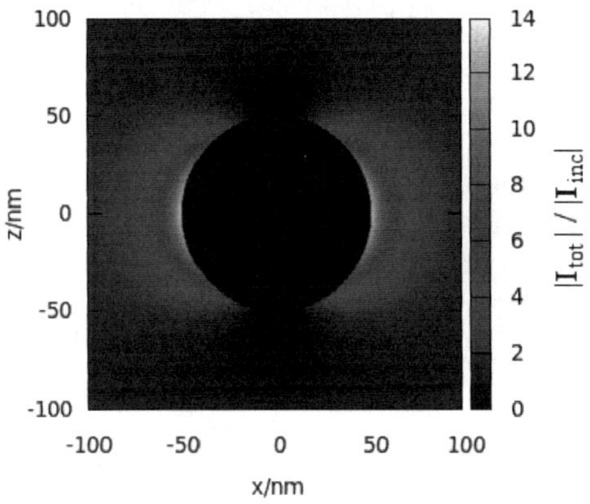

Figure 1. Total field enhancement near a 100 nm diameter gold nanosphere at x-z plane, driven by a 800 nm plane wave propagating in z-direction and polarized in x-direction. Maximum enhancement is found along polarization direction.

We model the gold conduction band in terms of a spherical square well potential. Based on these results, we performed a quantum-mechanical simulation of attosecond streaking from gold nanospheres [3]. Figure 2 shows calculated streaking spectra, where Fig. 2(a) includes plasmonically enhanced field, while in Fig. 2(b) only incident pulse is considered. Both are calculated from a single initial state taken at Fermi level of the gold conduction band. Our results show an enhanced oscillation amplitude, phase shift, and asymmetry of the streaking traces, as a result of the near-field enhancement.

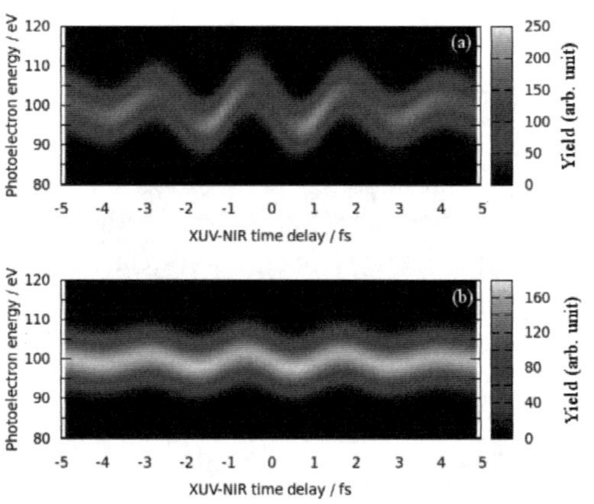

Figure 2. Calculated streaking spectra from a single initial state of a spherical square well potential (at the Fermi level with zero angular momentum) representing a 10 nm diameter gold nanosphere. (a) With plasmonically enhanced near field and (b) without near field enhancement.

References

[1] J. A. Stratton 2007 *Electromagnetic theory*, vol. 33, Wiley-IEEE Press.

[2] E. D. Palik 1985 *Handbook of optical constants of solids*, Elsevier Science.

[3] U. Thumm *et al* 2015 *Handbook of Photonics 1: "Attosecond physics"*, Chapter XIII: *"Attosecond streaking spectroscopy of atoms and solids"* (Wiley)

[1]E-mail: jasonlee@phys.ksu.edu
[2]E-mail: thumm@phys.ksu.edu
*Supported by NSF and the US DoE.

Content from this work may be used under the terms of the Creative Commons Attribution 3.0 licence. Any further distribution of this work must maintain attribution to the author(s) and the title of the work, journal citation and DOI.
Published under licence by IOP Publishing Ltd

Intracluster Coulombic decay following intense NIR ionization of clusters

Bernd Schütte*†[1], Mathias Arbeiter,‡ Thomas Fennel‡, Ghazal Jabbari§, Kirill Gokhberg§, Alexander I. Kuleff§, Marc J. J. Vrakking*, Arnaud Rouzée*

* Max-Born-Institut, Max-Born-Strasse 2A, 12489 Berlin, Germany
† Department of Physics, Imperial College London, SW7 2AZ London, United Kingdom
‡ Institut für Physik, Universität Rostock, Universitätsplatz 3, 18055 Rostock, Germany
§ Theoretische Chemie, PCI, Universität Heidelberg, Im Neuenheimer Feld 229, 69120 Heidelberg, Germany

Synopsis We report on the observation of a novel intracluster Coulombic decay process following Rydberg atom formation in clusters ionized by intense near-infrared fields. A new decay channel emerges, in which a Rydberg atom relaxes to the ground state by transferring its excess energy to a weakly bound electron in the environment that is emitted from the cluster. We find evidence for this process in the electron spectra, where a peak close to the corresponding atomic ionization potential is observed. For Ar clusters, a decay time of 87 ps is measured, which is significantly longer than in previous time-resolved studies of interatomic Coulombic decay.

When an excited atom or molecule is located in an environment, novel decay channels can emerge that are absent for isolated particles. Relaxation of the excited species can take place by a transfer of its excess energy to an atom or molecule in the environment that gets ionized, a process known as interatomic (intermolecular) Coulombic decay [1]. Thus far, the observations of such energy transfer mechanisms have been limited to dimers [2] and clusters [3] ionized by extreme-ultraviolet (XUV) pulses.

Here we present highly surprising results on a novel intracluster Coulombic decay (ICD) process that is of universal nature, as it is not restricted to a certain system or wavelength regime. In the experiment, we observe ICD following electron-ion recombination and extensive Rydberg atom formation in clusters interacting with intense near-infrared (NIR) laser pulses [4, 5]. When a highly excited Rydberg atom decays by transferring its excess energy to a second Rydberg atom in the environment [6], an electron is emitted whose initial kinetic energy is close to the atomic ionization potential. In Fig. 1, an electron spectrum from Ar clusters ionized by intense NIR pulses shows evidence for this process, where in addition to a thermal distribution, a peak is visible just below the ionization potential of Ar that can clearly be attributed to ICD. The novel ICD mechanism was observed in all systems under investigation, including Ar, Kr, Xe, O_2 and CH_4 clusters. We have further measured the decay time of the ICD contribution as 87 ps [7], which is significantly longer than the decay time of 150 fs recently reported for ICD in Ne dimers ionized by XUV pulses [8].

Intracluster Coulombic decay is expected to be important in a large variety of systems, including biomolecules interacting with intense X-ray pulses. The novel findings offer unprecedented opportunities in the future for an improved investigation of correlated electronic decay processes.

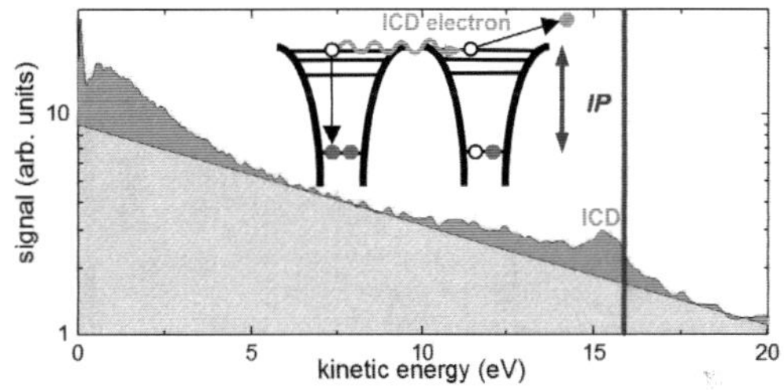

Figure 1. Electron kinetic energy spectrum from Ar clusters with an average size of $\langle N \rangle = 1000$ atoms ionized by NIR pulses at an intensity of 9×10^{13} W/cm^2. The peak just below the ionization potential of Ar (15.76 eV) is attributed to ICD between two Rydberg atoms (see inset). The gray area represents thermal electron emission.

References

[1] L. S. Cederbaum *et al* 1997 *Phys. Rev. Lett.* **79** 4778

[2] T. Jahnke *et al* 2004 *Phys. Rev. Lett.* **93** 163401

[3] S. Marburger *et al* 2003 *Phys. Rev. Lett.* **90** 203401

[4] B. Schütte *et al* 2014 *Phys. Rev. Lett.* **112** 073003

[5] B. Schütte *et al* 2014 *Phys. Rev. Lett.* **112** 253401

[6] A. I. Kuleff *et al* 2010 *Phys. Rev. Lett.* **105** 043004

[7] B. Schütte *et al submitted*

[8] K. Schnorr *et al* 2013 *Phys. Rev. Lett.* **111** 093402

[1]E-mail: schuette@mbi-berlin.de

Content from this work may be used under the terms of the Creative Commons Attribution 3.0 licence. Any further distribution of this work must maintain attribution to the author(s) and the title of the work, journal citation and DOI.
Published under licence by IOP Publishing Ltd

Time-resolved investigation of transient charges in laser-produced nanoplasmas

Bernd Schütte[*†1], **Marc J. J. Vrakking**[*], **and Arnaud Rouzée**[*]

[*] Max-Born-Institut, Max-Born-Strasse 2A, 12489 Berlin, Germany

[†] Department of Physics, Imperial College London, SW7 2AZ London, United Kingdom

Synopsis We report on the observation of a transient C^{4+} ion charge state in nanoplasmas produced by the interaction of intense near-infrared (NIR) laser pulses with CH_4 clusters. The underlying dynamics are studied by pump-probe spectroscopy, which reveals that the ion charge states are lowered by electron-ion recombination. Furthermore, we present direct evidence that autoionization of multiply-excited ions plays an important role in expanding nanoplasmas, in contrast to models that neglect quantum phenomena.

In recent experiments, transient charge states were observed following the interaction of intense extreme-ultraviolet (XUV) pulses with rare-gas clusters using fluorescence spectroscopy [1, 2]. The results were explained by the generation of highly charged ions that recombine with electrons in the nanoplasma after the XUV laser pulse has ended. In order to reveal the underlying processes involved in the nanoplasma relaxation dynamics, we have developed an intense XUV source based on high-order harmonic generation, and we have shown that highly excited Rydberg atoms are formed in the nanoplasma by charge recombination [3]. We were able to time-resolve these dynamics, and we demonstrated that similar processes take place following XUV and NIR ionization of clusters [4, 5].

Here we study the interaction of intense 32 fs NIR laser pulses with CH_4 clusters, where we observe the formation of a transient C^{4+} ion contribution in pump-probe experiments that is hardly visible in the ion time-of-flight (TOF) spectrum obtained by the pump pulse only (Fig. 1(a)). In the electron spectrum, we find a peak at ≈ 1 eV (not shown) that is explained by the reionization of excited states following charge recombination using the time-delayed probe pulse.

When using 400 fs NIR pump pulses at the same intensity, the cluster becomes strongly ionized, and exhibits a dominant C^{3+} ion contribution (Fig. 1(b)). Nevertheless, the C^{4+} ion yield using the pump pulse only remains small. It is strongly enhanced though by the time-delayed weak probe pulse, demonstrating that it results from reionization of a C^{3+} Rydberg state that has been formed by charge recombination of C^{4+} that must have been transiently formed during the cluster excitation. Furthermore, the electron spectrum in Fig. 1(c) exhibits a peak at an energy of 7 eV that is surprisingly identified as electron emission following autoionization of doubly-excited C^+ and C^{2+} ions.

Our results clearly demonstrate that charge recombination is responsible for the observation of transient charges in nanoplasmas. Moreover, the sharp ion charge state distribution that is dominated by a C^{3+} peak can be explained by extensive autoionization of multiply-excited atoms [6] and ions. The results are in stark contrast to existing models of nanoplasmas that treat electrons and ions as classical particles.

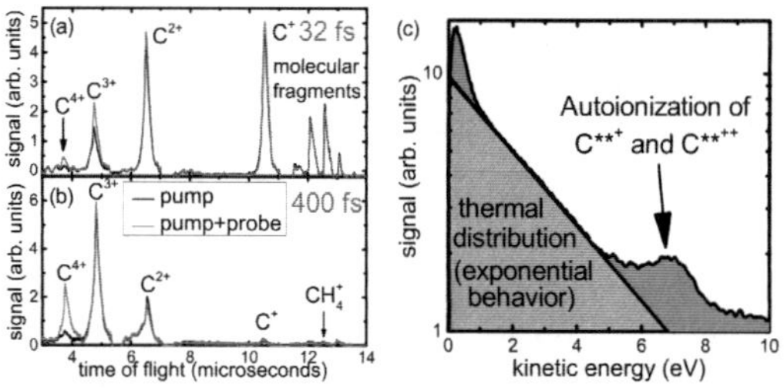

Figure 1. Ion TOF spectra from CH_4 clusters obtained using a 1×10^{14} W/cm^2 NIR pulse with a pulse length of (a) 32 fs and (b) 400 fs. When using an additional weak NIR probe pulse ($I=5 \times 10^{12}$ W/cm^2) at a time delay of 50 ps, the C^{4+} ion signal is significantly increased. (c) The corresponding electron spectrum using the NIR pump pulse exhibits a clear autoionization peak.

References

[1] L. Schroedter *et al* 2014 *Phys. Rev. Lett.* **112** 183401

[2] H. Iwayama *et al* 2015 *Phys. Rev. A* **91** 021402(R)

[3] B. Schütte *et al* 2014 *Phys. Rev. Lett.* **112** 073003

[4] B. Schütte *et al* 2014 *Phys. Rev. Lett.* **112** 253401

[5] B. Schütte *et al* 2015 *New J. Phys.* in press

[6] B. Schütte *et al* 2015 *Phys. Rev. Lett.* in press

[1]E-mail: schuette@mbi-berlin.de

Content from this work may be used under the terms of the Creative Commons Attribution 3.0 licence. Any further distribution of this work must maintain attribution to the author(s) and the title of the work, journal citation and DOI.
Published under licence by IOP Publishing Ltd

Journal of Physics: Conference Series **635** (2015) 102006 doi:10.1088/1742-6596/635/10/102006

Ab initio study of thermodynamically consistent physical properties of warm dense plasma

Xiang Gao* [1], **Jia-Ming Li** [†, ‡], **Peihong Zhang** [+]

[*] Beijing Computational Science Research Center, Beijing, 100084, China
[†] Department of Physics and Center for Atomic and Molecular Nanosciences, Tsinghua University, Beijing, 100084, China
[‡] Key Laboratory for Laser Plasmas (Ministry of Education) and Department of Physics, Shanghai Jiao Tong University, Shanghai, China, 200240
[+] Department of Physics, University at Buffalo, SUNY, Buffalo, NY 14260, USA

Synopsis Thermodynamically consistent equation of state (EOS) of aluminum across a wide range of parameter space is investigated using density functional theory (DFT) based first-principles electronic structure methods. Our results can serve as a benchmark for developing new EOS models and database, and can be conveniently used for modeling important properties and processes of high energy density systems. The interesting electronic structure variations with the increasing electronic temperature are observed and will also be presented for discussion.

The equation of state (EOS), i.e., the relation among state variables (e.g., pressure, volume, temperature, etc.) and Opacities, are the most important quantities for understanding thermodynamic and radiative properties of warm dense matter. In this work, thermodynamically consistent EOS and Opacities of dense plasma are investigated using density functional theory (DFT) based first-principles electronic structure methods [1] with the assumption that the Helmholtz free energy of a two-temperature plasma can be decomposed into three terms, i.e.,

$$F(T_e,T_{ion},V) = F^0(T_e = T_{ion} = 0,\{\vec{R}_i^{eq}\},V) + \Delta F_e(T_e,T_{ion} = 0,\{\vec{R}_i^{eq}\},V) + \Delta F_{ion}(T_e,T_{ion},V). \quad (1)$$

Both the first and the second terms can be calculated using DFT methods within the Mermin formalism [2]. At low temperatures, the ionic contribution to the free-energy, i.e., $\Delta F_{ion}(T_e,T_{ion},V)$, can be accurately evaluated using the quasi-harmonic approximation. At elevated temperatures, we propose that this term be further decomposed into two terms, i.e.,

$$\Delta F_{ion}(T_e,T_{ion},V) = \Delta\bar{F}_{e-ion}(T_e,T_{ion},V,\{\vec{R}_i\}) + \Delta F_{ion\_kinetic}(T_e,T_{ion},V) \quad . (2)$$

Using Al as an example, we demonstrate the applicability of this approach over a wide-range of electronic temperatures and densities, spanning from solid state to dense plasma regime. The interesting electronic structure variations with the increasing electronic temperature are observed and their influences to the opacities will also be presented for discussion.

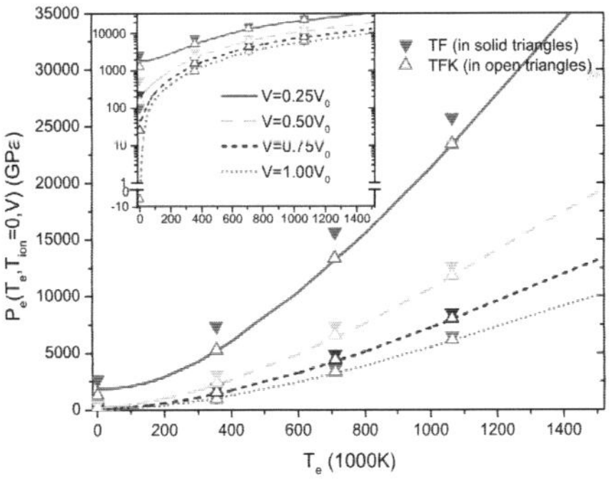

Figure 1. Electronic pressure of Al as a function of temperature (up to 1,500,000 K) and volume (up to a compression ratio of 4): Comparison between DFT (lines), TF (filled symbols) and TFK (unfilled symbols). The inset shows the results in logarithmic scale.

References

[1] P. Giannozzi *et al* 2009 *J. Phys.: Condens. Matter* 21 395502
[2] N. D. Mermin 1965 *Phys. Rev.* 137 A1441

[1] E-mail: xgao@csrc.ac.cn

Content from this work may be used under the terms of the Creative Commons Attribution 3.0 licence. Any further distribution of this work must maintain attribution to the author(s) and the title of the work, journal citation and DOI.
Published under licence by IOP Publishing Ltd

XXIX International Conference on Photonic, Electronic, and Atomic Collisions (ICPEAC2015) IOP Publishing
Journal of Physics: Conference Series **635** (2015) 102007 doi:10.1088/1742-6596/635/10/102007

Methyl formate ice stimulated by X-rays in protoplanetary discs

Edgar Mendoza[*,+ 1], **Wania Wolff**[† 2], **Maria Luiza Rocco**[‡] **and Heloisa Boechat-Roberty**[* 3]

[*] Observatório do Valongo, Universidade Federal do Rio de Janeiro, RJ, 20080-090 Rio de Janeiro, Brazil
[+] Instituto de Astronomia, Geofísica e Ciências Atmosféricas, Universidade de São Paulo, SP, 05508-090, São Paulo, Brazil
[†] Instituto de Física, Universidade Federal do Rio de Janeiro, RJ, 21941-909, Rio de Janeiro, Brazil
[‡] Instituto de Química, Universidade Federal do Rio de Janeiro, RJ, 21941-909, Rio de Janeiro, Brazil

Synopsis We study experimentally the processes of photoionization, photodissociation and photodesorption in methyl formate ice due to interaction of soft X-rays at two energy ranges, from 284 to 298 eV and from 528 to 546 eV, around the C 1s→π* and O 1s →π* resonances. The photodesorption ion yield was obtained for some of the desorbed species of the ice surface, like CH^+, HCO^+ and H_3CO+. The extrapolation of these results to an astrophysical environment, as the protoplanetary disc of TW Hydrae star, is presented

Methyl formate, HCOOCH3, is a molecule widely detected in the circumstellar medium. In regions of protoplanetary discs at low temperatures, molecules freeze forming a mantle of ice on surfaces of dust grains. The interaction of radiation emitted by the central protostar with this mantle stimulates ionization, dissociation and desorption favoring the formation of more complex species. The abundance of a given species, ionized or neutral, in these environments depends on the rate of formation, destruction and desorption from the surface of the grains.

We have studied experimentally the photodissociation, photoionization and photodesorption processes of methyl formate ices (130 K) stimulated by X-rays at two energy ranges, from 284 to 298 eV and from 528 to 546 eV, around the C 1s → π* and O 1s → π* resonances, respectively. The experiments were performed at the Brazilian Synchrotron Light Laboratory (LNLS) in the single bunch mode, using the Spherical-Grating Monochromator beamline and applying the time-of-flight mass spectrometry technique. Fifteen ionic fragments were identified in mass spectra, and the more abundant species were H^+, CH^+, CH_3^+, CO^+ and CH_3O^+, corresponding to around 85% of all fragments produced by the irradiation with 270 and 535 eV X-rays; the dissociation pathways were proposed.

These experimental data were then applied to determine the values of relevance to the astrochemistry in the conditions of the protoplanetary disc of the TW Hydrae star. This star has a dense disc and presents X-ray luminosity integrated from 0.2 to 2 keV of 10^{30} erg s^{-1}. The X-ray photon flux of the TW Hydrae disc was evaluated based on the disc density structure, the luminosity of the protostar and considering molecular hydrogen attenuation. For example, a photon flux of 6×10^8 photons cm^{-2} s^{-1} was estimated at 537 eV at a distance of 40 au from the central star [1]. The photodesorption ion yields of the relevant ionic species, H^+, CH^+, CH_3^+, HCO^+, CH_3O+, CH_3OH^+ and $COOH^+$, were extracted, as shown in figure 1. The column densities of these species considering both ions production and destruction rates, were determined and, for instance, the H^+ and HCO^+ column densities compared with values from other protoplanetary discs [2]. The half-life of methyl formate as function of the distance from the central star was determined and, in this context, the survival of the molecule in the circumstellar medium will be discussed.

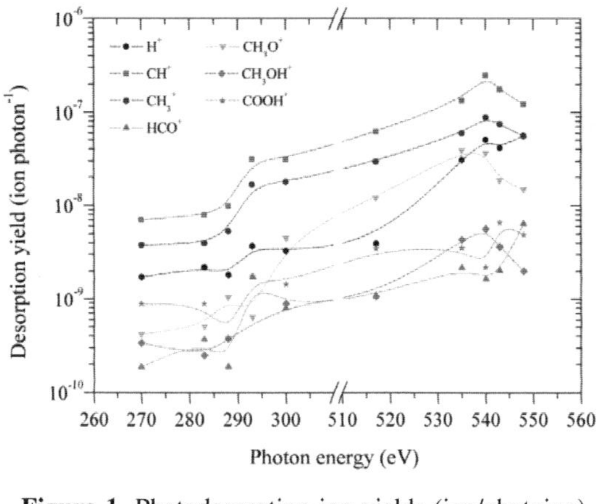

Figure 1. Photodesorption ion yields (ion/photoion) as function of the X-ray energy (eV) around the C1s and O1s edges for H^+, CH^+, CH_3^+, HCO^+, CH_3O+, CH_3OH^+ and $COOH^+$ ions.

References

[1] F. Fantuzzi *et al* 2011, MNRAS **417** 2631
[2] W F Thi *et al* 2011, A&A **530** L2

[1]E-mail: emr1918@gmail.com
[2]E-mail: wania@if.ufrj.br
[3]E-mail: heloisa@astro.ufrj.br

Content from this work may be used under the terms of the Creative Commons Attribution 3.0 licence. Any further distribution of this work must maintain attribution to the author(s) and the title of the work, journal citation and DOI.
Published under licence by IOP Publishing Ltd

Journal of Physics: Conference Series **635** (2015) 102008 doi:10.1088/1742-6596/635/10/102008

Spatial beam profile-induced effects in x-ray scattering pattern at high intensity

Malik Muhammad Abdullah[*,†,‡ 1], **Zoltan Jurek**[*,†], **Sang-Kil Son**[*,†] **and Robin Santra** [*,†,‡]

[*] Center for Free-Electron Laser Science, DESY, Notkestrasse 85, 22607 Hamburg, Germany

[†] The Hamburg Center for Ultrafast Imaging, Luruper Chaussee 149, 22761 Hamburg

[‡] Department of Physics, University of Hamburg, Jungiusstrasse 9, 20355 Hamburg, Germany

Synopsis Simulation of high-intensity x-ray scattering from a crystal that is larger than the transverse x-ray beam profile is presented. We consider a crystal with a size of a micron, when irradiated by an x-ray beam having a focus of 100 nm. We discuss the methodological challenges that had to be overcome in order to calculate the x-ray scattering pattern. Our calculations enable us to identify the characteristic features of the spatial beam profile imprinted in the scattering pattern.

X-ray free-electron lasers (XFEL) deliver intense femtosecond pulses that promise to yield high-resolution diffraction data of nanocrystals before the destruction of the sample by radiation damage. The characteristics of XFEL radiation and associated sample environments have triggered the development of new data collection methods, such as serial femtosecond crystallography (SFX). In SFX, a complete data-set can be obtained by exposing thousands of randomly oriented, individual crystals of proteins (Figure 1) to the x-ray beam. A bottleneck for x-ray structure determination in biology is radiation damage manifested by ionization and sample deterioration during irradiation. From the reconstruction aspect, that means the changes of scattering form factors [1] and atomic positions are strongly fluence dependent. A crystal with a size of a micron is large as compared to the 100 nm focus of an XFEL pulse. Thus it is not valid to consider a homogeneous fluence distribution throughout the whole crystal under the influence of such a focused beam. As a consequence of different fluence regions in the crystal, one might see the characteristics of the beam imprinted in the scattering pattern.

In realistic simulations of SFX experiments, it is mandatory to take into account the fluence dependent dynamics of different crystal regions as well as their different contributions to the scattering pattern. In our theoretical study we consider a micron-size protein crystal in a 100 nm focus beam. The number of atoms under the influence of the beam is, thus, of the order of 10^9. To follow the time evolution of the crystal we developed an extension of XMDYN [2], which is a complex simulation tool for modeling dynamics of molecular systems (e.g. biological molecules) irradiated by an intense hard x-ray pulse. Using the original XMDYN code, simulation of that many particles is too expensive. We present a model, employing our extension of XMDYN, that is capable of describing this challenging situation. We use, XATOM [3] toolkit to calculate all atomic x-ray scattering form factors required. To this end, we extend XATOM in order to enable the calculation of x-ray scattering form factors for many fluences for a large number of particles. Based on the sample dynamics and the scattering parameters we construct realistic scattering patterns for an inhomogeneous spatial pulse profile and compare them with the ideal ones. Moreover, we also investigate the effect of the shape of the beam profile by considering Gaussian and trapezoidal envelopes.

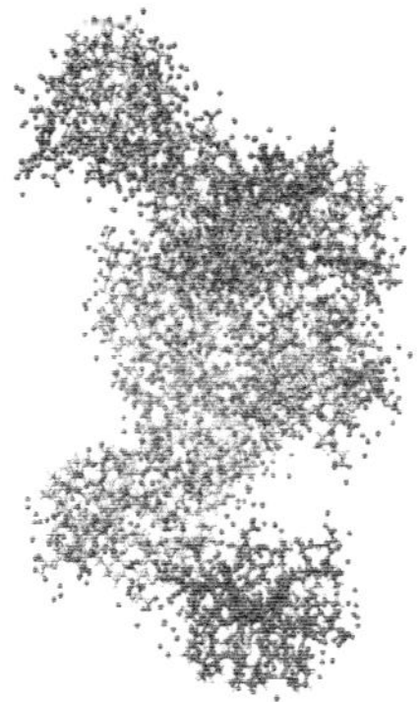

Figure 1. Unsolvated Ferredoxin unit-cell, containing 8 Ferredoxin molecules with a total of 9500 atoms.

References

[1] S.-K. Son, H. N. Chapman, and R. Santra, 2011 *Phys. Rev. Lett.* 107, 218102.

[2] B. F. Murphy *et al.*, 2014 *Nat. Commun.* 5, 4281.

[3] B. Rudek *et al.*, 2012 *Nat. Photonics* 6, 858.

[1]E-mail: muhammad.abdullah.malik@desy.de

Content from this work may be used under the terms of the Creative Commons Attribution 3.0 licence. Any further distribution of this work must maintain attribution to the author(s) and the title of the work, journal citation and DOI.
Published under licence by IOP Publishing Ltd

XXIX International Conference on Photonic, Electronic, and Atomic Collisions (ICPEAC2015) IOP Publishing
Journal of Physics: Conference Series **635** (2015) 102009 doi:10.1088/1742-6596/635/10/102009

X-ray two-photon absorption with high fluence XFEL pulses

J. Hoszowska* [1]**, J. Szlachetko**[†a]**, J.-Cl. Dousse*, W. Błachucki*, Y. Kayser**[†]**, Ch. Milne**[†]**, M. Pajek**[a] **, S. Boutet**[b]**, M. Messerschmidt**[b]**, G. Williams**[b]**, and C.T. Chantler**[c]

[*] Department of Physics, University of Fribourg, CH-1700 Fribourg, Switzerland
[†] Paul Scherrer Institut, CH-5232 Villigen PSI, Switzerland
[a] Institute of Physics, Jan Kochanowski University, 25-406 Kielce, Poland
[b] Linac Coherent Light Source (LCLS), SLAC National Accelerator Laboratory, 94025 California, USA
[c] School of Physics, University of Melbourne, Parkville, Victoria 3010, Australia

Synopsis. We report on nonlinear interaction of solid Fe with intense femtosecond hard x-ray free-electron laser (XFEL) pulses. The experiment was performed at the CXI end-station of the Linac Coherent Light Source (LCLS) by means of high-resolution x-ray emission spectroscopy. The focused x-ray beam provided extreme fluence of ~10^5 photons/Å^2. Two-photon absorption leading to K-shell hollow atom formation and to single K-shell ionization of solid Fe was investigated.

X-ray free electron laser (XFEL) facilities, with unprecedentedly high peak power densities reaching ~10^{20} W/cm^2, have paved the way to study nonlinear phenomena in the x-ray regime [1-6]. In this work we explored nonlinear interaction of high-fluence hard x-ray femtosecond pulses with solid Fe. Single and double K-shell electron ionization processes resulting from two-photon absorption were observed.

The experiment was carried out at the CXI end-station of the Linac Coherent Light Source (Menlo Park, USA) XFEL by means of the high energy resolution x-ray emission technique. The XFEL beam of ~5×10^{11} x-rays/pulse and pulse energy of 0.6 mJ was focused on a metallic Fe sample. The ultra-focused x-ray beam provided extreme fluence of ~10^5 photons/Å^2. Moving the sample out of the focus along the beam allowed varying the fluence. For the Fe Kα ($K^{-1}\rightarrow L^{-1}$) and Kα^h ($K^{-2}\rightarrow K^{-1}L^{-1}$) radiative transitions measurements the bent crystal von Hamos x-ray spectrometer of PSI [7] installed at CXI and equipped with the CSPAD detector developed at SLAC was employed. The K x-ray emission spectra were collected at photon beam energies below the Fe K-shell single- and double-ionization thresholds for the two-photon single ionization and double ionization processes, respectively.

For illustration, the probability of double K-hole formation *via* sequential absorption of two photons *versus* x-ray fluence is shown in figure 1. We observe a ~60-fold increase in the production probability of Fe hollow-atoms as compared to single-photon double ionization mediated by K-shell electron-electron correlations [8]. The cross-sections for double-K-hole formation and two-photon single K-shell ionization were derived from the x-ray fluence dependence of the measured x-ray emission intensities. For the two-photon single ionization process a square dependence of the Kα signal was found. This is the first observation of K-shell double core-hole creation following sequential photon absorption, and two-photon single K-shell ionization for metallic Fe.

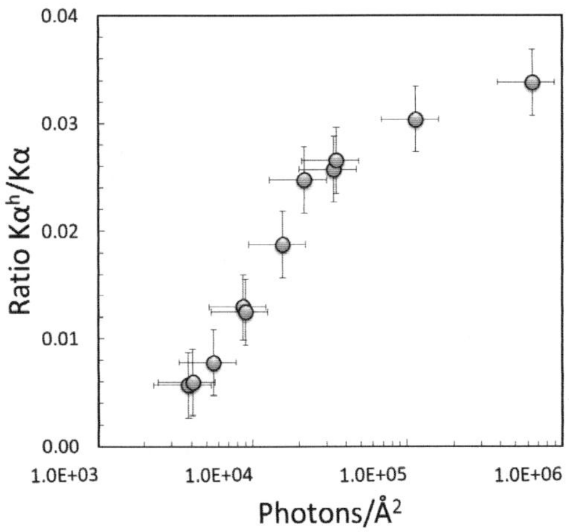

Figure 1. Probability of double K-hole formation *via* sequential absorption of two photons for Fe as a function of x-ray fluence. The data were derived from the Kα^h ($K^{-2}\rightarrow K^{-1}L^{-1}$) to Kα ($K^{-1}\rightarrow L^{-1}$) intensity ratios. The x-ray pulse energy was 7.6 keV and the duration 30 fs.

References

[1] L. Young *et al* 2010 *Nature* **466** 56
[2] G. Doumy *et al* 2011 *Phys. Rev. Lett.* **106** 083002
[3] B. Rudek *et al* 2012 *Nat. Photon.* **6** 858
[4] K. Tamasaku *et al* 2013 *Phys. Rev. Lett.* **111** 043001
[5] K. Tamasaku *et al* 2014 *Nat. Photon.* **8** 313
[6] H. Yoneda *et al* 2014 *Nat. Commun.* **5** 5080
[7] J. Szlachetko *et al* 2012 *Rev. Sci. Instrum.* **83** 103105
[8] J. Hoszowska *et al* 2009 *Phys. Rev. Lett.* **102** 073006

[1] E-mail: joanna.hoszowska@unifr.ch

Content from this work may be used under the terms of the Creative Commons Attribution 3.0 licence. Any further distribution of this work must maintain attribution to the author(s) and the title of the work, journal citation and DOI.
Published under licence by IOP Publishing Ltd

Journal of Physics: Conference Series **635** (2015) 102010 doi:10.1088/1742-6596/635/10/102010

High-Order Harmonic Generation from the Cu(111) surface

Néstor F. Aguirre*[1], Sergio Díaz-Tendero*, Fernando Martín*†‡

* Departamento de Química, Módulo 13,Universidad Autónoma de Madrid, 28049, Madrid, Spain

† Instituto Madrileo de Estudios Avanzados en Nanociencias (IMDEA-Nanociencia), 28049 Madrid, Spain

‡ Condensed Matter Physics Center (IFIMAC), Universidad Autónoma de Madrid, 28049 Madrid, Spain.

Synopsis We investigate theoretically the origin behind the formation of the plateau and cutoff structures in the high-harmonic spectra produced after the interaction of a near-infrared laser pulse with a metal surface. We use a wave packet propagation scheme and a one dimensional description of the Cu(111) surface.

It has been demostrated that solid-state samples can be used as generators of high-order-harmonic (HHG) radiation. The interest of this phenomenon resides in the fact that HHG represents one of the most reliable way to generate coherent ultraviolet to extreme ultraviolet light, with the possibility to observe strong-field physics with modest laser-fields strengths, in contrast with atoms (see for example [1]).

Image-potential states are quantized electronic states that appear near the vacuum level and are present in metal surfaces containing a band gap. The coupling of these states with bulk electronic states is the key point to understand a variety of dynamical processes, as for example the desorption of adsorbates induced by electronic excitations [2]. We are interested in the theoretical description of high-order harmonic generation when electrons are trapped in such image states.

In this work we present a theoretical study of HHG from a Cu(111) surface, by using a quantum-mechanical model within the single-active-electron (SAE) approximation. For the description of this surface, we use a model potential reported by Chulkov *et. al.* in ref. [3]. The time-dependent Schrödinger equation is solved numerically on a one-dimensional grid of equidistant points in the position coordinate, where the time evolution of the electronic wave function is described with a modified version of the split-operator technique [4]. We investigate the formation of the plateau and cutoff as a function of the field intensity and the screening length. For example, Fig. 1 shows some results for a given intensity and two values of the screening length, corresponding to the penetration of laser field inside the material in one and two atomic layers.

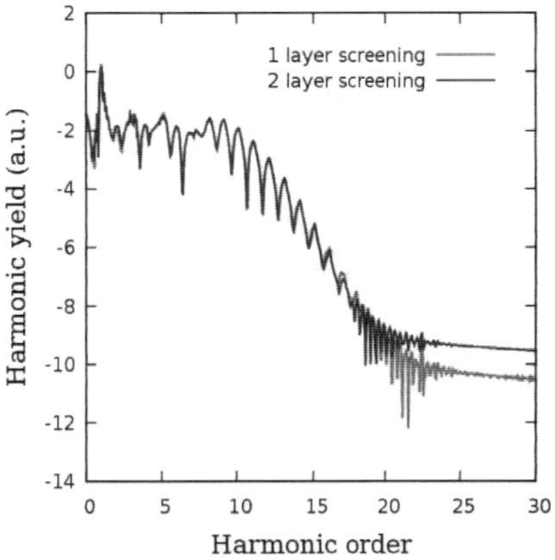

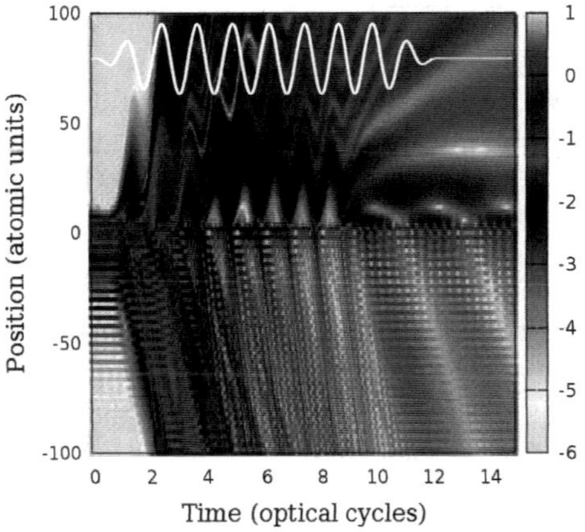

Figure 1. Top panel: harmonic spectra for a 1D Cu(111) surface irradiated by a 26.7 fs laser pulse (I_0 = 50 TW/cm^2, λ = 800 nm). Lower panel: time-dependent change in density on a log. scale.

References

[1] G. Wachter *et al* 2012 *Phys. Rev. B* **86** 035402

[2] J. Güdde *et al* 2006 *Chem. Rev.* **106** 4261

[3] E.V. Chulkov *et al* 1999. *Surf. Sci.* **437** 330

[4] C. M. Dion *et al* 2014. *Comput. Phys. Comm.* **185** 407

[1]E-mail: nestor.aguirre@uam.es

Content from this work may be used under the terms of the Creative Commons Attribution 3.0 licence. Any further distribution of this work must maintain attribution to the author(s) and the title of the work, journal citation and DOI.
Published under licence by IOP Publishing Ltd

Journal of Physics: Conference Series **635** (2015) 102011 doi:10.1088/1742-6596/635/10/102011

Slow ground state molecules from matrix isolation sublimation

C. L. Cesar[*1], A. N. Oliveira[†], R. L. Sacramento[*], B. X. Alves[*], B. A. Silva[*], and W. Wolff[*2]

[*] Instituto de Física, Universidade Federal do Rio de Janeiro, 21941-972 Rio de Janeiro, RJ, Brazil
[†] INMETRO, Av. Nossa Senhora das Graças, 50, Duque de Caxias - RJ, Brazil

Synopsis We describe a cryogenic beam of $^7Li_2$ dimers from sublimation of a neon matrix where Li atoms have been implanted via laser ablation of solid precursors of LiH. Laser absorption spectroscopy measured: T~7 K, Trot ~ 6K, drift velocity of 130 ms^{-1} with molecular density of 10^9 cm^{-3}. The formation of molecules in a matrix offers new possibilities.

The production of cold samples or beams of molecules is an ongoing research field with many applications [1], from quantum information to basic physics tests, such as the search for a permanent electric dipole moment of the electron with parity and time reversal symmetries violations. Some molecular species are complex to make in an environment compatible with low temperatures [2], requiring the use of hot ovens.

We describe the generation and properties of a cryogenic beam of $^7Li_2$ dimers from sublimation of a neon matrix where lithium atoms have been implanted via laser ablation of solid precursors of metallic lithium or lithium hydride (LiH). The sublimation regime can be controlled to lead to different parameters, such as temperature and forward velocities.

With laser absorption spectroscopy these parameters were measured using the molecular $^7Li_2$ (R) transitions $A^1\Sigma^+_u$ (v′ = 4, J′ = J″ + 1) ←$X^1\Sigma^+_g$ (v″ = 0, J″ = 0, 1, 3). In a typical regime, sublimating a matrix at 16 K, translational temperatures of 6–8 K with a drift velocity of 130 m s^{-1} in a free expanding pulsed beam with molecular density of 10^9 cm^{-3}, averaged along the laser axis, were observed. Rotational temperatures around 5–7 K were obtained. By monitoring the atomic Li signal – in the D2 line – concomitantly with the molecular signal, we see saturation in both the atomic and molecular signal [3].

Based on the data and a simple model, we discuss the possibility that a fraction of these molecules are being formed in the matrix, by mating atoms from different ablation pulses, which would open up the way to formation of other more interesting and exotic molecules to be studied at low temperatures. Such a source of cryogenic molecules have possible applications encompassing fundamental physics tests, quantum information studies, cold collisions, chemistry, and trapping.

A typical spectrum for the $A^1\Sigma^+_u$ (v′ = 4, J′ = 2) ←$X^1\Sigma^+_g$ (v″ = 0, J″ = 1) transition, at 665.927 nm, are shown in Fig. 1.

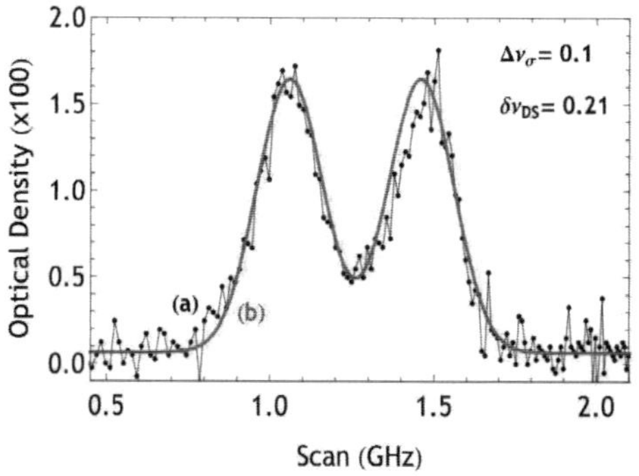

Figure 1. (a) Typical spectrum of Li_2 transition at 665.927 nm, black line; (b) best fit to the measured data, red line. The measured parameters $\Delta\nu_\sigma$ and $\delta\nu_{DS}$ are related to temperature and forward velocity.

Details of the experimental setup for Matrix Isolation Sublimation (MISu): a neon matrix is grown onto a cryogenic sapphire substrate, and atoms of Li and H are implanted via laser ablation of a solid LiH precursor; atoms and molecules are liberated from the matrix into vacuum with a sublimation heat pulse on the NiCr film resistor on the sapphire, will be presented. Also, details how to generate and characterize the molecular beam will be given.

References

[1] L. Carr *et al* 2009 New J. Phys. **11** 055049
[2] K. Walker *et al* 1993 Mol. Phys. **78** 577
[3] A. N. Oliveira *et al* 2014 J. Phys. B: At. Mol. Opt. Phys. **47** 245302

[1] E-mail: lenz@if.ufrj.br

[2] E-mail: wania@if.ufrj.br

Content from this work may be used under the terms of the Creative Commons Attribution 3.0 licence. Any further distribution of this work must maintain attribution to the author(s) and the title of the work, journal citation and DOI.
Published under licence by IOP Publishing Ltd

XXIX International Conference on Photonic, Electronic, and Atomic Collisions (ICPEAC2015) IOP Publishing
Journal of Physics: Conference Series **635** (2015) 102012 doi:10.1088/1742-6596/635/10/102012

Theoretical and experimental investigation for SO_3 production in SO_2-rich astrophysical environments

Víctor de Souza Bonfim* [1], **Roberto B. Castilho**†, **Leonardo Baptista**‡, **Sergio Pilling*** [2]

* Universidade do Vale do Paraíba – UNIVAP, São José dos Campos, Brazil.
† Universidade Federal do Amazonas – UFAM, Manaus, Brazil.
‡ Universidade Estadual do Rio de Janeiro (UERJ), Resende, Brazil.

Synopsis This work presents the results for the irradiation of pure SO_2 sample that was condensed in a pre-evacuated chamber, from Laboratório de Astroquímica e Astrobiologia (LASA/UNIVAP), at low temperature (12 K) and irradiated by ionizing photons which simulate Solar photons in the vacuum ultraviolet (VUV) and soft X-rays range. The infrared spectra of irradiated sample have presented the formation of SO_3. Experimental formation cross section was determined. Theoretical investigations were performed at Second-order Møller-Plesset perturbation theory (MP2) level and indicate the most likely SO_3 formation channels vary with the reaction supporting medium.

This work presents an experimental and theoretical research on the formation of SO_3 molecule during the photolysis of a pure SO_2 ice sample by synchrotron radiation at 12 K. SO_2 and its related ions, radicals and fragmentation products are important reagents when it comes to prebiotic synthesis, since they can lead to the production of sulfur-containing amino acids.

Experiments were performed using a portable high vacuum chamber from LASA/UNIVAP at the National Synchrotron Light Laboratory (LNLS), in Campinas, Brazil. Photons used were from the VUV to soft X-rays (near 6 eV to 1200 eV). In situ analysis was performed using an infrared Fourier transform spectrometer (FTIR) at different fluences. FTIR spectra of the irradiated SO_2 sample have presented formation of SO_3, a new chemical species (Figure 1). In sequence, exponential fits were employed and cross-sections were determined, of both SO_2 dissociation and SO_3 production.

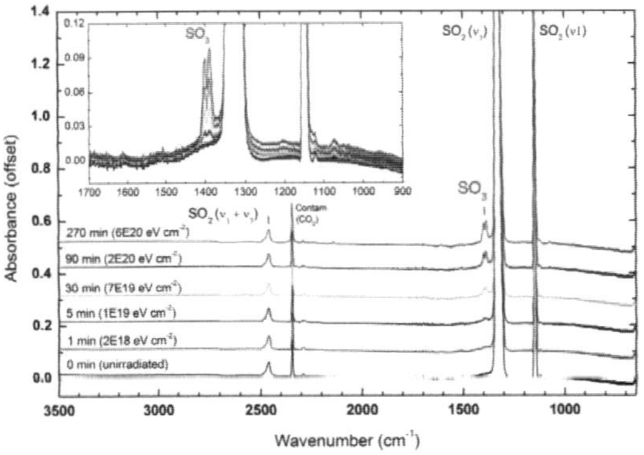

Figure 1. SO_2 ice FTIR spectra at 12 K, before and after different irradiation fluences (bottom to top).

The experimental results led to studies in Theoretical Chemistry. Calculations were performed at MP2 theory level with cc-pVTZ basis set, which is consistent with the studied systems size [1]. All species had the equilibrium geometry confirmed and the enthalpy value obtained at 12 K by applying standard statistical thermodynamics equations. Computations were carried out initially in the gaseous phase and then the solid (SO_2 ice). For calculations in the solid phase, the Polarized Continuous Model (PCM) has been chosen [2]. In the gas phase, the "daughter" chemical species that were most important for the SO_3 production were, in sequence, O^+, O_2 and SO, while for the SO_2 ice it were sequentially SO, O_2, and O species.

Through the calculation of reaction enthalpies with or without the solvent influence, one may infer that the presence of the solid SO_2, as described by its dielectric constant, decreases the reactivity of the O^+ species regarding the formation of SO_3. However, it should be noted that SO, O^+, O_2, and O are products of prior SO_2 dissociation and therefore more information would still be needed about the SO_2 molecule fragmentation pattern. The cross sections experimentally determined and the theoretically identified reaction channels in this study allow one to better understand the chemical evolution of certain astrophysical environments, such as the surface of Jupiter icy moons.

The authors would like to thank the FAPESP agencies (JP2009 / 18304-0) and CAPES, as well as the LNLS.

References

[1] S. Dobrin *et al* 2000 *J. Phys. Chem. A* **46** 10695
[2] S. Miertuš *et al* 1981 *Chem. Phys.* **55** 117

[1] E-mail: victordsb@gmail.com
[2] E-mail: sergiopilling@yahoo.com.br

Content from this work may be used under the terms of the Creative Commons Attribution 3.0 licence. Any further distribution of this work must maintain attribution to the author(s) and the title of the work, journal citation and DOI.
Published under licence by IOP Publishing Ltd

Journal of Physics: Conference Series **635** (2015) 102013 doi:10.1088/1742-6596/635/10/102013

Possible evidence for a liquid-gas phase transition in C_{60} systems

D. B. Qian[* 1], **X. J. Zhang**[†], **F. D. Shi**[*], **Z. Chen**[*], **D. C. Zhang**[*], **X. L. Zhu**[*], **J. Yang**[*], **X. Ma**[* 2]

[*] Institute of Modern Physics, Chinese Academy of Sciences, Lanzhou, 730000, China

[†]School of Physics and Engineering, Henan University of Science and Technology, Luoyang, 471003, China

Synopsis. We demonstrate that C_{60} fragmentation may be associated with a liquid-to-gas phase transition by analyzing the fragment size distributions in the C_{60} multifragmentation regime. The results will provide useful information to understand the formation mechanism of primary C_{60} in interstellar environments.

Recent studies have confirmed the presence of C_{60} in interstellar environments [1,2]. However, several aspects regarding C_{60} in space are still under debate, such as its formation in interstellar medium [3,4]. It is known that carbon is formed by fusion reactions in the cores of stars, and in the late stages of stellar evolution, massive stellar winds expel it into interstellar space. The primary C_{60} may be formed along with the expeled carbon vapor cools from thousands of degrees Celsius to an appropriate condensation temperature. Studying the Phase Transition (PT) in fragmenting C_{60} will provide useful information to understand its formation mechanism because of the thermodynamic reversibility of PT.

Our previous work has shown that a PT in C_{60} multifragmentation induced by nanosecond laser occurs at almost constant temperature covering a wide range of laser fluency [5]. Here, the yields of ionic fragments (IFs) resulting from the multifragmentation are measured within the PT region and are plotted in Fig. 1. By excluding two small IFs and magic IFs due to their abnormal behavior, the data for residual IFs are used to estimate the size distributions of primary intermediate-mass IFs. The distributions are found to obey power laws $n^{-\tau}$. Furthermore, the exponent τ values have sensitive dependence on lower laser fluency and converge to a constant for larger fluencies. These observations are in good agreement with an explanation based on the Fisher droplet model, offering the tantalizing possibility of a liquid-to-gas PT in C_{60} [6]. The physical picture is as follows:

The C_{60} on irradiation are firstly heated to the melting temperature (about 4000 K) at certain time during the laser pulse, causing its structural transition from a solid-like phase to a liquid-like phase. After a further photo-absorption from later part of the laser pulse, the instability of the melted C_{60} is encountered when the temperature reaches the fragmentation temperature (about 6000 K), in which various carbon clusters as fragments are formed. Such high temperature implies that, in the preliminary stage of fragmentation, each fragment has also an liquid-like structure.

If it's true, a model of liquid-gas PT can be used to describe the C_{60} formation in a cooling carbon vapor. The fragmentation temperature proposed by the previous work [5] provide a reference for the starting temperature of C_{60} formation in the stellar evolution. The two-stage PT proposed by another work [7] provide a physical picture for C_{60} formation, i.e. small carbon clusters form initially and then progress into C_{60} by incorporation of atomic and diatomic carbon.

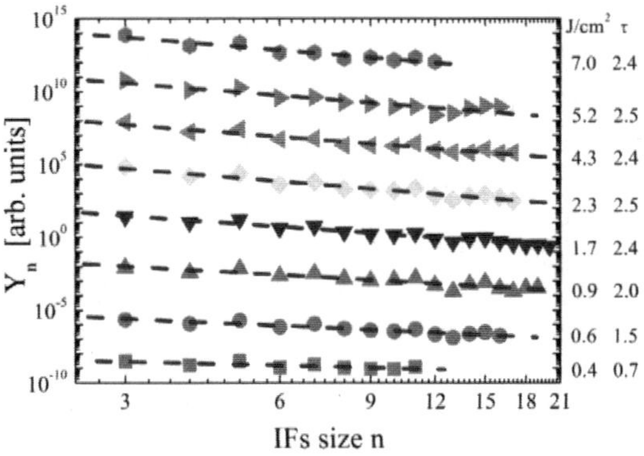

Figure 1. Yield distributions of IFs C_n^+ (n = 3–20) as a function of the size with different laser fluencies.

This work was partly supported by the NSFC (Grant Nos. U1232122 and 10904152).

References

[1] J. Cami *et al* 2010 *Science* **329** 1180
[2] K. Sellgren *et al* 2010 *Astrop. J. Lett.* **722** L54
[3] P. W. Dunk *et al* 2012 *Nat. comm.* **3** 855
[4] O. Berné *et al* 2012 *PNAS* **109** 401
[5] D.B. Qian *et al* 2011 *Phys.Chem.Chem.Phys.* **13** 3328
[6] D.B. Qian *et al* 2014 *J. Chem. Phys.* **141** 054307
[7] D.B. Qian *et al* 2013 *Phys. Rev. A* **87** 063201

[1] E-mail: qiandb@impcas.ac.cn
[* 2] E-mail: x.ma@impcas.ac.cn

Content from this work may be used under the terms of the Creative Commons Attribution 3.0 licence. Any further distribution of this work must maintain attribution to the author(s) and the title of the work, journal citation and DOI.
Published under licence by IOP Publishing Ltd

Journal of Physics: Conference Series **635** (2015) 102017 doi:10.1088/1742-6596/635/10/102017

Formation of hydrocarbon species and carbon clusters through femtosecond laser ablation of cryogenic solidified methane

Reza Karimi†, Asif Zaidi‡, Yiming Cai*, Mojtaba Hajialamdari* [1], Joseph Sanderson* [2], and Walter Duley*

* Physics and Astronomy, University of Waterloo, 200 University Avenue West, Waterloo, Ontario, N2L 3G1, Canada
† Energy Engineering and Physics, Amirkabir University of Technology, 424 Hafez Avenue, Tehran, Iran, 15875-4413
‡King Saud University, P. O. BOX 2454, Riyadh 11451, Saudi Arabia

Synopsis Femtosecond laser ablation of cryogenically solidified CH_4 using a peak intensity 8×10^{14} W/cm^2 at 800 nm has been studied using time of flight mass spectroscopy. Spectra show the presence of hydrocarbon species with molecular mass higher than the original target molecule.

The laser ablation of graphite has been extensively studied because of graphite's unique properties. It is well known that the laser ablation of graphite with nanosecond pulses produces pure carbon clusters and fullerenes such as C50, C60 and C70 as well as more complex species including nanotubes, depending on ablation conditions. It is then important to investigate the ablation of molecular systems containing C-H sigma bonds using femtosecond laser pulses in order to see the effect of hydrogenation of the target molecule. The use of femtosecond, rather than nanosecond laser pulses eliminates heating of the target. Alkanes are the simplest fully hydrogenated C-H molecular systems that occur naturally and methane (CH4), containing only C-H sigma bonds, is the simplest member of this family. A study of the ablation of solid methane is then important in understanding ablation in more complicated alkane systems.

Experiments were carried out using a 5×10^{-7} torr vacuum chamber attached to a time of flight (TOF) spectrometer. By flowing methane gas onto a cold finger made of copper plate in the vacuum chamber, a 0.5mm-thick film of solid methane (a van der Waals solid with a density of 0.522 gm cm^{-}3) was obtained on the cold finger. Five to eight successive laser pulses chopped from a 100-fs 1-KHz 800nm 300-μJ mode-locked Ti:sapphire laser system were focused on the solid methane by a 10-cm focusing lens. Each focused laser pulse has a peak intensity 8×10^{14} W/cm^2. A laser plume was generated which was extended from the cold finger surface up to a few millimeters above it. Positively charged particles generated in the laser plume were accelerated by a 4kV linear accelerator toward a micro-channel plate (MCP) detector. Signals from the MCP detector were recorded by a scope and sent to the PC

Fig. 1 shows the TOF spectrum recorded from solid methane. The peak assignments were derived from predictions from a model of the TOF apparatus. The presence of high kinetic energy fragment species in the plume makes precise identification difficult for higher mass to charge values and so only approximate bands are indicated. Nevertheless there is clearly a range of higher mass peaks the characteristics of which can be identified as a function of parameters such as laser intensity, and compared to previous work [1].

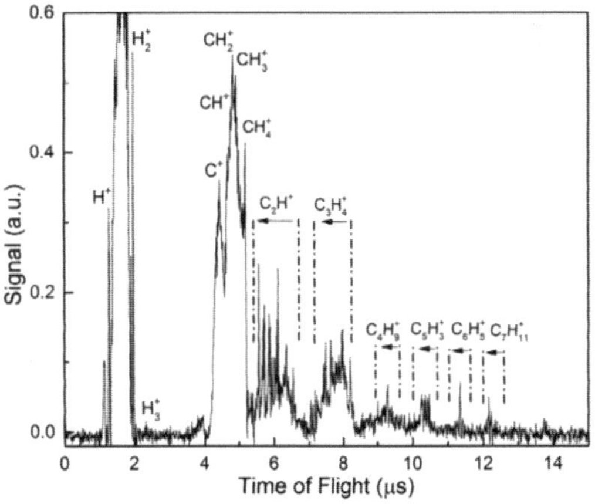

Figure 1. TOF spectrum resulted from the ablated solid methane.

References

[1] A.A. Zaidi *et al* 2015 *Int. J. Mass. Spec.* 376 32

E-mail: r3karimi@uwaterloo.ca

E-mail: j3sander@uwaterloo.ca

Content from this work may be used under the terms of the Creative Commons Attribution 3.0 licence. Any further distribution of this work must maintain attribution to the author(s) and the title of the work, journal citation and DOI.
Published under licence by IOP Publishing Ltd

Journal of Physics: Conference Series **635** (2015) 102018 doi:10.1088/1742-6596/635/10/102018

Gold nanoparticle electron and photon emissions after X-ray absorption

Romain Casta*[1], Jean-Philippe Champeaux*, Martine Sence*,Patrick Moretto-Capelle*, Pierre Cafarelli*

* LCAR - UMR 5589 IRSAMC Universit Paul Sabatier 118 route de Narbonne 31062 Toulouse, France

Synopsis In the last few years, numerous papers have been dedicated to nanoparticle radiotherapy enhancement. However, the biological and physical mechanisms leading to this enhancement are not well known, and can be improved by a better knowledge of the electron and photon emissions of nanoparticles after X-ray absorption. Experimental and theoretical results about these emissions are presented and discussed.

Gold nanoparticle have been increasingly used as radiosensitizer for cancer therapy in the last years and a lot of experimental work has been devoted to explaining why they effectively enhance the effect of radiotherapy treatments but the answer to this question remains elusive. Nevertheless, it seems that nanoparticle electron and photon emissions play a major role in the radiosensitization process: the absorption by the nanoparticles of the radiotherapy photon energy and its re-emission in the form of electrons and/or photons of lower energies causing damages to the cancerous cells is one the most admitted theories.

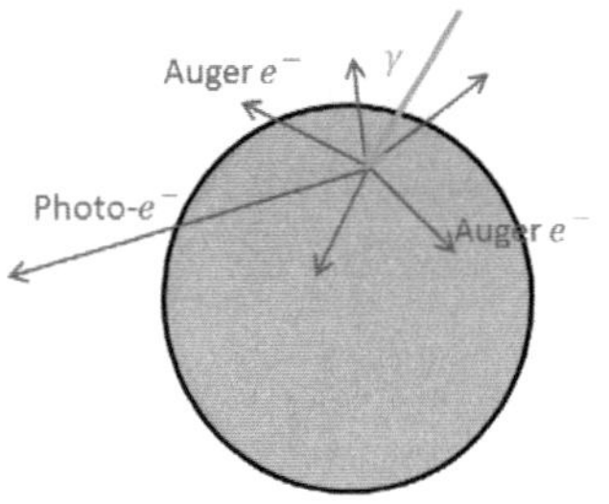

Figure 1. Electron emission of nanoparticle irradiated by X-ray photon (schematic diagram).

One of the key features to understand this radiosensitizing effect is the energy spectra of the electrons and photons from the gold nanoparticles. To do so, we present both experimental and theoretical studies : from the experimental point of view, we performed X-ray spectroscopy experiments on a commercial XPS apparatus but also on our home-made XPS allowing to reach the 1-10eV energy range. From the theoretical point of view, we present a model developed to evaluate nanoparticle photon and electron emission from Auger transition rates, stopping power and photoionisation subshell cross-sections.

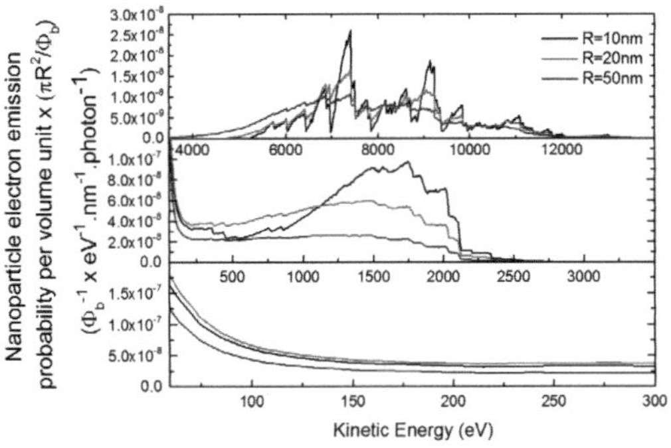

Figure 2. Example of electron emission spectra obtained with our theoretical model for three gold nanoparticles with different sizes and irradiated by $60keV$ X-ray photons.

References

[1] R. Casta *at al* 2014 *J.Nanoparticle Res.* **16**:2348.

[2] R. Casta *at al* 2014 *J.Nanoparticle Res.* **16**:2480.

[3] R. Casta *at al* 2015 *J.Nanoparticle Res.* **17**:3.

[1]E-mail: romain.casta@irsamc.ups-tlse.fr

Content from this work may be used under the terms of the Creative Commons Attribution 3.0 licence. Any further distribution of this work must maintain attribution to the author(s) and the title of the work, journal citation and DOI.
Published under licence by IOP Publishing Ltd

Journal of Physics: Conference Series **635** (2015) 112001 doi:10.1088/1742-6596/635/11/112001

Effect of Nuclear Mass in Strong-Field Ionization of Hydrogen Molecules and Dissociation of Hydrogen Molecular Ions

Igor V. Litvinyuk[i], Xiaoshan Wang, Han Xu, D. Kielpienski, R.T. Sang

Centre for Quantum Dynamics, Griffith University, Nathan, Queensland 4111, Australia
E-mail: i.litvinyuk @griffith.edu.au

Synopsis We measure isotope effect in strong-field ionization and dissociation of molecular hydrogen using a mixed-gas target in a Reaction Microscope (REMI). We compare experimental results with theoretical predictions.

In intense-field physics, a quantitative comparison of theoretical predictions with results of experimental measurements – necessary to validate any theory – often remains problematic. The difficulties arise most commonly from imprecise knowledge of laser (intensity, pulse duration, focal size), detector (acceptance angle, efficiency, resolution) and target (density, beam size) parameters. Even when all those parameters are known with sufficient precision, a proper comparison requires also careful accounting for focal and target geometry in combination with a realistic modeling of the detector. Most of those difficulties could be avoided by comparing measurements performed on different isotopes, i.e. light and heavy hydrogen, under exactly the same conditions. The relative isotope effect could be largely insensitive to variations of unknown or unstable parameters and may be then readily compared with predictions of various theoretical models.

Here we present two specific examples of experimental measurement of isotope effects observed in light and heavy hydrogen molecules – one validating a theory and another one revealing inadequacy of a model. To ensure that the measurements were performed under exactly the same conditions for H_2 and D_2, we used a 50:50 gas mixture as a target with REMI detection, which allowed us to mass-select the two isotopes.

In the first example, we measured single ionization yields for H_2 and D_2 for a number of different intensities. It has been recently predicted theoretically by Tolstikhin, Worner and Morishita [1] that tunneling ionization rate for neutral hydrogen molecules should exhibit a significant isotope effect, which becomes more pronounced at lower intensities. Since a significant fraction of singly ionized molecules also undergo dissociation (and also double ionization), which could itself exhibit an isotope effect, proper accounting for all final channels was necessary for a quantitative comparison. We performed the measurements using circularly polarized pulses and coincident REMI detection of protons, deuterons and electrons. We used ion detection to ascertain the relative ionization yields and coincident ion-electron detection to measure angular anisotropy of ionization for both H_2 and D_2. We found excellent agreement with predictions of Tolstikhin et al. with hydrogen more likely to be ionized (up to 1.4 times more likely at the lowest intensity) than deuterium.

In the second example, we measured CEP-dependent asymmetry of proton (deuteron) ejection in dissociation of H_2^+ (D_2^+) as a function of the fragment kinetic energy. We measured the highest asymmetry reported to date – 0.5 for H_2 and 0.4 for D_2. We compared our experimental results to calculations of a two-state model. While taken separately, both hydrogen and deuterium show excellent agreement with model calculations up to the unknown absolute CEP calibration. However, a side-by-side comparison reveals a significant discrepancy between the experimentally observed and calculated relative phase of the CEP dependence of asymmetry. Clearly, the theoretical model, while adequately reproducing the magnitude and energy dependence of CEP-dependent asymmetry, fails to properly account for its phase.

References

1. O.I. Tolstikhin, H.J. Warner and T. Morishita 2013 *Phys. Rev. A* **87** 041401(R)

[i]E-mail: i.litvinyuk@griffith.edu.au

Content from this work may be used under the terms of the Creative Commons Attribution 3.0 licence. Any further distribution of this work must maintain attribution to the author(s) and the title of the work, journal citation and DOI.
Published under licence by IOP Publishing Ltd

Detection of HCO^+, NO^+, $CNOH^+$ and CH_3^+ by multiphoton dissociation of the Nitromethane

Denhi Martínez[1]*, Ignacio Álvarez*, Alfonso Guerrero* and Carmen Cisneros*

*Instituto de Ciencias Físicas, Universidad Nacional Autónoma de México, Cuernavaca México

Synopsis. Using VUV radiation several ionic fragments were observed from the multiphoton dissociation of the Nitromethane. The detection of the molecular ions was focused on: HCO^+, NO^+, $CNOH^+$ and CH_3^+ due to their importance in reactions involving the formation of molecules in the interstellar clouds and the surrounding regions of the stars.

The interaction between the sample of gaseous Nitromethane (CH_3NO_2) with laser light at 532nm and 355nm wavelengths in a multiphotonic absorption regime was perform. The spectra were obtained by TOF in a Reflectron apparatus and at power range of ~10^9-10^{10} W/cm^2.

The attention was focused on the production of HCO^+, NO^+, $CNOH^+$ and CH_3^+ ions for their relevant role in the interstellar reactions involved on molecular clouds, stars and comets [1].

Nitromethane sample was introduced into the ionization chamber through a supersonic expansion process without carrier gas, by a pulsed valve. The sample gas and the laser light interact at 90°. The pressure inside the chamber was around $1x10^{-6}$ Torr. The formed ions, in the nanosecond regime, were accelerated to the free field region of the R-TOF spectrometer at 1.5 keV maintained at $1x10^{-8}$ Torr pressure and detected with a microchannel plate and processed with picosecond analyser.

The ions of interest in astrophysics: CH_3^+, $CNOH^+$, COH^+ were detected at both wavelengths with good intensity, increasing with the photon energy and also as a function of the radiation intensity. The fragment NO^+ was identified only at 355 nm. At both wavelengths the parent ion was not detected.

Figure 1 show the TOF spectra for three different radiant energies at 532nm wavelength, where the production of ions with higher m/z ions was favored.

Figure 2 shows TOF spectra at the wavelength of 355nm under the previous experimental conditions. At this photon energy the formation of small m/z ions was favored.

With the results obtained we conclude that the formation of fragment ions came for the process of dissociation - ionization of the Nitromethane and that the production of the selected ions, due to the proximity with its characteristic absorption bands, -198nm and 270nm-[2], were favored at 355nm, and increase with the radiant energy.

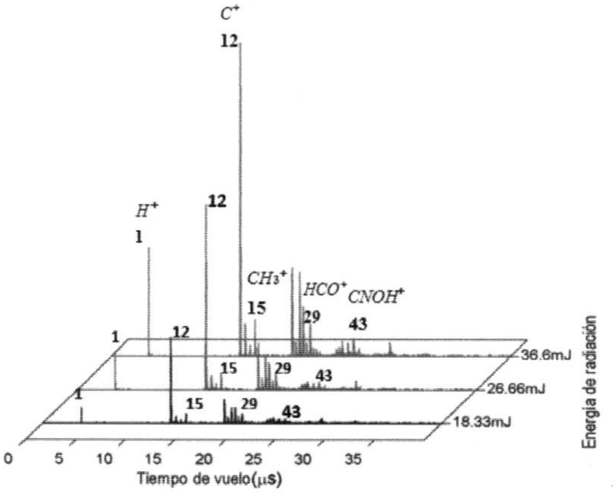

Figure 1. TOF spectra for 532nm at different radiant energies.

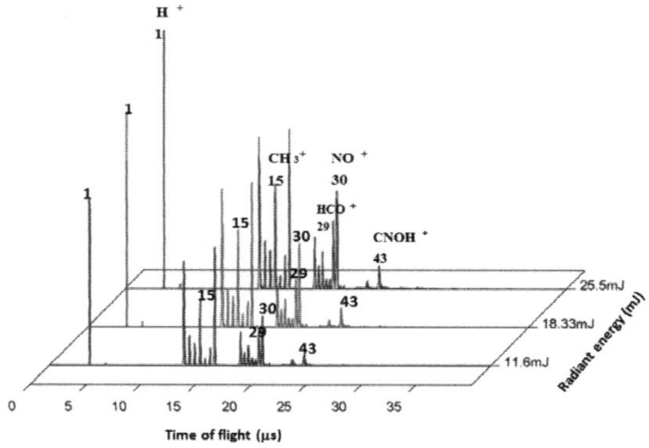

Figure 2. TOF spectra for 355nm at different radiant energies.

This work was supported by DGAPA Grant IN102613 and IN101215 CONACYT grant 165410.

References

[1] R. L. Pulliam *et al* 2011 *ApJ* **743** 36

[2] Nagakura, S. 1960 *Mol. Phys.* **3** 15

[1] E-mail: denhi27.10@gmail.com

Content from this work may be used under the terms of the Creative Commons Attribution 3.0 licence. Any further distribution of this work must maintain attribution to the author(s) and the title of the work, journal citation and DOI.
Published under licence by IOP Publishing Ltd

Shape of Model Potentials for a C_{60} Shell

Manson S. T.[†1], **Baltenkov A. S.***[2], **and Msezane A. Z.**[‡3]

[†] Georgia State University, Atlanta, Georgia 30303, USA
[*] Institute of Ion-Plasma and Laser Technologies, Tashkent 100125, Uzbekistan
[‡] Center for Theoretical Studies of Physical Systems, Clark Atlanta University, Atlanta, Georgia 30314, USA

Synopsis The spatial distribution of electric charges forming a square well potential has been analyzed. It has been shown that two concentric spheres each with a double layer of charges create the potentials of this type. It has been demonstrated that the phenomenological potentials simulating the potential of C_{60} shell belong to a family of potentials with non-flat bottoms. A C_{60} shell potential has been calculated under the assumption that it is formed by the averaged charge density of a neutral atom. It has been shown that this potential has a cusp-shaped form. Two model potentials of that type are proposed and their parameters have been calculated.

Phenomenological approaches on the basis of a rectangular (in the radial coordinate r) potential $U(r)$ for a C_{60} shell are widely used to describe fullerene-like molecular systems [1]. The C_{60} shell is formed by a superposition of the positive charge of carbon atomic nuclei and the negative charge of the tightly bound $1s^2$-electrons and the collectivized $2s^2 2p^2$-electrons. The spatial distribution of these charges is defined by the Poisson equation.

$$\frac{d^2}{dr^2}[rU(r)] = 4\pi r\rho(r).$$

The solutions of this equation were analyzed for a family of spherically symmetric stepwise potentials (with a flat bottom and parallel potential walls). It was shown that the density of electric charges $\rho(r)$ for these potentials is equal to zero in all space except the epsilon neighborhoods of the inner r_o and outer r_i radii, i.e. at the points where the function $U(r)$ changes discontinuously from zero to $-U_0$ and *vise versa*. Thus, the positive and negative charges of C_{60} shell can form the rectangular potential well only if one assumes that they are smeared over the surfaces of two concentric spheres and create a two-layer charge sandwich with a gap $\Delta = r_i - r_o$ between layers. This does not appear to correspond to the real structure of the C_{60}.

To obtain more detailed information on the behavior of the function $\rho(r)$ near the borders of potential walls, an exponential-power-law potential [2] was considered. This allows the potential shape to be changed continuously from a cusp-shaped type to a square-well type. It was shown that the potential well with the cusp-shaped bottom corresponds to real molecular structure of C_{60}, namely the atomic nuclei of the carbon atoms are localized on the sphere with radius R, while the electronic clouds border on the positively charged sphere inside and beyond it, i.e., the C_{60} shell charges form a three-layer charge sandwich. Here R is the distance between the center of the fullerene cage and nuclei of the carbon atoms.

The electrostatic potential created by the neutral atom charge density averaged over the sphere with radius R was calculated. It was shown that, calculated in such a way, the function $U(r)$ is very well approximated by the Lorentz-like curve, i.e., the potential $U(r)$ has a cusp-shape profile.

This analysis demonstrates that the phenomenological potentials simulating the C_{60} shell potential should belong to a family of potentials with non-flat bottoms.

Two model potentials for the C_{60} shell with a cusp-shaped bottom have been considered. The parameters of these potentials were chosen so that the bound state of the negative ion of C_{60} with affinity energy $\varepsilon \approx 2.65$ eV (as it follows from the data of UV spectroscopy) could exist in these potential wells.

It should be noted that while the potentials with flat bottoms (particularly the square well potential) remain usable from a practical viewpoint as a fitting potential, it is necessary to keep in mind that the actual potential of C_{60} has a cusp-shaped form.

Work was supported by Uzbek Foundation Award Ф2-ФА-Ф164 (ASB) and U.S. DOE, Basic Energy Sciences, Office of Energy Research (STM; AZM).

References

[1] V. Dolmatov 2009 *Adv. Quantum Chem.* **58** 13
[2] C. Lin and Y. Ho 2012 *J. Phys. B* **45** 155203

[1]E-mail: smanson@gsu.edu
[2]E-mail: arkbalt@mail.ru
[3]E-mail: amsezane@cau.edu

Content from this work may be used under the terms of the Creative Commons Attribution 3.0 licence. Any further distribution of this work must maintain attribution to the author(s) and the title of the work, journal citation and DOI.
Published under licence by IOP Publishing Ltd

Journal of Physics: Conference Series **635** (2015) 112006 doi:10.1088/1742-6596/635/11/112006

Photofragmentation of Formic Acid

Denhi Martínez*[1], Teonanacatl Bautista†, Jair Zarza†, Carmen Cisneros*[2], Ignacio Alvarez* and Alfonso Guerrero*

* Instituto de Ciencias Físicas, Universidad Nacional Autónoma de México, Cuernavaca México
† Facultad de Ciencias, Universidad Autónoma del Estado de México, Toluca, México

Synopsis Dissociation and ionization of formic acid was investigated by multiphotonic absorption regime using laser pulse at 355nm in conjunction with TOF mass spectrometry in reflectron mode (R-TOF). From the different type of detected fragment ions is inferred the formation of some small clusters of formic acid.

Different ions formation from formic acid (HCOOH) has been previously reported using synchrotron radiation. Experiments from Tabayashi et al. and Arruda et al. reported different ions coming from formic acid clusters [1,2]. Our findings show multiphoton dissociation using laser radiation of 355nm in the range of 10^9-10^{10} W/cm^2 intensity. Mass spectra were obtained using time of flight spectroscopy.

A beam was produced with formic acid 96% pure from Sigma-Aldrich. Two different gases (Argon and Helium) were used as carrier gases in order to investigate the influence of such carrier gases on the ion yield. The beam was collimated with a 0.5 mm skimmer and allowed to enter the main chamber of the apparatus, where it was crossed at 90° with a laser beam of 355nm in the ionization region of the R-TOF mass spectrometer. Ions were accelerated by a constant field of 800 V/cm into the R-TOF spectrometer and detected with a MCP. The signals were digitized and the mass spectra were obtained. Around 35 different ions were identified. Some of the most abundant were H^+, C^+, CH_3^+,H_2O^+, and HCO^+.

Table 1 shows the ion yield values in arbitrary units obtained with different carrier gas. Some ionic fragments included in the graph can only arise from small formic acid clusters.

Table 1. Ionic Yield at 4.11x10^9 W/cm^2.

Ion	No gas	Carried gas	
		Argon	Helium
$^{a,b,c}H^+$	1.11	0.74	2.36
$^{c}C^+$	5.57	4.26	5.74
$^{a,c}CH_3^+$	0.75	0.59	1.03
$^{a,b,c}H_2O^+$	0.25	0.19	0.77
$^{a,b,c}HCO^+$	2.85	2.15	6.38
$^{c}[CO_2H_2]H_2^+$	0.05	0.02	0.11

[a]Ions reported in [1], [b]Ions reported in [2] and [c]Ions reported in our work.

From Table 1, it is possible to observe the influence of the carrier gas which is not very evident when synchrotron radiation is used.

It can be concluded that the appearance of ions with mass/charge ratio of 15 and 48 can be explained as they are formed from parent molecular clusters. Such ions can be identified as a characteristics of proton transfer reactions within the clusters.

Figure 1 shows the dependence between the relative intensity and the radiant energy for some of the resulting ionic fragments listed in Table 1.

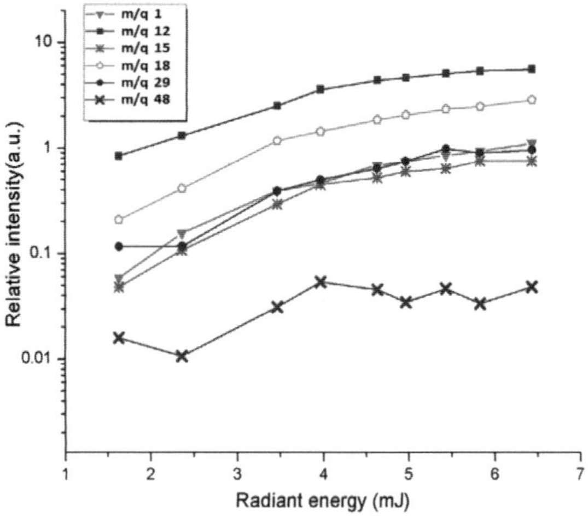

Figure 1. Relative intensity as a function of the radiant energy, the box in the graph related a symbol/color with the m/q ratio.

This work was supported by DGAPA Grants IN102613 and IN101215 and CONACYT Grant 165410.

References

[1] Kiyohiko Tabayashi *et al* 2006 *J. Chem. Phys.* **125** 194307

[2] Manuela S. Arruda *et al* 2012 *J. Phys. Chem. A* **116** 6693-67

[1]E-mail: denhi27.10@gmail.com
[2] E-mail: carmen@fis.unam.mx

Content from this work may be used under the terms of the Creative Commons Attribution 3.0 licence. Any further distribution of this work must maintain attribution to the author(s) and the title of the work, journal citation and DOI.
Published under licence by IOP Publishing Ltd

Doubly-charged negative ion of C_{60}

Baltenkov A. S.* [1], **Msezane A. Z.**† [2]

* Institute of Ion Plasma and Laser Technologies, Tashkent 100125, Uzbekistan
† CTSPS, Clark Atlanta University, Atlanta, Georgia 30314, USA

Synopsis An electronic structure of the negative ion C_{60}^{2-} has been studied. It is found that in the first approximation of the variational method (when a trial wave function of the extra electrons is a product of one-electron functions) the total energy of the system is negative and that the second electron affinity (EA) of the C_{60} molecule within the utilized model potentials is about 1 eV. The photodetachment cross sections of the C_{60}^{2-} ion have been calculated as well; they are found to exhibit two different threshold behaviors with their values being of the same order as the photodetachment cross sections of atomic negative ions with the same EA.

An electronic structure of the doubly-charged negative ion C_{60}^{2-} has been studied by a variational method. It was assumed that the closed-shell C_{60} molecule is spherically symmetric, and the interaction of the added electron with the molecule is described by a potential $U(r)$ having in the jellium model [1] the Lorentz-like profile

$$U(r) = -U_0 /[(r-R)^2 + d^2]$$

The parameters of $U(r)$ (the fullerene radius R, well depth U_0 and thickness d) are such that upon solving the wave equation with this potential for a single electron the first EA of the C_{60} molecule agrees with the measured value of ε_1=2.65 eV [2]. At the limit $d \to 0$ this potential is transformed into the Dirac-bubble potential [2]. It is assumed that in the $1s^2$ ground state of the system C_{60}^{2-} both electrons are in a state with zero orbital angular momentum and antiparallel spins. Firstly, the energy of the ground state of the system C_{60}^{2-} was estimated within the framework of perturbation theory, with the wave function of the electron pair represented as a product of one-electron functions. The obtained total energy of the electrons E, confined within the potential well $U(r)$ with thicknesses varying from d=0 to 1.5 atomic units (au), was found to be negative. Consequently, the existence of the doubly-charged negative ion of C_{60} within the Lorentz- and Dirac-bubble potential models is quite acceptable.

Secondly, the total energy E and the electronic wave function φ of the ground state of the doubly-charged negative C_{60}^{2-} ion were obtained by a direct variational method. A trial wave function of the extra electrons was written as a product of one-electron functions of the C_{60}^{-} negative ion having the asymptotically corrected behavior inside and beyond the fullerene bubble [2].

It has been shown that the total energy $E(\beta)$ as a function of the variational parameter β reaches its minimum at β_0=0.4269 au, which corresponds to the energy $E(\beta_0) \approx$ -1.59 eV. So, the second EA of the C_{60} molecule is equal to $\varepsilon_2 = \varepsilon_1 + E(\beta_0)$=1.06 eV, which is rather close to the results of other estimations [3].

Note that the long-lived doubly-charged carbon cluster anions C_{60}^{2-} in the gas phase have been reported from both sputtering of a graphite surface by Cs^+ ions [4] and the laser desorption from a surface covered with the neutral C_{60} molecules [3].

The photodetachment cross-sections $\sigma(\omega)$ of C_{60}^{2-} have also been calculated using the wave functions $\varphi(\beta_0)$. Near threshold $\sigma(\omega)$ is found to exhibit peculiar and interesting behavior. The first cross section accompanied by the transformation of the doubly-charged negative ion into a singly-charged one is exponentially small near the process threshold. The second cross section corresponds to the photodetachment of a singly-charged ion; it increases at the threshold as a power function of the kinetic energy of the photoelectron. These cross sections are of the same order as the photodetachment cross sections of atomic negative ions with the same EA.

Acknowledgments
Research was supported by the Uzbek Foundation Award Ф2-ФА-Ф164 (ASB) and U.S. DOE, Basic Energy Sciences, Office of Energy Research (AZM).

References

[1] V. Ivanov *et al* 2001 *J. Phys. B* **34**, L669
[2] L. Lohr and S. Blinder 1992 *Chem. Phys. Lett.* **198**, 100
[3] R. Hettich *at al* 1991 *Phys. Rev. Lett.* **67**, 1242
[4] N. Stephen *et al* 1990 *Phys. Rev. Lett.* **65**, 625

[1] E-mail: arkbalt@mail.ru
[2] E-mail: amsezane@cau.edu

Content from this work may be used under the terms of the Creative Commons Attribution 3.0 licence. Any further distribution of this work must maintain attribution to the author(s) and the title of the work, journal citation and DOI.
Published under licence by IOP Publishing Ltd

$e + C_{60}$ and $e + A@C_{60}$ elastic scattering versus the parameters of the C_{60}-model-square- well potential

V. K. Dolmatov [1], **M. B. Cooper and M. E. Hunter**

Department of Physics and Earth Science, University of North Alabama, Florence, Alabama 35632, U.S.A.

Synopsis Modelling of the C_{60} cage by a square-well potential $U_c(r)$ is a popular approximation. In the literature, some inconsistency is present in choosing the magnitudes of parameters of $U_c(r)$. In the present study, $e+C_{60}$ and $e + A@C_{60}$ elastic scattering is scrutinized versus the parameters of $U_c(r)$ in order to identify $U_c(r)$ which is best suited for studying *electron-fullerene* scattering and how the latter can be controlled by tuning the potential.

The work [1] demonstrated a reasonable usability of modelling the C_{60} cage by an attractive square-well (in the radial coordinate) potential $U_c(r)$ for the study of $e + C_{60}$ elastic scattering. This is because the most prominent features of $e + C_{60}$ scattering, predicted by the model [figure 1(a), solid line], agree qualitatively and even semi-quantitatively with calculated data obtained in the framework of a sophisticated *ab* - *t o* molecular-Hartree-Fock (MHF) approximation [figure 1(a), dash line] [1]. Later [2], the model was applied to $e+A@C_{60}$ elastic scattering as well. In [1, 2], $U_c(r)$ [to be labelled as $U_{1c}(r)$ in the paper] was defined as follows (in atomic units): the width $\Delta = 2.9102$, the inner radius $R_0 = 5.262$, the depth $U_0 = 0.2599$. The binding strength of $U_{1c}(r)$ (into a 2p state) matches the electron affinity of C_{60}, $EA = -2.65$ eV. However, in the literature, square-well potentials with other magnitudes of the parameters but the same binding strength as $U_{1c}(r)$ are used as well: $U_{2c}(r)$ with $R_0 = 5.8$, $\Delta = 1.9$, $U_0 = 0.302$ [3] and $U_{3c}(r)$ with $R_0 = 6.01$, $\Delta = 1.25$ and $U_0 = 0.422$ [4] (their binding strength matches the electron affinity of C_{60} relative to the 1s state). It is the ultimate aim of the present study (a) to clarify which one of these potentials is best suited for studying *electro -fullere e* scattering and (b) to learn how *electro -fullere e* scattering could be controlled by tuning the parameters of $U_c(r)$. The latter is an interesting in and of itself study from a point of view of basic science. Calculated electron elastic scattering cross sections $\sigma_{el}^{C_{60}}$ and $\sigma_{el}^{Ba@C_{60}}$ are depicted in figure 1. The results identify how, in principle, parameters of the resonances seen in $\sigma_{el}^{C_{60}}$ and $\sigma_{el}^{Ba@C_{60}}$, as well as even the very existence of some of them, can be modified by changing parameters of $U_c(r)$. Furthermore, results, obtained with the use of the potentials U_{2c} and U_{3c}, appear to differ noticeably from data calculated both with the use of U_{1c} and in *ab* *t o* MHF. Thus, the potential U_{1c} is a better potential than U_{2c} and U_{3c} for the study of *electro -fullere e* scattering, providing that MHF serves as the reference approximation.

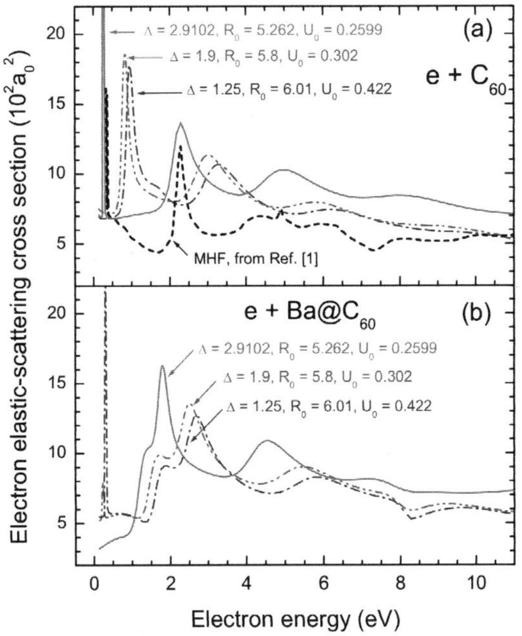

Figure 1. Electron elastic scattering cross sections $\sigma_{el}^{C_{60}}$ and $\sigma_{el}^{Ba@C_{60}}$ calculated with the use of different square-well potentials matching the electron affinity of C_{60}, $EA = -2.65$ eV, as well as $\sigma_{el}^{C_{60}}$ calculated in *ab* *t o* MHF [1], as marked.

This work was supported by the NSF grant No. PHY-1305085.

References

[1] C. Winstead and V. McKoy 2006 *Phys. Rev.* A **73** 012711

[2] V. K. Dolmatov *et al* 2014 *J. Phys.* **47** 115002

[3] P. C. Deshmukh *et al* 2014 *Phys. Rev.* A **89** 053424

[4] V. K. Dolmatov *et al* 2012 *J. Phys.* **45** 105102

[1] E-mail: vkdolmatov@una.edu

Content from this work may be used under the terms of the Creative Commons Attribution 3.0 licence. Any further distribution of this work must maintain attribution to the author(s) and the title of the work, journal citation and DOI.
Published under licence by IOP Publishing Ltd

Dynamics of dissociation in inner-valence excited 1,3-*trans*-Butadiene probed by 3-D multiple ion momentum imaging

S. Oghbaie[i], M. Gisselbrecht, A. Sankari, E. Mansson, J. Laksman, and S. L. Sorensen

Department of Physics, Lund University, S-221 00 Lund, Sweden

The study focuses on rapid nuclear rearrangement/isomerization taking place in an early stage of butadiene fragmentation using multiple ion momentum imaging spectroscopy and tunable synchrotron radiation. In addition, *ab initio* calculations are performed to disentangle the nuclear dynamics preceding dissociation.

Butadiene is the simplest conjugated diene, but its photochemistry is the most complex. Rapid H-transfer across the π-conjugated carbon chain can induce large-scale deformation of the molecules and bond rearrangement. Many studies on fragmentation of valence-excited butadiene have been carried out using laser and synchrotron radiation [1-2], however there is a lack of studies of dissociation dynamics of highly excited and ionized butadiene. Here, we use 3D-multiple ion momentum techniques [3] together with monochromatic synchrotron radiation to study the fragmentation dynamics of fixed in space localized-core-valence excited butadiene excited at different photon energies in the double-ionization continuum.

The sequence of dissociation channels in the ion-ion correlation map for ionization just above the double ionization threshold at 25.43 eV represents the early metastable dication dissociation where H-transfer to a terminal carbon is seen in the $CH_3^+/C_3H_3^+$ pair. Calculation of the dication potential energy surface finds a stable dication ground state with a 3 eV potential barrier for direct charge separation, but almost no potential barrier when a hydrogen adjacent to the twisted C=C bonds transfer to a terminal carbon. Comparison of calculation and experimental values of KER and AE indicates that the terminal C=C bond dissociates after H-transfer in the dictation ground state. Central C-C charge separation is initiated in excited states, where the bond dissociation is somewhat faster than H-migration and vibrational relaxation to the ground state. At 35 eV photon energy, three-body dissociation channels are found in the correlation map. The excess energy first deposits into the translational energy of H-atoms and is followed by CC charge separation in the dication species.

The fragment angular distribution in direct photoionization is determined first by the alignment selectivity in the transition, and is preserved in the fragmentation kinematics if the dissociation time scale is rapid compared to the rotational period of the molecule.

In a case where the axial-recoil approximation holds for a transition with π-symmetry the molecular plane is perpendicular to the polarization plane, and the C-H and C-C bonds are perpendicular resulting in $\beta = 1$. For parallel transitions (σ-symmetry) the carbon backbone will be aligned parallel to the polarization but C-H bonds will be perpendicular. This results in anisotropy values (β) with opposite signs for C-C and C-H bond dissociation.

In the double-ionization region the β parameter has a negative value for the early C-H bond dissociation and positive value for the secondary C-C charge separation fragments. According to the model described above the asymmetry suggests that transitions in the 35-45 eV range are dominated by alignment where the molecular backbone is oriented parallel to the polarization vector.

The fragment angular distribution after excitation of the inner-shell C1s→π* transition exhibits a different angular distribution. There we find that the molecular plane is perpendicular to the polarization vector of the light, which in fact explicitly reflects the early out-of-plane dissociation dynamics. For instance, the asymmetry parameter of H-atom in the early stage of symmetric charge separation depicts the primary torsional motion around the central bond with CC-CC dihedral angle ranges from 85° to 100° on that time scale.

References

[1]. H. Xu, Phys. Chem. Chem. Phys, 12 12939, (2010).
[2]. D. N. Neumark, J. Am. Chem. Phys. Soc, **124** 10211, (2002).
[3]. J. Laksman, Rev.Sci.Inst, **84** 123113, (2013).

[i]Shabnam.Oghbaiee@sljus.lu.se

Content from this work may be used under the terms of the Creative Commons Attribution 3.0 licence. Any further distribution of this work must maintain attribution to the author(s) and the title of the work, journal citation and DOI.
Published under licence by IOP Publishing Ltd

Journal of Physics: Conference Series **635** (2015) 112010 doi:10.1088/1742-6596/635/11/112010

Time-resolved and XUV spectroscopy of helium nanodroplets

M. Mudrich[1] *, J. von Vangerow[1], A. LaForge[1], F. Stienkemeier[1], A. Ciavardini[2], P. O'Keeffe[2], Y. Ovcharenko[3], T. Möller[3], M. Drabbels[4], P. Piseri[5], O. Plekan[6], P. Finetti[6], M. Coreno[7], C. Grazioli[6], R. Richter[6], K. C. Prince[6], C. Callegari[6]

[1] Physikalisches Institut, Universität Freiburg, Germany, [2] CNR - ISM, Monterotondo (RM), Italy, [3] Institut für Optik und Atomare Physik, TU-Berlin, Germany, [4] Federal Institute of Technology Lausanne (EPFL), Switzerland, [5] CIMAINA and Dipartimento di Fisica, Universitá di Milano, Italy, [6] Elettra-Sincrotrone Trieste, Basovizza, Trieste, Italy, [7] CNR-ISM, UOS Trieste, Basovizza, Trieste, Italy

Synopsis The ultrafast dynamics of pure and doped helium nanodroplets is studied using VIS and XUV femtosecond pump-probe spectroscopy in combination with ion and electron imaging detection.

Helium (He) nanodroplets currently attract considerable interest mainly for two reasons. 1) Due to their quantum fluid nature He nanodroplets feature extraordinary properties such as microscopic superfluidity. 2) He nanodroplets can be considered as a nearly ideal matrix for spectroscopy of embedded molecules and aggregates due to their transparency for light up to the extreme ultraviolet (XUV) spectral range, and their ability to efficiently aggregate and cool embedded species.

However, the short-time dynamics of He nanodroplets is still largely unexplored. Recently, strong perturbations of the rotational motion of embedded molecules induced by the He droplet environment was observed for the case of impulsive excitation [1]. In this contribution we study the ultrafast response to laser excitation of a prototype system consisting of an alkali metal atom residing at the surface of He nanodroplets. Using femtosecond pump-probe spectroscopy in combination with ion and electron imaging detection we follow in real-time the desorption of the excited atom off the He droplet surface (see Fig. 1) as well as the formation of alkali-He exciplex molecules.

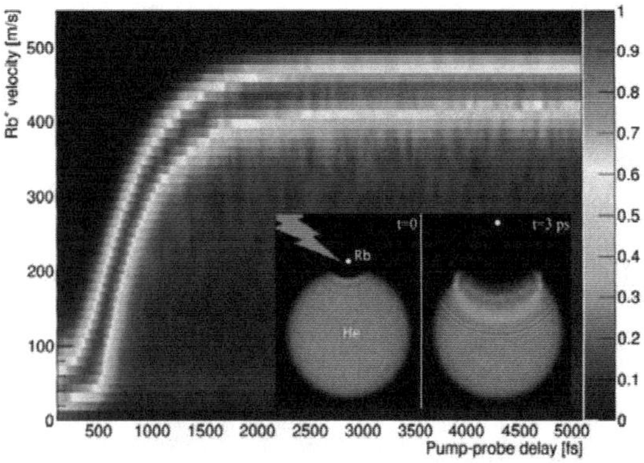

Figure 1. Time-evolution of the velocity of an excited Rb atom ejected off the surface of a He nanodroplet. Inset: TDDFT simulation [2].

Upon irradiation of He droplets with XUV light at $h\nu \geqslant 21$ eV, where He droplets are strongly absorbing, a complex photo dynamics is initiated by the excitation or ionization of He atoms inside the droplets [3]. The dynamic of relaxation, fragmentation, as well as indirect ionization processes are studied using synchrotron and free-electron-laser (Fermi@Elettra) radiation. In particular, the real-time dynamics of pure He nanodroplets excited by tunable femtosecond XUV-pulses is studied in a pump-probe experiment. The time-resolved photoelectron spectra reveal ultrafast intra- as well as interband relaxation dynamics, see Fig. 2.

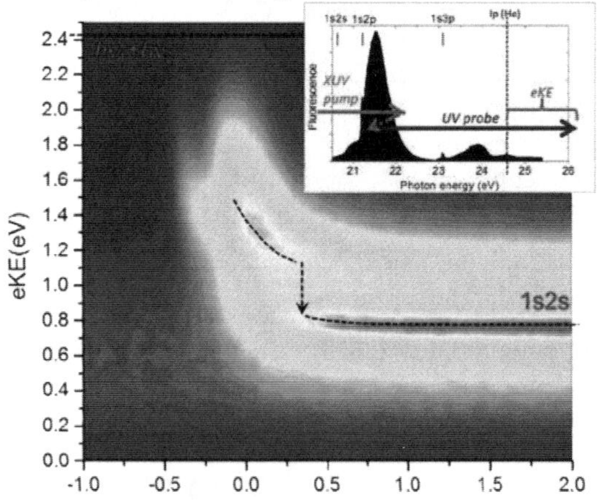

Figure 2. Pump-probe delay dependent photoelectron spectrum of pure He nanodroplets measured by XUV (22.2eV)/UV (4.8eV) fs pump-probe R2PI.

References

[1] D. Pentlehner *et al.*, Phys. Rev. Lett. **110**, 093002 (2013)
[2] J. von Vangerow *et al.*, J. Phys. Chem. A **118**, 6604 (2014)
[3] M. Mudrich and F. Stienkemeier, Int. Rev. Phys. Chem. **33**, 301 (2014)

* E-mail: mudrich@physik.uni-freiburg.de

Content from this work may be used under the terms of the Creative Commons Attribution 3.0 licence. Any further distribution of this work must maintain attribution to the author(s) and the title of the work, journal citation and DOI.
Published under licence by IOP Publishing Ltd

Valence-shell Photoelectron Circular Dichroism (PECD) on Chiral Halogenated Molecules in the Gas Phase

Maurice Tia,*[1] Gustavo Garcia,* Laurent Nahon* and Ivan Powis†

**Synchrotron SOLEIL, L'Orme des Merisiers, St. Aubin, B.P.48, 91192 Gif sur Yvette, France*
†School of Chemistry, University of Nottingham, Nottingham NG7 2RD, United Kingdom

Synopsis We measured and calculated the PECD of three different halogenated chiral molecules in order to study the role of the localization of the initial orbital of the outgoing electron with respect to the chiral center.

Photoelectron Circular Dichroism (PECD) is observed as a forward/backward asymmetry with respect to the photon axis, of the photoelectron angular distribution resulting from the CPL-induced photoionization of gas phase pure enantiomers of chiral species. This spectacular orbital-dependent chiroptical effect with intensities reaching up to the few tens of %, shows a rich photon energy dependence, and appears as a very sensitive probe of molecular conformations and of the chemical environment. [1] After the study of numerous organic compounds, we are interested here on unexplored questions, regarding to the role of the localization of the initial orbital of the outgoing electron with respect to the chiral center. Will the scattering phase of the outgoing electron be sensitive only to the nearby presence of a chiral center or is PECD a more global probe of the whole (chiral) molecular potential ?

Experiments were conducted on DESIRS beamline (Synchrotron SOLEIL, France) using the new electron/ion coincidence imaging spectrometer DELICIOUS 3, [2] which records Angle-Resolved photoelectron spectroscopy on mass-selected samples to extract the PECD. By alternating right- and left-handed circularly polarized radiation, the photoelectron circular dichroism (PECD) in the angular distribution was extracted from the photoelectron images obtained by coincidence with the ions corresponding to the parent and fragments, rejecting any spurious contribution from N_2 or H_2O for example.

Analysis of the photoelectron images of three well-chosen chiral halogenated molecules (Bromopropionic acid and Bromophenylethylamine isomers) on which the 4p Br orbitals are clearly localized (HOMO and HOMO-1 in the case of BPA, HOMO-3 and HOMO-4 in the case of BrPhEtA), allows us to observe and quantify the corresponding forward/backward asymmetry of the photoelectrons. Bromopropionic acid (BPA) possesses the Br atom directly bound to the chiral center whereas Bromophenylethylamine (BrPhEtA) isomers possess the Br atom localized far from the chiral center. Interestingly, the highest asymmetry (on the HOMO orbital) is found for the 1,4-BrPhEtA molecule although it possesses the Br atom localized far from the chiral center. Sensitivity of the PECD to isomerism is also observed especially at 9 and 15.5 eV photon energies. Indeed, figure 1 shows the different experimental PECD values observed for the HOMO orbital on the two BrPhEtA isomers as well as a very good mirroring between the 1,3- BrPhEtA enantiomers. Recent Continuum Multiple Scattering (CMS-Xα) calculations are then compared to experimental values at different photon energies and enable in some cases to determine the absolute configuration of the chiral molecule.

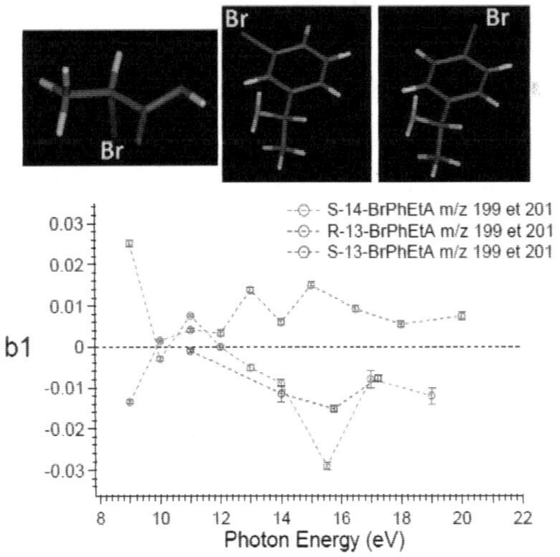

Figure 1. (up) Molecules corresponding to BPA and 1,3- and 1,4-BrPhEtA isomers. (bottom) Dichroic parameter extracted on the HOMO orbital as a function of the photon energy for the two isomers of BrPhEtA.

References

[1] A. Zehnacker (ed.), *Chiral recognition in the gas phase* (CRC Press - Taylor & Francis, New York, 2010).
[2] G. Garcia, B. Cunha de Miranda, M. Tia, S. Daly, and L. Nahon, 2013 Review of Scientific Instruments 84, 053112.

[1]E-mail: tia@atom.uni-frankfurt.de

Content from this work may be used under the terms of the Creative Commons Attribution 3.0 licence. Any further distribution of this work must maintain attribution to the author(s) and the title of the work, journal citation and DOI.
Published under licence by IOP Publishing Ltd

Journal of Physics: Conference Series **635** (2015) 112012 doi:10.1088/1742-6596/635/11/112012

Electron dynamics in the core-excited CS_2 molecule revealed through resonant inelastic x-ray scattering spectroscopy

T. Marchenko[*,†,1], S. Carniato[*,†], L. Journel[*,†], R. Guillemin[*,†], E. Kawerk[*,†], M. Žitnik[‡], M. Kavčič[‡], K. Bučar[‡], R. Bohinc[‡], M. Petric[‡], V. Vaz da Cruz[§], F. Gel'mukhanov[§] and M. Simon[*,†]

[*] CNRS, UMR 7614, Laboratoire de Chimie Physique-Matière et Rayonnement, F-75005 Paris, France
[†] Sorbonne Universités, UPMC Univ Paris 06, UMR 7614, Laboratoire de Chimie Physique-Matière et Rayonnement, F-75005, Paris, France
[‡]Jožef Stefan Institute, Jamova cesta 39, SI-1000 Ljubljana, Slovenia
[§] Theoretical Chemistry & Biology, School of Biotechnology, Royal Institute of Technology, SE-106 91 Stockholm, Sweden

Synopsis We present an experimental and theoretical study of resonant inelastic x-ray scattering (RIXS) in the CS_2 molecule near the S 1s edge. We show that localization of the S 1s core-hole occurs in CS_2 during the RIXS process due to the orientational dephasing of interference between the waves scattering on the two sulfur atoms. Strong evolution of the RIXS profile with the excitation energy far below the first absorption resonance reflects the onset of electron dynamics triggered by a coherent excitation of multiple electronic states..

Excitation of a molecule with a high-energy photon leads to an ultrafast dynamic response involving nuclear and electronic degrees of freedom. Alternatively to the time-resolved measurements using ultrafast laser sources, the nuclear and electron dynamics can be successfully studied in the x-ray energy domain using a technique known as 'core-hole clock' spectroscopy based on core-level excitation and decay [1, 2]. Our previous studies in gas-phase molecules excited at the Iodine L shell [3] show that resonant inelastic x-ray scattering (RIXS) provides information on sub-femtosecond nuclear dynamics in a molecule upon core-hole excitation. To our knowledge, so far there have been no attempts to reveal the electron dynamics in isolated molecules using core-hole clock spectroscopy.

In this study we report on a strong evolution of the RIXS profile with detuning below the first absorption resonance in CS_2 excited at the S 1s shell. The S K_β RIXS spectra of the CS_2 molecule, near the lowest unoccupied molecular orbital resonance $1s \rightarrow 3\pi_u$ at $\hbar\omega$=2470.4 eV, were measured at the ID26 beamline of the European Synchrotron Radiation Facility (ESRF) with the high-resolution X-ray spectrometer developed at the Jožef Stefan Institute in Ljubljana [4].

First, we have elucidated the mechanism responsible for the dominant dipole- and symmetry-forbidden transition from the *gerade* Σ_g^+ electronic ground state to the *ungerade* $|2\pi_g^{-1}3\pi_u^1\rangle\Sigma_u\,,\Delta_u$ final state. Violation of the symmetry selection rules occurs due to localization of the S 1s core-hole in CS_2 during the RIXS process. The reason for the core-hole localization is the orientational dephasing of Young double-slit interference patterns measured in randomly oriented molecules, which is specific to the hard x-ray regime.

Second, a strong evolution of the RIXS profile with the excitation energy tuned down to about 10 eV below the first resonance, is caused by a relative variation of the partial contributions from the different intermediate core-excited states.

Finally, the temporal evolution of the multi-electron wave packet launched by a coherent excitation of multiple electronic states is responsible for the excitation energy dependence of RIXS upon further detuning. Electron dynamics occurs on the characteristic time scale in the attosecond range, determined by the spacing of the coherently excited intermediate states. The RIXS profile is determined by the interplay of the effective scattering duration and the characteristic time of the electron wave packet evolution.

Direct time-resolved measurements of electron dynamics in core-excited molecules require using a short hard x-ray pulse with duration below the core-hole lifetime (~ 1fs), which is so far unavailable. Our study demonstrates the potential of RIXS and core-hole spectroscopy in general, as a promising alternative in application to the studies of electron dynamics in molecules excited at the deep core shells.

References

[1] P. Morin *et al* 1986 *Phys. Rev. Lett.* **56** 1913
[2] O. Bjőrneholm *et al* 1992 *Phys. Rev. Lett.* **68** 1892
[3] T. Marchenko *et al* 2011 *J. Chem. Phys.* **134** 144308
[4] M. Kavčič *et al* 2012 *Rev. Sci. Instrum.* **83** 033113

[1] E-mail: tatiana.marchenko@upmc.fr

Content from this work may be used under the terms of the Creative Commons Attribution 3.0 licence. Any further distribution of this work must maintain attribution to the author(s) and the title of the work, journal citation and DOI.
Published under licence by IOP Publishing Ltd

Journal of Physics: Conference Series **635** (2015) 112013 doi:10.1088/1742-6596/635/11/112013

Adiabatic treatment for vibrational predissociation of water dimers with channel interactions

H. Mineo*[1], S.H. Lin[2], and Y. Fujimura[2, 3]

[1] Institute of Atomic and Molecular Science, Academia Sinica, Taipei 115, Taiwan,
[2] Department of Applied Chemistry, Institute of Molecular Science and Center for Interdisciplinary Molecular Science, National Chiao Tung University, Hsinchu 30010, Taiwan
[3] Department of Chemistry, Tohoku University, Sendai, Japan

Synopsis The results of calculated vibrational predissociation (VPD) of water dimers are presented. The VPD processes considered here are the followings: initially closed channel of symmetric stretching mode is excited by one quantum, and dissociate to two open channels of the bending mode excitation by one quantum in the donor and acceptor units. The VPD rate constants are calculated in the quantum mechanical approach under the adiabatic approximation. The channel interactions were taken into account in and between the initial and final states. The results obtained by quantum-mechanical approach are compared with both experimental and semi-classical trajectory calculation results.

Infrared-induced vibrational predissociation (VPD) of water is considered to be an important reaction dynamics of water, but due to the complicated structure of the water cluster, direct measurement of the IR induced VPD rate constant is difficult. Therefore, the water dimer is expected to play an important role as a smallest water cluster to study the dynamics of VPD. However, the vibrational predissociation mechanism of hydrogen bonding water cluster is less understood at present stage.

The vibrational modes of the donor and accepter H_2O units of water dimer were grouped to high-frequency degrees of freedom which are localized in the donor or acceptor units, and the intermolecular modes were grouped to the low frequency-degrees of freedom in the adiabatic description.

We consider three VPD processes of water dimers $(H_2O)_2$ and $(D_2O)_2$ from the initial channel of the one-quantum excitation of the totally symmetric OH (OD) vibration in the donor water unit, $(1\ 0\ 0)_d + (0\ 0\ 0)_a$ to the two open channels of the bending-mode one-quantum excitation in the donor and acceptor water units, $(0\ 1\ 0)_d + (0\ 0\ 0)_a$ and $(0\ 0\ 0)_d + (0\ 1\ 0)_a$, respectively, and the other is to the open channel of the ground-state water dimer, $(0\ 0\ 0)_d + (0\ 0\ 0)_a$. Subscripts "a" and "d" denote the acceptor and donor water units respectively. $(l\ m\ n)$ specifies the quantum numbers of the totally symmetric stretching, bending, and anti-symmetric stretching modes respectively.

The corresponding VPD rate constants were calculated in the non-Condon scheme beyond the so-called Condon approximation. The VPD is induced by the breakdown of the vibrational adiabatic approximation and the two types of the nonadiabatic coupling matrix elements are involved: the VPD induced by the low-frequency dissociation mode and VPD of the other one through the channel interactions induced by the low-frequency vibrational modes. The channel interactions in the initial and/or final stages were taken into account.

The calculated VPD rates of the $(H_2O)_2$ are slow compared with those of the $(D_2O)_2$ for the case in which the channel interactions were omitted, while these are comparable order for the case in which the channel interactions were taken into account. The results are also compared with the results of the quasi-classical calculations and experimental results [1, 2].

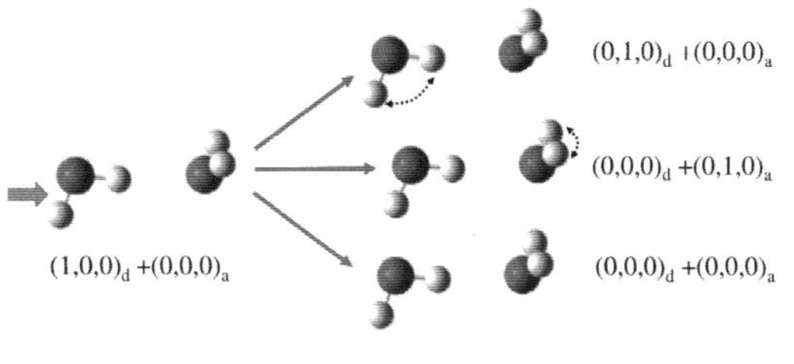

Figure 1. Vibrational VPD of water dimers from the initial channel $(1\ 0\ 0)_d + (0\ 0\ 0)_a$ to the final channels: $(0\ 1\ 0)_d + (0\ 0\ 0)_a$, $(0\ 0\ 0)_d + (0\ 1\ 0)_a$ and $(0\ 0\ 0)_d + (0\ 0\ 0)_a$, are depicted.

References

[1] L. C. Ch'ng, A. K. Samanta, G. Czakó, J. M. Bowman, H. Reisler, 2012 *J. Am. Chem. Soc.* **134** 15430.

[2] G. Czakó, Y. Wang, and J. M. Bowman, 2011 *J.Chem. Phys.* **135**, 151102.

* E-mail: mineo@gate.sinica.edu.tw

Content from this work may be used under the terms of the Creative Commons Attribution 3.0 licence. Any further distribution of this work must maintain attribution to the author(s) and the title of the work, journal citation and DOI.
Published under licence by IOP Publishing Ltd

Journal of Physics: Conference Series **635** (2015) 112014 doi:10.1088/1742-6596/635/11/112014

Resonance parameter calculations for molecular orbitals of H_2O exposed to strong DC electric fields.

Susana Arias Laso and Marko Horbatsch

Department of Physics and Astronomy, York University, M3J 1P3, Toronto, Canada

Synopsis Molecular orbitals for the water molecule obtained from a potential model are represented on a three-dimensional Cartesian mesh. The Schrödinger equation describing the molecule exposed to a strong DC electric field is solved to obtain the resonance parameters (position and line width, *i.e.*, lifetime) as a function of field orientation and for field strengths for which the outer classical turning point is within the computational grid.

The calculation of collisional ionization of water molecules on the basis of time-dependent Schrödinger equation calculations has reached a level of maturity [1,2], such that it warrants further studies, such as the exposure of H_2O vapor to laser fields.

Before proceeding with the program it is worthwhile to follow a path that was employed successfully in the case of H_2^+, i.e., to compute first the resonance parameters of the orbitals in a DC electric field [3], and follow this up with the AC Stark case [4]. In the former case one deals with a stationary Schrödinger problem with complex eigenvalues representing the energy shift (real part) and the decay width (imaginary part).

The problem of the H_2O molecular orbital response to fields is much more complicated than for H_2^+. Due to the multi-center nature of the molecule no variable transformation is known to allow one to deal with the Coulomb singularities. Thus, we resort to an approach which is based upon a discrete variable representation on a Cartesian mesh which employs derivatives on the basis of discrete Fourier transforms (pseudospectral method).

The periodic boundary conditions of the method are tempered by the use of a complex absorbing potential (CAP), which is needed for the resonance calculations [3]. One removes the absorber artefacts by considering the complex-plane trajectory of the eigenvalue as a function of the strength parameter of the CAP. Padé extrapolation to zero strength is complemented with the Riss-Meyer (RM) perturbative scheme at orders 0 (=stabilization) and 1.

The Coulomb potentials are regulated consistent with the lattice spacing of the Fourier method discretization. One needs to repeat the calculations for different lattice spacings to investigate the regulator effect. In order to learn about the prospects of the methodology we employ a potential model introduced by Illescas, *et al* [2].

In Fig. 1 complex-energy trajectories that depend parametrically on the CAP strength parameter are shown for the bonding state $|1b_1\rangle \sim |O(2p_y)\rangle$ exposed to a field $\mathbf{E} = F_0\,(\hat{\mathbf{e}}_x \cos(\alpha) + \hat{\mathbf{e}}_z \sin(\alpha))$ with α=30°, and $F_0 = 4 \times 10^{-2}$ au. The Padé extrapolated result is between the 1st and 2nd order RM trajectories and the width agrees reasonably with the RM-1 stabilized value, but the 2nd order RM results do not stabilize. The data demonstrate considerable, yet surmountable challenges faced by the method. Variation of the field orientation leads to factor-of-two changes in the lifetimes.

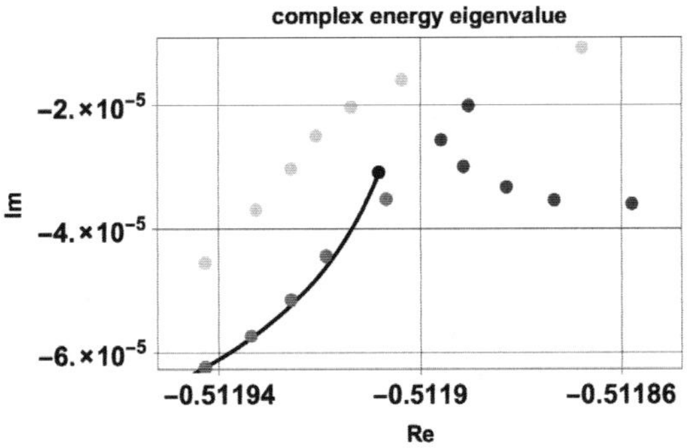

Figure 1. Complex energy trajectories in atomic units (au). Red dots represent the computed complex eigenvalues using a 64^3 Cartesian grid, the blue and green dots are the 1st- and 2nd-order RM energy trajectories, while the black dot represents a 5th order Padé extrapolated trajectory result.

References

[1] M Murakami, *et al* 2012 *Phys. Rev.* **A 85** 052713
[2] C Illescas, *et al* 2011 *Phys. Rev.* **A 83** 052704
[3] Ts Tsogbayar, M. Horbatsch 2013 *J. Phys. B: At. Mol. Opt. Phys.* **46** 085004
[4] Ts Tsogbayar, M. Horbatsch 2013 *J. Phys. B: At. Mol. Opt. Phys.* **46** 245005

Acknowledgment. Work supported by the Natural Sciences and Engineering Research Council, Canada

E-mail: marko@yorku.ca

Content from this work may be used under the terms of the Creative Commons Attribution 3.0 licence. Any further distribution of this work must maintain attribution to the author(s) and the title of the work, journal citation and DOI.
Published under licence by IOP Publishing Ltd

Journal of Physics: Conference Series **635** (2015) 112015 doi:10.1088/1742-6596/635/11/112015

Action spectroscopy of chlorophyll and other coordination complexes.

Mark H Stockett*[1], Jørgen Houmøller*, Christina Kjær*, Bruce F Milne†‡, Lihi Musbat§, Angel Rubio†¶, Yoni Toker§, and Steen Brøndsted Nielsen*

*Department of Physics and Astronomy, Aarhus University, Aarhus, Denmark
†Department of Materials Physics, University of the Basque Country, Donostia, Spain
‡Centre for Computational Physics, Department of Physics, University of Coimbra, Coimbra, Portugal
§Institute of Nanotechnology and Advanced Materials, Bar-Ilan University, Ramat-Gan, Israel
¶Max Planck Institute for the Structure and Dynamics of Matter, Hamburg, Germany

Synopsis Action spectroscopy provides key insights into the nature of electronic transitions of coordination complexes such as porhyrin-containing biochromophores like chlorophyll or transition metal complexes such as tris(bipyridine)ruthenium.

Coordination complexes where a metallic ion core is surrounded by organic ligands are common to important biochromophores as well as synthetic transition metal complexes. These molecules are strong light absorbers but the nature of their electronic transitions, as well as the role of the protein or solvent micro-environment, is not always obvious. Studying these molecules in the gas phase or in nano-solvated complexes significantly reduces these complexities and allows more direct comparison with theoretical calculations. Here, we provide two recent examples of how gas-phase action spectroscopy can provide clarity to otherwise intractable problems.

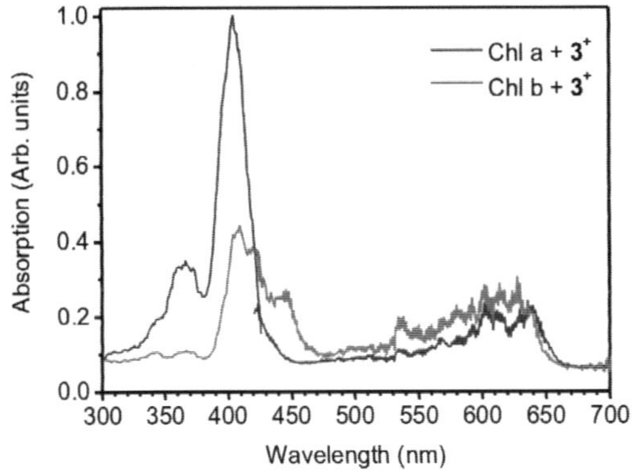

Figure 1. Photodissociation action spectra of Chl *a* and *b* tagged with acetylcholine.

Chlorophyll (Chl) *a* and *b* are the light-absorbers of plants and are composed of a porphyrin macrocyle with a divalent magnesium ion in the center. It is nontrivial to predict the absorption spectra of these molecules or what effect the protein environment may have. As a starting point, we have measured the absorption spectra of *gas-phase* Chl *a* and *b* using photodissociation action spectroscopy of Chl molecules tagged with quaternary ammonium ions. Chl *a* and *b* differ only on one peripheral substituent, but their absorption band maxima are shifted relative to each other by more than 30 nm. These shifts were reproduced in our gas-phase data, shown in Figure 1, meaning that they are intrinsic effects and not due to interactions with the micro-environment. On the other hand, the bands of both Chl *a* and *b* were significantly blue-shifted relative to plant-phase spectra. Interactions with the protein and solvent environments are thus crucial to understanding the absorption of chlorophylls.

Tris(bipyridine)ruthenium, $Ru(bipy)_3{}^{2+}$, has become a textbook case for the chemistry of coordination complexes over several decades of extensive study. The lowest-energy electronic transition involves a metal-to-ligand charge transfer (MLCT) from the divalent ruthenium core to the bipyridine ligands. The character of the MLCT transition has been an important, long-standing question. The permanent dipole moment of $Ru(bipy)_3{}^{2+}$ is zero in the ground state, but in the excited state depends on whether the symmetry of the electronic structure is broken by localization of the photoexcited electron to just one of the bipyridine ligands, rather than delocalization over all three. We attempt to elucidate this unresolved debate by comparing the photodissociation action spectra of gas phase $Ru(bipy)_3{}^{2+}$ with that of the monosolvated $[Ru(bipy)_3{+}CH_3CN]^{2+}$ complex. We find that the addition of a single solvent molecule (in this case acetonitrile CH_3CN) has no effect on the position of the band maximum. We conclude that, in the gas phase, the transition is delocalized and that a single solvent molecule is insufficient to break the symmetry of the transition.

[1]E-mail: stockett@phys.au.dk

Content from this work may be used under the terms of the Creative Commons Attribution 3.0 licence. Any further distribution of this work must maintain attribution to the author(s) and the title of the work, journal citation and DOI.
Published under licence by IOP Publishing Ltd

Journal of Physics: Conference Series **635** (2015) 112016 doi:10.1088/1742-6596/635/11/112016

Angular correlation measurements of a pair of Lyman-α photons emitted in the photodissociation of H_2

K. Hosaka[a1], Y. Nakanishi[a], R. Kogo[a], P. Schmidt[b], Y. Torizuka[a], K. Shigemura[a], T. Odagiri[d], A. Knie[b], K. Jänkälä[c], A. Ehresmann[b], M. Nakano, Y. Kumagai, K. Shiino, M. Kitajima[a] and N. Kouchi[a]

[a] Department of Chemistry, Tokyo Institute of Technology, Meguro-ku, Tokyo 152-8551, Japan
[b] Institute of Physics, University of Kassel, D-34132 Kassel, Germany
[c] Department of Physical Sciences, University of Oulu, PO Box 3000 90014 Oulu, Finland
[d] Department of Materials and Life Sciences, Sophia University, Chiyoda-ku, Tokyo 102-8554, Japan

Synopsis The angular correlation functions of a pair of Lyman-α photons in the photodissociation of H_2 are measured more precisely than in the early experiments [Nakanishi *et. al., Phys. Rev. A.* **90**, 043405(2014)]. The linearly polarized light is used to excite the photodissociation via the doubly excited $Q_2\ ^1\Pi_u(1)$ state.

A pair of $H(2p)$ atoms produced in the photodissociation of H_2 [1] is potentially entangled [2,3],

$$H_2(X^1\Sigma_g^+)+\gamma_{ex}\rightarrow H_2(Q_2\,^1\Pi_u(1))\rightarrow H(2p_0)+H(2p_{\pm1})$$
$$\rightarrow H(1s)+H(1s)+\gamma_{Ly\text{-}\alpha}+\gamma_{Ly\text{-}\alpha}, \quad (1)$$

where γ_{ex} is the incident photon and $\gamma_{Ly\text{-}\alpha}$ the Lyman–α photon. The subscripts of 0 and ±1 attached to $2p$ are the magnetic quantum numbers m with respect to the internuclear axis. The angular correlation function (ACF) of the Lyman-α photon-pair shows the features due to the entanglement in the $H(2p)$ pair [2,3]. Nakanishi *et. al.* [4] recently measured the ACF of the Lyman-α photon-pair in process (1) mediated by the linearly polarized light. They found that the ACFs at pressures of hydrogen gas lower than 1 Pa were free from the reactions $H(n = 2) + H_2$ and were able to be compared with the theoretical predictions. Their experiments were carried out at the undulator beamline BL28B and the bending beamline BL20A of the Photon Factory, KEK. The undulator line is much more advantageous than the bending line. In this paper we report the ACFs of the Lyman-α photon-pair in process (1) at the undulator line BL28B at hydrogen gas pressures of 0.1 and 1 Pa. The ACF at 1 Pa is taken from [4].

In Fig.1 the experimental ACFs at ~1 Pa[4] and 0.12 Pa are shown. As expected the period of 180° is seen, indicating the good alignment during the present experiments. The present ACFs are in good agreement with those in [4].

The blue solid [2] and red dashed [3] curves show the theoretical ACFs for the entangled $H(2p)$ pair. Miyagi *et. al.* [2] considered the possibility that the entanglement in $H(2p)$ pair is violated during the dissociation and $H(2p)$ pairs with definite magnetic quantum number of each atom are produced. The ACF for such a case is calculated and shown by the black dash-dotted curve [2]. Those three curves represent the ACFs convoluted with the angular resolution. Though the $H(2p)$ pair-state in process (1) still remains unknown in detail, we could exclude the possibility that the pair of $H(2p)$ atoms is in the definite states of magnetic quantum number of each atom relative to the internuclear axis, i.e. the possibility of the black dash-dotted curve.

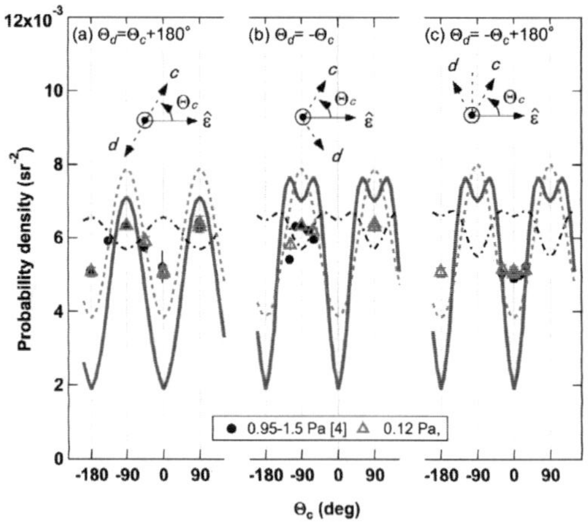

Figure 1. The Experimental and theoretical ACFs of the Lyman-α photon-pair. The pressure indicates H_2 gas pressure in the gas cell. See the text for details.

References

[1] T. Odagiri *et al* 2004 *J. Phys. B* **37** 3909
[2] H. Miyagi *et al* 2007 *J. Phys. B* **40** 617
[3] K. Jänkälä *et al* 2010 *J. Phys. B* **43** 065104
[4] Y. Nakanishi *et al* 2014 *Phys. Rev. A.* **90** 043405

[1] E-mail: hosakak@chem.titech.ac.jp

Content from this work may be used under the terms of the Creative Commons Attribution 3.0 licence. Any further distribution of this work must maintain attribution to the author(s) and the title of the work, journal citation and DOI.
Published under licence by IOP Publishing Ltd

XXIX International Conference on Photonic, Electronic, and Atomic Collisions (ICPEAC2015) IOP Publishing
Journal of Physics: Conference Series **635** (2015) 112017 doi:10.1088/1742-6596/635/11/112017

Field free alignment of hexapole state selected methyliodide

W.G. Roeterdink[1], **L. He**[2], **D. Ding**[3]

Institute of Atomic and Molecular Physics, Jilin University, Changchun 130012, PR China

Synopsis In this contribution we will discuss the field free laser induced spatial alignment of hexapole quantum state selected methyl iodide. We will demonstrate that we have control over the spatial alignment and can chose $\langle P_2(\cos\theta)\rangle$ between 0.7 and -0.1 coupled to $\langle P_4(\cos\theta)\rangle$ between 0.3 and 0.0. The effect of the relative orientation between a static electric field and a strong laser field on the structure of the field free rotational alignment revivals will be discussed.

In molecular frame dynamics studies the molecule needs to be oriented or aligned in the laboratory frame to a sufficient degree. Spatial alignment of molecules with intense laser pulses has been demonstrated for cold molecules in a molecular beam expansion [1]. However, the inherent problem of thermal averaging has a lowering effect on the maximum degree of alignment that can be achieved. Hexapole state selection can be used to prepare (a)symmetric top molecules in their lowest rotational state and the observed alignment increases to the max/min values of $\langle P_2(\cos\theta)\rangle$ 0.7 and -0.1 and $\langle P_4(\cos\theta)\rangle$ between 0.3 and 0.0.

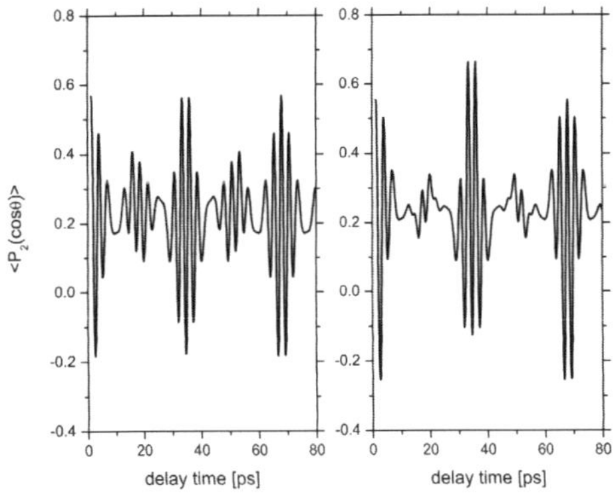

Figure 1. Revival structure of CH_3I, $|JMK\rangle = |1 \pm 1 \mp 1\rangle$ for the parallel configuration (left panel) and perpendicular configuration (right panel).

In the experimental configuration for ion (velocity map) imaging the extraction field of the time-of-flight detector is perpendicular to the laser polarization in order to preserve the cylindrical symmetry needed for Abel inversion of the obtained projected images. The state selected molecules have a spatially non-isotropic distribution and will be oriented and aligned along the direction of the static electric extraction field. The rotational wave packet revival transient is strongly dependent on the angle between the static electric extraction field and the polarization of the laser, see figure 1 [2]. The simulated revival structure is in good agreement with the experimental observed transient in the perpendicular geometry. Velocity map ion images are recorded at the peak positions at the half revival time (33.3 ps) and the full revival time (66.5 ps), see figure 2. In principal the cylinder symmetry around the laser polarization is broken due to the static electric field. However, our simulations show that at sufficient high laser intensities the symmetry is restored and that the requirements for the inverse Abel transform are fore filled.

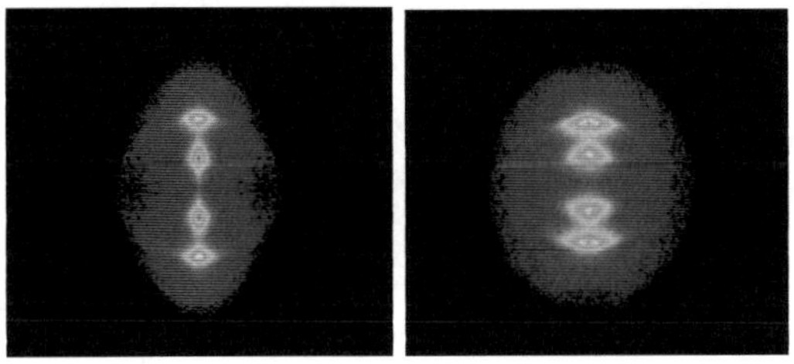

Figure 2. Velocity map ion images recorded at the half revival time where reaching a maximum alignment (left image) and the full revival time (right image).

The angular distribution observed in the ion images is a convolution of the recoil dynamics of the fragments and the spatial alignment of the parent molecule. The experimental obtained results will be discussed and compared to the simulated values.

References

[1] E. Hamilton *et al* 2005 *Phys. Rev. A* **72** 043402

[2] L.He *et al* 2015 *In preparation.*

[1]E-mail: wim@jlu.edu.cn
[2]E-mail: helh13@mails.jlu.edu.cn
[3]E-mail: dajund@jlu.edu.cn

Content from this work may be used under the terms of the Creative Commons Attribution 3.0 licence. Any further distribution of this work must maintain attribution to the author(s) and the title of the work, journal citation and DOI.
Published under licence by IOP Publishing Ltd

Journal of Physics: Conference Series **635** (2015) 112019 doi:10.1088/1742-6596/635/11/112019

H_n^+ Ejection from Aligned CH_3Cl Molecules in Intense *fs* Laser Fields

Sizuo Luo[1*], Lanhai He[*], Xiaokai Li[*], Wenhui Hu[*], Pan Ma[*], Chuncheng Wang* and Dajun Ding[2*]

[*] *Institute of Atomic and Molecular Physics, Jilin University, Changchun 130012, P R China*

Synopsis We report experimental results on strong-field laser interact with aligned CH_3Cl molecules, the ejection of fragments H_n^+ dependent on alignment of CH_3Cl molecules has been studied. The mechanism of fragmentation builds on the alignment sensitive ionization and dissociation from inner and outer valence orbitals or the excitation process of molecular have been discussed.

The study on the ejection of H_n^+ from Hydrocarbon molecules is very important which is related to many photophysical processes such as proton migration, deformation, and other ultrafast processes [1]. This phenomenon has been studied for different molecules in intense laser fields, such as CH_3OH, C_mH_n, and CH_3Cl [2]. For further understanding of the underlying mechanism, a deep investigation with elaborated measurement is required.

We have studied the different fragment ejections, using CH_3Cl as an example, and focus on the effect of orbitals or the excitation process on the fragment ejection of polyatomic molecules in intense laser initiate dynamics. We performed the experiment for field-free aligned CH_3Cl molecules from *fs* laser impulsive interaction. We align the rotation state selected CH_3Cl from a hexapole Stark selector by using a linearly polarized laser, with a stretched pulse of 300 *fs*, at the intensity of 3×10^{12} W/cm^2. The aligned molecules were ionized or dissociated by a 50 *fs* laser as a probe in the intensity range from 1×10^{13}- 4×10^{14} W/cm^2, and the alignment dependence of strong field ionization and dissociation yields were measured, as shown in Figure 1.

The alignment dependence of fragment ions Cl^{2+}, H_n^+ (n=1,2,3) are given in Fig. 1 (a)-(d), respectively. It is interesting to notice that H^+ and H_2^+ show the same alignment dependence as the case of Cl^{2+}, giving a larger yield in the direction of the probe laser polarization parallel to the molecular axis. But H_3^+ gives a prefer yield in the direction of the probe laser polarization perpendicular to the molecular axis, just like the parent ion. The ion yields H_n^+ of alignment and anti-alignment as a function of the laser intensity were also measured and the results are given in Figure 1(e)-(g). For H^+ and H_2^+, the ion yield for the aligned molecules is larger than that for anti-aligned molecules almost for the whole laser intensity region. The signal of H_3^+ is almost the same when laser intensity below 1.6×10^{14} TW, and as increasing the laser intensity the signal is stronger when probe laser is perpendicular than parallel to the molecular axis. Futhermore, H_n^+ ejection was controlled by manipulating the molecular alignment. The same dependence of the ejected H_3^+ and CH_3Cl^+ on the alignment observed indicates that H_3^+ is ejected from CH_3Cl^{2+}, produced by sequence double ionization.

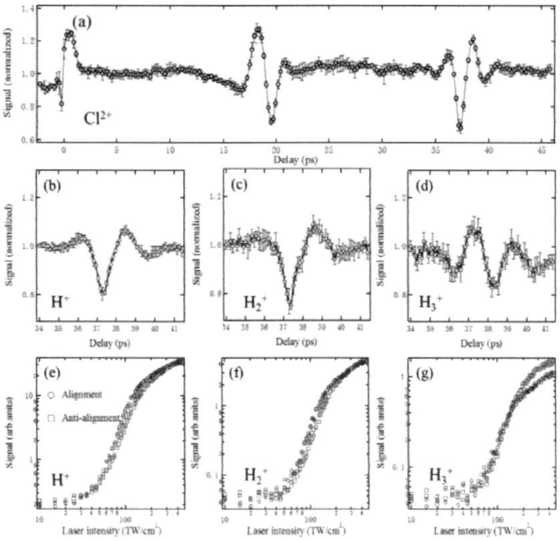

Figure 1. The delay dependent of the yields of signal produced from different fragmentation channels Cl^{2+}, H_n^+ (n=1,2,3) (a)-(d). (e)-(g) is ion intensities of H_n^+ (n=1,2,3) dissociated from aligned and anti-aligned molecules as a function of laser intensity.

References

[1] Y. H. Jiang *et al* 2010 *Phys. Rev. Lett.* 105 263002.

[2] E. Wells *et al* 2013 *Nat. Comms* 4 2895.

[1]E-mail: luoszl2@mails.jlu.edu.cn

[2]E-mail: dajund@jlu.edu.cn

Content from this work may be used under the terms of the Creative Commons Attribution 3.0 licence. Any further distribution of this work must maintain attribution to the author(s) and the title of the work, journal citation and DOI.
Published under licence by IOP Publishing Ltd

XXIX International Conference on Photonic, Electronic, and Atomic Collisions (ICPEAC2015) IOP Publishing
Journal of Physics: Conference Series **635** (2015) 112020 doi:10.1088/1742-6596/635/11/112020

Diffraction effects in the Recoil-Frame Photoelectron Angular Distributions of Halomethanes

Cédric Bomme*[1], Denis Anielski*,†, Evgeny Savelyev*, Rebecca Boll*,†, Benjamin Erk*, Sadia Bari•, Jens Viefhaus*, Mauro Stenerʃ, Piero Declevaʃ and Daniel Rolles*,$[2]

* Deutsches Electronen-Synchrotron(DESY), Hamburg, Germany
† Max-Planck-Institut f. Kernphysik, Heidelberg, Germany
•European XFEL GmbH, Hamburg, Germany
ʃ Universita' di Trieste, Trieste, Italy
$J. R. Macdonald Laboratory, Department of Physics, Kansas State University, Manhattan, KS, USA

Synopsis We have measured the Recoil Frame - Photoelectron Angular Distributions (RF-PADs) for inner-shell photoionization of CH_3F, CH_3I and CF_3I halomethane molecules for photoelectron energies up to 300 eV detected within a 4π solid angle in the gas-phase. For high kinetic energies, the RF-PADs are dominated by diffraction effects that encode information on the molecular geometry.

Recoil frame - photoelectron angular distributions supply deep insights into the molecular photoionization process and provide access to an unparalleled level of detailed information such as phases of photoelectron waves [1, 2], localization of core holes [3], and double-slit interference [4, 5]. RF-PADs can also be interpreted in terms of photoelectron diffraction [6, 7] and direct information on the geometric and electronic structure of the molecule can be obtained, e.g., by comparing the measured diffraction patterns and RF-PADs to single and multiple scattering calculations [8, 9, 10, 11, 12].

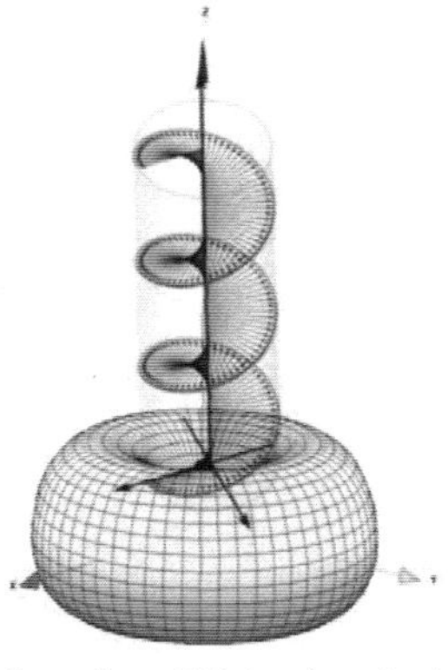

Figure 1: RF-PADs after F(1s) photoionization of CH_3F at 875eV photon energy, i.e. 180eV photoelectron energy. Light propagation axis and molecular F-C axis are parallel.

For high kinetic energies of 50eV and above, the RF-PADs contain diffraction effects that encode information on the molecular geometry in the RF-PADs. In order to illustrate this in more detail, we are comparing our experimental results to single and multiple scattering calculations similar to those used in earlier studies [10, 11, 12] and to DFT calculations [13, 14].

References:

[1] O. Geßner *et al.*, Phys. Rev. Lett. **88**, 193002
[2] S. Motoki *et al.*, Phys. Rev. Lett. **88**, 063003
[3] M. Schöffler *et al.*, Science **320**, 929
[4] D. Rolles *et al.*, Nature **437**, 711
[5] D. Akoury *et al.*, Science **318**, 949
[6] A. Landers *et al.*, Phys. Rev. Lett. **87**, 013002
[7] B. Zimmermann *et al.*, Nature Phys. **4**, 649
[8] D. Dill *et al.*, J. Chem. Phys. **65**, 3158
[9] R. Díez Muiño *et al.*, J. Phys. B **35**, L359
[10] A.V. Golovin *et al.*, J. Phys. B **38**, L63
[11] A.V. Golovin *et al.*, J. Phys. B **38**, 3755
[12] D. Rolles *et al.*, Proc. Of SPIE, Vol. 9198
[13] C. Bomme *et al.*, J. Phys. B **45**, 194005
[14] M. Stener *et al.*, J. Chem. Phys. **140**, 044305

[1]E-mail: cedric.bomme@desy.de
[2]E-mail: daniel.rolles@desy.de

Content from this work may be used under the terms of the Creative Commons Attribution 3.0 licence. Any further distribution of this work must maintain attribution to the author(s) and the title of the work, journal citation and DOI.
Published under licence by IOP Publishing Ltd

Journal of Physics: Conference Series **635** (2015) 112021 doi:10.1088/1742-6596/635/11/112021

Ionization of the hydrogen molecular ion H_2^+ by twisted light

A. A. Peshkov*[1], S. Fritzsche*,† and A. Surzhykov*

* Helmholtz-Institut Jena, Fröbelstieg 3, 07743 Jena, Germany

† Universität Jena, Institut für Theoretische Physik, 07743 Jena, Germany

Synopsis The ionization of the hydrogen molecular ion H_2^+ by twisted Bessel light has been theoretically investigated. Special attention is paid to scenario where the twisted radiation interacts with a macroscopic target which consists of randomly distributed but aligned H_2^+ molecules. We demonstrate for this case that the angle-differential photoionization cross section is sensitive to the transverse momentum of the incident light and alignment angle of molecules.

Interference effect, similar to what was observed in classical Young's double-slit experiment, occurs also in photoionization of diatomic molecules [1]. Electrons can be ejected from different atomic centres, and difference in their path length to detector leads to interference pattern of the photoionization cross section. Up to the present, Young-type interference was studied mainly for a plane wave incident radiation. However, due to recent developments in optics, ionization of molecular targets may be explored also for twisted (or vortex) light. In contrast to the plane wave photons, the twisted beams carry a non-zero projection of the orbital angular momentum (OAM) onto their propagation direction. One can expect that such a non-zero OAM may affect the ionization cross section and the emission pattern of the photoelectrons.

We have recently studied the ionization of the hydrogen molecular ion H_2^+ by twisted Bessel light. Such a beam propagates along the quantization axis and has well-defined longitudinal momentum k_z and absolute value of the transverse momentum κ, which define the so-called opening angle $\theta_k = \arctan(\kappa/k_z)$ [2, 3]. Theoretical analysis is performed within the non-relativistic framework and first-order perturbation approach. An experimental scenario was considered in which twisted light collides with a macroscopic target of aligned molecules. The angle-differential photoionization cross section $d\sigma/d\Omega$, obtained for this case, is presented in Fig. 1. Here, we calculated $d\sigma/d\Omega$ as a function of the photon energy, for the photoelectron emission perpendicular to the incident light direction and within the molecular plane. As seen from the figure, the cross section is very sensitive to the opening angle θ_k of the twisted light. For small opening angle $\theta_k = 5^o$, $d\sigma/d\Omega$ exhibits clear oscillatory behavior, similar to what is expected for the plane-wave case. In contrast, for $\theta_k = 30^o$ the energy behavior of the differential photoionization cross section becomes much smoother, especially for $\hbar\omega > 1000$ eV. This effect occurs due to the fact that twisted light has an inhomogeneous transverse field distribution, and therefore each atomic center interacts with a different intensity of the beam.

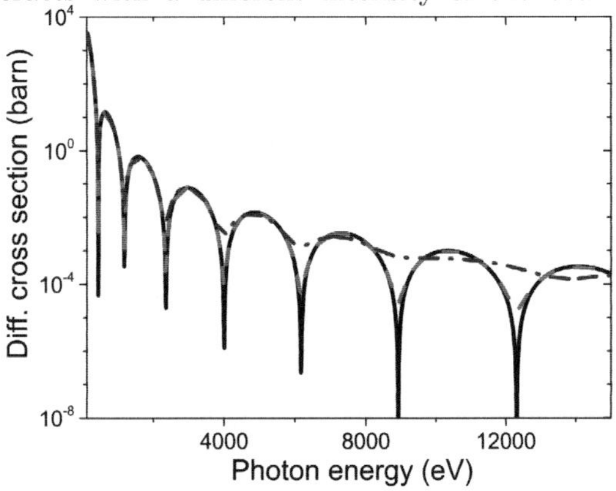

Figure 1. Differential photoionization cross section as a function of incoming photon energy. Black solid line corresponds to the plane-wave case; red dashed and blue dot-dashed lines describe the results obtained for the twisted light with opening angles $\theta_k = 5^o$ and $\theta_k = 30^o$ respectively. The target molecule is aligned along the axis, tilted by the angle $\alpha = 45^o$ with respect to the wave propagation direction.

References

[1] H. D. Cohen and U. Fano 1966 *Phys. Rev.* **150** 30

[2] O. Matula, A. G. Hayrapetyan, V. G. Serbo, A. Surzhykov and S. Fritzsche 2013 *J. Phys.* **46** 205002

[3] A. Surzhykov, D. Seipt, V. G. Serbo and S. Fritzsche 2015 *Phys. Rev. A* **91** 013403

[1]E-mail: antvor1991@gmail.com

Content from this work may be used under the terms of the Creative Commons Attribution 3.0 licence. Any further distribution of this work must maintain attribution to the author(s) and the title of the work, journal citation and DOI.
Published under licence by IOP Publishing Ltd

XXIX International Conference on Photonic, Electronic, and Atomic Collisions (ICPEAC2015) IOP Publishing
Journal of Physics: Conference Series **635** (2015) 112022 doi:10.1088/1742-6596/635/11/112022

Interatomic Coulombic decay in polyatomic rare-gas clusters.

N. Sisourat* [1]

* Laboratoire de Chimie Physique - Matière et Rayonnement, Université Pierre et Marie Curie, Paris, France

Synopsis A new method for a complete theoretical description of Interatomic Coulombic decay (ICD) in large polyatomic rare-gas clusters will be presented. This original method combines the projection-operator formalism of resonant scattering theory, the diatomics-in-molecules technique and a surface hopping algorithm. Such combined approach has fairly low computational costs and constitutes an efficient tool for studying ICD in polyatomic clusters. Benchmark examples will be given to illustrate the method. Results on ICD in large rare-gas clusters will finally be reported.

Interatomic (molecular) Coulombic decay (ICD) is an ultrafast non-radiative electronic decay process for excited atoms or molecules embedded in a chemical environment [1]. Via ICD, the excited system can get rid of the excess energy, which is transferred to one of the neighbors and ionize it. ICD produces two charged particles next to each other and thus leads to Coulomb explosion (see Fig.1).

In the last decade, ICD has attracted considerable and rising attention [2]. One of the reasons for the current broad interest in ICD is the possibility of analytical applications. For example, Auger effect is widely used to probe the atomic composition and the environment of the atoms of materials via Auger electron spectroscopy. ICD is clearly superior to the Auger effect in its sensitivity to the chemical environment. Indeed, here the decay itself is possible only due to the effect of the environment, even if the interaction of the ionized atom or molecule with the environment is very weak.

In order to develop such ICD spectroscopy a better knowledge of this effect in large polyatomic systems is necessary. From the theoretical point of view general quantum mechanical equations for describing the decay processes and the subsequent fragmentations are known but are only applicable for rather small systems. Approximate, but accurate, methods are therefore needed to go beyond benchmark systems.

To this end, an original approach for a complete theoretical description of ICD in polyatomic rare-gas clusters will be presented. This approach combines the projection-operator formalism of resonant scattering theory, the diatomics-in-molecules technique and a surface hopping algorithm. It thus involves a classical treatment for the nuclear motion while retaining a quantum description for the electron dynamics [3]. The potential energy and the decay widths surfaces are computing on-the-fly using the diatomics-in-molecules technique. Such combined approach has low computational costs and can be used to study much larger systems. Some examples to illustrate the method will be given. Results on ICD in large rare-gas clusters will finally be reported.

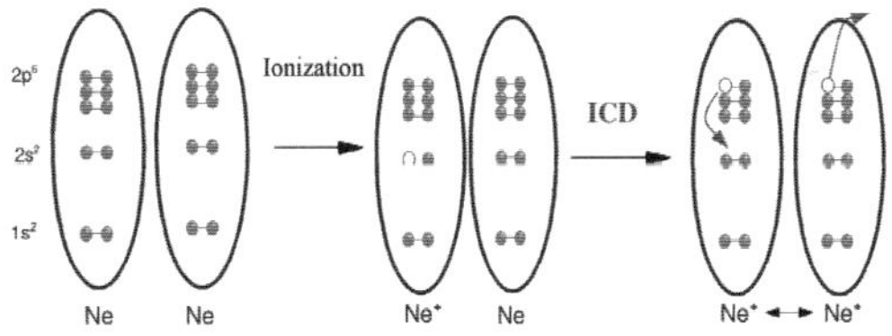

Figure 1. Scheme of the ICD process in Neon dimer.

References

[1] L. S. Cederbaum, J. Zobeley and F. Tarantelli 1997 *Phys. Rev. Lett.* **79** 4778

[2] see http://www.pci.uni-heidelberg.de/tc/usr /icd/ICD.refbase.html for a complete list of ICD publications

[3] N. Sisourat 2013 *J. Chem. Phys.* **139** 074111

[1]E-mail: Nicolas.Sisourat@upmc.fr

Content from this work may be used under the terms of the Creative Commons Attribution 3.0 licence. Any further distribution of this work must maintain attribution to the author(s) and the title of the work, journal citation and DOI.
Published under licence by IOP Publishing Ltd

XXIX International Conference on Photonic, Electronic, and Atomic Collisions (ICPEAC2015) IOP Publishing
Journal of Physics: Conference Series **635** (2015) 112023 doi:10.1088/1742-6596/635/11/112023

Auger-intercoulombic hybridized decay resonances in Kr@C_{60}

Maia Magrakvelidze* [1], Ruma De*, Steve Manson†, Himadri Chakraborty† [2]

* Department of Natural Sciences, D.L. Hubbard Center for Innovation and Entrepreneurship, Northwest Missouri State University, Maryville, Missouri 64468, USA

† Department of Physics and Astronomy, Georgia State University, Atlanta, Georgia 30303, USA

Synopsis We predict resonant hybridization of Auger and intercoulombic decay (ICD) processes in the photoionization of Kr@C_{60} using the time-dependent local density approximation (TDLDA) for the calculation.

For a combined binary system, AB, consider a precursor excitation forming an inner-shell vacancy created by promoting an inner electron of system A to an excited state by either electromagnetic radiation or charged-particle impact. This vacancy can then relax non-radiatively and the released energy can emit an outer electron of A itself in a regular resonant Auger decay (RAD) process. If, however, this released energy is transferred through the coulomb interaction and ionizes an outer electron of the companion system B, then this exotic decay process is known as the resonant intercoulombic decay (RICD). RICD signatures were measured in a number of experiments on dimers and clusters systems over the last decade. One most recent measurement detects a strong enhancement of the $HeNe^+$ yield, as He resonantly couples with the radiation [1].

If the inner-shell vacancy of the system AB decays resulting in the emission from an outer level which as a hybrid level between A and B, then the process entails a coherent superposition of RAD and RICD. This process was first predicted recently in the photoionization of Ar@C_{60} endohedral molecule [2], and was called resonant hybrid Auger-intercoulombic decay or RHA-ICD. In this work we report the prediction of RHA-ICD in the Kr@C_{60} molecule, suggesting the general abundance of this energy-transfer decay mode for atom-cluster or atom-nanostructure binary systems.

A jellium based time-dependent local density approximation (TDLDA) formulation [3], with the Leeuwen and Baerends exchange-correlation functional [4] to produce accurate asymptotic behavior for ground and continuum states, is employed to calculate the dynamical response of the system to the incident photon. Significant ground-state hybridization of Kr $4p$ orbital with C_{60} $3p$ is found to produce ($4p$ Kr) $\pm$ ($3p$ C_{60}) bonding and anti-bonding hybrid orbitals of the compound. Photoionization cross sections from these orbitals are shown in Figure 1 which show two hybrid resonances (RHA-ICD features) corresponding to the decay of the holes resulted from Kr $4s@ \rightarrow 5p@$, $6p@$ inner excitations; "@" denotes states in the combined system. Other structures in these two curves are RHA-ICDs from C_{60} inner-shell excitations. The corresponding pure RADs for $4p$ ionization of free Kr and $3p$ ionization of empty C_{60} are also shown. Comparisons indicate (i) a systematic red-shift in RHA-ICDs and (ii) a substantial increase in resonance strengths.

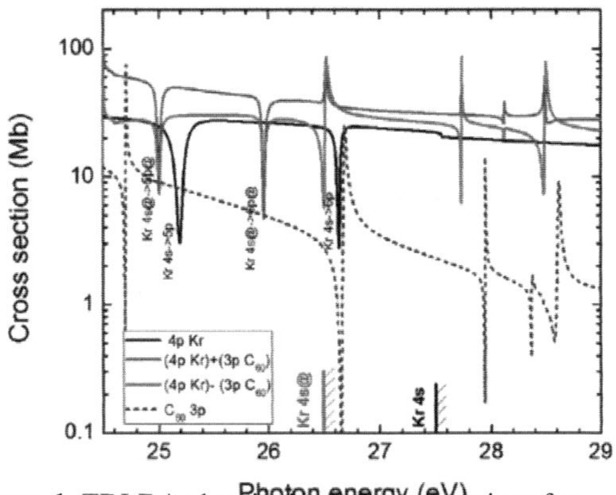

Figure 1. TDLDA photoionization cross sections for two hybrid states in Kr@C_{60}. Free Kr $4p$ and empty C_{60} $3p$ cross sections are also shown.

The work is supported by NSF and DOE, Basic Energy Sciences, USA.

References

[1] F. Trinter *et al* 2013 *Phys. Rev. Lett.* **111** 233004
[2] M. Javani *et al* 2014 Phys. Rev. A **89** 063420
[3] M. Madjet *et al* 2010 *Phys. Rev. A* **81** 013202
[4] Leeuwen & Baerends 1994 *Phys. Rev. A* **49** 2421

[1] E-mail: maia@nwmissouri.edu [2] E-mail: himadri@nwmissouri.edu

Content from this work may be used under the terms of the Creative Commons Attribution 3.0 licence. Any further distribution of this work must maintain attribution to the author(s) and the title of the work, journal citation and DOI.
Published under licence by IOP Publishing Ltd

Photo-induced ultrafast dissociation following deep-core-electron excitation

Oksana Travnikova[*◇ 1], **Tatiana Marchenko**[*◇], **Gildas Goldsztejn**[*◇], **Renaud Guillemin**[*◇], **Loïc Journel**[*◇], **Denis Céolin**[‡], **Kari Jänkälä**[∀], **Ralph Püttner**[†], **Hiroshi Iwayama**[∪], **Eiji Shigemasa**[∪], **Stéphane Carniato**[*◇], **Maria Novella Piancastelli**[*◇≀] **and Marc Simon**[*◇‡]

[*] CNRS, UMR 7614, Laboratoire de Chimie Physique-Matière et Rayonnement, F-75005 Paris, France
[◇] Sorbonne Universités, UPMC Univ Paris 06, UMR 7614, Laboratoire de Chimie Physique-Matière et Rayonnement, F-75005 Paris, France.
[∀]Department of Physics, University of Oulu, P. O. Box 3000, 90014 Oulu, Finland
[†] Institut für Experimentalphysik, Freie Universität Berlin, D-14195 Berlin, Germany
[‡] Synchrotron SOLEIL, l'Orme des Merisiers, Saint-Aubin, F-91192 Gif-sur-Yvette Cedex, France
[≀] Department of Physics and Astronomy, Uppsala University, SE-75120 Uppsala, Sweden
[∪] Ultraviolet Synchrotron Orbital Radiation Facility, Institute for Molecular Science, Okazaki 444-8585, Japan

Synopsis Creation of deep core holes leads to extensive nuclear dynamics on a few femtosecond timescale despite the very short ($\tau \leq 1$ fs) lifetime of such states. This is because the 1st steps of the relaxation processes (*i.e.* both radiative and non-radiative decays) generate intermediate states with one and multiple holes in core orbitals. As an example, ultrafast dissociation is observed in three well-distinguishable LVV Auger decay channels for HCl following Cl$1s \rightarrow \sigma^*$ excitation.

Absorption of an X-ray photon by a molecule may lead to the excitation of a localised core electron to a specific unoccupied valence orbital if the energy of the photon matches exactly the difference between the involved core and valence electronic levels. The created core-hole states are highly unstable and eventually decay on a very short timescale emitting a photon (radiative decay) or a so-called *Auger* electron (non-radiative or *Auger* decay).

Hard X-ray photons (>1keV) may reach deeper-lying core electrons. The lifetime (τ) of deep-core-hole states is very short – of the order of 1 fs or below. Nevertheless, the signature of nuclear dynamics was observed for Cl$1s \rightarrow \sigma^*$ ($\tau \approx 1$ fs) core-excited states of HCl [1] and even on a sub-femtosecond timescale ($\tau \approx 200$ attoseconds) in the case of I$2p \rightarrow 15a_1^2$ core electron transitions in CH_3I [2].

The electronic relaxation dynamics of deep-core-hole states is very rich. At variance with that, the very short lifetimes of these states do not allow for extensive nuclear dynamics to take place before electronic relaxation occurs. However, the dominant channels of the 1st step relaxation processes (both radiative and *Auger* decays) lead to intermediate states which bear 1 or 2 holes in core-electron shells. In the HCl molecule, the leading relaxation decays of the Cl$1s^{-1}\sigma^*$ state are Auger KLL and radiative Kα channels, which create Cl$2p^{-2}\sigma^*$ and Cl$2p^{-1}\sigma^*$ intermediate states, respectively. The latter one can be created by direct soft X-ray absorption and has been well characterised [3]. It is known to undergo ultrafast dissociation (UFD) within the Cl$2p^{-1}$ lifetime of ~8 fs [3, 4]. The former double core-hole Cl$2p^{-2}\sigma^*$ states are yet exotic and can be also created as so-called *"super"*-satellites [5] of direct $2p^{-2}$ double core-hole ionisation. Very recent theoretical studies [5] show that the energy gradients of the *core*$^{-2}$V states can be very large (3 times larger compared to the *core*$^{-1}$V state in the case of H_2O [5]). Therefore, nuclear dynamics is correspondingly faster in *core*$^{-2}$V *"super"*-satellites.

Our experimental measurements reveal UFD phenomena in Cl$2p^{-2}\sigma^*$ state of HCl with τ only about 3 fs. The measured LVV *Auger* decay spectrum of HCl clearly shows 3 possible decay channels following Cl$1s \rightarrow \sigma^*$ excitation. In fact, ultrafast dissociation is observed in *every* step of these LVV decay channels before the next electronic relaxation takes place. The observed results are supported by *ab initio* theoretical calculations.

References

[1] M. Simon *et al* 2006 *Phys. Rev. A* **73** 020706(R)

[2] T. Marchenko *et al* 2011 *J. Chem. Phys.* **134** 144308

[3] O. Björneholm *et al* 1997 *Phys. Rev. Lett.* **79** 3150

[4] O. Travnikova *et al* 2013 *J. Phys. Chem. Lett.* **4** 2361

[5] S. Carniato *et al* 2015 *J. Chem. Phys.* **142** 014307

[1]E-mail: oksana.travnikova@upmc.fr

Content from this work may be used under the terms of the Creative Commons Attribution 3.0 licence. Any further distribution of this work must maintain attribution to the author(s) and the title of the work, journal citation and DOI.
Published under licence by IOP Publishing Ltd

Attosecond time delay in the valence photoionization of C_{240} versus C_{60}

Kele Shi*, Maia Magrakvelidze* [1], Dylan Anstine*, Mohamed Madjet†, Himadri Chakraborty* [2]

* Department of Natural Sciences, D.L. Hubbard Center for Innovation and Entrepreneurship, Northwest Missouri State University, Maryville, Missouri 64468, USA

† Qatar Environment and Energy Research Institute, Qatar Foundation, P.O. Box 5825, Doha, Qatar

Synopsis We investigate effects of electron correlations on the attosecond time delay of the photoionization from HOMO and HOMO-1 electrons in C_{240}. A comparison with earlier C_{60} results assesses the molecular size effect.

Advances in technology to produce attosecond pulses or pulse trains made it possible to resolve electron dynamics in laser-matter interactions on ultrafast timescale. For atomic and bulk systems, photoemission time delays have been measured [1, 2]. Fullerenes structurally fall between atoms and condensed matters, and therefore is a prototype for attoscience interests in the field of clusters and nanostructures [3]. We present a theoretical study of the quantum phases and time delays of photoemissions from the valence orbitals of C_{240}. Effects of the electron correlation and the fullerene size are identified by comparing with the results of C_{60} [4].

The time-dependent local density approximation (TDLDA) [5] method with Leeuwen and Baerends (LB94) exchange-correlation functional is employed to calculate phases and corresponding Wigner-Smith delays. C^{4+} ions are kept fixed in space and smeared into a spherical jellium shell. The Kohn-Sham equations are solved for a delocalized collection of $2s^2 2p^2$ valence electrons from all carbon atoms.

Figure 1 shows the TDLDA phases for dipole allowed channels from HOMO and HOMO-1 levels of C_{240} and compares with those for C_{60}. Phase shifts of π (up or down) are observed at the cavity minima where electron correlations are dominant. Based on a geometry-dependent diffraction of the photoelectron waves, these minima arise from the nodes in the continuum wavefunctions. For HOMO-1, both in C_{60} and C_{240}, the jumps are roughly synchronized between the two dipole channels, in contrast to the results for HOMO. Also for HOMO, results differ between the two fullerenes. The series of jumps for C_{240} are seen to occur in smaller photon-energy period due to its larger molecular cavity size than C_{60}.

We calculate corresponding Wigner-Smith delays (not shown) by taking the energy differential of the total HOMO and HOMO-1 phases. Results point significant variations in the photoelectron speed at the cavity minima.

Current predictions open a new research direction for exploring dynamical correlations not only in fullerenes, but also in clusters and nanostructures (for which similar minima in photospectra are expected), using attosecond interferometry, such as, RABITT.

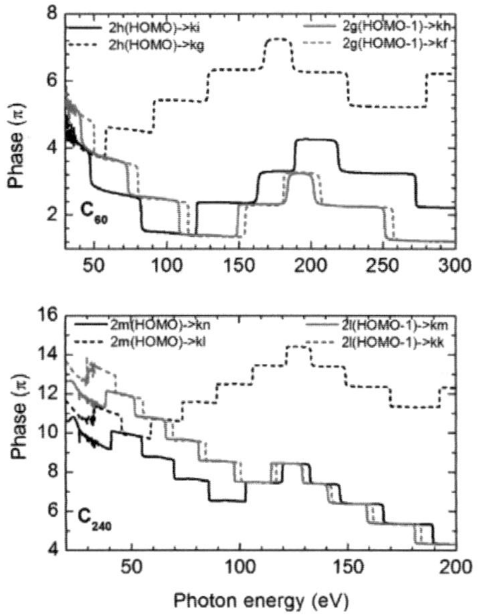

Figure 1. TDLDA phases as a function of photon energy for HOMO and HOMO-1 channels of C_{60} (top) and C_{240} (bottom).

This work was supported by the U.S. National Science Foundation.

References

[1] D. Guénot *et al* 2012 *Phys. Rev. A.* **85** 053424
[2] S. Neppl *et al* 2012 *Phys. Rev. Lett.* **109** 87401
[3] G. Dixit *et al* 2013 *Phys. Rev. Lett.* **111** 203003
[4] M. Magrakvelidze *et al* 2015 (*submitted*)
[5] M. Madjet *et al* 2008 *J. Phys. B.* **41** 015101

[1] E-mail: maia@nwmissouri.edu [2] E-mail: himadri@nwmissouri.edu

Content from this work may be used under the terms of the Creative Commons Attribution 3.0 licence. Any further distribution of this work must maintain attribution to the author(s) and the title of the work, journal citation and DOI.
Published under licence by IOP Publishing Ltd

Journal of Physics: Conference Series **635** (2015) 112026 doi:10.1088/1742-6596/635/11/112026

Symmetry forbidden vibronic transitions near the S K edge excitation of the SF_6 molecule

A.H. A. Gomes * [1], R. R. Oliveira Jr †, A. B. Rocha, W. Wolff, K. F. Alcantara, G. M. Sigaud#, and A. C. F. Santos*1

*Instituto de Física, Universidade Federal do Rio de Janeiro, PO 68528, 21941-972 Rio de Janeiro,RJ, Brazil

† Instituto de Química Universidade Federal do Rio de Janeiro, Ilha do Fundão, CT, Bloco A, Rio de Janeiro, RJ, CEP 21941-909, Brazil

In memorian

Synopsis Near sulfur K edge photoionization of the SF_6, molecule has been studied for X-ray photons. Total and partial photoion yields for the fragmentation products have been recorded as a function of the incident photon energy. The atomic fluorine ion, F^+, dominates the mass spectra in the energy range studied in this paper. The intensity of absorption for the dipole-forbidden but vibronically allowed transition S K ($1a_{1g} \rightarrow 6a_{1g}$) was calculated for the first time. This transition is shown to play a part in the fragmentation profile.

The dynamic interaction between the electronic and nuclear degrees of freedom in molecules, the so-called vibronic coupling becomes noticiable most evidently when the symmetry forbidden electronic transitions become observed due to the concomitant excitation of some asymmetric vibrations. It can ''appropriate'' intensities from further intermediate symmetry allowed transitions via some specific normal modes of a given symmetry, i.e. the reputed ''promoting mode''.

In this work, near sulfur K edge photoionization of the SF_6, molecule has been studied for X-ray photons. Total and partial photoion yields for the fragmentation products have been recorded as a function of the incident photon energy. The atomic fluorine ion, F^+, dominates the mass spectra in the energy range studied in this paper. The intensity of absorption for the dipole-forbidden but vibronically allowed transition S K ($1a_{1g} \rightarrow 6a_{1g}$) was calculated for the first time.

The SF_6 molecule is singly and multiply ionized by energy selected X-ray light in a region of strong uniform electric field (750 V/cm) which pulls the photoelectrons to a nearby microchannel plate (MCP) detector and the positively charged photoions into the flight path of a time-of-flight (TOF) mass spectrometer. The flight times of the moieities, timed from detection of at least one photoelectron by a multi-stop time-to-digital converter, are accumulated. The experiment was carried out on the Soft X-ray Spectroscopy beamline (SXS) at the Laboratório Nacional de Luz Síncrotron (LNLS), Campinas, São Paulo, Brazil.

Transition moments were calculated at inner-shell state-averaged multiconfigurational self consistent field. The configurations in this MCSCF calculation includes the Hartree–Fock (HF) configuration for the ground state and as many configuration as needed for a HF-like description of the core–hole excited state. In the present case one for $1a_{1g} \rightarrow 6a_{1g}$ and three for $1a_{1g} \rightarrow 6t_{1u}$. This method may also be considered as a state-averaged Hartree–Fock. Relativistic corrections, which are very important in S K-shell, are partially taken into account at the Douglas-Kroll method corrected to third order. All calculations have been done with aug-cc-pVTZ basis set for S and aug-cc-pVDZ for F.

Mass spectra measured at 2482 eV photon energy is presented in Fig. 1. The assignment of each peak in the time-of-flight mass spectra to its corresponding ion species is indicated on the figure. Some general observation can be made about the mass spectra of SF_6 around Si 1s edge.

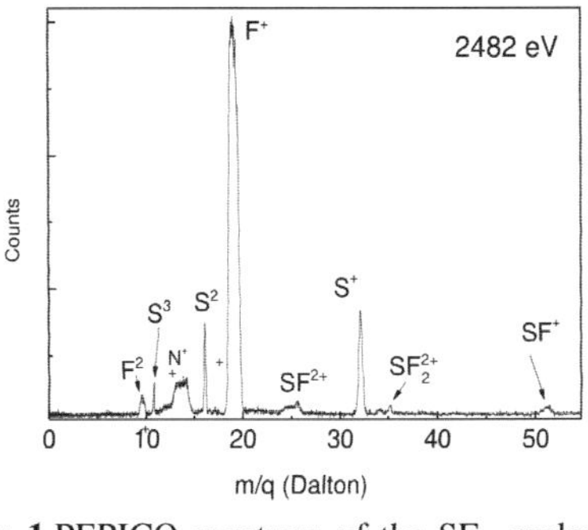

Figure 1.PEPICO spectrum of the SF_6, molecule at 2482 eV.

References

[1] A.H.A. Gomes, W. Wolff, N. Ferreira, K. F. Alcantara, H. Luna, G.M. Sigaud, A.C.F. Santos, Int. J. of Mass Spectr. **319– 320**, 1 (2012).

[1] E-mail: toni@if.ufjr.br

Content from this work may be used under the terms of the Creative Commons Attribution 3.0 licence. Any further distribution of this work must maintain attribution to the author(s) and the title of the work, journal citation and DOI.
Published under licence by IOP Publishing Ltd

Journal of Physics: Conference Series **635** (2015) 112027 doi:10.1088/1742-6596/635/11/112027

Fragment anisotropy on dissociative ionization of NO in two-color intense laser fields

Tomoyuki Endo, Ayaka Ishihara, Akitaka Matsuda, Mizuho Fushitani, Akiyoshi Hishikawa[1]

Department of Chemistry, Graduate school of Science, Nagoya University, Nagoya, Aichi 464-8602, Japan

Synopsis Dissociative ionization of NO, $NO \rightarrow NO^+ + e^- \rightarrow N^+ + O + e^-$, in ω-2ω two-color phase-controlled intense laser fields was investigated by three-dimensional ion momentum imaging. The momentum image of the N^+ ion exhibited anisotropic distributions peaked at 35° against the laser polarization, which show clear left-right asymmetry depending on the ω-2ω relative phase. The present study demonstrates that the molecular orbital can be retrieved by distinguishing the difference between the N and O side from the fragment anisotropy by intense laser pulses.

Visualization of molecular orbitals has been discussed in a number of studies in the last decades by using high-order harmonics generation [1] and Coulomb explosion [2]. The approach based on Coulomb explosion (or dissociative ionization) utilizes the fact that the spatial anisotropy of laser tunneling ionization rate in molecular frame depends on the shape of the outermost molecular orbital, which allows a direct read-out of the shape of molecular orbital from the spatial anisotropy of fragment ion. However, as demonstreted for our previous studies on NO [3], the asymmetry of molecular orbitals cannot be retrieved by a one-color laser pulse having a symmetric electric-field amplitude. To address this problem here, we employed ω-2ω two-color laser fields, where the tunneling ionization becomes an orientation selective process. In the present study, we investigate the dissociative ionization of NO, $NO \rightarrow NO^+ + e^- \rightarrow N^+ + O + e^-$, in ω-2ω two-color intense laser fields using three-dimensional (3D) ion momentum imaging.

The output of a Ti:Sapphire laser system (800 nm, 50 fs) was introduced into β-BBO crystal to obtain 2ω pulses (400 nm, 150 fs). A birefringent α-BBO crystal was placed to compensate the time-delay between ω and 2ω pulses. Both pulses were passed through a dual-wavelength waveplate ($\lambda/2$ for ω, λ for 2ω) to parallel the polarization axis of two beams. The relative phase between ω and 2ω was controlled by a feedback loop using a pair of fused silica wedges. The absolute value of the relative phase was determined by energy distribution of a back-scattered electron [4]. Two-color laser pulses were focused on the molecular beam of NO using a concave mirror (f = 300 mm) in an ultrahigh vacuum chamber. The fragment ions were accelerated to a position sensitive detector by four electrodes.The position (x, y) and the flight time (t) were recorded for each ion to calculate the 3D momentum vector $\mathbf{p} = (p_x, p_y, p_z)$.

The angular distribution of N^+ ion produced by dissociative ionization via $c^3\Pi$ state in two-color intense laser fields ($I_\omega = 4.3 \times 10^{13}\,\mathrm{W/cm^2}$, $I_{2\omega} = 3.4 \times 10^{12}\,\mathrm{W/cm^2}$, $E_{2\omega}/E_\omega = 0.28$) is peaked at 35° with respect to the laser polarization direction reflecting the nodal pattern of the 2π outermost molecular orbital of NO as observed in one-color laser fields (Fig. 1). On the other hand, a clear left-right asymmetry was observed, which depends strongly on the ω-2ω relative phase. The obtained result shows that the asymmetry of fragment distribution reflects the difference between N and O sides in the molecular orbital, demonstrating that the asymmetry of the molecular orbital can be retrieved using by two-color intense laser fields.

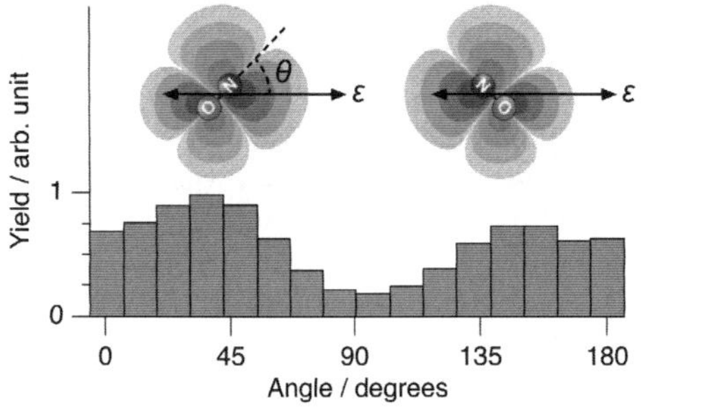

Figure 1. Angular distribution of the N^+ ion produced by dissociative ionization in two-color intense laser fields. The laser polarization denoted by ε.

References

[1] J. Itatani *et al* 2004 *Nature* **432** 867

[2] A.S. Alnaser *et al* 2004 *Phys. Rev. Lett.* **93** 113003

[3] T. Endo *et al* 2014 *J. Phys.:Conf. Ser.* **488** 032025

[4] D. Ray *et al* 2011 *Phys. Rev. A* **83** 013410

[1]E-mail: hishi@chem.nagoya-u.ac.jp

Content from this work may be used under the terms of the Creative Commons Attribution 3.0 licence. Any further distribution of this work must maintain attribution to the author(s) and the title of the work, journal citation and DOI.
Published under licence by IOP Publishing Ltd

Journal of Physics: Conference Series **635** (2015) 112028 doi:10.1088/1742-6596/635/11/112028

Let the trajectories tell a quantum story: Post-entangling the SHARC scheme

Pablo Sampedro Ruiz* [1], **Ignacio R. Sola*** [2], **Jesús González-Vázquez**† [3]

* Facultad de CC. Químicas,Departamento de Química Física I, UCM, Madrid, Spain

† Departamento de Químicas, Facultad de Ciencias, UAM, Madrid, Spain

Synopsis A new method is proposed to perform Quantum Wave Packet Nuclear Dynamics on large systems, by making use of information obtained by Semi-Classical Quantum Dynamics. The key of the method resides in expressing the nuclear wave function of the system in a basis set determined by the positions of multiple trajectories at each time.

Many approaches exist that take into account the nuclear motion in Quantum Dynamics simulations, from grid-based quantum dynamics techniques, discrete variable representation (DVR)[1] or multi-configurational time dependent Hartree (MCTDH)[2] methods, to mixed quantum classical dynamics (MQCD)[0, 0]. In particular, assuming the MQCD approach, we have recently developed the SHARC (Surface Hopping in the Adiabatic Representation including Random Couplings) scheme[0].

In the latter approach, the nuclear motion is determined classically from the electronic force field of a single representative reference electronic state, which is propagated by quantum dynamics. Electronic coherences and quantum interence effects (due to non-adiabatic couplins or laser-induced transitions) are therefore incorporated at the electronic level. Stochastic jumps from the chosen reference state (surface hopping) are accounted by Tully fewest schitches method. In order to statistically characterize the system dynamics, many trajectories must be propagated independently, but a clever choice of the quantum representation of the electronic basis allows to reduce the number of trajectories needed.

Although the MQCD approach (and the SHARC scheme in particular) was shown to reproduce many experimental results, any quantum effect at the nuclear level (such as quantum tunneling, zeroth point effects, symmetry and interference due to nodal patterns, etc.) are out of its reach. In order to deal with these problems, we have developed a method to recycle the information obtained by performing trajectory-based calculations to hugely reduce the complexity of full-dimensional quantum treatment of the nuclear motion.

In our method, the nuclear wave function is expressed as a linear combination of *trajectory based* basis functions. Each basis function consists on a product of monodimensional Gaussians, each of them centered at the correspondent position for a trajectory at each time. For each classical step a new set of orthonormal basis is obtained and the coefficients rotated accordingly. In this way, our basis set moves following the trajectories, allowing them to populate only the relevant regions of the potential energy surface (PES).

One great advantage of the method is the simplicity of the Hamiltonian expression and the economy of the calculation. Since MQCD methods require the calculation of the molecule's energy and its gradient at each step, the same information is used to obtain the potential energy operator, so that an analytical description of the PES is not required.

To show the performance of the method, we will present results for simple models for which it is affordable to compare them with full quantum dynamical simulations.

References

[1] R. Wyatt 1996 *Dynamics of Molecules and Chemical Reactions, 1st ed.*

[2] H. Meyer *et al.* 1990 *Chem. Phys. Lett.* **165**, 73

[3] J. C. Tully, 1990 *J. Chem. Phys.* **93** 1061.

[4] S. Hammes-Schiffer, J. C. Tully 1994 *J. Chem. Phys.* **101** 4657

[5] J. J. Bajo *et al.* 2012, *J. Phys. Chem. A* **116**, 2800

[1]E-mail: psampedro@ucm.es
[2]E-mail: isola@quim.ucm.es
[3]E-mail: jesus.gonzalezv@uam.es

Content from this work may be used under the terms of the Creative Commons Attribution 3.0 licence. Any further distribution of this work must maintain attribution to the author(s) and the title of the work, journal citation and DOI.
Published under licence by IOP Publishing Ltd

XXIX International Conference on Photonic, Electronic, and Atomic Collisions (ICPEAC2015) IOP Publishing
Journal of Physics: Conference Series **635** (2015) 112029 doi:10.1088/1742-6596/635/11/112029

Intense near-IR laser induced electron-ion scattering experiment on hydrocarbon molecules

Yuta Ito[1], Misaki Okunishi[1], Wang Chuncheng[2], Robert R. Lucchese[3], Toru Morishita[4], Oleg I. Tolstikhin[5], Lars B. Madsen[6], Kiyoshi Ueda[1]

[1] Institute of Multidisciplinary Research for Advanced Materials, Tohoku University, Japan
[2] Institute of Atomic and Molecular Physics, Jilin University, China
[3] Department of Chemistry, Texas A&M University, USA
[4] Department of Engineering Science, University of Electro-Communications, Japan
[5] National Research Center "Kurchatov Institute", Russia
[6] Department of Physics and Astronomy, Aarhus University, Denmark

Synopsis We have measured angle-resolved rescattering photoelectron spectra induced by ultrashort intense near-IR laser pulses at 1200-1650 nm. From the spectra, we have extracted field-free electron-ion differential scattering cross sections and compared them with theoretical calculations.

When atoms or molecules are exposed to linearly-polarized intense near-IR laser fields, a part of the electrons released by tunneling ionization will be driven back by the oscillating electric field into recollisions with their parent ions. Elastically scattered electrons have structural information of the parent ions. The rescattering process completes within one optical cycle of the laser (about 2.7 fs at 800 nm), which may enable us to observe ultrafast structural changes of molecules with a femtosecond temporal resolution. In our laboratory, we have measured rescattering photoelectron spectra (RPS) of atoms and molecules using laser pulses at 800 nm [1]. Electron-ion differential scattering cross sections (DCSs) extracted from the experimental RPS are well reproduced by theoretical calculations. In the case of 800 nm laser radiation, however, the collision energy of the rescattering electrons is limited due to the depletion of the targets by the strong field ionization. In this study, we measured RPS of C_2H_4 and C_2H_6 molecules using IR laser pulses at longer wavelengths to increase the recolliding momentum of the electrons. Also we have extended the detection range to smaller angles of the rescattering electron emission.

IR laser pulses at 1200-1650 nm are generated by an optical parametric amplifier pumped by Ti:Sapphire laser pulses at 800 nm (100 fs, 1.5 mJ, 1 kHz). The IR pulses are focused onto a sample gas effusively introduced in a vacuum chamber. Angular distributions of photoelectrons are obtained by rotating the polarization direction of the optical fields using a half-wave plate.

Figure 1 shows an angle-resolved RPS of C_2H_4 measured using the laser light at 1650 nm and a comparison between experimental and theoretical DCSs. Rescattering electrons having recolliding momentum around 3.0 a.u. are observed, while only up to 1.0 a.u. in the case of 800 nm. The electron momentum of 3.0 a.u. corresponds to 122 eV in the energy and 1.1 Å in the de Broglie wavelength, which is shorter than 1.34 Å, the C=C bond length of C_2H_4 molecules. In DCSs comparison, we have found that DCSs extracted from several RPS measured with different laser conditions agree well. Additionally, we have obtained fairy good agreement between experimental and theoretical DCSs for the rescattering at higher collision momenta.

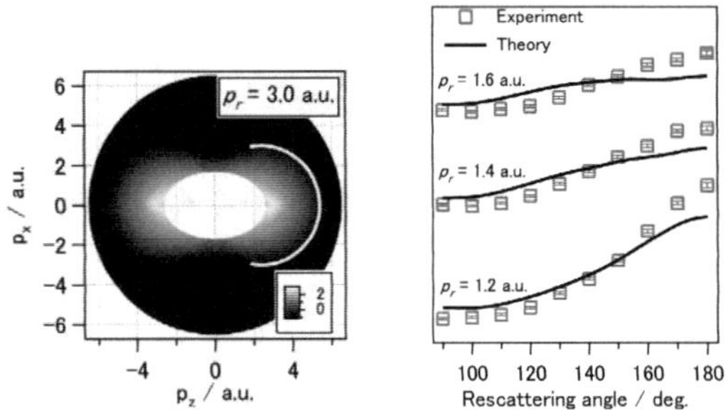

Figure 1. (left) angle-resolved RPS of C_2H_4, (right) comparison between experimental and theoretical DCSs of C_2H_4.

References

[1] M. Okunishi *et al.* 2008 *Phys. Rev. Lett.* **100** 143001; M. Okunishi *et al.* 2011 *Phys. Rev. Lett.* **106** 063001; C. Wang *et al.* 2012 J. *Phys. B: At. Mol. Opt. Phys.* **45** 131001

[1] E-mail: y-ito@mail.tagen.tohoku.ac.jp

Content from this work may be used under the terms of the Creative Commons Attribution 3.0 licence. Any further distribution of this work must maintain attribution to the author(s) and the title of the work, journal citation and DOI.
Published under licence by IOP Publishing Ltd

Journal of Physics: Conference Series **635** (2015) 112030 doi:10.1088/1742-6596/635/11/112030

Photodissociation of protonated Leucine-Enkephalin peptide in the VUV range

M. Lj. Ranković[1*], V. Cerovski[1], F. Canon[2], L. Nahon[3], A. Giuliani[3,4] and A. R. Milosavljević[1]

[1]Institute of Physics Belgrade, University of Belgrade, Pregrevica 118, 11080 Belgrade, Serbia
[2]INRA, UMR1324 Centre des Sciences du Goût et de l'Alimentation, F-21000 Dijon, France
[3]SOLEIL, l'Orme des Merisiers, St Aubin, BP48, 91192 Gif sur Yvette Cedex, France
[4]INRA, UAR1008, CEPIA, Rue de la Géraudière, BP 71627, 44316 Nantes, France

Synopsis We present results of VUV action spectroscopy of gas-phase protonated Leucin-Enkephalin peptide. The experiment has been performed by coupling a linear quadrupole ion trap with a synchrotron radiation VUV beamline.

Recent development of experimental methods that use electrospray ion source (ESI) to bring macromolecules into gas phase and collect the ions in an ion trap, which is coupled to bright synchrotron radiation sources, has allowed the investigation of large biopolymers (proteins and nucleotides) under well-defined conditions. These studies can provide more detailed insight into fundamental properties of biopolymers and their interaction with high energy photons.

We present the results of VUV photon action spectroscopy of Leucin-Enkephalin (Leu-Enk) peptide in 5-15 eV energy domain. This peptide has been studied with wide range of spectrometry techniques and has become a standard biomolecular target [1].

The experiment has been performed at the beamline DESIRS [2] of the synchrotron radiation facility SOLEIL near Paris, France. More details about the experimental setup can be found in recent publications [3, 4]. Briefly, precursor ions $[\text{Leu-Enk+H}]^+$ were produced by ESI and injected into the ion trap from the front side. Ions were irradiated with monochromatized VUV photon beam, during 500 ms from the back side of the ion trap. After irradiation all ion fragment species were ejected from the ion trap and tandem mass spectra (MS^2) as a function of the photon energies were recorded. Leu-Enk molecules were provided from Sigma Aldrich and dissolved in water/acetonitrile at 10 μM.

Photodissociation ion yields were extracted from the MS^2 and normalized to the total ion current and the photon flux. The ion yields show spectroscopic structure allowing the studying of absorption-like bands of the peptide. Figure 1 shows the relative ion yield of one of the backbone fragments. The present experimental results are in good agreement with our time dependent density functional theory (TD DFT) calculations.

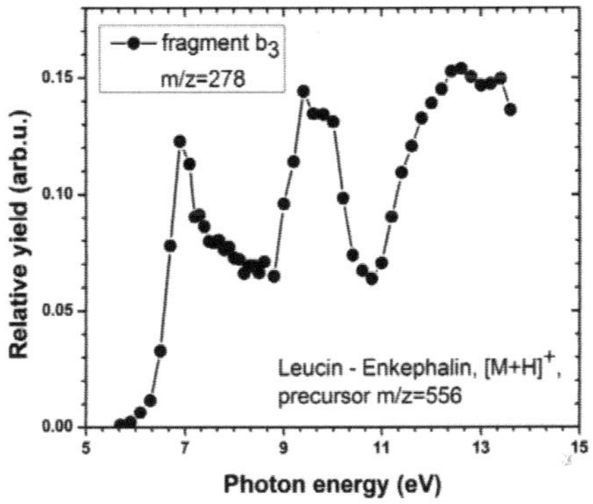

Figure 1. Relative ion yield of fragment b_3 from the photodissociation of precursor $[\text{Leu-Enk+H}]^+$.

We acknowledge the support by the ANR, France, under project ANR-08-BLAN-0065. A.R.M. and M.LJ.R. acknowledge support by the MESTD RS (Projects No. 171020). The SOLEIL synchrotron radiation facility is acknowledged for providing beamtime under project 20130388.

References

[1] S. Bari et al 2011, J. Chem. Phys. 18 **134**, 024314.
[2] L. Nahon et al 2012, J. Sync. Rad. **19**, 508 – 520.
[3] A. Milosavljević et al 2011, Phys. Chem. Chem. Phys. **13**, 15432–15436.
[4] A. Milosavljević et al 2012, J. Sync. Rad. **19**, 174.

* E-mail: mrankovic@ipb.ac.rs

Content from this work may be used under the terms of the Creative Commons Attribution 3.0 licence. Any further distribution of this work must maintain attribution to the author(s) and the title of the work, journal citation and DOI.
Published under licence by IOP Publishing Ltd

XXIX International Conference on Photonic, Electronic, and Atomic Collisions (ICPEAC2015) IOP Publishing
Journal of Physics: Conference Series **635** (2015) 112031 doi:10.1088/1742-6596/635/11/112031

Three-electron collective Auger decay in CH_3F

P. Kolorenč† [1], V. Averbukh*, J. Eland‡, R. Feifel◇, and F. Tarantelli♮

†Institute of Theoretical Physics, Charles University in Prague, Czech Republic
*Department of Physics, Imperial College London, UK
‡Physical & Theoretical Chemistry Laboratory, South Parks Road, Oxford University, UK
◇Department of Physics and Astronomy, Uppsala University, Sweden
♮Dipartimento di Chimica, Universitá degli Studi di Perugia and CNR I.S.T.M., Italy

Synopsis We report on experimental and theoretical study of collective Auger decay of double inner-valence vacancy in fluoromethane, in which two electrons simultaneously fill the vacancies and a third electron is ejected to continuum. Calculations predict the respective lifetime of the double vacancy state to be in the femtosecond range, typical for two-electron Auger decay. Measured electron spectra indirectly support this rather surprising result.

Inner-shell ionization of atoms and molecules leads to highly-excited ionic states that can relax by electron emission, the Auger effect being the most common example of such a process. The advent of high-intensity free-electron lasers brought into focus study of the non-radiative decay processes under conditions of multiple ionization. Here, we address so-called collective decay, a rather exotic three-electron Auger-like process. In collective decay, two vacancies recombine simultaneously and excess energy is carried away by a third electron that is ejected to continuum.

First such three-electron process reported was probably Auger LL-MMM transition in Ar [1]. More recently, related effects were observed in highly charged carbon atoms [2], three-electron interatomic Coulombic decay in rare-gas systems was studied both theoretically and experimentally [3, 4]. Some evidence that three-electron transitions do not occur exclusively in multiply ionized systems was given by Lee et al on the example of resonantly excited krypton atom [5].

Although the three-electron processes are usually considered to be extremely weak, under favorable conditions they can become dominant decay channels. In the present work, we focus on one such example which is the decay of doubly F(2s) ionized CH_3F molecule. Excitation energy of this state is well above triple-ionization threshold. However, calculations using the extended second-order algebraic diagrammatic construction (ADC) [6] show that all triply-ionized decay channels with at least one F(2s) vacancy are closed which is the condition we seek for the study of collective decay process. The figure shows triple-ionization spectrum of the $CH_3F(2s^0)^{2+}$ metastable state obtained by coincidence measurement of electrons emitted in a two step cascade initiated by C(1s) core ionization. The spectrum is compared to the theoretical spectra calculated using Fano-ADC method with and without the nuclear dynamics along C-F bond taken into account. Excellent agreement between the theoretical and experimental spectra supports the calculated lifetime is 2.5-3.0 fs, which is in the range of typical two-electron Auger lifetimes.

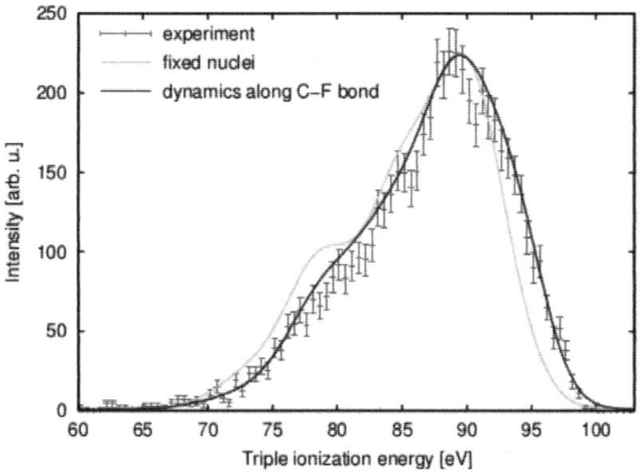

Figure 1. Measured and calculated spectrum of the $CH_3F(2s^0)^{2+}$ metastable state.

References

[1] V. V. Afrosimov *et al* 1975 *JETP Lett.* **21** 249

[2] E. De Filippo *et al* 2008 *Phys. Rev. Lett.* **100** 233202

[3] V. Averbukh and P. Kolorenč 2009 *Phys. Rev. Lett.* **103** 183001

[4] T. Ouchi *et al* 2011 *Phys. Rev. Lett.* **107** 053401

[5] I. Lee *et al* 1993 *J. Phys. B: At. Mol. Phys.* **26** L41

[6] J. Schirmer and A. Barth 1984 *Z. Phys. A* **317** 267

[1] E-mail: kolorenc@mbox.troja.mff.cuni.cz

Content from this work may be used under the terms of the Creative Commons Attribution 3.0 licence. Any further distribution of this work must maintain attribution to the author(s) and the title of the work, journal citation and DOI.
Published under licence by IOP Publishing Ltd

Imaging the predissociation dynamics of the methyl radical from the $3p_z$ Rydberg state

G. Balerdi*, J. Woodhouse*, R. de Nalda† [1], M.L. Senent‡, A. García-Vela§, L. Bañares* [2]

* Departamento de Química Física, Facultad de Ciencias Químicas (Unidad Asociada I+D+i al CSIC), Universidad Complutense de Madrid, 28040 Madrid, Spain
† Instituto de Química Física Rocasolano, CSIC, Serrano 119, 28006 Madrid, Spain
‡ Dpto. de Química y Física Teóricas, Instituto de Estructura de la Materia, CSIC, Serrano 121, Madrid 28006, Spain
§ Instituto de Física Fundamental, CSIC, Serrano 123, 28006 Madrid, Spain

Synopsis Direct, time-domain measurements are reported for the lifetimes of the lower-lying vibrational levels of the $3p_z$ Rydberg state of the methyl radical. Lifetimes measured are in the region of hundreds of femtoseconds, and show a decrease with increasing vibrational activity. The mechanism is interpreted with the aid of *ab initio* calculations.

The relevance of the methyl radical spans across both applied and fundamental areas of Chemistry, but crucial aspects of its spectroscopy and dynamics are still unknown today.

Due to its planar geometry, many states are spectroscopically dark, since one-photon transitions from the ground state are forbidden. REMPI (resonance-enhanced multiphoton ionization) spectroscopy allowed to obtain information about excited states previously unobserved [1]. Of those, the Rydberg-character $3p_z$ and $4p_z$ states have become widely used in 2+1 REMPI schemes for the detection of the methyl radical. However, it was soon realized that obtaining direct quantitative data on product state distributions of the methyl radical using these schemes was not possible due to predissociation processes that compete favorably with ionization, thus distorting line profiles in the REMPI spectra [2]. There have been no previous attempts to measure predissociation lifetimes in the time domain, or to propose a mechanism for predissociation based on high-level calculations. In this work, we report direct measurements of vibrational-level selective predissociation rates in the $3p_z$ state of the methyl radical. The work is complemented with *ab initio* calculations.

In the experiment, a "synthesis" pulse dissociates CH_3I in a molecular beam, giving rise to methyl fragments in the ground electronic state $CH_3(X, \nu)$. An ultrashort "excitation" laser pulse around 330 nm excites these methyl fragments by two-photon transitions to the $3p_z$ Rydberg state, $CH_3(3p_z, \nu')$. A third ultrashort "ionization" laser pulse at 400 nm removes an electron from the Rydberg orbitals. Velocity map images of the resulting CH_3^+ cation are aquired in a time-of-flight spectrometer. Tuning the excitation laser changes the distribution of excited vibrational states, and channel distinction is possible through energy-resolved detection in the velocity map images. Thus, for a given channel, we obtain a signal that is proportional to the instantaneous population in each vibrational level of the $3p_z$ state, which we map by changing the delay between the excitation and ionization pulses. This is shown in Figure 1, where the inset shows an example image. The results will be discussed under the framework of new *ab initio* calculations on this system.

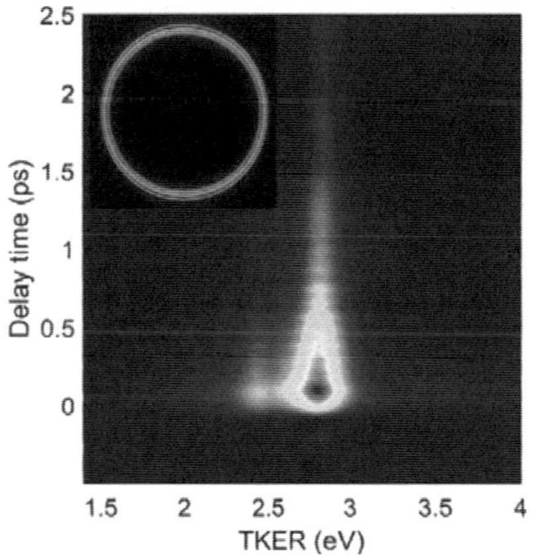

Figure 1. False-color map of the temporal evolution of the total kinetic energy distribution measured for the CH_3 fragment born after CH_3I dissociation. The inset shows a typical Abel-inverted velocity map images.

References

[1] J.W. Hudgens *et al* 1983 *J. Chem. Phys.* **79** 571
[2] J. Zhou *et al* 2003 *J. Chem. Phys.* **119** 2538

[1]E-mail: r.nalda@iqfr.csic.es
[2]E-mail: lbanares@quim.ucm.es

Content from this work may be used under the terms of the Creative Commons Attribution 3.0 licence. Any further distribution of this work must maintain attribution to the author(s) and the title of the work, journal citation and DOI.
Published under licence by IOP Publishing Ltd

Journal of Physics: Conference Series **635** (2015) 112033 doi:10.1088/1742-6596/635/11/112033

Laser pulse duration can control the breakage of multiple chemical bonds

X. Xie[1], E. Lötstedt[2,3,4], S. Roither[1], S. Erattupuzha[1], S. Larimian[1], D. Kartashov[1], M. Schöffler[1], K. Midorikawa[2,3], A. Baltuška[1], K. Yamanouchi[4], M. Kitzler[1]

[1] Photonics Institute, Vienna University of Technology, Gusshausstrasse 27, A-1040 Vienna, Austria
[2] Laser Technology Laboratory, RIKEN, Hirosawa, Wako, Saitama 351-0198, Japan
[3] RIKEN Center for Advanced Photonics, Hirosawa, Wako, Saitama 351-0198, Japan
[4] Department of Chemistry, The University of Tokyo, Hongo, Bunkyo-ku, Tokyo 113-0033, Japan

Synopsis We show that the interplay of electron removal and nuclear dynamics on the few femtosecond range, controlled by the duration of few-cycle laser pulses, determines the ratio of $C_2H_4^{3+}$ fragmentation into two respectively three moieties.

Using ethylene (C_2H_4) as an example, we experimentally demonstrate control of complex fragmentation reactions that involve the breakage of more than one chemical bond using ultrashort intense laser pulses with pulse durations from 4.5 fs to 25 fs. We show that the dynamics of removing three electrons by strong-field ionization determines the ratio of fragmentation of the molecular trication into two respectively three moieties.

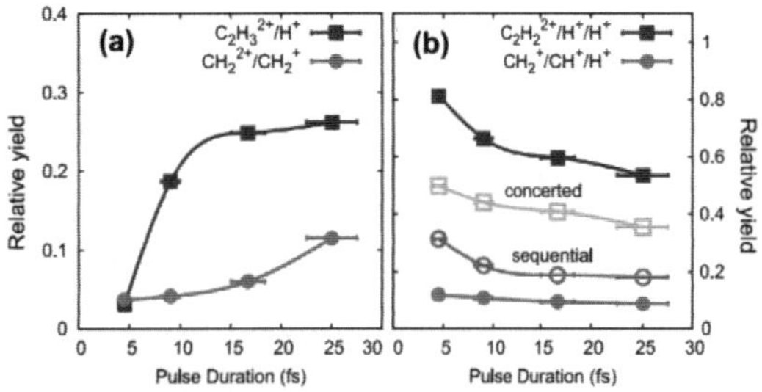

Figure 1. Normalized yield of (a) two-body and (b) three-body fragmentation channels (as indicated) as a function of laser pulse duration. Normalization is such that the sum of the yields of all four channels is 1 at each pulse duration. In (b) the yield of the three-body channel resulting in two protons (blue squares) is divided into the yields due to concerted and sequential fragmentation dynamics, respectively.

In our experiments we measured ionic fragments from $C_2H_4^{3+}$ in coincidence using the COLTRIMS technique [1] as a function of pulse duration from 4.5 fs to 25 fs (FWHM) with a constant peak intensity of 8×10^{14} W/cm^2. From all detected fragments we extracted the relative yield of in total four two- and three-body fragmentation channels, see Fig. 1. It can be seen that the probability of two-body fragmentation via breaking of a C-H bond dramatically increases as the pulse duration is increased from 4.5 fs to about 12 fs, and levels off for longer pulse durations. Fig. 1(a) furthermore shows that also the relative yield of two-body fragmentation via breaking of the center C-C bond monotonically increases with pulse duration. However, the slope is much smaller than in the case of the C-H bond, and the yield is still increasing at the longest measured pulse duration. In contrast to the two-body channels, the relative yields of the three-body fragmentation channels, for both the concerted and sequential fragmentation reactions, decrease with pulse duration [Fig. 1(b)].

Based on the dependence of the kinetic energy release and proton spectra as well as the angular momentum distributions on the pulse duration, and by performing intricate quantum dynamical simulations we explain the findings of Fig. 1 as follows. During the triple-ionization process the intermediate doubly charged ion is reached on an electronically excited PES as a result of removing (at least) one of the two electrons from an inner-valence shell. For sufficiently long laser pulse durations the C-H and C-C bonds do have enough time to significantly stretch on this excited PES. In contrast, for short pulse durations (4.5 fs) the nuclei stay essentially frozen. Thus, for short pulses the third ionization step prepares the triply charged molecular ion in a configuration close to the neutral's equilibrium configuration. For longer pulses (25 fs), however, the triply charged molecular ion is reached with significantly stretched C-H and also slightly stretched C-C bonds. Thus, with longer pulses the triply charged ion is prepared energetically further down the dissociative PES, i.e., on a position with lower potential energy as compared to short pulses. As a consequence, for longer pulses the potential energy is too small to overcome the energy barrier towards triple fragmentation and only two-body fragmentation is possible. In effect, the two-body fragmentation probability becomes strongly enhanced with increasing pulse duration, as observed in the experiment [Fig. 1]

References

[1] X. Xie *et al* 2012 *Phys. Rev. Lett.* **109** 243001

Content from this work may be used under the terms of the Creative Commons Attribution 3.0 licence. Any further distribution of this work must maintain attribution to the author(s) and the title of the work, journal citation and DOI.
Published under licence by IOP Publishing Ltd

Journal of Physics: Conference Series **635** (2015) 112034 doi:10.1088/1742-6596/635/11/112034

Two-Pulse Control over Double Ionization Pathways in CO_2

Sonia Erattupuzha, Seyedreza Larimian, Andrius Baltuška, Xinhua Xie, Markus Kitzler

Photonics Institute, Vienna University of Technology, Gusshausstrasse 27, A-1040 Vienna, Austria

Synopsis Using a double-pulse scheme we visualize and control molecular dynamics taking place on intermediate states populated during different sequential double ionization pathways of CO_2. Exchanging the pulse-sequence can almost completely switch the pathway.

When an atom or a molecule interacts with a strong laser field multiple ionization may take place. In this process the final ionic state may be reached via different pathways, which in the case of molecules may give rise to strongly different nuclear dynamics on the intermediately populated potential energy surfaces. We report on experiments that employ a sequence of two delayed pulses for controlling and visualizing the sequential double ionization pathway in CO_2. In the experiments, we measured electrons and ions in coincidence from the ionization and dissociation of CO_2 molecules using the COLTRIMS technique [1]. Measurements are done with two 795 nm, 25 fs laser pulses, with peak intensities of about 1×10^{14}W/cm^2 and 2×10^{14}W/cm^2, respectively. The time delay between the two pulses was varied in steps of 4 fs using a delay stage.

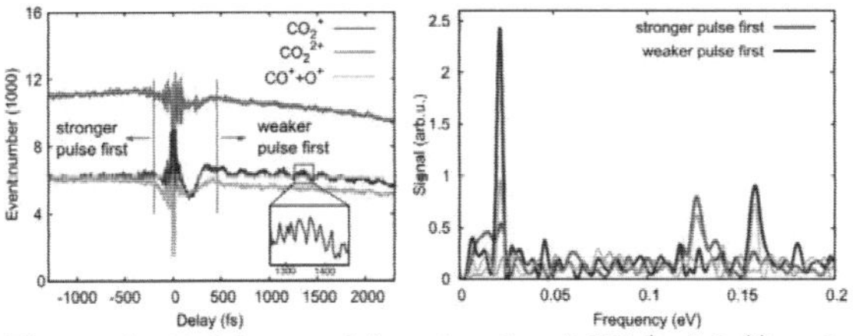

Figure 1. (a) Measured ion signals of CO_2^+, CO_2^{++} and CO^+/O^+ over time delay. Positive time delay means the weaker pulse comes first. (b) Fourier-transformed spectra of the yields from (a), separately for positive and negative time-delays for CO_2^{++} (blue and red, thicker lines) and CO^+/O^+ (orange and light-blue, thinner lines).

The measured signals of CO_2^+, CO_2^{++} and CO^+/O^+ are shown in Fig. 1 (a) as a function of the time delay between the two pulses. It can be clearly seen that the signals oscillate over time delay. There are at least two types of oscillations, a slow oscillation and a fast oscillation, as shown in the inset. To get insight to the oscillation pattern in CO_2^{++} and CO^+/O^+, we performed Fourier transforms of the signals over time delay. To avoid the region where the two pulses temporally overlap, we separated the scanned signal into left and right parts, as indicated in Fig. 1 (a). Fourier transforms are shown in Fig. 1 (b). The lower frequency peak at 0.022 eV (177 cm^{-1}) corresponds to the slow oscillation in the signal with a period of 190 fs. There are two higher frequency peaks at 0.126 eV (1016 cm^{-1}) and 0.158 eV (1274 cm^{-1}), which correspond to oscillation periods of 33fs and 26 fs, respectively. We assign the frequency peak at 0.158 eV to the symmetric stretch vibrational mode on the X-state of CO_2^+, while the peak at 0.126 eV is assigned to dynamics on the excited A-state [2].

We found that the higher vibrational frequency peak in CO_2^{++} and CO^+/O^+ moves from 0.158eV to 0.126 eV when we exchange the sequence of the two laser pulses. We explain this observation as follows. When the first pulse is weak, it singly ionizes the CO_2 molecule by removing a HOMO electron and the molecule ends at the X-state of CO_2^+. The second strong pulse further ionizes the cation to the dication. In such a scenario, the measured peak at 0.158 eV reflects the vibrational dynamics on the X-state of CO_2^+. However, when the strong pulse comes first, it can remove an electron from the HOMO-1 orbital and prepares CO_2^+ in its A-state. Because of its low intensity the following pulse can only further ionize CO_2^+ on the A-state, which explains why only the vibrational frequency at 0.126 eV from the A-state is present. From the measured frequency signal it can be seen that we can almost completely switch one pathway (through X-state) to the other path (through A-state) by simply exchanging the sequence of the two laser pulses. Thus, using a double-pulse scheme we could visualize and control molecular dynamics taking place on intermediately populated states during different sequential double ionization pathways of CO_2. During the conference, we will also discuss the origin of the low frequency peak.

References

[1] X. Xie *et al* 2012 *Phys. Rev. Lett.* **109** 243001
[2] A. Gharaibeh and D. J. Clouthier 2010 J. Chem. Phys. **132** 114307

Content from this work may be used under the terms of the Creative Commons Attribution 3.0 licence. Any further distribution of this work must maintain attribution to the author(s) and the title of the work, journal citation and DOI.
Published under licence by IOP Publishing Ltd

Journal of Physics: Conference Series **635** (2015) 112035 doi:10.1088/1742-6596/635/11/112035

Dynamical coupling of electrons and ions in Xray-induced dynamics

Ulf Saalmann[1], Abraham Camacho & Jan-Michael Rost

Max Planck Institute for the Physics of Complex Systems

Nöthnitzer Straße 38 · 01187 Dresden · Germany

Synopsis Photo-absorption from short and intense Xray pulses by a molecule or a cluster triggers a complicated electron and ion dynamics. Whereas the excitation process concerns largely core-shell electrons, there are various subsequent relaxation channels like electronic decays and ionic Coulomb explosion. We will discuss the interplay of those processes for molecular clusters and fullerenes.

Multiple ionization of a polyatomic system in an intense X-ray pulse lead — in contrast to narrow photo lines observed for atoms — to broad photo-electron spectra [1, 2], since the charge built-up modifies the overall potential, as shown in the figure below for a C_{60} molecule. Typically such spectra are difficult to interpret. Fullerenes offers a way to separate out the electron dynamics since the cage structure confines spatially the origin of photo and Auger electrons. Together with the sequential nature of the photo processes at intensities available at X-ray free electron lasers, this allows for a remarkably detailed interpretation of the photo-electron spectra [3]. The general features derived can serve as a paradigm for situations where an ion dynamics becomes relevant. For C_{60} this may occur already for pulses as short as 10 fs.

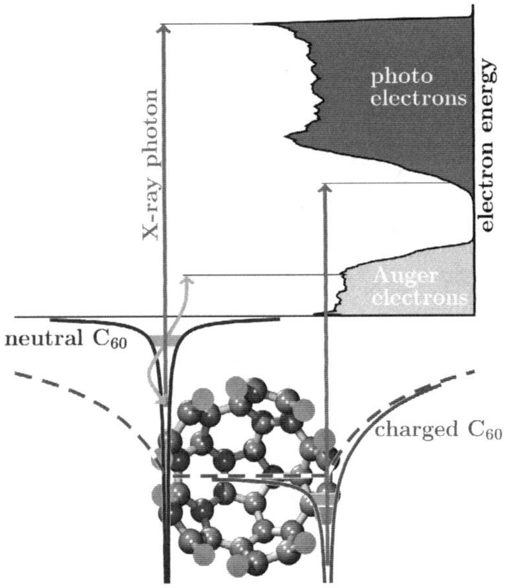

Figure: Sketch of the formation of broad electron spectra. The highest energy is observed for photo-ionization of core electrons from neutral C_{60} (blue solid line on the left). For a charged C_{60} (red solid line on the right) the energy is reduced due to the background potential (red dashed line).

The situation becomes more involved for clusters consisting of small molecules. We study systematically the electronic sequence of methane CH_4, ammonia NH_3, and water H_2O clusters, augmented by the "atomic limit" of neon clusters. Those containing hydrogen do eject fast protons when illuminated by short X-ray pulses. A suitable overall charging of the cluster controlled by the X-ray intensity induces electron migration from the surface to the bulk leading to efficient segregation of the protons. This "dynamical segregation" hinders globally the explosion of the heavy atoms even outside the screened volume. In contrast to core-shell systems where the outer shell is sacrificed to reduce radiation damage, the intricate proton dynamics of hydride clusters allows one to keep the entire backbone of heavy atoms intact [4]. Further studies on large molecules — being neither as homogeneous nor as spherical as the molecular cluster above — confirm this segregation effect [5].

References

[1] C. Bostedt *et al.*, New J. Phys. **12**, 083004 (2010).

[2] C. Gnodtke, U. Saalmann, and J.-M. Rost, Chem. Phys. **414**, 65 (2013).

[3] A. Camacho Garibay, U. Saalmann, and J. M. Rost, Phys. Rev. Lett. **113**, 083001 (2014).

[4] P. Di Cintio, U. Saalmann, and J.-M. Rost, Phys. Rev. Lett. **111**, 123401 (2013).

[5] A. Camacho Garibay, U. Saalmann, and J. M. Rost, unpublished.

[1]E-mail: us@pks.mpg.de

Content from this work may be used under the terms of the Creative Commons Attribution 3.0 licence. Any further distribution of this work must maintain attribution to the author(s) and the title of the work, journal citation and DOI.
Published under licence by IOP Publishing Ltd

XXIX International Conference on Photonic, Electronic, and Atomic Collisions (ICPEAC2015) IOP Publishing
Journal of Physics: Conference Series **635** (2015) 112037 doi:10.1088/1742-6596/635/11/112037

Control of ionization and dissociation of H_2^+ by elliptically polarized ultra-short VUV laser pulses.

Alejandro de la Calle* [1], **Alicia Palacios**†, **Fernando Martín**†, **Daniel Dundas***

* Atomistic Simulation Centre, School of Mathematics and Physics, ueen's University Belfast, Belfast BT7 1NN, Northern Ireland, UK.

† Departamento de uímica, Módulo 13,Universidad Autónoma de Madrid, Cantoblanco 28049, Madrid, Spain

Synopsis Resonance-enhanced multiphoton ionization of H_2^+ exposed to elliptically polarized VUV laser pulses is investigated. Differential cross sections for nuclei and electron are obtained using numerical solutions of the time-dependent Schrödinger equation. In this work in progress, we explore the dependence of the dissociative ionization observables with the polarization of the light.

The hydrogen molecular ion represents the simplest molecule in which correlations between electronic and ionic motion can be studied in a full quantum mechanical fashion. Understanding the interplay between electronic and ionic motion is fundamental in a wide range of areas such as condensed matter, plasma physics, medicine and biochemistry. Studying the interaction between the hydrogen molecular ion and the ultrashort VUV pulses therefore provides important insights into how these pulses interact with more complex molecules.

Since nowadays ultrafast intense radiation from sources such free-electron lasers and high-order harmonic generation are currently available in many laboratories around the world, the interest of photoionization of atoms and molecules by femto and sub-femtosecond laser pulses in the VUV and XUV frequency regime, has grown in the past years. Differential (in proton and electron energy) ionization cross sections have been theoretically obtained for H_2^+ [1] and H_2 [2] exposed to radiation polarized parallel to the molecular axis. Those studies show that, taking into account the vibrational motion of the molecule, it is possible to control the ratio between dissociative and non-dissociative ionization. In addition, photo-angular distributions (PAD) has been studied recently for linear [3] and elliptically polarized laser pulses [4, 5]. For fixed in space molecules, linearly polarized radiation oriented parallel to the molecular axis can only excite σ_u/σ_g states ($\Delta M = 0$), whereas perpendicular oriented one will populate π_u in one-photon absorption ($\Delta M = \pm 1$), and subsequent photon absorptions may reach σ_g/δ_g and so on. Elliptically polarized light is viewed as a combination of parallel and perpendicular oriented fields, therefore can simultaneously excite all those symmetries. Recently, it has been shown that the rotational symmetry of PAD with respect to the molecular and the polarization axis are strongly dependent of the interplay between intermediate resonant states [6].

All the available work using circular or elliptical polarizations has been performed within the fixed nuclei approximation. In our study, the electronic and vibrational degrees of freedom are described exactly through the solution of the Time-Dependent Schrödinger Equation (TDSE), assuming that the laser light can have elliptical polarization. In such case, as there is no longer a symmetry around the internuclear axis, the TDSE is solved in cartesian coordinates, with all the degrees of freedom treated using scaled finite-difference meshes [7, 8, 9]. This approach has been implemented in a massively parallel code called `RHYthMIC` (vibrating Hydrogen Molecular Ion in Cartesians) that allows the first full-dimensional studies of the hydrogen molecular ion in circularly polarized laser pulses for hundreds of femtoseconds in large grids.

References

[1] A. Palacios *at al* 2005 *Phys. Rev. A* **71** 063405

[2] A. Palacios *at al* 2006 *Phys. Rev. ett.* **96** 143001

[3] X. Guan *at al* 2012 *Phys. Rev. A* **85** 043419

[4] X. Guan *at al* 2013 *Phys. Rev. A* **87** 053410

[5] K-J. Yuan *at al* 2012 *Phys. Rev. A* **85** 053419

[6] K-J. Yuan *at al* 2014 *Chem. Phys. ett.* **592** 334-340

[7] D. Dundas 2012 *J. Chem. Phys.* **136** 194303

[8] D. Dundas *at al* 2000 *J. Phys.* **33** 3261

[9] D. Dundas 2002 *Phys. Rev. A* **65** 023408

[1]E-mail: a.delacalle@qub.ac.uk

Content from this work may be used under the terms of the Creative Commons Attribution 3.0 licence. Any further distribution of this work must maintain attribution to the author(s) and the title of the work, journal citation and DOI.
Published under licence by IOP Publishing Ltd

XXIX International Conference on Photonic, Electronic, and Atomic Collisions (ICPEAC2015) IOP Publishing
Journal of Physics: Conference Series **635** (2015) 112038 doi:10.1088/1742-6596/635/11/112038

Multiple-photon absorption in cooled molecular beams of acenes compounds at 266 and 355 nm laser radiation.

Juan Carlos Poveda* [1], **Ignacio Álvarez**[†], **and Carmen Cisneros**[† 2]

[*] Escuela de Química, Laboratorio de Espectroscopia Atómica Molecular, Universidad Industrial de Santander, Bucaramanga, Colombia

[†] Laboratorio de Colisiones Atómicas Moleculares, Instituto de Ciencias Físicas, Universidad Nacional Autónoma de México, Cuernavaca, Morelos, México

Synopsis: The cooled molecular beams of acenes compounds as naphthalene $C_{10}H_8$, anthracene $C_{14}H_{10}$, and benzantracene $C_{18}H_{12}$, were produced by adiabatic expansion in high vacuum chamber, 10^{-8} torr. Molecular beams interacted with laser radiation of 266 and 355 nm, from a Ng:YAG laser at intensities up to 10^{10} W·cm^{-2}. At higher intensities the multiple photon absorption was possible. The number of photons was calculated using Keldysh approach. The dominant photophysics of photon-molecule interactions were photodissociation and photoionization processes depending on the wavelength used during the experiment.

Photophysics of polycyclic aromatic compounds are a relevant subject of study given its ubiquitous presence and its toxicity[1]. In the multiple-photon absorption regimen the dissociation pattern is strongly linked to photon wavelength and intensity, and the molecule's possibility of resonant and non-resonant photon absorption.

In order to obtain cooled molecular beams, vapors of acenes were adiabatically expanded. Molecules interacted with radiation of 266 and 355nm at energies per pulse (0.1-10.0 mJ/p at 266 nm, and 5.0-30.0 mJ/p at 355nm) reaching intensities close to 10^9 and 10^{10} W·cm^{-2}, respectively. Charged ions coming from ionization process where detected using a R-ToF mass analyzer.

Spectra where analyzed in order to find the relation between ion-current and intensities of radiation, number of photons needed to allow the formation of a particular ion were calculated using the Keldysh approach.

The absorption of two photons was needed to ionize each molecule at 266 nm and molecular ions were detected. The dissociation was notorious as the energies per pulse increased. The main dissociation channels allowed the formation of $C_6H_4^{\cdot+}$ and $C_8H_6^{\cdot+}$ from naphthalene; $C_8H_6^{\cdot+}$ and $C_{10}H_8^{\cdot+}$ from anthracene; and $C_{14}H_{10}^{\cdot+}$, $C_{10}H_8^{\cdot+}$ from benzanthracene. From detected ions we were able to conclude that neutral fragments (C_2H_2, C_2H_2, C_2H_4, C_4H_2, C_8H_4) were eliminated.

At 355 nm, molecular parent ions were not detected. The photophysics was dominated by molecular dissociation followed by ionization. The ToF spectra showed the presence of ions with one to four atoms of carbon.

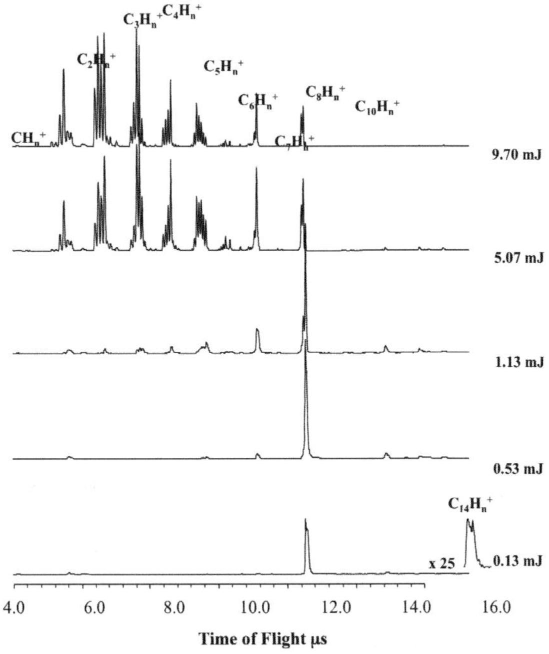

Figure 1. Photoionization spectra of anthracene at 266 nm.

This work was supported by CONACYT Project 165410 , PAPIIT Projects IN101215 and IN102613, and Vicerectoria de Investigaciones-Universidad Industrial de Santander.

References

[1] M.L. Lee et al., *Analytical Chemistry of Policyclic Aromatic Compounds*, Academic Press, New York, 1981.

[2] J.C. Poveda, *Ph.D. Thesis*, National Autonomous University of Mexico, 2010

[1] E-mail: jcpoveda@uis.edu.co

Content from this work may be used under the terms of the Creative Commons Attribution 3.0 licence. Any further distribution of this work must maintain attribution to the author(s) and the title of the work, journal citation and DOI.
Published under licence by IOP Publishing Ltd

Journal of Physics: Conference Series **635** (2015) 112039 doi:10.1088/1742-6596/635/11/112039

Molecular interferometer using XUV attosecond pulses to unravel electron and nuclear dynamics

Alicia Palacios*[1], Alberto González-Castrillo*, Fernando Martín*†

* Departamento de Química, Módulo 13,Universidad Autónoma de Madrid, Cantoblanco 28049, Madrid, Spain
† Instituto Madrileño de Estudios Avanzados (IMDEA) en Nanociencia, Cantoblanco, 28049 Madrid, Spain

Synopsis Two identical XUV attosecond pulses interact with the hydrogen molecule creating an interferometer resulting from the direct and sequential two-photon absorption paths reaching the same final ionized states. The dependence of the ionization yields with the time delay between the pulses allows to reconstruct the pumped vibronic (electronic and vibrational) wave packet created in the singly excited states of the molecule. The use of XUV pulses avoids a laser-induced distortion of the molecular potential, ensuring the characterization of the intrinsic behaviour of the system.

Monitoring and controlling electron dynamics in atoms and molecules requires the use of short and intense enough pulses. Significant progresses have been made using XUV-pump/IR-probe schemes, whose applications range from creating inteferometers in atoms to reconstruct an electron wave packet [1], to capturing time-resolved images of hole migration in molecules [2]. Despite these unquestionable achievements, the use of IR fields usually distorts the molecular potential and shorter-wavelength pulses are desirable to avoid the modification of the intrinsic behaviour of the system under study.

In the present theoretical work, we propose the use of a XUV-pump/XUV-probe scheme in the hydrogen molecule that allows for a full characterization of the pumped vibronic wave packet [3, 4]. The molecule is illuminated with identical XUV pulses with a given delay, creating an interferometric signal due to the electron ejection through both two-photon direct and sequential absorption processes reaching the same final state. We demonstrate that such a scheme allows us to unravel the coupled electron and nuclear dynamics, extracting both amplitudes and phases of the pumped vibronic wave packet. Figure 1 shows the dissociative ionization probability as a function of the time delay between two identical pulses of 2 fs duration and centered at 14 eV energy. A short-time Fourier transform of the ionization signal, with a proper choice of the time window to obtain a good balance in energy (y-axis) and time (x-axis) resolution, allows to disentangle two well defined components in the pumped wave packet. The probability localizes around 12 eV and around 14 eV of total absorbed energy revealing two different electronic states, each containing a nuclear wave packet (NWP). The averaged vibrational periods associated to each NWP (around 20 and 30 fs, respectively) are also captured. Finally, we also show that while total ionization yields reveal the entangled electronic and nuclear dynamics associated to the singly excited states, doubly differential yields (differential in both electronic and nuclear energies) exhibit in addition the dynamics of autoionization, i.e., of electron correlation in the ionization continuum.

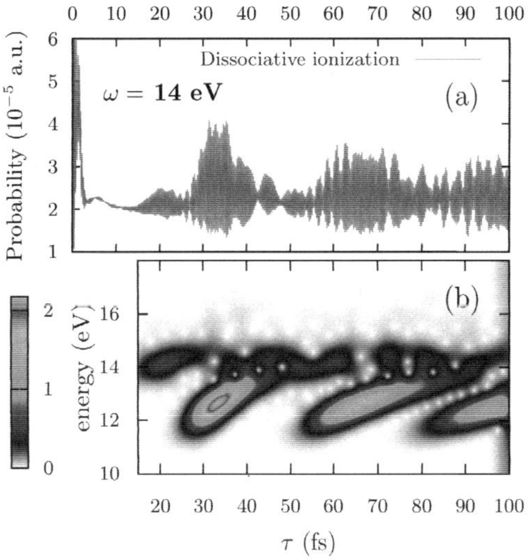

Figure 1. (A) Dissociative ionization probability as a function of the time delay between two XUV pulses. (B) Short-time Fourier transform of the ionization probability in A.

References

[1] T. Remetter *et al* 2006 *Nat. Phys.* **2** 323

[2] F. Calegari *et al* 2014 *Science* **346** 336

[3] A. Palacios *et al* 2014 *Proc. Natl. Acad. Sci.* **111** 3963

[4] P. A. Carpeggiani *et al* 2014 *Phys. Rev. A* **89** 023420

[1]E-mail: alicia.palacios@uam.es

Content from this work may be used under the terms of the Creative Commons Attribution 3.0 licence. Any further distribution of this work must maintain attribution to the author(s) and the title of the work, journal citation and DOI.
Published under licence by IOP Publishing Ltd

XXIX International Conference on Photonic, Electronic, and Atomic Collisions (ICPEAC2015) IOP Publishing
Journal of Physics: Conference Series **635** (2015) 112040 doi:10.1088/1742-6596/635/11/112040

Electron streaking in the autoionization region of H_2

Alicia Palacios*[1], Alberto González-Castrillo*, Fernando Martín*†

* Departamento de Química, Módulo 13,Universidad Autónoma de Madrid, Cantoblanco 28049, Madrid, Spain

† Instituto Madrileño de Estudios Avanzados (IMDEA) en Nanociencia, Cantoblanco, 28049 Madrid, Spain

Synopsis We use a UV-pump/IR-probe scheme, combining a single attosecond UV pulse and a 750 nm IR pulse, to explore laser-assisted photoionization of the hydrogen molecule in the autoionization region. The electron energy distributions exhibit unusual streaking patterns that are explored for different angles of the electron ejection with respect to the polarization vector and the molecular axis. Moreover, by controlling the time delay between the pulses, we observe that one can suppress the autoionization channel.

We present a theoretical study of H_2 ionization in the region of the doubly excited states by using a combination of a single attosecond UV pulse and a femtosecond IR pulse [1]. The IR field has a relatively low intensity and can not ionize the molecule by itself, simplifying the physics to a laser-assisted photoionization problem. Moreover, a well-known picture arises when a large-bandwidth UV pulse ionizes the target in a structureless region of the electronic continuum spectrum: the attosecond streaking phenomenom [2, 3]. If the UV pulse is significantly shorter than the IR pulse, the photoelectron is emitted at a well-defined time with respect to the IR field and the latter will then modify the electron motion in the continuum. We will explore how the presence of both autoionization and nuclear motion modifies this simple picture. We use an ab initio theoretical method to obtain total and energy differential probabilities for both dissociative ($H+H^++e^-$) and non-dissociative ($H_2^++e^-$) channels. We first investigate the behavior of the nuclear kinetic energy distributions and the electron-proton coincidence maps as a function of the time delay between the pulses. Most features can be explained in terms of perturbative transitions involving a UV photon and up to three IR photons. We found that autoionization is supressed when the IR pulse reaches the target after the UV absorption and with a delay shorter than 3-4 fs. This is due to the IR field depopulating the doubly excited states before they have time to decay into their background continua. At longer time delays, autoionization occurs before an the IR pulse arrives, and thus leaves its signature in the energy-differential distributions [4]. We later study the streaking patterns where the photoelectron is expected to be driven by the IR field, shifting its momentum in time following the vector potential ($A_{IR}(t)$). This simple picture, initially developed for atomic targets, is based on a single-active-electron approximation [2, 3]. We observe that when electrons ejected in both dissociative and non-dissociative ionization channels are not resolved (left panel in the figure), the distributions follow this simple pattern. However, this is no longer true for dissociative ionization alone (right panel in the figure), where the interferences with the autoionization process ($H_2^{**} \rightarrow H + H^++e^-$) destroy this simple picture. This opens the question about how to extract ionization times from a streaking signal where non-direct processes such as autoionization are present.

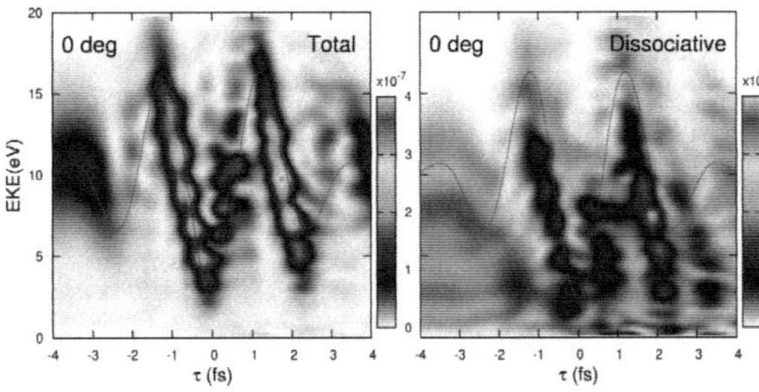

Figure 1. Electron energy distributions as a function of the time delay for ejection along the laser polarization axis. Left panel: total ionization. Right panel: dissociative ionization, which accounts for less than 10% of the total ionization (notice the different scales). Consequently, autoionization will strongly modify this channel and its streaking pattern. Red line: $-A_{IR}(t)$.

References

[1] A. Palacios *et al.* 2014 *J. Phys. B* **47** 124013

[2] J. Itatani *et al.* 2002 *Phys. Rev. Lett.* **88** 173903

[3] M. Kitzler *et al.* 2002 *Phys. Rev. Lett.* **88** 173904

[4] A. Palacios *et al.* 2013 *ChemPhysChem* **14** 1456

[1]E-mail: alicia.palacios@uam.es

Content from this work may be used under the terms of the Creative Commons Attribution 3.0 licence. Any further distribution of this work must maintain attribution to the author(s) and the title of the work, journal citation and DOI.
Published under licence by IOP Publishing Ltd

Journal of Physics: Conference Series **635** (2015) 112041 doi:10.1088/1742-6596/635/11/112041

Selectivity in the photofragmentation of halo-pyrimidines

P. Bolognesi*[1], A. Kettunen†, A. Cartoni&, R. Richter#, S. Tosic§, S. Maclot$, P. Rousseau$, R. Delaunay$, A. Domaracka$, L. Avaldi*

* CNR-Istituto di Struttura della Materia, Area della Ricerca di Roma1, Monterotondo Scalo, Italy
† Department of Physics, University of Oulu, Finland
&Dipartimento di Chimica, Sapienza Università di Roma, Roma, Italy
#Elettra-Sincrotrone Trieste, Basovizza, Italy
§nstitute of Physics, University of Belgrade, Belgrade, Serbia
$CIMAP, CEA/CNRS/ENSICAEN, Université de Caen Basse-Normandie, Caen, France

Synopsis The fragmentation of 2Br-, 2Cl- and 5Br-pyrimidine following direct valence photoionization or inner shell excitation and decay has been studied by electron-ion coincidence experiments. The results show that the fragmentation is strongly selective on the final singly charged ion state and the dominant fragmentation patterns correlate to the nearest appearance potential.

Pyrimidines are an important class of organic molecules because the pyrimidine ring forms the base structure of three nucleic acids (uracil, cytosine and thymine) and their halogenated substituted bases have found applications as radiosensitizers in radiotherapy.

In a previous work on pyrimidine [1] it has been observed that the resonant Auger spectrum shows a selective population of the final states of the cation and the site and state selected fragmentation patterns appear to depend only on these final states. In this work we have studied the fragmentation of 2Br-, 2Cl- and 5Br-pyrimidinc following direct valence shell photoionization as well as inner shell excitations. In the case of valence ionization the detection of the fragment ions in coincidence with energy selected photoelectrons allows a state selected study on the different cation states involved in the process. In the case of inner shell excitation the fragment ions are detected in coincidence with the resonant Auger electrons. In this way the selectivity on the site of the initial energy deposition is added, too.

The experiments have been performed at the Gas Phase Beamline at the Elettra Synchrotron (Trieste).

In Figure 1 the electron-ion coincidence yield for several 2Br-pyrimidine fragments as a function of the binding energy of the cation state measured either in direct photoionization at 100 eV or following the inner shell excitation are shown. Regardless of the initial process the coincidence mass spectra show a very similar behavior as long as the same final ionic state is selected. This behavior of the fragmentation can be explained by the fact that the fragmentation occurs on a time scale longer than the non-radiative relaxation of the inner shell vacancy. It is also clear that a certain ionic state preferentially correlates to the nearest fragmentation threshold, even though other fragmentation channels are energetically open.

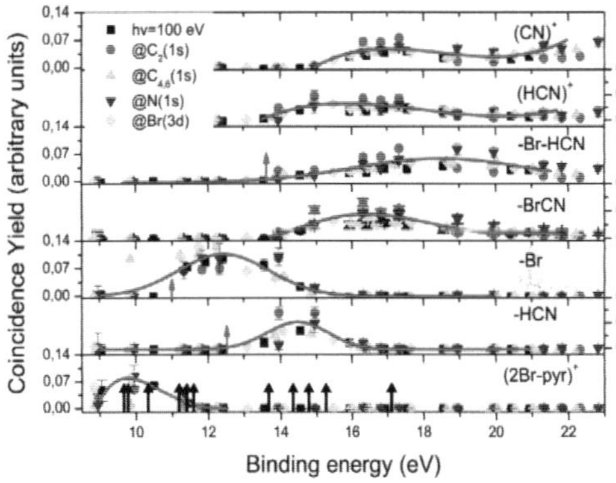

Figure 1. The PEPICO coincidence yields as a function of the binding energy of the cation state. The black arrows in the bottom panel indicate the position of the ion states , while the red ones the measured appearance energy of the fragments.

Acknowledgments Work partially supported by the Serbia – Italy Joint Research Project "Nanoscale Insight in the Radiation Damage".

References

[1] P. Bolognesi *et al* 2012 in"*Radiation Damage in Biomolecular Systems*", G. Garcia Gomez-Tejedor and M.C. Fuss eds.,Springer, 165-176

[1]E-mail: Paola.Bolognesi@cnr.it

Content from this work may be used under the terms of the Creative Commons Attribution 3.0 licence. Any further distribution of this work must maintain attribution to the author(s) and the title of the work, journal citation and DOI.
Published under licence by IOP Publishing Ltd

Journal of Physics: Conference Series **635** (2015) 112042 doi:10.1088/1742-6596/635/11/112042

Finite element DVR method for molecular single and double ionization by strong laser pulses

Denis Jelovina* [1], **Johannes Feist**◇, **Fernando Martín***†, **Alicia Palacios*** [2]
* Departamento de Química, Módulo 13,Universidad Autónoma de Madrid, Cantoblanco 28049, Madrid, Spain
◇ Departamento de Física Teórica de la Materia Condensada and Condensed Matter Physics Center (IFIMAC), Universidad Autónoma de Madrid, Cantoblanco 28049, Madrid, Spain
† Instituto Madrileño de Estudios Avanzados en Nanociencia, Cantoblanco, 28049 Madrid, Spain

Synopsis We are developing an *ab initio* numerical method to compute molecular single and double ionization processes induced by intense laser pulses. We present our first time-dependent simulations in the H_2^+ molecular ion including both electronic and nuclear degrees of freedom. We compare different methods for the extraction of excitation and ionization probabilities from the numerical time-dependent wave function, which will be applied to more complex targets in the future.

We solve the time-dependent Schrödinger equation (TDSE) for the H_2^+ molecule in full dimensionality, with three spatial degrees of freedom for the electron, and nuclear motion restricted to one dimension (neglecting its angular dependence). The method is closely related to previous approaches employed in atoms [1, 2], employing a single-center expansion for the electronic degrees of freedom, with spherical harmonics describing the angular dependence. The radial component of both the electronic and nuclear motion is described using a finite-element discrete variable representation (FEDVR) basis.

In order to avoid unphysical reflections and to impose pure outgoing boundary conditions in the driven equations (see below), we apply exterior complex scaling (ECS) for both the electronic and internuclear radial coordinates [3].

As an initial benchmark result, we compute the (pure) dissociation and ionization cross sections for single-photon absorption, using three distinct methods. First, we obtain the total absorption (dissociation+ionization) from the optical theorem, $\sigma = \mathrm{Im}\langle\psi_0|\hat{\mu}\frac{1}{\hat{H}-\omega}\hat{\mu}|\psi_0\rangle$, where $|\psi_0\rangle$ is the ground state, $\hat{\mu}$ is the dipole operator (along the internuclear axis), $\hat{H}$ is the Hamiltonian, and ω is the driving frequency. Second, we propagate the system under the influence of an ultrashort laser pulse (which also gives access to multiphoton processes). We follow the formalism developed in [1] to implicitly integrate to infinite times for extracting the final observables. The dissocation and dissociative ionization cross sections are obtained by either integrating the probability flux through properly chosen surfaces, or by extracting the energy- and angle-resolved cross sections using surface integrals over testing functions [3, 1].

Figure 1 shows the cross sections for single-photon absorption using the latter implementation, extracted from the time-dependent wave packets obtained after interaction with laser fields of varying central frequency and duration. We obtain perfect agreement with ref. [4].

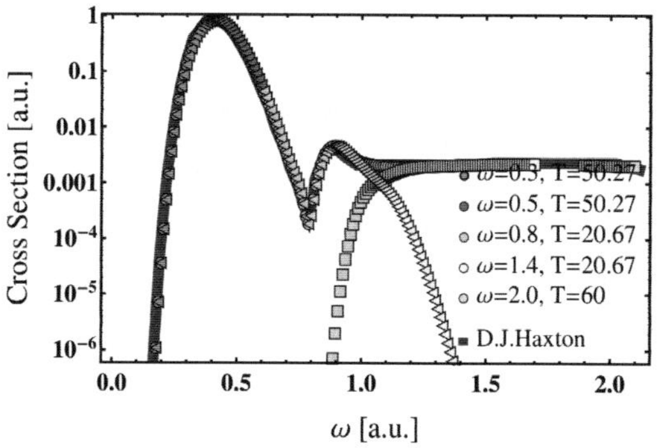

Figure 1. Absorption cross section for molecular ion for polarization parallel to the bond axis as a function of the photon energy. Full line: Optical theorem. Rectangles: dissociative ionization. Triangles: dissociative excitation. Circles: total cross section. Each color stands for a different field use in the solution of the TDSE; the central frequency and pulse duration in atomic units are indicated.

References

[1] A. Palacios, T. N. Rescigno and S. W. McCurdy 2007 *Phys. Rev. A* **76** 043420

[2] J. Feist *et al.* 2008 *Phys. Rev. A* **77** 043420

[3] T. N. Rescigno and C. W. McCurdy 2000 *Phys. Rev. A* **62** 032706

[4] D. J. Haxton 2013 *Phys. Rev. A* **88** 013415

[1]E-mail: denis.jelovina@uam.es
[2]E-mail: alicia.palacios@uam.es

Content from this work may be used under the terms of the Creative Commons Attribution 3.0 licence. Any further distribution of this work must maintain attribution to the author(s) and the title of the work, journal citation and DOI.
Published under licence by IOP Publishing Ltd

The importance of Rydberg orbitals in dissociative ionization of small hydrocarbon molecules in intense few-cycle laser pulses

Bethany Jochim*[1], R. Siemering†, M. Zohrabi*, A. Voznyuk, J. B. Mahowald**, D. G. Schmitz**, K. J. Betsch*, Ben Berry*, T. Severt*, Nora G. Kling*‡ T. G. Burwitz**, U. Ablikim*, K. D. Carnes*, M. F. Kling*‡, I. Ben-Itzhak*, E. Wells**[2], R. de Vivie-Riedle†**

*J.R. Macdonald Laboratory, Department of Physics, Kansas State University, Manhattan, Kansas 66506 USA
†Department für Chemie, Ludwig-Maximilians-Universität München, D-81377 München, Germany
**Department of Physics, Augustana College, Sioux Falls, South Dakota 57197 USA
‡Department of Physics, Ludwig-Maximilians-Universität München, D-85748 Garching, Germany

Synopsis We demonstrate the importance of ionization from Rydberg orbitals via experimental and theoretical work focusing on the strong-field dissociative single ionization of small hydrocarbons. Our findings suggest that Rydberg states should be routinely considered when studying polyatomic molecules in intense laser fields.

Much of our intuition about strong-field processes is built upon studies of diatomic molecules, which have electronic states that are typically well separated in energy. In polyatomic molecules, however, the electronic states are closer together, leading to the involvement of more states. We studied a specific process initiated by intense few-cycle laser pulses in C_2D_2, C_2D_4, and C_2D_6, namely single ionization of the parent molecule followed by hydrogen elimination. We measure the momentum of the remaining $C_2D_{n-1}^+$ fragment in each case using velocity map imaging (VMI). Our measurements, along with angle-dependent ionization probability calculations [1], demonstrate that Rydberg orbitals are readily populated for certain orientations of the molecules relative to the laser polarization. Moreover, these commonly-neglected Rydberg orbitals make a key contribution to strong-field ionization.

As an example, the comparison between measured and calculated $C_2D_3^+$ angular distributions from C_2D_4, shown in Fig. 1, clearly illustrates the need to include contributions from Rydberg states. The general four-lobed structure of $C_2D_3^+$ from dissociative single ionization is independent of pulse intensity and duration (at least up to 45 fs [2]). Calculations that do not include ionization from Rydberg orbitals result in an approximately isotropic angular distribution, as illustrated by Fig. 1(c), while including contributions from Rydberg orbitals yields the four-lobed structure shown in Fig. 1(d), which qualitatively matches the experimental results. This improved understanding of the links between photofragment images and the electronic dynamics in the molecule can improve endeavors towards image-based strong-field coherent control.

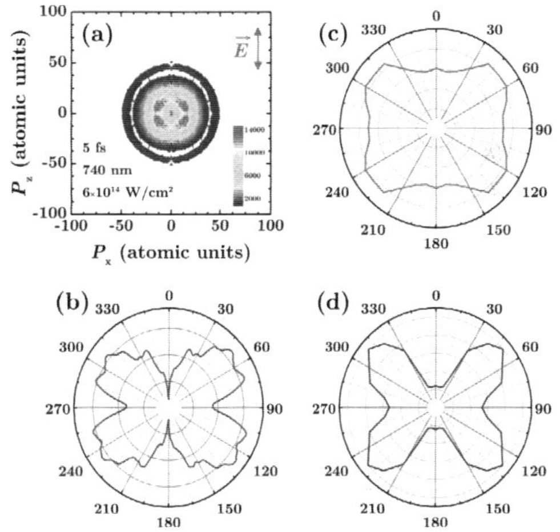

Figure 1. (a) Momentum distribution of $C_2D_3^+$. The inner lobes come from dissociative single ionization of C_2D_4, and the weaker outer ring is due to dissociative double ionization. (b) Measured yield of the inner lobes as a function of the angle between the $C_2D_3^+$ and the laser polarization. Calculated angular distribution for $C_2D_3^+$ (c) without including Rydberg orbitals and (d) with ionization from Rydberg orbitals included.

JRML is supported by the Chemical Sciences, Geosciences, and Biosciences Division, Office of Basic Energy Sciences, Office of Science, U.S. Department of Energy. Augustana College is supported by NSF, EPSCoR, and the South Dakota Space Grant. BJ is also supported by DOE-SCGF. MFK, R dV-R, and RS are supported by DFG via the Cluster of Excellence: Munich Center for Advanced Photonics.

References

[1] P. von den Hoff *et al* 2010 *Appl. Phys. B* **98** 659
[2] E. Wells *et al* 2013 *Nat. Comm.* **4** 2895

[1]E-mail: bjochim@phys.ksu.edu
[2]E-mail: eric.wells@augie.edu

Content from this work may be used under the terms of the Creative Commons Attribution 3.0 licence. Any further distribution of this work must maintain attribution to the author(s) and the title of the work, journal citation and DOI.
Published under licence by IOP Publishing Ltd

XXIX International Conference on Photonic, Electronic, and Atomic Collisions (ICPEAC2015) IOP Publishing
Journal of Physics: Conference Series **635** (2015) 112044 doi:10.1088/1742-6596/635/11/112044

Strong-field dissociation dynamics of molecular dications

Bethany Jochim*[1], T. Severt*[2], M. Zohrabi*, U. Ablikim*, Ben Berry*, B. Gaire*, K.J. Betsch*, F. Anis*, Tereza Uhlíková†, K.D. Carnes*, E. Wells‡, B.D. Esry*, I. Ben-Itzhak*

*J.R. Macdonald Laboratory, Department of Physics, Kansas State University, Manhattan, Kansas 66506 USA
†Department of Analytical Chemistry, Institute of Chemical Technology, Prague Technická, Czech Republic
‡Department of Physics, Augustana College, Sioux Falls, South Dakota 57197 USA

Synopsis We focus on the dissociation of metastable molecular dications induced by intense, ultrafast laser pulses. In particular, we demonstrate the dominant role of commonly-neglected permanent-dipole transitions and drive dissociation via a pump-dump-like mechanism within a single laser pulse.

Many doubly-charged molecules have ground electronic states with lifetimes spanning from picoseconds to seconds [1], thus enabling unique studies. We focus on dissociation of several-keV beams induced by intense, ultrafast laser pulses. Specifically, we demonstrate that permanent-dipole (PD) transitions can play a key role and that the structure of the metastable electronic ground state can delay stimulated emission, leading to pump-dump-like transitions.

In NO^{2+}, dissociation parallel to the laser polarization is prominent at high intensities, as shown in Fig. 1(a). Contrary to common intuition dictating that electronic transitions always prevail, we find that permanent-dipole transitions are extremely important in this system. We demonstrate this with time-dependent Schrödinger equation (TDSE) calculations [2], shown in Fig. 1(b), together with the most likely number of photons involved in the transition, derived from fitting the measured angular distribution in Fig. 1(a) [3]. Specifically, we find that the dominant pathways leading to the aligned feature at high kinetic energy release (KER$\geq$8.9 eV) are multiphoton permanent-dipole-driven vibrational excitations on the electronic ground state leading to its continuum.

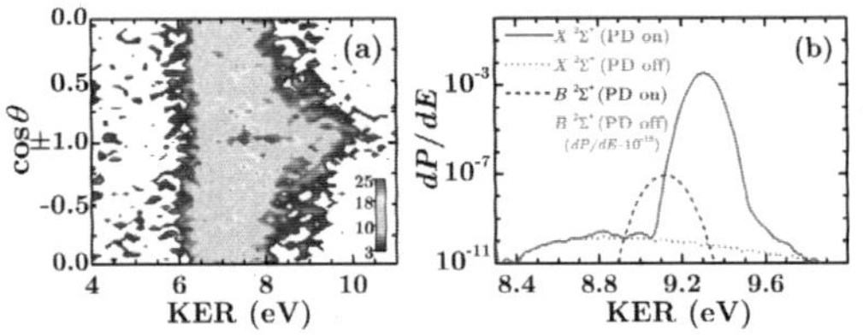

Figure 1. (a) Density plot of $NO^{2+} \rightarrow N^{+}+O^{+}$ dissociation as a function of KER and $\cos\theta$ for 5×10^{15} W/cm^2 laser intensity. (b) Calculated KER spectra for transitions starting from $X\,^2\Sigma^+(v{=}3)$ and dissociating on the indicated state.

In CS^{2+}, photodissociation via $X\,^3\Pi \rightarrow A\,^3\Sigma^-$ transitions yield the main KER peak, marked as 1ω in Fig. 2(a). In addition, a more intriguing peak, marked 0ω, appears at about one-photon energy lower KER. This 0ω peak can be understood as a pump-dump process driven by one laser pulse, where the same $X\,^3\Pi \rightarrow A\,^3\Sigma^-$ excitation ("pump") is followed by delayed stimulated emission ("dump"). This delay, on the order of a few tens of fs, is due to the potential barrier of the electronic ground state, which makes stimulated emission energetically forbidden until the dication has stretched far enough, as indicated in Fig. 2(b). Once the Condon point is reached, stimulated emission is most likely, and the pump-dump-like process results in $C^+ + S^+$ fragments with a KER value reflecting no net energy exchanged with the laser field. Finally, we also observe this "pump-dump-like" mechanism in $CO^{2+}(v{=}0)$.

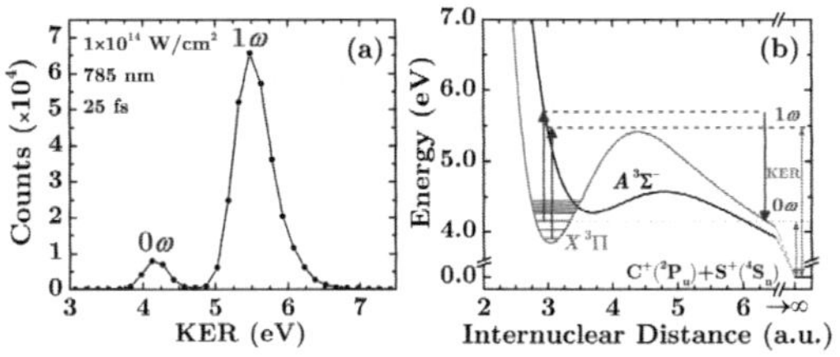

Figure 2. (a) Measured KER spectrum for $CS^{2+} \rightarrow C^+ + S^+$. (b) Potential energy curves for CS^{2+} [4] with dissociation pathways for the one-photon (1ω) and pump-dump (0ω) transitions.

This work was supported by the Chemical Sciences, Geosciences, and Biosciences Division, Office of Basic Energy Sciences, Office of Science, U.S. Department of Energy. BJ is also supported by DOE-SCGF (DE-AC05-06OR23100). TU is supported by GACR and MetaCentrum.

References

[1] D. Mathur 2004 *Physics Reports* **391** 1
[2] F. Anis and B. D. Esry 2008 *Phys. Rev. A* **77** 033416
[3] A. M. Sayler *et al* 2007 *Phys. Rev. A* **75** 063420
[4] T. Šedivcová *et al* 2006 *J. Chem. Phys.* **125** 164308

[1]E-mail: bjochim@phys.ksu.edu
[2]E-mail: tsevert@phys.ksu.edu

Content from this work may be used under the terms of the Creative Commons Attribution 3.0 licence. Any further distribution of this work must maintain attribution to the author(s) and the title of the work, journal citation and DOI.
Published under licence by IOP Publishing Ltd

Journal of Physics: Conference Series **635** (2015) 112045 doi:10.1088/1742-6596/635/11/112045

Carrier-envelope phase control over fragmentation of H_2^+ and D_2

M. Zohrabi* [1], **B. Berry***, **Nora G. Kling***, **Bethany Jochim***, **T. Severt***, **U. Ablikim***, **K. J. Betsch***, **S. Zeng***, **F. Anis***, **Z. Wang***, **M. Kübel**†, **M. F. Kling***†, **K. D. Carnes***, **B. D. Esry***, **and I. Ben-Itzhak***

* J. R. Macdonald Laboratory, Physics Department, Kansas State University, Manhattan, Kansas 66506, USA
† Max-Planck-Institut für Quantenoptik, 85748 Garching, Germany

Synopsis We demonstrate control over fragmentation of H_2^+ and D_2 molecules via the carrier-envelope phase of sub-5 fs laser pulses. Moreover, we attribute our findings to interferences between different pathways involving different net numbers of photons [1], revealing "high-order" pathways and the importance of the bandwidth.

The carrier-envelope phase (CEP) of ultrashort laser pulses is a known control knob for molecular fragmentation [2]. Recently, we performed a detailed study of CEP dependences in H_2^+ dissociation [3] and Rydberg D^* production upon fragmentation of D_2.

In our first experiment, an H_2^+ ion-beam target, produced in an ion source, interacts with sub-5 fs laser pulses. The CEP of every laser shot is tagged using a stereo-ATI CEP meter [4]. We observe CEP-dependent asymmetries in the emission direction of H^+ fragments relative to the laser polarization. The CEP dependence of these asymmetries match that predicted by the general CEP theory [1] and are due to interference of odd and even photon number pathways. To further our understanding, the measurements are quantitatively compared to state-of-the-art *ab initio* calculations that include nuclear rotation and intensity averaging [3].

In the second experiment, the CEP dependence in the fragmentation of D_2 molecules into D^* has been studied using ultrashort linearly-polarized laser pulses. The total dissociation yield of D^* fragments exhibits an oscillatory behavior as a function of CEP, as shown in Fig. 1(a). These oscillations result from interference between photon pathways differing by net two photons (Δn=2). The difference in the emission direction (asymmetry) of these D^* fragments along the laser polarization, shown in Fig. 1(b), reveals a similar CEP dependence to those seen in H_2^+ [3]. It is noted that these CEP dependences vary with the specific band of excited states of the D^* fragments that were measured.

The asymmetry map possesses higher-order oscillations related to a higher net number of photons involved in the interference pathways. We employ a Fourier analysis method to extract information linked to these pathways, as shown in Fig. 1(c). Lastly, for the first time, the amplitude and phase of "higher-order" pathways, specifically those differing by net three photons, are revealed and plotted as a function of kinetic-energy release (KER) in Fig. 1(d).

The broad spectral bandwidth of the laser pulses plays a crucial rule in these CEP dependences. This importance is demonstrated by utilizing chirped pulses of the same spectral bandwidth as our 5 fs FTL laser pulses. Clear CEP dependences persist for pulses chirped up to 20 fs.

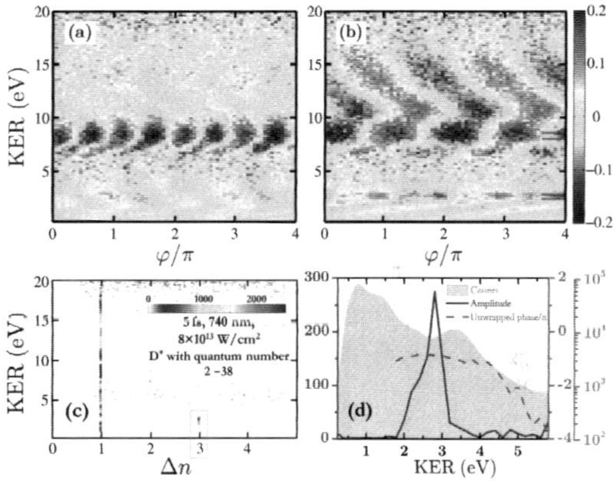

Figure 1. D^* fragment (a) normalized yield and (b) asymmetry as a function of CEP and KER. (c) Fourier transform of the asymmetry map shown in (b) as a function of Δn and KER. (d) Amplitude and phase extracted from (c) for Δn=3 (Gray: the D^* yield averaged over CEP) as a function of KER.

This work was supported by the Chemical Sciences, Geosciences, and Biosciences Division, Office of Basic Energy Sciences, Office of Science, U.S. Department of Energy. BJ is also supported by DOE-SCGF.

References

[1] V. Roudnev and B. D. Esry 2007 *Phys. Rev. Lett.* **99** 220406

[2] M. F. Kling *et al* 2006 *Science* **312** 246

[3] Nora G. Kling *et al* 2013 *Phys. Rev. Lett.* **111** 163004

[4] A. M. Sayler *et al* 2011 *Opt. Lett.* **36** 1

[1] E-mail: mza@phys.ksu.edu

Content from this work may be used under the terms of the Creative Commons Attribution 3.0 licence. Any further distribution of this work must maintain attribution to the author(s) and the title of the work, journal citation and DOI.
Published under licence by IOP Publishing Ltd

XXIX International Conference on Photonic, Electronic, and Atomic Collisions (ICPEAC2015) IOP Publishing
Journal of Physics: Conference Series **635** (2015) 112046 doi:10.1088/1742-6596/635/11/112046

tSURFF — *ab initio* solutions to the time-dependent non-perturbative half-scattering problem

A. Zielinski, V. P. Majety, and A. Scrinzi,[†1]

Department of Physics, Ludwig Maximilians University, Munich, Germany

Synopsis A new approach to the half-scattering problem is introduced and demonstrated on several of the challenging applications in laser-matter interaction: precision single photo-electron spectra, fully differential double-emission spectra at near IR wave-length, and emission from multi-electron systems. A long-standing discrepancy between theory and measurement in the orientation dependence of CO_2 ionization is resolved. Application to many-body systems will be demonstrated and extension to the full scattering problem will be discussed.

The *ab initio* computation of particle emission spectra from systems that undergo strong, time-dependent perturbations require that the solutions are followed until the perturbation ceases and final energies and emission channels can be established. For determining spectra, the asymptotics of the solution must be computed and analyzed. Only in the case of potential scattering, spectral decomposition suffices.

In a series of papers we have introduced a new approach to this problem. It combines mathematically and computationally perfect absorption [1] with the time-dependent analysis of outgoing flux [2] and the time-evolution of the emission channels including double continuum channels [3]. The power of this "time-dependent surface flux" (tSURFF) approach has allowed us to solve key problems in strong laser-matter interaction: (1) the computation of precision photoelectron spectra at elliptic polarization (attoclock, [4]) and at wave-length up to $2\mu m$, (2) accurate fully differential (angle-resolved) double emission spectra from the He atom and from H_2 with polarization parallel to the molecular axis, (3) single photo-emission from the CO_2 molecule in dependence of molecular orientation. In all cases we can provide dependable accuracy estimates.

The attoclock results strictly show that the measured deflection angles in elliptically polarized near-infrared pulses are not compatible with solutions of the single-electron Schrödinger equation, implying that an effect of either exchange or multi-electron dynamics is observed.

An example of precision near-IR double emission calculations is shown in Fig. 1.

For computing emission from molecules, strong field dynamics is combined with quantum chemical structure (COLUMBUS, [5]) including full interchannel coupling and exchange [6]. We find that including exchange is crucial: it explains the observed orientation dependence of ionization of CO_2. Multi-electron effects can clearly be detected, but have no influence on the peak emission angle at 45°.

In addition to these examples, we will present extensions of the method to many-body physics and discuss its potential for applications in standard scattering problems.

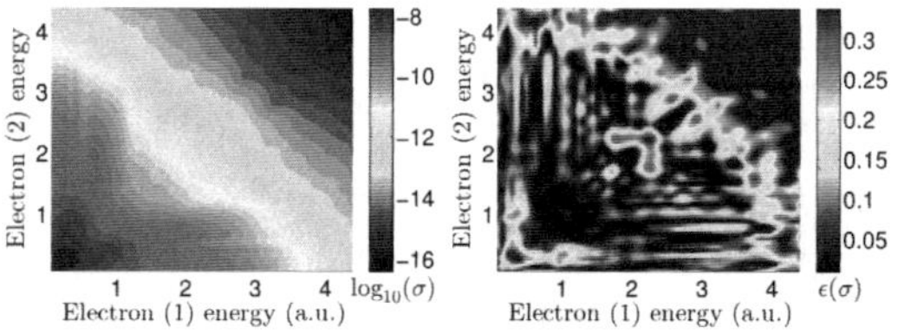

Figure 1. Double emission spectra of He by a 3 cycle 730nm pulse at intensity $3\times10^{14}W/cm^2$ (left) and estimate of the relative error (right). Sequential and non-sequential emission is clearly distinguishable. The fully 6-dimensional calculation was performed during two days on 96 CPUs. The error is $\leq 10\%$ in large part of the spectrum.

References

[1] A. Scrinzi 2010, *Phys. Rev. A* **81**, 1.

[2] Liang Tao and A. Scrinzi 2012, *New J. Phys.* **14**, 013021.

[3] A. Scrinzi 2012, *New J. Phys.* **14**, 085008.

[4] C. Hofmann, A. S. Landsman, A. Zielinski, C. Cirelli, T. Zimmermann, A. Scrinzi, and U. Keller 2014, *Phys. Rev. A* **90**, 043406.

[5] H. Lischka, Th. Mueller, P. G. Szalay, I. Shavitt, R. M. Pitzer, and R. Shepard 2011, *Wiley Interdiscip. Rev.-Comp. Mol. Science* **1**, 191.

[6] V. P. Majety, A. Zielinski, and A. Scrinzi 2014, *New. J. Phys.*, accepted for publication; *arXiv:1412.3666 [physics.comp-ph]*.

[1]E-mail: armin.scrinzi@lmu.de

Content from this work may be used under the terms of the Creative Commons Attribution 3.0 licence. Any further distribution of this work must maintain attribution to the author(s) and the title of the work, journal citation and DOI.
Published under licence by IOP Publishing Ltd

Journal of Physics: Conference Series **635** (2015) 112047 doi:10.1088/1742-6596/635/11/112047

Multicenter effect in high-order above-threshold ionization of polyatomic molecules

C. Wang[1], X. Hao[3], M. Okunishi[2], J. Chen[4,5*], Y. Yang[1], M. Zhang[1], B. Yan[1], D. Ding[1†] and K. Ueda[2‡]

[1]*Institute of Atomic and Molecular Physics, Jilin University, Changchun 130012, PR China*
[2]*Institute of Multidisciplinary Research for Advanced Materials, Tohoku University, Sendai 980-8577, Japan*
[3]*Institute of Theoretical Physics, Shanxi University, PR China*
[4]*Institute of Applied Physics and Computational Mathematics, P. O. Box 8009, Beijing 100088, China*
[5]*HEDPS, Center for Applied Physics and Technology, Peking University, Beijing 100084, China and*

Synopsis The multicenter nature of molecules can affect the strong field ionization. By studying the resonance-like enhancement structure in the high-order ATI spectrum of C_2H_4 and C_2H_6, we illustrate how the multicenter effect contributes to the molecular tunneling ionization. We find the hydrogen contributions lead to the weak resonance-like structure of C_2H_6 but have no affect for C_2H_4, this target dependence origins from their different HOMO structures. Those conclusions are supported by Strong Field Approximation calculation.

Ionization of rare atoms is frequently described by the tunneling model, which works well for the predicate ionization rate and high-order above-threshold ionization. However, laser tunneling of molecule is more complex because of its structure complexity and the extra nuclear degrees of freedom and this issue still needs further research. Hydrocarbons are proper targets to test the multicenter interference and the well known resonance-like enhanced (RLE)[1] structures in high-order ATI (HATI) photoelectron spectrum can be used to show this effect. When the laser peak intensity changes only very slightly, the magnitudes of HATI peaks, located at $6Up$ and $8Up$, can exhibits a significant enhancement by up to an order, so called RLE.

Fig.1 depicts the measured ATI spectra of C_2H_4 and C_2H_6 evolving as function of the laser intensities. To construct 2D image of fig.1, the ATI spectrum was normalized at each laser intensity to clearly emphasize the overall evolution of the spectra with the laser intensity. For C_2H_4, singly strong RLE pattern has been found around 30 eV at 45 TW/cm^2, and then suppressed as increase laser intensity. But for C_2H_6, First RLE patterns appear around 25 eV at 35 TW/cm^2, as increase the laser intensity, another RLE structures appear at higher energy region (around 35 eV), further increase the laser intensity, weaker RLE structure still can be observed for higher energy region. To explore the origin of this difference, we implement the numerical calculation in the framework of the Strong Field Approximation. The simulation for C_2H_4 and C_2H_6 gives qualitatively same results as our experiment. C_2H_4 has a HOMO with lobes perpendicular to the molecular plane, the multicenter interference are expected quite weak, then C_2H_4 should have strong RLE structure because of constructive interference of long "quantum orbits" at channel closing. For C_2H_6, only the hydrogen atoms contribute to constructive interference, which produced weak multiple RLE structures. Those results clearly show the multicenter effect in the strong field tunneling ionization of polyatomic molecules.

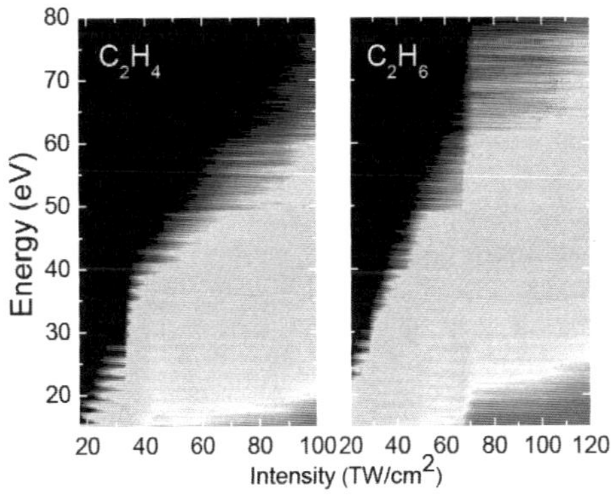

Figure 1. The measured ATI spectra of C_2H_4 and C_2H_6. The horizontal axis reflects the laser intensity and the vertical axis gives the kinetic energy of ATI electrons. The false colour scale reflects the ionization yields. The bright yellow horizontal lines indicate the RLE structures as discussed in the text.

References

[1] W. Quan *et al* 2013 *Phys. Rev. A.* **88** 021401(R)

*E-mail: chen_jing@iapcm.ac.cn

†E-mail: dajund@jlu.edu.cn

Content from this work may be used under the terms of the Creative Commons Attribution 3.0 licence. Any further distribution of this work must maintain attribution to the author(s) and the title of the work, journal citation and DOI.
Published under licence by IOP Publishing Ltd

Dynamical effects in the vibrationally resolved C $2s^{-1}$ photoionization cross section ratios of Methane

Saikat Nandi* [1]**, Minna Patanen*, Catalin Miron*§, John D. Bozek*, Etienne Plésiat†, Fernando Martín†, Daniele Toffoli‡ and Piero Decleva‡** [2]

*Synchrotron SOLEIL, L'Orme des Merisiers, Saint-Aubin, BP 48, 91192 Gif-sur-Yvette Cedex, France
†Departamento de Química, Módulo 13, Universidad Autónoma de Madrid, Cantoblanco 28049, Madrid, Spain
‡ Dipartimento di Scienze Chimiche e Farmaceutiche, Universita di Trieste, Via L. Giorgieri 1, 34127 Trieste, Italy
§Extreme Light Infrastructure - Nuclear Physics (ELI-NP), 'Horia Hulubei' National Institute for Physics and Nuclear Engineering, 30 Reactorului Street, RO-077125 Măgurele, Jud. Ilfov, Romania

Synopsis The vibrationally resolved C 2s photoionization cross-section of methane was investigated both theoretically and experimentally. When compared to that of C 1s photoionization, a rather different pattern has been observed, suggesting a strong interplay between the electron diffraction and interference effects.

Interference and diffraction effects in molecular photoionization have been recently revealed and subjected to a number of studies [1-5]. The experimental observation of such effects is very difficult when measuring absolute cross sections, but they can be evidenced in a much easier way by monitoring the photon energy dependence of the appropriate ratios of partial cross sections. One such possibility is the study of v-ratios that is cross section ratios of different vibrational levels in the same electronic channel. In absence of any diffraction and interference effects, the v-ratios should be essentially energy-independent, a constant equal to the Franck-Condon ratios. However, the experimentally studied v-ratios of valence [1] and core ionized [4,5] show apparent energy-dependent oscillations, which can be qualitatively predicted from the Cohen-Fano model [1,3] and simple diffraction effects [2,4,5]. The method to study v-ratios has proven to be a very sensitive probe, as it can reveal even the tiny diffraction associated with the light hydrogen atoms, as has been shown in the core ionization of CH_4 [2].

A key feature of the oscillations is their relationship to the molecular geometry, and in particular the bond length between emitting and scattering centers [5]. While the observed patterns are quite accurately reproduced by high quality ab-initio theoretical calculations, a full understanding of all factors involved is still incomplete. To this end we have compared the behavior of v-ratios in the photoionization of the inner-valence $2a_1$ orbital of methane which is a linear combination of C 2s and H 1s orbitals, with the completely localized C 1s core orbital. The vibrationally resolved photoionization cross section ratios of the symmetric stretching mode in the C 1s and C 2s photoionization of CH_4 were measured at PLÉIADES beam-line [6] using a VG-Scienta R4000 spectrometer. Unexpectedly, both experimental and theoretical v-ratios of $2a_1$ photoionization exhibit a quite different pattern compared to the C1s photoionization, indicating that in case of the $2a_1$ state the relative cross section oscillations are also modulated by other effects in addition to the diffraction of an electron issued from a localized emitter. A possible explanation lays in the more delocalized and spatially extended nature of the initial $2a_1$ (C 2s) orbital, although additional complications (e.g. apparent predissociation effects in the experimental spectra) may also come into play. A better understanding of the behavior of the v-ratios of an inner-valence orbital may give us not only structural information (e.g., the C-H distance in this case) but also reveals information of other contributing factors such as the molecular orbital composition (i.e. electron density).

References

[1] S. E. Canton, E. Plésiat, J. D. Bozek, B. S. Rude, P. Decleva, and F. Martín 2011 *Proc Natl Acad Sci USA* **108** 7302
[2] E. Plésiat, L. Argenti, C. Miron, E. Kukk, K. Ueda, P. Decleva and F. Martín 2012 *Phys. Rev. A* **85** 023409
[3] R. K. Kushawaha M. Patanen, R. Guillemin, L. Journel, C. Miron, M. Simon, M. N. Piancastelli, P. Decleva 2013 *Proc Natl Acad Sci USA* **110** 15201
[4] K. Ueda *et al* 2013 *J. Chem Phys.* **139** 124306
[5] E. Kukk *et al* 2013 *Phys. Rev. A* **88** 033412
[6] http://www.synchrotron-soleil.fr/Recherche/LignesLumiere/PLEIADES

[1] E-mail: saikat.nandi@synchrotron-soleil.fr
[2] E-mail: decleva@units.it

Content from this work may be used under the terms of the Creative Commons Attribution 3.0 licence. Any further distribution of this work must maintain attribution to the author(s) and the title of the work, journal citation and DOI.
Published under licence by IOP Publishing Ltd

Journal of Physics: Conference Series **635** (2015) 112049 doi:10.1088/1742-6596/635/11/112049

Hydrogen migration in photodissociation of the pyridine molecules

Tomasz J. Wasowicz*[1], Antti Kivimäki[†,‡], Marcello Coreno[‡,§] and Mariusz Zubek*

[*] Department of Physics of Electronic Phenomena, Gdańsk University of Technology, 80-233 Gdańsk, Poland
[†] CNR-IOM, Laboratorio TASC, 34149 Trieste, Italy
[‡]Gas Phase beamline@Elettra, Basovizza Area Science Park, 34149, Trieste, Italy
[§]CNR-IMIP, Monterotondo, 00016 Roma, Italy

Synopsis Photodissociation of the pyridine molecules has been studied experimentally over the 16-70 eV photon energy range. Formation of the excited NH($A^3\Pi$) radicals has been observed among several other atomic and diatomic fragments, using the photon-induced fluorescence spectroscopy. Since the pyridine molecules do not contain the NH structural units, this is an indication of the bond rearrangement processes associated with migration of the hydrogen atoms prior to dissociation. The plausible mechanisms for the NH formation will be discussed.

In the migration process the hydrogen (proton) atom moves from one location to another within a molecule [1]. This movement takes place in a chemical bond rearrangement leading to isomeric deformation of the initial molecular geometrical structure of the molecule and may open new dissociation reaction pathways [2-5]. It usually occurs on a femtosecond time scale [3] and is faster than the molecular bond breaking in dissociation. It may therefore be pertinent for controlling the chemical bond scission (or formation), for example, in combustion [6], catalytic [7] or the biological radiation damage [5] processes.

Pyridine, C_5H_5N, (Figure 1) is a simple six-membered azine, containing one nitrogen N atom that replaces methine group CH in benzene. Pyridine occurs in nature within the molecular structure of many compounds that play important roles in a wide range of the biological media.

In the present study, the H atom migration was observed, for the first time, in the photodissociation processes of the pyridine molecules in the gas-phase applying photon-induced fluorescence spectroscopy (PIFS) [8]. The excited NH ($A^3\Pi$) radicals were detected by observing its $A^3\Pi \rightarrow X^3\Sigma^-$ fluorescence (Figure 1). Emission of other excited dissociation fragments was also identified [9]. NH is not a structural unit of the pyridine molecule and its fluorescence gives a clear evidence of the hydrogen migration prior to photodissociation. The fluorescence yields from the excited NH($A^3\Pi$) and H(n=5) (H_γ line) fragments were measured as a function of the incident photon energy in the region 16-70 eV (Figure 2). The observed yields rise at about 10 eV above the first ionization potential of pyridine indicating that the superexcited states may play a role in producing the excited NH($A^3\Pi$) radicals. The possible mechanism for NH formation based on recent *ab initio* molecular calculations will be discussed.

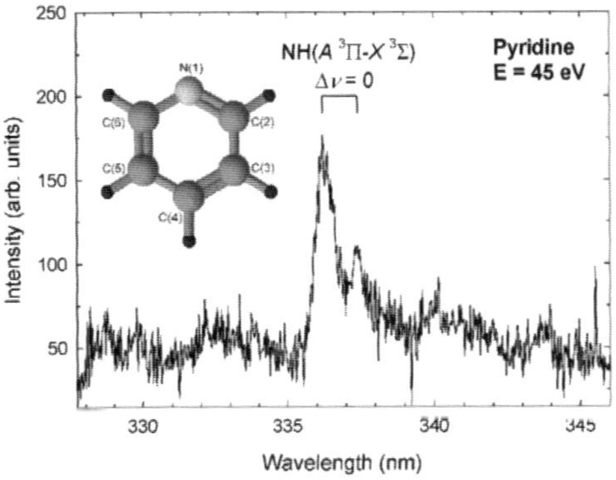

Figure 1. Emission spectrum showing the $A^3\Pi \rightarrow X^3\Sigma^-$ bands at 336.2 and 337.4 nm.

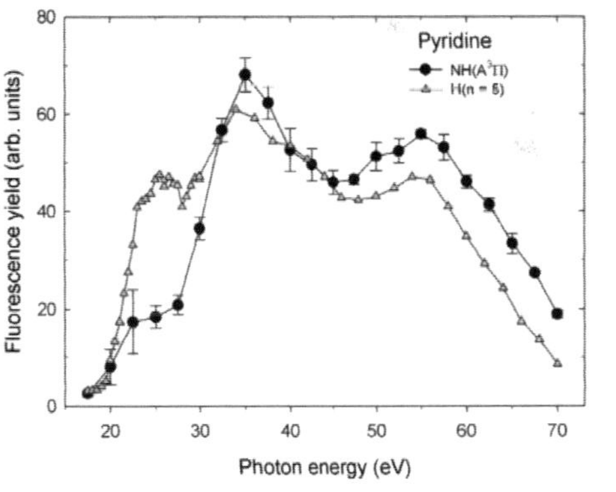

Figure 2. The NH($A^3\Pi$) and H(n=5) fragmentation yields obtained in the 16-70 eV photon energy range.

References

[1] H. F. Schaefer III 1979 *Acc. Chem. Res.* **12** 288
[2] K. M. Morgan *et al* 2001 *J. Org. Chem.* **66** 1600
[3] T. Osipov *et al* 2003 *Phys. Rev. Lett.* **90** 233002
[4] Y. H. Jiang *et al* 2010 *Phys. Rev. Lett.* **105** 263002
[5] M. Zubek *et al* 2014 *J. Chem. Phys.* **141** 064301
[6] A. Ratkiewicz 2013 *Reac. Kinet. Mech. Cat.* **108** 545
[7] D. R. Killelea *et al* 2008 *Science* **319** 790
[8] T.J. Wasowicz *et al* 2011 *Phys. Rev.A* **83** 033411
[9] T.J. Wasowicz *et al* 2014 *J. Phys. B* **47** 055103

[1]E-mail: twasowicz@mif.pg.gda.pl

Content from this work may be used under the terms of the Creative Commons Attribution 3.0 licence. Any further distribution of this work must maintain attribution to the author(s) and the title of the work, journal citation and DOI.
Published under licence by IOP Publishing Ltd

Resonant Auger-decay induced fragmentation of glycine molecule

L. Fang [1, a], H. Xiong [2], T. Osipov [3], V. S. Petrovic [4] and N. Berrah [2, b]

[1]Center for High Energy Density Science, University of Texas, Austin, TX 78712, USA
[2]University of Connecticut, Physics Department, Storrs, CT 06269, USA
[3]LCLS, SLAC, Menlo Park, CA 94025, USA
[4]PULSE Institute, SLAC, Menlo Park, CA 94025, USA

Synopsis We investigated the fragmentation of glycine molecule induced by resonant Auger-decay following core-shell ionization at different atomic sites by synchrotron light. We show that the dissociation pathways as well as the charge state distributions are modified due to core-level photoionization followed by Auger-decay at the different target atomic sites. The initial states for the dissociation are associated with vacancies in valence or inner-valence shells that are not localized as in core shells. We also compare the molecular dissociation following resonant and non-resonant Auger decay processes to gain insight into the influence on bond cleavage by different ionization mechanisms.

Molecular dissociation has been one of the important topics in studies of molecular interaction with photons. Fragmentation patterns and the final charge distributions bear information on the electronic properties of the initial states of the molecule and even the temporal information of the interaction [1]. We investigated fragmentation of glycine molecule induced by resonant Auger decay processes of core-shell vacancies at different atomic sites. By comparing the scenarios for different target sites of the initial core-level photoexcitation, we are able to gain insight into the various electronic states that leads to the asymptotic ionic outcomes. Moreover, we compare the dissociation pattern and charge distribution for cases with photon energy above and on resonance of K-shell electrons at particular atomic sites. This allows us to gain information regarding the roles the population mechanisms play on determining the final fragment outcomes.

The experiment was carried at the Advanced Light Source (ALS) at the Lawrence Berkley National Laboratory (LBNL). The photon energies were judiciously chosen to core -excite or -ionize the carbon and oxygen atoms in glycine molecule. Photon energy both above and at resonance of the K-edge of the above atomic sites were used. Ion signals were recorded by a velocity map imaging (VMI) spectrometer with a hexanode delay-line detector which enables there-dimensional momentum measurement.

In figure 1, we show the ion time-of-flight (TOF) spectra taken at various photon energies. As seen in these figures, resonant Auger-decay processes lead to different ion yields branching ratios with certain peak enhanced, compared to that by normal Auger-decay. Resonant Auger-decay of core vacancies at C and O results in different charge distribution at particularly the single atomic C, N, O ions with or without hydrogen atoms attached. We will also present the coincidence measurements showing various dissociation channels at different photon energies.

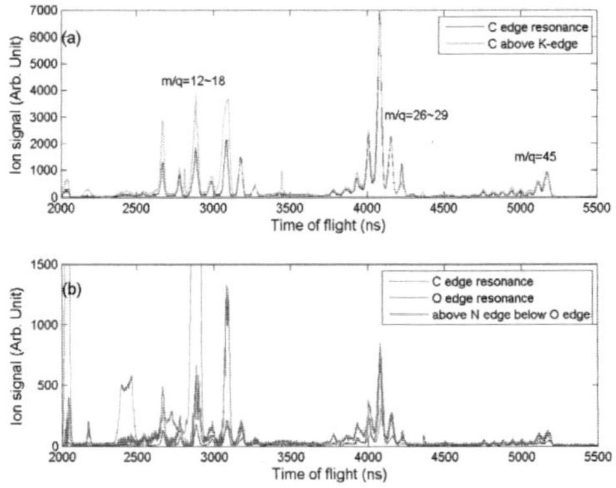

Figure 1. (a) Ion TOF spectra with photon energies at (a) C K-edge resonance and above C K-edge (normalized to the m/q=~26 group) and at (b) O and C K-edge resonance (normalized to the m/q=~26 group); also shown is the spectrum with a photon energy above N but below O K-edge (normalized to N^+ peak in the O resonance spectrum without background subtraction).

This work was funded with DOE-BES, No. DE-FG02-92ER14299.

References

[1] L. Fang *et al.* 2012 *Phys. Rev. Lett.* **109** 263001

[a] E-mail: lifang@austin.utexas.edu [b] E-mail: Nora.Berrah@uconn.edu

Content from this work may be used under the terms of the Creative Commons Attribution 3.0 licence. Any further distribution of this work must maintain attribution to the author(s) and the title of the work, journal citation and DOI.
Published under licence by IOP Publishing Ltd

XXIX International Conference on Photonic, Electronic, and Atomic Collisions (ICPEAC2015) IOP Publishing
Journal of Physics: Conference Series **635** (2015) 112051 doi:10.1088/1742-6596/635/11/112051

Direct Signature of Light-Induced Conical Intersections

Ágnes Vibók*[1], Gábor J. Halász†[2]

* Department of Theoretical Physics, University of Debrecen, H-4010 Debrecen, PO Box 5, Hungary
† Department of Information Technology, University of Debrecen, H-4010 Debrecen, PO Box 12, Hungary

Synopsis By analyzing the photodissociation process of the D_2^+ molecule, we found a robust effect in the angular distribution of the photofragments which serves as a direct signature of the light-induced conical intersection (LICI) providing undoubted evidence for its existence.

Nonadiabatic effects are ubiquitous in physics, chemistry and biology. They are strongly amplied by conical intersections (CIs) which are degeneracies between electronic states of triatomic or larger molecules. A few years ago it has been revealed that CIs in molecular systems can be formed both by running and standing laser light even in diatomics. In this situation the rotational degree of freedom comes into play and serves as a proper second degree of freedom so as to form the branching space where the degeneracy is lifted. It has been demonstrated in several former theoretical works that due to the prevailing strong nonadiabatic couplings, the LICIs have strong impact on the dynamical properties of molecular systems. Recently, we have made an attempt to provide and analyze a physical event which may serve as an undoubted evidence of the LICI, giving a direct observable signature of the presence of this intersection [1].

It is known in the field of nonadiabatic molecular dynamics that due to the extreme breakdown of the Born-Oppenheimer approximation, conical intersections are responsible for ultrafast radiationless processes, typically on the femtosecond time scale. They provide pathways for extremely fast population transfer between electronic states. This latter effect is probably the most important inherent feature of the CIs. Nevertheless, until now one could not find an unambiguous experimentally measurably quantity which reflects directly this population transfer between electronic states for a LICI.

Recently a physical process has been found, where an ultrafast population transfer takes place between the electronic states of a diatomic molecule, providing direct evidence for the existence of the LICI. The photodissociation dynamics of the D_2^+ molecule serves as a show case physical example. We used the exact time dependent Hamiltonian and calculated the dissociation rate which provides a direct and experimentally measurable impact of the LICI. One and two dimensional calculations (1d, 2d) have been performed. In the 1d calculations the Hamiltonian depends only parametrically on the rotational degree of freedom (θ), therefore the LICI is not taken into account. The structure and magnitude of the 2d ($\nu = 5$) dissociation rates close to $\theta = \pi/2$ undoubtedly demonstrate the strong nonadiabatic effects due to the presence of the LICI.

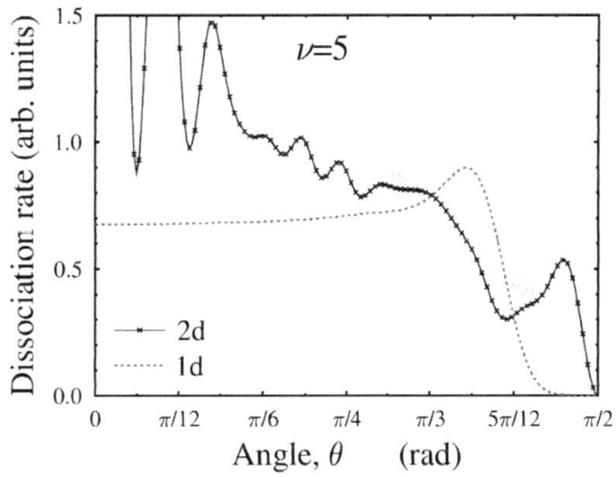

Figure 1. Fragment angular distributions of the dissociating D_2^+ molecule for initial vibrational states $\nu = 5$. Curves are presented both for one dimensional (1d) and for two dimensional (2d) cases. The applied field intensity is $1 \times 10^{14}\ W/cm^2$.

References

[1] Halász, G. J. *at al J. Phys. Chem. Lett.*, 6, 348, (2015)

[1]E-mail: vibok@phys.unideb.hu
[2]E-mail: halasz@inf.unideb.hu

Content from this work may be used under the terms of the Creative Commons Attribution 3.0 licence. Any further distribution of this work must maintain attribution to the author(s) and the title of the work, journal citation and DOI.
Published under licence by IOP Publishing Ltd

XXIX International Conference on Photonic, Electronic, and Atomic Collisions (ICPEAC2015) IOP Publishing
Journal of Physics: Conference Series **635** (2015) 112052 doi:10.1088/1742-6596/635/11/112052

Femtosecond x-ray induced fragmentation of $Ho_3N@C_{80}$

H. Xiong*[1], B. Murphy[2], L. Fang[3], T. Osipov[4], E. Kukk[5], M. Guehr[6], R. Feifel[7], V. S. Petrovic[6], K. R. Ferguson[4, 8], J. D. Bozek[4], C. Bostedt[4], L. J. Frasinski[9], P. H. Bucksbaum[6], J. C. Castagna[4] and N. Berrah[1]

[1]University of Connecticut, Physics Department, Storrs, CT 06269, USA
[2] Western Michigan University, Physics Department, Kalamazoo, MI 49008, USA
[3]University of Texas, Center for High Energy Density Science, Austin, TX 78712, USA
[4]LCLS, National Accelerator Laboratory, Menlo Park, CA 94025, USA
[5] University of Turku, Department of Physics and Astronomy, FI-20014 Turku, Finland
[6] Stanford PULSE Institute, SLAC National Accelerator Laboratory, Menlo Park, CA 94025 USA
[7] University of Gothenborg, Physics Department, SE-412 96 Gothenburg, Sweden
[8]Department of Applied Physics, Stanford University, Stanford CA 94305, USA
[9]Imperial College London, Department of Physics, UK

Synopsis - We report on new data regarding on the photoionization and fragmentation dynamics of gas phase endohedral fullerenes $Ho_3N@C_{80}$ obtained with intense femtosecond X-ray pulses from the Linac Coherent Light Source (LCLS) free electron laser (FEL).

Investigations of ionization and fragmentation dynamics of nano-size fullerenes subjected to ultrashort strong x-ray laser field [1] are very important for understanding the mechanisms of radiation damage and will benefit the development of bio-imaging techniques and radiotherapies. Recently, we reported an investigation on photoionization and fragmentation of C_{60}. Although the dynamics were complicated because of the number of electrons and atoms involved, the experimental results could be well reproduced by molecular dynamics (MD) modeling with electronic structure and nuclear motion taken into account [2]. To understand the dynamics further, we extended our work to endohedral fullerenes, $Ho_3N@C_{80}$.

The experiment was carried out at the AMO hutch at the Linac Coherent Light Source (LCLS) at Stanford. Gas phase $Ho_3N@C_{80}$ molecules were evaporated from a heated oven and interacted with strong x-ray laser pulses. The photon energy of the x-rays was chosen at 1530 eV to optimize the ionization from the Ho atoms, which were encapsulated in C_{80} cages along with a nitrogen atom in the center. The pulse duration of the FEL was set to 80 fs and the pulse energy was nominally 2.2 mJ. The x-rays were focused to a spot approximately 5μm×5μm. A typical mass spectrum is shown in Figure 1. Parent ions up to quintuply charged $Ho_3N@C_{80}^{5+}$ were observed, suggesting a stable structure even with 5 charges on the parent molecule. Atomic Ho^+ ions, originating from the encapsulated Ho_3N molecule, were observed at the highest yield compared to other fragments and parent ions.

In this contribution, we will also present the kinetic energies of fragments and compare them with our prior C_{60} experiment. Furthermore, the correlation between ions will be examined by the covariance mapping technique.

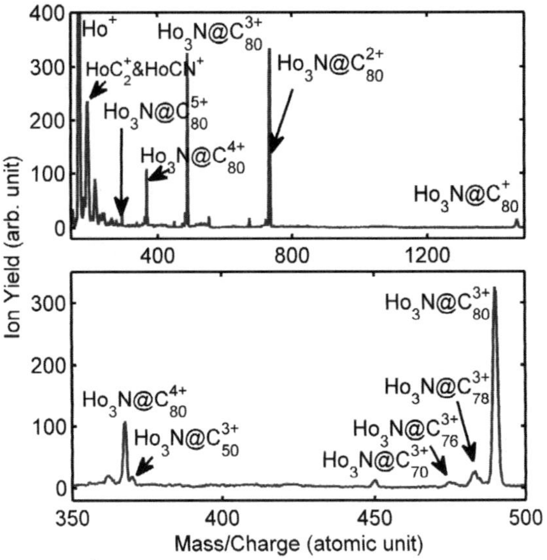

Figure 1. Fragmentation pattern of gas phase $Ho_3N@C_{80}$ exposed to an intense femtosecond X-ray laser beam at 1530eV. The top panel displays the ion fragments from Ho^+ to parent ions while the bottom panel focuses on the triply charged ions.

References

[1] N. Berrah, P.H. Bucksbaum 2014 *Scientific American* **310** 64
[2] B. Murphy *et al* 2014 *Nature Comm.* **5** 4281

* E-mail: hui.xiong@uconn.edu

Content from this work may be used under the terms of the Creative Commons Attribution 3.0 licence. Any further distribution of this work must maintain attribution to the author(s) and the title of the work, journal citation and DOI.
Published under licence by IOP Publishing Ltd

XXIX International Conference on Photonic, Electronic, and Atomic Collisions (ICPEAC2015) IOP Publishing
Journal of Physics: Conference Series **635** (2015) 112053 doi:10.1088/1742-6596/635/11/112053

Nondipole Effects in Chiral Molecules Measured with Linearly Polarized Light

K.P. Bowen*[1], O. Hemmers*, R. Guillemin†, W.C. Stolte‡, M.N. Piancastelli§, D.W. Lindle*

*Department of Chemistry & Biochemistry, University of Nevada Las Vegas, Las Vegas, NV, USA
† Laboratoire de Chimie Physique – Matière et Rayonnement, Université Pierre et Marie Curie, CNRS UMR 7614, Paris Cedex 05, France
‡National Security Technologies, LLC (NSTec), Livermore Operations, Livermore, CA, USA
§ Department of Physics and Astronomy, Molecular and Condensed Matter Physics, Uppsala Universitet, Uppsala, Sweden

Synopsis: We present the first-ever measurement of nondipole chiral angular distribution parameters for C 1s photoemission from each enantiomer of camphor in the photon energy range 296-343eV using linearly polarized light. The angular distribution parameters are determined to be enantiomer-specific, suggesting a new form of linear dichroism.

Over the past two decades, it has been demonstrated that higher-order corrections to the dipole approximation are necessary for the description of light-matter interactions in the soft x-ray range. These so-called 'nondipole effects' present themselves as asymmetries in the angular distributions of photoelectrons. Chiral molecules, known to have asymmetries in photoelectron angular distributions when exposed to circularly polarized light, have been proposed to demonstrate a chiral-specific nondipole effect when exposed to linearly polarized light[1]. This chiral-specific contribution to the angular distribution of photoelectrons is characterized by the parameter ε, which introduces a "twisting" of the angular distribution in the plane containing the polarization and perpendicular to the incoming photon propagation. This study presents ε(hν) values, as well as the conventional angular distribution asymmetry parameter β(hν), for C 1s photoemission from both (R)-Camphor and (S)-Camphor, in the range of 5-50eV above the C 1s threshold.

The measurements were performed using electron time-of-flight (TOF) spectroscopy at the Advanced Light Source (ALS) at Lawrence Berkeley National Lab (LBNL) in Berkeley, CA.

The individual parameter values with respect to photon energy are presented in figures 1 and 2, respectively. The large difference in ε(hν) between enantiomers implies a new form of linear dichroism.

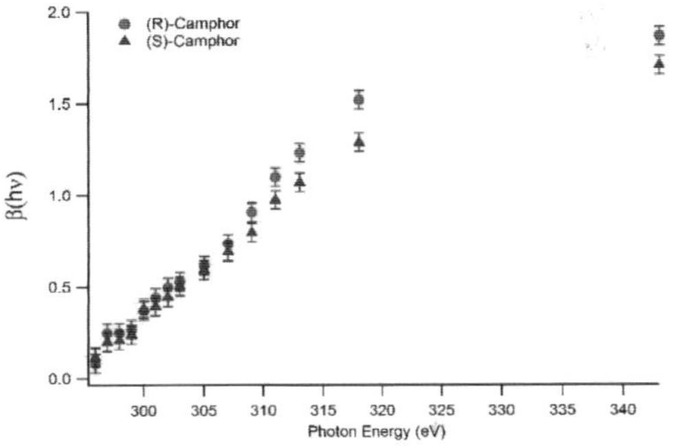

Figure 1. The anisotropy parameter β(hν) for C 1s photoemission from (R)-Camphor and (S)-Camphor

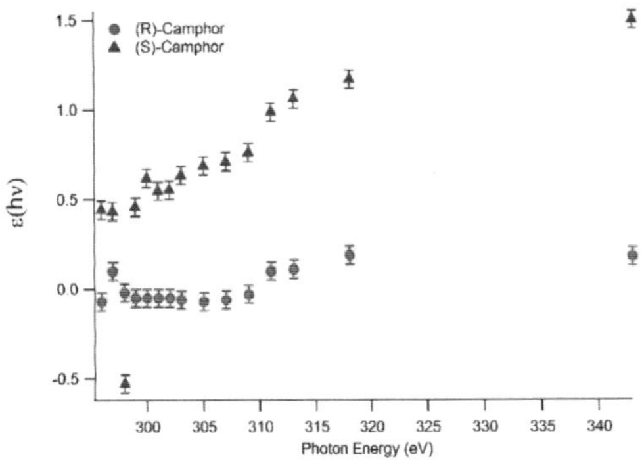

Figure 2. The chiral nondipole paremeter ε(hν) for C 1s photoemission from (R)-Camphor and (S)-Camphor

References

[1] A.N. Grum-Grzhimailo 2003 *J. Phys. B* **36** 2385

[1] Email: bowenk4@unlv.nevada.edu

Content from this work may be used under the terms of the Creative Commons Attribution 3.0 licence. Any further distribution of this work must maintain attribution to the author(s) and the title of the work, journal citation and DOI.
Published under licence by IOP Publishing Ltd

Journal of Physics: Conference Series **635** (2015) 112054 doi:10.1088/1742-6596/635/11/112054

Laser-induced delayed electron emission of Co_4^- anions

Sebastian George*[1], Klaus Blaum*, Christian Breitenfeldt*†, Jürgen Göck*, Jonas Karthein*, Thomas Kolling§, Christian Meyer*, Jennifer Mohrbach§, Gereon Niedner-Schatteburg§, Dirk Schwalm*, Lutz Schweikhard†, Andreas Wolf*

* Max-Planck-Institut für Kernphysik, Saupfercheckweg 1, 69117 Heidelberg, Germany

† Institut für Physik, Ernst-Moritz-Arndt Universität, Felix-Hausdorff-Str. 6, 17487 Greifswald, Germany

§ Fachbereich Chemie, Technische Universität Kaiserslautern, Erwin-Schrödinger-Straße 52, 67663 Kaiserslautern, Germany

Synopsis The cryogenic trap for fast ion beams (CTF) located at the Max-Planck-Institut für Kernphysik has been used to study laser-induced delayed electron emission of various molecular and metal-cluster anions. We will discuss recent experiments investigating the radiative cooling respectively heating of Co_4^- anions produced in hot and cold ion sources. The experiments will be continued at the cryogenic storage ring (CSR), which is presently being commissioned, to disentangle laser-induced electron emission from atom or cluster dissociation.

In the CTF [1] stored ions with an energy of some keV are reflected back and forth between two electrostatic mirrors. Einzel lenses focus the ion beam towards the trap center ensuring radial confinement. The CTF can be operated at any temperature between 300 K and 4 K. Excellent vacuum conditions allow storage times up to seconds at room temperature and minutes under cryogenic conditions. Confined particles can interact with a laser beam in cross beam configuration within a field free region between the mirrors. Neutralized particles leaving the trap through holes in the electrodes of one of the mirror are detected by a microchannel plate detector. The setup is well suited to investigate dynamical processes of stored ion beams. Vibrational electron autodetachment (also called delayed electron detachment) is observed by measuring the rate of neutral particles escaping from the EIBT as a function of storage time as well as a function of the time after laser excitation.

In a recent measurement campaign the electron emission of cobalt cluster anions consisting of four atoms has been investigated. The ions were produced in two different types of ion sources: (A) a cesium sputter ion source, producing ions ro-vibrational temperatures around 1000 Kelvin, (B) a laser vaporization ion source with helium expansion to produce Co_4^- with ro-vibrational excitation levels corresponding to cryogenic temperatures. The ions were stored in the CTF at room temperature and the cooling respectively heating of the ions was visible in the change of the delayed electron detachment rate. The photo excitation measurements were performed by a pulsed, tunable laser system (OPO) as a function of storage time and wavelength of the laser. In figure 1 the counts of delayed electron emission for Co_4^- produced with source (A) are displayed in dependence of storage time (top) and photon energy (bottom); they clearly reflect the change of the clusters internal energy distribution by radiative cooling. These results together with recent findings using the cold ion source (B) will be discussed.

After successful commissioning of the CSR at cryogenic temperatures these experiments will be extended towards the observation of the competing photofragmentation processes.

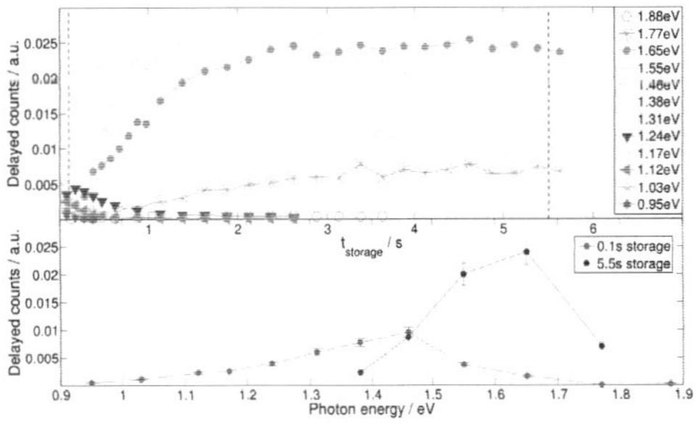

Figure 1. Top panel: The counts of delayed neutralized Co_4^- clusters are plotted for different photon energies for excitation (legend) as a function of storage time. Bottom panel: Delayed electron count-rate in dependence of the laser excitation energy for two storage times indicated by dashed lines in the upper panel. (All data taken at room temperature)

References

[1] M. Lange et al., Rev. Sci. Instrum. **81** (2010) 055105.

[1]E-mail: Sebastian.George@mpi-hd.mpg.de

Content from this work may be used under the terms of the Creative Commons Attribution 3.0 licence. Any further distribution of this work must maintain attribution to the author(s) and the title of the work, journal citation and DOI.
Published under licence by IOP Publishing Ltd

Journal of Physics: Conference Series **635** (2015) 112055 doi:10.1088/1742-6596/635/11/112055

Temperature dependence on VUV-absorption cross sections of CO_2 molecule in the energy region of 10.6 – 11.8 eV

M. Hoshino[1,†], M. Matsui[1], Y. Mochizuki[1], T. Odagiri[1], A. Kondo[2], K. Shigemura[2], M. Kitajima[2], N. Watanabe[3], J. Adachi[4], P. Limão-Vieira[5] and H. Tanaka[1]

[1] Department of Material and Life Sciences, Sophia University, Tokyo 102-8554, Japan.
[2] Department of Chemistry, Tokyo Inst. of Tech., Tokyo 152-8551, Japan.
[3] Institute of Multidisciplinary Research for Advanced Materials, Tohoku University, Sendai 980-8577, Japan.
[4] Photon Factory, Institute of Materials Structure Science, Tsukuba 305-0801, Japan.
[5] Laboratório de Colisões Atómicas e Moleculares, CEFITEC, Departamento de Física, FCT-Universidade Nova de Lisboa, P-2829-516 Caparica, Portugal.

Synopsis We have developed a new setup to measure temperature dependence on vacuum ultraviolet absolute photo-absorption cross sections for linear triatomic molecules from room temperature (~300 K) up to 700 K. The setup consists of a stainless steel absorption-cell wrapped by sheath wires of 1.2φ in diameter and 2 m length. For the first set of measurement on CO_2 in the photon energy range 10.6 – 11.8 eV, we found that the cross section values at room temperature are in very good agreement with previous measurements as well as the temperature dependence on the cross sections are clearly observed in the measured temperature range.

The vibrational wave function of the vibratioanlly-excited molecule has a larger spatial extent than that of the vibrational-ground state. Therefore, spectroscopic studies of vibrationally-excited molecules can probe different regions of potential energy surfaces of electronically excited states based on the Franck-Condon principle. It is expected that molecules in which non-totally symmetric vibrations are excited, may exhibit new properties due to symmetry breaking. Indeed the dramatically enhanced vibronic-coupling effects were clearly observed in electron impact excitation [1] reported recently as well as our previous x-ray absorption spectra of the vibrationaly-excited CO_2 molecules [2]. In this work, in order to observe a temperature effect of valence-electron excitations, we have developed a setup to measure the temperature dependence of the vacuum ultraviolet (VUV) photoabsorption cross sections.

Measurements were performed on the 3-m normal incidence monochromator of the BL-20A beam line at the Photon Factory synchrotron facility. A 1200 lines/mm grating was used to achieve a resolution of 3 meV (FWHM) with entrance and exit slit widths of 50 μm. The setup consists of a LiF window, a stainless steel absorption-cell with an effective length of 220-mm wrapped by sheath wires of 1.2φ in diameter and 2-m length, a window coated by sodium salicylate and a photomultiplier tube with a CsI-coated photocathode assembled after the exit slit of the absorption-cell. The absorption-cell, which has been newly built at Sophia University can be heated from room temperature (~300 K) up to 700 K by resistive heating.

The pressure of the target molecule in the absorption-cell is monitored with a capacitance manometer (Baratron 626A, MKS Co. Ltd) kept at room temperature. The target pressure was corrected by the thermal transpiration effect with the empirical expression developed by Takaishi and Sensui [3] for each set of temperature measurements.

VUV absorption spectra of the target gases recorded as the photocurrent, were converted to absolute photoabsorption cross sections from target gas densities at a certain temperature and length of absorption-cell by applying the normal Lambert-Beer law. From the first set of measurements using the new setup in photon energy range of 10.6 – 11.8 eV, we found very good agreement between our cross sections at room temperature (310 K) and previous data by Stark et al. [4] at 295 K. Temperature dependence on the cross sections for the transition to the $^1\Sigma_u^+$ and $^1\Pi_u$ from the electronic ground state has been clearly observed in the measured temperature range.

References

[1] N. Watanabe *et al* 2013 *J. Chem. Phys.* **138** 184311
[2] T. Tanaka *et al* 2005 *Phys. Rev. Lett.* **95** 203002
[3] T. Takaisi and Y. Sensui 1963 *Trans. Faraday Soc.* **59** 2503
[4] G. Stark *et al* 2007*J. Quant, Spectr. and Rad. Trans.* **103** 67.

[1] E-mail: masami-h@sophia.ac.jp

Content from this work may be used under the terms of the Creative Commons Attribution 3.0 licence. Any further distribution of this work must maintain attribution to the author(s) and the title of the work, journal citation and DOI.
Published under licence by IOP Publishing Ltd

Journal of Physics: Conference Series **635** (2015) 112056 doi:10.1088/1742-6596/635/11/112056

Ionization of Kr.CF_4 and Xe.CF_4 van der Waals clusters: from face to vertex geometry

R. Püttner*[1], V. A. Alekseev†[2], G. A. Garcia‡, R. Kevorkyants†, L. Nahon‡

* Institut für Experimentalphysik, Freie Universität Berlin, Arnimallee 14, D-14195, Germany
† Institute of Physics, St.Petersburg State University, Uljanovskaja St.1, Peterhof, 198504 Russia
‡ Synchrotron Soleil, Orme des Merisiers, St. Aubin BP 48, 91192 Gif sur Yvette cedex, France

Synopsis We report on photoionization studies of Xe.CF_4 and Kr.CF_4 van-der-Waals (vdW) clusters produced in a supersonic expansion and detected using synchrotron radiation and photoelectron- photoion coincidence techniques. Complementary *ab initio* calculations of the potential energy surfaces for the ground state and the first ionized state are performed and show generally good agreement with the experimental results. In particular, the calculation reveals a change in the equilibrium configuration from face to vertex geometry upon ionization. This observation is fully in line with the lineshapes of the Xe.CF_4 and Kr.CF_4 clusters in the photoelectron spectra.

Recent studies of Xe-CF_4 and Kr-CF_4 quasi-molecules (collision pairs) in static gas cell experiments revealed an interesting coupling effect induced by an interaction of transient dipole of IR active CF_4 ($\nu_3 = 0 \leftrightarrow \nu_3 = 1$) vibrational transition with transient dipoles of IR transitions between excited states of the Rg atom. This coupling manifests itself via intense satellite bands at energies of the Rg atom states increased by the energy of one ν_3 quantum of CF_4 ([1] and references therein).

In this paper, we report on photoionization studies of the rare gas (Rg) vdW clusters Xe.CF_4 and Kr.CF_4. The experiments were performed on the DESIRS VUV beamline located at the French national synchrotron facility, SOLEIL. The ions and electrons were detected in coincidence using the DELICIOUS III double imaging photoelectron-photoion coincidence (i^2PEPICO) spectrometer [2]. In this way, both conventional photoelectron spectra at fixed photon energies and threshold photoelectron spectra were recorded. For the interpretation of the experimental results complementary *ab initio* calculations for the ground and the first ionized state of the Rg.CF_4 vdW clusters are performed. These calculations show that the vertex geometry is energetically most favorable for the ionized states, which differs from the well-known face geometry for the ground state of Rg.CH_4 and Rg.CF_4 clusters.

For both clusters, the lineshapes of the photoelectron spectra show an asymmetric structure with a main peak approximately 0.05 eV below the $Rg^+ + CF_4$ dissociation asymptote as well as a shoulder with an onset approximately 0.2 eV below the asymptote. Based on the calculations the main peaks can be explained with vertical transitions in the face geometry while the shoulders are due to adiabatic transitions to vibrational states which are mainly located around the vertex geometry.

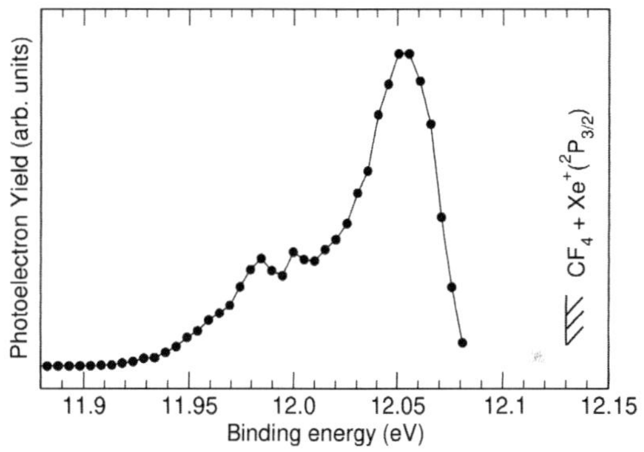

Figure 1. Threshold photoelectron spectrum for Xe.CF_4. The dissociaton limit Xe^++CF_4 is also indicated.

For Xe.CF_4 also the energy range above the first ionization threshold of Xe^+ was probed. However, in this region no photoelectron - Xe.CF_4^+ ion coincidence signal was found. This suggests a fast decay of the Xe.CF_4^+ clusters above the first dissociation limit, in agreement with previous PEPICO measurements for Kr_2 and Xe_2 dimers [3].

References

[1] V. A. Alekseev and N. Schwentner 2011 *J. Chem. Phys.* **135** 044313.

[2] G. A. Garcia at al 2013 *Rev. Sci. Instrum.* **84** 053112.

[3] K. Norwood*at al* 1989 *J. Chem. Phys.* **90** 4689.

[1]E-mail: puettner@physik.fu-berlin.de
[2]E-mail: v.alekseev@spbu.ru

Content from this work may be used under the terms of the Creative Commons Attribution 3.0 licence. Any further distribution of this work must maintain attribution to the author(s) and the title of the work, journal citation and DOI.
Published under licence by IOP Publishing Ltd

Double core-hole states in SiX_4 (X = F, Cl, Br, and CH_3) molecules derived by photoelectron and KLL Auger spectroscopy

R. Püttner*[1], T. Marchenko†,$, R. Guillemin†,$, L. Journel†,$, G. Goldsztejn†,$, J.-P. Rueff¶, D. Céolin¶, D. W. Lindle§, A. Lago#, K. Ueda%, O. Takahashi&, M. N. Piancastelli†,$,‡ M. Simon†,$,¶

* Institut für Experimentalphysik, Freie Universität Berlin, Germany
† CNRS, UMR 7614, Laboratoire de Chimie Physique-Matière et Rayonnement, France
$ Sorbonne Universités, UPMC Univ Paris 06, UMR 7614, LCPMR, France
¶ Synchrotron SOLEIL, Gif-sur-Yvette, France
§ Department of Chemistry, University of Nevada, Las Vegas, USA
# Centro de Ciências Naturais e Humanas, Universidade Federal do ABC (UFABC), Santo André-SP, Brazil
% Institute of Multidisciplinary Research for Advanced Materials, Tohoku University, Sendai, Japan
& Department of Chemistry, Hiroshima University, Japan
‡ Department of Physics and Astronomy, Uppsala University, Sweden

Synopsis In recent years double core-hole states are intensively studied since their chemical shifts provide detailed information about initial-state and relaxation effects in a molecule. We derived the Si $1s^{-1}$, $2s^{-1}$, and $2p^{-1}$ binding energies as well as the Si $2s^{-2}$, $2s^{-1}2p^{-1}$, and $2p^{-2}$ double-core hole binding energies of different SiX_4 systems in order to derive the chemical shifts. Based on these results we created Wagner plots, which give insight in the initial state and the relaxation effects in the different molecules.

In recent years double core-hole states are intensively studied since their chemical shifts provide detailed information about initial-state and relaxation effects in a molecule. Most of these measurements are done either by coincidence spectroscopy and single-photon absorption using synchrotron radiation (e.g. [1]) or by two-photon sequential absorption using free electron lasers as light (e.g. [2]) sources. However, some of these double core-hole states can also be obtained by other means. This holds in particular for $2s^{-2}$, $2s^{-1}2p^{-1}$, and $2p^{-2}$ core-hole states as a result of KLL Auger decay, which can readily be measured using conventional photoelectron spectroscopy.

Using the new experimental setup for high-resolution HAXPES (HArd X-ray PhotoElectron Spectroscopy) in gas phase, which is mounted at the GALAXIES beamline of the synchrotron SOLEIL, we measured the Si KLL Auger spectra of the molecules SiX_4 with X = F, Cl, Br, and CH_3. Complementary photoelectron spectra of these molecules were recorded in order to derive the Si $1s^{-1}$, $2s^{-1}$, and $2p^{-1}$ single core-hole binding energies. Using these binding energies, the chemical shifts of the Si $2s^{-2}$, $2s^{-1}2p^{-1}$, and $2p^{-2}$ double core-hole binding energies were derived from the Auger spectra. The obtained binding energies are compared with theoretical results [3] revealing good agreement. Moreover, the results were used to create Wagner plots (i.e. plotting single core-hole energies versus double core-hole energies) for the Si $1s$, Si $2s$, and Si $2p$ core hole state, which give a detailed insight in the initial state and relaxation effects [4]. In particular, they show that the various core holes provide similar information. In addition, the Wagner plots reveal that the initial state effects depend only on the electron-accepting (F,Cl,Br) or electron-donating (H,CH_3) character of the ligands. Within these groups differences in the ionization energies are due to varying relaxation energies.

Recently it has been shown for argon that the Ar $1s^{-1}2s,p^{-1}$ double core-hole shake-up states are accessible with conventional photoelectron spectroscopy [5]. Here, we present similar Si $1s^{-1}2s,p^{-1}$ spectra for all SiX_4 molecules. In summary, we show that KLL Auger and photoelectron spectroscopy are well suited to derive information about double core-hole states.

References

[1] P. Linusson *et al* 2011 *Phys. Rev. A* **83** 022506 and references therein; P. Lablanquie *et al* 2013 *Phys. Rev. Lett.* **110** 163001 and references therein.

[2] N. Berrah *et al* 2011 *PNAS* **108** 16912; P. Salen *et al* 2012 *Phys. Rev. Lett.* **108** 153003, and references therein.

[3] O. Takahasi *et al* 2011 *Chem. Phys.* **384** 28.

[4] T. D. Thomas 2012 *J. Phys. Chem. A* **116** 3856.

[5] R. Püttner *et al* 2015 *Phys. Rev. Lett.* **114** 093001.

[1]E-mail: puettner@physik.fu-berlin.de

Content from this work may be used under the terms of the Creative Commons Attribution 3.0 licence. Any further distribution of this work must maintain attribution to the author(s) and the title of the work, journal citation and DOI.
Published under licence by IOP Publishing Ltd

Core-excitation from excited triplet states of organic molecules: X-ray absorption spectra and doubly excited potential curves

Sena Nagashima[*], **Kazufumi Yamamoto**[*], **Osamu Takahashi**[**], **and Atsunari Hiraya**[*,† 1]

[*]Department of Physical Science, [*]Hiroshima University, Higashi-Hiroshima 739-8526, Japan
[**]Institute for Sustainable Science and Development, Hiroshima University, Higashi-Hiroshima 739-8526, Japan
[†] Hiroshima Synchrotron Radiation Center, Hiroshima University, Higashi-Hiroshima 739-0026, Japan

Synopsis Soft X-ray absorption spectra of the lowest excited triplet states of several organic molecules (quinones, benzaldehydes, and other carbonyls) and potential curves of valence- and core- doubly excited states are calculated for the O 1s region.

Resonant core-excitation from valence excited organic molecules has attracted attentions from the view point of the valence and core interactions and dynamics of valence- and core-doubly excited molecules [1].

The x-ray absorption spectrum of the lowest excited triplet state (T_1-XAS) were calculated by using the similar procedure for that of the ground state (S_0-XAS) [2] with modifying the initial state of the resonant core-excitation to the lowest excited triplet state. Potential energy curves of the valence- and core- doubly excited states (PEC-VCDES) for a specific bond are calculated by changing the bond distance with fixing other geometrical parameters [3].

T_1-XAS of para-cyanobenzaldehyde (PCBA) exhibits a lowest energy peak (Fig. 1) of transition to the hole created at the highest occupied molecular orbital (HOMO). PCBA and other mono-carbonyls show almost the same intensity for HOMO and LUMO (lowest unoccupied) peak. However HOMO peaks of quinones, symmetric di-carbonyls, are very weak that would indicate a new type of valence-core interaction.

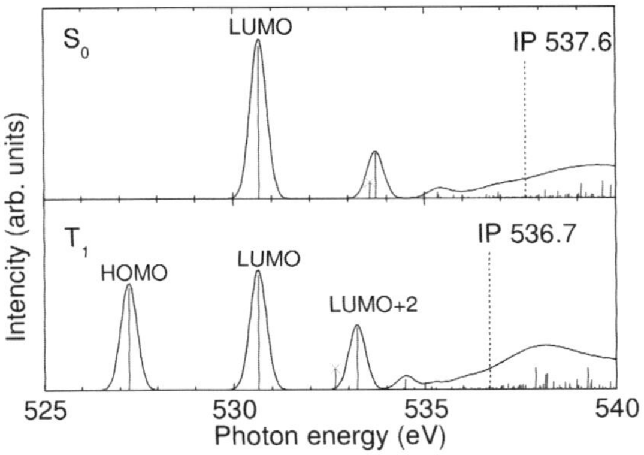

Figure 1. XAS from the S_0 (top) and the T_1 (bottom) of para cyanobenzaldehyde. Solid vertical lines are calculated transition moments including very weak transitions to the LUMO+1 multiplied by 100 and designated by × mark. Dotted vertical lines indicates the calculated ionization potential.

The PEC-VCDES from the T_1 have negative gradient for some bonds in several molecules. For the C≡N bond of PCBA all PEC-VCDES ($O1s^{-1}$) are repulsive at the equilibrium bond distance in the T_1 state, especially for the $O1s^{-1}$ $HOMO^1LUMO^2$ (-77 eV/nm), while all PEC from the S_0 are almost flat regardless of the repulsive nature of the excited orbitals. It is noteworthy that strongly repulsive nature that would result in the site-selective bond scission is found at the most far C≡N bond from the core-hole oxygen atom across the benzene ring.

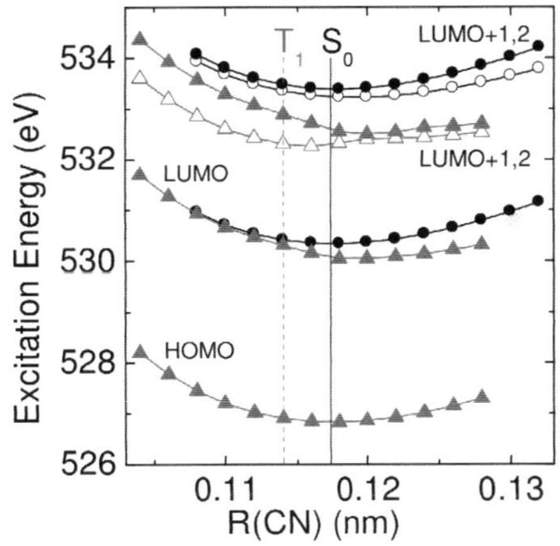

Figure 2. Potential curves along C≡N bond of para-cyanobenzaldehyde in the O 1s core-excited states from the S_0 (circle) and the T_1 (triangle). Open symbol designates the states with very weak transition moment. Vertical lines indicate the C-N equilibrium distances in the S_0 (solid) and the T_1 (broken) states.

References

[1] T. Privalov, F. Gel'mukhanov and H. Ågren, 2003 *J. Electron Spectrosc. Relat. Phenom.* **129** 43
[2] C. Kolczewski, R. Püttner, O. Plashkevych, H. Ågren, V. Staemmler, M. Martins, G. Snell, A.S. Schlachter, M.S. Anna, G. Kaindl, L.G.M. Pettersson, 2001 *J. Chem. Phys.* **115** 6426
[3] O. Takahashi, T. Matsui, A. Kawano, K. Tabayashi, K. Yamasaki, 2007 *J. Mol. Struct.* **808** 35

[1] E-mail: hiraya@sci.hiroshima-u.ac.jp

Content from this work may be used under the terms of the Creative Commons Attribution 3.0 licence. Any further distribution of this work must maintain attribution to the author(s) and the title of the work, journal citation and DOI.
Published under licence by IOP Publishing Ltd

Journal of Physics: Conference Series **635** (2015) 112059 doi:10.1088/1742-6596/635/11/112059

Coupling of collective excitation in proton and photon interaction with PAHs

P M Mishra*[1], L Avaldi[†], P Bolognesi[†], K C Prince[δ], J Rajput[Σ], R Richter[δ], C P Safvan[ϑ], S Vig[λ], and U R Kadhane[φ]

[*] Max-Planck Institut für Kernphysik, Heidelberg 69117, Germany
[†] CNR-Instituto di Metodologie Inorganiche e dei Plasmi, CP10-00016, Monterotondo, Italy
[δ] Elettra-Sincrotrone Trieste, Strada Statale 14, km 163.5, Area Science Park, I-34149 Basovizza, Italy
[Σ] Department of Physics and Astrophysics, University of Delhi, Delhi 110007, India
[ϑ] Inter-University Accelerator Centre, Aruna Asaf Ali Marg, New Delhi 110067, India
[λ] Department of Earth and Space Science, Indian Institute of Space Science and Technology, Trivandrum 695547, India
[φ] Department of Physics, Indian Institute of Space Science and Technology, Trivandrum 695547, India

Synopsis Collective excitation in polycyclic aromatic hydrocarbon molecule is studied in collision experiments with intermediate velocity (50-180 keV) proton and FUV (15-40 eV) photon energy. In proton impact experiments a clear comparison between pure ionization and capture ionization (or pure capture) process is done on the basis of ionization, evaporation, fragmentation process as a function of impact energy. Two mechanisms: large impact parameter collective excitation mode and closer encounters with higher amount of electronic energy loss leading to fragmentation were observed in proton collision process. The photon impact results were used to understand the origin of evaporation processes.

The structural and dynamical study for polycyclic aromatic hydrocarbons (PAHs) is important in astrophysics and in medical science. A typical photoabsorption spectra of PAH molecules is dominated by two main feature, the π-π* resonance at about 6 eV and a very broad, strong σ-σ* at approximately 20 eV leading to collective excitation with large oscillator strength. This excitation is difficult to isolate from other physical processes due to broader energy deposition in ion-molecule collision. We have studied the contribution of collective excitation in various post collisional processes like fragmentation, ionization, and collision dynamics of PAHs in two independent ways such as pure ionization process and capture process (pure as well as capture ionization). The collision experiments of proton (ranging 50 to 180 keV) with a PAH molecule naphthalene (D_{2h} symmetry) was performed using electro cyclotron resonance source. It is observed that the coupling of this excitation is to pure ionization as well as evaporation processes like loss of neutral H, C_2H_2 (see Figure 1) [1]. The photoelectron spectroscopy experiments using synchrotron source with pyrene (D_{2h}) and fluorene (C_{2v}) showed that collective excitation is coupled to inner valence shell [2] and hence a good amount of excitation energy goes in heating which leads to above evaporation process. In addition to this the energetic of H, 2H and C_2H_2 loss, proton emission is also studied. The fragmentation dynamics is studied in both the process as a function of projectile impact energy.

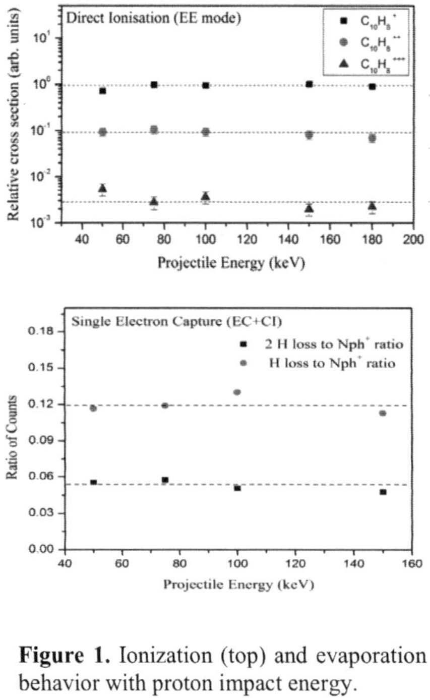

Figure 1. Ionization (top) and evaporation (bottom) behavior with proton impact energy.

References

[1] P. M. Mishra *et al* 2013 *Phys. Rev. A* **88** 052707
[2] P. M. Mishra *et al* 2014 *J. Phys. Chem. A* **118** 3128

[1] E-mail: pmishra@mpi-hd.mpg.de

Content from this work may be used under the terms of the Creative Commons Attribution 3.0 licence. Any further distribution of this work must maintain attribution to the author(s) and the title of the work, journal citation and DOI.
Published under licence by IOP Publishing Ltd

XXIX International Conference on Photonic, Electronic, and Atomic Collisions (ICPEAC2015) IOP Publishing
Journal of Physics: Conference Series **635** (2015) 112060 doi:10.1088/1742-6596/635/11/112060

Probing O_2^+ potential curves with an XUV–IR pump–probe experiment

Philipp Cörlin*[1], Andreas Fischer*, Michael Schönwald*, Alexander Sperl*, Tomoya Mizuno*, Uwe Thumm†, Thomas Pfeifer*, Robert Moshammer*[2]

* Max-Planck-Institut für Kernphysik, Saupfercheckweg 1, 69117 Heidelberg, Germany
† Kansas State University, Manhattan, Kansas 66506, USA

Synopsis Upon ionization of ground state O_2 molecules in a short XUV pulse, we observe a time-dependent vibrational wave packet in the potential of the binding $O_2^+(a\,^4\Pi_u)$ state. Our pump–probe delay dependent kinetic-energy-release (KER) spectra are in qualitative agreement with the results of coupled-channel simulations that are based on calculated Born-Oppenheimer potential-energy curves (PECs). Using a Morse potential adjusted to the experimental data most features of the experimental spectra are reproduced quantitatively.

In an XUV–IR pump–probe experiment, an extreme ultraviolet (XUV) pulse initiates an oscillating nuclear vibrational wave packet in the binding $O_2^+(a\,^4\Pi_u)$ potential by single-photon ionization. After a variable time delay population is transferred to the weakly repulsive $O_2^+(f\,^4\Pi_g)$ state by an infrared (IR) pulse [(Duration (FWHM) $\approx$ 12 fs, Intensity $\approx$ 3×10^{12} W/cm^2] (Fig. 1). The initial three-dimensional momenta of the created photoions and photoelectrons are determined using a reaction microscope.

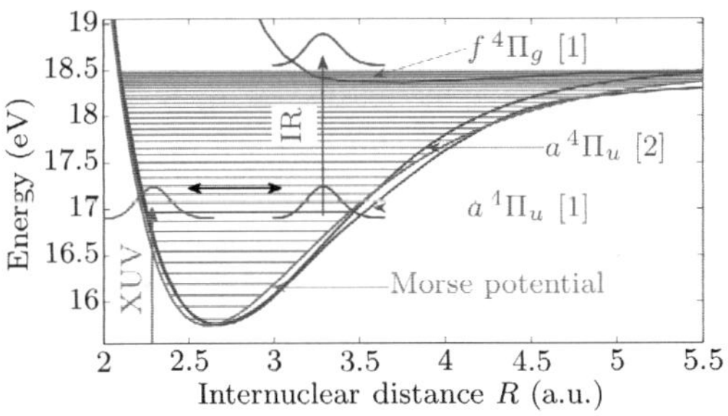

Figure 1. O_2^+ PECs calculated in Refs. [1] and [2] as well as a Morse potential adjusted to the experimental data.

The experimental yield of O^+ ions with KER < 0.08 eV is plotted as a function of the pump–probe delay in Fig. 2. Clearly visible is an oscillation with a period of 40 fs. The anharmonicity of the $a\,^4\Pi_u$ PEC causes the originally well localized wave packet to dephase over a period of approximately 640 fs before it revives after 1270 fs.

With a coupled-channel calculation, we simulate the wave-packet oscillation and the probing process for different sets of PECs. The results are shown in the three lower graphs of Fig. 2. The revival time of the wave packet is overestimated in the simulations using the PECs calculated in Refs. [1] and [2]. If the wave-packet is propagated within a Morse potential adjusted to the experimental data, the revival time is in very goo

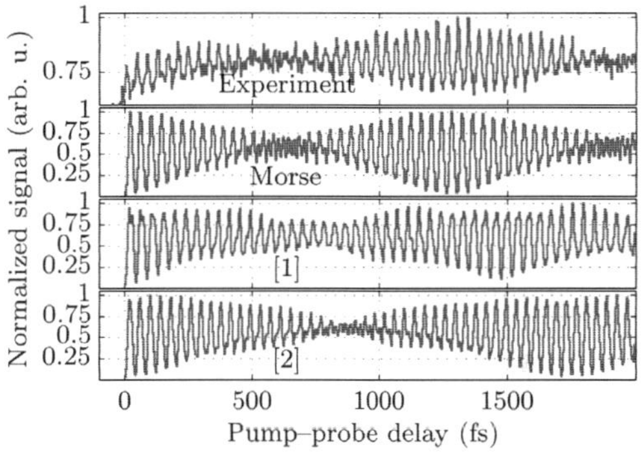

Figure 2. O^+ yield as a function of the pump–probe delay. Experimental data (top) and three simulated spectra using a Morse potential as well as the PECs calculated in Refs. [1] and [2].

The close resemblance between the Morse potential and the predicted PECs in Fig. 1 demonstrates that the revival time and our experimental method is sensitive to small variations in the shape of the binding PEC. Additionally, the KER distribution of the O^+ ions contains information about the shape of the repulsive PEC. This allows us to distinguish between the $O_2^+(f\,^4\Pi_g)$ PECs calculated in Refs. [1] and [2].

References

[1] C. M. Marian, *et al.*, 1982 *Mol. Phys.* **46** 779

[2] M. Magrakvelidze, C. M. Aikens, and U. Thumm, 2012 *Phys. Rev. A* **86** 023402

[1]E-mail: philipp.coerlin@mpi-hd.mpg.de
[2]E-mail: robert.moshammer@mpi-hd.mpg.de

Content from this work may be used under the terms of the Creative Commons Attribution 3.0 licence. Any further distribution of this work must maintain attribution to the author(s) and the title of the work, journal citation and DOI.
Published under licence by IOP Publishing Ltd

Journal of Physics: Conference Series **635** (2015) 112061 doi:10.1088/1742-6596/635/11/112061

An ion source platform of the cryogenic storage ring (CSR)

P M Mishra* [1], K Blaum*, C Breitenfeldt*,†, S George*, J Göck*, M Grieser*, R von Hahn*, J Karthein*, T Kolling§, C Krantz*, H Kreckel*, S Kumar*, J Mohrbach§, G Niedner-Schatteburg§, C Meyer*, R Repnow*, L Schweikard†, and A Wolf*

* Max-Planck-Institut für Kernphysik, D-69117 Heidelberg, Germany

† Institut für Physik, Ernst-Moritz-Arndt Universität Greifswald, D-17487 Greifswald, Germany

§ Fachbereich Chemie, Technische Universität Kaiserslautern, Erwin-Schrödinger-Straβe 52, 67663 Kaiserslautern, Germany

Synopsis A network of ion sources is being developed on the 300-kV acceleration platform of the cryogenic storage ring (CSR) at the Max-Planck-Institut für Kernphysik. It consists of several types of sources like a metal ion sputtering source (MISS), a Penning source, a laser vaporization (LVAP) source, and an electrospray ionization (ESI) source to produce a large variety of ions which can be studied for photon and electron interaction in a ro-vibrationally cold environment. Furthermore a storage device such as a radiofrequency quadrupole (RFQ) is foreseen for internal state cooling and accumulation of rarely produced species.

In order to determine ground state properties of small molecules and clusters and to understand their structure as well as various quantum dynamical interactions, it is crucial to produce and accumulate sufficient amount of state-selected ions of interest. The ions will be studied in the cryogenic storage ring (CSR) [1], currently under commissioning at the Max-Planck-Institut für Kernphysik in Heidelberg, by miscellaneous experimental probes such as laser, cold electron, and neutral atom collisions. Photodissociation, electron-ion recombination, and ion-atom interaction taking place in the extremely background free environment of the CSR will provide a unique insight into the ground state properties and interaction of various ion species ranging from small diatomic molecules to larger molecules like clusters, biomolecules or PAHs.

In this contribution we describe the various ion sources and traps in the first phase of the CSR. Figure 1 shows the layout of the sources to be installed on the high-voltage platform of the CSR enabling various ion beams of energies up to 300 keV. Standard sources installed for the commissioning of the CSR are a Penning ion source (positive Heinecke type) and a Middleton type sputter source (MISS) for anionic species. In addition to this the development of a laser vaporization source (LVAP) as well as an electrospray ion (ESI) source for complex biomolecules has been triggered. A radiofrequency quadrupole (RFQ) trap will be added for ion accumulation and internal state cooling using buffer gas. To transport the ions between the different sources and the RFQ trap as well as further towards injection into the CSR (via an optional mass analyzing magnet) a purely electrostatic ion optical beam line has been designed. Ion optical simulations performed using SIMION 8.1 and the entire layout of the source facility will be presented.

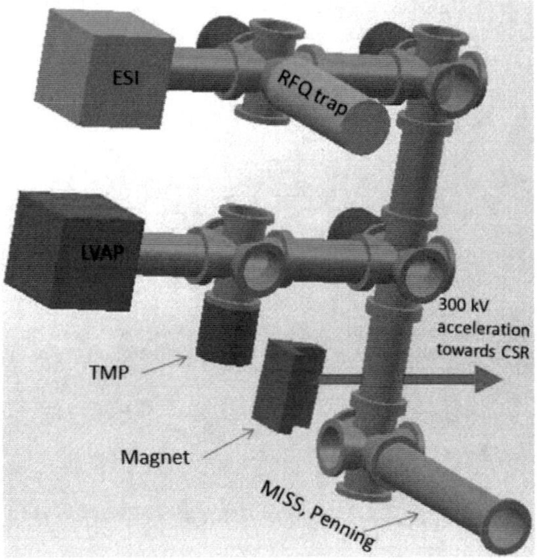

Figure 1. Phase-I layout of the ion source network for the CSR

References

[1] R. von Hahn *et al* 2011 *Nucl. Instrum. Meth. B* **269** 2871

[1] E-mail: pmishra@mpi-hd.mpg.de

Content from this work may be used under the terms of the Creative Commons Attribution 3.0 licence. Any further distribution of this work must maintain attribution to the author(s) and the title of the work, journal citation and DOI.
Published under licence by IOP Publishing Ltd

XXIX International Conference on Photonic, Electronic, and Atomic Collisions (ICPEAC2015) IOP Publishing
Journal of Physics: Conference Series **635** (2015) 112062 doi:10.1088/1742-6596/635/11/112062

A study of the dynamical energy flow in uracil

P. Bolognesi [a], P. O'Keeffe [a], T. Mazza[b], J. Bozek[c], R. Coffee[c], C. Bostedt[c], S. Schorb[c],S. Carron[c], R. Feifel[d], M. Mucke[d], M. Guehr[e], E. F. Sistrunk[e], J. Grilj[e], B. K. McFarland[e], M. Koch[e], M. Larsson[f], P. Salem[f], N. Berrah[g], L. Fang[g], T. Osipov[g], B. Murphy[g], R. R. Lucchese[m], M. Meyer[b], M.N. Piancastelli[d], K. Ueda[h], S. Mondal[h], C. Miron[i], R. Richter[l], K.C. Prince[l] , O. Takahashi[n] and L. Avaldi [1a]

[a]CNR-Istituto di Struttura della Materia, UOS- Montelibretti, CP10,Area della Ricerca di Roma1, Italy
[b]European XFEL GmbH,D-22761 Hamburg, Germany
[c]LCLS, SLAC National Accelerator Laboratory, Menlo Park, CA 94025, USA
[d]Department of Physics and Astronomy, Uppsala University, SE-751 20 Uppsala, Sweden
[e]PULSE, SLAC National Accelerator Laboratory, Menlo Park, CA 94025, USA
[f]Phys. Dept., Stockholm University, Sweden
[g] Phys. Dept, University of Connecticut, Storrs, CT 06269-3046, USA
[h]Department of Physics, Tohoku University, Sendai 980-8577, Japan
[i]Synchrotron SOLEIL, Saint-Aubin, BP 48, 91192 - Gif-sur-Yvette Cedex, France
[l]Elettra-Sincrotrone Trieste, Basovizza I-34149, Italy
[m]Dept. of Chemistry, Texas A&M University, College Station, TX 77843-3255,USA
[n]Dept. Chemistry, Hiroshima University, Hiroshima, Japan

Synopsis The time resolved photoionization of C 1s in uracil following excitation of the neutral molecule by 260 nm pulses has been studied at LCLS.

Uracil, is one of the building blocks of RNA and is closely related to thymine, one of the letters in the DNA alphabet. A fundamental question in ultrafast photodynamics of these molecular subunits is: how do they resist damage due to the strong absorption of UV radiation? Pyrimidine and purine derivatives, the main components of DNA bases, can dissipate dangerous electronic energy before it causes bond breakage and consequent errors in the DNA code. Current theories focus on conical intersections of the excited state, and the roles of allowed and dark excited states.

We have investigated the dynamics of core ionized uracil by time-resolved pump-probe techniques, using a femtosecond laser system as the pump and the X-ray pulse of the LCLS as probe, combined with an efficient electron detector: a magnetic bottle. By laser pumping the ground state of the molecule and probing the excited states with core level spectroscopy, one can shed light on the nature of internal electronic energy conversion after UV photoexcitation. In the experiments the photoionization of C1s at 350 eV has been studied at different delay times between the UV excitation (260 nm, $\pi\pi^*$ excitation) and the X-Fel ionisation pulses. The operating conditions of LCLS have been chosen in order to avoid non-linear processes. The difference between the spectra with laser on and off is shown in figure (1b). In the figure a fast sub 100 fs dynamics and a slower one can be easily identified.

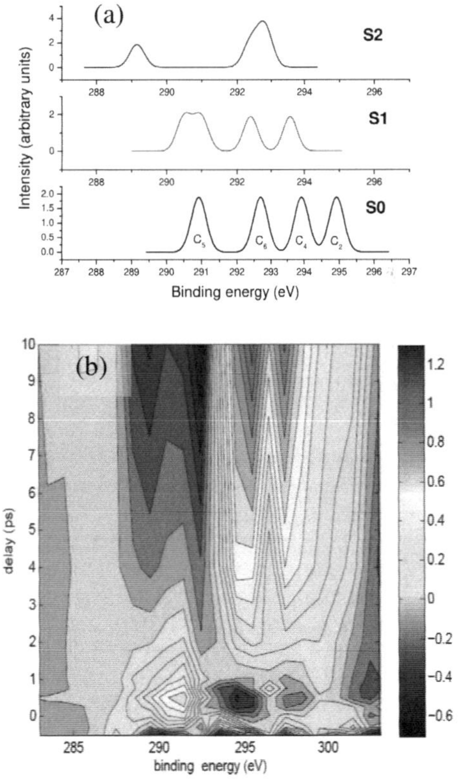

Figure 1. (a) Calculated XPS spectra of the ground and two excited states of uracil. (b) Difference between the measured photoelectron spectra with and without ultraviolet excitation as a function of the delay

[1]E-mail: lorenzo.avaldi@ism.cnr.it

Content from this work may be used under the terms of the Creative Commons Attribution 3.0 licence. Any further distribution of this work must maintain attribution to the author(s) and the title of the work, journal citation and DOI.
Published under licence by IOP Publishing Ltd

Journal of Physics: Conference Series **635** (2015) 112063 doi:10.1088/1742-6596/635/11/112063

Einstein–Bohr recoiling double-slit gedanken experiment performed at the molecular level

X.-J Liu[ab1], Q. Miao[cd], F. Gel'mukhanov[c†], M. Patanen[a], O. Travnikova[ae], C. Nicolas[a], H. Ågren[c], K. Ueda[f] and C. Miron[ag2]

[a] Synchrotron SOLEIL, Saint Aubin BP48, 91192 Gif-sur-Yvette Cedex, France ; [b] School of Physics, Beihang University, 100191, Beijing, China; [c] School of Biotechnology, Royal Institute of Technology, S-10691 Stockholm, Sweden; [d]Shandong University of Science and Technology, 266590, Shandong, China; [e] CNRS, Sorbonne Universités, UPMC Univ Paris 06, UMR 7614, Laboratoire de Chimie Physique-Matiére et Rayonnement, F-75005 Paris, France; [f] Institute of Multidisciplinary Research for Advanced Materials, Tohoku University, Sendai, 980-8577, Japan; [g] ELI-NP, Horia Hulubei National Institute for Physics and Nuclear Engineering, RO-077125, Romania.

Synopsis Double-slit experiments illustrate proof for wave–particle complementarity. The essence of Einstein–Bohr's debate about wave–particle duality was whether the momentum transfer between a particle and a recoiling slit could mark the path, thus destroying the interference. We realized this recoiling double-slit gedanken experiment by resonant X-ray photoemission from molecular oxygen for geometries near equilibrium (coupled slits) and in a dissociative state far away from equilibrium (decoupled slits). Interference is observed in the former case, while the electron momentum transfer quenches the interference in the latter case.

Einstein–Bohr recoiling double-slit gedanken experiment [1,2] reveals one of the most impressive deviations from classical concepts. In a double-slit experiment one can infer that achieving information about the particle path via momentum transfer erases the interference pattern characteristic of the wave behavior [3]. However, when using macroscopic slits, momentum transfer from the particle to the slit is too small to be measured. For this reason, coherence and decoherence in double slit experiments have also been studied using ultra-light atomic slits.

Our experiment, based on resonant X-ray ionization of the O_2 molecule (Fig. 1) using synchrotron radiation and a state-of-the-art coincidence set-up, offers the possibility to distinguish the momenta imparted to each of the two slits materialized by identical atomic sites: during the X-ray-induced ultrafast dissociation, the symmetry is broken owing to a site-dependent momentum exchange between the ejected electron and the dissociating molecule.

We demonstrate the occurrence of symmetry breaking in the dissociating O_2^+ cation by a direct measurement of the momenta of co- and counter-propagating O+ ions with respect to the ejected electron momentum. Which-path information (WPI) about the recoiling atomic slit is thus revealed, and the interference fringes are washed out through the measurement of the opposite Doppler shifts of two paths—the 'left' and 'right' dissociating oxygen ions.

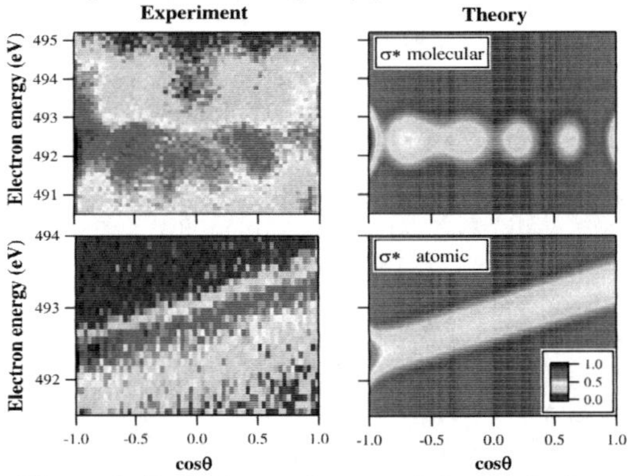

Figure 1. By both experiment and theory, the interference patterns show up when O_2 molecular geometry near equilibrium (upper part), while they disappear when O_2 dissociate far away from equilibrium(lower part).

We thus provide a consistent experimental proof and corresponding theoretical support showing a Doppler marker eliminates the interference pattern, in full agreement with Bohr's complementary principle [4].

References

[1] N. Bohr, in *Quantum Theory and Measurement* 9–49 (Princeton Univ. Press, 1983).
[2] W. K. Wootters *et al* 1979 *Phys. Rev. D* **19**, 473
[3] P. Bertet et al 2001 *Nature* **411**, 166
[4] Liu, X.-J et al 2015 *Nat. Photon.* **9** 120

xjliuxjliu@gmail.com
catalin.miron@synchrotron-soleil.fr

Content from this work may be used under the terms of the Creative Commons Attribution 3.0 licence. Any further distribution of this work must maintain attribution to the author(s) and the title of the work, journal citation and DOI.
Published under licence by IOP Publishing Ltd

New insight on the photofragmentation of CH_2I_2

A. R. Casavola*, A. Cartoni†[1], P. Bolognesi *[2], S. Borocci#, L. Avaldi*

*Istituto di Struttura della Materia, Area della Ricerca Roma 1, 00016, Monterotondo Scalo, Italy
† Dip. di Chimica, Università di Roma Sapienza, p.le Aldo Moro 5, 00185 Roma, Italy
#Dip. per l' Innovazione nei sistemi Biologici, Agroalimentari e Forestali (DIBAF), Università della Tuscia, L.go dell'Università, s.n.c., 01100 Viterbo, Italy

Synopsis An experimental and theoretical investigation of the VUV photofragmentation of CH_2I_2 molecule is reported. The work focuses on the I-loss channel. The experimental results demonstrate that at energies lower than 12 eV diiodomethane radical cation dissociates only into the CH_2I^+ ion and an I atom while the theoretical calculations show that isomerisation of $[CH_2I_2]^{\cdot+}$ into iso-diiodomethane $[CH_2I\text{-}I]^{\cdot+}$ intermediates can occur before the loss of iodine atom.

Diiodomethane is a halomethane widely studied for its role played in the terrestrial atmosphere as well as in solution chemistry. In the environment CH_2I_2 is a source of iodine atom which affects the ozone chemistry and contributes to the formation of the ultrafine marine aerosol particles, while in photoreaction chemistry this molecule is used for cyclopropanation of alkene. Different studies [1] have demonstrated that in condensed phase diiodomethane molecule under photon irradiation isomerizes into iso-diiodomethane and becomes a source of "CH_2" species. Moreover in the recent literature it has been reported that in the gas phase halomethane can isomerize into isohalomethane before loosing the halogen molecule [2]. In a previous study [3] we have investigated the photofragmentation of dihalomethanes (CH_2X_2, X=F, Cl, Br, I) and in particular we found that diiodomethane generates different fragment ions and the I-loss channel is the one opening at the lowest energy. In this work we report an experimental and theoretical investigation of VUV photofragmentation of CH_2I_2 molecule with the main focus on I-loss channel, which occurs from the lower electronic ionic states as showed in our PEPICO experiments.

Theoretical calculations have been performed at the MP2/CCSD(T) level of theory to obtain the structures and the energy of each stationary point found for the $[CH_2I_2]^{\cdot+}$ ion. The results show that the isomerization of diiodomethane radical cation $[CH_2I_2]^{\cdot+}$ into iso-diiodomethane intermediates $[CH_2I\text{-}I]^{\cdot+}$ is an energetically competing channel with respect to the direct C-I bond breaking (Figure 1).

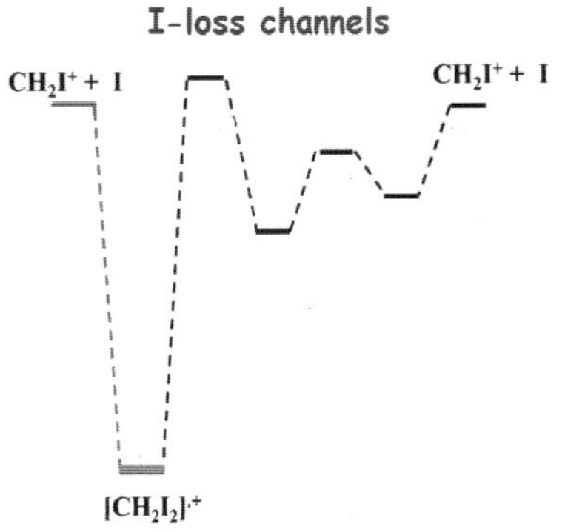

Figure 1. Schematic representation of energetic paths to I atom and CH_2I^+ ion from $[CH_2I_2]^{\cdot+}$ through $[CH_2I\text{-}I]^{\cdot+}$ intermediates (black pattern on the right).

Acknowledgment

This work is supported by the MIUR FIRB RBFR10SQZI project 2010.

References

[1] Phillips, D. L.; Fang, W.-H.; Zheng, X. J. Am. Chem. Soc. 2001, 123, 4197-4203.
[2] Kalume, A.; George, L.; Reid, S. A. J. Phys. Chem. Lett. 2010, 1, 3090–3095.
[3] Cartoni, A.; Bolognesi, P.; Fainelli, E.; Avaldi L. J. Chem. Phys. 2014, 140, 184307-1--13

[1] E-mail: antonella.cartoni@uniroma1.it
[2] E-mail: paola.bolognesi@cnr.it

Content from this work may be used under the terms of the Creative Commons Attribution 3.0 licence. Any further distribution of this work must maintain attribution to the author(s) and the title of the work, journal citation and DOI.
Published under licence by IOP Publishing Ltd

Direct Determination of Molecular Handedness via Coulomb Explosion Imaging

M. Pitzer*, M. Kunitski*, G. Kastirke*, T. Jahnke*, T. Bauer*, C. Goihl*, F. Trinter*, C. Schober*, K. Henrichs*, J. Becht*, S. Zeller*, H. Gassert*, M. Waitz*, A. Kuhlins*, F. Wiegandt*, R. Wallauer*, L. Ph. H. Schmidt*, M. Mazenauer°, B. Spenger°, S. Marquardt†, H. Schmidt-Böcking*, J. Stohner°, R. Dörner*, M.S. Schöffler*[1] R. Berger†,

* Institut3 for Nuclear Physics, Goethe-Universität, Max-von-Laue-Str. 1, 60438 Frankfurt, Germany
† Fachbereich Chemie,Philipps-Universität Marburg,Hans-Meerwein-Str. 4, 35032 Marburg, Germany
° Institute of Chemistry and Biological Chemistry, Zurich University of Applied Sciences, Campus Reidbach, Einsiedlerstr. 31, 8820 Wädenswil, Switzerland

Synopsis In this work, we show that the direct determination of a chiral molecule's absolute configuration can be achieved with Coulomb Explosion Imaging in a C LTRIMS reaction microscope. We compare the results after ionisation with femto-second laser pulses and with X-ray synchrotron light. Different fragmentation pathways of the prototypical chiral molecule CHBrClF are identified that carry information on the handedness. The applicabilty of the technique towards bigger molecules is discussed.

Symmetries and asymmetries have always been intriguing phenomena for mankind. In molecular physics and chemistry, the notion of chirality (handedness) plays an important role: Many molecules exist in two forms that are mirror images of each other, like our hands. These two forms of the molecules are commonly called enantiomers. Apart from intellectual curiosity, there is practical interest for the investigation of chiral molecules in pharmacology.
Whereas many techniques exist to distinguish enantiomers, e.g. via optical or vibrational spectroscopy, the determination of the absolute molecular configuration, i.e. the geometry of the underlying structure model, is still challenging. This is especially the case for substances that usually exist in the gas phase.
In the present work, we show that absolute configuration can be determined on a single-molecule level via Coulomb Explosion Imaging (see also [1]). For this purpose, the chiral methane derivative CHBrClF is expanded into a supersonic jet and crossed with laser or synchrotron radiation. After ionisation and fragmentation, the resulting ions are recorded with a COLTRIMS setup [2]. Ionisation by an intense femtosecond laser pulse leads to five-fold ionisation and subsequent fragmentation into atomic ions. From the momenta of these ions, the absolute configuration can be inferred for individual molecules [3]. Figure 1 shows the momenta in the molecular frame of reference. As complete fragmentation is becoming less likely with increasing complexity of the molecules, the possibility of partial fragmentation for the determination of absolute configuration needs to be considered. Results show that fragmentation pathways involving molecular ions can be used to distinguish enantiomers.

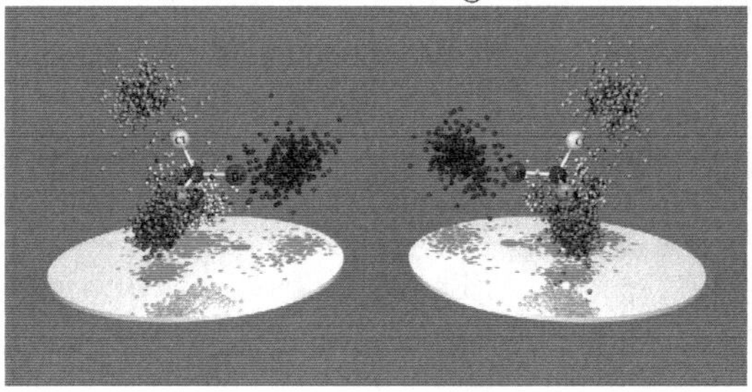

Figure 1. Momenta in the molecular frame for the fragmentation of CHBrClf into Br^+ (red), Cl^+ (yellow), F^+ (green), H^+ (white), C^+ (black arrow) after ionisation with a femtosecond laser pulse.

In a subsequent experiment, it could be demonstrated that also ionisation by a single X-ray photon from a synchrotron source leads to five-fold ionisation and complete fragmentation, as well as to various fragmentation patterns involving neutral fragments. Even though the neutral dissociation products cannot be detected by spectroscopic methods, these break-up channels carry a sign of handedness, increasing the possibility to extend the method towards larger molecules.

References

[1] Ph. Herwig *et al* 2013 *Sc e ce* **342** 1084

[2] R. Dörner *et al* 2000 *Phys cs Re orts* **330** 95

[3] M. Pitzer *et al* 2013 *Sc e ce* **341** 1096

[1]E-mail: schoeffler@atom.uni-frankfurt.de

Content from this work may be used under the terms of the Creative Commons Attribution 3.0 licence. Any further distribution of this work must maintain attribution to the author(s) and the title of the work, journal citation and DOI.
Published under licence by IOP Publishing Ltd

Journal of Physics: Conference Series **635** (2015) 112066 doi:10.1088/1742-6596/635/11/112066

Photoinduced Ultrafast Electronic Dynamics in Polyatomic Molecules: A time-dependent computational study

Astrid Nikodem* [1], **Françoise Remacle*** [2],

* Theoretical Physical Chemistry, Department of Chemistry, University of Liège
B4000 Liège, Belgium

Synopsis We report on the dynamics of photoexcitation and photoionization induced by ultrashort strong optical pulses in two medium sized polyatomic molecules, 1-azabicyclo[2,2,2]octane and caffeine. The time-dependent Schrödinger equation is solved for an electronic Hamiltonian that includes the interaction with the strong pulse. The partitioning technique is applied to include both the neutral and the ionized subspaces.

The interaction between middle sized polyatomic molecules and ultrashort optical pulses with XUV or IR carrier frequencies can lead to photoexcitation or photoionization. To describe the dynamics of these processes, we solve the time-dependent Schrödinger equation for an electronic Hamiltonian that includes the interaction with the strong pulse within the partitioning approach, so that both the neutral and the ionized subspaces are taken into account. As a basis set, we use the field free stationary states of the neutral and the cation at frozen nuclei geometry, which can be gotten by methods like averaged CASSCF or TD-DFT. When the pulse is over one gets a coherent superposition of electronic states [1].

We report on electronic photoexcitation and photoionization dynamics in two medium sized molecules: 1-azabicyclo[2,2,2]octane and caffeine. While both molecules are of comparable size they have different symmetries, which influences their interaction and their resonse to the strong optical pulse. We are especially interested in the charge beating of the coherent superposition of states after the pulse. The charge transfer between different molecular moieties is reflected in the oscillations of the time dependent dipole moment, an observable that can be probed experimentally.

Not only the spatial distribution of the density of the bound electrons is of interest but also the one of the unbound electron after the ionization. Figure 1 shows preliminary results on the electronic dynamics in both the photoelectron spectrum and the angular distribution computed at two values of the kinetic energy for the caffeine molecule. The dynamics is computed for a EUV pulse with a carrier frequency of 7.6 eV, a width of 2fs and a strength of 0.005 a.u., polarized in the z direction of the molecular frame. At the end of the pulse, mainly the GS and the 8th excited state of the cation are populated, as can be seen in the inset of Figure 1a). In Figure 1b) one sees that the spatial distribution for the first peak in the photoelectron spectrum is dominated by the angular distribution of the ground state of the cation, while that of the second peak is dominated by that of the 8th state.

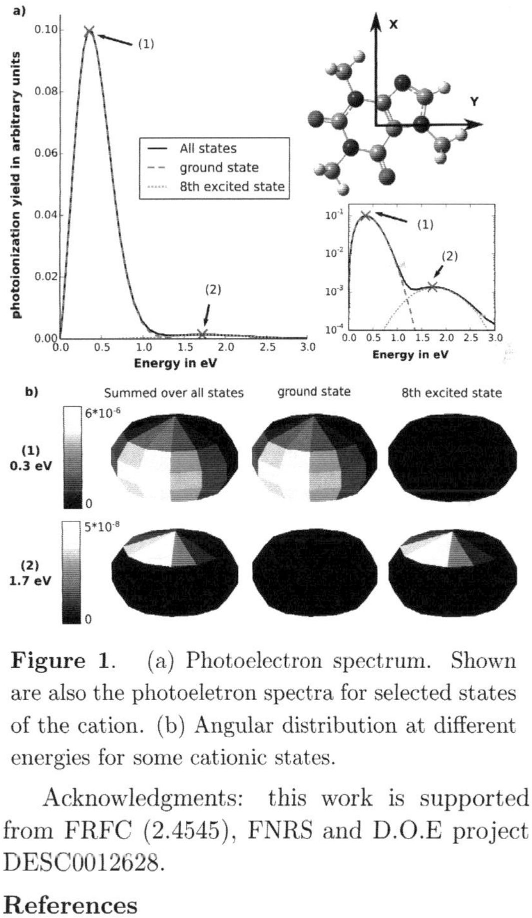

Figure 1. (a) Photoelectron spectrum. Shown are also the photoeletron spectra for selected states of the cation. (b) Angular distribution at different energies for some cationic states.

Acknowledgments: this work is supported from FRFC (2.4545), FNRS and D.O.E project DESC0012628.

References

[1] B. Mignolet *at al* 2014 *J. Phys. B* **47** 124011

[1]E-mail: anikodem@ulg.ac.be
[2]E-mail: fremacle@ulg.ac.be

Content from this work may be used under the terms of the Creative Commons Attribution 3.0 licence. Any further distribution of this work must maintain attribution to the author(s) and the title of the work, journal citation and DOI.
Published under licence by IOP Publishing Ltd

Interatomic Coulombic Decay Processes after Multiple Valence Excitations in Ne Clusters

D. Iablonskyi[a] [1], K. Nagaya[b], H. Fukuzawa[a] K. Motomura[a], Y. Kumagai[a], S. Mondal[a], T. Tachibana[a], T. Takanashi[a], T. Nishiyama[b], K. Matsunami[b], P. Johnsson[c], P. Piseri[d], G. Sansone[e], A. Dubrouil[e], M. Reduzzi[e], P. Carpeggiani[e], C. Vozzi[e], M. Devetta[e], M. Negro[e], D. Faccialà[e], F. Calegari[e], A. Trabattoni[e], M. Castrovilli[e], Y. Ovcharenko[f], T. Möller[f], M. Mudrich[g], F. Stienkemeier[g], M. Coreno[h], M. Alagia[i], B. Schütte[j], N. Berrah[k], C. Callegari[l], O. Plekan[l], P. Finetti[l], C. Spezzani[l], E. Ferrari[l], E. Allaria[l], G. Penco[l], C. Serpico[l], G. De Ninno[l], B. Diviacco[l], S. Di Mitri[l], L. Giannessi[l], K. C. Prince[i,l], M. Yao[b], and K. Ueda[a]

[a] Institute of Multidisciplinary Research for Advanced Materials, Tohoku University, 980-8577 Sendai, Japan
[b] Department of Physics, Graduate School of Science, Kyoto University, 606-8502 Kyoto, Japan
[c] Department of Physics, Lund University, P. O. Box 118, 22100 Lund, Sweden
[d] Dipartimento di Fisica, Università degli Studi di Milano, 20133 Milano, Italy
[e] CNR-IFN, Piazza Leonardo da Vinci 32, 20133 Milan, Italy
[f] Institut für Optik und Atomare Physik, TU Berlin, 10623 Berlin, Germany
[g] Physikalisches Institut, Universität Freiburg, 79104 Freiburg, Germany
[h] CNR-ISM, Area della Ricerca di Roma 1, Monterotondo Scalo, Italy
[i] CNR-IOM, Area Science Park, 34149 Trieste, Italy
[j] Max-Born-Institut, Max-Born-Strasse 2 A, 12489 Berlin, Germany
[k] Department of Physics, University of Connecticut, Storrs, CT 06269, USA
[l] Elettra-Sincrotrone Trieste, Area Science Park, 34149 Trieste, Italy

Synopsis We present a comprehensive analysis of autoionization processes in Ne clusters (~5000 atoms) after multiple valence excitations by free electron laser radiation. The evolution from 2-body interatomic Coulombic decay (ICD) to 3-body ICD is demonstrated when changing from surface to bulk Frenkel exciton excitation. Super Coster-Kronig type 2-body ICD is observed at Wannier exciton which quenches the main ICD channel.

Previously, Yase *at al* [1] investigated multiple excitation of Wannier type excitons (corresponding to the $2p \rightarrow 3d$ atomic resonance) at 20.26 eV in Ne clusters by the intense extreme ultraviolet free electron laser (EUV-FEL) at SCSS (SPring-8 Compact SASE Source, Japan) and found that the electron emission is dominated by low energy electron emission that originates from a nanoplasma.

In the present experiment, we have extended our observations to multiple excitations of surface and bulk Frenkel type excitons (corresponding to the $2p \rightarrow 3s$ atomic resonance) at 17.12 eV and 17.65 eV, respectively, using the new seeded EUV-FEL, FERMI (Trieste, Italy) [2].

At the lowest surface Frenkel exciton we can clearly see the pure 2-body ICD peak at ~11.5 eV, predicted by Kuleff *at al* [3], with its multistep ICD tail which is similar to direct multistep ionization in Ar clusters [4]. The situation changes for the bulk Frenkel exciton, where the broad structure around 5 eV is identified as 3-body ICD of knock-off type (also known as collective autoionization [5]) and becomes dominant over 2-body ICD at high FEL intensities. For the Wannier exciton we can see complete quenching of the main 2-body ICD by super Coster-Kronig type ICD in which one 3d electron relaxes to a 3s orbital and another 3d electron is ejected with ~1.8 eV kinetic energy (see Fig. 1).

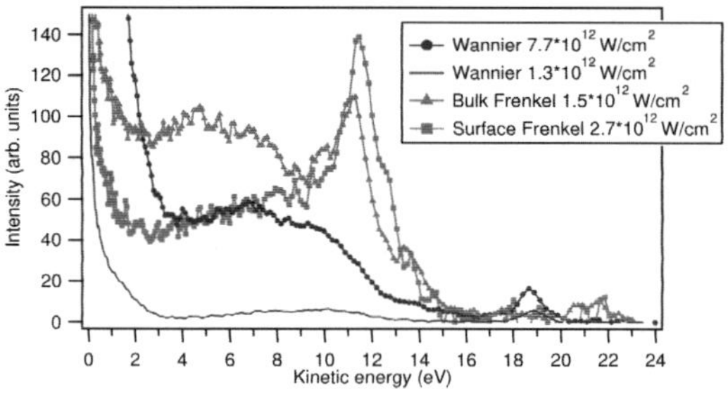

Figure 1. Electron emission spectra for excitation of different excitons and selected FEL intensities.

References

[1] S. Yase *et al.* 2013 *Phys. Rev. A* **88** 043203

[2] E. Allaria *et al.* 2012 *Nat. Photonics* **6** 699

[3] A. I. Kuleff *et al.* 2010 *Phys. Rev. Lett.* **105** 043004

[4] C. Bostedt *et al.* 2008 *Phys. Rev. Lett.* **100** 133401

[5] Y. Ovcharenko *et al.* 2014 *Phys. Rev. Lett.* **112** 073401

[1] E-mail: denys@tagen.tohoku.ac.jp

Content from this work may be used under the terms of the Creative Commons Attribution 3.0 licence. Any further distribution of this work must maintain attribution to the author(s) and the title of the work, journal citation and DOI.
Published under licence by IOP Publishing Ltd

Journal of Physics: Conference Series **635** (2015) 112068 doi:10.1088/1742-6596/635/11/112068

Ultrafast dynamics in the bifunctional PENNA neutral and cation molecule induced by ultrashort photoexcitation

Shoutian Sun*[1], Benoît Mignolet*, Françoise Remacle*[2]

* Theoretical Physical Chemistry, Department of Chemistry, University of Liege, Belgium

Synopsis We report on a theoretical investigation on ultrafast dynamics induced in the neutral and cationic states of the PENNA molecule by a short, strong optical pulse. The ultrafast, short time, electronic dynamics is computed by solving the time-dependent Schroedinger equation for the time-dependent Hamiltonian that includes the coupling to the strong field. The partitioning technique is implemented to take into account the coupling to the ionization continua. Preliminary results on the onset of nuclear dynamics that follows the excitation by the pump pulse will be discussed.

The recent developments in the generation of optical attopulses suggest that it will soon become experimentally feasible to induce and subsequently directly probe ultrafast charge transfer between the end moieties of the modular molecule. One ultrafast pulse creates a non-stationary state of the neutral or of the cation and a second one ionizes it. Such experiments would allow characterizing a purely electronic time scale, before the coupling to the nuclei takes place. This is a pre Born-Oppenheimer regime where the electronic states are not stationary.

We will report on the simulation of realistic pump probe experiments that monitor the ultrafast electronic dynamics in the medium size bifunctional molecule PENNA ($C_{10}H_{15}N$)[1] (Fig.1) using a coupled equation scheme that includes the ionization continua and field effects. Examples using short UV and IR pulses will be reported, both for the cation and the neutral. The electronic density motion between the two ends of a molecular system is probed by the anisotropy ionization parameter computed as the normalized difference between the ionization yields at the two moieties. Heatmaps of the ionization anisotropy parameter as a function of the delay time between the two pulses and the kinetic energy of the photoelectron exhibit oscillations that reflect the beating periods of the electron density.

Preliminary results on the onset of nuclear dynamics that follows the excitation by the pump pulse will be discussed, both in the cation state and in the neutral.

Acknowledgments: this work is supported from FRFC (2.4545), FNRS and D.O.E project DESC0012628.

[1]E-mail: shoutian.sun@ulg.ac.be

[2]E-mail: fremacle@ulg.ac.be

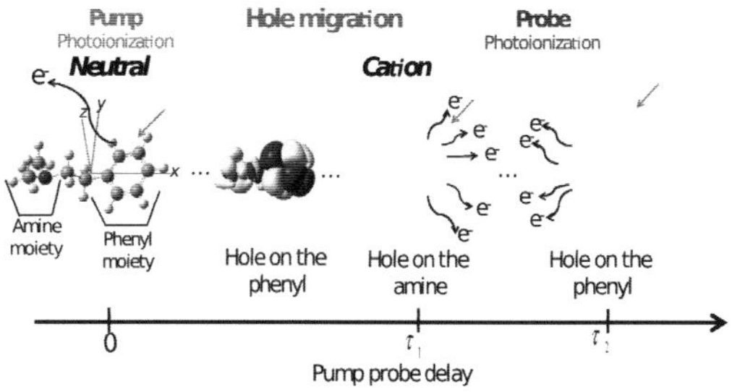

Figure 1. Pump-probe scheme for the simulation of electronic dynamics in the PENNA cation.

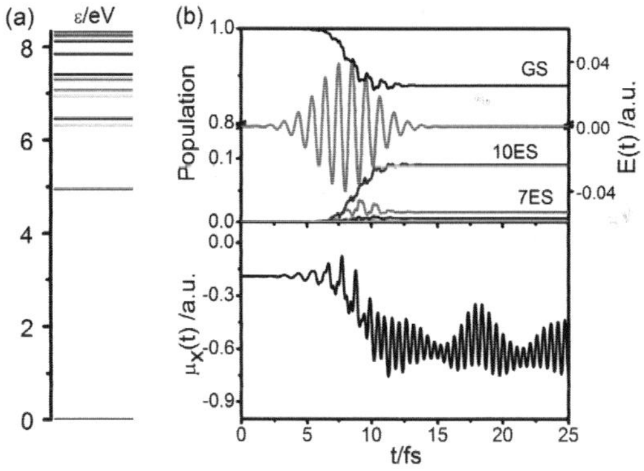

Figure 2. a) Level structure of the neutral excited states of PENNA. b) populations and dipole moment following excitation by a short pulse (ω=3.9eV, pulse width=5fs, field strength=0.04 a.u, polarization along *x*.

References

[1] Mignolet, B.; Levine, R. D.; Remacle, F., 2014 J. Phys. B: At. Mol. Opt. Phys. 47, 124011.

Content from this work may be used under the terms of the Creative Commons Attribution 3.0 licence. Any further distribution of this work must maintain attribution to the author(s) and the title of the work, journal citation and DOI.
Published under licence by IOP Publishing Ltd

Journal of Physics: Conference Series **635** (2015) 112069 doi:10.1088/1742-6596/635/11/112069

"Slow " Molecular Fragmentation after Ultrafast Interaction

Seyedreza Larimian*, **Sonia Erattupuzha***, **Erik Lötstedt**†, **Tamas Szidarovszky**†, **Raffael Maurer***, **Stefan Roither***, **Markus Schöffler***, **Daniil Kartashov***, **Andrius Baltuška***, **Kaoru Yamanouchi**†, **Markus Kitzler***, **Xinhua Xie***[1]

* Photonics Institute, Vienna University of Technology, Gusshausstrasse 27, A-1040 Vienna, Austria
† Department of Chemistry, School of Science, The University of Tokyo, 7-3-1 Hongo, Bunkyo-ku, Tokyo 113-0033, Japan

Synopsis We report experimental observation of "slow" fragmentation from ethylene molecules on nanosecond or even microsecond timescale after their interaction with femtosecond laser pulses. ur results indicate that such fragmentation process originates from meta-stable high-lying vibrational states of ethylene ions.

Molecular fragmentation from dissociative states in molecules is a common process after their interaction with strong laser fields [1]. In general, such process happens on a very short, *coseco d* or even *femtoseco d*, timescale. In this contribution, we present experimental observation of "slow" fragmentation on *a osec- o d* or even *m croseco d* timescale from ethylene molecules which are ionized within a few femtosecond by few-cycle laser pulses.

In the experiments, we recorded ionic products in coincidence from the interaction of ethylene molecules with strong near-infrared femtosecond laser pulses (with duration of 4.5 fs and peak intensity from a few 10^{14} W/cm^2 up to 2×10^{15} W/cm^2) in a reaction microscope apparatus [2, 3]. During the strong field interaction, the molecules can be doubly ionized and may be further broken into two moieties. We observed stable dications of ethylene and three two-body fragmentation channels from dications, as indicated in the photoion-photoion coincidence (PIPICO) spectrum in Figure 1.

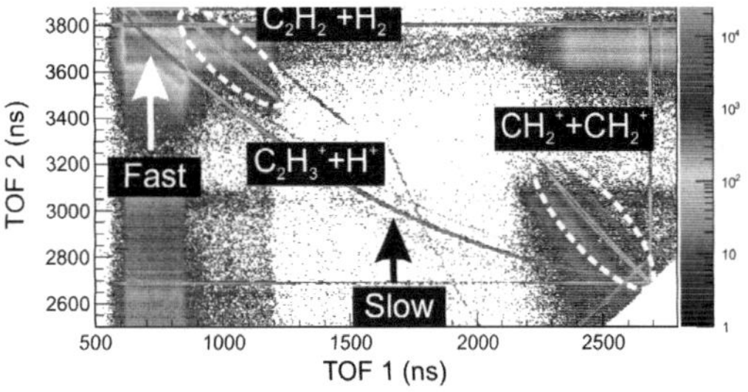

Figure 1. Measured PIPICO spectrum of ethylene molecule. The long tail of slow deprotonation of ethylene dication can be distinguished from the fast fragmentation process.

The two-body fragmentation channels are presented as parabolic lines in the PIPICO spectrum due to momentum conservation during the fragmentation process. Since the kinetic energy released (KER) of a two-body fragmentation from ethylene is on the order of a few electron volt, the corresponding PIPICO line is normally rather short. We found there is a long tail attached to the deprotonation PIPICO line, which is formed by a slow fragmentation process happening on nanosecond up to microsecond timescale after the interaction with ultrafast laser pulses. The KER of the slow fragmentation process and the lifetime of the corresponding ethylene dication is retrieved from the measurements for studying the origin of such process.

In the previous work, we found that the electronic ground state of ethylene dication is meta-stable [4]. Our observation indicates that the slow fragmentation of ethylene dication happens from meta-stable states with lifetime on the nanosecond timescale. In the quantum chemical simulations, we found that there are high-lying vibrational states on the electronic ground state of ethylene dication with nanosecond timescale lifetime. During the strong field interaction, such states might be populated through coupling with higher electronic excited states. After the laser pulse, dications can be trapped in such long lifetime meta-stable states which later lead to the observed slow fragmentation.

References

[1] K. Yamanouchi 2002 *Sc e ce* **295** 1659

[2] R. Dörner *et al* 2000 *Phys cs Re orts* **330** 95

[3] X. Xie *et al* 2012 *Phys. Rev. ett.* **108** 193004

[4] X. Xie *et al* 2014 *Phys. Rev. X* **4** 021005

[1]E-mail: xinhua.xie@tuwien.ac.at

Content from this work may be used under the terms of the Creative Commons Attribution 3.0 licence. Any further distribution of this work must maintain attribution to the author(s) and the title of the work, journal citation and DOI.
Published under licence by IOP Publishing Ltd

Attosecond transient absorption spectroscopy of molecular hydrogen

Fernando Martín[1,2,3,+], Yan Cheng[4], Michael Chini[4], Xiaowei Wang[4,5], Alberto González-Castrillo[1], Alicia Palacios[1], Luca Argenti[1], and Zenghu Chang[4]

[1]Departamento de Química, Módulo 13, Universidad Autónoma de Madrid, Madrid 28049, Spain
[2]Condensed Matter Physics Center (IFIMAC), Cantoblanco, Madrid 28049, Spain
[3]Instituto Madrileño de Estudios Avanzados en Nanociencia, Cantoblanco, Madrid 28049, Spain
[4]Institute for the Frontier of Attosecond Science and Technology, CREOL and Department of Physics, University of Central Florida, Orlando, Florida 32816, USA
[5]Department of Physics, National University of Defense Technology, Changsha, Hunan 41000, China

Synopsis We extend attosecond transient absorption spectroscopy (ATAS) to the study of hydrogen molecules, demonstrating the potential of the technique to resolve – simultaneously and with state resolution – both the electronic and nuclear dynamics.

Attosecond pulses promise to allow new forms of quantum control in which a broadband isolated attosecond pulse excites a wave packet, which is a coherent superposition of multiple excited electronic states. In molecules, this electronic excitation triggers nuclear motion on the molecular manifold of coupled potential energy surfaces and can result in permanent rearrangement of the constituent atoms. The ability to steer reaction outcomes by controlling the exciting light field will enable attosecond photochemistry [1,2]. In this communication, we demonstrate attosecond transient absorption spectroscopy as a viable probe of the coupled electronic and vibrational wave packet dynamics initiated by an isolated attosecond pulse in the vacuum ultraviolet (VUV) spectral region in neutral hydrogen molecules. The technique allows high-resolution and state-resolved measurements of electronic and nuclear dynamics without the need for inducing subsequent ionization [3] or dissociation [4]. The addition of a moderately intense few-cycle laser field allows control over the subsequent evolution of the molecule. An illustration of the ATAS for the H_2 molecule obtained from experiment and a solution of the time-dependent Schrödinger equation is shown in Figure 1. The origin of the different structures will be discussed at the conference.

* E-mail: fernando.martin@uam.es

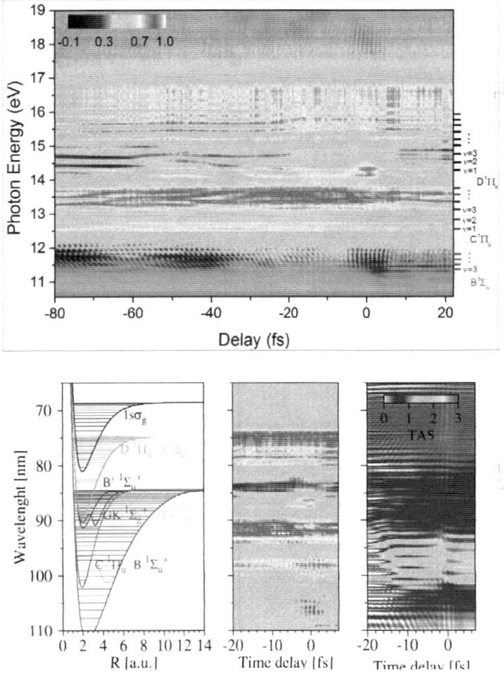

Figure 1. Measured (central panel) and calculated (right panel) ATAS for the H_2 molecule. The time-resolved absorption spectra uniquely encode the molecular dynamics in the vicinity of the B $^1\Sigma_u^+ \leftarrow$X $^1\Sigma_g^+$, C $^1\Pi_u\leftarrow$X, and D $^1\Pi_u\leftarrow$X transition bands of H_2. The corresponding potential energy curves are shown on the left panel.

References

[1] F. Lépine, G. Sansone, and M. J. J. Vrakking, *Chem. Phys. Lett.* **578**, 1 (2013).
[2] F. Lépine, M. Y. Ivanov, and M. J. J. Vrakking, *Nat. Photon.* **8**, 195 (2014).
3. P. Ranitovic, *et al.*, *Proc. Natl. Acad. Sci.* **111**, 912 (2014).
4. G. Sansone, *et al.*, *Nature* **465**, 763-766 (2010).

Content from this work may be used under the terms of the Creative Commons Attribution 3.0 licence. Any further distribution of this work must maintain attribution to the author(s) and the title of the work, journal citation and DOI.
Published under licence by IOP Publishing Ltd

XXIX International Conference on Photonic, Electronic, and Atomic Collisions (ICPEAC2015) IOP Publishing
Journal of Physics: Conference Series **635** (2015) 112071 doi:10.1088/1742-6596/635/11/112071

Tunneling Ionization of Aligned CH_3X (X=I, Br, Cl) Molecules in Intense Femtosecond Laser Fields

Sizuo Luo, Yujun Yang, Fuming Guo, Chuncheng Wang, Dajun Ding[1]

Institute of Atomic and Molecular Physics, Jilin University, Changchun 130012, P. R. China

Synopsis We perform an experimental investigation on tunneling ionization of CH_3X (X=I, Br, Cl) molecules in intense femtosecond laser fields (10^{13} to 10^{15} W/cm^2). The angular distribution of the parent and fragment ions from the well aligned molecular targets has been measured. Compared with theoretical calculation, the results show some new insights into the ionization mechanism of these molecules in intense laser fields.

Tunneling ionization is a fundamental process in interaction of molecules with intense femtosecond (*fs*) laser fields. There are still some ambiguous questions in understanding and controlling the ionization of molecules, originating from the complicated quantum many-body interaction [1]. We investigate the tunneling ionization angular distribution of CH_3X (X=I, Br, Cl) molecules under intense laser fields in molecular frame.

Aligned or oriented molecules can be obtained by the interaction with an intense laser field. However, the obtained degree of alignment or orientation is limited. It depends crucially on the initial rotational states of the molecules. Experimentally, the molecules are selected in the |111> rotational state by hexapole focusing, and a target of nonadiabatically aligned molecules is created by a aligning laser beam, with the pulse duration stretched to 300 *fs* and the intensity of 10^{12} W/cm^2. The aligned molecules are ionized by a 50 *fs* intense laser with the intensity ranging from 10^{13} to 10^{15} W/cm^2. The angular distributions of parent and fragment ions are measured at a certain delay time with respect to the aligning pulse by a VMI method. The technique used in the present work enables us to achieve accurate control of orientation or alignment for CH_3X molecules [2] and reveal the information of the molecular oriented dependence of the ionization.

The yields of the parent and fragment ions as the delay time between pump and probe laser are measured experimentally, the evolutions directly give the angular distributions of the parent and fragment ions from ionizing the aligned molecule. The measurement show off-phase feature for parent and fragment ions, *i.e.*, that the yield of the fragment ions X^+ or X^{2+} peaks at parallel alignment, while the yield of the parent ions CH_3X^+ peaks at perpendicular alignment. The measured angular distributions of parent ions CH_3I^+, CH_3Br^+, and CH_3Cl^+ are shown in Figure 1, in which the angular distribution of parent ions from diatomic molecules N_2, is also given for compared. It is noteworthy that the maximum of the ionization yield of CH_3X molecules is observed with the laser polarization perpendicular to the molecular axis. This behavior has been seen previously in triatomic molecules OCS and attributed to the contribution from molecular excited states [3]. We perform the strong-field molecular theoretical calculations based on MO-ADK and MO-SFA for a detailed understanding. In the presentation, we will give the details of discussion.

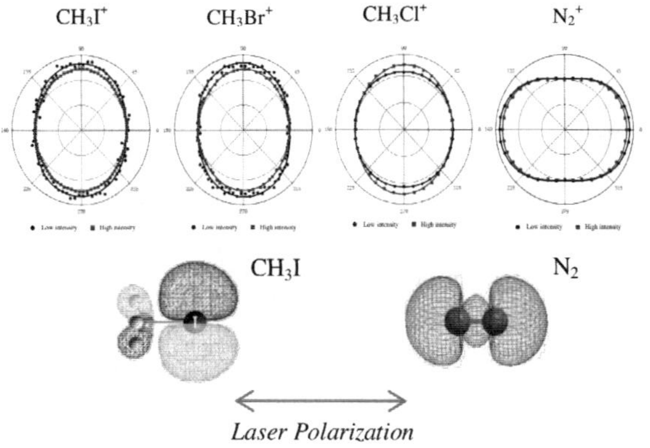

Figure 1. Angular distribution of tunneling ionization yields for CH_3X (X=I, Br, Cl) and N_2. The laser intensity is 2 x 10^{13} W/cm^2 (black), 1 x 10^{14} (blue). The electron density distribution of the HOMO orbital of CH_3I and N_2 are also shown for reference.

References

[1] R. Murray *et al* 2010 Phys. Rev. A **81** 023413; D. Pavičić *et al* 2007 *Phys. Rev. Lett.* **98** 243001
[2] S. Luo *et al* 2015 *Phys. Rev.* A, accepted; W. Roeterdink *et al* 2014 *Chem. Phys. Lett.* **598** 96.
[3] J. Hansen *et al* 2012 *J. Phys.* B **45** 015101

[1] E-mail: dajund@jlu.edu.cn

Content from this work may be used under the terms of the Creative Commons Attribution 3.0 licence. Any further distribution of this work must maintain attribution to the author(s) and the title of the work, journal citation and DOI.
Published under licence by IOP Publishing Ltd

Journal of Physics: Conference Series **635** (2015) 112072 doi:10.1088/1742-6596/635/11/112072

Toward laser-induced vibrational emission spectroscopy of C_{60}^{+}

Sunil Kumar S*[1], Peter Bizenberger†, Klaus Blaum*, Christian Breitenfeldt*,‡, Jürgen Göck*, Ulrich Grözinger†, Thomas Henning†, Jonas Karthein*, Bastian Kern§, Christian Meyer*, Gaël Rouillé†, Dmitry Strelnikov§, Andreas Wolf*, Sebastian George*, Holger Kreckel*[2]

* Max-Planck-Institut für Kernphysik, Saupfercheckweg 1, 69117 Heidelberg, Germany
† Max-Planck-Institut für Astronomie, Königstuhl 17, 69117 Heidelberg, Germany
‡ Institut für Physik, Ernst-Moritz-Arndt-Universität Greifswald, Felix-Hausdorff-Straße 6, 17489 Greifswald, Germany
§ Institut für Physikalische Chemie II, Karlsruher Institut für Technologie, Kaiserstraße 12, 76131 Karlsruhe, Germany

Synopsis C_{60}^{+} has been proposed to be responsible for two of the *diffuse interstellar bands* (DIBs), the absorption features observed in the visible-to-near-infrared spectra of the interstellar medium. However, a confirmation requires laboratory gas-phase spectra, which are so far not available. We plan to develop a novel spectroscopy technique that will allow us to obtain *the first gas-phase spectra* of C_{60}^{+}, and that will be applicable to other complex organic molecules such as polycyclic aromatic hydrocarbons. The current status of the experimental setup, the ideas behind the measurement scheme and the preparatory work toward its implementation will be presented.

The *diffuse interstellar bands* (DIBs) [1] are absorption features that can be observed in the visible to near-infrared spectra of the interstellar medium. Despite the observation of several hundred lines over the last few decades, none of them has been unequivocally attributed to a known chemical species. Still, astronomical observations and laboratory spectra of C_{60}^{+} [2] isolated in neon matrix suggest that this molecular ion could be responsible for two DIB features at 957.7 nm and 963.2 nm. However, a definitive conclusion will have to be based on laboratory gas phase spectra, which are not available to date. Using the Cryogenic Trap for Fast ion beams (CTF) [3], we intend to carry out *the first gas-phase spectroscopy* of C_{60}^{+}. The CTF allows us to prepare conditions that have close resemblance to the interstellar medium, where temperatures go down to ∼10 K and background gas densities are extremely low. We will store the molecular ions inside the CTF and excite them with laser light at the relevant wavelengths, spanning the absorption features attributed to C_{60}^{+}. At resonance, the excitation is expected to be followed by the delayed emission of infrared radiation through vibrational transitions in the mid-infrared [4, 5]. A fraction of the emitted light will be guided onto an array of highly sensitive blocked impurity band (BIB) detectors, which feature an extremely low dark current. The delayed emission intensity will be observed as a function of the excitation wavelength, and this signal will be used for a novel type of extremely sensitive, almost background-free gas-phase spectroscopy.

After the proof-of-principle experiments with C_{60}^{+}, this spectroscopic scheme can be applied to a wide range of other complex organic molecular ions such as polycyclic aromatic hydrocarbons. The current status of the experimental setup, the measurement scheme and the preparatory work toward its implementation will be presented.

[1]E-mail: sunil@mpi-hd.mpg.de
[2]E-mail: kreckel@mpi-hd.mpg.de

References

[1] Herbig G H 1995 *Annu. Rev. Astron. Astrophys.* **33** 1

[2] Foing B H and Ehrenfreund P 1994 *Nature* **369** 296

[3] Lange M *et al* 2010 *Rev. Sci. Instrum.* **81** 055105

[4] Allamandola L J *et al* 1989 *Astrophys. J. Suppl. Scr.* **71** 733

[5] Berné O *et al* 2013 *Astron. Astrophys.* **550** L4

Content from this work may be used under the terms of the Creative Commons Attribution 3.0 licence. Any further distribution of this work must maintain attribution to the author(s) and the title of the work, journal citation and DOI.
Published under licence by IOP Publishing Ltd

Journal of Physics: Conference Series **635** (2015) 112073 doi:10.1088/1742-6596/635/11/112073

Sturmian approach to study photoionization of molecules

C. M. Granados–Castro*,†1, L. U. Ancarani*, X. Assfeld*, G. Gasaneo†,‡, D. M. Mitnik§,‡

* Équipe TMS, SRSMC UMR 7565, Université de Lorraine, 57078 Metz, France

† Departamento de Física, Universidad Nacional del Sur, 8000 Bahía Blanca, Argentina

§ Instituto de Astronomía y Física del Espacio (IAFE) and Universidad de Buenos Aires, Argentina

‡ Consejo Nacional de Investigaciones Científicas y Técnicas (CONICET), Argentina

Synopsis We explore the implementation of generalized Sturmian functions for the study of ionization of molecules by photon or electron impact. A first part is devoted to finding a good enough description of the molecular model scattering potential and imposing accurate asymptotic boundary conditions to the molecular continuum wave functions. A second part deals with the calculation, within a generalized Sturmian approach, of single ionization cross sections for small molecules such as CH_4, H_2O and NH_3. Issues related to the random molecular orientation are addressed.

It is generally not easy to find an accurate representation of continuum states for molecular systems. The Sturmian approach, using Generalized Sturmian Functions (GSF), has been applied successfully for the study of several atomic ionization processes [1], but the extension of the method to molecular systems is under development. In the present contribution, we present our results of the implementation of GSFs for studying photoionization of CH_4, NH_3 and H_2O molecules, and preliminary applications to $(e, 2e)$ processes.

In our study of ionization processes we use (i) the single-electron approximation; (ii) a non-central molecular model potential [2] as scattering potential, and (iii) the one-electron wave functions in the one-center expansion calculated by Moccia [3]. As in most experiments the orientation of the molecule in the laboratory frame is unknown, an angular average over all possible spatial orientations of the molecule must be performed. This can be done either with an angular average of the molecular model potential (pre-averaged calculation), or using a non-central model potential and then evaluating the angular average of the calculated cross sections (post-averaged). In figure 1 we show the effective charge associated to both potentials for H_2O.

In order to obtain the required transition amplitudes, we solve the time-independent, first-order perturbative, Schrödinger equation, expanding the scattering wave function in a GSFs basis set. Their adequate asymptotic behavior allows us to extract directly such transition amplitudes from the expansion coefficients [1]. When the noncentral model potential is used we have to solve a set of angular-coupled of equations. The calculated cross sections (partially presented in [4] for CH_4) are compared with theoretical and experimental data.

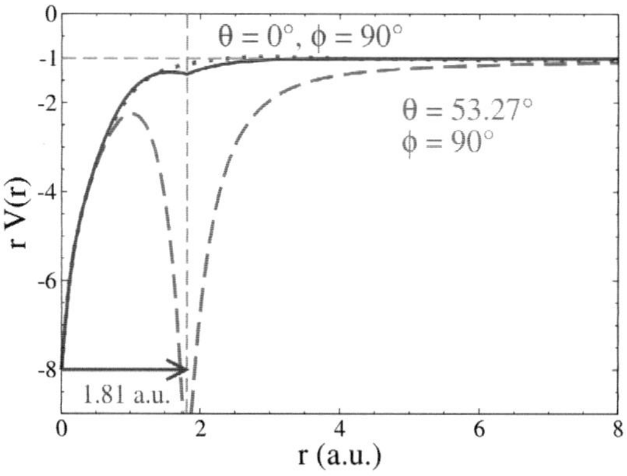

Figure 1. Effective charge associated with the used molecular model potentials for H_2O: non-central potential (red dashed; green dotted) at indicated angles, and the angular averaged potential (blue, solid). The equilibrium distances of each H respect to the O atom is shown by the vertical dashed line.

We acknowledge the CNRS (PICS project No. 06304) and CONICET for funding our French-Argentinian collaboration.

References

[1] G. Gasaneo *et al* 2013 *Adv. Quantum Chem.* **57** 153

[2] L. Fernández-Menchero and S. Otranto 2010 *Phys. Rev. A* **82** 022712

[3] R. Moccia 1964 *J. Chem. Phys.* **40** 2164; **40** 2176; **40** 2186

[4] C. M. Granados-Castro *et al* 2014 *Few-Body Syst.* **55** 1029

[1] E-mail: carlos.mario-granados.castro@univ-lorraine.fr

Content from this work may be used under the terms of the Creative Commons Attribution 3.0 licence. Any further distribution of this work must maintain attribution to the author(s) and the title of the work, journal citation and DOI.
Published under licence by IOP Publishing Ltd

Journal of Physics: Conference Series **635** (2015) 112074 doi:10.1088/1742-6596/635/11/112074

Attosecond time delays in C_{60} valence photoemissions at the giant plasmon

T. Barillot*, M. Magrakvelidze†, V. Loriot*, C. Bordas*, P-A. Hervieux[1], M. Gisselbrecht[2], P. Johnsson[2], J. Laksman[2], E. P. Mansson[2], S. Sorensen[2], S. E. Canton[3], J. M. Dahlström[4], G. Dixit[5], M. E. Madjet[6], F. Lépine* [1], H. S. Chakraborty† [2]

* Institut Lumiére Matière, UMR5306, Université Lyon 1-CNRS, 10 rue Ada Byron, 69622 Villeurbanne cedex, France
† Department of Natural Sciences, D.L. Hubbard Center for Innovation and Entrepreneurship, Northwest Missouri State University, Maryville, Missouri 64468, USA
[1] Institut de Physique et Chimie des Matériaux de Strasbourg, 23 rue du Loess, BP 43, 67034 Strasbourg Cedex 2, France
[2] Department of Physics, Lund University, Box 118, 221 00 Lund, Sweden
[3] Department of Synchrotron Radiation Instrumentation, Lund University, Box 118, 221 00 Lund, Sweden
[4] Department of Physics, Stockholm University, Alba Nova University Centrum, S-106 91, Stockholm, Sweden
[5] Max Born Institute, Max-Born-Strasse 2A, 12489 Berlin, Germany
[6] Qatar Environment and Energy Research Institute, Qatar Foundation, P.O. Box 5825, Doha, Qatar

Synopsis We perform time-dependent local density functional calculations of the time delay in C_{60} HOMO and HOMO-1 photoionization at giant plasmon energies. A semiclassical model is used to develop further insights.

Oscillations of the correlated electron density in nanosize systems stimulated by external radiation can quantize, forming many-body quasi-partcle, the plasmon. For nanoclusters, the direct consequence of the plasmon is the formation of a giant resonance in the response spectra that holds promise in optical applications. For the C_{60} molecule, a giant dipole plasmon can be ionized by XUV photons around 20 eV which therefore can serve as a spectral laboratory to study the effects of the collective motions on the emission time delay.

We compute the phases of HOMO and HOMO-1 photoamplitudes of C_{60}, using the time-dependent local density approximation (TDLDA) [1] that includes many-particle correlation effects. Previous calculations [2] were improved by including a scheme that angle-integrates the dipole amplitudes to calculate the total phase, whose energy-derivatives gave the Wigner-Smith time delays of emission. A semiclassical description is utilized to interpret some aspects of the result.

Numerical results for HOMO and HOMO-1 delay are presented in Figure 1(a). The comparison with single-particle LDA predictions, that omit correlations, clearly reveals the role of the plasmon to affect the delay. A roughly opposite delay pattern between the growing versus waning part of the plasmon resonance (peaked at 20 eV) is noted.

The complex interaction between the 240 valence electrons of C_{60} and the incident photon can be mimicked without solving the full time-dependent many-body problem, by introducing an effective Hamiltonian. Based on a semi-classical model, we show that the plasmon acts as a screening effect that modifies the phase of the electron wavepacket. It induces a local variation of the angular distribution which is accompanied by a variation of the photoemission delays on the attosecond timescale.

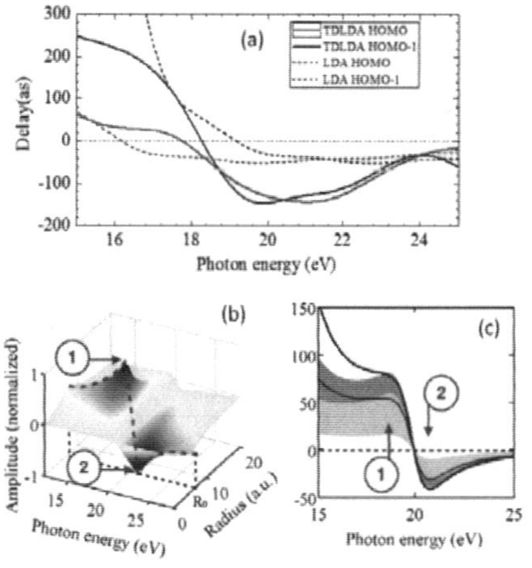

Figure 1. (a) TDLDA and LDA delays for HOMO and HOMO-1 emissions. (b) Real part of the screening potential. (c) Delays from semi-classical model.

This work was supported by NSF (USA) and ANR-10-BLAN-0428-01 "MUSES".

References

[1] M. Madjet *et al* 2010 *Phys. Rev. A* **81** 013202
[2] T. Barillot *et al* 2015 *Phys. Rev. A (accepted)*

E-mail: franck.lepine@univ-lyon1.fr

Content from this work may be used under the terms of the Creative Commons Attribution 3.0 licence. Any further distribution of this work must maintain attribution to the author(s) and the title of the work, journal citation and DOI.
Published under licence by IOP Publishing Ltd

Controlling high harmonic generation in diatomic molecules by phase shaped laser pulses

Manuel Lara-Astiaso* [1], **Christoph Meier**† [2], **Fernando Martín***,‡ [3],

* Departamento de Química,Módulo 13, Universidad Autónoma de Madrid, Cantoblanco 28049, Madrid, Spain
‡ Instituto Madrileño de Estudios Avanzados en Nanociencia (IMDEA-Nanociencia) Cantoblanco, E-28049 Madrid, Spain
† Laboratoire de Collisions Agrégats Réactivité, CNRS UMR 5589, IRSAMC, Université de Toulouse (UPS), 118 Route de Narbonne, 31062 Toulouse CEDEX 4, France

Synopsis In this work we explore the high-order harmonic generation in the H_2^+ molecule under the action of phase shaped pulses. To this end, quantum mechanical wavepacket calculations are performed within a collinear model of H_2^+. We analyze the effect of a chirp to the plateau extension of the HHG spectrum, observing significant differences with respect to the sign of the chirp parameter. This result is found to originate from a subtle electron recollision dynamics within the chirped laser pulses, as substantiated by classsical trajectory calculations. Extension of this work focuses on including the nuclear motion of H_2^+.

The process of high harmonic generation has proven to be an effective source of extreme utraviolet and soft x-ray light with a crucial application to generate attosecond light pulses, which can be used in time-resolved pump-probe experiments. Coherent control over the HHG spectrum has been studied by many authors, commonly in atoms, for various purposes such as the selective enhancement of harmonics or the temporal shape harmonic emission.

The aim of the present work is to analyze the effect of modifying the spectral phase of the IR laser onto the generation of high harmonics. For that purpose, we first analyze the dependence of the cutoff energy with respect to the chirp parameter. In addition to an extension of the plateau region for chirped pulses, we have found significant differences with respect to the sign of the chirp parameter depending also of the peak intensity of the laser pulse which determines the tunnel rate.

For the usual laser intensities, negative (down) chirp rates are found to be more efficient than positive (up) chirps. Nevertheless, when using strong pulses, the efficiency of HHG is reversed. An increase of the ionization yield penalizes down-chirped pulses due to significant groundstate depletion. Since the up/down chirped pulses have the same spectrum, the observed asymmetry cannot be attributed to purely spectral arguments, using the three-step model, but is the result of a complex electron recombination dynamics. Quantum results are substantiated by classical trajectory calculations.

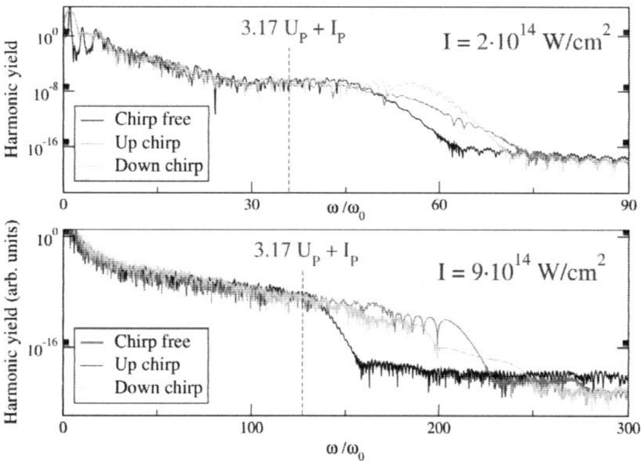

Figure 1. Chirp effects on harmonic generation. The harmonic yield is greatly enhanced near the cutoff frequency, especially for the down-chirped pulse (green) when normal intensities are applied. In the high-intensity regime, up chirps (red) become more efficient to extend the cutoff.

Founded on the obtained results, the effect on nuclear motion for those pulses is analyzed by 2-dimensional wavepacket calculations. Indeed, recent calculations have shown the importance of correlated electron-nuclear dynamics in the high-order harmonic generation process. [1]

References

[1] Xue-Bin Bian, André D. Bandrauk 2014 *Phys. Rev. Lett.* **113** 193901

[1]E-mail: m.lara@uam.es
[2]E-mail: chris@irsamc.ups-tlse.fr
[3]E-mail: fernando.martin@uam.es

Content from this work may be used under the terms of the Creative Commons Attribution 3.0 licence. Any further distribution of this work must maintain attribution to the author(s) and the title of the work, journal citation and DOI.
Published under licence by IOP Publishing Ltd

XXIX International Conference on Photonic, Electronic, and Atomic Collisions (ICPEAC2015) IOP Publishing
Journal of Physics: Conference Series **635** (2015) 112078 doi:10.1088/1742-6596/635/11/112078

Stability of Auger final dicationic states of core-ionized N_2 molecules studied by Auger-electron ion coincidence measurements

Hiroshi Iwayama and Eiji Shigemasa

† UVSOR Facility, Institute for Molecular Science, Okazaki 444-8585, Japan

Synopsis We present experimental results for AEPICO measurements following core photoionization of N2 molecules, with using an AEPICO spectrometer in which a toroidal electron energy analyzer and an ion momentum spectrometer are installed.

Absorption of a soft x-ray creates a core hole in a molecules. A core hole state is unstable and usually leads to Auger decay for light elements. This Auger decay process produces various valence-hole configuration. Although the dicationic states are generally unstable and come apart into ionic fragments, quasi-bound or metastable dicationic states are known to exist. In this work, we investigate a stability of Auger final states of core ionized N2 molecules with Auger-electron ion coincidence measurements.

The Auger-electron-ion coincidence measurements for core ionized N2 molecules were carried out on the undulator beamline BL6U at UVSOR. The radiation from an undulator was monochromatized by a variable included angle varied line-spacing plane grazing monochromator. The electrons ejected at 54.7° with respect to the electric vector of the ionization light were analyzed in energy by double toroidal analyzer (DTA), while ions were extracted from the interaction region into the momentum spectrometer by a pulsed electric field according to the electron detection. Arrival positions on the detector and time-of-flights of ions were recorded for every event. The pass energy of the DTA was set to 200 eV for observing resonant Auger electrons. The energy resolution was about 1.2 eV. All signals from electron and ion detectors were recorded with an 8ch TDC board. We used photon energies of 450eV. Here we note that ion time-of-flight can't be used to distinguish two ion species of N2++ and N+ due to the same mass-to-charge ratio. Since fragmentation process give large kinetic energy, we can identify ion species with their kinetic energies of ions.

Figure 2 (a) shows an Auger-electron ion coincidence map for core-ionized N2 molecules. The vertical and horizon axes of the map correspond to the kinetic energy of ion and Auger electron, respectively. Figure 2(b) and (c) shows total ion kinetic distributions and Auger electron spectrum with a binding energy scale. As shown in Fig. 2(b), ion kinetic energy distribution has two peaks at 0 eV and 6 eV. This represents that ions with nearly zero kinetic energy are N2++ while energetic ions are N+ fragment ions.

For (N+, N+) fragmentation process, we found that strong correlation between Auger-electron and ion kinetic energies, as shown in Fig.2(a). The slope of -1 means that these fragmentations reach a same fragmentation limit. From NIST Atomic Database, we found that they correspond to N+(3P) + N+(3P) fragmentations.

Auger final states ranging from 50 to 54 eV in binding energy scale are unstable and immediately lead to fragmentations. From the theoretical works[1], these Auger final states mainly comes from the 1Δg(1πu-2) configuration. Conversely Auger final states ranging from 46 to 50 eV and from 54 to 56 eV are stable. These Auger final states mainly comes from 3σg-2 and 2σg-11πu-1 configurations. In conclusion, we found that two 1πu bonds of N2 molecules play an important role on a stability of dicationic states.

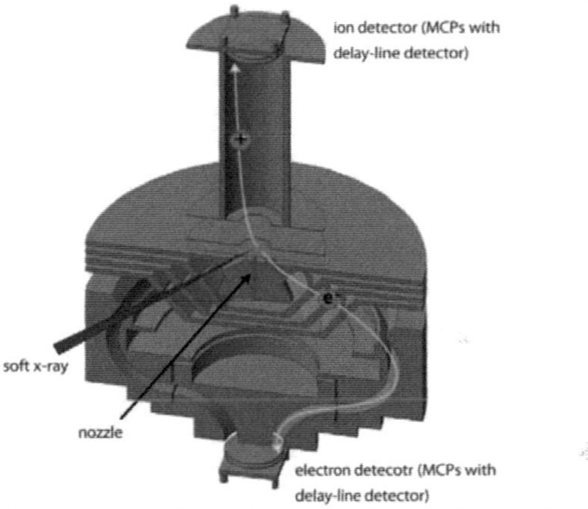

Figure 1. A schematic view of Auger electron-ion coincidence spectrometer.

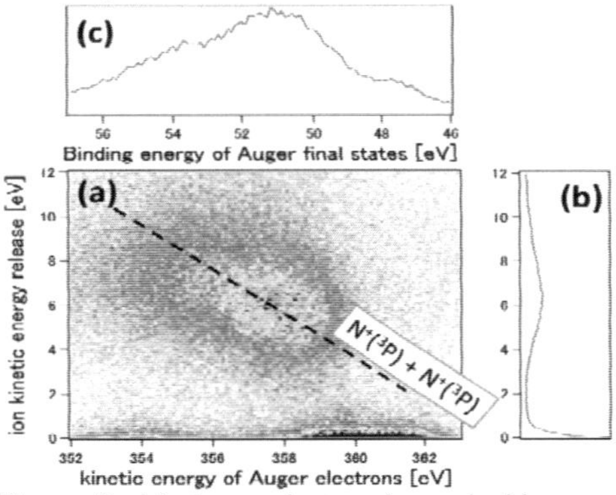

Figure 2. (a) Auger-electron ion coincidence map for core-ionized N_2 molecules. The spectra (b) and (c) shown at the right and top panels represents the total ion kinetic distribution and Auger electron spectra in binding energy scale, respectively.

[1] H. Agren, J. Chem. Phys. **75** (1982) 1267

Content from this work may be used under the terms of the Creative Commons Attribution 3.0 licence. Any further distribution of this work must maintain attribution to the author(s) and the title of the work, journal citation and DOI.
Published under licence by IOP Publishing Ltd

Towards XUV-pump XUV-probe experiments with attosecond pulses at the Lund Laser Centre

S. Maclot*[1], H. Coudert-Alteirac*, F. Campi*, P. Rudawski*, L. Rading*, B. Manschwetus*, C.M. Heyl*, M. Miranda*, B. Farkas†, F. Brizuela*, B. Kim*, A. L'Huillier* and P. Johnsson*[2]

* Department of Physics, Lund University, Box 118, SE-22100 Lund, Sweden

† ELI-HU Non-Profit Ltd., Dugonics tér 13, 6720 Szeged, Hungary

Synopsis The relaxation dynamics of an ionized/excited molecular system leads to different fundamental processes occurring at sub-femtosecond timescales. Time-resolved XUV-pump XUV-probe experiments should allow deeper insight to the key roles of these processes. Such studies should be feasible soon thanks to experimental schemes similar to those at the high-intensity high-order harmonic generation beamline at the Lund Laser Centre.

The stability of ionized and excited complex molecular systems depends on fundamental processes as electron dynamics, charge transfer/migration and nuclear motion [1, 2, 3], leading to chemical reactions and specific dissociations through, for instance, intramolecular rearrangements [4]. That is why, in this context of radiation damage, it is of prime importance to understand and characterize the effects of these processes on the behaviour of the molecule.

In order to follow the dynamics of these processes occurring on attosecond to femtosecond timescales, specific time-resolved tools are required [5]. An efficient technique involves pump-probe experiments using high-order harmonic pulses (XUV) produced after the focusing of an intense infrared (IR) femtosecond laser pulse in a gas. Most of the recent works have employed two color XUV/IR pump-probe experiments giving sub-femtosecond resolution. The limitation of this method lies in the fact that the IR pulse, most of the time, controls the dynamics rather than making an observation of the dynamics intrinsic to the molecule. A solution to this problem, would be to use XUV-pump XUV-probe experiments.

At the Lund Laser Centre, using the high-order harmonic generation (HHG) technique with a Ti:Sapphire laser at 800 nm and a repetition rate of 10 Hz [6], we are developing a new device based on a split mirror (splitting/delaying unit) to perform XUV-pump XUV-probe experiments. This method requires high intensities of the XUV pulses which was recently measured by two photons experiments performed at our beamline [7]. The target will be produced in the gas phase by an Even-Lavie pulsed valve and the products of the interaction will be analyzed by a double-sided velocity map imaging detector (VMI) allowing to have access to electron and ion information. Thereby, this technique allow us to study time-dependent fragmentation of small molecules and attosecond relaxation dynamics in atoms.

In more detail, the split mirror is composed of two silica plates with IR anti-reflection coatings. The XUV beam impinges equally on the two plates at grazing incidence, and is thus split in two pulses. One of the plates is driven by piezoelectric actuators, allowing a very fine control of the translation and the relative tilt, thus of the delay between the two pulses and of the beam pointing. The piezo actuators can be varied with nanometric precision, which leads to a sub-femtosecond control of the delay. The performance of the split mirror assembly has been benchmarked using the output from an OPCPA system to perform broadband all-reflective autocorrelation measurements, demonstrating an excellent stability and repeatability over its total delay range of 40 fs, with a resolution below 50 attoseconds.

References

[1] Cederbaum L S *et al* 1999 *Chem. Phys. Lett.* **307** 205

[2] Remacle F *at al* 2006 *Proc. Natl. Acad. Sci.* **103** 6793

[3] Calegari F *et al* 2014 *Science* **346** 336

[4] Jiang Y H *et al* 2010 *Phys. Rev. Lett.* **05** 263002

[5] Lépine F *et al* 2014 *Nat. Photonics* **8** 195

[6] Rudawski P *et al* 2013 *Rev. Sci. Instrum.* **84** 073103

[7] Manschwetus B *et al* Manuscript in preparation

[1]E-mail: sylvain.maclot@fysik.lth.se
[2]E-mail: per.johnsson@fysik.lth.se

Content from this work may be used under the terms of the Creative Commons Attribution 3.0 licence. Any further distribution of this work must maintain attribution to the author(s) and the title of the work, journal citation and DOI.
Published under licence by IOP Publishing Ltd

Journal of Physics: Conference Series **635** (2015) 112080 doi:10.1088/1742-6596/635/11/112080

Mapping ultrafast dynamics of highly excited D_2^+ by ultrashort XUV pump - IR probe radiation

R. Bello*[1], L. S. Martin†, C. W. Hogle†, A. Palacios*, J. L. Sanz-Vicario§, X. M. Tong‡, F. Martín*◇, M. Murnane†, H. C. Kapteyn† and P. Ranitovic†[2]

* Departamento de Química, M-13, Universidad Autónoma de Madrid, Cantoblanco 28049, Madrid, Spain
† JILA and Department of Physics, University of Colorado and NIST, Boulder, CO 80309-0440, USA
§ Grupo de Física Atómica y Molecular, Instituto de Física, Universidad de Antioquia, Medellín, Colombia
◇ Instituto Madrileño de Estudios Avanzados (IMDEA) en Nanociencia, Cantoblanco, 28049 Madrid, Spain
‡ Institute of Materials Science, University of Tsukuba, 1-1-1 Tennodai, Tsukuba, Ibaraki 305-8573, Japan

Synopsis An ultrashort XUV laser pulse ionizes the D_2 molecule creating an electronic and nuclear wave packet, with the dominant contributions from the $2s\sigma_g$ and $2p\pi_u$ ionic states. A delayed interaction with a 780 nm IR field ejects the second electron, leading to the Coulomb explosion of the molecule, whose nuclear fragments, recorded in coincidence, map the dynamics associated to those two ionic excited states. By varying the orientation of the light polarization, one can control the molecular dynamics by modifying the ratio between the ionic states. Experimental and ab initio theoretical data are jointly reported.

The ultimate goal of attosecond science is to achieve control of electron dynamics in matter. Our aim is to trace and control nuclear and electronic wave packets in excited [1] or ionized molecules [2]. In spite of its simplicity, the D_2 molecule is an ideal benchmark for coherent control mechanisms combining accurate experimental and theoretical methodologies. In our experiment, a single XUV pulse (energy 42.7 eV, 7 fs pulse duration) ionizes the D_2 molecule, creating a wave packet involving the four lowest states of D_2^+. After some delay, a 780 nm IR field ejects the second electron, leading to Coulomb explosion. All fragments are measured in coincidence using the well-known COLTRIMS technique, which allows for a complete reconstruction of the fragment momenta, and thus for identification of the molecular orientation. The nuclear kinetic energy (NKE) distributions map the evolution in the more highly excited ionic states, $2s\sigma_g$ and $2p\pi_u$, which is usually hard to obtain. The coherent superposition of nuclear wave packets in the two electronic states $2s\sigma_g$ and $2p\pi_u$ can be manipulated by varying the XUV pulse polarization with respect to the molecular axis. Moreover, the nuclear wave packet associated to each ionic state accumulates a different momentum since they follow different potential energy curves, therefore leading to nuclear fragments with different energies in the Coulomb explosion channel. Additionally, we observe interference between these two channels when identical final states are reached after the multiphoton IR absorption. This is shown in figure 1, where we plot the theoretical data for the double ionization probabilities as a function of the NKE and the time delay between the pulses for perpendicular (left panel) and parallel (right panel) orientations. In both cases, we observe the coherent superposition of the $2s\sigma_g$ and $2p\pi_u$ states. The time-dependent simulations are carried out using a $\mathcal{L}^2$ close-coupling method [3, 4], including nuclear motion and electron-electron interactions. The wave packets created by the XUV pulse are obtained from the ab initio calculation, while the Coulomb explosion induced by the IR probe pulse is described as a sudden transition in a two-step model. A very good agreement is bound between experiment and theory.

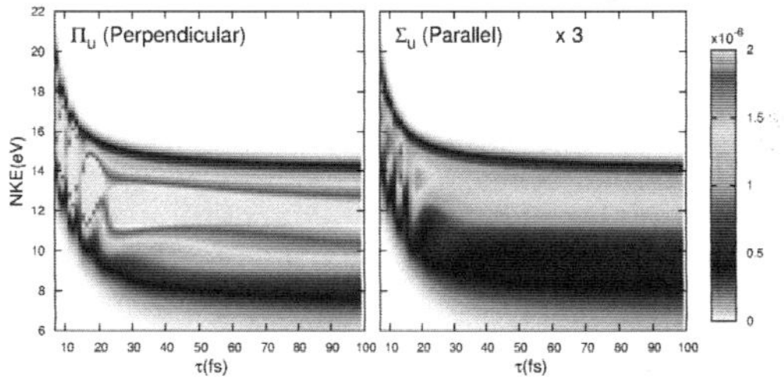

Figure 1. Calculated nuclear kinetic energy (NKE) distributions associated to the Coulomb explosion of D_2 as a function of the pump-probe time delay τ, for perpendicular ($^1\Pi_u$) and parallel ($^1\Sigma_u^+$) molecular orientations with respect to the polarization axis.

References

[1] P. Ranitovic *et al.* 2014 *PNAS* **111** 912

[2] F. Calegari *et al.* 2014 *Science* **346** 336

[3] A. Palacios *et al.* 2007 *PRA* **75** 013408

[4] J. L. Sanz-Vicario *et al.* 2006 *PRA* **73** 033410

[1]E-mail: roger.bello@uam.es
[2]E-mail: pranitovic@lbl.gov

Content from this work may be used under the terms of the Creative Commons Attribution 3.0 licence. Any further distribution of this work must maintain attribution to the author(s) and the title of the work, journal citation and DOI.
Published under licence by IOP Publishing Ltd

Electronic excitation and ionic dissociation of the vanillin molecule: photoionization with flavor

A. Moreno Betancourt[1†], R. B. Bernini[2], L. H. Coutinho[3], G. G. B. de Souza[1*]

[1] Instituto de Química, Universidade Federal do Rio de Janeiro (UFRJ) 21949-900 Rio de Janeiro – RJ, Brazil
[2] Instituto Federal de Ciência e Tecnologia do Rio de Janeiro (IFRJ), 25050-100 Duque de Caxias – RJ, Brazil
[3] Instituto de Física, Universidade Federal do Rio De Janeiro (UFRJ) 21941-972 Rio de Janeiro – RJ, Brazil

Synopsis: The electronic properties of vanillin molecule were investigated by different excitation and ionization techniques. The Ionic fragmentation mechanisms were studied using time-of-flight mass spectrometry as a function of the energy of the incident radiation in the valence energy region and the inner and core levels. It was stablished as a general feature that the ruptures related to the oxygen atoms are predominant at all energies. The resonances and ionization potential of the oxygen 1s was also determined by NEXAFS spectra.

The major component of natural vanilla is 4-hydroxy-3-methoxybenzaldehyde better known as vanillin (Fig. 1), one of the most used and important flavoring material worldwide. The vanilla is extracted from the bean or the pod of the tropical Vanilla orchid (mainly *Vanilla planifolia* Andrews, syn. *V. fragrans* (Salisb. Ames)).[1] Our research group has recently discussed the electronic excitation and ionic fragmentation of volatile natural products, such as limonene,[2] carvone,[3] camphor[4] among others.

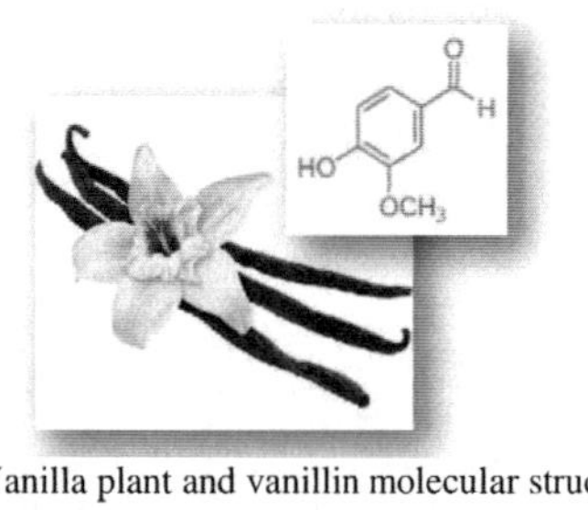

Fig 1. Vanilla plant and vanillin molecular structure.

In the present work, the core electronic excitation and fragmentation of the vanillin molecule has been studied using the NEXAFS and PEPICO techniques. Soft X-rays from the TGM and SGM beamlines at the Brazilian Synchrotron Light Laboratory were used to ionize the sample, and the cations resulting from the photofragmentation process, were characterized according to their mass/charge ratios using a time-of-flight mass spectrometer. For the sake of comparison, TOF mass spectra were also obtained using a He(I) lamp, to access the valence region. An electron gun, tuned at an energy of 800 eV was also employed with the same purpose. Details concerning our experimental methodology may be found in the afore mentioned papers.

The NEXAFS spectrum of the vanillin molecule (Fig. 2) has been studied around the O 1s edge (525 – 555 eV). Since the vanillin molecule features oxygen atoms in three different chemical environments (C-O-H, C-O-CH_3 and C=O) several bands would be expected. However, within our experimental resolution, a single resonance was observed at 531 eV, being assigned mainly to a O 1s → $\pi^*_{C=O}$ transition.

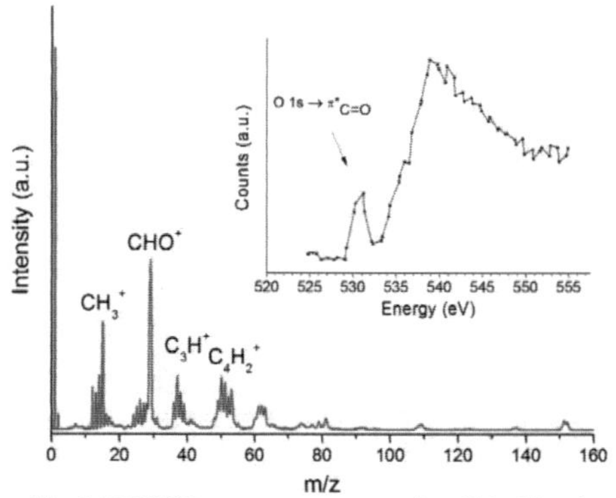

Fig 2. PEPICO spectrum measured at 531 eV and NEXAFS (insert) spectrum around the O 1s energy region for the vanillin molecule.

The TOF-MS spectra, measured either in a PEPICO using a He I Lamp (12 and 21.20 eV), synchrotron radiation (130, 310, 531 and 550 eV) or using an electron gun (800 eV) showed as a common feature ruptures involving the oxygen atoms. For example, the main fragments observed at all energies are H^+, CH_3^+, CHO^+, $C_6H_5O^+$, $C_8H_7O_3^+$ and $C_8H_8O_3^+$. Although the degree of molecular fragmentation increases steadily with increasing photon energy, essentially the same fragments are observed. A preference for site-specific dissociation of the oxygen atom is suggested.

References

[1] J. Nicholas *et al* 2003 *Phytochem. 63, 505.*
[2] R. B. de Castillo *et al* 2007 *J. Elect. Spectr. Rel. Phenom. 155, 77.* [3] R. B. de Castillo *et al* 2014 *J. Elect. Spectr. Rel. Phenom. 192, 61.* [4] R. B. de Castillo *et al* 2014 *Rapid Commun. Mass Spectrom. 28, 1769.*

*E-mail: gerson@iq.ufrj.br
†E-mail: angelicam9@hotmail.com

Content from this work may be used under the terms of the Creative Commons Attribution 3.0 licence. Any further distribution of this work must maintain attribution to the author(s) and the title of the work, journal citation and DOI.
Published under licence by IOP Publishing Ltd

Journal of Physics: Conference Series **635** (2015) 112082 doi:10.1088/1742-6596/635/11/112082

Designing a COLTRIMS Reaction-Microscope for multi-hit coincident measurements with the SQS instrument at European XFEL

Gregor Kastirke* [1]**, Alexander Hartung*, Lothar P.H. Schmidt*, Till Jahnke*, Markus S. Schoeffler*, Robert Moshammer**†**, Michael Meyer**‡ **and Reinhard Dörner***

* Institut für Kernphysik, Goethe-Universität Frankfurt, 60438 Frankfurt am Main, Germany
† Max-Planck Institut für Kernphysik, 69117 Heidelberg, Germany
‡ European XFEL GmbH, 22761 Hamburg, Germany

Synopsis Using intense photon beams from the XFEL free electron laser, the COLTRIMS Reaction Microscope will be a versatile tool for examination of the dynamics in small quantum systems.

The new free electron laser at the European XFEL facility will provide photon beams of extremely high density consisting of very short light pulses. We designed a highly modular REaction MIcroscope (REMI), based on the COLTRIMS- (COLd Target Recoil Ion Momentum Spectroscopy) technique to investigate correlated dynamics of electrons and ions in atomic, molecular and cluster systems [1], [2].

In a COLTRIMS-setup containing three different UHV-recipients (expansion-, target- and jetdump-chamber (Figure 1)), the target is prepared in a supersonic gas expansion. This gas-jet can be shaped with up to four differential pumping stages, high precision motorized collimators and a slotted disk velocity selector. Varying parameters as the driving pressure and the temperature of the nozzle (12 K – 500 K, using a He-flow cryostat or heating) makes it possible to generate an extremely well defined and localized atomic or molecular gas-jet which can selectively consist of e.g. superfluid He-clusters, molecular clusters or organic molecules. The X-ray beam interacts with the molecular gas-jet, allowing photoionization with subsequent Auger cascades to take place. It is possible to measure up to 10 ions and electrons per X-ray pulse in coincidence.

The SQS-Instrument at the SASE3 beamline of the European XFEL will provide a photon beam with a timing structure of 27 kHz repetition rate and microscopic bunch spacing of 220 ns. Each pulse with a duration of less then 2 fs will contain $2{\cdot}10^{14}$ photons focused on a spot of 1µm. Thus, already a single pulse is sufficient to saturate the K-Shell ionization e.g. in carbon or nitrogen. However, in order to measure all charged fragments from an ionization event in coincidence, it is necessary to have less than one reaction per pulse. This requires low density targets and correspondingly very good vacuum, i.e. less than $2{\cdot}10^{-11}$ mbar. This is achieved by using a tempered and electro-polished recipient with including NEG- and TSP-pumps.

To investigate time resolved processes in molecules with pump-probe experiments, SQS additionally provides a tunable femtosecond laser. This laser can be injected collinear to the X-ray beam. With this setup, different pump-probe scenarios are possible: X-ray pump-laser probe, laser pump-X-ray probe, laser pump-laser probe and X-ray pump-X-ray probe.

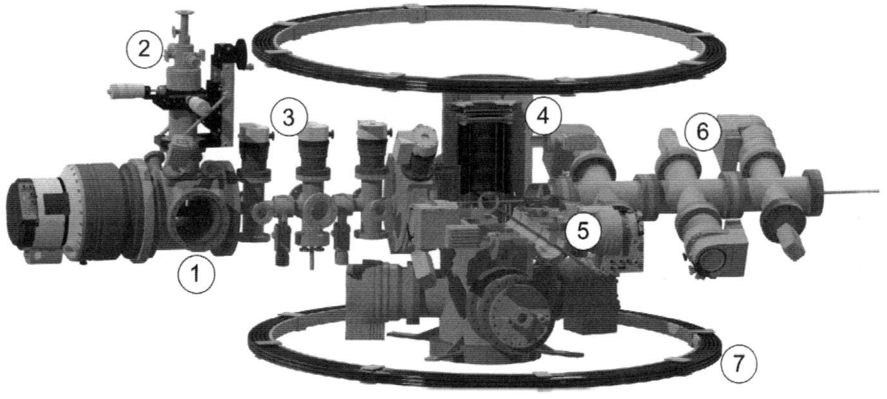

Figure 1. CAD-Model of the XFEL-REMI. Blue line: FEL-beam, yellow line: gas-jet, red circle: interaction-zone. (1) expansion-chamber, (2) cryostat, (3) differential jet-stages, (4) target-chamber with spectrometer and detectors, (5) differential beamline pump-stages, (6) jetdumps, (7) Helmholtz-coils.

References

[1] R. Dörner *et al* 2000 *Phys. Rep.* **330** 96-192

[2] J. Ullrich *et al* 2003 *Rep. Prog. Phys.* **66** 1463-1545

Acknowledgement

Supported by BMBF

[1] E-mail: kastirke@atom.uni-frankfurt.de

Content from this work may be used under the terms of the Creative Commons Attribution 3.0 licence. Any further distribution of this work must maintain attribution to the author(s) and the title of the work, journal citation and DOI.
Published under licence by IOP Publishing Ltd

An intense electrospray ionization source for soft X-ray photoionization of gas phase protein ions

Dmitrii Egorov*[1], Sadia, Bari[†], Ronnie Hoekstra*, Arkadiusz Lawicki[‡], Konstantin Hirsch[‡], Vicente Zamudio-Bayer[‡], Tobias Lau[‡], Bernd von Issendorf[§], Thomas Schlathölter*

* Zernike Institute for Advanced Materials, University of Groningen, Nijenborgh 4, 9747AG Groningen, Netherlands
[†] European XFEL GmbH, Notkestrasse 85, 22607 Hamburg, Germany
[‡] Helmholtz Zentrum Berlin, Albert-Einstein-Str. 15, 12489 Berlin, Germany
[§] Physikalisches Institut, Universität Freiburg, Stefan-Meier-Strasse 21, 79104 Freiburg

Synopsis For gas-phase experiments on complex biomolecular ions, the target density is a most critical issue. We developed a novel high-current electrospray ionization source, delivering 10 times higher currents than conventional systems.

Complementary to conventional biomolecular spectrocopy in the liquid phase, over the past few years the investigation of large and complex biomolecules in the gas phase has gained momentum. Such studies for instance allow to distinguish intrinsic molecular properties from environment effects, to investigate fundamental processes such as nanosolvation, or to benchmark quantum chemical calculations.

The first requirement for experimental studies is non-destructive transfer of fragile biomolecular species (e.g. proteins, DNA) from the liquid phase into the gas phase. Electrospray ionization (ESI) has proven to be a powerful technique for production of beams of gas-phase biomolecular ions [1]. Despite its versatility, ESI has the major drawback of delivering notoriously low ion fluxes. Typical mass selected biomolecular ion currents obtained with commercial instruments are of the order of few pA - much too low for many applications.

We developed an ESI source, optimized for high ion flux. The source is equipped with a radiofrequency (RF) ion funnel and a housing optimized for efficient pumping. Using a roots pump system then allows the use of a heated capillary with a much larger than usual inner diameter. First tests with the small protein melittin proved, that the source can deliver more than 1 nA of total ions and more than 100 pA of mass selected protein ions.

In a first test, the high flux ESI source was interfaced with the NanoClusterTrap [2] setup at the Helmholtz Zentrum Berlin (HZB). Here, electrosprayed melittin cations were RF guided, mass selected and eventually accumulated in a cryogenic linear RF ion trap. The trapped cations were then exposed to soft X-rays from the HZB UE52_PGM beamline.

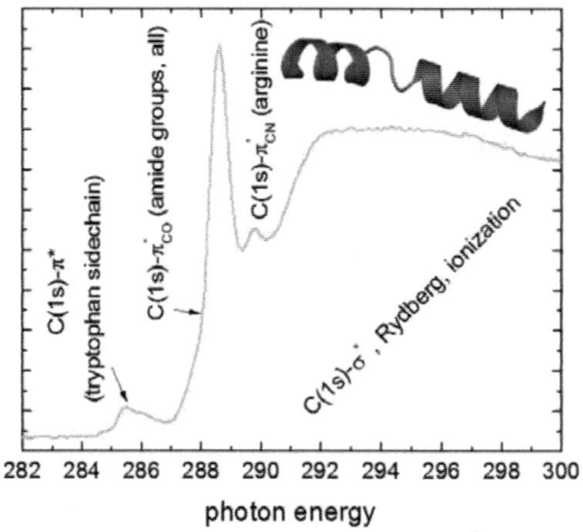

Figure. Fragment yield for (melittin+2H)$^{2+}$ after soft X-ray absorption at the C K-edge.

Preliminary results for photoionization of (melittin+2H)$^{2+}$ are displayed in figure. The quality of the data renders for instance circular dichroism studies feasible.

References

[1] Photophysics of Ionic Biochromophores, Physical Chemistry in Action, Ed. S. Brøndsted Nielsen, J.A. Wyer, Springer Berlin Heidelberg (2013)
[2] J. T. Lau, J. Rittmann, V. Zamudio-Bayer, M. Vogel, K. Hirsch, Ph. Klar, F. Lofink,T. Möller, B. v. Issendorff, Phys. Rev. Lett. **101** (2008) 153401

[1] E-mail: d.egorov@rug.nl

Content from this work may be used under the terms of the Creative Commons Attribution 3.0 licence. Any further distribution of this work must maintain attribution to the author(s) and the title of the work, journal citation and DOI.
Published under licence by IOP Publishing Ltd

The time-dependent two-particle reduced density matrix method: LiH in strong laser fields

Fabian Lackner*[1], Iva Březinová*[2], Takeshi Sato†, Kenichi L. Ishikawa†, Joachim Burgdörfer*

* Institute for Theoretical Physics, Vienna University of Technology, Wiedner Hauptstraße 8-10/136, 1040 Vienna, Austria, EU

† Photon Science Center, Graduate School of Engineering, The University of Tokyo,7-3-1 Hongo, Bunkyo-ku, Tokyo 113-8656, Japan

Synopsis We present a time-dependent many-body theory that is based on the propagation of the two-particle reduced density matrix (2-RDM) [1]. We apply our method to simulate the non-linear response of one-dimensional lithium hydride in femtosecond mid-IR laser pulses.

With the availability of laser pulse intensities competing with the internal Coulomb force the electron dynamics in atoms and molecules can be perturbed far beyond the linear regime. Such strong perturbations give rise to fascinating non-linear effects such as above threshold ionization, high harmonics generation and multiple ionization. Along with these advanced experimental techniques the development of sophisticated numerical simulations is required to describe the correlated many-body systems.

We have developed a new approach, the time-dependent two-particle reduced density matrix (TD-2RDM) method, to describe the time-dependent many-body response. As a prototypical molecular many-body system we apply it to the one-dimensional model for lithium hydride (LiH) in strong infrared laser fields. We solve the equations of motion for 2-RDM by approximating the three-particle reduced density matrix (3-RDM) in terms of lower order RDMs via a new reconstruction functional that preserves norm, energy, and spin symmetries during time propagation. We show that approximately enforcing N-representability during time evolution is essential for achieving stable solutions. For the induced dipole moment (Fig. 1) our results are in very good agreement with numerically exact solutions obtained from the multiconfigurational time-dependent Hartree-Fock method. Further we find that the time-dependent 2-RDM method (TD-2RDM) performs consistently better than time-dependent Hartree-Fock (TDHF) and time-dependent density functional theory (TDDFT) within the adiabatic local-density approximation over the entire time interval. This advance opens up the intriguing possibility of solving the many-body problem within the two-particle subspace previously only studied for the ground state of complex atoms and molecules [2, 3].

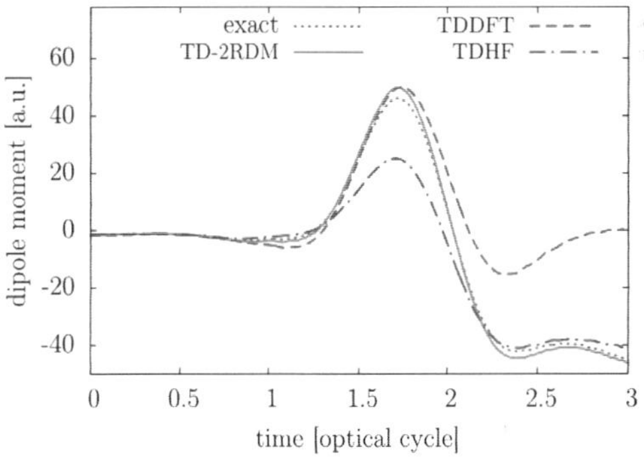

Figure 1. Dipole moment of LiH subject to a laser pulse with $I = 8 \times 10^{14}\mathrm{W/cm^2}$ within the TD-2RDM method compared with the exact (MCTDHF) reference, the TDHF and the TDDFT calculations.

Acknowledgement

This work has been supported by the FWF doctoral school Solids4Fun, FWF SFB-041 ViCoM, and FWF SFB-049 Next Lite, and in part by KAKENHI, the Photon Frontier Network Program of MEXT (Japan), and Center of Innovation Program from JST (Japan).

References

[1] F. Lackner *et al* 2015 *Phys. Rev. A* **91** 023412

[2] D. Mazziotti 1998 *Phys. Rev. A* **57** 4219

[3] Z. Zhao *et al* 2004 *J. Chem. Phys.* **120** 2095

[1]E-mail: fabian.lackner@tuwien.ac.at
[2]E-mail: iva.brezinova@tuwien.ac.at

Content from this work may be used under the terms of the Creative Commons Attribution 3.0 licence. Any further distribution of this work must maintain attribution to the author(s) and the title of the work, journal citation and DOI.
Published under licence by IOP Publishing Ltd

Journal of Physics: Conference Series **635** (2015) 112085 doi:10.1088/1742-6596/635/11/112085

Absence of entanglement effect on the decay dynamics of H(2p) pairs produced by VUV photodissociation of H_2

X Urbain*[1], A Dochain*, C Lauzin* and B Fabre†

* Institute of Condensed Matter and Nanosciences, Université catholique de Louvain, B-1348 Louvain-la-Neuve, Belgium

† Université de Bordeaux-CNRS-CEA, CELIA, UMR5107, F33405 Talence, France

Synopsis The effects of entanglement on the spontaneous decay of a pair of H(2p) previously observed by Tanabe *et al.* (*Phys. Rev. A* **82**, 040101 (2010)) could not be reproduced despite careful determination of emission times with respect to the synchrotron bunch.

The entanglement of a pair of H(2p) generated by VUV photodissociation of H_2 along the $Q_2\ ^1\Pi_u$ doubly excited state has been discussed in a series of publications [1, 2, 3]. In an attempt to reproduce the effects attributed to entanglement, namely the shortening of the apparent lifetime and the stronger modulation of the angular distribution at lower pressure, we have studied this reaction at the DESIRS beamline of the SOLEIL Synchrotron facility operating in single-bunch mode.

In the present experiment, the synchrotron beam is injected into an interaction cell filled with H_2 between 2×10^{-4} and 10^{-2} mbar. On either side of the cell, a CsI-coated multichannel plate assembly located behind a 1 mm-thick MgF_2 window detects Lyman-α photons emitted by H(2p) at time t_1 and t_2. The synchrotron clock signal t_0, and the photon arrival times t_1 and t_2 are recorded by a multichannel TDC (120 ps binwidth) receiving the triple coincidence signal as a common stop. This setup differs from that used by Tanabe *et al* [2] who had only access to $t_2 - t_1$. It also allows us to measure the emission time of the first incoming photon $t_<$ and that of the second photon $t_>$ independently.

The histogram of $t_1 - t_0$, $t_2 - t_0$ (Fig.1(a)), can be fitted by an exponential function giving lifetimes $\tau_1 = 1.65(5)$ ns and $\tau_2 = 1.57(5)$ ns respectively, both compatible with the lifetime of an isolated H(2p) atom. The histogram of $t_< - t_0$ and $t_> - t_0$ (Fig.1(b)) gives the temporal distribution for the first and second photon, respectively. Those follow the functional dependence discussed by Sanjo and Plaja [4]. All fits are consistent with a single decay time affecting photon emission. We have systematically varied the experimental conditions, i.e. the pressure of H_2, the incoming photon energy and polarization, the latter giving access to the angular distribution of the emitted photons.

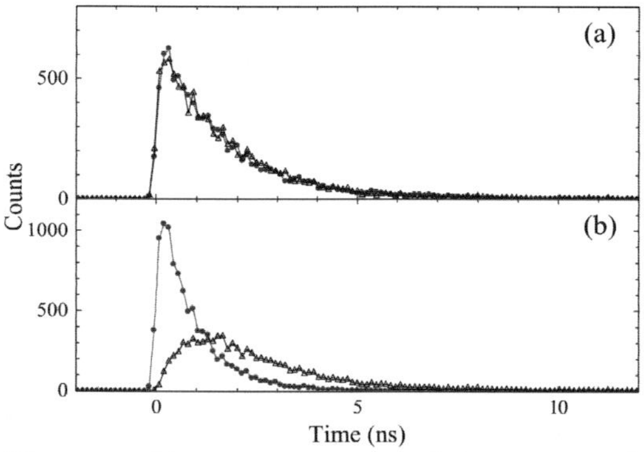

Figure 1. Time spectrum recorded at 3×10^{-3} mbar - (a) detector 1 (dots) and 2 (triangles), (b) first (dots) and second photon detected (triangles).

The invariability of the decay time throughout the measurements, determined to be identical to the natural lifetime of the 2p state, rules out any modification of the decay rate due to entanglement of the atom pair and its possible decoherence upon collision with surrounding molecules. Superradiance or concerted emission [5] is not expected since the atoms are ~90 μm apart at the time of their radiative decay. A similar conclusion was reached independently by Nakanishi *et al* [6] who attributed earlier findings to a flaw of the experiment.

References

[1] Tanabe T , Odagiri T, Nakano M, Suzuki I H and Kouchi N 2009 *Phys. Rev. Lett.* **103** 173002

[2] Tanabe T *et al* 2010 *Phys. Rev. A* **82** 040101

[3] Odagiri T, Tanabe T and Kouchi N 2012 *J. Phys. Conf. Ser.* **388** 012024

[4] Sancho P and Plaja L 2011 *Phys. Rev. A* **83** 066101

[5] Gross M and Haroche S 1982 *Phys. Rep.* **93** 301

[6] Nakanishi Y *et al* 2014 *Phys. Rev. A* **90** 043405

[1]E-mail: xavier.urbain@uclouvain.be

Content from this work may be used under the terms of the Creative Commons Attribution 3.0 licence. Any further distribution of this work must maintain attribution to the author(s) and the title of the work, journal citation and DOI.
Published under licence by IOP Publishing Ltd

XXIX International Conference on Photonic, Electronic, and Atomic Collisions (ICPEAC2015) IOP Publishing
Journal of Physics: Conference Series **635** (2015) 112086 doi:10.1088/1742-6596/635/11/112086

Implementation of fixed-nuclei polyatomic MCTDHF capability and the future with nuclear motion

Daniel J. Haxton*,# Keith V. Lawler† François-Henry Rouet° Jeremiah R. Jones+ Khaled Z. Ibrahim° Samuel Williams° Eugene Vecharynski° Chao Yang° Xiaoye Li° Brant Abeln*,#,⊗ Thomas N. Rescigno* C. William McCurdy*,#,⊗

*Chemical Sciences Division, Lawrence Berkeley National Laboratory, Berkeley 94720, California, USA
#Ultrafast X-ray Science Laboratory, Lawrence Berkeley National Laboratory, Berkeley 94720, CA, USA
†Department of Chemistry, University of Nevada-Las Vegas, Las Vegas 89154, Nevada, USA
+School of Mathematical and Statistical Sciences, Arizona State University, Tempe, Arizona 85281
°Computing Sciences Divison, Lawrence Berkeley National Laboratory, Berkeley 94720, California, USA
⊗Departments of Applied Science and Chemistry, Davis 95616, California, USA

Synopsis We discuss the implementation (https://commons.lbl.gov/display/csd/LBNL-AMO-MCTDHF) of Multiconfiguration Time-Dependent Hartree-Fock for polyatomic molecules using a Cartesian product grid of sinc basis functions, and present absorption cross sections and other results calculated with it.

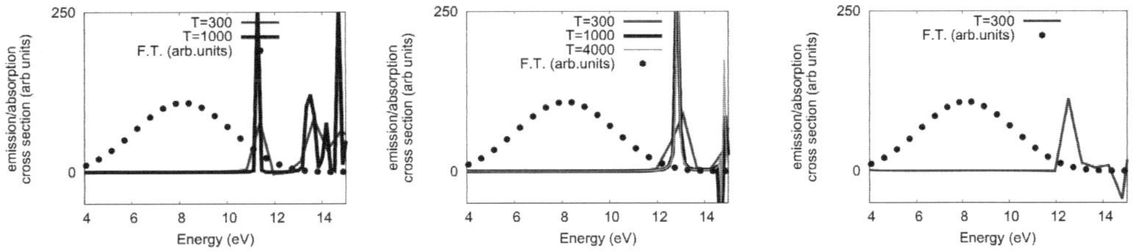

Figure: Absorption spectra for methane obtained from calculations with (left) grid spacing $\Delta = 0.4a_0$, $n = 31$ points on a side; (middle) $\Delta = 0.4a_0$, $n = 31$; (right) $\Delta = 0.2a_0$, $n = 63$.

The implementation of MCTDHF for polyatomic molecules using Cartesian product sinc basis functions will be discussed. There are two key components to the polyatomic representation [1]:

(1) A diagonal expression for the Coulomb matrix elements is given by a resolution-of-the-identity approximation based upon the interpolating property of the sinc basis. This expresson is obtained through a generalization of Ref. [2] and yields surprisingly accurate results. Interesting questions remain about the application of Ref. [2] to arbitrary generalized discrete variable representations in 3D.

(2) The mean fields and two-electron matrix elements are constructed using established techniques [3] for triple Toeplitz matrix algebra, in parallel, using an OpenMP/MPI three-dimensional fast Fourier transform, requiring $O(N \log N)$ multiplication operations; $N = n^3$ for a cube.

Absorption spectra and other properties of methane and other polyatomics, calculated using n =16-399, will be presented. Applications of the implementation including the decomposition of nonlinear wave mixing signals [4] will be discussed.

The method fixes the nuclei at the electron grid points, does not include complex scaling for ionization, and the nuclear kinetic energy terms are not yet implemented. A method for arbitrary, complex-valued deformations of the grid will permit a polyatomic implementation with nonadiabatic nuclear motion and ionization. Generalizations of Ref. [2] involving not the kinetic energy operator, but a Dirac operator, may provide the way forward.

References

[1] J. R. Jones *et al*, "An efficient basis set representation for calculating electrons in molecules," in preparation.

[2] C. W. McCurdy, M. Baertschy, and T. N. Rescigno 2004 *J. Phys. B* **37 R137**

[3] J. Demmel, P. Koev, and X. Li in *Templates for the Solution of Algebraic Eigenvalue Problems: A Practical Guide*, edited by Z. Bai, J. Demmel, J. Dongarra, A. Ruhe, and H. van der Vorst. SIAM, Philadelphia (2000)

[4] L. Seidner, G. Stock, and W. Domcke 1995 *J. Chem. Phys.* **103 3998**

Content from this work may be used under the terms of the Creative Commons Attribution 3.0 licence. Any further distribution of this work must maintain attribution to the author(s) and the title of the work, journal citation and DOI.
Published under licence by IOP Publishing Ltd

Journal of Physics: Conference Series **635** (2015) 112088 doi:10.1088/1742-6596/635/11/112088

Experimental measurement of Interatomic Coulombic Decay rates in Argon dimers

J. Rist*† 1, M. Schöffler* 2, A. Moradmand‡, A. Landers‡, B. Gaire†, T. Weber†, P. Kolorence§, K. Gokhberg‖, T. Jahnke* and R. Dörner*

* Institut für Kernphysik, Goethe-Universität, Max-von-Laue-Str.1, 60438 Frankfurt am Main, Germany
† Chemical Sciences Division, Lawrence Berkeley National Laboratory, Berkeley, CA 94720, USA
‡ Department of Physics, Auburn University, AL 36849, USA
§ Institute of Theoretical Physics, Charles University in Prague, Prague 116 36, Czech Republic
‖ Theoretical Chemistry Group, Heidelberg University, 69117 Heidelberg, Germany

Synopsis We present a method by which the internuclear distance-dependent decay width Γ(r) of Interatomic Coulombic Decay can be obtained using a measured energy distributions and a classical nuclear dynamics model.

Interatomic Coulombic Decay (ICD) is a fundamental decay process, which can occur in weakly bound systems such as the Argon dimer after excitation. It was predicted in 1997 by L. S. Cederbaum *et al* [1] and confirmed experimentally in 2004 [2].

The decay widths of ICD depend strongly on the internuclear distance of the two atoms involved in the decay. Here we present a method employing a classical nuclear dynamics model in order to extract the decay width Γ(r) from measured experimental data.

As an experimental setup we used a Cold Target Recoil Ion Momentum Spectroscopy (COLTRIMS) [3] apparatus located at the Advanced Light Source (ALS) in Berkeley, a 3rd gen. synchrotron, to populate a singly ionized and excited Argon dimer state. As ICD takes place the system breaks up into the three fragments $Ar^+ + Ar^+ + e_{ICD}$. These fragments and the emitted photoelectron are measured in coincidence. From the measured momenta the state excited prior to ICD can be determined, as shown in Figure 1.

Since the different initial states can be analyzed individually and the potential curves of the ground state and excited states are known, the nuclear dynamics occurring during the decay can be modelled in order to deduce a decay time and internuclear distance from the measured kinetic energy release (KER).

Figure 2 depicts the decay width in dependence of the internuclear distance of the $(^1D)4d(^2S)$ state marked in Fig. 1.

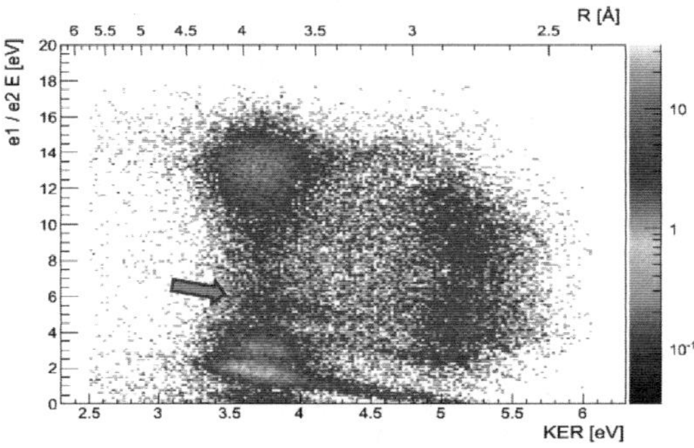

Figure 1. Electron energy distribution in dependence of the recoil kinetic energy release in Argon dimers.

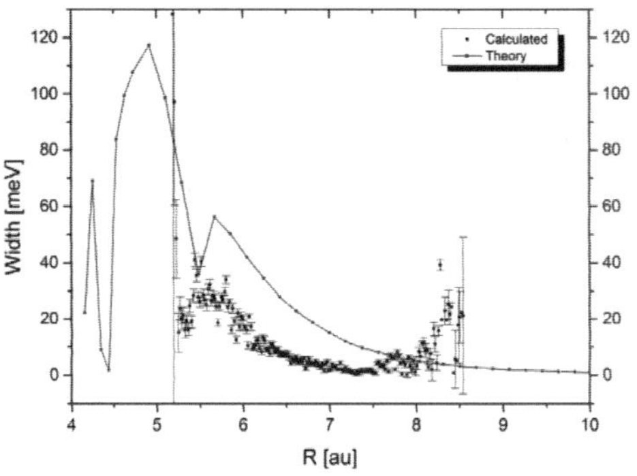

Figure 2. Calculated Γ(r) (dots) compared with the theory [4] (line) for the $(^1D)4d(^2S)$ state.

Comparing the values obtained from the experiment to theoretical predictions yields good qualitative agreement for internuclear distances greater than 5.5 au.

References

[1] L. Cederbaum *et al* 1997 *Phys. Rev. Lett.* 79 4778
[2] T. Jahnke *et al* 2004 *Phys. Rev. Lett.* 93 163401
[3] R. Dörner *et al* 2000 *Phys. Rep.* 330 95-192
[4] Private Communication. (P. Kolorence)

1 E-mail: rist@atom.uni-frankfurt.de
2 E-mail: schoeffler@atom.uni-frankfurt.de

Content from this work may be used under the terms of the Creative Commons Attribution 3.0 licence. Any further distribution of this work must maintain attribution to the author(s) and the title of the work, journal citation and DOI.
Published under licence by IOP Publishing Ltd

XXIX International Conference on Photonic, Electronic, and Atomic Collisions (ICPEAC2015) IOP Publishing
Journal of Physics: Conference Series **635** (2015) 112089 doi:10.1088/1742-6596/635/11/112089

Double Photoionization of confined atoms in n-walled fullerenes

J M Randazzo[*,†,1], **A L Frapiccini**[‡,†], **D Mitnik**[§,†], **G Gasaneo**[‡,†], **F D Colavecchia**[*,†]

[*] División Física Atómica, Molecular y Optica, Centro Atómico Bariloche, 8400 S. C. de Bariloche, Río Negro, Argentina.

[§] Instituto de Astronomía y Física del Espacio, FCEyN Universidad de Buenos Aires, Argentina.

[‡] Departamento de Física, Universidad Nacional del Sur, 8000 Bahía Blanca, Buenos Aires, Argentina.

[†] Consejo Nacional de Investigaciones Científicas y Técnicas, CONICET, Argentina.

Synopsis In this work we study the effect of fullerene confinement on the triple differential cross section (TDCS) for double ionization of atoms. We compare the results corresponding to the free case with single cage C_{60} and two cage $C_{60}@C_{240}$ confinement.

Fullerene cages have been intensively studied because of their utility in nano-science and development of materials with unique properties [1]. The carbon spheres C_n have different sizes depending on the number n of carbon atoms and they can be arranged one inside the other in a configuration which could be named "bucky-onions" , and single photo ionisation of atoms inside these structures has been studied [2, 3]. In this work we deal with the double photo ionisation processes, first with free atoms, and then of atoms inside these structures, each onion layer modeled with a square well potential [3]:

$$V_n(r) = \begin{cases} U_n & if \quad R_n \leq r \leq R_n + \Delta_n \\ 0 & otherwise \end{cases}$$

for each electron.

In order to obtain the TDCS we search first for the initial state Ψ_0, solution of the system's Schrdinger equation with hamiltonian H_0. Once we have the state we prepare a non-homogeneous driven equation for the scattering wave function Ψ^+_{sc}:

$$[H_0 - E]\,\Psi^+_{sc} = -W\Psi_0, \qquad (1)$$

where W is the field operator both in velocity or length gauge. Both wave functions are obtained through a Generalized Sturmian Function expansion [4] which allow imposition of bounded or outgoing-flux boundary conditions. Once the solution is obtained, TDCS is evaluated according to the formulae presented in [5]. An example of the results are shown in figure (1), where we show experimental TDCS results for the free He atom, together with our results and Exterior Complex Scaling Calculations [5], to which we found an excellent agreement. The blue line shows the strong modification of the TDCS by the presence of the fullerene potential model corresponding to the C_{60}, where $R_{60} = 5.8\ a.\ u.$, $\Delta_{60} = 1.9\ a.\ u.$, and $U_{60} = 8.22\ eV$ [3].

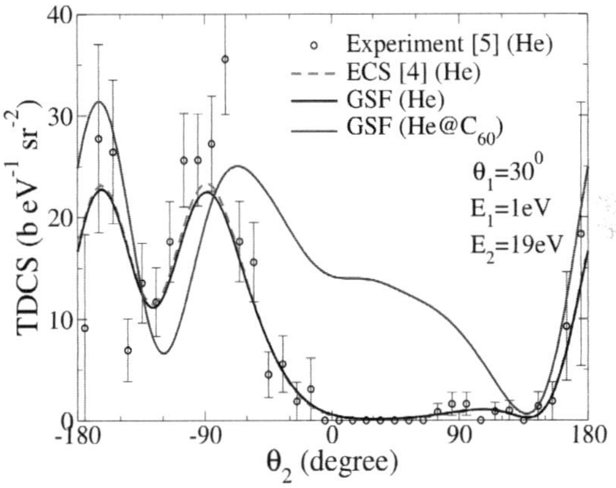

Figure 1. Triple differential cross section (TDCS) for double ionization by single photon absorption of He and $He@C_{60}$. Circles: He Experimental data from [6]. Red dashed: He ECS calculations. Continuous line He GSF (black) and $He@C_{60}$ GSF (blue) [7].

We will analyze the effect of combining more potentials such as the C_{240}, C_{540}, $C_{60}@C_{240}$ and $C_{60}@C_{240}@C_{540}$ over the triply differential cross sections.

References

[1] H. Shinohara 2000 *Rep. Prog. Phys.* **63** 843

[2] F. D. Colavecchia *J. At., Mol. Opt. Phys.* 2011 **2011** 817034

[3] Dolmatov VK *et al* 2008 *Phys Rev A* **78** 013415

[4] G. Gasaneo *et al* 2013 *Adv. Quantum Chem.* **87** 153

[5] C. W. McCurdy et al 2004 *Phys. Rev. A* **69** 032707

[6] H. Bräuning *et al* 1998 *J. Phys. B* **31** 5149

[7] J. M. Randazzo *et al* submitted to *Eur. Phys J. D*

[1]E-mail: randazzo@cab.cnea.gov.ar

Content from this work may be used under the terms of the Creative Commons Attribution 3.0 licence. Any further distribution of this work must maintain attribution to the author(s) and the title of the work, journal citation and DOI.
Published under licence by IOP Publishing Ltd

New process to describe radiation damage at the molecular level. Application to ^{125}I seeds in water

R. Colmenares*, K. Krupa°, A. Williart†, L. Ellis-Gibbings§, Ali Traore Dubuis§, A. Verkhovtsev§ and G. García§1

* Unidad de Radiofísica Clínica, Hospital Ramón y Cajal, Madrid
° Faculdade de Ciencias e Tecnologias - Universidade nova de Lisboa
§ Instituto de Física Fundamental, CSIC, Madrid
† Departamento de Física de los Materiales, UNED, Madrid

Sinopsis We present the first step of a general study on induced damage by ^{125}I in water at the molecular level. Requirements for such a study are introduced and preliminary results on the emission spectra of ^{125}I seeds are presented.

The main goal of this work is to study the biological damage induced by the radiation in a nanometric target volume (nanodosimetry), which is related to molecular dissociation.

The Monte Carlo codes used in this study includes all relevant interactions (LEPTS, Low Energy Particle Track Simulation [1], for energies < 10 keV with GEANT4 for >10keV) to obtain not just absorbed doses but also the number of dissociations in any nanovolume (~nm^3). According to the nanodosimetry concept, the MC code should track the particles with low energy, in order to take into account some interactions which are very efficient in this energy range. The input parameters for the MC are the cross sections (elastic and inelastic) and the electron energy loss spectra which have been previously measured experimentally or calculated depending on the energy range [2].The final goal is to improve the calculations used in clinical dosimetry based on these new concepts, or at least to define where the absorbed dose is low but some molecular damage is actually calculated by our model during treatment.

The first case to be studied are the low dose rate prostate treatments with ^{125}I seeds. The first measurements are based on quantifying the ionization generated by the ^{125}I seeds (Figure 1) in liquid water and comparing the experimental results with those obtained by the MC code. In addition, the absorbed dose calculated by MC will be checked by clinical planning system based on the TG43 protocol (using a water medium).

For this purpose, the ^{125}I seed is placed in a water phantom with dimensions of 50cmx50cmx50cm. The ionization chamber (model CC 01 of IBA) with an active volume of 0,01 cm^3 is used. The electrometer MAX-4001 of Standart Imaging is connected to the chamber and operated in Low Range module with a resolution of 0,01 pC. The main difficulty of these measurements consist in the determination of the reference point between the chamber and the irradiation source, as other variables are well controlled. This uncertainty was kept below 1 mm.

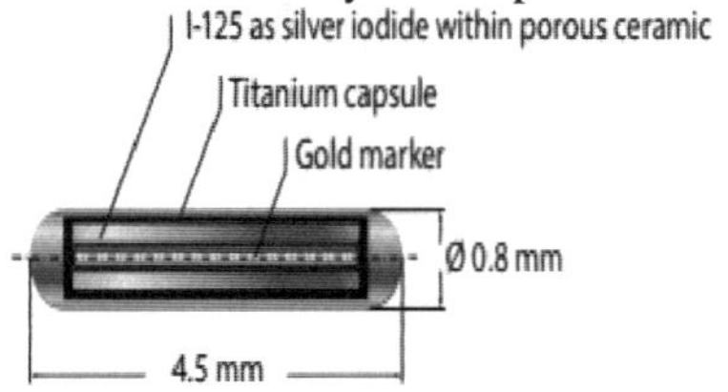

Figure1. IsoSeed®I25.S061-1

All the measurements are corrected for radioactive decay (S_k=2.483 U, $T_{1/2}$ = 59.41 d), pressure and temperature to the standard conditions of normalised radioactive activity, 20ºC, and 1013 mbar.

In the Table 1 the ionization data obtained at different geometries is showed.

z(mm)	x(mm)	y(mm)	Mean (pC)	sd (pC)	v.i (%)
2	0	0	19.84	0.36	1.8
4	0	0	10.31	1.67	16.2
7	0	0	4.30	0.54	12.6
10	0	0	2.97	0.32	10.9
20	0	0	1.04	0.12	11.2
30	0	0	0.63	0.08	12.3
10	0	5	2.65	0.17	6.3
10	0	10	2.02	0.22	11.1
10	0	20	1.28	0.10	7.7
10	0	30	0.45	0.05	11.6
10	14	0	1.11	0.11	9.5

x: coordinate on longitudinal axis on the seed; sd: standard deviation; i.v.: variation index.

The spectrum of the ^{125}I seeds was measured with a Si(Li) detector and so all the required information is available to begin MC simulations as future work.

Aknowlegments

This work was supported by MINECO (FIS2012-31230) and FCT/MEC, RaBBiT, PD-F,PD/00193/2012); UID/Multi/04378/2013 (UCIBIO); UID/FIS/00068/2013 (CEFITEC); and grant number SFRH/BD/52536/2014.

References

[1] G. García *et. al.* 2012 *Springer*
[2] F. Blanco *et. al.* 2013 *Eur. Phys. J. D*, **67**, 199.

1 E-mail: g.garcia@csic.es

Content from this work may be used under the terms of the Creative Commons Attribution 3.0 licence. Any further distribution of this work must maintain attribution to the author(s) and the title of the work, journal citation and DOI.
Published under licence by IOP Publishing Ltd

XXIX International Conference on Photonic, Electronic, and Atomic Collisions (ICPEAC2015) IOP Publishing
Journal of Physics: Conference Series **635** (2015) 112091 doi:10.1088/1742-6596/635/11/112091

Electron production by noble metal nanoparticles in collision processes ith photons and fast ions

Alexey Verkhovtsev[*,†1], **Andrei Korol**[†], **and Andrey Solov'yov**[†]

[*] CSIC, Instituto de Fisíca Fundamental, Serrano 113-bis, 28006 Madrid, Spain

[†] MBN Research Center, Altenhöferallee 3, 60438 Frankfurt am Main, Germany

Synopsis We analyzed electron production by noble metal nanoparticles irradiated by photons and fast ions. For that, photoabsorption spectra of several gold clusters were calculated in a broad photon energy range by means of time-dependent density-functional theory (TDDFT). Two dominant peaks appearing in the spectra of electron emission of noble metal clusters were assigned to two distinct types of collective electron excitations. This study revealed that the decay of the collective excitations formed in the nanoparticles is an important mechanism of generation of low-energy electrons, which are known to act as important agents of biodamage.

It is currently acknowledged that low-energy secondary electrons (having the kinetic energy from a few eV to several tens of eV) and other species, which are produced when fast ions propagate through a biological medium, largely cause the biological damage [1, 2]. As was proposed recently, cancer treatments with ionizing radiation can be improved by injecting metal (noble metal, in particular) nanoparticles which may increase the number of secondary species in the target region. However, the nanoscale mechanisms leading to an improvement of existing techniques are still to be thoroughly substantiated.

We performed a theoretical and numerical analysis of electron production by noble metal (gold, platinum, silver) nanoparticles irradiated by photons and fast ions. For that, we calculated the photoabsorption spectra of Au_N ($N = 18 - 42$) clusters in a broad photon energy range by means of the TDDFT approach. Two prominent features of the spectra were assigned to the two distinct types of collective electron excitations (Fig. 1). Plasmons, i.e. collective excitations of delocalized valence electrons, dominate the spectra of electron emission from metallic nanoparticles in the energy range of about $1 - 10$ eV. For higher electron energies (of a few tens of eV), the main contribution to the electron yield arises from the atomic giant resonance associated with the collective excitation of $5d$ electrons in individual atoms of a nanoparticle.

Similar to the photoionization, the two distinct types of collective electron excitations appear in the process of impact ionization. We demonstrated that noble metal nanoparticles, especially those made of gold and platinum, significantly enhance the electron yield due to the collective response to an external electric field of a charged projectile. As a result of these effects, the number of the low-energy electrons generated by small noble metal nanoparticles significantly exceeds that produced by an equivalent volume of water [3, 4]. Thus, decay of the collective electron excitations formed in the nanoparticles represents an important mechanism of generation of the low-energy electrons.

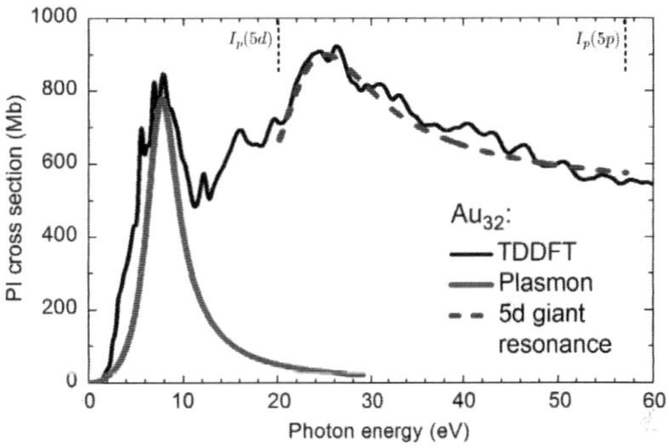

Figure 1. Photoabsorption cross section of the Au_{32} cluster calculated within TDDFT (thin curve). Thick solid and dashed curves represent the contributions of the plasmon-type and the localized (giant resonance) excitations, respectively [3, 4]. Vertical lines mark the $5d$ and $5p$ ionization thresholds in the atom of gold.

References

[1] B. Boudaïffa *et al* 2000 *Sc e ce* **287** 1658

[2] E. Surdutovich and A. V. Solov'yov 2014 *Eur. Phys. J. D* **68** 353

[3] A. V. Verkhovtsev, A. V. Korol and A. V. Solov'yov 2015 *Phys. Rev. ett.* **114** 063401

[4] A. V. Verkhovtsev, A. V. Korol and A. V. Solov'yov *J. Phys. Chem. C* (in print); DOI: 10.1021/jp511419n

[1]E-mail: verkhovtsev@iff.csic.es

Content from this work may be used under the terms of the Creative Commons Attribution 3.0 licence. Any further distribution of this work must maintain attribution to the author(s) and the title of the work, journal citation and DOI.
Published under licence by IOP Publishing Ltd

Journal of Physics: Conference Series **635** (2015) 112092 doi:10.1088/1742-6596/635/11/112092

Ultrafast Dynamics And Electronic State - Lifetime Interferences In Chlorine-Containing Molecules

G. Goldsztejn[1,2*], T. Marchenko[1,2], R. Guillemin[1,2], L. Journel[1,2], R.K. Kushawaha[1,2], R. Püttner[3], D. Céolin[4], M.N. Piancastelli[1,2,5] and M. Simon[1,2,4]

[1] *CNRS, UMR 7614, Laboratoire de Chimie Physique-Matière et Rayonnement, F-75005, Paris, France*
[2] *Sorbonne Universités, UPMC Univ Paris 06, UMR 7614, Laboratoire de Chimie Physique-Matière et Rayonnement, F-75005, Paris, France*
[3] *Institut für Experimentalphysik, Freie Universität Berlin, Arnimallee 14, D-14195 Berlin, Germany*
[4] *Synchrotron SOLEIL, l'Orme des Merisiers, F-91192, Gif-sur-Yvette Cedex, France*
[5] *Department of Physics and Astronomy, Uppsala University, SE-75120, Uppsala, Sweden*

Synopsis In order to have access to fine phenomena such as ultrafast nuclear dynamics and electronic state - lifetime interferences (ELI) at high energies (few keV), one has to rely on great experimental resolution. The new experimental HAXPES endstation at the GALAXIES beamline of the French synchrotron SOLEIL provides photons in the 2-12 keV energy range along with a total instrumental resolution below 500 meV. In this work we show experimental evidences of ultrafast nuclear dynamics and ELI on HCl and CH_3Cl excited around the Cl 1s resonance. We show also simulations that allow to disentangle the contribution of nuclear dynamics and ELI in our experimental spectra.

Ultra-fast photodissociation in chlorine-containing molecules has been known for more than twenty years [1,2].

While the first observations were made after photoexcitation of the Cl L shell, the study of nuclear motion in these molecules has been recently extended to deep shells by means of Resonant Inelastic X-ray Scattering (RIXS) in the tender x-ray region (2–12 keV) [3,4,5].

A new experimental station [6] dedicated to HArd X-ray PhotoElectron Spectroscopy (HAXPES), has been mounted on the GALAXIES beamline of the synchrotron SOLEIL.

Using this setup, we have studied the electronic relaxation of HCl and CH_3Cl [7] after Cl 1s excitation under resonant Raman conditions, i.e. with total instrumental resolution below the natural line width of the decay features.

In particular, we were able to observe nuclear dynamics on the 1-femtosecond time scale using the 'core-hole clock approach' [8].
These measurements also allowed us to unravel subtle phenomena as electronic state-lifetime interferences, which were also observed in RIXS [9] and in Argon using HAXPES [10]. We present a 'simple model' simulation, initially based on a theoretical model [11], which reproduces well the electronic state-lifetime interferences in both HCl and CH_3Cl.

References

[1] H. Aksela, *et al* 1990, *Phys. Rev. A.*, 41 6000
[2] O. Bjorneholm *et al* 1997, *Phys. Rev. Lett.*, 79 3150
[3] M. Simon *et al* 2006 *Phys. Rev. A*, 020706
[4] T. Marchenko *et al* 2011, *J. Chem. Phys.*, 144308
[5] L. El Khoury *et al* 2012, *J. Chem. Phys*, 024319
[6] D. Céolin *et al* 2013, *J. of Elect. Spectro. and Rel. Ph.*, **190** 188
[7] M. N. Piancastelli *et al* 2014, *J. Phys. B*, **47** 124031
[8] F. Gel'mukhanov *et al* 1999, *Phys. Rep.*, **312** 87
[9] M. Kavčič *et al* 2010, *Phys. Rev. Lett.*, **105** 113004
[10] D. Céolin *et al.*2015, *Phys. Rev. A*, **91** 022502
[11] G. B. Armen *et al* 1996, *Phys. Rev. A.*, **53** 772

* E-mail: gildas.goldsztejn@upmc.fr

Content from this work may be used under the terms of the Creative Commons Attribution 3.0 licence. Any further distribution of this work must maintain attribution to the author(s) and the title of the work, journal citation and DOI.
Published under licence by IOP Publishing Ltd

XXIX International Conference on Photonic, Electronic, and Atomic Collisions (ICPEAC2015) IOP Publishing
Journal of Physics: Conference Series **635** (2015) 112093 doi:10.1088/1742-6596/635/11/112093

Single photon core ionization with core excitation: a new spectroscopic tool.

Penent F.*[1], Carniato S. *[2], Lablanquie P.*, Selles P.*, Palaudoux J.*, Andric L.*, Žitnik M.[&], Bučar K.[&], Shigemasa E.[£], Nakano M.[†], Hikosaka Y.[#] and Ito K.[†]

[*] Sorbonne Université, UPMC and CNRS, LCP-MR (UMR 7614), 11 rue P. & M. Curie, 75231 Paris Cedex 05, France
[&] Jozef Stefan Institute, P.O. Box 3000, SI-1001 Ljubljana, Slovenia
[£] UVSOR Facility, Institute for Molecular Science, Okazaki 444-8585, Japan
[†] Institute of Materials Structure Science, Photon Factory, Tsukuba, Ibaraki 305-0801, Japan
[#] Department of Environmental Science, Niigata University, Niigata 950-2181, Japan

Synopsis . The simultaneous core ionization and core excitation process (or $K^{-2}V$ process) induced by absorption of a single photon provides the basis of a new spectroscopy that offers both advantages of X-ray Photoelectron Spectroscopy (XPS) and near-edge x-ray absorption fine structures (NEXAFS) spectroscopy

In the last few years, double-core-hole states (DCHs) have been extensively studied by single photon double-core ionization at synchrotron centers [1] and, in relation to X-ray two photon double-core spectroscopy [2], at X-ray free electron laser facilities (XFELs). Closely related to double core ionization, the single photon process where a core electron is ionized while another core electron is excited simultaneously, reveals interesting properties on the formation mechanisms of $K^{-2}V$ states and provides a new spectroscopic insight that offers the advantages of both XPS and NEXAFS spectroscopies.

The creation of $K^{-2}V$ states and their subsequent decay are the following:

$h\nu + AB \rightarrow AB^{+}(K^{-2}V) + e^{-}_{ph} \rightarrow AB^{++} + e^{-}_{A1} \rightarrow AB^{3+} + e^{-}_{A2}$.

Since one photoelectron and two Auger electrons are emitted, by detecting in coincidence these three electrons, it is possible to filter out the specific photoelectron spectrum associated with $K^{-2}V$ states. This can be done very efficiently by using a magnetic bottle time-of-flight electron spectrometer under single bunch operation at SOLEIL (France) and Photon Factory (Japan).

The study of the above process for the C_2H_{2n} (n=1-3) series [3] has revealed a NEXAFS signature in the spectra (Fig.1). Two competing channels appear with comparable intensities:

-The first one corresponds to the dipolar ionization of a core electron accompanied by monopolar excitation (direct shake-up) of the remaining core electron and can be interpreted as a "Super" K^{-1} satellite state.

The second one corresponds to the dipolar excitation of a core electron to a vacant orbital accompanied by monopolar shake-off of the other core electron (conjugate shake-up process) which has a NEXAFS-like character.

Close to the K^{-2} threshold, these two processes appear with comparable intensities. This can be understood as resulting from aborted double-core ionization where one of the outgoing electrons is recaptured in a vacant orbital. At higher photon energy, the conjugate shake-up process is unfavorable since the sudden approximation leads only to the direct process.

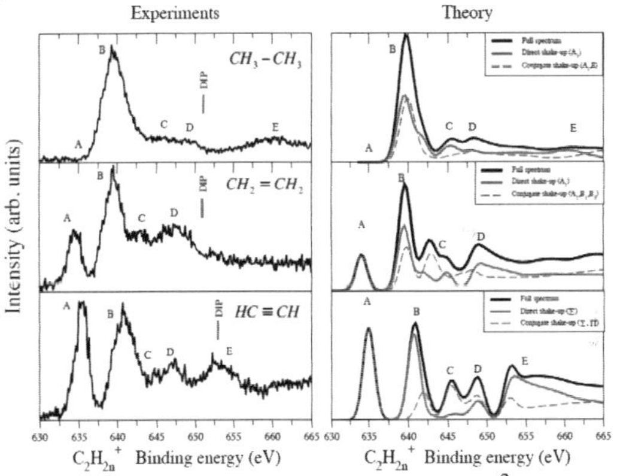

Figure 1: Experimental and theoretical $K^{-2}V$ spectra in C_2H_{2n} series at a photon energy $h\nu$ = 770eV.

Examples on small molecules: H_2O [4], N_2 [5] and many others, have been obtained, which show how powerful this new spectroscopy is and allows also a better interpretation of resonances appearing in NEXAFS spectrum.

References

[1] Lablanquie *et al* 2011, *Phys. Rev. Lett.* **106**, 063003
[2] Santra *et al* 2009, *Phys. Rev. Lett.* **103**, 013002
[3] Nakano *et al* 2013, *Phys. Rev. Lett.* **111**, 123001
[4] Carniato et al 2015, *J. Chem. Phys.* **142**, 014307
[5] Carniato et al 2015, *J. Chem. Phys.* **142**, 014308

[1] E-mail: francis.penent@upmc.fr
*[2] E-mail: stephane.carniato@upmc.fr

Content from this work may be used under the terms of the Creative Commons Attribution 3.0 licence. Any further distribution of this work must maintain attribution to the author(s) and the title of the work, journal citation and DOI.
Published under licence by IOP Publishing Ltd

XXIX International Conference on Photonic, Electronic, and Atomic Collisions (ICPEAC2015) IOP Publishing
Journal of Physics: Conference Series **635** (2015) 112094 doi:10.1088/1742-6596/635/11/112094

X-ray induced fragmentation dynamics of doubly charged L-alanine in gas phase

Yang Wang[*,†,1], **Helena Levola**[§], **Estefanía Rossich**[*], **Dariusz Piekarski**[*], **Sergio Díaz-Tendero**[*], **Edwin Kukk**[§], **Manuel Alcamí**[*], **and Fernando Martín**[*,†,¶]

[*] Departamento de Química, Módulo 13, Universidad Autónoma de Madrid, 28049 Madrid, Spain
[†] Instituto Madrileño de Estudios Avanzados en Nanociencia (IMDEA-Nanociencia), 28049 Madrid, Spain
[§] Department of Physics and Astronomy, University of Turku, FIN-20014 Turku, Finland
[¶] Condensed Matter Physics Center (IFIMAC), Universidad Autónoma de Madrid, 28049 Madrid, Spain

Synopsis The molecular fragmentation of doubly charged L-alanine in gas phase was studied in radiation synchrotron experiments. In this presentation, we summarize our theoretical study on the dynamics of this fragmentation, using various computational methods. We show that in practice the ground state MD simulations are able to statistically reproduce the experimental results of the photo-fragmentation initiated at the excited state.

The study of fragmentation dynamics of charged biomolecules is of great interest, aiming at helping us to better understand the processes of biological damage by irradiations. In recent experiments, L-alaline molecules were ionized into doubly charged excited states, as a result of the Auger decay induced by N 1s core photoionization. The excited molecules then dissociated into fragments. The Auger electron spectra were recorded in coincidence with time-of-flight mass spectra of the cationic fragments using the Auger Electron Photo-ion Photo-ion Coincidence (AEPIPICO) technique (see Fig. 1) [1]. The probabilities of various dissociation channels were also measured as functions of the kinetic energy of the Auger electron, which implied that the fragmentation of the parent dication would follow different mechanisms, depending strongly on the internal energy deposited on the molecule.

Here, we summarize our theoretical study on the fragmentation dynamics of doubly charged L-alaline molecule in gas phase by using various computational methods, including the time-dependent density functional theory (TD-DFT) Ehrenfest Molecular Dynamics (MD), the Born-Oppenheimer MD simulations, the Car-Parrinello MD, the self-consistent charge density functional tight-binding (SCC-DFTB) MD and the atom-centered density matrix propagation MD simulations. As one of the simplest amino acids, L-alanine is an excellent benchmark model to test the performance of these methods, By comparing with the experimental results, we show that the more accurate TD-DFT method for excited state dynamics is too time-consuming to afford a statistical study. In practice, one can apply ground state MD simulations by depositing the vibrational internal energy in place of the excitation energy onto the molecule. As shown in Fig. 1, the simulated AEPIPICO map by the SCC-DFTB method gives an overall agreement with experiment. We also show how the fragmentation dynamics depends on the spin state and the initial conformation of the molecule.

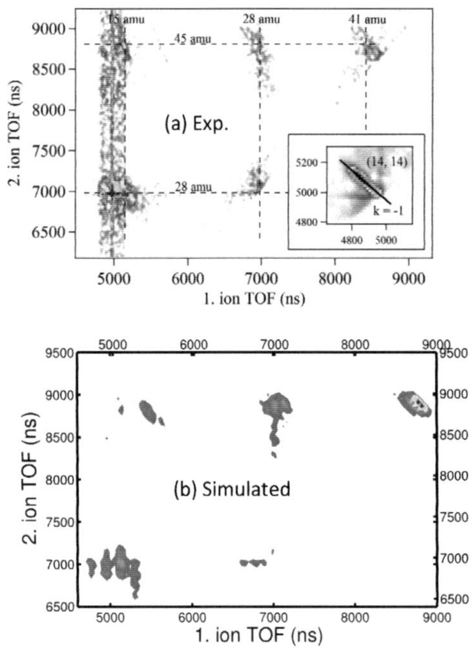

Figure 1. Comparison between (a) the experimental AEPIPICO map and (b) that obtained from the SCC-DFTB MD simulations.

References

[1] D. T. Ha *et al* 2014 *J. Phys. Chem. A* **118** 1374

[2] S. Maclot *et al* 2013 *J. Phys. Chem. Lett.* **4** 3903

[3] M. Capron *et al* 2012 *Chem. Eur. J.* **18** 9321

[1]E-mail: yang.wang@uam.es

Content from this work may be used under the terms of the Creative Commons Attribution 3.0 licence. Any further distribution of this work must maintain attribution to the author(s) and the title of the work, journal citation and DOI.
Published under licence by IOP Publishing Ltd

Controlling charge migration in molecules

Nikolay V. Golubev[1], **Alexander I. Kuleff**[2]

Theoretische Chemie, Universität Heidelberg, Im Neuenheimer Feld 229, 69120 Heidelberg, Germany

Synopsis Due to electronic many-body effects, the ionization of a molecule can trigger ultrafast electron dynamics appearing as a migration of the created hole charge throughout the system. Here we propose a scheme for control of the charge migration dynamics with a single ultrashort laser pulse. We demonstrate by fully *ab initio* calculations on a molecule containing a chromophore and an amine moieties that simple pulses can be used for stopping the charge-migration oscillations and localizing the charge on the desired site of the system. We argue that this control may be used to predetermine the follow-up nuclear rearrangement and thus the molecular reactivity.

In recent years, the rapid development of the attosecond pulse techniques opened the door for studying and eventually controlling the electronic motion. Due to the coupling between the electronic and the nuclear motion, the control over the pure electronic step offers the extremely interesting possibility to steer the succeeding chemical reactivity by predetermining the reaction outcome at a very early stage.

One example of physical phenomena, where an electronic dynamics significantly affect on reactivity is the process of an ultrafast *charge migration* [1]. The positive charge created upon ionization of a molecule can migrate throughout the system on a few-femtosecond time scale solely driven by the electron correlation and electron relaxation. Charge migration triggered by ionization appeared to be a general phenomenon taking place both after inner- [2] and outer-valence [3] ionization of molecules.

We present a scheme for controlling the many-body electron dynamics of the charge migration process by a single ultrashort laser pulse. We show that by appropriately tailored [4] infrared pulses one can stop the pure electronic, few-femtosecond oscillation of the charge, localizing it on the desired site of the molecule 3-methylen-4-penten-N,N-dimethylamine (MePeNNA) [5]. To describe the correlated motion of 69 electrons (the number of electrons in MePeNNA cation) in the presence of a laser field we use *ab initio* methods only.

We would like to emphasize that the scheme presented in this work is general and not restricted to only stopping the charge migration oscillations. Through the appropriately chosen laser pulses it is possible to drive the system to any combination of the two electronic states, and thus bring the ion to the optimum initial condition for the desired follow-up nuclear motion.

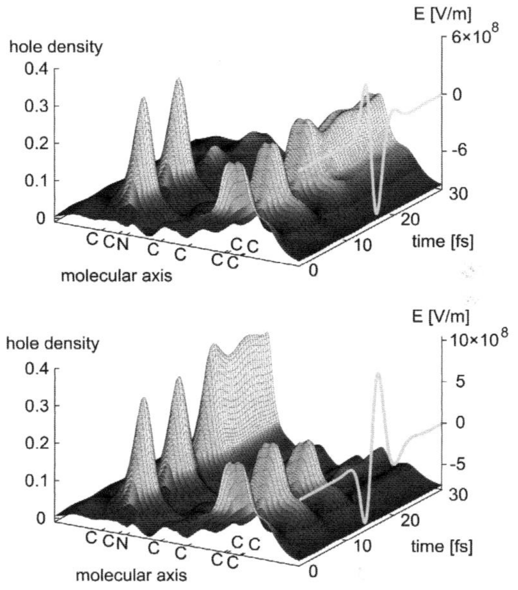

Figure 1. Time evolution of the hole density along the molecular axis of the molecule MePeNNA, after an initial localized ionization of the chromophore, controlled with a laser pulse (shown on the right) centered at 15 fs. Upper panel: the laser pulse is designed to achieve localization of the charge on the chromophore. Lower panel: the laser pulse is designed to achieve localization of the charge on the amine site.

References

[1] L. S. Cederbaum and J. Zobeley, Chem. Phys. Lett. **307**, 205 (1999).

[2] A. I. Kuleff, S. Lünnemann, and L. S. Cederbaum, J. Phys. Chem. A **114**, 8676 (2010).

[3] S. Lünnemann, A. I. Kuleff, and L. S. Cederbaum, J. Chem. Phys. **129**, 104305 (2008).

[4] N. V. Golubev and A. I. Kuleff, Phys. Rev. A **90**, 035401 (2014).

[5] N. V. Golubev and A. I. Kuleff, arXiv:1502.05613

[1]E-mail: nikolay.golubev@pci.uni-heidelberg.de
[2]E-mail: alexander.kuleff@pci.uni-heidelberg.de

Content from this work may be used under the terms of the Creative Commons Attribution 3.0 licence. Any further distribution of this work must maintain attribution to the author(s) and the title of the work, journal citation and DOI.
Published under licence by IOP Publishing Ltd

XXIX International Conference on Photonic, Electronic, and Atomic Collisions (ICPEAC2015) IOP Publishing
Journal of Physics: Conference Series **635** (2015) 112096 doi:10.1088/1742-6596/635/11/112096

Observation of the Efimov state of the helium trimer

Maksim Kunitski* [1], **Stefan Zeller***, **Jörg Voigtsberger***, **Anton Kalinin***, **Lothar Ph. H. Schmidt***, **Markus Schöffler***, **Achim Czasch***, **Wieland Schöllkopf**[†], **Robert E. Grisenti***, **Till Jahnke***, **Dörte Blume**[#] **and Reinhard Dörner*** [2]

[*] Institut für Kernphysik, Goethe-Universität Frankfurt am Main, Max-von-Laue-Str. 1, 60438 Frankfurt/M, Germany
[†] Department of Molecular Physics, Fritz-Haber-Institut, Faradayweg 4-6, 14195 Berlin, Germany
[#] Department of Physics and Astronomy, Washington State University, Pullman, WA 99164-2814, USA

Synopsis We report experimental observation of the long predicted but experimentally elusive Efimov state of ^{4}He trimer by means of Coulomb explosion imaging. We show spatial images of an Efimov state, confirming the predicted size and a typical structure where two atoms are close to each other while the third is far away.

In 1970 Vitali Efimov predicted remarkable counterintuitive behavior of a three-body system made up of identical bosons. Namely, a *weakening* of pair interaction in such a system brings about in the limit appearance of *infinite* number of bound states of a huge spatial extent [1]. The helium trimer has been predicted to be a molecular system having an excited state of the Efimov character under natural conditions without artificial tuning of the attraction between particles by an external field. Though many theoretical works predict the existence of this state in the helium trimer, it has not been observed experimentally so far. Main reasons for that are the tiny binding energy and a huge spatial extent that makes it fragile for preparation and elusive for detection.

Here we report experimental observation of the Efimov state of $^4He_3$ by means of Coulomb explosion imaging of mass-selected clusters. Helium trimers were prepared under supersonic expansion of the gaseous helium through a 5 μm nozzle. The clusters were selected from the molecular beam by means of matter wave diffraction [2]. Each atom of a trimer was singly ionized by a strong ultrashort laser field resulting in Coulomb explosion of the cluster. The momenta, the ions acquired during Coulomb explosion, were measured by COLTRIMS. These momenta were utilized for reconstruction of the initial spatial geometry of the neutral trimer at the instant of ionization using Newton's equation of motion.

The spatial extent of the excited Efimov state of the $^4He_3$ (Figure 1A) is about eight times larger than that of the ground state (Figure 1B). Whereas the ground state corresponds to an almost randomly distributed cloud of particles [3], the excited Efimov state is dominated by configurations in which two atoms are close to each other and the third one further away (Figure 1C).

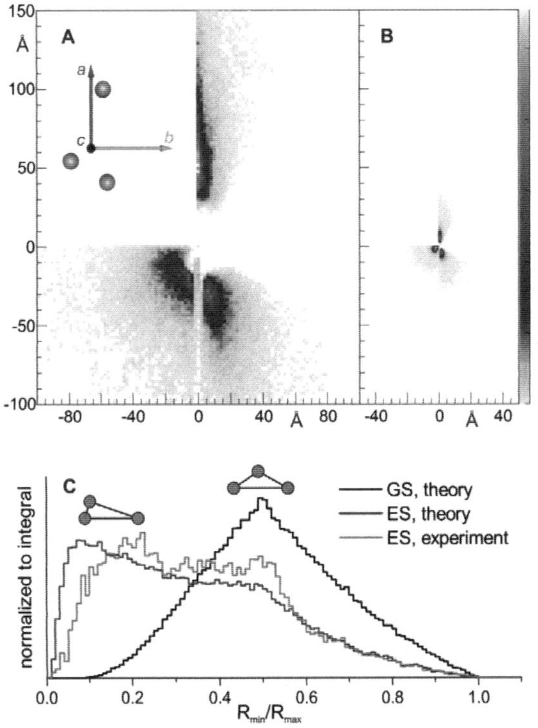

Figure 1. Structures of the helium trimer: A – excited state, experimental, B – ground state, theoretical. Three helium atoms of each trimer are plotted in the principal axis frame *abc*. C: Distributions of the ratio of the shortest interparticle distance to the longest interparticle distance for ground state structures (black), theoretical excited state structures (blue) and experimental excited state structures (red).

References

[1] V. Efimov 1970 *Phys. Lett. B* **33** 563–564
[2] W. Schöllkopf, J. P. Toennies 1994 *Science* **266** 1345–1348
[3] J. Voigtsberger *et al* 2014 *Nat. Comm.* **5** 5765

[1] E-mail: kunitski@atom.uni-frankfurt.de
[2] E-mail: doerner@atom.uni-frankfurt.de

Content from this work may be used under the terms of the Creative Commons Attribution 3.0 licence. Any further distribution of this work must maintain attribution to the author(s) and the title of the work, journal citation and DOI.
Published under licence by IOP Publishing Ltd

XXIX International Conference on Photonic, Electronic, and Atomic Collisions (ICPEAC2015) IOP Publishing
Journal of Physics: Conference Series **635** (2015) 112097 doi:10.1088/1742-6596/635/11/112097

CORRELATIONS OF RYDBERG ORBITALS OF DIATOMIC MOLECULES

Hyun-Jin Kim[1] and Chun-Woo Lee[2,*]

[1]Department of Energy Systems Research, Ajou University, Suwon, 443-749, Korea

[2]Department of Chemistry, Ajou University, 206 Worldcup-ro, Yeongtong-Gu, Suwon, 443-749, Korea

Synopsis Calculations of the potential energy curves of diatomic molecules are performed for the entire range of internuclear separation in order to correlate Rydberg orbitals at the united atom and separated atoms which are needed to properly assign highly excited Rydberg states.

The assignment of highly excited Rydberg states is difficult because of the absence of the knowledge on the behaviors of potential energy curves (PECs) at small internuclear separation *R* for most of cases whereby the correlation between the united atom (UA) limit and separated atoms (SA) limit cannot be given. We performed all electron multi-reference configuration interaction calculations of PECs for H_2, He_2, Li_2 [1], HeH, LiH [2], BeH and BH to obtain the correlation diagrams between Rydberg orbitals at both limits.

Calculations show that the behaviors of quantum defect curves (QDCs) are dominated by electrons in Rydberg orbitals in contrast to those of PECs which are dominated by core electrons (see Fig. 1), indicating that QDCs are suitable for the study of correlations between Rydberg orbitals at the UA and SA limits. Note that usual UA and SA designations of Rydberg orbitals are derived from the studies of one-electron molecular systems and cannot be used for many-electron molecular systems. There may be several reasons for this failure. One of the reasons comes from the strong influence of core electrons in the neighborhood of united atoms. Their influence can be ignored in the Franck-Condon region so that groupings of Rydberg states into series akin to that of the molecular ion are attained. There is another reason coming from the incompatibility between Rydberg orbitals at both limits which causes distortion to the behaviors of Rydberg series. Diversities in the correlation patterns are observed for different molecules.

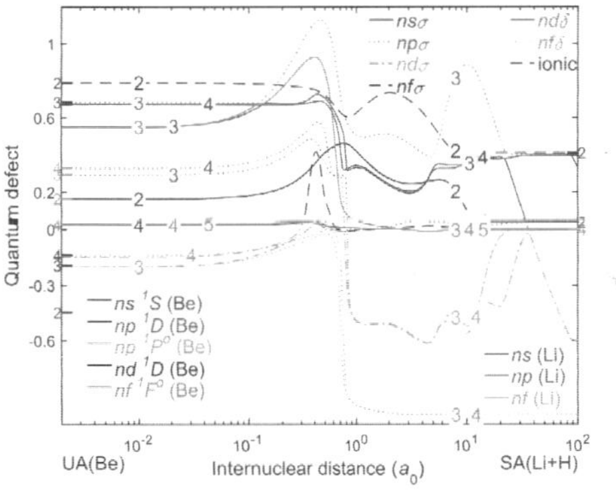

Figure 1. Correlations of Rydberg orbitals of a LiH molecule at the united atoms and separated atoms.

References

[1] H-J Kim and C-W Lee 2015 *J. Phys. B* **48** 015101
[2] Y Gim and C-W Lee 2014 *J. Chem. Phys.* **141** 144313

* E-mail: clee@ajou.ac.kr

Content from this work may be used under the terms of the Creative Commons Attribution 3.0 licence. Any further distribution of this work must maintain attribution to the author(s) and the title of the work, journal citation and DOI.
Published under licence by IOP Publishing Ltd

XXIX International Conference on Photonic, Electronic, and Atomic Collisions (ICPEAC2015) IOP Publishing
Journal of Physics: Conference Series **635** (2015) 112098 doi:10.1088/1742-6596/635/11/112098

Excitation of vibrational modes in the ionization of water molecule by XUV/X-ray radiation

Selma Engin[† 1], Jesús González-Vázquez[†], Inés Corral[†], Alicia Palacios[†], David Ayuso[†], Piero Decleva[**], and Fernando Martín[†‡]

[†] Departamento de Química, Módulo 13, Universidad Autónoma de Madrid, Cantoblanco 28049 Madrid, Spain
[‡] Instituto Madrileño de Estudios Avanzados en Nanociencia, Cantoblanco 28049 Madrid, Spain
[*] Dipartimento di Scienze Chimiche e Faramaceutiche, Università di Trieste, 34127 Trieste, Italy
[⋆] CNR-IOM, Trieste, Italy

Synopsis We present a theoretical study of the vibrationally resolved core photoionization of the water molecule up to high photon energies. In order to understand the role of the coupled electron-nuclear motion in polyatomic molecules, we thus have implemented a new methodology to describe all vibrational modes of a polyatomic molecule. We show our preliminary results on the O(1s) photoionization, with special focus on the vibrationally resolved cross sections in a large range of photon energies, reaching up to 1500 eV.

Most theoretical studies of coupled electron and nuclear dynamics in photoionization problems have been restricted to diatomic molecules [1] or to a single vibrational mode of polyatomic molecules [2]. In some cases, a reasonable approximation is to reduce the nuclear degrees of freedom to one principal normal mode. This was shown in photoionization problems for highly symmetric targets such that CH_4 or BF_3 [2], whose symmetric stretching mode was the main active mode upon the photon absorption. However, such simplification is not valid in most problems.

The aim of this study is to extend the methodology previously developed in those molecular targets, which used a one-dimensional vibrational wave function, to include an arbitrary number of nuclear degrees of freedom. We thus seek to achieve a complete description of vibrational motion, although, for simplicity our first implementation uses Jacobi coordinates, which are particularly suitable for non-linear triatomic molecules. In the present study, we report results for the O(1s) photoionization of the water molecule induced by synchroton radiation, where all normal modes are simultaneously accounted for : bending, symmetric stretching and asymmetric stretching. Accurate vibrationally resolved cross sections require the computation of dipole matrix elements for the transition from the ground state of the neutral to the core-ionized water molecule. These dipole elements are obtained for the full grid of nuclear coordinates using the static-exchange DFT method developed by P. Decleva and collaborators [3]. The reliability of our approach is tested by comparing our calculated O(1s) vibrationally resolved photoelectron spectrum with available experimental data for a photon energy of 590 eV [4]. We find a good agreement with the vibrational progression at 590 eV, where the dominant contribution is that of the bending mode of the core-ionized water molecule. We further analyze the v-ratios as a function of photon energy, i.e. the ratios between the photoionization cross sections for individual vibrational states. For all normal modes, we observe oscillations around the Franck-Condon factor value, similar to those found in previous works for other diatomic and polyatomic targets (CO, CH_4, BF_3, etc) [1, 2], whose origin comes from diffraction of the ejected electron by the surrounding H atoms.

References

[1] E. Kukk *et al.*, *Phys. Rev. A* **88** 033412 (2013) D. Ayuso *et al.*, *J. Elec. Spec. Rel. Phen.* **195** 320-326 (2014)

[2] E. Plésiat *et al.*, *Phys. Rev. A* **85** 023409 (2012) K. Ueda *et al.*, *J. Chem. Phys.* **139** 124306 (2013)

[3] M. Stener *et al.*, *J. Chem. Phys.* **112** 10871 (2000) M. Stener *et al.*, *Chem. Phys. Lett.* **351** 469 (2002) D. Toffoli *et al.*, *Chem. Phys.* **276** 2543 (2002)

[4] R. Sankari *et al.*, *Chem. Phys. Lett.* **380** 647-653 (2003)

[1]E-mail: selma.engin@uam.es

Content from this work may be used under the terms of the Creative Commons Attribution 3.0 licence. Any further distribution of this work must maintain attribution to the author(s) and the title of the work, journal citation and DOI.
Published under licence by IOP Publishing Ltd

Journal of Physics: Conference Series **635** (2015) 112099 doi:10.1088/1742-6596/635/11/112099

A Single Atom Antenna

Florian Trinter* [1], **Joshua B. Williams*, Miriam Weller*, Markus Waitz*, Martin Pitzer*, Jörg Voigtsberger*, Carl Schober*, Gregor Kastirke*, Christian Müller*, Christoph Goihl*, Phillip Burzynski*, Florian Wiegandt*, Robert Wallauer*, Anton Kalinin*, Lothar Ph. H. Schmidt*, Markus S. Schöffler*, Ying-Chih Chiang†, Kirill Gokhberg†, Till Jahnke*, and Reinhard Dörner***

* Institut für Kernphysik, J. W. Goethe-Universität, Max-von-Laue-Strasse 1, D-60438 Frankfurt am Main
† Theoretische Chemie, Universität Heidelberg, Im Neuenheimer Feld 229, D-69120 Heidelberg, Germany

Synopsis Here we demonstrate the smallest possible implementation of an antenna-receiver complex which consists of a single (helium) atom acting as the antenna and a second (neon) atom acting as a receiver.

In radio technology antennas are used to efficiently collect energy from the electromagnetic field. In the optical regime nano-scale antennas have been developed [1] and in nature specialized antenna molecules efficiently collect the visible light for light harvesting. Schematically all these systems consist of the antenna itself which couples to the radiation field, a receiver which uses the energy and a route to transport the energy between them.

Here we demonstrate the smallest possible implementation of such an antenna-receiver complex which consists of a single (helium) atom acting as the antenna and a second (neon) atom acting as a receiver [2,3]. We investigate the ionization of HeNe from below the He 1s3p excitation to the He ionization threshold. We observe $HeNe^+$ ions with an enhancement by more than a factor of 60 when the He side (the antenna atom) of the antenna couples resonantly to the radiation field. The energy transfer occurs via Interatomic Coulombic Decay (ICD) [4]. The $HeNe^+$ ions are an experimental proof of a two-center resonant photoionization mechanism predicted by Najjari et al. [5]. Furthermore, our data provide electronic and vibrational state resolved decay widths of interatomic Coulombic decay in HeNe dimers. We find that the interatomic Coulombic decay lifetime strongly increases with increasing vibrational state.

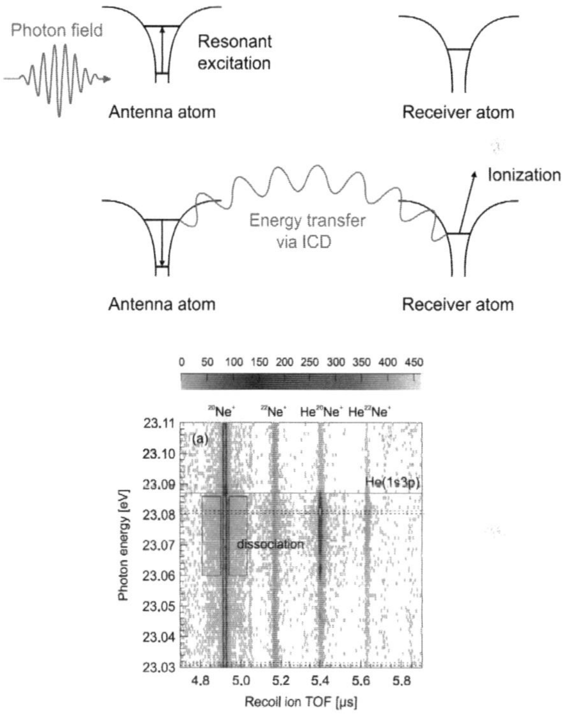

Figure 1. Schematic of the antenna mechanism and experimental results: Ion time of flight versus photon energy in the vicinity of the He(1s3p) resonance.

References

[1] L. Novotny and N. van Hulst 2011 *Nat. Photonics* 5 83
[2] F. Trinter *et al* 2013 *Phys. Rev. Lett.* **111** 233004
[3] G. Jabbari *et al* 2014 *J. Chem. Phys.* **140** 224305
[4] L. S. Cederbaum, J. Zobeley, and F. Tarantelli 1997 *Phys. Rev. Lett.* **79** 4778
[5] B. Najjari, A. B. Voitkiv, and C. Müller 2010 *Phys. Rev. Lett.* **105** 153002

[1] E-mail: trinter@atom.uni-frankfurt.de

Content from this work may be used under the terms of the Creative Commons Attribution 3.0 licence. Any further distribution of this work must maintain attribution to the author(s) and the title of the work, journal citation and DOI.
Published under licence by IOP Publishing Ltd

XXIX International Conference on Photonic, Electronic, and Atomic Collisions (ICPEAC2015) IOP Publishing
Journal of Physics: Conference Series **635** (2015) 112100 doi:10.1088/1742-6596/635/11/112100

A molecular movie of Interatomic Coulombic Decay in NeKr

Florian Trinter* [1], **Tsveta Miteva**[†], **Miriam Weller***, **Sebastian Albrecht***, **Alexander Hartung***, **Martin Richter***, **Joshua Williams***, **Averell Gatton**[#], **Bishwanath Gaire**[#], **Thorsten Weber**[#], **James Sartor**[§], **Allen Landers**[§], **Ben Berry**[&], **Vasili Stumpf**[†], **Kirill Gokhberg**[†], **Reinhard Dörner***, and **Till Jahnke***

[*] Institut für Kernphysik, Goethe-Universität, 60438 Frankfurt am Main, Germany
[†] Theoretische Chemie, Physikalisch-Chemisches Institut, Universität Heidelberg, 69120 Heidelberg, Germany
[#] Lawrence Berkeley National Laboratory, Chemical Sciences Division, Berkeley, California 94720, USA
[§] Department of Physics, Auburn University, Auburn, Alabama 36849, USA
[&] J. R. Macdonald Laboratory, Department of Physics, Kansas State University, Manhattan, Kansas 66506, USA

Synopsis Interatomic Coulombic Decay in mixed NeKr dimers has been measured time-resolved and the nuclear dynamics of the decay have been investigated.

During the last 15 years a novel decay mechanism of excited atoms has been discovered and investigated. This so called "Interatomic Coulombic Decay" (ICD) [1,2] involves the chemical environment of the electronically excited atom or molecule: the excitation energy is transferred to a neighbor of the initially excited particle usually ionizing that neighbor. It turned out that ICD is a very common decay route in nature as it occurs across van der Waals and hydrogen bonds. The time evolution of ICD is predicted to be highly complex, as its efficiency strongly depends on the distance of the atoms involved and this distance typically changes during the decay.

Here we present a direct measurement of the temporal evolution of ICD using a novel experimental approach [3] as well as theoretical results. The results show the evolution of the vibrational wavepacket of a NeKr dimer during the decay. So we gain insight into the complex behavior of ICD in the time domain.

Figure 1 shows the norm of the wavepacket as a function of the IC-decay time, i.e. temporal behavior of the survival probability of the intermediate state prior to ICD. Two photon energies as well as theory are included. The data points are extracted from two datasets employing two different photon energies in order to access different decay times: for the black points a photoelectron kinetic energy of 120 meV was chosen while the red points belong to a dataset with a photoelectron energy of 200 meV.

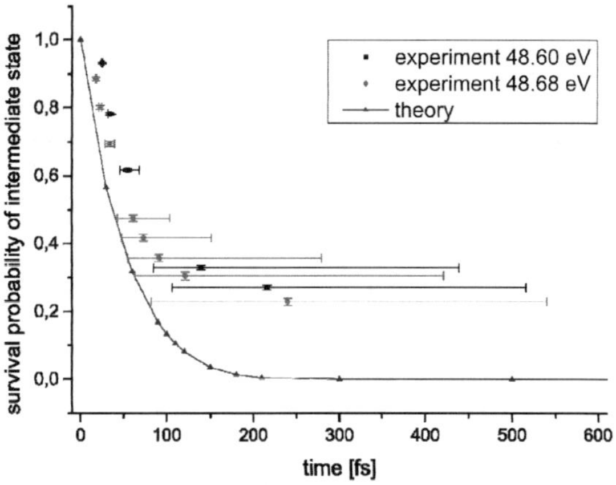

Figure 1. Results: Norm of the wavepacket as a function of the IC-decay time, i.e. temporal behavior of the survival probability of the intermediate state prior to ICD.

References

[1] L. S. Cederbaum, J. Zobeley, and F. Tarantelli 1997 *Phys. Rev. Lett.* 79 4778
[2] T. Jahnke *et al* 2004 *Phys. Rev. Lett.* **93** 163401
[3] F. Trinter *et al* 2013 *Phys. Rev. Lett.* **111** 093401

[1] E-mail: trinter@atom.uni-frankfurt.de

Content from this work may be used under the terms of the Creative Commons Attribution 3.0 licence. Any further distribution of this work must maintain attribution to the author(s) and the title of the work, journal citation and DOI.
Published under licence by IOP Publishing Ltd

XXIX International Conference on Photonic, Electronic, and Atomic Collisions (ICPEAC2015) IOP Publishing
Journal of Physics: Conference Series **635** (2015) 112101 doi:10.1088/1742-6596/635/11/112101

Mapping the dissociative ionization dynamics of molecular nitrogen with attosecond resolution

M. Klinker[1 1], A. Trabattoni[2], J. González-Vázquez[1], C. Liu[3], G. Sansone[2], R. Linguerri[4], M. Hochlaf[4], J. Klei[5], M. J. J. Vrakking[5], F. Martin[1,6,7], M. Nisoli[2,8], F. Calegari[8]

[1] Departamento de Química, Módulo 13,Universidad Autónoma de Madrid, Cantoblanco 28049, Madrid, Spain. [2] Dipartamento di Fisica di Milano, Piazza Leonardo da Vinci 32, I-20133 Milano, Italy. [3] State Key Laboratory of High Field Physics, Shanghai Institute of Optics and Fine Mechanics, Chinese Academy of Sciences, Shanghai 201800, China. [4] Université Paris-Est, Laboratoire Modélisation et Simulation Multi Echelle, MSME UMR 8208 CNRS, 5bd Descartes, 77454 Marne-La-Vallée, France. [5] Max-Born-Institut, Max Born Straße 2A, D-12489 Berlin, Germany. [6] Instituto Madrileño de Estudios Avanzados en Nanociencia (IMDEA-Nano), Ciudad Universitaria de Cantoblanco, 28049 Madrid, Spain. [7] Condensed Matter Physics Center (IFIMAC), Universidad Autónoma de Madrid, 28049 Madrid, Spain. [8] Institute for Photonics and Nanotechnologies, IFN-CNR, Piazza Leonardo da Vinci 32, I-20133 Milano, Italy.

Synopsis We wish to understand the processes underlying the ionization dynamics of N_2 as experimentally induced and studied by recording the kinetic energy release (KER) in a XUV-pump/IR-probe setup. To this end a theoretical model was developed describing the ionization process using Dyson Orbitals and, subsequently, the dissociation process using a large set of diabatic potential energy surfaces (PES) on which to propagate. From said set of PES, a small subset is extracted allowing for the identification of one and two photon processes chiefly responsible for the experimentally observed features.

N_2 is the most abundant element is the earth's atmosphere and as such is of integral importance to processes induced by solar radiation, in particular the dissociation of N_2 induced by solar XUV light in the upper atmosphere. Recent advances in atto second technology have made it possible to gain new insights into ultra fast electron dynamics following sudden removal of an electron [1, 2], providing the perfect tools to study the photo induced dissociation of N_2.

We present the experimental and theoretical investigation of the ionization and dissociation dynamics of N_2. Isolated attosecond pulses (energy range: 16-50 eV, duration: 300 as) were used to ionize N_2 molecules, through a single photon transition. After a variable delay, 4-fs NIR/VIS CEP controlled pulses (peak intensity: $8 \cdot 10^{12}$W/cm^2) were used to probe the subsequent dissociation dynamics. The angularly resolved momentum distribution of the produced N^+ fragments was measured as a function of the pump-probe delay using a velocity map imaging (VMI) spectrometer. We observed a depletion of a quasi-bound state of N_2^+, 8 fs after zero time delay, together with sub-cycle modulation of the N^+ yield.

To understand the origin of the dynamics a model was developed: The ionization process was simulated assuming an instantaneous transition of the nuclear ground state wave function of N_2 to a set of diabatic PESs of the N_2^+ ion, obtained using *ab initio* multiconfigurational methodology. The relative initial populations of said set of ionic PESs were approximated using Dyson orbitals in first order perturbation theory [3, 4] and modelling the leaving electron by a Coulomb wave. The dissociating ion subjected to the IR pulse was modelled by solving the TDSE using Split Operator [5] and Fast Fourier [6] techniques.

Careful investigation of the data produced by aforementioned techniques yielded the $F^2\Sigma_g$, $3^2\Sigma_g$, $C^2\Sigma_u$ and $5^2\Sigma_u$ states of N_2^+ as being capable of describing most features observed experimentally; most importantly a two photon transition from $F^2\Sigma_g$ to $3^2\Sigma_g$ via $5^2\Sigma_u$ was found to be responsible for the sub-femtosecond dynamics, as it causes the wave packets initially in $F^2\Sigma_g$ and $3^2\Sigma_g$ to interfere.

References

[1] M. Nisoli, G. Sansone 2009 *Prog. Quantum Electronics* **33** 17-59

[2] G. Sansone, et al. 2010 *Nature* **465** 763-766

[3] C.M. Oana, A.I. Krylov 2007 *J. Chem. Phys.* **127** 234106

[4] M. Spanner et al. 2012 *Phys. Rev. A* **86** 053406

[5] M.D. Feit, J. Fleck, A. Steiger 1983 *J. Comp. Phys.* **47** 412-433

[6] R. Kosloff 1994 *Ann. Rev. Phys. Chem.* **45** 145-178

[1]E-mail: m.klinker@uam.es

Content from this work may be used under the terms of the Creative Commons Attribution 3.0 licence. Any further distribution of this work must maintain attribution to the author(s) and the title of the work, journal citation and DOI.
Published under licence by IOP Publishing Ltd

XXIX International Conference on Photonic, Electronic, and Atomic Collisions (ICPEAC2015) IOP Publishing
Journal of Physics: Conference Series **635** (2015) 112102 doi:10.1088/1742-6596/635/11/112102

Near-Threshold Photoionization of Molecular Nitrogen

G. VanGyseghem,*[1] T. W. Gorczyca,*[2] and C. P. Ballance†

*Department of Physics, Western Michigan University, Kalamazoo, MI 49008-5252, USA
†Department of Physics, Auburn University, Auburn, AL 86301, USA

Synopsis Photoabsorption of molecular nitrogen (N_2) is investigated near the first ionization threshold using an R-matrix, multi-channel quantum defect (MQDT) approach.

Photoabsorption of molecular nitrogen (N_2) is investigated near the first ionization threshold using an R-matrix, multi-channel quantum defect (MQDT) approach. Building on an existing fixed-nuclei R-matrix photoionization model [1], single and multiple configuration molecular descriptions have been considered using the UKRmol suite of codes. Photoionization cross sections are first computed over a wide energy range to study the behavior of the strong, broad shape-resonance that affects the near-threshold energy-dependence in scattering quantities (see Fig. 1).

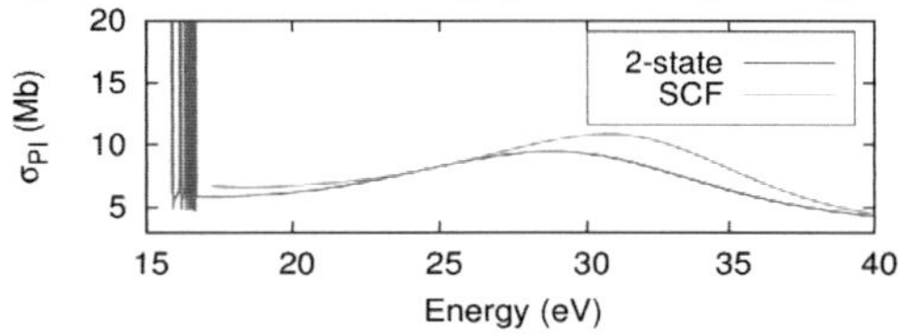

Figure 1. N_2 photoionization cross section, showing the broad shape resonance behavior above threshold.

The near-threshold region, expanded in Fig. 2, shows prominent resonance series converging to the $A^2\Pi_u$ threshold.

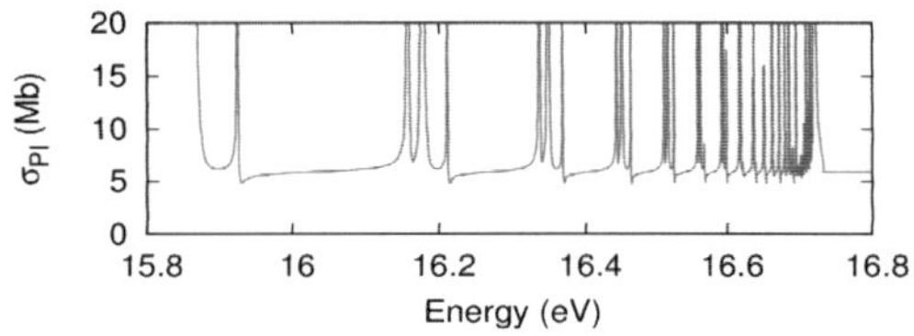

Figure 2. N_2 photoionization cross section, showing the fixed-nuclei $A^2\Pi_u nl$ Rydberg series.

This complicated resonance structure is governed by smoother eigenquantum defects μ_α, defined via diagonalization of the fixed-nuclei, unphysical reactance matrices:

$$K_{ij}(R) \;=\; \sum_\alpha U_{\alpha i} \tan(\pi\mu_\alpha(R))\, U_{\alpha j} \;.$$

Eigenquantum defects for a simpler 2-state (X, A), 4-channel case, for each final photoionized symmetry, $^1\Sigma_u$ and $^1\Pi_u$, are seen in Fig. 3 to be slowly-varying in energy except for the broad shape resonance feature.

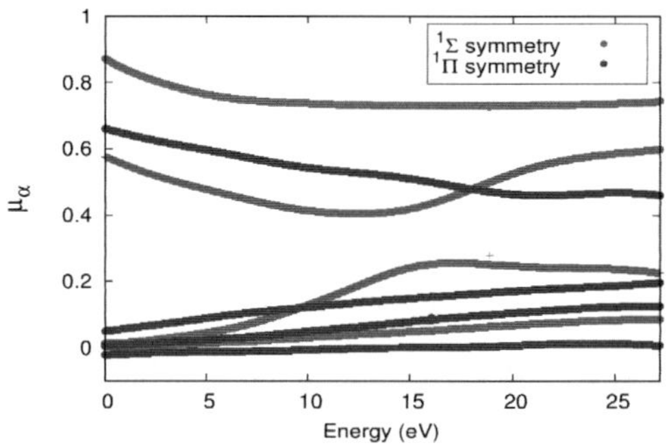

Figure 3. N_2 eigenquantum defects for the two-state, four-channel case in $^1\Sigma$ and $^1\Pi$ symmetries.

The smooth eigenquantum defects shown in Fig. 3 are found to have threshold values in excellent agreement with the empirical MQDT treatments of [2, 3], based on below-threshold spectra. The computed eigenquantum defects are then used in further MQDT vibrational frame transformation calculations via the transformed, unphysical vibrationally-resolved reactance matrix

$$K_{iv,jv'}(R) \;=\; \int dR\, \langle v|R\rangle\, K_{ij}(R)\, \langle R|v'\rangle \;,$$

and similarly for the dipole matrices. Comparison with high-resolution experimental photoionization cross sections is made near threshold, a region complicated by multiple vibrationally-resolved, interacting Rydberg series.

References

[1] Tashiro M 2010 em J. Chem. Phys. **132** 134306

[2] Huber K P and Jungen C 1990 *J. Chem. Phys.* **92** 850

[3] Jungen C *et al* 2003 *J. Chem. Phys.* **118** 4517

[1]E-mail: gaetan.l.vangyseghem@wmich.edu
[2]E-mail: gorczyca@wmich.edu

Content from this work may be used under the terms of the Creative Commons Attribution 3.0 licence. Any further distribution of this work must maintain attribution to the author(s) and the title of the work, journal citation and DOI.
Published under licence by IOP Publishing Ltd

Journal of Physics: Conference Series **635** (2015) 112103 doi:10.1088/1742-6596/635/11/112103

A VMI setup to study prompt and delayed electron emission from trapped cluster anions

K. Saha*[1], A. Prabhakaran*[2], M. L. Rappaport*, O. Heber*, D. Schwalm*†, D. Zajfman*

* Department of Particle Physics and Astrophysics, Weizmann Institute of Science, Rehovot 7610001, Israel

† Max-Planck-Institut für Kernphysik, Saupfercheckweg 1, Heidelberg 69117, Germany

Synopsis We have designed a velocity map imaging (VMI) setup and integrated it with an Electrostatic Ion Beam Trap (EIBT) to study photon induced prompt and delayed electron emission from trapped cluster anions. The operational details are discussed here. The performance of the VMI is also reported for photoionization of trapped O^- ions.

We report the performance of a velocity map imaging (VMI) setup which is designed to study photon induced prompt as well as delayed electron emission from trapped cluster anions in an event by event mode. The trapping of anionic species (produced in a cesium sputter source) is done with an Electrostatic Ion Beam Trap (EIBT) [1]. In the EIBT, 4.2 keV anions are kept oscillating along the trap axis in a field-free region between two electrostatic mirrors for a certain period of time (a few milliseconds to a few seconds). Our VMI setup is mounted perpendicular to ion beam direction in the field-free region of the EIBT.

Electron emission from the anions is induced by photons from a Nd-YAG laser with an OPO (Ekspla NT342 B-SH). The laser beam is introduced in an orthogonal direction to both the trapped ions and the VMI axis. The emitted electrons are extracted by electrostatic fields of the VMI and are detected by an MCP with a resistive anode encoder at its end. Electrons with different velocities will hit the detector at different radii, thus, the energy of the emitted electrons can be deduced. As the trapped anions are oscillating back and forth along the trap axis, the emitted electron experience an initial velocity along the motion of the anion which produces two shifted electron images. To overcome this, the ion beam is bunched within the trap by applying an RF potential to one of the trap mirror electrodes such that all the ions move in a particular direction during interaction with the laser beam. Electrons are now emitted from anions having the same initial velocity direction and thus we observe a single electron image on the detector albeit with a shift. Since the initial ion velocity is known, the energy of the emitted electrons can be reconstructed [2].

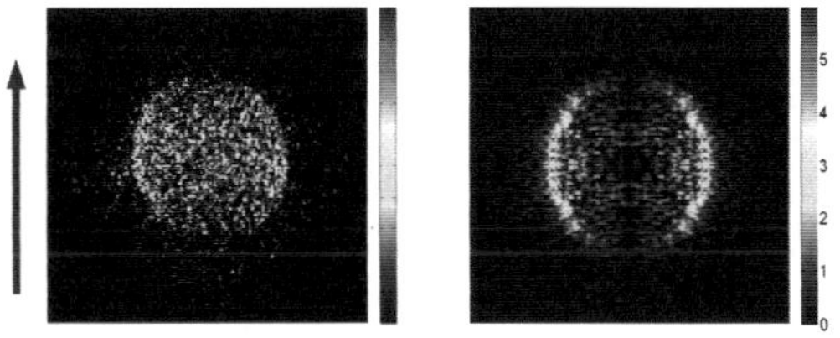

Figure 1. Photoelectron image upon ionization by 600 nm laser. Left: Raw image. Right: After inverse Abel transformation. The arrow shows the polarization direction of the laser.

We have tested the performance of our setup for trapped O^- beam. Various laser wavelengths (500 nm – 675 nm) were used to get photoelectrons of different energies for calibration purposes. The raw electron images were inverse Abel transformed using the BASEX code developed by Dribinski *et al.* [3]. Figure 1 shows the raw image and the inverse Abel transformed image of electrons emitted upon ionization by photons of 600 nm wavelength. The energy resolution was found to be 0.017 eV for 0.606 eV electrons emitted at 600 nm wavelength. We have also computed the asymmetry parameter from the angular distribution of the electrons. The value for asymmetry parameter at 600 nm wavelength was found to be $-0.84(10)$ which is in agreement with the value of -0.89 reported by Cavanagh *et al.* [4] for photoionization of O^- by photons of 532 nm wavelength. Experiments with hydrocarbons and cluster anions are in progress.

References

[1] M. Dahan *et al.* 1998 *Rev. Sci. Instrum.* **69** 76

[2] C. J. Johnson *et al.* 2011 *Rev. Sci. Instrum.* **82** 105105

[3] V. Dribinski *et al.* 2002 *Rev. Sci. Instrum.* **73** 2634

[4] S. J. Cavanagh *et al.* 2007 *Phys. Rev. A* **76** 052708

[1]E-mail: koushik.saha@weizmann.ac.il

[2]Present address: Thermo Fisher Scientific GmbH, Bremen 28199, Germany

Content from this work may be used under the terms of the Creative Commons Attribution 3.0 licence. Any further distribution of this work must maintain attribution to the author(s) and the title of the work, journal citation and DOI.
Published under licence by IOP Publishing Ltd

XXIX International Conference on Photonic, Electronic, and Atomic Collisions (ICPEAC2015) IOP Publishing
Journal of Physics: Conference Series **635** (2015) 112104 doi:10.1088/1742-6596/635/11/112104

Triggering photochemical processes in frozen extraterrestrial worlds by soft X-rays

Sergio Pilling*[1], Alexandre Bergantini *, Fredson A. Vasconcelos*, Will R. M. Rocha*

* Laboratório de Astroquímica e Astrobiologia (LASA), Instituto de Pesquisa & Desenvolvimento (IP&D), Universidade do Vale do Paraíba (UNIVAP), São Jose dos Campos, Brazil

Synopsis We presents and experimental investigation on the effects produced by broad band soft X-rays (combined with a small fraction of vacuum ultra violet photons and possibly secondary electrons) on the surface of three moons of giant planets: Europa, Titan and Enceladus. Such environments are constantly exposed to space ionizing agents (UV and soft X-rays photons, electrons and ions) allowing photodissociation processes, surface photochemistry and prebiotic chemistry. The processing of such spatial ices have promoted an enhancement in the chemical complexity, similar what may have happened in the early earth triggering the arising of life.

In this work, we investigate the effects produced mainly by broad band soft X-rays (and X-ray-induced secondary electrons) in ice mixtures that simulates the surface of three moons of giant planets: Europa, Titan and Enceladus. Such environments are constantly exposed to space ionizing agents (UV and soft X-rays photons, electrons and ions) allowing photodissociation processes, surface photochemistry and prebiotic chemistry.

The experiments have been performed using a high vacuum portable chamber from the Laboratório de Astroquímica e Astrobiologia (LASA/UNIVAP) coupled to the SGM beamline in the Brazilian Synchrotron Light Source (LNLS) at Campinas, Brazil. The beamline was operated in off-focus and white beam mode, which produces a wide band spectral range of photons, mainly from 6 eV up to 2000 eV, with the total average flux at the sample of about 1E14 photons cm^{-2} s^{-1}.

Briefly, the samples were produced by the adsorption of a gaseous mixture at 12 K and following by slowly heating to the temperatures in which the irradiation occur, simulating this way the frozen surface of an specific moon of a given giant planet: a) $H_2O:CO_2:NH_3:SO_2$ (10:1:1:1) at 90 K for Europa, moon of Jupiter; b) $N_2:CH_4$ (19:1) at 12 K for Titan, moon of Saturn, and c) ($H_2O:CO_2:CH_4:NH_3$) (10:1:1:1) at 80 K for Enceladus, another moon of Saturn. For experiment b, we simulate the effect of incoming radiation in the aerosols in the upper atmosphere of Titan. *In-situ* sample analyses were performed by Fourier transform infrared (FTIR) spectrometer. Complete experimental description is given elsewhere [1].

The spectral analysis of the processed samples shows several new bands associated with the formation of organic molecules, including nitriles, hydrocarbons, acids and other organic compounds. The dissociation cross sections of parental species were in the order of 10^{-18}-10^{-19} cm^2. Half-lives of the parental species extrapolated to the astrophysical scenario was determined. In case of Europa´s experiment, we observe the presence of H_2O_2, H_3O^+, SO_3, CO, and OCN^- among the new species produced. For the experiment simulation aerosols in Titan, we identify the formation of HCN, HNC, CCCN, NH_3 and C_2H_2. Such molecules possibly will be deposited into the ground and increase the chemical complexity of the surface with time. The irradiation of Enceladus like surface shows as daughter species, OCN^-, H_2CO and CO among others

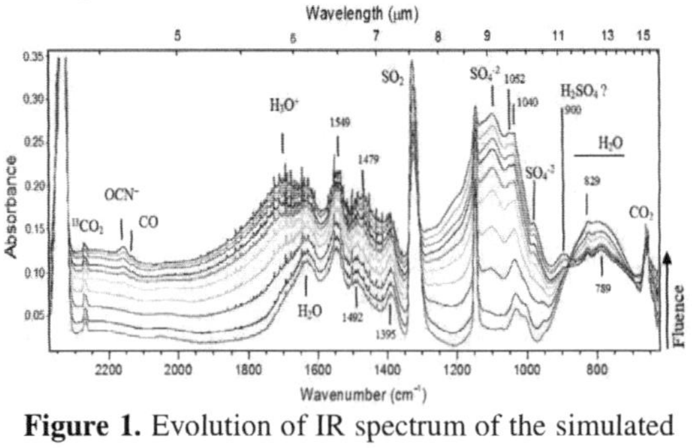

Figure 1. Evolution of IR spectrum of the simulated Europa´s surface at 90 K during the irradiation. Bottommost curve is the unirradiated spectrum. Adapted from [1].

References

[1] Pilling S. & Bergantini A., 2015, *MNRAS, Submitted.*

[1] E-mail: sergiopilling@pq.cnpq.br

Content from this work may be used under the terms of the Creative Commons Attribution 3.0 licence. Any further distribution of this work must maintain attribution to the author(s) and the title of the work, journal citation and DOI.
Published under licence by IOP Publishing Ltd

Journal of Physics: Conference Series **635** (2015) 112105 doi:10.1088/1742-6596/635/11/112105

Molecular dynamics of photodissociation: towards more complex systems

E. Kukk[1*], D. T. Ha[2], Y. Wang[2,3], D. G. Piekarski[2], S. Diaz-Tendero[2], K. Kooser[1], E. Itälä[1], H. Levola[1], M. Alcamí[2], E. Rachlew[4] and F. Martín[2,3,5§]

[1]Department of Physics and Astronomy, University of Turku, FI-20014 Turku, Finland
[2]Departamento de Quimica, Modulo 13, Universidad Autonoma de Madrid, Cantoblanco 28049 Madrid, Spain
[3]Instituto Madrileno de Estudios Avanzados en Nanociencia, Cantoblanco, 28049 Madrid, Spain
[4]KTH, Physics, Stockholm, Sweden
[5]Condensed Matter Physics Center (IFIMAC), Universidad Autonoma de Madrid, 28049 Madrid, Spain

Synopsis We present a combined experimental and theoretical study of the photodissociation of thiophene molecule using energy-resolved electron-ion-ion coincidence technique and self-consistent charge density functional tight-binding theory combined with a statistical approach. The observed complex molecular dynamics with many internal-energy-dependent fragmentation pathways is successfully described by the theoretical simulations.

Following the dynamics of molecular reactions as they proceed has been a topic of keen interest for chemists as well as physicists, from both the fundamental and applied points of view. Unimolecular reactions that follow the deposition of energy into the system by photoabsorption is a particular category that has recently spurred significant theoretical and experimental efforts [1,2]. Despite of active studies, our understanding of the molecular dynamics in radiation-induced reactions is still very limited, particularly when it comes to medium- and large-sized systems.

Here, we present a combined experimental and theoretical study of thiophene molecule – a ring structure of 9 atoms. We employed energy-resolved electron-ion-ion coincidence (PEPIPICO) technique using synchrotron radiation at MAX-lab, Sweden and observed the molecular fragmentation pathways depending on the amount of the internal energy deposited into the molecule during the sulphur 2p photoionization and the subsequent Auger relaxation processes.

Theoretical simulations were performed by using the semi-empirical self-consistent charge density functional tight-binding (SCC-DFTB) method. By running thousands of these simulations, the variability of the initial conditions encountered in the experiment were taken into account, including the systematic dependencies on the internal energy.

Our chosen molecule exhibits a rich pattern of molecular dynamics with numerous ring rupture processes accompanied by hydrogen migration and/or ejection (see Figure 1). Both the ring ruptures and the hydrogen dynamics are highly internal-energy-dependent and the branching ratios into all available dissociation pathways show major variations: in experiment as a function of the Auger electron kinetic energy and in simulations as a function of the initial total vibrational energy of the molecule.

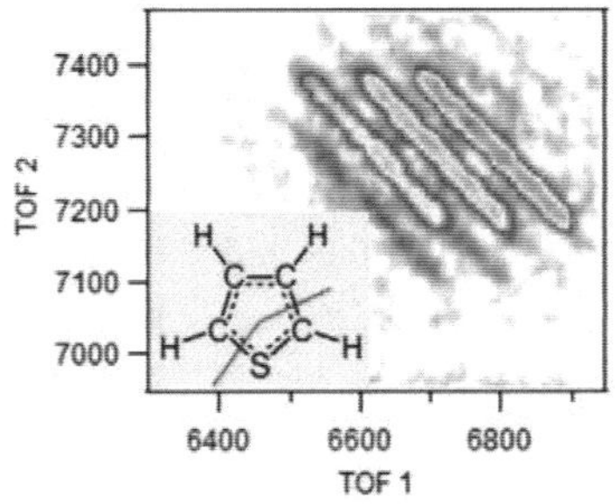

Figure 1. A small region of the PEPIPICO map following S 2p photoionization and Auger decay in thiophene. Illustration shows the ring bond ruptures and the sequence of patterns is due to additional ejection of hydrogen(s).

This study demonstrates that a computationally feasible theoretical model can successfully reproduce the complex dissociation pattern and its energy-dependent variations in a medium-sized molecule. The approach can be extended to larger systems in future.

References

[1] J. Ullrich, A. Rudenko, and R. Moshammer, Annual Review of Physical Chemistry, 63, 635–(2012).

[2] P. Casavecchia, Reports on Progress in Physics, 63, 355 (2000).

---

*E-mail: edwin.kukk@utu.fi

§E-mail: fernando.martin@uam.es

Content from this work may be used under the terms of the Creative Commons Attribution 3.0 licence. Any further distribution of this work must maintain attribution to the author(s) and the title of the work, journal citation and DOI.
Published under licence by IOP Publishing Ltd

Journal of Physics: Conference Series **635** (2015) 112106 doi:10.1088/1742-6596/635/11/112106

Molecular dynamics observed with nonlinear autocorrelation in the VUV range

Dimitris Rompotis, Thomas Gebert, Oliver Schepp, Arne Baumann, Marek Wieland, Markus Drescher[1]

Institut für Experimentalphysik, Universität Hamburg, Luruper Chaussee 149, 22761 Hamburg, Germany
Center for Free-Electron-Laser-Science, Luruper Chaussee 149, 22761 Hamburg, Germany
Center for Ultrafast Imaging, Luruper Chaussee 149, 22761 Hamburg, Germany

Synopsis 160 nm vacuum-ultraviolet pulses generated as the 5th harmonic of an 800 nm femtosecond laser are sufficiently energetic to drive multiphoton transitions in molecules. 2nd order intensity- as well as interferometric autocorrelation were used as time-metrology tools for studying the photodissociation dynamics in ethylene and oxygen molecules. An imaging ion spectrometer enables the separate capture of dynamics in different fragmentation channels. Excitation in a non-collinear beam geometry is shown to deliver temporal resolution already in a single shot.

Driving multiphoton transitions at short wavelengths reduces the order of nonlinearity and necessary field-strengths, thereby allowing an observation of molecular targets closer to a weak-field case. An efficient 5th harmonic frequency-conversion scheme generates sub-20 fs pulses at 160 nm (7.8 eV photon energy) [1], falling into an interesting spectral range for triggering photo-processes in many molecules. Pulse energies exceeding 1 μJ support nonlinear transitions even under loose focusing conditions. A transfer of the nonlinear autocorrelation technique, regularly serving for time-metrology in the ultraviolet, visible, and infrared range, into the vacuum-ultraviolet (VUV) regime requires novel concepts for the optical layout. We have realized both intensity- as well as interferometric autocorrelation at 160 nm. Fig. 1 shows a fringe-resolved autocorrelation (FRIAC) trace obtained from two-photon ionization of krypton gas, from which duration as well as linear chirp of the VUV pulses can be derived [2].

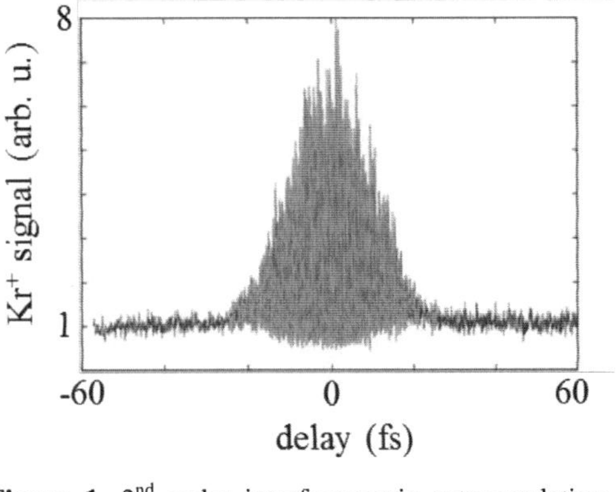

Figure 1. 2nd–order interferometric autocorrelation trace taken at 160 nm wavelength in krypton. The analysis yields a pulse duration of 18.4 ± 0.7 fs.

The simultaneous capture of this instantaneously responding reference target together with the molecular fragments under study is essential for a reliable deconvolution of the autocorrelation trace to extract information on molecular fragmentation times. Of similar importance are a high degree of modulation and good signal-to-noise ratio. The trace in Fig. 1 demonstrates a contrast approaching the 1:8 ratio expected for an ideal collinear FRIAC.

Upon non-collinear beam propagation the modulation vanishes and the contrast reduces to 1:3, but such geometry enables a very efficient detection of all delays at once. This single-shot capability is particularly important for light sources like FELs, known to be subject to strong fluctuations of the pulse properties. But even for the limited energies available with the laser based source we approach the single-shot case.

Both intensity- and interferometric autocorrelators were used for studying the photo-fragmentation of oxygen and ethylene molecules. For the $C_2H_4^+$ molecular ion of the latter we find a time-constant of 20 fs, while the formation of the $C_2H_3^+$ fragment takes considerably longer. The preliminary analysis of autocorrelation traces taken in oxygen indicates a dynamical component of less than 10 fs.

References

[1] *D. Rompotis et al. 2015 Optics Letters, accepted*
[2] *T. Gebert et al. 2014 New J. Phys.* ***16*** *07304*

E-mail: markus.drescher@uni-hamburg.de

Content from this work may be used under the terms of the Creative Commons Attribution 3.0 licence. Any further distribution of this work must maintain attribution to the author(s) and the title of the work, journal citation and DOI.
Published under licence by IOP Publishing Ltd

XXIX International Conference on Photonic, Electronic, and Atomic Collisions (ICPEAC2015) IOP Publishing
Journal of Physics: Conference Series **635** (2015) 112107 doi:10.1088/1742-6596/635/11/112107

Molecular Resolvent Operator for H_2^+ molecule

R. E. F. Silva* [1], **F. Catoire**†, **P. Rivière***, **H. Bachau**†, **and F. Martín***,**

* Departamento de Química, Módulo 13,Universidad Autónoma de Madrid, Cantoblanco 28049, Madrid, Spain
† Centre des Lasers Intenses et Applications CNRS-CEA-Univ. Bordeaux I, 351 Cours de la Libération, Talence F-33405, France
** Instituto Madrileño de Estudios Avanzados en Nanociencia (IMDEA-Nanociencia), Cantoblanco, 28049 Madrid, Spain

Synopsis We present a theoretical method to extract photoelectron and nuclear kinetic energy spectra of molecules under intense ultrashort laser pulses. The method is applied to H_2^+ photoionization and dissociation within a 3D model of the H_2^+ molecule [1]. This method is an extension to molecules of the resolvent technique originally proposed for atoms [2], which was recently implemented to extract absolute values of photoelectron and nuclear kinetic energy spectra [3, 4].

One of the most valuables informations that we can extract from a theoretical solution of the Time-Dependent Schrödinger equation (TDSE) is the spectra of the fragments. In a molecule we can have different fragments and the usually calculated spectra are the photoelectron spectra and the nuclear kinetic energy release. These two spectra can provide more information in a correlated spectra [5]. In this work we have extended the resolvent technique to molecules within a 3D model of the H_2^+ molecule [1]. To achieve that we propose a factorization of the resolvent operator into an electronic resolvent operator and a nuclear resolvent operator, where the first one selects a given electronic state and the second one extracts the probability associated to a given vibronic energy [4].

$$\hat{R} = \hat{R}_N \hat{R}_{ele} \qquad (1)$$

This method allows to extract differential probability densities for the different ionization and dissociation channels in the Born-Oppenheimer approximation.

We have studied the multiphoton ionization of the H_2^+ molecule with low intensities of the laser field and we are able to obtain fully differential spectra both in nuclear and electronic energy and also angular distributions. With a fully differential spectra we are capable of providing a much more complete description of the dynamics behind photoionization and photodissociation. Our results have been successfully compared with those previously obtained in the literature [6].

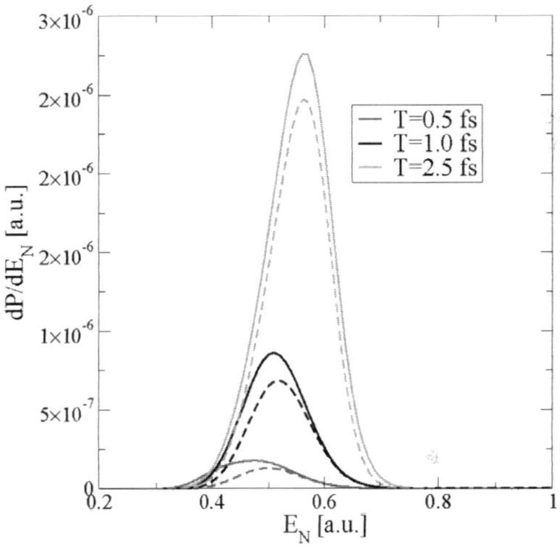

Figure 1. KER spectra resulting from a pulse with central frequencies $\omega = 0.6$ a.u. The pulse duration is indicated in the figure. We have used relatively low intensities (10^{12} W/cm^2). We also show the results from Ref. [6] (dashed lines).

References

[1] T. Niederhausen, U. Thumm and F. Martín, *Journal of Physics B* **45**, 105602-1 105602-9 (2012).

[2] K. J. Schafer and K. C. Kulander, *Phys. Rev. A* **42**, 5794 (1990).

[3] F. Catoire and H. Bachau, *Phys. Rev. A* **85**, 023422 (2012).

[4] F. Catoire, R. E. F. Silva, P. Rivière, H. Bachau and F. Martín, *Phys. Rev. A* **89**, 023415 (2014).

[5] R. E. F. Silva, F. Catoire, P. Rivière, H. Bachau and F. Martín, *Phys. Rev. Lett.* **71**, 063405 (2005).

[6] A. Palacios, S. Barmaki, H. Bachau, and F. Martín, *Phys. Rev. A* **110**, 113001 (2013).

[1] E-mail: rui.silva@uam.es

Content from this work may be used under the terms of the Creative Commons Attribution 3.0 licence. Any further distribution of this work must maintain attribution to the author(s) and the title of the work, journal citation and DOI.
Published under licence by IOP Publishing Ltd

Radiative stabilization of C_6^-

V. Chandrasekaran,[*] B. Kafle,[*] A. Prabhakaran,[*] O. Heber,[*,1] M. L. Rappaport,[*] H. Rubinstein,[*] D. Schwalm,[*,+] Y. Toker,[*,#] and D. Zajfman[*]

[*]Department of Particle Physics and Astrophysics, Weizmann Institute of Science, Rehovot, 7610001, Israel
[+]Max-Planck-Institut für Kernphysik, D-69117 Heidelberg, Germany
[#]Department of Physics, Bar-Ilan University, Ramat-Gan 5290002, Israel

Synopsis Absolute rate coefficients were determined for the recurrent fluorescence (RF) process in C_6^- anions at excitation energies above the adiabatic electron attachment energy. The RF process is an important ingredient in understanding the formation and stabilization of anions in the interstellar medium.

In collision-free environments like the interstellar medium (ISM), electron attachment to a neutral molecule having a positive electron affinity can yield a temporary negative ion (TNI), where the total electronic excitation energy has been rapidly converted into vibrational excitation due to non-adiabatic coupling between the electronic and vibrational degrees of freedom (intramolecular vibrational redistribution (IVR)). Since the total energy of the TNI is above the electron affinity, it can either decay back to the neutral molecule and a free electron (vibrational auto-detachment (VAD)), or it can stabilize by emitting infrared (IR) photons. In anions exhibiting excited low-lying electronic configurations, however, much faster radiative stabilization can occur by the emission of an optical photon (recurrent or Poincare fluorescence (RF) [1]). Here, we report about a measurement of absolute RF rates in C_6^- as an example for the importance of the RF process in the radiative stabilization of small, open-shell, cluster anions.

The experiment was performed by extracting C_6^- ions from a sputter ion source and storing them in a bent electrostatic ion beam trap [2]. After 1 s of storage, during which the anions cooled down to temperatures close to room temperature, they were excited by a short laser pulse to excitation energies above the adiabatic electron attachment energy of EA=4.18 eV, and the neutralization rate due to VAD was measured as a function of time at several laser wavelengths. Due to the different energy dependence of the two competing decay processes, RF and VAD their contributions to the measured total decay rate coefficients $k_{tot}(E)$ could be disentangled [3]. For energies $E \leq 4.7$ eV, the decay is found to be dominated by the RF process with decay rate coefficients on the order of $5 \times 10^4\ s^{-1}$.

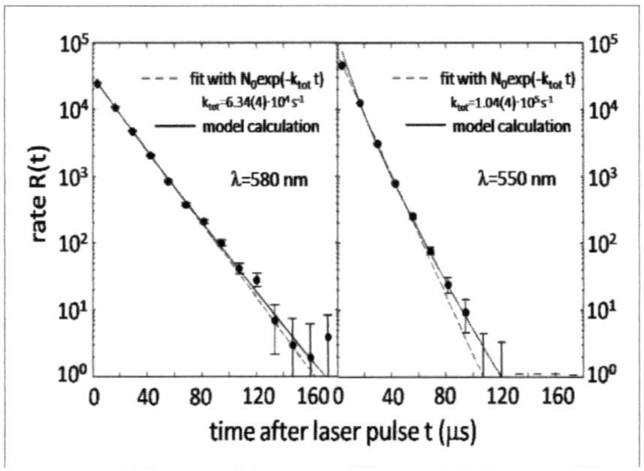

Figure 1. Neutralization rates observed as a function of time after two-photon excitation of C_6^- with a short laser pulse at two wavelengths. The laser was fired 1 s after the injection of the ions into the trap.

Our experiment constitutes the first quantitative measurement of RF rate coefficients in a small polyatomic molecule. The result can serve as a benchmark for the validity and accuracy of the statistical approach in describing the RF process even in small molecules, a process indispensable for understanding the formation and abundance of anions in the interstellar medium.

Acknowledgements: This research was supported by the Israel Science Foundation (Grant No. 1242/09), by a Yeda Sela Grant, and by a Weizmann Institute Senior Grant. D.S. acknowledges support by the Weizmann Institute of Science through the Joseph Meyerhoff program.

References

[1] A. Leger *et al.* 1988 *Phys. Rev. Lett.* **60**, 921
[2] O. Aviv *et al.* 2008 *Rev. Sci. Instrum.* **79**, 083110.
[3] V. Chandrasekaran *et al.* 2014 *J. Phys. Chem. Lett.* 5 4078.

E-mail: oded.heber@weizmann.ac.il

Content from this work may be used under the terms of the Creative Commons Attribution 3.0 licence. Any further distribution of this work must maintain attribution to the author(s) and the title of the work, journal citation and DOI.
Published under licence by IOP Publishing Ltd

XXIX International Conference on Photonic, Electronic, and Atomic Collisions (ICPEAC2015) IOP Publishing
Journal of Physics: Conference Series **635** (2015) 112109 doi:10.1088/1742-6596/635/11/112109

Resonant X-ray Scattering of carbonyl sulfide at the sulfur K edge

Loïc Journel[*1], **Tatiana Marchenko**[*], **Renaud Guillemin**[*], **Elie Kawerk**[*], **Matjaž Kavčič**[†], **Matjaž Žitnik**[†], **Klemen Bučar**[†], **Rok Bohinc**[†] **and Marc Simon**[*]

[*]Sorbonne Universités, UPMC Université Paris 06 and CNRS, UMR 7614, Laboratoire de Chimie Physique-Matière et Rayonnement, F-75005, Paris, France
[†] J. Stefan Institute, Jamova 39, SI-1000, Ljubljana, Slovenia

Synopsis New results on free OCS molecules have been obtained using Resonant X-ray Inelastic Scattering spectroscopy. A deconvolution algorithm has been applied to improve the energy resolution spectra of which we can extract detailed information on nuclear dynamics in the system.

We present results obtained using Resonant Inelastic X-ray Scattering (RIXS) spectroscopy on isolated OCS molecules after sulfur K-shell photoexcitation. This technique has been demonstrated to be sensitive to both nuclear dynamics on a femtosecond timescale [1, 2] (the core-hole lifetime is used here as a characteristic time to probe nuclear rearrangements) and to the chemical environment of the molecule [3]. We present the first results obtained on free OCS molecules which is the most abundant sulfur compound naturally present in the atmosphere and is a significant compound of the global sulfur cycle [4].

We have recorded OCS KL RIXS map where the intensity is plotted as a function of incident (γ_1) and scattered (γ_2) photon energies (top of figure 1.). This map is composed of 87 consecutive spectra recorded at excitation energies ranging from 2467 to 2480 eV with 150 meV step. KL emission lines correspond to the filling of the S 1s hole by an electron of the 2p orbitals. The spectra were recorded at the ID26 beamline of the ESRF synchrotron using the X-ray spectrometer developed at the J. Stefan institute, Ljubljana [5]

In core-shell spectroscopies, like RIXS, image analysis is complicated due to the broadening induced by a number of physical or instrumental processes. The experimental broadening is mainly due to the limited bandwidth of the incident photon beam and energy resolution of the spectrometer. The measured image I as a function γ_1 and γ_2 is then the result of the convolution of the true 'physical' image O with a function P representing the effects of experimental broadening, i.e. I = P * O where * represents the convolution symbol. The problem of extracting the original image O from I has been the subject of many papers in different experimental fields (including deblurring images from Hubble space telescope [6]) and different methods have been applied to remove intrinsic broadening due to very short core-hole lifetime of heavy elements [7]. We have implemented iterative deconvolution algorithms (Richardson – Lucy method) and we will show that such algorithms provide an effective way to deconvolve instrumental broadening typical of RIXS data (see bottom of figure 1).

The main molecular features of x-ray emission characteristic of the OCS system obtained at high energy resolution after deconvolution procedure will be presented.

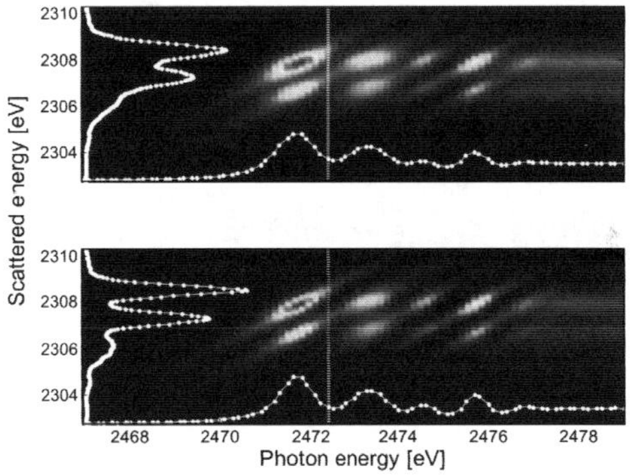

Figure 1. KL emission map before (top) and after deconvolution (bottom). X-ray emission spectra for an excitation energy indicated by the white dashed line are plotted on the left side of the maps.

References

[1] M. Simon *et al* 2006 *Phys. Rev. A* 73 020706(R).
[2] R. Guillemin et al 2012 Phys. Rev. A 86, 013407.
[3] S. Carniato *et al, 2012 J. Chem. Phys.* 137, 144303.
[4] R. P. Wayne 1991 *Chemistry of Atmospheres,* Oxford University Press, New York, USA.
[5] M. Kavcic *et al* 2012 *Rev. Sci. Instrum.* 83, 033113
[6] H. M. Adorf et al 1995, *Int. J. Imaging Syst. Technol.* 6, 339.
[7] T. Fister *et al* 2004 *Phys. Rev. B* 75, 174106.

[1] E-mail: loic.journel@upmc.fr

Content from this work may be used under the terms of the Creative Commons Attribution 3.0 licence. Any further distribution of this work must maintain attribution to the author(s) and the title of the work, journal citation and DOI.
Published under licence by IOP Publishing Ltd

Journal of Physics: Conference Series **635** (2015) 112110 doi:10.1088/1742-6596/635/11/112110

Describing ionization of small molecules with a Gaussian and B-Splines Mixed Basis (GABS)

Jesús González-Vázquez[*1], **Markus Klinker**[*], **Carlos Marante**[*], **Luca Argenti**[*], **Inés Corral**[*], **Fernando Martín**[*†]

[*] Departamento de Química, Módulo 13,Universidad Autónoma de Madrid, Cantoblanco 28049, Madrid, Spain
[†] Instituto Madrileo de Estudios Avanzados en Nanociencia (IMDEA-Nanociencia), Cantoblanco 28049, Madrid, Spain

Synopsis An interface between quantum chemistry electronic structure and scattering electronic theory was developed for representing the ionization of molecules. This description is achieved by creating a mix of Gaussian and B-spline functions (GABS) for the ejected electron and standard multiconfigurational methodology for the remaining bounded ones

The extraordinary evolution of experiments with time resolution at atomistic level requires, in general, the use of two laser pulses in a pump-probe scheme [1]. Although it is possible to use fluorescence or stimulated emission to probe the time evolution of the system, it is more common to ionize the molecule. In this frame, the knowledge of the ionization channel is crucial to understand the experimental outcome. Moreover, the possibility to synthesize ultrashort light pulses [2], belonging to extreme UV and X-ray regimes, has opened the way to create the so-called autoionizing states where one electron is spontaneously emitted without a probe pulse.

The use of Gaussians as basis functions to describe electronic bound states using multiconfigurational methods is nowdays a well-established technique. However, ejected electrons are difficult to describe using his type of methodology. In our group, an alternative is presented using a mix of Gaussian and B-splines functions (GABS) to describe the electron ejected in the ionization process [3].

In this contribution we expand this methodology to be used in polielectronic system by dividing the system in two different parts. On the one hand, the method takes into account the neutral molecule and the cationic eigenfunctions using Complete Active Space Self Consistent (CASSCF) methodology [4] with Gaussian basis centered in the atoms. On the other hand the ejected electron is described using the GABS methodology. The interface between the electron in the continuum and the ones in the bound states is calculated using a modified version of the MOLCAS suite of programs [5, 6]. As a proof of principle, we show the first results in small molecules.

[1]E-mail: jesus.gonzalezv@uam.es

References

[1] A. H. Zewail 1998 *Science* **242** 1645.

[2] A. Scrinzi, M. Y. Ivanov, R. Kienberger, and D. M. Villeneuve 2006 *J. Phys. B* **39** R1.

[3] C. Marante, L. Argenti, and Fernando Martín 2014 *Phys. Rev. A* **90** 012506.

[4] B. O. Roos and K. P. Lawley 1987 *Advances in Chemical Physics* **69** 399.

[5] Karlström *et al.* 2003 *Computational Materials Science* **28** 222.

[6] Aquilante *et al.* 2013 *Wiley Interdisciplinary Reviews: Computational Molecular Science* **3** 143.

Content from this work may be used under the terms of the Creative Commons Attribution 3.0 licence. Any further distribution of this work must maintain attribution to the author(s) and the title of the work, journal citation and DOI.
Published under licence by IOP Publishing Ltd

XXIX International Conference on Photonic, Electronic, and Atomic Collisions (ICPEAC2015) IOP Publishing
Journal of Physics: Conference Series **635** (2015) 112111 doi:10.1088/1742-6596/635/11/112111

An ab initio multiconfigurational description of core hole and shake up excited states in small molecules

Inés Corral* [1], Jesús González-Vázquez* [2], and Fernando Martín*[3]

* Departamento de Química, Módulo 13, Universidad Autónoma de Madrid, Cantoblanco 28049, Madrid Spain

Synopsis In this contribution, equilibrium geometries and potential energy curves computed at the CASSCF level of theory for the core hole and shake up superexcited states of small size molecules will be presented. The effect of dynamic correlation on the description of these species will be discussed based on preliminary multireference configuration interaction energetics.

Recent experimental progresses in the development of sub-femtosecond extreme ultraviolet light sources have opened the way to monitor the ultrafast dynamics of nuclei and electrons in atomic and molecular systems in their natural attosecond time regimes [1].

Extreme UV radiation induces inner-shell ionization and results in core hole states or shake up excited states, if valence electron excitation follows the hole creation. The embedding of these electronic states in the electronic continuum and their complex open shell electronic structure, with close lying near degenerate states, hamper the electronic structure description from a theoretical viewpoint. Furthermore, the characterization of these superexcited electronic states is subject to the variational collapse of the wavefunction, invalidating standard quantum chemical procedures designed for the description of low-lying excited states.[2]

With the long-term aim of achieving a quantitatively reliable description of photoionization processes, here we present a multiconfigurational approach based on CASSCF[3] wavefunctions, and which relies on the control of the occupancy of the core orbital for the electronic characterization of the superexcited parent ions.

The performance of this method will be illustrated by comparing the equilibrium geometries, harmonic frequencies, anharmonicity constants, ionization potential and potential energy curves along selected coordinates computed with our approach and those obtained with other methods or extracted from experiments. Preliminary energetics including dynamic correlation by means of the Multireference Configuration Interaction method will be also presented.

Finally, the bonding in these superexcited electronic states will be analyzed on the basis of the Atoms in Molecules Theory of Bader [4] and will be compared to that of their valence cations, ground and valence excited state analogues.

References

[1] F. Krausz, M. Ivanov 2009, *Rev. Mod. Phys.* **81** 163.

[2] A. V. Carravetta, H. Ågren 2012 Computational X-ray Spectroscopy in Computational Strategies for Spectroscopy: from Small Molecules to Nano Systems, Ed. V. Barone, John Wiley & Sons, Inc. Hoboken, New Jersey.

[3] B. O. Roos 1987 in Ab initio Methods in Quantum Chemistry II, ed. K. P. Lawley, Wiley, Chichester.

[4] R. F. W. Bader 1990 Atoms in Molecules. A Quantum Theory, Clarendon Press, Oxford.

[1] E-mail: ines.corral@uam.es

*[2] E-mail: jesus.gonzalezv@uam.es

*[2] E-mail: fernando.martin@uam.es

Content from this work may be used under the terms of the Creative Commons Attribution 3.0 licence. Any further distribution of this work must maintain attribution to the author(s) and the title of the work, journal citation and DOI.
Published under licence by IOP Publishing Ltd

Development of a table-top electrostatic ion storage ring (μE-ring)

Jun Matsumoto[1], Kodai Gouda, Natsumi Kondo, and Haruo Shiromaru

Department of Chemistry, Tokyo Metropolitan University, Hachioji, Tokyo 192-0397, Japan

Synopsis A table-top electrostatic ion storage ring (μE-ring) is being developed. All the electrodes of the ring were mounted on a single rectangle plate of 480 mm × 200 mm, and 4 keV Ar^+ ions were injected for a test of the ring. In the presentation, we will show the detailed design and the status of the μE-ring.

An electrostatic ion storage ring is a powerful device for investigation of slow reactions such as metastable decay and radiative cooling of molecular ions including clusters and biomolecules. We are developing a table-top electrostatic ion storage ring (μE-ring), of which the size is reduced ten times from that of the existing ring at Tokyo Metropolitan University (TMU E-ring) [1], keeping the basic idea for beam manipulation unchanged. That is, the μE-ring consists of two 160 degree deflectors, four 10 degree deflectors and four focusing and defocusing electrostatic quadrupole doublets (Q-lens). It is designed for storage of either positive or negative ions with the energy range from several to 20 keV with the storage time up to several seconds under ultra-high vacuum. Compact size of the ring has a great advantage in the time resolution of the measurements detecting neutral particles, which is limited to the residence time of the ion beam in the straight section. The portability enables us to bring it to various ion beam facilities. For example, use of the ring in a synchrotron radiation facility will definitely expand the field explored by the ion-storage ring.

First, we reconsidered transport property of an ion beam in order to check the effect of the miniaturization. Based on the equation of motion of ions in the ring, transfer matrices were produced, and the Twiss parameters were calculated. Then the stable conditions of betatron oscillations were derived by changing voltages of the quadrupole doublets. Second, actual electric fields generated by the electrodes were calculated by a simulation code, SIMION. The simulated ion beam trajectories show considerable deviation from those expected from the transport property. Therefore, the ion optics was refined with field clamps recursively. The dimensions of each electrode were determined with considering accuracies of machining and assemblage due to the miniaturization. In the final design, the circumference of an ion beam trajectory is about 0.8 m.

All the electrodes are made of aluminum, mounted on a single rectangle plate of 480 mm × 200 mm, for precise alignment of the electrodes and efficient bake-out of the ring, as shown in figure 1. Voltages of all the electrodes, the deflectors and the Q-lenses, are provided from high-voltage modules controlled by PC. Presently, the vacuum pressure is ≈ 1 × 10^{-5} Pa with a 400 L/s turbo molecular pump.

Ion storage tests were carried out by injection of Ar^+ ions at an energy of 4 keV extracted from an electron impact ion source. On the basis of the ion currents measured at various places in the ring, we confirmed that the ion beam was properly controlled on the whole, in the first revolution by the steady voltages close to the designed values. For further revolutions, optimization of all the voltages including the pulse profile of the voltage applied to the switching electrodes is under way.

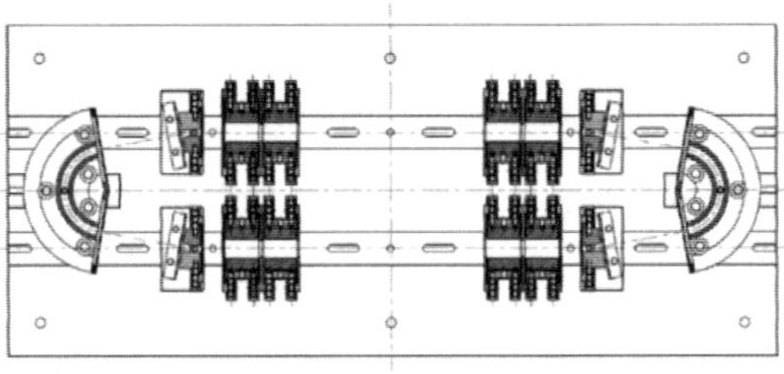

Figure 1. Schematic drawing of μE-ring . All electrodes are mounted on a single rectangle plate of 480 mm x 200 mm

References

[1] S. Jinno *et al* 2007 *Nucl. Instrum. Meth. A* **572** 568

[1] E-mail: junmatsu@tmu.ac.jp

Content from this work may be used under the terms of the Creative Commons Attribution 3.0 licence. Any further distribution of this work must maintain attribution to the author(s) and the title of the work, journal citation and DOI.
Published under licence by IOP Publishing Ltd

Potential energy curves of the ground and excited states of molecular hydrogen calculated by quasicomplete bases

Xian-Fang Yue* ‡, Xiang Gao* [1], and Jia-Ming Li† § ǂ [2]

* Beijing Computational Science Research Center, Beijing 100084, China
‡ Department of Physics and Information Engineering, Jining University, Qufu 273155, China
† Key Laboratory for Laser Plasmas (Ministry of Education) and Department of Physics and Astronomy, Shanghai Jiao Tong University, Shanghai 200240, China
§ Department of Physics and Center for Atomic and Molecular Nanosciences, Tsinghua University, Beijing 100084, China
ǂ Collaborative Innovation Center of Quantum Matter, Beijing 100084, China

Synopsis The "quasi-complete" basis calculation scenario for atomic excited states were extended to molecular system. We successfully construct the "quasi-complete" bases which consist of the spectroscopy orbitals determined by the key physical states and the pseudo-orbitals determined by the import electron correlations for the hydrogen molecule of H_2. By using such "quasi-complete" bases, the potential energy curves of the ground and highly excited states of the H_2 has been calculated and obtained in large scale $0 \leq R_{H\text{-}H} \leq 120$ Bohr. This will be very useful for further studies on spectroscopy, photoionization and photo-dissociation processes.

Molecular hydrogen of H_2 plays a vital role in forming and powering stars. It is also the simplest diatomic molecule, and had been a benchmark molecule to study the spectroscopy, photoionization, photo-dissociation, and so on. A lot of astronomical, physical and chemical researchers had paid attention to this molecule system and carried out both the experimental and theoretical studies. In theoretical aspects, the potential energy curves (PECs) of molecules are very important and the basis for the studies of the collisions, spectroscopy, photoionization, photo-dissociation, etc. Sharp [1] had summarized some early PECs of the H_2. However, most of previous calculations on PECs of the H_2 are focused on the ground and lower excited states. As for the higher excited state PECs, few studies had been reported, because the calculation of highly excited state PECs is a rather difficult and challenge work.

In the present work, the "quasi-complete" basis calculation scenario for atomic excited states [2] were extended to molecular system. Applying this scenario, we carried out calculations on the higher excited states of H_2 potential energy curves using the Molpro program [3]. The definition of the "quasi-complete" basis, here, is similar to that in atomic systems we have used [2]. It consists of spectroscopic orbitals and pseudo-orbitals, and can satisfies the desired accuracy and speed of calculations. Specifically, the construction of our present calculations can be divided into the following three main steps. The first step is to prepare the spectroscopy orbitals by optimizing the main physical states. The second step is to prepare the pseudo-orbitals based on the fixed spectroscopic orbitals. Note that the character of the spectroscopy orbitals influences the optimization of the pseudo-orbitals. The pseudo-orbitals are used to deal with the important electron correlations, involving the dipole and some quadrupole dynamical correlations. The effects of the limited pseudo-orbitals in the Configuration Interaction (CI) calculation are very close to the full CI result using the infinite configurations of the whole Hilbert space. Therefore the pseudo-orbitals can be viewed as a specific linear combination of infinite-bound-type orbitals and continuum orbitals. Finally, the third step is a large-scale CI calculation, i.e., calculating the energy levels based on the "quasi-complete" bases. Another technical issue one must pay attention to is the choice of Gaussian type orbitals (GTO) used in different interatomic distances, which we will discuss.

With the "quasi-complete" bases, we have obtained not only the ground state but also the higher excited PECs of the molecular hydrogen in large scale of $0 \leq R_{H\text{-}H} \leq 120$ Bohr, which will provide the helpful facilities for further studies on spectroscopy, photoionization and photo-dissociation processes.

References

[1] T. E. Sharp 1971 *Atomic Data* **2** 119
[2] X. Han *et al* 2014 *Phys. Rev. A* **89** 042514
[3] MOLPRO, version 2010.1, a package of ab initio programs, H.-J.Werner, P. J. Knowles, et al.

[1] E-mail: xgao@csrc.ac.cn

[2] E-mail: jmli@Tsinghua.edu.cn

Content from this work may be used under the terms of the Creative Commons Attribution 3.0 licence. Any further distribution of this work must maintain attribution to the author(s) and the title of the work, journal citation and DOI.
Published under licence by IOP Publishing Ltd

Photoelectron imaging of interstellar medium anions

Nrisimhamurty Madugula [1], Roby Chacko, Pranawa C. Deshmukh, and Aravind Gopalan [2]

Department of Physics, Indian Institute of Technology Madras, Chennai-600036, India

Synopsis A state-of-the-art photoelectron spectroscopy (PES) experimental setup is designed and built to study the structure and dynamics of interstellar medium anions, in particular polyaromatic hydrocarbon chain (PAH) anions.

Study of fundamental processes like photoionization, photodetachment, photodissociation, electronic excitation, and electron attachment, help us to understand the structure and dynamics of interstellar medium (ISM) anions, including clusters. Studies on anions are of paramount importance as they are the best-suitable systems to study the role of electron correlations.

Towards this study, a state-of-the-art photoelectron spectroscopy (PES) experimental setup is newly designed and constructed in our laboratory. This setup will be employed to probe the electronic structure of molecules (particularly, PAHs), which are proposed constituents of ISM. The constituents will then be identified by comparing our results with diffuse interstellar bands (DIBs), obtained by very high-resolution astronomical telescopes.

The experimental setup consists of a pulsed supersonic expansion discharge ion-source (one includes a solenoid valve and the other consists of piezo valve) [1] and is coupled with Wiley-McLaren type [2] time-of-flight (TOF) mass spectrometer, for mass selecting the desired anions. This source produces cold ions and thus helps in minimizing the spectral congestion. We have also constructed cold-cathode plasma source for the production of negative ions. TOF spectra obtained from our newly constructed experimental setup are presented in Fig. 1 and Fig. 2.

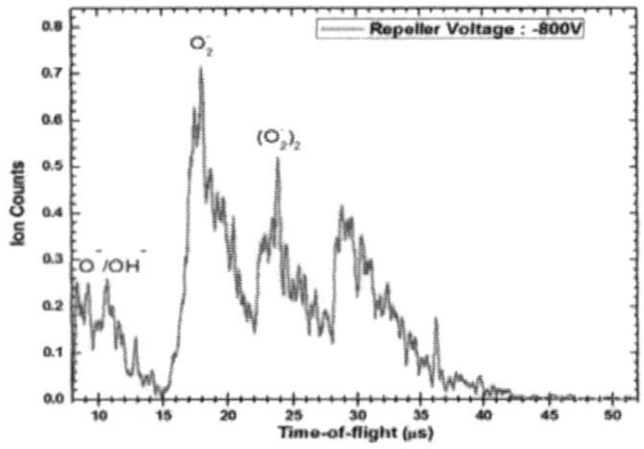

Figure 1. TOF spectra for the precursor, oxygen, is presented. In this case, we have used pulsed supersonic discharge ion source (solenoid valve). Supersonic expansion through 150μm nozzle has yielded cluster anion.

The mass-selected negative ions are photodetached using a Nd: YAG laser. The photoelectron kinetic energy released and angular distributions are simultaneously measured using a home-built velocity-map imaging (VMI) [3-5] spectrometer. For a given anion, the photoelectron angular distributions will be characteristic of the electronic state, from which they are photodetached. Our results on molecular oxygen anion (O_2^-) clusters [6, 7] and SF_6^- , will be presented in comparison with the reported experimental and theoretical data.

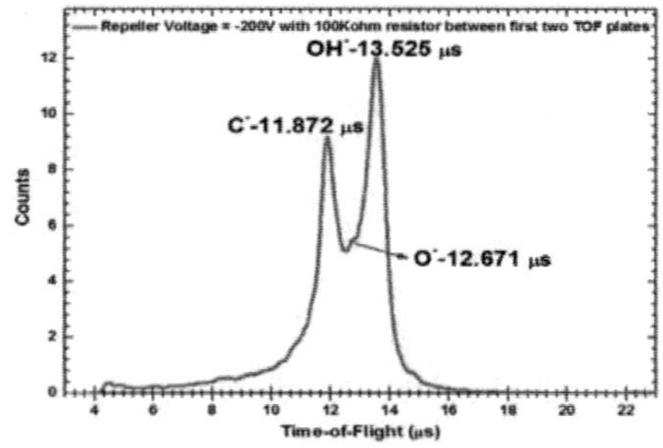

Figure 2. TOF spectra with air as the discharge medium. In this case, we have used cold-cathode plasma source. Unlike the supersonic source, this source does not yield cluster anions.

References

[1] D. L. Osborn *et al* 1996 *J. Chem. Phys.* **104** 5026

[2] W. C. Wiley and I. H. McLaren 1955 *Rev. Sci. Instrum.* **26** 1150

[3] A. T. J. B. Eppink and D. H. Parker 1997 *Rev. Sci. Instrum.* **68** 3477

[4] E. Surber *et al* 2003 *J. Phys. Chem. A* **107** 8215

[5] G. Aravind *et al* 2009 *Phys. Rev. A* **79** 043411

[6] Wenyun Lu *et al* 1995 *J. Photochem. Photobiol., A* **92** 13

[7] D. Hern Paik *et al* 2006 *J. Chem. Phys.* **125** 133408

[1] E-mail: snmurty@physics.iitm.ac.in
[2] E-mail: garavind@iitm.ac.in

Content from this work may be used under the terms of the Creative Commons Attribution 3.0 licence. Any further distribution of this work must maintain attribution to the author(s) and the title of the work, journal citation and DOI.
Published under licence by IOP Publishing Ltd

Core-Level Photoelectron Spectroscopy of Metal Clusters at FLASH

F. Martinez [1*], P. Oelßner*, M. Köther*, J. Tiggesbäumker*, K. H. Meiwes-Broer*, D. Bauer*, G. Ganteför[†], B. v. Issendorff[‡], M. Meyer[$], T. Möller[§], E. Rühl[×]

* University of Rostock, Germany; [†] University of Konstanz, Germany; [‡] University of Freiburg, Germany; [$] European XFEL, Hamburg, Germany; [§] TU Berlin, Germany; [×] FU Berlin, Germany

Synopsis Free size-selected metal clusters are investigated by core-level photoelectron spectroscopy at the free electron laser FLASH. The binding character of the clusters, as well as a multitude of relaxation channels are indicated by size-dependent core-level shifts and low-energy secondary electron emission, respectively. A new apparatus including a cryogenic ion-trap is currently built to extend the studies towards the smallest cluster sizes (less than 10 atoms), where low photo-absorption cross sections have to be matched by increased cluster density.

XUV radiation of high brilliance from the free electron laser facility FLASH (Hamburg) allows to access electronic core levels of free metallic clusters by photo-electron spectroscopy (PES). Recent experiments on lead cluster anions revealed a systematic shift of 5d and of 4f level energies as a function of cluster size [1,2]. For small clusters, a variation of the electron binding energy from the metallic sphere model is observed. It results from a reduced screening of the core-hole by the electrons, thus indicating a transition from a metallic to a non-metallic binding character in the cluster [2]. Moreover, a considerable contribution of low-energetic electrons in the PE-spectrum is due to electronic relaxation processes of the residual, highly excited cluster.

With increasing photon energies, but also with decreasing cluster size, the photoionization cross sections decrease rapidly, thus limiting the photo-electron yield. To increase the target density of size-selected clusters, and hence the electron yield, a new apparatus with a cryogenic, linear radiofrequency ion trap is currently being constructed (Fig.1). Additionally, such an ultra cold environment will prepare target clusters, which are close to their electronic and geometric ground states. With this setup many-electron dynamics due to x-ray photon absorption and highly-correlated phenomena on ultrashort timescales in clusters will be addressed in future experiments.

In this contribution, recent results on correlation effects in the photoemission of core-excited, mass selected metal clusters will be discussed, and the new ion trap setup will be introduced. The work is funded by the bmbf (FSP 302), and supported by the DFG (SFB 652).

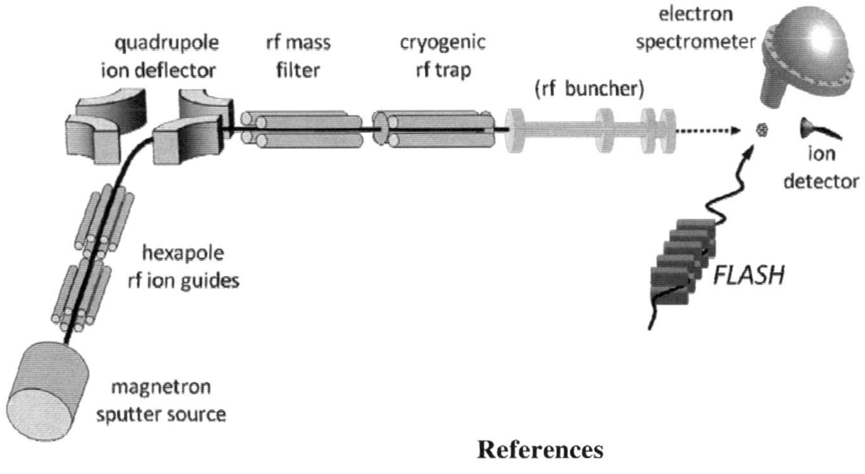

Figure 1. New experimental setup for core-level PES at free, small size-selected metal clusters.

References

[1] V. Senz *et al* 2009 *Phys. Rev. Lett.* **102** 138303

[2] J. Bahn *et al* 2012 *New J. Phys.* **14** 075008

[1] E-mail: franklin.martinez@uni-rostock.de

Content from this work may be used under the terms of the Creative Commons Attribution 3.0 licence. Any further distribution of this work must maintain attribution to the author(s) and the title of the work, journal citation and DOI.
Published under licence by IOP Publishing Ltd

Journal of Physics: Conference Series **635** (2015) 112117 doi:10.1088/1742-6596/635/11/112117

Asymmetric electron emission in laser-assisted photoionization of diatomic molecules

Diego I.R. Boll*[1], Omar A. Fojón*[2]

*Laboratorio de Colisiones Atómicas, IFIR, CONICET-UNR, Rosario, Argentina

Synopsis We study theoretically the different kinds of asymmetries that may arise in the laser-assisted photoionization of diatomic molecules.

The study of the electron dynamics now available within its natural time scale gave place to the birth of the so called attosecond physics [1]. Extreme ultraviolet (XUV) light pulses with durations in the range of hundreds of attoseconds (as, 1 as $= 1 \times 10^{-18}$ s) in combination with a weak or moderately strong near-infrared laser field (NIR), allow to monitor and/or control the electron dynamics.

The asymmetric electron emission is closely related to the control of dynamical properties in laser-assisted reactions. The electron localization in dissociating molecular states [2] and the orbital parity mix [3] have been proved to be valid tools to control the reactions.

In this work, we study theoretically the laser-assisted photoionization of diatomic molecules by means of an analytical model [4, 5], with emphasis in the asymmetric electron emission. In this model, the initial molecular wavefunctions are described as linear combination of Slater type orbitals variationally optimized, whereas the final states are represented by separable Coulomb-Volkov type wavefunctions. Multiple differential cross sections (MDCS) obtained by means of our analytical model [4] are in qualitative agreement with ab-initio calculations [6].

When the reaction is assisted by a NIR, strong asymmetries in the MDCS are obtained compared to the completely symmetric MDCS produced by the XUV solely. Our model provides an interpretation of this effect based on interferences of the different channels opened by the NIR presence. For heteronuclear diatomic molecules, the aforementioned effects, are superimposed to the corresponding asymmetric emission patterns for the one-photon ionization processes.

Finally, the effect of the XUV and NIR pulses durations are analyzed. In Fig. 1, we show the angle-energy photoelectron spectra for H_2^+ molecules ionized with XUV pulses of different duration in the presence of a NIR. As can be seen, asymmetric electron emission is obtained for XUV pulses with smaller durations (Fig. 1, first row), corresponding to the streaking regime. On the contrary, almost symmetric electron emission distributions are obtained when the XUV pulse duration is comparable to the period of the NIR (Fig. 1, second row).

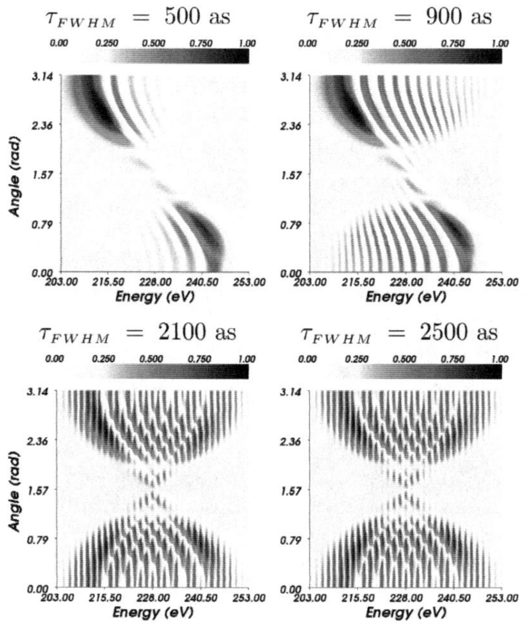

Figure 1. Angle-energy photoelectron spectra for H_2^+ ionized with XUV pulses of different durations, in the presence of a moderately intense NIR.

References

[1] F. Krausz, M. Ivanov 2009 *Rev. Mod. Phys.* **81** 163

[2] F. Kelkensberg *et al* 2011 *Phys. Rev. Lett.* **107** 043002

[3] G. Laurent *et al Phys. Rev. Lett.* **109** 083001

[4] D. Boll, O. Fojón 2014 *Phys. Rev. A* **90** 053414

[5] G. Yudin *et al* 2007 *J. Phys. B: At., Mol. Opt. Phys.* **40** F93

[6] X. Guan *et al* 2013 *Phys. Rev. A* **87** 053410

[1]E-mail: boll@ifir-conicet.gov.ar

[2]E-mail: fojon@ifir-conicet.gov.ar

Content from this work may be used under the terms of the Creative Commons Attribution 3.0 licence. Any further distribution of this work must maintain attribution to the author(s) and the title of the work, journal citation and DOI.
Published under licence by IOP Publishing Ltd

XXIX International Conference on Photonic, Electronic, and Atomic Collisions (ICPEAC2015) IOP Publishing
Journal of Physics: Conference Series **635** (2015) 112118 doi:10.1088/1742-6596/635/11/112118

Time-resolved photoemission from endohedral C_{60} probed by attosecond streaking

S. Nagele*[1], G. Wachter*[2], M. Wais*, R. Pazourek†, and J. Burgdörfer*

* Institute for Theoretical Physics, Vienna University of Technology, Vienna, Austria, EU

† Department of Physics and Astronomy, Louisiana State University, Baton Rouge, Louisiana, USA

Synopsis We theoretically investigate the time-resolved photoionization of endohedral buckminsterfullerenes as probed by attosecond streaking. We show that the time delays of electrons emitted from the central atom contain the signature of confinement resonances but are also sensitive to molecular near-fields due to enhancement and screening of the probing streaking field by the C_{60} molecule.

Recent experiments, using either attosecond streaking or the complementary interferometric RABBIT technique, have allowed to study photoemission from rare gas atoms or surfaces in the time-domain with attosecond precision. In this contribution we will study the time-resolved photoionization of endohedral buckminsterfullerenes (A@C_{60}) where a central guest atom A is surrounded by a cage of carbon atoms (C_{60}). Endohderal fullerenes have been the focus of recent time-integrated experiments [1] and *confinement resonances*, due to the interaction of the emitted electrons from the central atom with the quantum well formed by the C_{60} atoms, have been identified.

It can be expected that the confinement resonances will also leave their signature in the time domain when the ionization process from A@C_{60} is probed by the attosecond streaking technique [2]. However, in this complex system additional effects due to the interaction of the probing IR field and transport through the cage have to be accounted for. In particular, the 240 valence electrons in the C_{60} shell feature a high dipole polarizability resulting in an effective screening of the streaking field inside the fullerene and an enhanced dipolar near-field outside the "polar caps" of C_{60} (Figure 1). The near-field alters the readout of timing information two-fold: the photoelectron departing from the central atom is effectively subject to the probing streaking field only after passing through the C_{60} shell. Arrival in the streaking field is thus delayed compared to the ionization of a free atom by the travel time to the surface of the shell. Furthermore, after leaving the shell, the photoelectron experiences the locally dipole-enhanced field instead of a spatially homogenous streaking field.

In our theoretical approach, we simulate the ionization of A@C_{60} by solving the time-dependent Schrödinger equation (TDSE) using a suitable model potential representing the static C_{60} cage [2]. The response of the fullerene to the IR field is calculated using time-dependent density functional theory (TDDFT).

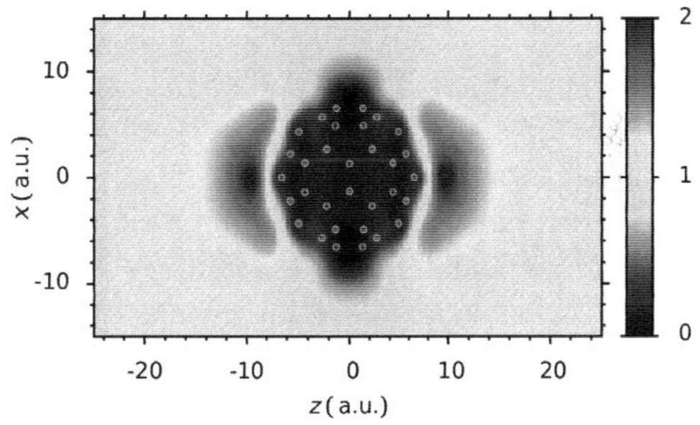

Figure 1. Local streaking field near C_{60} calculated using TDDFT. The positions of the C atoms are indicated by the grey circles.

The resulting time dependent near field is then used in the solution of the TDSE for the emitted electron. We show that the time delays extracted by attosecond streaking contain the temporal signature of confinement resonances but contain also information on the electron transport through the cage and on the near-field enhancement. We identify the separate contributions with attosecond precision.

This work was supported by the FWF-Austria (SFB NEXTLITE, SFB VICOM), by the Max Planck Society (IMPRS-APS), by the NSF through XSEDE resources (TG-PHY090031), and through computational resources at the Vienna Scientific Cluster (VSC).

References

[1] A. Kilcoyne *et al* 2010 *Phys. Rev. Lett.* **105** 213001

[2] S. Nagele *et al* 2014 *J. Phys.: Conf. Ser.* **488** 012004

[1] E-mail: stefan.nagele@tuwien.ac.at

[2] E-mail: georg.wachter@tuwien.ac.at

Content from this work may be used under the terms of the Creative Commons Attribution 3.0 licence. Any further distribution of this work must maintain attribution to the author(s) and the title of the work, journal citation and DOI.
Published under licence by IOP Publishing Ltd

Storing keV negative ions for hours: Lifetime measurements in new time domains

M. Kaminska*‡[1], E. Bäckström*, D. Hanstorp†, O. M. Hole*, R. F. Nascimento*, M. Blom*, M. Björkhage*, A. Källberg*, P. Löfgren*, P. Reinhed*, S. Rosèn*, R. D. Thomas*, S. Mannervik*, H. T. Schmidt*[2], and H. Cederquist*.

* Department of Physics, Stockholm University, Alba Nova University Centre, SE-106 91 Stockholm, Sweden
‡ Institute of Physics, Jan Kochanowski University, 25-369 Kielce, Poland
† Department of Physics, University of Gothenburg, SE-412 96 Göteborg, Sweden

Synopsis We have used one of the cryogenic ion storage rings of DESIREE to measure the lifetime of the $^2P^o_{1/2}$ level in the sulfur anion to be 503 ± 43 seconds. This is orders of magnitude longer than any previously measured lifetime in a negatively charged ion.

We have measured the radiative lifetime of the $^2P^o_{1/2}$ level in $^{32}S^-$ using the cryogenic electrostatic ion-storage ring DESIREE (Double ElectroStatic Ion Ring ExpEriment) [1, 2, 3]. With its extremely good vacuum of a few times 10^{-14} mbar, DESIREE is capable of storing negative ions for very long times. Further, the cold environment makes it possible to study weakly bound systems, which rapidly would be photodetached by the black-body radiation at ambient temperatures. The double-ring feature of DESIREE allows for studies of interactions between positive and negative ions down to very low center-of-mass energies. Here, however, we focus on laser experiments with ions stored in one of the rings.

The $^2P^o_{1/2}$ fine structure level in S^- is metastable and decays to the $^2P^o_{3/2}$ ground state through a slow magnetic dipole (M1) transition. The theoretically predicted lifetime of the upper state is 437 s [4]. Until now this has not been challenged by direct measurements as it has not been possible to store negative ions for sufficiently long times before. We utilized a laser probing technique photo-detaching small fractions of the ions in a stored beam by a series of laser pulses. By using two different laser wavelengths we monitored the neutral yield due to detachment of ions in the metastable state only and due to detachment of ions in both the ground- and metastable states. The inverse of the longer (1/e)-lifetime when both states were detached was taken as a measure of the residual-gas density. We repeated these steps at slightly elevated ring temperatures of 15-17 K, which gave higher residual-gas densities than at 13 K. In figure 1 we show the decay rates for different residual-gas densities. The intercept with the vertical axis gives the inherent spontaneous lifetime of the $^2P^o_{1/2}$ level (τ=503±43 s) [3]. This is by far the longest lifetime ever measured for a negatively charged ion. The difference from the theoretical prediction is 1.3σ showing that the multi-configuration Dirac-Fock method applied in [4] may be appropriate to describe the features of excited finestructure levels in atomic anions at this level of precision. The present results demonstrate the power of the new method and opens up for a broad survey of lifetimes of excited states in atomic ions in the time range from some tens of milliseconds to tens of minutes.

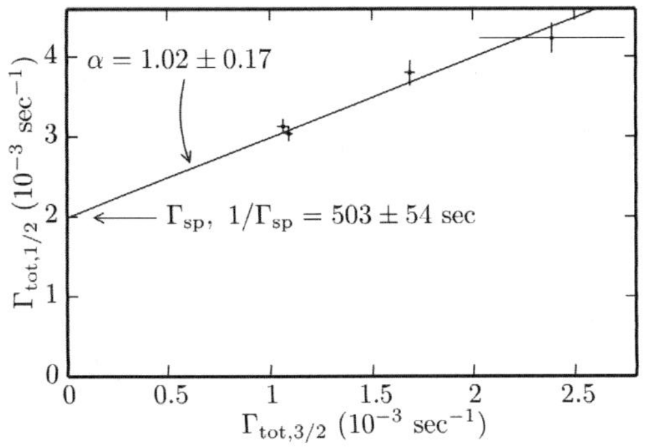

Figure 1. The measured decay rate of the $^2P^o_{1/2}$ level (ordinate) plotted with measured decay rate of the $^2P^o_{3/2}$ level (abscissa) in S^- [3].

References

[1] R. D. Thomas *et al* 2011 *Rev. Sci. Instrum.* **82** 065112

[2] H. T. Schmidt *et al* 2013 *Rev. Sci. Instrum.* **84** 055115

[3] E. Bäckström *et al* 2015 *Phys. Rev. Lett.* accepted

[4] P. Andersson *et al* 2006 *Phys. Rev. A* **73** 032705

[1]E-mail: magdalena.kaminska@fysik.su.se
[2]E-mail: schmidt@fysik.su.se

Content from this work may be used under the terms of the Creative Commons Attribution 3.0 licence. Any further distribution of this work must maintain attribution to the author(s) and the title of the work, journal citation and DOI.
Published under licence by IOP Publishing Ltd

Journal of Physics: Conference Series **635** (2015) 112120 doi:10.1088/1742-6596/635/11/112120

Photoionization and photofragmentation of multiply charged $Lu_3N@C_{80}$ ions

J. Hellhund[1], A. Borovik Jr.[1], K. Holste[1], S. Klumpp[2], M. Martins[2], S. Ricz[1,3], S. Schippers[1], and A. Müller[1]

[1] Institut für Atom- und Molekülphysik, Justus-Liebig-Universität Gießen, 35392 Gießen, Germany
[2] Institut für Experimentalphysik,Universität Hamburg, Luruper Chaussee 149, 22761 Hamburg, Germany
[3] Institute for Nuclear Research of the Hungarian Academy of Sciences (ATOMKI), Bem tér 18/c, H-4026 Debrecen, Hungary

Synopsis Relative cross sections for photoionization and photofragmentation of endohedral fullerene $Lu_3N@C_{80}^{q+}$ (q=1,2,3) ions have been measured employing the new photon-ion spectrometer PIPE at PETRA III. Prominent structures related to the carbon K-shell ionization threshold were observed in the energy range 280 to 330 eV.

We present experimental cross sections for photoionization of endohedral fullerene ions $Lu_3N@C_{80}^{q+}$ (q=1,2,3) extending recent studies on endohedral fullerenes [1, 2, 3] to a heavier and more complex species.

Photoreactions $Lu_3N@C_{80}^{q+} \rightarrow Lu_3N@C_{80-2r}^{p+}$ [q=1,2,3; p=2,3,4,5,6; r=0,1,3,4] have been investigated. The photon-ion merged-beams technique was employed at the PIPE endstation [4] of beamline P04 of the PETRA III synchrotron at DESY in Hamburg. The cross-section measurements focused on photon energies around the carbon K-shell threshold (280 - 330 eV), the nitrogen K-Shell threshold (380 - 435 eV) and the lutetium M-shell threshold (1500 - 1700 eV).

In every reaction channel investigated between 280 and 330 eV, we could identify a group of resonances. Since this group of resonances seems to make up for the most prominent structures in all spectra, we made an attempt to model each spectrum as a sum of seven peak features (Fano profile resonances, dark yellow lines), a threshold feature (blue line) and a constant background (grey line). The relative strengths of the resonance peaks P1 through P7 are given by the vertical blue bars in the insets of the figure. Additionally, in the double and triple ionization channels, a threshold can be observed at about 294 eV. We observe a shift of the ionization threshold towards higher photon energies when comparing double ionization of $Lu_3N@C_{80}^{2+}$ with double ionization of $Lu_3N@C_{80}^{+}$. We can determine the size of the endohedral fullerene ion based on this threshold shift. The cross sections measured at higher photon energies (380-435 eV and 1500-1700 eV) show no structures and decrease with increasing photon energy.

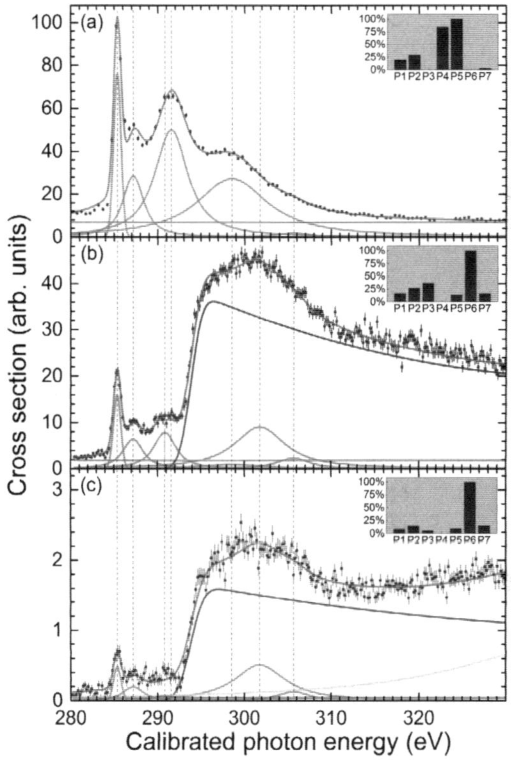

Figure 1. Relative cross sections for photoreactions of $Lu_3N@C_{80}^{q+}$: (a) single, (b) double, and (c) triple ionization of $Lu_3N@C_{80}^{1+}$. Dark grey squares with error bars represent the experimental data. Red lines correspond to fitting results. The blue and yellow lines indicate individual cross-section contributions.

References

[1] A. Müller *et al.* 2008 *Phys. Rev. Lett.* **101** 133001
[2] A.L.D. Kilcoyne *et al.* 2010 *Phys. Rev. Lett.* **105** 213001
[3] R.A. Phaneuf *et al.* 2013 *Phys. Rev. A* **88** 053402
[4] S. Schippers *et al.* 2014 *J. Phys. B* **47** 115602

[1]E-mail: Jonas.Hellhund@physik.uni-giessen.de

Content from this work may be used under the terms of the Creative Commons Attribution 3.0 licence. Any further distribution of this work must maintain attribution to the author(s) and the title of the work, journal citation and DOI.
Published under licence by IOP Publishing Ltd

XXIX International Conference on Photonic, Electronic, and Atomic Collisions (ICPEAC2015) IOP Publishing
Journal of Physics: Conference Series **635** (2015) 112121 doi:10.1088/1742-6596/635/11/112121

Synchrotron excitation - field ionization studies of high-Rydberg fragments produced after inner-shell ionization of small molecules

A. Kivimäki* [1]**, J.A. Kettunen**[†]**, C. Stråhlman**[+]**, A. Sankari**[#]**, J. Álvarez Ruiz**[‡]**, and R. Richter**[°]

* Consiglio Nazionale delle Ricerche – Istituto Officina dei Materiali, Laboratorio TASC, 34149 Trieste, Italy
[†] Department of Physics, University of Oulu, P.O. Box 3000, 90014 Oulu, Finland
[+] MAX IV Laboratory, Lund University, P.O. Box 118, 22100 Lund, Sweden
[#] Department of Physics, Lund University, P.O. Box 118, 22100 Lund, Sweden
[‡] Colegio Los Naranjos, Fuenlabrada, 28941, Madrid, Spain
[°] Elettra – Sincrotrone Trieste, Area Science Park Basovizza, 34149 Trieste, Italy

Synopsis We have studied the production of neutral high-Rydberg (HR) fragments in small molecules after inner-shell excitation and ionization. Such fragments were ionized using a pulsed electric field and resulting ions were mass-analyzed with an ion time-of-flight (TOF) spectrometer. As an example, the results obtained at the C 1s ionization threshold of the methane molecule will be discussed.

A small molecule usually fragments if one of its inner-shell (or core) electrons is promoted to an unoccupied molecular or Rydberg orbital or is completely removed from the molecule. This is because electronic states with an inner-shell hole typically decay through Auger transitions, whose two-hole final states are unstable in most molecules. The molecule then breaks up, yielding in general both ionic and neutral fragments.

In the present study, we have constructed an experimental set-up that allows us to detect neutral fragments in high-Rydberg (HR) states and determine their identities. It combines soft x-ray excitation with pulsed field ionization (PFI) and ion TOF spectroscopy. The experiment exploits PFI with high electric fields (of the order of kV/cm). A special feature of our apparatus is that field ionization of HR fragments occurs in a region that is separated from their place of creation. Therefore the measurements can be performed in the usual multi-bunch operation mode of a synchrotron light source.

Previous neutral particle-photoion coincidence experiments of the CH_4 molecule [1] revealed that the production of metastable H atoms hugely increases just above the C 1s ionization potential. There the so-called post collision interaction between a slow photoelectron and a subsequently emitted fast Auger electron can induce recapture processes, where the slow C 1s photoelectron is pushed back to a HR orbital of the molecular ion. The present experiments (figure 1) show that all possible HR fragments – not only H(HR) – can be created after subsequent dissociation processes. The C(HR) and H(HR) fragments dominate the spectrum, but molecular fragments CH_n(HR), n=1-3, and H_2(HR) were also observed. The findings are rationalized in comparison with the results of Auger electron-ion coincidence experiments [2,3]. The large intensity of the C(HR) fragments can be explained if they result from recapture processes after double Auger decay, whereas CH_3(HR), CH_2(HR) and H_2(HR) could mostly be created in recapture processes after single Auger decay.

We have performed similar experiments at the C 1s and O 1s edges of the CO_2 molecule, and at the S 2p edge of the SF_6 molecule.

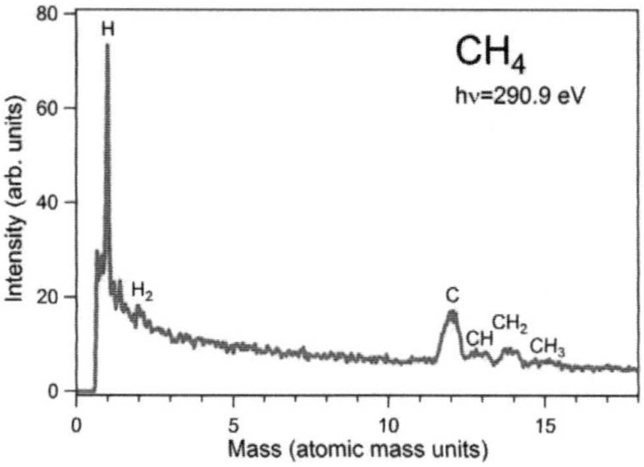

Figure 1. The ion-mass spectrum of field-ionized HR fragments of the CH_4 molecule measured just above the C 1s ionization potential.

References

[1] Kivimäki A *et al* 2013 *Phys. Rev.* A **88** 043412
[2] Kukk E *et al* 2007 *J. Phys. B* **40** 3677
[3] Flammini R *et al* 2009 *New J. Phys.* **11** 083006

[°] E-mail: kivimaki@iom.cnr.it

Content from this work may be used under the terms of the Creative Commons Attribution 3.0 licence. Any further distribution of this work must maintain attribution to the author(s) and the title of the work, journal citation and DOI.
Published under licence by IOP Publishing Ltd

Journal of Physics: Conference Series **635** (2015) 112122 doi:10.1088/1742-6596/635/11/112122

Strong field ionization of small hydrocarbon chains with full 3D momentum analysis

Claus Peter Schulz[1], **Sascha Birkner, Federico J. Furch, Alexandria Anderson, Jochen Mikosch, Felix Schell, Marc J.J. Vrakking**

Max Born Institute, Max-Born-Str. 2a, 12489 Berlin, Germany

Synopsis Strong field ionization of small hydrocarbon chains is studied in a kinematic complete experiment using a reaction microscope. By coincidence detection of ions and electrons different ionization continua populated during the ionization process are identified. In addition, photoelectron momentum distributions from laser-aligned molecules allow to characterize the electron wavepackets emerging from different Dyson orbitals.

Strong field driven spectroscopy of molecules is a very active field of research which promises to bring unprecedented time- and spatial resolution to molecular dynamics studies. It has become clear in recent years that a number of the initial simplifying assumptions, such as the adiabatic and the single-electron approximation, can break down and have to be replaced by a multi-electron picture. In particular the appearance of multiple ionization continua in Strong-Field Ionization and High Harmonic Generation of polyatomic molecules is being increasingly investigated theoretically and experimentally. The interpretation of the strong field ionization is largely facilitated by a complete experiment, where the full 3-D momentum vectors of all outgoing particles created during the photoionization process are registered. A so-called reaction microscope, which has already been developed more than a decade ago, enables this type of complete experiments by detecting the energy and angular distribution of photo electrons and ions in coincidence. Recently, we have set-up a laboratory where we combine a high repetition rate laser with a reaction microscope [1].

In a first experimental run we have studied small hydrocarbon chains like n-Butane and Butadiene. These molecules have already been investigated in similar experiments [2]. By correlated ion-electron detection it was shown that different ionization continua are excited during strong field ionization. In our experiment the hydrocarbon chains are ionized with a sub-7 fs, 4 μJ pulses at 800 nm. These pulses are focused by an 80 mm focal length spherical mirror onto a supersonic molecular beam. The intensity in the interaction region is on the order of 10^{13} to 10^{14} W/cm^2. Our experiments confirm the population of different excited ionic states. In addition, we were able to determine a complete energy and angular distribution of the photo electrons emerging from different populated states.

Presently, this study is extended by using laser-aligned hydrocarbon molecules [3]. In these experiments we are studying how multiple strong-field ionization continua manifest themselves in electron recollision driven by the strong laser field. Electron diffraction patterns for different fragments promise to characterize ionization channel resolved continuum electron wavepackets, emerging from different Dyson orbitals. This might evolve into an ultrafast dynamical probe of electronic coherences in molecules.

References

[1] F. J. Furch *et al.* 2013 *Opt. Express* **21** 22671

[2] A. E. Boguslavskiy *et al.* 2012 *Science* **335** 1336

[3] J. Mikosch *et al.* 2013 *Phys. Rev. Lett.* **110** 023004

[1]E-mail: cps@mbi-berlin.de

Content from this work may be used under the terms of the Creative Commons Attribution 3.0 licence. Any further distribution of this work must maintain attribution to the author(s) and the title of the work, journal citation and DOI.
Published under licence by IOP Publishing Ltd

Negative-ion/positive-ion coincidence spectroscopy with a novel spectrometer

Christian Stråhlman* [1], **Rami Sankari***, **Antti Kivimäki**†, **Robert Richter**‡, **Marcello Coreno**# **and Ralf Nyholm***

* MAX IV Laboratory, Lund University, Box 118, 221 00 Lund, Sweden
† CNR-IOM, Laboratorio TASC, 34149 Trieste, Italy
‡ Elettra–Sincrotrone Trieste, Area Science Park, 34149 Trieste, Italy
# CNR-ISM, S.S. 14 Km 163.5, Basovizza, I-34149 Trieste, Italy

Synopsis A novel instrument for negative-ion/positive-ion coincidence spectroscopy (NIPICO) is presented. The instrument consists of a double TOF spectrometer with a common extraction region. Electrons are deflected by means of a magnetic field. We demonstrate detections of double (NIPICO) and triple (NIPIPICO) coincidences.

Core-excited states usually decay by emitting electrons or much less probably by emitting soft x-ray photons. The yield of negative ions typically increases at these resonances. Since negative ions only will be created in negative-ion/positive-ion pairs, these fragments can be detected in coincidence. A negative-ion/positive-ion coincidence technique (NIPICO) allows us to study how yields of different fragmentation channels vary when the photon energy is tuned to different resonances and associated kinetic energy releases.

We have designed a negative-ion time-of-flight spectrometer to be used together with a positive-ion time-of-flight spectrometer provided by the Gas Phase beamline at the Elettra storage ring [1]. The two spectrometers are mounted facing each other with a common interaction region. Ions and electrons are extracted by means of an electric field. The electrons are deflected by a magnetic field, similar to pioneering experiments by Schermann *et al* [2], in the negative spectrometer flight tube, reducing efficiently the detection of electron/negative-ion coincidences. The instrument could be operated either using a pulsed electric potential in the interaction region or using a constant field. In the latter case, the identities of the positive and negative ions are deduced from the unique time between the two ion detections.

As a first target for our instrument we chose the SF_6 molecule, known to produce negative ions at resonances close to the S 2p and F 1s edges [3,4]. Figure 1 shows a measurement at two resonances above and below the S 2p ionization edge. They were measured with a constant extraction field and show three main NIPICO channels S^-/F^+, F^-/F^+ and F^-/S^+. The pronounced doublet structure of the two former indicates a significant kinetic energy release in the corresponding dissociation channels. A change in branching ratios between the S^-/F^+ and F^-/F^+ coincidence channels is also seen. We have also detected several three-particle coincidence channels (NIPIPICO), such as $F^-/F^+/F^+$ and $F^-/S^+/F^+$.

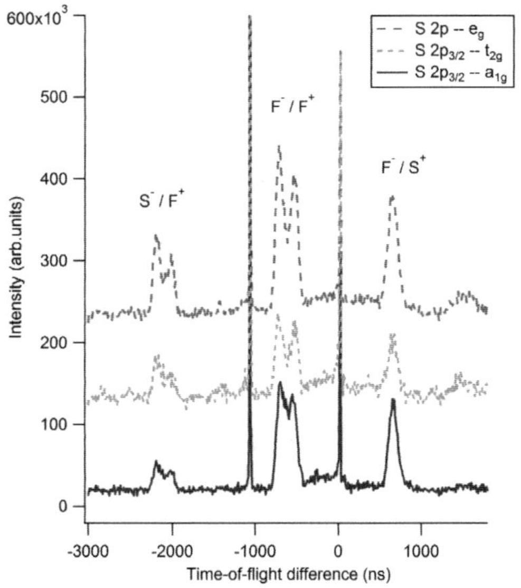

Figure 1. NIPICO spectrum for three resonances close to the S 2p IP. The t-scale indicates the differences between the positive ion TOF and the negative ion TOF. The main coincidence channels are indicated. The sharp peak at t = 0 arise either from cross-talk between the two detectors or fluorescence emission. The second sharp peak at t = -1100 is a computing artefact.

References

[1] O. Plekan *et al.*, *Phys. Scr.* **78**, 058105 (2008)
[2] C. Schremann *et al.*, J. Phys. E: Sci. Instrum., **11**, 746 (1978)
[3] S. W. J. Scully *et al*, *J. Phys. B: At. Mol. Opt. Phys.* **35**, 2703-2711 (2002)
[4] M. N. Piancastelli, *et al.*, *J. Chem. Phys.* **122**, 094312 (2005)

[1] E-mail: Christian.Strahlman@maxlab.lu.se

Content from this work may be used under the terms of the Creative Commons Attribution 3.0 licence. Any further distribution of this work must maintain attribution to the author(s) and the title of the work, journal citation and DOI.
Published under licence by IOP Publishing Ltd

Journal of Physics: Conference Series **635** (2015) 112124 doi:10.1088/1742-6596/635/11/112124

Radiative cooling of hot C_n^- and C_nH^- molecules

M. Kaminska*†[1], R. F. Nascimento*, M. H. Stockett, E. K. Anderson*, R. Delaunay‡, V. Vizcaino‡, P. Rousseau‡, L. Adoui‡, B. A. Huber‡, R. D. Thomas*, M. Gatchell*, K. Hansen*♯, H. Zettergren*, H. T. Schmidt*, and H. Cederquist*[2]**

* Department of Physics, Stockholm University, Alba Nova University Centre, SE-106 91 Stockholm, Sweden
† Institute of Physics, Jan Kochanowski University, 25-369 Kielce, Poland
⋆ Department of Physics and Astronomy, Aarhus University, Ny Munkegade, DK 8000 Aarhus C, Denmark
‡ CIMAP, UMR 6252, CEA/CNRS/EINSCAEN/Université de Caen Basse-Normandie, bd Henri Becquerel, BP 5133, F-14070 Caen cedex 05, France
♯ Department of Physics, University of Gothenburg, SE-412 96 Göteborg, Sweden

Synopsis We have measured the rates of neutrals produced from 10 keV C_n^- or C_nH^- (n=2, 4, 6, 8, and 10) ion beams stored in one of DESIREE's 14 K storage rings. For n=4, 6, and 8 we observe marked differences between C_n^- and C_nH^- cooling rates as inverse internal conversion [cf. S. Martin *et al* (2013) *Phys. Rev. Lett.* **110**, 063003] processes are effective for the C_n^- ions only. Knowledge of the cooling rates of these ions are important for estimates of their formation and destruction rates in cold interstellar environments.

We investigate the cooling of internally hot C_n^- and C_nH^- ions with n=2, 4, 6, 8, and 10 stored in the electrostatic ion storage ring DESIREE [1, 2] operating at 14 K. We record the signal of neutrals detected after a straight section of the ring as a function of time, t, after producing internally hot C_n^- or C_nH^- ions in a sputter ion source. This production method gives wide internal temperature distributions. If the ions do not stabilize radiatively very fast, the hottest ones will decay first giving overall rates of neutrals following t^{-p} power laws [3].

In Fig. 1 we show data for n=4 (the results for n=6 and n=8 are qualitatively similar). While the C_4^--signal is small and disappears rapidly (sub-milliseconds), the C_4H^-- signal follows a power law for tens of milliseconds. Similar to C_6^- [4, 5], C_4^- and C_8^- are open-shell systems that cool very rapidly as they can convert (part of) their internal vibrational energies to low-lying electronically excited states, which decay through fast radiative transitions. The C_nH^- ions, on the other hand, are closed shell systems and have much higher HOMO-LUMO gaps. This means that inverse internal conversions to electronically excited states become highly unlikely [4, 5] and internal cooling must proceed through much slower vibrational transitions. While C_2H^- and $C_{10}H^-$ have similar decay behaviors as C_4H^-, C_6H^- and C_8H^-, the decays for C_2^- and C_{10}^- are much slower than those for C_4^-, C_6^- and C_8^-. The C_{10}^- decay is also more complex possibly reflecting the presence of linear and ring isomers in the stored beam. For C_n^- and C_nH^- anions, knowledge of their cooling rates are considered to be very important for estimates of their interstellar abundances [6].

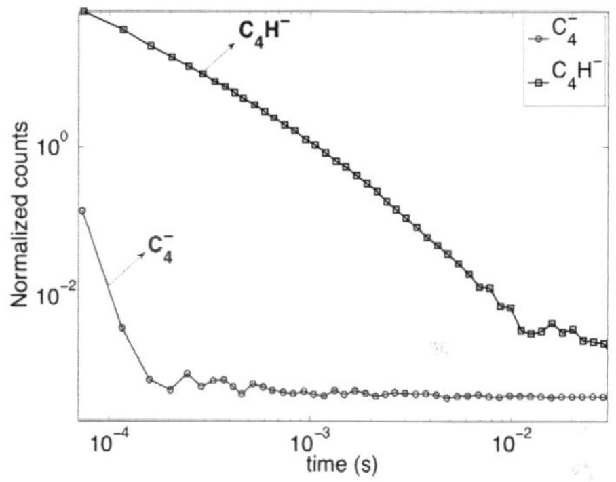

Figure 1. Rates of detected neutral particles normalized to the injected ion currents for the two molecular ions C_4^- and C_4H^-. The rapid disappearance of the signal for C_4^- is due to fast cooling through inverse internal conversion [4, 5, 7].

References

[1] R. D. Thomas *et al* 2011 *Rev. Sci. Instrum.* **82** 065112

[2] H. T. Schmidt *et al* 2013 *Rev. Sci. Instrum.* **84** 055115

[3] K. Hansen *et al* 2001 *Phys. Rev. Lett.* **87** 123401

[4] G. Ito *et al* 2014 *Phys. Rev. Lett.* **112** 183001

[5] V. Chandrasekaran *et al* 2014 *Phys. Chem. Lett.* **5** 4078

[6] H. Brünken *et al* 2007 *Astrophysical J.* **664** L43

[7] S. Martin *et al* 2013 *Phys. Rev. Lett.* **110** 063003

[1]E-mail: magdalena.kaminska@fysik.su.se
[2]E-mail: cederq@fysik.su.se

Content from this work may be used under the terms of the Creative Commons Attribution 3.0 licence. Any further distribution of this work must maintain attribution to the author(s) and the title of the work, journal citation and DOI.
Published under licence by IOP Publishing Ltd

Journal of Physics: Conference Series **635** (2015) 112125 doi:10.1088/1742-6596/635/11/112125

Generation of polyyne and methylpolyyne molecules from toluene by intense femtosecond laser pulse irradiation

Ali Ramadhan*[1], Michal Wesolowski*, Tomonari Wakabayashi†, Haruo Shiromaru‡, Tatsuya Fujino‡, Takeshi Kodama‡, Walter Duley*, Joseph Sanderson*[2]

* Department of Physics, University of Waterloo, 200 University Ave. W, Waterloo, Ontario, Canada N2L 3G1
† Department of Chemistry, Kinki University, Kowakae 3-4-1, Higashi-Osaka 577-8502, Japan
‡ Department of Chemistry, Tokyo Metropolitan University, Minami-Osawa 1-1, Tokyo 192-0397, Japan

Synopsis Hydrogen-capped and methyl-capped carbon chains (polyynes) have been generated by intense femtosecond laser irradiation of pure liquid toluene. UV-Vis and Raman spectroscopy were used to confirm the presence of polyynes in the irradiated samples, and high performance liquid chromatography (HPLC) was used to separate polyynes up to $C_{18}H_2$ and $HC_{13}CH_3$.

Polyynes are linear carbon chain clusters with an even number of carbon atoms alternating triple and single sp-hybridized C–C bonds, usually terminated by hydrogen atoms. They have been detected in the interstellar medium, and could be important precursor molecular components in the formation of fullerenes and carbon nano-tubes [1]. Polyynes have been used as nano-conductors and can form novel coaxial nano-wires by insertion into single wall carbon nano-tubes.

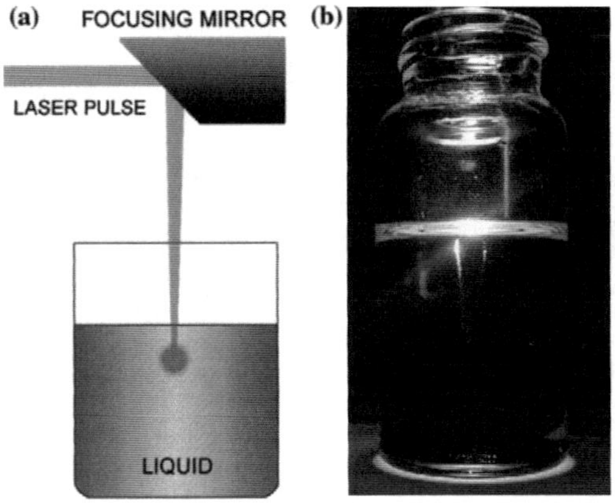

Figure 1. (**a**) A schematic of the experimental setup. (**b**) An image of the laser-liquid interaction during irradiation.

Toluene in small vials was irradiated by 100 fs pulses from a 240 μJ regeneratively amplified Ti:Sapphire tabletop laser with a 1 kHz repetition rate and central wavelength of 800 nm. The irradiation time was varied between 1-4 hours. A schematic of the irradiation setup is shown in Fig. (1a). The laser beam was focused directly below the meniscus of the liquid using a 50 mm focal length mirror. Consequently, a very intense ($> 10^{13}$ W/cm2), highly luminous, dissociation and ionization region formed, as shown in Fig. (1b).

After irradiation, Raman spectroscopy was performed on irradiated samples, indicating the presence of polyynes as large as $C_{20}H_2$ when compared to density functional simulations [2]. Fig. 2 shows that HPLC confirms the presence of polyynes up to $C_{18}H_2$ and $HC_{12}CH_3$. Hexane experiments suggest that formation via single step rearrangement of the parent molecule is not likely [3]. Formation mechanisms will be discussed including the significance of laser filamentation.

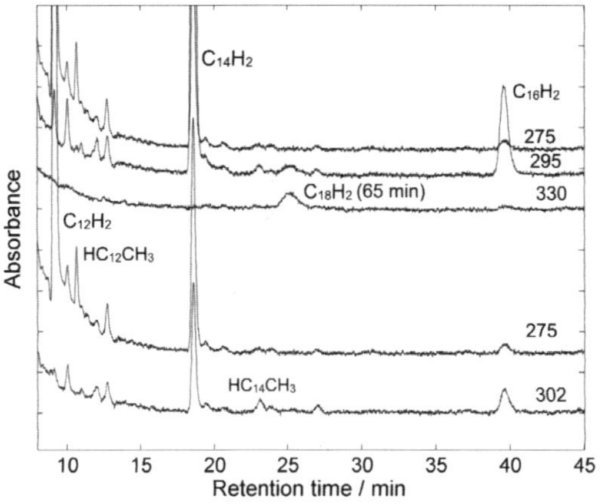

Figure 2. Analytical HPLC chromatograms of the irradiated samples, each labelled on the extreme right by its corresponding absorption wavelength in nm.

References

[1] T. Wakabayashi *et al* 2009 *Eur Phys J D* **52** 79

[2] A. K. Ott *et al* 1998 *J. Chem. Phys.* **109** 9652

[3] M. J. Wesolowski *et al* 2011 *Carbon* **49** 625

[1]E-mail: ali.ramadhan@uwaterloo.ca
[2]E-mail: j3sanderson@uwaterloo.ca

Content from this work may be used under the terms of the Creative Commons Attribution 3.0 licence. Any further distribution of this work must maintain attribution to the author(s) and the title of the work, journal citation and DOI.
Published under licence by IOP Publishing Ltd

NO_2 molecular frame photoelectron angular distributions for a range of geometries using the R-matrix method

Danilo S Brambila*, **Alex G Harvey***, **Zdeněk Mašín**†, **Olga Smirnova***

* Max-Born-Institut, Max-Born-Str 2A, 12489 Berlin, Germany
† Department of Physical Sciences, The Open University, Walton Hall, Milton Keynes, MK7 6AA, United Kingdom

Synopsis We present R-matrix calculations of photoionization from NO_2, resolved in energy, angle, and both neutral and ionic state, for a range of molecular geometries, including in the vicinity of the $^2A_1/^2B_2$ conical intersection.

The advent of new XUV light sources such as free electron lasers, and the development of high harmonic emission both as a light source and as a direct probe, coupled with advances in detector technology and the laser alignment of molecules, has opened up the possibility of observation, initiation and control of the fastest molecular processes. Exposing aligned molecules to these short, femtosecond to sub-femtosecond pulses leads to ionization, and the angular distribution of ejected electrons encodes information about the parent molecule [1, 2]. High harmonic emission from molecules can also encode information about molecular structure and dynamics [3, 4].

The theoretical descriptions of both of these types of experiment need high quality dipole matrix elements. They are used to calculate molecular frame photoelectron angular distributions and in the recombination step of current, sophisticated, models of HHG [5]. With this in mind we have converted the UK R-matrix scattering codes [6] to treat photoionization (or recombination) from aligned molecules [7].

The molecular R-matrix method is a powerful, multichannel, ab initio technique for treating electron-molecule interactions. It works by dividing the configuration space of the molecule into an inner and outer region allowing the optimal approximation and computational method to be used in each. The inner region rigorously takes into account electron-electron effects such as exchange, correlation and polarization and in the outer region the electron moves in the long range, multipole potential of the parent molecule. Matching to known asymptotic boundary conditions gives the wavefunction representing the ionized molecule plus continuum electron. Transition dipole matrix elements between this wavefunction and the initial bound state of the molecule are produced, from which the desired photoionization observables may be obtained.

We have applied the R-matrix approach to the photoionization of NO_2 resolved in energy, angle, and both neutral and ionic state, for a range of geometries, including in the vicinity of the $^2A_1/^2B_2$ conical intersection. These result represent the first application of our new R-matrix photoionization codes to both non-equilibrium molecular geometries, and excited states of the neutral.

References

[1] for a recent review see K. L. Reid, Molecular Physics, 3, 131, (2012).

[2] A. Rouze, A. G. Harvey, F. Kelkensberg, D. Brambila, W. K. Siu, G. Gademann, O. Smirnova and M. J. J. Vrakking, J. Phys. B, 47, 124017, (2014).

[3] O. Smirnova, Y. Mairesse, S. Patchkovskii, N. Dudovich, D. Villeneuve, P. Corkum and M. Yu. Ivanov, Nature 460, 972977, (2009).

[4] S. Haessler, J. Caillat, W. Boutu, C. Giovanetti-Teixeira, T. Ruchon, T. Auguste, Z. Diveki, P. Breger, A. Maquet, B. Carr, R. Taeb and P. Salires, Nature Physics 6, 200-206, (2010).

[5] see e.g. A. T. Le, R. R. Lucchese, S. Tonzani, T. Morishita, and C. D. Lin, Phys. Rev. A 80, 013401, (2009).

[6] J. M. Carr, P. G. Galiatsatos, J. D. Gorfinkiel, A. G. Harvey, M. A. Lysaght, D. Madden, Z. Man, M. Plummer, J. Tennyson and H. N. Varambhia, Euro. J. Phys. D, 66, 58 (2012).

[7] A. G. Harvey, D. S. Brambila, F. Morales, O. Smirnova J. Phys. B, 47, 215005, (2014)

Content from this work may be used under the terms of the Creative Commons Attribution 3.0 licence. Any further distribution of this work must maintain attribution to the author(s) and the title of the work, journal citation and DOI.
Published under licence by IOP Publishing Ltd

XXIX International Conference on Photonic, Electronic, and Atomic Collisions (ICPEAC2015) IOP Publishing
Journal of Physics: Conference Series **635** (2015) 112127 doi:10.1088/1742-6596/635/11/112127

Ultrafast time-resolved photoelectron spectroscopy of solvated systems

Inga Jordan*[1], **Martin Huppert***, **Sebastian Hartweg***, **Hans Jakob Wörner***[2]

* Laboratorium für physikalische Chemie, ETH Zürich, Vladimir-Prelog-Weg 2, 8093 Zürich, Switzerland

Synopsis Photoionization delays reveal electron dynamics on ultrashort time scales. Here, we present attosecond time-resolved experimental results of gas-phase and liquid water and we discuss the analysis methods to retrieve molecular ionization delays therefrom.

We describe the design and operation of an experiment based on an actively-stabilized interferometer and a magnetic-bottle-time-of-flight spectrometer which allows studying liquid systems by means of photoelectron spectroscopy with attosecond time resolution. Subjects of interest for our pump-probe studies are water, solvated salts and transition metal complexes as well as nanoparticles.

Using an attosecond pulse train generated in a semi-infinite gas cell and an infrared dressing field we are able to perform RABBIT [1] (reconstruction of attosecond beating by interference of two-photon transitions) experiments. The phases of the sideband oscillations are extracted from the RABBIT scan in order to calculate the ionization delays between the molecular orbitals as a function of the harmonic energy.

Ionization delays for non-overlapping sidebands are readily retrieved from a RABBIT trace by integrating the sideband photoelectron peak along the delay axis and performing a sine fit or Fourier transform. Such a situation can be met for gas-phase water by using a tin filter which isolates harmonic orders 11, 13 and 15 (Fig.1). The resulting ionization delay between HOMO (1b1) and HOMO-1 (3a1) for sideband order 14 is (8±6) as in our data. Using chrome or titanium metal filters, higher harmonic orders can be chosen in order to gain the ionization delays for more sideband orders. Additionally, we are working on a global fitting procedure for the disentanglement of more complicated photoelectron spectra generated by ionizing several valence orbitals with a high number of harmonic energies. This is expected to enable the analysis of sideband oscillations from overlapping photoelectron bands.

We will present photoelectron spectra from liquid and gas-phase water, solvated sodium iodide as well as RABBIT scans of gas-phase and liquid water. We expect these experiments to provide a more detailed understanding of the ultrafast electron dynamics in the liquid phase.

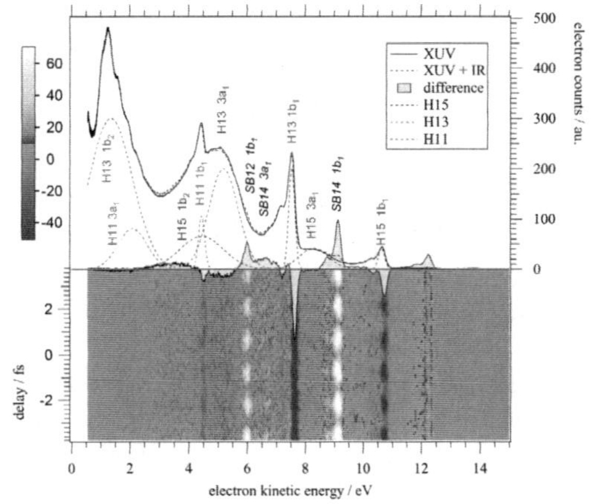

Figure 1. RABBIT trace and the photoelectron spectrum of gas-phase water generated with harmonic orders 11, 13 and 15. Ionization delays between HOMO and HOMO-1 are retrieved from the sideband oscillations.

References

[1] P. M. Paul *et al.* 2001 *Science* **292** 1689-1692

[1]E-mail: inga.jordan@phys.chem.ethz.ch
[2]E-mail: hansjakob.woerner@phys.chem.ethz.ch

Content from this work may be used under the terms of the Creative Commons Attribution 3.0 licence. Any further distribution of this work must maintain attribution to the author(s) and the title of the work, journal citation and DOI.
Published under licence by IOP Publishing Ltd

Tabletop imaging using 266nm femtosecond laser pulses, for characterization of structural evolution in, single molecule, chemical reactions

Heide Ibrahim[*1], Benji Wales[†], Samuel Beaulieu[*], Bruno E. Schmidt[*], Nicolas Thiré[*], Éric Bisson[*], Christoph T. Hebeisen[‡,♦], Vincent Wanie[*], Mathieu Giguére[*], Jean-Claude Kieffer[*], Michael Spanner[♦], André D. Bandrauk[♠], Michael S. Schuurman[‡], Joseph Sanderson[†2], François Légaré[*]

**Institut National de la Recherche Scientifique, Centre Énergie Matériaux et Télécommunications, 1650 Boulevard Lionel-Boulet, Varennes, Qc, J3X1S2, Canada*

[†]Department of Physics and Astronomy, University of Waterloo, 200 University Avenue West, Waterloo, ON N2L 3G1, Canada

[‡]National Research Council of Canada, 100 Sussex Dr., Ottawa, On, K1A 0R6, Canada

[♦]Department of Physics, University of Ottawa, 150 Louis Pasteur, Ottawa On, K1N 6N5, Canada

[♠]Laboratoir De Chimie Théorique, Université de Sherbrooke,Qc, Canada

Synopsis. We have demonstrated a generally applicable tabletop approach utilizing a 266nm femtosecond laser pulse pump, 800nm pulse probe, coupled with Coulomb explosion imaging (CEI). We have investigated two simple chemical reactions in $C_2H_2^+$ simultaneously: proton transfer and C=C bond-breaking, triggered by multiphoton ionization to excited states. Too and fro proton migration results are in excellent agreement with new ab initio trajectory simulations which predict isomerization timescales and pathways.

Sources of photon with high energy, such as the Free Electron Laser (FEL), [1] have been used to efficiently launch dynamics in excited states of molecules with a large energy gap between their ground and excited states. In our work [2] we show that multiphoton ionization with ultrashort UV laser pulses provides a powerful alternative to such VUV sources.

Acetylene (C_2H_2) has been serving as a model system for decades, because of its vinylidene isomer, nevertheless proton transfer dynamics in its cation is not yet fully elucidated and has recently attracted attention of both experimental and theory groups. Given its small size, it is an ideal candidate for the application of quantitative electronic structure theory, which can be brought to bear on its ground and excited states.

Experiments were carried out at the Advanced Laser Light Source ALLS (INRS-EMT, Varennes, Canada). We employ pump-probe CEI: A pump-pulse (266 nm, 32 fs, 3.2 μJ) ionizes the system to the cation and launches proton migration dynamics. A time delayed probe-pulse (800 nm, 40 fs, 44 μJ) further ionizes it to higher charged states. Its electric field stripes-off electrons almost immediately, leaving positively charged fragments behind which undergo Coulomb explosion. This gives CEI the ability to directly indicate geometrical structures. Ionic fragments resulting from CE are collected with a uniform-electric-field ion imaging spectrometer. Their full 3D momenta are retrieved using a time- and position-sensitive delay line detector at the end of the 23 cm ion time of flight. Fig. 1 shows how the species yield varies as a function of pump probe delay, indicating an isomerization time of 41fs.

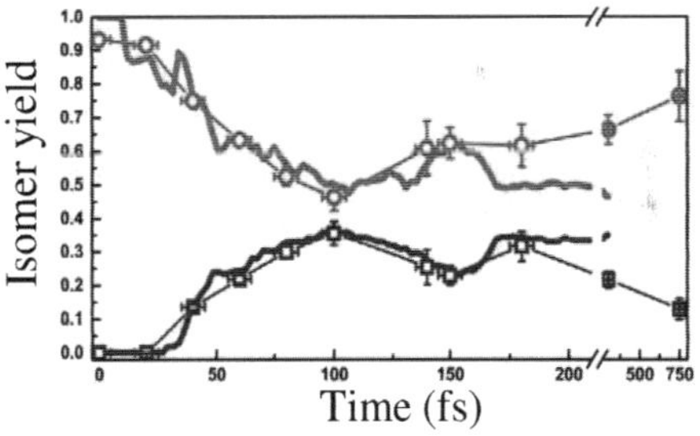

Figure. 1 Comparison of theory (thick lines) and experiment (open symbols) for acetylene (red) and vinylidene (blue) yield: Theoretical curves show the superposition of acetylene and vinylidene yield in both the ground state and excited states of the cation as a function of time

References

[1] Y. H. Jiang *et al* 2010 *Phys. Rev. Lett.* **105**, 263002.

[2] H. Ibrahim *et al* 2014 *Nature Communications* **5**, 4422

[3] B. Wales et al 2012 *J. Phys. B.* **45** 045205

E-mail: ibrahim@emt.inrs.ca

E-mail: j3sander@uwaterloo.ca

Content from this work may be used under the terms of the Creative Commons Attribution 3.0 licence. Any further distribution of this work must maintain attribution to the author(s) and the title of the work, journal citation and DOI.
Published under licence by IOP Publishing Ltd

Journal of Physics: Conference Series **635** (2015) 112129 doi:10.1088/1742-6596/635/11/112129

Theoretical Study of Ultrafast Electron Dynamics in Amino Acids

D. Ayuso*[1], A. Palacios*, P. Decleva†, F. Martín*‡[2]

* Departamento de Química, Módulo 13,Universidad Autónoma de Madrid, Cantoblanco 28049, Madrid, Spain
† Dipartimento di Scienze Chimiche e Farmaceutiche, Università di Trieste and CNR-Istituto Officina dei Materiali, 34127 Trieste, Italy
‡ Instituto Madrileño de Estudios Avanzados en Nanociencia, Cantoblanco 28049 Madrid, Spain

Synopsis Prompt ionization of large biological molecules may induce ultrafast charge migration along the molecular skeleton, preceding any nuclear rearrangement[1, 2]. This phenomenon has been recently observed in the amino acid phenylalanine [3] in a two-color pump probe experiment, where the production of ionic fragments was measured as a function of the time delay between the two pulses and charge fluctuations manifested as sub-4.5 fs oscillations in the quantum yield of a specific doubly charged fragment. We present our latest results in glycine, and compare with previous findings in phenylalanine [3]. We seek to perform a systematic study including larger aminoacids such as tryptophan.

The development of attosecond technology has enabled the real-time observation of electron motion in atoms, molecules and condensed matter [4]. In the case of large biological systems, such as amino acids, sudden ionization from a molecular orbital may trigger an ultrafast electronic response which is solely driven by electron correlation, as predicted in various theoretical works [1, 2]. However, due to the large bandwidth of attosecond pulses employed in state-of-the-art experiments [3, 4], electrons are usually removed not from one but from many orbitals, thus leading to a coherent superposition of ionic states. Therefore, an accurate description of the ionization process becomes crucial in the description of these experimental observations.

We have performed a theoretical study of the field-free evolution of the electronic wave packet generated in amino acids (glycine, phenylalanine and tryptophan) upon the interaction with with attosecond pulses. The ionization amplitudes have been quantitatively determined for all open channels using the static-exchange DFT method [5] developed by Decleva and collaborators, which makes use of the Kohn-Sham DFT to describe molecular bound states and of the Galerkin approach to evaluate continuum wave functions in the field of the corresponding Kohn-Sham density. The evolution of the electronic wave packet has then been described by using a standard time-dependent density matrix formalism. The results of the numerical simulations clearly show the production of an ultrafast electronic response occurring in a sub-femtosecond time scale and are in good agreement with recent experimental observations [3]. We notice that, in contrast with most previous works, most valence and inner-valence electrons are efficiently ionized and therefore the observed dynamics is that of a delocalized hole. We have further analyzed the role of the emitted electron in the hole migration mechanism and observed that even faster charge fluctuations (faster than the experimental resolution) might arise as a consequence of the interplay between the ground and continuum states.

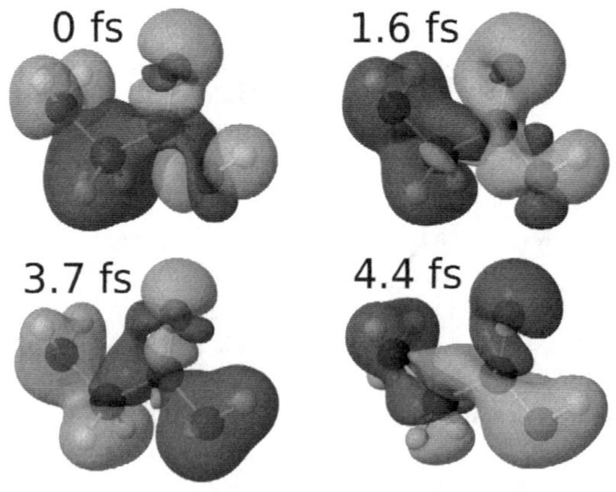

Figure 1. Relative variation of the hole density generated by an attosecond XUV pulse in the amino acid glycine with respect to its time-averaged value (yellow: 10^{-4}a.u., purple: -10^{-4}a.u.). Time is refereed to the end of the pulse.

References

[1] L. S. Cederbaum *et al* 1999 *Chem. Phys. Lett.* 307, 205-210.

[2] F. Remacle *et al* 2006 *P.N.A.S.* 103, 6793-6798.

[3] F. Calegari *et al* 2014 *Science* 346, 6207 336-339.

[4] F. Krausz *et al* 2009 *Rev. Mod. Phys.* 81, 163.

[5] M. Stener *et al* 2002 *Chem. Phys. Lett.* 351 469.

[1]E-mail: david.ayuso@uam.es
[2]E-mail: fernando.martin@uam.es

Content from this work may be used under the terms of the Creative Commons Attribution 3.0 licence. Any further distribution of this work must maintain attribution to the author(s) and the title of the work, journal citation and DOI.
Published under licence by IOP Publishing Ltd

Excitation-energy resolved fluorescence spectra of hydrogen molecules in the regime of singly excited molecular states

Philipp Schmidt*, Andreas Hans*, Christian Ozga*, Philipp Reiß*, Ltaief Ben Ltaief*, Kouichi Hosaka*, Masashi Kitajima†, Noriyuki Kouchi†, André Knie*, and Arno Ehresmann* [1]

* Institut für Physik and CINSaT, Universität Kassel, Heinrich-Plett-Straße 40, 34132 Kassel, Germany
† Department of Chemistry, Tokyo Institute of Technology, Meguro-ku, Tokyo 152, Japan

Synopsis We have measured the emission spectrum of gaseous hydrogen molecules in the far ultraviolet spectral range over the large interval of excitation energies between 11 eV and 17 eV. This regime is governed by singly excited electronic states below the ionization threshold. While these features have been already investigated separately in the past, the completeness of the present measurements allows a very intuitive representation of the participating molecular effects and their respective correlations. It also contains the first complete investigation of the Condon diffraction bands visible for hydrogen through transitions into the ground state continuum.

The hydrogen molecule is the most abundant molecule in the universe and has always been one of the most fundamental prototype systems in the field of molecular physics and spectroscopy in general. Precise knowledge of its characteristics is essential to most astrophysical observations and calculations [1].

At the same time, even the regime of singly excited states of this diatomic molecule is still a field of active work due to its complicated crossing of a variety of individual states [2,3,4]. By applying the method of photon-induced fluorescence spectroscopy (PIFS), where dispersed fluorescence is measured as a function of photon energy upon excitation by monochromatized synchrotron radiation, over a large interval of excitation energies, we were able to provide a complete compilation of the emission features in this regime. These include molecular fluorescence with vibrational progressions, predissociation resonances through various coupled electronic states as well as continuous emissions into the continuum of the X ground state. While these effects are already known individually, this kind of overview allows a much deeper understanding of their respective relations. It is also the first complete investigation of those continuum emissions and their intensity oscillations known as Condon diffraction bands [5, 6].

The comparison of the present data to readily available and ever improving theoretical data allows a concise evaluation of the corresponding theoretical methods.

In the future the present experimental technique can be used to deliver benchmark data for potential energy landscape calculations for species where such theoretical data is not available from first principles. Due to the excitation state selective nature of the method, it can separate the individual features already with only modest resolving power.

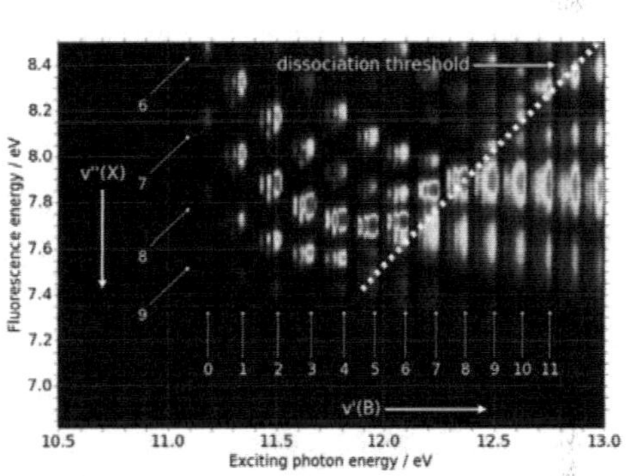

Figure 1. Part of the excitation-energy resolved fluorescence spectra showing vibrational progressions from the B state of hydrogen to the ground state (Lyman bands) from various upper and lower vibrational levels as well as continuum emissions beyond the dissociation threshold.

Acknowledgement: This work was supported by the Hesse-State Initiative for the Development of Scientific and Economic Excellence (LOEWE) in the LOEWE-Focus project "ELCH".

References

[1] A. Broadfoot *et al 1979 Science* ***204*** *979*
[2] L. Wolniewcz *et al 2006 J Mol Spec* ***238*** *118*
[3] M. Glass-Maujean *et al 2007 J Chem Phys* ***126****, 144303*
[4] M. Glass-Maujean *et al 2007 J Phys B* ***40*** *F19*
[5] E. Condon *1928 Phys Rev* ***32*** *858*
[6] H. Schmoranzer *et al 1978 Phys Rev A* ***18*** *1472*

[1] E-mail: Ehresmann@physik.uni-kassel.de

Content from this work may be used under the terms of the Creative Commons Attribution 3.0 licence. Any further distribution of this work must maintain attribution to the author(s) and the title of the work, journal citation and DOI.
Published under licence by IOP Publishing Ltd

Journal of Physics: Conference Series **635** (2015) 112131 doi:10.1088/1742-6596/635/11/112131

XUV induced hydrogen migration in 5-halouracil

M.C. Castrovilli[1,*], A. Trabattoni[†], P.Bolognesi[§], P. O'Keeffe[§], L. Avaldi[§], M. Nisoli[†,*], F. Calegari[*] and R. Cireasa[#]

[*]CNR-IFN, Piazza Leonardo da Vinci 32, 20133 Milano, Italy
[†]Dipartimento di Fisica, Politecnico di Milano, Milano, Italy
[§]CNR-ISM, Area della Ricerca di Roma 1, Monterotondo Scalo, Italy
[#]Institut des Sciences Moléculaires d'Orsay, Université Paris XI, Orsay, France

Synopsis This work presents a study of the XUV-induced ultrafast molecular dynamics of 5XU conducted with unprecedented temporal resolution. A clear dynamics occurring in 35 fs for both 5FU and 5BrU has been identified and assigned to the hydrogen migration from the C6 site to the C5 site of the molecule. Since 5XU are routinely used as radio sensitizers, the present work provides new important insights in the XUV induced structural changes responsible for the DNA breaking when these molecules are embedded in the DNA chain for radiotherapy.

The interaction of high-energy photons with complex biomolecules is responsible for many chemical changes the molecule undergoes leading to mutations and ultimately damage. The substitution of the H atom at position 5 of the uracil ring with a halogen atom allows to obtain the 5-Halouracils (5XU). When the DNA of a tumor cell is modified by replacing the thymine nucleobase with 5XU, the cell becomes more sensitive to lethal effects of XUV, X-ray and proton radiation [1]. For this reason 5XU molecules such as 5FU are routinely used as radiosensitizers in tumor radiotherapy.

Here we present a time-resolved study of the XUV-induced ultrafast molecular dynamics of 5XU conducted with unprecedented temporal resolution. Charge dynamics in 5XU was initiated by a short train (sub-2fs) of attosecond pulses, with photon energies in the spectral range between 17 eV and 35 eV, obtained by high-order harmonic generation in krypton. The subsequent dynamics was probed by 4-fs, waveform-controlled visible (NIR-VIS) pulses [2,3].

Time-of-Flight (TOF) spectra for the photofragments of 5FU and 5BrU were recorded using a Velocity Map Spectrometer as a function of the delay between the XUV pump pulse and the NIR-VIS probe pulse. The time dependent yield of most of the fragments displayed clear femtosecond dynamics. In the case of 5FU, the XUV pulse ionizes the molecule catalysing the most probable hydrogen migration from C6 to C5 at the first step of the reaction, leading to the fragment m/z=44, C6C5FH (Fig. 1a) and m/z=32 whose unique assignment is the $HC5F^+$ structure. The pump-probe measurement also reveals that at early times, the NIR-VIS pulse inhibits the hydrogen migration process thus resulting in a sudden reduction of fragments m/z=44 and m/z=32, and a consequent increase of fragments m/z=43 C6C5F (Fig. b) and m/z=31 (C5H). As can be observed in Fig. 1, the rising (decaying) dynamics associated to fragment m/z=44 (m/z=43) occurs with a time constant of about 35 fs which can be related to the time required by the hydrogen atom to migrate from the C6 site to the C5 site of the molecule. Similar behaviours were observed for the equivalent 5BrU fragments.

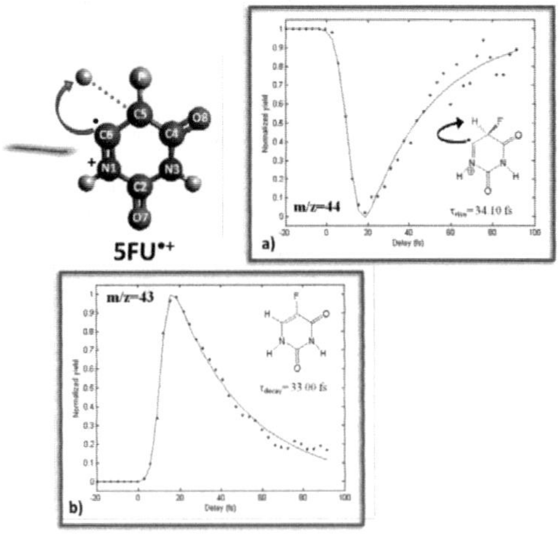

Figure 1. Time dependent yield for fragments a) m/z=44 and b) m/z=43 of 5FU.

References

[1] T.J. Kinsella et al., *Jr., Int. J. Radiat. Oncol., Biol., Phys.* 13, 733 (1987).
[2] L. Belshaw et al., *J. Phys. Chem. Lett.* 3, 3751 (2012).
[3] F. Calegari et al., *Science* 346, 336-339 (2014).

[1] E-mail: matteac@gmail.com

Content from this work may be used under the terms of the Creative Commons Attribution 3.0 licence. Any further distribution of this work must maintain attribution to the author(s) and the title of the work, journal citation and DOI.
Published under licence by IOP Publishing Ltd

XXIX International Conference on Photonic, Electronic, and Atomic Collisions (ICPEAC2015) IOP Publishing
Journal of Physics: Conference Series **635** (2015) 112132 doi:10.1088/1742-6596/635/11/112132

Experimental study of the ion-molecule reaction $H_2^+ + H_2 \rightarrow H_3^+ + H$ at low collision energies

P. Allmendinger [1], **J. Deiglmayr, F. Merkt**

Laboratory of Physical Chemistry, ETH Zurich, Switzerland

Synopsis We present a new method to study ion-molecule reactions at low temperatures or low collision energies and illustrate its application to the study of the $H_2^+ + H_2 \rightarrow H_3^+ + H$ reaction.

H_2 molecules in a skimmed supersonic beam are excited to nonpenetrating $n = 30$, $M_J = 3$ Rydberg-Stark states belonging to series converging to a selected rotational level ($N^+ = 0, 1, 2$ or 3) of the $X^+\,{}^2\Sigma_g^+$ ($v^+ = 0$) ground state of H_2^+. The beam of Rydberg molecules is then deflected and velocity selected using a bent surface decelerator [1] and merged with another supersonic beam of ground state H_2. In $n = 30$, $M_J = 3$ Rydberg Stark states, the Rydberg electron is completely decoupled from the H_2^+ ion core motion and acts as a spectator when the H_2^+ ion core reacts with an H_2 molecule located within the Rydberg-electron orbit [2]. The experimental apparatus (see Figure 1) has been conceived so that the relative velocities of the two beams can be adjusted, making the study of the reaction at collision energies below 10 meV possible.

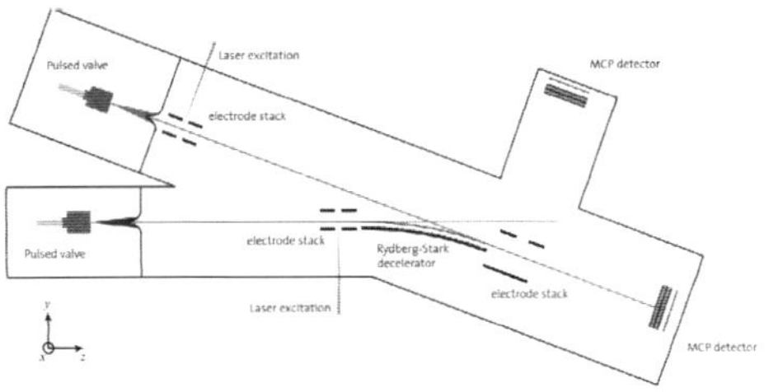

Figure 1. Schematic representation of the experimental arrangement used to study the $H_2^+ + H_2 \rightarrow H_3^+ + H$ reaction at low and tunable collision energies.

The reaction products are monitored on an imaging multichannel plate detector following extraction with a pulsed electric field. The information contained in the ion time of flight (TOF) distributions and in the distributions of H_3^+ product ions on the images are used to determine the relative cross section in dependence of the collision energy and the selected rotational state of the H_2^+ ion core. A typical TOF spectrum showing H_3^+ ions formed from the reaction of H_2^+ ($v^+ = 0$, $N^+ = 0$) at a collision energy of 6 meV is displayed in Figure 2 as illustration.

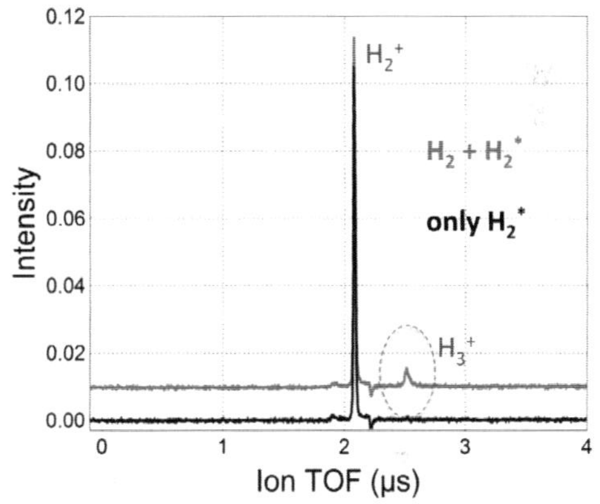

Figure 2. Upper trace: TOF spectrum showing H_2^+ ions (main peak) produced by pulsed field ionization of the initially prepared $n = 30$, $M_J = 3$ Rydberg states and H_3^+ ions resulting from the $H_2^+ + H_2 \rightarrow H_3^+ + H$ reaction at 6 meV collision energy.

The results of a systematic investigation of the $H_2^+ + H_2$ reaction at low collision energies will be presented.

References

[1] P. Allmendinger, J. Deiglmayr, J. A. Agner, H. Schmutz, and F. Merkt, 2014 *Phys. Rev. A.* **90** 043403

[2] P.M. Dehmer, W. A. Chupka, 2005 *J. Phys. Chem.* **99** 1686

[1] E-mail: p.allmendinger@phys.chem.ethz.ch

Content from this work may be used under the terms of the Creative Commons Attribution 3.0 licence. Any further distribution of this work must maintain attribution to the author(s) and the title of the work, journal citation and DOI.
Published under licence by IOP Publishing Ltd

Lyman series emission of valence and inner-shell excited gaseous H_2O

Andreas Hans*, Philipp Schmidt*, Marko Förstel†, Uwe Hergenhahn†, Ltaief Ben Ltaief*, Henning Huckfeldt*, Christian Ozga*, Philipp Reiß*, André Knie*, and Arno Ehresmann*[1]

*Universität Kassel, Institut für Physik, Heinrich-Plett-Straße 40, 34132 Kassel, Germany
† Max-Planck-Institut für Plasmaphysik, Wendelsteinstraße 1, 17491 Greifswald, Germany

Synopsis We present emission cross section functions for Lyman series fluorescence emission after photoexcitation of water molecules with monochromatized synchrotron radiation as function of the exciting-photon energy. Both the valence and inner-shell excitation energy range of the water molecule are covered. The results are compared to literature and contribute to complete the model of dissociative relaxation dynamics of excited small molecules.

Its unique role in nature made the water molecule always subject to intense investigations of theoretical and experimental research. Of particular interest, e.g. in atmosphere physics and chemistry, radiation biology and fundamental photochemistry, is the interaction with radiation and subsequent decay processes such as photoexcitation, -ionization, and -dissociation.

The tracking of neutral fragments is challenging, because they cannot be guided by electric or magnetic fields as it is the case for ionic particles. However, if they are excited, fluorescence spectrometry is suitable for their investigation. The Balmer emission series as well as Ly_α after photoexcitation of water molecules have been measured as functions of the exciting-photon energy in previous experiments and absolute emission cross sections are available for some exciting-photon energy ranges [1,2,3].

However, for the completion of models of dissociation dynamics, measurements of the emission of higher Lyman transitions ($\geq Ly_\beta$) are essential, as the Lyman decay is the dominant channel for all excited hydrogen fragments.

Here, we present experimentally determined cross sections for Lyman series emission as function of the exciting-photon energy. The experiments were performed with tunable synchrotron radiation in the range of valence excitations of the water molecule (15 – 34 eV) as well as for excitation in the oxygen inner-shell (1s) electron range (533 – 542 eV). In the valence energy range, thermodynamical limits of production of excited hydrogen fragments are measured and agree with previous results. Above those limits, absolute emission cross sections are determined by calibrating to literature values (see Fig. 1). Oscillator strengths at emission maxima are calculated and branching ratios between Lyman and Balmer transitions are determined and compared to theoretical values.

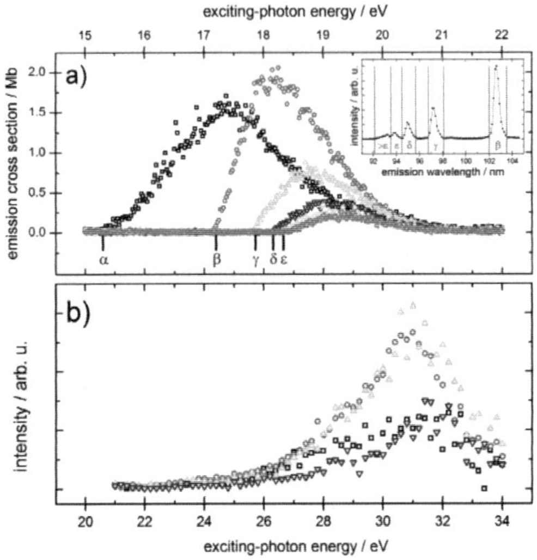

Figure 1. a) Exciting-photon energy dependent fluorescence emission cross sections: a) Absolute Lyman series emission cross sections after excitation of $1b_2^{-1}$ states of the water molecule. Black squares: Ly_α, red circles: Ly_β, green triangles: Ly_γ, blue triangles: Ly_δ, cyan diamonds: Ly_ε, magenta triangles $Ly_{>\varepsilon}$. b) Relative Lyman series emission cross sections after excitation into doubly excited and $2a_1^{-1}$ states.

In the inner-shell excitation range, the emission function is found to be strongly dependent on the character of the excitation, as observed before for the Balmer series [3].

Acknowledgement: This work was supported by the Hesse-State Initiative for the Development of Scientific and Economic Excellence (LOEWE) in the LOEWE-Focus project "ELCH".

References

[1] M. Kato *et al* 2004 *J. Phys. B* **37** 3127
[2] A. Kivimäki *et al* 2006 *J. Phys. B* **39** 1101
[3] E. Melero García *et al* 2006 *Phys. Rev. Lett.* **96** 063003

[1] E-mail: ehresmann@physik.uni-kassel.de

Content from this work may be used under the terms of the Creative Commons Attribution 3.0 licence. Any further distribution of this work must maintain attribution to the author(s) and the title of the work, journal citation and DOI.
Published under licence by IOP Publishing Ltd

Vibrationally resolved B 1s photoionization cross section of BF_3

D. Ayuso*[1], M. Kimura†, K. Kooser‡, M. Patanen¶, E. Plesiat*⊥, L. Argenti*, S. Mondal†, O. Travnikova¶§, K. Sakai†, A. Palacios*, E. Kukk‡, P. Decleva△, K. Ueda†, F. Martín*□, C. Miron¶

* Departamento de Química, Módulo 13, Universidad Autónoma de Madrid, Cantoblanco 28049, Madrid, Spain
† Institute of Multidisciplinary Research for Advanced Materials, Tohoku University, Sendai 980-8577, Japan
‡ Department of Physics and Astronomy, University of Turku, 20014 Turku, Finland
¶ Synchrotron SOLEIL, L'Orme des Merisiers, Saint-Aubin, BP 48, 91192 Gif-sur-Yvette Cedex, France
§ CNRS, UMR 7614, UPMC Univ Paris 06, LCPMR, 75005 Paris, France
⊥ Institut fur Physik, Humboldt-Universitat zu Berlin, Newtonstrasse 15, 12489 Berlin, Germany
△ Dipartimento di Scienze Chimiche e Farmaceutiche, Università di Trieste, 34127, and CNR-IOM, Trieste, Italy
□ Instituto Madrileño de Estudios Avanzados en Nanociencia, Cantoblanco, 28049 Madrid, Spain

Synopsis We present a study of the vibrationally resolved B 1s photoionization cross section of the BF_3 molecule. A combination of high-resolution photoelectron spectroscopy measurements and of state-of-the-art calculations shows the evolution of the photon energy dependence of the cross section from a complete trapping of the photoelectron wave (low energies) to oscillations due to intramolecular scattering [1, 2]. These diffraction patterns allow to access structural information of both the neutral molecule and the core-hole species generated upon photoabsoption [3].

The advent of third-generation synchrotron radiation facilities, in combination with high energy-resolution detection techniques, has opened the way for the investigation of vibrationally-resolved inner-shell photoionization in small molecules, where an electron is emitted from a 1s orbital of a first-row atom. We present a study on B 1s photoionization of BF_3, comparing experimental results with full first-principle calculations and showing that the most relevant features can be understood by means of simple models.

In order to calculate vibrationally-resolved cross sections, we have evaluated (bound and continuum) electronic wave functions using the static-exchange and the time-dependent DFT methods [4], developed by Decleva and collaborators, for different molecular geometries along the totally symmetric stretching mode. This mode is the most affected by the structural rearrangement accompanying core ionization [1, 2]. Our theoretical results are in good agreement with experimental measurements from SOLEIL synchrotron and, at high energies, qualitatively agree with a first-Born approximation model.

We have found that the relative cross sections show clear oscillations in the high-energy region as a function of photoelectron momentum which are due to an intramolecular scattering mechanism: in its way out of the molecule, the photoelectron is diffracted by the surrounding atomic centers, encoding the geometry of the molecule [1, 2, 3]. Very close to the photoionization threshold a complete trapping of the photoelectron is observed, manifesting itself as an emission angle dependent shape resonance feature [1, 2].

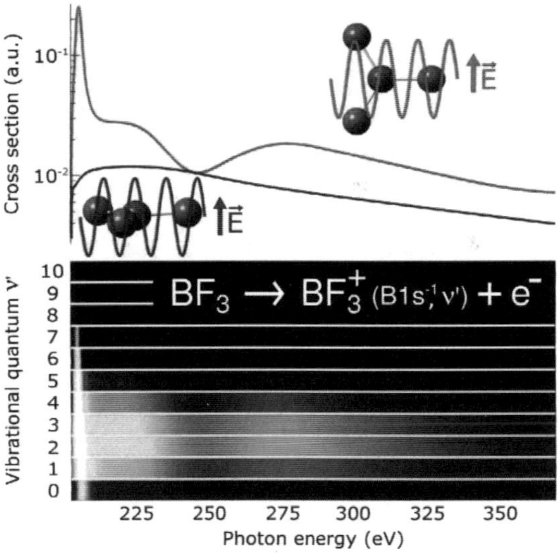

Figure 1. Calculated cross section (upper panel) for different molecular orientations and vibrationally resolved cross section in the totally symmetric stretching mode (lower panel) for the case of randomly oriented molecules.

References

[1] D. Ayuso *et al* 2015 *J. Phys. Chem. A* (Accepted)
[2] K. Ueda *et al* 2013 *J. Chem. Phys.* 139, 124306
[3] E. Kukk *et al* 2013 *Phys. Rev. A* 88, 033412
[4] H. Bachau *et al* 2001 *Rep. Prog. Phys.* 64, 1815

[1] E-mail: david.ayuso@uam.es

Content from this work may be used under the terms of the Creative Commons Attribution 3.0 licence. Any further distribution of this work must maintain attribution to the author(s) and the title of the work, journal citation and DOI.
Published under licence by IOP Publishing Ltd

Journal of Physics: Conference Series **635** (2015) 112135 doi:10.1088/1742-6596/635/11/112135

Resolving the dynamics of valence-shell electrons and nuclei through laser-induced diffraction and holography

Samuel G. Walt*[1], N. Bhargava Ram*, Aaron von Conta*, Denitsa Baykusheva*, Marcos Atala*, Hans Jakob Wörner*[2]

* Laboratorium für Physikalische Chemie, ETH Zürich, 8093 Zürich, Switzerland

Synopsis We have studied a coupled electronic-nuclear wave packet in nitric oxide using time-resolved strong-field photoelectron holography and rescattering. We show that the electronic dynamics mainly appears in the holographic structures whereas nuclear motion strongly modulates the angular distribution of the rescattered photoelectrons.

The interaction of intense laser pulses with molecules has opened several new avenues for imaging ultrafast dynamics. Strong-field photoelectron holography has been used to measure static atoms [1, 2] and molecules [3] whereas rescattering has additionally revealed first evidence of nuclear motion [4]. Here, we transpose the powerful methods of photoelectron holography and rescattering to observing electronic dynamics and its coupling to nuclear dynamics.

First, we show that the electronic dynamics modulate the contrast of the holographic pattern which is found to probe the lateral-momentum distribution of the photoelectron wave packet at the instant of ionization. Second, the electronic dynamics is shown to neither modify the structure of the holographic pattern nor the angular distribution of the rescattered electrons, providing a clean separation of nuclear dynamics occurring simultaneously to the electronic dynamics. Third, owing to this separation, the nuclear dynamics is recorded by rescattering in a background-free manner.

We use impulsive stimulated Raman scattering (ISRS) to prepare a coupled electronic and rotational wave packet in the neutral NO molecule. The wave packet is probed by strong-field ionization and the photoions and photoelectrons are detected by a velocity-map-imaging spectrometer. An overview of the experimental results is given in Fig. 1. Figure 1a shows the normalized NO^+ yield as a function of the pump-probe delay. The rapid regular oscillation originates from the electronic wave packet whereas the two features around 5 and 10 ps correspond to the quarter and half rotational revivals, respectively. Figure 1b shows a cut through the three-dimensional photoelectron momentum distribution recorded at a delay of 1.56 ps. Figures 1c and 1d show normalized differences of photoelectron momentum distributions recorded at maxima and minima of the photoelectron (or, equivalently, NO^+) yield. The electronic dynamics (Fig. 1c) appears as a pronounced modulation of a holographic pattern whereas the nuclear dynamics manifests itself by modifying the angular distribution of the rescattered photoelectrons.

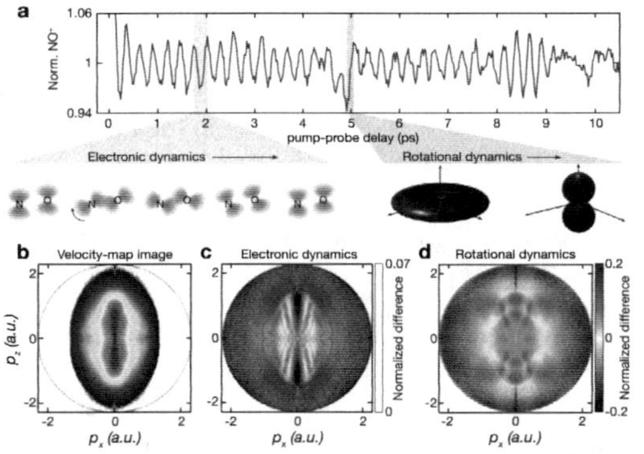

Figure 1. Resolving electronic and nuclear dynamics through holography and rescattering. (**a**) Time-dependent normalized NO^+ signal and illustration of the prepared electronic and rotational dynamics, (**b**) photoelectron momentum distribution of excited NO molecules recorded at a delay of 1.56 ps, (**c**) normalized difference of momentum distributions at maxima and minima of the NO^+ signal dominated by electronic dynamics (1.55 and 1.72 ps, respectively), (**d**) normalized difference of momentum distributions at maxima and minima of the NO^+ signal dominated by rotational dynamics (4.92 and 1.50 ps, respectively).

References

[1] Y. Huismans *et al.* 2011 *Science* **331.** 61

[2] D. D. Hickstein *et al.* 2012 *Phys. Rev. Lett.* **109.** 073004

[3] M. Meckel *et al.* 2014 *Nat. Phys.* **10.** 594

[4] C. I. Blaga *et al.* 2012 *Nature* **483.** 194

[1]E-mail: samuel.walt@phys.chem.ethz.ch
[2]E-mail: hansjakob.woerner@phys.chem.ethz.ch

Content from this work may be used under the terms of the Creative Commons Attribution 3.0 licence. Any further distribution of this work must maintain attribution to the author(s) and the title of the work, journal citation and DOI.
Published under licence by IOP Publishing Ltd

XXIX International Conference on Photonic, Electronic, and Atomic Collisions (ICPEAC2015) IOP Publishing
Journal of Physics: Conference Series **635** (2015) 112136 doi:10.1088/1742-6596/635/11/112136

Attosecond charge migration and its laser control

P. M. Kraus[1*], B. Mignolet[2], D. Baykusheva[1], A. Rupenyan[1], L. Horný[1], E. F. Penka[3], O. I. Tolstikhin[4], J. Schneider[1], F. Jensen[5], L. B. Madsen[6], A. D. Bandrauk[3], F. Remacle[2] and H. J. Wörner[1†]

[1]Laboratorium für Physikalische Chemie, ETH Zürich, 8093 Zürich, Switzerland
[2]Department of Chemistry, B6c, Université de Liège, B4000 Liège, Belgium
[3]Laboratoire de Chimie Théorique, Université de Sherbrooke, Québec, Canada, J1K 2R1
[4]Moscow Institute of Physics and Technology, Dolgoprudny 141700, Russia
[5]Department of Chemistry, Aarhus University, 8000 Aarhus C, Denmark
[6]Department of Physics and Astronomy, Aarhus University, 8000 Aarhus C, Denmark

Synopsis: We advance high-harmonic spectroscopy to resolve molecular charge migration in time and space and simultaneously demonstrate extensive control over the process. A multidimensional approach enables us to reconstruct both quantum amplitudes and phases with a resolution of better than 100 attoseconds and to separately reconstruct field-free and laser-driven charge migration. Our techniques make charge migration in molecules measurable on the attosecond time scale and open new avenues for laser control of electronic primary processes.

The spatial migration of an electron hole following excitation or ionization is a fundamental process in nature and engineering. Charge migration controls the regioselectivity of chemical reactions, energy transport in biological systems and underlies molecular electronics. Moreover, charge migration following ultrafast ionization has been predicted to provide direct information on electron correlation in molecules [1]. The extreme spatio-temporal scales of charge migration pose major challenges to the latest techniques of time-domain spectroscopy.

Here, we experimentally study charge migration on the attosecond time scale and demonstrate laser control over the dynamics. We develop high-harmonic spectroscopy (HHS) further to perform a direct reconstruction of time-dependent populations and phases of electronic eigenstates from measured amplitudes and phases of the harmonic radiation with a resolution of about 100 as. Photorecombination dipole moments and the angular variation of strong-field ionization rates of the relevant eigenstates of the cation are the necessary input in the reconstruction. This procedure allows us to characterize field-free charge migration in spatially-oriented iodoacetylene (ICCH) molecules as depicted in the picture below.

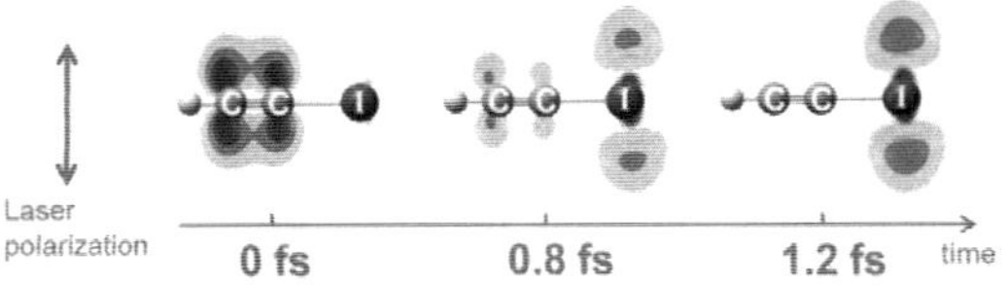

We further demonstrate extensive laser control over the charge migration by controlling the orientation of the molecules [2,3], which enables us to separately reconstruct field-free and laser-driven charge migration. The laser-driven charge migration fundamentally differs from the field-free process. We find significant population transfer between the two relevant electronic eigenstates for both investigated wavelengths (800 nm and 1300 nm) as depicted below. This is mediated by a strong dipole coupling between the two lowest-lying eigenstates of the cation.

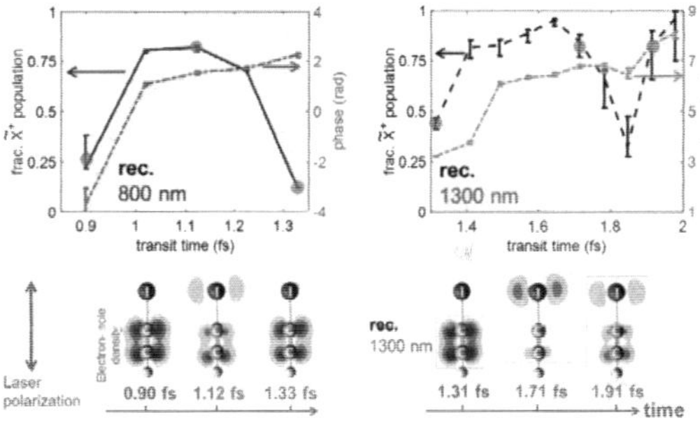

Our results enable the direct reconstruction of electronic state populations. They thus constitute a unique experimental access to the rich variety of field-free charge migration in molecules. Moreover, laser-control over charge migration in molecular systems opens new perspectives for controlling chemical reactions on the electronic time scale [4].

References

[1] L. Cederbaum, J. Zobeley, *Chemical Physics Letters* **307**, 205 (1999).
[2] P. M. Kraus, A. Rupenyan, H. J. Wörner, *Phys. Rev. Lett.* **109**, 233903 (2012)
[3] P. M. Kraus, D. Baykusheva, H. J. Wörner, *Phys. Rev. Lett.* **113**, 023001 (2014)
[4] F. Remacle, R. D. Levine, *PNAS* **103**, 6793 (2006).

*E-mail: krauspe@ethz.ch
†E-mail: hwoerner@ethz.ch

Content from this work may be used under the terms of the Creative Commons Attribution 3.0 licence. Any further distribution of this work must maintain attribution to the author(s) and the title of the work, journal citation and DOI.
Published under licence by IOP Publishing Ltd

Journal of Physics: Conference Series **635** (2015) 112137 doi:10.1088/1742-6596/635/11/112137

Dissociative ionization dynamics of triatomic molecules induced by soft X-rays

Ali Ramadhan[*1], **Benji Wales**[*], **Isabelle Gauthier**[†], **Mike MacDonald**[†], **Lucia Zuin**[†], **Joe Sanderson**[*2]

[*] Department of Physics, University of Waterloo, 200 University Avenue West, Waterloo, Ontario, Canada N2L 3G1
[†] Canadian Light Source, 44 Innovation Boulevard, Saskatoon, Saskatchewan, Canada S7N 2V3

Synopsis Soft X-rays (50eV to 170eV) have been used in conjunction with a triple coincidence time and position sensitive detection apparatus to observe the Coulomb explosion of OCS and CS_2. By varying the X-ray energy we can access ionization channels which result in two or three body break up, for 3+, such as $O^++C^++S^+$ and for 4+ such as $S^++C^++S^{2+}$. We can compare the kinetic energy release to the value calculated for a purely Coulombic break up from near equilibrium. The extent to which channels are concerted or stepwise is also determined.

The single ionization of molecules by X-ray excitation of core electrons can lead to multiple ionization of valence electrons through the processes of Auger decay [1]. The resultant highly charged molecule will dissociate into predominantly ionic fragments due to ionic repulsion. The momentum imparted to the fragments is measurable and carries with it an imprint of the molecular geometry, provided the interaction potential is known. Recent studies of OCS have investigated the nature of molecular ionization via high-intensity few cycle femtosecond laser pulses [2] and highly charged ion impact [3]. In both cases time and position sensitive detection techniques were used along with TOF (Time of Flight) in order to measure ion fragments in coincidence. In this work the same method is used to study OCS and CS_2 ionized by a range of low energy X-rays (50-173eV) from the VLS-PGM beamline at the Canadian Light Source (CLS) [4].

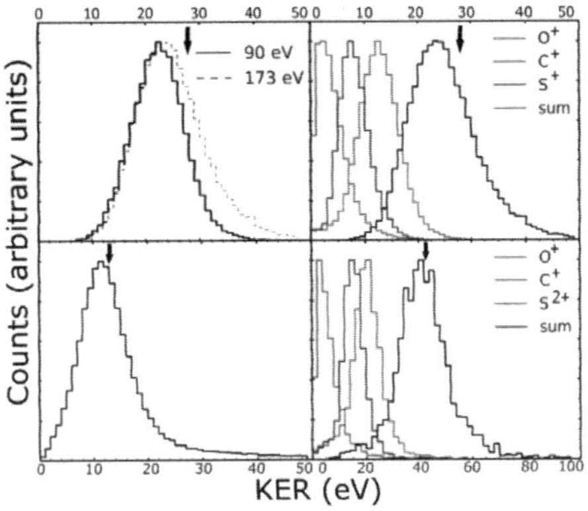

Figure 1. Energy distributions of OCS^{3+} fragmenting into (1,1,1) at 90eV and 173eV (top left), each ion of OCS^{3+} fragmenting into (1,1,1) at 173eV (top right), OCS^{2+} fragmenting into S^{2+}+CO^+ at 173eV (bottom left), and of each ion of OCS^{4+} fragmenting into (1,1,2) at 173eV. Coulombic energies produced by each fragmentation are indicated by arrows.

The resulting fragmentation dynamics of the OCS^{3+} and $CS_2{}^{3+}$ channels show dependence on the photon energy, as shown in Figure 1 with 90eV giving rise to a limited energy release in the (1,1,1) channel for OCS, compared to 173eV. We can interpret this in terms of the production of photo electron or Auger electron followed by knock off of further electrons.

By presenting our data in a Dalitz plot in Figure 2 we can see the contributions from the concerted process (x=0.1, y=-0.3) and the two stepwise break-up channels for OCS^{3+}.

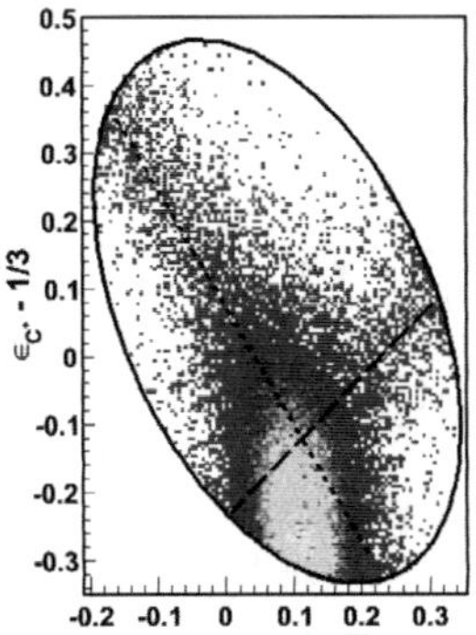

Figure 2. Dalitz plot of OCS^{3+} fragmenting into (1,1,1) channel via 173eV X-ray ionization. Concerted fragmentation makes up the bulk of the image while stepwise processes into metastable CO^{2+} and CS^{2+} are indicated by dotted and dashed lines respectively.

References

[1] T. Masuoka *et al* 1993 *Phys. Rev. A* **47** 278
[2] B. Wales *et al* 2014 *J. Elec. Spec. Rel. Phen.* **195** 332
[3] B. Wales *et al* 2012 *J. Phys. B* **45** 045205
[4] Y.F. Hu *et al* 2007 *Rev. Sci. Instrum.* **78** 083109

E-mail: ali.ramadhan@uwaterloo.ca
E-mail: j3sander@uwaterloo.ca

Content from this work may be used under the terms of the Creative Commons Attribution 3.0 licence. Any further distribution of this work must maintain attribution to the author(s) and the title of the work, journal citation and DOI.
Published under licence by IOP Publishing Ltd

XXIX International Conference on Photonic, Electronic, and Atomic Collisions (ICPEAC2015) IOP Publishing
Journal of Physics: Conference Series **635** (2015) 112138 doi:10.1088/1742-6596/635/11/112138

Towards a complete dynamical description of D_2^+ in strong laser fields

A. Tóth* [1], **S. Borbély**†, **G. J. Halász**‡, **and Á. Vibók***

* Department of Theoretical Physics, University of Debrecen, H-4010 Debrecen, PO Box 5, Hungary
† Faculty of Physics, Babeş-Bolyai University, Cluj-Napoca, Romania
‡ Department of Information Technology, University of Debrecen, H-4010 Debrecen, PO Box 12, Hungary

Synopsis In the framework of the present work we investigate the ionization and dissociation dynamics of D_2^+ irradiated by strong laser fields. We propose a simple three-level model, where the nuclear dynamics on the ground, dissociative, and ionization channels are considered. Our attention was focused on the calculation of ionization probability densities.

In previous studies it was shown that light-induced conical intersections (LICIs) can be formed in diatomic molecules either by standing or by running laser waves [1, 2]. The position of these LICIs is directly controlled by the laser frequency, while the strength of the nonadiabatic couplings is defined by the field intensity. The presence of the LICI has a strong impact on the physical properties of diatomic molecules. This impact was shown in the angular and energy differential dissociation probabilities of the D_2^+ target [3, 4, 5] calculated for different laser pulse parameters.

In order to obtain a more complete picture regarding the influence of LICIs on the molecular dynamics, beside dissociation, the ionization should also be considered. We have extended our previous model [3, 4] by including these ionization channels. The ionization probabilities of the discretized continuum states are obtained as the norm of the nuclear wave packets on the ionization channels.

Here we present our first results on the ionization and dissociation of D_2^+ by linearly polarized ultrashort laser pulses. In our calculations we considered only the vibrational degree of freedom of the molecule, while keeping its orientation parallel to the laser polarization direction, and monitored the ionization probability densities as a function of the energy and ejection angle of the continuum electrons.

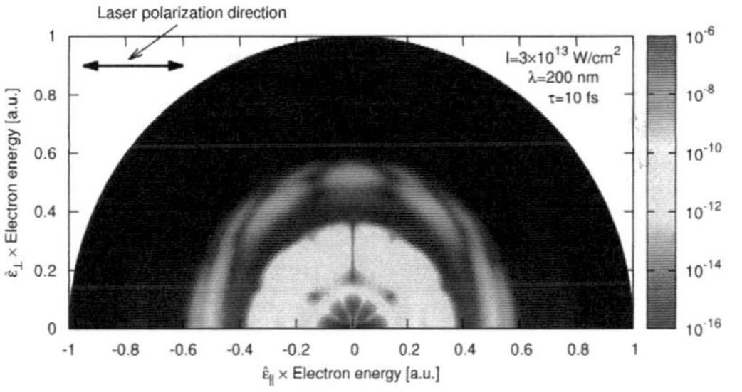

Figure 1. Ionization probability density as a function of electron energy and ejection direction ($\hat{\varepsilon}$).

The present work was supported by the OTKA (NN103251) project, the Hungarian Academy of Science, and by a grant of the Romanian National Authority for Scientific Research (CNCS UEFIS-CDI) project number PN-II-ID-PCE-2011-3-0192.

References

[1] N. Moiseyev *at al* 2008 *J. Phys. B* **41** 22101

[2] M. Sindelka *at al* 2011 *J. Phys. B* **44** 045603

[3] G. J. Halász *at al* 2013 *J. Phys. Chem. A* **117** 8528

[4] G. J. Halász *at al* 2013 *Phys. Rev. A* **88** 043413

[5] G. J. Halász *at al* 2014 *J. Phys. Chem. A* **118** 11908

[1]E-mail: attila.toth@ubbcluj.ro

Content from this work may be used under the terms of the Creative Commons Attribution 3.0 licence. Any further distribution of this work must maintain attribution to the author(s) and the title of the work, journal citation and DOI.
Published under licence by IOP Publishing Ltd

Journal of Physics: Conference Series **635** (2015) 112139 doi:10.1088/1742-6596/635/11/112139

Valence-Shell-Photoelectron Imaging of Controlled Biomolecules

Joss Wiese[a 1], **Sebastian Trippel**[a], **Jochen Kpper**[a,b,c]

[a] Center for Free-Electron Laser Science, DESY, Hamburg, Germany
[b] The Hamburg Center for Ultrafast Imaging, Hamburg, Germany
[c] Department of Physics, University of Hamburg, Germany

A molecules chemical behavior is governed by its electronic properties. Thus, a view at the evolution of the involved molecular orbitals during a reaction lures with insight into the fundamentals of chemistry.

In our experiments molecules are investigated by means of molecular-frame photoelectron imaging. State-selected molecular ensembles are laser aligned or mixed-field oriented inside a velocity map imaging spectrometer. The resulting velocity maps display the projected three-dimensional distributions of both, the photoelectrons kinetic energy and their release angle, in the molecular frame (MFPADs). MFPADs of selected complex molecules and molecular clusters will be discussed in terms of the molecular orbitals which are involved in the ionization process. Three-dimensional MFPADs will be presented, which were obtained through tomographic reconstruction of the projected 2D images. Theoretical analyses of the experimental data will be presented.

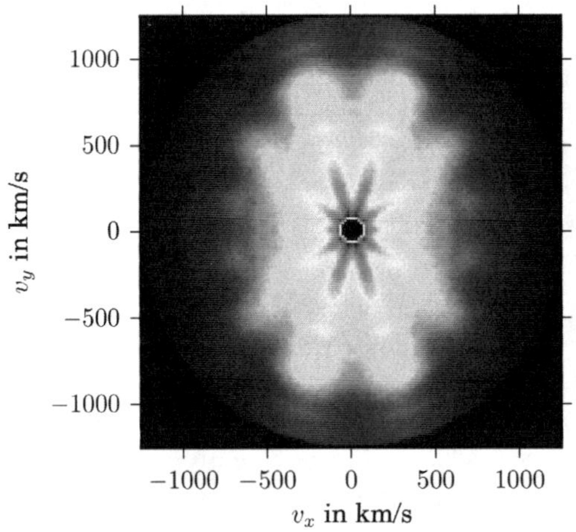

[1]E-mail: joss.wiese@desy.de

Content from this work may be used under the terms of the Creative Commons Attribution 3.0 licence. Any further distribution of this work must maintain attribution to the author(s) and the title of the work, journal citation and DOI.
Published under licence by IOP Publishing Ltd

Molecular frame photoemission: a sensitive probe of the complete polarization state of high harmonic generation

K. Veyrinas* [1], **V. Gruson** †, **S.J. Weber** †, **L. Barreau** †, **T. Ruchon** †, **J.C. Houver***, **B. Carré**†, **R.R. Lucchese** §, **P. Salières** †, **D. Dowek** *

*Institut des Sciences Moléculaires d'Orsay (UMR 8214 CNRS-Université Paris-Sud), Bâtiment 350, Université Paris-Sud, 91405 Orsay, France
†CEA-Saclay, IRAMIS, Laboratoire Interactions, Dynamique et Lasers, 91191 Gif-sur-Yvette, France
§ Department of Chemistry, Texas A&M University, College Station, TX 77840, USA

Synopsis Molecular polarimetry, based on the determination of molecular frame photoemission in dissociative photoionization (DPI) studied by electron-ion coincident momentum spectroscopy, allows us to measure the complete state of elliptically polarized light. Here, we demonstrate the first complete characterization of high-order harmonic polarization state for three benchmark generation processes with inherent symmetry breaking based on the polarization of the generating beam (elliptical or counter-rotating fields) or on anisotropic targets e.g. aligned molecules. The ability to disentangle circular and unpolarized components of the XUV pulses is of particular interest to qualify the HHG sources.

The strong motivation for the control of the polarization state of XUV coherent light sources such as high-order harmonic generation (HHG) or Xray free electron lasers (XFELs) is two-fold. Firstly polarization is a fundamental property of electromagnetic waves: its vectorial nature reveals geometrical or symmetry properties of matter in the generation process. Secondly, there is a strong interest for well characterized elliptically polarized ultrashort sources in order to perform e.g., dynamical studies of circular dichroism in dilute [1] or condensed [2] matter.

The molecular polarimetry method [3] relies on the fact that molecular frame photoelectron angular distributions (MFPADs) in photoionization (PI) of molecules encode both the complex dipole matrix elements describing the dynamics of PI *and* the three normalized Stokes parameters (S_1, S_2 and S_3) characterizing the complete polarization state of the ionizing light. The determination of the (S_1, S_2, S_3) vector is equivalent to that of (ψ, ε, P), the orientation and signed ellipticity of the polarization ellipse, *and* the degree of polarization, respectively. Molecular polarimetry has been bench-marked combining electron-ion vector correlation and fully controlled VUV elliptically polarized light on the beamline DESIRS (SOLEIL).

The reported experiments were performed on the PLFA beamline at CEA Saclay [4], where the laser delivers 10 mJ, 50 fs, 800 nm pulses at 1 kHz. The generated harmonics were focused at the center of a COLTRIMS set-up [4] where dissociative ionization of the NO molecule through the $NO^+(c^3\Pi)$ ionic state [3] was recorded leading to MFPADs for harmonics H15-17-19-21. The measured ellipse orientation Ψ obtained from (S_1,S_2) will be compared to previously reported results, obtained in optical polarimetry using a rotating polarizer [1,5]. The signed ellipticity ε derived from (S_1, S_2, S_3), measured for the first time, is significantly smaller than the expected values previously reported for aligned N_2, due to depolarization of the harmonic light (see Fig. 1).

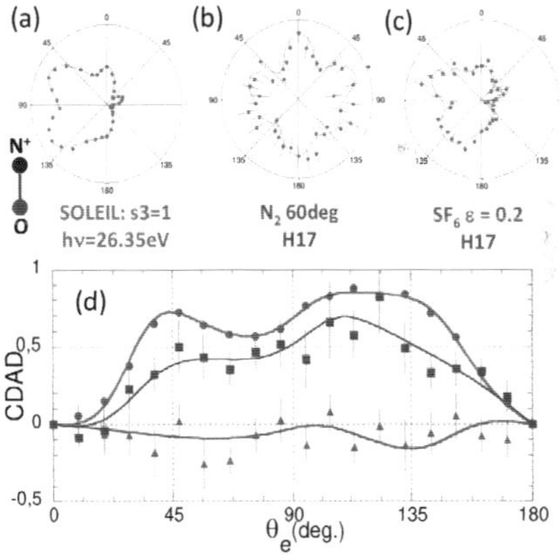

Figure 1. MFPADs for NO perpendicular to the **k** light vector, measured at SOLEIL for hν=26.35 eV and $S_3 \approx 1$ (a), for H17 generated from (b) N_2 aligned at 60 deg. from the IR linear polarization and (c) SF_6 using elliptical IR light with $\varepsilon \approx 0.2$. (d) CDAD parameter [3] for cases (a), (b) and (c) characterizing circular dichroism in the MF, proportional to S_3.

References

[1] A. Ferré *et al* 2015 *Nat. Photon.* 9 93
[2] O. Kfir *et al* 2015 *Nat. Photon.* 9 99
[3] K. Veyrinas *et al* 2013 *Phys. Rev. A.* 88 063411
[4] S. Weber *et al* 2015 *RSI in press*
[5] X. Zhou *et al* 2009 *Phys. Rev. Lett.* 102 073902

[1] E-mail: kevin.veyrinas@u-psud.fr

Content from this work may be used under the terms of the Creative Commons Attribution 3.0 licence. Any further distribution of this work must maintain attribution to the author(s) and the title of the work, journal citation and DOI.
Published under licence by IOP Publishing Ltd

XXIX International Conference on Photonic, Electronic, and Atomic Collisions (ICPEAC2015) IOP Publishing
Journal of Physics: Conference Series **635** (2015) 112141 doi:10.1088/1742-6596/635/11/112141

Dissociation Dynamics of Doubly Ionized Water Using 57 eV Photons

D. Reedy*[1], B. Gaire†, J. Williams^δ, M. Weller†δ, A. Gatton*†, T. Bauer^δ, K. Henrichs^δ, P. Burzynski^δ, J. Sartor* B. Berry^ϵ R. Doerner^δ, Th. Weber†, A. L. Landers*[2],

* Department of Physics, Auburn University, Auburn, AL, USA
† Lawrence Berkeley National Lab, Berkeley, CA, USA
δ Experimental Atomic Physics Group, Goethe-Universität Frankfurt, Frankfurt, Germany
ϵ Department of Physics, Kansas State University, Manhattan, Kansas, USA

Synopsis We have measured the photoelectron emission and dissociation of water in the gas phase following double photoionization by a 57 eV photon. The momenta of the resulting dication ion fragments and electrons were measured in coincidence with Cold Target Recoil Ion Momentum Spectroscopy (C LTRIMS) 1 . The dissociation dynamics following the photoionization process show a marked dependence on both bond angle and kinetic energy release distribution among the fragments.

We have focused our attention the following final state following the photionization of H_2O by 57 eV photons:

$$\gamma + H_2O \rightarrow H^+ + H^+ + O + e^- + e^- \qquad (1)$$

Because of the innate asymmetry of H_2O, fixing the molecular body frame to investigate electronic dynamics has proven to be a challenge. Because the kinetic energy release (KER) is not always equally partitioned between the two ions (in addition to the unmeasured neutral oxygen), the axial recoil approximation breaks down in general. Furthermore, because water has a vibrational mode that corresponds to the bending of the molecule, interpreting KER with a 1-dimensional potential model is inadequate.

For example, Figure 1a) shows the energy correlation between the two measured protons in the isolated dissociation channel above. The distribution of energy sharing as well as total ion energy is quite broad, demonstrating the challenge of isolating particular dissociation pathways from energy measurements alone.

Figure 1b) shows the momentum space histogram in a frame defined by the momentum sum P_{par} of the ion fragments (horizontal axis) vs. the corresponding perpendicular momentum. We note that the magnitude of the ion fragment momentum has a clear correlation with the relative angle between the two momentum vectors. This seems plausible, as one might expect a higher KER for smaller initial bond angles. Also evident in Figure 1b) is the presence of two yet unidentified distinct dissociateve states.

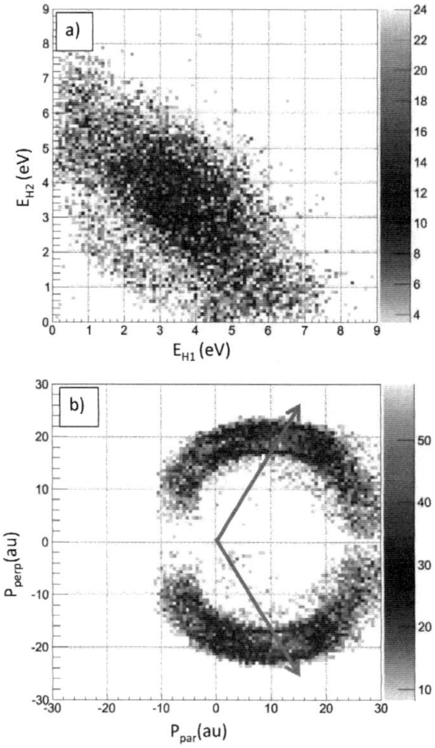

Figure 1. a) Energy correlation map for H^+ ions. b) Molecular frame momenta for both hydrogen fragments, H1 top half, H2 bottom half. The red arrows indicate the average bond angle for neutral water.

Analysis is underway to isolate the fixed-body-frame of the molecule in order to investigate molecular frame electronic correlation effects.

References

[1] T. Jahnke, *et al.* , J. Electron Spectrosc. Relat. Phenom. 141, 229 (2004).

[1]E-mail: dpr0008@auburn.edu
[2]E-mail: landeal@auburn.edu

Content from this work may be used under the terms of the Creative Commons Attribution 3.0 licence. Any further distribution of this work must maintain attribution to the author(s) and the title of the work, journal citation and DOI.
Published under licence by IOP Publishing Ltd

Journal of Physics: Conference Series **635** (2015) 112142 doi:10.1088/1742-6596/635/11/112142

Ultracold hydrogen and deuterium production via Doppler-cooled molecules

Ian C. Lane* [1]

* School of Chemistry and Chemical Engineering, Queen's University Belfast, UK

Synopsis A scheme to produce ultracold hydrogen or deuterium atoms via photo-fragmentation, even allowing direct formation of a quantum gas

A counterintuitive scheme [1] to produce ultracold hydrogen via fragmentation of laser cooled diatomic hydrides is presented (figure 1) where the final atomic H temperature is inversely proportional to the mass of the molecular parent. In addition, the critical density for formation of a Bose-Einstein Condensate (BEC) at a fixed temperature is reduced by a factor $\left(\frac{m_H}{m_{MH}}\right)^{3/2}$ over directly cooled hydrogen atoms.

cooled to a Doppler temperature of ~26 μK with just two rovibronic transitions (figure 2), the simplest molecular cooling scheme identified to date. Preparation of a hydrogen atom gas below the critical BEC temperature Tc is feasible with present cooling technology, with optical pulse initiation of the entire condensation process. As isotopic substitution does not alter the feasibility of laser cooling the parent molecule, this method can be used to produce ultracold deuterium atoms as well, a species yet to be brought to quantum degeneracy.

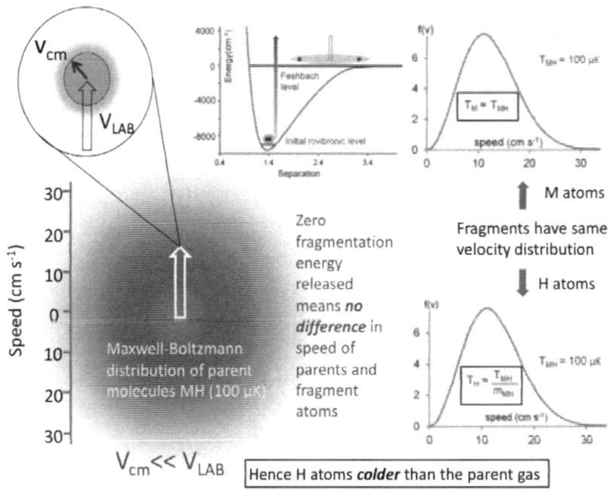

Figure 1. Outline of the production scheme.

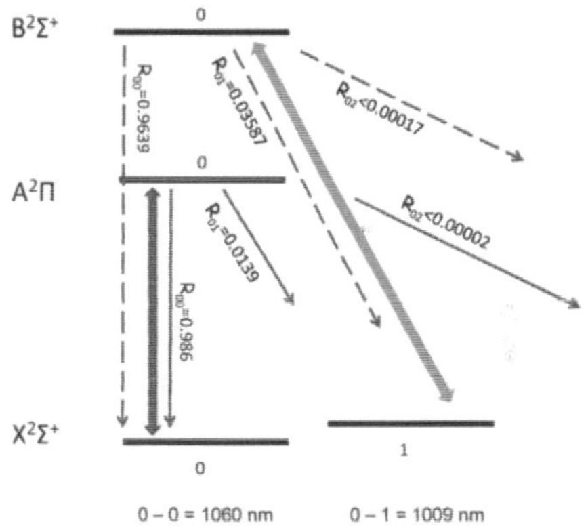

Figure 2. Two color laser scheme for BaH.

The narrow Feshbach resonances between a $^1S_0$ atom and hydrogen are well suited to a tiny center of mass energy release necessary during fragmentation. With the support of *ab initio* quantum chemistry, it is demonstrated that BaH is an ideal diatomic precursor that can be laser

References

[1] I. C. Lane, arXiv:1311.7081.

[1] E-mail: i.lane@qub.ac.uk

Content from this work may be used under the terms of the Creative Commons Attribution 3.0 licence. Any further distribution of this work must maintain attribution to the author(s) and the title of the work, journal citation and DOI.
Published under licence by IOP Publishing Ltd

Resonant Compton scattering of photons by He($1s^2\ ^1S$) in astrophysical plasmas

Sabyasachi Kar* [1], **Yang Wang*** [1], and **Y. K. Ho**[† 2]

[*]Center for Theoretical Atomic and Molecular Physics, Academy of Fundamental and Interdisciplinary Sciences, Harbin Institute of Technology, Harbin 150080, People's Republic of China

[†] Institute Atomic and Molecular Sciences, Academia Sinica, PO Box 23-166, Taipei, Taiwan 106, Republic of China

Synopsis The nonthermal effect on resonant Compton scattering of photons by the He ($1s^2\ ^1S$) ground state is investigated in Lorentzian astrophysical plasmas using exponential correlated basis functions and with a variational technique. The nonthermal character of the Lorentzian plasma enhances the resonant Compton scattering cross sections.

In the space environments, plasmas are generally observed to possess a non-Maxwellian high-energy tail that can be well modeled by a generalized Lorentzian (or kappa) distribution function containing the spectral index κ. The kappa distribution function has been well described by several research groups [1-7] with the following form:

$$f_\kappa(v) = \left(\frac{m}{2\pi E_\kappa}\right)^{3/2} \frac{\Gamma(\kappa+1)}{\Gamma(\kappa-1/2)} \left(1+\frac{mv^2}{2\kappa E_\kappa}\right)^{-(\kappa+1)},$$

where m and v are the mass and velocity of the plasma electron, respectively, $E_\kappa (\equiv (1-\frac{3}{2\kappa})E_T)$ is a characteristic energy in generalized Lorentzian plasmas, and κ (>3/2) is a spectral index, $E_T = \kappa_B T_e$, κ_B is the Boltzmann constant, T_e is electron temperature. We have considered the effective screening potential of the form $\varphi_{eff}(r) = \frac{Ze}{r}\exp(-\frac{r}{\lambda_\kappa})$, where $\lambda_\kappa (= \mu_\kappa^{1/2}\lambda_D)$ is the effective screening length, with λ_D is the standard Debye length in Maxwellian plasmas and $\mu_\kappa = \mu_D\sqrt{\alpha^\kappa}$, $\alpha^\kappa = \sqrt{\frac{\kappa-1/2}{\kappa-3/2}}$, $\mu_D (\equiv \lambda_D^{-1})$ is Debye screening parameter.

In this work, we have carried out calculations to investigate the effects of Lorentzian astrophysical plasmas on the resonant Compton scattering of photons by the ground state electrons of helium in the framework of Ritz variational method. Total resonance Compton scattering cross sections σ_C^T [1,3] can be written as $\sigma_C^T = (\frac{3}{2\pi})\Lambda_{fi}^2$, where $\Lambda_{fi} (\equiv 2\pi c/\omega_{fi})$ is the resonant wavelength between the final atomic state f and the initial state i, ω_{fi} is the frequency between the atomic states i and f. Detail calculations can be found from the earlier works Li and Kar ([3], references therein). We solve the Schrödinger equation for helium atom in generalized Lorentzian plasma in the framework of Ritz variational principle using highly correlated exponential basis set [8, 9] of the form:

$$\{x(\delta_1^{(i)}r_1)x(\delta_2^{(i)}r_2)x(\delta_{12}^{(i)}r_{12})\} \text{ with } x(\delta r) = e^{-\delta r}.$$

The nonlinear variational parameters in the exponents are generated by a pseudorandom process. It is well-known that resonances can occur when the energy difference between the incoming and outgoing photons is equal to transition energy for target atoms or ions. Resonant scattering cross sections σ_C^T between $1s^2\ ^1S$ and $1s2p\ ^1P$ states is presented in figure 1 as a function of the spectral index for different screening parameter.

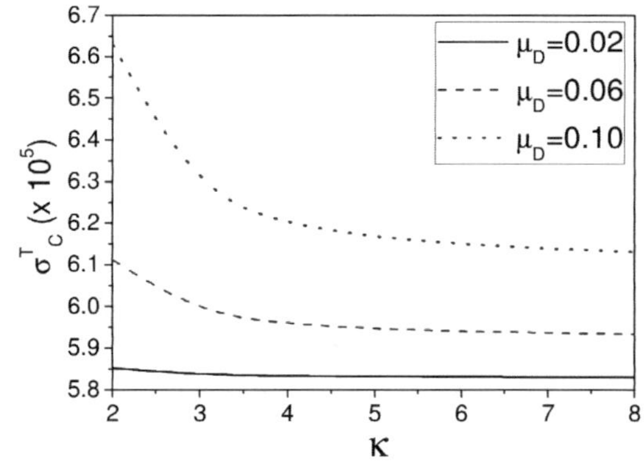

Figure 1. Resonant scattering cross section σ_C^T as a function of κ for three different values of μ_D.

Acknowledgments: YKH is supported by MOST of Taiwan and YW by NNSF of China.

References

[1] Y. D. Jung 2009 *Ap. J.* **695** 917
[2] V. Pierrard, M. Lazar 2010 *Sol. Phys.* **267** 153
[3] H.W. Li, S. Kar 2012 *Euro. Phys. J. D* **66** 304
[4] P. Jiang *et al* 2013 *Phys. Plasmas* **20** 012126
[5] P. Jiang *et al* 2013 *Few-Body Syst.* **54** 1911
[6] G. Livadiotis et al 2013 *Space Sci. Rev.* **175** 183
[7] S. Kar *et al* 2013 *ICPEAC Book of Abstracts*
[8] S. Kar, Y.K. Ho 2008 *JQSRT* **109** 445
[9] S. Kar 2012 *Phys. Rev. A* **86** 062516

[1] E-mail: skar@hit.edu.cn
[2] E-mail: ykho@pub.iams.sinica.edu.tw

Content from this work may be used under the terms of the Creative Commons Attribution 3.0 licence. Any further distribution of this work must maintain attribution to the author(s) and the title of the work, journal citation and DOI.
Published under licence by IOP Publishing Ltd

Journal of Physics: Conference Series **635** (2015) 122002 doi:10.1088/1742-6596/635/12/122002

Low-order harmonic generation in a ZnS laser ablation plasma

M. Oujja, A. Benítez-Cañete, I. López-Quintás, M. Martín, R. de Nalda [1], M. Castillejo [2]

Instituto de Química Física Rocasolano, CSIC, 28006 Madrid, Spain

Synopsis Low-order (3rd to 9th) harmonic generation of a near-infrared driving laser (1064 nm) is reported in a laser ablation plasma of ZnS. Temporal analysis shows two distinct components with respect to the ablation event. The late temporal component exhibits high conversion efficiency for the highest harmonic orders observed. This is attributed to a dramatic modification of the plasma medium with the driving laser.

Harmonic generation has become a standard technique for the synthesis of coherent radiation in the high frequency regions of the electromagnetic spectrum, from the UV through VUV and XUV, to X-ray. Low-order harmonic generation is possible with moderately intense lasers of nanosecond duration, and is used as a tool for VUV spectroscopy [1]. In contrast, high-order harmonic generation is only possible in the high laser intensity regime (typically above 10^{14} W/cm^2) only easily attained in systems that deliver ultrashort laser pulses of femtosecond duration.

Atomic and molecular gas jets have been the most common nonlinear media employed for harmonic generation, but the universality of the phenomenon makes alternative sources viable. Laser ablation plasmas offer interesting avenues for research, and have been explored as nonlinear media for harmonic generation in recent years in a number of works [2,3] both in the low and high intensity regimes. Particularly attractive is the possibility to tailor the characteristics of the medium through the control of the ablation laser characteristics. In particular, control of the density, composition, spatial extent or degree of ionization are possible with a wise choice of ablation laser parameters like wavelength, intensity, pulse duration or spatial properties of the laser beam on the surface of the target.

In this work, we report on the generation of low-order harmonics (from the 3rd to the 9th) of a fundamental NIR laser source of nanosecond duration at 1064 nm in a NIR laser ablation plasma of the wide bandgap semiconductor ZnS. The ablation laser irradiated the target at normal incidence, and the driving laser propagated parallel to the target at a controllable distance (typically 1 mm). The delay between the ablation event and the harmonic generation event was fully controllable electronically, and the focal position of the driving laser could be placed on-axis with respect to the ablation plasma propagation direction, or out-of-axis. This permitted full freedom of spatiotemporal exploration.

Harmonic generation was found to maximize at two distinct temporal delays with respect to the ablation event, as shown in Figure 1. The first peak was found around 250 ns, and the second, around 10 μs. Interestingly, the early component dies off quickly with increasing harmonic order (becoming absent for the 9th harmonic), whereas the late component is extremely intense for the 7th harmonic and is still clearly visible with the 9th harmonic. This evidence points to the idea that the driving laser induces a secondary plasma in a precursor population composed of ZnS aggregates and nanoparticles, and this secondary plasma is characterized by a particularly high conversion efficiency towards higher harmonic orders. The control of the properties of the secondary plasma using this method opens new routes for finding efficient nonlinear media for nonlinear wavelength conversion.

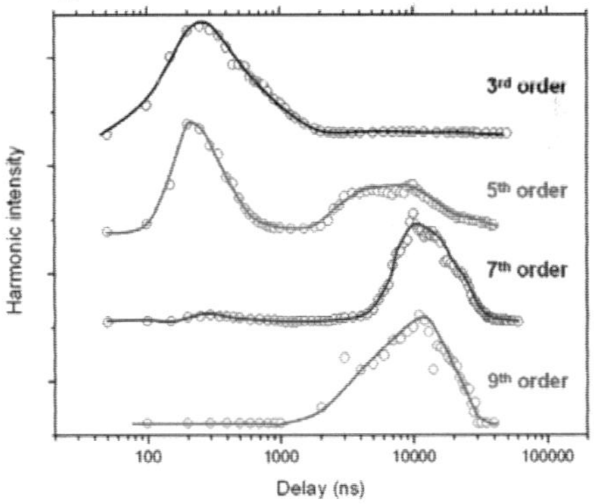

Figure 1. Normalized harmonic emission at orders 3, 5, 7 and 9 of the fundamental NIR driving laser in a ZnS laser ablation plasma as a function of delay with respect to the ablation event.

References

[1] C.Y. Ng 2002 *Annu. Rev. Phys. Chem.* 53 101
[2] R. Ganeev 2014 *J. Opt. Soc. Am. B* 31 2221
[3] R. de Nalda *et al* 2011 *Phys. Chem. Chem. Phys.* 13 10755

[1]E-mail: r.nalda@iqfr.csic.es
[2]E-mail: martacastillejo@iqfr.csic.es

Content from this work may be used under the terms of the Creative Commons Attribution 3.0 licence. Any further distribution of this work must maintain attribution to the author(s) and the title of the work, journal citation and DOI.
Published under licence by IOP Publishing Ltd

Femtosecond Electron Delocalization in Polymer:Fullerene Blend Films

Yunier Garcia-Basabe*, Cleber F. N. Marchiori†, Lucimara S. Roman†, and Maria Luiza Rocco*,[1]

* Institute of Chemistry, Federal University of Rio de Janeiro, Rio de Janeiro, 21941-909, Brazil
† Departament of Physics, Federal University of Paraná, Curitiba, 81531-990, Brazil

Synopsis Ultrafast electron dynamics in the low-femtosecond regime was evaluated for a low band-gap polymer and its blend with fullerene by resonant Auger spectroscopy. Remarkable changes developed by tuning the photon energy along the sulfur and silicon 1s absorption edges. The effect of thermal annealing and molecular orientation on the charge transfer delocalization times were probed and correlated with solar cells performance.

Ultrafast electron dynamics is a relevant topic for fundamental and applied areas like femtochemistry, surface photochemistry, molecular electronics, solar energy, and so on. It is possible to apply two different spectroscopic approaches to pursue it. Ultrashort pulse laser pump-probe spectroscopies have been used widely, where delocalization of excited electrons in the presence of delocalized valence hole states are probed. Besides laser-based pump-probe methods, resonant Auger spectroscopy following core excitation emerges as an alternative with major advantages. First, the core hole lifetime probed by core level spectroscopy can be used as a fast internal clock and insofar very low time scales in the range of femtoseconds down to hundred attoseconds can be achieved, the so-called core-hole clock (CHC) method. Second, the inherent atomic specificity of core levels, which deals with localized hole states, can be explored.

A fundamental aspect regarding charge transfer dynamics is related to the organizational level of the polymeric films and its morphology. In this regard, X-ray absorption spectroscopy is a powerful technique which gives information about the orientation of molecular films with respect to the surface while probing unoccupied density of states. One way in pursing this organization is by means of thermal annealing treatments, which changes the polymer morphology and possibly enhances its charge transport properties.

In the present work, we have selected the internal donor/acceptor copolymer poly[2,7-(9,9-bis(2-ethylhexyl)-dibenzosilole)-alt-4,7-bis(thiophen-2-yl)benzo-2,1,3-thiadiazole] (PSiF-DBT) and its blend with fullerene (PSiF-DBT:PCBM) to investigate how the thermal annealing treatment affects molecular orientation and consequently the charge delocalization time (τCT) [1,2].

Near edge X-ray absorption fine structure (NEXAFS) and Resonant Auger spectra were measured at the soft X-ray spectroscopy (SXS) beamline at the Brazilian Synchrotron Light Source (LNLS), in Campinas, São Paulo. It was demonstrated a decrease of the τCT after thermal annealing which may be associated with more ordered polymeric films, as probed by NEXAFS, with the increase of orbital overlapping, and consequently electron delocalization. Further reduction of the τCT was observed for the polymer:fullerene blends and the subsequent annealing. These results will be presented and discussed in the context of solar cells performance.

Work supported by LNLS, CNPq, FAPERJ and CAPES.

References

[1] Y. Garcia-Basabe *et al* 2014 *J. Appl. Phys.* **115** 134901
[2] Y. Garcia-Basabe *et al* 2014 *J.Phys. Chem. C* **118** 23863

[1]E-mail: luiza@iq.ufrj.br

Content from this work may be used under the terms of the Creative Commons Attribution 3.0 licence. Any further distribution of this work must maintain attribution to the author(s) and the title of the work, journal citation and DOI.
Published under licence by IOP Publishing Ltd

Dissociation mechanisms for the doubly-ionized DMDS (CH_3-SS-CH_3) molecule: use of a multivariate normal distribution method.

L. R. Varas* [1], **A.C.F. Santos**† [2], **L. H. Coutinho**†[2], **G.G.B. de Souza*** [3]

* Instituto de Química, Universidade Federal do Rio de Janeiro, Cidade Universitária, Ilha do Fundão, 21949-900, Rio de Janeiro, RJ, Brazil.

† Instituto de Física, Universidade Federal do Rio de Janeiro, Cidade Universitária, Ilha do Fundão, 21949-900, Rio de Janeiro, RJ, Brazil.

Synopsis The use of Multivariate Normal Distribution was used to represent the whole spectra as combination of parameters. Fragmentation and ionic dissociation of dimethyl disulfide (CH_3SSCH_3 or DMDS) at 800 eV is studied for single and multiple coincidence. The proposed methodology allows to obtain dissociation parameters when there is an overlap of the ion-ion data spectrum.

Double ionization of molecules usually leads to highly dissociative states. Time-of-Flight mass spectrometry coupled to ion-ion coincidence techniques allows for the characterization of these unstable ionic states [1]. The principal focus in this work is to represent the correlation between charged ions associated with a given dissociation process using an alternative way, based on a multivariate distribution. We thus, recover some of the initial Frasinsky's covariance approach [2]. Card et al. also used the covariance idea in the description of ammonia clusters subjected to femtosecond lasers pulses [3], but neither Fransinsky nor Card went further in order to represent the whole coincidence spectra as a combination of statistical parameters.

All the bivariate density functions can be expressed as the combination of two matrices, μ and $\sum$:

$$f = \frac{1}{\sqrt{2\pi|\Sigma|}} e^{-\left(\frac{1}{2}(x-\mu)^T\Sigma^{-1}(x-\mu)\right)} \qquad (1)$$

Where μ is the mean vector and $\sum$ is the covariance matrix. As an example, considering the dissociation of $DMDS^{2+}$ into CH_3SS^+ and CH_3^+, following electron beam irradiation of the neutral molecule (Fig. 1a), we can show that these matrixes can be defined as:

$$\mu = \begin{pmatrix}231\\2611\end{pmatrix},\ \Sigma = \begin{pmatrix}206{,}5 & -107{,}9\\ -107{,}9 & 180{,}6\end{pmatrix}$$

Considering also the $[SS]^+$ / $[CH_3]^+$ ion pair coincidence (Fig.1b), the matrix are represented as:

$$\mu = \begin{pmatrix}231\\2189\end{pmatrix},\quad \Sigma = \begin{pmatrix}240{,}7 & -114{,}47\\ -114{,}47 & 125{,}2\end{pmatrix}$$

We can now use the previous information to disentangle the compounded coincidence between an ion and a set of related ionic species, as for instance, $[CH_nSH_n]^+$ / $[CH_3]^+$ (Fig.1c)

The latter ion-ion coincidence spectrum shows infinite slope possibilities. Using the presently proposed methodology we can nonetheless obtain the dissociation parameters for ion pairs such as $[CHS]^+$ / $[CH_3]^+$ (Fig 1d).

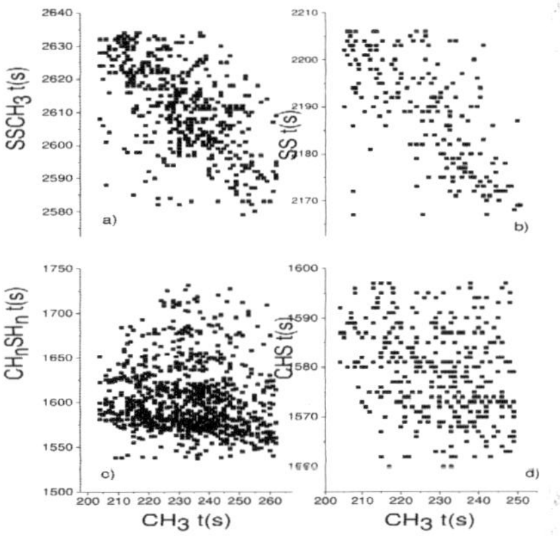

Figure 1. a) $[CH_3SS]^+$ / $[CH_3]^+$ b) $[SS]^+$/ $[CH_3]^+$ c) [CHnSHn]$^+$ / $[CH_3]^+$ d) $[CHS]^+$/ [CH3]$^+$ coincidence.

Dissociation of the ionic pair $[SS]^+$ / $[CH_3]^+$ can also be represented in a three-dimensional plot, as shown below:

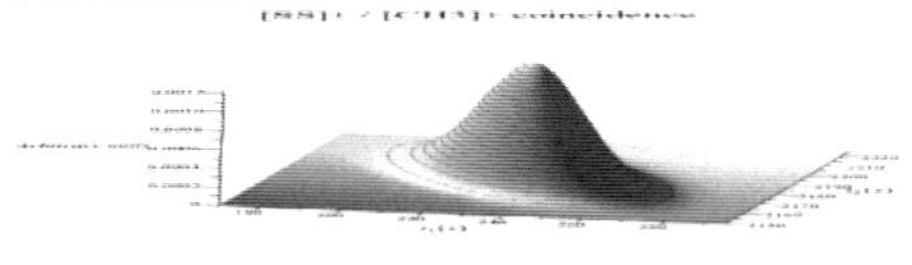

Figure 2. $[SS]^+$/ $[CH_3]^+$ coincidence represented.

References

A.C.F. Santos et al. 2002. Journal of Physics B. Atomic and Molecular Physics, vol. 19, pp. L819 -L824, 1986.
L, J, Frasinsky, et al. 1986. 1986, Journal of Physics B: Atomic and Molecular Physics, Vol. 19, pp. L819 -L824.
D, A, Card, et al. 2003. 2003, International Journal of Mass Spectrometry , Vols. 223-224, pp. 355-36

[1] E-mail: lautaro.ramirezvaras@ucr.ac.cr
[2] E-mail: gerson@iq.ufrj.br

Content from this work may be used under the terms of the Creative Commons Attribution 3.0 licence. Any further distribution of this work must maintain attribution to the author(s) and the title of the work, journal citation and DOI.
Published under licence by IOP Publishing Ltd

Harmonic-emission from Rydberg states in the vicinity of metallic nanostructures

I. Yavuz* [1], **Y. Tikman***, **and Z. Altun*** [2]

* Dep. of Physics, Marmara University, 34722 Ziverbey, Istanbul, TURKEY

Synopsis We theoretically demonstrate that harmonic emission could be achieved from Rydberg atoms without deteriorating the nanostructures.

The localized fields enhanced by surface plasmons often degrade the metallic nanostructure itself, since high field-intensities are in turn needed to achieve harmonic emission from ground-state atoms. [1]

We theoretically investigate harmonic emission from bound states of target atoms exposed to locally inhomogeneous electric fields. We numerically solve three-dimensional time-dependent Schrödinger equation for hydrogen atom in the vicinity of metallic nanostructure. The interaction between the nanostructure-enhance incident field and the target atom is taken as $z(1+\beta z)E(t)$ [2]. As we go up in Rydberg states of the hydrogen atom, we n-scale the input parameters for harmonic emission. Classical assessments have shown that incident field intensity and frequency and the field-free ionization potential (I_p) of hydrogen-like atoms scales as n^{-8}, n^{-4} and n^{-2}, respectively. Thus, for higher Rydberg states the incident field intensity becomes so weak that the nanostructure would not be damaged [3].

For field intensities on the order of $100TW/cm^2$, the tunnel ionization distance I_p/E_0 plus the excursion amplidute of the continuum electron E_0/ω_0^2 becomes $\sim 20-30a.u.$ and scales as n^2. Thus, due to electron confinement in the nano element, within a dimension of $\sim 200a.u.$ and the chances of absorbtions from metallic surfaces, one cannot speak of the likelyhood of recombination above a certain n-state. Therefore, tweaking the input parameters for an efficient HHG process in the vicinity of metallic nanostructure is essential.

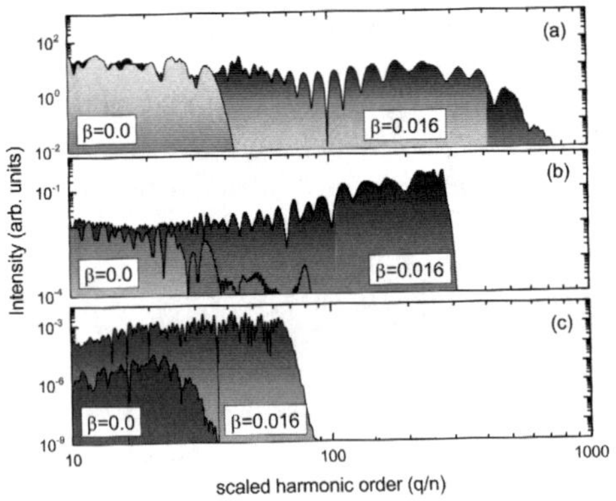

Figure 1. (a) n=1, (b) n=2, (c) n=4. $I_0 = 200TW/cm^2$

References

[1] Sivis, M., *et al.* 2013 *Nature Phys.* **9** 304-309.

[2] Yavuz, I., *et al.* 2012 *Phys. Rev. A* **85** 013416.

[3] Bleda, Erdi A., *et al.* 2013 *Phys. Rev. A* **88** 043417.

[1] E-mail: ilhan.yavuz@marmara.edu.tr
[2] E-mail: zikalt@marmara.edu.tr

Content from this work may be used under the terms of the Creative Commons Attribution 3.0 licence. Any further distribution of this work must maintain attribution to the author(s) and the title of the work, journal citation and DOI.
Published under licence by IOP Publishing Ltd

Journal of Physics: Conference Series **635** (2015) 122006 doi:10.1088/1742-6596/635/12/122006

Interrogation and fabrication of nm scale hot alkali vapour cells

K. A. Whittaker*[1], J. Keaveney*, I. G. Hughes*, A. Sargysyan†, D Sarkisyan†, B. Gmeiner,‡ V. Sandoghdar,‡ C. S. Adams*

* Joint Quantum Centre (JQC) Durham-Newcastle, Department of Physics, Durham University, South Road, Durham, DH1 3LE, United Kingdom

† Institute for Physical Research, National Academy of Sciences, Ashtarak 2, 0203, Armenia

‡Max Planck Institute for the Science of Light,Günther-Scharowsky-Str. 1, Building 24, 91058, Erlangen Germany

Synopsis Here we describe our use of cells with nm scale thickness ('nanocells') to investigate inter-atomic, atom-light and atom-surface interactions. Such cells allow us to access high density regimes previously unexplored in vapour cell spectroscopy, and to probe the interaction of atoms within the near field of surfaces, where atoms interact with their own virtual image. We also discuss the future fabrication of such cells in Durham.

Spectroscopy of alkali vapour cells is a well established and long lived field of physics [1, 2] which still offers new and exciting discoveries; providing a medium to probe fundamental properties of atomic interactions, such as the cooperative Lamb shift [3], superluminal propagation [4] and atom-surface interactions [5].

In order to probe such media, we use nm scale sapphire vapour cells ('nanocells') with a thickness ranging from 2 μm to 40 nm. By tuning the temperature, we can move through over 6 orders of magnitude in density (up to 10^{17}cm^{-3}), enabling us to investigate dipole-dipole interactions at small atomic separation. Furthermore, spectra from nanocells have greatly reduced linewidths [2], resulting in single-beam spectroscopy with individually resolved hyperfine lines and huge negative group refractive indices reaching up to -2.4$\times 10^5$, resulting in superluminal pulse propagation [4].

We have also exploited the tight atomic confinement to investigate the interaction of atoms with their virtual images in nearby surfaces. The van der Waal atom-surface (AS) force induces a shift of the atomic transitions, resulting in a characteristic asymmetric line-shape. We detect off-axis scattered light using single photon counting modules to measure absorption spectra from nanocells over the thickness range 40-250 nm. We fit these absorption spectra using a model of the vapours susceptibility to extract information about the AS interaction. A plot comparing experimental data (black line) and fit (red line) for cell length ℓ =38 nm shown in Fig. 1 (a). As a result, we have confirmed the form the AS interaction takes over these length scales [5].

Fabrication of our own nanocells will allow us to customize the cell interior, affecting how the atoms interact with each other and the cell walls. Our approach uses optical contacting of quartz plates, one etched with a region 500 nm thick. After heat treatment a permanent hermetic bond is formed that has almost 100% transmission with none of the common problems associated with glued cells, e.g. outgassing and expansion under thermal stresses. An example of a cell produced with this process is shown in Fig. 1 (b).

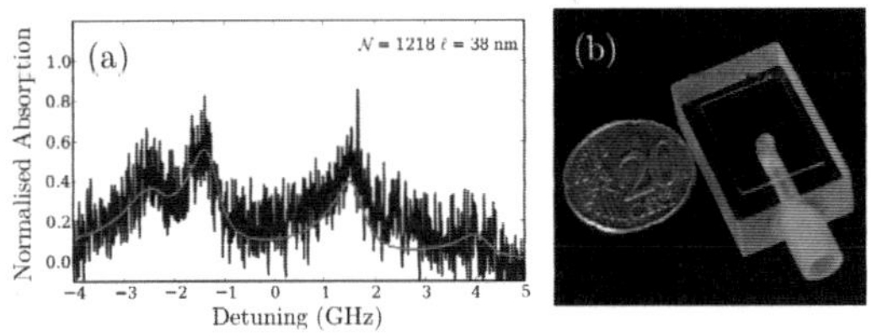

Figure 1. (a) AS shifted Rb D1 spectrum from a cell length of 38 nm. (b) Partially fabricated cell made in Durham.

References

[1] D Sarkisyan *et al.* 2001 *Opt. Comm.* **200** 201

[2] G. Dutier *et al* 2003 *Europhys. Lett.* **63** 35

[3] J. Keaveney *et al* 2012 *Phys. Rev. Lett.* **108** 173601

[4] J. Keaveney *et al* 2012 *Phys. Rev. Lett.* **109** 233001

[5] K. A. Whittaker *et al* 2014 *Phys. Rev. Lett.* **112** 253201

[1]E-mail: k.a.whittaker@durham.ac.uk

Content from this work may be used under the terms of the Creative Commons Attribution 3.0 licence. Any further distribution of this work must maintain attribution to the author(s) and the title of the work, journal citation and DOI.
Published under licence by IOP Publishing Ltd

Institute of Physics
Dirac House, Temple Back
Bristol BS1 6BE UK

ISSN: 1742-6588
ISBN 978-1-5108-1197-3

CURRAN ASSOCIATES INC.
proceedings.com